Photosynthesis: Mechanisms and Effects

Photosynthesis: Mechanisms and Effects

Volume V

Proceedings of the XIth International Congress on Photosynthesis,
Budapest, Hungary, August 17–22, 1998

edited by

G. GARAB
Biological Research Center,
Hungarian Academy of Sciences,
Szeged, Hungary

KLUWER ACADEMIC PUBLISHERS
DORDRECHT / BOSTON / LONDON

A C.I.P. Catalogue record for this book is available from the Library of Congress.

ISBN 0-7923-5542-3 (Vol. 1)
ISBN 0-7923-5543-1 (Vol. 2)
ISBN 0-7923-5544-X (Vol. 3)
ISBN 0-7923-5545-8 (Vol. 4)
ISBN 0-7923-5546-6 (Vol. 5)
ISBN 0-7923-5547-4 (set)

Published by Kluwer Academic Publishers,
P.O. Box 17, 3300 AA Dordrecht, The Netherlands.

Sold and distributed in North, Central and South America
by Kluwer Academic Publishers,
101 Philip Drive, Norwell, MA 02061, U.S.A.

In all other countries, sold and distributed
by Kluwer Academic Publishers,
P.O. Box 322, 3300 AH Dordrecht, The Netherlands.

Printed on acid-free paper

Printed in the Netherlands.

GENERAL CONTENTS

GENERAL CONTENTS

CONTENTS TO VOLUME V

17. CO_2 entry, concentration and fixation

CONGRESS ON PHOTOSYNTHESIS

XIth INTERNATIONAL

BUDAPEST ✳ 1998

RUBISCO CATALYSIS *IN VITRO* AND *IN VIVO*

T. John Andrews, Susanne von Caemmerer, Zhili He, Graham S. Hudson and Spencer M. Whitney. Molecular Plant Physiology, Research School of Biological Sciences, Australian National University, PO Box 475, Canberra ACT 2601, Australia

Key words: antisense suppression, chloroplast transformation, gas exchange, genetic manipulation, photorespiration, Rubisco activase

1. Introduction

Studies of the mechanisms and regulation of photosynthesis in higher plants can benefit from detailed comparisons between the properties of isolated proteins and inferences about them derived from physiological measurements with whole leaves. This approach allows the increasing understanding of molecular details provided by the reductionist approach to be tested and integrated so that the relative importance of the various details can be assessed.

Rubisco is a particularly favourable candidate for this kind of comparison. Its activity and properties can be measured *in vitro* like any other enzyme and many of its properties *in vivo* can be inferred quite reliably from leaf gas-exchange measurements (1). When the two approaches agree, we can be confident that our understanding of the molecular details is sound and relevant to the functioning of the whole system. When they don't agree, then we know that something is missing - some part of the molecular details has not been discovered or understood correctly and needs to be investigated further.

Two factors have combined to make these kinds of comparative studies interesting currently. First, both biochemical and leaf gas-exchange techniques have improved enough that meaningful comparisons between them are possible. Second, our increasing ability to make precise molecular genetic interventions into both nuclear and chloroplast genomes now allows us to perturb the physiology experimentally.

Here we review two examples drawn from our recent research. The first uses nuclear transformation to effect antisense suppression of Rubisco's regulatory protein, Rubisco activase. This case has revealed quite a gulf between current biochemical understanding of the function of activase and physiological observations with activase-deficient tobacco. The second case involves transformation of the chloroplast genome to mediate site-direct mutagenesis of Rubisco's large subunit. It has confirmed that current ideas about the dual role of Rubisco in photosynthesis and photorespiration are soundly based.

G. Garab (ed.), Photosynthesis: Mechanisms and Effects, Vol. V, 3307–3312.
© 1998 *Kluwer Academic Publishers. Printed in the Netherlands.*

2. Procedure

2.1 Growth and analysis of activase-deficient tobacco

Transgenic tobacco (*Nicotiana tabaccum* L. cv. W38) with a nuclear antisense gene directed against activase (2) were grown, together with wildtype controls, in an atmosphere enriched in CO_2 to 0.15% in a growth cabinet with a 14-h photoperiod (450 μmol quanta m^{-2} s^{-1}, 25/18 °C). The plants were the progeny of primary transformant A52 which had two independent copies of the antisense gene (2). Photosynthetic gas exchange (1000 μmol quanta m^{-2} s^{-1}, 350 μbar CO_2, 25 °C), activase content, Rubisco content, Rubisco carbamylation and Rubisco activity immediately after extraction and after full *in vitro* activation were measured as described (3,4).

2.2 Site-directed mutagenesis of the large subunit of tobacco Rubisco

The transforming plasmid, pL335V (Fig. 1), was derived by site-directed mutagenesis (5) from pLEV1, which contained a fragment of the large-single-copy region of the tobacco plastid genome comprising *rbc*L and its flanking regions. pLEV1 also contained a promoterless *aad*A gene (conferring resistance to aminoglycoside antibiotics) inserted downstream of *rbc*L (Fig. 1). This construct was homologously recombined into the plastid genome of tobacco (*Nicotiana tabaccum* L. cv. Petit Havana) using the biolistic procedure (6). Spectinomycin-resistant plantlets were screened for homoplasmicity and the presence of the introduced *Eco*RV site by Southern analysis with leaf DNA (7). One plantlet which became homoplasmic and remained so through several cycles of regeneration was cloned, transferred to soil, and grown, together with wildtype control regenerants, in an atmosphere of 0.3% CO_2 as described in Section 2.1. Its gas-exchange and Rubisco parameters were measured as described in Section 2.1. Rubisco was extracted from leaves of mutant and control plants and purified by polyethyleneglycol fractionation and sucrose density gradient centrifugation (8). These preparations were used to determine the Michaelis constants for the substrates (9) and the CO_2/O_2 specificity (10).

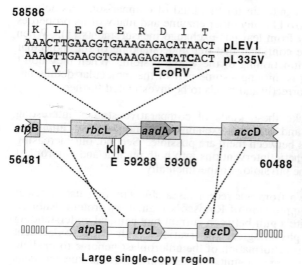

FIG. 1. The pL335V plasmid used for transformation of the tobacco plastid genome. This plasmid was based on pLEV1 with the nucleotide substitutions (in bold) required to convert codon 335 of *rbc*L from Leu to Val and silent substitutions to introduce an *Eco*RV site for use in selecting mutants (5). The numbers refer to the sequence of the tobacco plastid genome (11). K, *Kpn*I; N, *Nru*I; E, *Eco*RV; T, *rps*16 terminator sequence.

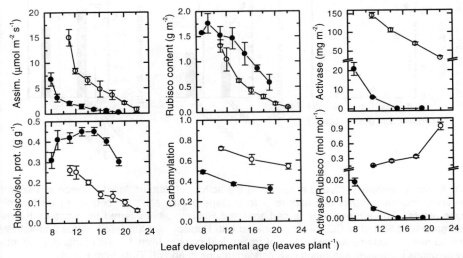

FIG. 2. CO_2 assimiliation rate, activase content and Rubisco content and carbamylation during leaf development in wildtype (O) and anti-activase (●) plants grown in 0.15% CO_2. The same leaf was sampled throughout its development. To compensate for differences in growth rate between the genotypes, the developmental age of a leaf is expressed in terms of the total number of leaves that the plant had at sampling. Data from Ref. 4.

3. Results and Discussion

3.1 A new role for Rubisco activase?

Biochemical studies with activase suggest that it is a motor protein, powered by ATP hydrolysis, that manipulates Rubisco's active sites so that inhibitors bound to them are released. It releases non-reactive or poorly reactive analogs of the substrate or catalytic intermediates from carbamylated Rubisco sites and it releases the substrate, ribulose-1,5-bisphosphate (RuBP), from its unproductive complex with uncarbamylated sites (12,13). Hitherto, it has not been thought to influence the catalytic process itself.

Plants with 5% or less of the wildtype's activase content required CO_2 supplementation to grow. Under these conditions they developed normally, but more slowly than the wildtype. The leaves contained more Rubisco than the wildtype and they retained it longer during development (Fig. 2). Therefore, the older anti-activase leaves had much higher ratios of Rubisco to soluble protein. The activase content, which was very low even in the young anti-activase leaves, soon dropped to undetectable levels and the activase to Rubisco ratio moved in opposite directions in the anti-activase and control plants as the leaves aged. Rubisco carbamylation was reduced in the anti-activase plants, consistent with expectations based on current ideas about the function of activase.

In the wildtype controls, CO_2-assimilation rates were consistent with the content of carbamylated Rubisco throughout leaf development. However, in the anti-activase plants, the average activity of carbamylated Rubisco sites was lower initially and declined markedly during leaf development so that, in the oldest leaves, carbamylated Rubisco turned over catalytically less than one-tenth as fast in the activase-deficient

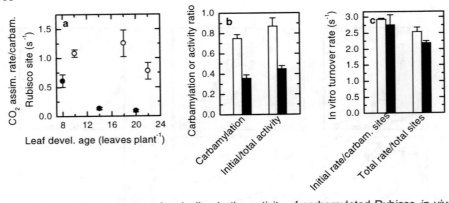

FIG. 3. Activase deficiency caused a decline in the activity of carbamylated Rubisco *in vivo* but not *in vitro*. Hollow symbols and bars refer to wildtype plants, solid symbols and bars to anti-activase plants. The turnover rate of carbamylated Rubisco sites inferred from gas-exchange measurements at 350 μbar CO_2 (a) is compared with Rubisco carbamylation status and activity measured after instantaneous extraction of frozen leaf powder from 18-leaf plants directly into assay solutions containing saturating CO_2 concentrations (b and c). Data from Ref. 4.

leaves as in the controls (Fig. 3a). This inhibition might be explained by an accumulation on carbamylated sites of tight-binding inhibitors that activase normally releases. However, inhibitors that bind tightly enough to need the assistance of activase to be released must be released slowly when unaided. *In vitro* measurements show no signs of lingering inhibition. The ratio of initial activity immediately after rapid extraction to fully activated activity was close to that predicted from the carbamylation status (Fig. 3b) and the turnover rates after extraction were similar in both genotypes (Fig. 3c). This was true even when the interval between extraction and assay was eliminated by direct addition of freeze-clamped, powdered leaf tissue to activity and carbamylation assays.

This strong inhibition that is readily apparent *in vivo* but vanishes instantly upon extraction is inconsistent with current models of Rubisco regulation and activase function. It is not a result of RuBP limitation or product inhibition (4) and is a specific response to activase deficiency, not seen in other photosynthetically impaired transgenic tobacco. It indicates that activase is required to stimulate catalytic turnover by carbamylated Rubisco *in vivo*, but not *in vitro*. The molecular details of this unsuspected role of activase will be interesting topics for future research. Perhaps activase functions to correct some kind of conformational error that Rubisco is prone to in the concentrated stromal environment but escapes from upon extraction. An alternative, more in keeping with current understanding of activase function, would be a requirement *in vivo* for activase to release an unknown inhibitor that is instantly destroyed by extraction. Revelation of these unsuspected complexities demonstrates the usefulness of comparing protein function *in vitro* and *in vivo*.

3.2 *Manipulation of photorespiration in tobacco by directed mutation of* rbcL

So far, no eukaryotic Rubisco has been expressed in properly folded, assembled and active form in any foreign host. This has hampered mutagenic study of the higher-plant enzyme. Chloroplast transformation (6) provides a way of circumventing this frustration and it has the additional advantage of providing the mutant enzyme within the whole-plant context for physiological study.

We aimed to induce as large a change in Rubisco's CO_2/O_2 specificity as possible without disabling the carboxylating capacity seriously. The latter requirement obviously precludes alterations to catalytically essential residues, such as Lys-201 (the site of carbamylation), Lys-175 (essential for RuBP enolisation), or Lys-334 (which stabilises the transition state for carboxylation of the enediol) (14). We chose Leu-335 which shares with Lys-334 a position at the apex of a flexible loop that closes over the bound substrate and makes contact with the phosphate group attached to C5 (15). We changed it to Val (see Section 2.2), guided by reasoning that this should induce subtle changes in the critical positioning of the neighbouring Lys-334 side chain and by previous mutagenesis with prokaryotic Rubiscos (16,17).

Chloroplast transformation depends on a poorly understood process by which the transformed plastome completely supplants the wildtype plastome. An intermediate phase occurs during which both genomes coexist and recombination between them can occur. This caused us considerable difficulty. Many spectinomycin-resistant plantlets that had the introduced EcoRV site (see Section 2.2) initially lost it during this phase, presumably through secondary crossovers between mutant and wildtype plastomes that restored the wildtype rbcL sequence but retained the aadA gene. This leads us to suspect that impaired Rubisco function is deleterious even during growth on sucrose-containing medium. Nevertheless, one plantlet retained the EcoRV site throughout the sorting out process. This plant and its clones retained the mutation and remained homoplasmic through a full generation. Sequencing of rbcL confirmed the presence of the intended changes. When transferred to soil, these plants grew very slowly, even in 0.3% CO_2, but they have recently managed to set seed.

Photosynthetic CO_2 assimilation was grossly impaired in the mutants and the CO_2-compensation point was increased approximately four-fold, implying that the mutant

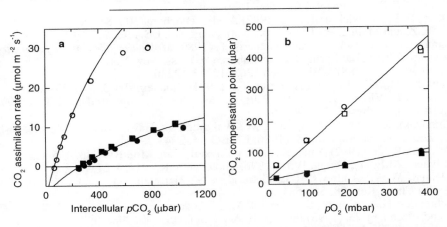

FIG. 4. Gas-exchange measurements with wildtype (O,□) and Val-335 (●,■) mutant plants. **a**, response of CO_2 assimilation to intercellular pCO_2 at 200 mbar pO_2; **b**, response of the CO_2-compensation point to pO_2.

Rubisco's CO_2/O_2 specificity was impaired (Fig. 4a). Consistent with this, the slope of the O_2 dependence of the CO_2 compensation point increased four fold in the mutant (Fig. 4b). Biochemical measurements with the isolated and purified mutant protein were consistent with these observations. Mutant plants had a similar Rubisco content to the wildtype but the maximum carboxylase activity and the CO_2/O_2 specificity of the mutant Rubisco were both reduced by over 70 percent. The Michaelis constants for CO_2, O_2 and RuBP were all reduced by the mutation. The carbamylation status of the mutant Rubisco was nearly twice that of the wildtype under the high-CO_2 growth conditions. In general, the parameters measured *in vitro* agreed well with those inferred from gas exchange. Further details are given in the contribution by S.M. Whitney, S. von Caemmerer, G.S. Hudson and T.J. Andrews elsewhere in these proceedings.

This work demonstrates the feasibility of site-directed mutagenesis with higher-plant Rubisco. In this case, the observations *in vivo* were entirely concordant with those *in vitro*. The impaired CO_2/O_2 specificity led to the expected increase in the CO_2 compensation point of the leaves. This engenders confidence in the current understanding that the linkage between photosynthesis and photorespiration is mediated by a common dependence on Rubisco. It also provides evidence that the stoichiometry between CO_2 release and O_2 uptake in photorespiration must be close to 0.5, as predicted by the conventional formulation of the photorespiratory glycolate pathway.

References

1   von Caemmerer, S., Evans, J.R., Hudson, G.S. and Andrews, T.J. (1994) Planta 195, 88-97
2   Mate, C.J., Hudson, G.S., von Caemmerer, S., Evans, J.R. and Andrews, T.J. (1993) Plant Physiol. 102, 1119-1128
3   Mate, C.J., von Caemmerer, S., Evans, J.R., Hudson, G.S. and Andrews, T.J. (1996) Planta 198, 604-613
4   He, Z., von Caemmerer, S., Hudson, G.S., Price, G.D., Badger, M.R. and Andrews, T.J. (1997) Plant Physiol. 115, 1569-1580
5   Deng, W.P. and Nickoloff, J.A. (1992) Analyt. Biochem. 200, 81-88
6   Svab, Z. and Maliga, P. (1993) Proc. Natl. Acad. Sci. USA 90, 913-917
7   Sambrook, J., Fritsch, E.F. and Maniatis, T. (1989) *Molecular Cloning. A Laboratory Manual*, 2 Ed., Cold Spring Harbor Laboratory Press, New York
8   Andrews, T.J. (1988) J. Biol. Chem. 263, 12213-12220
9   Paul, K., Morell, M.K. and Andrews, T.J. (1991) Biochemistry 30, 10019-10026
10 Kane, H.J., Viil, J., Entsch, B., Paul, K., Morell, M.K. and Andrews, T.J. (1994) Aust. J. Plant Physiol. 21, 449-461
11 Shinozaki, K., Ohme, M., Tanaka, M., Wakasugi, T., Hayashida, N., Matsubayashi, T., Zoutu, N. and Sugiura, M. (1986) EMBO J 5, 2043-2049
12 Portis, A.R., Jr. (1992) Annu. Rev. Plant Physiol. Plant Mol. Biol. 43, 415-437
13 Salvucci, M.E. and Ogren, W.L. (1996) Photosynth. Res. 47, 1-11
14 Cleland, W.W., Andrews, T.J., Gutteridge, S., Hartman, F.C. and Lorimer, G.H. (1998) Chem. Rev. 98, 549-561
15 Schreuder, H.A., Knight, S., Curmi, P.M.G., Andersson, I., Cascio, D., Brändén, C.-I. and Eisenberg, D. (1993) Proc. Natl. Acad. Sci. USA 90, 9968-9972
16 Terzaghi, B.E., Laing, W.A., Christeller, J.T., Petersen, G.B. and Hill, D.F. (1986) Biochem. J. 235, 839-846
17 Lee, G.J., McDonald, K.A. and McFadden, B.A. (1993) Protein Sci. 2, 1147-1154

PHOSPHORIBULOKINASE:
3-DIMENSIONAL STRUCTURE & CATALYTIC MECHANISM

Henry M. Miziorko, Jennifer A. Runquist, and David H.T. Harrison
Medical College of Wisconsin, Milwaukee, Wisconsin 53226

Key words: Calvin cycle; phosphoribulokinase; R.. Sphaeroides; site-directed mutagenesis; stromal enzymes; X-ray diffraction

Introduction

Phosphoribulokinase (PRK) catalyzes production of the Calvin cycle's CO_2 acceptor, ribulose 1,5-bisphosphate. As might be anticipated for a reaction so crucial to carbon assimilation, PRK is highly regulated in both the prokaryotic and eukaryotic organisms in which it functions. In eukaryotes, activity is modulated by a thioredoxin-mediated thiol/disulfide exchange. The prokaryotic enzyme is allosterically regulated, with NADH functioning as a positive effector while AMP and PEP function as negative effectors.

Enzyme function has been investigated in work employing both eukaryotic PRKs, which are typically dimers of 40 kDa subunits, and also prokaryotic PRKs, which are often octamers of 32 kDa subunits. Protein modification studies have implicated eukaryotic PRK's C16 and C55 in thiol/disulfide exchange (1). Neither residue is conserved in prokaryotic PRKs and these cysteines do not greatly influence catalysis (2). Mutagenesis work, directed either by affinity labeling results or by the observation of residue conservation in the sequence alignment of diverse PRK proteins, has implicated other active site residues. For example, prokaryotic PRK's H45 and R49 influence binding of the substrate, ribulose 5-phosphate (3) and the invariant acidic residues D42, E131, and D169 clearly support catalysis (4). To better establish a framework in which to understand these observations and in an attempt to eliminate ambiguity in functional assignments of residues involved in regulation or catalysis, we have determined the structure of R. sphaeroides PRK (5) and find that the enzyme adopts a fold typical of the nucleotide monophosphate (NMP) kinases.. A concise interpretation of available PRK solution structure data in the context of the X-ray structure is presented in this report. Additionally, structure/function assignments for PRK prompt predictions of functionally significant regions for other enzymes that exhibit NMP kinase folds.

Procedure

R. sphaeroides PRK was over expressed in E. coli and isolated as previously described (4). PRK crystallizes (5) in cubic space group (P432; a=129.6Å). A solvent-flattened MIR map was generated based on diffraction data from native enzyme, mercury and platinum derivatives. Iterative model building and refinement produced a final model with an R-value

G. Garab (ed.), Photosynthesis: Mechanisms and Effects, Vol. V, 3313–3318.
© 1998 Kluwer Academic Publishers. Printed in the Netherlands.

3314

of 21.3% and a free R-value of 28.4% at 2.5 Å resolution (6).

Results and Discussion

<u>Quaternary Structure and Subunit Contacts</u>. As anticipated, *R. sphaeroides* PRK has been demonstrated by X-ray diffraction analysis to consist of eight 32 kDa subunits. The octamer has two layers, four subunits on top of four two-fold rotated subunits on the bottom, displaced by about one eighth of a turn (Fig. 1). Thus, the octamer resembles a "pinwheel" in which each blade represents a two-fold symmetric dimer. This subunit pair appears to be the dimer conserved throughout evolution (Fig. 2). Contact between these two subunits (1760 Å^2) is much larger than that between a top layer subunit and the other adjacent bottom layer subunit (450 Å^2) or between two subunits in the same layer (640 Å^2).

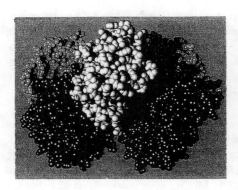

Fig. 1 (above). A view of the PRK octamer.

Fig.2 (right). Ribbon drawing of the PRK dimer believed to be conserved throughout evolution.

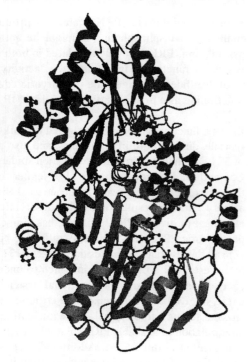

<u>Structure of the PRK Subunit</u>. Each subunit folds into a seven member mixed β sheet surrounded by α helices (Fig. 3); there is also an auxiliary antiparallel pair of β strands. Precedent with enzymes that exhibit the NMP kinase fold suggests the boundaries of the active site region. One side is defined by the C-terminal ends of the five parallel strands found in the β sheet; the opposite face is formed by the "lid" comprised of the helix immediately following the first of the parallel β strands as well as the helix-loop-helix region

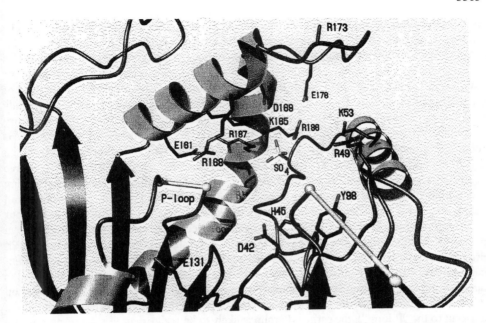

Fig. 3. Active site structure of PRK. One side of the active site cavity is defined by the edge of the central beta sheet. This region includes a Walker A or "P loop" motif depicted with a break in the alpha carbon backbone between residues 17 and 18. Glu-131 is situated at the end of a beta strand and falls within a Walker B consensus sequence. Asp-42, located at the C-terminus of the adjacent beta strand, is proposed to function as a general base in deprotonation of Ru5P's C1 hydroxyl. The opposite side of the active site cavity is formed by a mobile lid that includes Arg-168, Arg-173, Arg-186, Arg-187, and the catalytically essential, Asp-169.

that follows the fourth parallel strand. In the current PRK structure, there is no occupancy of ATP or sugar phosphate binding sites, merely an immobilized sulfate which may bind where Ru5P's phosphoryl group would normally be situated during turnover. In this unoccupied state, NMP kinases normally exhibit an "open" conformation, with a large span between the β sheet and the "lid" region; such an open conformation is evident in the PRK structure. Upon binding of substrate(s) or analogs, it has been well established that the space between the lid and the β sheet is diminished and active site residues become properly juxtaposed to support catalysis.

<u>Active Site Residues and Consensus Binding Motifs in PRK</u>. While lack of substrate site occupancy in the currently available structure precludes any definitive assignment of the active site apparatus, the results of site-directed mutagenesis work as well as the observation

of consensus nucleotide binding motifs allow some confidence in developing a self-consistent prediction of how certain structural elements are likely to contribute to this site. It has long been known that the N-terminal region of both prokaryotic and eukaryotic PRKs harbors a glycine-rich stretch of residues that fit the consensus sequence for a "P-loop" or "Walker A" motif. The "P-loop" of PRK is situated at the C-terminal end of the third of the five parallel β strands. Upon binding of the nucleotide substrate, the "P loop" is typically positioned over the β and γ phosphoryl groups. In eukaryotic PRKs, one of the regulatory cysteines (C16) is found in this "P loop". If a change in conformation of the "P loop" is required for successive substrate turnovers, it is now clear why tethering C16 to C55 in a disulfide linkage would impair catalysis in the eukaryotice enzyme. Recently, we have identified a second consensus sequence in PRK, namely a "Walker B" motif. The striking invariance of the EGLH sequence (residues 131-134 or 125-128 in prokaryotic and eukaryotic PRKs, respectively) contrasts with the lack of apparent conservation in the four preceding residues. However, invariably those four preceding residues are hydrophobic in nature. We predict that this qualifies the region as a "Walker B" nucleotide binding motif, prompting the assignment of the invariant acidic residue (E131) as a ligand to the divalent cation of the Mg^{++}-ATP substrate. The two orders of magnitude effect on k_{cat} that results upon substitution of this invariant glutamate supports this assignment. Moreover, based on this residue's position at the C-terminus of the second of the parallel β strands (and, thus, adjacent to the "P-loop"), the carboxyl group is likely to be well oriented with respect to the cation complexed with ATP's γ-phosphoryl. While recognition of consensus sequences aids in understanding the involvement of components in or near the second and third parallel β strands, the importance of the C-terminus of the first parallel β strand is suggested by mutagenesis results. D42 (D57 in eukaryotic PRK) is situated two residues after the end of this β strand. This residue is well positioned to function as a general base catalyst. The carboxyl group would be juxtaposed to Ru5P's C1 hydroxyl; after hydroxyl deprotonation, the resulting anion would be ideally positioned to attack ATP's γ-phosphoryl group. In addition to qualifying for such an assignment on a structural basis, D42 meets the functional criteria expected for such a key chemical step. Elimination of the carboxylate at this position results in diminution of activity by more than five orders of magnitude. Such an observation is even more impressive in view of the structural integrity of the D42 mutants, demonstrated by their ability to stoichiometrically bind ATP analogs in a fashion indistinguishable from wild-type PRK. Thus, the evidence for assigning the C-terminal edge of the β sheet as one side of PRK's active site is compelling.

Despite the substantial distance that separates the β sheet from the "lid" regions, there remains good reason to believe that these structural elements follow the NMP kinase fold precedent and move substantially closer to the β sheet during catalysis. Mutagenesis data clearly implicate H45 and R49 as interacting with Ru5P, since elimination of basicity at these positions correlates with substantial increases in $K_{m\ Ru5P}$. These residues are situated in the stretch of sequence between the first parallel β strand and the following helix. Given the strong rationale for D42's interaction with Ru5P's C1 hydroxyl, it is reasonable to assign

H45 and R49 to additional interactions with this substrate. The helix-loop-helix region that follows the fourth parallel β strand has also been implicated by mutagenesis work as important to catalysis. D169 is clearly involved in catalysis, since elimination of a carboxyl group from residue 169 diminishes k_{cat} by more than four orders of magnitude. The precise role of this residue awaits clarification by future experimental work. Basic residues in this region have also been implicated as active site components. Elimination of basicity contributed by the invariant arginines, R168 and R173 results in a >300-fold drop in k_{cat} in the case of R168Q and a 100-fold increase in $K_{m\ Ru5P}$ in the case of R173Q. Precise interpretation of these effects and unambiguous functional assignments to residues in the "lid" region may only be possible after additional structural results become available. Nonetheless, the argument that supports inclusion of this region as part of the active site is quite clear. Thus, while all elements of the PRK active site may not yet be identified, the general boundaries of the catalytic apparatus have been defined.

The elucidation of the PRK structure has been coincident with the reports of structures of several enzymes that catalyze metabolically unrelated but mechanistically similar reactions, namely, estrogen sulfotransferase, fructose 6-phosphate 2-kinase, thymidine kinase, and shikimate kinase; all of these enzymes exhibit the NMP kinase fold. Based on an analysis of the structures and any available functional data, a consensus is emerging. In all cases, a "P-loop" is found at the C-terminal end of the third parallel strand of the β sheet. Additionally, a "Walker B" acidic residue at the C-terminal tip of the second parallel strand has been assigned for several of these enzymes and a candidate residue for this assignment could be identified in other cases. Finally, while NMP kinases require no general base catalysis of their reactions, the transferases listed above are likely to require such assistance to facilitate deprotonation of the alcohol groups of their acceptor substrates. In the sequence following the C-terminus of the first parallel strand of these enzymes, a general base catalyst candidate commonly appears. A clear functional assignment is most obvious in the case of PRK but persuasive assignments, based on limited published mutagenesis results, can also easily be made for thymidine kinase (7) and estrogen sulfotransferase (8). Candidate residues are also suitably located in shikimate kinase (9). While no assignment of a general base has been offered for fructose 6-phosphate 2-kinase (10), the general base consensus emerging from these mechanistically related enzymes argues that this issue merits further scrutiny.

In summary, a combination of structural work and mechanistic investigation allows functional assignments and predictions for both prokaryotic and eukaryotic PRKs. Moreover, when the PRK results are interpreted in the context of the NMP kinase folding paradigm, a consensus emerges that suggests functional assignments for a wider array of enzymes that fold similarly but catalyze diverse transferase reactions.

(Supported in part by USDA's NRICRG-Photosynthesis program (HMM) and the Herman Frasch Foundation (DHTH))

References

1. Porter, M.A., Stringer, C.D., and Hartman, F.C. (1988) J. Biol. Chem. 263, 123-129
2. Milanez, S. Mural, R.J., and Hartman, F.C. (1991) J. Biol. Chem. 266, 10694-10699
3. Sandbaken, M.G., Runquist, J.A., Barbieri, J.T., and Miziorko, H.M. (1992) Biochemistry 31, 3715-3719
4. Charlier, H.A., Runquist, J.A., and Miziorko, H.M. (1994) Biochemistry 33, 9343-9350
5. Roberts, D.L., Runquist, J.A., Miziorko, H.M., and Kim, J.J.P. (1995) Prot. Sci. 4, 2442-2443
6. Harrison, D.H.T., Runquist, J.A., Holub, A., and Miziorko, H.M. (1998) Biochemistry 37, 5074-5085
7. Wild, K., Bohner, T., Folkers, G., and Schulz, G.E. (1997) Prot. Sci. 6, 2097-2106
8. Kakuta, Y. Pedersen, L.G., Carter, C.W., Negishi, M., and Pedersen, L.C. (1997) Nat. Struct. Biol. 4, 904-908
9. Krell, T., Coggins, J.R., and Lapthorn, A.J. (1998) J. Mol. Biol. 278, 983-997
10. Hasemann, C.A., Istvan, E.S., Uyeda, K., and Deisenhofer, J. (1996) Structure 4, 1017-1029

THREE-DIMENSIONAL STRUCTURE OF PHOSPHOENOLPYRUVATE CARBOXYLASE FROM *Escherichia coli* AT 2.8Å RESOLUTION

Yasushi Kai*, Hiroyoshi Matsumura*, Tsuyoshi Inoue*,
Kazutoyo Terada[‡], Yoshitaka Nagara*, Takeo Yoshinaga[†],
Akio Kihara[‡], Katsura Izui[‡ ¶]

*Department of Materials Chemistry, Graduate School of Engineering, Osaka University, Suita, 565-0871, Japan
[†] Department of Public Health, Graduate School of Medicine,
[‡] Department of Chemistry, Graduate School of Science, and
[¶] Division of Applied Biosciences, Graduate School of Agriculture, Kyoto University, Sakyo-ku, Kyoto, 606-8501, Japan

Key words: PEPC, C4, CO_2 concentration, single crystals, X-ray diffraction, elevated CO_2

1. Introduction

Phosphoenolpyruvate carboxylase (PEPC; EC 4.1.1.31) irreversibly catalyzes the carboxylation of phosphoenolpyruvate to form oxaloacetate and inorganic phosphate using Mg^{2+} as a cofactor.

$$HOCO_2^- + \ ^-OOC\text{-}\underset{O^-\ PO_3^{2-}}{C}=CH_2 \xrightarrow{Mg^{2+}} \ ^-OOC\text{-}\underset{O}{C}\text{-}CH_2\text{-}CO_2^- + HPO_4^{2-}$$

The enzyme is widespread in all plants and many kinds of bacteria, and plays anaplerotic functions by replenishing C_4-dicarboxylic acids for the syntheses of various cellular constituents and for the maintenance of the citric acid cycle. Higher plants have several isoforms of PEPC with different kinetic and regulatory properties which correlate with their respective roles in cellular metabolism. Much attention has been paid on this enzyme, since in C_4 plants such as maize and sugarcane, and in CAM plants such as pineapple and cactus, one of the isoform PEPCs is abundantly expressed and plays a key role in the C_4- and CAM-photosynthesis. Molecular structural studies on PEPC are thus expected to provide clues for the development of innovative techniques for the augmentation of productivity of photosynthetic organisms and for the conversion of this greenhouse gas into useful organic compounds.

PEPCs from various sources are usually composed of four identical subunits whose molecular masses are 95 to 110 kD. The primary structure was first deduced from a cloned DNA of *Escherichia coli* in 1984 (1). Ever since more than twenty molecular species of PEPC have been elucidated for their primary structures including the ones from maize, cyanobacteria and extreme thermophile. Alignment of all amino acid sequences available in 1994 and construction of phylogenetic tree by neighbor-joining method revealed that PEPC had evolved from the same ancestral origin and that the amino acid identities and

G. Garab (ed.), Photosynthesis: Mechanisms and Effects, Vol. V, 3319–3324.
© 1998 *Kluwer Academic Publishers. Printed in the Netherlands.*

similarities among them were more than 31% and 52%, respectively (2). Especially *E. coli* PEPC is very similar to plant PEPC in the primary structure except the extra residues on N-terminus in the latter. Thus, the three-dimensional structure of *E. coli* PEPC can be directly applied to plant PEPC.

Although X-ray crystallographic analysis of PEPC is indispensable together with functional analysis by site-directed mutagenesis (3) for studies on reaction mechanism and regulation by allosteric effectors or covalent modification, it has been hampered mainly due to the unavailability of a large amount of PEPC of high purity. This obstacle had been overcome for *E. coli* PEPC by establishing a novel and simple method of preparation (4) and first crystallization was reported in 1989 (5). After the crystallization report, however, we met another big obstacles on the way to the three-dimensional structure of PEPC. They were the polymorphism in the crystallization process and the unstability of the crystals obtained. The obstacles were finally overcome by the fine modifications on the precipitant and the additives of the crystallization solution together with the use of synchrotron radiation. Here we report the three-dimensional structure of PEPC complexed with the allosteric inhibitor, L-aspartate, determined by X-ray diffraction method at 2.8 Å resolution.

2. Procedure

Crystallization and diffraction data.
Single crystals of PEPC from *E.coli* belong to orthorhombic space group of *I*222 with unit cell parameters of $a = 117.6$, $b = 248.4$, and $c = 82.7$ Å. The asymmetric unit contains one PEPC monomer. Thus, the tetrameric PEPC molecule has D_2 crystallographic symmetry. The crystals for structure determination were obtained by the hanging-drop vapor diffusion method of the following conditions. The mother solution in the 6 μl droplet contains 10 mg/ml protein in 50 mM Tris-HCl (pH 7.4) containing 6 mM Na-L-Asp, 45 mM $CaCl_2$, 0.6 mM DTT, and 10% of PEG300. The droplet was equilibrated against the 500 μl reservoir solution containing 2.5 mM Na-L-Asp, 90 mM $CaCl_2$, 0.25 mM DTT, and 15% of PEG300 in the same buffer.

X-ray diffraction intensities were measured at station BL6B of the Photon Factory, Tsukuba, Japan, using the Sakabe's Weissenberg camera for macromolecular crystallography and imaging plates as a detector. The data were processed using DENZO and scaled with the program SCALEPACK. The crystal structure was determined by multiple isomorphous replacement method. Three useful heavy atom derivatives were obtained by soaking crystals in the presence of 1 mM methylmercuric chloride, 1 mM mersalyl acid, and 1 mM EMTS, respectively. Heavy atom parameters including positions, occupancies and temperature factors were refined using program MLPHARE (figure of merit of 0.47 at 2.8 Å), and solvent-flattening and histogram matching were performed with the program DM.

Model building and refinement.
The electron density map phased by three mercuric derivatives had enough quality to locate most of the secondary structures included in the enzyme. Interpretation of the map and building of the atomic model were proceeded by using the graphics program O. The initial structure model including partially located secondary structures was refined with X-PLOR, and the phases obtained from the refined structure were combined with the original phase probability distributions using program SIGMAA. The refinements using X-PLOR and REFMAC, and manual modification of the model structure were repeatedly carried out up to an R-factor of 21.9% for the significant 26,242 reflections between 10.0 and 2.8 Å resolutions. Using a 5% reflection test set (1,409 reflections) the R_{free} value is 25.9%. The r.m.s. deviation from standard values of bond lengths and angles are 0.011 Å and 1.4 °,

respectively. Monomer of PEPC has the α/β barrel motif with eight β-strands connected by the chains including one to several α-helices. Whereas the whole monomer structure was fitted well on the electron density map, it is interesting that a loop including seven residues (Lys702- Gly708) near the probable active site is missing. The high mobility of these residues might contribute to the catalytic function in its active form as will be described later. The final model includes 6,898 non-hydrogen protein atoms and 39 solvent molecules.

3. Results and Discussion

Tetramer structure.

The overall structure of PEPC is shown in Fig. 1. Four monomer subunits are related by the crystallographic 222 symmetry resulting the molecular symmetry of D_2 for the enzyme. Two of four subunits, I and II or III and IV, are related by rather close contacts in contrast with alternative pairs of subunits. The contact surface area between I and II subunits or I and III subunits was calculated to be 1,610 or 520 $Å^2$, respectively, by program GRASP based on a probe radius of 1.4 Å. In the center of the tetramer of PEPC, vacant holes are found. So, the tetramer structure of PEPC is described as "dimer of dimers". From the studies on the site-directed mutagenesis hitherto proceeded, the conserved Arg438 is found as the residue essential to maintain tetrameric structure of the enzyme. By the replacement of Arg438 with Cys, the dissociation of tetramer structure to dimer is caused. After completed the X-ray structure determination, the Arg438 was searched in the tetrameric structure of the enzyme. Then, the Arg was found just on the boundary of two subunits with rather loose contact, I and III or II and IV subunits. The intersubunit interactions through Arg438 are the salt bridges between the residue and Glu433 of the neighboring subunit. By the replacement of Arg438 with Cys, the salt bridges may be broken resulting the dissociation of tetrameric enzyme into two dimers. The revealed tetrameric structure of PEPC readily explains the inherent tendency of the enzyme to dissociate into rather stable dimer (6).

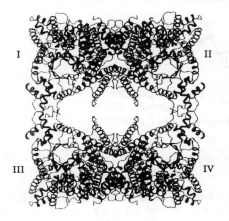 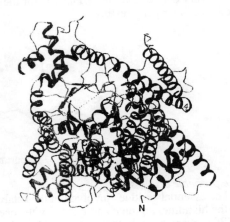

Fig. 1. Tetramer structure of PEPC   Fig. 2. Monomer structure of PEPC

Momoner structure.

The monomer structure of PEPC from *E. coli* is shown in Fig. 2. The characteristic features of the structure are an eight-stranded β-barrel and abundant numbers of α-helices.

Among the secondary structures included in the enzyme, no β-strand is found except eight included in a barrel. In contrast to the limited number of β-strand, a total of 41 α-helices are found. A total number of residues included in the α-helices is 572, which is 65% of the whole amino acid residues of the enzyme, while the corresponding numbers are 40 residues and 5% for β-strands. Many of the α-helices come together on the C-terminus side of the β-barrel, while scarce of them are on the N-terminus side of the barrel. The aspartate, the regulatory effector molecule, was found among the α-helices on the C-terminus side of the barrel.

The aspartate binding site.

L-aspartate is one of the allosteric effector molecules for PEPC and causes the inhibition of the catalytic activity. Four amino residues, Lys773, Arg832, Arg587, and Asn881, participate in the binding of the aspartate (Fig. 3). Lys773 is one of the two lysine residues strictly conserved in all PEPCs hitherto reported. When Lys773 was replaced with alanine, enzymatic activity was completely lost and the accumulation of the mutant enzyme was severely decreased (3). Lys773 is hydrogen bonded to the carboxyl group in the side chain of aspartate by the full extension of its side chain. To the same carboxyl group Arg832, which is highly conserved in PEPC (2), is also hydrogen bonded. Arg587 is the second arginine in the unique sequence of GRGGXXGRGG (XX; TV, SI, or SV), which is definitely conserved in PEPC (2). The loop with six glycine residues serves a flexible long arm for functional two arginine residues in this sequence. Arg587 is hydrogen bonded to the carboxyl group of the aspartate. The replacement of Arg587 by Ser causes a virtual loss of the catalytic activity to form oxaloacetate (7). The hydrogen bond between Arg587 and the asparatic acid is full of suggestiveness to show that the Arg587 indispensable for enzymatic activity is trapped by the inhibitory effector molecule of aspartate. Therefore, the active site of the enzyme may be limited within the region swept by the flexible long arm of Arg587 supported on the flexible loop with the unique amino acid sequence in PEPC. The fourth residue supporting the aspartate is Asn881, which is the second next residue to the C-terminus and is completely conserved in all PEPCs (2). The participation of the very end residue of the enzyme for the recognition and/or binding of the inhibitor molecule is very interesting.

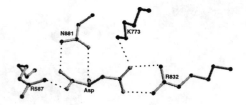

Fig. 3. Aspartate binding site.

The probable active site.

The essential residues involved in the catalytic function of PEPC have been suggested to be the histidine, arginine and lysine by the chemical modifications of these residues (8). In PEPCs, there are only two conserved histidines, His138 and His 579. By the site-directed mutagenesis on these histidines revealed the essential role of these residues on the catalytic activities (9, 10). The locations of two conserved histidines are shown in Fig.4 together with the important residues for catalytic activity and inhibitory effector of aspartate. These residues and aspartate are close together on the C-terminus side of the β–barrel. Just after

His579, unique amino sequence of GRGGSIGRGG was found including two arginine residues important for catalytic activity of the enzyme. This glycine-rich loop may essentially be mobile even though the aspartate deprives the loop from the catalytic site and restricts its mobility. Arg587 is stretched toward the aspartate and away from two histidines in this inactive form of PEPC.

The amino acid residues untraced in the crystal structure are only limited, three on the N-terminus and seven of the loop from Lys702 to Gly708. Therefore, this missing loop is very suggestive for the active site of the enzyme. In the sequence of the loop, Arg703 is completely conserved in all the PEPCs, and Lys702 and Arg704 are well conserved in the forms of Lys or Arg. These functional residues are sandwiched by Ala701 or Ser701 and Gly707Gly708, which make this loop flexible. In the aspartate complex of PEPC, this missing loop may have no role to fix its structure. The importance of this mobile loop in catalytic activity was shown by several observations. Firstly, when PEPC was treated with trypsin, the enzyme was readily inactivated concomitant with the single cleavage at carboxyl side of Arg703 (Yoshinaga et al. to be published). Secondly, when both Arg703 and Arg704 were replaced to Gly, the k_{cat} value decreased about 20-fold but the k_{cat} value for the unfavorable reaction, bicarbonate-dependent PEP hydrolysis reaction increased 10-fold. Furthermore, K_m value for bicarbonate increased more than 60-fold (Tsumura et al. to be published). When substrate molecules come close to interact with the active site, the mobile loop may serve the side chains of the functional residues together with those on the loop supporting Arg581 and Arg587. Figure 4 also includes some conserved residues, Arg396 essential for PEPC function (11) and Lys546 associated with bicarbonate-binding (12), together with the conserved Arg residues, Arg581 and Arg699. So, the active site of PEPC may probably be in the region around the C-terminus side of the β-barrel.

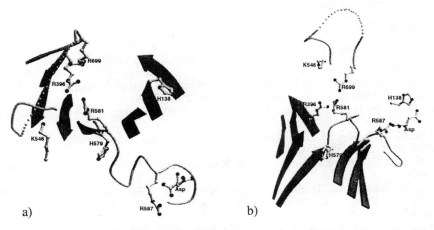

Fig. 4. Probable active site. a) Top view, and b) Side view.

4. Conclusion and perspective

The first three-dimensional structure of PEPC was determined by X-ray diffraction method for the enzyme obtained from *E. coli* at 2.8 Å resolution in the form of L-aspartate complex. The tetrameric structure of PEPC is well interpretable as "dimer of dimers" with the total molecular symmetry of D_2. Monomer of PEPC has the α/β barrel motif with eight β-strands surrounded by many helices. The active site of PEPC is assigned on the C-terminus side of the β-barrel, since most of the catalytically essential residues identified by site-directed mutagenesis are located in this region. A missing loop crossing over this side of the barrel was also shown to be involved in the catalytic function. L-Aspartate, the inhibitory effector molecule, is bound close to the probable active site fixing catalytically essential Arg587, which is on the conserved glycine-rich sequence, GRGGXXGRGG, unique to PEPC. The logical implication of this finding is that the allosteric inhibition by L-aspartate is exerted at least in part by trapping the flexible loop away from the catalytic site. The C-terminus chain of PEPC forms highly hydrophobic α-helix and is embedded in the hydrophobic pocket of the enzyme. The very end of C-terminus chain is fixed through the hydrogen bond to the complexed L-aspartate.

The detailed molecular mechanism for the enzymatic function of PEPC depends on the determination of three-dimensional structure of the active form of PEPC, which will be achieved by the complexation of PEPC with a substrate analogue, e.g. 3,3-dichloro-2-dihydroxyphosphinoylmethyl-2-propenoate, Mg^{2+}, and allosteric activators such as acetyl-CoA and fructose 1,6-bisphosphate (13) . The crystallization of the active form of PEPC is in progress.

Among ca. 900 amino acid residues of PEPCs from a wide range of sources, 99 and 139 residues are identical and similar to one another and molecular evolutionary analysis suggested the divergence from the common ancestral origin. Thus the basic folding structure is supposedly very similar to one another. The functional differentiation such as kinetic parameters and the kind of regulatory effectors may have been realized by minor substitutions of amino acid residues or by additions of extra segments. In fact, the N-terminus extension which carries the site of regulatory phosphorylation is ubiquitously found in higher plant PEPCs but is lacking in bacterial PEPCs. To elucidate the molecular mechanism of regulatory phosphorylation the X-ray crystallographic analysis of maize PEPC is also now in progress.

References

1. Fujita, N., Miwa, T., Ishijima, S., Izui, K. & Katsuki, H. (1984). *J. Biochem. (Tokyo)* **95**, 909-916.
2. Toh, H., Kawamura, T. & Izui, K. (1994). *Plant, Cell and Environment* **17**, 31-43.
3. Yano, M. & Izui, K. (1997). *Eur. J. Biochem.* **247**, 74-81.
4. Terada, K., Fujita, N., Katsuki, H. & Izui, K. (1995). *Biosci. Biotechnol. Biochem.* **59**, 735-737.
5. Inoue, M., Hayashi, M., Sugimoto, M., Harada, S., Kai, Y. & Kasai, N. (1989). *J. Mol. Biol.* **208**, 509-510.
6. Yoshinaga, T., Teraoka, H., Izui, K. & Katsuki, H. (1974). *J. Biochem. (Tokyo)* **75**, 913-924.
7. Yano, M., Terada, K., Umiji, K., & Izui, K. (1995). *J. Biochem.* **117**, 1196-1200.
8. Chollet, R., Vidal, J. & O'Leary, M. H. (1996). *Annu. Rev. Plant Physiol. Plant Mol. Biol.* **47**, 273-298.
9. Terada, K. & Izui, K. (1991). *Eur. J. Biochem.*, **202**, 797-803.
10. Terada, K., Murata, T. & Izui, K. (1991). *J. Biochem.*, **109**, 49-54.
11. Gao, Y. & Woo, K. C. (1996). *FEBS Lett.* **392**, 285-288.
12. Gao, Y. & Woo, K.C. (1995). *FEBS Lett.*, **375**, 95-98.
13. Izui, K., Taguchi, M., Morikawa, M. & Katsuki, H. (1981). *J. Biochem. (Tokyo)* **90**, 1321-1331.

Induction of CO$_2$ and Bicarbonate Transport in a Green Alga

Brian Colman[1], Gale G. Bozzo[1] and Yusuke Matsuda[2]

[1] Department of Biology, York University, Toronto, Ontario
Canada. M3J 1P3
[2] Department of Chemistry, Kwansei-Gakuin University,
1-1-155 Uegahara, Nishinomiya, Japan 662-8501

Key words: Acclimation, *Chlorella kessleri*, C metabolism, elevated CO$_2$, gene regulation, signal transduction.

1. Introduction

A number of microalgae and cyanobacteria have been shown to have active transport systems for the uptake of CO$_2$ and bicarbonate from the external medium. Both transport systems have been found to be repressed when these phototrophs are grown on high concentrations of CO$_2$.

The acclimation of high CO$_2$-grown cells to air results in an increased affinity of the cells for dissolved inorganic carbon (DIC). In a detailed study of the acclimation of *Chlorella ellipsoidea* to air, it was shown that the affinity for both CO$_2$ and bicarbonate increased rapidly and that these changes were correlated with a decrease in the CO$_2$ compensation point and an increased capacity to accumulate intracellular pools of acid-labile inorganic carbon (1). This induction of transport systems was inhibited by cycloheximide indicating the requirement for *de novo* cytoplasmic protein synthesis. Similar changes have been demonstrated in other *Chlorella* species (2).

Protein synthesis appears to be initiated by some signal triggered by the limitation of inorganic carbon supply to the cells, but the nature of the exact critical conditions and the signal produced are not known. During the adaptation period the available DIC in the medium drops to very low levels, causing a change in the CO$_2$/O$_2$ ratio and thus increasing the rate of oxidation of RuBP and the production of glycolate. It has been suggested that the increase in the internal concentration of glycolate or some other photorespiratory pathway intermediate may trigger the protein synthesis necessary for the induction of the DIC transport systems (3). However, if this is the signal, then induction of transport would only occur in the light and, while light has been reported to be required

G. Garab (ed.), Photosynthesis: Mechanisms and Effects, Vol. V, 3325–3330.
© 1998 *Kluwer Academic Publishers. Printed in the Netherlands.*

for induction in *Chlamydomonas* (4), it has been shown that induction occurs in the dark in *C.ellipsoidea* (5). The cells of *C.ellipsoidea* appear to respond to the CO_2 concentration in the medium and induction occurs only when the CO_2 concentration falls below a critical level (5). In support of this idea, mutants of *C.ellipsoidea*, obtained by X-ray irradiation, were found to be insensitive to CO_2 and had the capacity to actively transport DIC even when grown on high CO_2 (6). It was further demonstrated that a closely-related species, *C.saccharophila*, was also CO_2-insensitive (7).

It is possible that algae in which active DIC transport is repressed by high CO_2 during growth have a CO_2 'sensor' at the cell membrane. Occupation of the active site of this sensor by CO_2 might then produce a signal to repress the production of proteins of the active DIC transport systems.

In an effort to demonstrate the generality of the conditions for the repression of transport, we have examined the induction of active transport during acclimation to air of *Chlorella kessleri*, a species which, when grown on air, has high rates of bicarbonate transport.

2. Materials and Methods

Chlorella kessleri (UTEX 1808) was grown on Bold's basal medium at pH 6.6 and bubbled with air supplemented with 5% CO_2. To examine changes in the photosynthetic parameters of the cells during adaptation to air, cells were harvested and resuspended in growth medium at pH 6.6 and bubbled with air, without supplementary CO_2 or with CO_2-free air. The photosynthetic oxygen evolution rate of cells was determined with a Clark-type O_2 electrode at pH 7.8, a dissolved inorganic carbon (DIC) concentration of 50 µM, a light fluence of 400 µmol $m^{-2} s^{-1}$ and at $25°C$; the stimulation of O_2 evolution on the addition of 10 µg mL^{-1} bovine carbonic anhydrase (CA) was used as a measure of active CO_2 transport while the rate without added CA were used as a measure of HCO_3^- uptake.

3. Results and Discussion

Changes in the rate of photosynthetic O_2 evolution were determined over a period of 12h during the adaptation of high CO_2-grown cells to air (0.035% CO_2) or to CO_2-free air. The cells initially had a low photosynthetic rate both with and without added bovine CA, but both rates increased rapidly with time, without a perceptible lag, and the maximum photosynthetic rate, i.e. complete induction of both CO_2 and bicarbonate transport was reached over 5 to 6 h (Fig.1). The rate of induction was the same for cells acclimating to air and to CO_2-free air, although in the latter case the maximum rate of photosynthetic O_2 evolution was higher than that on air. This result indicates that the CO_2 concentration in the suspending medium during the period of adaptation has a determining effect on the rate of photosynthesis, presumably through its effect on the DIC transport systems.

The induction of DIC transport in *C.kessleri* (Fig.1) occurred much more rapidly than was observed in *C.ellipsoidea* where induction was not complete after 8 h adaptation and was acheived after an initial lag of 2 to 3 h. The results indicate clearly that the photosynthetic rate, after one hour of adaptation, exceeds the spontaneous rate of CO_2 production and is therefore dependent on bicarbonate uptake while the rate of CO_2 transport (in the presence of added bovine CA) increases rapidly and quickly surpasses that of bicarbonate transport. It should be noted that this species does not have an external CA.

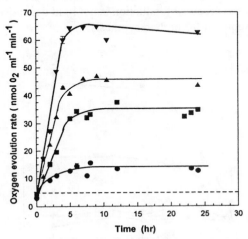

Figure 1.   Rates of photosynthetic O_2 evolution of high CO_2-grown cells during adaptation to air or to CO_2-free air determined at 50 μM DIC, pH 7.8 and 25°C. Rates of O_2 evolution in air-adapted cells without (●) and with added CA (■) and those of CO_2 -free air-adapted cells without (▲) and with added CA (▼). The calculated rate of spontaneous CO_2 formation from bicarbonate (------).

The induction of DIC transport was found to be inhibited when the cells were allowed to adapt to air in the presence of 5 μg mL^{-1} cycloheximide but not in the presence of 400 μg mL^{-1} chloramphenicol. This result indicates that induction was dependent on *de novo* cytoplasmic protein synthesis as in previous studies (1,2).

During the course of adaptation the DIC concentration in the growth medium declined markedly from 5 mM to 30 μM in air-bubbled cultures and to almost zero in the CO_2-free air-bubbled cultures. Because the high CO_2-grown cells have a limited capacity for DIC uptake, this drop in external DIC concentration would cause a profound change in the CO_2 / O_2 ratio so that it would favour the the production of glycolate and products of the photorespiratory pathway. However, cells grown on, and maintained on high CO_2 when incubated with 5 mM aminooxyacetate or with 10 mM isonicotinyl hydrazide, both of which inhibit the photorespiratory pathway, did not cause any induction of active DIC transport, although it is known that isonicotinyl hydrazide causes a significant glycolate build up and release in *C.kessleri* (unpublished results).

If the signal for DIC transport induction is due to an increase in an intermediate of the photorespiratory pathway then induction should only occur when cells are allowed to adapt in the light. To further test this possibility, high CO_2-grown cells were transferred to air and allowed to adapt in the dark (Fig.2). The time course of adaptation of cells in the dark was similar to that of cells in the light except that the maximum rate of photosynthetic O_2 evolution in dark-adapted cells was lower than that of cells adapted in light (Fig.2). Clearly the process of induction is not light dependent but may depend on light for the energy required for protein synthesis.

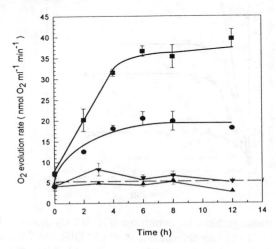

Fig.2. Rates of photosynthetic O_2 evolution of high CO_2-grown cells during adaptation to CO_2-free air in the dark. Photosynthetic rates were determined at pH 7.8, 50 µM DIC and 25°C. Cells maintained on CO_2-free air with (■) and without added bovine CA (●): cells maintained on high CO_2 with (▼) and without added bovine CA (▲).

The finding that induction of DIC transport can take place in the dark and that it occurs independently of the production of metabolites in the light suggests that the alga, like *C.ellipsoidea* responds directly to its immediate external environment, that is to the concentration of total DIC, or the bicarbonate concentration or the concentration of CO_2.

The DIC and CO_2 concentrations at which the induction of DIC transport occurred were determined by transferring high CO_2-cells to media of precisely defined pH and DIC concentration and allowed to adapt for 5.5 h. This period of adaptation was chosen because DIC transport was found to be fully induced after this period of time (Fig.1). During adaptation the CO_2 and DIC concentrations in the medium were closely monitored and maintained constant by adjusting the pH and the inflow CO_2 concentration. After the adaptation period the photosynthetic rate of the cells, with and without added bovine CA, was determined as measures of the presence of CO_2 and bicarbonate transport, respectively. Critical CO_2 concentrations were determined at 10 to 12 concentrations over the range of zero to 125 μM CO_2 at pH 6.6 and pH 7.5. The parameters of CO_2 and bicarbonate transport induction are given in Table 1, with comparable data for *C.ellipsoidea*.

Table 1. Concentrations of CO_2 at which half maximum DIC transport occurs in the presence of added CA and at which bicarbonate transport is first observed in high CO_2-grown cells after 5.5 h adaptation at various pHs.

pH	5.5	6.6	7.5
	CO_2 Concentration (μM)		
[CO$_2$] at half maximum DIC transport:			
C.kessleri	-	21	25
C.ellipsoidea	71	72	73
[CO$_2$] at which HCO$_3^-$ first detected:			
C.kessleri	-	76	70
C.ellipsoidea	60	63	59

The initial point at which HCO_3^- transport was recorded, was that CO_2 concentration at which the rate of photosynthesis (without added CA) exceeded the calculated rate of spontaneous CO_2 production in medium at a concentration of 50 μM DIC. Since the stimulation of photosynthesis by added CA is due to the an increased supply of CO_2 to the CO_2 transporter, the CO_2 concentration for half maximum induction is a good measure of the induction of CO_2 transport.

These results indicate that while the total DIC in the medium increases with pH, the concentration of CO_2 at which induction occurs remains relatively constant with an increase in pH. In *C.kessleri*, HCO_3^- transport is induced at a slightly higher CO_2 concentration than in *C.ellipsoidea*. This is consistent with the finding that the affinity of *C.kessleri* cells for HCO_3^- is much higher than that of *C.ellipsoidea* (8).

The results with *C.kessleri* are consistent with previous results obtained with *C.ellipsoidea* in that, both CO_2 and HCO_3^- transport appear to be induced at low and constant CO_2 concentrations in the suspending medium, irrespective of the pH. This strongly suggests that the concentration of free CO_2 in the medium is the critical criterion for induction and constitutes the initial trigger for the induction of DIC transport. It also suggests that the cells can monitor the concentration of free CO_2 in the external medium by means of some type of sensor and it is possible that membrane proteins in these *Chlorella* species might change conformation by interaction with CO_2 to transmit a signal to the nucleus which triggers the induction of DIC transport proteins.

This work was supported by grants from the Natural Sciences and Engineering Research Council of Canada.

4. References

1. Matsuda,Y. and Colman, B. (1995) Plant Physiol. 108, 247-252
2. Shiraiwa, Y. and Miyachi, S. (1985) Plant Cell Physiol. 26, 543-549
3. Marcus, Y., Harel, E. and Kaplan, A. (1983) Plant Physiol. 71, 208-210
4. Spalding, M.H., and Ogren, W. (1982) FEBS Lett. 145, 41-44
5. Matsuda, Y. and Colman, B. (1995) Plant Physiol. 108, 253-260
6. Matsuda, Y. and Colman, B. (1996) Plant Physiol. 110, 1283-1291
7. Matsuda, Y. and Colman, B. (1996) J.Exp.Botany 47, 1951-1956
8. Matsuda, Y., Williams, T.G. and Colman, B. Plant, Cell & Environ. In Press

OXYGEN-INSENSITIVE GROWTH OF ALGAE WITH AND WITHOUT CO2-CONCENTRATING MECHANISMS

Raven J A (1), Kübler J E (1), Johnston A M (1,2), Poole L J (1), Taylor R (1) and McInroy S G (1)
(1) Department of Biological Sciences, University of Dundee, Dundee DD1 4HN, Scotland
(2) Scottish Crop Research Institute, Invergowrie, Dundee DD2 5DA, Scotland

Key words: active oxygen, anoxia, C_3, CO_2 concentrating mechanisms, O_2 evolution

1. Introduction

The present atmosphere contains some 0.208 moles of O_2 per mole of total gas. This normoxic situation corresponds to some 21 kPa O_2 with the standard sea-level atmospheric pressure of 101.3 kPa, which is equivalent to an equilibrium concentration in solution of between 0.1 mol O_2 m^{-3} for very saline water at close to the upper temperature limit for eukaryotic O_2-evolvers (e.g. the Dead Sea) and 0.4 mol O_2 m^{-3} for freshwater close to the freezing point (1,2). Physical phenomena such as decreasing atmospheric pressure with altitude and pressure from breaking waves and from formation of ice from snow followed by melting can respectively produce dissolved O_2 which are respectively higher and lower than the equilibrium value. However, it is biological activity which produces the most extreme O_2 concentration in the biosphere. Hypoxia and thence anoxia develop when the rate of organic C supply for growth and maintenance of organisms exceeds the O_2 supply needed for aerobic performance of these functions. Hyperoxia of up to several times the air-equilibrium value can occur in (i) water bodies containing O_2-evolvers with abundant inorganic C and restricted O_2 exchange with the atmosphere (ii) cells and tissues of organisms with CO_2 concentrating mechanisms, including CAM and PEPck- and NADme-C_4 plants and (iii) a combination of (i) and (ii) (3). Thus, the photosynthetic apparatus of many of the 325,000 or so currently described species of O_2-evolving organism (4) spend at least part of their life cycle under anoxic, hypoxic or hyperoxic conditions. There does not seem to be conclusive evidence that any photosynthetic eukaryote can complete its life cycle in the absence of O_2, using fermentative chemoorganotrophy in the dark, or photosynthetic partial reactions other than PsII for photoorganotrophy or photolithotrophy with H_2 or H_2S as electron donor, in the light (5). However, photolithotrophy with H_2S as electron donor can support photolithotrophic growth in anoxia in some cyanobacteria (6). By contrast, many O_2-evolvers (e.g. some aquatic organisms, PEPck- and NADme-C_4 plants and CAM plants) can complete their life cycle with some or all of their photosynthetic apparatus exposed to hyperoxia (3).

G. Garab (ed.), Photosynthesis: Mechanisms and Effects, Vol. V, 3331–3337.
© *1998 Kluwer Academic Publishers. Printed in the Netherlands.*

One influence of O_2 concentration on the growth of O_2-evolvers is on the essential O_2 assimilation by cytochrome oxidase related to aerobic ATP synthesis and by the oxygenases involved in essential biosyntheses (2, cf.6). The cytochrome oxidase requirement for O_2 is fulfilled by less than 10 mmol O_2 m^{-3} unless there are O_2 uptake sites separated from the O_2 source by a transport limitation (2,7), while some oxygenases have a very low affinity for O_2 (8). However, few oxygenases have a half-saturation O_2 concentration in excess of 1 mol m^{-3} (8) and such enzymes could generally compensate for a low O_2 affinity by a higher enzyme concentration per unit biomass which would permit biosynthesis to occur at low O_2 levels (2).

The other influence of O_2 concentration on the growth of O_2-evolvers is inhibition by high O_2 levels. Leaving aside diazotrophy (3), O_2 can inhibit growth via the oxygenase activity (competitive with CO_2) of RUBISCO, and by generating active oxygen species ($O_2^{\bullet-}$, H_2O_2 and organic peroxides and, especially, $^1O_2$ and $^{\bullet}OH$). In the case of RUBISCO, organisms (algae, C_3 higher plants) with diffusive CO_2 entry show growth responses to O_2 in the range 1-40 kPa and normal air CO_2 in the gas phase which are at least in semiquantitative agreement with RUBISCO kinetics *in vitro* for the organism and with effects of O_2 on short-term photosynthesis (3). By contrast, when organisms with a CO_2 pump are investigated, or with very high CO_2 levels when CO_2 movement to RUBISCO is wholly by diffusion, there is little or no growth response to O_2 variation in the range 1-40 kPa O_2 (3). Any cost (in terms of growth rate) of unrepaired damage by active oxygen species, or of the avoidance or repair of damage, is not detectable in the range 1-40 kPa O_2 (3). However, at higher O_2 levels there are effects on growth rate which may be attributable to costs of avoidance, tolerance or repair of damage by active oxygen species. Raven *et al.* (3) review data on *Chlorella sorokiniana* grown on 5 kPa CO_2 where the wild type has growth rate unaffected by O_2 up to 70 kPa, while a mutant with higher superoxide dismutase levels maintains its growth rate up to 95 kPa O_2. While there are relatively few data on $^{18}O_2$ uptake in the light (and on $^{16}O_2$ uptake in the dark) at O_2 levels in excess of 21 kPa (200-300 mmol O_2 m^{-3} in solution under most experimental conditions) it appears that RUBISCO oxygenase is the only major O_2 uptake process whose activity continues to increase as O_2 levels increase above 21 kPa (7,9). This means that any O_2 uptake process which is specifically involved in generating damaging active oxygen species at high O_2 levels corresponds to a small fraction of the O_2 flux; such uptake processes could involve a low-affinity oxygenase (8) or a non-enzymic reaction with (for example) UQ$^{\bullet-}$ or PQ$^{\bullet-}$ (10) with a linear relationship to O_2 concentration.

The work described here addresses O_2 sensitivity of growth of two marine macroalgae. The green (ulvophycean) alga *Enteromorpha intestinalis* grows intertidally on rock platforms and in rock pools; high intertidal rockpools are only flushed at spring tides and can achieve O_2 levels of several times air-equilibrium during neap tides as net O_2 evolution over 24 h (with inorganic C acquired by a CO_2 pump from the 2 mol m^{-3} in sea water as oxidant) which is not balanced by O_2 evasion to the atmosphere (11,12). Populations of *E. intestinalis* from rock pools and rock platforms (where there is no external build-up of O_2) were grown in the laboratory at various O_2 levels to see if the rock pool and rock platform algae differed in O_2 sensitivity. The red (floridiophycean) alga *Lomentaria articulata* grows intertidally on the north-facing (shaded) side of rocks on north-eastern Atlantic shores where it is not subject to external O_2 build-up. *L. articulata* behaves like a C_3 plant on the basis of a number of physiological characteristics, e.g. inorganic C affinity, CO_2 compensation concentration based on direct measurements on emersed algae and from the compensation pH value for submersed algae, the lower CO_2 compensation concentration in 1-2 kPa O_2 than in 21 kPa O_2, and the low (i.e. negative) $\delta^{13}C$ of algal organic C relative to the $\delta^{13}C$ of CO_2 in seawater (13-18) although O_2 sensitivity of photosynthesis (other than the CO_2 compensation concentration) and growth has not thus far been measured. The work reported here involves an examination of the effects on growth of several combinations of inorganic C and O_2 concentrations.

2. Materials and Methods

Algae were collected from St Andrews (rock shelf population of *E. intestinalis*) and Fife Ness (rock pool population of *E. intestinalis*; *L. articulata*), Fife, Scotland. Growth under controlled CO_2 and O_2 concentrations involved Provasali-enriched sea water at 10°C in 40-50 μmole photon m^{-2} s^{-1} (12L:12D) at 10°C for *L. articulata* and at 10°C in 200 or 500 μmole photon m^{-2} s^{-1} (16L:8D) for *E. intestinalis* using a gas mixer of the type described in Parsons, Raven & Sprent (19). The gas mixtures used were 350 μmole CO_2 mole^{-1} and 0.2, 0.4, 0.6 or 0.8 mole O_2 mole^{-1}, and 220 or 370 μmole CO_2 mole^{-1} with 0.2 or 0.4 mole O_2 mole^{-1} for *E. intestinalis* (Poole, in preparation) and all combinations of 235, 350, 700, 1750 μmole CO_2 mole^{-1} with 0.02, 0.1, 0.2 or 0.4 mole O_2 mole^{-1} for *L. articulata* (Kübler, Johnston & Raven, in preparation).

3. Results and Discussion

The highest gas phase O_2 level (0.8 mole O_2 mole^{-1}) used for growth of *E. intestinalis* was close to the value in equilibrium with the highest values that were ever recorded in the rock pools at Fife Ness (0.86 mole O_2 m^{-3}) and was much higher than those around the rock platform populations. There was no significant effect of any of the increased O_2 concentrations tested on the growth of either population of *E. intestinalis*. These findings are consistent with the occurrence of a CO_2 concentrating mechanism in *E. intestinalis* (11-13) which restricts the effect of external O_2 on carboxylation by RUBISCO by maintaining a high internal CO_2 concentration (3). The presence of a CO_2 concentrating mechanism in *E. intestinalis* is consistent with the absence of a significant effect of equilibrating the growth medium with CO_2 at less than the present atmospheric level under normoxic conditions. Thus, the specific growth rate of *E. intestinalis* at 500

μmole photon m^{-2} s^{-1} is 0.143 ± 0.012 (standard error of the mean; n=3) d^{-1} in 370 μmole CO_2 mole^{-1}, and 0.158 ± 0.019 (s.e.m., n=3) d^{-1} in 220 μmole CO_2 mole^{-1} (cf. work on *L. articulata* mentioned below). The data are also consistent with a negligible impact on growth of any less effective resource acquisition, or resource allocation and use, resulting from avoidance, tolerance or repair of damage resulting from any increased production of active oxygen species in the higher O_2 treatments. Since this absence of effect on growth was found for both populations it appears that there is no greater genetic adaptation to tolerance of high O_2, in the range tested, for the rock pool than the rock platform population.

Another O_2-related source of active oxygen species which may have ecological relevance to *E. intestinalis* is during recovery from anoxia (5). The rock-pool population studied here could suffer anoxia during ice encasement in winter, with the rock platform population being much less likely to suffer in this way (22). While no comparative studies of anoxia tolerance have been performed on these populations, Taylor and McInroy (unpublished) has found generally high tolerance (assayed as post-anoxia growth rate) of prolonged (months) anoxia in isolates of *E. intestinalis* and other *Enteromorpha* spp. from the Ythan estuary (Aberdeenshire, Scotland) for conditions mimicking *in situ* winter survival in sediment, albeit with some inter-isolate differences (cf.23).

The results (Kübler, Johnston & Raven, in preparation) on *L. articulata* show no significant effect of O_2 on growth within the range tested, with a significant growth stimulation by doubled CO_2 relative to lower or higher CO_2 concentrations. While the growth response to doubled CO_2 is quantitatively compatible with C_3 physiology as judged from work on higher plants the absence of an O_2 effect is not (3). No convincing mechanistic explanation for these data is yet available (Kübler, Johnston & Raven, in preparation). Thus, the CO_2/O_2 selectivity factors for RUBISCO in bangiophycean (24) and floridiophycean (25) red algae are high, and cyanidiophycean red algae have the highest selectivity factors so far reported (26,27), there would still be significant RUBISCO oxygenase activity in an alga with C_3 physiology in air-equilibrated solutions, as is indicated for *L. articulata* by CO_2 compensation data (14). The interpretation of gas exchange in *L. articulata* is further complicated by our relative ignorance of the pathway(s) of glycolate metabolism in red algae (11,28). At all events the suggestion that *L. articulata* shows classic C_3 physiology seems to be too simplistic.

4. Conclusions

The absence of effect of hyperoxia on growth of *E. intestinalis* is consistent with the occurrence of a CO_2 concentrating mechanism, and an insignificant impact in terms of growth rate of whatever mechanism permits tolerance of enhanced production of active oxygen species during hyperoxia. There is no difference in hyperoxia tolerance between isolates which normally encounter hyperoxia and isolates which do not.

The absence of effect of hypo- or hyperoxia on growth of *L. articulata* is not consistent with other data suggesting that this alga has C_3 physiology.

5. Acknowledgements

AMJ was, and JEK and LJP are, supported by the Natural Environment Research Council (UK), and AMJ and RT are supported by the Scottish Office Agriculture, Environment and Fisheries Department.

6. References

1 Gilmour, D.J. (1982) PhD Thesis, University of Glasgow.
2 Raven, J.A. (1984) Energetics and Transport in Aquatic Plants. A R Liss, New York.
3 Raven, J.A., Johnston, A.M., Kübler, J. and Parsons, R. (1994) Biol. Revs. 69, 69-94.
4 Falkowski, P.G. and Raven, J.A. (1997) Aquatic Photosynthesis. Blackwell Science, Malden, Mass.
5 Vartapetian, B.B. and Jackson, M.B. (1997) Ann. Bot. 79, supplement A, 3-20.
6 Padan, E. and Cohen, Y. (1982) in The Biology of Cyanobacteria (Carr, N.G. and Whitton, B.A., eds) pp.215-235, Blackwell, Oxford.
7 Raven, J.A. and Beardall, J. (1981) in Physiological Bases of Phytoplankton Ecology (Platt, T., ed.) pp.55-92, Canadian Bulletin of Fisheries and Aquatic Science No.210.
8 Schomburg, D. and Stephen, D. (eds) (1994) Enzyme Handbook 8, Class 1.13-1.97, Oxidoreductases. Springer-Verlag, Berlin.
9 Badger, M.R. (1985) Ann. Rev. Plant Physiol. 36, 27-53.
10 Skulachev, V.P. (1996) FEBS Lett. 397, 7-10.
11. Raven, J.A. (1997) Adv. Bot. Res. 27,85-209.
12. Poole, L.J. and Raven, J.A. (1997) Progr. Phycol. Res. 12, 1-148.
13 Maberly, S.C. (1990) J. Phycol. 26, 439-449.
14 Johnston, A.M., Maberly, S.C. and Raven, J.A. (1992) Oecologia 92, 317-326.
15 Maberly, S.C., Raven, J.A. and Johnston, A.M. (1992) Oecologia 91, 481-492.
16 Kübler, J.E. and Raven, J.A. (1994) Mar. Ecol. Progr. Ser. 110, 203-208.
17 Kübler, J.E. and Raven, J.A. (1996a) Hydrobiologia 326/327, 401-406.
18 Kübler, J.E. and Raven, J.A. (1996b) J. Phycol. 32, 963-964.
19 Parsons, R., Raven, J.A. and Sprent, J.I. (1992) J. exp. Bot. 43, 595-604.
20 Poole, L.J. and Raven, J.A. (1997) Progr. Phycol. Res. 12, 1-148.
21 Raven, J.A. (1997) Adv. Bot. Res. 27, 85-209.
22 Raven, J.A. and Scrimgeour, C.M. (1997) Ann. Bot. 79, Supplement A, 79-86.
23 Kamermans, P., Malta, E.-j., Verschure, J.M., Lentz, L.F. and Schrijvers, L. (1998) Mar. Biol. 131, 45-51.
24 Read, B.A. and Tabita, F.R. (1994) Arch. Biochem. Biophys. 312, 210-218.
25 Uemura, K., Suzuki, Y., Shikanai, T., Wadano, A., Jensen, R.G., Chmara, W. and Yokota, A. (1996) Plant Cell Physiol. 37, 325-331.

26 Uemura, K., Anwaruzzaman, Miyachi, S. and Yokota, A. (1997) Biochim. Biophys.
    Res. Communs. 233, 567-571.
27 Whitney, S.M. and Andrews, T.J. (1998) Austr. J. Plant Physiol. 25, 131-138.
28 Raven, J.A., Johnston, A.M. and MacFarlane, J.J. (1990) in The Biology of the Red
    Algae (Sheath, R.G. and Cole, K.M., eds) pp.171-202, Cambridge University Press.

CRYSTAL STRUCTURE OF RUBISCO FROM A THERMOPHILIC RED ALGA, GALDIERIA PARTITA

Hajime Sugawara[*], Hiroki Yamamoto[*], Tsuyoshi Inoue[*], Chikahiro Miyake[+], Akiho Yokota[+] and Yasushi Kai[*]
[*]Department of Materials Chemistry, Graduate School of Engineering, Osaka University, Suita 565-0871, Japan; [+]Graduate School of Biological Sciences, Nara Institute of Science and Technology, Ikoma 630-0101, Japan

Key words: X-ray diffraction, CO_2 uptake, H-bond, photorespiration, single crystals, high specificity factor

1. Introduction

Ribulose 1,5-bisphosphate carboxylase/oxygenase (RuBisCO, EC 4.1.1.39) is the initial step enzyme in the Calvin-Benson cycle of photosynthesis. It catalyzes the addition of gaseous CO_2 to ribulose 1,5-bisphosphate (RuBP) and produces two molecules of 3-phosphoglycerate (3-PGA) (1,2). However, this enzyme also catalyzes O_2 addition for RuBP as the primary reaction of photorespiration. The reaction yields one molecule each of 3-PGA and 2-phosphoglycolate from RuBP. The oxygenation reaction impairs the photosynthetic efficiency by up to 60 % (3). Thus the improvement of carboxylation/ oxygenation ratio by genetic engineering has been attempted to increase the productivity of crop plants.

From phylogenetic analysis, RuBisCO genes are devided into three groups, α-purple bacterial, β-purple bacterial, and γ-bacterial-cyanobacterial group (4). Even though the structural difference amongst these groups was presumed, crystal structure of β-purple bacteria group was not determined.

The gaseous substrate CO_2 and O_2 are the competitive inhibitor of oxygenation and carboxylation, respectively. The carboxylation/oxygenation ratio is defined as CO_2/O_2 relative specificity factor $(\tau), V_c K_o/V_o K_c$, where V_c and V_o are maximum velocity of carboxylation and oxygenation, respectively, and K_o and K_c are the Michaelis constants for O_2 and CO_2 (5). High τ value means the RuBisCO with the activity for effective carbon fixation. RuBisCO from *R. rubrum* has low τ value of 15. τ value is 93 for higher plants spinach (6). Moreover, RuBisCOs from marine algae (brown algae, red algae *et al.*) have been found to exhibit high τ values over 100 (7). RuBisCO genes from marine algae belong to β-purple bacterial group. Especially, RuBisCOs from thermophilic red algae have higher τ values. RuBisCO from *Galdieria partita* indicates τ value of 238, which is the highest among the RuBisCOs hitherto reported (8).

In this report, crystallographic analysis of the complex of activated RuBisCO from a thermophilic red alga *Galdieria partita* and reaction intermediate analogue of 2-CABP at 2.4 Å resolution is presented.

2. Procedure

Galdieria RuBisCO was purified by the same method as described for that of red alga *Prophyra yezoensis* (6). Crystallization was carried out using the hanging-drop vapor

3339

G. Garab (ed.), Photosynthesis: Mechanisms and Effects, Vol. V, 3339–3342.
© *1998 Kluwer Academic Publishers. Printed in the Netherlands.*

diffusion method. The drops consisted of 2μl of protein solution of 10.0 mg/ml comprising 20 mM $MgCl_2$, 20 mM $NaHCO_3$ and 2 mM 2-CABP, and 2 μl of precipitating solution on a siliconized glass coverslip and suspended over a 0.5 ml reservoir containing the same precipitating solution. The hexagonal crystals were obtained at 293 K in 9 % PEG 8,000 in the presence of 4 % MPD and 50 mM Hepes pH 7.5. Crystals grew to a typical size of 0.4 x 0.3 x 0.3 mm^3 in 4-5 days.

Data collection was performed on beamline BL18B of Photon Factory, Tsukuba, Japan. The space group of RuBisCO was determined to be hexagonal $P6_4$ with unit cell parameters of $a = b = 117.07$ and $c = 319.63$ Å. Assuming L_4S_4 of RuBisCO in the assymmetric unit, the value of the Matthews constant (9) gives $V_M = 2.55$ Å^3/Da corresponding to a solvent content of 51.5 %. The data was processed by using *DENZO* and *SCALEPACK* (10). A total of 277,331 measurements of 83,303 unique reflections were recorded and R_{merge} was 9.7 % for the data between 40 and 2.4 Å resolutions, the completeness of which is 86.4 %.

Crystal structure of *Galdieria* RuBisCO was solved by the molecular replacement method using *AMoRe* (11). Search model was based upon L_1S_1 of the activated *Synechococcus* RuBisCO with 2-CABP (12). The model was refined by the use of *X-PLOR* with 2- and 4-fold non crystallographic symmetry (NCS) restraints (13). After one round of simulated annealing was applied, multiple cycles of model fitting and refinements were alternated. NCS restraints were performed throughout the refinement process. At the final stage of refinement, bulk solvent correction was applied. The final R and R_{free} factors for all reflections between 20.0 and 2.4 Å resolutions were 0.180 and 0.209, respectively.

3. Result and Discussion

3.1. *Active site structure*
Compared with *Synechococcus* and spinach RuBisCO's active site structures (12,14), *Galdieria* RuBisCO has the significant structural differences around P5 phosphate of 2-CABP. Especially, side chain of H327 and R295 have large structural differences (Fig. 1, 2). Around H327, terminal oxygen of Y346 forms hydrogen bond with carbonyl oxygen of G329 (Fig.1). This bond may move the imidazole ring position of *Galdieria* RuBisCO. In higher plants and *Synechococcus*, valine is conserved at residue 346. Asn, Ser and Ala mutants on H327 in *R. rubrum* RuBisCO decrease less than 1/9 in the 2-CABP exchange rate, however, K_m for RuBP of these mutants increase only 4-fold (15). As the consequence, it is presumed that H327 preferentially binds the transition state of carboxylate reaction. In other words, imidazole ring orientation of H327 appears to influence K_m for CO_2 and may play a role of decreasing it in *Galdieria* RuBisCO. The side chains of H327 and S379 in *Galdieria* RuBisCO translate to approach ca. 0.5 Å each other in comparison with that in spinach RuBisCO. Thr and Cys mutants on S379 of RuBisCO from cyanobacterium *Anacystis nidulans* are almost devoid of oxygenation activity (16). The imidazole position of H327 in *Galdieria* RuBisCO may influence the decrease of oxygenation activity through S379.

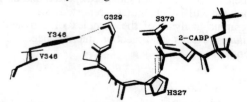

Fig. 1. Structure around H327 (Thick: *Galdieria* RuBisCO, Thin: spinach RuBisCO)

On the other hand, H298 is completely conserved except RuBisCOs from non-green algae. This residue is close to R295 in the crystal structure of spinach or *Synechococcus* RuBisCO. However, residue 298 in the RuBisCOs from non-green algae is asparagine, the nitrogen atom of which makes hydrogen bond to terminal oxygen of Y301. Residue 301 is Val, Ile or Leu except in β-purple bacterial RuBisCO. D302 of spinach RuBisCO makes ion pair with H298, which residue 302 of *Galdieria* RuBisCO is Ser. These two different types of interactions may cause the structural difference around R295. Leu and Lys mutants of R295 in RuBisCO from cyanobacterium *Anacystis nidulans* are almost devoid of carboxylation activity (17). Side chain orientation of R295 during the enzyme reaction seems to be important.

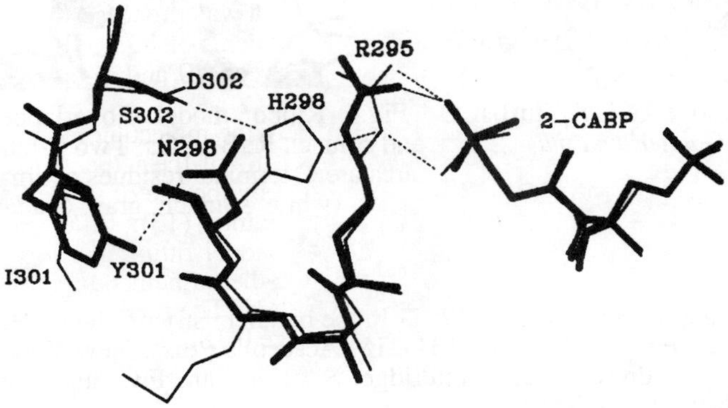

Fig.2. Structure around R295 (Thick: *Galdieria* RuBisCO, Thin: spinach RuBisCO)

Side chains of H327 and R295 make ionic interaction with P5 phosphate oxygen atoms. So these two residues appear to influence the K_m for RuBP. It has not been determined in *Galdieria* RuBisCO, but RuBisCO from red alga *Porphyridium* has the K_m value of 3.7 µM (7). This is 1/8 of the K_m of spinach RuBisCO. *Porphyridium* RuBisCO has the highly sequence identity with *Galdieria*'s and seems to have similar K_m value. Supposing that, the position of two side chains may affect to reduce the K_m for RuBP in *Galdieria* RuBisCO.

3.2. *C-terminal structure of small subunit*
Along the non-crystallographic 4-fold axis, there is a solvent channel passing through the center of RuBisCO complex, In green algae, *Euglena* and higher plants, genes of small subunit are nuclear-encoded and have 12-18 amino acid insertions, which are not present on plastid-encoded genes. RuBisCOs from red algae have more than thirty unique amino acid residues on C-terminus (18). The extra amino acid residues in *Galdieria* RuBisCO make up hairpin-loop structure (Fig. 3). So, the solvent channel of *Galdieria* RuBisCO is narrower than spinach RuBisCO. C-terminal region of small subunit makes a lot of interactions with adjacent small subunits. When structures of *Galdieria* and spinach RuBisCOs are superimposed, hairpin region of neighboring small subunit and apical portion of insertion residues are almost overlapped to each other (Fig.4). However, this insertion part has a few interactions with large and small subunits constituted of L_1S_1. There is no interaction with neighboring small subunit. C-terminal residues of small subunit in *Galdieria* RuBisCO may play a role of making the complex structure more rigid.

3342

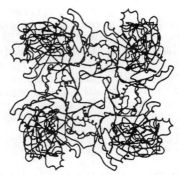 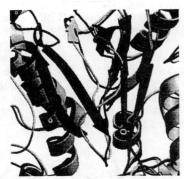

Fig.3. Cα model of L_4S_4 RuBisCO from *Galdieria partita*

Fig.4. Ribbon model of overlapped *Galdieria* and spinach RuBisCOs. Two β-hairpins of this area are C-terminal residues of small subunits (white: spinach, gray: *Galdieria*)

References

1 Andrews, T.J. and Lorimer, G.H. (1987) In the biochemistry of plants (Hatch, M.D., Boardman, N.K., eds.) vol. 10, pp.131-218, Academic Press, New York

2 Cleland, W.W., Andrews, T.J., Gutteridge, S., Hartman, F.C. and Lorimer, G.H. (1998) Chem. Rev. 98, 549-561

3 Zelitch, I. (1975) Science 188, 626-633

4 Assali, N.-E., Mache, R. and Loiseaux-de Goer, S. (1990) Plant Mol. Biol. 15, 307-311

5 Liang, W.A., Ogren, W.L. and Hageman, R.H. (1974) Plant Physiol. 54, 678-685

6 Uemura, K., Suzuki, Y., Shinkai, T., Wadano, A., Jensen, R.G., Chmara, W. and Yokota, A. (1996) Plant Cell Physiol. 37, 325-331

7 Read, B.A. and Tabita, F.R. (1994) Arch. Biochem. Biophys. 312, 210-218

8 Uemura, K., Anwaruzzaman, Miyachi, S. and Yokota, A.(1997) Biochem. Biophys. Res. Commun. 233, 568-571

9 Otwinowski, Z. (1993). Proceedings of CCP4 Study Weekend 1993 (Sawyer, L., Issacs, N. and Bailey, S., eds) pp. 56-62. Daresbury Laboratory, Warrington, England

10 Matthews, B. W. (1968). J. Mol. Biol. 33, 491-497

11 Navaza, J. (1994). Acta Cryst. A50, 157-163

12 Newman, J. and Gutteridge, S. (1993) J. Biol. Chem. 268, 25876-25886

13 Brunger, A.T., Kuriyan, J. and Karplus, M. (1987) Science 235, 458-460

14 Shibata, N., Inoue, T., Fukuhara, K., Nagara, Y., Kitagawa, R., Harada, S., Kasai, N., Uemura, K., Kato, K., Yokota, A. and Kai, Y. (1996) J. Biol. Chem 271, 26449-26452

15 Harpel, M.R., Larimer, F.W. and Hartman, F.C., (1991) J. Biol. Chem., 266, 24734-24740

16 Lee, G. J. and McFadden, B. A. (1992) Biochemistry 31, 2304-2308

17 Haining, R. L. and McFadden, B. A., (1990) J. Biol. Chem., 265, 5434-5439

18 Assali, N.-E., Martin, W.F., Sommerville, C.C. and Loiseaux-de Goer, S. (1991) Plant Mol. Biol. 17, 853-863

DIFFERENT EFFECTS OF MERCAPTOETHANOL ON CYANOBACTERIAL RECOMBINANT RUBISCO AND ITS SITE - DIRECTED MUTANTS

Alla K. Romanova[1], Zhen - qi Cheng[2], Bruce A. McFadden[2]

[1]Institute of Basic Bioligical Problems RAS, Pushchino Moscow region, 142292, Russia, [2]Department of Biochemistry and Biophysics, Washington State University, Pullman, WA, USA

Key words: CO_2 binding, enzymes, disulphide bonds, genetic manipulation

1. Introduction

Ribulose-1,5-bisphosphate carboxylase/oxygenase (rubisco) is the main enzyme responsible for primary productivity of the overwhelming majority of photoautotrophs [1].

High molecular (Mr 540 kD) form of rubisco most widely spread among the phototrophs consists of eight 57 kD large subunits (LS) and eight small 14 kD subunits. Either of two LS is organised as a pair, L_2, and each of the two catalytic sites is shared between two adjacent LS. Conservative amino acid residue *Asn*123 (spinach numeration) localised at N-domain of one of the rubisco LS is believed to contribute to the active centre of the adjacent second LS. Barrel-shape structure of the C-terminal domain of LS consists of 8 alternating β-strands and α-helixes connected with loops. Loop 6 and its adjacent secondary structures are of special interest because its flexibility during the catalysis.

The activity of native rubisco, isolated from the cells of cyanobacteria ranged between 3.3 s^{-1} and 5.0 s^{-1} [2]. The published values of carboxylase activity for the purified recombinant rubisco from *A. nidulans* are ranged much more widely: from approximately 1.7 s^{-1} to about 10 s^{-1}[3, 4]. It was difficult to explain these differences because the details of the experiments were not described in any case.

The necessity of SH-group stabilisers at rubisco *in vitro* assay was known earlier, but neither the degree of their effect on activation level nor the mechanism of action were studied. It was supposed that *Cys* residues could be significant in maintenance of the enzyme conformation or structural integrity.

The purpose of this study was to assay carboxylase activity of rubisco and its sensitivity to mercaptoethanol (ME) in recombinant rubisco and some site-directed mutants of the enzyme from *A. nidulans*.

2. Procedure

Mutagenesis and gene expression were accomplished as described in details earlier [5]. Two types of mutants were constructed. Conservative amino acid *Asn*123 was changed in one case. The main motive for construction of the other mutants was

G. Garab (ed.), Photosynthesis: Mechanisms and Effects, Vol. V, 3343–3346.
© 1998 *Kluwer Academic Publishers. Printed in the Netherlands.*

substitution of the nonconservative elements of strand β6, loop6 and helix α6 rubisco LS from *A. nidulans* for the analogous ele

ments specified in rubisco from diatom alga *Cylindrotheca* sp. N^o1. [6]. *Purification of rubisco* by sucrose density gradient centrifugation and its carboxylation activity assay were carried out as described earlier [5].

3. Results and Discussion

3.1 *Wild type (WT) rubisco*

From Fig. 1 it is seen, that without ME or in the presence 5 mmol ME the time courses of $^{14}CO_2$ fixation were linear, and the enzyme activities were close to each other. In presence of 130 mM ME after 1 min preincubation substantial increase of velocity of carboxylation is seen. The distinctly manifested activation of rubisco continued during the catalytic process.

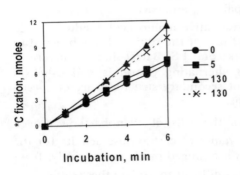 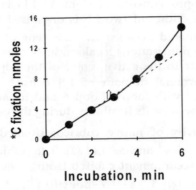

Figure 1. Influence of ME on time course of carboxylation catalysed by the wild type recombinant rubisco from *A. nidulans.* Activation in the presence $NaH^{14}CO_3$ and $MgCl_2$ - 1 min. Concentrations of ME added in course of preincubation are shown in the inserted legend. Dotted line is approximation to the linearity .
Figure 2. Influence of addition of ME on the carboxylation time course of the recombinant rubisco from *A. nidulans.* Preincubation and the first 3 min of the catalytic reaction were carried out with ME omitted. The arrow shows the time point of addition 275 mM ME (final concentration). Dotted line as in Fig. 1.

When rubisco was activated with Mg^{2+} plus $NaH^{14}CO_3$ without additional ME, the velocity was evidently less than after addition of 280 mmol ME during time-course of the reaction (Fig. 2). In other case rubisco was preliminary activated by 3 min preincubation in presence of the excess ME. Further increase of carboxylation velocity was not observed whatever more ME was added, while catalysis was going on (not shown). Three min preincubation resulted the linear time-course of the reaction was used routinely as indispensable and satisfactory time to get complete activation of the enzyme in the presence ME.

The maximal turnover numbers around 13.0 s^{-1} [cf. 3,4] were obtained for the WT recombinant rubisco A. *nidulans*, when the freshly thawed preparations have been diluted with buffer mixture containing 320 mmol ME.

When preincubation of WT rubisco was carried out in presence of 130 mmol ME, the carboxylation activity on average was about 40% higher than in the presence of 5 mM ME or with ME omitted.

3.2 Mutant rubisco

The activity of mutant rubisco having one conservative amino acid residue changed at the N-domain of LS, *Asn123His*, with ME omitted from reaction mixture was rather high, 78.6% that of WT rubisco. Being the highest in the presence 130 mmol ME (Fig. 3) it reached about 70% activity of completely activated WT rubisco.

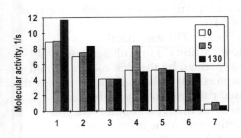 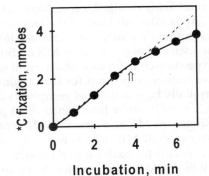

Figure 3. Effect of ME on carboxylation activity of the wild type and some mutant types of rubisco from A. *nidulans*.
Concentrations of ME added are shown in the inserted legend. 1 - Wild type. Mutants: 2 - *N123H*; 3 - *L326I*; 4 - *S328A*; 5 - *T342I*; 6 - *L343K*, and 7 - *KASTL* (339--343)*PLMIK*.

Figure 4. Influence of ME addition on the carboxylation time-course of the mutant rubisco from A. *nidulans* with five aminoacids substituted *KASTL*(339 -- 343) *PLMIK*. Preincubation and the first 4 min the reaction was carried out with ME omitted. The arrow shows the time-point of 275 mM ME (final concentration) addition. Dotted line as in Fig. 1.

Substitutions of non conservative amino acid residues in the strand β6, loop 6, and helix α6 of the LS were made, and the influence of ME on rubisco activity was assayed (Fig. 3).

The mutant enzyme with substitution *Leu326Ile* (strand β6) was half as active as WT rubisco. This mutant rubisco was not sensitive to ME, as a result its activity in the presence 130 mmol ME retained only 35% WT rubisco velocity at the same ME concentration.

Mutant *Ser328Ala* (the very beginning of loop 6) displayed as low level of the activity without ME addition as the preceding one, but it was sensitive to 5 mmol ME: its activity became 62% higher, and approached to WT rubisco velocity at the same ME concentration. However, the carboxylation velocity of the mutant decreased again at 130 mmol ME. That was the first example of inhibition of the rubisco activity with excess of ME.

Two substitutions close to the middle of the helix α6, namely, *Thr342Ile* and *Leu343Lys*, also decreased the carboxylase activity approximately to a half of the WT species. Both of them were not sensible to ME presence in activation medium, hence finally their activity was about 40% that of WT rubisco.

Mutant rubisco with five amino acid residues changed *Lys,Ala,Ser,Thr,Leu*(339--343)*Phe,Lys,Met,Ile,Lys* disposed at the beginning through the middle of the helix α6 was the lowest one: with ME omitted or in the presence of 5 mmol ME the mutant was about 10 times less active, than WT rubisco (Fig. 3). However the mutant rubisco activity was decreased in the presence 130 mmol ME. As a result this rubisco species exhibited only about 5% fully activated WT rubisco.

From Fig. 4 it is seen, indeed, that carboxylation velocity of the mutant rubisco with five amino acids changed, on opposite to that of wild type enzyme, began to decrease immediately after addition of the large excess of ME.

Finally, mutant rubisco *Val,Asp,Leu*(346--348)*Tyr,His,Thr* was inactive.

The necessity of high excess of ME addition (low "affinity") for full activation of wild type rubisco from cyanobacterium can be explained by the lack of accessibility of *Cys* residues, important for the obtaining the maximal velocity values. Perhaps the exceptionally high activity of completely activated rubisco from cyanobacterium [3,4, and this work] can be interpreted as an peculiarity of the recombinant enzyme. Insensibility to ME and especially decrease of activity of some mutants as influenced by high excess of ME suggest necessity to keep a favourable ratio -SH and S-S groups specific for each kind of rubisco. Some of disulphide bonds which can be accessible for ME in WT molecule become inaccessible in some mutants.

This work was supported by the CAST-93 Program of the National Academy of Sciences USA and partially by the International Science Foudation, Supplementary Grant SDH 000.

References

1. Andrews T.J. and Lorimer G.H. (1987) In: The Biochemistry of Plants (Hatch MD and Boardman N.K., eds.), Vol. 10, pp. 131--218. New York, London, Academic Press.

2. Badger M.R. (1980) Arch. Biochem. Biophys. 201, 247--254

3. Paul K. , Morell M.K., and Andrews T.J. (1991) Biochemistry 30, 10019--10026.

4. Gutteridge S., Rhoades D.F., and Herrmann C. (1993) J. Biol. Chem. 268, 7818--7824

5. Romanova A.K., Cheng Zh.-qi, and McFadden B.A. (1997) Biochem. Molec. Biol. Internat. 42, 299--307

6. Read B.A., and Tabita F.R. (1992) Biochemistry 31, 5553--5560

Expression of *Chromatium vinosum* rubisco in the carboxysome deficient mutant of cyanobacterium *Synechococcus* PCC7942

Kojima, K., Haranoh, K., [1]Kobayashi, K, Iwaki, T and Wadano, A
Department of Applied Biochemistry, Osaka Prefecture University, Sakai, Osaka 599-8531, Japan and [1]Laboratory of Plant Cell Technology, University of Shizuoka, Prefecture, 52-1 Yada, Shizuoka, 422, Japan

C metabolisms, Calvin-cycle, CO_2 concentration, light regulation, molecular biology

1. Introduction

The control coefficient is an important index to show how an enzyme regulates a metabolic pathway. On higher plants, antisence technique decrease RuBisCO protein for the elucidation of a regulatory role of the enzyme in photosynthetic inorganic carbon anabolism (1). The technique can not be applicable to cyanobacteria because of their high possibility of homologous recombination. On the other hand, the control coefficient could be estimated dependent on the increase of the enzyme concerned. As a trial for the estimation, RuBisCO protein of purple sulfur photosynthetic bacterium *Chromatium vinosum* was expressed in the cyanobacterium *Synechococcus* PCC7942 by the transformation with bi-directional expression vector. In this transformant MT1, we confirmed that introduced expression vector replicated stably and *C. vinosum* RuBisCO was expressed in the cytosol in active L8S8 form. But the expression level of *C. vinosum* RuBisCO was around 30% of total RuBisCO activity. The low expression level makes it difficult to analyze the effect of the foreign enzyme on the photosynthetic physiology of the transformant. Thus, in this paper, we try to increase its expression level after checking promoters derived from cyanobacterium and *E. coli* with luciferase as a reporter.

2. Materials and Methods

2.1 *Construction of a plasmid for expressing C. vinosum RuBisCO gene in Synechococcus PCC7942*

The plasmid pUC303 contains a chloramphenicol resistance gene and streptomycin resistance gene and also contains the *Synechococcus* PCC7942 origin of replication (2). The 3.0 kb PvuII fragment from plasmid pCV23 containing a *lac* promoter and *C. vinosum* RuBisCO promoter upstream of the *C. vinosum* RuBisCO gene was ligated into the XbaI-XhoI site in the plasmid pUC303. This produced 11.2 kb *E. coli /Synechococcus* shuttle vector, plasmid pUCV1. The plasmid DNA was transformed into *Synechococcus* PCC7942 by the method of Williams (3). The transformant MT1 was grown in the presence of 5 g chloramphenicol in 1 l of BG-11 medium.

2.2 *Luciferase assay of transformants*

Cells of transformants were grown at 30°C in BG-11 medium with 10 μg/ml streptomycin.

G. Garab (ed.), Photosynthesis: Mechanisms and Effects, Vol. V, 3347–3350.
© 1998 *Kluwer Academic Publishers. Printed in the Netherlands.*

At logarithmic phase of growth (OD$_{730}$=1.5), cells were harvested by centrifugation, washed twice with ice-cold phosphate buffered saline, and resuspended in 200 μl of ice-cold 50 mM Tris-H$_3$PO$_4$ (pH 7.8). Cells were disrupted by sonic treatment at minimal power for 1 min at 0°C, then added equal volume of cell disruption buffer (25 mM Tris-H$_3$PO$_4$ (pH 7.8), 2 mM DTT, 2 mM 1,2-cyclohexanediaminetetraacetic acid, 10% (v/v) glycerol, 1% (v/v) tritonX-100), and incubated for 15 min at room temperature. The lysate were clarified by centrifugation at 15,000 x g for 10 min at room temperature. The chlorophyll content of each sample was adjusted about 2.5 x 10^{-2} mg/ml with cell disruption buffer containing 1 mg/ml BSA. Twenty μl of cell extracts were added in 100 μl of luciferin mixture containing 20 mM Tricine-NaOH (pH7.8), 1.07 mM (MgCO$_3$)$_4$Mg(OH)$_2$•5H$_2$O, 2.67 mM MgSO$_4$, 0.1 mM EDTA, 33.3 mM DTT, 0.27 mM CoenzymeA and 0.47 mM luciferin. Luciferase activity is measured as light-units per μg chlorophyl by a luminometer (a machine capable of detecting light at 562 nm).

2.3 *Enzyme assay*
Cells were grown in BG11 medium until OD$_{730}$ of the medium reached around 1.0. Preparation of cytosol and carboxysome fraction were mainly according to the method of Price et. al (4). RuBisCO activity was measured spectrophotometrically (5) on these fractions. The assay solution contained 50 mM Hepes-NaOH, pH 7.6, 20 mM MgCl$_2$, 50 mM NaHCO$_3$, 1 mM ATP, 5 mM phosphocreatine, 0.15 mM NADH, 22.2 units/ml yeast 3-phosphoglycerate kinase, 6.8 units/ml rabbit muscle glycerardehyde-3-phosphate dehydrogenase, and 5 units/ml rabbit muscle creatine kinase.

3. Results
3.1 *Immunochemical differentiation of Synechococcus PCC7942 RuBisCO and C. vinosum RuBisCO proteins*
RuBisCO of *C. vinosum* was expressed in cyanobacterium *Synechococcus* PCC7942 by the transformation with the bi-directional expressing vector. The immunological analysis showed that *C.vinosum* RuBisCO protein was expressed in cytosol in active and the same molecular mass as cyanobacterilal L8S8 enzyme.

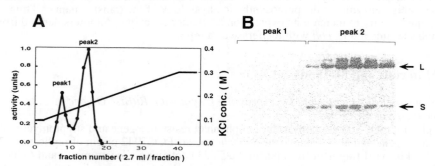

Fig. 1. *Differentiation of C. vinosum RuBisCO L- and S-subunit proteins in the cytosol fractions on an anion exchange column*
A, Elution profile of RuBisCO activity (•) from DEAE TOYO PEARL 650M packed in FPLC column. The column (1.0 x 1.5 cm) was equilibrated with a buffer consisting of 10 mM HEPES-NaOH (pH 7.6), 1 mM EDTA, 50 mM KCl and 10 mM 2-mercaptoethanol. The active fraction from the Superdex 200 was loaded onto the column, and RuBisCO was eluted as shown with 0.1 to 0.3 M KCl linear gradient. B, SDS-PAGE of RuBisCO

purified from MT1.   Twenty μl of fractions at the peak of RuBisCO activity was separated in 15% acrylamide gel and electroblotted onto PVDF membrane.   After blocking with skim milk, the subunits were proved with rabbit antibodies to *C. vinosum* RuBisCO followed by goat-rabbit immunoglobulin conjugated to horseradish peroxidase (6). Immuno-reactive proteins were visualized by peroxidase staining using diaminobenzidine and H_2O_2 as a developer.

3.2   *Promoter search for cyanobacterium Synechococcus PCC7942*

In order to search strong promoters for *Synechococcus* PCC7942, *Synechococcus* PCC7942 *psbAI* and *rbc* promoter, *Synechosystis* PCC6803 *psbAII* and *rpoD1* promoter were

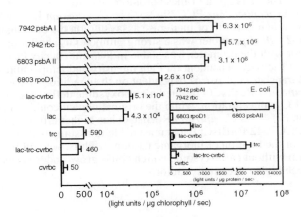

Fig. 2.   *Luciferase activity in transformed cyanobacteria and E. coli*
Cells of transformed cyanobacteria were grown in BG-11 medium containing 10 μg/ml streptomycin at 30°C and harvested at the mid-logarithmic phase of growth.   Luciferase activity was measured by using a luminometer.   Inset; the result of luciferase assays in transformed *E. coli* JM109.   Cells were grown in 2xTY medium (pH7.0) containing 100 μg/ml streptomycin for 17 hrs at 37°C.

checked by luciferase reporter system. The HindIII-BamHI fragment of the plasmid PicaGene cassette vector containing luciferase gene from *Photunis pyralis* was ligated into the various promoter.   The results of luciferase assay showed that the activity of *Synechococcus* PCC7942 *psbAI* promoter was more that 10 times as strong as that of *lac* promoter (Fig. 2).

3.3 *Expression of RuBisCO protein of C. vinosum in Synechococcus PCC7942*

Synechococcus PCC7942 *psbAI* promoter was used for expressing RuBisCO protein of *C. vinosum* in *Synechococcus* PCC7942, and the promoter made a high activity of *C. vinosum* RuBisCO in the cyanobacterium, as confirmed with the Western blotting and measurement of RuBisCO activity.   Stability of the shuttle vector in *Synechococcus* PCC7942 was checked by the plasmid rescue method.   Southern analysis showed that the vector was not incorporated into the cyanobacterial genome. The highest activity of RuBisCO in a transformant was about 6 times larger than that of MT1. The enzyme activity was increased at 24 hr after shifting the cells to either high light intensity or low temperature. The transformation rescued a carboxysome deficient mutant to grow under 0.6% of CO_2 at least on the plate culture.

4. Discussion

The *tac* promoter which is origin of *trc* promoter was known as strong promoter in another strain *Synechocystis* PCC6803. From result of luciferase assays, it was shown that this promoter can not function in *Synechococcus* PCC7942, although *lac* promoter functioned well. The *trc* promoter combines the -35 region from *trp* promoter and the -10 region from *lacUV5* promoter. The -10 region of these promoter would function but *trp*'s -35 region could not work in *Synechococcus* PCC7942. The *C. vinosum* RuBisCO promoter alone could not work as shown in result of luciferase assays. On the other hand, *Synechococcus* PCC7942 *psbAI* promoter was a strong promoter as shown previously (7,8), and a shuttle vector with the promoter was stable in *Synechococcus* PCC7942. The regulation of the RuBisCO activity dependent on light intensity and temperature resemble to that of transcription and translation of *psbAI* dependent on the condition (7,8). The high level expression of *C. vinosum* RuBisCO rescued a carboxysome deficient cyanobacterium to grow under 0.6% of CO_2 concentration, while the original mutant could not survive under the same condition (9). As *C. vinosum* RuBisCO has higher affinity for CO_2 than that of *Synechococcus* PCC7942 RuBisCO, it is reasonable to assume that the expressed RuBisCO worked well for the fixation of inorganic carbon. Thus, we will be able to get a mutant of cyanobacterium without carboxysome which could grow under normal air by using a RuBisCO of much higher specificity factor.

5. References

1 Stitt, M., Quick, W.P., Schurr, U., Schulze, E.D., Rodermel, S.R. and Bogorad, L. (1991) Planta, 183, 555-566
2 Kuhlemeier, C.J., Thomas, A.A., van der Ende, A., van Leen, R.W., Borrias, W.E., van den Hondel, C.A. and van Arkel, G.A. (1983) Plasmid, 10,156-163
3 Williams, J.G.K. (1988) Method in Enzymology, 167, 766-778
4 Price, G.D., Coleman, J.R. and Badger, M.R. (1992) Plant Physiol., 100, 784-793
5 Lilley, R.M. and Walker, D.A. (1974) Biochim. Biophys. Acta, 358, 226-229
6 Viale, A.M., Kobayashi, H., Takabe, T. and Akazawa, T. (1985) FEBS Lett., 192, 283-288
7 Kulkarni, R. D. and Golden, S. S. (1994) J. Bacteriology, 176, 959-965
8 Campbell, D., Zhou, G., Gustafsson, P., Öquist, G. and Clarke A. (1995) EMBO J. 14, 5457-5466
9 Price, G.D. and Badger, M.R. (1991) Can. J. Bot., 69, 963-973

ENGINEERING RUBISCO IN HIGHER PLANT CHLOROPLASTS

Ivan Kanevski, Pal Maliga, Dan Rhoades and Steven Gutteridge.
Waksman Institute, Rutgers, State University of New Jersey
Piscataway, NJ 08855 and CR&D Department, DuPont Company
present address: Agricultural Products, Stine Research Labs.,
DuPont Company, Newark, DE 19714
Keywords: chloroplast transformation, Rubisco

1. Introduction

Attempts to alter the characteristics of higher plant Rubisco by applying the usual molecular biological procedures of construction and over-production from plasmids in the usual bacterial hosts is fraught with difficulties. A major problem, so far not surmounted is the inability of the recombinant protein to assemble into any aggregate that remotely resembles a competent functioning entity. Most examples of successful manipulation of an L8S8 form of the enzyme in *E.coli* have been forthcoming using the dicistronic gene of cyanobacteria (see refs cited in 1). Even plant - cyanobacteria chimeric constructions resulted in only limited success in recovering active enzyme.

We have therefore exploited the recent developments in plastid genome manipulation (2, 3) to demonstrate that the *rbcL* gene of the plastome can be replaced by homologous recombination in high enough frequency for transgenic plants to be regenerated and supply enough Rubisco for functional analysis. A further advantage of using what is essentially a binary approach to construct then express and produce the hybrid enzyme, is that the effect of Rubisco variants on whole plant photosynthesis and development can be assessed with this system.

2. Procedure

2.1 Modifying rbcL *genes*

The process of constructing Rubisco L-subunits variants and the organisation of the plastid targetted vector are shown in Figure 1. The steps involved introducing usable restriction sites within and outside of the *rbcL* genes from sunflower and the cyanobacterium *Synechococcus* PCC6301. These restriction sites allowed fusion in-frame of most of the subunit except for the first 30 or 24 bases of tobacco *rbcL* ORF, respectively. These chimeras were first constructed and amplified in *E.coli* before the modified L-subunit was introduced into a plant compatible vector containing a spectinomycin selectable cassette (*aadA* gene) 200 bases from the 3' end of the L-subunit gene. The vector also contained parts of the *atpB* and *accD* genes from the plastid genome to enhance the probability of homologous recombination.

G. Garab (ed.), Photosynthesis: Mechanisms and Effects, Vol. V, 3351–3354.
© *1998 Kluwer Academic Publishers. Printed in the Netherlands.*

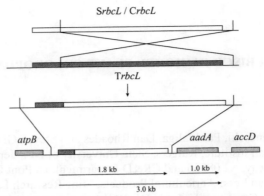

Figure 1. The vector used to replace tobacco *rbcL* with foreign L-subunit genes and the sizes of the transcripts detected in transformed chloroplasts (see figure 3).

3. Results

3.1 Regeneration and Analysis of Transformants

The plant vectors were introduced into leaves by biolistic means and transformed plants were regenerated in sucrose supplemented medium containing spectinomycin. A number of stable transplastomic lines were eventually isolated. Figure 2 shows two lines (both green) which contained all or part of the S *rbcL* gene (S*rbcL*/R*rbcL*), the third (yellow) contained the L-subunit gene for *Synechococcus* Rubisco (C*rbcL*).

Figure 2. Regenerated plants expressing foreign rbcL genes compared to wild type.

Figure 3 shows that all lines expressed message (Fig. 3A) and produced protein (Fig. 3B) from the L- subunit gene. In the case of the C*rbcL* line containing the cyanobacterial gene (C) only message was detectable. Note that two different populations of RNA were observed. The smaller message of 1.8 kb is from the coding region of the L-subunit gene, the larger 3.0 kb message encompasses both the L-subunit and *aadA* genes, presumably as a result of incomplete transcription termination.

The presence of significant amounts of S-subunit with the L-subunit from the S*rbcL* (S) lines (Fig. 3B) and that regenerated plants were phenoptypically indistinguishable from the wild type plants on sucrose-containing media, suggested strongly that the heterologous L-subunit was folding correctly, assembling and associating with the indigenous S-subunit population to produce functional enzyme.

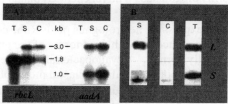

Figure 3. A) Transcripts detected with either *rbcL* or *aadA* probes (T – tobacco, S – sunflower, C – cyanobacteria lines). B) Rubisco L- and S-subunit antibodies.

Rubisco activity was determined by measuring ribulose-P_2 dependent CO_2 fixation of leaf extracts from all three lines. The S*rbcL* lines had significant carboxylase activity about 15% of that of wild type plants. Since *Synechococcus* specific L-subunit antibodies did not detect cyanobacterial L-subunit protein and only a trace amount of the plant S-subunit, it was not surprising to find that CO_2 fixation by C*rbcL* extracts was undetectable. When *Synechococcus* S-subunits separately expressed and isolated from bacteria were added to the assay for Rubisco activity, again no CO_2 fixation was evident.

3.2 Chloroplast structure

Rubisco can be as much as 80% of the stromal contents of chloroplasts so it was of interest to see if altering the *rbcL* gene had had any effect on the morphology of the plastids of these various lines. Electron micrographs (Figure 4) were taken of chloroplasts from leaf thin sections of regenerated transformants. The left panel is a chloroplast from tobacco that expresses the L8s/S8t hybrid and looks almost indistinguishable from a plastid from wild type plants in terms of membrane distribution, stacking and density. In the two right-hand panels are chloroplasts from a tobacco plant (mid) that has no L-subunit gene (4) and hence cannot produce Rubisco compared to those plants that do express *Synechococcus* L-subunit message but do not accumulate any Rubisco L-subunit protein (right). In both cases the internal structue of the plastid is not well developed and its organisation atypical.

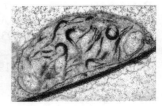

Figure 4. Electron micrographs of plastids from transplastomic plants.

4. Discussion

Homologous recombination in transformed tobacco plants can be used to generate chimeric and hybrid forms of Rubisco. The regenerated plants had the following features:

- Transplastomic plants expressing heterologous plant Rubisco L-subunit genes engineered in bacteria, produce active hybrid forms of the enzyme and their plastid morphology is essentially normal. A cyanobacterial *rbcL* gene produced only message and L-subunit did not accumulate. Plastid morphology was undeveloped.
- The amount of active Rubisco that accumulated in the regenerated lines using these vectors is only about 30% of wild type levels.
- The hybrid Rubisco composed of sunflower L-subunits and tobacco S-subunits had at most 20% of the activity of tobacco Rubisco.
- The Km for CO_2 and ribulose-P_2 substrates was higher by about a factor of five and Vmax for carboxylation lower by a similar amount.
- Activation with CO_2 was compromised only slightly as was the relative specificities of the carboxylase and oxygenase reactions compared to the sunflower enzyme.
- Plants could be maintained on sucrose supplemented media and develop seed by grafting onto wild type stock.
- Unlike other heterologous expression systems, plant transformation allows the compatible folding machinery to generate engineered forms of Rubisco.
- Enough enzyme can be isolated for analysis and potentially crystallography.
- Plants can be investigated for the effects of altered Rubisco on photosynthesis.
- Engineered Rubisco might provide insight into that essential yet elusive interaction with activase.
- Eventually and maybe in combination with tandem expression of L- and compatible S-subunits from the plastid genome, the effects of superior forms of Rubisco through enhanced carbon fixation might now be investigated.

5. Acknowledgements

We wish to thank Zora Svab for the *aadA* gene, Benoit Ranty for the sunflower *rbcL* gene, Tim Bourett for electron micrography and Charles Herrmann for technical assistance. Research was supported by the NSF, and a DuPont Grant to P.M.

References

1. Gutteridge S, Gatenby A. A. (1995) Plant Cell 7: 809-819
2. Svab Z, Maliga P (1993) Proc Natl Acad Sci USA 90: 913-917
3. Rochaix J-D (1997) Trends in Pl Sci 2: 419-425
4. Kanevski I, Maliga P (1994) Proc Natl Acad Sci. USA 91: 1969-1973

LEAF PHOTOSYNTHESIS IN RICE PLANTS TRANSFORMED WITH THE ANTISENSE OR SENSE GENE TO THE SMALL SUBUNIT OF RUBISCO.

Amane Makino, Hiromi Nakano, Tadahiko Mae, Takiko Shimada, Makoto Matsuoka, Ko Shimamoto, Naoki Yamamoto
Graduate School of Agricultural Science, Tohoku University, Sendai 981-8555, Japan (A.M., H.N., T.M.); The Ishikawa Agricultural College, Ishikawa 921, Japan (T.S.); Bioscience Center, Nagoya University, Nagoya 464, Japan (M.M.); Graduate School of Biological Sciences, Nara Institute of Science and Techonology, Ikoma 630, Japan (K.S.); National Institute of Agrobiological Resources, Tsukuba 305, Japan (N.Y.).

Key words: biomass, crop improvement, elevated CO_2, gas exchange, transgenic plants

Introduction

Rubisco is both the key enzyme of photosynthesis/photorespiration and the most abundant leaf protein.   Since Rubisco has a low rate of catalysis, this enzyme is a rate-limiting factor for the rate of light-saturated photosynthesis at present atmospheric CO_2 levels (1, 2) and its amount accounts for 15 to 35% of total leaf N in C_3 species (3).

Genetic engineering using antisense technology provided model plants with decreased Rubisco protein.   These plants have been used as new experimental materials to evaluate the contribution of Rubisco itself to leaf photosynthesis and plant growth (4, 5, 6).   However, such antisense technology has not been undertaken for a major crop.   In addition, Rubisco has not been taken as a target of sense transformation.

In the present work, we transformed rice plants with the antisense or sense gene to rbcS, and examined how these transgenic plants show leaf photosynthesis and plant growth. Some of the data have been already published (7).

Procedure

A cultivar of rice (Oryza sativa L. cv Notohikari) was used, and transformed with the antisense or sense gene to the rice rbcS cDNA and the rice rbcS promoter.   Both genes were introduced by particle bombardment.   The antisense transformants were screened for the presence of the antisense gene verified by PCR with two synthetic oligonu-cleotides corresponding to the rbcS promoter and the rbcS antisense cDNA sequence. The sense transformants were screened for the absence of the interveing sequence between the promoter and the sense rbcS gene.   The selected transformants were transferred to a 3.5-L pot containing paddy soil in a greenhouse under natural sunlight conditions, and allowed to self-fertilize.

The R_1 progeny of the selected transgenic plants were used for the following experiments.   All plants were grown hydroponically with three N concentrations (0.5 mM, 2.0 mM and 8.0 mM N) in an environmentally controlled growth chamber (8).   The uppermost, fully expanded leaves of the 70- to 80-d-old plants were used.

Gas exchange was determined with an open gas exchange system (7).   Measurements were made at an irradiance of 1800 $\mu mol\ m^{-2}\ s^{-1}$, a leaf temperature of 25°C, and

3355

a leaf-to-air vapor difference of 1.0-1.2 kPa.. The amounts of Rubisco and total leaf N were determined according to Makino et al. (7).

Results and Discussion

The primary antisense- and sense-transformants obtained were screened for the Rubisco to leaf N ratio versus leaf N content in young, fully expanded leaves (Fig. 1). For the antisense transformants, the Rubisco to leaf N ratio was decreased by up to 35%, compared with wild-type levels. On the other hand, this ratio from the sense transformants was enhanced by up to 135% wild-type Rubisco. This means that the overproduction of the Rubisco-small subunit leads to an increase in the amount of the holoenzyme protein. However, all sense plants with enhanced Rubisco were completely sterile, whereas even the antisense plants with severely decreased Rubisco were fertile. The A:Ci curve studies with the primary transformants showed the initial slope of the sense plant was enhanced and the CO_2-saturation point was around atmospheric CO_2 levels (S-12 in Fig. 2). In contrast, the initial slope of the antisense plants decreased and the CO_2-saturation point was above 80 Pa of Ci (AS-77 and AS-71). Interestingly, the photosynthetic rate at elevated CO_2 levels in the antisense plants with moderately decreased Rubisco (AS-77) was higher than in the wild-type plants.

Among the primary antisense-transformants, we selected two transformants with 65% (AS-77) and 40% (AS-71) wild-type Rubisco, respectively (indicated by the arrows in Fig. 1). Both transformants were allowed to self-fertilize, and the seeds were

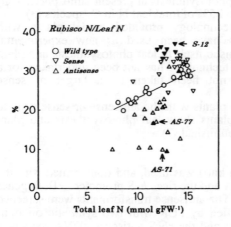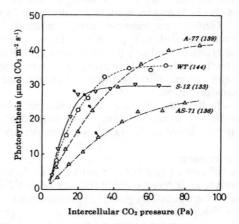

Fig. 1 Ratio of Rubisco N to total leaf N content versus total leaf N content in leaves from the primary transformants screened for the presence of the transgenes. (○) wild type plants, and (▽) sense and (△) antisense Ro transformants. Closed symbols show the sterile transformants. The arrows indicate the selected transformants.

Fig. 2 Photosynthesis as a fuction of Ci (A:Ci curve) in the leaves of the wild-type plant (WT) and the sense (S-12) and antisense (AS-77 and AS-71) Ro transformants. The arrows indicate the points obtained at an external [CO₂] of 36 Pa. The values in parentheses show leaf N content (mmol m⁻²). Measurements were made at 25°C and 1800 μmol quanta m⁻² s⁻¹.

obtained. The segregation of the Rubisco content of the R_1 population from the both transformants was 3:1. The respective R_1 plants with decreased Rubisco had also 65% and 40% wild-type Rubisco.

We examined gas exchange characteristics at normal CO_2 (36 Pa) and elevated CO_2 (100 Pa) levels (Fig. 3). At 36 Pa CO_2, the photosynthetic rates from the both antisense plants were lower than that from the wild-type plants. At 100 Pa CO_2, however, the antisense plants with 65% wild-type Rubisco (AS-77) showed 5-15% higher rates of photosynthesis than the wild type plants for the same leaf N content, whereas the antisense plants with 40% wild-type Rubisco had lower photosynthesis. These results indicated that N was optimally distributed between Rubisco and components limiting CO_2-saturated photosynthesis in the antisense plants with moderately decreased, whereas N was not optimally distributed in the plants with severely decreased Rubisco (see the detailed discussion in ref 7). Thus, the plants with 65% wild-type Rubisco (AS-77) may have a higher N-use efficiency of photosynthesis under the conditions of the elevated CO_2.

To analyze the growth at the level of the whole plant, we next selected homo-like segregants among R_2 progeny from the respective antisense plants. Fig. 4 shows total plant biomass including roots on the 77th d after germination. When the plants were grown in 36 Pa CO_2, the total biomass of the antisense plants was much smaller than that of the wild-type plants. These results indicated that the growth of the whole plant was correlated with Rubisco content in the leaf. When they were grown in 100 Pa CO_2, the difference in the total biomass between the antisense and wild-type plants was very small, but the biomass of the antisense plants with 65% wild-type Rubisco was not greater than that of the wild-type plants. Thus, the antisense plants with higher N-use efficiency of leaf photosynthesis at the elevated CO_2 levels do not necessarily perform

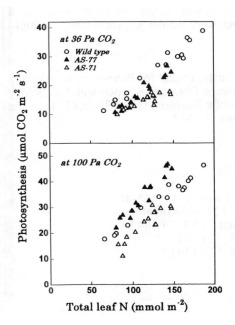

Fig. 3 **Rates of photosynthesis at 36 Pa (upper panel) and 100 Pa (lower panel) CO_2 versus total leaf N content.** (O) wild type, (▲) AS-77 and (Δ) AS-71 R_1 antisense plants. Measurements were made at a leaf temperature of 25°C and 1500 1800 μmol quanta m^{-2} s^{-1}.

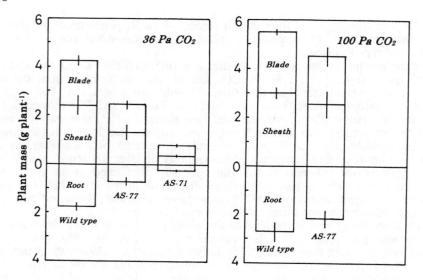

Fig. 4   Plant biomass of wild-type plants and AS-77 and AS-71 R_2 antisense plants on the 70th d after germination.   The vertical bars indicate the SE ($p < 0.05$, n = 3 - 4). Plants were grown hydroponically at a PPFD of 1000 µmol quanta m^{-2} s^{-1} (14-h photo-period), a day/night temperature of 25/20°C and the indicated CO_2 partial pressures.

better in elevated CO_2 environments.   This was mainly caused by lower N uptake of the antisense plants (data not shown).   Further work on the growth analysis in high-CO_2 environments is in progress.

Acknowledgements
This work was supported by the Special Coordination Fund for Promoting Science and Technology, Enhancement of Center-of-Excellence from the Science and Technology Agency, Japan, and by Grants-in-Aid for Scientific Research (No. 09660061) from the Ministry of Education, Science and Culture, Japan to AM.

References
1.   Evans, J.R. (1986) Planta 167, 351-358
2.   Makino, A. et al. (1985) Planta 166, 414-420
3.   Evans, J.R. (1989) Oecologia 78, 9-19
4.   Stitt, M. and Schulze, D. (1994) Plant Cell Environ. 17, 465-487
5.   Andrews, T.J. et al. (1995) J. Exp. Bot. 46, 1293-1300
6.   Furbank, R.T. and Taylor, W.C. (1995) Plant Cell 7, 797-807
7.   Makino, A. et al. (1997) Plant Physiol. 114, 483-491
8.   Makino, A. et al. (1994) Plant Physiol. 105, 173-179

DIRECTED MUTAGENESIS OF THE LARGE SUBUNIT OF TOBACCO RUBISCO ASSESSED *IN VIVO*.

Spencer M. Whitney, Susanne von Cammerer, Graham S. Hudson and T. John Andrews. Molecular Plant Physiology Group, Research School of Biological Sciences, The Australian National University, PO Box 475, Canberra, ACT 2601, Australia.

Key words: chloroplast transformation, gas exchange, mutational analysis, photorespiration, transgenic plants

1. Introduction.

Mutational analyses of the CO_2-fixing enzyme Rubisco from higher plants has been hindered by the inability to fold and assemble the enzyme correctly in heterologous expression systems. Nevertheless, a wealth of information exists about Rubisco's structure and function (1,2). Development of a method for transforming the chloroplast genome (plastome) (3) provides an avenue for using this knowledge to manipulate the higher plant enzyme. Chloroplast transformation occurs via homologous recombination, thereby allowing site directed mutagenesis of chloroplast genes, such as the *rbc*L gene for the large subunit of Rubisco.

Using a new transformation vector, we mutated codon Leu-335 of *rbc*L and conservatively changed it to Val. Following insertion of the gene into the tobacco plastome, the consequences of the mutation were analysed by measuring the kinetic properties of the altered Rubisco and by leaf gas-exchange measurements with the mutant plant.

2. Procedure.

The transformation plasmid, pLEV1, contained 4kb of the tobacco plastome encompassing *rbc*L and its flanking regions (nucleotides 56481 to 60488; (4)). A promoterless *aad*A gene, that confers resistance to spectinomycin, was inserted downstream of *rbc*L. The Leu-335 codon of *rbc*L was mutated to Val and an *Eco*RV site introduced 15 nucleotides downstream by silent changes . This plasmid (pL335V) was used to transform in *Nicotiana tabacum* L. cv Petit Havana, via biolistic bombardment (3,5) and resistant plantlets selected on RMOP media (3) containing spectinomycin (500 µg/ml). Resistant plantlets were screened for the introduced *Eco*RV site by Southern analysis of digested leaf DNA. Homoplasmic (according to Southern analysis) mutant

3359

G. Garab (ed.), Photosynthesis: Mechanisms and Effects, Vol. V, 3359–3362.
© *1998 Kluwer Academic Publishers. Printed in the Netherlands.*

plants and wildtype controls were grown in 10 l pots of soil in an artificially lit (400 μmol quanta m^{-2} s^{-1}) growth cabinet in air supplemented with 0.3% CO_2 using a 14 h photoperiod (25°C /18°C). Gas exchange measurements were made on young, fully expanded leaves of mutant and wildtype plants as previously described (6). Kinetic parameters of the mutant and wildtype Rubiscos measured *in vitro* included; Rubisco content (CABP-binding) (6); substrate-saturated activity (7); Michaelis constants (at pH 8.3) for CO_2, O_2 and RuBP (8); and the CO_2/O_2 specificity (9). For the latter assays, Rubisco was purified by centrifuging 8 to 18% polyethylene glycol fractions (10) on sucrose density gradients (11).

3. Results and Discussion

3.1 *Transformation of the tobacco chloroplast genome.*

Guided by previous mutagenesis of prokaryotic Rubiscos (12,13) and crystallographic information about Rubisco's active site (14), we reasoned that mutation of Leu-335 to Val might decrease the CO_2/O_2 specificity without seriously compromising activity (see the contribution by T.J. Andrews, S von Caemmerer, Z. He, G.S. Hudson and S.M. Whitney in these proceedings for more details). A new vector, pLEV1, was engineered for transforming the plastome of tobacco (Fig 1a). Codon 335 of *rbc*L was converted from Leu to Val by directed mutagenesis to produce the transformation vector pL335V. To identify transformants with the introduced mutation, a nearby *Eco*RV restriction site was introduced. The vector was transformed into the tobacco plastome and, after two rounds of selection, resistant plantlets were screened for the introduced *Eco*RV site by Southern analysis of digested leaf DNA. One plantlet (from eight transformants) contained the *Eco*RV site. Subclones of this plantlet were subjected to two additional rounds of regeneration on the selective medium, after which they were homoplasmic (Fig 1b). The *rbc*L gene was amplified by PCR and sequenced to confirm that only the

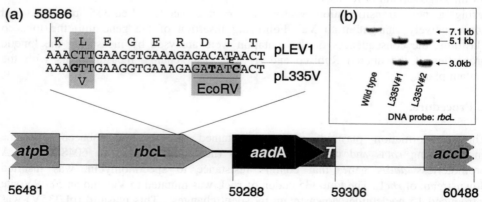

Figure 1. (a) The pL335V plasmid used for transforming the tobacco plastome and (b) Southern analysis of EcoRV-digested leaf DNA from wildtype and two homoplasmic transformants. Numbers refer to the tobacco plastome (4). T, rps16 terminator sequence.

intended changes had been introduced. Five mutants were transferred to soil and grown in air supplemented with CO_2 (0.3%). Even under these growth conditions, mutant plants grew much slower than wildtype plants.

3.2 In vivo *analyses.*

Gas-exchange characteristics of leaves from mutant and control plants were compared (see Fig 4a in Andrews *et al.* manuscript). In the mutant plants, the CO_2-compensation point in air was increased four-fold relative to wildtype. Assimilation in the mutant plants remained limited by Rubisco activity at CO_2 concentrations up to 1200 μbar. In contrast, assimilation in wildtype leaves became limited by light-dependent RuBP regeneration above 300 μbar, as expected. CO_2-response curves were measured at different O_2 concentrations. Consistent with the poorer CO_2/O_2 specificity of the mutant Rubisco, the CO_2 compensation point increased with increasing O_2 concentration four times as steeply with mutant leaves than with control leaves (see Fig 4b, Andrews *et al.* manuscript).

3.3 *Comparing kinetic parameters measured* in vivo *and* in vitro

Kinetic parameters of the wildtype and mutant Rubiscos were calculated from the *in vivo* analyses using a model for photosynthetic gas exchange (15). These were compared with *in vitro* measurements with the isolated enzymes (Table 1). Data from the two methods are in general agreement. The CO_2/O_2 specificity and maximum carboxylase activity of the mutant enzyme were both reduced over 70%. Likewise the Michaelis constants of the mutant enzyme for were also reduced by 50 to 90 % compared to the wildtype enzyme. In contrast to these reductions, the Rubisco content of mutant plants was similar to wildtype; however the carbamylation status was almost twice that of the wildtype under

Parameter	in vivo		in vitro	
	Leu-335*	Val-335	Leu-335[†]	Val-335
Rubisco content (μmol m^{-2})	30.3 ± 1.6	33.0 ± 5.6	-	-
% carbamylation[‡]	48.1 ± 5.6	89.9 ± 5.4	-	-
V_{cmax} (s^{-1})	3.53	1.02	6.2 ± 0.4	1.8 ± 0.1
$K_m(CO_2)$ (μM)	$8.6^\S - 13.5^\$$	$8.6^\S - 10.6^{\$¼}$	11.5 ± 0.9	5.0 ± 1.1
$K_i(O_2)$ (μM)	$226^\S - 313^\$$	$55^\S - 70^\$$	222 ± 51	40.9 ± 12.9
$K_m(RuBP)$ (μM)	-	-	18.8 ± 3.2	2.1 ± 0.2
$S_{c/o}$	102.0	27.0	81.1 ± 1.6	20.1 ± 1.5

Table 1. Kinetic parameters for Leu-335 (wildtype) and Val-335 Rubiscos in vivo *and* in vitro.* **Measurements with anti-rbcS N. tabacum L. cv. W38 (16).* [†]*Measurements with wildtype regenerants.* [‡]*Measured with leaves sampled under the growth conditions.* § *Assuming a value of 0.3 mol m^{-2} s^{-1} bar^{-1} for the conductance for CO_2 transfer between the intercellular air spaces and the sites of carboxylation.* $^\$$*Assuming infinite conductance for CO_2 transfer between the intercellular air spaces and the sites of carboxylation.*

growth conditions. The reduced $K_m(CO_2)$ for the mutant enzyme measured *in vitro* was the only parameter not replicated closely in the *in vivo* data.

3.4 *Conclusions*
This work demonstrates that chloroplast transformation enables directed mutagenesis of higher-plant Rubisco and that it provides the mutant Rubisco in the context of the whole plant for physiological as well as biochemical analysis. The agreement between the kinetic parameters of the mutant and wildtype Rubiscos calculated from gas exchange measurements with those measured *in vitro* engenders confidence in current understanding of the linkage between photosynthesis and photorespiration in terms of a common dependence on Rubisco.

References

1. Hartman, F.C. and Harpel, M.R. (1994) Annu. Rev. Biochem. 63, 197-234
2. Cleland, W.W., Andrews, T.J., Gutteridge, S., Hartman, F.C. and Lorimer, G.H. (1998) Chem. Rev. 98, 549-561
3. Svab, Z., and Maliga, P. (1993) Proc Natl Acad Sci. 90, 913-917
4. Shinozaki, K., Ohme, M., Tanaka, M., Wakasugi, T., Hayashida, N., Matsubayshi, T., Zoutu, N. and Sugiura, M (1986) EMBO J 5, 2043-2049
5. Bock, R (1998) Methods. 15, 75-83
6. He, Z., von Caemmerer, S., Hudson, G.S., Price, D.G., Badger, M.R. and Andrews, T.J. (1997) Plant Physiol. 115, 1569-1580
7. Mate, C.J., Hudson, G.S., von Caemmerer, S., Evans, J.R., and Andrews, T.J. (1993) Plant Physiol. 102, 1119-1128
8. Paul, K., Morell, M.K. and Andrews, T.J. (1991) Biochemistry 30, 10019-10026
9. Kane, H.J., Viil, J., Entsch, B., Paul, K., Morell, M.K. and Andrews, T.J. (1994) Aust J Plant Physiol. 21, 449-461
10. Servaites, J.C. (1985) Arch. Biochem. Biophys. 238, 154-160
11. Andrews, T.J. (1988) J. Biol. Chem. 263, 12213-12220
12. Lee, G.J., McDonald, K.A., and McFadden, B.A. (1993) Protein Science 2, 1147-1154
13. Terzaghi, B.E., Laing, W.A., Christeller, J.T., Petersen, G.B., and Hill, D.F. (1986) Biochem. J. 235, 839-846
14. Andersson, I. (1996) J. Mol. Biol. 259, 160-174
15. Farquhar, G.D., von Caemmerer, S. and Berry, J.A. (1980) Planta 149, 78-90
16. von Caemmerer, s., Evans, J.R., Hudson, G.S. and Andrews, T.J. (1994) Planta 195, 88-97

RUBISCO SSMT AND LSMT: RELATED ᵅN- AND ᵉN-METHYLTRANSFERASES THAT METHYLATE THE LARGE AND SMALL SUBUNITS OF RUBISCO

Zhentu Ying[a], R. Michael Mulligan[b], Noel Janney[a], Malcolm Royer[a], and Robert L. Houtz[a], [a]University of Kentucky, Lexington, KY 40546-0091 and [b]University of California, Irvine, CA 92697-2300

Keywords: molecular biology, processing enzymes, protein synthesis, stromal enzymes, protein methyltransferases

1. Introduction

Ribulose-1,5-bisphosphate carboxylase/oxygenase (Rubisco) experiences several types of post-translational modifications during the expression, import and assembly of the protein. The large subunit (LS) from many plant species contains a trimethyllysyl residue at Lys-14 (1-3). The small subunit (SS) is post-translationally modified by removal of the targeting presequence, and the N-terminal methionine residue subjected to ᵅN-methylation (4). The methylation of Lys-14 in the LS is catalyzed by S-adenosyl-L-methionine (AdoMet): Rubisco LS (lysine) ᵉN-methyltransferase (Rubisco LSMT, Protein Methylase III, EC 2.1.1.127) (5). In spinach a homologue of Rubisco LSMT exists that encodes an enzyme that methylates the SS of Rubisco (Rubisco SSMT). Recombinant tobacco Rubisco LSMT expressed in *E. coli*, and both recombinant and purified pea Rubisco LSMT, also catalyze the methylation of the α amino group of the processed form of the SS.

2. Procedures

Two *rbcMT-S* cDNAs were obtained by RT-PCR and 5' and 3' RACE using degenerate oligonucleotide primers corresponding to conserved peptide sequences between pea (5) and tobacco (6) Rubisco LSMTs. Full-length S-38 and S-40 spinach cDNAs (accession #'s AF071544 and AF071545) were obtained by ligation of partial clones. S-38 and S-40 were cloned into pET-23d for expression in pLysS cells. A subclone of spinach Rubisco SSMT was transcribed with T3 RNA polymerase in the presence of ^{32}P-rATP (800 Ci/mMol) to produce a radiolabeled antisense transcript for RNase protection assays. The activity of Rubisco LSMT was determined in thawed cell lysates as described previously (7). Rubisco SSMT activity was determined by incubation of purified S-38 with lysates from spinach chloroplasts isolated as described in reference (8). After incubation at 30°C for 30 min, reactions were terminated by addition of SDS-PAGE sample preparation buffer, and samples electrophoresed on 15% acrylamide gels. After transfer of proteins to a PVDF membrane, radioactivity was visualized using a ^{3}H phosphor-imager screen (Molecular Dynamics Model 425) and polypeptides visualized by staining with Coomassie Brilliant Blue R-250.

3. Results and Discussion

LSMT-specific primers were designed to conserved sequences of pea and tobacco Rubisco LSMT (Fig. 1, arrows) and used to amplify a 786-bp fragment from first-strand cDNAs.

G. Garab (ed.), Photosynthesis: Mechanisms and Effects, Vol. V, 3363–3366.
© 1998 *Kluwer Academic Publishers. Printed in the Netherlands.*

Sequence analysis of the 786-bp fragment indicated that the spinach cDNA had a large extent of identity with pea and tobacco LSMT (Fig. 1). PCR amplification resulted in the identification of two 5′ RACE products (836-bp and 848-bp fragments). The 848 bp product (designated as S-40′) had a 12-bp insertion relative to the 836 bp fragment (designated as S-38′). The *rbcMT-S* cDNAs encode polypeptides of 495-aa (S-40) and 491-aa (S-38) with predicted molecular masses of 55.5 kD and 55.0 kD, respectively, similar to the masses of pea (55.0 kD) and tobacco (56.0 kD) Rubisco LSMT (Fig. 1), and also included five imperfect copies of a leucine-rich repeat (LRR), similar to pea and tobacco LSMT (Fig. 1, underlined regions) (9). The amino acid sequence from the *rbcMT-S* cDNAs shows 60% and 62% identity with the amino acid sequences of pea and tobacco Rubisco LSMT, respectively. A 3.1 kb genomic clone that included the entire coding region of the *rbcMT-S* gene was obtained and sequenced (GenBank accession #AF071543). The *rbcMT-S* gene has similar genomic organization with the tobacco Rubisco LSMT gene (*rbcMT-T*). The 12-bp insertion in the *rbcMT-S* S-40 cDNA corresponds to the 3′ 12 nt of intron III. Examination of the DNA sequence of this intron and flanking regions suggested that either of two 3′ splice sites (separated by the 12-bp sequence) could be utilized during splicing of the *rbcMT-S* transcripts. RNase protection studies were performed to analyze the relative expression of the S-38 and S-40 mRNAs in spinach leaves. The results revealed that S-40 and S-38 transcripts were present in approximately equal abundance (Fig. 2, Lanes 3-5). Thus, two transcripts of a single *rbcMT-S* gene are expressed in spinach leaves that differ by a 12-bp insertion as a

```
Spinach40    MATLFTLIPS-SNSTFLNPFKTTQHSKLHFATPSPTFKNPLSIRCFRPPETDTPPEIQKFWGW   62
Spinach38    MATLFTLIPS-SNSTFLNPFKTTQHSKLHFATPSPTFKNPLSIRCFRPPETDTPPEIQKFWGW   62
Pea          MATIFSGG---SVSPFLFHTNKGTSFTPKAPILHLKRSFSAKSVASVGTEPSLSPAVQTFWKW   60
Tobacco      MASVFSVHPLPSSSFLCPLKTTKSRTKHHQTFYTYQKTILINSLQLTELDPKIPQPVQTFWQW   63

Spinach40    LSDKGIISPKCPVKPGIVPEGLGLVAQKDISRNEVVLEVPQKFWINPDTVAASEIGSVCNGLK 125
Spinach 38   LSDKGIISPKCPVKPGIVPEGLGLVAQKDISRNEVVLEVPQKFWINPDTVAASEIGSVCNGLK 125
Pea          LQEEGVITAKTPVKASVVTEGLGLVALKDISRNDVILQVPKRLWINPDAVAASEIGRVCSELK 123
Tobacco      LCKEGVVTTKTPVKPGIVPEGLGLVAKRDIAKGETVLQVPKRFWINPDAVAESEIGNVCSGLK 126

Spinach40    PWVSVALFLMREKKLGNSSSWKPYIDILPDSTNSTIYWSEEELSELQGSQLLNTTLGVKELVA 188
Spinach40    PWVSVALFLMREKKLGNSSSWKPYIDILPDSTNSTIYWSEEELSELQGSQLLNTTLGVKELVA 188
Pea          PWLSVILFLIRER-SREDSVWKHYFGILPQETDSTIYWSEEELQELQGSQLLKTTVSVKEYVK 185
Tobacco      PWISVALFLLREK-WRDDSKWKYYMDVLPKSTDSTIYWSEEELSEIQGTQLLSTTMSVKDYVQ 188

Spinach40    NEFAKLEEEVLVPHKQLFPFDVTQDDFFWAFGMLRSRAFTCLEGQSLVLIPLADLWVQQANHS 251
Spinach38    NEFAKLEEEVLVPHKQLFPFDVTQDDFFWAFGMLRSRAFTCLEGQSLVLIPLADL----ANHS 247
Pea          NECLKLEQEIILPNKRLFPDPVTLDDFFWAFGILRSRAFSRLRNENLVVVPMADL----INHS 244
Tobacco      NEFQKVEEEVILRNKQLFPFPITLDDFFWAFGILRSRAFSRLRNQNLILVPFADL----TNHN 247

Spinach40    PDITAPKYAWEIRG-AGLFSRELVFSLRNPTPVKAGDQVLIQYDLNKSNAELALDYGLTESRS 313
Spinach38    PDITAPKYAWEIRG-AGLFSRELVFSLRNPTPVKAGDQVLIQYDLNKSNAELALDYGLTESRS 209
Pea          AGVTTEDHAYEVKGAAGLFSWDYLFSLKSPLSVKAGEQVYIQYDLNKSNAELALDYGFIEPNE 307
Tobacco      ARVTTEDHAHEVRGPAGLFSWDLLFSLRSPLKLKAGDQLFIQYDLNKSNADMALDYGFIEPSS 310

Spinach40    ERNAYTLTLEIPESDSFYGDKLDIAESNGMGESAYFDIVLEQPLPANMLPYLRLVALGGEDVF 376
Spinach38    ERNAYTLTLEIPESDSFYGDKLDIAESNGMGESAYFDIVLEQPLPANMLPYLRLVALGGEDVF 372
Pea          NRHAYTLTLEISESDPFFDDKLDVAESNGFAQTAYFDIFYNRTLPPGLLPYLRLVALGGTDAF 370
Tobacco      ARDAFTLTLEISESDEFYGDKLDIAETNGIGETAYFDIKIGQSLPPTMIPYLRLVALGGTDAF 373

Spinach40    LLESIFRNSIWGHLDLPISPANEELICQVIRDACTSALSGYSTTIAEDEKLLAEGDIDPRLEI 439
Spinach38    LLESIFRNSIWGHLDLPISPANEELICQVIRDACTSALSGYSTTIAEDEKLLAEGDIDPRLEI 435
Tobacco      LLESIFRNSVWGHLGLPVSRANEELICKVVRDACKSALSGYHTTIEEDEKLMEEGNLSTRLQI 436
Pea          LLESLFRDTIWGHLELSVSRDNEELLCKAVREACKSALAGYHTTIEQDREL-KEGNLDSRLAI 432

Spinach40    AITIRLGEKKVLQQIDEEFKEREMELGGYEYYQERRLKDLGLAGEQGEKLPWIGEV        495
Spinach38    AITIRLGEKKVLQQIDEEFKEREMELGGYEYYQERRLKDLGLAGEQGEKLPWIGEV        491
Pea          AVGIREGEKMVLQQIDGIFEQKELELDQLEYYQERRLKDLGLCGENGDILGDLGKFF       489
Tobacco      AVGIRLGEKRVLKQIDDIFRERELELDELEYYGERRLKDLGLVGEQGDIIFWEPK         491
```

Figure 1

result of 3′ alternative splice site selection. Polyclonal antibodies were prepared against purified S-38 protein and used to detect translation products. A single immunoreactive polypeptide was detected in spinach chloroplast lysates (~50.6 kD, Fig. 3). The presence of chloroplast-localized translation products for S-38 and S-40 with homology to both pea and tobacco Rubisco LSMT suggested that these proteins may have a different, but related, type of methyltransferase activity. Pea Rubisco LSMT catalyzes incorporation of methyl groups from [^3H-methyl]AdoMet into Rubisco isolated from unmethylated species such as spinach. All attempts to measure methyltransferase activity towards isolated and purified spinach Rubisco were negative. A recent report of α-methylation of the N-terminus of the processed form of the SS of Rubisco (4) lead us to examine the possibility that the S-38 and S-40 cDNAs coded for an ^αN-methyltransferase for the SS of spinach Rubisco. Cell lysates from bacterial expression of S-38 and S-40 constructs were assayed for methyltransferase activity with spinach chloroplast lysates as a proteinaceous substrate. Tritium radiolabel incorporation

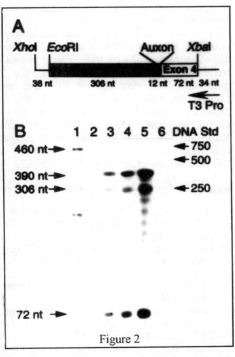

Figure 2

from [^3H-methyl]AdoMet into spinach chloroplast polypeptides was assessed by SDS-PAGE, electrophoretic transfer to PVDF membranes, and phosphor image analysis. Incorporation of radiolabel was detected into a region corresponding to the SS of Rubisco which was dependent on the addition of chloroplast lysates and purified S-38 (Fig. 4). Similar results were obtained with purified S-40. These results suggest that a small pool of free SSs exists in the stromal fraction from spinach chloroplasts as has been previously demonstrated in chloroplasts from *Chlamydomonas* (10) and peas (11). That purified recombinant S-38 catalyzed SS methylation at the N-terminal Met of processed SS was documented by Edman degradation of the SS with simultaneous determination of radioactivity. During the first cycle

of Edman degradative sequencing, 92% of the radiolabel incorporated into the small subunit region of the PVDF blot was released (Fig. 5). The amino acid residue in this cycle did not correspond to any of the standard amino acids, but migrated as an unknown with a retention time just slightly greater than DPU, similar to that reported for methylated Met in other work (4). The next 11 amino acid

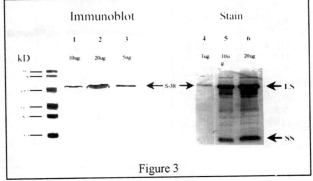

Figure 3

residues corresponded exactly to the known N-terminal residues of the SS of spinach Rubisco (12). Given that a thorough chemical and physical structural analysis of the N-terminal residue of the SS of spinach Rubisco has already identified this residue as N-methylmethionine, the results presented here demonstrate that the two gene products from the *rbcMT-S* gene code for enzymes capable of methylating the N-terminal Met residue of the processed form of the small subunit of Rubisco. We propose that these enzymes be referred to as Rubisco small subunit ᵅN-methyltransferase or Rubisco SSMT.

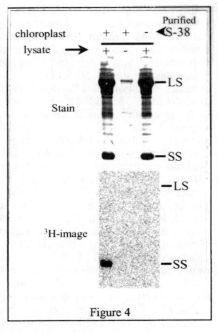

Figure 4

References

1 Houtz, R.L., Stults, J.T., Mulligan, R.M. and Tolbert, N.E. (1989) Proc. Natl. Acad. Sci. USA 86, 1855-1859

2 Houtz, R.L., Royer, M. and Salvucci, M.E. (1991) Plant Physiol. 97, 913-920

3 Houtz, R.L., Poneleit, L., Jones, S.M., Royer, M. and Stults, J.T. (1992) Plant Physiol. 98, 1170-1174

4 Grimm, R., Grimm, M., Eckerskorn, C., Pohlmeyer, K., Rohl, T. and Soll, J. (1997) FEBS Letters 408, 350-354

5 Klein, R.R. and Houtz, R.L. (1995) Plant Mol. Biol. 27, 249-261

6 Ying, Z., Janney, N. and Houtz, R.L. (1996) Plant Mol. Biol. 32, 663-671

7 Wang, P., Royer, M. and Houtz, R.L. (1995) Prot. Expr. Pur. 6, 528-536

8 Mills, W.R. and Joy, K.W. (1980) Planta 148, 75-83

9 Kobe, B. and Deisenhofer, J. (1994) Trends Biochem. Sci. 19, 415-421

10 Schmidt, G.W. and Miskind, M.L. (1983) Proc. Natl. Acad. Sci. USA 80, 2632-2636

11 Hubbs, A. and Roy, H. (1992) Plant Physiol. 100, 272-283

12 Martin, W., Mustafa, A.-Z., Henze, K. and Schnarrenberger, C. (1996) Plant Mol. Biol. 32, 485-491

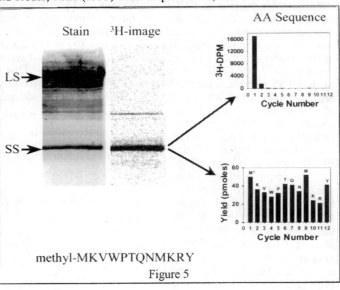

methyl-MKVWPTQNMKRY

Figure 5

ACTIVE OXYGEN SPECIES TRANSIENTLY SUPPRESS THE SYNTHESIS OF THE CHLOROPLAST ENCODED RUBISCO LSU

Michal Shapira and Vered Irihimovitch
Ben-Gurion University of the Negev, Beer-Sheva 84105, Israel

Key words: Chlamydomonas , chloroplast genes, light stress. Rubisco, active oxygen, molecular biology

Background

Excess light intensities are the source of oxidizing activity in the chloroplast (1). Photoinhibitory effects can also be observed upon immediate shifts from low to high light intensities, that create an imbalance between the antenna size and the excess amount of light. Most plants have the capacity to recover from light stress through the process of photoacclimation which involves increased synthesis of D1 and the CO_2 fixing enzyme Rubisco, the most abundant, but slowly turned-over chloroplast protein (2).

The chloroplast redox state has been implicated in control of the photosynthetic apparatus when plants or algal cells are shifted from low to high light. This has been demonstrated for transcription of the nuclear encoded *cab* genes (3) and for translation of the chloroplast *psbA* gene which encodes for D1 (4). Redox is assumed to play a key role in regulation of gene expression, and efforts have been directed to identify the target which senses it and sends out the regulatory signals (5).

The Ribulose-1,5-biphosphate carboxylase/oxygenase (Rubisco) holoenzyme is the key enzyme in CO_2 fixation (6). This protein complex consists of 8 large subunits (LSU) encoded by the chloroplast *rbcL* gene, and 8 small subunits (SSU), encoded by the *rbcS* gene in the nucleus. We previously demonstrated that shifting low light (LL) adapted cells of *C. reinhardtii* to high light (HL) results in a dramatic but transient downregulation in the synthesis of Rubisco LSU (7). We assume that active oxygen species (AOS) which accumulate during the shift from LL to HL, have a key role in control of LSU synthesis.

Methodology

LL grown *C. reinhardtii* (CC-125) were pulse labeled with [^{35}S] H_2SO_4 in LL and after transfer to HL under autotrophic growth conditions, as previously described (7), in the presence or absence of herbicides and ascorbic acid. Mutants of electron transfer were grown on acetate containing media. Transfer of these mutants from LL to HL and their metabolic labeling were performed after depletion of the acetate during 24 h. Labeling was performed in whole cell cultures and the labeled proteins were separated on SDS-PAGE and further analyzed by autoradiography.

Results and Discussion

Transfer of Chlamydomonas reinhardtii *from LL to HL leads to a transient arrest in synthesis of Rubisco LSU.* Transfer of the cells from LL (70 µmol m^{-2} s^{-1}) to HL (700 µmol m^{-2} s^{-1}) or to medium light (ML, 300 µmol m^{-2} s^{-1}), led to a transient but almost

3367

G. Garab (ed.), Photosynthesis: Mechanisms and Effects, Vol. V, 3367–3370.
© 1998 *Kluwer Academic Publishers. Printed in the Netherlands.*

complete shut down in LSU synthesis and a concomitant increase in D1 synthesis after 15 min and 2h. Recovery of the capacity to synthesize LSU was evident at 4-6 h and occurred sooner if the cells were exposed to ML rather than to HL (Fig. 1). Synthesis of LSU had fully resumed 12 h after transfer to ML or HL.

Experiments in which D1, LSU as well as other photosynthetic proteins were pulse labeled and then immunoprecipitated confirmed the clearly differential responses in the synthesis of these two proteins following transfer of LL grown cells to HL, and further indicated that the drastic downregulation was unique to Rubisco LSU, and was not observed with other tested photosynthetic genes (7).

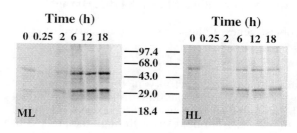

Fig. 1. *Effects of shifting LL grown cells of* C. reinhardtii *cells to HL on synthesis of the Rubisco LSU and PSII D1 proteins.* LL grown cells were transferred to ML or HL for 0.25 to 18h and then pulse labeled for 15 min with 500 μCi of [^{35}S] H_2SO_4 in the presence of anisomycin. Labeled proteins were separated by LDS-PAGE and quantified by phosphorimaging.

The observed changes in D1 and LSU synthesis could not be correlated with changes in the steady state levels of their corresponding mRNAs, implying that translational regulation was involved. Primer extension analysis of *rbcL* mRNA isolated from LL grown and HL (2h) cells, revealed the presence of two transcripts which differed in their 5' ends, mapping to positions -93 and -168 nts upstream of the ATG initiation codon. Although both mRNAs were always present, the ratio between them varied under different light conditions, and the higher abundance of the -93 transcript correlated with active translation of LSU (7).

The effects of light shifts on chlorophyll a fluorescence parameters, chlorophyll content, and cell division. A rapid increase in the redox state of the electron transport pathway is indicated by the increase in the reduction state of Q_A, as inferred from changes in 1-qP seen immediately following transfer of LL grown cells to HL (Fig. 2A). This was followed by only a slight decline in the highly reduced state over the next 24 h. In contrast, photosynthetic efficiency measured as Fv/Fm decreased sharply during the first hour in HL and then gradually recovered by 6 h (Fig. 2B). Consistent with the cells' acclimation following initial photoinhibition, total chlorophyll content per biomass declined by 60% during the first 6 h at HL and then stabilized (Fig. 2C). During this photoacclimation process, cell number barely increased during the first 6h, rose 2.3 fold during the next 6 h and then increased 16 fold between 12 and 24 h. Concomitantly cell size (A_{750} per 10^8 cells) and total protein content (mg per 10^8 cells) increased continuously during the first 12 hours and then dropped back to the values typical of LL grown cells as the cells divided. The foregoing pattern is consistent with the LL->HL shift serving to synchronize temporarily the growth and division of the logarithmic population of LL grown cells. It also indicated that HL acclimation is associated with cell growth and subsequent cell division.

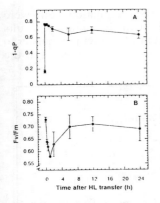 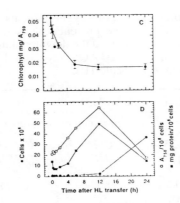

Fig. 5. *Effect of shifting LL grown cells of C. reinhardtii to HL on reduction state of QA, photoinhibition, photoacclimation, cell size, protein content and cell division.* A. Increase in reduction state of QA indicated by 1-qP. B. Initial photoinhibition following transfer to HL, indicated by the decline of Fv/Fm, followed by recovery during photoacclimation. C. Decline in total chlorophyll content per unit biomass following transfer to HL. D. Increase in cell number (X10⁶ cells), cell biomass (A750/10⁸ cells) and total cell protein (mg/10⁸ cells) following transfer to HL. All data points represent means of three or more independent determinations.

Effects of herbicides that inhibit the electron transport on LSU downregulation. To examine how blocking of the electron transport at specific points affected the downregulation of LSU synthesis, wildtype *C. reinhardtii* (CC-125) cells were labeled in LL and after transfer to HL (2h) in the presence of DCMU and DBMIB. DCMU inhibits oxidation of QA, and blocks reduction of plastoquinone whereas DBMIB, a PQ analog, inhibits its oxidation. Preliminary experiments revealed that downregulation of LSU synthesis was inhibited selectively by DCMU, while no such inhibition was observed with DBMIB, suggesting that the redox state of PQ could play a role in LSU regulation (Table 1A). A similar indication was obtained from labeling of *C. reinhardtii* mutants defective in their electron transport. The analysis included mutants of PSII deficient of D1 [*FUD7*, (8)] or defective in that protein due to an Ala251->Arg point mutation [*A251R*, (9)], and a PSI mutant including a frameshift mutation in the *psaB* gene [*ac-u-g-2-3*, (10)]. However, since the downregulation of LSU synthesis is transient, and the PQ is reduced immediately after transfer to HL and remains in a reduced state [(7) and Fig. 2], a direct role for redox was ruled out. Thus, DCMU could exert its effect by preventing the accumulation of certain AOS.

Potential signaling for LSU downregulation by AOS. A possible explanation for the downregulation of LSU is that its synthesis is regulated by AOS. A signal could be generated by the imbalance that occurs when cells grown in LL with their large chlorophyll antenna complexes are shifted to HL, where the large antenna size traps too many light quanta for optimal photosynthetic electron transport. As the antenna size is adjusted downward in response to redox, the imbalance would be dissipated and LSU synthesis increases once again. AOS, whose levels increase during the period of imbalance, could be involved in initiating the signal that leads to the translational arrest of LSU.

Changes in the level of AOS within the cells were evaluated using the oxidatively sensitive fluorophore dichlorofluorescin (DCFH). Shifting the cells from LL to HL lead to a transient increase of 1.7 fold in the level of AOS, which returned to the basal level after 2.5 h. This transient increase in the level of AOS correlated with the transient nature of LSU downregulation, and could signal its translational arrest. In the presence of ascorbic acid, a scavenger of AOS, this increase was inhibited, and basal AOS levels were obtained after less than 90 minutes.

To examine the potential role of oxygen radicals in LSU regulation, cells were labeled at LL and at HL (2h) in the presence of ascorbic acid. Preliminary results demonstrated that ascorbic acid prevented the downregulation of LSU after the transfer to HL, indicating that the increase in AOS level could initiate the signal for the downregulation of LSU synthesis (Table 1B).

Table 1. *HL/LL ratio of LSU synthesis in:*

A. C. reinhardtii *mutants defective in the electron transport pathway*

Strain	CC-125 (w.t.)	CC-741 (FUD 7)	CC-3376 (A251R)	CC-703 (ac-u-g-2-3)
HL/LL	0.1-0.5	1.1	0.9	0.1

B. Wildtype C. reinhardtii *labeled in the presence of herbicides and AOS scavengers*

Herbicides and Ascorbate	DCMU (10^{-7}M)	DBMIB (10^{-6}M)	Ascorbic acid (10mM)
HL/LL	1.0	0.5	0.9

CC-125-w.t. *C. reinhardtii;* CC-741 *(FUD7)*-a *psbA* deletion mutant; CC-3376 *(A251R)*-a *psbA* mutant with an Ala->Arg substitution at position 251 of D1; CC-703 *(ac-u-g-2-3)* -a *psaB* frame shift mutant.

The potential physiological significance of LSU downregulation. The possible significance of the short term downregulation in LSU synthesis is that it could lead to a transient arrest of Rubisco assembly. Thus chaperones such as groEL and groES could temporarily be released from the assembly process (11) and become available for other functions, such as overcoming the damaging effects of AOS to proteins and membranes.

Acknowledgment
Experiments presented in Figures 1&2 were performed in the laboratory of Prof. J.E. Boynton and Prof. N.W. Gillham.

References
1. Barber, J. and Andersson, B. (1992) Trends Biochem. Sci. 17, 61-66.
2. Anderson, J. M. & Osmond, C. B. (1987). In Shade-sun responses: Compromises between acclimation and photoinhibition. Elsevier, Amsterdam. pp. 1-38.
3. Escoubas, J.-M., Lomas, M., LaRoche, J. & Falkowski, P. G. (1995) Proc. Natl. Acad. Sci. USA, 92, 10237-10241.
4. Danon, A. & Mayfield, S. P. (1994) Science, 266, 1717-1719.
5. Allen, J. F., Alexciev, K. & Hakansson, G. (1995) Curr. Biol., 5, 869-872.
6. Spreitzer, R. J. (1993) Annu. Rev. Plant. Physiol. Plant. Mol. Biol., 44, 1-49.
7. Shapira, M., Lers, A., Heifetz, P., Irihimovitz, V., Osmond, B. C., Gillham, N.W. & Boynton, J.E. (1997) Plant Mol. Biol., 33, 1001-1011.
8. Bennoun, P., Spierer-Herz, M., Erickson, J., Girard-Bascou, J., Pierre, Y., Delosme, M., & Rochaix, J.-D., (1986) Plant Mol. Biol. 6, 151-160.
9. Lardans, A.,Gillham, N. W., & Boynton, J. E.(1997) J. Biol. Chem. 272, 210-216.
10. Grant, D.M., Gillham, N.W., & Boynton, J.E. (1980), Proc. Natl. Acad. Sci. USA, 77, 6067-6071.
11. Hemmingsen, S.M., Woolford, C., van der Vies, S.M., Tilly, K., Dennis, D.T., Georgopoulos, C.P., Hendrix, R.W. & Ellis, R.J. (1988) Nature, 333, 330-334

RUBISCO ACTIVATION IS IMPAIRED IN TRANSGENIC TOBACCO PLANTS WITH REDUCED ELECTRON TRANSPORT CAPACITY

Ruuska, S.A.[1], Andrews, T.J.[1], Badger, M.R.[1], Lilley, R.McC.[2], Price, G.D.[1] and von Caemmerer, S[1]. [1]Research School of Biological Sciences, Australian National Univ., PO Box 475, Canberra ACT 2601, Australia; [2]Biological Sciences, Univ. of Wollongong, Northfields Av., Wollongong 2522, Australia.

Keywords: Calvin-cycle, cytochrome complexes, electron transport, Rubisco, transgenic plants

1. Introduction

Rubisco can be catalytically competent only after a specific lysyl residue within the active site has been carbamylated. Before carbamylation can occur, any inhibitory ligands bound at the site must be released, and this process is facilitated by another enzyme, Rubisco activase. It has been shown *in vitro* that Rubisco activase needs to hydrolyse ATP to function and is inhibited by ADP, and so presumably is sensitive to the chloroplast ATP/ADP ratio (1). However, there are indications that activase is also regulated by transthylakoid pH gradient (ΔpH) and electron transport through PSI (2).

We studied the connection between electron transport, adenylates and Rubisco activation in two different transgenic tobacco types. Plants with reduced amounts of the cytochrome *bf* complex (anti-FeS plants) (3) have impaired electron transport and low ΔpH, which should decrease ATP synthesis. On the other hand, plants with low activity of chloroplast glyceraldehyde-3-P dehydrogenase (anti-GAPDH plants) (4) have a decreased capacity to use ATP in carbon assimilation, which should lead to high ATP/ADP ratio and ΔpH.

2. Procedure

Anti-GAPDH plants were grown in a glasshouse with a peak irradiance of 700-900 µmol quanta m^{-2} s^{-1}. Anti-FeS plants were grown in a growth cabinet at 70-100 µmol quanta m^{-2} s^{-1} and a 20-h photoperiod. Leaf gas exchange was measured at 350 µbar CO_2 in air, 1000 µmol quanta m^{-2} s^{-1} and a leaf temperature of 25 °C in a chamber attached to a rapid-kill apparatus. After gas exchange measurements leaf discs were rapidly freeze-clamped (5). Rubisco initial and total activity, site concentration and carbamylation level

G. Garab (ed.), Photosynthesis: Mechanisms and Effects, Vol. V, 3371–3374.
© 1998 *Kluwer Academic Publishers. Printed in the Netherlands.*

3372

were measured from the samples as described (6). RuBP, ATP and ADP contents were measured by a luminometric method (7) after heat-treatment of the acid extracts (8). Activation state of NADP-malate dehydrogenase was measured as described (9). The leaves were characterised by measuring the total activity of GAPDH (6) or the amount of the cytochrome *f* protein by quantitative Western blotting (3).

3. Results and Discussion

3.1 *RuBP content and Rubisco characteristics* (Fig. 1)

The reduction in either cytochrome *bf* complex or GAPDH activity decreased CO_2 assimilation rates (3,4). This was accompanied by lowered RuBP contents (Fig.1A-B). In anti-GAPDH plants, the RuBP regeneration was limited because the activity of GAPDH was reduced, whereas in anti-FeS plants the supply of NADPH and ATP to the Calvin cycle was restricted.

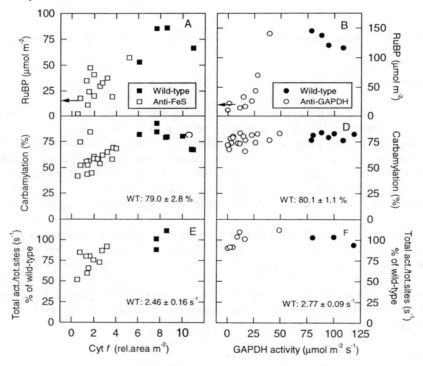

Figure 1. RuBP content and Rubisco characteristics of anti-FeS and anti-GAPDH plants. The arrows in panels A&B indicate the average Rubisco site concentration and the numbers within panels C-F the average ± S.E. of carbamylation state and total *in vitro*-activity per total Rubisco sites of the control plants. Samples were collected by freeze-clamping after leaves were stabilised at 350 μbar CO_2 in air, irradiance of 1000 μmol quanta m^{-2} s^{-1} and 25 °C.

The carbamylation state of Rubisco was high (80%) in all wild-type and anti-GAPDH plants whereas it decreased in anti-FeS plants (C-D). In addition, the total *in vitro*-activity per total sites declined in these plants, suggesting the presence of tightly-binding inhibitors in uncarbamylated sites (E-F). The catalytic activity of carbamylated sites (measured as initial *in vitro*-activity per carb. sites) was comparable to control plants in both anti-FeS and anti-GAPDH plants (data not shown), which indicates that carbamylated sites function normally *in vitro*, as was seen *in vivo* (von Caemmerer et al, this issue).

The reason for the reduced Rubisco carbamylation state and the presence of tightly-binding inhibitors in uncarbamylated sites may be that the functioning of Rubisco activase is impaired in anti-FeS plants. This could be mediated by reduced electron transport rate and ATP synthesis.

3.2. ATP/ADP ratios and NADP-malate dehydrogenase (MDH) activation level

We examined whether the decreased Rubisco carbamylation in anti-FeS plants could be attributed to low ATP (Fig. 2). However, the whole-leaf ATP/ADP in anti-FeS plants was similar to control plants. As expected, ATP/ADP increased as the GAPDH activity reduced, indicating that the consumption of ATP was restricted due to limited carbon assimilation.

Another measure of the chloroplast energy status is the $NADP^+/NADPH$ ratio. We measured the activation state of NADP-MDH and used it as an indicator of the reduction state of NADP-pool (10). MDH activation decreased dramatically as the amount of cytochrome *bf* complex reduced, confirming that electron transport and NADPH synthesis are severely impaired in anti-FeS plants. In anti-GAPDH plants MDH activation did not change, indicating that they can avoid overreduction of $NADP^+$ pool by high nonphotochemical quenching, as shown earlier (4).

Conclusions

Low RuBP content alone does not cause decarbamylation of Rubisco in anti-FeS plants since carbamylation remained high in anti-GAPDH plants.

We suggest that the decline in carbamylation and the total *in vitro*-activity per total Rubisco sites in anti-FeS plants were a consequence of impaired functioning of Rubisco activase. This could be caused by reduced electron transport rate and low ΔpH and/or low ATP, but does not appear to correlate with whole-leaf adenylate content or ATP/ADP ratio.

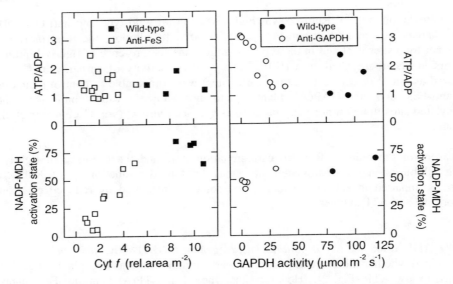

Fig. 2. Whole-leaf ATP/ADP ratios and NADP-malate dehydrogenase activation levels in anti-FeS and anti-GAPDH plants. Measuring conditions and leaf sampling are as in Fig. 1.

References

1 Robinson S.P and Portis A.R, Jr. (1989) Arch. Biochem. Biophys. 268,93-989.

2 Campbell, W.J and Ogren, W.L (1990) Plant Phys. 94, 479484

3 Price, G.D., Yu, J-W., von Caemmerer, S., Evans, J.R., Chow, W.S., Anderson, J.M., Hurry, V. and Badger, M.R. (1995a) Austr. J. Plant Phys. 22, 285-297

4 Price, G.D., Evans, J.R., von Caemmerer, S., Yu, J.W. and Badger, M.R. (1995b) Planta 195, 369-378

5 Badger, M.R., Sharkey, T.D. and von Caemmerer, S. Planta 160,305-13

6 He, Z., von Caemmerer, S., Hudson, G.S., Price, G.D., Badger, M.R. and Andrews, T.J. (1997) Plant Phys. 115,1596-80

7 Lilley, R. McC., Grahame, P.G. and Ali, S.R.M. (1985) Anal. Biochem. 148,282-87

8 Biotto, P.A. and Siegenthaler, P.A. (1994) Potato Res. 37,151-160

9 Scheibe, R. and Stitt, M. (1988) Plant Phys Biochem. 26,473-81

10 Harbinson, J., Genty, B. and Foyer, C. (1990) Plant Phys. 94,545-53

MEASUREMENTS OF RIBULOSE 1, 5-BISPHOSPHATE CARBOXYLASE/OXYGENASE ACTIVITIES BY NMR

Z.-Y. Wang, S. Luo, K. Sato, M. Kobayashi and T. Nozawa
Department of Biomolecular Engineering, School of Engineering,
and Center for Interdisciplinary Science, Tohoku University,
Sendai 980-8579(Japan)

Key words: Calvin-cycle, CO_2, enzymes, metabolic processes, Rubisco, spinach

1. Introduction

Ribulose 1,5-bisphosphate carboxylase/oxygenase (Rubisco; EC 4.1.1.39) is a bifunctional enzyme that catalyzes both photosynthetic carbon assimilation and photorespiration. The former reaction involves incorporation of one molecule of CO_2 into ribulose bisphosphate (RuBP) to form two molecules of 3-phosphoglycerate (PGA), while the latter reaction incorporates one molecule of O_2 into RuBP to generate one equivalent each of PGA and 2-phosphoglycolate (PG) (Scheme I). Establishment of a sensitive, reliable and straightforward method for evaluation of both carboxylase and oxygenase specificities is an important part of the effort to improve photosynthetic efficiency of the Rubisco and to investigate its CO_2/O_2 discriminating mechanism.

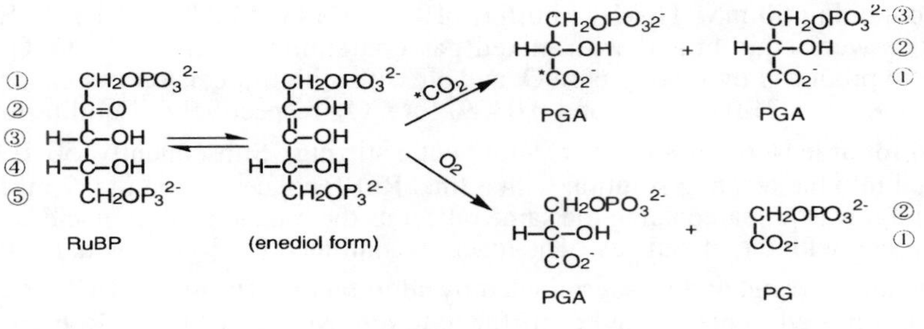

Scheme I

In the present study, we demonstrate that both carboxylase and oxygenase activities, as well as specificity factor, can be determined by high-resolution nuclear magnetic resonance (NMR) spectroscopy with similar sample preparation to that used in the conventional assay. The carboxylase reaction can be monitored in situ by each of the ^1H-, ^{31}P-, and ^{13}C-NMR measurements. In contrast to the early ^1H-NMR experiments where D_2O was used as a solvent to avoid the gross H_2O signals or to facilitate observation of the exchange between C-3 proton of RuBP and deuterium of the solvent, all experiments of this study were conducted using H_2O as a solvent, therefore a more accurate evaluation of the compound quantity is possible, while the problem of the giant H_2O peak can be overcome by adopting the advanced water-suppression technique combined with the application of field-gradient.

G. Garab (ed.), Photosynthesis: Mechanisms and Effects, Vol. V, 3375–3378.

2. Experimental

2.1 *Materials*
A detailed description of the purification of three Rubisco enzymes from higher plant (spinach) and photosynthetic bacteria (*Chromatium vinosum*, and *Rhodospirillum rubrum*) with different reported τ values was given in a previous paper[1].

2.2 *Carboxylase assay by NMR*
Purified Rubisco enzymes were preincubated at 23 °C and purged with N_2 gas for 30 min in a 5 mm NMR tube with a solution volume of 0.5 ml. The incubating mixture consisted of 70 ~ 150 µg enzyme, 50mM Tris-HCl, 10mM $MgCl_2$, 5mM DTT, 4mM EDTA and 100mM $NaHCO_3$. Unless otherwise stated, pH was adjusted to 7.8. In the ^{13}C-NMR experiments, $NaHCO_3$ was replaced by $NaH^{13}CO_3$. The carboxylase reaction was initiated by injection of 100 µl 50 mM RuBP into the preincubated solution, and monitored immediately after inserting the NMR tube into the probe. NMR experiments were performed at 23 °C. Field-frequency locking was achieved by using a commercially available inner tube filled with D_2O as described previously[1]. Proton spectra were obtained using Bruker 400 MHz instrument (DRX400) equipped with a z-axis field-gradient module. Water peak suppression was achieved using a spin echo sequence, WATERGATE[2]. A total of 8 transients were accumulated. ^{31}P-NMR spectra were recorded with Bruker DPX400 spectrometer, using a 60° pulse, and 100 scans. No proton spin-decoupling was applied during the measurements. ^{13}C-NMR spectra were collected on a Bruker DRX500 spectrometer with low-power ^{1}H-decoupling.

2.3 *Assay of Specificity factor by NMR*
The reactions were carried out in flat-bottom glass tubes (20 mmϕ , 10 ml). Rubisco samples were preincubated in the tubes at concentrations of 2.0~2.5 mg/ml in a solution containing 50 mM Tris-HCl buffer, pH8.5, 10 mM $MgCl_2$ and 1 mM EDTA. The samples were equilibrated with mixed gas at the desired CO_2/O_2 mole fraction ratio produced by mixing pure O_2 and air with a high-precision gas mixer. The air composition was 21.0% for O_2 and 0.035% for CO_2, respectively. Equilibrium was maintained for at least 60 min at 23 °C with gentle stirring. Subsequently, 50 µl RuBP was injected into the reaction solution with a final RuBP concentration of 0.7 mM. The RuBP solution was prepared using the same buffer as the reaction solution and was also pre-equilibrated with the mixed gas. The reaction continued for 30 min with continuing gas purging and stirring, and was terminated by addition of 40 µl of 2M HCl. Reaction mixture was adjusted to pH 8.5 and centrifuged to remove the denatured Rubisco. The supernatant was used directly for NMR measurement. NMR measurements were similar to that for the carboxylase reaction except for different scan numbers and that inverse-gated ^{1}H-decoupling was applied during the ^{31}P-NMR measurements[3]. The specificity factors τ were calculated by the following equation given by Kane *et al.*[4].

$$\tau = \left(\frac{v_c}{v_o}\right)\left(\frac{[O_2]_{gas}}{[CO_2]_{gas}}\right) \times 0.0375 \qquad (1)$$

where v_c and v_o are carboxylase and oxygenase rates at the gas-phase concentrations of CO_2 and O_2.

3. Results and Discussion

3.1 *In situ detection of carboxylase reaction*

Figure 1 shows a time-variation of the proton spectra for the Rubisco from *R. rubrum*. The resolution is sufficient for quantitative determination of the specific activity. Two signals, H-3 of RuBP at 4.34 ppm and H-2 of PGA at 4.06 ppm, were chosen as indicators for tracking the reaction as these signals had clear shapes and well-separated from other resonances. Other signals of RuBP were severely obscured with residual H_2O peak or due to an exchange with its enol form (see Scheme 1). The signal for one of the H-3 protons of PGA at 3.87 ppm, that was more tightly coupled with ^{31}P nucleus as identified by ^{31}P-decoupling experiment, also showed a clear multiplet, but partially overlapped with the resonances of RuBP. It can be seen that the H-3 signal of RuBP decreased as the carboxylase reaction proceeded, while the H-2 signal of PGA increased constantly.

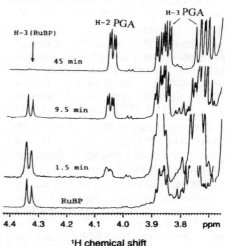

Fig. 1 ^{1}H-NMR spectra for carboxylase reaction of *R. rubrum*.

Figure 2 shows the time-variation of ^{31}P-NMR spectra for the Rubisco from *R. rubrum*. Compared to ^1H spectra, the ^{31}P spectra exhibited a much simpler pattern. All signals were well separate and appeared as triplets due to coupling with the two protons attached to the neighboring carbon atoms. Two signals at 4.2 and 4.8 ppm correspond to P-5 and P-1 phosphoruses of RuBP, respectively. With progress of the carboxylation reaction, intensities of RuBP decreased and PGA peak increased continuously. Advantages of using ^{31}P-NMR are (1) simple requirements for hardware and software because no solvent-suppression and proton decoupling are involved and only a single pulse is needed for the measurement; (2) no overlap between the signals over a wide pH range; and (3) the availability of evaluation of the degree of reaction at any time during the reaction. By using the data of early stage, specific activities for the three Rubiscos were determined to be 3.3, 2.3 and 2.1 U/mg·protein for *R. rubrum*, spinach and *C. vinosum*, respectively. These values were consistent with those reported by other methods

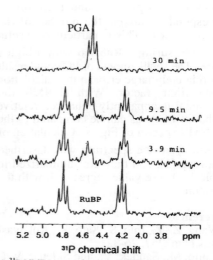

Fig. 2 ^{31}P-NMR spectra for carboxylase reaction of *R. rubrum*.

3.2 Specificity factor measurements by NMR

Figure 3 shows typical ^{31}P-NMR spectra of the carboxylase/oxygenase reaction products by three Rubiscos. The τ values can be evaluated by Eq. (1) from the individual experiment. In addition to the advantages mentioned before for the ^{31}P-NMR, an important feature in the τ measurement is that a small amount of inorganic phosphate can readily be detected if using any non-phosphate buffer as a reaction solution. This is very useful for monitoring the activity of phosphatases as a trace amount of PG phosphatase could cause an overestimate of the τ values. With ^{31}P-NMR, the PG phosphatase is not necessarily removed completely from the Rubisco preparation, because even in this case, the τ can still be calculated by taking into account the inorganic phosphate if no other phosphate sources exist in the reaction system.

Figure 4 shows the ^{1}H-NMR spectra for the Rubisco from *C. vinosum* under different O_2/CO_2 mixes. An attractive feature of ^{1}H-NMR is that the PG signal intensity represents *two* protons, and therefore it corresponds to *double* the number of the PG molecules. This is particularly useful when measuring a Rubisco with a high τ value. The small amount of PG could result in a relatively large error in the calculation of specificity factor. With ^{1}H-NMR, the PG signal is apparently enhanced relative to that of PGA, as can be compared to the ^{31}P-NMR spectra of Fig. 3. Calculation of τ value from the spectrum was described elsewhere[3], and they are summarized in Table 1. These values agree well with the literature.

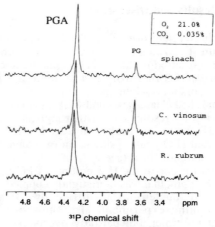

Fig. 3 ^{31}P-NMR spectra for CO_2/O_2 reaction.

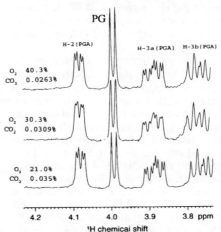

Fig. 4 ^{1}H-NMR spectra for CO_2/O_2 reaction.

Table 1 Specificity factors

	τ
R. rubrum	12.9
C. vinosum	34.2
spinach	86.0

References

1 Wang, Z.-Y., Luo, S., Sato, K., Kobayashi, M. and Nozawa, T. (1998) Anal. Biochem. 257, 26-32.
2 Piotto, M., Saudek, V., and Sklenar, V. (1992) J. Biomol. NMR 2, 661-665.
3 Wang, Z.-Y., Luo, S., Kobayashi, M. and Nozawa, T. Photosyn. Res. (in press).
4 Kane, H. J., Viil, J., Entsch, B., Paul, K., Morell, M. K. and Andrews, T. J. (1994) Aust. J. Plant Physiol. 21, 449-461.

RUBISCO: KINETIC CONSTANTS OF PARTIAL REACTIONS AND THEIR RELATION TO V_{MAX} AND K_M

Juta Viil
Institute of Experimental Biology
Harku, Harjumaa, Estonia

Key words: CO2 uptake, enzymes, modelling, regulatory processes, Rubisco.

Carboxylation of ribulose-1,5-bisphosphate (RuBP) catalyzed by ribulose-1,5-bisphosphate carboxylase/oxygenase (Rubisco) may be studied as a sequence of three partial reactions:

$$CO_2$$
$$k_1 \quad \downarrow k_2 \quad k_3$$
$$E_f + R_f \text{-----> } ER \text{-----> } ERC \text{-----> } 2\,C_3 + E_f$$

where k_1, k_2, k_3 are the apparent rate constants;

E_f is the concentration of free reaction centers;

R_f is the concentration of free RuBP;

ER is the concentration of 2,3-enediol;

ERC is the concentration of the C6 intermediate

C_3 is 3-phosphoglyceric acid;

The total concentration of competent active centers (E_t) may be expressed:

$$E_t = E_f + ER + ERC,$$

if oxygen concentration in the medium is very low.

A change in the activity of Rubisco indicates that one or several of the kinetic constants: E_t, k_1, k_2, or k_3 must have changed. Conventionally, activity of an enzyme is estimated by V_{max} and K_M. However, from a mathematical model of Rubisco (1) it may be calculated that different combinations of values of the rate constants of partial reactions and E_t result in identical values of V_{max} and K_M. In this work, an attempt is made to study, by means of the model, the influence of changes in the rate constants and E_t upon V_{max} and K_M. The values of the parameters used in calculations, which are given below, have been estimated in intact barley leaves at 25 °C (2; 3):

E_t	4.18 nmol cm^{-2}
k_1	0.45 mM^{-1} s^{-1}
k_2	63.0 mM^{-1} s^{-1}
k_3	3.0 s^{-1}

G. Garab (ed.), Photosynthesis: Mechanisms and Effects, Vol. V, 3379–3382.
© 1998 Kluwer Academic Publishers. Printed in the Netherlands.

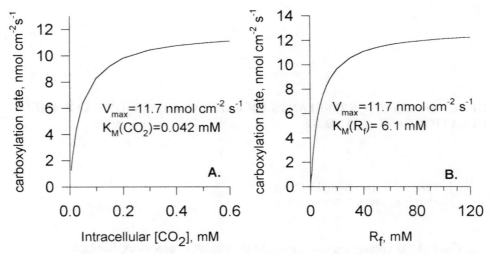

Figure 1. Substrate dependence of the carboxylation rate calculated from the model (1). The constants applied: E_t=4.18 nmol cm^{-2}, k_1=0.45 mM^{-1} s^{-1}, k_2=63.0 mM^{-1} s^{-1}, k_3=3.0 s^{-1}. A: CO2 dependence calculated for the saturating RuBP concentration R_f=50 mM. B: RuBP dependence calculated for the saturating CO$_2$ concentration 1 mM.

Fig. 1A shows the dependence of the calculated carboxylation rate on the CO$_2$ concentration. $K_M(CO_2)$ was 0.042 mM. This is about 5 times higher than estimated experimentally. V_{max} is 11.7 nmol cm^{-2} s^{-1} which is about 6 fold higher than values estimated experimentally for barley populations (4). Fig. 1B shows the R_f dependence of carboxylation at 1 mM CO$_2$. $K_M(R_f)$ was 6.1 mM.
which is also 5 to 6 times higher than measured experimentally (5).

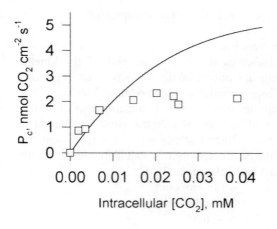

Figure 2. Comparison of the carboxylation rate (P_c) calculated from the model (line) with the rate measured in intact barley leaves in the saturating irradiance (1800 µE m^{-2}), 1.5% O$_2$ (points).

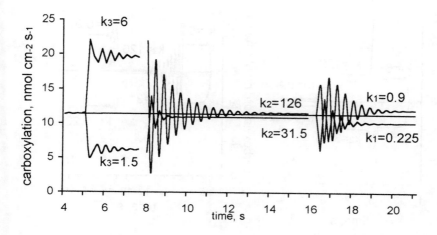

Figure 3. Influence of changes in the values of rate constants upon the carboxylation rate at R_f 50 mM and CO_2 1mM. Horizontal line with the ordinate of 11.7 nmol cm^{-2} s^{-1} corresponds to steady state equal to V_{max} calculated from the model applying the values of constants as in Fig. 1. From these values, shifts of rate constants have been carried out as shown next to the curves in the figure. At the shifts of k_2, k_3 was transferred back to the initial level. At the shifts of k_1, k_2 and k_3 were transferred to their initial level.

Fig. 2 shows that carboxylation in the leaf is saturated with CO_2 at the level markedly lower than that calculated from the model, although the rates at concentrations below 0.01 mM are equal. There are several reasons for the earlier saturation of the leaf carboxylation. The leaf may become R_f-limited at higher carboxylation rates. In the model R_f remains saturating. Moreover, when the CO_2 concentration increases, a proportion of reaction centers of Rubisco is blocked (3). This means a decrease in the concentration of the competent centers in the leaf. The model does not take that into account because of the lack of sufficient amount of quantitative data.

Fig. 3 demonstrates that a four fold shift in k_2 has almost no influence upon V_{max}. A significant change in k_1 has only a minor influence. Nearly proportional change in the carboxylation rate is brought about by a change in k_3. It may be calculated from the model that even a ten fold change in k_2 has no significant influence upon V_{max} (1) because at saturating CO_2 carboxylation of enediol has only a weak limiting effect on the rate of the whole carboxylation reaction.

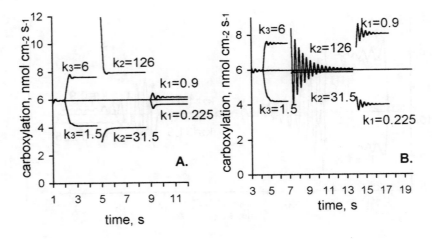

Figure 4. Influence of changes in the values of rate constants upon the carboxylation rate at rate limiting CO_2 and R_f concentrations. Horizontal line with the ordinate of 5.9 nmol cm^{-2} s^{-1} represents the half-saturated steady state rate of carboxylation. with the values of rate constants as in Fig. 1. A: CO_2 0.042mM, R_f 50 mM. B: CO_2 1 mM, R_f 6.1 mM. Shifts in rate constants are shown next to the curves.

Fig.4A shows that at the half-saturating concentration of CO_2 the carboxylation rate is very sensitive to changes in k_2 and k_3. Changes in k_1 have minor influence. When R_f is half-saturating (Fig. 4B), changes in k_2 have transient influence, but do not affect the steady state level. k_1 and k_3 have a significant effect on the steady state level of carboxylation.

References

1 Viil, J. (1995) in From Light to Biosphere (Mathis P., ed.) V. 5, pp. 215-218, Kluwer Academic Publishers, Dordrecht, The Netherlands
2 Viil, J. and Pärnik, T (1995) J. Experim. Botany 46, 1301-1307
3 Viil, J. and Pärnik, T. (1989) in Current Research in Photosynthesis (Baltscheffsky M, ed.) v. 3, pp 415-418, Kluwer Academic Publishers, Dordrecht, The Netherlands
4 Rinehart, C.A., Tingey, S.V. and Andersen, W.R.(1983) Plant Physiol. 72, 76-79
5 Li, L.R., Sisson, V.A. and Kung, S.D. (1983) Plant Physiol. 71, 404-408

RIBULOSE-1,5-BISPHOSPHATE CARBOXYLASE/OXYGENASE (RUBISCO) CONTENT IN LEAVES, ASSIMILATORY CHARGE AND MESOPHYLL CONDUCTANCE

Hillar Eichelmann and Agu Laisk, Tartu Ülikooli Molekulaar- ja Rakubioloogia Instituut, Riia tn 23, Tartu, EE2400, Estonia

Key words: CO_2 uptake, gas diffusion, intercellular CO_2 concentration, light acclimation, mesophyll conductance, Rubisco.

Introduction

All factors, which finally influence the photosynthetic rate, can do this by influencing the activity of Rubisco (EC 4.1.1.39) and its substrate, CO_2 and RuBP concentrations. The CO_2 concentration is usually well below the $K_M(CO_2)$ of the enzyme and it is the initial slope of the kinetic curve $V_M/K_M(CO_2)$, termed carboxylation conductance, which becomes important. The reciprocal of the conductance, carboxylation resistance must be smaller than other resistances in series, the gas phase resistance (stomatal and boundary layer) and liquid phase diffusion resistance (r_{md}). The carboxylation conductance in intact leaves *in vivo* may be found as the initial slope μ of the A *vs* C_c graph (A, net CO_2 uptake rate, C_c, CO_2 concentration at the reaction sites) at low C_c values. If C_c cannot be calculated because the mesophyll diffusion resistance r_{md} is unknown, the closest approximation is A *vs.* C_w or A *vs* C_i plot (C_w, dissolved CO_2 concentration in cell walls, C_i, intercellular gas space CO_2 concentration). True parameters of the carboxylase can be found only from experiments carried out in nonphotorespiratory conditions (1-2% O_2), otherwise the competing oxygenase reaction consumes a part of RuBP and partially inhibits the carboxylase activity. In the present work we investigate the relationship between Rubisco content and leaf mesophyll conductance in plants grown under different light intensities and allowing leaves to adapt to changed light intensity.

Materials and methods

Sunflower (*Helianthus annuus* L.) was grown in commercial fertilized peat-soil mixture at 16/8 h 28/18 °C day/night cycle at PFD of 250 to 300 (further referred to as high-light) or 40 to 60 (further referred to as low-light) μmol m^{-2}s^{-1}. A rapid-response leaf gas exchange measurement system was used (leaf chamber $4.4\times4.4\times0.3$ cm^3, gas flow rate 20 cm^{3}s^{-1}), that consisted of two similar open gas channels 1 and 2, which allowed independent gas preconditioning and which were equipped with laboratory-made psyhrometers for water vapour measurements and IR CO_2 analysers (LI 6262, LiCor, Lincoln, NE). The leaf chamber could be rapidly switched from one channel to the other, what made it possible to produce rapid changes in CO_2 concentration and to start gas-exchange recording already 3 s after switching. As the calculated C_w is presented in μM, dissolved in water, we also present the external CO_2 concentration C_{wo} in μM, as such which would be in water in equilibrium with that in gas. The integral postillumination CO_2 uptake (assimilatory charge, AC), was measured in 1.5% O_2 and $C_{wo}=3.25$ μM (C_w from 1.1 to 3.3 μM)

3383

G. Garab (ed.), Photosynthesis: Mechanisms and Effects, Vol. V, 3383–3386.
© 1998 *Kluwer Academic Publishers. Printed in the Netherlands.*

independent of the previous conditions of steady-state photosynthesis. For the AC measurements, light was switched off and the leaf chamber was simultaneously switched to channel 2, where CO_2 and O_2 concentration were as specified above. The catalytic rate constant k_{cat} was calculated as $k_{cat}=\mu K_M/E_t$ where E_t is the concentration of Rubisco active sites. Equation binds the μ with enzyme characteristics. The content of Rubisco was determined on the basis of the large subunit band in SDS gel electrophoresis (1).

Results

In leaves used in experiments Rubisco content varied between 2 and 75 µmol active sites m^{-2} (0.14 to 5 g Rubisco protein m^{-2}). The measured μ vs. E_t relationships were not linear but saturated at E_t of 30 µmol active sites m^{-2} (2 g Rubisco protein m^{-2}, Fig. 1). The k_{cat} in Fig. 1 was 8. s^{-1}. In low-light grown sunflowers the maximum, Rubisco-saturated μ was only about 5 mm s^{-1}, compared with 9-13 mm s^{-1} in high-light plants. From this difference the importance of light adaptation in the Rubisco kinetics is evident. When the low-light plants where transferred to high light, the content of Rubisco in existing leaves dropped

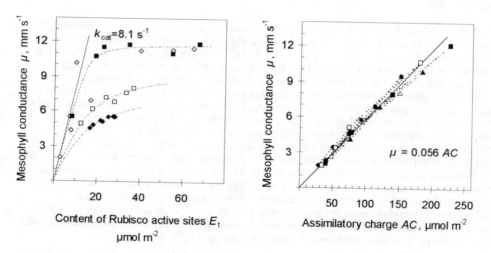

Figure 1. Relationships between Rubisco concentration and μ in leaves. Adaptation to different light intensities. Filled diamonds - 2.5 weeks old low light-grown plants; open diamonds - low light-grown plants adapted at high light intensity for 2 weeks; filled squares - 2 weeks old high light-grown plants; open squares - high light-grown plants adapted 2 weeks at low light intensity

Figure 2. Relationship between mesophyll conductance and assimilatory charge for five different-aged leaves of one and the same plant. Different assimilatory charges were generated with decreasing light intensity to different lower levels for 10 s (dotted lines join the data points obtained with one leaf, initial PAD = 1300 µmol m^{-2}s^{-1}), then μ and AC were measured at C_{w0} of 3.25 µM. Rubisco special activity (SCE) was 0.056 mm s^{-1} per µmol m^{-2} (the slope of the solid line).

drastically during the two-week adaptation time, to the extent that the Rubisco content became to limit the μ (empty diamonds on the slope of the relationship in Fig. 1). New leaves grown under high light behaved like typical high-light leaves, having high μ and capability to synthesise large amounts of Rubisco up to 65 μmol m^{-2} (empty diamonds on the plateau area of the relationship in Fig. 1). In a reverse experiment high-light plants were readapted under low light. Leaves initially having high μ (11 mm s^{-1}) and large amounts of Rubisco (up to 68 μmol m^{-2} in fully expanded leaves, filled squares) were converted into leaves reminding typical low-light leaves, with low maximum μ (6-7.5 mm s^{-1}) and limited range of Rubisco content (15-40 μmol active sites m^{-2}, empty squares). Mesophyll conductance of a leaf depends on the Rubisco activity on one hand, but on the pool of RuBP on the other hand. The latter can be estimated from the postillumination CO_2 uptake (assimilatory charge, AC) in intact leaves. The dependence of μ on AC was checked in experiments where light intensity was varied and AC measured promptly after that (Fig. 2). Data for five different-aged leaves of the same plant all present a proportional dependence with the slope (Specific Carboxylation Efficiency, SCE) of 0.056 mm s^{-1} per μmol AC m^{-2}. However SCE seems not to be a basic constant of Rubisco, because in some leaves, mostly in low light-grown leaves, which had small AC, SCE was significantly higher, up to 0.066-0.076 $\mu M^{-1}s^{-1}$. However, as far as SCE remains constant, μ proportionally depends on AC.

Discussion

By varying growth conditions we obtained leaves that had a wide range of Rubisco content. We related Rubisco content to the mesophyll conductance μ in nonphoto-respiratory conditions. μ was not proportional to Rubisco content E_t but saturated at higher E_t values. This saturation caused a $2 - 3$-fold decrease in the apparent k_{cat}. Similar relationship was typical also for cotton, bean, potato and English spinach (1). The saturation of the μ vs. E_t relationship may be caused by physical and chemical reasons. Physically, mesophyll diffusion conductance may become limiting at very high carboxy-lation conductance. Usually diffusional component is between 0.02 and 0.04 s mm^{-1}, not exceeding 10 - 20% of the total mesophyll resistance (5). Therefore, diffusion-limited μ is expected to be about 5 to 10 times greater than the actual maximum μ. Thus, μ was always carboxylase-limited, independent of the content of Rubisco.

Carboxylation resistance has two components: one determined by the content and activity of Rubisco and the other determined by inhibiting stroma metabolites, PGA and free P_i. Our conditions were chosen to minimize PGA and P_i, thus, we may neglect the metabolite effects here. Data in Fig. 1 show that in high-light leaves at low Rubisco contents μ increases proportionally with Rubisco content. The calculated k_{cat} is 8 s^{-1}, which is only slightly higher than the highest values obtained in Rubisco assays *in vitro*, 5-6 s^{-1} (2,3). Thus, in leaves that had the highest k_{cat} values we may suppose that all Rubisco was completely active. When the Rubisco content exceeded 15 μmol reaction sites m^{-2} and μ approached 11 mm s^{-1}, the relationship abruptly saturated. There was a ceiling limiting the amount of active Rubisco to 15-20 μmol reaction sites m^{-2}. In low-light leaves the saturation occurred more or less at the same Rubisco content as in the high-light leaves, but the maximum μ was much lower. It looks like in low-light leaves the same maximum number of Rubisco molecules could be activated as in high-light leaves, but each molecule had a smaller number of active sites. Thus, this model defines two different processes that determine the Rubisco activity *in vivo*, one, which determines the number of activated molecules and the other that determines the number of active reaction sites per molecule. The second process is sensitively regulated by growth light, while the first is relatively insensitive to this parameter. It may be speculated that the maximum number of activated molecules may be determined by the number of sites where Rubisco activase can bind or be connected with thylakoid membranes, i.e., not the content and activity of Rubisco activase but its binding sites at the PS I region of thylakoids may govern the activation process of Rubisco, in parallel with the limited diffusibility of the protein at its high concentration. Whatever the mechanistic reasons for it, our results clearly demonstrate that a large part of Rubisco is not active for carboxylation and plays the role of storage protein.

The mesophyll conductance surprisingly well correlated with assimilatory charge, which is a gas exchange measure of mainly RuBP pool (1,4,6,7). It is not easy to explain why the μ vs. AC relationships were so linear up to the RuBP concentrations of 10-20 mM, considering that K_M(RuBP) for Rubisco is about 40 µM in *in vitro* experiments (8). More or less linear kinetics have been modelled to describe the postillumination CO_2 fixation process during which RuBP pool decreases and PGA pool increases, considering that PGA is a competitive inhibitor (4). The proportionality constant relating μ to AC, SCE was 0.04-0.065. The product-inhibition model predicts that the true uninhibited mesophyll (carboxylation) conductance that may be related to Rubisco content can be measured only under conditions where PGA and P_i pools are minimal, i.e., under saturating light and limiting CO_2 concentrations and low O_2 concentrations.

Acknowledgement. This work was supported by grant No. 1808 from Estonian Science Foundation.

References

1 Eichelmann, H. and Laisk, A. (1990) Fiziol. Rastenii 37, No.6, 1053-1064.
2 Evans, J.R. and Seemann, J.R. (1984) Plant Physiol. 74, 759-764.
3 Flachmann, R., Zhu, G., Jensen, R.G. and Bohnert, H.J. (1997) Plant Physiol. 114, 131-136.
4 Laisk, Á., Kiirats, O., Eichelmann, H. and Oja, V. (1987) in Progress in Photosynthesis Research (Biggins J., ed.), pp. 245-252, Martinus Nijhoff Publishers, Dordrecht, the Netherlands.
5 Laisk, A. and Loreto, F. (1996) Plant Physiol 110, 903-912.
6 Laisk, A., Oja, V. and Kiirats, O. (1984) Plant Physiol. 76, 723-729.
7 Oja, V. and Laisk, A. (1998) Biochim. Biophys. Acta submitted.
8 Servaites, J.C., Shieh, W.-J. and Geiger, D.R. (1991) Plant Physiol. 97, 1115-1121.

ACTIVATION STATE OF RUBISCO UNDER DIFFERENT PARTIAL PRESSURES OF CO_2 DURING LEAF AGING OF RICE

H. FUKAYAMA, N. UCHIDA, T. AZUMA, T. YASUDA
The Graduate School of Science and Technology, Kobe University,
Rokko, Nada, Kobe 657-8501, Japan

Key words : gas exchange, higher plant, light activation, O_2 evolution, Rubisco, Rubisco activase.

1. Introduction

Rubisco activase catalyzes the light dependent activation of Rubisco activity. The amount of Rubisco activase in rice leaves has shown high correlation with the carboxylation efficiency under low partial pressure of CO_2 (1) and the maximum evolution rate of O_2 under saturated partial pressure of CO_2 (2). These investigations suggest that Rubisco activase is a limiting factor of photosynthetic rate, especially in young rice leaves as they have a lower rate of synthesis of Rubisco activase than of Rubisco (Activase/Rubisco ratio). Investigations on Rubisco activation states during leaf aging are needed to verify this hypothesis. However, it is apparent that the activation mechanism of Rubisco is a complicated process and differs among plant species (3).

The aim of this study was to investigate the activation state of Rubisco in young leaves under ambient and saturated partial pressures of CO_2 and to examine the relationship between activation state of Rubisco and Activase/Rubisco ratio during leaf aging of rice. The states of Rubisco activation were analyzed in terms of the activity and the number of carbamylated catalytic site.

2. Procedure

2.1 Plant growth and sampling
The plants of Japonica rice (*Oryza sativa* cv. Nipponbare) were grown in a soil media under natural conditions. The 10th leaf blades on the main stems were sampled once in every 10 days from 9.5 leaf stage (the numbering of leaves on the main stem is based on phyllotactic order). The leaves were kept under following conditions before the sampling was done: photon flux densities (PFD) of 60, 500 or 1800 $\mu mol\ m^{-2}\ s^{-1}$, gas mixture of 36 Pa CO_2 and 21 kPa O_2 (ambient CO_2), or PFD of 2000 $\mu mol\ m^{-2}\ s^{-1}$, gas mixture of 5 kPa CO_2 and 21 kPa O_2 (saturated CO_2 ; condition for the measurement of maximum O_2 evolution rates as described previously (2)). After 1 h (ambient CO_2) or 15 min (saturated CO_2) of treatments, the leaves were sampled and immediately frozen in liquid nitrogen and stored at -80°C.

2.2 Biochemical assays
Initial activity of Rubisco was measured using $^{14}CO_2$ by the method of (4) with some

3387

G. Garab (ed.), Photosynthesis: Mechanisms and Effects, Vol. V, 3387–3390.
© 1998 *Kluwer Academic Publishers. Printed in the Netherlands.*

modifications. Rice leaves were homogenized in extraction buffer (50 mM Bicine-NaOH, 5 mM $MgCl_2$, 5 mM DTT, 0.1 mM EDTA, pH 7.8) within 30 s at room temperature. Then, An aliquot of the homogenates containing cell debris was removed and directly applied to the assay mixture (100 mM Bicine-NaOH, 25 mM $MgCl_2$, 5 mM DTT, 0.5 mM RuBP, 20 mM $NaH^{14}CO_2$, pH 8.2) to measure activity of Rubisco. After 1 min, the assays were terminated by adding formic acid. The solution was dried, and acid stable ^{14}C-products were estimated by liquid scintillation counting. The numbers of initial carbamylated sites of Rubisco were determined by the stoichiometric binding of (^{14}C)CABP (5). Another aliquot of the homogenates was removed and incubated with the buffer (50 mM Bicine-NaOH, 5 mM $MgCl_2$, 1 mM EDTA, 25 μM (^{14}C)CABP, pH 7.8) on ice for 1 h. Then, an excess amount of (^{12}C)CABP was added to the solution and precipitated by adding PEG and $MgCl_2$ to a final concentration of 20% (w/v) and 25 mM, respectively. After centrifugation, the ^{14}C retained within the pellet was determined by liquid scintillation counting.

The amounts of Rubisco and Rubisco activase were determined by SRID and ELISA, respectively (2).

3. Results and Discussion

The initial activity of Rubisco under ambient partial pressure of CO_2 and PFD of 500 μm ol m^{-2} s^{-1} and 1800 μmol m^{-2} s^{-1} tended to increase with leaf aging (Fig.1A). In contrast, these values under PFD of 60 μmol m^{-2} s^{-1} were almost constant during leaf aging. Photosynthetic rates in rice leaves are largely limited by the activity of Rubisco under ambient partial pressure of CO_2 and high light intensity (6). During the analysis of flux control, however, it was apparent that the coefficient of Rubisco is lower under low light intensity when compared that with high light intensity (7). Thus, these results may be due to the influence of different limiting factors between relatively high and low PFDs. The number of initial carbamylated catalytic site of Rubisco under PFD of 500

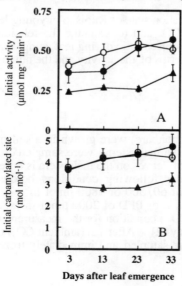

Figure 1. Changes in the initial activity (A) and the number of initial carbamylated catalytic site (B) of Rubisco during leaf aging of rice under ambient partial pressure of CO_2 at PFD of 60 μmol m^{-2} s^{-1}(▲), 500 μmol m^{-2} s^{-1} (●) and 1800 μmol m^{-2} s^{-1} (■). Bars in the figure represent standard error of three indipendent estimation.

μmol m^{-2} s^{-1} and 1800 μmol m^{-2} s^{-1} increased with leaf aging (Fig. 1B). These results show that the activity of Rubisco depends on the number of carbamylated catalytic site under ambient partial pressure of CO_2.

The number of initial carbamylated catalytic sites at a gradient of Activase/Rubisco ratios showed linear correlation under ambient CO_2 and relatively high PFD (Fig. 2). These results suggest that there is a lower initial activity of Rubisco under these conditions in young leaves because of low number of initial carbamylated site through the Activase/Rubisco ratio. However, suppression of photosynthesis was not recognized in transgenic tobacco plant without a severe reduction of Rubisco activase

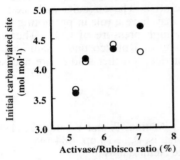

Figure 2. Relationship between the ratio of the amounts of Rubisco activase to Rubisco (Activase/Rubisco ratio) and the number of initial carbamylated site of Rubisco under ambient partial pressure of CO_2 at PFD of 500 μmol m^{-2} s^{-1} (●) and 1800 μmol m^{-2} s^{-1} (O).

(8). In contrast, antisense Rubisco activase with a moderate low concentration of Rubisco activase gave significantly suppressed photosynthetic rates in *Arabidopsis thaliana* (9). These reports imply that the extent of limitation of photosynthesis by Rubisco activase varies with the plant species. Therefore, it is suggested that *in vivo* activities of Rubisco activase are lower in rice leaves than in the leaves of other plant species and causes the limitation of photosynthesis under ambient CO_2 in the developing leaves.

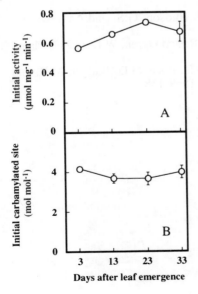

Figure 3 Changes in the initial activity (A) and the number of initial carbamylated catalytic site of Rubisco under 5 kPa CO_2 at PFD of 2000 μmol m^{-2} s^{-1} during leaf aging of rice. Bars in the figure represent standard error of five indipendent estimation.

The initial activities tended to increase with leaf aging in the developing stage under the conditions of saturated CO_2 and PFD (Fig. 3A). The trend of this is similar to that of under ambient CO_2 and relatively high PFD. These findings imply that the Activase/Rubisco ratio affects the activity of Rubisco even under saturated CO_2. However, the trend of initial carbamylated site did not coincide with that of initial activity (Fig. 3B). These results indicate that the initial carbamylated site does not always reflect the activity of Rubisco under saturated CO_2. In this case, changes in the rates of catalytic turn over should be taken into consideration in order to investigate the activity of Rubisco. Recent investigations in transgenic tobacco plants with reduced amounts of Rubisco activase revealed that the *in vivo* catalytic turn over rate of Rubisco was repressed by the deficiency of Rubisco activase (10). These results with the results of the present study suggest that Rubisco activase may have a role in promoting the rate of catalytic turnover of Rubisco, especially under high pressure of CO_2. Therefore, the Activase/Rubisco ratio may determine the activity of Rubisco under saturated CO_2 as well as under ambient CO_2 during leaf aging of rice. Further studies are necessary for the confirmation of this hypothesis.

References

1. Uchida, N., Fukayama, H., Koshimura, H., Azuma, T. and Yasuda, T. (1995) in Photosynthesis : From Light to Biosphere (Mathis, P., ed.) pp. 684-690, Kluwer Academic Publishers, Dordrecht, The Netherlands
2. Fukayama, H., Uchida, N., Azuma, T. and Yasuda, T. (1996) Jpn. J. Crop Sci. 65 : 296-302
3. Seemann, J.R., Kobza, J. and Moore, B.D. (1990) Photosynth. Res. 23 : 119-130
4. Lorimer, G.H., Badger, M.R. and Andrews, T.J. (1977) Anal. Biochem. 78 : 66-75
5. Butz, N.D. and Sharky, T. D. (1989) Plant Physiol. 89 : 735-739
6. Makino, A., Mae, T. and Ohira, K. (1984) Plant Cell Physiol. 25 : 511-521
7. Stitt, M. (1994) Flux Control in Biological System (Schulze, E.D., ed.) pp 13-36. Academic Press, New York
8. Mate, C.J., von Caemmerer, S., Evans, J. Hudson, G.S. and Andrews, T.J. (1996) Planta. 198 : 604-613
9. Eckardt, N.A., Snyder, G.W., Portis, A.R. and Ogren, W.L. (1997) Plant Physiol. 113 : 575-586
10. He, Z., von Caemmerer, S., Hudson, G.S. Price, G.D. Badger, M.R. and Andrews T.J. (1997) Plant Physiol. 115 : 1569-1580

THE RESPONSES OF RUBISCO PROTEIN TO LONG-TERM EXPOSURE TO ELEVATED CO_2 IN RICE AND BEAN LEAVES

Hiromi Nakano[1], Amane Makino, Tadahiko Mae
Department of Applied Biological Chemistry, Faculty of Agriculture,
Tohoku University, Tsutsumidori-Amamiyamachi, Sendai 981-8555, Japan
[1]*Research Fellow of the Japan Society for the Promotion of Science*

Key words: acclimation, C3, carbohydrate, Chl, nitrogen nutrition

1. Introduction

For a number of C_3 species, a photosynthetic acclimation to elevated CO_2 involves a decline in the Rubisco activity/content in leaves (see 1, 2).   Stitt (1) has suggested that this phenomenon may be one of the photosynthetic regulatory mechanisms mediated through an accumulation of carbohydrate(s) under the sink-source imbalance. In addition, many studies have also shown that CO_2 enrichment induces a decrease in the total leaf N content (3, 4).   The change in total leaf N is directly related to changes in the amounts of photosynthetic proteins including Rubisco, because 70-80% of total leaf N is allocated into chloroplasts in C_3 plants (5, 6).   Furthermore, the ratio of Rubisco to total leaf N is also altered with changing total leaf N content in several C_3 species (7, 8). Nevertheless, there are few studies discussing an elevated-CO_2-induced decline in the Rubisco activity/content in relation to a change in the total leaf N (2, 9).   The nature of the response of Rubisco to elevated CO_2, and its species-dependent difference are still unclear.

In our previous studies with rice, the decline in Rubisco content by CO_2 enrichment could be explained by a decrease in total leaf N (10).   Although CO_2 enrichment led to a decline in total leaf N content, Rubisco content for a given leaf N content was unaffected by growth CO_2 pressures.   To examine whether such a tendency is found for other plants, we investigated the long-term effects of CO_2 enrichment on Rubisco, Chl and total leaf N contents and starch and sucrose contents in bean, and compared them with rice plants.

2. Materials and Methods

2.1 *Plant materials and cultural conditions*

Rice (*Oryza sativa* L.) and bean (*Phaseolus vulgaris* L.) were grown hydroponically in a controlled-environmental growth chamber (10).   The maximum PPFD was about 850 µmol quanta m^{-2} s^{-1}.   The basal nutrient solutions for rice and bean plants were the same as described by Mae and Ohira (11) and by Makino and Osmond (5), respectively.   Both plants were grown with different N concentrations and at normal (36 Pa) or elevated CO_2 (100 Pa) partial pressures.   After over 2 weeks, the youngest, fully expanded leaves of rice and the central leaflets of the 2nd trifoliate leaves of bean were used for the experiments.

G. Garab (ed.), Photosynthesis: Mechanisms and Effects, Vol. V, 3391–3394.

2.2 *Rubisco, Chl and total leaf N determinations*

Rubisco, Chl and total leaf N contents were determined as described by Makino et al. (12). Leaves were homogenized in 50 mM Na-phosphate buffer (pH 7.0) containing 2 mM Na-iodoacetate, 120 mM 2-mercaptoethanol and 5% (v/v) glycerol. A portion of the leaf homogenates was centrifuged after addition of Triton X-100 (final conc. 0.1% [v/v]) to solubilize membrane-bound Rubisco (5). The amount of Rubisco in the supernatant was measured spectrophotometrically after formamide extraction of Coomassie brilliant blue R-250-stained subunit bands separated by SDS-PAGE. A calibration curve was obtained with rice-purified Rubisco. Chl content was measured from an 80% (v/v) acetone extraction of the leaf homogenates according to Arnon (13). Total leaf N content was measured with Nessler's reagent after digestion of the leaf homogenate in H_2SO_4/H_2O_2.

2.3 *Carbohydrates determinations*

The procedure used for measuring starch and sucrose contents was as previously described (14). Briefly, a portion of the leaf homogenates was extracted at 80°C with 80% (v/v) ethanol. After evaporation of the ethanol, the residue was solubilized with 0.02 M NaOH. The sucrose content in the preparations was measured by the method of Jones et al. (15). Starch in the 80% ethanol-insoluble fraction was extracted with 0.5 M KOH, and digested with amyloglucosidase after neutralization with 0.5 M $HClO_4$. The starch content was taken to be 0.9 times the amount of glucose determined by Somogyi-Nelson method (16).

3. Results and Discussion

3.1 *Changes in Rubisco, Chl and total leaf N contents in rice and bean leaves*

Rubisco content in leaves increased with increasing growth N levels in both rice and bean plants (Fig. 1). However, for both plants, CO_2 enrichment induced a significant reduction in Rubisco content in the low N treatment. Similar results are also reported by Wong (4). Although Chl content decreased in rice grown at elevated CO_2 pressures, no effect on Chl content was found in bean (Fig. 1). These results indicated that the decline in the Rubisco content was selective in the elevated-CO_2-grown bean. On the other hand, the decreases in Rubisco and Chl in rice grown at elevated CO_2 pressures were correlated with the decrease in the total leaf N content. Thus, the decline in Rubisco in bean grown at elevated CO_2 pressures could not be explained by a decrease in total leaf N, because there was no significant change in the total leaf N content by CO_2 enrichment (Fig. 1). Actually, the ratio of Rubisco N to total leaf N contents was reduced in bean, whereas this ratio in rice was unaffected by growth CO_2 pressures (Fig. 2). These results suggested that there was a difference in the response of N allocation to Rubisco to elevated CO_2 between the two species.

3.2 *Carbohydrate accumulations by CO_2 enrichment*

In rice and bean, starch and sucrose contents increased significantly at the elevated CO_2 pressures (Fig. 3). In addition, the amount of starch increased with a decrease in growth N levels. However, the absolute amount of accumulated starch in bean is much greater than in rice. By contrast, the sucrose accumulation by CO_2 enrichment was

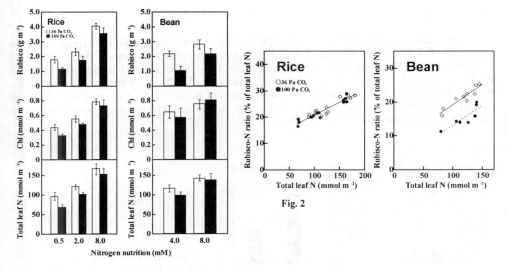

Fig. 1

Fig. 1 Rubisco, Chl, and total leaf N contents in leaves of rice (left panels) and bean (right panels) grown hydroponically under two CO_2 partial pressures of 36 (open columns) and 100 (closed columns) Pa CO_2 at different N concentrations. The vertical bars indicate SE (n = 3-6).

Fig. 2 Ratio of Rubisco N to total leaf N versus total leaf N content in leaves of rice (left panel) and bean (right panel) grown hydroponically under two CO_2 partial pressures of 36 (○) and 100 (●) Pa CO_2. Data are taken from Fig. 1. Regression analysis was performed using first-order kinetics. For rice, y = 0.097x + 11, r = 0.945; for bean grown at 36 Pa CO_2, y = 0.121x + 7.1, r = 0.930; for bean grown at 100 Pa CO_2, y = 0.118x + 1.2, r = 0.859.

greater in rice. Thus, this large accumulation of starch in bean apparently was correlated with the selective reduction in Rubisco. However, although the suppression of some photosynthetic gene expressions by soluble sugar(s) has been reported (17), it is still unknown how starch regulates the amount of Rubisco.

3.3 *The implication of the selective reduction in Rubisco protein at elevated CO_2*

The carboxylation capacity by Rubisco becomes relatively in excess of the capacities by the other photosynthetic processes under the conditions of elevated CO_2 pressures (2). Therefore, the selective reduction in Rubisco by CO_2 enrichment has been thought to be related to an optimal N reallocation among photosynthetic components. Sage et al. (2), however, suggested that such a Rubisco reduction did not necessarily lead to an idealized acclimation to elevated CO_2. Similarly, our results show that no significant increase in Chl content or cytochrome *f* content was observed in bean instead of the reduction of Rubisco (data not shown). Therefore, it is possible that N from decreased Rubisco is reallocated into non-photosynthetic components in bean grown at elevated CO_2 pressures. Thus, although Rubisco is the largest sink for N, the selective reduction in Rubisco by CO_2 enrichment may not mean an optimal N allocation at elevated CO_2 lev-

3394

els.

   In summary, our results indicate that there is a species-dependent difference in the response of Rubisco protein to elevated CO_2 between rice and bean. Bean showed a reduction of N allocation to Rubisco protein by CO_2 enrichment, but rice did not. In addition, absolute amount of carbohydrate accumulated as well as its partitioning under the conditions of CO_2 enrichment was also different between the two species. In bean grown at elevated CO_2, the selective reduction in Rubisco protein may be related to an excessive accumulation of starch in the leaves.

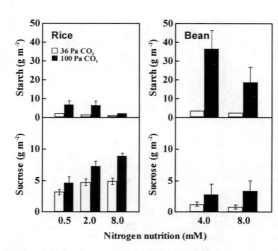

Fig. 3   Starch and sucrose contents in leaves of rice (left panels) and bean (right panels) grown hydroponically under two CO_2 partial pressures of 36 (open columns) and 100 (closed columns) Pa CO_2 at different N concentrations. The vertical bars indicate SE (n = 3-5).

References

1   Stitt, M. (1991) Plant Cell Environ. 14, 741-762
2   Sage, R.F., Sharkey, T.D. and Seemann, J.R. (1989) Plant Physiol. 89, 590-596
3   Rowland-Bamford, A.J., Baker, J.T., Allen, L.H., Jr and Bowes, G. (1991) Plant Cell Environ. 14, 577-583
4   Wong, S.C. (1979) Oecologia 44, 68-74
5   Makino, A. and Osmond, B. (1991) Plant Physiol. 96, 355-362
6   Morita, K. (1980) Ann. Bot. 46, 297-302
7   Makino, A., Sakashita, H., Hidema, J., Mae, T., Ojima, K. and Osmond, B. (1992) Plant Physiol. 100, 1737-1743
8   Evans, J.R. (1989) Oecologia 78, 9-19
9   Makino, A. (1994) J. Plant Res. 107, 79-84
10   Nakano, H., Makino, A. and Mae, T. (1997) Plant Physiol. 115, 191-198
11   Mae, T. and Ohira, K. (1981) Plant Cell Physiol. 22, 1067-1074
12   Makino, A., Nakano, H. and Mae, T. (1994) Plant Physiol. 105, 173-179
13   Arnon, D.I. (1949) Plant Physiol. 24, 1-15
14   Nakano, H., Makino, A. and Mae, T. (1995) Plant Cell Physiol. 36, 653-659
15   Jones, M.G.K., Outlaw, W.H.Jr and Lowry, O.H. (1977) Plant Physiol. 60, 379-383
16   Somogyi, M. (1952) J. Biol. Chem. 195, 19-23
17   Sheen, J. (1994) Photosynth. Res. 39, 427-438

TEMPERATURE ACCLIMATION OF THE PHOTOSYNTHETIC APPARATUS: BALANCING REGENERATION AND CARBOXYLATION OF RIBULOSE BISPHOSPHATE

Hikosaka K., Murakami A., Hirose T.
Biological Institute, Graduate School of Science,
Tohoku University, Aoba, Sendai 980-8578, Japan

Key words: adaptation, gas exchange, modelling, temperature acclimation, temperature dependence, nitrogen use

1. Introduction

Temperature dependence of photosynthesis varies even in plants of the same species grown under different temperature conditions (1). In many species, leaves acclimated to lower temperatures tend to have lower temperature optimum of photosynthesis. However, although several properties in relation to the photosynthetic apparatus are known to change depending on growth temperature, it is still unclear how temperature dependence of photosynthesis changes.

It is well known that changes in organisation of the photosynthetic components are involved in sun-shade acclimation and that it contributes to efficient use of nitrogen in leaves. Nitrogen to construct photosynthetic proteins is generally limited in natural environments. Leaves grown under the shade invest more nitrogen into light-harvesting function, i. e., chl-protein complexes, whilst those grown under the sun invest more nitrogen into energy conversion and CO_2 fixation, i. e., electron carriers, ATPase, and Calvin cycle enzymes (2, 3).

We have hypothesised that the similar context may be applicable to the temperature acclimation of photosynthesis. In this article, we summarise our recent studies on the temperature acclimation. First, using a theoretical model, we show that changes in organisation of photosynthetic components can alter temperature dependence of the photosynthetic rate. Second, we test the model prediction in an evergreen tree, *Quercus myrsinaefolia*.

2. Theoretical study

Hikosaka & Terashima (3) proposed a model for nitrogen partitioning among photosynthetic components. In this model, photosynthetic components were categorised into 5 groups according to their function. Amount of each group was responsible to the light-saturated rate and/or the initial slope of light-response curve of photosynthesis. Optimal nitrogen partitioning among these groups maximising daily photosynthesis was calculated. However, in this model, temperature response was ignored. Extending this model, Hikosaka (4) proposed a new model incorporating temperature dependence of each component. In the new model, the photosynthetic rate is limited either regeneration or carboxylation of ribulose bisphosphate (RuBP). In order to have higher capacity of RuBP carboxylation, leaves should invest more nitrogen in RuBP carboxylase, while large investment of nitrogen in electron carriers, ATPase, and other Calvin cycle enzymes

3395

G. Garab (ed.), Photosynthesis: Mechanisms and Effects, Vol. V, 3395–3398.

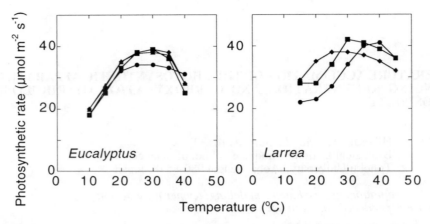

Figure 1. Model simulation of temperature dependence of light-saturated rates of photosynthesis of leaves optimised at various temperature regimes. Left and right show results where temperature dependence of RuBP regeneration was based on *Eucalyptus pauciflora* and on *Larrea divaricata*, respectively. Triangles, diamonds, squares, and circles denote leaves optimal at 10°C, 20°C, 30°C, and 40°C, respectively. Leaf nitrogen content is assumed to be fixed at 0.18 mol m^{-2}.

makes the capacity of RuBP regeneration higher. Temperature dependence of these two processes was formulated according to literature. Temperature dependence of RuBP carboxylation was deduced from that of RuBP carboxylase of spinach determined in vitro (5), which has temperature optimum at 30°C. For temperature dependence of the light-saturated rate of RuBP regeneration, two contrasting curves were used. One is obtained with gas exchange method for *Eucalyptus pauciflora*, which has temperature optimum at 30°C (6). The other is obtained as the electron transport rate of thylakoid membrane purified from *Larrea divaricata*, which has temperature optimum at 40°C (7). For simplification, temperature dependence of each process was assumed to be independent of growth temperature. Optimal nitrogen partitioning among the groups maximising daily photosynthesis at different temperature conditions was calculated.

Figure 1 shows calculated temperature dependence of the light-saturated rate of photosynthesis of leaves optimised at various temperatures. In *E. pauciflora*, differences in the temperature dependence of the photosynthetic rate among the leaves are small. On the other hand, in *L. divaricata*, the temperature optima clearly shift toward the temperature at which leaf photosynthesis is optimised. Figure 2 illustrates how the change in temperature dependence of photosynthesis occurs. The photosynthetic rate (circles) is limited by the lower rate of two processes, regeneration and carboxylation of RuBP. If the investment of nitrogen in the process that has a lower optimal temperature (carboxylation) is reduced and that in the other (regeneration) is increased, the photosynthetic rate has a lower optimal temperature. Inversely, if investment in RuBP regeneration is reduced, the optimal temperature of the photosynthetic rate increases. Therefore, If the temperature dependence of the two limiting processes are different from each other, temperature dependence of photosynthesis can change by altering the organisation of photosynthetic components. In *E. pauciflora*, temperature dependence of regeneration was relatively similar to that of carboxylation. Therefore, we predict that changes in nitrogen partitioning cause little change in temperature dependence of

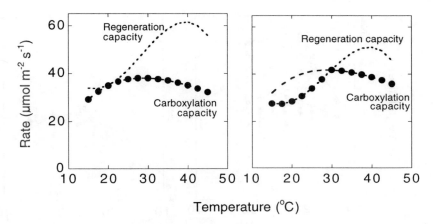

Figure 2. Model simulation of photosynthetic rates and capacities of regeneration and carboxylation of RuBP in *Larrea divaricata* leaves optimal at 20°C (left) and at 30°C (right). Dotted lines, capacity of RuBP regeneration at 35 Pa CO_2; broken lines, capacity of RuBP carboxylation at 35 Pa CO_2; circles, photosynthetic rate at 35 Pa CO_2.

photosynthesis. This prediction is consistent with the fact that the change in temperature optimum of photosynthesis of *E. pauciflora* is larger than that of *L. divaricata*.

3. Temperature acclimation in *Quercus myrsinaefolia*

To test the prediction that the change in the balance between regeneration and carboxylation of RuBP contributes to temperature acclimation, we investigated gas exchange characteristics of an evergreen tree, *Quercus myrsinaefolia*, grown under two temperature regimes (15 and 30°C) (Hikosaka K., Murakami A. and Hirose T., submitted). It is known that the capacity of RuBP carboxylation tends to limit photosynthesis at low CO_2 concentrations while that of RuBP regeneration does so at high CO_2 concentrations. We determined temperature dependence of photosynthesis at various CO_2 levels and, using a kinetic model, estimated the capacity of these process at 35 Pa CO_2 in order to know which limits the photosynthetic rate at 35 Pa CO_2 (8).

    In Fig. 3, the estimated capacities of carboxylation and regeneration of RuBP at 35 Pa CO_2 are compared with the photosynthetic rate at 35 Pa CO_2. For 15°C-grown leaves, the capacity of RuBP carboxylation was always lower than that of RuBP regeneration. The photosynthetic rate at 35 Pa CO_2 nearly corresponded to the capacity of RuBP carboxylation at all temperatures. This suggests that the photosynthetic rate at 35 Pa CO_2 is always limited by RuBP carboxylation. On the other hand, for 30°C-grown leaves, the capacity of RuBP carboxylation was higher than that of RuBP regeneration below 22°C but was lower above this temperature. The photosynthetic rate at 35 Pa CO_2 nearly corresponded to the lower value of the capacities of the two processes, suggesting that the photosynthetic rate is limited by RuBP regeneration below 22°C and is limited by RuBP carboxylation above 22°C. These results suggest that the balance between regeneration and carboxylation changes depending on growth temperatures, as predicted by the theoretical study.

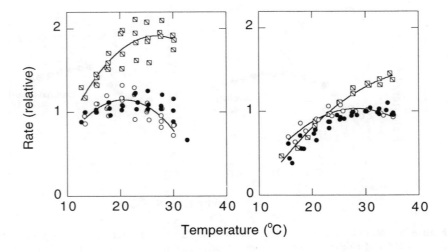

Temperature (°C)

Figure 3. Comparison of photosynthetic rates and capacities of carboxylation and regeneration of RuBP at 35 Pa CO_2 for *Quercus myrsinaefolia* leaves grown at 15 (left) and 30°C (right). Closed circles, photosynthetic rate at 35 Pa; open circles, capacity of RuBP carboxylation at 35 Pa; squares, capacity of RuBP regeneration at 35 Pa. Relative values normalised at growth temperature are shown. Fitted curves are two-dimensional polynomials.

However, it is still unclear how the balance between the two processes changes. As presented above, we hypothesized that the balance changes due to changes in nitrogen partitioning among photosynthetic components. Therefore, further studies on the amount of photosynthetic components are still needed. On the other hand, it is also obvious that temperature dependence of each process changes with growth temperatures (Fig. 3). Temperature dependence of RuBP regeneration was large in leaves grown high temperature. Such a change is known for the electron transport rate at thylakoid membrane (7), which may cause the difference in the temperature dependence of the RuBP regeneration capacity. Temperature dependence of the RuBP carboxylation capacity also differed: although 30°C-grown leaves showed increases in the capacity at higher temperatures, the change was small in the capacity of 15°C-grown leaves. This may be due to difference in heat stability of RuBP carboxylase in vivo (9). Such changes in each process may also contribute to temperature acclimation of photosynthesis.

References
1. Berry, J.A. and Björkman, O. (1980) Ann. Rev. Plant Physiol., 31, 491-543.
2. Evans, J.R. (1989) Aust. J. Plant Physiol., 16, 533-548.
3. Hikosaka, K. and Terashima, I. (1995) Plant Cell Environ., 18, 605-618.
4. Hikosaka, K. (1997) Ann. Bot., 80, 721-730.
5. Jordan, D.B. and Ogren, W.L. (1984) Planta, 161, 308-313.
6. Kirschbaum, M.U.F. and Farquhar, G.D. (1984) Aust. J. Plant Physiol., 11, 519-538.
7. Armand, P.A., Schreiber, U.S. and Björkman, O. (1978) Plant Physiol., 61, 411-415.
8. Farquhar, G.D., von Caemmerer, S. and Berry, J. (1980) Planta, 149, 78-90.
9. Badger, M.R., Björkman, O. and Armand, P.A. (1982) Plant Cell Environ., 5, 85-99.

PHOSPHORIBULOKINASE:
MUTAGENESIS OF THE MOBILE LID AND "P-LOOP"

Jennifer A. Runquist, Hanane A. Koteiche,
David H.T.Harrison, & Henry M. Miziorko
Medical College of Wisconsin, Milwaukee, WI 53226

Key words: ATP; Calvin cycle; kinase; phosphoribulokinase; R. Sphaeroides; site-directed mutagenesis

Introduction

Phosphoribulokinase (PRK) catalyzes the in-line transfer of ATP's γphosphoryl to the C1 hydroxyl of ribulose 5-phosphate (Ru5P) forming ribulose 1,5-bisphosphate, the Calvin cycle CO_2 acceptor. Recently several acidic residues crucial to catalysis, D42 and D169, were identified in *R. spheroides* PRK (1). The mutation of these two residues to alanines profoundly reduced catalysis, with a 10^5-fold diminution in rate for D42 and a 10^4-fold diminution in rate for D169; E131 also has a significant influence on the catalytic efficiency. These acidic amino acids could function as activator cation ligands or as a catalytic base. A specific function in sugar phosphate substrate binding has been proposed for prokaryotic PRK's R49, based on the large $K_{m\,Ru5P}$ effect that is observed upon mutagenesis of this residue to glutamine (2).

Elucidation of a high resolution structure for *R. spheroides* PRK (3) places PRK within the adenylate kinase (AK) family of proteins, all of which have a "mobile lid". This observation stimulated our investigation of the remaining invariant PRK arginines (R168, R173, R187 as well as R186 which is conserved in prokaryotic PRKs) all situated in PRK's mobile lid domain. In addition, the consensus "P-loop" is known to be important to ATP binding in the AK family of proteins. As a first step in pursuing an understanding of the dynamics of this loop, an A16C enzyme was studied. In eukaryotic PRKs there is a regulatory cysteine at this P-loop position.

Procedure

Arginine and Cysteine Mutagenesis. For construction of arginine mutants R168Q, R173Q, R186Q, R187Q, cassette mutagenesis was utilized (4). Cysteine mutants were produced using PCR technology. For C194A single-stage PCR was utilized. The TGC codon was changed to GCC with a synthetic oligonucleotide containing this mutation as well as a unique *Nsi*I site 26 bp upstream of the mutation site. A 610 bp PCR fragment generated by use of this primer and the universal T_7 terminator primer

G. Garab (ed.), Photosynthesis: Mechanisms and Effects, Vol. V, 3399–3402.
© 1998 *Kluwer Academic Publishers. Printed in the Netherlands.*

was digested with *Nco*I and *Nsi*I. The purified 486 bp fragment was ligated with a 1298 bp *Nco*I/*Pst*I and 3906 *Nsi*I/*Pst*I fragments to generate the mutant plasmid. The A16C mutation was constructed using PCR over-lap extension methodology. First a *Sac*I restriction site was created by a silent mutation at DNA position 239 in the prkA gene. The 4 necessary oligonucleotides included the two mutagenic oligonucleotides, the T_7 promoter primer and a synthetic primer. The 690 bp PCR piece was digested with *Nco*I and *Sac*II producing a 520 bp fragment which was ligated to 3340 bp *Nco*I/*Pst*I and 1831 bp *Pst*I/*Sac*II fragments. The A16C mutation was implemented using the C194A, *Sac*I plasmid. A 440 bp PCR over-lap extension piece was constructed employing 4 oligonucleotides as described above. After digestion with *Sac*I and *Nco*I, the 240 bp piece was ligated with 3340 bp *Pst*I/*Nco*I and 2000 bp *Pst*I/*Sac*I fragments to complete the mutant plasmid.

Methods. Procedures followed for expression and purification of PRK mutants have been described previously (1). The assay of mutant PRKs involved trapping the PRK reaction product RuBP via the RuBP carboxylase-dependent incorporation of $^{14}CO_2$. NADH and TNP-ATP stoichiometries were determined using fluorescence titrations, and ATPγSAP (adenosine 5'-O-(S-acetamidoproxyl 3-thiotriphosphate) stoichiometries were determined from ESR titrations (5,6).

Results and Discussion

Structural and Kinetic Characterization of Arginine Mutants. The arginine mutant proteins were expressed and isolated to homogeneity. The structural integrity of the mutants was evaluated by fluorescence titrations using the allosteric activator NADH and trinitrophenyl-ATP, a fluorescent nucleotide triphosphate that functions as an alternative PRK substrate. Stoichiometries determined from these titrations are presented in Table 1. These mutants show a full complement of functional substrate and effector binding sites, thus confirming their substantial structural integrity. Table 2 displays the kinetic characterization of the arginine mutants.

TABLE 1: SUBSTRATE AND EFFECTOR BINDING BY PRK MUTANTS

ENZYME	$n_{TNP-ATP}$	n_{NADH}
WILD-TYPE	1.0 ± .03	0.97 ± .03
R168Q	1.1 ± .03	0.88 ± .05
R178Q	0.90 ± .05	0.87 ± .05
R186Q	0.90 ± .04	1.1 ±.04
R187Q	1.1 ± .04	1.1 ± .04

TABLE 2: STEADY-STATE KINETIC PARAMETERS OF WILD-TYPE AND ARGININE MUTANT PRKS

ENZYME	$K_{m\,RuSP}$ (mM)	n_H	$S_{1/2\,ATP}$ (mM)	n_H	V_{max} (units/mg)
WT	0.096	(h)	0.55	2.0	338
R168Q	4.6	(h)	2.4	1.9	0.98
R173Q	10.5	(h)	1.5	2.0	22.0
R186Q	0.048	(h)	0.17	1.0	273
R187Q	0.55 ($S_{1/2}$)	2.9	3.2	2.7	150

Structural and Kinetic Characterization of Cysteine Mutants. The cysteine mutant proteins were expressed and isolated to homogeneity. The structural integrity of the mutants was evaluated by CD scans from 195-260 nm, which showed the secondary structure to be similar to that of wild-type. Gel filtration on Superose 6 suggests that wild-type, C194A, and A16C exhibit comparable quaternary structure. Structural integrity of mutants was also indicated by maintenance of activation by the allosteric effector NADH as seen in Table 3 below. ESR titrations using the spin-labeled probe ATPγSAP, an ATP analogue, gave stoichiometries close to wild-type for both C194A and A16C demonstrating that the mutants have a full complement of functional ATP binding sites.

TABLE 3: KINETIC AND BINDING PARAMETERS FOR WILD-TYPE AND CYSTEINE MUTANT PRKS

ENZYME	K_{mRuSP} (mM)	$S_{1/2ATP}$ (mM)	V_{max} (units/mg)	NADH ACTIVATION (-fold)	$n_{ATPγSAP}$
WT	0.082	0.568	322	~80	0.83
C194A	0.080	0.716	383	~100	ND
A16C	0.518	3.2	0.25	>30	0.73

Interpretation of Mutagenesis Results. Precedent for the significant function of arginines in phosphotransferase reactions prompted substitution of glutamine for each of these three invariant lid arginines. Substantial k_{cat}(R168Q), $K_{m\,RuSP}$(R173Q) and cooperativity effects (R186Q, R187Q) confirm the importance of the mobile lid to catalysis. The rate reduction of >300-fold in k_{cat} for R168Q, a residue next to the catalytically essential D169, suggests a role in the catalysis process. A role in transition state stabilization during phosphoryl transfer is possible if the mechanism is associative. For a dissociative process this basic residue might assist the leaving group. In this context, the 300-fold reduction in catalysis rate for R168Q is comparable to that seen upon replacement of E131, a Walker B residue which typically contributes a ligand to the cation of the M-ATP substrate (1). For R173Q it seems reasonable to correlate the 100-fold reduction in $K_{m\,Ru5P}$ with diminished substrate affinity. This 10^2-fold effect is comparable in magnitude to the the effect previously reported for R49Q (2). However, there would have to be considerable movement of the lid domain to close the active site cavity and bring these two residues into proximity for support of Ru5P binding. The cooperative effects on the saturation curves of R186Q and R187Q could imply that these residues play a role in communication between the effector and active site.

For the cysteine mutants it is clear that C194A resembles wild-type PRK. Surprisingly, A16C has a greatly reduced catalysis rate in spite of showing a full complement of ATP binding sites as indicated by stoichiometric binding of ATPγSAP. Additionally, the CD, gel filtration, substrate binding, and NADH activation results indicate that structural features are similar to wild-type for A16C. Since eukaryotic PRKs have cysteine in this P-loop position, the significant k_{cat} effect for A16C was unexpected. Additional work will be required in order to explain the observed effect.

References

1. Charlier,H.A., Runquist,J.A., & Miziorko, H.M. (1994) Biochemistry 33, 9343-9350
2. Sandbaken,M.G., Runquist,J.A., Barbieri,J.T. & Miziorko,H.M. (1992) Biochemistry 31, 3715-3719
3. Harrison,D.H.T., Runquist,J.A., Holub,A. & Miziorko,H.M. (1998) Biochemistry 37, 5074-5085
4. Runquist,J.A.,Harrison,D.H.T., & Miziorko,H.M. (1998) Biochemistry 37, 1221-1226
5. Koteiche,H.A., Narasimhan,C., Runquist,J.A. & Miziorko,H.M.(1995) Biochemistry 34, 15068-15074
6. Runquist,J.A.,Narasimhan,C.,Wolff,C.E., Koteiche,H.A., &Miziorko,H.M. (1996) Biochemistry 35, 15049-15056

CHARACTERIZATION OF HIGH AND LOW MOLECULAR MASS ISOFORMS OF PHOSPHO*ENOL*PYRUVATE CARBOXYLASE FROM THE GREEN ALGA *SELENASTRUM MINUTUM*

Jean Rivoal[1], Stacy Trzos[3], Douglas A. Gage[3], William C. Plaxton[1,2] and David H. Turpin[1] Depts. of [1]Biology and [2]Biochemistry, Queen's University, Kingston, ON, K7L 3N6, Canada. [3]Dept. of Biochemistry, Michigan State University, East Lansing, MI 48824 USA.

Key words: PEPC, green algae, protein-protein interaction, C metabolism, N metabolism, phosphorylation/dephosporylation

1. Introduction

Phospho*enol*pyruvate carboxylase (PEPC) is a ubiquitous cytosolic enzyme in plants and is also found in a wide range of prokaryotes (1). This enzyme has been extensively studied and characterized in higher plants, in particular in C4 and CAM leaves. In contrast, lower plant PEPCs have only received a modest amount of attention. However, physiological studies in C3 green algae have shown that, in these organisms, PEPC is a key enzyme involved in anaplerotic C fixation during N assimilation (2,3). This is an important enzyme to study because approximately 70% of global N assimilation occurs in algal cells. We have recently purified PEPC from two green algae *Selenastrum minutum* (4) and *Chlamydomonas reinhardtii* (5) and found multiple PEPC isoforms in these unicellular organisms. In this paper, we present an overview of our current knowledge on the kinetic and structural properties and of green algal PEPCs and propose a model for the structural organization of PEPC in green algae.

2. Procedures

2.1 *PEPC extraction, assay and purification*
PEPC was extracted from N-sufficient *S. minutum* cells (6). Unless otherwise stated, the extraction buffer contained 25 mM Hepes-KOH, pH 7.5, 20% (v/v) glycerol, 1 mM EDTA, 1 mM EGTA, 10 mM $MgCl_2$, 0.1% (v/v) Triton X-100, 25 mM NaF, 5 mM malate, 2 mM DTT, 2 mM PMSF, 2 mM 2,2'-dipyridyl, 10 µg/ml pepstatin, 10 µg/ml chymostatin. PEPC assay and purification were as previously described (4).

2.2 *Native gel electrophoresis and activity staining*
PEPC isoforms were analyzed by native gel electrophoresis as previously described (5). PEPC activity bands were visualized by incubating the gel in a buffer containing 50 mM Bis-Tris-propane pH 8.4, 15 % (v/v) glycerol, 20 mM $MgCl_2$, 5 mM $KHCO_3$, 0.2 mM NADH, 5 mM PEP, 10 units/ml malate dehydrogenase. PEPC

G. Garab (ed.), *Photosynthesis: Mechanisms and Effects, Vol. V,* 3403–3406.
© 1998 *Kluwer Academic Publishers. Printed in the Netherlands.*

activity bands appeared dark on a fluorescent background when the gel was illuminated with UV light.

2.3 *Protein phosphorylation and dephosphorylation assays*

Protein kinase assays were conducted as described (7). Dephosphorylation of purified PEPC was performed using bovine protein phosphatase 2A (PP2A), in a buffer containing 50 mM Hepes-KOH (pH 7.5), 5 mM $MgCl_2$, 1 mM EDTA, 50 nM microcystin LR, 20% (v/v) glycerol, 2 mM DTT.

3. Results and Discussion

3.1 *S. minutum contains two classes of PEPC isoforms*

Four isoforms of PEPC have been purified from *S. minutum* (4). These isoforms fall into two classes. Class 1 (low M_r) is represented by PEPC1 and Class 2 (high M_r) contains PEPC2, PEPC3 and PEPC4. Table 1 summarizes the properties of the two PEPC classes. These data clearly indicate that the two PEPC classes have very different structural and kinetic properties.

Table 1: Summary of the structural, physical and kinetic properties of the two classes of PEPC isoforms.

Parameter	Class 1	Class 2	Comments
Native M_r	400 kDa	1000-1600 kDa	Determined by SEC on Superose 6
Subunit M_r	102 kDa	130 kDa, 102 kDa, several peptides at 60-75 kDa	Determined by SDS/PAGE
Structure	Homotetramer	Protein complexes	
Thermostability	Heat labile (denaturation at T > 25°C)	Heat stable (denaturation at T > 53°C)·	Determined with pure enzyme preparations
Sensitivity to:			
DHAP	Activation	No effect	Sensitivity to activators
Gln	Potent activation	Weak activation	and inhibitors is
Glu	Inhibition	No effect	markedly increased at
2-oxoglutarat	Inhibition	Weak inhibition	physiological pH (7.4)
Asp	Inhibition	No effect	compared to optimal
malate	Potent inhibition	No effect	pH 8.4)
Affinity for PEP	Low ($S_{0.5}$ = 1.6 mM)	High (K_m = 0.3 mM)	Values obtained at pH 8.4
PEP binding kinetics	Cooperative	Michaelis-Menten	

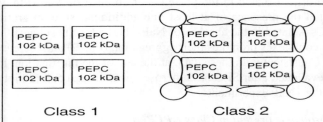

Figure 1: A model for the structural organization of the two classes of PEPC isoforms in *S. minutum*. The round shaped boxes represent APs. Note that the stoichiometry of the various Class 2 subunits is not known.

Evidence from CNBr peptide mapping and immunoblotting of the 102 kDa polypeptide present in all *S. minutum* PEPC isoforms indicate that the two classes of isoform contain the same catalytic subunit (4,5). The other polypeptides associated with Class 2 PEPCs are not immunologically related to the catalytic subunit (4). We therefore propose that, in the Class 2 PEPCs, association of the 102 kDa PEPC catalytic subunit with immunologically unrelated associated proteins (APs) is responsible for the striking differences in physical and kinetic properties observed when comparing the two PEPC classes (Table 1). These data are consistent with the model for the structural organization of *S. minutum* PEPCs presented in Figure 1.

3.3 *Involvement of protein phosphorylation in the regulation of the structural organization and activity of S. minutum Class 2 PEPCs*

Table 2: The effect of NaF on the extractable PEPC activity and on the ratio of the two classes of isoforms.

Parameter	+ NaF	- NaF
Extractable PEPC (units/mg protein)	0.018	0.030
Ratio Class 1/Class 2	1.15	0.13

S. minutum PEPC was extracted in the absence or presence of 25 mM NaF. The presence of this phosphatase inhibitor lowered the yield of extractable PEPC activity (Table 2). Purification of the two classes of PEPCs +/- NaF revealed that the yield of Class 2 PEPCs was greatly depressed when 25 mM NaF was present in chromatographic buffers (Table 2).

The overall yield of PEPC1 was not significantly affected by the presence of NaF.

A native PAGE analysis followed by activity staining was performed

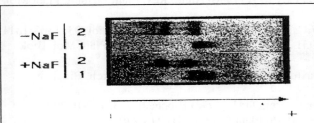

Figure 2: Native PAGE analysis of PEPC isoforms purified in the presence or absence of NaF in chromatographic buffers. 1: Class 1, 2: Class 2

on the two classes of isoforms isolated in the presence or absence of NaF (Fig. 2). The activity stain revealed the presence of at least one additional, slow-migrating band in the Class 2 PEPCs isolated in the absence of NaF. These results suggest that dephosphorylation of the complexes promotes their aggregation (Fig. 2) and increases their recovery (Table 2). Consistent with these data, at the end of the purification, the yield of Class 2 PEPCs, evaluated after SEC on Superose 6 was at least 5-fold higher in the -NaF compared to the +NaF experiment.

3.4 A 130 kDa phosphoprotein is present in Class 2 PEPCs

The 130 kDa polypeptide co-purifying with Class 2 PEPCs, p130, is the substrate for a protein kinase present in *S. minutum* crude extracts. This protein kinase, however, does not copurify with the Class 2 complexes. p130 is also a substrate for bovine protein kinase A. In protein kinase assays where Class 2 PEPCs were used as a substrate, p130 was always the major protein labeled with both the endogenous kinase and protein kinase A. This polypeptide is therefore a good candidate to mediate the phosphorylation-dependent change in aggregation state of Class 2 PEPCs. We are currently sequencing p130 tryptic peptides using MALDI-MS peptide mapping and post source decay analysis to identify this protein and localize its phosphorylation site.

4. Conclusions

Two classes of PEPC isoforms with contrasting physical and kinetic properties exist in the unicellular green alga *S. minutum*. PEPC1 has a homotetrameric structure whereas PEPC2, PEPC3 and PEPC4 are complexes containing the PEPC catalytic subunit and immunologically unrelated associated proteins. Circumstantial evidence indicates that *S. minutum* PEPCs are regulated by protein-protein interactions. We have also obtained evidence that the structural organization of the large M_r isoforms is influenced by phosphorylation, with dephosphorylation promoting aggregation of these complexes. Work is underway to identify the protein(s) involved in this process using MALDI-MS and microsequencing.

Acknowledgements

We thank T. F. Moraes, A. E. McDonald, J. R. Moustgaard and A. Messenger for their help with this manuscript.

References

1. Chollet R., Vidal J. and O'Leary M.H (1996) Annu. Rev. Plant Physiol. Plant Mol. Biol. 47, 273-298
2. Schuller K.A., Plaxton W.C. and Turpin D.H. (1990) Plant Physiol. 93, 1303-1311.
3 Huppe H.C. and Turpin D.H. (1994) Annu. Rev. Plant Physiol. Plant Mol. Biol. 45, 577-607
4. Rivoal J., Dunford R., Plaxton W.C. and Turpin D.H. (1996) Arch. Biochem. Biophys. 332, 47-57
5. Rivoal J., Plaxton W.C. and Turpin D.H. (1998) Biochem. J. 331, 201-209
6. Elrifi I.R. and Turpin D.H. (1986) Plant Physiol. 81, 273-279
7. Law R.D. and Plaxton W.C. (1997) Eur. J. Biochem. 247, 642-651

REGULATORY PHOSPHORYLATION OF MAIZE C4-FORM PEP CARBOXYLASE (PEPC): ANTIBODY DETECTION SYSTEM AND PURIFICATION OF PEPC-PROTEIN KINASE

Ueno Y.,[1] Emura J.,[2] Kumagaye K. Y.,[2] Nakajima K.,[2] Inami K.,[2] Shiba T.,[2] Hata S.[1] and Izui K.[1]

[1]Div. Applied Biosciences, Grad. Sch. Agriculture, Kyoto Univ., Kyoto 606-8502, Japan, and [2]Peptide Institute, Inc., 4-1-2 Ina, Osaka 562-0015, Japan

Key words: chemiluminescence, circadian rhythm, enzymes, light regulation, signal transduction, synthetic oligopeptide

1. Introduction

Activity of phospho*enol*pyruvate carboxylase (PEPC) for C_4 photosynthesis is upregulated by light (reviewed in [1, 2]). The regulation is mediated by a specific protein kinase (PEPC-PK) that phosphorylates Ser near N-terminus (Ser-15 in the maize C_4-form PEPC). This phosphorylation is dependent on light intensity in C_4-plants [3]. Hartwell *e al* [4] reported that PEPC-PK was regulated by the level of translatable mRNA. No evidence, however, has been obtained to indicate how PEPC-PKs or its mRNAs are regulated and there has been no report about the gene for PEPC-PK.

In order to study this signal transduction cascade we undertook two lines of approach. 1) For the analysis of phosphorylation of PEPC *in vivo* under various physiological conditions, we developed specific antibodies to detect phosphorylated PEPC. With these antibodies, time course of phosphorylation of PEPC in leaves upon light-dark transition could be followed semiquantitatively. 2) Since it is now most plausible that major PEPC-PKs are Ca^{2+}-independent PK with Mrs of 30 and 37 kD, we purified one of these kinase to homogeneity in order to clone its cDNA.

2. Procedure

2.1 The specific antibodies to phosphorylated PEPC (anti-PNP antibodies)

A phosphorylated synthetic oligopeptide PNP (GCPGEKHHS(*P*)IDAQLR) was conjugated with keyhole limpet hemocyanin and injected to rabbit as antigen. The obtaine antiserum was purified with successive use of NNP (non-phosphorylated synthetic peptide: GCPGEKHHSIDAQLR)- and PNP-immobilized affinity column chromatography. The specificity of purified antibodies was confirmed by ELISA and Western blots.

G. Garab (ed.), Photosynthesis: Mechanisms and Effects, Vol. V, 3407–3410.
© 1998 *Kluwer Academic Publishers. Printed in the Netherlands.*

2.2 Purification of PEPC-PK

Maize leaves were harvested repeatedly on sunny mornings (7:00-11:00) and disrupted within 30 min with Polytron after removal of main veins (2.6 kg fresh weight without main veins in total). Crude soluble extract was fractionated with 25-60% saturation of $(NH_4)_2SO_4$. The obtained suspension was dissolved to a final concentration of $(NH_4)_2SO_4$ to 1 M and purified as follows: Step 1, Centrifugation at 10^5 x g; step 2, TOYOPEARL HW-55 F (TOSOH) hydrophobic chromatography; step 3, Phenyl-Sepharose FF (Pharmacia) hydrophobic chromatography; step 4, Blue-Sepharose CL-6B (Pharmacia) affinity chromatography; step 5, Superose 12 (Pharmacia) size exclusion chromatography; step 6, Chitopearl SU-01 (FUJIBO) cation exchange chromatography; step 7, Hydroxyapatite (KOKEN) HPLC chromatography; step 8, MiniQ (Pharmacia) SMART anion exchange chromatography.

3. Result and Discussion

3.1 Diurnal change of the phosphorylation status of PEPC in vivo

Anti-PNP antibodies we developed in this study were highly specific to phosphorylated form of PEPC. As shown in Figure 1, these antibodies reacted with recombinant wild-type or K12N mutant PEPCs phosphorylated with PEPC-PKs and PEPC prepared from illuminated maize leaves. In contrast, the antibodies did not react with wild-type and K12N mutant PEPC [5] which had not been phosphorylated nor S15D mutant PEPC. It was also revealed that the antibodies did not react with root-form PEPC irrespective of its phosphorylation status, PEPCs from illuminated leaves of *Echinochloa utilis*, *Echinochloa crus-galli*, *Mesembryanthemum crystallinum*, *Flaveria pringlei*, *Flaveria ramosissima*, and *Flaveria trinervia* but cross-reacted with PEPCs from Pearl Millet and *Sorghum bicolor* (data not shown).

Anti-PNP antibodies were useful to monitor dynamic changes of the phosphorylation status of PEPC in maize leaves. In 2 weeks-old young plants, the phosphorylation and dephosphorylation were found to be completed within 2 h (data not shown). In this experiment, no indication of circadian rhythm for this phosphorylation was observed. These results were consistent with the report by Bakrim *et al* [3]. Figure 2, however, shows that the phosphorylation and dephosphorylation were triggered prior to the onset and offset of illumination, respectively, in 13 weeks-old mature maize plants grown in the field. And this phosphorylation status changed closely in parallel with the activity of PEPC-PKs. These results suggest that there are another trigger(s) to initiate phosphorylation of PEPC other than light intensity in mature maize leaves.

3.2 Purification of PEPC-PK

PEPC-PKs were highly hydrophobic on chromatographic behavior. We employed TOYOPEARL resin as a support of hydrophobic chromatography in the first step. The later purification was conducted according to conventional procedures. Each of three active fractions in the final step (MiniQ) gave a single band of about 30 kD on SDS-PAGE

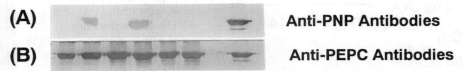

1 2 3 4 5 6 7 8

(A)   **Anti-PNP Antibodies**

(B)   **Anti-PEPC Antibodies**

Figure 1. Western blot analysis of various maize C_4-form PEPCs

Each membrane was probed with anti-PNP antibodies (A) or anti-PEPC antibodies. Recombinant wild-type (1 and 2), K12N mutant (3 and 4), S15D mutant (5 and 6) and without (7) PEPCs were incubated with Mg-ATP and with (2, 4, 6 and 7) or without (1, 3 and 5) partially purified PEPC-PKs before SDS-PAGE. PEPC prepared from illuminated maize leaves was loaded on lane 8.

(A)

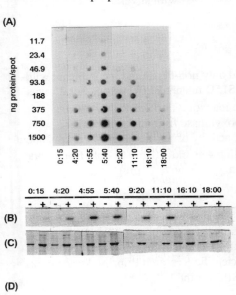

(B)

(C)

(D)

Figure 2. Diurnal change of the phosphorylation status of PEPC and the activity of PEPC-PKs

Each leaf extract from 13 weeks-old mature maize plant grown in the field was subjected to dot blot analysis and the assay of PEPC-PK activity. Light intensity was measured at each sampling time and the data are given in panel (D). Dot blot was probed with anti-PNP antibodies (A) and anti-PEPC antibodies. Each sample gave almost the same intensity of signal probed with anti-PEPC antibodies (data not shown). For the assay of PEPC-PKs in each sample, the samples were freed from endogenous PEPC by hydrophobic chromatography on TOYOPEARL resin. The obtained samples were incubated with Mg-[γ-^{32}P]ATP and with (+) or without (-) recombinant PEPC [5]. After the reaction the mixtures were subjected to SDS-PAGE. Gels were stained with Coomassie Brilliant Blue (C) and ^{32}P incorporation was analyzed by autoradiography (B) and BAS 2000 (Fuji) (D).

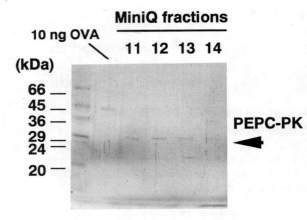

Figure 3. Purified 30 kDa PEPC-PK The active fractions of MiniQ were subjected to 12.5 % SDS-PAGE. Molecular weight markers and 10 ng chicken egg white albumin (OVA) were loaded on the left side. Gel was stained with silver.

by silver staining (Figure 3) and phosphorylated only non-denatured recombinant wild-type and K12N mutant C$_4$-form PEPC but not S15D mutant PEPC whose Ser-15 was replaced by Asp. Ca^{2+} was not required for the reaction. Purified PEPC-PK phosphorylated neither boiled PEPC, PEP carboxykinase from maize nor other conventional substrates. The estimated specific activity of purified PEPC-PK was about 100-500 nmol/min/mg and the attained purification was more than 10^5-fold from crude extract. Amino acid sequence analysis of this enzyme is now on progress.

4. Acknowledgement

Authors thank to Furumoto, T. for his generous gift of recombinant maize PEP carboxykinase. This work was supported in part by grants from "Research for the Future" program JSPS-RFTF 96L00604 and Scientific Research Priority Areas Grants (09274214, 10170217). Y. U. thanks to the JSPS for the fellowship.

References

1 Chollet, R., Vidal, J. and O'Leary, M. H. (1996) Annu. Rev. Plant Physiol. Plant
 Mol. Biol. 47, 273-298
2 Vidal, J. and Chollet, R. (1997) Trends in Plant Sci. 2, 230-237
3 Bakrim, N., Echevarria, C., Cretin, C., Arrio-Dupont, M., Pierre, J. N., Vidal, J.,
 Chollet, R. and Gadal, P (1992) Eur. J. Biochem. 204, 821-830
4 Hartwell, J., Smith, L. H., Wilkins, M. B., Jenkins, G. I. and Nimmo, H. G. (1996)
 Plant J. 10, 1071-1078

MOLECULAR EVOLUTION OF C4-FORM PEPC IN MAIZE: COMPARISON OF PRIMARY SEQUENCES AND KINETIC PROPERTIES WITH A NEWLY CLONED RECOMBINANT ROOT-FORM PEPC

Dong L.-Y., Masuda T., Hata S. and Izui K
Division of Applied Biosciences, Graduate School of Agriculture,
Kyoto University, Sakyo-ku, Kyoto 606-8502, Japan

Key words: C metabolism, enzymes, isoforms, cDNA cloning, allosteric effectors, phosphorylation/dephosphorylation.

1. Introduction

Phosphoenolpyruvate carboxylases (PEPC, EC4.1.1.31) are encoded by a small mutigene family in plants. Each isoform PEPC is thought to play a specific physiological role. Three isoforms have been reported in C4 plants such as maize [1, 2], sorghum [3] and *Flaveria trinervia* [4]; the C4-form PEPC involved in C4 photosynthesis, the C3-form PEPC for housekeeping and the root-form PEPC prevalent in roots. Only for sorghum full-length cDNAs for the three isoforms were cloned and sequenced [3]. Although we previously cloned and sequenced cDNAs for C4-form [5, 6] and C3-form PEPCs [7] in maize, the third one has remained unknown. We report here the cloning and prokaryotic expression of a full-length cDNA for the root-form PEPC and the comparison of its kinetic and regulatory properties with the C4-form PEPC [8]. Structural features of C4-form PEPCs acquired during evolution of C4 photosynthesis are discussed.

2. Procedure

2.1 *Cloning of a root-form PEPC cDNA*

Maize (*Zea mays* H84) seedlings were grown in vermiculite for 7d at 25°C. Poly(A)-rich RNA from roots was used to construct an oligo(dT)-primed cDNA library using λgt10 as a vector. Partial cDNA fragments for the maize root-form PEPC previouly obtained [2] was used as probes [2].

2.2 *Prokaryotic expression and purification of recombinant fusion root-form and C4-form PEPCs*

The entire coding sequence was cloned into a pET32a vector and expressed in *E. coli* BL21(DE3). Active fusion PEPC in soluble fraction was purified to homogeneity with a His-Bind resin, and then the tag peptide was removed by enterokinase [9].

2.3 *Assay of PEPC activity and in vitro phosphorylation*

PEPC activity was assayed by a coupled spectrophotometric method [8] and the regulatory phosphorylation was carried out as described [10].

G. Garab (ed.), Photosynthesis: Mechanisms and Effects, Vol. V, 3411–3414.
© *1998 Kluwer Academic Publishers. Printed in the Netherlands.*

3412

3. Result and Discussion

3.1 *Cloning and sequence analysis of a full-length cDNA for the maize root-form PEPC*

The longest cDNA insert was 3,319 bp in size, and consisted of 197 bp of 5' non-coding, 2,880 bp of coding and 242 bp 3' non-coding regions, respectively. When the amino acid sequences of the three isoforms of PEPC are aligned, the identities between root-form/C3-form, root-form/C4-form and C3-form/C4-form are 85%, 81% and 77%, respectively. Comarison of hydropathy plots among the three PEPCs revealed 5 regions different from one another (Fig.1). In the regions B, D, and E the profiles are similar in root-form and C3-form PEPC, but distinct from C4-form PEPC. In addition in the regions A and C the profiles are characteristic to each isoform. The regions characteristic to C4-form PEPC seem to be resposible for its characteristic enzymological properties.

A	root	1 M-----P---ERHQS*IDAQLR	**D**	root	560 DLEAAPAAVA
	C3	1 .AAL-G.-KM..LS........		C3	566 ....L.
	C4	1 .ASTKA.GPG.KHH........		C4	570 S.QLR..S.E

B	root	111 HRRR-IKLKRGDFADEASAP	**E**	root	875 PYLKQRLRLRES
	C3	117 Y...-....K......N..I		C3	881 L....|.....DA
	C4	119 ....NS...K.G....G..T		C4	885 PF...G.V..NP

C	root	330 LRIRADELH-RSSRKAAKH
	C3	336 ..M...V..-L.TK.D...
	C4	339 V.V..E...SS.GS.VT.Y

Figure 1. The local sequences of maize PEPC isoforms. For the regions B, D and E, hydropathy profiles for the root-and C3-form PEPC are similar but different from the C4-form PEPC. For the regions A and C, the profiles are characteristic to each isoform. Serine residue marked with * in region A denotes the site of regulatory phosphorylation. The boxed residues in region E denotes one of the general replacements unique to C4-form PEPC according to [11].

3.2 *Expression and purification of the root-form PEPC*

Both active root-form and C4-form PEPCs could be expressed in *E. coli* cells. Interestingly, when the crude cell extracts were centrifuged at 20,000xg for 30 min, majority (c.a. 80%) of the root-form PEPC was found in the pellet presumably as inclusion body, while the majority (c.a. 90%) of the C4-form PEPC remained in the supernatant. This suggests that the C4-form PEPC can fold more efficiently than the root-form PEPC at least in *E. coli* cells. Both enzymes in the soluble fractions were purified to homogeneity and tag-peptide was removed for kinetic studies.

3.3 *Comparison of kinetic and regulatory properties*

Table 1 summarizes kinetic properties of both isoform enzymes. The Vmax value of root-form PEPC was about 1.2-fold larger than C4-form PEPC. Notably the Km values of root-form PEPC for PEP, Mg^{2+} and HCO_3^- at pH 7.3 were about 30, 10 and 2-fold lower than C4-form PEPC, respectively. The Ki (L-malate) values at pH 7.3 obtained by Dixon plot were 0.12 and 0.43 mM for root- and C4-form PEPC, respectively. As shown in Table 2, root-form PEPC was neither sensitive to glucose 6-P nor to glycine which are activators of C4-form PEPC. Furthermore, the inhibition by malate was markedly reduced by these activators in C4-form PEPC. These results indicate that there are significant functional differences between the two isoform enzymes.

Similarly to the case with C4-form PEPC, root-form PEPC was also susceptible to the regulatory phosphorylation and about 2-fold increase in $I_{0.5}$ for malate was brought about by the phosphorylation (data not shown). Serine residue for the phosphorylation was conserved also in root-PEPC (see Fig.1).

Table 1. Kinetic parameters of purified recombinant root-form and C4-form PEPCs. Each value represents the average of three separate measurements. The standard errors were within 5 to10% for Km (PEP) and Km (Mg^{2+}), and within 25% for Km (HCO_3^-).

Parameter	Root-form PEPC		C4-form PEPC	
	pH 8.0	pH 7.3	pH 8.0	pH 7.3
Vmax (unit/mg)	28.3	21.8	23.0	18.2
Km (PEP) (mM)	0.04	0.05	0.59	1.48
Km (Mg^{2+}) (mM)	0.06	0.17	0.58	1.79
Km (HCO_3^-) (mM)	0.05	0.06	0.10	0.12

Table 2. Comparison of regulatory properties between the root- anc C4-form PEPCs. The assay was performed at pH 7.3 in the presence of 50 mM Hepes-NaOH, 5 mM $MgSO_4$ and 1 mM $KHCO_3$. The concentrations of both glucose 6-phosphate (G-6-P) and glycine were 10 mM. The concentrations of PEP were 2-fold of each Km (PEP), being 0.1 and 3.0 mM for root- and C4-form PEPCs, respectively (see Table 1). Each value is the average of three separate measurements. Standard error was within 10%.

Isoform	Activation (fold)			$I_{0.5}$ (L-malate) (mM)		
	None	+G-6-P	+Glycine	None	+G-6-P	+Glycine
Root-PEPC	1	1.08	0.79	0.24	0.57	0.28
C4-PEPC	1	2.44	2.33	0.82	10.0	13.1

3.4 *Residues presumed to be responsible for characteristic enzymatic properties*
Westhoff and his colleagues reported that there is a *Ppc* gene in *Flaveria pringlei* (C3) orthologous to the *Ppc* gene for C4 photosynthesis in *Flaveria trinervia* (C4) with the amino acid identity of 94.7% [11, 12]. Although the number of different amino acid residues were only 51 in total of 966, the kinetic and regulatory properties are markedly different from each other. Km (PEP) snd I_{50} (malate) of C4-form PEPC were about 10-fold higher than those of C3-form PEPC also in *Flaveria* . They pointed out 5 residues which might be of functional significance for all C4-form PEPCs, but the alignment of

increasing number of PEPC sequences ([13] and data available at present) revealed that only one residue among them still holds true as shown in Fig. 2.

Zea mays	C4	772	LRAIPWIFSWTQTRFHLPVWL
Sorghum vulgare	C4	763	....
F. trinervia	C4	767	....
A. hypochondriacus	C4	766	....
Zea mays	C3	768	........A............
	root	762	........A............
Sorghum vulgare	C3	762	........A............
F. pringlei	C3	766	........A............
A. thaliana	C3	767	........A............
G. max	C3	768	........A............

Figure 2. An amino acid residue specific to C4-form PEPC

Inspection of the corresponding residue on the tertiary structure of *E. coli* PEPC revealed that the residue is located near the putative catalytic site (courtesy of Drs. Kai and Matsumura, Osaka Univ.). Site-directed mutagenesis on this residue with root-form and C4-form PEPCs will provide a chance to test this notion.

4. Acknowledgements

This work was supported in part by grants from "Research for the Future" program JSPS-RFTF 96L00604 and Scientific Research Priority Areas Grants (09274214, 10170217).

References

1 Ting, I.P. and Osmond, C.B. (1973) Plant Physiol. 51, 448-453
2 Kawamura, T., Shigesada, Yanagisawa, S. and Izui, K. (1990) J. Biochem. 107, 165-168
3 Lepiniec, L., Keryer, E., Philippe, H., Gadal, P. and Cretin, C. (1993) Plant Mol. Biol. 21, 487-502
4 Ernst, K. and Westhoff, P. (1997) Plant Mol. Biol. 34, 427-443
5 Izui, K., Ishijima, S., Yamaguchi, Y., Katagiri, F., Murata, T., Shigesada, K., Sugiyama, T. and Katsuki, H. (1986) Nucl. Acids Res. 14, 1615-1628
6 Yanagisawa, S., Izui, K., Yamaguchi, Y., Shigesada, K. and Katsuki H. (1988) FEBS Lett. 229, 107-110
7 Kawamura, T., Shigesada, K., Toh, H., Okumura, S., Yanagisawa, S. and Izui, K. (1992) J. Biochem. 112, 147-154
8 Dong, L.-Y., Masuda, T., Kawamura, T., Hata, S. and Izui, K. (1998) Plant Cell Physiol. 39 (8) in press
9 Dong, L.-Y., Ueno, Y., Hata, S. and Izui, K. (1997) Biosci. Biotech. Biochem. 61, 545-546
10 Ueno, Y., Hata, S. and Izui, K. (1997) FEBS Lett. 417, 57-60
11 Hermans, J. and Westhoff, P. (1992) Mol. Gen. Genet. 234, 275-284
12 Svensson P., Blasing, O.E. and Westhoff, P. (1997) Eur. J. Biochem. 246, 452-460
13 Toh, H., Kawamura, T. and Izui, K. (1994) Plant Cell Environ. 17, 31-43

ENHANCED AFFINITY OF C_4 PEP CARBOXYLASE TO BICARBONATE ON ILLUMINATION IN *AMARANTHUS HYPOCHONDRIACUS* LEAVES

K. Parvathi, A.S. Bhagwat and A.S. Raghavendra
Department of Plant Sciences, School of Life Sciences,
University of Hyderabad, Hyderabad 500 046

Key words: Carbonic anhydrase, light activation, malate sensitivity, pH, phosphorylation/dephosphorylation, protein synthesis

1. Introduction

Phospho*enol*pyruvate carboxylase (PEPC, EC 4.1.1.31) mediates the primary carbon assimilation in C_4 and CAM plants (1). The activity of PEPC is highly regulated by light. Illuminated leaves exhibit two- to three- fold higher activities than that of dark-adapted ones. Besides an increase in the activity, there is a marked decrease in the sensitivity to L-malate in the light-form of PEPC (2), mainly due to the light-induced phosphorylation of the enzyme. Most of the attention was focused on the effect of external factors such as pH or temperature on regulatory properties of PEPC, i.e., the sensitivity of the enzyme to either malate or Glc-6-P (3,4). We report for the first time that the affinity of PEPC to HCO_3^- in C_4 leaves markedly increases upon illumination.

2. Materials and Methods

2.1. Extraction, assay and light activation
Plants of *Amaranthus hypochondriacus* L. (cultivar AG-67) were raised from seeds and grown in field (approximate photoperiod of 12 h and temperature of 30-40 ^{0}C day/25-30 ^{0}C night). Leaves were harvested from three to four-week old plants.

Light activation of PEPC in leaf discs by illumination at 1000 μmol m^{-2} s^{-1} for 30 min as well as feeding with cycloheximide (CHX) were carried out, as described earlier (5).

Leaf discs were extracted with a medium containing 100 mM HEPES-KOH, pH 7.3, 10 mM MgCl$_2$, 2 mM K$_2$HPO$_4$, 1 mM EDTA, 10% (v/v) glycerol, 10 mM β-mercaptoethanol, 2 mM PMSF and 10 μg/ml leupeptin and 100 μg/ml chymostatin were also present. The homogenate was cleared by centrifugation and the supernatant was used for desalting on a Sephadex G-25 column (1 x 3 cm).

G. Garab (ed.), Photosynthesis: Mechanisms and Effects, Vol. V, 3415–3418.
© 1998 *Kluwer Academic Publishers. Printed in the Netherlands.*

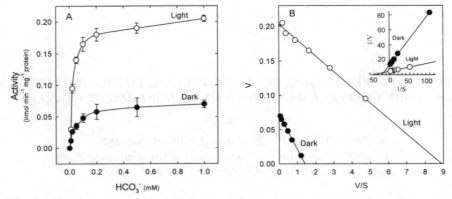

Figure 1. A: The response of PEPC to externally added HCO_3^- (0.01-1 mM).
B: The Eadie-Hofstee plots of PEPC activity (V) against PEPC activity (V)/added
bicarbonate (S). The inset shows the double reciprocal plots of the same data.

The activity of PEPC was determined by the coupled enzyme reaction in a Shimadzu
UV-Vis Spectrophotometer at 30 ^{0}C (6). The response of PEPC to varying
concentrations of malate (0 - 2.5 mM) was monitored at pH 7.3 and 2.5 mM PEP.
The K_i (malate) values were determined by using computer program developed by
Brooks (7). Precautions were taken to minimize the level of dissolved bicarbonate in
the assay media. Boiled distilled water was used and the media were bubbled with N_2
to make the assay media CO_2-free. The very small volume of the extract ensured that
the carry over of dissolved CO_2 into the assay medium was minimized.

2.2. *In vitro* phosphorylation of PEPC

PEPC was purified from dark-adapted leaves of *Amaranthus hypochondriacus*, by a
modified procedure of Iglesias et al. (8), and phosphorylation was performed, using a
procedure of Duff et al. (9). The phosphorylation reaction was allowed to proceed for
up to 120 min at 30 ^{0}C. K_m (HCO_3^-) and K_i (malate) of the enzyme were determined
at the specified times after the start of incubation.

3. Results and Discussion

Illuminated leaves exhibited 2- to 5-fold more activity than that of the dark-adapted
leaves. The activity of light-form of PEPC was saturated at lower concentration of
HCO_3^- than that of dark-form (Fig. 1A). The K_m (HCO_3^-) of PEPC from illuminated
leaves was less than half of that from dark-adapted ones (Fig. 1B). K_m (HCO_3^-) in
illuminated extracts was 0.02 mM and V_{max} was 0.20 µmol min^{-1} mg^{-1} protein while
in dark-adapted extracts K_m (HCO_3^-) is 0.05 mM and V_{max} is 0.07 µmol min^{-1} mg^{-1}
protein. Despite the precautions, it was impossible to remove completely the
dissolved bicarbonate. We have evaluated our data by subtracting the back ground
activity of PEPC without any externally added bicarbonate.

Table 1. Effect of cycloheximide and ethoxyzolamide on the activity and kinetic characteristics of PEPC in extracts prepared from dark-adapted or illuminated leaves.

Exposure/Test compound	Parameter		
	V_{max}	K_i (malate)	K_m (HCO_3^-)
	$\mu mol\ min^{-1}\ mg^{-1}$ protein	mM	mM
Dark			
None (control)	0.24 ± 0.019	0.07 ± 0.01	0.05 ± 0.002
Light			
None (control)	0.83 ± 0.035	0.22 ± 0.02	0.02 ± 0.002
+ 5 μM Cycloheximide*	0.48 ± 0.015	0.06 ± 0.01	0.03 ± 0.001
+ 500 μM Ethoxyzolamide**	0.82 ± 0.030	0.29 ± 0.02	0.02 ± 0.002

* preincubation of leaf discs   ** included during the assay.

The stimulation by light of PEPC activity was more when assayed at supra-optimal concentrations of HCO_3^- compared to higher concentrations. As bicarbonate concentration increased during the assay of PEPC, there was a marked decrease in the ratio of PEPC activity in light-exposed leaves to that in dark-adapted ones (Light/Dark ratio) (Table 1). We therefore suggest that illumination of leaves leads to the sensitization of PEPC to low levels of bicarbonate.

CHX, an inhibitor of cytoplasmic protein synthesis, blocks the PEPC-protein kinase synthesis and the light activation of PEPC (10). Feeding of 5 μM CHX inhibited the light activation and light-induced increase in the K_i (malate) and decrease in K_m (HCO_3^-) of PEPC (Table 2). In contrast, ethoxyzolamide (an inhibitor of carbonic anhydrase) (11) had no effect on either the extent of light activation or the decrease in K_m (HCO_3^-) of light-activated PEPC (Table 2). Thus, we conclude from these results that the light-induced increase in the affinity of PEPC to bicarbonate is due to the involvement of light-activated PEPC-PK but not carbonic anhydrase. Further, we add bicarbonate directly during the assay of PEPC. Therefore, the involvement of carbonic anhydrase was remote during light-induced regulatory changes of PEPC.

Table 2. Effect of HCO_3^- on the extent of light activation (indicated by ratio) of PEPC. PEPC was assayed at suboptimal (0.5 mM) and optimal (2.5 mM) PEP.

HCO_3^-	0.5 mM PEP	2.5 mM PEP
mM	Light/Dark	
0.05	3.0	5.0
0.1	3.0	4.7
1.0	2.8	4.2
10.0	2.6	2.1

3418

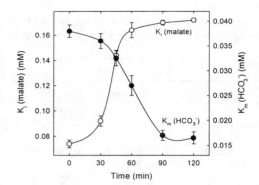

Figure 2. The time course of changes in K_i (malate) and K_m (HCO$_3^-$) of purified PEPC from dark-adapted leaves of *A. hypochondriacus* during *in vitro* phosphorylation by incubation with mammalian PKA and Mg.ATP.

There was a two-fold increase in K_i (malate) and two-fold decrease in K_m (HCO$_3^-$) of PEPC on incubation with protein kinase A (Fig. 2). These results confirm that phosphorylation of PEPC can lead to a decrease in K_m (HCO$_3^-$).

On illumination, there is a simultaneous increase in the availability of HCO$_3^-$ due to the rise in cytosolic pH (12), which can in turn lead to the increased affinity of PEPC to HCO$_3^-$. We propose that the combined effect of rise in dissolved HCO$_3^-$ and affinity of PEPC to bicarbonate can play a significant role, besides phosphorylation-dephosphorylation cascade during the light activation of PEPC in C$_4$ leaves.

Acknowledgements

This work is supported by a grant No. 4/13/95-R&D-II/709 from Department of Atomic Energy, Mumbai, India and KP is a UGC- Senior Research Fellow.

References

1. Andreo, C.S., Gonzalez, D.H. and Iglesias, A.A. (1987) FEBS Lett. 213, 1-8
2. Doncaster, H.D. and Leegood, R.C. (1987) Plant Physiol. 84, 82-87
3. Chollet, R., Vidal, J. and O'Leary, M.H. (1996) Annu. Rev. Plant Physiol. Plant Mol. Biol. 47, 273-298
4. Vidal, J. and Chollet, R. (1997) Trends Plant Sci. 2, 230-237
5. Rajagopalan, A.V., Devi, M.T. and Raghavendra, A.S. (1993) Photosynth. Res. 38, 51-60
6. Gayathri, J. and Raghavendra, A.S. (1994) Biochem. Mol. Biol. Intl. 33, 337-344
7. Brooks, S.P.J. (1992) Biotechniques 13, 906-911
8. Iglesias, A.A., Gonzalez, D.H. and Andreo, C.S. (1986) Planta 168, 239-244
9. Duff, S.M.G., Andreo C.S., Pacquit, V., Lepiniec, L., Sarath, G., Condon, S.A., Vidal, J., Gadal, P. and Chollet, R. (1995) Eur. J. Biochem. 228, 92-95
10. Jiao, J.A., Echevarria, C., Vidal, J. and Chollet, R. (1991) Proc. Natl. Acad. Sci. 88, 2712-2715
11. Badger, M.R. and Pfanz, H. (1995) Aust. J. Plant Physiol. 22, 45-49
12. Hatch, M.D. and Burnell, J.N. (1990) Plant Physiol. 93, 825-828

MODULATION BY CALCIUM OF PEP CARBOXYLASE AND PEPC-PROTEIN KINASE FROM LEAVES OF *AMARANTHUS HYPOCHONDRIACUS*, AN NAD-ME TYPE C$_4$ PLANT

A.S. Raghavendra, K. Parvathi and J. Gayathri
Department of Plant Sciences, School of Life Sciences,
University of Hyderabad, Hyderabad 500 046

Key words: Calmodulin, CO$_2$ uptake, inhibitors, light activation, phosphorylation/dephosphorylation, regulatory mechanism

1. Introduction

Phospho*enol*pyruvate carboxylase (PEPC; EC 4.1.1.31) is a principal enzyme in primary CO$_2$ fixation in C$_4$ plants. PEPC undergoes reversible phosphorylation by PEPC-protein kinase (PEPC-PK) in light. The phosphorylated form is more active and less sensitive to L-malate than the dephosphorylated form (1,2,3). The involvement of Ca^{2+} during the PEPC phosphorylation is a matter of debate. Two types of PEPC-PKs: Ca^{2+}-dependent and Ca^{2+}-independent are reported to occur in leaf extracts of C$_4$-species (1,2,3). Duff et al. (4) and Giglioli-Guivarc'h et al. (5) suggested that Ca^{2+} was involved in the *in situ* phosphorylation of PEPC in *Digitaria sanguinalis*, but an upstream level of regulation. Multiple forms of protein kinases could be involved in the regulation of PEPC phosphorylation (5). The present article is an attempt to re-evaluate the role of Ca^{2+} on PEPC-PK activity.

2. Materials and Methods

Plants of *Amaranthus hypochondriacus* L. (cultivar AG-67) were grown in field (approximate photoperiod of 12 h and temperature of 30-40 ^{0}C day/25-30 ^{0}C night). Leaves were harvested from 3 to 4 week-old plants. PEPC was purified from dark-adapted leaves to homogeneity by a modified procedure of Iglesias et al. (6).

Leaf discs were illuminated at 1000 µE m^{-2} min^{-1} for 30 min and extracted in chilled mortar and pestle with extraction medium containing 100 mM HEPES-KOH, pH 7.3, 10 mM MgCl$_2$, 2 mM K$_2$HPO$_4$, 1 mM EDTA, 10% (v/v) glycerol, 10 mM β-mercaptoethanol and 2 mM PMSF. The homogenate was centrifuged at 7000 g for 5 min. The supernatant was used as a PEPC-PK source, whereas purified PEPC from dark-adapted leaves was used as the substrate for PEPC-PK. The activity of PEPC

3419

3420

was determined by a coupled enzyme assay in a Shimadzu UV-Vis Spectrophotometer at 30^0C (7).

In vitro phosphorylation of PEPC was performed for 60 min at 30 °C according to Jiao and Chollet (8). After the phosphorylation reaction was completed the protein was precipitated with anti-PEPC antiserum and resolved on SDS-PAGE and autoradiographed on X-ray film. Various concentrations of Ca^{2+}, EGTA or BAPTA or inhibitors of different protein kinases were included in the reaction mixture, as required for the experiments. The activity of PEPC-PK was determined by measuring the intensity of PEPC band on the autoradiogram, using a computer programme "Image tools".

3. Results and Discussion

High levels of Ca^{2+} (>100 μM) inhibited PEPC activity and such inhibitory effects of calcium depended on the pH. The extent of inhibition by Ca^{2+} of PEPC was more at pH 7.8 than that at pH 7.3 (data not shown). The inhibition by Ca^{2+} of PEPC activity was due to the competition with Mg^{2+}. K_m for Mg^{2+} increased from 1.48 mM (in absence of Ca^{2+}) to 2.85 mM (in presence of 100 μM Ca^{2+}), while V_{max} (1.45 μmol min^{-1} mg^{-1} protein) did not change (Fig. 1).

The phosphorylation of PEPC was dependent on the concentration of Ca^{2+} included in the assay mixture (Fig. 2A). The rate of phosphorylation was enhanced by 10 μM Ca^{2+}. The involvement of Ca^{2+} was further confirmed by the reversal of the BAPTA-inhibition of PEPC phosphorylation by Ca^{2+}, but not by Mg^{2+} (Fig. 2B). Phosphorylation of PEPC thus appears to need low concentration of Ca^{2+}.

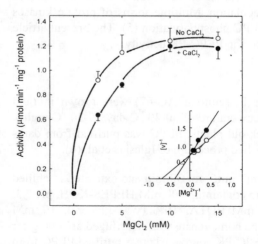

Figure 1. The response of PEPC to varying levels of $MgCl_2$ in crude extracts from dark-adapted leaf discs. The enzyme was assayed in absence or presence of 100 μM $CaCl_2$. The inset is a reciprocal plot of PEPC activity vs concentration of $MgCl_2$.

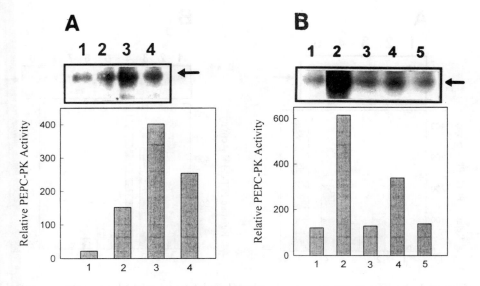

Figure 2. A: The effect of $CaCl_2$ on the phosphorylation of PEPC. Lanes 1 to 4: 0, 5, 10 and 20 µM $CaCl_2$ **B:** The effect of BAPTA on PEPC-PK activity. Lane1: control (no $CaCl_2$), Lane 2: 10 µM $CaCl_2$, Lane 3: 1 mM BAPTA, Lane 4: 1 mM BAPTA + 10 µM $CaCl_2$, Lane 5: 1 mM BAPTA + 10 µM $MgCl_2$.

Phosphorylation of PEPC was stimulated by Ca^{2+} and was further enhanced by CaM (Fig. 3A). Diacylglycerol or phosphotidylserine had only marginal effect on Ca^{2+}-induced stimulation of PEPC-PK activity (Fig. 3B). We tested also the effects of H_7, staurosporine (Protein kinase C inhibitors) and CaM antagonists, W_7 and TFP and ML_7 (inhibitor of myosin light chain kinase, belonging to Ca^{2+}-CaM-dependent protein kinase family). All the above inhibitors restricted the Ca^{2+}-induced stimulation of PEPC-PK activity (Fig. 3B).

These results provide a strong evidence for the involvement of Ca^{2+} and CaM during the phosphorylation of PEPC. The inhibition of PEPC-PK activity by ML_7 suggests that PEPC-PK is related to Ca^{2+}-CaM-dependent protein kinase family. There are only few reports on the involvement of Ca^{2+} during *in vitro* phosphorylation of PEPC (9,10). Yet there is no agreement on the involvement of Ca^{2+} or CaM. Our results suggest that phosphorylation of PEPC in *Amaranthus hypochondriacus* leaves occurs in a Ca^{2+}-CaM-dependent manner and the fine tuning of Ca^{2+} is needed for the optimal activity of both PEPC and PEPC-PK.

Acknowledgements

This work is supported by a grant No. 4/13/95-R&D-II/709 from Department of Atomic Energy, Mumbai, India and KP is a UGC- Senior Research Fellow.

3422

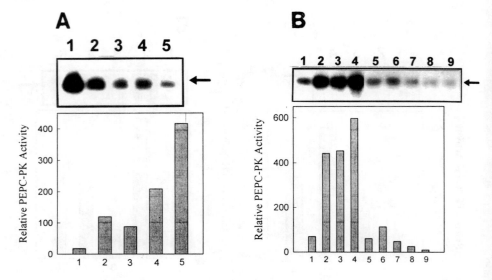

Figure 3. A: The response of PEPC phosphorylation to calmodulin. Lane 1: control (no $CaCl_2$ or no CaM), Lanes 2 to 5: 0, 2.5, 5, 10 μg CaM, all with 10 μM $CaCl_2$. **B:** The effect of activators and inhibitors of different protein kinases on PEPC phosphorylation. Lane 1: control (no $CaCl_2$), Lane 2: 10 μM $CaCl_2$, Lane 3: 10 μM $CaCl_2$ + Phosphotidylserine (20 μg/ml) + diacylglycerol (200 μM), Lane 4: 10 μg CaM, Lane 5: 500 μM H_7, Lane 6: 50 μM staurosporine, Lane 7: 500 μM TFP, Lane 8: 500 μM W_7 and Lane 9: 100 μM ML_7.

References

1. Rajagopalan, A.V. Devi, M.T. and Raghavendra, A.S. (1994) Photosynth. Res. 39, 115-135
2. Chollet, R., Vidal, J. and O'Leary, M.H. (1996) Annu. Rev. Plant Physiol. Plant Mol. Biol. 47, 273-298
3. Vidal, J. and Chollet, R. (1997) Trends Plant Sci. 2, 230-237
4. Duff, S.M.G., Giglioli-Guivarc'h, N., Pierre, J-N., Vidal, J., Condon, S.A. and Chollet, R. (1996) Planta 199, 467-474
5. Giglioli-Guivarc'h, N., Pierre, J-N., Brown, S., Chollet, R., Vidal, J. and Gadal, P. (1996) Plant Cell 8, 573-586
6. Iglesias, A.A., Gonzalez, D.H. and Andreo, C.S. (1986) Planta 168, 239-244
7. Gayathri, J. and Raghavendra, A.S. (1994) Biochem. Mol. Biol. Intl. 33, 337-344
8. Jiao, J-A. and Chollet, R. (1992) Plant Physiol. 98, 152-156
9. Ogawa, N. and Izui, K. (1992) in Research in Photosynthesis (Murata N., ed.) Vol. III, pp. 831-834, Kluwer Academic Publishers, Dordrecht, The Netherlands
10. Ogawa, N., Okumura, S. and Izui, K. (1992) FEBS Lett. 302, 86-88

HIGH LEVEL EXPRESSION OF C4 ENZYMES IN TRANSGENIC RICE PLANTS

Sakae Agarie[1,2], Hiroko Tsuchida[2], Maurice S.B. Ku[3], Mika Nomura[4],
Makoto Matsuoka[5], Mitsue Miyao-Tokutomi[2]
[1]Saga Univ., Saga 840-8502, JAPAN, [2]Natl. Inst. Agrobiol. Resources,
Tsukuba 305-8602, JAPAN, [3]Washington State Univ., Washington 99164-
4238, USA, [4]Kagawa Univ., Kagawa 761-0795, JAPAN, [5]Nagoya Univ.,
Nagoya 464-8601, JAPAN.

Key words: C3, C4, CO_2 uptake, overexpression, PEPC

1. Introduction

In C4 plants, the C4 photosynthetic pathway acts as a "CO_2 pump" and confers a number of advantages, including enhanced photosynthetic capacity, high growth rate, and high nutrient and water use efficiencies. Several attempts have been made to transfer these features to C3 plants using recombinant DNA techniques (1-4). However, the activities of C4 photosynthetic enzymes in transgenic C3 plants were much lower than those of C4 plants, and no significant impact on photosynthetic characteristics was observed.

We have designed gene constructs for high level expression of C4 photosynthetic enzymes in a C3 plant, rice, based on our previous studies of the expression of promoters of the C4 photosynthesis genes in C3 plants (5, 6). We report here that each of three key enzymes of maize C4 photosynthesis pathway, namely, PEPC, PPDK and NADP-ME, can be expressed in rice at levels comparable to or even higher than those in maize.

2. Materials and Methods

Genes used for transformation of rice were as follows: the intact gene encoding maize C4-specific PEPC (7), the intact gene and a cDNA encoding maize C4-specific PPDK (8), the intact gene and a cDNA encoding maize NADP-ME (unpublished), and a cDNA encoding rice NADP-ME (unpublished). These genes were cloned into pIG121Hm and introduced into calli derived from rice (cv. *Kitaake*) using *Agrobacterium*-mediated transformation (9). Transgenic plants were regenerated from hygromycin-resistant calli, planted in soil, and grown in a naturally illuminated greenhouse maintained at 27 °C/22 °C (day/night).

Abbreviations: PEPC, phospho*enol*pyruvate carboxylase (EC 4.1.1.31); PPDK, pyruvate orthophosphate dikinase (EC 2.7.9.1); NADP-ME, NADP-malic enzyme (EC 1.1.1.40).

G. Garab (ed.), Photosynthesis: Mechanisms and Effects, Vol. V, 3423–3426.
© 1998 Kluwer Academic Publishers. Printed in the Netherlands.

Levels of expression of the introduced C4 photosynthesis genes in transgenic rice were determined by assaying the activity of the corresponding C4 enzyme in leaf protein extracts. Approximately 0.1 g of leaf tissue was harvested in the light, immediately frozen in liquid nitrogen, and ground in 1.0 ml of extraction buffer containing 50 mM Tris-HCl (pH 7.0), 10 mM MgCl$_2$, 1 mM EDTA, 5 mM dithiothreitol, 5% insoluble polyvinylpolypyrrolidone and 10% glycerol, using a cold mortar and pestle. For assaying PPDK activity, the extraction buffer was supplemented with 2 mM pyruvate and 2.5 mM phosphate. After total maceration, the homogenate was centrifuged at 15,000 x g for 10 min and the resulting supernatant was used for enzyme assays. PEPC activity was determined by the method of Slack and Hatch (10), PPDK activity was assayed as described by Aoyagi and Bassham (11), and NADP-ME activity was assayed according to Hatch and Mau (12). All enzyme assays were carried out at 30 °C, and activities were calculated on a protein basis. Polypeptide composition of leaf extracts was analyzed by SDS-polyacrylamide gel electrophoresis.

3. Results

Our previous studies demonstrated that the promoters of the maize genes encoding C4-specific PEPC and PPDK can drive high level expression of reporter genes in transgenic C3 plants (5, 6). Based on these results, we initially introduced intact maize genes into rice. The maize PEPC gene used was an 8.8-kb fragment containing all exons, introns, and promoter (1.2 kb) and terminator (2.5 kb) sequences (7). As expected, introduction of the intact gene was effective in increasing the PEPC activity of rice plants. Approximately 10% of transgenic rice plants exhibited PEPC activity ranging from 40 to 120 times that of non-transgenic rice (Fig. 1A). The PEPC activity in these plants was 1-4 times that found in maize.

To confirm the effectiveness of this strategy, we also introduced the intact gene for maize PPDK into rice. The gene used was a 7.2-kb fragment containing all exons, introns, and promoter (1.3 kb) and terminator (0.4 kb) sequences (8). In this case, a 4-kb part of the first intron was deleted. The activity of PPDK in transgenic rice plants was increased to as much as 40 times that of non-transgenic rice (Fig. 1B), or the equivalent of about 50% of maize activity. In contrast, the introduction of a maize PPDK cDNA fused to the rice Cab promoter led to at most a five-fold increase in PPDK activity in transgenic rice plants (data not shown).

We also introduced an intact maize NADP-ME gene into rice. However, the NADP-ME activity of transgenic rice plants did not increase appreciably (data not shown). Introduction of a maize cDNA encoding NADP-ME fused to the rice Cab promoter was effective in conferring high level expression, and the NADP-ME activity in transgenic rice plants was enhanced by as much as 22-fold over that of non-transgenic rice (Fig. 1C). In contrast, the introduction of a rice NADP-ME cDNA fused to the rice Cab promoter did not lead to increased activity (data not shown).

In all cases, the activity of the introduced C4 enzyme correlated with an increase in the amount of the corresponding protein (data not shown), suggesting that the maize enzyme remained functional in transgenic rice plants.

4. Discussion

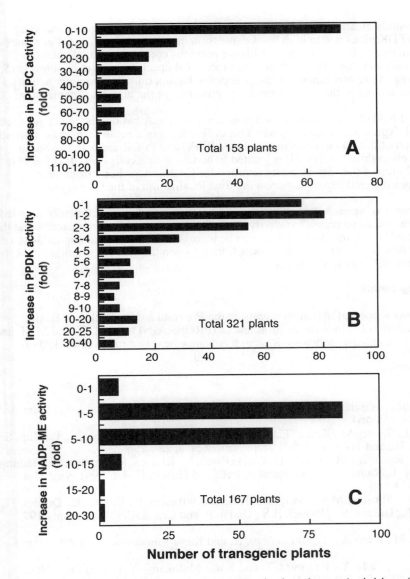

Figure 1. Expression of C4 enzymes in transgenic rice plants. Activities of the introduced C4 enzymes in primary transgenic rice plants were determined and expressed as -fold increases relative to those in non-transgenic plants. (A) PEPC activities of transgenic plants carrying an intact maize PEPC gene. (B) PPDK activities of transgenic plants carrying an intact maize PPDK gene. (C) NADP-ME activities of transgenic plants carrying a maize NADP-ME cDNA fused to the rice *Cab* promoter. Activities of PEPC, PPDK and NADP-ME in maize were 30-40, 80-100, and 20-30 times higher than those of non-transgenic rice, respectively.

The present study clearly shows that the introduction of intact maize genes encoding PEPC and PPDK gives rise to high level expression of the corresponding C4 enzymes in a C3 plant, rice. This is consistent with our previous conclusion that C3 plants possess the necessary genetic machinery to express C4-specific genes at high levels (5, 6). According to our preliminary study, it appears that not only the promoter activity but also other factors of the intact gene are responsible for the high level expression.

In the case of NADP-ME, the introduction of a maize cDNA but not the intact gene led to high level expression of the enzyme. This difference might result from the difference in localization of the C4 enzymes in C4 plants: PEPC and PPDK are both located in mesophyll cells, while NADP-ME is located in bundle sheath cells (13). It is possible that the evolution of the various genes required for C4 photosynthesis from their ancestor genes differed depending upon the final localization of the C4 enzymes.

Earlier attempts to increase the activity of C4 enzymes in transgenic C3 plants have led to only 0.5- to 3-fold increases in activity (1-4). The present study demonstrates that the activity of the C4 enzymes in C3 plants can be increased to the levels comparable to or even higher than those in C4 plants. These findings open the way for installing the C4 photosynthetic pathway in C3 plants.

Acknowledgements

This work was supported in part by a grant at the Program for Promotion of Basic Research Activities for Innovative Biosciences (PROBRAIN) to M.M.-T. and M.M. and from the Bio-Oriented Technology Research Advancement Institution (BRAIN) of Japan.

References

1 Hudspeth, R.L., Grula, J.W., Dai, Z., Edwards, G.E. and Ku, M.S.B. (1991) Plant Physiol. 98, 458-464
2 Kogami, H., Shono, M., Koike, T., Yanagisawa, S., Izui, K., Sentoku, N., Tanifuji, S., Uchimiya, H., and Toki, S. (1994) Transgenic Res. 3, 287-296
3 Gehlen, J., Panstruga, R., Smets, H., Merkelbach, S., Kleines, M., Porsch, P., Fladung, M., Becker, I., Rademacher, T., Hausler, R.E. and Hirsch, H.-J. (1994) Plant Mol. Biol. 32, 831-848
4 Gallardo, F., Maslow, M.M., Sangwan, R.S., Decottignies, P., Keryer, E., Dubois, F., Bismuth, E., Galvez, S., Norreel, B.S., Gadal, P. and Cretin, C. (1995) Planta 197, 324-332
5 Matsuoka, M., Kyozuka, J., Shimamoto, K. and Kano-Murakami, Y. (1994) Plant J. 6, 311-319
6 Matsuoka, M., Tada, Y., Fujimura, T. and Kano-Murakami, Y. (1993) Proc. Natl. Acad. Sci. USA 90, 9586-9590
7 Matsuoka, M. and Minami, E. (1989) Eur. J. Biochem.181, 593-598
8 Matsuoka, M. (1990) J. Biol. Chem. 265, 16772-16777
9 Toki, S. (1997) Plant Mol. Biol. Reporter 15, 16-21
10 Slack, C.R. and Hatch, M.D. (1967) Biochem. J. 103, 660-665
11 Aoyagi, K. and Bassham, J.A. (1983) Plant Physiol. 73, 853-854
12 Hatch, M.D. and Mau, S.L. (1977) Arch. Biochem. Biophys. 179, 361-369
13 Hatch, M.D. (1988) Biochim. Biophys. Acta 895, 81-106

MASSIVE INORGANIC CARBON CYCLING IN CYANOBACTERIA GEARED TO THE ENERGY TRANSFER TO PSII.

Tchernov, D., Keren, N. Hess, M., Ronen-Tarazi, M. Luz, B. Kaplan A.
The Avron-Evenari Minerva Center, The Hebrew University of Jerusalem

Key words: CO_2-concentrating mechanism, Cyanobacteria

1. Introduction

Light energy-dependent mechanisms which raises the concentration of CO_2 at the carboxylating sites (CCM) have been recognized in many photosynthetic microorganisms [1]. We have followed CO_2 and HCO_3^- fluxes associated with the CCM in several photosynthetic organisms. Sustained net CO_2 efflux was observed during net photosynthesis in certain marine organisms including the cyanobacterium *Synechococcus* sp. WH7803 [2] and *Nannochloropsis* sp. [3]. This efflux resulted from HCO_3^- uptake and interacellular conversion to CO_2, part of which leaked and raised its concentration in the cell vicinity. On the other hand, the steady state $[CO_2]$ near photosynthesizing microorganisms, including the fresh water *Synechococcus* sp. PCC 7942, is often lower than expected at CO_2/HCO_3^- chemical equilibrium [4], a phenomena attributed to CO_2 uptake and fixation during photosynthesis.

Changes in the CO_2 concentration measured during light/dark transitions, are related to variations in the Ci fluxes leading to the accumulation of Ci internaly. Energization of these fluxes is likely to depend on the energy transfer from the harvesting complex to the reaction center [5]. Variable fluorescence increased significantly when Ci was provided to CO_2 depleted cells [6]. We report sustained net CO_2 uptake by *Synechococcus* sp. PCC 7942, even in the absence of net CO_2 fixation, suggesting massive light-dependent HCO_3^- efflux. The CO_2 fluxes were strongly affected by changes in the efficiency of energy transfer during light-dark transitions.

2. Procedure

Cells of *Synechococcus* sp. PCC 7942 were grown in medium BG11. Gas exchange and fluorescence parameters were measured simultaneously on a 3 ml cell suspension held under constant, controlled temperature (30C) and light intensity (varied as required). A membrane inlet quadrupole mass spectrometer (Balzers QMG 421) was used to follow CO_2 and O_2 concentrations. Simultaneous measurements of argon concentration (not shown) were used to assess and correct the stability of the system. PAM101 (Waltz, Germany) was used to assess the fluorescence parameters. Modulated frequency was 1.6 kH, flash duration was 1 s, applied at various intervals.

3. Results and Discussion

Figure 1 shows a typical experiment where CO_2 and O_2 concentrations were measured in a *Synechococcus* sp. PCC 7942 culture. The fast initial decline in CO_2 concentration, on illumination, has been attributed to "filling up" of the internal Ci pool [4]. Following a short photosynthetic induction, indicated by the rise in the slope of O_2 evolution, CO_2 concentration declined slowly due to its utilization in photosynthesis.

G. Garab (ed.), Photosynthesis: Mechanisms and Effects, Vol. V, 3427–3430.

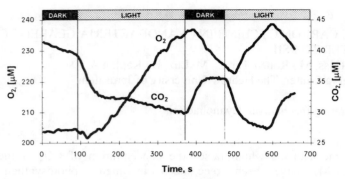

Figure 1
[CO₂] and [O₂] levels in the medium of *Synechococcus* sp. PCC 7942 as affected by light/dark cycles. 20 mM Hepes-NaOH pH-7.8, 26°C, 160 μmol quanta*m⁻²*s⁻¹.

Darkening resulted in a fast rise of CO_2 concentration to the level expected at chemical equilibrium (sometime transiently even above this level due to leak of the internal Ci pool). It was widely accepted that the CO_2/HCO_3^- disequilibrium in the light resulted from CO_2 fixation in photosynthesis. A strong indication that this may not be the case came from experiments on high-CO_2-requiring mutants bearing defective carboxysomes showing massive CO_2 uptake even in the absence of net photosynthesis (Fig. 2). Raising the Ci concentration to 5 mM enabled CO_2-dependent O_2 evolution by the mutant (Fig. 2B). Nevertheless, CO_2 concentration was held below equilibrium, indicated by the fact that darkening (or addition of CA, not shown) resulted in a fast rise in CO_2 concentration to the equilibrium level. Studies with these mutants enabled distinction between CO_2 uptake followed by fixation and other means of dissipation of CO_2. Inhibition of CO_2 fixation by iodoacetamide [7] enabled similar distinction in the wild type (Fig. 3). The CO_2 concentration was held below equilibrium (indicated by the fast rise following addition of CA) in the absence of CO_2 fixation. Similar results were obtained by Canvin and colleagues [7]. We conclude that light-dependent reduction of CO_2 concentration below equilibrium did not result from CO_2 fixation only. In the absence of CO_2 fixation, the net rate of CO_2 uptake is similar to the net rate of formation from HCO_3^- in the medium. Maintenance of a steady state CO_2 concentration below that expected at equilibrium, without significant CO_2 fixation (O_2 evolution, Figs. 2 and 3), is only possible if net HCO_3^- efflux (at a rate approximately equal to CO_2 formation from HCO_3^- in the medium) occurred at the same time. The other alternative that the cells continuously accumulated the "missing" CO_2 in the internal pool is unlikely. Efflux of HCO_3^- at a rate approximately equal to CO_2 uptake minus CO_2 fixation contributed to the maintenance of CO_2/HCO_3^- concentration ratio below equilibrium.

Measurement of the fluorescence parameters [8] showed very low Fm in dark-adapted cells (Fig. 4), suggesting low efficiency of energy transfer from the harvesting complex (the phycobilisomes) to RCII. Fm (and Fs) increased with light intensity and reached maximum at light levels where maximal rate of O_2 evolution was observed. At higher light intensities, Fm declined suggesting reduction of energy transfer efficiency. The extent of CO_2/HCO_3^- disequilibrium, an indicative of CO_2 uptake, fixation and conversion to HCO_3^- and subsequently HCO_3^- efflux) also increased with light level.

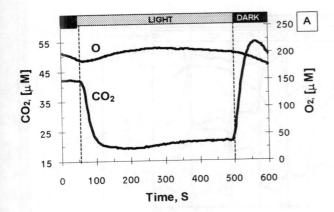

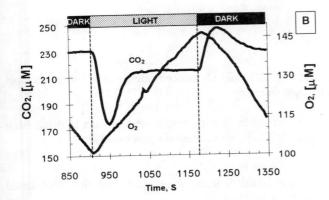

Figure 2
CO_2 and O_2 exchange in a *Synechococcus sp.* PCC 7942 mutant with aberrant carboxysomes. Cells were exposed to 1 mM Ci (A) where they do not perform net photosynthesis (O_2 evolution) or 5 mM Ci (B). Uptake of CO_2 lowered the external steady state concentration in the light below that expected at chemical equilibrium with HCO_3^-, even in the absence of net CO_2 fixation (net O_2 evolution).

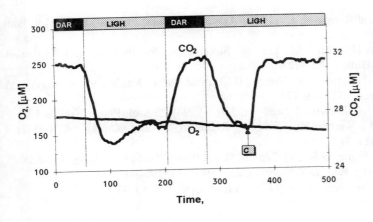

Figure 3
CO_2 and O_2 exchange in *Synechococcus* sp. PCC 7942. CO_2 fixation was inhibited by iodoacetamide (3.3 mM).

3430

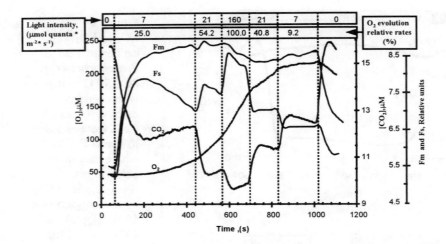

Figure 4
Fluorescence parameters (Fm and Fs), CO_2 and O_2 concentrations in a *Synechococcus*
PCC 7942 culture as affected by light intensity.

Comprehensive discussion of the significance of such results with respect to dissipation
of excess light energy by recycling of Ci is being discussed elsewhere [1].

Acknowledgement
This research was supported by the Israel Science foundation and the USA-Israel
Binational Science Foundation (BSF), Jerusalem.

References
[1]   Kaplan, A. and Reinhold, L. (1999) Annu. Rev. Plant Physiol. Plant Mol. Biol.
       50, (in press).
[2]   Tchernov, D., Hassidim, M., Luz, B., Sukenik, A., Reinhold, L. and Kaplan, A.
       (1997) Curr. Biol. 7, 723-728.
[3]   Sukenik, A., Tchernov, D., Huerta, E., Lubian, L.M., Kaplan, A. and Livne, A.
       (1997) J. Phycol. 33, 969-974.
[4]   Badger, M.R., Palmqvist, K. and Yu, J.W. (1994) Physiol. Plant. 90, 529-536.
[5]   Hassidim, M., Keren, N., Ohad, I., Reinhold, L. and Kaplan, A. (1997) J.
       Phycol. 33, 811-817.
[6]   Miller, A.G., Espie, G.S. and Canvin, D.T. (1988) Plant Physiol. 86, 655-658.
[7]   Salon, C. and Canvin, D.T. (1997) Can. J. Bot. 75, 290-300.
[8]   Schreiber, U. and Krieger, A. (1996) FEBS Lett. 397, 131-135.

IS A METABOLITE IN THE GLYCOLATE PATHWAY THE SIGNAL FOR ACCLIMATION TO CHANGING AMBIENT CO_2 CONCENTRATION ?

Shochat S., Zer H., Rosenbaum A. Kaplan A.
The Avron-Evenari Minerva Center, The Hebrew University of Jerusalem

Key words: Acclimation, CO_2-concentrating mechanism, Cyanobacteria

1. Introduction

Photosynthetic microorganisms acclimate to large changes in the ambient CO_2 concentrations. The cells undergo a syndrome of changes in response to the concentration of CO_2 experienced during growth [1]. The most prominent aspect of this syndrome but certainly not the only one, is the modulation of the activity of the inorganic carbon (Ci)-concentrating mechanism (CCM) operating in these organisms.

· The nature of the signal which induces the syndrome of changes, characteristic of acclimation to changing ambient CO_2 concentration is still not resolved. The rate and extent of acclimation to low CO_2 in cyanobacteria and in *Chlamydomonas* is strongly affected by the CO_2/O_2 concentration ratio rather than by CO_2 concentration as such. The lower this ratio the faster is the acclimation [2,3]. Earlier studies [2] demonstrated that exposure of cyanobacteria to a low CO_2/O_2 ratio resulted in phosphoglycolate (a product of the oxygenase activity of Rubisco) accumulation. It was suggested [2], but never tested experimentally, that the accumulation of phosphoglycolate may serve as the cellular signal for the process of acclimation to low CO_2. This possibility was strenghten by the demonstration that small metabolites, such as acetyl phosphate, in signalling starvation in bacteria. However, while acetyl phosphate can directly phosphorylate a response regulator without the participation of a sensor kinase [4], the free energy in the phosphate bond in phosphoglycolate is probably too low to phosphorylate a response regulator.

Attempts to modulate the level of phosphoglycolate in *Synechococcus* PCC7942 independently of the CO_2/O_2 concentration ratio (by partly permeabilizing the cells) were not successful (Marcus and Kaplan, unpublished). The identification of cbbZ encoding phosphoglycolate phosphatase (PGPase) in Alcaligenes (Ralstonia) [5] opened the way to *in situ* modification of the level of phosphoglycolate.

2. Procedure

Synechococcus sp. PCC 7942 were grown under 5% CO_2 in air (high CO_2) or low CO_2 (1:1 mixture of air with CO_2-free air) as described elsewhere [6]. Standard DNA procedures were used to clone the *cbbZ* in an *E. coli*/cyanobacterial shuttle vector used to transfect *Synechococcus* PCC7942 (mutant VNPK). Activity of phosphoglycolate phosphatase was measured as described [6] using cell-free extracts. *De-novo* synthesis of proteins was assessed after labelling (of high-CO_2-grown and cells exposed to low CO_2 for various duration) with ^{35}S-methionine for 15 min. The level of protein phosphorylation was determined in 50mM HCl-Tris buffer pH 8.0 containing 10 mM $MgCl_2$ and 1 μCi/nmole ^{32}P-ATP, 20 μg protein in a final volume of 50 μl. Incubation was for 1 hour at 30°C.

G. Garab (ed.), Photosynthesis: Mechanisms and Effects, Vol. V, 3431–3434.
© *1998 Kluwer Academic Publishers. Printed in the Netherlands.*

3. Results and Discussion

Figure 1 shows that in the wild type, the activity of PGPase increased significantly following exposure to low CO_2 and decreased thereafter to the level observed in high CO_2 grown cells. On the other hand, PGPase activity in the mutant (VNPK) was high in both high and low-CO_2-grown cells.

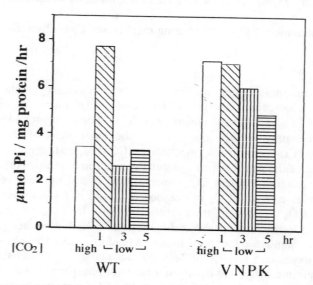

Figure 1: PGPase activity in *Synechococcus* sp. PCC 7942 and mutant VNPK nearing *cbbZ* from *Alcaligenes*. Cells were grown under high CO_2 and exposed to low CO_2 for various duration

The apparent photosynthetic $K_{1/2}$ for external Ci decline during acclimation from high to low CO_2. This was used to assess the extent and rate of acclimation (Fig. 2) following exposure of the wild type and mutant VNPK to low CO_2. The $K_{1/2}(HCO_3^-)$ decreased slower in mutant VNPK suggesting slower acclimation to low CO_2 in the mutant.

Analysis of the net rate of polypeptide synthesis by *Synechococcus* sp. PCC 7942 cells (Fig. 3) showed marked changes in the level of several polypeptides of as yet unknown nature (some of them are marked in the figure), following transfer from a high- to a low level of CO_2. The rate of synthesis of several polypeptides increased transiently during acclimation of the wild type to low CO_2. In contrast and regardless of the concentration of CO_2, these polypetides were constitutively synthesized at a high rate in VNPK.

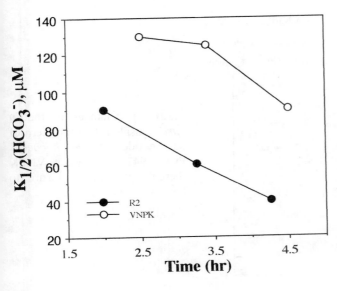

Figure 2: The apparent photosynthetic affinity for extracellular Ci in *Synechococcus* PCC 7942 and mutant VNPK thereof. Cells were grown under high CO_2 and transferred to low CO_2 for the indicated duration.

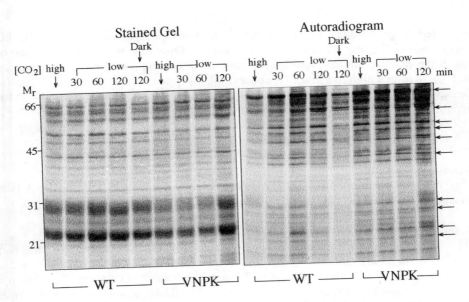

Figure 3: Electrophoretic profiles of the stained and ^{35}S-labeled polypeptide patterns from the soluble fractions of *Synechococcus* PCC 7942 and mutant VNPK thereof.

Phosphorylation plays a critical role in the signal transduction pathway, regulation of gene expression and post translational modifications [7]. The data presented in Figure 4

3434

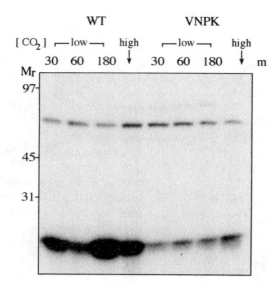

Figure 4: Changes in the *in vitro* phosphorylation pattern of polypeptides of *Synechococcus* PCC 7942 and mutant VNPK thereof following exposure to low CO_2 conditions

suggested large changes in the extent of *in vitro* phosphorylation of a 20 kDa polypeptide (following transfer from one level of CO_2 to another) in the wild type but not in the mutant. Similar results were obtained following transfer of *Synechococcus* PCC 7942 from a high to a low CO_2 concentration [8]. N-terminal sequencing identified the 20 kDa polypeptide as the beta-phycocyanin core protein (Dr. N. Mann unpublished). The nature of the 60 kDa polypeptide is not known.

Our results indicate possible involvement of the glycolate pathway in the acclimation of *Synechococcus* PCC 7942 to low ambient CO_2 concentration.

Acknowledgement

This research was supported by the Israel Science foundation and the USA-Israel Binational Science Foundation (BSF), Jerusalem.

References

[1]   Kaplan, A. and Reinhold, L. (1999) Annu. Rev. Plant Physiol. Plant Mol. Biol. 50, (in press).

[2]   Kaplan, A., Schwarz, R., Ariel, R. and Reinhold, L. (1990) in: Regulation of Photosynthetic Processes., Vol. 2, pp. 53-72. (R., K., Katoh, S. and Miyachi, S., Eds.) Botanical Magazine, special issue.

[3]   Marek, L.F. and Spalding, M.H. (1991) Plant Physiol. 97, 420-425.

[4]   Nystrom, T. (1994) Mol. Microbiol. 12, 833-843.

[5]   Schaferjohann, J., Yoo, J.G., Kusian, B. and Bowien, B. (1993) J. Bacteriol. 175, 7329-40.

[6]   Kaplan, A. et al. (1998) Can. J. Bot. (in press).

[7]   Sultemeyer, D., Klughammer, B., Badger, M.R. and Price, G.D. (1998) Plant Physiol. 116, 183-192.

[8]   Bloye, S.A., Silman, N.J., Mann, N.H. and Carr, N.G. (1992) Plant Physiol. 99, 601-606.

NETWORK OF INORGANIC CARBON UPTAKE REGULATION IN THE CYANOBACTERIUM *SYNECHOCYSTIS* PCC 6803.

S. Bédu, M. Hisbergues, R. Jeanjean, N. Tandeau de Marsac* and F. Joset, LCB-CNRS, 31 Ch. J. Aiguier, 13402 Marseille cedex 20, France; *Unité de Physiologie Microbienne, Institut Pasteur, 75724 Paris cedex 15, France

Keys words : bicarbonate, C metabolism, CO_2 uptake, environment, phosphorylation /dephosphorylation, regulatory processes.

1. Introduction

Inorganic carbon (Ci) in cyanobacteria is modulated by external factors such as the Ci regime or light intensity. Up to now, determination of kinetic parameters for Ci uptake has brought the only direct evidence for a participation of several Ci transport systems in the adaptation of these bacteria to variations of external Ci concentration.

The strain *Synechocystis* PCC6803 exhibits a low affinity (300 to 400 µM) HCO_3^- uptake activity when grown under high Ci regime, whereas under Ci limiting conditions (LC), a high affinity transport activity (60 µM) is present (1). The low affinity Km, measured on whole cells, is close to the affinity of the RUBISCO for CO_2; the latter could be the limiting factor for Ci entry under these trophic conditions. The high affinity uptake activity measured under LC conditions is more likely representative of the transport activity *per se*; the HCO_3^- pump would be, in this case, the limiting factor.

Regulation of Ci metabolism is essential for cell survival and is obviously connected with regulation of other physiological processes, such as supply of energy or of other nutritional elements. Earlier works (2) have shown, in *Anacystis nidulans*, an interaction between CO_2 and ammonium assimilation, *via* the modulation of nitrate uptake (see 3 for review). More recently, an implication of the regulatory protein PII in these two metabolic processes has been demonstrated in *Synechococcus* PCC7942: PII is either phosphorylated or dephosphorylated depending on the ammonium and inorganic carbon assimilation activities (4). Moreover, the unphosphorylated protein is involved in nitrate and nitrite uptake inhibition (5).

The *glnB* gene, coding for protein PII, has been identified in *Synechocystis* PCC6803, and a null mutant constructed. We present here evidences for the mediation of this protein in Ci uptake modulation.

2. Procedures

Strains and growth conditions : Wild-type (WT) *Synechocystis* PCC6803 and mutant were grown in modified Allen's medium (1). This standard medium (SM) contains 12mM bicarbonate. The low Ci medium (LC) is totally devoid of bicarbonate, the only

G. Garab (ed.), Photosynthesis: Mechanisms and Effects, Vol. V, 3435–3438.

Ci source being the CO_2 present in air. The high Ci condition (HC) was obtained by bubbling cultures in standard medium with 5% CO_2 in air.

A *PII null mutant*, ΔPII, was constructed by deletion of a 30 bp *HincII-HincII* fragment within the *glnB* coding region, and insertion of a spectinomycin/streptomycin resistance cassette.

$H^{14}CO_3^-$ *uptake* was measured according to (1).

P700 redox changes were measured by absorbance changes at 820 nm, in whole cells illuminated with white light, at the indicated intensity, as in (6). PSI cyclic activity was measured in the presence of 10 μM DCMU.

3. Results and Discussion

Growth rate of mutant ΔPII, as a function of the Ci regime.

Growth of ΔPII was analysed in the three Ci regimes, SM, LC and HC (Table 1). Under all three conditions, the mutant growth rate was affected : the higher the Ci concentration in the medium, the more the mutant generation time increased, compared to that of the wild type.

Table 1: Generation times (hour) for wild type and ΔPII strains, grown under different inorganic carbon regimes.

Growth conditions	WT	ΔPII
air CO_2 as unique carbon source (LC)	16	24
standard medium : 12 mM HCO_3^- (SM)	10	20
standard medium + 5% CO_2 (HC)	6	14

Role of the PII protein in Ci uptake.

When WT cells are grown under LC, a high affinity bicarbonate transport activity is detected, which is not present in cells grown under higher Ci concentrations (1). Ci uptake kinetic parameters determined on the ΔPII mutant were similar to those of WT cells when grown under LC conditions (data not shown). In contrast, a dramatic change was observed when mutant cells were grown under SM conditions: a high affinity transport activity was present, in spite of the 12 mM HCO_3^- in the culture medium (Table 2). The Km of this transport system was identical to that of the WT one. The ΔPII mutant permanently shows a high affinity transport activity, e.i. reacts as if the cells received a LC signal whatever the Ci concentration in the medium. It is noteworthy, also, that the same profile was obtained whatever the nature of the nitrogen source, indicating that the control process involved here is independent of that through which PII regulates nitrogen metabolism

Forschhammer *et al.* (4) have demonstrated that, in *Synechococcus* PCC7942, PII is modified by phosphorylation, instead of uridylylation in *E. coli*. The protein is phosphorylated when cells are grown under nitrate, and it is dephosphorylated when growth is in the presence of ammonium. The protein is also dephosphorylated under LC regime (4). Identical properties have been demonstrated for the *Synechocystis* PCC6803 protein (data not shown). From these results and those presented in Table 2, it may be concluded that the phosphorylation level of PII is not implicated in the control of the high affinity transport activity: this activity is undetectable under standard conditinswith NH_4^+ as nitrogen source, when PII is dephosphorylated, while it is present under LC regime, condition under which PII is also dephosphorylated. However, PII itself is clearly involved in the regulation of Ci transport, since the high affinity transport activity is constitutive in the ΔPII mutant, even under SM regime. In other words, PII, which regulates nitrate transport *via* its posttranlationnal modification, is implicated in Ci uptake regulation *via* another process.

Table 2 : Kinetics parameters (Km, μM) of bicarbonate uptake, for wild type and ΔPII strains grown under standard medium, with nitrate or ammonium.

nitrogen source strain	NO_3^-	NH_4^+
WT	300 $\pm$ 60	320 $\pm$ 55
ΔPII	49 $\pm$ 12	59 $\pm$ 12
	337 $\pm$ 20	264 $\pm$ 50

Correlation between the cellular energetic state and PII control activity.
A commun feature to Ci and nitrogen metabolism is their strict dependence on the energetic capacity of the cells, dependent mainly on photosynthetic activity. Particularly, under LC and/or nitrate regimes, ATP and reducing power are required for uptake and assimilation into organic molecules. This raises the possibility of a interaction between the mode of control of PII and the energetic state of the cells. To challenge this hypothesis, P700 redox changes were measured on WT and ΔPII mutant, *via* either linear or cyclic electron transfer, as function of light intensity. A typical curve for redox change of P700 is shown in Fig. 1A. The level of P700 oxydation was not different in the mutant and the WT, under linear electron flow. It was, however, notably reduced under high light intensities (above 30 μEinstein.m^{-2}.s^{-1}) in mutant cells incubated with DCMU: under saturating light, P700 oxidation capacity was reduced by 50% after cultivation in

3438

ammonium (Fig. 1B); under nitrate regime, reduction of this activity was about 30% (data not shown).
The deficiency in light - energy conversion in the mutant may explain the permanent LC phenotype of the cells. The implication of the PII protein in Ci uptake regulation would be directed not toward the transport system itself but rather to the energization of the transport process. The question remains of how the PII protein modulates electron transfert activity.

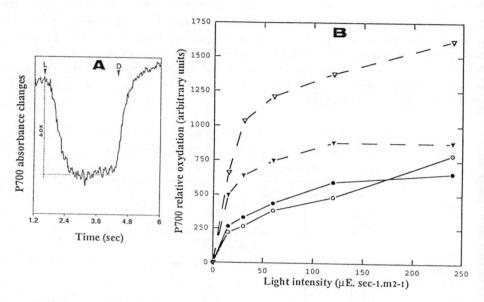

Fig. 1 : P700 redox changes on the WT and ΔPII mutant cells. A : typical measure of P700 absorbance changes (Δox), at 830 nm. B : P700 absorbance changes are plotted as a function of light intensity, as indicated. empty symbols : wt cells; full symbol : mutant ΔPII cells; continuous lines : linear electron transfert; dotted lines : cyclic electron transfert, when cells are incubated with 10 μM DCMU.

References
1. Bédu S., Pozuelos P. Cami B. and Joset F. (1995) . *Molec Microbiol* **18** : 559-568.
2. Romero J.M., Lara C. and Guerrero M.G. (1985) Arch. Microbiol. **237** : 396-401.
3. Lara C. and Guerrero G. (1997) Rai A.K. ed., Springer-Verlag, Cyanobacterial Nitrogen Metabolism and Environmental Biotechnology, pp. 131-153.
4. Forchhammer K. and Tandeau de Marsac N. (1995) *J Bact* **177** : 2033-2040.
5. Lee H.M., Flores E., Herrero A., Houmard J. and Tandeau de Marsac N. (1998) *FEBS letters* **427** : 291-295.
6. Schreiber U., Klughammer C. and Neubauer C. (1988) *Z. Naturforsh.* **43c** : 686-698.

PHOTOSYNTHETIC SYSTEM OF A THERMOPHILIC CYANOBACTERIUM GROWN UNDER HIGH-CO_2

Sachio Miyairi, Nat. Inst. Biosci. Human-Technol.,
Tsukuba, Ibaraki 305-8566, Japan

Key words: adaptation, cell structure, environmental stress, pH-dependence, photosynthetic activity, pigments.

1. Introduction

Cyanobacteria is a very primitive photosynthetic organism which survived the high CO_2 environment and temperature of ancient earth and then adapted to our present environment. This makes one imagine that cyanobacteria might be an adequate subject to study how organisms adapt to a CO_2 environment. This report describes some effects of CO_2 concentrations on the growth and photosystem of the thermophilic cyanobacterium, *Synechococcus elongatus*, isolated from a hot spring (1).

2. Materials and Methods

S. elongatus was cultivated at 52°C in an inorganic medium (2), supported with CO_2 of various concentrations (200cc/min) and illuminatied with tungsten lamps at an intensity of 1,400 lux. Growth was followed by absorbance at 750 nm. The growth vessel was a 650 ml flat bottle with a wide of 6 cm. 5% CO_2-grown cells were used as a seed and cultivations were started at an OD_{750} of 0.02, unless otherwise stated. The cells cultivated for 1 day under various conditions were used for every assay.

For electron microscopy, the cells were fixed with glutaraldehyde and post-fixed with OsO_4. After washing and dehydration, the fixed samples were infiltrated in embedding media and polymerized. Thin sections were stained with uranyl acetate and lead citrate and examined on Hitachi H-4000 transmission electron microscope.

Chlorophyll-a and phycobiliproteins extracted from the cells were measured using the methods as previously reported (3, 4), respectively.

Photosynthetic oxygen evolution of the cells was measured with a Clark-type electrode at 45°C in a 25 mM buffer containing 5 mM $NaHCO_3$. MES-NaOH was used in the range pH 4.5-7.0 and HEPES-NaOH in the range pH 7.0-9.1. Illumination was provided from a 250 W halogen lamp through a water filter at an intensity of 80 mW/cm^2. The maximum rate of activity was calculated of each 3 min experiment.

3. Results and Discussion

The time course of the Synechococcus growth was measured under various CO_2

G. Garab (ed.), Photosynthesis: Mechanisms and Effects, Vol. V, 3439–3442.
© 1998 *Kluwer Academic Publishers. Printed in the Netherlands.*

conditions (Fig. 1A). Doubling times of the cells at logarithmic growth phase were 3.5, 3.2 and 9.8 h under 0.04% (air), 5% and 80% CO_2, respectively. Although the cells scarcely grew under 100% CO_2 under the ordinary condition, when the cells were inoculated into a medium preliminarily alkalinized with NaOH, they grew with a doubling time of 8.3 h. And when large amounts of cells were inoculated into the normal medium, they grew under 100% CO_2 with a doubling time of 8.1 h. While the pH of the medium was initially lowered to a level depending on CO_2 concentration, it gradually rose as the cells grew (Fig. 1B). This might partly be ascribed to NO_3^- uptake by the cells from the medium which originally contained KNO_3 and $NaNO_3$. The results led to follow up

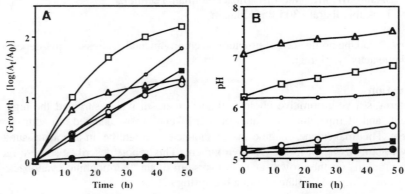

Figure 1. Time course of growth of *S. elongatus* under various concentrations of CO_2 (A) and pH change of the medium during growth (B). The cells were grown under 0.04% (△), 5% (□), 80% (■) and 100% (●) CO_2. The cells were also grown in a medium preliminarily added with NaOH (○) and grown with a large inoculum starting at an OD_{750} of 0.12 (◌), under 100% CO_2.

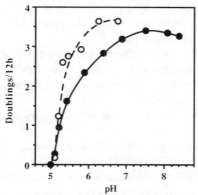

Figure 2. Relationship between the growth rate and pH in the medium. pH was adjusted with NaOH or HCl (—●—), or defined by CO_2 concentrations (---○---).

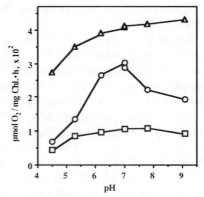

Figure 3. Photosynthetic O_2-evolving activity in the cells. Cells were grown under 0.04% (△), 5% (□) and 100% (○) CO_2.

Table 1. Pigment contents in the cells grown under different concentration of CO_2.

Pigments	Contents (mg/g of dry cells)		
	0.04% CO_2	5% CO_2	100% CO_2
Chlorophyll-a	12	19	12
Phycobiliproteins	83	143	72
Phycocyanin	50	84	35
Allophycocyanin	28	51	34
Phycoerythrin	5	8	3

experiments on the effect of medium pH on the growth.

When the cells were cultivated under 0.04% CO_2 in the mediums of various pHs prelinarily adjusted with NaOH or HCl, they showed active growth in slightly alkaline environments, but a steep decline of growth at acidic levels leading to no growth at pH 5.0 (Fig. 2). When growth rates were plotted against pH in the medium defined by CO_2 concentrations, similar descending curve on the acidic side leading to critical pH of about 5.0 were obtained (Fig. 2). The results strongly suggest that high CO_2 levels repress the *Synechococcus* growth through lowering the pH in the medium.

Pigment contents in 0.04% CO_2-grown cells and 100% CO_2-grown cells were much less than those in 5% CO_2-grown cells (Table 1). In the 100% CO_2-grown cells, phycobiliprotein content was only a half of that in the 5% CO_2-grown cells, and contents of phycocyanin and allophycocyanin became almost equal.

0.04% CO_2-grown cells were the largest in photosynthetic oxygen evolving activity and followed by 100% CO_2-grown cells and 5% CO_2-grown cells (Fig. 3). The differences of the activity among these cells become less significant when calculated as per weight of cells. Optimum pH of the activity was in the neutral range for 100% CO_2-grown cells but it shifted to the slightly alkaline range as environmental CO_2 concentration became lesser. As environmental CO_2 concentration became higher, pH dependence of the activity became more eminent.

Fig. 4 shows the electron micrographs of cells grown under CO_2 of 0.04% (4-A), 5% (4-B) and 80% (4-C) in the normal medium, and cells grown under 0,04% CO_2 in pH 5.4 medium adjusted with HCl (4-D). Many carboxysome-like bodies can be seen in the 0.04% CO_2-grown cells compared to the 5% CO_2-grown cells. Carboxysome-like bodies are also seen, in no small numbers, both in the 80% CO_2-grown cells and in the cells grown under 0.04% in the low pH medium, though the bodies seems relatively small compared with those in 0.04% CO_2-grown cells. Some contraction in length and distortion can be seen in the 80% CO_2-grown cells and in the cells grown under 0.04% CO_2 in the low pH medium. Only in the micrographs of 80% CO_2-grown cells were cell fragments frequently observed, suggesting the envelopes of such cells may be fragile to the fixation procedures.

As CO_2 molecules are liable to exist as un-ionized form in the acidic medium, only minute quantities may be transferred through the cell envelope and cytoplasm to carboxysomes,

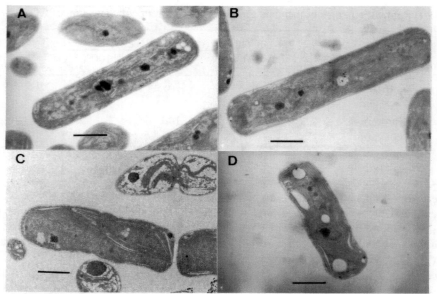

Figure 4. Electron microscopy of the cells. The cells were grown under 0.04% (A), 5% (B) and 80% (C) CO_2 in the normal medium, and grown under 0.04% CO_2 in pH 5.4 medium (D). Bar = 1 μm.

as suggested by present knowledge on inorganic carbon-concentrating mechanism of cyanobacteria (5, 6). Thus, the cells might be starved with inorganic carbon under very high CO_2, which would be related to contents of carboxysomes and pigments (7).

Acknowledgements
I am grateful to Ms. Rena Kawai for preparing the electron micrographs.

References
1 Yamaoka, T., Satoh, K. and Katoh, S. (1978) Plant Cell Physiol. 19, 943-945
2 Dyer, D.L. and Gafford, R.D. (1961) Science 134, 616-617
3 Mackinney, G. (1941) J. Biol. Chem. 140, 315-322
4 Tandeau DE Marsac, N. and Houmard, J. (1988) in the Methods in Enzymology Vol. 167 (Packer L. and Glazer A.N. eds.) pp.318-328, Academic Press, New York, U.S.A.
5 Kaplan, A., Schwarz, R., Lieman-Hurwitz, J., Ronen-Tarazi, M. and Reinhold, L. (1994) in The Molecular Biology of Cyanobacteria (Bryant D.A., ed.) pp.469-485, Kluwer Academic Publishers, Dordrecht, The Netherlands
6 Price, G.D., Coleman, J.R. and Badger M.R. (1992) Plant Physiol. 100, 784-793
7 Grossman, A.R., Schaefer, M.R., Chiang, G.G. and Collier J.L. (1994) in The Molecular Biology of Cyanobacteria (Bryant D.A., ed.) pp.641-675, Kluwer Academic Publishers, Dordrecht, The Netherlands

ADAPTATION OF *CHLAMYDOMONAS REINHARDTII* TO LIMITING CO$_2$ CONDITIONS

James V. Moroney, Mark D. Burow, Zhi-Yuan Chen, Olga N. Borkhsenious, Catherine B. Mason and Aravind Somanchi, Department of Biological Sciences, Louisiana State University, Baton Rouge, LA 70803, USA

Key words: CO$_2$ concentrating mechanism, *Chlamydomonas,* gene regulation, pyrenoid

1. Introduction

The unicellular green alga *Chlamydomonas reinhardtii* can adapt to limiting CO$_2$ conditions through the action of a CO$_2$ concentrating mechanism (CCM). When grown on elevated CO$_2$ (10 to 100 times ambient levels of CO$_2$), *C. reinhardtii* exhibits relatively poor affinity for external inorganic carbon. However if cells grown on elevated CO$_2$ are switched to low CO$_2$ conditions (ambient levels of CO$_2$) the alga will adapt to the limiting CO$_2$ environment within four hours. During this adaptation a number of new proteins are synthesized and significant morphological changes occur. We have found that over 90% of the Rubisco within the cell is localized to the pyrenoid when cells are grown on low CO$_2$ conditions. In contrast, in cells grown on elevated CO$_2$ a majority of the Rubisco (60%) is found in the chloroplast stroma. In addition to the redistribution of Rubisco, messages encoding three isoforms of carbonic anhydrase, a cyclophilin and three novel membrane proteins increase under low CO$_2$ conditions. The sequence of one of these novel cDNAs named *lci2*, will be reported here.

2. Procedure

2.1 *Growth and Maintenance of Algal Cultures*
C. reinhardtii line 137 cells were obtained from Robert Togasaki, Indiana University, Bloomington, IN, USA. Cells were grown in Minimal Media as previously described (1) bubbled with either air (low CO$_2$) or air supplemented with 5% CO$_2$ (high CO$_2$). For construction of the cDNA library, northern blot analysis, and electron microscopy, cells were grown on high CO$_2$ and switched to low CO$_2$ for the indicated times.

2.2 *Electron microscopy and immunolocalization*
For transmission electron microscopy, various cell strains were prepared according to Henk et al. (2) with the modifications described by Borkhsenious et al. (3). The

3443

G. Garab (ed.), Photosynthesis: Mechanisms and Effects, Vol. V, 3443–3446.
© 1998 *Kluwer Academic Publishers. Printed in the Netherlands.*

immunocytochemical procedure was similar to the method of Schroeder et al. (4) with the modifications described by Borkhsenious et al. (3). The sections were incubated for 1 h with diluted primary antibody (1 to 50 dilutions of anti-Rubisco) or preimmune serum diluted similarly as a control. In order to detect bound antibodies, Protein A conjugated with 15 nm colloidal gold particles diluted 1:50 was used. The quantitation of immunogold particles was described previously (3).

2.3 Synthesis of the cDNA library and mRNA preparation

Total RNA was isolated from cells adapting to low CO_2 conditions as described previously (5). For northern blots, 5 μg of total RNA were probed with *Lci2* DNA labeled by random priming. Hybridization and washing of the blots was done by standard procedures (5). mRNA was purified from the total RNA using Poly AT tract beads (Promega) as described by the manufacturer. For the synthesis of the cDNA library 5 μg of mRNA from cells transferred to low CO_2 for 2.25 h were reverse-transcribed into cDNA and ligated into the Stratagene uni-ZAP II vector according to the manufacturer's instructions.

3. Results and Discussion

3.1 The distribution of Rubisco changes in response to the CO_2 level in the environment

There has been a number of studies in the literature concerning the location of Rubisco in *C. reinhardtii*. Everyone agrees that Rubisco is localized to the chloroplast and that the pyrenoid is composed largely of Rubisco. However, the actual percentage of Rubisco found in the pyrenoid versus the chloroplast stroma has been controversial. The published estimates of the percent of Rubisco in the pyrenoid range from 5% to 99% (See 6 for a review). We have found that one reason for this controversy is that the amount of Rubisco in the pyrenoid varies with growth condition. If cells are grown on elevated CO_2 concentrations about 60% of the Rubisco is located in the chloroplast stroma and 40% is in the pyrenoid (Table I). However if cells are grown on low CO_2 concentrations about 90% of the Rubisco is in the pyrenoid and 10% in the chloroplast stroma (Table I). Morita et al. (7) claimed that 99% of the Rubisco was localized to the pyrenoid and they grew their cells on air levels of CO_2. This indicates that Rubisco is packaged in the pyrenoid when the CO_2 concentration mechanism is functioning.

3.2 Identification of low CO_2 inducible cDNAs in C. reinhardtii

A cDNA library was constructed from cells adapting to low CO_2 conditions and screened with labeled cDNA from either high CO_2-grown cells or low CO_2-grown cells. About 450 clones were initially screened as being up-regulated under low CO_2 conditions. Many of these cDNAs have now been sequenced and identified. Ten classes of cDNAs that are up-regulated under low CO_2 conditions are summarized in Table II.

Immunogold density (particles•mm^{-2})

Growth conditions	Pyrenoid	Stroma	Cytoplasm	% Rubisco in pyrenoid
High CO$_2$	70 ± 14	9.0 ± 2.1	2.0 ± 0.9	40
Low CO$_2$	65 ± 15	2.5 ± 1.2	2.0 ± 0.8	89

Table I. To obtain the fraction of Rubisco in the pyrenoid, the average density of particles in the cytoplasm was subtracted from the average density of particles in the stroma or pyrenoid. The net particle density of the pyrenoid or stroma was then multiplied by the average volume of that compartment which is 2.4 mm^3 and 35.6 mm^3, respectively. The number of particles in the pyrenoid was then divided by the combined number of particles in the pyrenoid and stroma to obtain the %Rubisco in the pyrenoid.

cDNA	Gene product(s)	High CO$_2$	Low CO$_2$
Cah1	periplasmic carbonic anhydrase	-	+++
Mca1 Mca2	mitochondrial carbonic anhydrase	+/-	+++
Ccp1 Ccp2	chloroplast envelope carrier protein	-	+++
Aat1	alanine aminotransferase 1	+	++
Cyp	cytosolic cyclophilin	+	++
PsaE	photosystem I subunit	+	++
Lca1* family	photosystem I antenna protein	+	++
Lcb1* family	photosystem II antenna protein	+	+++
Lci1	membrane protein	-	+++
Lci2	membrane protein	-	+++

Table II. A summary of 10 classes of cDNAs that have been found using the differential screen. The last two columns indicate the relative abundance of the mRNAs corresponding to these genes in cells grown under high CO$_2$ and low CO$_2$ conditions.

3.3 Sequence of Lci2

Among the low CO$_2$-inducible cDNAs identified were some that had no significant similarity with genes deposited in GenBank. These genes have been given the designation Lci for Low CO$_2$-Inducible gene (8). Lci2 is the second cDNA identified that shows very significant up-regulation under low CO$_2$ but has not been assigned a physiological function. Clone Lci2 is 962 nucleotides long and has an open reading frame that would encode a protein of 131 amino acids (Genbank accession number AF081461). The sequence of Lci2 is shown in Figure 1. The mRNA corresponding to Lci2 is about 0.9 kb on northern blots in agreement with the cDNA size. Its message abundance increases dramatically when C. reinhardtii is grown on low CO$_2$ (referred to as clone 2I5 in reference 8). While the function of Lci2 is still unknown the sequence does show some homology with previously described thylakoid-bound ascorbate peroxidases. However the sequence similarity is limited to the C-terminal region or thylakoid membrane anchoring part of the known ascorbate peroxidases (Fig. 1). The

N-terminal portion of *Lci2* does not appear to be similar to these other proteins and likely has a different metabolic function. Future studies include raising antibodies to the Lci2 protein to determine its intracellular location and attempting to obtain mutants of this gene to determine its metabolic role.

Lci2 full-length sequence:

```
  1 MALRQAVRVPGAGLRMSAVAPVRPVRSVLARAEPPSGAKVTREYREDTGE    50
 51 VTAPGASQTTRRESDGLYVNADGPRPVPRKDNMSKEMKARLRQEYTGLGG   100
101 AENKAMSNNYFLYISIFVAILAIMSKAIGAI                      131
```

Comparison:

```
Lci2   79  RKDNMSKEMKARLRQEYTGLGGAENKAMSNNYFLYISIFVAILAIMSKAIG 129
S. o.  364 KRSELSDSMKEKIRAEYEGFGGSPNKPLPTNYFLNIMIVIGVLAVLSYLAG 414
M. c.  379 KKELSDSMRQKIRAEYEGFGGSPNNPLPTNYFLNIMIVVAVLAVLTYLTG  428
Con.       ++   +S  M+ ++R EY G GG+ N+ +  NYFL I I + +LA+++   G
```

Figure 1. The sequence of *Lci2* and an alignment of the C-terminal half of *Lci2* with the C-terminal sequence of thylakoid-bound ascorbate peroxidase from *Spinacia oleracea* (9), *Mesembryanthemum crystallinum* (10), and consensus sequence (Con.).

4. Acknowledgments

This work was supported by the National Science Foundation grant IBN-9632087.

References

1. Chen Z.-Y., Lavigne L.L., Mason C.B. and Moroney J.V. (1997) Plant Physiol. 114:265-273
2. Henk M.C., Rawat M., Hugghins S.Y., Lavigne L.L., Ramazanov Z., Mason C.B. and Moroney J.V. (1995) in Photosynthesis: from Light to Biosphere, Vol. V. (Mathis P., ed.) pp. 595-598, Kluwer Academic Press, Dordrecht, The Netherlands
3. Borkhsenious O.N., Mason C.B. and Moroney J.V. (1997) Plant Physiol. 116:1585-1591
4. Schroeder M.R., Borkhsenious O.N., Matsuoka K. and Raikhel N.V. (1993) Plant Physiol. 101:451-458
5. Sambrook J., Fritsch E.F. and Maniatis T. (1989) Molecular Cloning: a laboratory Manual. Cold Spring Harbor Laboratory press, Cold Spring Harbor, NY
6. Moroney J.V. and Chen Z.-Y. (1998) Can. J. Bot. (in press)
7. Morita E., Kuroiwa H., Kuroiwa T. and Nozaki H. (1997) J. Phycol. 33:68-72
8. Burow M.D., Chen Z.-Y., Mouton T.M. and Moroney J.V. (1996) Plant Mol. Biol. 31:443-448
9. Ishikawa T., Sakai K., Yoshimura K., Takeda T. and Shigeoka S. (1996) FEBS Lett. 384: 289-293
10. Michalowski C.B., Quigley-Landreau F. and Bohnert, H.J. (1998) Accession AF069315

MAIN PRINCIPLES TO BUILD UP THE EFFECTIVE CO_2 CONCENTRATING MECHANISM IN CYANOBACTERIA AND MICROALGAE

L.E. Fridlyand
Institute of Experimental Botany, National Academy of Sciences,
Scorina 27, 220072, Minsk, Belarus

Key words: bicarbonate, carbonic anhydrase, CO_2 concentration, modelling

1. Introduction

Many microalgae and cyanobacteria, when growing under low dissolved inorganic carbon concentrations, acquire the capacity to efficient utilization CO_2. This is caused by the induction of a specific CO_2 concentrating mechanism (CCM). However, CCM action in these cells is not completely understood. Therefore, an attempt has been made to elaborate the theoretical concept of the CCM action. This concept uses mathematical models and takes into account the complex structure of the cells (1,2). This approach can result in clarification of the possible mechanisms that may be used by the cells for the creation of an efficient CCM and explain the contradictory data accumulated so far.

2. Results and Discussion

CO_2 is the only form of dissolved inorganic carbon fixed by Rubisco. Because of this, the goal of the CCM operation should be mainly the increase of the CO_2 concentration in the place of Rubisco localisation. However, the cells have to overcome some difficulties to achieve this aim.

2.1. *CO_2 evolution*
Bicarbonate is the only possible source of CO_2 evolution in chloroplasts or carboxysomes if the C4 mechanism is absent. However, alkaline medium exists inside the chloroplasts and cyanobacteria cells in light, and thus the interconversion of CO_2 and bicarbonate is shifted mainly in the direction of bicarbonate. Therefore, a sufficient rate of CO_2 evolution should be provided in unfavourable alkaline medium.
According to (1) the cyanobacterial cell achieves this objective through the accumulation of bicarbonate up to such a big concentration that CO_2 evolution proceeds spontaneously inside carboxysomes where carbonic anhydrase is localised. Apparently, in microalgae the CO_2 evolution takes place in thylakoids inside pyrenoids (2). In this case CO_2 evolution can occur to use the acidic medium inside thylakoids

2.2. *CO_2 diffusion*
CO_2 diffusion away from the site of its concentration diminishes the efficiency of CCM. Thus, a decrease of the outward CO_2 flow should also be a goal of a cell. However, the

G. Garab (ed.), Photosynthesis: Mechanisms and Effects, Vol. V, 3447–3449.
© 1998 *Kluwer Academic Publishers. Printed in the Netherlands.*

big CO_2 diffusion coefficient was obtained by Gutknecht et al. (3) for a double lipid layer membrane. The question arises as to whether a cell can reach the big CO_2 diffusion resistance between the place of CO_2 concentrating and a surrounding medium in such conditions.

The diffusion law for the spherical layer differs essentially from it for flat layer. Permeability coefficient (P) for the spherical layer can be written as (see for example (4))

$$P = \frac{4\pi\ D}{1/\ R_{in}\ -\ 1/\ R_{out}}$$

where D is the diffusion coefficient for a substance that penetrates through layer, R_{in} and R_{out} are the inner and outer radii of a layer.

This means that the permeability coefficient through a whole spherical layer diminishes sharply with the decrease of inner radius and it tends to zero when R_{in} is close to zero even though the layer thickness remains invariable. Such peculiarities can be explained by the necessity for the molecules to penetrate through the area at the diffusion. However, it is evident that this area diminishes sharply with decrease of the radii. So, to decrease the CO_2 leakage, a cell has to try to localise the CO_2 evolution inside a small space and surround it by layers with low CO_2 diffusion coefficient

Apparently, this principle of the cell construction was used by cyanobacteria, where carbonic anhydrase and Rubisco are localised in carboxysomes and great CO_2 diffusion resistance can be explained by the existence of spherical protein layer around the centre of carboxysomes (1).

Such structures as pyrenoid, starch sheath and concentric thylakoid system around it may be used to explain the low diffusion losses of CO_2 from the place of its evolution in pyrenoid of microalgae (2).

2.3. "short-circuited" fluxes

"short-circuited" inorganic carbon fluxes between the external medium and the cytoplasm under active bicarbonate transport through the plasmalemma and in the presence of carbonic anhydrase in the cytoplasm can lead to a decrease of the efficiency of CCM, since the accumulated bicarbonate can transform spontaneously to CO_2 that leaves a cell.

Cyanobacteria cells are able to prevent such possibility because of the absence of carbonic anhydrase in cytoplasm (1).

Present study shows that a microalgae cell can eliminate "short-circuited" inorganic carbon fluxes between external medium and cytoplasm under active bicarbonate transport through plasmalemma and in the presence of carbonic anhydrase in cytoplasm by forcing out a cytoplasm from the space between the chloroplast envelope and plasmalemma upon the microalgae adaptation to low concentration of the dissolved inorganic carbon (2).

2.4. Scavenging mechanisms

To diminish the losses of energy and to increase CCM efficiency a cell have to catch the CO_2 that leaves the site of its concentrating.

It was supposed that cyanobacteria cell can have special scavenging mechanism to perform this (1). This mechanism can be localised near the plasmalemma inside a cell and it can use some additional energy to produce bicarbonate from CO_2.

In microalgae this role can be performed by carbonic anhydrase on plasmalemma and in cytoplasm. Calculations show considerable decrease of diffusion fluxes of CO_2 from the cell as results of these processes (2).

It should be pointed out that the procedure was developed to assess bicarbonate and CO_2 fluxes associated with cyanobacteria and green algae during steady-state photosynthesis (5). Supposition was used that there is the same CO_2 flux from cell in dark and in light. However, according to (1,2) the effective scavenging mechanisms act in cells in light and CO_2 flux from the cell can exist mainly in dark. For this reason, the procedure developed in article (5) and used widely now can lead to incorrect results.

2.5. *Mechanisms of apparent CO_2 absorption*
Models including all these mechanisms can be useful for the explaining some incomprehensible data. For example, a set of experimental data show the existence of CO_2 flux into cell while the CCM action. There are suppositions about the existence of the active CO_2 transport through plasmalemma, too. However, it is likely that this active transport through membrane is possible only for bicarbonate. No experimental evidence is available on the possibility of an active transport of neutral gas molecules such as CO_2.

Proof of the active CO_2 transport can be interpreted in another way based on the developed models. In cyanobacterial model the apparent active CO_2 transport is attributed to the passive diffusion of CO_2 through plasmalemma followed by its active interconversion to bicarbonate by the action of scavenging mechanism localised near the plasmalemma inside a cell (1).

Calculations demonstrate also a possibility a net flux of CO_2 into the microalgae cell if the bicarbonate active transport mechanism on plasmalemma is somehow restricted (2). An apparent flux of CO_2 into a cell is linked to passive diffusion of CO_2 into the cytoplasm, interconversion of CO_2 to bicarbonate and the following active transport of bicarbonate into the chloroplast. By this means the cytoplasmic carbonic anhydrase which usually serves to diminish the CO_2 flow from pyrenoid is displayed as an activator of the visible CO_2 absorption if bicarbonate active transport is inhibited.

In any event the works (1,2) show that different CCMs are possible for cyanobacteria and microalgae and that the complex structure of cells has to be taken into account when developing mathematical models of CCMs.

References

1. Fridlyand, L.E., Reinhold, L. and Kaplan, A. (1996) BioSystems 37, 229-238
2. Fridlyand, L.E. (1997) BioSystems 44, 41-57
3. Gutknecht, J., Bisson, M.A. and Tosteson, D.C. (1977) J. Gen. Physiol. 69, 779-794
4. Jacobs, M.H. (1935) in:Ergebnisse der Biologie. 12, pp. 1-159
5. Badger, M.R., Palmqvist, K. and Yu, J-W. (1994) Physiologia Plantarum 90, 529-536

A PSII-ASSOCIATED CARBONIC ANHYDRASE FACILITATES CO_2 SUPPLY TO CARBON ASSIMILAITON IN *CHLAMYDOMONAS REINHARDTII*

Y.-I. Park, J. Karlsson, I. Rojdestvenski, N. Pronina[*], V.V. Klimov[+], G. Öquist and G. Samuelsson   *Dept. of Plant Physiol., Umeå Univ. S90187, Umeå, Sweden; [*]Inst. of Plant Physiol., Moscow 127276 and [+]Inst. of Soil Science and Photosynthesis, Russian Academy of Science, Puschcino 14292, Russia*

Key words: bicarbonate, Calvin-cycle, pH-gradient, Rubisco, thylakoid membranes

1. Introduction

CO_2 limitations in the environment are overcome by CO_2-concentrating mechanisms (CCM) in photosynthetic organisms such as freshwater algae and C_4 plants. Intracellular carbonic anhydrases (CA) which catalyze the interconversion of $CO_2 + H_2O$ and $HCO_3^- + H^+$ have been suggested to play a crucial role in CCM [1]. Recently, we identified a constitutively expressed CA which is located in the chloroplast thylakoid membranes and is responsible for the viability of *Chlamydomonas* cells under ambient CO_2 growth condition [2]. Further, this thylakoid membrane-bound CA appears to be in a close connection with PSII and localized on the lumenal side of the thylakoid membranes (Park et al., submitted). In order to understand the physiological role of PSII-CA, we compared cell wall-less mutant of *Chlamydomonas reinhardtii*, CW92 with the *Cia3*/CW15 mutant which can not grow at low CO_2 condition due to its presence of inactive PSII-CA [2].

2. Materials and Methods

Cell wall-less strains of *Chlamydomonas reinhardtii* CW92 and a mutant lacking active CA (*Cia3*/CW15) were grown in batch culture at 25°C under a continuous irradiance of 150 µmol m^{-2}s^{-1} with 5% CO_2 [2]. Cells were harvested and washed three times in a CO_2-free buffer (10 mM Mes-KOH , pH 5.3). Chl *a* fluorescence and O_2 evolution were simultaneously measured using a pulse amplitude modulation fluorometer (PAM, Walz, Germany) and a PAM-compatible system of cuvette, magnetic stirrer, oxygen electrode and Björkman type actinic lamp (Hanstech, King's Lynn, England) [2].

3. Results and Discussion

Cessation of net O_2 evolution is induced by bicarbonate-deficiency. Upon illumination by non-saturating light (100 µmol m^{-2}s^{-1}), CW95 cells showed three distinct phases of oxygen evolution; a lag phase that lasted for a few minutes, followed by a linear phase with a duration of several minutes, and then finally a stationary phase (Fig. 1A). This final phase is due to the similar rates of respiratory O_2 consumption and photosynthetic O_2 evolution. Chl fluorescence yield, F and Fm' were simultaneously recorded to determine the PSII quantum efficiency (1 - F/Fm') as well as the dissipation of excitation energy (1/Fm'). Unlike many higher plants, the dark adapted Fm is lower than that obtained under actinic light condition, indicating that state transition, or photoactivation of PSII (estimated as 1 - F/Fm'), may occur during the induction period of photosynthesis in *Chlamydomonas* cells. When net O_2 evolution ceased in the final phase (Fig. 1A), F and Fm' started to decrease rapidly down to the steady state level. Upon addition of 5 µM DCMU at the quenched level, the Chl fluorescence yields recovered very quickly to values somewhat higher than

3451

those observed under the preceeding phase of steady net O_2 evolution. In order to check whether this cessation of net O_2 evolution and hence decline in PSII activity is due to depletion of bicarbonate, we added 50 µmol bicarbonate 2 min before it reached the final cessation phase. As shown in Fig. 1B, addition of extra bicarbonate allowed net O_2 evolution for an extended long period (cf Fig 1A and 1B). Further, the period required to reach the final cessation phase was inversely correlated with the bicarbonate concentration added and the intensities of actinic light (not shown).

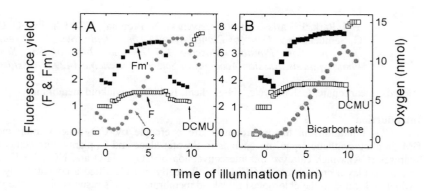

Figure 1. Changes in Chl fluorescence yields (F and Fm') and O_2 evolution in CW92 cells treated with (B) or without (A) 50µM bicarbonate. 5 µM DCMU was added as indicated by arrow.

Limitation of CO_2 supply to the Calvin-cycle is responsible for the final cessation phase. The final stage induced by bicarbonate-deficiency may be caused by either a malfunction of PSII photochemistry or a reduced carboxylation efficiency since bicarbonate is required for proper function of both PSII [3] and Rubisco [1]. Upon addition of an artificial electron acceptor of PSII, 2 mM 2,5-dichloro-1,4-benzoquinone (DCBQ) in the presence of 5 mM $K_3Fe(CN)_6$, O_2 evolution resumed immediately (Fig. 2A). This rapid recovery of O_2 evolution by allowing PSII to keep electrons flow from water to DCBQ suggests that the inorganic carbon concentration under the final stage is enough for optimal PSII photochemistry. We, therefore, consider that the efficiency of carboxylation is responsible for the cessation of net O_2 evolution. Furthermore, under partially uncoupled conditions, bicarbonate would be utilized continuously but with slower rates than under coupled conditions. This view is supported by the fact that treatment with an uncoupler, 1 µM nigericin, did only extend the time to reach the cessation stage, but failed to prevent cells reaching the final stage (Fig. 2B). The rapid declines in Fm' is probably due to a high pH-gradient created when the consumption of ATP ceases due to CO_2 limitation of carboxylation at the cessation phase. Thus, the final stationary phase seen by bicarbonate depletion is attributable to the limitation of CO_2 supply to the Calvin-cycle via Rubisco rather than improper functioning of PSII at donor or acceptor side.

Similar photosynthetic responses of CW92 cells treated with CA inhibitor and the *Cia3* mutant. CW92 cells treated with an internal carbonic anhydrase inhibitor, ethoxyzolamide (EZ), showed much more gradual depletion of bicarbonate (Fig. 3A) and hence there was no pronounced "cut-off" effects of O_2 evolution (cf Fig. 1A). Further, F and Fm' levels decreased slowly relatively in the presence of EZ and the PSII photochemical efficiency remained higher in cells with than without EZ (*cf* Fig. 1A and 3A), suggesting that PSII-associated CA is likely to be involved in the consumption of bicarbonate.

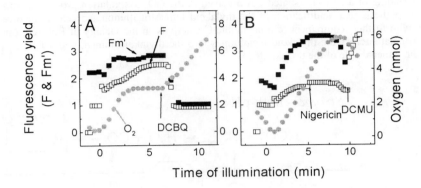

Figure 2. Changes in Chl fluorescence yields (F and Fm') and O_2 evolution in CW92 cells treated with 2 mM DCBQ (A) and 1 μM nigericin (B).

As shown in Fig. 3B, the behaviors of O_2 evolution and Chl *a* fluorescence of the *Cia3* cells with inactive PSII-CA [2] are very similar to CW92 cells treated with EZ. The reduced rates of utilization of bicarbonate observed in the *Cia3* mutant with inactive PSII-CA, or in WT cells when PSII-CA was inhibited, suggest strongly the direct involvement of PSII-CA in controlling the Calvin cycle activity.

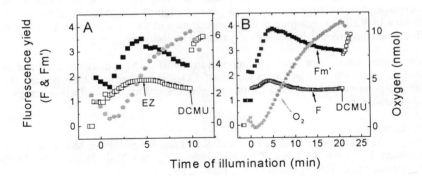

Figure 3. Changes in Chl fluorescence yields (F and Fm') and O_2 evolution in CW92 (A) treated with 10 μM ethoxyzolamide (EZ) and *Cia3* mutant (B).

Given that the only difference between CW92 and *Cia3* mutants is the presence of an active PSII-CA in the former, PSII-CA appears to help provide a steady supply of CO_2 for the assimilation

3454

process. As expected, the rate of maximal photosynthetic O_2 evolution of CW92 cells are about 50% higher than those of *Cia3* mutants or CW92 cells treated with EZ (Park et al., submitted).

This finding of a PSII-associated CA involved in the CO_2 assimilation process is consistent with the hypothesis that acidification of the thylakoid lumen of illuminated cells is involved in CCM [4], with further requirement of a negligible CA activity in the vicinity of Rubisco, transport of a bicarbonate across the thylakoid membranes into the thylakoid lumen, and a CA activity on the acidic lumenal side of the thylakoids [5].

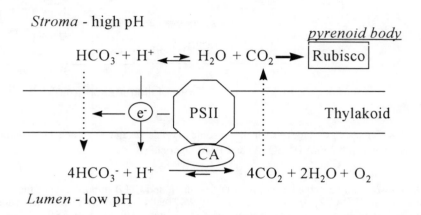

Figure 4. Model of a PSII-CA driven CO_2 pump. In light, photosynthetic electron transport develops a pH gradient across the thylakoid membrane, resulting in alkalization of stroma and acidification of lumen. Bicarbonate is transported across the thylakoid membrane at the expense of part of the pH gradient. In the acidic lumen, bicarbonate is dehydrated into CO_2 as catalyzed by PSII-CA, resulting in high concentration of CO_2 in the lumenal compartment, which diffuses into the stroma where it is finally trapped by Rubisco.

In conclusion, we suggest that a the proposed thylakoid CCM is driven by PSII and catalyzed by PSII-CA, thus providing an ample supply of CO_2 for the carboxylation process (Fig. 5). The proposed pump would be fueled by the pH gradient created by the PSII-driven photosynthetic electron transport.

References
1 Badger, M.R. and Price, G.D. (1994) Ann. Rev. Plant Physiol. Plant Mol. Biol. 45, 369-392
2 Karlsson, J., Clarke, A., Chen, Z.Y., Hugghins, S.Y., Park Y.-I., Husic, H.D., Moroney, H.D. and Samuelsson, G. (1988) EMBO J. 17, 1208-1216
3 Klmov, V.V., Allakhverdiev, S.I., Baranov, S.V. and Feyziev, Y.M. (1995) Photosynth. Res. 46, 219-225
4 Pronina, N.A. and Semenenko, V.E. (1990) in Current Research in Photosynthesis (Baltscheffsky, M., ed.) vol. 4, pp. 489-492, Kluwer Academic Publishers, Dordrecht, The Netherlands
5 Raven, J.A. (1997) Plant, Cell and Environ. 20, 147-154

EFFECT OF REDUCTION IN CARBONIC ANHYDRASE ACTIVITY CAUSED BY ZINC DEFICIENCY ON CO_2 TRANSFER RESISTANCE

Haruto Sasaki, Tatsuro Hirose, Yoshito Watanabe and Ryu Ohsugi
Graduate School of Agriculture and Life Science, University of Tokyo, 1-1-1 Yayoi, Bunkyo-ku, Tokyo 113, Japan (H.S.); National Institute of Agrobiological Resources,(T.H., R.O.); and National Institute of Radiological Sciences (Y.W.)

Key words: C3, carbon isotope, CO_2 concentration, gas diffusion, gene expression, Rubisco

Introduction

Carbonic anhydrase (CA; EC.4.2.1.1) catalyzes the reversible conversion of CO_2 to HCO_3^-. CA has been recognized as an important enzyme that is closely associated with photosynthesis and the activity of CA can be specifically inhibited by zinc deficiency. In a previous study, we succeeded in estimating the magnitude of the CO_2 transfer resistance (r_t) from the intercellular space of mesophyll cells to the site of CO_2 fixation in the chloroplasts (1). The purpose of the present study was to elucidate the relationship between the activity of CA and each component of mesophyll resistance and the mechanism of the specific inhibition of CA activity by using zinc deficient rice plants.

Materials and methods

Cultivation of plants-Seedlings of rice (Oryza sativa L.) were transplanted to 40-liter containers filled with culture solution without zinc prepared as described by Yoshida (2). The plants were treated with 150 nM $ZnCl_2$ from 4 weeks after germination (+Zn), with 15 nM $ZnCl_2$ from 11 weeks after germination (-/+Zn), and without zinc (-Zn) by addition of the salt to the culture solution as appropriate. The plants were grown in a naturally illuminated green house.

Measurements of photosynthesis and the levels of enzymes-Leaf photosynthesis (LPS), transpiration, the CA activity and the level of RubisCO were measured simultaneously for the uppermost fully expanded leaf on the main stem, in 13 weeks after germination. The measurements of LPS and transpiration were made at a CO_2 concentration of 340 µl l^{-1}, at an irradiance above 1,400 µmol $m^{-2}s^{-1}$ photon flux density under artificial light. The CA activity and the level of RubisCO were determined by the method of Sasaki et al. (1). The activity of

3455

RubisCO was measured in terms of the initial activity at 30°C.

Measurement of $\delta\,^{13}C$ values of soluble sugars-Plants were placed in darkness at 1200h to make the leaves starved for the photosynthetic products. The next morning, the leaves were exposed to artificial light at an irradiance above 1,400 $\mu mol\ m^{-2}s^{-1}$ photon flux density from 0900h to 1200h, and then they were cut off from the plant. The soluble sugars were extracted and $\delta\,^{13}C$ was determined with mass spectrometer as reported previously (1).

Calculation of CO_2 concentrations and resistances-The concentrations of CO_2 in the stomatal cavity ([CO_2]stc) and at the CO_2 fixating site of RubisCO in the chroloplast ([CO_2]cht), and the stomatal (r_s), transfer (r_t) and fixation resistance (r_c) were estimated by measuring LPS, transpiration and $\delta\,^{13}C$ values of soluble sugars (1).

Isolation of RNA and Northern hybridization- Total RNA was fractionated on a 1.15% agarose gel, transferred to a nylon membrane and allowed to hybridize with radiolabelled DNA probes. The signals due to the radiolabelled probe were visualized and quantified with a Bio-Imaging Analyzer (BAS 2000; Fujix, Tokyo, Japan). A 0.64-kbp cDNA fragment of cDNA for rice chloroplastic CA (Suzuki and Burnell, unpublished; Gene Bank Accession No. U08404) and a full-length clone of rice Ssu ((3); Gene Bank Accession No. D00644) were used as probes.

Results

Levels of zinc and enzymes and rates of photosynthesis-The zinc contents per unit leaf area of the -Zn and -/+Zn plants were as low as 0.19 and 0.23 mg m^{-2}, respectively, as compared to 0.51 mg m^{-2} in the +Zn plants (Table 1). In spite of the great reduction in the zinc content of leaves, we observed no change in SLW or in chlorophyll content, which is considered to be a critical indicator of zinc deficiency. Thus, —Zn and -/+Zn plants were only moderately stressed. By contrast, while the CA activity in -Zn plants decreased dramatically to as little as 14% of that in +Zn plants, the RubisCO activity decreased only to 89% of that in +Zn plants. The levels of soluble protein and also of RubisCO increased slightly. Moreover, little change in the rate of photosynthesis was observed in the leaves with a reduced zinc content. These results indicated that zinc deficiency resulted in the specific inhibition of CA activity.

Stomatal, transfer and fixation resistance-Although no significant difference was found in r_c and r_s in leaves of +Zn, -Zn and -/+Zn plants, we observed a 2.3-fold increase in r_t, which increased from 1.2 $mol^{-1}CO_2m^2sec$ in +Zn plants to 2.7 $mol^{-1}CO_2m^2sec$ in -Zn plants (Table 2). This increase corresponded to an increase from 10% to 21% when r_t was calculated as a percentage of total resistance. This indicates that the zinc deficiency affected only the CO_2 transfer step in the CO_2 assimilation process.

The concentration of CO_2 in leaves-In all leaves examined, [CO_2]stc was maintained at about 220 $\mu l\ l^{-1}$. By contrast, [CO_2]cht in +Zn leaves was 195±4 $\mu l\ l^{-1}$, and [CO_2]cht decreased with a reduction in the zinc content of leaves to 171±6 $\mu l\ l^{-1}$ and 166±5 $\mu l\ l^{-1}$ in -/+Zn and -Zn leaves,

Table 1. Effects of zinc deficiency on zinc content, LPS, δ ^{13}C value, chlorophyll content, levels of CA and RubisCO and the expression of mRNAs in uppermost fully expanded leaves.

Parameter		Plants		
		+Zn	-/+Zn	−Zn
Zinc content	(mg m^{-2})	0.51±0.05	0.23±0.03	0.19±0.04
SLW	(g m^{-2})	19.3±2.0	20.2±2.6	19.9±3.6
Chlorophyll content	(g m^{-2})	0.59±0.09	0.49±0.15	0.45±0.14
LPS	(μmol CO$_2$m^{-2}sec^{-1})	22.9±1.7	19.7±3.6	20.4±2.3
Soluble protein	(g m^{-2})	6.5±0.3	7.7±0.3	7.5±0.2
CA activity	($\times10^3$U m^{-2})	473±57	126±23	68±36
RubisCO content	(g m^{-2})	3.5±0.3	4.3±0.3	4.8±0.4
activity	(μmol CO$_2$m^{-2}sec^{-1})	29.4±2.6	23.5±4.3	26.3±5.2
δ ^{13}C value	(‰)	-25.83±0.24	-24.05±1.27	-23.55±1.91
mRNA expression	(%)			
CA		100	26±4	13±4
RubisCO Ssu		100	147±20	65±24

Levels of mRNAs in +Zn plants are set at 100%.

Values are expressed as means ± se.

Table 2. Effects of zinc deficiency on CO$_2$ diffusion resistance and concentration of CO$_2$ in the stomatal cavity and at the site of CO$_2$ fixation.

Parameter	Plants		
	+Zn	-/+Zn	−Zn
Resistance (mol^{-1}CO$_2$m^{2}sec)			
r$_s$	4.7±0.5	5.7±0.6	5.2±0.5
	(40)	(42)	(40)
r$_r$	1.2±0.3	2.2±0.9	2.7±0.7
	(10)	(16)	(21)
r$_c$	5.9±0.4	5.8±0.8	5.3±0.6
	(50)	(42)	(40)
CO$_2$ concentration (μl l^{-1})			
[CO$_2$]stc	224±7	215±8	225±6
[CO$_2$]cht	195±4	171±6	166±5

Values are expressed as means ± se. Numbers in parentheses are percentages relative to the total resistance.

respectively (Table 2). The gradient of the CO_2 concentration between the stomatal cavity and the site of CO_2 fixation increased from 29 µl l^{-1} to 59 µl l^{-1}. The resistance in CO_2 flux with zinc deficiency and our findings support the hypothesis that CA plays a role in facilitating the supply of CO_2 to the sites of carboxylation.

Northern hybridization analysis-The level of the mRNA for CA decreased with the reduction in zinc content of the leaf, falling to 26% (-/+Zn) and 13% (-Zn) of that in control plants (+Zn) (Table 1). By contrast, the level of the mRNA for Ssu showed no consistent trend.

Discussion

Major reductions in CA activity were observed without any change in r_s or r_c. Thus, we were able to detect a correlation between the transfer of CO_2 from the stomatal cavity to the site of CO_2 fixation and CA activity. It appeared, then, that r_t increased with the reduction in CA activity (Tables 1 and 2). While $[CO_2]stc$ was maintained at about 220 µl l^{-1} during photosynthesis, $[CO_2]cht$ decreased with the reduction in CA activity, and the gradient of the concentration of CO_2 between the stomatal cavity and the site of CO_2 fixation increased from 29 µl l^{-1} to 59 µl l^{-1} with the reduction in CA activity. Consequently, we can conclude that a reduction in CA activity tampers the transfer of CO_2 from the stomatal cavity to the site of RubisCO and causes a low $[CO_2]cht$. CA is mostly localized in chroloplasts in C_3 plants. Thus, our findings support the hypothesis that carbonic anhydrase has a role in facilitating the supply of CO_2 to the sites of carboxylation within the chloroplast.

Zinc is necessary for catalysis by CA (4). In this study, expression of the mRNA for CA decreased to 34% (-/+Zn) and 13% (-Zn) of that in control plants. This decrease resembled that in the CA activity in these plants with the reduction in the zinc content of leaves (Table 1). Thus, we suggest that the reduction in CA activity due to zinc deficiency was a result not of failure to activate CA but of a decrease in the level of CA. Although modification of transcript abundance does not always indicate transcriptional regulation, it is possible that a feedback system balances the amount of CA synthesized with the available zinc in the cell.

References

1 Sasaki, H., Samejima, M. and Ishii, R. (1996) Plant Cell Physiol. 37, 1161-1166

2 Yoshida, S., Forno, D.A. Cock, J.H. and Gomes, K.A. (1976) in Laboratory Manual for Physiological Studies of Rice , The International Rice Research Institute, Philippines, pp.61-66

3 Matsuoka,M., Kano-Murakami, Y., Tanaka, Y., Ozeki, Y. and Yamamoto, N. (1988) Plant Cell Physiol. 29, 1015-1022

4 Silverman, D.N. (1991) Can. J. Bot. 69, 1070-1078

MORPHOLOGICAL, ULTRASTRUCTURAL AND PHOTOSYNTHETIC FEATURES OF TWO FRESHWATER ANGIOSPERMS

RASCIO N., DALLA VECCHIA F. AND LA ROCCA N.
Dipartimento di Biologia, via U. Bassi 58/b, I-35131 ,Padova, Italy

Key words: Aquatic ecosystems, carbonic anhydrase, CO_2 uptake, electron microscopy, O_2 evolution, leaf anatomy,

1. Introduction

Colonization of the aquatic environment by flowering plants has required the evolution of morphological and physiological adaptative mechanisms useful to face the numerous limiting factors which can make it difficult for a macrophyte to live in water.

Of particular interest are the systems carried out by the different species for supplying photosynthetic cells with inorganic carbon (IC) (1), whose concentration as CO_2 can be greatly limited in the aquatic medium (2). For a plant survival and growth in any stressing environment, in fact, depend on the possibility to maintain a photosynthetic gain in terms of dry matter production.

In this paper the adaptative systems and photosynthetic strategies of two freshwater macrophytes have been analyzed. The plants studied were *Ceratophyllum demersum* L., a totally submerged dicotyledon, and *Sparganium emersum* L., a monocotyledon with partially floating leaves. Morphological, ultrastructural and biochemical investigations were carried out.

2. Materials and Methods

The plants were collected in early summer from a stream of the natural oasis "Mulino di Cervara" (Treviso) in North Italy.

Leaf samples were processed for light and electron microscopy as described in (3). Chlorophyll contents were analyzed according to (4). Oxygen release from leaf tissues was measured according to the method of Ishii *et al.*(5), modified by Rascio *et al.* (3). These latter analyses were carried out in water media with different pH values, containing 5 mM $KHCO_3$ as IC source and with or without addition of AZA (5-acetamido-1,3,4-thiadiazole-2-sulphonamide).

3. Results and Discussion

In the submerged species *C. demersum* the leaves are finely divided into thin laciniae each one having an epidermis, a parenchyma with large intercellular spaces and a central

3459

G. Garab (ed.), Photosynthesis: Mechanisms and Effects, Vol. V, 3459–3462.
© 1998 *Kluwer Academic Publishers. Printed in the Netherlands.*

vascular bundle (Fig.1). As in other submerged macrophytes (3, 5-7), the epidermis cells are rich in chloroplasts with electron-dense stroma, well organized thylakoid system and several starch grains (Fig.2). A feature of these cells, also noticed in other plants living under water (3, 5, 8), is a peculiar ultrastructure of the outer cell wall. In *C. demersum* this wall (Fig.3) shows a compact, cutinized outermost layer and, below it, a loose region which extends along the cell sides (Fig.4).

C. demersum is a "HCO_3^--user" species, able to utilize environmental HCO_3^-, besides CO_2, as IC source for photosynthesis (1). This is confirmed by the persistence of oxygen emission from leaf tissues (Fig.5) at 9, 9.5 pH values, leading to CO_2 concentrations in the experimental medium close to those of the compensation point calculated for other aquatic plants (9).

Addition to the water medium of AZA, an apoplastic inhibitor of carbonic anhydrase (CA), causes a striking decrease in photosynthetic activity (Fig.6), suggesting that in the mechanism of IC uptake an important role is played by the extracellular activity of this enzyme, which accelerates HCO_3^- conversion to CO_2 in the cell wall environment. The involvement of this kind of AC in supplying the photosynthetic cells with CO_2 has also been found in leaves of other submerged dicotyledons, such as *Ranunculus penicillatus* (10) and *Miriophyllum spicatum* (personal comunication).

Thus, the peculiar organization of the epidermis cell wall of *C. demersum* leaf seems to assume a precise adaptative significance. The loose inner region might form an environment enriched in aquatic medium in which IC availability would be increased by the enzymatic conversion of HCO_3^- to CO_2. The compact outermost layer might act as a barrier for hampering the CO_2 back diffusion and, as a consequence, for favouring its entry into the photosynthetic cells.

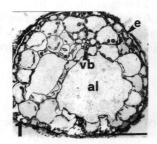  

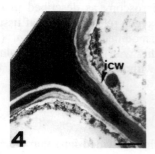

"Figures 1-4." Leaf of *C. demersum*.
Figure 1. Transversal section of a leaf lacinia (al = aerenchymatic lacuna, e = epidermis, vb = vascular bundle) (bar = 100 μm). *Figure 2.* A well differentiated chloroplast of an epidermis cell (t = thylakoids, s = starch) (bar = 2 μm). *Figure 3.* The surface wall of an epidermis cell. Note the compact and cutin (cu) covered outer layer (ocw) and the loose inner region (icw) (bar = 2 μm). *Figure. 4.* Cell wall between adjacent epidermis cells showing the loose region (icw) along the cell sides (bar = 2 μm)

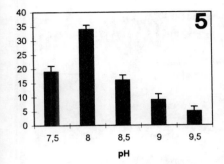

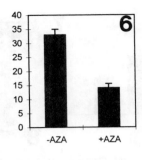

"Figures 5, 6." Oxygen emission (μmol O_2 /h/mg chl (a+b)) from *C. demersum* leaves.
Figure 5. At increasing pH values.
Figure 6. At pH 8, without (-) or with (+) 250 μM AZA in the water medium.

S. emersum, on the contrary, is a partially emerged species, whose long ribbon-like leaves float with their apical region. This plant belongs to the amphibious species, unable to use HCO_3^- as IC source for photosynthesis (1). The analyses carried out at increasing pH values on the floating leaf tissues, in fact, show that in the experimental conditions adopted, the O_2 emission ceases at pH 9, when CO_2 concentration in the water medium approaches the compensation point (Fig.7). The survival and growth of *S. emersum* are rather related to the leaf structural adaptations for CO_2 uptake from the atmosphere and for gas circulation in the submerged tissues. In the floating apical region, the leaf is flattened, with huge inner cavities delimited by cellular septa in which vascular bundles run (Fig.8). Well differentiated stomata (Fig.9) occur in both the epidermises, but are more frequent in the upper one, exposed to the air. The leaf shows an asymmetrical inner organization, with a mesophyll which is multilayered towards the adaxial surface and monolayered towards the abaxial one. This photosynthetic tissue, rich in chloroplasts, is very loose (Fig.10), so most of the cell surfaces are in contact with the gaseous environment of the aerenchymatic cavities. The presence of stomata and the absence of chloroplasts (Fig.10) in the epidermises, also noticed in other amphibious species (7), are terrestrial features, maintained by this kind of leaves which are still connected to the aerial environment. In *S. emersum* the leaves show a marked gradient of pigmentation along their longitudinal axis, acquiring a deep green colour only at the apical region. This can be due to the fact that the leaf growth depends on the activity of a basal meristem and therefore tissues of increasing age occur from the base to the leaf tip. However, the striking colour gradient suggests an ontogenetic slowness. In fact, in the middle zone of the leaf, below the floating region, the stomata of the lower epidermis are still indifferentiated, while those of the upper epidermis show early differentiation stages, with different degrees of guard cell separation (Figs.11-14). Moreover, in the mesophyll cells, the chloroplasts, besides thylakoids, contain large prolamellar bodies (Fig.15), which are characteristic of young organelles (11). Thus, only near the floating region of the leaf does the photosynthetic apparatus reach a complete differentiation.

Most of the leaf under water, therefore, seems rather to act as a petiole, sustaining the emerged photosynthetic tissues and favouring O_2 circulation in the submerged ones by means of the aerenchymatic cavities which are distributed throughout the leaf length.

3462

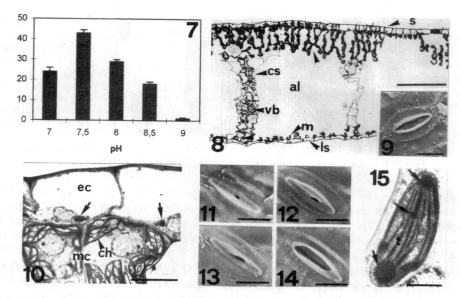

"*Figures 7-15.*" Leaf of *S. emersum*. *Figure 7.* Oxygen emission (μmol O_2/h/mg chl (a+b)) from the floating apical region at increasing pH values. *Figure 8.* Transversal section of the floating apical region showing the huge aerenchymatic lacunae (al) and the mesophyll (m) thickness (cs = cellular septum, ls = lower surface, us = upper surface, vb = vascular bundle) (bar = 100 μm). *Figure 9.* Stomate of the upper epidermis (bar = 10 μm). *Figure 10.* Epidermis cells (ec) with rudimentary chloroplasts (arrows) and mesophyll cells (mc) rich in well differentiated chloroplasts (ch) in the floating apical region (bar = 5 μm). *Figures 11-14.* Differentiating stomata in the upper epidermis of a submerged middle region (bar = 10 μm). *Figure 15.* Young chloroplast with thylakoids (t) and prolamellar bodies (arrows) in the submerged middle region (bar = 1 μm)

4. References

1 Prins, H.B.A. and Elzenga, J.T.M. (1989) Aquatic Botany 34, 59-83
2 Sand-Jensen, K. (1983) J. Exp. Bot. 34, 198-210
3 Rascio, N., Mariani, P., Tommasini, E., Bodner, M. and Larcher, W. (1991) Planta 185, 297-303
4 Moran, R. And Porath, D. (1980) Plant Physiol. 65, 478-479
5 Hishii, R., Yamagishi, T.Y. and Murata, Y (1977) Jap. J. Crop Sci. 46, 53-57
6 Rascio, N., Mariani, P., Dalla Vecchia, F., Zanchin, A., Pool, A. and Larcher, W. (1994) J. Plant Physiol. 144, 314-323
7 Dalla Vecchia, F., Zuppini, A., Baldan, B., Mariani, P. and Rascio, N. (1995) in Photosynthesis: From Light to Biosphere (Mathis P., ed.), vol. 5, pp. 575-578, Kluwer Academic Publishers, Dordrecht, The Netherlands
8 Madsen,T.V. and Maberly, S.C. (1991) Freshwater Biol. 26, 175-187
9 Sültemeyer, D., Schmidt, C. and Fock, H.P. (1993) Physiol. Plant. 88, 179-190
10 Newman, J.R. and Raven, J.A. (1993) Plant Cell Environm. 126, 491-500
11 Casadoro, G. and Rascio, N. (1979) J. Ultrastruct. Res. 69, 307-315

CARBON DIOXIDE DIFFUSION INSIDE C₃ LEAVES

Evans, John R. Research School of Biological Sciences,
Australian National University, GPO Box 475 Canberra
ACT 2601 Australia

Key words: C3, chloroplast, CO_2, gas diffusion, intercellular CO_2 concentration, leaf anatomy

1. Introduction

CO_2 reaches Rubisco by diffusion down a concentration gradient. The gradient forms during photosynthesis as a result of restrictions to free movement imposed by stomata and leaf anatomy. Recent techniques have enabled the internal conductance, g_i, of leaves, that is, from sub-stomatal cavities to the sites of carboxylation, to be measured (1). A strong correlation has been found between g_i and the rate of CO_2 assimilation under high irradiance and ambient CO_2 partial pressure, C_a (2,3,4). Unfortunately, the title of the paper by Lloyd et al. (3), 'low conductances for CO_2 diffusion from stomata to the sites of carboxylation in leaves of woody species' has been taken to mean that there is a correspondingly low partial pressure of CO_2 in the chloroplast, which contributes to the low CO_2 assimilation rate of sclerophyllous leaves. The claim was repeated by Epron et al. (5) 'high internal resistances to CO_2 transfer may account for the low net CO_2 assimilation rates that often characterise tree leaves.

However, leaves with lower photosynthetic capacities have lower g_i. When all the data currently available is examined together, the drawdown in CO_2 from substomatal cavities, C_i, to the sites of carboxylation, C_c, is similar between leaves with high and low photosynthetic capacities. That is, a low g_i generally does not mean that photosynthetic efficiency is reduced because it is matched by a low photosynthetic capacity. The drawdown from substomatal cavities to the sites of carboxylation within the chloroplast averages 80 μbar for an actively photosynthesising leaf.

Internal conductance, g_i, can be factored into two components, namely diffusion restrictions through the intercellular airspaces and the liquid path across cell walls into the chloroplasts. For leaves with large g_i, intercellular airspaces apparently offer little resistance and g_i correlates strongly with the surface area of chloroplasts exposed to intercellular airspace.

G. Garab (ed.), Photosynthesis: Mechanisms and Effects, Vol. V, 3463–3466.
© 1998 *Kluwer Academic Publishers. Printed in the Netherlands.*

2. Procedure

Internal conductance has been estimated by combining gas exchange measurements with either stable carbon isotope or chlorophyll fluorescence measurements. As the fluorescence method becomes unreliable when internal conductances exceed 0.3 mol m^{-2} s^{-1} bar^{-1}, only data collected using the isotopic method were used for the mesophytic species. For the sclerophytic species, data collected by either method were used as direct comparisons between the two methods agreed closely when gi was less than 0.3 mol m^{-2} s^{-1} bar^{-1} (4). Surface area of chloroplasts exposed to intercellular airspace per unit leaf area can be quantified from transverse sections of leaves (6).

3. Results and Discussion

A total of 70 estimates of g_i were available in the literature that satisfied the procedure (2,3,4,5,7,8,9). Each species was assigned to either a mesophytic or sclerophytic category, with the sclerophytes generally having lower photosynthetic capacity per unit leaf area or mass. A strong relationship was observed between g_i and CO_2 assimilation rate measured under high irradiance and ambient CO_2 partial pressure (Figure 1). The

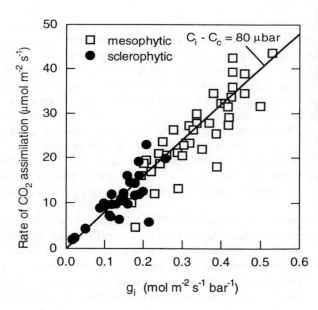

Figure 1 Relationship between CO_2 assimilation rate and internal conductance for sclerophytic (solid) and mesophytic (open) leaves. CO_2 assimilation rates were measured under high irradiance and ambient CO_2 partial pressures. CO_2 drawdown from substomatal cavities to sites of carboxylation (C_i - C_c) is given by the slope of the diagonal line.

average drop in CO_2 partial pressure between substomatal cavities and sites of carboxylation in the chloroplast can be calculated from the ratio A / g_i. The slope of the diagonal line was 80 µbar, irrespective of leaf category. There was no indication that the low internal conductance of sclerophytic leaves reduced CO_2 assimilation to any greater extent than in mesophytic leaves. The mean and standard errors for CO_2 assimilation rate, stomatal and internal conductance to CO_2 for each leaf type are given in the Table, along with the drop in CO_2 partial pressure across stomatal pores and within leaves. Sclerophytic leaves had CO_2 assimilation rates only 40% of the mean of mesophytic leaves, reflecting a smaller biochemical capacity (e.g. less Rubisco per unit leaf area) rather than Rubisco having a lower catalytic rate due to lower CO_2 partial pressures in chloroplasts of sclerophytic leaves. Stomatal conductances were slightly less than internal conductances for both leaf types. While the drop in CO_2 partial pressure within leaves was similar for both leaf types, the diffusive barrier imposed by stomata in sclerophytic leaves was proportionately greater than in mesophytic leaves. Consequently, the CO_2 partial pressure at the sites of carboxylation in sclerophytic leaves was lower than in mesophytic leaves.

Leaf Type	Sclerophytic (n=30)	Mesophytic (n=40)
CO_2 assimilation rate, A (μmol m^{-2} s^{-1})	10.9 ± 0.9	25.6 ± 1.4
stomatal conductance, g_s (mol CO_2 m^{-2} s^{-1})	0.10 ± 0.01	0.30 ± 0.02
internal conductance, g_i (mol CO_2 m^{-2} s^{-1})	0.14 ± 0.01	0.34 ± 0.02
$C_a - C_i$ (A/g_s) (μbar)	121 ± 5	90 ± 3
$C_i - C_c$ (A/ g_i) (μbar)	81 ± 4	77 ± 3

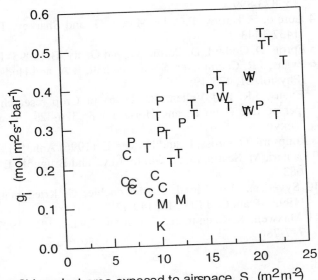

Figure 2 Relationship between internal conductance and chloroplast surface area exposed to intercellular airspace per unit leaf area.
T - tobacco;
W - wheat,
P - peach,
C - Citrus,
M - Macadamia;
K - Kalanchoe.

The measurement of chloroplast surface area exposed to intercellular airspace is laborious. Consequently there is limited data from which to generalise (Figure 2). A similar relationship appears to exist for the sclerophyllous leaves of *Citrus* to that of the mesophytic peach, tobacco and wheat, despite considerable variation in leaf thickness and porosity. This is surprising given that the thick hypostomatous leaves of *Citrus* would be expected to impose considerably greater restriction to CO_2 diffusion through intercellular airspaces which would then have to be offset by a more permeable cell wall and liquid path for diffusion. The thick hypostomatous leaves of *Macadamia* have impermeable bundle sheath extensions which prevent free lateral diffusion through the intercellular airspaces in these leaves. In this case, the lower internal conductance per unit exposed chloroplast surface could be due to diffusion restrictions through intercellular airspaces. This could be tested experimentally through the use of helox (air in which helium replaces nitrogen to increase the diffusivity of CO_2 2.3 fold). The succulent leaves of *Kalanchoe* exceed 1 mm in thickness and have limited airspaces which may also account for the low internal conductance per exposed chloroplast surface. The internal conductance per exposed chloroplast surface declined in wheat as the leaves aged while the ratio A/ g_i remained constant (7). This suggests that another factor is involved in determining the internal conductance which remains to be elucidated.

References

1 Evans, J.R. and Caemmerer, S.von (1996) Plant Physiol. 110, 339-346

2 Caemmerer, S.von and Evans, J.R. (1991) Aust. J. Plant Physiol. 18, 287-305

3 Lloyd, J. Syvertsen, J.P. Kriedemann, P.E. and Farquhar, G.D. (1992) Plant Cell Env. 15, 873-899

4 Loreto, F. Harley, P.C. Di Marco, G. and Sharkey, T.D. (1992) Plant Physiol. 98, 1437-1443

5 Epron, D. Godard, D. Cornic, G. and Genty, B. (1995) Plant Cell Env. 18, 43-51

6 Evans, J.R. Caemmerer, S.von Setchell, B.A. and Hudson, G.S. (1994) Aust. J. Plant Physiol. 21, 475-495

7 Evans, J.R. and Vellen, L. (1996) in Crop research in Asia: Achievements and perspective (Ishii R. and Horie T.) pp.326-329, Asian Crop Science Association, Tokyo

8 Roupsard, O. Gross, P. and Dreyer, E. (1996) Ann. Sci. For. 53, 243-254

9 Lauteri, M. Scartazza, A. Guido, M.C. and Brugnoli, E. (1997) Funct. Ecol. 11, 675-683

10 Syvertsen, J.P. Lloyd, J. McConchie, C. Kriedemann, P.E. and Farquhar, G.D. (1995) Plant Cell Env. 18, 149-157

11 Maxwell, K. Caemmerer, S.von and Evans, J.R. (1997) Aust. J. Plant Physiol. 24, 777-786

GROWTH LIGHT INTENSITY AFFECTS PHOTOSYNTHETIC CARBON METABOLISM IN SPINACH

A. Battistelli [1], S. Moscatello, S. Proietti, D. Salvini, A. Scartazza, and A. Augusti.
CNR, Istituto per l'Agroselvicoltura, Porano, Italy. a.battistelli@ias.tr.cnr.it

Key words: C partitioning, carbohydrate, leaf physiology, light acclimation, sun and shade leaves, sucrose phosphate synthase.

1. Introduction

Growth of plants at different light intensities causes the appearance of large differences in the leaf physiology, particularly at the level of photosynthetic metabolism. Leaves grown at low light have lower light and CO_2 saturated photosynthesis, a lower amount of ribulose-1,5-bisphosphate carboxylase (Rubisco), a reduced capacity of electron transport, a lower Mg^{2+}-dependent ATPase activity than leaves grown at high light intensity [1]. Despite a large number of studies carried out on the effect of growth light intensity on the modification of leaf photosynthesis, very little effort has been spent to study the effect of growth light intensity, on the control of photosynthetic end product synthesis and on carbohydrate partitioning.

The aim of this study was to evaluate the effect of growth at different light intensities on the accumulation of leaf sugars, on the activities of ADPglucose pyrophosphorylase (AGP) and sucrose phosphate synthase (SPS), two key enzymes on the starch and sucrose synthesis pathways respectively, and on photosynthesis and growth of spinach plants.

2. Procedure

Spinach plants were grown in a growth chamber (Sanyo Gallencamp U.K.) in 1l pots in garden soil, watered daily and supplied with full nutrient solution twice a week. Light cycle was 10 hours light and 14 hours dark, temperature was controlled at $25 \pm 0.5°C$ and $20 \pm 0.5 °C$ during the day and night respectively. Relative humidity was set constant at $70\% \pm 5\%$. Light intensity was 800 ± 40 and 200 ± 20 μmol quanta $m^{-2} s^{-1}$ (400-700 nm) for the high and low light treatment respectively, from seeding to emergence of cotyledons, all pots were kept at the low light intensity. Measurements were made 4 weeks after germination, on the third true leaf, after full expansion. For growth analysis, plant material was dried at 105°C, to constant weight. Gas exchange were measured with an open gas exchange system (ADC LCA2 U.K.) at 25°C and ambient CO_2. Samples, for carbohydrate and enzymes measurements, were taken with a cork borer ($1.1 cm^2$), immediately frozen in liquid N_2, and then extracted with a glass potter. Carbohydrates extraction and measurements were performed as in [2]. Enzymes extraction and SPS maximum activity *via* UDP formation were made as in [3], AGP was measured spectrophotometrically in the pyrophosphorolytic direction as in [4]. Statistical

3467

G. Garab (ed.), Photosynthesis: Mechanisms and Effects, Vol. V, 3467–3470.

significance was tested by ANOVA, at the 5% (*) and 1% (**) significance level for the F test.

3. Results and Discussion

Growth of spinach plants at low light halved the plant growth rate, decreased the leaf area of the 20%, reduced the specific leaf dry weight and increased the shoot/root ratio with respect to growth at high light (Table 1). These results, indicate that plant acclimation to low light was attained with an optimisation of carbon partitioning at the plant level in the attempt of increase the light interception at the expenses of investment of carbon on the

Table 1 Plants characteristics, and photosynthetic parameters at the time of sampling for enzymes and sugars analysis.

	Growth light intensity		Statistical significance
	High	Low	
Plant leaf area (cm^2)	80.3	66	*
Number of expanded leaves	6	6	n. s.
Specific leaf dry weight of expanded leaves ($mg \cdot cm^{-2}$)	5.1	2.8	**
Plant dry matter ($g \cdot plant^{-1}$)	0.49	0.22	**
Shoot/root	5.56	6.41	*
CO_2 assimilation rate ($\mu mol \cdot m^{-2} s^{-1}$)	17.2	12.4	*
Stomatal conductance ($mol \cdot m^{-2} s^{-1}$)	0.46	0.25	*
Sub-stomatal CO_2 partial pressure ($\mu bar \cdot bar^{-1}$)	242.7	233.1	n.s.

root system and by decreasing leaf thickness, similar results were obtained for example in rice (5). The fact that growth was dramatically decreased in low light grown plants, indicates that the strategy adopted by the plant could only partially mitigate the decreased availability of fixed carbon caused by the drop of the CO_2 assimilation rate. Lower photosynthesis of low light grown plant was obviously caused primarily by the low growth light intensity. In addition, when photosynthesis was measured at high light, leaves of the low light grown plants showed a 30% decrease of assimilation rate, a 45% decrease of stomatal conductance and a similar sub-stomatal CO_2 partial pressure with respect to leaves grown at high light. This indicates that the photosynthetic capacity of low light grown plants was decreased by non-stomatal components. Such lower photosynthetic capacity could have contributed to the low plant carbon gain under low light.
A decrease of photosynthetic capacity of low light grown plants might be caused by a decrease of Rubisco activity connected to a lower leaf nitrogen content (5) and to a reduced capacity for ribulose-1,5-bisphosphate regeneration (6). The soluble protein content of low light grown leaves was 37% lower than that of the high light grown leaves (Table 2) and this is indeed indicative of a reduction of Rubisco activity (1). Similarly, the decrease of chlorophyll content of low light grown leaves, might be taken as an indication of a reduction of the component of the light harvesting and electron transport chain that is known to occur in shade leaves in association with a decrease of the Mg^{2+}-specific ATPase activity (1). Measurements of the maximum extractable activity of two key enzymes, of the sucrose (SPS) and of the starch (AGP) pathways, showed that, when expressed on the area basis, SPS activity was decreased by a significant 40% and AGP by a 20%, although the AGP change was not statistically significant. Due to the high control exerted by each of these two enzymes on their respective pathways (7), it is possible to

infer that in low light grown leaves, the total capacity of the photosynthetic end products synthesis was decreased. A limitation of the photosynthetic end products synthesis, might cause a decrease of the maximum photosynthetic capacity (8). Our results indicate that this type of limitation might contribute to the decrease of photosynthetic capacity, of low light grown leaves, when compared with high light grown ones, particularly when measurements are made under photosynthetic flux unusually high for low light grown leaves.

Taken together, the results on growth and photosynthesis of spinach plants grown under low light, indicate that these plants experienced a severe reduction of the availability of photosynthetic products. Measurements of carbohydrate in leaves during the day, showed that the limitation of photosynthetic carbon supply caused a 75% decrease of leaf carbohydrate accumulation (Table 2) and a significant increase of the relative starch accumulation. Photosynthetic products have to be properly partitioned, at the leaf level, in order to support the requirements of the photosynthetic cell, and the plant requests of carbon supply. It is known that a sufficient amount of photosynthetic products has to be

Table 2 Leaf characteristics, and carbohydrate metabolism parameters in spinach leaves.

	Growth light intensity		Statistical significance
	High	Low	
Total leaf chlorophyll content (mg $\cdot$ m^{-2})	48	37	**
Soluble leaf protein content (g $\cdot$ m^{-2})	12.8	8	**
SPS Max. activity (μ mol product $\cdot$ cm^{-2} h^{-1})	0.46	0.27	**
AGP (μ mol product $\cdot$ cm^{-2}h^{-1})	0.92	0.74	n.s.
AGP/SPS	2.05	3.09	*
Total leaf carbohydrate accumulation during the day (μ mol hexose $\cdot$ cm^{-2})	3.14	0.77	**
% of carbohydrate accumulated as starch	49.9	62.3	*

accumulated in leaves, during the day, to support the cell metabolism at night. Starch-less mutants of *Arabidopsis*, for example, can only survive under a long day photoperiod, since the very low amount of starch accumulated, is insufficient to support cell metabolism under long nights (9) It is obvious, that under a severe limitation of photosynthetic carbon fixation, carbohydrate availability at the leaf level drops dramatically (see for example ref. 5). Our results demonstrate that the severe reduction of photosynthesis, in low light grown spinach plants, caused a large decrease of the amount of sugars in the leaf, at the end of the day, that was accompanied by a significant switch of partitioning toward starch. Partitioning of photosynthetic products at the leaf level depends on many factors, and particularly on the relative strength of the sucrose *vs* the starch biosynthetic pathways. SPS exerts a relevant control over the photosynthetic sucrose synthesis and part of this control is at the level of the enzyme maximum activity (7). A reduction of SPS maximum activity would undoubtedly cause a restriction on the potential for sucrose synthesis of the cell. The activity of AGP exerts a crucial role in controlling starch synthesis in photosynthetic cells, and a decrease of this activity would restrict starch synthesis at low and at high light (10). In this study, we found that growth of spinach plants at light intensity that was sufficiently low to determine shade-like acclimation of plants and leaves, but was sufficient for sustained growth, not only causes a decrease of SPS maximum activity but determined also a smaller, and not statistically significant, decrease of AGP, that lead to a significant increase in the AGP/SPS ratio. Changes of gene expression, protein level and SPS activity were already found to occur in spinach, upon change from moderate light to light intensity lower than the light compensation point, in close link with changes in the Rubisco small sub-unit (11). Makino *et al.* (5) found no differences in SPS activity, on rice leaves, grown at different light intensity, when the activity was expressed as function of the leaf nitrogen content. Our results confirm, at the activity level, and at more growth compatible low light intensity, the decrease of SPS per leaf area unit, found by Klein *et al.* (11). Furthermore, in accordance with Makino *et al.* (5), when SPS activity was expressed as function of soluble proteins content (data not shown), changes between low and high light grown plants were much smaller and not statistically significant, showing that changes in SPS were related to changes in soluble leaf proteins an hence nitrogen and Rubisco (1). However, the AGP/SPS ratio was obviously not affected by the reference scale used for enzymes activity and the significant increase of this indicator of the relative strength of the starch over the sucrose synthetic pathways, remains significant, even taking into account the general reduction of the protein content and of the photosynthetic machinery, represents a good biochemical basis for the relative increase of starch accumulation.

In conclusion, with this work we obtained evidences that growth of spinach plants in low light affects not only the photosynthetic light driven reactions and the capacity of the Calvin cycle, but also the starch and sucrose synthesis pathways, in a way that most likely provides a means to increase, in relative terms, the accumulation of starch during the day and hence the retention of photosynthetic products in leaves. In the view of the importance of the accumulation of leaf carbohydrate to support leaf metabolism during the dark period, and of the relevance of the night turn-over of starch for plant growth and development (12), the shown changes in carbohydrate metabolism of spinach plants, grown under low light, may play an important acclimatory role for plant survival, under shade conditions.

4. Addendum

This work was supported by the Special Project "Biologia e Produzione Agraria per una Agricoltura Sostenibile" of the CNR-National Institute "Biologia e Produzione Agraria"

References

1 Terashima, I. and Evans, J. R. (1988) Plant Cell Physiol. 29, 143-155.
2 Massacci, A., Battistelli, A. and Loreto, F. (1996) Aust. J. Plant Physiol. 23, 331-340
3 Stitt, M., Wilke, I., Feil, R., and Held, H. W. (1988) Planta 174, 217-230
4 Kleczkowski, L. A., Villand, P., Lüthi, E., Olsen, O.-A. and Preiss, J. (1993) Plant Physiol. 101, 179-186
5 Makino, A., Sato, T., Nakano, H., and Mae T. (1997) Planta 203, 390-398
6 Anderson, J. M. and Osmond, C.B. (1987) In: Photoinhibition. (Kyle D.J., Osmond C.B. and Arntzen C.J.eds) pp.1-38, Elsevier Science Publishers, Amsterdam.
7 Stitt, M. and Sonnewald, U. (1995) Annu. Rev. Plant Physiol. Plant Mol. Biol. 46, 341-368.
8 Foyer, C. H. (1988) Plant Physiol. Biochem. 26, 483-492.
9 Lin, T-P, Caspar, T., Somerville, C. and Preiss J. (1988) Plant Physiol. 86, 1131-1135.
10 Neuhaus, H. E. and Stitt, M. (1990) Planta 182, 445-454.
11 Klein, R. R., Crafts-Brandner, S. J. and Salvucci, M. E. (1993) Planta 190, 498-510.
12 Schulze, W., Stitt, M., Schulze, E.-D., Neuhaus, H. E. and Fichtner, K. (1991) Plant Physiol. 95, 890-895

THE INFLUENCE OF LEAF THICKNESS ON THE CO_2 TRANSFER CONDUCTANCE AND LEAF $\delta^{13}C$ FOR SOME EVERGREEN TREE SPECIES.

Yuko T. Hanba, Shin-Ichi Miyazawa, and Ichiro Terashima
Department of Biology, Graduate School of Science, Osaka University,
Toyonaka 560-0043, Japan

Key words: CO_2, gas diffusion, gas exchange, leaf anatomy, stable isotopes

1. Introduction

It has been shown that carbon isotope discrimination during photosynthesis linearly correlated to the ratio of internal (P_i) to external (P_a) partial pressure of CO_2 of the leaves (1). Carbon isotopic composition in plants ($\delta^{13}C$), therefore, can be used as an index of various plant physiological features that are influenced by P_i/P_a such as water use efficiency. However, for thick hypostomatous leaves, besides the variation in P_i/P_a, the large resistance of CO_2 diffusion inside leaves may affect the leaf $\delta^{13}C$. So it would be dangerous to use the leaf carbon isotopic composition as an simple index of leaf physiological status in such leaves.

The aim of this work is to examine effects of leaf thickness on the CO_2 transfer conductance inside leaves, and to clarify whether variations in leaf thickness affect leaf dry matter $\delta^{13}C$ through changing CO_2 transfer conductance.

2. Materials and methods

2.1 *Plant materials*

Two to three-year-old seedlings of evergreen tree species, *Quercus glauca* and *Castanopsis sieboldii* , were grown under field condition in 5-litter vinyl pots from March 1997. Shoots of *Quercus phillyraeoides*, *Cinnamomum camphora*, *Ligustrum lucidum*, and *Camellia japonica* were collected from the sunny sides of the tree crowns in November and December 1997 and treated as (2).

2.2 *Gas exchange and carbon isotope measurements*

Two to three leaves of the plants were enclosed in an chamber, and gas exchange measurements were made before and after the gas collection. CO_2 gas was collected after (3) with some modification. The carbon isotope ratio of CO_2 and leaf dry matter were measured with a mass-spectrometer (MAT252, Finnigan MAT, Bremen, Germany). The CO_2 transfer conductance from the substomatal cavities to the sites of carboxylation (g_w) was calculated by the equations given in (4).

2.3 *Light microscopy*

G. Garab (ed.), Photosynthesis: Mechanisms and Effects, Vol. V, 3471–3474.
© 1998 *Kluwer Academic Publishers. Printed in the Netherlands.*

Leaf pieces were embedded in Spurr's resin (5). Microphotographs of the sections of 0.8 μm thick were taken with a microscope (BX50/PM30, OLYMPUS, Tokyo, Japan) and mesophyll thickness and porosity were measured.   Surface area of mesophyll cells exposed to intercellular air spaces (S_{mes}) were measured by the method of (6).

3. Results and discussion

The CO_2 transfer conductance (g_W) were tended to be larger when leaf mesophyll thickness were larger (Fig. 1). The positive correlation between g_W and mesophyll thickness would be due to the variation in leaf anatomical characteristics with leaf thickness.

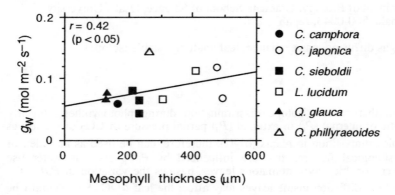

Fig. 1 The dependence of g_W on mesophyll thickness.

The species that have thicker mesophyll tended to have more developed palisade tissues, therefore, they have larger wall surfaces exposed to intercellular air spaces (S_{mes}), and smaller porosity among the evergreen species (Fig. 2a, b).

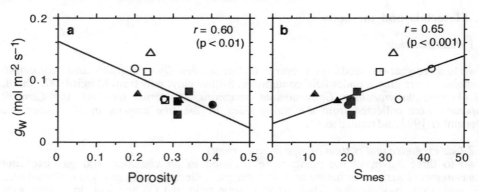

Fig. 2 The correlations between g_W and leaf anatomical characteristics. a, Porosity and g_W, b, surface area of mesophyll cells exposed to intercellular air spaces (S_{mes}) and g_W. Symbols are as shown in Fig. 1.

CO_2 diffuses from the substomatal cavities to the carboxylation site in two different media (7): CO_2 diffusion within intercellular air spaces occurs in gas phase (g_{ias}), and that across the cell wall and chloroplasts in liquid phase (g_{liq}).

The g_{ias} would be proportional to porosity (8), so if g_{ias} determine the g_w, g_w would be positively correlated to porosity. However, the opposite was the case (Fig. 2a), therefore, the difference of CO_2 transfer conductance within intercellular air spaces may not be a major limitation to g_w among the evergreen tree species examined in this study.

The surface area of mesophyll cells exposed to intercellular air spaces was positively correlated to g_w (Fig. 2b). If liquid phase diffusion is a major limitation, surface area of chloroplasts exposed to intercellular air spaces should be almost proportional to g_w (9). If the density of chloroplasts layers covered the mesophyll cell surface is almost similar between species, surface area of chloroplasts exposed to intercellular air spaces is strongly related to surface area of mesophyll cells exposed to intercellular air spaces (S_{mes}). The positive correlation between g_w and S_{mes} (Fig. 2b), therefore, suggests that liquid phase diffusion may be a major determinant of g_w for the evergreen tree species used in this study.

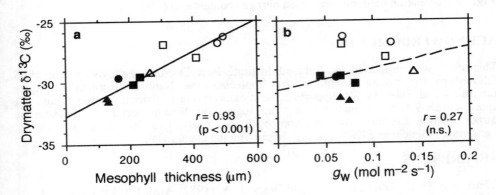

Fig. 3 The relationships between stable carbon isotope ratio ($\delta^{13}C$) in leaf dry matter and leaf characteristics. **a**, Mesophyll thickness; **b**, CO_2 transfer conductance (g_w). Symbols are as shown in Fig. 1.

Leaf dry matter $\delta^{13}C$ was positively correlated with mesophyll thickness (Fig. 3a). However, no significant correlation was obtained between leaf dry matter $\delta^{13}C$ and g_w (Fig. 3b). These results suggest that variation in g_w among the evergreen tree species in this study could not affect the variation in leaf $\delta^{13}C$ among these species. This result is coincident with (10) that thicker leaves of chestnut had positive $\delta^{13}C$ and larger g_w. More positive $\delta^{13}C$ in the species of thicker leaves which have higher leaf nitrogen contents on an area basis (Fig. 4) also suggests that variation in long-term photosynthetic capacity is the most important determinant of the difference in leaf $\delta^{13}C$ in our study, as some previous works (11, 12). The variation in long-term photosynthetic capacity probably results in the variation of long-term P_i. We propose that among the similar leaf functional types (e.g. evergreen), variation in long-term photosynthetic capacity rather than that in g_w would be an important determinant of leaf $\delta^{13}C$.

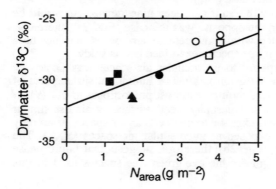

Fig. 4 Correlation between area-based nitrogen content and $\delta^{13}C$.

ACKNOWLEDGMENTS

This study was supported by National Institute Post Doctoral Fellowship and JSPS fellowships for Young Scientists. We appreciate the National Institute of Agro-Environmental Sciences for supporting the measurements of carbon isotope ratios. We appreciate Dr S. von Caemmerer for her kind suggestion about the technique of concurrent carbon isotope discrimination. We appreciate the constructive comments from Drs S. Funayama-Noguchi and K. Noguchi to improving the manuscript.

REFERENCES

1 Farquhar, G. D., O'Leary, M. H. and Berry, J. A. (1982) Aust. J. Plant Physiol. 9, 121-137.
2 Koike, T. (1986) J. Jap. Forest. Soci. 68(10), 425-428.
3 von Caemmerer, S. and Evans, J. R. (1991) Aust. J. Plant Physiol. 18, 287-305.
4 Evans, J. R., Sharkey, T. D., Berry, J. A. and Farquhar, G. D. (1986) Aust. J. Plant Physiol. 13, 281-292.
5 Spurr, A. R. (1969) J. Ultrastructure Res. 26, 31-43.
6 Thain, J. F. (1983) J. Exp. Bot. 34, 87-94.
7 Evans, J. R. and von Caemmerer, S. (1996) Plant Physiol. 110, 339-346.
8 Parkhurst, D. F. (1994) New Phytol. 126, 449-479.
9 Evans, J. R., von Caemmerer, S., Setchell, B. A. and Hudson, G. S. (1994) Aust. J. Plant Physiol. 21, 475-495.
10 Lauteri, M., Scartazza, A., Guido, M. C. and Brugnoli, E. (1997) Funct. Ecol. 11, 675-683.
11 Ehleringer, J. R., Field, C. B., Lin, Z.-f. and Kuo, C.-y. (1986) Oecologia 70, 520-526.
12 Hanba, Y. T., Mori, S., Lei, T. T., Koike, T. and Wada, E. (1997) Oecologia 110, 253-261.

GAS EXCHANGE AND ENVIRONMENTAL PARAMETERS AT FBPASE ANTISENSE POTATOES

Michael Muschak, Lothar Willmitzer and Joachim Fisahn
Max-Planck-Institut f. Molekulare Pflanzenphysiologie,
Karl-Liebknecht-Str. 25, D-14476 Golm, Germany

Key words: air humidity, Calvin-cycle, CO_2 concentration, ligth sensing, stomata, transgenic plants

Introduction

It has been frequently shown that stomatal conductance responded to changes in ambient CO_2 concentration (1,2). On the other hand insensitivity was observed under special conditions (1,2). Reduction of leaf conductance and stomatal aperture in response to an elevation in ambient $[CO_2]$ is commonly indicated as "normal" response (2). Scarcely, the opposite reaction was observed and indicated as "exceptional" response. (2) The following findings could provide an explanation for these contradictory results. There is evidence that endogeneous plant hormones like ABA or cytokinins could influence the quality and extent of stomatal response to changes in $[CO_2]$ (2). Furthermore, it was suggested that stomatal response is linked by these hormons to other parameters like plant water status or photosynthetic activity of mesophyll cells (1,2).

To study stomatal conductance as a function of intercellular CO_2 levels we tried to use a new approach. Transgenic plants have been described that exhibited increased levels of intercellular $[CO_2]$ at ambient CO_2 levels due to an inhibition in Calvin cycle activity (3,4). These transgenic plants expressed an antisense mRNA targeted to the chloroplastic FBPase. As a consequence assimilation rates were severly reduced (4).

The scope of the present work was to investigate leaf conductance and assimilation rate in response to light at different levels of air humidity and CO_2. Transgenic plants were used as a model system to study the effects of elevated intercellular CO_2 levels, however without changing ambient $[CO_2]$. Thus gas exchange measurements will be presented in an attempt to provide new aspects to the field of stomatal control.

Materials and methods

Plants. Transgenic potato plants were generated as described previously (3); these plants expressed a FBPase-antisense mRNA targeted to the chloroplastic FBPase. Two independently generated lines of these plants were used in the experiments described here. These different lines contained FBPase activities of 809 ± 130 Um^{-2} (100% Wt) and 103 ± 12 Um^{-2} (12%). Wildtype potato (*Solanum tuberosum* L. cv. Desirée) as well as transgenic lines were grown in the greenhouse. Three days prior to the measurements all

G. Garab (ed.), Photosynthesis: Mechanisms and Effects, Vol. V, 3475–3478.
© 1998 *Kluwer Academic Publishers. Printed in the Netherlands.*

plants were transferred to a growth chamber (Noske-Kaeser, Germany), where the gas exchange measurements were carried out.

Gas-exchange measurements were performed in an open system made in collaboration with Walz Inc. (Effeltrich, Germany). Special gas atmospheres, in particular air containing 500 ppm CO_2, were produced by two gas-mixing units (GMA-2 Walz, Effeltrich, Germany).

Results

Assimilation rate. Wildtype plants show normal saturation curves upon increasing photon flux density at 70–80% relative air humidity with a maximum of 14 μmol $m^{-2}s^{-1}$. At 60–70% r.h. a limitation by C_i emerged at a PFD of 400 μmol $m^{-2}s^{-1}$, which was caused by low leaf conductances. Maximum assimilation rate was 10.2 μmol $m^{-2}s^{-1}$ at a PFD of 600 μmol $m^{-2}s^{-1}$ under these conditions. αcp-FBP12 plants revealed reduced assimilation rates and a sigmoid curve shape. For these plants no significant differences were detected between the two levels of air humidity. Mean values of their assimilation rates did not exceed 7.2 μmol $m^{-2}s^{-1}$ (at a PFD of 1000 μmol $m^{-2}s^{-1}$).

At an ambient CO_2 concentration of 500 ppm maximal assimilation rates of wildtype plants were elevated up to a maximum of 17.7 μmol $m^{-2}sec^{-1}$ at PFDs of 800 and 1000 μmol $m^{-2}s^{-1}$. at 70–80% r.h. and up to 12.5 μmol $m^{-2}s^{-1}$ at 600 μmol $m^{-2}s^{-1}$ at 60–70% r.h. Assimilation rates of αcp-FBP12 plants remained the same as under atmospheric CO_2.

Transpiration rate. Transpiration rates exhibited a linear increase vs. PFD at 60-70% air humidity but a convex shape at 70–80% in both plant lines. At 70–80% transpiration rates were elevated compared to that at 60–70%. This difference was more pronounced in αcp-FBP12 plants. At 60–70% transpiration rates were equal for wildtype and αcp-FBP12 plants.

At 500 ppm CO_2 and 60–70% relative humidity same transpiration rates were found as at atmospheric CO_2 concentration. Transpiration at 70–80% air humidity did not reach the same maxima as at atmospheric CO_2 level. There was an increase of transpiration up to 1.6 mmol $m^{-2}s^{-1}$ at a PFD of 600 μmol $m^{-2}s^{-1}$ for wildtype and up to 1.5 mmol $m^{-2}s^{-1}$ at 400 μmol $m^{-2}s^{-1}$ for αcp-FBP12 plants. Beyond these PFDs transpiration declined.

Intercellular CO_2 concentration. At atmospheric CO_2 level intercellular CO_2 concentrations revealed the same patterns for wildtype and αcp-FBP12 plants. But in αcp-FBP12 intercellular CO_2 was elevated with a difference of about 80 ppm compared to wildtype plants under same conditions. Intercellular CO_2 concentration at 70–80% r.h. increased with PFD up to 220 ppm and 310 ppm at a PFD of 600 μmol $m^{-2}s^{-1}$ for wildtype and αcp-FBP12 plants respectively and remained constant up to a PFD of 1000 μmol $m^{-2}s^{-1}$. At 60–70% relative humidity intercellular CO_2 concentration decreased from 200 ppm at a PFD of 250 μmol $m^{-2}s^{-1}$ to 175 ppm at 1000 μmol $m^{-2}s^{-1}$ in wildtype and from 273 ppm to 241 ppm in αcp-FBP12 plants.

When an ambient CO_2 concentration of 500 ppm was applied intercellular CO_2 declined with increasing PFD in both plant lines and at both levels of air humidity. At a PFD of

50 µmol m^{-2}s^{-1} intercellular CO_2 level was at 450 ppm with no significant differences between plant lines or humidity levels. Differences emerged when PFD increased. At 1000 µmol m^{-2}s^{-1} wildtype plants show 202 ppm at 60–70% r.h. and 328 ppm at 70–80% r.h. αcp-FBP12 plants reached 352 ppm and 394 ppm respectively.

Leaf conductance. Leaf conductance of wildtype plants increased with increasing PFD at air humidities of 70–80% and atmospheric CO_2 concentration from 115 mmol m^{-2}s^{-1} at a PFD of 250 µmol m^{-2}s^{-1} up to 190 mmol m^{-2}s^{-1} at a PFD of 600 µmol m^{-2}s^{-1} and remained constant at the PFDs beyond. In αcp-FBP12 plants leaf conductance was 140 mmol m^{-2}s^{-1}, reached 270 mmol m^{-2}s^{-1} at a PFD of 600 µmol m^{-2}s^{-1} and increased further up to 310 mmol m^{-2}s^{-1} at a PFD of 1000 µmol m^{-2}s^{-1}. In contrast at 60–70% both plant lines kept their leaf conductances low between 100 and 110 mmol m^{-2}s^{-1}.

At 500 ppm ambient CO_2 and 70–80% relative humidity leaf conductance of wildtype plants increased with elevating photon flux density up to 170 mmol m^{-2}s^{-1} at a PFD of 400 µmol m^{-2}s^{-1} and declined for the PFDs above. In comparison, αcp-FBP12 plants show much lower leaf conductances at 70–80% air humidity than at atmospheric $[CO_2]$. They increased with rising PFD up to 200 mmol m^{-2}s^{-1} at a PFD of 200 µmol m^{-2}s^{-1}. Beyond this PFD they declined down to 130 mmol m^{-2}s^{-1} at a PFD of 1000 µmol m^{-2}s^{-1}. At 60–70% relative humidity wildtype plants exhibited constant leaf conductance of 100 mmol m^{-2}s^{-1} up to a PFD of 400 µmol m^{-2}s^{-1} followed by a decline down to 70 mmol m^{-2}s^{-1} at 1000 µmol m^{-2}s^{-1} PFD. Leaf conductance of αcp-FBP12 plants at a relative humidity of 60–70% resembled the curve shape of that of the wildtype plants but at a higher level. It was 152 mmol m^{-2}s^{-1} at a PFD of 50 µmol m^{-2}s^{-1} and 113 mmol m^{-2}s^{-1} at 1000 µmol m^{-2}s^{-1}. In comparison, this results in higher leaf conductances at PFDs below 600 µmol m^{-2}s^{-1} than at atmospheric CO_2 at the same PFDs.

Discussion

Transgenic plants with reduced assimilation rates were used to analyze effects of changes in photon flux density, different relative air humidities, and elevated intercellular CO_2 levels on stomatal conductance. Noteworthy, increases in intercellular $[CO_2]$ were obtained without a rise in ambient CO_2 levels. The results were compared to experiments performed at elevated ambient $[CO_2]$. Furthermore, antisense limitation of FBPase activity enabled to observe stomatal response to elevated levels of ambient CO_2 without elevating assimilation rates and changing mesophyll photosynthetic activity.

Comparison of leaf conductances of wildtype and transgenic plants exhibited some interesting aspects. A change in leaf conductance upon elevating photon flux density at atmospheric CO_2 levels was observed in wildtype and transgenic plant lines only when relative air humidity exceeded 70%. Furthermore, stomatal conductance increased in αcp-FBP12 plants under atmospheric CO_2 concentrations and 70-80% relative air humidity with rising photon flux density stronger than in wildtype plants. At 60-70% relative humidity the leaf conductances were the same in both plant lines. Therefore, it might be assumed that the stomatal response to air humidity remained unimpaired by FBPase inhibition. But the response to changes in photon flux density seems to be altered and hence to be linked to photosynthesis. Two possible suggestions about the underlying mechanism arise. First: Adaptation to light in guard cells is linked to mesophyll photosynthesis. ABA might be a suitable candidate for such a signal. Preliminary results suggest that ABA levels are reduced in these leaves. Second: Guard cell photosynthesis exerts major control on the light response of stomata (5) and FBPase antisense has changed guard cell photosynthesis. Since the FBPase antisense construct is expressed behind an unspecific promotor (CaMV 35S) the results presented herein do not allow to make a final decision between the two suggestions described above. In general, the results with respect to Calvin cycle activity in guard cells are not unique (6,7).

At an ambient CO_2 concentration of 500 ppm leaf conductances of strongly inhibited FBPase antisense plants showed a pronounced reduction at air humidities of 70–80% compared to atmospheric conditions. Since Calvin cycle activity is genetically limited in these plants assimilation rates were not altered by elevating ambient [CO_2]. At 60–70% relative humidity stomata of FBPase antisense plants displayed the opposite response i.e. an elevation of leaf conductance upon 500 ppm CO_2. It was demonstrated (8) that guard cells of *Vicia faba* showed an elevation of stomatal aperture with increasing [CO_2] if the guard cells were not functionally connected to mesophyll cells. If the assumption is correct that signals triggered by mesophyll photosynthesis are unaltered by elevated [CO_2] in FBPase antisense plants we should conclude that this "exceptional response" is a result of guard cell internal processes at high CO_2 concentrations in combination with low air humidities.

In summary, our results suggest that stomatal response to variations in photon flux density is linked to photosynthetic activity in the mesophyll. Noteworthy, elevated intercellular CO_2 levels have not induced lower leaf conductances in FBPase antisense plants under atmospheric conditions. These results suggest an involvement of photosynthesis in the CO_2 response of stomata. The responses to elevated CO_2 and low air humidity revealed a complicated interplay which should be subject of further investigations.

We thank Dr. J. Koßmann for his generous gift of the FBPase antisense plants. This work was supported by a DFG grant # Fi 571 1-1 to Joachim Fisahn.

References

1   Raschke K. (1975) Stomatal action. Annu. Rev. Plant Phys. 26: 309 - 340
2   Mansfield, T.A., Hetherington, A.M., Atkinson, C.J. (1990) Annu. Rev. Plant Physiol. Plant Mol. Biol. 41: 55 - 75
3   Koßmann, J., Sonnewald, U., Willmitzer, L. (1994) The Plant J. 6: 637 - 650
4   Muschak M., Hoffmann-Benning S., Fuss H., Koßmann J., Willmitzer L., Fisahn J. (1997) Photosynthetica (Prague) 33 (3-4): 455 - 465
5   Talbott L.D., Zeiger E. (1996) Plant Physiol. 111: 1051 – 1057
6   Raschke K., Dittrich P. (1977) Planta 134: 69 - 75
7   Shimazaki K.-I., Terada J., Tanaka K., Kondo N. (1989) Plant Physiol. 90: 1057 - 10643
8   Spence R.D., Sharpe P.J.H., Powell R.D. (1984) New Phytologist 97: 145 - 154

CHLOROPHYLL FLUORESCENCE AS AN INDICATOR OF PHOTOSYNTHETIC BEHAVIOUR OF *IN VITRO* CHESTNUT DURING ACCLIMATIZATION AT HIGH CO_2 AVAILABILITY

L. F. Carvalho,[1] M. L. Osório,[2] S. Amâncio[1]
[1]Dept de Botânica e Eng. Biológica. [2]Dept de Eng. Florestal. Instituto Superior de Agronomia. UTL. 1399 LISBOA. PORTUGAL

Key words: fluorescence, elevated CO_2, photoinhibition, quantum yield, light acclimation

1.Introduction

In vitro propagation takes place in controlled environments regarding physical (temperature and light), hormonal, nutritional and sanitarian conditions. One of the main applications is plant clonal propagation; nevertheless, the survival rate of *in vitro* plantlets when transferred to *in vivo* can be very low. An acclimatization period, defined as the natural process where plants can adjust their physiological processes to the changing man-controlled environmental conditions is normally required. It has been reported that the manipulation of the irradiance regime gives rise to a decrease of the acclimatization period and an increase of survival rates (1). It is also possible to stimulate growth through the increase in the CO_2 concentration (2, 3, 4). In the present study the use of two CO_2 concentrations (350 ppm and 700 ppm) at a light intensity of 300 µmol quanta $m^{-2}.s^{-1}$ during the acclimatization of *in vitro* chestnut was tested. Treatments were compared through growth analysis and fluorescence parameters.

Chlorophyll fluorescence is an indicator of photosynthetic reactions that take place in the chloroplasts of green plants (4) and can be used as an indicator of photoinhibition damage (5, 6). Chlorophyll fluorescence *in vivo* originates primarily from chlorophyll **a** in PSII. Its yield is controlled by factors linked to properties of the photosynthetic apparatus like antenna organization, primary photochemistry and electron transport (7). Therefore, chlorophyll fluorescence yield reflects in the first place the excitation and energy conversion occurring at the PSII level (8). However, due to the functional connection of the PSII with other components of the photosynthetic apparatus, fluorescence yield can be considered as an indicator of the process as a whole.

2. Procedure

2.1. *Plant material and culture conditions*
In vitro plantlets of one hybrid clone of *Castanea sativa* x *C. crenata* were used. Stock shoot multiplication cultures had a four weeks multiplication cycle on Greshoff and Doy

G. Garab (ed.), *Photosynthesis: Mechanisms and Effects, Vol. V,* 3479–3482.

(9). Cultures were kept at 25°C by day and 20°C by night, with 16 h photoperiod at a PPFD of 45±5 μmol.m^{-2}.s^{-1} provided by cool-white fluorescent lamps. After 5 days in *in vitro* rooting induction medium, added of the synthetic auxin IBA at 3 mg.L^{-1}, plantlets were transferred to acclimatization.

2.2. *Acclimatization*

Root expression took place *ex vitro*, during the 6 weeks acclimatization period. This was obtained by placing the induced microcuttings in 6 cm $\varnothing$ pots containing a sterilised mixture of hydrated peat and perlite (1:1, v/v). The pots were placed in a 450L glass chamber (500E, Aralab, Portugal), with control of CO_2 concentration by an IRGA system, light and humidity. During the first two weeks and at night CO_2 concentration was the atmospherical (350 ppm) on both treatments. Thereafter, CO_2 concentratiom was 350 ppm on treatment 1 (control) and 700 ppm on treatment 2. Light at the intensity of 300 μmol quanta m^{-2}.s^{-1} for 16 hours.day^{-1} was provided by 16 fluorescent lamps placed over the transparent top of the chamber. Relative humidity was controlled by an hygrometer, with a sensor placed inside the glass chamber, through an ultrasonic fog system. RH was decreased daily from 98% until ambient RH was reached.

2.3. *Chlorophyll fluorescence*

Chlorophyll fluorescence parameters were measured in freshly-detached leaf disks from *in vitro* plantlets and weekly during acclimatization, from both persistent and newly expanded leaves. Plants were kept for 30 minutes in the dark prior to the disks were cut. Modulated fluorescence was measured with a pulse amplitude modulation system (PAM 101, 103, Walz, Effeltrich, Germany) chlorophyll fluorimeter. The values of maximal variable fluorescence (Fv = Fm - F_0) and the quantitative measure of the photochemical efficiency of PSII (Fv/Fm) were calculated for dark-adapted leaves. Using both light and dark fluorescence parameters it was also possible to calculate the variable fluorescence in light-adapted leaves (F'v = F'm - F'$_0$), PSII excitation capture efficiency in the light-adapted leaves (F'v/F'm) (7), the proportion of PSII reaction centres open, photochemical quenching, q_p=(F'm-Fs)/F'v and ϕe = (F'm - Fs)/F'm (8).

3. Results and Discussion

3.1 *Growth*

Fig. 1 A shows the biomass of plantlets submitted to both CO_2 treatments when elevated CO_2 was applied (day 14 of acclimatization), at day 28 and at the end of the acclimatization phase (day 42). Treatment 2 presents higher biomass in comparison to treatment 1, reaching a difference of 100% on the 42nd day, 48.8 mg on treatment 1 versus 103.01 on treatment 2. This difference in biomass can be explained by the results shown in fig. 1 B: the earlier appearance of new leaves in treatment 2, on day 21 of acclimatization against day 28 in treatment 1; the higher dry weight of new leaves in treatment 2.

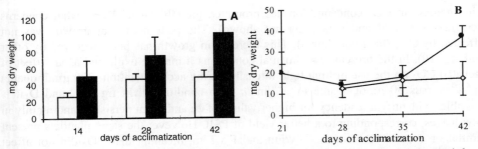

Figure 1. Total biomass (A, treatment 1, white, and treatment 2, gray) and dry weight of the 1st and 2nd new expanded leaves (B, treatment 1, ◊; treatment 2, •), during the acclimatization period.

3.2 *Chlorophyll fluorescence*

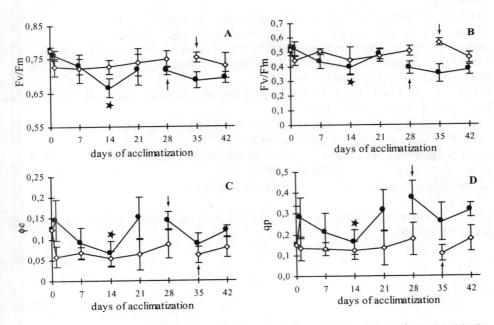

Figure 2. Fv/Fm (A), F'v/F'm (B), φe (C) and qp (D) on treatments 1 (◊) and 2 (•). (→: new leaves; ★: elevated CO_2).

From the chlorophyll fluorescence parameters presented in fig 2 A-D it is possible to verify that both the photochemical efficiency of PSII (Fv/Fm) and the PSII excitation capture efficiency in light adapted leaves (F'v/F'm) present lower values at the end of treatment 2 when compared to treatment 1, (fig. 2 A-B) although the quantum yield of PSII (φe) was higher on treatment 2 (fig 2 C). The values of photosynthetic quenching (qp) were also higher on treatment 2 (fig. 2, D). Apparently PSII yield can be higher on treatment 2.

The increase in CO_2 concentration has promoted growth, especially in what concerns new leaves weight and area (data not shown). The pattern of root growth was not affected by CO_2 (data not shown). This increase in growth has been observed by other authors (10). In the present experimental conditions it may be explained as an effect of elevated CO_2 on the use of high light intensity during acclimatization (300 µmol quanta $m^{-2}.s^{-1}$ versus 50 µmol quanta $m^{-2}.s^{-1}$ in *in vitro* conditions). In fig. 2 C and D it is possible to identify a tendency for higher values of ϕe and qp in persistent and mainly in new leaves, corresponding to a better yield of PSII. However the same plantlets present similar or even lower values of Fv/Fm and F'v/F'm, meaning that CO_2 did not affect neither the maximal capacity for PSII to accept electrons nor the electron flux through it. From the results taken as a whole it is possible to suggest that plantlets acclimatized at higher CO_2 concentration are able to make a better use of the available light, as the consequence of an increase in PSII yield.

References

1 Donnelly, D. J. and Vidaver, W. E. (1984) J. Amer. Soc. Hort. Sci. 109(29)172-176.

2 Desjardins, Y.; Gosselin, A. and Yelle, S. (1987). J. Amer. Soc. Hort. Sci. 112(5):846-851.

3 Lakso, A.; Reisch, B.; Mortensen, J. and Roberts, M. (1986) J. Amer. Soc. Hort. Sci. 111(4):634-638.

4 Schreiber, U.; Endo, T.; Mi, H. & Asada, K. (1995). Plant Cell Physiol. 36(5): 873-882.

5 Krause, G. H. (1988) Physiol. Plant. 74:566-574.

6 Bjorkmann and Demmig-Adams (1994) in Ecophysiology of Photosynthesis (Schulze E. D., Caldwell, M. M., ed.) pp17-47, Springer-Verlag, Berlin.

7 Schreiber, U.; Bilger, W. and Neubauer, C. (1994) in Ecophysiology of Photosynthesis (Schulze E. D., Caldwell, M. M., ed.) pp49-70, Springer-Verlag, Berlin.

8 Genty, B.; Briantais, J-M. and Baker, N. R. (1989). Biochim. Biophys. Acta 990: 87-92.

9 Greshoff, P. M. and Doy, C. H. (1972). Planta. 107:161-170.

10 Usuda, H, and Shimogawara, K. (1998) Plant Cell Physiol. 39(1):1-7.

This work was supported by the Grant PRAXIS XXI BD/3097/96 to L.C. and PRAXIS XXI BIO/1064/95.

STUDIES ON CO_2 EXCHANGE OF SCOTS PINE NEEDLES USING CUVETTE FIELD MEASUREMENTS AND A 3-DIMENSIONAL STOMATAL MODEL INCLUDING GASEOUS PHASE AND LEAF MESOPHYLL

T. Aalto, T. Vesala, T. Mattila, P. Simbierowicz, P. Hari*, E. Juurola*
Dept. of Physics, P.O.Box 9, FIN-00014 University of Helsinki, Finland
*Dept of Forest Ecology, P.O. Box 24, FIN-00014 University of Helsinki, Finland

Key words: modelling, gas diffusion, intercellular air spaces, CO_2 concentration, CO_2 uptake, temperature dependence

1. Introduction

The uptake of CO_2 by plants can be divided into several subprocesses. The molecules move in the gaseous phase through the boundary layer adjacent to the leaf surface and diffuse through the stomatal pore, the substomatal cavity and intercellular air spaces onto the surrounding cell walls. In the surfaces of the mesophyll cells molecules dissolve in the water film and diffuse in the liquid phase into the stroma of the chloroplast. There, after series of complex biochemical reactions, CO_2 or its ionized forms are finally fixated into carbohydrates essential for plant functioning.

A three-dimensional model was constructed for studying the diffusion of CO_2 molecules from air into stomatal air spaces and further into leaf mesophyll cells. The model consists of two different phases, mesophyll and air with appropriate boundary conditions. Sinks for CO_2 molecules were obtained from a well-known biochemical photosynthesis model parametrized for Scots pine ([1],[2]). In addition, the results of the biochemical model combined with simple conductance calculations were compared to results of cuvette field measurements designed for long-time monitoring of CO_2 exchange of Scots pine shoots.

2. Materials and methods

In the diffusion model for CO_2 the stomatal geometry was assessed by constructing a graphical object including air and leaf mesophyll in the vicinity of one stoma. Fig. 1 shows the stomatal construction where the surface of the object represents the air-mesophyll interface. The geometry is quite rough in comparison to the fine structure of the real leaf, but it captures the essential requirement of large amount of cell surface available for CO_2 assimilation.

To solve the diffusion equations in the stomatal geometry boundary conditions must be determined for each limiting surface. Flux across outer boundaries of the system was inhibited, because each stoma was assumed to provide CO_2 only for its own mesophyllic volume and stomata were assumed to function identically. A constant ambient CO_2 concentration was set into top of the boundary layer above the leaf surface. Carbon dioxide was assimilated only by irregular substomatal surfaces (except walls of the capillary tube and the upper part of the sub-stomatal cavity) which are indicated in Fig. 1. The boundary condition for them is $Hc_{g,s} = c_{l,s}$, where $c_{g,s}$ and $c_{l,s}$ refers to gas and mesophyll concentrations at the interface and

3483

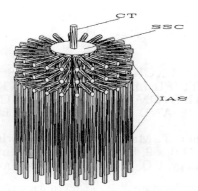

Figure 1: A 3-dimensional description of the stomatal object, i.e. the gaseous volume inside a portion of leaf containing one stoma. CT = capillary tube from the leaf surface, SSC = substomatal chamber (only the upper end is visible) and IAS = intercellular air spaces [5].

H is the Henry's law absorption equilibrium constant. Sinks for CO_2 were distributed continuously throughout the mesophyll. Finally, concentrations and fluxes were solved numerically in a 3-D calculation grid, which was fine in the absorbing boundaries of the air-leaf system where gradients were large and coarse in distant areas of lesser importance.

Cuvette field measurements considered in this study were performed in two boreal measuring stations; Värriö (67°46'N, 29°35'E) in northern Finland and Hyytiälä (61°51'N, 24°17'E) in southern Finland. The measuring systems consisted of trap-like cuvettes, CO_2 and H_2O infrared gas analyzers and supporting controlling equipment ([3],[4]). A pine shoot (150-250 needles) was installed inside the cylindrical cuvette and a fan stirred the air while the lid of the cuvette was closed for a measuring period of 70 s once in every 10-30 minutes. Temperature and PAR (photosynthetically active radiation) were also measured from the cuvette.

3. Results and discussion

The effects of the varying stomatal geometry on the CO_2 flux and intercellular-air-space (IAS) and mesophyll concentrations were examined. The net CO_2 flux was sensitive to the size of the IAS-mesophyll interface when the mesophyll transport coefficent was low equalling the diffusivity of CO_2 in water, while using a 100 times higher diffusion coefficient the length of the intercellular bristles did not practically have any effect on the flux. The simulations also revealed an optimum between the volume of the assimilating mesophyll and the absorbing interface. The highest CO_2 fluxes were produced by a relatively low air-mesophyll interface area. A more irregular interface with larger area would decrease the mesophyll volume that assimilates CO_2 making the exchange rate decrease. With exact description of the chloroplasts as sinks for CO_2 in the vicinity of the cell walls the result might be slightly different leading to higher optimum of the interface area.

The temperature maximum of the CO_2 flux was a couple of degrees lower than the biochemical model parameters would indicate. This shift is due to the temperature and pH dependent solubility of CO_2 molecules into water in the cell surfaces (taken

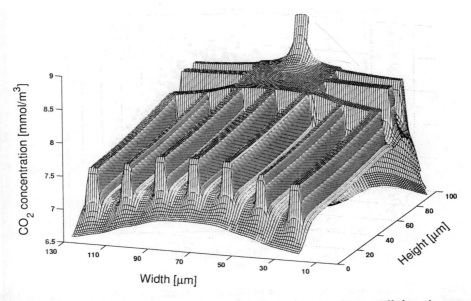

Figure 2: Concentration of CO_2 in substomatal chamber and intercellular air spaces and in mesophyll. $D_l=1.7\times10^{-4}$cm^2/s ($10\times D_{CO_2}$ in water) and $T=20°$C [6].

into account using the factor H in the boundary condition for absorbing surfaces). Increasing solubility from the value indicated by the physical solubility (pH $\sim$ 6) leads to a gradual change of temperature maximum into higher temperatures, but also to unrealistically high CO_2 concentrations in the mesophyll. The role of the solubility obviously deserves more attention since it has a clear effect on the CO_2 flux and concentrations. Typical CO_2 concentrations in substomatal chamber and mesophyll are shown in fig 2. The values are calculated using only the physical solubility. Decrease in concentrations from the cell walls into mesophyll is clearly seen. Lower diffusion coefficient in mesophyll would cause sharper drawdown.

Field measurements of Scots pine shoot CO_2 exchange were compared to results of the biochemical model [1] adjusted for Scots pine grown in southern Finland conditions [2]. The results were generally in a good agreement, the correlation coefficients varied from 0.74 to 0.95. Some discrepancies were also found. The model followed more intensively changes in temperature. This could be seen in northern Finland measurements at low temperatures ($<18°$C). The modelled temperature response indicated low fluxes at low temperatures, but measurements did not show any decrease (Fig. 3). The data points shown in Fig. 3 were chosen so that low irradiance and large water vapor deficit were not limiting. The irradiation response was relatively similar in both measuring sites and according to the model. Cuvette measurements showed slightly smaller quantum yields (0.025-0.035) as a result from shading of the needles. The temperature dependences of the biochemical model parameters J_{max} and $V_{c(max)}$ were re-evaluated from the field measurements. The results for $V_{c(max)}$ agreed well with earlier estimations, while the results for J_{max}

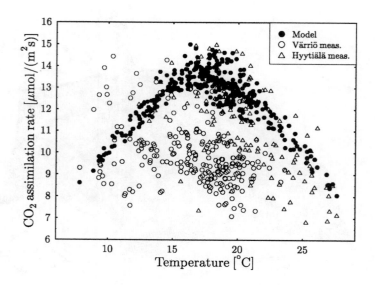

Figure 3: Temperature dependence of the daily maximum CO_2 assimilation rate according to biochemical model and measurements at Värriö and Hyytiälä during July -96 [7].

suggested relatively high values at low temperatures especially in northern Finland indicating adaptations to colder climate. Processing of the measurement results produced also substomatal concentrations of CO_2, which agreed quite well with the model. The daily minimum of substomatal/ambient concentration ratio varied from 0.4 to 0.8.

References

1. Farquhar G.D., von Caemmerer S. and Berry J.A. (1980) Planta, 149,78-90.
2. Wang K.Y. (1996) *Effects of long-term CO_2 and temperature elevation on gas exchange of Scots pine*. PhD thesis, University of Joensuu, Finland.
3. Haataja, J. and Vesala, T. (1997) Station for Measuring Forest Ecosystem - Atmosphere Relations:SMEAR II. University of Helsinki, Department of Forest Ecology Publications 17, Orivesi: Aseman kirjapaino.
4. Ahonen, T., Aalto, P., Rannik, Ü., Kulmala, M., Nilsson ED, Palmroth, S., Ylitalo, H. and Hari, P. (1997) Atm. Env. 31(20), 3351-3362.
5. Aalto, T. (1998). Gas exchange of Scots pine shoots: Stomatal modelling and field measurements. PhD thesis, University of Helsinki, Finland, In press in Rep. Ser. in Aerosol Sci.
6. Aalto, T., Vesala, T., Mattila, T., Simbierowicz, P. and Hari, P. (1998). Submitted to J. of Theor. Biol.
7. Aalto, T. (1998). Submitted to Silva Fennica.

EFFECT OF LOW CO$_2$ GROWTH ON *PSA E* AND *DES 6* cDNAS IN *CHLAMYDOMONAS REINHARDTII*

Aravind Somanchi, Eric R. Handley and James V. Moroney, Department of Biological Sciences, Louisiana State University, Baton Rouge, LA 70803 USA

Key words: *Chlamydomonas reinhardtii*, CO$_2$ concentrating mechanism, *des 6*, *Psa E*.

1. Introduction

A CO$_2$ concentrating mechanism in *Chlamydomonas reinhardtii*, increases the efficiency of photosynthesis. This is seen from the apparent increase in affinity for CO$_2$ when the alga is grown in low CO$_2$ conditions. In *C. reinhardtii* this concentrating mechanism is induced only under low CO$_2$ conditions. This adaptation to low CO$_2$ occurs in concurrence with the synthesis of a number of new proteins. To identify the components involved in this mechanism a cDNA library has been constructed. This library has been differentially screened for clones upregulated under low CO$_2$ conditions (1). Several putative cDNAs such as those encoding carbonic anhydrases, LHCs, and cyclophilins have been scored as upregulated under low CO$_2$ conditions. Here we discuss the identification and the possible roles played by the clones 49 I 10 (*Lci 4*) and 17 I 1 (*Lci 6*).

2. Procedure

2.1 Cell Culture
Wild type cells of *Chlamydomonas reinhardtii* (CC 137), obtained from Togasaki, R. K., Indiana University, USA, were grown in minimal media as described previously (1), bubbled with high CO$_2$ (5% CO$_2$ in air) or low CO$_2$ (0.035% CO$_2$ in air). High CO$_2$ grown cells were switched to low CO$_2$ for adaptation. RNA extracted from these low CO$_2$ adapted cells was used in the construction and screening of the cDNA library and for northern blot analyses.

2.2 Northern blot analyses and cDNA library synthesis
Total RNA extracted from cells adapting to low CO$_2$ was isolated as described previously (1). Ten µg of total RNA was probed with ^{32}P labeled fragments from the appropriate cDNAs. Hybridization and washing were done following standard

G. Garab (ed.), Photosynthesis: Mechanisms and Effects, Vol. V, 3487–3490.
© 1998 Kluwer Academic Publishers. Printed in the Netherlands.

procedures (1). mRNA was extracted from total RNA using Poly AT tract beads (Promega) as described by the manufacturer. The cDNA library was synthesized using five µg of mRNA from low CO_2 adapted cells. This mRNA was reverse transcribed into cDNA and ligated into the Stratagene uni-ZAP II vector following the manufacturer's instructions.

2.3 Sequencing and homology searches
Purified plasmids containing the cDNA inserts were sequenced from the 3' and 5' ends using M13 forward and reverse primers at the automated sequencing facility at Colorado State University. The sequence information was used in homology searches using the BLAST software from NCBI.

3. Results

3.1 Identification of cDNAs upregulated under low CO_2 conditions
A cDNA library, constructed from low CO_2 adapting cells, was screened with labeled cDNAs from high CO_2 grown cells and low CO_2 adapting cells. This differential screening yielded about 425 putative clones upregulated under low CO_2 conditions. A number of these have been sequenced and characterized to fall into different classes (1,2). Differential screening of the remaining clones showed that there are several other clones that show upregulation under low CO_2 conditions. Two such clones, Lci 4 and Lci 6, are characterized in this report.

3.2 Northern blot analyses
Northern hybridization was done using RNA from high CO_2 grown cells and low CO_2 adapting cells. Figure 1 shows that there is a distinct increase in abundance of the transcript for the clones Lci 4, Lci 6, Lhca1*2 and Lhcb1*2 under low CO_2. A clone encoding glyceraldehyde 3 phosphate dehydrogenase (G3PDH), which is constitutively expressed in high CO_2 and low CO_2, was used as control.

3.3 Sequencing and Homology
The primary clone of Lci 4 was about 0.6 kb and the size of the mRNA transcript was about 2.2 kb. This short 600 bp fragment was used to rescreen the cDNA library to pull out longer clones. A database search using the longer sequence of Lci 4 shows high homology to the previously identified C. reinhardtii cDNA des 6 that encodes for an omega 6 desaturase (3) as seen in Fig. 2. The size of the Lci 4 message agrees with the cDNA size reported by Sato et al.(3). From the sequence alignment it is clear that Lci 4 is identical to des 6.

The sequence of Lci 6 shows that is 623 nucleotides long and has an open reading frame of 120 amino acids (Fig. 3). Database searches showed that this cDNA is identical to the C. reinhardtii Psa E protein previously reported (4).

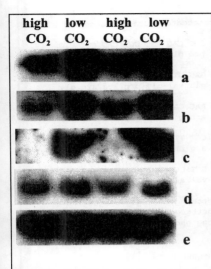

high	low	high	low
CO₂	CO₂	CO₂	CO₂

Fig.1. Northern blot analysis using total RNA from high CO₂ grown cells and low CO₂ adapted cells probed with *Lci 4* (a), *Lci 6* (b), *Lhca1*2* (c), *Lhcb1*2* (d) and G3PDH (e)

4. Discussion

We have discovered that the transcripts of the cDNAs *des 6* and *Psa E* show increase in abundance in response to low CO_2. PsaE is a peripheral photosystem I subunit. It has been shown to play a role in light harvesting and transfer of excitation energy to PSI. Extrapolating from the cyanobacterial model, it may be speculated that PsaE plays a role in cyclic electron transport around PSI for the production of ATP which might be used for the CO_2 concentrating mechanism. The enzyme omega 6 desaturase introduces unsaturation in membrane lipids. This increases the fluidity of the membrane. It is interesting to note that some of the proteins that are upregulated by low CO_2 are membrane proteins such as LHCs and LIP-36. In *Synechocystis,* omega 6 desaturase was shown to play an important role in the assembly of the D1 polypeptide after photodamage (5). The function of *des 6* in *C. reinhardtii* may be similar. Mutants of *C. reinhardtii* with an inactive omega 6 desaturase show poor photosynthetic rates (6) suggesting that membrane unsaturation is important for photosynthesis. This may play a direct role enabling the newly induced proteins to associate with the membrane or

```
des 6        TGCCACGTGTACGACGAGAAGGTCAACTACAAGCCCTTTGACTACAAGAAGGAG
             :::::::::::::::::::::::::::::::::::::::::::::::::::::::
Lci 4        TGCCACGTGTACGACGAGAAGGTCAACTACAAGCCCTTTGACTACAAGAAGGAG

des 6        GAGGCCCTGTTTGCCGTGCAGCGCCGCGTCCTGCCCGACTCCGCCGCCTTCTAAACTAGC
             :::::::::::::::::::::::::::::::::::::::::::::::::::::::::::::::
Lci 4        GAGGCCCTGTTTGCCGTGCAGCGCCGCGTCCTGCCCGACTCCGCCGCCTTCTAAACTAGC

des 6        TGGCTACAACTGGCTATCGCGCTAGTGAGCTAATCGGAGCTGAGCAAGCTTCTCGGAGCC
             :::::::::::::::::::::::::::::::::::::::::::::::::::::::::::::::
Lci 4        TGGCTACAACTGGCTATCGCGCTAGTGAGCTAATCGGAGCTGAGCAAGCTTCTCGGAGCC

des 6        TTGCGAGAGCTTCTGATGTTGGCTTTATAGGGCTGGCTAAAACAGCTAGCAGCTAGGTAG
             :::::::::::::::::::::::::::::::::::::::::::::::::::::::::::::::
Lci 4        TTGCGAGAGCTTCTGATGTTGGCTTTATAGGGCTGGCTAAAACAGCTAGCAGCTAGGTAG

des 6        TGTGGCTGCGCCGGCCGCTTGCGCGGCCGCGCGAGAGAGGGAGGGGCGTCAGGCATGTGC
             :::::::::::::::::::::::::::::::::::::::::::::::::::::::::::::::
Lci 4        TGTGGCTGCGCCGGCCGCTTGCGCGGCCGCGCGAGAGAGGGAGGGGCGTCAGGCATGTGC
```

Fig.2. Sequence alignment of *Lci 4* with the *Chlamydomonas reinhardtii des 6* gene.

```
                                gtgacacttcgcctttcatctccccaccaaaaccggaaa

ATG CAG GCC CTG TCG TCT CGC GTG AAC ATC GCG GCC AAG CCC CAG CGC GCT CAG CGC CTG
Met gln ala leu ser ser arg val asn ile ala ala lys pro gln arg ala gln arg leu

GTG GTC CGC GCC GAG GAG GTT AAG GCT GCC CCC AAG AAG GAG GTC GGC CCC AAG CGC GGC
val val arg ala glu glu val lys ala ala pro lys lys glu val gly pro lys arg gly

TCG CTG GTG AAG ATC CTG CGC CCC GAG TCC TAC TGG TTC AAC CAG GTC GGC AAG GTC GTC
ser leu val lys ile leu arg pro glu ser tyr trp phe asn gln val gly lys val val

TCC GTC GAC CAG AGC GGC GTC CGC TAC CCC GTC GTT GTC CGT TTC GAG AAC CAG AAC TAC
ser val asp gln ser gly val arg tyr pro val val val arg phe glu asn gln asn tyr

GCT GGT GTC ACG ACG AAC AAC TAC GCT CTG GAT GAG GTT GTT GCC GCC AAG TAA
ala gly val thr thr asn asn tyr ala leu asp glu val val ala ala lys Stop

atggctggcttcgcgcagcatcacggaatgccgagaggagcggcaagtgtcgttacgggacctagagaagagaagatt
agcgggctggccttagggcaagtcgcgattttgtgagctagttttgggcgcgcttgcgcgaccccagccaagcataga
cttggcgatcggacacatacaggtgtcggacactgaccatgagctaaggagcggtgacgagggagcaggcgggtgtaa
tactaaatgctgaaaaaaaaaaaaaaa
```

Fig. 3. Sequence of *Lci 6* along with the deduced amino acid sequence

some secondary role. To characterize the specific roles played by *Psa E* and *des 6* in *C. reinhardtii* cells during low CO_2 adaptations requires further investigation.

5. Acknowledgments

This work was funded by the National Science Foundation grant IBN-9632087.

References

1. Burow M.D., Chen Z.-Y., Mouton T.M. and Moroney J.V. (1996). Plant Mol. Biol. 31: 443-448.
2. Chen Z.-Y., Lavigne L.L., Mason C.B. and Moroney J.V. (1997). Plant Physiol. 114: 265-273.
3. Sato N., Fujiwara S., Kawaguchi A. and Tsuzuki M. (1998). J. Biochem (Tokyo) 122: 1224-1232.
4. Franzen L.-G., Rochaix J.-D. and von Heijne G. (1989) FEBS Lett. 260: 165-168.
5. Kanervo E., Tasaka Y., Murata N. and Aro E.M. (1997). Plant Physiol. 114: 841-849.
6. Sato N., Tsuzuki M., Matsuda Y., Ehara T., Osafune T. and Kawaguchi A. (1995). J. Biochem. 230: 987-993.

EFFECT OF ELEVATED AIR CO_2 CONCENTRATION ON THE CO_2 ASSIMILATION OF WINTER WHEAT

N. Harnos[1], K. Szente[2], Z. Tuba[2]

[1]Agricultural Research Institute of the Hungarian Academy of Sciences, Martonvásár, Hungary;
[2]Dept. of Botany and Plant Physiology, University of Agricultural Sciences, Gödöllõ, Hungary

Key words: acclimation, rubisco capacity, modelling, CO_2 assimilation

1. Introduction

Increasing number of experiments have been carried out since the seventies to analyse the effect of elevated air CO_2 concentration on wheat (1, 2, 3). It was found that the photosynthetic rate of plants grown at higher CO_2 concentrations was more intense, while the transpiration rate decreased compared to plants grown in our present day CO_2 level, leading to a better water use efficiency (2), a greater amount of biomass and, in many cases, to higher yields (4). Although elevated atmospheric CO_2 increases the photosynthesis and growth of C_3 plants, the magnitude of this increase is very difficult to predict. Simulation models would be able to describe the effects of high CO_2 on photosynthesis, but the contradictory results obtained in a vast number of studies nevertheless raises the question of how photosynthesis can be predicted without having to carry out the measurements.

In the present work leaf net CO_2 assimilation was simulated according to the biochemical model described by Farquhar et al. (5) and Farquhar and von Caemmerer (6) and compared with measured data obtained from our winter wheat experiment under present and elevated CO_2 levels. This model has a small number of parameters which can be obtained from CO_2 gas exchange measurements.

2. Procedure

2.1 Growth conditions

The experiment was carried out at the Global Climate Change and Plant Research Station in Gödöllõ, Hungary. The climate of the region is temperate-continental with hot dry summers, the annual mean temperature is 11 °C and the annual precipitation is 500 mm. The soil is a light moderately calcareous sand soil. NPK fertilisers were applied at rates of 100, 50 an 50 kg ha^{-1} before sowing. The seeds of a Hungarian winter wheat variety Emma were sown on 10 October. The air CO_2 concentration in the elevated (700 µmol mol^{-1}) and present day (350 µmol mol^{-1}) open top chambers were maintained

G. Garab (ed.), Photosynthesis: Mechanisms and Effects, Vol. V, 3491–3494.
© *1998 Kluwer Academic Publishers. Printed in the Netherlands.*

as described elsewhere (2). The plants were occasionally irrigated. For gas exchange measurements the flag leaves of the plants were used at the beginning of flowering (16 May).

2.2 Gas exchange measurements

Intercellular CO_2 dependence of light saturated net CO_2 assimilation rates (A/C_i curves) was measured on the flag leaves using an LCA2-type IRGA system (ADC Co. Ltd., Hoddesdon, U.K.), operated in differential mode and a Parkinson LC-N leaf chamber with an air flow of 300 mL min^{-1}. Ambient CO_2 concentrations of 30, 100, 200, 330, 540, 730 and 900 μmol mol^{-1} were produced by a gas diluter (GD 600, ADC Co. Ltd., Hoddesdon). Photosynthetic active radiation (PAR) and leaf temperature were kept constant at 1000 μmol m^{-2} s^{-1} and 20 ± 0.5 °C, respectively (2). Gas exchange parameters were calculated according to von Caemmerer and Farquhar (7).

2.3 Modelling approach

The response of the net leaf CO_2 assimilation rate (A) to intercellular CO_2 concentration (C_i) shows two distinct regions predicted by the theory of Farquhar et al. (5): one in which A responds almost linearly to increasing C_i, where A is limited by ribulose-1,5-biphosphate carboxylase/oxygenase (Rubisco) capacity (activity and/or amount). The dependence of A on C_i is given by

$$A = V_{max} \frac{C_i - \Gamma *}{C_i + K_c \cdot (1 + \frac{O}{K_o})} - R_d \qquad [1]$$

where V_{max} is the Rubisco activity at saturating CO_2 concentration, K_c and K_o are the Michaelis constants for CO_2 and O_2, respectively, O is the O_2 concentration, $\Gamma *$ is the CO_2 compensation point and R_d is dark respiration. In the second region A responds less rapidly to increasing C_i, because the rate is limited by electron transport (J) and by the regeneration of ribulose-1,5-biphosphate (RuBP). In this phase

$$A = \frac{J(C_i - \Gamma *)}{4.5C_i + 10.5\Gamma *} - R_d \qquad [2]$$

It is assumed (8) that CO_2 assimilation will be the minimum of either Eq. [1] or [2], whichever is more limiting. Equations [1] and [2] were used for analysing the gas exchange data and biochemical parameters. The parallel measured environmental parameters (temperature, photosynthetically active radiation, C_a) were also inserted in the model. Other parameters needed for the model were calculated as described in (5,6,8). The Rubisco content was estimated from the modelling results.

3. Results and Discussion

The observed values of net leaf CO_2 assimilation rate (A) differed for winter wheat grown at current and high CO_2 concentrations (Fig. 1). The difference in CO_2 assimilation rate between the treated (high CO_2) and control (present day CO_2) plants' leaves was small in the initial slope of the A/C_i curves (C_i<350 μmol mol⁻¹). The net leaf CO_2 assimilation rate of plants grown at high CO_2 increased almost linearly until C_i reached 500 μmol mol⁻¹. For higher values of C_i the CO_2 assimilation rate remained constant for both the treated and control plants, in agreement with earlier results (2,9). The values of saturated net CO_2 assimilation rate were approximately 12.5 μmol m⁻²s⁻¹ for plants grown at the present CO_2 level and 20 μmol m⁻²s⁻¹ for plants grown at 700 μmol mol⁻¹.

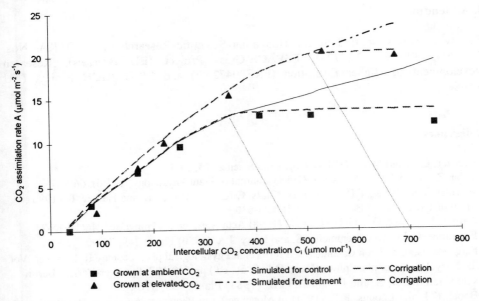

Figure 1. Measured and simulated responses of light-saturated rates of net flag leaf CO_2 assimilation rate (A) to intercellular CO_2 concentrations (C_i) for winter wheat grown at present (350 μmol mol⁻¹) and at elevated (700 μmol mol⁻¹) CO_2 concentrations. Simulated values were calculated from equations [1] and [2] (see before: 2.3 Modelling approach) using observed environmental parameters. The dotted lines interconnecting axis X and A/C_i curves indicate the supply function, i.e. the C_i obtained for a given C_a with varying A. The broken lines illustrate simulated values of A with incorporated RuBP regeneration.

It is not yet known, whether acclimation to long-term elevated CO_2 causes higher or lower Rubisco capacity in the leaves (9). Our results show a small but statistically not significant increase in Rubisco capacity calculated by the model (Eq. 1) in response to long-term elevated air CO_2 exposure, supporting the findings of some earlier reports (10, 11) but contradicting others (12).

Comparing the simulated and observed values for net CO_2 assimilation rate it can be seen (Fig. 1) that the model used here (Eqs. 1 and 2) simulates the net CO_2 assimilation rate response to C_i and acclimation to long-term high CO_2 exposure reasonable until it reaches the second phase, i.e. the RuBP regeneration phase, without any change in the parameters involved in Rubisco activity. When A reaches the second phase, these model parameters should be changed. The question was arising is how the exact C_i value and Rubisco activity can be predicted when CO_2 assimilation becomes RuBP limited?

4. Addendum

This work was supported by the Hungarian Scientific Research Fund (OTKA, Nos. F020704, T18595, T22723), ESPACE-Grass Project (EU Brussels), Research Development of Higher Education (FKFP0472/97) and MEGARICH Project (EU Brussels).

References

1 Krenzer, E.G. and Moss, D.N. (1975): Crop Science, 15, 71-74
2 Tuba, Z., Szente, K. and Koch, J. (1994): Journal of Plant Physiology, 144(6), 669-678
3 Wheeler, T.R., Hong, T.D., Ellis, R.H., Batts, G.R., Morison, J.I.L. and Hadley, P. (1996): Journal of Experimental Botany, 47(298), 623-630
4 Harnos, N., Veisz, O. and Tischner, T. (1998) Acta Agronomica Hungarica, 46(1), 15-24
5 Farquhar, G.D., von Caemmerer, S. and Berry, J.A. (1980) Planta 149, 78-90
6 Farquhar, G.D. and von Caemmerer, S. (1982) in Physiological plant ecology II. New Ser. Vol. 12B. Encyclopedia of plant phys. (Lang O.L. et al., ed.) pp.549-587, Springer-Verlag, Berlin
7 von Caemmerer, S. and Farquhar, G.D. (1981) Planta 153, 376-387
8 Boote, K.J. and Loomis, R.S. (1991) in Modeling Crop Photosynthesis -from Biochemistry to Canopy. CSSA Special Publication no. 19. (Boote, K.J. and Loomis, R.S., Eds) pp. 109-140
9 Long, S.P. and Drake, B.G. (1992) in Crop Photosynthesis: spatial and temporal determinants (Baker N.R., Thomas H., eds) pp.69-103, Elseviel Sci. Publishers B.V.
10 Campbell, W.J., Allen Jr., L.H. and Bowes, G. (1990) J. Exp. Bot. 41, 427-433
11 Vu J.C.V, Allen Jr., L.H. and Bowes, G. (1989) Environ. Exp. Bot. 29, 141-147
12 Rowland-Bamford, A.J., Baker, J.T., Hartwell, A. and Bowes, G. (1991) Plant Cell Environ. 14, 557-587

ON IMPORTANCE OF PHYSICAL PHENOMENA IN THE TEMPERATURE DEPENDENCE OF PHOTOSYNTHESIS -A SENSITIVITY ANALYSIS

E. Juurola[1], T. Vesala[2], T. Aalto[2], T. Markkanen[2], P. Hari[1]
[1] Department of Forest Ecology, University of Helsinki, Finland
[2] Department of Physics, University of Helsinki, Finland

Key words: CO_2 uptake, gas diffusion, dissolution, modelling, stomata

1. Introduction

Photosynthesis is a temperature-dependent process. On the whole, it is determined by biochemistry together with a stomatal component. In ecological and plant physiological research the temperature dependence is studied mainly according to detailed biochemical modelling with varying or constant conductance (1). In this kind of approach inherent physical processes, namely transport phenomena, have only been implicitly included in concepts of conductance and intercellular partial pressure of CO_2. However, transport is a highly temperature dependent phenomenon and each component process, transfer in the air, absorption on air-mesophyll interface and transport in the mesophyll, has its own temperature response. Thus in order to understand the temperature dependence of photosynthesis thoroughly, it is necessary to distinguish the solely physical processes.

In a stoma, like in most transfer systems, there is an interface with concentration gradients on both sides. CO_2 molecules are transported down the gradient through the laminar boundary layer, stomatal pore, substomatal cavity and intercellular air spaces (IAS) to the surface of a mesophyll cell and further across cell walls through the extracellular water to a site of carboxylation. The temperature dependence of mass transport arises from those of diffusion coefficients and absorption equilibrium concentration. In the gaseous part of the route (from the boundary layer to the mesophyll surface) the transport is governed by the binary diffusion in a carrier gas (air), the temperature dependence of which is well-known (2). In the liquid part of the route (mesophyll), the actual transport mechanism is not well known and still remains under discussion (3). However, the temperature dependence of this transport cannot differ much from that in the pure liquid phase, and can be reasonably estimated from the hydrodynamical or Eyring theories (4). When surface-active species are not present on the mesophyll surface, the interface resistance is negligible and the gas and liquid CO_2 concentrations adjacent to the air-mesophyll surface lie on the equilibrium curve. The equilibrium is expressed in terms of Henry's law coefficient, the temperature dependence of which is well-known (5).

Introducing physical phenomena into existing biochemical models is not possible since usually they are implicitly incorporated into the models in the values of different parameters. Thus, for the sensitivity analysis we have utilised a three dimensional diffusion model which is ideal for investigating the physico-biochemical phenomena separately. The aim of the study was to identify the physcial processes involved in the temperature

G. Garab (ed.), Photosynthesis: Mechanisms and Effects, Vol. V, 3495–3498.

dependence of CO_2 exchange and to clarify their role in CO_2 exchange of non-interacting stomata on leaf scale. With this kind of approach we wanted to emphasise the importance of physical phenomena in CO_2 exchange, which has often been omitted in physiological research. Thus, the model used here is not intended to be a complete model of photosynthesis.

2. Model Description

2.1 Overall Structure

The above-mentioned physical processes are incorporated into a three-dimensional cylindrically symmetric model that offers a rigorous way to investigate the roles of physics and geometrical structure in the stomatal gas exchange (detailed description in (6)). The transport phenomena are described by ordinary diffusion equations, the use of which is emphasised by Parkhurst (7), rather than by a sequential, one-dimensional resistance model. The model includes some idealisations: IAS are not explicitly included, biochemistry is described by the homogeneous first-order reaction (i.e. CO_2 uptake is limited by low CO_2 concentration) and transpiration feedback on the stomatal pore aperture is not included.

2.2. Temperature-dependent Variables

The temperature dependence of the binary diffusion coefficient of CO_2 in the air is estimated according to (2). The temperature dependence of the binary diffusion coefficient of CO_2 in the mesophyll is estimated according to (4). The viscosity of a liquid decreases with increasing temperature and thus the temperature dependence of the diffusion coefficient of CO_2 in the mesophyll is stronger than first power. For the realistic functioning of the model, the sub-stomatal and mesophyllic CO_2 concentration at light saturation should be significantly lower than the ambient value but not near zero (6). To achieve this, a diffusion coefficient about ten times greater was used as in (6). The use of greater diffusion coefficient is assumed to be justified also by the fact that the intercellular air spaces are omitted in the model.

The equilibrium between gaseous and dissolved CO_2 can be expressed by Henry's law coefficient (8). The temperature dependence of the coefficient is given by (5). In general, Henry's law coefficients increase in value as temperature decreases, reflecting greater solubility at lower temperatures. The values used here reflect only the physical solubility of CO_2 in pure water at the equilibrium pH 5.6 regardless of the subsequent fate of the dissolved gas.

Dark respiration, determined experimentally earlier, followed Q_{10}-rule ($Q_{10} = 2.2$). Photorespiration (determined by the CO_2 compensation concentration in the absence of dark respiration) was assumed to follow the temperature dependence of (9) in liquid phase. Other biochemistry of photosynthesis is considered here in a simplified form in terms of the overall reaction rate constant, following Q_{10}-rule ($Q_{10} = 2$). Possible direct inhibition of photosynthesis at high temperatures is omitted here. As a result, the temperature optimum is not reached. This does not, however, mask the effects of physical processes.

Light intensity was at saturating level (850 $\mu mol\ m^{-2}\ s^{-1}$) and the ambient CO_2 concentration was 350 $\mu mol\ mol^{-1}$.

3. Results and Discussion

Sensitivity of modelled CO_2 exchange to changes in the temperature-dependent parameters was tested by applying the parameter value at 20 °C for the whole temperature range, one parameter at a time (Figure 1).

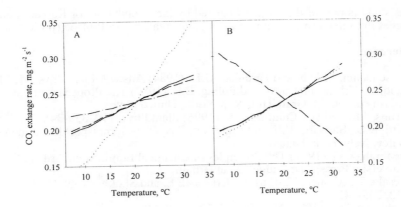

Figure 1. The sensitivity of the model (——) to changes in six temperature-dependent variables. (A) The physical variables: diffusion coefficient of CO_2 in air (– – –), diffusion coefficient of CO_2 in the mesophyll (— · —) and Henry's law coefficient (-----), and (B) the biochemical variables: the parameter for overall biochemistry (– – –), CO_2 compensation concentration (— · —) and dark respiration (-----), all applied to the model as constants determined at 20 °C.

Of the physical parameters, the dissolution of CO_2 in water was clearly most important factor controlling the modelled temperature dependence. The dissolution of CO_2 in water, as well as the other physiological parameters, has a clear physical background and well-defined temperature behaviour. Hence, this result emphasises the importance of dissolution on plant physiology. This, in turn, might affect e.g. the temperature optimum of CO_2 exchange (see also Aalto et al. this book). The diffusion of CO_2 in the air had virtually no effect on the modelled temperature dependence. Furthermore, the diffusion of CO_2 in the mesophyll only slightly affected the modelled temperature dependence (Figure 1A).

Of the biochemical parameters, the parameter for the overall biochemistry had the strongest effect on the modelled CO_2 exchange, as was expected (Figure 1B). This does not, however, affect the results shown in Figure 1A, and it is exchangeable to a parameter with different temperature behaviour (e.g. modified Farquhar parametrisation (1)). Compared to this, the dark respiration and the CO_2 compensation concentration only slightly affected the modelled CO_2 exchange.

The physical processes involved in the temperature dependence of photosynthesis are, for the most part, similar in every plant species; but the biochemistry varies from species to species. The parameter values most often used in biochemical models, however, do not describe the biochemical phenomena as such, but implicitly include other phenomena, namely physical, as well. Therefore, it is essential to construct a model that merges the temperature behaviour of the biochemical processes as well as the physical phenomena. Furthermore, the mechanisms for the diffusion of CO_2 in the mesophyll and also the role of pH in the dissolution of CO_2 in water are still unclear and need to be considered in more detail.

Acknowledgements

This study is part of the project financed by the Foundation for Research on Natural Resources in Finland (project no. 1521/98) and Academy of Finland (project no. 33687).

References

1 Kirschbaum, M.U.F. and Farquhar, G.D. (1984) Austr. J. Plant Phys. 11, 519-538
2 Reid, R.C., Prausnitz, J.M. and Poling, B.E. (1987) The Properties of gases and liquids. Fourth edition, McGraw-Hill, New York., United States
3 Evans, J.R. and von Cammerer, S. (1996) Plant Phys. 110, 339-346
4 Bird, R.B., Stewart, W.E. and Lightfoot, E.N. (1960) Transport Phenomena. John Wiley, New York, United States
5 Denbigh, K.G. (1971) The Principles of Chemical Equilibrium. 3rd ed. Cambridge University Press, Cambridge, Great Britain
6 Vesala, T., Ahonen, T., Hari, P., Krissinel, E. and Shokhirev, N. (1996) New Phytol. 132, 235-245
7 Parkhurst, D.F. (1994) New Phytol. 126, 449-479
8 Seinfeld, J.H. (1986) Atmospheric Chemistry and Physics of Air Pollution. John Wiley, New York, United States
9 Brooks, A. and Farquhar, G.D. (1985) Planta 165, 397-406

STOMATAL REGULATION IN TRANSGENIC TOBACCO PLANTS WITH REDUCED LEVELS OF b6f COMPLEX

Katharina Siebke*, Murray Badger, Dean Price, Susanne von Caemmerer
*Institut für Botanik, Schloßgarten 3, D-48149 Münster, Germany
Research School of Biological Sciences, Australian National University
GPO Box 475, Canberra , ACT 2601, Australia

Key words:
blue light, CO_2 concentration, electron transport, modelling, red light

1. Introduction

The control of stomatal aperture is multi-factorial. The signal perception and transduction chains within the guard cells are still unknown. In transgenic tobacco plants (*Nicotiana tobaccum*) with a reduced amount of b6f complexes the capacity of the linear electron transport rate is decreased. Because of it's role in the control of stomatal opening, it should be expected, that the transgenic plants have a lower stomatal opening. Different signals were given to induce short term opening and responses were examined. Although the aim of the experiments was primarily to show a different response of wildtype and B6F plants to opening signals, which involve the linear electron transport chain, we also want to draw your attention to the differences in the kinetic response of stomata depending on the quality of the triggering signal regardless of the plants.

2. Plant Material and Methods

T2 seeds from the B6F antisense line B6F2.2-513 [1], were germinated on soil and fertilised with a complete nutrient solution. Plants were grown in a growth cabinet at a photon flux density of 120-150 mmol m^{-2} s^{-1} and a daily light period of 20 hours at a temperature of 25 °C/20 °C day night. Plants were selected on the basis of chlorophyll fluorescence characteristics and repotted into 5 L pots containing sterilised garden soil. 10 to 14 weeks post-germination upper fully expanded leaves were used for the experiments.

The upper fully expanded leaves were placed into a dual channel rapid gas exchange system and brought to steady state photosynthesis. To the first set of plants 200 µmol m^{-2} s^{-1} of white light was given until the steady state of photosynthesis was reached. The

3499

G. Garab (ed.), Photosynthesis: Mechanisms and Effects, Vol. V, 3499–3502.

CO_2 partial pressure in air was adjusted to 21 ± 1 Pa intercellular CO_2 concentration and the H_2O partial pressure was 1.9 kPa. The stomatal reaction was recorded after, (a) a 5 min pulse of increased white light (770 μmol m^{-2} s^{-1}), (b) a 5 min pulse of low CO_2 concentration (6.3 ± 0.5 Pa). The second set of plants obtained 150 μmol m^{-2} s^{-1} red light (red Hansatech LED, peak at 610 nm) to reach the steady state of photosynthesis. The stomatal reaction was subsequently recorded after, (c) a 5 min pulse of increased red light (625 μmol m^{-2} s^{-1} Schott RG 610 filter with halogen lamp), (d) a 5 min pulse of blue light (175 μmol m^{-2} s^{-1}, B51 Balzers plus 9782 Corning filters with halogen lamp) replacing the red light with a constant assimilation rate, and, (e) a 5 min pulse of low CO_2 concentration (*see* Abb.1). The difference in the steady state and the peak value of stomatal conductance was assessed. During the 5 min pulse of increased white or red light the external CO_2 concentration was slightly increased to compensate for the expected drop in the intercellular CO_2 concentration due to increased photosynthesis. The values of signals were chosen to result in a similar slope during the 5 min period, but the steady state, which was achieved if the signal was given continuously, was very different between the different treatments. The steady state increase of stomatal conductance was in low CO_2: 110, with increased white light: 65, with increased red light: 45, and with the change to blue light: 35 mmol m^{-2} s^{-1} in a wild type plant.

3. Results and Discussion

There was a slightly lower response to a pulse of increased white or red light in transgenic plants in comparison to wild type plants (Table 1). This should be expected, if the linear electron transport is involved in the stomatal opening reaction. There was no difference in the response to a blue light or low CO_2 pulse between wildtype and trangenic plants supporting the idea, that the blue light response and the low CO_2 response is independent from the contribution of the linear electron transport within the chloroplasts [*for a review see*: 2].

Fig. 1 shows examples of transients recorded during a short term disturbance of B6F (C, D) and wildtype plants (A, B). The examples were chosen to show the whole variety of responses. The responses to a pulse of increased red or white light are very similar to each other and lay in contrast to the responses to a blue light or a low CO_2 pulse. They are faster and have a smaller lag phase. Especially the closing reaction occurs far earlier. In red light sometimes a short oscillation was observed (see Fig. 1,D), the stomates seemed to be less stable in red than in white light. The opening reaction to blue light and low CO_2 was often very similar, but the closing reaction was usually slower after the low CO_2 pulse.

A model similar to the one developed by Kirschbaum et al. [3] was used to simulate the obtained transients. It contains four mangnitudes: disturbance, biological signal, osmotic potential and stomatal conductance, which change proportianal to the difference of their actual state and the state of their prevailing magnitude (not shown). The observed faster response to red and white light in comparison to the response to low CO_2 or blue light can be modeled by coupling the biochemical signal directly to the disturbance, what is equivalent to a shortened signal transduction chain. The difference in the closing reaction after the blue light in comparison to the low CO_2 pulse can be explained with the fact that the low CO_2 signal was stronger in that respect, that a higher opening

Table 1

Stomata reaction to a 5 min pulse of changed environmental conditions as described *in Material and Methods.*

The external CO_2 concentration was chosen different for B6F and WT plants to compensate for different assimilation rates and to achieve the same intercellular CO_2 concentrations. Values are values±standard deviation. Significance (T-test), probability of error: * $\leq 10\%$, ** $\leq 1\%$, *** $\leq 0.1\%$.

first set of plants		WT (3 plants)	B6F (9 plants)	significance
qP at 200 μmol m^{-2} s^{-1} white light		0.86±0.03	0.44±0.23	***
Assimilation at steady state (μmol m^{-2} s^{-1})		5.8±0.5	2.9±1.8	**
Conductance at steady state (mmol m^{-2} s^{-1})		70±25	65±21	n.s.
Peak Increase of Conductance	a) white light increased	10.5±1.8	6.5±2.8	*
(mmol m^{-2} s^{-1}) after a change fo 5 min	b) c_i decreased	13.2±3.6	14.8±3.6	n.s.
second set of plants		(4 plants)	(6 plants)	
qP at 150 μmol m^{-2} s^{-1} red light		0.89±0.02	0.71±0.13	*
Assimilation at steady state (μmol m^{-2} s^{-1})		4.0±0.6	3.2±0.8	*
Conductance at steady state (mmol m^{-2} s^{-1})		50±16	40±14	n.s.
Peak Increase of Conductance	c) red light increased	11±2	8.5±2.1	*
(mmol m^{-2} s^{-1}) after a change	d) red light exchanged with blue light	13.4±2.9	13.0±3.5	n.s.
for 5 min	e) c_i decreased	15.6±2	13.1±3.6	n.s.

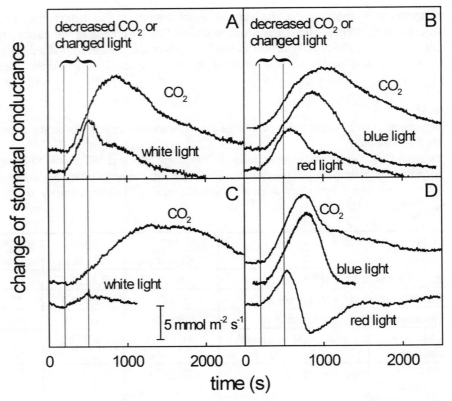

Fig. 1 Examples for the change of stomatal conductance to CO_2 in wildtype (A, B) and trangenic (C, D) tobacco plants after a 5 min pulse of changed environmental conditions as described *in Material and Methods*.

status would be achieved in a continuous change. A closing reaction with two phases, as observed in Fig. 1 D after the disturbance with low CO_2 can be achieved, if a lag phase is introduced into the model. An undershooting oscillation as observed in Fig. 1 D after the increased red light pulse can only be obtained with feed back reactions in addition to the lag phase.

References
1 Price GD, Yu J-W, von Caemmerer S, Evans JR,Chow WS, Anderson JM, Hurry V, Badger MR (1995) Australian Journal of Plant Physiology 22, 285-297
2 Assmann SM,(1993) Annu. Rev. Cell. Biol. 9, 345-75
3 Kirschbaum MUF, Gross LJ, Pearcy RW (1988) Plant, Cell and Environment 11, 111-121

18. Metabolic pathways

CONGRESS ON PHOTOSYNTHESIS

XIth INTERNATIONAL

BUDAPEST ✴ 1998

REGULATION OF PLANT METABOLISM BY PROTEIN PHOSPHORYLATION. Possible Regulation of Sucrose-Phosphate Synthase by 14-3-3 Proteins.

Huber, S.C., Toroser, D., Winter, H., Athwal, G.S. and Huber, J.L.
USDA Plant Science Research, North Carolina State University, Raleigh, NC 27695-7631 (U.S.A.)

Key words: C-metabolism, enzymes, N-metabolism, phosphorylation/dephosphorylation, protein kinase, sucrose.

1. Introduction

Reversible protein phosphorylation is known to be an important regulatory mechanism for several plant enzymes involved in primary carbon and nitrogen metabolism (1,2). Our laboratory has been studying two enzymes of sucrose metabolism, sucrose-phosphate synthase (SPS) and sucrose synthase (SuSy), and an enzyme of N-metabolism, NADH:nitrate reductase (NR), which are known to be phosphorylated in vivo. Interestingly, the effect of phosphorylation is somewhat different for the three enzymes. With SuSy, phosphorylation of the single, major phosphorylation site (Ser-15; ref. 3) has little effect on activity but appears to be one of the factors that controls its association with the plasma membrane (4) and actin cytoskeleton (5). Thus, phosphorylation affects intracellular localization of the enzyme. With NR, phosphorylation of Ser-543 has no direct effect on activity but completes the motif required for binding of a 14-3-3 inhibitor protein (6,7). Thus, phosphorylation affects enzyme activity indirectly by affecting protein:protein interactions. With SPS, phosphorylation appears to directly affect enzyme activity, at least in some cases. Current work suggests that SPS may have three regulatory phosphorylation sites: i) Ser-158, which is largely (or solely) responsible for light/dark modulation (8); ii) Ser-424, which is thought to be responsible for osmotic stress activation of SPS (9), and iii) Ser-229, which may constitute a binding site for a 14-3-3 inhibitor protein (10).

The possible regulation of SPS by 14-3-3 proteins is still rather preliminary. Previous work in our laboratory (10) has established that the two proteins can be associated based on co-elution during gel filtration and co-immunoprecipitation. Moreover, the content of 14-3-3 proteins associated with SPS was inversely related to SPS activity suggesting that 14-3-3s function to inhibit activity. The 14-3-3-binding site on SPS has not been unequivocally established, but available evidence is consistent with involvement of Ser-

G. Garab (ed.), Photosynthesis: Mechanisms and Effects, Vol. V, 3505–3510.
© *1998 Kluwer Academic Publishers. Printed in the Netherlands.*

229. Briefly, a synthetic phosphopeptide based on the Ser-229 sequence was shown to bind to 14-3-3 proteins in vitro, and the phosphorylated SPS-229 synthetic peptide was shown to stimulate SPS activity, presumably by disrupting the SPS:14-3-3 complex. In order to further characterize the SPS:14-3-3 interaction, we are currently determining whether phosphorylation of Ser-229 is required for 14-3-3 binding and if so, whether leaves contain a protein kinase capable of phosphorylating this residue in the native SPS molecule.

2. Procedures

2.1 Materials
Spinach (*Spinacia oleracea* L. cvs. Tyee and Bloomsdale) were grown in soil in the greenhouse. Highly activated (dephospho-Ser-158) SPS was obtained by D-mannose feeding to excised leaves (11). Leaf tissue was harvested directly into liquid nitrogen prior to extraction (12). The SPS-158 peptide (GRJRRISSVEJJ; J=norleucine), designated as the "SP2" peptide in Fig. 2, was obtained from Dr. Jan Kochansky (USDA, Beltsville, MD). The phosphorylated SPS-229 peptide (CRVDLLTRQVphosphoSAPGVDK) was purchased from Genemed Synthesis, Inc. (San Francisco, CA). The unphosphorylated peptide (designated "SP42" in Fig. 2) was synthesized in-house on a Synergy Model 32A Peptide Synthesizer (Perkin Elmer), and differs only in the lack of a Cys residue at the N-terminus.

2.2 Assay of SPS in immune complexes
SPS was immunopurified from mannose-treated spinach leaf extracts using spinach leaf SPS-specific monoclonal antibodies (Mab) and Immunoprecipitin (Gibco-BRL) as the precipitating agent as described (9). The Immunoprecipitin-MAb-SPS complex (immune complex) was washed several times to remove adhering proteins and preincubated at room temperature for 30 min with no additions (control); with 1 mM ATP plus the PK_I kinase (see below); or with alkaline phosphatase (14 units per mL). Following preincubation, the immune complexes were washed and assayed for SPS activity using the selective assay (9).

2.3 Assay and partial purification of leaf protein kinases
Spinach leaf tissue was extracted in a buffer containing 50 mM Mops-Na, pH 7.5, 2 mM EDTA, 5 mM DTT, 0.5 mM PMSF, 1 mM ε-amino-*n*-caproic acid, 1 mM benzamidine, 25 mM NaF, 0.25 μM microcystin-LR, and 0.1% (v/v) Triton X-100, as described (9). The proteins precipitating between 3 and 25% (w/v) polyethylene glycol(PEG)-8000 were then fractionated by Fast Protein Liquid Chromatography (FPLC), using a Resource Q anion exchange resin. Bound proteins were eluted with a linear gradient from 0 to 500 mM NaCl. Fractions were assayed for peptide kinase activities as described (12).

3. Results and Discussion

3.1 Ser-229 is conserved across species

Regulatory sites that are important in the control of an enzyme are generally conserved across species. In the special case of phosphorylation sites, not only must the Ser or Thr residue be conserved, but also the positive recognition elements surrounding the phosphorylatable residue that constitute the motif targeted by the requisite protein kinase(s).

For example, in the case of Ser158, all deduced sequences available to date for leaf enzymes contain a homologous seryl residue. When the seryl residues are aligned, it is clear that although there is considerable variation among the sequences, there is absolute conservation of basic residues at -3, -6 and -8 and hydrophobic residues at -5 and $+4$, relative to the Ser at position 0) (13). Several of these conserved residues are essential for recognition by protein kinases in vitro (12).

If Ser-229 is also an important regulatory phosphorylation site, it is reasonable to expect that it would also be conserved across species. As shown in Table 1, this indeed is the case. All of the deduced sequences available to date for the leaf enzyme contain a homologous Ser residue and several residues surrounding it are similarly conserved. In particular, the Pro at $+2$ and the Arg at -3 are probably important for 14-3-3 binding.

Table 1 Sequence alignment of amino acids surrounding the putative 14-3-3 protein binding site in leaf SPS from a diverse group of species.

Species	Sequence
Spinacia oleracea	DLLTRQVSAPGV
Vicia faba	DLLTRQVSSPEV
Craterostigma plantagineum	DLLTRQVSAPGV
Solanum tuberosum	DLLTRQVSSPEV
Zea mays	DLFTRQVSAPGV
Saccharum officinarum	DLFTRQVSSPDV
Oryza sativa	DLFTRQVSSPEV
Musa acuminata	DLLTRQ I SAPDV
Actinidia deliciosa	DLLTRQVSSPEV
Concensus	DLxTRQxSxPxV

3.2 Phosphorylation enhances inhibition of SPS activity by 14-3-3 proteins

It was demonstrated previously (10) that a synthetic phosphopeptide ("pSPS-229 peptide"), based on the putative binding site sequence, increased SPS activity in vitro apparently by disrupting the SPS:14-3-3 complex. Because the unphosphorylated SPS-229 peptide did not increase SPS activity, it was suggested that 14-3-3 binding required phosphorylation of Ser-229 (10). Similar observations have been made for disruption of the 14-3-3 complex with the phosphorylated form of NR by synthetic peptides in vitro(6,7).

To test this notion further, we used immunopurified SPS that was depleted of bound proteins by extensive washing in vitro. In these experiments, SPS was extracted from mannose-treated leaf tissue in order to obtain protein that was relatively dephosphorylated at the outset (11). The SPS in the washed immune complex was then subjected to phosphorylating or dephosphorylating conditions prior to addition of 14-3-3 proteins and assay of activity. As shown in Fig. 1, the addition of 14-3-3s without pretreatment resulted in about 16% inhibition of SPS activity. Pretreatment with alkaline phosphatase reduced the inhibition by 14-3-3s to 8% whereas pretreatment under phosphorylating conditions (ATP + PK$_I$, see below) increased inhibition to 32%. These results are consistent with previous findings and suggest that phosphorylation of Ser-229 is necessary for binding of the 14-3-3 inhibitor protein.

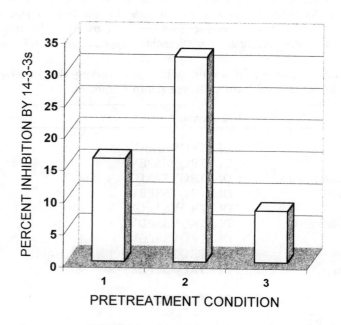

Figure 1. Phosphorylation is required for inhibition of SPS activity by 14-3-3 proteins. A washed SPS immune complex was preincubated with no additions (control), with ATP + PK$_I$ kinase, or with alkaline phosphatase (pretreatment conditions 1,2 and 3, respectively).

A mixture of 14-3-3 proteins were then added and SPS activity was assayed with the selective assay (12). Values shown are the means of two separate experiments.

3.3 Identification of an endogenous protein kinase that phosphorylates the SPS-229 synthetic peptide

If phosphorylation of Ser-229 is necessary for binding of 14-3-3s, it follows that an endogenous leaf kinase(s) must be present to phosphorylate this residue. We used the unphosphorylated SPS-229 synthetic peptide to identify kinase(s) that might be able to phosphorylate Ser-229 in the native SPS molecule. Leaf proteins were precipitated with PEG-8000 (3 to 25%) and fractionated by FPLC-anion exchange chromatography.

A single peak of kinase activity was observed that coeluted with a kinase activity previously designated as PK_I (6,8,12). The SPS-229 peptide was not phosphorylated by other kinases that readily phosphorylated the SPS-158 synthetic peptide (see Fig. 2). The major difference between the two synthetic substrates is that the SPS-229 peptide lacks a basic residue at the −6 position, which is almost essential for phosphorylation by many kinases in leaves (14) and and other tissues such as cauliflower florets (12). Apparently, even though the PK_I kinase shows some preference for substrates with a basic residue at −6 (or a prolyl residue at −7), the dependence on this element is not as strict as the other kinases. We are presently further purifying the SPS-229 peptide kinase(s) for the purpose of characterizing the kinase(s) and for further investigations into the phosphorylation of Ser-229 in the native SPS molecule.

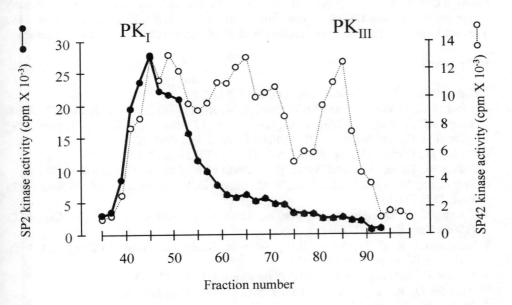

Fig. 2. Resolution of peptide kinase activities by FPLC-ResourceQ anion exchange chromatography of spinach leaf proteins. Extracts were prepared in the absence of Mg^{2+} and protein from a 3—25% (w/v) PEG-8000 precipitation was fractionated as described under Procedures. Peptide kinase activities were determined by the

incorporation of ^{32}P in synthetic peptides: SP2 (the SPS-158 peptide; o); and SP42 (the SPS-229 peptide; •). The positions of the PK$_I$ and PK$_{III}$ kinases (15) are identified.

3.4 Concluding Remarks

The regulation of SPS activity by reversible protein phosphorylation continues to grow in complexity with the recent suggestion that Ser-229 may be involved in binding of a 14-3-3 inhibitor protein (10). The results presented herein extend this notion by showing that Ser-229, and the sequences surrounding it, are conserved among species (Table 1), and that phosphorylation of SPS is required for the interaction to occur (Fig. 1). We presume that Ser-229 is at least one of the residues phosphorylated under conditions that increase interaction with the 14-3-3s. Lastly, we have tentatively identified an endogenous leaf protein kinase that can phosphorylate the SPS-229 peptide in vitro, and thus, is a candidate for the kinase that may function to regulate the interaction with 14-3-3s in vivo.

4. Addendum

Cooperative investigations of the U.S. Department of Agriculture, Agricultural Research Service, and the North Carolina Agricultural Research Service. The research was supported in part by funds from U.S. Department of Energy (DE-AI05-91ER20031). Mention of a trademark or proprietary product does not constitute a guarantee or warranty of the product by the USDA or the North Carolina Agricultural Research Service and does not imply its approval to the exclusion of other products that might also be suitable.

References

1 Krebs, E.G. (1985) Biochem. Soc. Trans. 13, 813-820

2 Huber, S.C., Huber, J.L., and McMichael, R.W. Jr., (1994) Int. Rev. Cytol. 149: 47-98

3 Huber, S.C., Huber, J.L., Liao, P.-C., Gage, D.A., McMichael, R.W. Jr., Chourey, P.S. Hannah, L.C., and Koch, K. (1996) Plant Physiol. 112: 793-802

4 Winter, H., Huber, J.L., and Huber, S.C. (1997) FEBS Lett. 420: 151-155

5 Winter, H., Huber, J.L., and Huber, S.C. (1998) FEBS Lett. 430: 205-208

6 Bachmann, M., Huber, J.L., Liao, P.-C., Gage, D.A. and Huber, S.C. (1996) FEBS Lett. 387: 127-131

7 Moorhead, G., Douglas, P., Morrice, N., Scarabel, M., Aitken, A. and MacKintosh, C (1996) Curr. Biol. 6: 1104-1113

8 McMichael, R.W. Jr., Klein, R.R., Salvucci, M.E., and Huber, S.C. (1993) Arch Biochem. Biophys. 307: 248-252

9 Toroser, D. and Huber, S.C. (1997) Plant Physiol. 114: 947-955

10 Toroser, D., Athwal, G.S., and Huber, S.C. (1998) FEBS Lett, in press

11 Huber, J.L.A., and Huber, S.C. (1992) Biochem. J. 283: 877-882

12 Toroser, D. and Huber, S.C. (1998) Arch. Biochem. Biophys. 355: 291-300

13 Huber, S.C. and Huber, J.L. (1996) Annu. Rev. Plant Physiol. Plant Mol. Biol. 47 431-444

14-3-3 PROTEINS: FROM PLANT NITRATE REDUCTASE TO WIDER ROLES IN PLANT RESPONSES TO HORMONES, STRESSES, AND NUTRIENTS

Carol MacKintosh, MRC Protein Phosphorylation Unit, Biochemistry Department, University of Dundee, MSI/WTB Complex, Dow Street, Dundee, Scotland, U.K., DD1 5EH.

Key words: Arabidopsis, cytosolic enzymes, N metabolism, phosphorylation/dephosphorylation, protein-protein interactions, signal transduction.

1. Regulation of cytosolic enzymes in response to changes in photosynthesis and other factors

Photosynthesis generates the reduced ferredoxin, NADPH and ATP that is used to support assimilation of carbon (C) and nitrogen (N) into starch, sugars, and amino acids in leaves. Carbon in the form of triose-phosphates can either be used to make starch in the plastid, sucrose in the cytosol, or can be converted into organic acids by the activity of phosphoenolpyruvate carboxylase (PEPc) to provide the C for amino acid synthesis. N supplied as nitrate is reduced nitrite by nitrate reductase (NR) in the cytosol, and the nitrite is further reduced to ammonia for amino acid synthesis by the reduced ferredoxin-dependent nitrite reductase (NiR) in the chloroplast.

When photosynthesis is active, cytosolic sucrose synthesis, organic acids synthesis and nitrate assimilation are activated by the dephosphorylation and activation of sucrose-phosphate synthase (SPS) and NR, and the phosphorylation and activation of PEPc (1, 2). The nature of the signals that are generated by photosynthesis to control the activties of these three cytosolic enzymes are not yet known. SPS, NR and PEPc are not always regulated in parallel; their activities can be differentially regulated in response to water, O_2 and N availability, growth regulators, and developmental stage. For example, water stress causes activation of SPS (3) and inactivation of NR (4). Understanding the mechanisms that control SPS, NR and PEPc will be crucial to understanding what controls primary partitioning of photosynthate. My group have been dissecting the mechanism of regulation of NR, and it has been discovered that NR activity is not controlled by a 'simple' phosphorylation/dephosphorylation cycle.

2. Phosphorylated nitrate reductase in plants is inhibited by 14-3-3 proteins

NR in leaves is inactivated within minutes when photosynthesis is inhibited, or when plants suffer water loss (1, 2, 4). The inactivation of NR occurs in two steps (Figure 1). First serine-543 (in the spinach enzyme) is phosphorylated by an NR kinase (5). Phosphorylation alone has no effect on NR activity. However, the phosphorylated protein binds to a protein that we originally named NIP, for nitrate reductase inhibitor protein (6, 7). We purified NIP and discovered by amino acid sequencing that it comprises several 14-3-3 isoforms (8, 9). It is the binding of 14-3-3s to

G. Garab (ed.), Photosynthesis: Mechanisms and Effects, Vol. V, 3511–3516.
© 1998 Kluwer Academic Publishers. Printed in the Netherlands.

3512

phosphorylated NR that inhibits the enzyme. Reactivation in the light is by dephosphorylation of NR with a type 2A protein phosphatase and dissociation of NIP (7, 9, 10). The exact activation state of NR depends on the proportion of the enzyme that is phosphorylated and bound to 14-3-3 proteins at any given time. NR activity is also regulated under other physiological circumstances, and in other plant tissues.

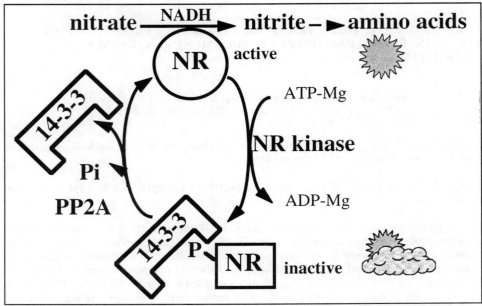

Figure 1. Post-translational regulation of plant leaf nitrate reductase by reversible phosphorylation and 14-3-3 proteins in response to changes in photosynthesis. Photosynthesis and NR activity state can change rapidly, as in this diagram where photosynthesis is inhibited by cloud cover on a sunny day.

3. Serine-543 on NR is phosphorylated *in vitro* by SNF1-like protein kinases and CDPKs

After fractionation of spinach leaf extracts by Mono Q anion-exchange chromatography, three major peaks of NR kinase activity were identified (termed PKI, PKII and PKIII) that each phosphorylated NR on serine-543 and rendered the enzyme sensitive to inhibition by 14-3-3 proteins (11, 12, 13), and phosphorylated and inactivated SPS (11, 12).

PKIII was identified as a plant SNF1-related enzyme by the following properties; inactivation by dephosphorylation with protein phosphatase 2A or 2C, requirement for a hydrophobic residue in the +4 position of the peptide substrate AMARAASAAALARRR, its inactivation of *Arabidopsis* 3-hydroyxyl-3-methyl glutaryl-CoA reductase-1 by phosphorylation of serine-577, a molecular mass of 140 kDa, its Ca^{2+}-independent activity, and recognition of its 58-kDa catalytic subunit by an antibody raised against the rye SNF1-related enzyme, RKIN1 (13, 14). Thus, the SNF1-kinase has the potential to co-ordinately regulate nitrate assimilation, sucrose synthesis, and isoprenoid biosynthesis in plants.

The term SNF1 comes from the <u>s</u>ucrose <u>n</u>on-<u>f</u>ermenting *S. cerevisiae* mutant which is unable to grow on carbon sources other than glucose because of a defect in this protein kinase (15). The mammalian homologue is the AMP-activated protein kinase that phosphorylates and inactivates several biosynthetic enzymes in response to stresses that deplete cells of ATP (16, 17).

It is intriguing to find that plant cytosolic C and N metabolism is regulated by familiy of kinases that are central components of sugar and energy-sensing signalling pathways in other organisms. Like the yeast and mammalian SNF1-like kinases, the plant enzyme is part of a protein kinase cascade and is only active when phosphorylated by an activating kinase kinase (13, 14). However, how the activities of the plant SNF1-related kinase cascades respond to changes in photosynthesis or other signals that regulate NR is not yet known.

We have also purified a calcium-dependent NR kinase that comigrates with PKI from spinach leaf extracts (18), and established by Edman sequencing of 80 amino acid residues that it is a calcium-dependent (calmodulin-domain) protein kinase (CDPK), with peptide sequences very similar to *Arabidopsis* CDPK6 (accession no. U20623; also known as CPK3). The spinach CDPK was recognised by antibodies raised against *Arabidopsis* CDPK. Nitrate reductase was phosphorylated at serine-543 by bacterially-expressed His-tagged CDPK6, and the phosphorylated NR was inhibited by 14-3-3 proteins.

Ca^{2+} is implicated in regulating a huge number of cellular responses in plants, and cytosolic Ca^{2+} concentrations have been reported to increase in leaf cells in the dark (19), a time when phosphorylation and inactivation of NR also occurs. However, we cannot yet say whether this means that the CDPK is operating on NR in the dark. Clearly, elucidating the physiological cues that control the expression and activity of the NR kinase CDPK (along with the SNF1 and any other NR kinases) will be important in order to understand which of these enzymes is actually responsible for phosphorylating NR in plants in response to different environmental stimuli.

4. 14-3-3 proteins bind to phosphopeptide motifs in target proteins

The 14-3-3 proteins were first identified as abundant brain proteins, and named by their elution position on ion-exchange columns and mobility on 2-dimensional gel electrophoresis (reviewed in 20). We now know that 14-3-3s are ubiquitous in eukaryotes. In most species there are many isoforms, for example, 10 isoforms in *Arabidopsis* (21). The 3-dimensional structures of 14-3-3 proteins show them to be homo- and hetero-dimers, whose L-shaped monomers come together to form a broad central groove that contains two sites for binding to target proteins.

Many 14-3-3-binding partners have been identified, particularly signalling components in mammalian cells (cdc25, Raf-1, p53 (the 'guardian of the genome'), insulin-like growth factor-1 receptor, and many others), and from the phenotypes of 14-3-3 mutants in *Drosophila* and yeasts, 14-3-3s are implicated in a wide variety of cellular functions, including vesicular trafficking, cell growth, and programmed cell death (apoptosis).

It is emerging that 14-3-3s recognise and bind to phosphopeptide motifs in many of their target proteins (9, 22, 23). For example, we noticed that the peptide sequence around the phosphorylated serine-543 on NR (RSTS*TP, where S* is phosphorylated serine-543) is strikingly similar to the sequences surrounding phosphoserines-259 (RSTS*TP) and -621 (RSAS*EP) in the mammalian protein kinase, Raf-1, that interact with 14-3-3s. We found that a phosphopeptide containing the 14-3-3-binding 'consensus' completely blocked the inhibition of phosphorylated NR by 14-3-3s, while a dephosphorylated peptide or unrelated phosphopeptides had no effect. These and other results show that 14-3-3s bind directly to the phosphorylation site of NR. More recently, we (Fiona Milne, Sarah Meek, and Carol MacKintosh, unpublished) found that the following peptides bind 14-3-3s:-

ARAAS*APA > QRSTS*TPN > ARSTS*TPA = ARSLS*VEA > AASTS*TPA > ARSTATPA
(best)   (no effect)

5. Functional effects of 14-3-3 proteins

The identification of 14-3-3s as regulators of NR in plants provided the first case where a functional effect (inhibition of NR) of 14-3-3s had been identified in a physiological context (in response to inhibition of photosynthesis, or water stress). 14-3-3 proteins are also involved in binding to and activating a plant plasma membrane proton-ATPase (for example, 24). The plasma membrane proton-ATPase generates a proton gradient that supports the import of nitrate into the cytosol by the plasma membrane nitrate transporter. It will be interesting, therefore, to discover whether the activities of NR and the proton-ATPase are in any way co-ordinated by reversible phosphorylation and 14-3-3 proteins. For example, do any of the NR kinases phosphorylated the proton-ATPase?

More recently, several regulatory functions have also been defined in mammalian cells. For example, binding of 14-3-3 proteins to phosphorylated BAD in response to interleukin 3 blocks programmed cell death (25). Binding 14-3-3s to phosphorylated cdc25 in DNA-damaged cells causes G2 arrest, thereby preventing potentially lethal cell division in cells with damaged DNA (26, 27). In normal cells p53 is phosphorylated on serines 376 and 378. In response to DNA-damaging irradiation, serine 376 on p53 is dephosphorylated and the remaining phosphorylated serine 378 provides a 14-3-3-binding target site. Binding of 14-3-3s to p53 increases the affinity of p53 for sequence-specific DNA (28). 14-3-3s maintain the mammalian protein kinase Raf-1 in an inactive state in the absence of GTP-bound Ras and also stabilize an active conformation of Raf produced during activation *in vivo* (29). 14-3-3 proteins, therefore, have crucial roles in regulation of mammalian cell growth and cell cycle.

6. What is the true significance of 14-3-3-binding to nitrate reductase?

The finding that NIP comprises 14-3-3s gives a new perspective to our thinking on NR regulation. In addition to our systematic study of the mechanism of NR control, we are faced with many new and difficult questions.

14-3-3s seem to be mostly associated with signalling components in mammalian cells, and with events important in control of mammalian 'cell survival'. Does this make NR the 'odd one out', or is NR binding to 14-3-3s part of a signalling pathway, or are 14-3-3s involved in many processes other than signalling? Should we be thinking of NR inhibition by 14-3-3s as a 'cell survival' role for 14-3-3s (a reasonable suggestion since inhibition of NR is thought to prevent build-up of toxic nitrite under conditions where the nitrite cannot be further reduced to ammonia)?

14-3-3s are dimers, so can they bind NR at one site and 'something else' at the other? If so, what is the 'something else'? How many other 14-3-3 targets exist in plants, and to what extent does their regulation (phosphorylation and 14-3-3 binding) coordinate with changes in NR activity? Do the same protein kinases that phosphorylate NR also phosphorylate other 14-3-3 binding partners, or are there '101' different kinases that operate on different 14-3-3-targets under different circumstances? Do different 14-3-3 targets compete with each other for binding to 14-3-3s, or is there an excess of 14-3-3s sufficient to bind all target proteins? Are the 14-3-3 proteins themselves subject to regulation? What are the cellular effects of 14-3-3-target interactions?

7. Affinity binding methods to identify novel 14-3-3-binding partners, and monitor how their binding changes in response stresses, nutrients, and hormones.

At the time of finding that NIP comprised 14-3-3s, we knew that mammalian Raf-1 and plant NR both contained similar phosphopeptide motifs that bound to 14-3-3s. By assuming that many or all other target proteins also contain phosphopeptide binding motifs, we have developed biochemical strategies to tackle many of the questions listed in the section above. Over the past year or so, we have developed sensitive and specific biochemical affinity procedures that allow us to identify potential new 14-3-3 binding partners; monitor how their phosphorylation status changes in response to hormones, nutrients, stresses and other cell stimuli; and identify the functional effects of 14-3-3 binding to individual target proteins.

During this work, we have discovered that several enzymes of plant C and N metabolism are regulated by reversible phosphorylation and interactions with 14-3-3 proteins (Greg Moorhead, Sarah Meek and Carol MacKintosh, unpublished). In future, these studies will allow us to dissect signalling pathways starting from the physiological changes in enzyme activity, and work backwards up the signalling pathways towards the receptors for extracellular stimuli.

References

1. Huber SC, Bachmann M, McMichael RW, Huber JL (1995) in Sucrose Metabolism, Biochemistry, Physiology and Molecular Biology (Pontis, H.G., Salerno, G.L. and Echeverria, E.J., eds), pp 6-13. American Society of Plant Physiologists Series, Vol. 14
2. MacKintosh, C. (1998) New Phytol. 139, 153-159
3. Toroser, D. and Huber, S.C. (1997) Plant Physiol. 114, 947-955
4. Kaiser, W.M. and Förster, J. (1989) Plant Physiol. 91, 970 974
5. Douglas, P., Morrice, N. and MacKintosh, C. (1995) FEBS Lett. 377, 113-117
6. Spill, D. and Kaiser, W.M. (1994) Planta 192, 183-188
7. MacKintosh, C., Douglas, P. and Lillo, C. (1995) Plant Physiol. 107, 451-457
8. Bachmann, M., Huber, J.L., Liao, P.-C., Gage, D.A. and Huber, S.C. (1996) FEBS Lett. 387, 127-131.
9. Moorhead, G., Douglas, P., Morrice, N., Scarabel, M., Aitken, A. and MacKintosh, C. (1996) Curr. Biol. 6, 1104-1113
10. Bachmann, M., Huber, J.L., Athwal, G.S., Wu, K., Ferl, R.J. and Huber, S.C. (1996) FEBS Lett. 398, 26-30
11. McMichael RW, Bachmann M. and Huber SC (1995) Plant Physiol. 108, 1077-1082
12. Huber, S.C., Bachmann, M. and Huber, J.L. (1996) Trends Plant Sci. 1, 432-438

13. Douglas, P., Pigaglio, E., Ferrer, A., Halford, N.G. and MacKintosh, C. (1997) Biochem. J. 325, 101-109
14. Ball, K.L., Barker, J. Halford, N.G. and Hardie, D.G. (1995) FEBS Lett 377, 189-192
15. Celenza, J.L. and Carlson, M. (1986) Science 233, 1175-1180
16. Carling, D., Aguan, K., Woods, A., Verhoeven, A.J., Beri, RK., Brennan, C.H., Sidebottom, C., Davison, M.D. and Scott, J. (1994) J. Biol. Chem. 269, 11442-11448
17. Hardie, D.G., Carling, D. and Halford, N. (1994) Seminars in Cell Biol. 5, 409-416
18. Douglas, P., Moorhead, G., Hong, Y., Morrice, N. and MacKintosh, C. (1998) Planta (in press)
19. Miller, A.J. and Sanders, D. (1987) Nature 326, 397-400
20. Aitken, A. (1996) Trends Cell Biol. 6, 341-347.
21. Wu, K., Rooney, M.F. and Ferl, R.J. (1997) Plant Physiol. 114, 1421-1431.
22. Muslin, A.J., Tanner, J.W., Allen, P.M., and Shaw, A.S. (1996) Cell 84, 889-897
23. Yaffe, M.B., Rittinger, K., Volinia, S., Caron, P.R., Aitken, A., Leffers, H., Gamblin, S.J., Smerdon, S.J. and Cantley, L.C. (1997) Cell 91, 961-971
24. Jahn, T., Fuglsang, A.T., Olsson, A., Bruntrup, I.M., Collinge, D.B., Volkmann, D., Sommarin, M., Palmgren, M.G. and Larsson, C. (1997) Plant Cell 9, 1805-1814
25. Zha, J., Harada, H., Yang, E., Jockel, J. and Korsmeyer, S.J. (1996) Cell 87, 619-628
26. Peng, C.Y., Graves, P.R., Ogg, S., Thoma, R.S., Byrnes, M.J., Wu, Z., Stephenson, M.T. and Piwnica-Worms, H. (1998) Cell Growth Differ. 9, 197-208
27. Sanchez, Y., Wong, C., Thoma, R.S., Richman, R., Wu, Z., Piwnica-Worms, H. and Elledge SJ (1997) Science 277, 1497-1501
28. Waterman, M.J., Stavridi, E.S., Waterman, J.L. and Halazonetis, T.D. (1998) Nature Gen. 19, 175-178
29. Tzivion, G., Luo, Z. and Avruch, J. (1998) Nature 394, 88-93

MOLECULAR CLONING AND CHARACTERIZATION OF CYTOSOLIC ISOFORM OF RIBULOSE-5-PHOSPHATE 3-EPIMERASE FROM RICE.

Stanislav KOPRIVA, Anna KOPRIVOVA and Karl-Heinz SÜSS[1]
Institute of Plant Physiology, Altenbergrain 21, 3013 Bern, Switzerland
[1] Faculty of Science, University of Kuwait, Safat 13060, Kuwait

Key words: C-metabolismus, Calvin-cycle, cytosolic enzymes, gene expression, molecular biology

1. Introduction

Although the oxidative pentose phosphate pathway (OPPP) in plants partially contributes to respiration, it plays a much more important role as a source of NADPH for the reduction of metabolites in various biosynthetic pathways. In animal and yeast cells, OPPP is present in the cytosol and, therefore, it was generally accepted that the cytosol of plant cells also contains this pathway (1). After a CO_2 release from [1-^{14}C]glucose by isolated chloroplasts was demonstrated it was believed that two sets of isoenzymes of this pathway occur in chloroplasts and in cytosol (2,3). However, the quantitative distribution of OPPP enzymes among the cell compartments in green and non-green plant cell is still unclear. There remains no doubt about the existence of cytosolic and plastidic isoforms of glucose-6-phosphate dehydrogenase and 6-phosphogluconate dehydrogenase in leaf mesophyll and non-green plant cells (4). In contrast, cell fractionation experiments combined with chromatographic separation of enzyme isoforms did not provide evidence for the occurrence of cytoplasmic OPPP enzymes catalyzing the regenerative sequence of the pathway including transketolase, ribose-5-phosphate isomerase, and ribulose-5-phosphate 3-epimerase (RPE) (5). This led to the conclusion that plastids contain a complete OPPP, whereas an incomplete pathway is operating in the cytoplasm of leaf mesophyll cells.

2. Procedure

2.1 Sequence Analysis
The GeneBank and EMBL sequence databases were screened with the BLAST software. The sequences were further analysed by the PCGENE package (Intelligenetics). The cDNA clone corresponding to the GeneBank accession number D41947 has been obtained from the MAFF DNA Bank, Tsukuba, Japan and completely sequenced on both strands.

2.2 DNA and RNA Analysis
RNA was isolated from leaves, roots, and seedlings of rice by phenol extraction and LiCl precipitation. Southern analysis of genomic DNA from *A. thaliana*, and *O. sativa* and Northern blotting of the RNA isolated from rice tissues has been performed as described (6). The blots were hybridized either with ^{32}P labelled total cDNA of rice cytosolic RPE or chloroplastic RPE from *S. tuberosum* (7).

G. Garab (ed.), *Photosynthesis: Mechanisms and Effects, Vol. V,* 3517–3520.
© 1998 *Kluwer Academic Publishers. Printed in the Netherlands.*

2.3 Protein Analysis.

The cytosolic epimerase has been expressed in *E. coli* by the pET expression system (Novagen). The RPE coding region has been cloned into pET14b plasmid what ensured an addition of a 6-histidine tag at N-terminus of the expressed protein. The modified protein has been purified on Ni^{2+} columns according to the manufacturer's instructions (Novagen). The RPE activity was measured spectrophotometrically by coupling with transketolase, triose phosphate isomerase, and glycerol-3-phosphate dehydrogenase with ribulose-5-phosphate as a substrate (8). Bacterial extracts for activity measurements were prepared from overnight cultures by resuspending the bacteria in PBS buffer and sonication. Antibodies were raised in rabbits against the purified cytosolic RPE. SDS-PAGE and Western blotting were performed using standard procedures. The proteins were quantitated with the Bio-Rad kit.

3. Results and Discussion

3.1. Sequence Analysis

The previous cloning of chloroplast pentose-5-phosphate 3-epimerases from potato (7) and spinach (9) opened up the way for a search for cytoplasmic enzyme isoforms. The search of EMBL and GeneBank databases for epimerase-related sequences revealed several EST entries from *Arabidopsis thaliana* and rice similar to the RPE cDNA. The EST sequences from rice could be divided into two groups with 65 % and 45 % identity with the RPE from potato. From the sequences of the former group a 1 kb cDNA could be reconstructed which coded for a protein with amino acid sequence 90 % identical to the potato chloroplast RPE. This protein contained a 39 amino acid presequence carrying all characteristics of a chloroplast targeting peptide, implying that the cDNA codes for a chloroplastic isoform of ribulose-5-phosphate epimerase. The other cDNA clones, corresponding to ESTs OSS4145a, OSS4976a, and OSR28401a, were obtained from the MAFF DNA Bank, Tsukuba, Japan, found to be identical, and the OSS4976a clone was completely sequenced. The 942 base pairs long cDNA contained a single open reading frame coding for protein of 228 amino acids with molecular mass of 24.3 kDa. Compared to the plant chloroplast epimerases this protein does not contain the chloroplast targeting peptide. Its primary structure is 40 % identical with the potato and rice chloroplastic RPE isoforms and 27-35 % with 3 RPE isoenzymes from *E. coli*. Most importantly, comparison with RPE from *S. cerevisae* and human, both known to be localised in the cytosol, revealed 52 % identity, whereas human and rice chloroplastic RPEs possess only 36 % identical amino acid residues.

As the evidence that the cDNA codes for cytosolic isoform of RPE was still indirect, the protein was expressed in *E. coli*, purified, and used to produce polyclonal antibodies. Expression constructs of both isoforms of RPE in pET3a and pET14b vectors were transformed in *E. coli* BL21 strain, and the epimerase activity was measured in the bacterial extracts. As documented in Table 1, extracts from *E. coli* transformed with RPE constructs showed clearly much higher epimerase activity than extracts from *E. coli* transformed with pET14b alone. There were not substantial differences in the specific activity between both RPE isoforms or between the constructs in various vectors.

3.2. Expression Analysis

In order to investigate the tissue-specific distribution of the two RPE isoforms Northern analysis of RNA prepared from different tissues of rice was performed. The cytosolic RPE transcript is present mostly in roots and seedlings, whereas in adult leaves only a very small amount has been detected (Fig. 1). On the other hand, the chloroplastic isoform of RPE is present prevalently in rice leaves, and to much lesser extent in seedlings

Table 1: Ribulose-5-phosphate 3-epimerase activity in *E. coli* BL21 transformed with different expression constructs.

construct	RPE activity (mU/min/mg)
pET14b	1.08
chl.RPE / pET14b	6.18
chl.RPE / pET3a	6.06
cyt.RPE / pET14b	6.8
cyt.RPE / pET3a	5.81

and roots, similar to the distribution observed in potato (7). The Western analysis of proteins prepared from the same tissues corroborated the results of RNA analysis for the cytosolic epimerase (Fig.1) showing its presence in all organs with maximum in roots. Similar to Teige et al. (1995) we again could not detect the chloroplastic RPE in roots, although the corresponding mRNA is clearly present. A small amount of this RPE isoform was, however, located in seedlings germinated in dark, suggesting that a posttranscriptional mechanism, but not light, is responsible for the tissue-specific synthesis of RPE protein.

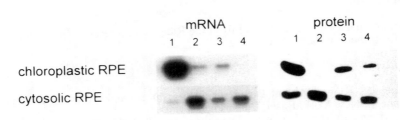

Figure 1 Tissue-specific expression of RPE isoforms in rice.   **A** Northern analysis of the cytosolic and chloroplastic RPE mRNA in RNA from leaves (1), roots (2), 2 weeks old seedlings (3), and seedlings germinated 2 weeks in dark (4). **B** Western analysis of the twoRPE isoforms in protein extracts from the tissues as indicated in A.

3.3. Southern Analysis
Southern analysis indicated presence of 1-2 copies of the cytosolic RPE gene in rice genome. More important, however, is the finding of hybridization signals with DNA from *A. thaliana* and maize, an evidence for presence of cytosolic RPE genes also in other plant species (Fig 2). We can conclude that the existence of cytosolic RPE isoform is not connected with a metabolic pathway specific for rice but that also other plant species possess this gene.

3.4. Conclusions
In conclusion, for the first time, a cytosolic isoform of any enzyme involved in the regenerative part of OPPP was characterized. Since the RPE activity not connected with pentose phosphate cycles is not of high significance for plant metabolism we can expect also other enzymes of this pathway to be present in cytosol. We can hypothetize, therefore, that higher plants possess two complete oxidative pentose phosphate cycles compartmentalized in chloroplasts and in the cytosol.

3520

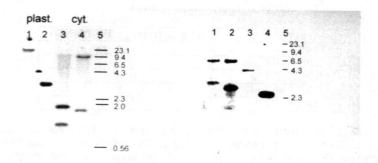

Figure 2. Southern analysis of genomic DNA. (A): The DNA from rice was cut with EcoRI (1,3) and HindIII (2,4) and hybridized with total cDNAs coding for plastidic and cytosolic RPE. (B): DNA from *A. thaliana* was cut with EcoRI (1), HindIII (2), XhoI (3),and SacI (4) and hybridized with total cDNA encoding cytosolic RPE from rice.

4. Acknowledgements

The authors thank Dr. Takuji Sasaki from the Rice Genome Research Program, Tsukuba, Japan, for providing the cDNA clones.

References

1. Lüttge, U., Kluge, M. and Bauer, G. (1994) in Botanik. WCH Verlagsgesellschaft Weinheim, New York, Basel, Cambridge, Tokio
2. Stitt, M. and ap Rees, T. (1980) Phytochemistry 18, 1905-1911
3. Dennis, D. T. and Turpin, D. H. (1990) in Plant Physiology, Biochemistry and Molecular Biology. Longman Scientific & Technical, Singapore
4. Schnarrenberger, C., Oeser A. and Tolbert N. E. (1973) Arch. Biochem. Biophys. 154, 438-448
5. Schnarrenberger, C., Flechner, A. and Martin, W. (1995) Plant Physiol. 108, 609-614
6. Sambrook, J., Fritsch, E. F. and Maniatis, T. (1989) Molecular Cloning: A Laboratory Manual. Cold Spring Harbor Laboratory Press, New York
7. Teige, M., Kopriva, S., Bauwe, H. and Süss, K.-H. (1995) FEBS Lett. 377, 349-352
8. Williamson, W. T. and Wood, W. A. (1966) Meth. Enzymol. 9, 605-608
9. Nowitzki, U., Wyrich, R., Westhoff, P., Hense K., Schnarrenberger, C. and Martin, W. (1995) Plant Mol. Biol. 29, 1279-1291

DISEQUILIBRIUM OF CHLOROPLASTIC PHOSPHOGLUCOSE ISOMERASE INFERRED FROM DEUTERIUM ISOTOPE DISCRIMINATION

J. Schleucher[1], P. Vanderveer[2], J.L. Markley[2], T.D. Sharkey[2]
(1) Dept. of Organic Chemistry & Medical Biophysics, Umeå Univ., S-90187 Umeå, Sweden. (2) Univ. of Wisconsin, Madison, WI 53706, USA.

Key words: Calvin cycle; deuterium; intramolecular isotope distribution; NMR; regulatory processes; stable isotopes.

1. Summary

Intramolecular deuterium (D) distributions of plant metabolites carry information whether an enzyme reaction *in vivo* is in equilibrium or not. We find that cytosolic phosphoglucose isomerase (PGI) is in equilibrium, but chloroplastic PGI is not.

This is done by recording the intramolecular D distribution of glucose isolated from sucrose and leaf starch, using deuterium NMR spectroscopy. The NMR spectra are interpreted in terms of enzyme isotope effects. This information can be obtained without any disturbance of the plant during growth.

Methods to study plant physiology without sampling during growth are useful for plant breeding and paleobotany. For plant breeding, time-integrated measures of enzyme kinetics can be obtained from a sample taken at a single time, because kinetic information is stored in the intramolecular D distribution. For paleobotany, the physiological information stored in the D distribution can be retrieved at any time after growth, e.g. from prehistoric samples. This opens a way to extract physiological information from ancient plant material and can help to understand how past climates and physiology interacted.

2. Why deuterium?

Deuterium is a stable isotope of hydrogen. In all water and organic material, a small fraction ($\approx 0.015\%$) of hydrogen is deuterium. The exact fraction (D/H ratio) is determined by physical and biochemical isotope discriminations during the formation of the sample. The D/H ratio therefore carries information about these physical and

G. Garab (ed.), Photosynthesis: Mechanisms and Effects, Vol. V, 3521–3524.
© 1998 *Kluwer Academic Publishers. Printed in the Netherlands.*

biochemical processes, i.e. information about environmental conditions and plant physiology.

In precipitation, the D/H ratio is correlated with temperature.

In plant matter, several influences interact (1):

[1] the D/H ratio of precipitation
[2] the water source of the plant
[3] D discrimination by evapotranspiration
D discrimination $\approx 10\%$

[4] biochemical D discrimination, by kinetic isotope effects of enzymes
D discrimination $\approx 10 - 90\%$!

How do these mechanisms influence the D abundance in the seven carbon-bound hydrogens of glucose, or of other metabolites? Mechanisms [1] and [2] influence all intramolecular positions of all metabolites in the same way, so they influence only the average D/H ratio. If only these mechanisms existed, all intramolecular positions would have equal D/H ratios, and average D/H ratios of plant material would correlate equally well with climate as do D/H ratios of precipitation. Mechanism [3] influences the average D/H ratio of metabolites formed in leaves. When metabolites get translocated, enzyme-mediated isotope exchange can still lead to non-random intramolecular D distributions. Mechanism [4] invariably leads to non-random intramolecular D distributions. Because of the size of kinetic isotope effects, this can be the overriding effect.

Measurements of the average D/H ratio by isotope ratio mass spectrometry (IRMS) are blind to the intramolecular D distribution. Mechanism [4] can therefore influence the average D/H ratio of plant material in a way that cannot be detected by IRMS. This can introduce uncertainties into correlations between D/H ratios and climate. Measurement of intramolecular D distributions allow to isolate the physiological mechanisms [3] and [4]. The combination of intramolecular D distributions and average D/H ratios should allow to infer climate and physiology from hydrogen isotopes.

3. Cytosolic and chloroplastic PGI

Figure 1 shows deuterium NMR spectra of the glucose derivative 3,6-anhydro-1,2-O-isopropyliden-α-D-glucofuranose isolated from bean sucrose and bean leaf starch. Experimental details can be found elsewhere (Schleucher *et al.*, submitted). Each of the seven signals represents one of the seven carbon-bound hydrogens of glucose, the integral of each signal is proportional to the D abundance of that position. The overall pattern is similar between the two spectra, because sucrose and starch are made from a common pool of chloroplastic triose phosphates. In contrast to previous work (3, 2), all seven carbon-bound hydrogen signals are well resolved.

A striking difference is the depletion of the C(2) position of glucose from leaf starch by ≈ 40%. This depletion was the same whether the bean plants were grown under 150 or 700 μmol photons s⁻¹ m⁻². From organic chemistry it is known that a 40% D depletion can only be caused if the C(2)-H bond is broken. In the biosynthetic pathway from triose

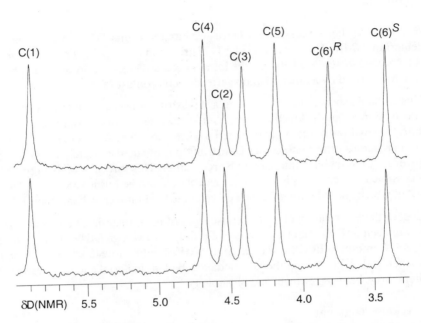

Figure 1: Deuterium NMR spectra of 3,6-anhydro-1,2-O-isopropyliden-α-D-gluco-furanose isolated from bean leaf starch and sucrose. The indices R and S denote the two C(6) methylene hydrogens.

phosphates to starch, this happens only in the PGI reaction. For an enzyme like PGI, two kinetic conditions can be distinguished:

a) high PGI activity, reaction in equilibrium, high [G6P], no D discrimination;

b) low PGI activity, reaction in disequilibrium, low [G6P], C(2) of G6P depleted in deuterium.

The D depletion of C(2) of chloroplast starch tells that chloroplast PGI has little access activity, compared to the metabolic flux of starch synthesis. If the chloroplastic PGI reaction was in equilibrium, a high chloroplastic G6P concentration would withdraw a large pool of reduced carbon from regeneration of RuBP.

The 40% depletion of C(2) of starch, averaged over seven carbon-bound hydrogens leads to a ≈ 6% reduction in the average D abundance of starch, relative to sucrose. This calculation can be done because most other hydrogens must have the same D abundance in starch and sucrose, because both are formed from triose phosphates. The D depletion of C(2) explains most of the low D/H ratio observed for starch (4). This shows that a single enzyme can have a large influence on D/H ratios, which underlines the importance of studies of intramolecular D distributions.

4. Outlook

A scheme has been worked out to measure intramolecular D distributions of glucose by deuterium NMR. The intramolecular D distribution of glucose is non-random, and there is a very big difference between the intramolecular D distributions of sucrose and starch, which is caused by a kinetic isotope effect of chloroplast PGI.

This example shows that intramolecular D distributions can be interpreted and yield information about the kinetics of individual enzyme reactions. Many enzyme reactions introduce hydrogen into organic material, each showing a distinct H/D isotope effect. Further studies of intramolecular D distributions of sucrose, starch and cellulose will therefore yield insights into metabolic regulation. This information is in principle accessible from ancient plant samples and should not be influenced by the D/H ratio of water, which should only influence the average D/H ratio (mechanisms [1], [2]).

Most of the information of hydrogen isotopes of plant material is contained in the intramolecular D distribution and is lost if only the average D/H ratio is measured by IRMS. Combining IRMS and deuterium NMR should allow to separate climatic and biochemical influences on the D abundance in plant material and to investigate interactions between climate and physiology.

5. Acknowledgment

This study made use of the National Magnetic Resonance Facility at Madison, which is supported by NIH (grant RR 02301), the University of Wisconsin, NSF and USDA. This study was supported by the US DOE (grant DE-FG02-87ER13785 to TDS). JS was supported in part by a scholarship from Deutsche Forschungsgemeinschaft.

6. Literature

1 Schleucher J. (1998) in Stable Isotopes and the Integration of Biological, Ecological and Geochemical Processes (Griffiths, H., ed.) pp. 63-73, Bios publishers, Oxford, UK.
2 Martin G.J., Zhang B.-L., Naulet N. & Martin M.L. (1986) J. Am. Chem. Soc. 108, 5116-5122.
3 Zhang B.-L., Quemarais B., Martin M.L. & Martin G.J. (1994) Phytochem. Anal. 5, 105-110.
4 Luo Y. & Sternberg L. (1991) Phytochemistry 30, 1095-1098.

INFORMATION TRANSFER IN GAPDH-PRK COMPLEX OF *CHLAMYDOMONAS REINHARDTII*

Gontero, B., Lebreton, S., Avilan, L. and Ricard, J.
Institut J. Monod. Universités Paris 6 & 7. Tour 43, 2 place Jussieu.
75 251 Paris Cedex 05. France.

Key words: Calvin cycle, complex, light regulation

Introduction
We have purified a bi-enzyme complex from a green alga, *Chlamydomonas reïnhardtii*. This complex is made up of two molecules of dimeric phosphoribulokinase (PRK) and of two molecules of tetrameric glyceraldehyde-3-phosphate dehydrogenase (GAPdH) [1]. These two enzymes become more active in the light [2-4]. The aim of this paper is mainly to show that interaction of PRK with GAPdH leads to a significant activity of PRK under oxidized state.

Results and Discussion

In Fig. 1 is shown the PRK activity of the oxidized complex. When this activity is continuously monitored (curve 1), a lag is displayed. It lasts for several minutes.

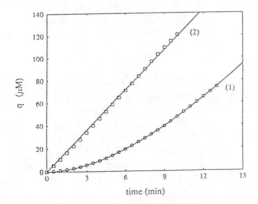

Figure 1. Progress curves of the phosphoribulokinase activity within the bi-enzyme complex. The total enzyme concentration in the reaction mixture, is 2.6 nM. The activity of the bi-enzyme complex is either followed immediately after dilution in the assay cuvette (curve 1) or after 15 mins of incubation of this complex in the assay cuvette in the absence of substrates (curve 2). The ATP and Rib5P concentrations in the reaction mixture are 1mM.

G. Garab (ed.), Photosynthesis: Mechanisms and Effects, Vol. V, 3525–3528.

The lag is apparently monoexponential and is due to the slow spontaneous dissociation as a consequence of a high dilution. This conclusion can be drawn by incubating the complex for 15 mins (Fig. 1, curve 2) since in that case no lag is detectable. The fact that the apparent steady-state of the reaction is higher than the initial velocity implies that the free dissociated form is more active than the form inserted in the complex. Nonetheless it cannot be excluded, at this stage, that the complex is inactive. The distribution of the residuals as indicated in the legend of Fig 2 allows one to clear cut this question. Indeed, the distribution of the residuals (Fig. 2A) is biased if one assumes the free form of the complex to be the only active species. Moreover, the residuals are more randomly dispersed if one assume the free and the inserted form of PRK to be active (Fig. 2B).

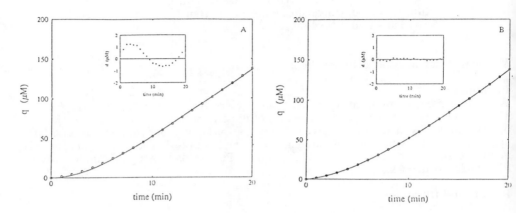

Figure 2. Analysis of progress curves. The PRK activity of the bi-enzyme complex is continuously monitored and the experimental results fitted assuming the free enzyme to be the only active (A) or assuming that both the complex and the free enzyme are active (B). Full lines are theoretical curves. In both inserts are shown the distribution of residual (d, expressed in μM). The total enzyme concentration in the reaction mixture, is 2.4 nM.

These results suggest that the oxidized phosphoribulokinase which interacts with glyceraldehyde-3-phosphate dehydrogenase in a bi-enzyme complex has an activity. Both forms of phosphoribulokinase, the one inserted in the complex and the one that just recently dissociates from the complex are active. They all follow Michaelis-Menten kinetic. The Km for Rib5P is for the PRK of the bi-enzyme complex and for the free metastable state respectively, 30 ± 2.3 μM and 59 ± 2.5 μM. The Km for ATP is for the PRK of the bi-enzyme complex and for the free metastable state respectively, 46 ± 4.6 μM and 48 ± 0.9 μM. One should notice that the form that dissociates from the complex is called metastable for its activity continuously varies to finally reach a stable state (data not shown).

A model has been proposed which accounts for the observed experimental results (Fig. 3).

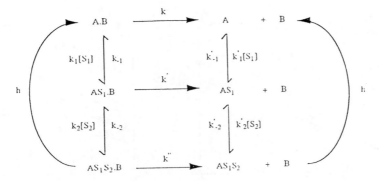

Figure 3. Theoretical model for dissociation of a bi-enzyme complex. k, k' and k" are the apparent rate constants for the dissociation of the various forms of the enzyme complexes (AB, AS_1B, AS_1S_2B). S_1 and S_2 are the two substrates (Rib5P or S_1 and ATP or S_2) of one (A) of the enzymes. Both A and B may represent one, or several, enzyme molecule(s). The two catalytic rate constants are h and h'.

In the model presented in Fig. 3 it is assumed that the binding of the substrates is ordered, ribulose-5-phosphate (Rib5P) being bound first. This result was supported by other studies [5-6]. The inhibition by the product ribulose bisphosphate (RuBP) shows that phosphoribulokinase inserted in the complex (Fig 4.), and the metastable form (Fig. 5.) have different behaviours.

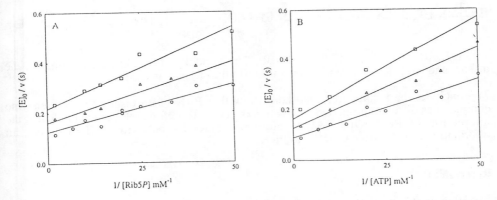

Figure 4 Inhibition by RuBP of oxidized phosphoribulokinase bound to glyceraldehyde-3-phosphate dehydrogenase. **A**. Non-competitive inhibition with respect to Rib5P . The ATP concentration is constant and equal to 1 mM. The RuBP concentrations are as follows : 0 (○), 1.86 mM (△) and 4.7 mM (□). The total enzyme concentration is 8.5 nM. **B**. Non-competitive inhibition with respect to ATP. The Rib5P concentration is constant and equal to 1 mM. The RuBP concentrations are as follows : 0 (○), 2 mM (△) and 4 mM (□). The total enzyme concentration is 10 nM.

3528

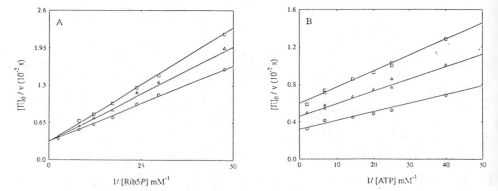

Figure 5 Inhibition by RuBP of oxidized metastable phosphoribulokinase **A**. Competitive inhibition with respect to Rib5*P* . The ATP concentration is constant and equal to 1 mM. The RuBP concentrations are as follows : 0 (O), 1.55 mM (Δ) and 3.1 mM (□). **B**. Non-competitive inhibition with respect to ATP. The Rib5*P* concentration is constant and equal to 1 mM. The RuBP concentrations are as follows : 0 (O), 1.8 mM (Δ) and 3.6 mM (□). The total enzyme concentration is 0.8 nM.

To explain the inhibition studies, a simple iso-model has been proposed for PRK bound to GAPdH:

$$E \xrightleftharpoons{R} ER \xrightleftharpoons{ATP} ER.ATP \xrightleftharpoons{} ER.ADP \underset{ADP}{\xrightleftharpoons{}} E'RP \underset{RP}{\xrightleftharpoons{}} E' \xrightleftharpoons{} E$$

where R, RP stand for Rib5*P* and RuBP. This model is different than the one for the dissociated enzyme where no isomerization (E' into E) step is required.

To conclude, the results presented here show that PRK when inserted in a bi-enzyme complex has different properties than when isolated. Therefore one can conclude that the interaction of PRK with GAPdH quite modulate its behaviour. There exists an information transfer from GAPdH to PRK.

References

1 Avilan, L., Gontero, B., Lebreton, S. and Ricard, J. (1997) Eur. J. Biochem 246, 78-84
2 Scheibe, R. (1991) Plant Physiol. 96, 1-3
3 Wolosiuk, R.A. and Buchanan, B.B. (1978) Arch. Biochem. Biophys. 189, 97-101
4 Wolosiuk, R.A. and Buchanan, B.B. (1978) Plant Physiol. 61, 669-671
5 Lebreton, S., Gontero, B., Avilan, L. and Ricard, J. (1997) Eur. J. Biochem 246, 85-91
6 Lebreton, S., Gontero, B., Avilan, L. and Ricard, J. (1997) Eur. J. Biochem 250, 286-295

DIFFERENTIAL REGULATION OF GLYCERALDEHYDE-3-PHOSPHATE DEHYDROGENASES IN THE GREEN ALGA *Chlorella fusca*

Federico Valverde, Manuel Losada and Aurelio Serrano
Instituto de Bioquímica Vegetal y Fotosíntesis, CSIC-Universidad de Sevilla, CICIC, c/Américo Vespucio s/n, E-41092 Sevilla, Spain

Key words: Calvin-cycle, C metabolism, gene regulation, glucose regulation, glycolysis, microalgae.

1. Introduction

Plants and eukaryotic microalgae posses three distinct glyceraldehyde-3-phosphate dehydrogenases (GAPDH) which are encoded by different nuclear genes, located in different cell compartments and expressed under different conditions (1). We have studied the regulation of these enzymes in a strain of the green alga *Chlorella fusca* that can use glucose for heterotrophic or mixotrophic growth. By using RT-PCR with RNA extracted from photoautotrophic cells, we cloned partial DNA fragments from the *gapN* gene encoding the NADP-dependent non-phosphorylating cytosolic enzyme (EC 1.2.1.9, GAPDHNP), the *gapA* gene for the chloroplastic NADP-dependent phosphorylating enzyme (EC 1.2.1.13, GAPDHA), and the *gapC* gene for the cytosolic NAD-dependent enzyme (EC 1.2.1.12, GAPDHC). These probes were used for gene expression studies by Northern blots in cells grown under different conditions and the results compared with activity and Western blot data obtained with cell-free extracts. Although glycolysis has been extensively studied in bacteria and vertebrates, there are marked differences with the structure, compartmentalization and regulation of this degradative pathway in plants and algae (2). Our results strongly suggest that, in photosynthetic eukaryotes, carbohydrate catabolism is carried out both in the chloroplast and the cytosol and that in the latter cell compartment two different degradative pathways may be operative, one involving the phosphorylating GAPDH and another implicating the non-phosphorylating one.

2. Procedure

2.1 *Growth conditions*

Chlorella fusca SAG 211-8b was grown in Sueoka medium at 25°C under continuous white light (Sylvania-daylight) and bubbled with a mixture of air and CO_2 (2% v/v). When chlorophyll concentration reached 10 μg/ml, cells were submitted to diverse changes in culture conditions, glucose was added at 0.5% (w/v, 17 mM) and cultures remained in either light or in the dark. Cells were usually monitored during a week cycle and aliquots for chlorophyll determination, RNA and protein were taken every 12 h, centrifuged, washed in Tris-HCl 50 mM (pH 7.5) and stored at -20°C.

2.2 *Protein techniques*

Frozen pellets of cells were resuspended in disruption buffer (3) at a ratio of 2 ml/g and disrupted by ultrasonic treatment in a Branson 25U sonifier at medium strength. After centrifugation for 20 min at 40,000g activities were measured in the supernatants as in (3), except for non-phosphorylating activity, in which no arsenate or phosphate was added to the assays. Protein was quantified in whole cells by a modification of the method of Lowry or in cell-free extracts by the Bradford method using ovoalbumine as a standard. Immunoblot experiments were as in (3) using 50 μg protein of extract per lane in 12%

G. Garab (ed.), Photosynthesis: Mechanisms and Effects, Vol. V, 3529–3532.

acrylamide SDS-page gels, incubated in TBS buffer with polyclonal antibodies (1:1000) raised in rabbit against the GAPDH2 from *Synechocystis* sp. PCC 6803 (3), against the GAPDH1 from *Anabaena* sp. PCC 7120 (4), and against the GAPDHNP from pea (4) for the detection of *Chlorella* GAPDHA, GAPDHC and non-phosphorylating enzymes, respectively. The developed filters were quantified with an analytical imaging instrument (Bio Image, Millipore, USA).

2.3 *Nucleic acid analysis*

Total DNA and RNA from *Chlorella fusca* frozen cells were extracted as described in (5). Partial cDNA clones for *gapA* and *gapC* genes were obtained by RT-PCR from total algal RNA using degenerate primers gap2 and gap4 as described in (3). A 700 bp fragment of *gapN* gene from *Chlorella fusca* was also obtained in a similar way using degenerate primers NPCO1 and NPCO2 (4) constructed from highly conserved aa sequences from plant *gapN* genes. All three DNA fragments were cloned in pGEMt vector (Promega) and sequenced in the two strands to completion either by manual or automatic DNA sequencing. Southern blot experiments were carried out as in (3) using 30 μg of *C. fusca* DNA restricted with different enzymes per lane and the blotted filters incubated with ^{32}P labeled *gapA*, *gapC* or *gapN* algal probes. Northern blot experiments were as in (3) with 30 μg of total RNA per lane, the filters incubated with the same radiolabeled probes as above and the resulting radioactivity quantified in an Instantimager Electronic Autoradiography apparatus (Packard Inst., USA).

3. Results and Discussion

Although there have been extensive studies on GAPDHA or GAPDHC of plants in the past, most data regarding regulation of these enzymes were contradictory, in particular those concerning activity measurements (6,7) and reports on microalgae were scarce (8). In this work we have tried to make a comparative study of the regulation of the three GAPDH encoding *gap* genes using different complementing approaches, namely, activity measurements and Western and Northern blots. *Chlorella fusca* SAG 211-8b was chosen because it is a green alga with very versatile trophic metabolism, able to grow photoautotrophically, mixotrophically or heterotrophically with glucose and other sugars as organic C-sources. By following algal cultures that were incubated in light/dark +/- glucose during a week growth cycle, a differential regulation of the three *gap* genes was detected that can give new information about algal and plant C-metabolism.

Fig. 1 shows Southern blots of total genomic DNA from *C. fusca* cells restricted with different enzymes and incubated with radiolabeled probes from *gapN* (Fig.1A), *gapA* (Fig. 1B) and *gapC* (Fig.1C) of this alga, obtained by RT-PCR as described above. Notice that there is a single copy of the three genes in the genome, and that the three GAPDH proteins are coded by nuclear genes as is also the case in plants and other algae (7). We have sequenced these partial RT-PCR clones that contain sequences from the structural genes. This is the first report of an algal *gapN* gene (although a partial sequence) that shows a 75 % identity in its deduced aa sequence with those from plants and ca. 60% with the NADP-dependent non-phosphorylating GAPDH found in the bacterium *Streptococcus mutans*. The *gapA* gene from *C. fusca* shares a 83% identity in its deduced aa sequence with that of the *Chlamydomonas reinhardtii* gene and 80% identity with those from *gapA* genes from plants, while the identity to homologues *gap2* genes from cyanobacteria is 75%. Thus, GAPDHA proteins are apparently more conserved in its deduced aa sequence than the GAPDHNP ones. The cytosolic *gapC* gene keeps an overall identity of 70% in its deduced aa sequence with those of *gapC* genes from plants.

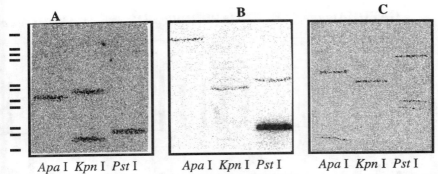

Fig. 1. Southern Blots of *gapA*, *gapN*, and *gapC* respectively with *C. fusca* genomic DNA. *Eco*R I/*Hin*d III-restricted lambda DNA fragments are used as markers.

It was previously reported by our group (8) that the addition of metabolizable sugars to autotrophically growing *C. fusca* cells produced a differential effect on enzyme activity of the three GAPDHs from this green alga. Metabolization of the sugar in light promoted a marked decrease in the activity levels of the chloroplastic GAPDHA, whereas the cytosolic GAPDHC and GAPDHNP were not affected or even increased their activity levels. This is also the case in *Chlorella kessleri*, where *gapC* transcripts increased during a short time period after the addition of glucose (6), a similar effect being detected in transcript levels of cytosolic GAPDHC from tobacco. Furthermore, it is well known that the addition of sugar to autotrophic cultures of microalgae promotes the activation of several genes implicated in C-catabolism and the inhibition of some photosynthetic reactions (glucose bleaching), as the cells adapt their metabolism from autotrophy to mixotrophy-heterotrophy (2). In order to find at which level of genetic regulation was the activation-repression of the components of the GAPDH system being observed we have used a multidisciplinary approach, measuring activity levels, protein amount by Western blot and transcript levels by Northern blot analysis. In this way, we could also get information on the involvement of the non-phosphorylating GAPDH in carbon catabolism, because no definite function has been assigned to this enzyme which is characteristic of the cytosol of photosynthetic eukaryotic organisms (1).

Fig.2 presents the results on the study of the regulation of the three GAPDH enzymes and *gap* genes of *Chlorella fusca* either in autotrophic cultures (above) or in glucose light (below) during a week time course with punctual samples every 24 h. The open bars represent specific activities of the three enzymes, namely GAPDHA, GAPDHNP, or GAPDHC. The stippled bars represent the quantitation of immunostained GAPDH proteins and the closed bars the quantitation of transcript levels of the *gap* genes in the same conditions. In all cases the 24h sample, in which glucose (0,5% w/v) is added, which correspond to autotrophic cells with ca. 10 µg/ml chlorophyll, is considered as the 100% value of the corresponding measured parameters. Note that while GAPDHC and GAPDHNP have a similar activation pattern, which is parallel to the culture growth, GAPDHA is inhibited by the end of the stationary phase. In the presence of glucose, the inhibition of *gapA* expression, estimated either by protein or transcript amounts is dramatic, reaching almost undetectable levels 48 h after the addition of the sugar. In this trophic condition, *gapC* increases its transcription levels until mid stationary phase, and eventually decreases, but unexpectedly, GAPDHNP activity levels increase due to a greater amount of protein, and a greater transcription of the *gapN* gene.

3532

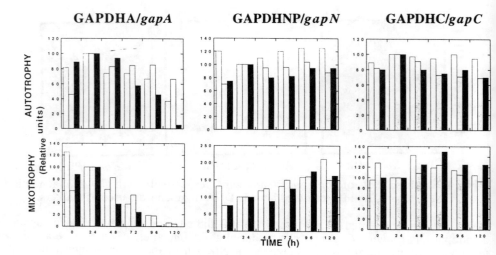

Fig.2. Differential regulation of the three components of the GAPDH system in *C. fusca*. (The inserts show Western blots analysis of GAPDHA during culture evolution).

According to these data, a different metabolic role for each *Chlorella* GAPDH is suggested. While GAPDHA should carry out the classical anabolic function (Calvin cycle) under autotrophic conditions and would be clearly repressed by exogenous organic C-sources, its presence in darkness +/- glucose may indicate also a possible catabolic function, acting in the degradation of either imported glucose or that coming from starch into G3P which would then be exported to the cytosol (data not shown). In this way, it could be acting in the chloroplast stroma as an amphibolic enzyme, rather like the homologous cyanobacterial NAD(P)-depending cytosolic GAPDH2 (3) which may resemble its molecular ancestor. In contrast to that situation, the *gapC* expression is enhanced in the presence of glucose, either in light or in darkness, but decreases at the late stationary phase, while the amount of GAPDHNP increases and remains high until the end of the culture growth. It has been recently described in plants a possible less regulated, non-phosphorylating glucose catabolic pathway, which would involve pyrophosphate-dependent enzymes (2) as well as the non-phosphorylating GAPDH. A similar non-phosphorylating pathway has been also described in some bacteria. This alternative route seems to be less strictly regulated than normal glycolysis, and could be the preferred pathway when ATP levels in the cytosol are too high as a consequence of mitochondrial metabolism, but the need for reducing power and intermediate anabolic metabolites is high due to cell growth.

Supported by grants PB94-033 from the DGICYT and Junta de Andalucía PAI CVI-198.

References
1 Mateos, M.I. and Serrano, A. (1992) Plant science 84:163-170
2 Plaxton, W.C. (1996). Annu. Rev. Plant Physiol. Plant Mol. Biol. 47:185-214
3 Valverde, F., Losada, M. and Serrano, A. (1997) J. Bacteriol. 179:4513-4522
4 Valverde, F., Losada, M. and Serrano, A. (manuscript in preparation)
5 Serrano, R.(1993) Methods in Biochemistry and Molecular Biology. EMBL-Heidelberg
6 Hilgarth, C., Sauer, N. and Tanner, W. (1991) J. Biol. Chem. 266: 24044-24047
7 Brown, J.R. and Doolittle, W.F. (1997) Microbiol. Mol. Biol. Rev. 61:456-502
8 Serrano, A., Mateos, M.I. and Losada, M. (1991) B.B.R.C. 181:1077-1083

REGULATION OF CARBOHYDRATE PARTITIONING IN WHEAT LEAVES

S.J.Trevanion.
IACR-Rothamsted, Harpenden. Herts., AL5 2JQ. U.K.

Key words: fructose 2,6-bisphosphate, metabolic processes, sucrose, sugar, starch.

1. Introduction

Extensive work over the past 15 years has examined the importance of fructose 2,6-bisphosphate (Fru-2,6-P_2), an inhibitor of cytosolic FBPase, in regulating the partitioning of fixed carbon between sucrose and starch. This work has mainly been done in spinach leaves. In this plant, changes in the amounts of Fru-2,6-P_2 regulate the activity of cytosolic FBPase, and under certain conditions this can regulate partitioning of fixed carbon between sucrose and starch [1]. Although this model almost certainly applies to many species other than spinach, its general applicability should not be taken for granted. In this poster I examine the significance of starch and sucrose as intermediary carbon stores in mature wheat leaves, and correlate this with measurements of Fru-2,6-P_2. I have addressed this using two experimental systems. 1) What are the changes in the amounts of sucrose, starch and Fru-2,6-P_2 during the normal day / night growth period. 2) How do the amounts of Fru-2,6-P_2 and the fluxes into sucrose and starch vary in plants photosynthesising under different light intensities. The results suggest that although changes in Fru-2,6-P_2 may regulate carbohydrate partitioning between the night and the day, the metabolite is not important in regulating partitioning when photosynthesis rates are varied by changing the light intensity. Further experiments using a wider range of experimental conditions and/or transgenic plants are required to examine in much greater detail the role of this metabolite in wheat leaves.

2. Procedure

Wheat plants (Triticum aestivum c.v. Bob White) were grown in a mixture of 50% compost, 50% perlite in a controlled environment cabinet (18°C 14h day, 14°C 10 h night; 150μmol.m^{-2}.s^{-1} irradiance) for four weeks. The third fully expanded leaf was selected for measurement at this stage. Gas exchange measurements and $^{14}CO_2$ feeding experiments with individual leaves were made using an infrared gas analyser (ADC, Hoddesdon, U.K.). Leaves were frozen in $N_{2(l)}$ and extracted in about 10 volumes 50mM KOH. Aliquots for measurement of sugars were heated at 70°C for 20 minutes and then assayed according to standard spectrophotometric techniques. Starch in the insoluble pellet was measured as glucose released after digestion with α–amylase and amyloglucosidase essentially as described in [2]. Fru-2,6-P_2 was measured in aliquots heated at 80°C for 10 minutes and then treated with activated charcoal (10mg/ml, 10 minutes on ice). Measurement of Fru-2,6-P_2 was essentially as

G. Garab (ed.), Photosynthesis: Mechanisms and Effects, Vol. V, 3533–3536.
© 1998 Kluwer Academic Publishers. Printed in the Netherlands.

described previously [3], except that the concentration of HCl used during acid treatment of the extract was reduced five fold. [14]C in the neutral components of the aqueous fraction (99% sucrose) was measured after ion exchange; ^{14}C in starch was measured as label solubilised after digestion α–amylase and amyloglucosidase. Chlorophyll was measured in 80% acetone as described in [4].

3. Results and Discussion

3.1 *Intermediary carbon storage during the diurnal cycle and the role of Fru-2,6-P$_2$.*

Changes in the amounts of sucrose and starch during a 2d normal growth period are shown in Figure 1. This clearly shows that both are used as intermediary carbon stores in this plant (about 80% of the fixed carbon was stored as sucrose and about 20% as starch). Interestingly, the rates of synthesis of both sugars did not vary much throughout this period, suggesting that feedback inhibition of sucrose synthesis does not occur in wheat leaves of this age from plants grown under the specific conditions used here. This is very different from the decrease in rates of sucrose synthesis seen during the day in plants such as spinach [5]. These wheat plants were grown under low light (150 $\mu E.m^{-2}.s^{-1}$) and it may be that even at the end of the day they do not exceed their capacity for sucrose storage. Similarly, rates of breakdown of both sucrose and starch were relatively constant at all times of the night period.

Amounts of Fru-2,6-P$_2$ were measured in samples from the second 24h period. As shown in Figure 1, there was a two-fold decrease in amounts within 1h of the light in the growth cabinet being switched on. Similarly, there was a two-fold increase in amounts of this metabolite within 1h of the plants going into the dark. These changes in Fru-2,6-P$_2$ correlate with differences in the rates of sucrose synthesis between the night and the day, and are consistent with a role for the metabolite in regulating sucrose synthesis. By contrast neither the rates of sucrose synthesis nor the amounts of Fru-2,6-P$_2$ changed dramatically during either the light or the dark periods. Again these results are entirely consistent with a role for Fru-2,6-P$_2$ in regulating the partitioning of carbohydrate between sucrose and starch in wheat leaves.

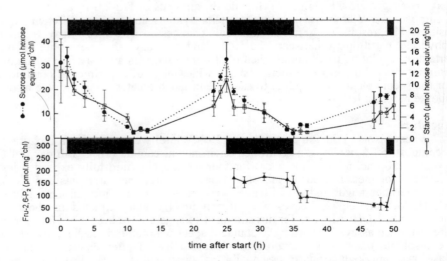

Figure 1 Changes in sugars, starch and Fru-2,6-P$_2$ in wheat leaves during a 48 h growth period

3.2 *Regulation of carbohydrate partitioning by Fru-2,6-P$_2$ during light induced changes in photosynthesis.*

In the second experiment I varied rates of photosynthesis by altering the light intensity, and examined relationships between amounts of Fru-2,6-P$_2$ and partitioning of $^{14}CO_2$ into sucrose and starch. Plants used were taken at the end of the photoperiod. Amounts of Fru-2,6-P$_2$ decreased as photosynthesis rates increased from 0.5 to 1.5 $\mu mol.mg^{-1}Chl$ (Figure 2a, curve fitting used a linear plus exponential fit, $r^2 = 0.49$). This decrease is less than, but consistent with those seen in spinach [6] and barley [7]. Interestingly, amounts of Fru-2,6-P$_2$ increased as photosynthesis rates increased above about 1.5$\mu mol.min^{-1}.mg^{-1}Chl$. This high light induced increase in Fru-2,6-P$_2$ has previously been observed with spinach leaf discs at 14°C [8].

The incorporation of $^{14}CO_2$ into all fractions examined (starch, sucrose, and the basic fraction) increased with increases in photosynthesis rates (not shown). Figure 2b shows the relationship between the rate of sucrose synthesis and amounts of Fru-2,6-P$_2$. This appears to shows two phases, depending on the rate of photosynthesis. At low rates of photosynthesis, a light induced increase in sucrose synthesis was accompanied by a decrease in the amount of Fru-2,6-P$_2$. However at higher rates of photosynthesis (above about 1.5$\mu mol.min^{-1}.mg^{-1}Chl$) further increases in sucrose synthesis were correlated with increases in the amount of Fru-2,6-P$_2$. Although potentially interesting, this correlation was observed for incorporation of label into all other fractions examined and its significance is uncertain.

There was apparently very little incorporation of label into starch at low rates of photosynthesis, leading to a higher sucrose:starch incorporation ratio at low light (Figure 2c) (the ratios at 0.5 and 0.9$\mu mol.min^{-1}.mg^{-1}Chl$ are significantly different by a student's t-test, assuming no variance in the photosynthesis measurements; p=0.012). The sucrose:starch incorporation ratio did not alter as photosynthesis rates were increased above 0.9 $\mu mol.min^{-1}.mg^{-1}Chl$. Figure 2d shows the relationship between amounts of Fru-2,6-P$_2$ and the sucrose:starch incorporation ratio in wheat leaves with different photosynthesis rates. These data allow me to assess the significance of Fru-2,6-P$_2$ in regulating carbohydrate partitioning in this system. (i) As photosynthesis was increased by raising the light intensity from 150 to 1150 $\mu E.m^{-2}.s^{-1}$ there were changes in Fru-2,6-P$_2$ but no changes in partitioning between sucrose and starch. The conclusion from this is that Fru-2,6-P$_2$ does not regulate carbohydrate partitioning in wheat leaves when photosynthesis is varied by altering light intensity. This could be because either (a) Fru-2,6-P$_2$ does not regulate carbohydrate partitioning in wheat leaves, (b) the changes in amount are too small to affect partitioning, or (c) compartmentation of Fru-2,6-P$_2$ between different cell types masks the relevant changes in amounts. (ii) As photosynthesis was increased by raising the light intensity from 75 to 150 $\mu E.m^{-2}.s^{-1}$ there was a decrease in Fru-2,6-P$_2$ and a possible *decrease* in the sucrose:starch incorporation ratio. This is not what would be expected if Fru-2,6-P$_2$ was regulating carbohydrate partitioning (a decrease in Fru-2,6-P$_2$ should lead to an *increase* in the sucrose:starch incorporation ratio), but can be explained as follows. At low photosynthesis rates 3-PGA will fall, activating PFKII. This will lead to a rise in Fru-2,6-P$_2$ and a partial inhibition of FBPase. However, the low 3-PGA:Pi ratio will inhibit chloroplastic ADPG-PPase, resulting in photosynthate being used for the synthesis of sucrose rather than starch. Previous work feeding $^{14}CO_2$ to tomato leaf discs [9] also demonstrated in increase in partitioning into sucrose at low light ($\leq$400$\mu E.m^{-2}.s^{-1}$), with no effects on partitioning at higher light. However amounts of Fru-2,6-P$_2$ were not measured in these experiments.

3536

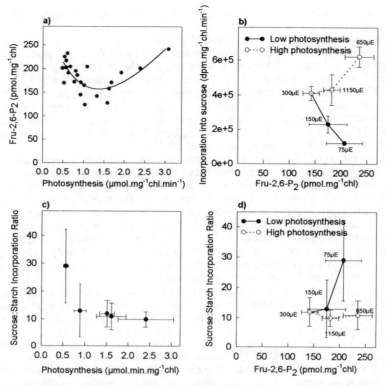

Figure 2 Changes in Fru-2,6-P₂ and carbohydrate partitioning with light induced changes in the rate of photosynthesis.

References

1 Stitt, M. (1990) Annu.Rev. Plant Physiol. Mol. Biol. 41, 153-185

2 Coombs, J., hind, G., Leegood, R.C., Tieszen, L.L. and Vonshak (1985) In Techniques in Bioproductivity and Photosynthesis 2nd edition (Coombs, J., Hall, D.O., Long, S.P. and Scurlock, J.M.O., eds.) pp. 219-228, Pergamon Press, London

3 Stitt, M. (1990) in Methods in Plant Biochemistry (Dey, P.M. and Harbourne, J.B., eds.) pp. 87-92, Academic Press Ltd., London

4 Porra, R.J., Thompson, W.A. and Kriedemann, P.E. (1989) Biochem. Biophys. Acta 975, 384-394

5 Stitt, M., Gerhardt, R., Kürzel, B. and Heldt, H.W. (1983) Plant Physiol. 72, 1139-1141

6 Neuhaus, H.E. and Stitt, M. (1989) Planta 179, 51-60

7 Sicher, R.C. and Bunce, J.A. (1985) Plant Physiol. Biochem. 25, 525-530

8 Stitt, M. and Grosse, H. (1988) J. Plant Physiol. 133: 392-400

9 Galtier, N., Foyer, C.H., Murchie, E., Alred, R., Quick, P., Voelker, T.A., Thépenier, C., Lascève, G. & Betsche, T. (1995) J. Exp. Bot. **46**: 1335-1344

SYNTHESIS OF FLORIDEAN STARCH IN THE RED ALGA *GRACILARIA GRACILIS* OCCURS VIA ADP-GLUCOSE.

Juliana I. Sesma and Alberto A. Iglesias. Instituto Tecnológico de Chascomús (INTECH, SECyT-CONICET), Camino Circunv. Laguna km 6, CC 164, Chascomús, 7130, Bs. As., Argentina.

Key words: algae, C partitioning, enzymes, polysaccharides, red algae, starch.

Introduction.
In higher plants and green algae starch is the major product of photosynthesis (1). The route conducting to polysaccharide synthesis in these organisms has been well characterized, taking place by utilizing ADPGlc as the glucosyl donor (for reviews see 1-3). Moreover, ADPGlc pyrophosphorylase (AGP; EC 2.7.7.27), the enzyme catalyzing the production of the sugar nucleotide in the presence of a divalent metal ion (Mg^{2+}):

$$Glc1P + ATP \Leftrightarrow ADPGlc + PPi$$

is the key regulatory enzyme in the biosynthetic pathway (1-4). AGP from higher plants, green algae and cyanobacteria is allosterically regulated by 3P-glycerate (3PGA, activator) and Pi (inhibitor). It has been clearly demonstrated that the amount of starch accumulated depends upon the fine regulation of the enzyme by these metabolites (4).
Rodophyta (red algae) is one of the major lineages of the algae, being a very distinct group of organisms, phylogenetically distant from green plants and comprising over 10,000 species (5-7). Many of these are economically important species, utilized in food industry and for the production of polysaccharides and phycocolloids (i.e. agar) (5). In red algae, photosynthetically fixed carbon is mainly partitioned into two products: agar and floridean starch (a reserve polysaccharide characteristically found in these organisms) (5). Studies on the characterization of the metabolic pathways for the formation and degradation of polysaccharides in red algae are scarce, thus being the understanding of the carbon partitioning in these organisms very limited.
In this work we study the process of starch biosynthesis in the red alga *Gracilaria gracilis*, determining the sugar nucleotide utilized as glucosyl donor and characterizing the properties of a key enzyme of the pathway.

Materials and Methods.
All specimens of *G. gracilis* were kindly provided by Dr. M.L. Piriz (Centro Nacional Patagónico, Chubut, Argentina). Freshly harvested material, collected at Golfo Nuevo (Mar Argentino) was exhaustively washed using sea and distilled water and then stored at -80°C until use. Frozen material was resuspended in a medium containing 25 mM Tris-HCl buffer pH 7.5, 1 mM EDTA, 0.5 mM PMSF and 10 mM 2-mercaptoethanol, then homogeneized using a mortar and pestle and centrifuged at 4°C. The resulting supernatant is referred as the crude extract.
Incorporation of [^{14}C]Glc into glycogen was carried out in a medium containing 10 µmol Mops-KOH pH 7.5, 5 µmol KCl, 1 µmol DTT, 0.1 mg glycogen, 0.2 µmol

G. Garab (ed.), Photosynthesis: Mechanisms and Effects, Vol. V, 3537–3540.

[^{14}C]sugar nucleotide (UDP- or ADP-Glc; about 150 cpm/nmol) and crude extract in a final volume of 150 µl. After 45 min incubation at 37°C the reaction was stopped by the addition of 2 ml 75% methanol containing 1% KCl. Amount of ^{14}C incorporated into glycogen was determined as described (8).

Purification of AGP was performed at 4°C in a medium identical to that utilized for obtaining the crude extract except for the absence of PMSF. Successive chromatographic steps include columns of DEAE-cellulose, Affi-gel Blue and Mono Q.

AGP activity was assayed in the ADPglucose synthesis (assay A) and pyrophosphorolysis (assay B) directions as previously described (9). For assay A, the standard reaction mixture (200 µl final volume) contained 20 µmol Tris-HCl buffer pH 7.5, 50µg BSA, 1 µmol MgCl$_2$, 0.4 µmol ATP, 0.1 µmol [^{14}C]Glc1P (about 1000 cpm/nmol) and 0.3 units of inorganic pyrophosphatase. Assays were initiated by the addition of enzyme and, after 60 min at 25°C, they were stopped by heating at 100°C during 30 s. For assay B, the standard assay milieu contained 10 µmol Mops-KOH buffer pH 7.5, 0.6 µmol MgCl$_2$, 0.6 µmol NaF, 25 µg BSA, 0.5 µmol DTT, 0.3 µmol ADPglucose, 0.2 µmol [^{32}P]PPi (about 4000 cpm/nmol) and enzyme in a final volume of 125 µl. The reaction was runned during 60 min at 25°C and stopped by the addition of 1 ml 5% TCA. For assays (A or B) under activated conditions, 1.5 mM 3PGA was included in the reaction mixtures.

Results and Discussion.

In crude extracts from the red alga *G. gracilis* the ratio UDPGlc:ADPGlc pyrophosphorylase activity was about 3. However, as shown in Table I, ADPGlc (but not UDPGlc) was effective as a substrate in the incorporation of [^{14}C]Glc into glycogen catalyzed by the extracts. The incorporation was proportional to time of incubation and amount of sample only when ADPGlc was added as the glucosyl donor. These results suggest that synthesis of starch in red algae takes place via ADPGlc, as occurs in higher plants and green algae (1-4).

To further study the occurrence and regulation of the metabolic pathway, AGP from *G. gracilis* was partially purified and its properties characterized. Purification procedure included ion exchange and affinity chromatographies rendering an about 40% pure enzyme according to SDS-PAGE. The native molecular mass of the enzyme was estimated in 170±5 kDa by size exclusion chromatography on FPLC (Superose 12 column) (Figure 1). Maximal activity of the enzyme was obtained at 25°C and pH 7.5 or 8.0, depending upon the buffer used (Mops or Tris, respectively) (data not shown).

Table II shows the effect of different metabolites on the pyrophosphorolytic activity of AGP from *G. gracilis*. As shown, the enzyme was activated by 3P-glycerate (3PGA), Glc6P, Fru6P, P-enolpyruvate (PEP) and, to a lesser extend by Fru1,6bisP. On the other hand, Pi behaved as a potent inhibitor of the enzyme, whereas pyruvate exhibited a slight inhibitory effect (Table II). Other compounds tested: NAD$^+$, NADP$^+$, AMP, and NH$_4$Cl showed no effect on the enzyme activity (data not shown). Results indicate that AGP from red algae is distinctively regulated by key compounds of carbohydrates metabolism in these organisms.

Table I. Incorporation of $[^{14}C]$Glc into glycogen catalyzed by crude extracts from *G. gracilis*.

Addition	nmol ^{14}C in glycogen / 10 min x mg
$[^{14}C]$ UDPGlc	< 0.4
$[^{14}C]$ ADPGlc	248

Table II. Effect of different metabolites on the activity (pyrophosphorolytic direction) of AGP from *G. gracilis*.

Compound *	nmol ATP/ 10 min x ml	Relative Activity
None	1.1	1.0
3PGA	15.1	14.0
Glc6P	12.9	12.3
Fru6P	10.6	10.1
PEP	10.4	9.9
Fru1,6bisP	2.5	2.4
Pyruvate	0.8	0.8
Pi	< 0.01	< 0.01

*All compounds were added at 3 mM concentration.

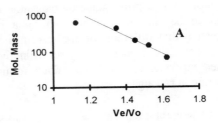

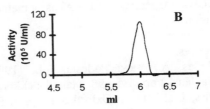

Figure 1. Molecular mass estimation of AGP from *G. gracilis*.
A. Calibration plot of the Superose 12 FPLC size exclusion column. Molecular mass markers utilized are: thyroglobulin 669 kDa, apoferritin 443 kDa, ß-amylase 200 kDa, alcohol dehydrogenase 150 kDa and BSA 66 kDa.
B. FPLC elution profile of AGP runned in the purification buffer stated under Materials and Methods supplemented with 150 mM KCl.

The very distinctive properties exhibited by red algae include metabolic and structural differences respect to those found in green plants (5-7). In this way, floridean starch, the characteristic reserve polysaccharide accumulated by red algae, possesses different physical and chemical properties than starches from green plants (5). Despite these differences, results shown in the present work indicate that synthesis of floridean starch occurs by using ADPGlc as the glucosyl donor, in a similar way to that found in higher plants and green algae (1-4). For the latter organisms, AGP is the key regulatory step of the biosynthetic pathway, the enzyme being allosterically regulated by metabolites strictly related with the main carbohydrate route of the organism (2,3). In the same sense, the regulatory properties exhibited by AGP from *G. gracilis*, suggest that it catalyzes a main regulatory step in polysaccharide biosynthesis. However, activity of AGP from red algae seems to be allosterically affected by a broader number of metabolites, this result suggesting that different metabolic routes for carbohydrates may be important in the whole metabolic scenario of these distinctive photosynthetic organisms. It is clear that further experimental work is nedded to understand the complex carbon partitioning occurring in red algae.

Acknowledgements. This work was supported, in part, by grants from Fundación Antorchas and Comisión de Investigaciones Científicas (CIC). JIS is a fellow from CONICET, Argentina, and AAI is a member of the Research Career from the same Institution.

References.
1. Iglesias, A.A. and Podestá, F.E. (1996) In: Handbook of Photosynthesis (M. Pessarakli, ed.) pp. 681-698. Marcel Dekker, New York.
2. Preiss, J. (1991) In: Oxford Surveys of Plant Molecular and Cell Biology (J. Miflin, ed.) pp. 59-114. Oxford Univ. Press, Oxford.
3. Iglesias, A.A. and Preiss, J. (1992) Biochem. Education 20, 196-203.
4. Stark, D.M., Timmerman, K.P., Barry, G.F., Preiss, J. and Kishore, G.M. (1992) Science 258, 287-292.
5. Yu, S. (1992) Ph Thesis, Uppsala University, Uppsala, Sweden.
6. Stiller, J.W. and Hall, B.D. (1997) Proc. Natl. Acad. Sci. USA 94, 4520-4525.
7. Bhattacharya, D. and Medlin, L. (1998) Plant Physiol. 116, 9-15.
8. Eidels, L. and Preiss, J. (1970) Arch. Biochem. Biophys. 140, 75-89.
9. Ghosh, H.P. and Preiss, J. (1966) J. Biol. Chem. 241, 4491-4504.

OVEREXPRESSION OF PYROPHOSPHATE-DEPENDENT PHOSPHOFRUCTOKINASE (PFP) IN TRANSGENIC TOBACCO

Monika M. Kuzma, Steven P. King, Maryse Chalifoux, Susan M. Wood, Stephen D. Blakeley, David T. Dennis. Performance Plants Inc., Biosciences Complex, Queen's University, Kingston, On, Canada K7L 3N6.

Key words: C-metabolism, C-partitioning, Transgenic plants, PFP

1. Introduction

PPi- dependent phosphofructokinase (PFP) catalyzes a reversible cytosolic interconversion of fructose-6-phosphate and fructose-1,6-bisphosphate. This reaction is also catalyzed in the glycolytic direction by ATP-dependent phosphofructokinase (PFK) and in the gluconeogenic direction by fructose-1,6- bisphosphatase (FBPase). The role of PFP in plants is not clear but many functions have been proposed: a role in glycolysis, a role in gluconeogenesis, regulation of PPi content in the cytosol, equilibration of hexose-P and triose-P pools and a role in adaptation to stress (eg. low Pi, low temp., anaerobiosis) (1). To elucidate the role of PFP in plants previous studies examined transgenic tobacco (2) and transgenic potato (1) with strongly reduced expression of PFP. However, no adverse effects of this reduction were found on plant growth and metabolism under both control and stress (low Pi, low N, low temp., anaerobiosis) conditions. To gain further insight into the role of PFP, the present study examined the effect of overexpression of unregulated PFP on plant metabolism in tobacco.

2. Materials and Methods

Nicotiana tabacum L. SR1 was transformed with a gene coding for the PFP from *Giardia lamblia* (not regulated by F-2,6-P$_2$) cloned in the sense direction behind a tandem cauliflower mosaic virus 35S promoter. Transgenic plants were screened using Southern analysis, western analysis and PFP enzyme assays. T$_3$ and T$_4$ generations of transgenics (PFP) and their siblings without the transgene (Controls) were grown in a randomized complete block design in a growth chamber or a greenhouse to assess their performance. The following parameters were measured: seedling vigor (seedling length), vegetative growth of plants (total leaf area), leaf photosynthesis at both unsaturating (300 umol m^{-2} s^{-1}) and saturating (1200 umol m^{-2} s^{-1}) light and ambient CO$_2$ (350ppm) and leaf respiration in the dark (CO$_2$ gas exchange), sucrose and starch contents of young fully expanded leaves, composition of mature and developing seeds (fiber, lipid, protein, carbohydrates) and structure of young developing seeds (light microscopy).

G. Garab (ed.), Photosynthesis: Mechanisms and Effects, Vol. V, 3541–3544.
© 1998 Kluwer Academic Publishers. Printed in the Netherlands.

3. Results and Discussion

Since levels of *Giardia* PFP expression in the leaves were similar in the 4 transgenic lines examined, data were pooled for all transgenic lines (PFP) and all control lines (Control). PFP transgenics were shown to have 12 to 17 fold higher PFP total extractable activity than Controls while other enzymes remained unaffected (Fig. 1).

Figure. 1. Enzyme activities in fully expanded photosynthetically active leaves of transgenic (PFP) and Control tobacco plants (n=12).

No differences were observed in the seedling length over the first two weeks of growth suggesting that overexpression of PFP in tobacco does not affect

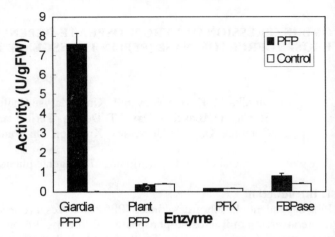

seedling vigor (data not shown). During early vegetative growth the total leaf area of the PFP plants was slightly higher than that of controls but this difference became insignificant as plants matured (data not shown). Lack of differences in the vegetative growth between transgenics and Controls was reflected by lack of significant differences in their photosynthetic rates, both at ambient light and CO_2 (Fig. 2) and at high light and ambient CO_2.

Similarly, leaf respiration in the dark was not significantly altered between transgenics and controls (Fig. 2).

Figure. 2. Leaf photosynthesis at ambient light (300 umol m^{-2} s^{-1}) and CO_2 (350ppm) and leaf respiration in the dark of transgenic (PFP) and control plants (n=11 for both).

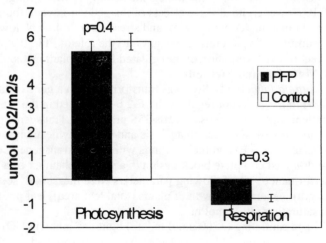

No significant change in the sucrose content of leaves was observed (data not shown) but a significant reduction in starch of transgenic leaves both at the beginning and 12 h into the photoperiod was detected (Fig. 3) suggesting changes in carbon partitioning. Seed composition of mature seed was not affected by PFP overexpression but in developing seeds differences were observed.

Figure 3. Starch content of fully developed leaves of PFP transgenic and Control tobacco plants at the beginning of photoperiod (7:00) and 12 hours into the photoperiod (n=16 for PFP, n=13 for control).

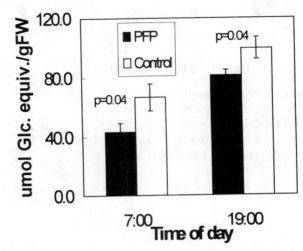

A range of *Giardia* PFP activities was found in developing seeds 9 days after anthesis (DAA) and this activity was positively correlated with soluble protein and lipid content while it was negatively correlated with fiber content (Fig. 4). Additional evidence for higher lipid content of transgenic young seeds was provided by micrographs of 9DAA seeds. The transgenic seeds were shown to contain more lipid droplets than the Control seeds (Fig. 5).

4. Conclusions
(a) Under optimal conditions overexpression of PFP in tobacco does not affect the growth and vegetative performance of the plants.
(b) Differential partitioning of imported carbon during transgenic tobacco seed development results in earlier accumulation of lipids and protein in young seeds.
(c) Fiber content of young seeds may be reduced in higher expressing transgenics.

5. Acknowledgements
(a) *Giardia* PFP DNA – M. Muller, Rockefeller University, (b) Transformations – B.L. Miki, Agriculture and Agrifood Canada, (c) NSERC Industrial Fellowships (MMK, SPK, SMW)

6. References
1) Hajirezaei M., Sonnewald U., Viola R., Carlisle S., Dennis D., Stitt M. (1994) Planta 192:16-30
2) Paul M., Sonnewald U., Hajirezaei M., Dennis D., Stitt M. (1995)Planta **196**: 277-283

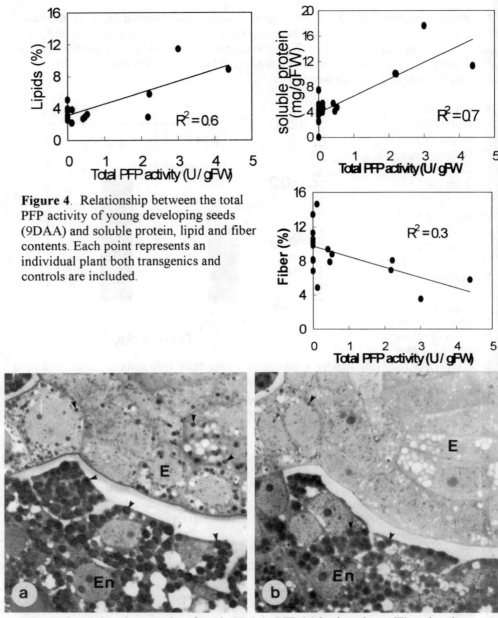

Figure 4. Relationship between the total PFP activity of young developing seeds (9DAA) and soluble protein, lipid and fiber contents. Each point represents an individual plant both transgenics and controls are included.

Figure 5. Light micrographs of seeds 9DAA. PFP (a) both embryo (E) and endosperm (En) cells are more densely packed with lipid droplets (small arrows) than in Control (b seeds. Mag. 1120x

MODULATION OF THIOREDOXIN AFFINITY FOR CHLOROPLASTIC FRUCTOSE-1,6-BISPHOSPHATASE BY ELECTROSTATIC INTERACTIONS

Santiago Mora-García, Paula D. Duek and Ricardo A. Wolosiuk
Instituto de Investigaciones Bioquímicas, Antonio Machado 151
1405, Buenos Aires, Argentina

Key words: thiol/disulfide exchange, protein-protein interactions, disulfide bonds, site-directed mutagenesis, enzymes

1. Introduction

Thioredoxins (Trx) are small proteins found in the cytoplasm of all living cells. They catalyze thiol/disulfide exchange reactions with other proteins by means of two cysteine residues that alternate between reduced and oxidized states (1). Biosynthesis of deoxyribonucleotides (2), DNA transcription (3) and carbon and sulfur assimilation are subject to redox modulation. In particular, Trxs play a central role in oxygenic photosynthetic metabolism, as a link between the redox state of photosystems and the activity of soluble enzymes. *In vitro*, the reduction of disulfide bonds in chloroplast fructose-1,6-bisphosphatase (CFBPase) enhances the capacity of the enzyme to hydrolyze fructose 1,6-bisphosphate at low concentrations of Mg^{2+} (4). Two Trxs coexist in plant chloroplasts: Trx-f, of eukaryotic lineage, is most efficient in the reductive activation of CFBPase, whereas Trx-m, related to prokaryotic Trxs, is ineffective. However, modification on non-covalent interactions in CFBPase causes drastic changes in its affinity for less proficient Trxs. Not only do protein perturbants devoid of redox activity (cosolvents, chaotropic anions, high pressure) accelerate the reduction of the enzyme (4), but also inefficient modulators such as *E. coli* Trx become indistinguishable from Trx-f when the activation is carried out in the presence of fructose 1,6-bisphosphate and Ca^{2+} (5). Electrostatic interactions are relevant for the interaction between proteins (6). If the electrostatic component played a role in the association between Trxs and CFBPase, it should be possible to change the association rate by altering the surface net charge on one, or both, partners. Also, the modulation of the enzyme should be sensitive to the shielding of electrostatic interactions by neutral salts (7). We describe herein the stimulatory effect of positive charges introduced near the active site of *E. coli* Trx. Moreover, affinities for the enzyme were alike when the activation of CFBPase was performed in the presence of high concentrations of KCl, irrespective the modulator is highly efficient (e.g Trx-f) or highly inefficient (e.g. Trx-m). These features are congruent with a crucial role of the electrostatic component in the affinity of Trxs for CFBPase.

3545

G. Garab (ed.), Photosynthesis: Mechanisms and Effects, Vol. V, 3545–3548.
© 1998 Kluwer Academic Publishers. Printed in the Netherlands.

2. Experimental Procedures

2.1 *Materials and General Methods:* Enzymes and solutions for DNA manipulations were used according to manufacturers (New England Biolabs, Promega). Oligonucleotides were purchased from NBI or GenSet, and biochemical reagents from Sigma-Aldrich. Protein concentrations were determined by the method of Lowry (8). A molar extinction coefficient of 13,700 $M^{-1}cm^{-1}$ (9) and 20,600 $M^{-1}.cm^{-1}$ were used to quantitate *E. coli* Trx and Trx-m, respectively. CFBPase and Trx-f were purified from fresh spinach leaves, according to Stein and Wolosiuk (10).

2.2 *Expression and Purification of Recombinant Proteins:* The entire sequences of *E. coli* Trx and mature *B.napus* Trx-m, flanked both by NdeI and EcoRI and BamHI, respectively, restriction sites, were cloned into the plasmid pET22b+, and transformed into the overexpression strain JM109/DE3. Details for the isolation and preparation of Trx-m will be given elsewhere. Recombinant protein expression was induced by the addition of IPTG to exponentially growing cultures. Homogeneous preparations of Trxs were obtained after cell lysis and chromatography through columns of Sephadex G-50 and DEAE-Sepharose Fast Flow.

2.3 *Site-directed Mutagenesis:* Site directed mutagenesis of *E. coli* Trx was accomplished using the "megaprimer" method described by Sarkar and Sommer (11). The primers used for the mutagenesis were (in bold type, the introduced mutations): E30K: TTTCTGGGCA**AA**GTGGTGCGGT; L94K: GACCTTTAGAC**TTT**GCACCCA.

2.4 *Protein Disulfide Reductase Activity:* The activity of Trxs as general protein disulfide reductant was assessed in the presence of a dithiol (DTT) by the novel fluorometric substrate di-fluoresceinthiocarbamil-insulin (di-FTC insulin) (12). The reduction process was monitored in a Jasco FP-770 spectrofluorometer by following the emission intensity at 519 nm when the fluorophore was excited at 495 nm. Maximum rate of change of emission was used for the calculation of the protein disulfide reductase activity.

2.5 *CFBPase activity:* The activity was analyzed by the two-stage assay (13).

3. Results and Discussion

Determinants in chloroplastic Trxs that contribute to the specific recognition of target proteins should be located at regions that do not participate in redox reactions, since both proteins share the same active site sequence. Analysis of the molecular surface of Trxs indicated that they all display an extended, non-charged surface in the immediate vicinity of the active site, but that large patches of negatively charged residues are exposed surrounding this region on Trx-m and *E. coli* Trx, whereas Trx-f displays a more balanced distribution. This observation matches the differences in pI found for both proteins (5.01 for Trx-m, 6.1 for Trx-f). On the other hand, cysteine residues involved in the reductive regulation in CFBPase are located in a negatively charged region, providing grounds to the hypothesis that electrostatic interactions might be involved in the initial steps of the heterocomplex formation (14). Therefore, we investigated whether novel positive charges on the surface of *E. coli* Trx could influence its affinity for CFBPase. We introduced two mutations (E30K and L94K) at positions exposed to the solvent, highly variable among Trxs and not involved in defined structures.

In our studies on chloroplast enzymes, we analyzed the reductive activation by separating the conversion of the enzyme to the catalytically active form (modulation), from the transformation of substrates (catalysis). In this scheme, $A_{0.5}$, the concentration of modulator required to obtain half of the maximal specific activity, is the kinetic constant that defines the modulation process. Table 1 shows that *E. coli* and chloroplast Trx-m are much less effective ($A_{0.5}=31$ µM and 82 µM, respectively) than chloroplast Trx-f ($A_{0.5}=0.9$ µM) in the reductive activation of CFBPase. However, the incorporation of two positively charged side chains on the surface of *E. coli* Trx brought about a marked reduction in the $A_{0.5}$ (2.5 µM). Congruent with the participation of electrostatic components in the formation of a non-covalent complex between Trx and CFBPase, high salt concentrations (0.5 M KCl) lowered the $A_{0.5}$ of *E. coli* Trx and Trx-m (10 and 12 µM respectively), but at the same time increased that of Trx-f and of E30K/L94K *E.coli* Trx (4 µM and 7 µM respectively). The convergence of $A_{0.5}$'s to a limiting value (*ca.* 9 µM) indicates that Trxs behave in much the same way when electrostatic interactions are screened out; i.e., all share similar surface features near the active site.

Table 2 illustrates that Trxs clearly showed different behaviour in the cleavage of cystines in a model protein as insulin. Both chloroplastic Trxs were significantly less efficient than wild-type or mutant *E. coli* Trx. These results evince the importance of the affinity between interacting proteins for the reductive step to take place.

Thus, a collision event leads to the formation of a non-covalent complex between Trx and CFBPase, prior to the thiol-disulfide exchange reaction. Our results, along with previous ones on spinach Trx-f and -m (15), suggest that the reductive activation of CFBPase is mostly driven by the rate of the association step, which in turn is dependent on long-range electrostatic components. These interactions should help in the discrimination between both chloroplastic Trxs.

4. Acknowkledgments

This work was supported by grants from University of Buenos Aires (UBA), Consejo Nacional de Investigaciones Científicas y Técnicas (CONICET) and Fondo Nacional para la Investigación Científica y Técnica (FONCYT). SMG and PDD are recipient of fellowships from CONICET, and RAW is a research member of the same institution.

REFERENCES
1.- Eklund, H., Gleason, F.K. and Holmgren, A. (1991) Proteins. Struct. Funct. and Gen. 11, 13-28.
2.- Laurent, T.C., Moore, E.C. and Reichard, P. (1964) J. Biol. Chem. 239, 3436-3444.
3.- Matthews, J.R., Wakasugi, N., Virelizier, J-L., Yodoi, J. and Hay, R.T. (1992) Nucleic Acid Res. 20, 3821-3830.
4.- Wolosiuk, R.A., Ballicora, M.A. and Hagelin, K. (1993) FASEB J. 7, 622-637.
5.- Mora-Garcia, S., Ballicora, M.A. and Wolosiuk, R.A. (1996) FEBS Lett. 380, 123-126.
6.- Stone, R.S., Dennis, S. and Hofsteenge, J. (1989) Biochemistry 28, 6857-6863.

7.- Schreiber, G., and Fersht, A.R. (1996) Nature Struct. Biol. 3, 427-431.

8.- Lowry, O.H., Rosebrough, N.J., Farr, A.L., and Randall, R.J. (1951) J. Biol. Chem. 193, 265-275.

9.- Reuntimann, H., Straub, B., Luisi, P.L., and Holmgren, A. (1981) J. Biol. Chem. 256, 6796-6803.

10.- Stein, M. and Wolosiuk, R.A. (1987) J. Biol. Chem. 262, 16171-16179.

11.- Sarkar, G., and Sommer, S. (1990) Biotechniques 8, 404-407.

12.- Heuck, A.P. and Wolosiuk, R.A. (1997) Anal.Biochem. 248, 94-101.

13.- Hertig, C.M. and Wolosiuk, R.A. (1983) J. Biol. Chem. 258, 984-989.

14.- Rodriguez-Suarez, R., Mora-Garcia, S. and Wolosiuk, R.A. (1997) Biochem. Biophys. Res. Commun. 232, 388-393.

15.- Geck, M, Larimer, F y Hartman, F (1996) J. Biol. Chem. 271, 24736-24740.

TABLE 1. Effect of KCl on the $A_{0.5}$ of Trx

Trx	Control	0.5 M KCl
chloroplast Trx-f	0.9	4
chloroplast Trx-m	82	12
wild type *E. coli*	31	10
E30K/L94K *E. coli*	2.5	7

The $A_{0.5}$ was measured by following the activation of CFBPase in the presence of varying concentrations of Trx (01. μM to 130 μM).

TABLE 2. Protein disulfide reductase activity of Trx

Trx	PDR activity (units. min^{-1})
chloroplast Trx-f	0.012
chloroplast Trx-m	0.003
wild type *E. coli*	0.029
E30K/L94K *E. coli*	0.038

The PDR activity of 1 μM Trx was estimated as described under Experimental Procedures.

CHEMICAL CROSS-LINKING BETWEEN FRUCTOSE-1,6-BISPHOSPHATASE AND THIOREDOXIN F.

Ocón, E., Hermoso, R. and Lázaro. J.J.
Biochemistry Department. Estación Experimental del Zaidín. CSIC.
Granada. Spain.

Key words: Calvin-cycle, complex formation, protein-protein interaction.

Introduction

Fructose-1,6-bisphosphatase (FBPase), a key enzyme in the regulation of the reductive pentose-phosphate cycle, is modulated by thioredoxin f (Trx f), that links photosynthetic electron transport to the enzyme (1). Both proteins form an associated complex that is important to the reductive activation. This protein-protein interaction is favoured at low ionic strength and a restricted pH range (2). Earlier results suggested that the regulatory site preceding region in the FBPase, which shows a high concentration of acid residues, is the docking point (3, 4) and that, conversely, amino residues from Trx f are likely to be involved in the complex formation (5).

In order to further investigate this interaction, we tried to stabilize the complex between the two proteins by chemical cross-linking. With this purpose we have used the water-soluble cross-linking agent 1-ethyl-3-[3-(dimethylamino)propyl] carbodiimide (EDC), which catalyzes the formation of an amide bond between carboxylates and neighbouring primary amines, the very residues that are likely to be involved in the electrostatic interaction. This reaction proceeds in two steps. The carboxyl group is activated by carbodiimide to form an o-acylisourea intermediate, which can react further in a subsequent step with the amino group displacing the urea derivative with the formation of a peptide bond between both proteins.

Materials and Methods

FBPase and Trx f were purified from cell lysates after expression in *E. coli* (6, 7). Trx f-dependent FBPase activity was determined according to published methods (8). Antisera against FBPase and Trx f were raised in rabbits, and specific test as well as ELISA analysis were carried out according to Hermoso et al (9). SDS-PAGE electrophoresis (10) was performed using 7% and 4% (w/v) acrylamide for the separating and stacking gel, respectively. Covalent cross-linking was carried out with both proteins in 10mM phosphate buffer, pH 7.9, in the presence of 0.4mM FBP and 5mM EDC; after incubation for 30min at 28°C the reaction was stopped by the addition of electrophoresis loading buffer and the samples were boiled before the application to SDS-electrophoresis gels. Chromatography of FBPase and Trx f after reaction with EDC was performed on a Pharmacia model FPLC system using a calibrated Superose 12 column preequilibrated with 30mM Tris-HCl pH 7.9, 200mM NaCl. Fractions (100µl) were immunologically determined by ELISA. The reaction

G. Garab (ed.), Photosynthesis: Mechanisms and Effects, Vol. V, 3549–3552.

was carried out as above, but was stopped by adding ammonium acetate to 0.1M final concentration.

Results and Discussion

Our results indicate that FBPase cross-links to Trx f in the presence of EDC, with the appearance by SDS-PAGE of a new molecular mass component of about 56kDa. This cross-linked product showed positive cross-reactivity to antibodies raised against both proteins (Fig. 1). We have also detected a component of higher molecular mass with the same positive cross-reactivity, probably a dimer of the monomeric complex. Subunits of FBPase cross-link by means of EDC, but this products decrease with the presence of Trx f.

There is a relationship between covalent complex and the ratio of FBPase and Trx f. With a fixed amount of one protein, the amount of the complex increased with the rise of the other protein (Fig. 2).

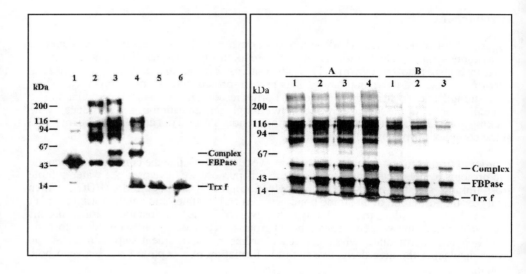

Fig. 1 (left). Covalent complex formation between FBPase and Trx f. Western-blotting developed with antibodies anti-FBPase (lanes 1, 2 and 3) and anti-Trx f (lanes 4, 5 and 6). Lane 1: FBPase. Lane 2: FBPase+EDC. Lanes 3 and 4: FBPase+Trx f+EDC. Lane 5: Trx f+EDC. Lane 6: Trx f. FBPase concentration was $1\mu M$, Trx f was $65\mu M$. Molecular mass markers (in kDa) were myosin (200), β-galactosidase (116), phosphorilase b (94), bovine serum albumin (67), ovoalbumin (43) and α-lactalbumin (14).

Fig. 2 (right). Dependance of the ratio of FBPase and Trx f on the cross-linking reactions. Reactions were carried out as described in Materials and Methods and analyzed by SDS-PAGE. A) The FBPase concentration was $1\mu M$ in all cases. The Trx f concentration was $8\mu M$ (lane 1), $24\mu M$ (lane 2), $40\mu M$ (lane 3), or $55\mu M$ (lane 4). B) The Trx f concentration was $55\mu M$ in all cases. The FBPase concentration was $0.5\mu M$ (lane 1), $0.3\mu M$ (lane 2) or $0.16\mu M$ (lane 3). Molecular mass markers (in kDa) were as in Fig. 1.

The efficiency of cross-linking was strongly dependent on the ionic strength (Fig. 3). The amount of cross-linked material decreased up to 200mM NaCl and was nil at 300mM NaCl. Even though the FBPase-Trx f complex did not retain enzyme activity, this was consistent with the Trx f-dependent FBPase activity (Fig. 3) and confirmed the strong contribution of electrostatic forces to the interaction. This suggests that formation of a specific functional electrostatic complex between these proteins is previously required for the covalent complex formation. Pea FBPase contains an exposed acid cluster which is involved in the electrostatic binding with Trx f (3). Furthermore, we have evidence that four available basic amino acids of pea Trx f are responsible for linkage with FBPase (11).

The amount of cross-linked product was also dependent on the pH and, as in the case of the ionic strength, there was a positive correlation with the Trx f-dependent FBPase activity (Fig. 4). For the three pHs tested (7.0, 7.5 and 7.9), the highest amount of covalent complex and activity was at pH 7.9, decreased at pH 7.5, whereas at pH 7.0 the activity was negligible and the amount of complex was the lowest. This is consistent with the "in vivo" rise in the stromal pH (from 7.0 to about 7.9) during the dark-light transition, and provides additional evidence that pH is important in this interaction (2).

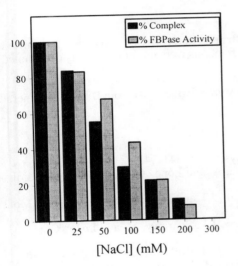
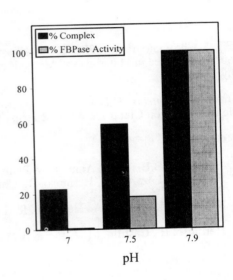

Fig.3 (left). Effect of ionic strength on the FBPase-Trx f covalent complex formation and on the Trx f-dependent FBPase activity. The percentage of complex corresponds to a densitometric measure of a gel stained with Coomassie Brilliant Blue R.

Fig. 4 (right). Effect of pH on the FBPase-Trx f covalent complex formation and on the Trx f-dependent FBPase activity. Conditions as in Fig. 3.

When FBPase and Trx f were chromatographied after reaction with EDC, four fractions appeared positive to Trx f antibodies (ELISA), two of them being also positive to FBPase. This elution profile suggests the presence of cross-linked products between FBPase (tetramer and dimer) and Trx f (Fig. 5).

3552

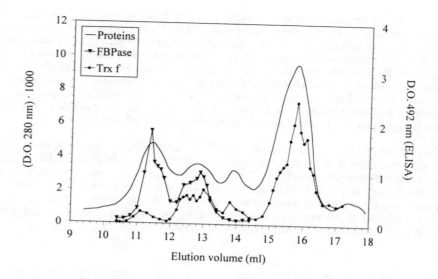

Fig. 5. Chromatography of FBPase and Trx f after reaction with EDC. Reaction was carried out as described in Materials and Methods and fractions were immunologically determined by ELISA.

Acknowledgements: This work was supported by grant PB95-0081 from Dirección General de Investigación Científica y Técnica (Spain).

References

1 Buchanan, B.B. (1980) Annu. Rev. Plant Physiol. 31, 341-347.
2 Reche, A., Lázaro, J.J., Hermoso, R., Chueca, A. and López Gorgé, J. (1997) Physiol. Plant. 101, 463-470.
3 Sahrawy, M., Chueca, A., Hermoso, R., Lázaro, J.J. and López Gorgé, J. (1997) J. Mol. Biol. 269, 623-630.
4 Hermoso, R., Castillo, M., Chueca, A., Lázaro, J.J., Sahrawy, M. and López Gorgé, J. (1996) Plant Mol. Biol. 30, 455-465.
5 Geck, R.K., Larimer, F.W., Hartman, F.C. (1996) J. Biol. Chem. 271, 24736-24740.
6 Carrasco, J.L., Chueca, A., Prado, F.E., Hermoso, R., Lázaro, J.J., Ramos, J.L., Sahrawy, M. and López Gorgé, J. (1994) Planta 193, 494-501.
7 Hodges, M., Miginiac-Maslow, M., Decottignies, P., Jacquot, J.P., Stein, M., Lepiniec, L., Crétin, C. and Gadal, P. (1994) Plant Mol. Biol. 26, 225-234.
8 Prado, F.E., Lázaro, J.J., Hermoso, R., Chueca, A. and López Gorgé, J. (1992) Planta, 188, 345-353.
9 Hermoso, R., Chueca, A., Lázaro, J.J. and López Gorgé, J. (1987) Photosynth. Res. 14, 269-278.
10 Laemmli, U.K. (1970) Nature 227, 680-685.
11 López-Jaramillo, J., Chueca, A., Jacquot, J.P., Hermoso, R., Lázaro, J.J., Sahrawy, M. and López Gorgé, J. (1997) Plant Physiol. 114, 1169-1175.

TWO CHLOROPLAST FRUCTOSE-1,6-BISPHOSPHATASES ARE EXPRESSED IN RAPESEED (*Brassica napus*)

Roberto J. Rodriguez-Suarez, Ana C. D'Alessio and Ricardo A. Wolosiuk. Instituto de Investigaciones Bioquimicas, Antonio Machado 151, 1405 Buenos Aires, Argentina.

Key words: fructose-1,6-bisphosphatase, chloroplast, isoforms, gene expression, Calvin cycle, enzymes.

1. Introduction

In leaf cells, the nuclear DNA codes for a large number of chloroplast proteins that are synthesized as precursors in the cytoplasm (1). After removal of the transit peptide at the chloroplast envelope, the mature polypeptide refolds and assembles as the functional entity. The paradigm for proteins imported by chloroplasts, the small subunit of Rubisco, exists as a multigene family in most of the plants examined; e.g. transcripts of RbcS1 and RbcS2 were found at high levels in bundle sheat cells of maize seedlings indicating that both genes are expressed in the C4-type plants (2). However, it remains unknown whether other important stromal proteins follow a similar trend; e.g. light-modulated enzymes of the Benson-Calvin cycle (3).

Another important issue regarding the expression of chloroplast proteins is that multiple homologous copies of genes are present in polyploid species (4). The expression of duplicated genes might lead to the multiplicity of proteins even though gene silencing could preclude the appearance of isoproteins (5). On this basis, we analyzed the expression of chloroplast fructose-1,6-bisphosphatase (CFBPase) in the allotetraploid *Brassica napus* (rapeseed). This species possesses the complete genomic complement of their parental diploids: *B. oleracea* and *B. campestris* (6). Here, we report a novel isoform of rapeseed CFBPase precursor whose primary structure is clearly different from that already described (7).

2. Experimental procedures

2.1 *Plant Material*: Rapeseed seeds (*Brassica napus* L., cv.Global) were purchased from Nidera S.A.(Buenos Aires, Argentina). Leaves, grown under field conditions, were harvested 30 days after germination, washed with deionised water and immediately processed for RNA preparation.

2.2 *cDNA library construction*: We followed the procedures described in (8) and (9) for the purification of total RNA and RNA-poly(A)+, respectively, and, we subsequently constructed the cDNA library between NotI/SalI sites of the vector pSPORT1 (SuperScript plasmid system, GIBCO-BRL, Bethesda, USA). Recombinant plasmids were electroporated in *Escherichia coli* DH5α cells by standard methods (9).

2.3 *Screening and DNA sequencing*: The cDNA library was screened for the expression of CFBPase with polyclonal antibodies raised in rabbits against the spinach enzyme and goat anti-rabbit IgG coupled to peroxidase (7). Nucleotide sequences

G. Garab (ed.), Photosynthesis: Mechanisms and Effects, Vol. V, 3553–3556.

were determined on both strands by the dideoxyribonucleotide termination method using "Sequenase version 2.0" (Amersham, UK).

3. Results and discussion

After screening for the expression of rapeseed CFBPase with an antibody raised against the spinach enzyme, the presence of CFBPase coding DNA was confirmed by Southern Blot using a probe derived from the wheat counterpart (kindly provided by Dr. T. Dyer) (11). These studies provided two full-length clones coding for the CFBPase precursor: BnpFBP1 and BnpFBP3, and we here report on the later clone because the former was previously characterized (7,12). The sequence of BnpFBP3 comprised 1370 nucleotides followed by a poly(A) tail. From the ATG triplet in the 5'-end started an open reading frame of 1254 bp (predicted polypeptide molecular mass 45038 Da) that left 14 bp and 102 bp in the 5'- and 3'-unstranslated regions, respectively (13). These data established that BnpFBP3 is 18 bp, 9 bp and 3 bp longer than BnpFBP1 in the open reading frame, the 5'- and the 3'-untranslated regions, respectively (7). The alignment of the deduced amino acid sequences of BnpFBP3 with counterparts of dicots - *B.napus* BnpFBP1 (7), *Arabidopsis thaliana* (14), *Solanum tuberosum* (15)- and monocots -*Triticum aestivum* (11)- clearly evinced two distinct regions in the CFBPase precursor: i.e., the transit peptide (Fig. 1) and the mature enzyme (Fig. 2) (16).

The 57-amino acid transit peptide lacked negatively charged amino acid residues but retained a high proportion of non-polar (47%), hydroxylated (26%) and positive (18%) residues (Fig. 1). The amino acid sequence was 77%, 71%, 21% and 40% identical to *B.napus* BnpFBP1, *A.thaliana, S.tuberosum,* and *T.aestivum* CFBPases, respectively. Moreover, the transit peptide of BnpFBP3 showed the conserved amino acid sequence –(G/N)V(R/K)CM- within the C-terminal region. This sequence, followed by –A(V/I)- in the mature enzyme, represents the site for cleavage by a specific protease within the chloroplast. In line with this feature, alanine was the unique N-terminus of CFBPases hitherto analyzed by Edman degradation (16).

The deduced amino acid sequence of the mature form of BnpFBP3 (molecular mass 39202 Da) not only showed high identity to BnpFBP1 (95%) but also to *A.thaliana* (94%) (Fig.2). This considerable homology in the primary structure of the functional polypeptide diminished when compared to *S.tuberosum* (87%) and *T.aestivum* (79%). Alignment of amino acid sequences of mature CFBPases from a variety of higher plants has shown that the N-terminal region and the sequence spanning from Cys-158 to Cys-176 -i.e., the 170's loop (17)- are highly variable with respect to BnpFBP1. BnpFBP3 CFBPase had insertions in these regions [Ala-13, Asp-162] and, except for one position [168], the 170's loop of BnpFBP3 is identical to *A.thaliana* but different from BnpFBP1. The high homology of both the transit peptide and the mature form of BnpFBP3 and *A.thaliana* is congruent with the phylogenetic proximity of *A.thaliana* to *Brassica* species (18); in fact, both group under the *Brassicaceae* (formerly *Cruciferae*). In this context, it was interesting to find that the expression of the

BnpFBP3 mature form in *E.coli* provides a recombinant enzyme with bisphosphatase activity (not shown).

Thus, rapeseed contains homologous genes, transcribed under field conditions, coding for enzymatically active isoforms of CFBPase. Further molecular studies are currently underway to establish whether *B.napus* sequences, BnpFBP1 and BnpFBP3, belong to homologous loci on different genomic complements, i.e. those from *B.oleracea* and *B.campestris,* or conform an already undescribed gene family in *Brassica* species.

4. Acknowledgments

This work was supported by grants from University of Buenos Aires (UBA.), Consejo Nacional de Investigaciones Científicas y Técnicas (CONICET) and Fondo Nacional para la Investigación Científica y Técnica (FONCYT). RRS and ACD are recipient of fellowships from UBA and RAW is a research member of CONICET.

REFERENCES

1.- Cline, K. and Henry, R. (1996) Annu. Rev. Cell Dev. Biol. 12, 1-26.
2.- Ewing, R.M., Jenkins, G.I. and Langdale, J.A. (1998) Plant Molec. Biol. 36, 593-599.
3.- Wolosiuk, R.A., Ballicora, M.A. and Hagelin, K. (1993) FASEB J. 7, 622-637.
4.- Lagercrantz, U. and Lydiate, D.J. (1996) Genetics 144, 1903-1910.
5.- Meyer, P. and Saedler, H. (1996) Annu. Rev. Plant Physiol. Mol. Biol. 47, 23-48.
6.- Kimber, D.S. and McGregor, D.I. eds. (1995) *Brassica* Oilseeds: Production and utilization. CAB International, University Press. Cambridge, UK.
7.- Rodriguez-Suarez, R. and Wolosiuk, R.A. (1993) Plant Physiol. 103, 1453-1454.
8.- Chomczynski, P. and Sacchi, N. (1987) Anal. Biochem. 162,156-159.
9.- Sambrook, J., Fritsch, E., and Maniatis, T. (1989) Molecular Cloning. A laboratory manual, 2nd Ed., Cold Spring Harbor Laboratory Press, Cold Spring Harbor, NY.
10.- Ausubel, F.M., Brent, R., Kingston, R., Moore, E.W., Seidman, J.G., Smith, J.A., and Struhl, K. (1995). Current Protocols in Molecular Biology, John Wiley & Sons, NY.
11.- Raines, C.A., Lloyd, J.C., Longstaff, M., Bradley, D. and Dyer, T. (1988) Nucleic Acids Res. 16, 7931-7942.
12.- Rodriguez-Suarez, R.J. and Wolosiuk, R.A. (1995) Photosynth. Res. 46, 313-322
13.- Joshi, C.P., Zhou, H., Huang, X. and Chiang, V.L. (1997) Plant Mol. Biol 35, 993-1001.
14.- Horsnell, P.R. and Raines, C.A. (1991) Plant Mol. Biol. 17, 185-186.
15.- Kossmann, J., Muller-Rober, B., Dyer, T.A., Raines, C.A., Sonnewald, U. and Willmitzer, L. (1992) Planta 188, 7-12.
16.- Rodriguez-Suarez, R.J. and Wolosiuk, R.A. (1995) *in* Photosynthesis: from light to biosphere (Mathis, P. ed.) vol.V, 167-170. Kluwer Academic Publishers, Dordrecht, The Netherlands.
17.- Villeret V., Huang, S., Zhang, Y., Xue, Y. and Lipscomb, W.N. (1995) Biochemistry 34, 4299-4306.
18.- Lagercrantz, U., Putterill, J., Coupland, G. and Lydiate, D. (1996) Plant J. 9, 13-20.

Fig. 1. Amino acid alignment of deduced BnpFBP3 CFBPase transit peptide with counterparts of other species. In all five sequences, totally and highly conserved residues were shaded in dark and light gray, respectively. Gaps are indicated by dashes. Total amino acid residues are between parenthesis. The numbering follows the largest sequence; i.e. *A.thaliana.*

Fig. 2. Amino acid alignment of deduced CFBPases. Shading and gaps were similar to Fig 1. The numbering follows the largest sequence; i.e. BnpFBP3 *B.napus.*

3556

Fig. 1.

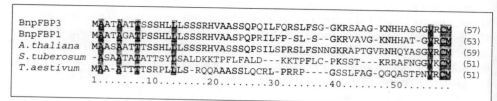

```
BnpFBP3      MAATAATTSSSHLLLSSSRHVAASSQPQILFQRSLFSG-GKRSAAG-KNHHASGGVRCM  (57)
BnpFBP1      MAATAGATPSSHLLLSSSRHVAASPQPRILFP-SL-S--GKRVAVG-KNHHAT-GVRCM  (53)
A.thaliana   MAASAATTTSSHLLLSSSRHVASSSQPSILSPRSLFSNNGKRAPTGVRNHQYASGVRCM  (59)
S.tuberosum  -ASAATATATTSYLSALDKKTPFLFALD---KKTPFLC-PKSST---KRRAFNGGVKCM  (51)
T.aestivum   MAA-ATTTTSRPLLLS-RQQAAASSLQCRL-PRRP----GSSLFAG-QGQASTPNVRCM  (51)
             1........10........20........30........40........50........
```

Fig. 2.

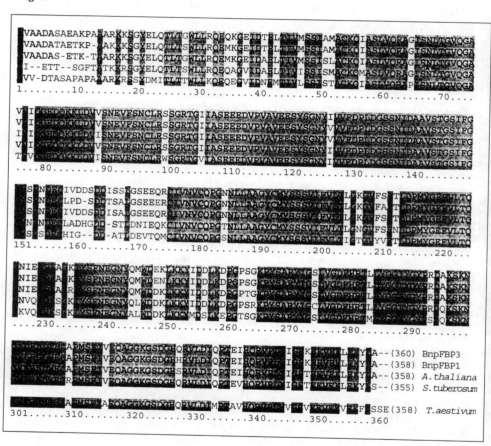

EFFECT OF ALTERING POD INTENSITY ON MONOCARPIC SENESCENCE IN COWPEA

SRIVALLI, B. & RENU KHANNA-CHOPRA
Water Technology Centre, I.A.R.I., New Delhi-110 012, India

Key Words: Carbohydrate, proteases, Rubisco, senescence, sink strength.

1. Introduction

Monocarpic senescence is a genetically programmed decline in physiological functions which are induced during fruit development following a single reproductive phase. Leaf senescence, is subject to regulation by many environmental and internal factors, where the internal factors include reproductive development (Gan and Amasino, 1997)[1]. In legumes, each reproductive node behaves as an independent unit, where the leaf at a particular reproductive node supports the fruits developing in its axil (Grover *et al.*, 1985)[2]. In the present investigation, it has been examined whether (I) the regulation of sink strength on monocarpic senescence is due to altered protease activity, using the in vivo substrate, Rubisco and a non-plant substrate, casein and (ii) the neutral protease activity has a role in nitrogen mobilisation or not. The results suggest that the reproductive sink regulates monocarpic senescence, the removal of which delays it, also making the source leaf as a temporary carbohydrate reserve. These observations are related to the previously observed (Khanna-Chopra and Reddy, 1988)[3] changes in leaf area, chlorophyll content and total soluble proteins.

2. Materials and Methods

A determinate cowpea (*Vigna unguiculata*) var. Komal was grown in the field following standard agronomic practices. During early pod development stage (3 days after fertilization), pod number was maintained as 2 pods or 1 pod or 0 pod per node respectively. The leaf and pod at sixth node were used for all parameters under study and sampling was done from the day of treatment up to pod maturity at 4 days interval. The plant materials were dried in an oven at 80^0C to obtain dry matter. Three replicates were maintained for all measurements. For the biochemical assays leaves were cut into small pieces after measuring their fresh weight. The biochemical parameters include chlorophyll content (Arnon, 1949)[4], total soluble proteins (Lowry *et al.*, 1951)[5], total sugars (Dubois *et al.*, 1956)[6], starch (Reed and Singletary, 1989)[7] and protease activity

3557

G. Garab (ed.), Photosynthesis: Mechanisms and Effects, Vol. V, 3557–3560.
© *1998 Kluwer Academic Publishers. Printed in the Netherlands.*

following the slightly modified method of Feller $(1979)^8$, using casein and Rubisco as substrates. One unit of proteolytic activity was defined as an increment of 0.01 in A_{340} in 1 hr. Photosynthesis rate was measured in the morning between 10-11 AM using Licor 6200 portable photosynthesis instrument (Licor Inc., USA). Total nitrogen in leaves and fruits was determined before dried samples by Kjeltech 1030 autoanalyser (Tecator, USA) after digestion in sulphuric acid.

3. Results & Discussion

The seed wt./node reduced from 2.06 g to 1.3 g in 2P to 1P and the decrease in seed N/node was from 71.2 mg to 53.3 mg in 2P to 1P leading to a final reduction of 38% and 25% respectively. There was a compensation of 14% and 20% in seed wt./seed and seed N/seed respectively. Decreasing the sink intensity reduced the rate of senescence as revealed by the loss of chlorophyll and total soluble proteins (*Figures 1A and 1B*). This is in agreement with the earlier report (3). The rate of loss of nitrogen was similar in 2P and 1P plants. The largest source of N remobilization is the Rubisco protein, the loss of which may be responsible for the high rate of loss of total soluble proteins (*Figure 1B*).

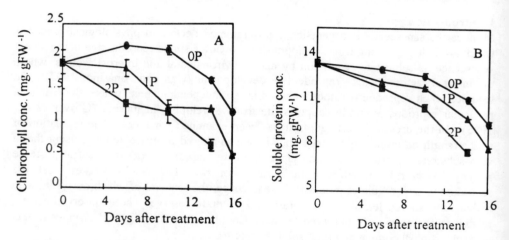

"Figure 1." Effect of altered sink treatment on chlorophyll concentration (A) and total soluble proteins (B) in cowpea leaves.

In 0P plants, protease activity remained unchanged during pod development (*Figures 2A & B*). This may be regulating the slower rate of degradation of proteins. So far, increased neutral protease activity has not been correlated with nitrogen mobilization (8). Our study shows a significant correlation between neutral protease activity with casein as substrate and loss in N from the leaf tissues (*Figure 3*).

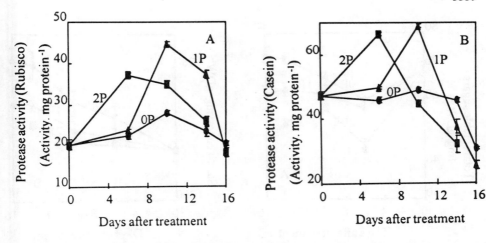

"Figure 2." Effect of altered sink treatment on protease activity with Rubisco (A) and casein (B) as substrates respectively in cowpea leaves.

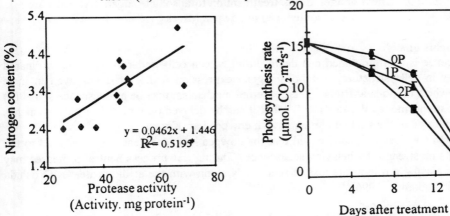

"Figure 3." Correlation between the protease activity at pH 7.5 and the loss in nitrogen (%) in the leaf tissue (*P<0.05* and r=0.721).

"Figure 4." Effect of altered sink treatment on photosynthesis rate in cowpea.

The rate of decline in the photosynthesis rate was slowed down on the removal of the sink (*Figure 4.*). This occured in spite of the accumulation of photosynthesis products in the leaves (*Figures 5A & 5B*). Thus, in cowpea, photosynthesis products do not show feedback inhibition on the photosynthesis rate as shown in studies in soybean (Miceli *et al.*, 1995)[9].

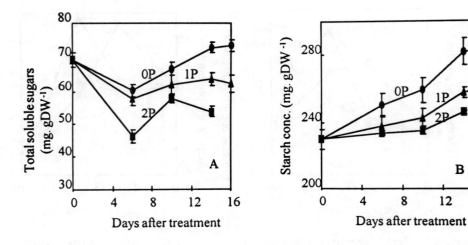

"Figure 5." Effect of altered sink treatment on total soluble sugars (A) and starch content (B) in cowpea leaves.

4. Conclusions

Senescence is a programmed cell death during which cells undergo highly co-ordinated changes in cell structure, metabolism gene expression and it proceeds even in male plants which do no bear fruits. The triggering mechanism may be a senescence signal (3) or a death hormone (Wilson, 1997)[10], but it can be delayed on removal of the sink as has been revealed in the study undertaken with cowpea var. Komal.

In conclusion, it can be said that the cowpea leaf senescence does respond to the altered sink strength by delaying senescence. During this process neutral proteases may also be playing a role in the loss of reduced N along with the acidic proteases (Yoshida and Minamikawa, 1996)[11].

References

1 Gan, S. and Amasino, R.M. (1997) Plant Physiol. 113, 313-319.

2 Grover, A. Koundal, K.R. and Sinha, S.K. (1985) Physiol. Plant. 63, 87-92.

3 Khanna-Chopra, R. and Reddy, P.V. (1988) Ann. Bot. 61, 655-658.

4 Arnon, D.I. (1949) Plant Physiol. 24, 1-15

5 Lowry, O.H. Roserbrough, N.J. Farr, A.L. and Randall, R.J. (1951) J. Biol. Chem. 193, 265-275.

6 Dubois, N. Gilles, K. Hamilton, J.K. Rebers, P.A. and Smith, F. (1956) Nature 168, 167.

7 Reed and Singletary, 1989 Reed, A.J. Singletary, G.W. (1989). Plant Physiol. 91, 986-992.

8 Feller, U. (1979) Plant & Cell Physiol. 20(8) 1577-1583.

9 Miceli, F. Crafts-Brandner, S.J. and Egli, D.B. (1995) Crop Sci. 35, 1080-1085.

10 Wilson, J.B. (1997) Physiol. Plant. 99, 551-516.

11 Yoshida, T. and Minamikawa, T. (1996) Eur. J. Biochem. 238, 317-324.

EFFECTS OF A MAIZE SPS GENE EXPRESSED IN RICE ON ITS ACTIVITY AND CARBON PARTITIONING

K. ONO, K. FURUKAWA[1], T. MURATA[1], K. ISHIMARU, K. OZAWA, Y. OHKAWA and R. OHSUGI

Natl. Inst. Agrobiol. Resources, Tsukuba, 305-8602, Japan, [1]Fac. Agric. Iwate Univ., Morioka, 020-0066, Japan

Key words: carbohydrate, enzymes, higher plants, transgenic plants

Introduction

It is generally thought that sucrose-phosphate synthase (SPS) can be a limiting enzyme of sucrose synthesis and photosynthesis. With soybean plants, Huber and Israel (1) showed that leaf starch contents were negatively correlated with the SPS activities (r = −0.71) and concluded that SPS is a key control point regulating the formation of not only sucrose but also starch. The partitioning of carbon between sucrose and starch is different among species (2). To our knowledge, transgenic plants over-expressing a maize SPS gene were made from only starch-formers such as tomato (3,4,5) and *Arabidopsis* (6). As demonstrated previously with those transgenic plants, over-expressing maize SPS increased the ratios of sucrose/starch accumulated in leaves (4,6). It is an open question whether the same effect is obtained in a sucrose-former, such as rice, or not. We investigated the effect of over-expressing a maize SPS gene on carbon partitioning in a sucrose-former, rice.

Materials and Methods

The maize SPS cDNA was cloned based on the sequence described by Worrell et al. (3). The pBS-SK containing a maize SPS cDNA linked to the cab promoter of rice (7) was introduced by electorpolation. T2 generations (indicated by number of T1 plant −number of individual plant, 54-3, 54-8, 54-9, 59-1, 69-6, 69-7, 100-4, 100-9, 139-4, 139-5 and 139-9) of transgenic plants and control (wild type) plants were grown under ambient condition. Transgenic rice plants were selected by the presence of the maize SPS cDNA verified by PCR or the expression of the transgene by northern hybridization (data not shown). Flag leaf blades were sampled between 2:00 p.m. and 2:30 p.m. of the day two days after flowering. Leaf blades were immediately frozen with liquid nitrogen and stored at −80°C until measurement. Frozen flag leaf blades were ground in liquid nitrogen to a fine powder with a moter and pestle, and soluble protein was extracted with a buffer containing 100 mM

3561

G. Garab (ed.), Photosynthesis: Mechanisms and Effects, Vol. V, 3561–3564.
© 1998 *Kluwer Academic Publishers. Printed in the Netherlands.*

sodium phosphate buffer (pH7.0), 1% (w/v) insoluble PVP, 1% (v/v) 2-mercaptoethanol and 1 mM PMSF. After centrifugation, 12 μg or 24 μg soluble leaf protein was subjected to SDS-PAGE on a 12.5% (w/v) polyacrylamide gel. After electrophoresis, proteins were transferred to a nitrocellulose and probed with a rabbit polyclonal antibody raised against maize SPS or rice SPS and alkaline phosphatase-conjugated goat anti-rabbit immunoglobulin G antibody. The amounts of SPS protein were quantified using an imazing analyzer. SPS activity was measured by a slight modification of Lunn and Hatch (8). For sucrose and starch measurement, frozen samples were ground with a morter and a pestle in liquid nitrogen into fine powder and suspended in 80% ethanol. Supernatants were combined, dried up, resuspended in water and used to determine the sucrose content. Sucrose was assayed enzymatically (9). The pellets were boiled in water, digested with amyloglucosidase and used to detect starch as newly generated glucoses enzymatically according to Keppler and Decker (10).

Results and Discussion
Western blotting
The quantities of maize SPS protein expressed in the transgenic rices differed among plants (Fig.1A). The antibody against maize SPS protein did not crossreacted with rice SPS protein. Some transgenic plants (100-9, 139-4, 100-4, 139-9 and 139-5) had almost no maize SPS protein and others (69-7, 54-9 and 54-8) had large amounts of maize SPS protein. Another analysis was also carried out with an anti-rice SPS antibody (Fig.1B).

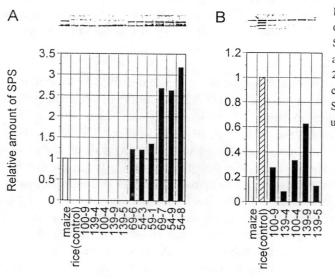

Fig.1 Western-blot analysis of SPS protein using maize SPS antibody (A) and rice SPS antibody (B). 12 μg (A) or 24 μg (B) protein was applied each lane. The amounts of SPS protein were quantified using an imazing analyzer.

Some transgenic rice plants (100-9, 139-4, 100-4, 139-9 and 139-5) had less amounts of rice SPS protein than control rice plants. Since rice SPS cDNA (11) has 78.6% homology to maize SPS cDNA (3), these decrease may be caused by co-suppression.

SPS activity

SPS activity of the control plant was close to saturating with 10 mM UDPG (Fig. 2). SPS activity was higher in transgenic rice plant with a large amount of SPS protein (TL) (54-8) than in the control plant and did not saturate with 20 mM UDPG. Under the same condition, SPS activity of the control or the transgenic rice plant with a small amount of SPS protein (TS) (100-9) was saturated. SPS activity of the TL (54-8) was higher than that of the control plant, even when measured at low substrate concentrations. The apparent Vmax values of SPS were calculated by Lineweaver-Burk plot (Fig. 3). SPS activity of TL (54-8) was about ten folds as high as those of the control plants. SPS activities in TS (100-9 and 139-4) were one-third of those of the control plants. The amounts of maize SPS protein were positively correlated with maximal SPS activities (r = 0.867).

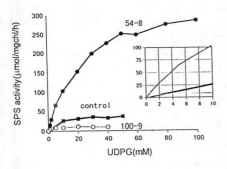

Fig.2　SPS activities in the transgenic rice plants. Flag leaves were harvested between 2:00 p.m. and 2:30 p.m. and the activity of SPS was measured with 8 mM F6P, 27.6 mM G6P and different concentration of UDPG in the reaction mixture. Control is a wild type rice plant. 54-8 is a TL and 100-9 is a TS. The values are the results of the one experiment using one of the flag leaf blades of the plant.

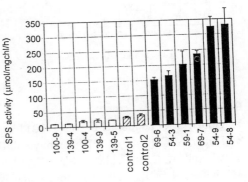

Fig.3　Maximal SPS activities in the flag leaf blades of the transgenic rice plants and the control plants. The values are calculated by Lineweaver-Burk plot. The values are the means of three different flag leaves of the same plants. Bars indicate standard error.

3564

Carbon partitioning

Sucrose and starch were measured in the flag leaves of the control and transgenic plants (Fig. 4). No marked changes were observed in the leaf sucrose contents among plants ($r^2 = 0.372$)(Fig. 4A). On the other hand, TL had lower starch contents than controls (Fig. 4B). The starch contents were negatively correlated with the SPS activities ($r^2 = 0.733$). The sucrose/starch ratios were positively correlated with the SPS activities ($r^2 = 0.898$) (Fig. 4C). Transgenic tomato plants with high SPS activities had more than twice values of sucrose/starch than control plants (4). Sucrose/starch ratio increased about 60% in one of the TL compared with that of the control rice (Fig. 4C). Our result suggested that sucrose/starch ratios increased with SPS activities even in a sucrose-former (Fig.4) and that the effect on carbon partitioning was greater in a starch-former, such as tomato than in a sucrose-former, such as rice.

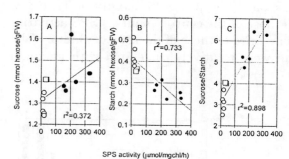

SPS activity (μmol/mgchl/h)

Fig.4 The relationship between the maximal SPS activity and the sucrose content (A), the starch content (B) and sucrose/starch ratio (C) of the flag leaves of the transgenic rice plants and the control plants. The values are the means of three different flag leaf blades of the same plants.
□; control, ○; TS, ●; TL.

References
1. Huber, S. C. and Israel D. W. (1982) Plant Physiol. 69, 691-696
2. Huber, S. C. (1981) Z. Pflanzenphysiol. Bd. 101, 49-54
3. Worrell, A. C. et al. (1991) Plant Cell 3,1121-1130
4. Galtier, N. et al. (1995) in Photosynthesis: from light to biosphere, Vol.V. (Mathis P., ed.) pp.301-304, Kluwer Academic Publishers, The Neherlands
5. Micallef, B. et al. (1995) Planta 196, 327-334
6. Signora, L. et al. (1998) J. Exp. Bot. 49, 669-680
7. Tada, Y. et al. (1991) EMBO J. 10, 1803-1808.
8. Lunn, J. E. and Hatch, M. D. (1997) Aust. J. Plant Physiol. 24, 1-8
9. Bergmeyer, H. U. and Bernt, E. (1974) in Methods of enzymatic analysis, 2nd ed, vlo3 (Bergmeyer, H. U. ed) pp. 1176-1179, Academic Press, Inc., New York
10. Keppler, D. and Decker, K. (1974) in Methods of enzymatic analysis, 2nd ed, vlo3 (Bergmeyer, H. U. ed) pp. 1127-1131, Academic Press, Inc., New York
11. Valdez-Alarcon, J. J. et al. (1996) Gene 170, 217-222.

EXPLOITATION OF NON-PHOTOSYNTHETIC ENZYMES TO OPERATE A C$_4$-TYPE CCM WHEN [CO$_2$] IS LOW

Srinath K. Rao, Julia B. Reiskind, Noël C. Magnin, and
George Bowes, Department of Botany, University of Florida,
Gainesville FL, 32611-8526, USA

Key words: aquatic ecosystems, cDNA cloning, CO$_2$ concentration, PEPC, phylogeny, sequence analysis, Kranz anatomy

1. Introduction

Hydrilla verticillata is a submersed monocot that typically functions as a C$_3$ plant. However when CO$_2$ is low, C$_4$-type photosynthesis is induced, but without Kranz anatomy (1). This is the only higher plant known to operate a C$_4$-type CO$_2$ concentrating mechanism (CCM) without separating into mesophyll and bundlesheath cells the initial carboxylation from the fixation by Rubisco. Instead, phosphoenolpyruvate carboxylase (PEPC) is located in the cytosol while the decarboxylase (NADP-malic enzyme: NADP-ME) and Rubisco are in the chloroplasts, where CO$_2$ is concentrated to 400 μM (2,3). All the C$_4$-cycle enzymes increase substantially in activity under low [CO$_2$], including PEPC, pyruvate Pi dikinase (PPDK), and NADP-ME. Western blots show that during C$_4$ induction the proteins for these enzymes are up-regulated (3).

Hydrilla is in the Hydrocharitaceae, a family in the Alismatidae, which can be traced back at least to the Paleocene using *rbc*L sequence analysis (4). Some fossil evidence places *Hydrilla* in the upper Eocene of 35-mya (5). In which case it may predate modern terrestrial C$_4$ monocots that became abundant in the Miocene of 7 to 10 mya. Terrestrial C$_4$ photosynthesis possibly arose in response to declining atmospheric CO$_2$ in the Cretaceous (6). However, in densely-vegetated waters daytime dissolved [CO$_2$] are not in air-equilibrium, and can decline to near zero even when atmospheric CO$_2$ is high (1). Thus, it is likely that submersed species were exposed to low CO$_2$ environments before terrestrial species. The common occurrence of CCMs among cyanobacteria and microalgae attests to this. These facts raise questions as to whether forms of C$_4$ photosynthesis arose in water before their advent on land, and whether non-photosynthetic enzymes were appropriated to operate the C$_4$ cycles.

G. Garab (ed.), Photosynthesis: Mechanisms and Effects, Vol. V, 3565–3570.
© 1998 *Kluwer Academic Publishers. Printed in the Netherlands.*

Because of the fundamental importance of PEPC to C_4 and CAM photosynthesis, PEPC sequences have been used for phylogenetic analysis, and results have been reported that correspond with those based on traditional methods (7,8). Such studies are consistent with terrestrial C_3 being ancestral to C_4 and CAM, and the latter two as polyphyletic in origin. CAM first appears in the Pteridophyta, but C_4 seems to have arisen after angiosperms appeared (7,8). However, we have evidence of a Kranz-less, C_4-type system operating in the Chlorophyta; in the marine, macroalga *Udotea* (9).

PEPC is encoded by a nuclear multigene family with isoforms that differ in function and location, and with species (10). It has been reported (8), using a partial, conservative sequence with the active site, that a housekeeping C_3 isoform is ubiquitous, and thus possibly ancestral to other isoforms. The monocot C_4 isoforms are clustered, distinct from C_3-monocot, C_4-dicot and CAM isoforms. The clustering of the C_4-monocot isoforms is consistent with an early origin for the genes by duplication (7,8,10).

The study reported here is an attempt to clarify the origins of aquatic C_4-type photosynthesis. Recent work with *Hydrilla* indicates it has at least three PEPC isoforms, two leaf forms and one root (11). Sequence homologies and a cladogram analysis using published PEPC sequences (8), paired to match those from *Hydrilla*, are used to explore the hypothesis that *Hydrilla* represents an early form of C_4-monocot photosynthesis.

2. Procedures

2.1 *Plant material, PEPC assays, and RNA isolation*
Hydrilla verticillata (L.f.) Royle in the C_3 state was collected from Lake Lochloosa, FL. Shoots were incubated to induce C_4-type photosynthesis (3). PEPC activity was monitored daily, and shoots, midway through the induction period, were harvested and homogenized (3). Filtered extracts were assayed spectrophotometrically (3). Soluble protein was determined with BSA as standard (12). Total RNA was extracted (Qiagen).

2.2 *PCR amplification, cDNA library construction, screening and cloning*
Two degenerate primers were designed (see Fig. 1). cDNAs were synthesized from total RNA, amplified with GEC119 and GEC120, and cloned into a pCR2.1 vector (Invitrogen). The cDNA library was constructed (Clontech), cloned in a λgt11 vector, and immunoscreened using antiserum against maize PEPC.

3. Results and Discussion

Two fragments of cDNA (1 kb and 0.9 kb) were amplified by PCR with the degenerate primers (GEC119 and GEC120). Their deduced amino acid sequences showed a 78 to 79% homology with the maize PEPC at the positions 602 to 964 (Fig.1).

A cDNA expression library was constructed, resulting in the identification of 16 positive primary clones. A PCR-based screening with primers (forward primer; 5'GCGAAGCAATATGGAGTGAAGTTGA3' and reverse primer; 5'TTGTACATTG TACCCTGGGTCCCTT3') specific to PEPC-I that amplified the 0.9 kb product resolved four positive clones. Restriction digestion with *Eco* RI followed by Southern

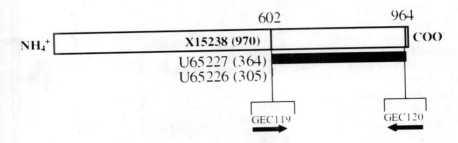

Figure 1. A representation showing the relative location of the conserved domains (shaded portions) with respect to the deduced amino acid sequence of *Zea mays* (accession number X15238) and the amplified portion of the two *Hydrilla* isoforms (accession numbers U65227 {PEPC-I} and U65226 {PEPC-II}) by RT-PCR. The arrows denote the direction of amplification with degenerate primers GEC119; 5'-TC(A/T/C) GA(C/T) TC(G/T/C) GG(C/T) AA(A/G) GA(C/T)-3'; and GEC120; 5'-GC(A/G/T) (G/C)(G/C)(A/G/T) A(C/T)(A/G) CCC TTC ATG GT-3'. The deduced amino acid number is in parenthesis.

analysis using a probe of the 0.9 kb fragment indicated a 5 kb insert of (data not shown). Three more clones, that did not give a PCR product with the above primers, did react positively to the 0.9 kb probe in Southerns, and showed a large insert (>12 kb). These clones were digested separately with *Eco* R1 and *Sal* 1 followed by Southern analysis (Fig. 2A). The pattern with *Eco* R1 showed 16 bands (4 kb to ~12 kb) and with *Sal* 1 two bands (6 and 12 kb). The 6 kb *Sal* I fragment was cloned in pBluescript (Stratagene) and analysed to confirm its identity (Fig. 2B).

The deduced amino acid sequence of the *Hydrilla* isoforms (PEPC-I, PEPC-II) were subjected to pairwise comparison together with eleven other organisms from different taxa (Table 1). The two *Hydrilla* PEPC sequences shared 97% homology, similar to that in *Flaveria*, as compared with 80% between two *Sorghum* leaf sequences. *Hydrilla* PEPC isoforms indicated good homology (79 to 86%) with other seed plants, as do the two seed plant CAM-isoforms (81 to 86%), which, however, showed only 73% with the *Isoetes* (Pteridophyte) CAM isoform. The bryophyte C_3 isoform was more homologous to the seed plants than to the Pteridophytes.

3568

Figure 2. Southern blot analysis of (A) three recombinant λgt11 clones digested with *Eco* R1 (lanes 2,3,4) and *Sal* I (lanes 5,6,7). Lane 1 is undigested recombinant clone. (B) Recombinant plasmid (pBluescript) with the released insert (arrow Fig. 2A) was cloned and analyzed further. Lane 1 is undigested supercoiled plasmid with insert, lanes 2 and 3 are *Eco* RI and *Sal* I digested recombinant plasmids, respectively. Lane 4 is the plasmid pCR2.1 with 1 kb cDNA encoding partial PEPC in *Hydrilla* (U65227).

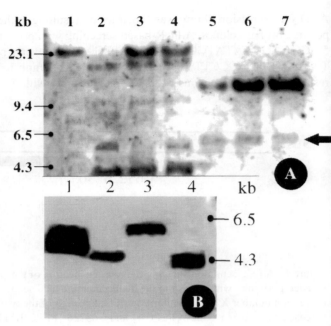

	Organisms (Accession numbers)	b	c	d	e	f	g	h	i	j	k	l	m
a	*Hydrilla verticillata* (U65227)	97	80	86	86	83	86	82	83	87	71	76	80
b	*Hydrilla verticillata* (U65226)	-	79	85	85	83	86	82	83	87	71	75	80
c	*Sorghum vulgare* (X63756)		-	80	82	78	78	76	80	78	68	73	76
d	*Sorghum vulgare* (X59925)			-	88	85	88	86	86	90	73	77	81
e	*Sorghum vulgare* (X65137)				-	81	84	85	86	88	71	77	82
f	*Flaveria trinervia* (X64143)					-	95	81	83	83	70	73	78
g	*Flaveria pringlei* (Z48966)						-	83	84	87	71	75	79
h	*Mesembryanthemum crystallinum* (X14587)							-	82	83	67	73	79
i	*Tillandsia usneoides* (X91406)								-	85	68	73	79
j	*Picea abis* (X79090)									-	70	76	78
k	*Equisetum hyemale* (X95855)										-	69	71
l	*Isoetes histrix* (X95854)											-	74
m	*Sphagnum* species (X95852)												-

Table 1. Identity matrix, expressed as a percentage of the total number of sites, occupied by identical amino acids in pairwise comparisons. Sequences from angiosperm, gymnosperm, pteriodyphyte and bryophyte representatives possessing C3 (d, e, g, j, k, m), C4 (c, f) or CAM (h, i, l) type PEPCs were compared with the 305 homologous residues of the *Hydrilla* PEPC isoforms.

An attempt, using parsimony analysis, was made to resolve the *Hydrilla* PEPCs with other PEPC sequences (Fig. 3). Higher plant PEPC isoforms grouped as largely unresolved polytomies.

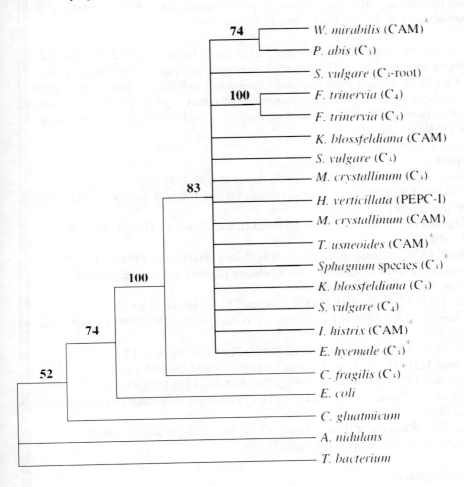

Figure 3. Cladogram obtained from a consensus tree drawn from PAUP with 21 PEPC sequences and 100 bootstraps. Only 377 amino acid residues relative to the partial PEPC of *Hydrilla* sequence were considered. The shaded box is to indicate the predominant polytomies. *The PEPC data obtained from Gehrig et al. (1998).

The identity matrix, which uses the conserved functional domain of the *Hydrilla* PEPC isoforms, demonstrates that there is a strong homology among all the seed plants. There seems to be too little variation within the domain to allow significant phylogenetic groupings. This is also apparent in the clades from parsimony analysis. There is no significant discrimination in the primary structure of the domain to categorize unequivocally *Hydrilla* PEPC, or any other, as C_3, C_4 or CAM type. The residual changes that have occurred in this part of the sequence do not seem to relate to the specialized functions of the various PEPC isoforms.

We conclude that phylogenetic or functional analyses based on only this conserved domain have limited value. The polyphyletic origins of C_4 and CAM photosynthesis may make PEPC alone a weak molecular marker to resolve the origins of the various photosynthetic systems. Full length sequence analyses, coupled with other parameters, might help resolve the differences among the C_3, C_4, and CAM isoforms, including that of *Hydrilla*.

References

1. Bowes, G. and Salvucci, M.E. (1989) Aquat Bot 34,233
2. Reiskind, J.B., Madsen, T.V., van Ginkel, L.C. and Bowes, G. (1997) Plant Cell Environ 2,211
3. Magnin, N.C., Cooley, B.A., Reiskind, J.B. and Bowes G. (1997) Plant Physiol 115,1681
4. Chase, M.W., Stevenson, D.W., Wilkin, P. and Rudall, P.J. (1995) in Monocotyledons, Systematics and Evolution (Rudall, P.J et al eds.) pp 685-730, Royal Botanic Gardens, Kew
5. Mai, D.H.and Walther, H. (1985) Abh Staal Mus Mineral Geol 33,1
6. Ehleringer, J.R., Sage, R.F., Flanagan, L.B. and Pearcy RW. (1991) Trends Ecol Evol 6,95
7. Toh, H., Kawamura, T. and Izui K. (1994) Plant Cell Environ 17,31
8. Gehrig, H.H., Heute, V. and Kluge, M. (1998) J Mol Evol 46,107
9. Reiskind, J.B. and Bowes, G. (1991) Proc Nat Acad Sci USA 88,2883
10. Chollet, R., Vidal, J. and O'Leary, M.H. (1996) Annu Rev Plant Physiol Mol Biol 47,273
11. Rao, S.K., Reiskind, J.B. and Bowes, G. (1998) Plant Physiol 117,s675
12. Bradford, M. (1976) Anal Biochem 72,248

Acknowledgements

This work was supported by the USDA/NRICG Photosynthesis and Respiration Program (93-37306-9386) and the National Science Foundation (IBN-9604518). We thank Dr. M. Whitten (Florida Museum of Natural History) and Dr. R. Ferl (Interdisciplinary Center for Biotechnological Research) for their assistance.

Contribution to the study of the Hatch and Slack photosynthetic cycle. Part II : Kinetic properties of the PEPcase-MDH-Malic enzyme multienzymatic complex purified from a C4 plant (sugarcane).

P. Baret[1], C. Queiroz[2], C. Rouch[1], J.C. Meunier[2] and F. Cadet[1].

1 - Faculté des Sciences et Technologies. Université de la Réunion. Laboratoire de Biochimie et Génétique Moléculaire. 15 avenue René Cassin. B.P. 7151. 97715 St-Denis Messag Cedex 9. Réunion France-Dom. cadet@univ-reunion.fr
2 - I.N.A.P.G. Chaire de Biochimie. Laboratoire de Chimie Biologique. 78850 Thiverval-Grignon.France.meunier@cardere.grignon.inra.fr

Introduction

We found out the existence in fresh young leaves of sugarcane of a PEPcase-MDH(NADP)-Malic enzyme multienzymatic complex. We have been able to purify this complex. Its functional advantage has been investigated by comparing the kinetic properties of the constitutive enzymes in their free state with those in their complexed state.

Results and Discussion

Free enzymes and the multienzymatic complex were purified following the protocol in figure 1.

PEPcase activity (table 1). The catalytic efficiency of PEPcase in the complexed form showed to be higher than in its free state form (116.10^3 s^{-1}M^{-1} for the complexed form and 62.10^3 s^{-1}M^{-1} for the free form) while Km with regard to PEP is decreased in the complex. The behaviour of the two PEPcase forms towards pH was found to be identical (optimum at pH 8.2). However it was noticed that the inhibitory effect of L-malate on PEPcase activity was higher in the complex with a Ki of 10 mM in the complexed form and a Ki of 18 mM in the free state.

G. Garab (ed.), Photosynthesis: Mechanisms and Effects, Vol. V, 3571–3574.

MDH activity (table 2). The comparative studies undergone with MDH in the complexed form and in the free form confirms that its association with PEPcase and malic enzyme leads to enhanced kinetic properties. An increase (3.5 fold) in NADP-MDH catalytic efficiency was observed (14.10^6 $s^{-1}M^{-1}$ for the complexed form against $3.9.10^6$ $s^{-1}M^{-1}$ for the free form). Moreover, higher concentrations of 2,3-butadiene inhibitor were needed in the complexed form than in the free state.

Malic enzyme activity (table 3). The results obtained with malic enzyme showed at the studied pH that the enzyme had a stronger affinity towards L-malate and had an improved catalytic efficiency in its complexed form than in its free form.

Table 1 : Activity of PEPcase as a function of PEP concentration

	Free PEPcase	Complexed PEPcase
k_c (s^{-1})	117	93
K_m (mM)	1.89	0.80
k_c/K_m ($s^{-1}.M^{-1}$)	$62x10^{+3}$	$116x10^{+3}$
$K_{i_{L-malate}}$ (mM)	18	10
pH optimum	8.2	8.2

Table 2 : MDH activity with OAA as substrat

	Free MDH	Complexed MDH
k_c (s^{-1})	276	1500
K_m (mM)	0.07	0.11
k_c/K_m ($s^{-1}.M^{-1}$)	$3.9x10^{+6}$	$14x10^{+6}$
50% inhibition (2,3-butadione)	30 mM	50 mM
pH optimum	8.0	8.0

Table 3 : Malic enzyme activity with L-malate as substrate

| | At pH 8.0 | |
	Free enzyme	Complexed
kc (s^{-1})	605	2000
$Km_{L\text{-malate}}$ (mM)	0.57	0.28
kc/Km (s^{-1}.M^{-1})	$1.1 \times 10^{+6}$	$7.2 \times 10^{+6}$
	Free enzyme	Complexed
pH optimum	7.6	7.4

Conclusion

The study of the kinetic properties of PEPcase, MDH and malic enzyme showed that their association into a complex induced important changes in their respective kinetic properties. This strongly suggests that the conformations of the proteins in the free form are different to those in the complexed form.

3574

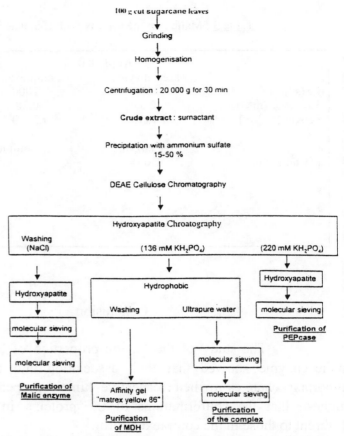

Figure 1 : Summary of first steps for extraction and Purification procedures for the complex and for each enzyme in their free state

References

- Gontero B., Giudici-Orticoni M.T. and Ricard J. – 1994 – The modulation of enzyme reaction rates within multi-enzyme complexes. 2 – Information transfer within a chloroplast multi-enzyme complex containing ribulose bisphosphate carboxylase-oxygenase – *Eur. J. Biochem.*, 226, 999-1006.
- Hermoso R., Castillo M., Chueca A. and Lazaro J.J. – 1996 – Binding site on pea chloroplast fructose-1,6-bisphosphatase involved in the interaction with thioredoxin – *Plant Molecular Biology*, 30, 455-465.
- Ricard J., Giudici-Orticoni M.T. and Gontero B. – 1994 – The modulation of enzyme reaction rates within multi-enzyme complexes. 1 – Statistical thermodynamics of information transfert through multi-enzyme complexes – *Eur. J. Biochem.*, 226, 993-998.
- Wang X., Tang X., Anderson L.E. – 1996 – Enzyme-enzyme interaction in the chloroplast : physical evidence for association between phosphoglycerate kinase and glyceraldehyde-3-phosphate dehydrogenase in vitro – *Plant Science*, 117, 45-53.

Contribution to the study of the Hatch and Slack photosynthetic cycle. Part I : Purification of a PEPcase-MDH-malic enzyme multienzymatic complex in a C4 plant (sugarcane).

P. Baret[1], C. Queiroz[2], C. Rouch[1], J.C. Meunier[2] and **F. Cadet**[1]

1 - Faculté des Sciences et Technologies. Université de la Réunion. Laboratoire de Biochimie et Génétique Moléculaire. 15 avenue René Cassin. B.P. 7151. 97715 St-Denis Messag Cedex 9.Réunion France-Dom. cadet@univ-reunion.fr
2 - I.N.A.P.G. Chaire de Biochimie. Laboratoire de Chimie Biologique. 78850 Thiverval-Grignon. France. meunier@cardere.grignon.inra.fr

Extraction and Purification

We found out the existence in fresh young leaves of sugarcane of a PEPcase-MDH(NADP)-Malic enzyme multienzymatic complex. The enzyme was purified as follows :

Extraction

100g of leaves were homogenised in a waring blendor with 300 ml of citrate-phosphate buffer 100 mM pH 6.8. The suspension was then filtered through a bluter cloth and the filtrate was centrifuged during 30 min at 20000 g.

Purification

Precipitation with ammonium sulfate. The crude extract was then fractionnated with ammonium sulfate in a range of 15% to 50% saturation before it was loaded onto a DEAE-Cellulose column (figure 1). Elution was carried out in a NaCl gradient increasing from 0 M to 1 M in the equilibration buffer (phosphate buffer 20 mM, pH 6.5). It was observed that the activities of PEPcase, MDH and Malic enzyme were eluted at the same salt concentration (NaCl 425 mM) and that no activity was detected at 1 M NaCl. The fractions that contained the activities of the three enzymes were pooled then concentrated.

G. Garab (ed.), Photosynthesis: Mechanisms and Effects, Vol. V, 3575–3578.

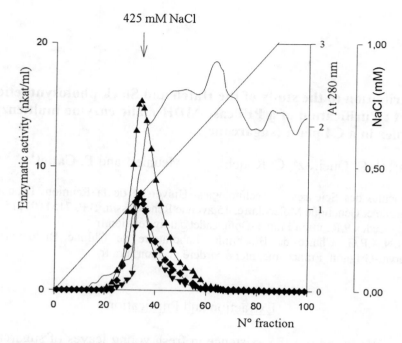

Figure 1 : DEAE-Cellulose Chromatography

Elution with a linear gradient from 0 to 1 M NaCl dissolved in the phosphate buffer (20 mM, pH 6.5). Flow rate : 4 ml/min. The volume of fractions was 8 ml.

▼ PEPcase activity (nkat/ml)
▲ NADP-MDH activity (nkat/ml)
◆ Malic enzyme activity (nkat/ml)

After chromatography on hydroxyapatite and hydrophobic chromatography, two succesive gel filtrations was performed on a TSK G 4000 SW column (7.5 x 300 mm) coupled to a Waters Associated Inc. HPLC system. The chromatograms obtained show that there was still superimposition of the peaks of the three enzyme activities. However peak PEPcase, MDH and malic enzyme activities could be detected in other fractions. This again suggests that the PEPcase-MDH-Malic enzyme complex tend to fragment.

The first molecular siewing obtained showed that the complex has a tendency to fragment into several proteins. The complex peak obtained was loaded again onto a second molecular siewing column in the same conditions so as to obtain a perfect superimposition of the protein peak with the complex peak activities (figure 2). Equilibration and elution were carried out at a flow rate of 0.5 ml/min and at 4°C with 20 mM phosphate buffer (pH 7). Eluted protein was detected at 280 nm in a Waters 486 spectrophometer.

Results and Discussion

We subsequently experienced difficulties in the determining the homogeneity of the complex by electrophoresis. The estimation of the apparent weight of the complex was hence established by molecular siewing chromatography. The complex has an estimated molecular weight of 711 kDa. The weight is close to the sum of the molecular weight of each enzyme in their free state : 736 kDa (PEPcase, 398 kDa; NADP-MDH, 69 kDa; malic enzyme, 268 kDa).

Two hypothesis could attempt to explain the origin of this complex. In first instance it is possible that the complex is formed during the extraction procedure since the enzymes are reported to be located in different subcellular compartments. The other hypothesis would suggest that the subcellular localisation of the enzymes in young sugarcane leaves could be different with that observed in aged leaves : this is supported by the fact that it has been shown that the photosynthetic type could change during the different stages of the developpement of leaves.

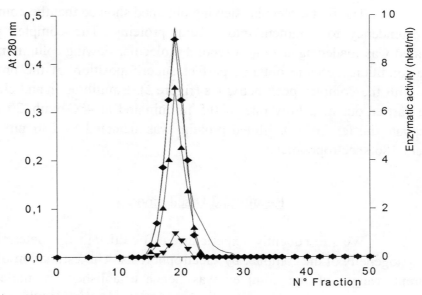

Figure 2 : Exclusion chromatography
The column was equilibrated with a 20 mM phosphate buffer, pH 7.0
The volume of fractions was 0.5 ml. Flow rate : 0.5 ml/min.
The absorbance at 280 nm was monitored.
▼   PEPcase activity (nkat/ml)
▲   NADP-MDH activity (nkat/ml)
◆   Malic enzyme activity (nkat/ml)

References

- Baret P., Rouch C., Meunier J.C. and Cadet F. - 1995 - A PEP/MDH enzymatic complex in sugar cane (C4 plant) - Photosynthesis Research: from Light to Biosphere, Kluwer Acad. Publish ed., 5, 309-312.
- Dai Z., Ku M.S.B., Edwards G.E. - 1995 - C4 Photosynthesis. The effects of leaf developpment on the CO2 concentration mechanism and photorespiration in Maize. Plant Physiol, 107, 815-825.
- Hatch M.D. and Slack C.R. - 1966 - Photosynthesis by sugar cane leaves. A new carboxylation reaction and the parthway of sugar formation - Biochem. J., 101, 103-111.
- Iglesias A.A. and Andreo C.S. - 1989 - Purification of NADP-Malic enzyme and phosphoenolpyruvate carboxylase from sugar leaves - Plant Cell. Physiol., 30, 399-405.
- Queiroz-Claret C. and Queiroz O. - 1992 - Malate Dehydrogenase forms a complex with and activates phosphoenolpyruvate carboxylase from Crassulacean Acid Metabolism Plants - J. Plant Physiol., 139, 385-389.

NADP-MALIC ENZYME FROM THE C_3-C_4 INTERMEDIATE SPECIE FLAVERIA FLORIDANA.

Carlos S. Andreo, Paula Casati, Analia Fresco and María F. Drincovich.
Centro de Estudios Fotosintéticos y Bioquímicos-Universidad Nacional de Rosario. Suipacha 531. 2000 Rosario. Argentina.

Key Words: C metabolisms, C3, C4, NADP-malic enzyme

1. Introduction

C_4 plants have fully differentiated mesophyll (MC) and bundle sheath cells (BSC) which cooperate to fix CO_2 by the C_4 pathway. C_4 plants are more efficient than C_3 under some environmental conditions due to the higher CO_2 concentration in BSC. C_3-C_4 intermediate species are thought to represent a stage in the evolutionary transition from the C_3 to the C_4 photosynthetic mechanism [1,2]. MC of C_3-C_4 species function, as in C_3 plants, fixing atmospheric CO_2 through RuBisCO and generating C-2 compounds for the photosynthetic oxidation cycle. Two mechanisms are proposed to account for the low apparent photorespiration in these intermediate species. In one of them, which may be common to all intermediates, metabolites generated as a consequence of the RuBisCO oxygenase reaction in MC are metabolized in BSC and the CO_2 released refixed by RuBisCO. In this way, reduced photorespiratory CO_2 evolution occurs without the operation of a C_4 cycle. In other class of intermediates, the operation of a limited C_4 cycle between MC and BSC contributes to the further reduction of photorespiration. *Flaveria floridana*, a C_3-C_4 intermediate specie, was used in the present report to characterize different isoforms of NADP-malic enzyme (NADP-ME) and to evaluate their expression in different photosynthetic cell types. In previous studies, we detected three isoforms of the enzyme in various *Flaveria* intermediate species [3]. One of them was found to be constitutively expressed in photosynthetic and non-photosynthetic tissues of the different species examined, while the other two isoforms were abundant only in photosynthetic tissues having partial or complete C_4 photosynthesis. A correlation between these proteins, their function and the cDNAs already cloned [4] is necessary to understand the evolution of this protein among the different *Flaveria* species.

2. Procedure

2.1. *Gel electrophoresis and western blot.* Protein samples, prepared from various tissues and cellular fractions, were analyzed by native isoelectrofocusing (IEF) and SDS-

G. Garab (ed.), Photosynthesis: Mechanisms and Effects, Vol. V, 3579–3582.

PAGE. After electrophoretic separation, proteins on gels were either stained for malic enzyme activity or transfered onto nitrocellulose membrane for western blotting (3). Purified antibodies against the 62 kDa NADP-ME from maize green leaves were used (3).

2.2. *Isolation and purification of MC and BSC.* MC and BSC were isolated using a procedure previously described (5). The MC and BSC were examined under light microscope for purity and analyzed for PEPC, NADP-MDH and NADP-ME activity and western blot analysis using antibodies against NADP-ME and RuBisCO (spinach large subunit). The reaction medium used for activity measurements were: *a) PEPC:* 50mM Tris-HCl pH 8.0, 5mM $MgCl_2$, 10mM $KHCO_3$, 4mM PEP, 0.15mM NADH and 10IU of MDH; *b) NADP-MDH:* (samples were preincubated in 100mM DTT for 2 hs) 50mM Tris-HCl pH 8.0, 1mM EDTA, 0.25 mM NADPH and 1mM OAA; *c) NADP-ME:* 50mM Tris-HCl pH 8.0, 10mM $MgCl_2$, 0.5mM NADP and 4mM L-malate.

2.3. *Purification of NADP-ME from leaves and measurement of kinetic parameters.* The protocol used for purification was similar to the previously described for the enzyme from maize (6) with the following modifications: (a) the crude extract was precipitated with 30-70 % $(NH_4)_2SO_4$; (b) the *affi-gel blue* column was omitted; (c) the enzyme eluted from the hidroxilapatite column was further purified by a *Superose 12 HR 12/30*. Enzyme activity was determined at 30°C by monitoring NADPH production at 340 nm. Initial velocity studies were performed by varying the concentration of one of the substrates around its K_m, while keeping the other substrates concentrations at saturating levels (0.5mM NADP, 4mM L-malate and 10mM $MgCl_2$). K_m values of the substrates were calculated in terms of free concentrations by linear regression in all cases.

3. Results

3.1. *Non-denaturing isoelectrofocusing and western blot analysis of extracts from F. floridana.* Fig 1A shows a native isoelectrofocusing gel revealed for NADP-ME activity. Three active bands were found in *F. floridana* (C_3-C_4) and *F. bidentis* (C_4) in contrast to a single band found in *F. pringlei* (C_3). Western blot analysis (Fig 1B) of crude extracts shows three immunoreactive bands in the case of *F. floridana*, two in *F. bidentis* and one in *F. pringlei*.

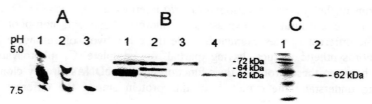

Figure 1. Native IEF revealed for NADP-ME activity (1mU) **(A)** and western blot revealed against NADP-ME antibodies (30 μg) **(B)** of crude extracts of *F. bidentis* (1), *F. floridana* (2), *F. pringlei* (3) and purified NADP-ME from *F. floridana* (1 μg) (4). **(C)** Coomassie blue staining of MW markers (1) and purified NADP-ME from *F. floridana* (5 μg) (2).

3.2. *Location of photosynthetic enzymes in MC and BSC of F. floridana.* Three purified type of cells were obtained in the Percoll gradient, as in the case of *F. ramossisima* (5), which were different when examined under light microscope. These fractions were called MC1, MC2 and BSC (when ordered by increasing density). The activities (IU/mg protein) of NADP-ME, PEPC and NADP-MDH measured in each fraction are indicated in Table 1:

Fraction	NADP-ME	PEPC	NADP-MDH
MC1	0.094	0.550	0.107
MC2	0.325	0.160	0.096
BSC	0.702	0.052	0.064

Fig 2 shows the western blot analysis of the three cell samples obtained (MC1, MC2 and BSC) revealed against RuBisCO and NADP-ME antibodies. The results show that, although RuBisCO is not restricted to a cell type, the three type of cells express different NADP-ME isoforms. In this way, the BSC and MC2 express the three isoforms of NADP-ME, while MC1 express only one.

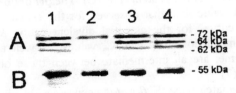

Figure 2. Western blot of protein samples (A: 30 μg; B: 5 μg) extracted from total extract (1), MC1 (2), MC2 (3) and BSC (4) revealed against NADP-ME **(A)** and RuBisCO large subunit **(B)** antibodies.

3.3. *Purification of NADP-ME from Flaveria floridana leaves.* NADP-ME was purified from *F. floridana* leaves. The purified protein presented a molecular mass of 62 kDa, as revealed by western blot and Coomassie Blue staining (Fig 1C). Analysis of the dependence of activity on pH revealed a broad maximun of activity centered at pH 7.5. The V_{max} was 15 IU/mg. The K_m values for the substrates are indicated in Table 2 (at pH 7.5). A biphasic double reciprocal plot was obtained in the case of Mg^{2+}, and two K_m values were calculated.

Substrate	K_m(mM)
L-malate	0.46
NADP	0.012
Mg^{2+}	0.005
	0.16

3582

4. Discussion

Three active forms of NADP-ME were found in *F. floridana* leaves which differ in their native isoelectric point (Fig 1A). Western blot analysis (Fig 1B) of extracts from *F. floridana* shows three immunoreactive bands. These results, in comparison to the results obtained with other *Flaveria* species (Fig. 1 and ref. 3), allow us to conclude that the 62 kDa form is correlated to C_4 photosynthesis, while the 72 kDa is a form found in all the species studied independently of the metabolism used for CO_2 fixation. The 64 kDa form is a form restricted to C_3-C_4 intermediate species, and may represent an intermediate form of the protein.

Protoplast fractions of mesophyll and bundle sheath cells were purified from *F. floridana* leaves. Activity measurement and western blotting of the three fractions obtained (MC1, MC2 and BSC) indicate that NADP-ME isozyme expression is cell-specific in this specie, with BSC presenting a higher activity (Table 2) and expressing the 64 and 62 kDa forms (Fig 2). The activity measured in MC1 may correspond to the 72 kDa form of the enzyme, which is the only form found by western blot analysis of this fraction. PEPC activity was also partially differentially distributed among the different cell types, as revealed by activity measurements of the three fractions obtained (Table 2). On the other hand, RuBisCO has an even distribution as revealed by western blot analysis (Fig 2).

The 62 kDa enzyme from *F. floridana* leaves was purified to homogeneity, as revealed by Coomassie Blue staining of the protein (Fig 1C). The pH optimum of activity was 7.5, which is a more acidic value than that observed for the enzyme from C_4 leaves. Non-hyperbolic saturation curves for the essential divalent cation Mg^{2+} indicate different binding sites for the metal cofactor. The K_m values, as well as the specific activity obtained for this enzyme, are all intermediate between those observed in NADP-ME from C_3 and C_4 plants (7).

In this way, the intermediate specie *F. floridana* express a key photosynthetic enzyme of the C_4 cycle with high specific activity and kinetic features between those of C_3 and C_4 plants. This isoform is also expressed in a cell-specific way, suggesting that the distribution of C_4 enzymes such as NADP-ME may be one of the first steps in the evolution to C_4 photosynthesis, which may occur even before RuBisCO compartmentalization.

References

1) Monson RK & Moore Bd (1989) Plant, Cell and Environment **12**, 689.
2) Rawsthorne S (1992) Plant J. **2**, 267.
3) Drincovich MF; Casati P; Andreo CS; Chessin SJ; Franceschi VR; Edwards GE & Ku MSB (1998) Plant Physiol., in press.
4) Marshall JS; Stubbs JD & Taylor WC (1996) Plant Physiol. **111**, 1251.
5) Moore Bd; Monson RK; Ku MSB & Edwards GE (1988). Plant Cell Physiol. **29**, 999.
6) Iglesias AA & Andreo CS (1989) Plant Cell Physiol. **30**, 399.
7) Edwards GE & Andreo CS (1992). Phytochemistry **31**, 1845.

IDENTIFICATION OF ONE PAIR OF DOMAIN-LOCKING, REDOX-SENSITIVE CYSTEINE RESIDUES IN MAIZE CHLOROPLAST MALATE DEHYDROGENASE

Elizabeth H. Muslin, Louise E. Anderson
Department of Biological Sciences, University of Illinois at Chicago,
Chicago, IL 60607-7060 USA

Key words: light-activation, maize, malate dehydrogenase, regulatory processes, site-directed mutagenesis, thiol/disulfide exchange.

1. Introduction

The NADP-linked malate dehydrogenase (EC 1.1.1.82) in the chloroplasts of higher plants is reductively activated in the light (1). We (2) predicted, on the basis of the orientation and separation of the residues in the modeled enzyme, that Cys-325 in the carbon substrate-binding domain forms an inhibitory, redox-sensitive disulfide with Cys-132 or Cys-139 in the nucleotide-binding domain. (Numbering throughout is according to the model enzyme used in those experiments, porcine cytosolic malate dehydrogenase, Protein Data Bank entry 4MDH. See Fig. 1.) We suggested that formation of the inter-domain disulfide bond would interfere with movement necessary for catalysis. Later experiments with the *Escherichia coli* enzyme (3) made it unlikely that Cys-139 is involved. Cys-325 had already been identified as a regulatory Cys (4), in site-directed mutagenesis experiments. Although the chloroplast enzyme is activated by DTT, it is not activated by the monothiol mercaptoethanol (5). Consistent with (6), we find that replacement of Cys-325 in the maize chloroplast malate dehydrogenase with an alanine results in an enzyme equally sensitive to mercaptoethanol and to DTT. Surprisingly, we find that mutation of Cys-132 renders the enzyme less sensitive to mercaptoethanol than to DTT.

SORGHUM	64	69	215	223	248	368	405	417
PIG	-20	-15	132	139	164	289	325	337

Figure 1. Numbering of Cys residues in the chloroplast malate dehydrogenase from sorghum according to (13) and the corresponding residues in pig heart cytoplasmic enzyme (Protein Data Bank entry 4MDH).

G. Garab (ed.), Photosynthesis: Mechanisms and Effects, Vol. V, 3583–3586.
© *1998 Kluwer Academic Publishers. Printed in the Netherlands.*

2. Materials and Methods

2.1 Mutation

The maize (*Zea mays* L.) malate dehydrogenase gene cloned into the expression vector pET-3b (7) (Novagen, Madison WI) was provided by A.R. Ashton. Cys codons were substituted into the DNA sequence by the method of (8). The mutations were confirmed by sequencing. *E. coli* strain BL21(DE3) was transformed (9) with the pET-3b plasmid carrying the appropriate construct.

2.2 Selection and testing for redox-sensitivity

Transformants were selected and cultured as described in (10) except that there was no kanamycin in the medium, the induction period was only 3 hours, and the lysis buffer was 50 mM Tris-HCl, pH 7.9, 1 mM potassium EDTA, 11.5 mM phenylmethylsulfonyl fluoride, 0.1 mg ml^{-1} chicken lysozyme. DTT activation was as described in (10). Mercaptoethanol treatment was essentially as described by (5). The lysate supernate was diluted with an equal volume of 25 mM potassium Hepes, pH 7.5, 200 mM KCl, 0.5 mM EDTA, allowed to stand for 1 hour at 25°C, and then made 100 mM in mercaptoethanol under N$_2$. Activity was determined 30 min later.

2.3 Activity assay and protein estimation

Malate dehydrogenase activity was routinely assayed in the direction of NADPH oxidation according to (11). The reference cuvette contained all of the components of the assay mixture except the carbon substrate. Protein was estimated by the method of Bradford (12).

3. Results and Discussion

In order to determine whether Cys-325 and Cys-132 are involved in the redox-sensitivity of chloroplast malate dehydrogenase we generated two single mutants, C325A and C132A, by site-directed mutagenesis. There is both chemical and kinetic evidence for at least two different disulfides which differ in redox-sensitivity in the maize NADP-linked malate dehydrogenase (5), and, as expected, the mutant enzymes were still redox-sensitive when tested in the standard activation assay with DTT (Fig. 2). The wild type enzyme is not mercaptoethanol-sensitive, but we found complete activation with the C325A mutant when we substituted mercaptoethanol for DTT in the activation assay. Clearly Cys-325 is involved in a regulatory disulfide. When we tested the C132A mutant we found that mercaptoethanol was not as effective for activation as DTT (Fig 2). The increase in activity after mercaptoethanol treatment was about 40% of that after DTT-treatment. Cys-132 is, then, involved in a disulfide with Cys-325. Because the C132A mutant is only partially activated by mercaptoethanol, there must a second Cys residue that can pair with 325. If, as is indicated by mutagenesis experiments with the *E. coli* enzyme (3,10), Cys-139 cannot form a disulfide with Cys-325, then the alternate Cys that forms a disulfide with Cys-325 must be one of the residues in the N- or C-terminal extension.

In a similar set of experiments Issakidis et al. (6) created a set of sorghum malate dehydrogenase mutants in which Cys residues were replaced by alanines. Consistent

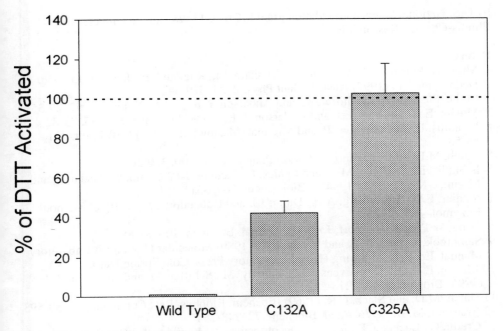

Figure 2. Effect of mercaptoethanol (100 mM) on the activity of recombinant wildtype and mutant maize malate dehydrogenases in crude bacterial extracts expressed as percent of activity after 10 mM DTT treatment. Results are means from replicate assays for 4 single colony transformants for the C325A mutant and for recombinant wildtype, and for 3 single colony transformants for the C132A mutant. After DTT-treatment activity values ranged from 84 to 137 units mg protein[-1] with a mean of 110 for the recombinant wildtype enzyme, from 218 to 657 units mg protein[-1] with a mean of 428 for the C132A mutant enzyme, and from 283 to 489 with a mean of 399 for the C325A mutant enzyme.

with our experiments, Cys-325 appeared to be a regulatory cysteine. They also identified Cys (-20), Cys (-15) and Cys-337 in the extensions as regulatory Cys residues. This is not inconsistent with our results, but they identified the bonding partners as the two N-terminal cysteines and the two C-terminal cysteines. Our experiments suggest, instead, that one (or more) of the cysteines in the extensions is the alternative to Cys-132 in cystine disulfide bond formation with Cys-325.

Acknowledgments

The oligonucleotides were generated by Paul Gardner at the University of Chicago. We thank Anthony Ashton, CSIRO, Canberra, for providing us with the cDNA coding for the maize NADP-malate dehydrogenase cloned into pET-3b. This work was supported by U.S. National Science Foundation grant MCB-9513498 and the University of Illinois-Chicago Research Board.

References

1 Miginiac-Maslow, M., Issakidis, E., Lemaire, M., Ruelland, E., Jacquot, J.P. and Decottignies, P. (1997) Aust. J. Plant Physiol. 24, 529-542
2 Li, D., Stevens, F.J., Schiffer, M. and Anderson, L.E. (1994) Biophys. J. 67, 29-35
3 Muslin, E.H., Stevens, F.J. and Anderson, L.E. (1998) Photosyn. Res. 55, 75-82
4 Issakidis, E., Decottignies, P. and Miginiac-Maslow, M. (1993) FEBS Lett. 321, 55-58
5 Hatch, M.D. and Agostino, A. (1992) Plant Physiol. 100, 360-366
6 Issakidis, E., Saarinen, M., Decottignies, P., Jacquot, J-P., Cretin, C., Gadal, P. and Miginiac-Maslow, M. (1994) J. Biol. Chem. 269, 3511-3517
7 Studier, F.W., Rosenberg, A.H., Dunn, J.J. and Dubendorff, J.W. (1990) Methods Enzymol. 185, 60-89
8 Deng, W.P. and Nickoloff, J.A. (1992) Anal. Biochem. 200, 81-88
9 Sambrook, J., Fritsch, E. and Maniatis, T. (1989) Molecular Cloning: A Laboratory Manual, Ed. 2, Cold Spring Harbor Laboratory Press, Cold Spring Harbor, USA
10 Muslin, E.H., Li, D., Stevens F.J., Donnelly, M., Schiffer, M. and Anderson, L.E. (1995) Biophys. J. 68, 2218-2223
11 Hatch, M.D. and Slack, C.R. (1969) Biochem. Biophys. Res. Commun. 34, 589-593
12 Bradford, M.M. (1976) Anal. Biochem. 72, 248-254
13 Crétin, C., Luchetta, P., Joly, C., Decottigenies, P., Lepiniec, L., Gadal, P., Sallantin, M., Huet, J-C. and Pernollet, J.C. (1990) Eur. J. Biochem. 192, 299-303

METABOLIC EFFECTS OF WATER DEFICIT IN CROP TROPICAL PLANTS

Castrillo M*, D.Fernandez,A.M.Calcagno,I.Trujillo
*Depto. Biologia Organismos. Universidad Simon Bolivar. Caracas 1080.
Venezuela

Key words : C3, C4, Rubisco, PEPC , Chl, water potential, RWC

1.Introduction

This work represents a compilation of research, undertaken in our laboratory in the past years, about the effects of water deficit on the photosynthetic metabolism. in tropical cultivars . In mesophyte plants the ability to adapt themselves metabolically to water deficit is an essential condition for survival.There have been many reports about the water deficit effects on decreased RBPC activity, it was reported (1) that the reversible decrease in the photsynthetic activity of the chloroplasts occurs at RCW values of 70 to 40% and this decrease is attributed to a reversible inhibitory effect on the enzymatic activity by the increase in cellular solute.It was reported   (2) that RBPC activity decreased sharply whereas the Rubisco protein is affected to a lesser extent in maize plants under water. However, it has been reported (3) that in two sunflower hybrids under water stress, the primary sit of limitation of net photosynthesis might be related to concentraion of chlorophyll , total soluble protein or Rubisco protein.The aim of this work was to study the effects of water deficit on some enzyme activities such as Ribulose-1,5-bis-phosphate carboxylase (RBPC) and Phosphoenolpyruvate carboxylase (PEPC), on total chlorophyll content and on soluble protein content and the relationship between these parameters and components of leaf water status such as water potential and relative water content.

Procedure

2.1 *Plant material*   Maize (*Zea mays*) : plants of two hybrids CPB2 and CPB8 of thirty days age were used, 30 plants were subjected to water deficit treatment and another 30 plants were wattered(control). Tomato (*Lycopersicon esculentum):* plants of two varieties Rio Grande (RG) and Pera Quibor (PQ) of 39 days old were used, 40 plants were subjected to water deficit treatment and another 40 plants were wattered (control). Bean (*Phaseolus vulgaris*): plants of 30 days of age for the cv Tacarigua and 42 days of age for the cv VUL-73-401, 40 plants were subjected to water deficit treatment and another 40 plants were wattered (control). The plants were kept in a greenhouse. For environmental greenhouse conditions see: (4)

G. Garab (ed.), Photosynthesis: Mechanisms and Effects, Vol. V, 3587–3590.
© *1998 Kluwer Academic Publishers. Printed in the Netherlands.*

2.2 *Assays and measurements* Water potential was measured with a HR-33T Dew Point Microvoltimeter using C-52 chambers (Wescor Inc., Logan Utah, USA). Relative water content (RWC) was determined (5).All determinations were performed at predawn time.RBPC activitiy was assayed (6). PEPC activity was assayed (4). The assays were performed at midday within 5 to 15 min after extraction.Chlorophyll determination was performed (7).Soluble protein estimation was performed (8).The measurements were taken at three days intervals during a period of two or three weeks.A set of four replicates, of the second and third expanded leaf from the apex (tomato and bean) and the sixth leaf from the base (maize) were used to measure all parameters.

3. Results and Discussion

Under water deficit the results of RBPC and PEPC activities, chlorophyll (Chl) content and protein content in maize, are shown in Table 1, at moderate stress the RBPC activity decrease, but is higher for CPB8 hybrid, whereas at severe stress the values are similar for both hybrids. The PEPC activity decreased being higher in hybrid CPB8 at both moderate and severe water deficit.Chl. content decreased at both moderate and severe water deficit. Hybrid CPB8 showed slightly high values of protein content than CPB2 hybrid at moderate stress whereas at severe stress both hybrids had similar soluble protein contents.In tomato plants, we observed lower values of water potential and RWC for PQ cv., these values, at both moderate and severe stress are reached at a slower rate in cultivar PQ(data not shown).In Table 2, are shown the values of RBPC activity, PQ cultivar showed slightly high values at both moderate and severe water deficit. It can be observed that the Chl. content showed less decrease in PQ cv. compared with RG cv.The protein content decreased but PQ cv showed higher values than RG cv. The protein content decrease is proportionally lower than the RBPC activity decrease for both cultivars. The comparison between cultivars allow us to observe how in cultivar PQ lower values of water potential and RWC are reached as compared with cultivar RG, but a greater number of days was required to reach these lower values. This gradual experimenting of the water stress in cultivar PQ could explain the higher RBPC activity chlorophyll and protein contents observed values. In bean, Table 3, the plants of variety VUL-73-401 (V) had higher age days than those of Tacarigua (T) plants.The RBPC activity decreased for both varieties.The Chl content decreased for both varieties, being slightly higher for T. The protein contents decreased and are similar for both varieties. In a comparison about the responses to water deficit among the three different studied species : maize, tomato and bean, we found :- The values of water potential were lower in tomato and the values RWC were lower in bean and maize - The lowest values obtained for RBPC activity were for bean- The lowest values obtained for Chl. content were for bean- The protein content values were lower for maize and higher for bean- A greater relationship between RWC and water potential and all parameter studied in all cases.It has been reported that the RWC could be a better water leaf parameter to explain the metabolic changes in the leaf (11) (Walter and Krebs 1970, mentioned by (9) and (10)..It was confirmed (that RWC or cell volume may be more appropiate indicators of the water leaf status (12). It has been evidenced that the mesophyll photosynthesis is affected at RWC values of 50 to 70% (1) (13)(14) which agrees with the values observed in the present work, although in our case lower RWC values were reached.. It has been reported that a slow dehydration (period of two or more weeks) could affect mesophyll photosynthesis at RWC relatively higher than those imposed by a rapid water deficit (12). Our experiments were undertaken in period between 8 and 24 days. We can point out that, although the applied water deficit was higher, it did not produce an intense water deficit in the plants and it could be due to the non-controlled environmental conditions in the greenhouse and/or intrinsic characteristics of each specie or cultivar. In C-4 plants (14) such as maize the non-stomatic components may be more sensitive to water stress thanC-3 plants. This agrees with our results in maize, over a water deficit period of 24 days, the water leaf state components were higher as compared with the two C-3 plants, whose water deficit period were shorter, although its intensity was higher and the responses of metabolic parameters decreased in all cases.These observations could lead us to presume that in C-4 plants like maize, under water deficit, there may be a greater control of leaf water state parameters associated to an increased metabolic sensitivity to water deficit.Recently however, it was reported (15) that in maize, the PEPC activity and PEPC protein were stabilized during drought, as it has been also reported (16).This difference with our results could be attributed to different age leaf used in the experimentation(17)

Table 1. Maize - Values of RBPC and PEPC activities, Chl. and protein contents at water deficit for hybrids CPB2 (2) abd CPB8 (8)

RWC %	PEPC act. $\mu mol\ CO_2\ min^{-1}\ g^{-1}\ DW$		RBPC act.		Chl. $mg\ g^{-1}\ DW$		Prot.	
	2	8	2	8	2	8	2	8
90-60	3.5-1.5	3.8-1.5	1.0-0.6	1.4-0.8	22-12	18-16	22-19	27-22
<60	2.0-1.2	2.8-2.0	0.9-0.5	1.2-0.5	15-12	15-12	19-15	19-16
ΨMpa								
-0.40 to-1.40	3.0-1.8	2.5-1.8	1.2-0.6	1.4-0.8	21-13	19-14	22-20	27-20
-1.41 to-2.40	2.0-1.5	3.2-2.0	0.8-0.6	1.2-0.6	13-12	16-13	20-15	25-16

Table 2 .Tomato - Values of RBPC and PEPC activities, Chl. and protein contents at water deficit for varieties PQ and RG

RWC %	RBPC act. $\mu mol\ CO_2\ min^{-1}\ g^{-1}\ DW$		Chl. $mg\ g^{-1}\ DW$		Prot.	
	PQ	RG	PQ	RG	PQ	RG
90-70	45-35	38-20	20-16	16-10	115-80	120-60
<70	30-18	35-15	18-15	16-10	138-80	90-60
ΨMpa						
-0.50 to-1.50	45-25	38-18	19-15	17-10	150-100	122-100
-1.50 to-2.52	28-20	26-20	18-15	16-13	125-80	90-60

Table 3. Bean -Values of RBPC and PEPC activities, Chl. and protein contents at water deficit for varieties T and V

RWC%	RBPC act. μmol CO_2 min^{-1} g^{-1} DW		Chl. mg g^{-1} DW		Prot.	
	T	V	T	V	T	V
90-60	45-28	40-15	24-10	18-8	180-138	250-100
<60	35-4	10-5	16-3	13-3	150-30	80-30
ΨMpa						
-0.50 to-1.20	35-18	42-13	24-8	18-7	180-110	280-100
-1.20 to-2.00	22-4	8-7	12-3	13-3	140-30	80-25

References

1 Kaiser W.M. 1987. Physiol. Plant. 71 : 142-49
2 Martinez-Barajas E., Villanueva-Verduzco C., Molina-Galan J. Loza-TalaveraH., Sanchez-Jimenez. 1992.Crop Sc. 32:718-22
3 Gimenez C., MitchellV.J.,Lawlor D.W.1992.Pl.Physiol. 98: 516-24
4 Castrillo M., Fernandez D.1990. Maydica 35: 67-72
5 Turner N.C. 1981 Pl. & Soil 58 : 339-66
6 Castrillo M. 1985. Photosynthetica 19:56-64
7 Bruinsma J. 1963. Photochem. Photobiol. 2 :241-49
8 Bradford M.M. 1976. Anal. Biochem.72 : 248-54
9 Schulze E.D.1986. Ann. Rev. Pl. Physiol. 37 : 247-74
10 Schulze E.D.,Steudle G.,GollanT.,Schurr V. 1988. Pl. Cell.Environ.11:573-6
11 Sinclair T.R., Ludlow M.M. 1985.Aust.J.Pl.Physiol. 12 : 213-17
12 Chaves M.M. 1991. J. Exp. Bot. 42 :1-16
13 Robinson S.P., Grant W.J. Loveys B.R.1988.Aust.J.Pl.Physiol.15:495-503
14 Cornic G., Le Gonallec J.L., Briantais J.M., Hodges M. 1989. Planta 177: 89-90
15 Foyer C., Valadier M.H., Migge A., Becker T.1998.Pl.Physiol 117:283-92
16 Saccardy K.,CornicG.,Brulfert J.,ReyssA.1996.Planta 199:589-595
17 Nelson T., Langdale J.A.1992. Ann.Rev.Pl.Physiol.Pl.Mol.Biol.43:25-47

IDENTIFICATION OF METAL-BINDING SITE OF MAIZE NADP-MALIC ENZYME BY AFFINITY CLEAVAGE BY Fe^{2+}-ASCORBATE.

S.R. Rao, B.G. Kamath and **A.S. Bhagwat**
Molecular Biology & Agriculture Division
Bhabha Atomic Research Centre,
Mumbai-400 085, India

Key words: active oxygen, active site, magnesium, C-4, enzymes, hydroxyl radicals.

1. Introduction

NADP-malic enzyme (EC 1.1.1.40) is found in bundle sheath chloroplasts of maize, a C-4 plant, and catalyses the divalent metal ion dependent decarboxylation of malate into pyruvate and CO_2. The pigeon liver malic enzyme has been well characterized and requires Mn^{2+} which provides a bridge between malate and the enzyme and functions as a second sphere complex with the substrate [1]. By metal catalyzed oxidation (MCO) and affinity cleavage ASP^{258}, ASP^{141}, ASP^{194} and ASP^{464} have been shown as the coordination sites for the metal binding in the pigeon liver enzyme [2,3]. By site directed mutagenesis, the catalytic role of metal ion in the pigeon liver malic enzyme and role of ASP^{258} as the metal coordinate was confirmed by Wei *et al.* [3]. Since the 3-D structure of maize NADP malic enzyme is not available, we thought that the MCO (non-enzymatic) system might be useful for identifying the metal binding site in maize NADP-malic enzyme. Such studies on other metal ion requiring enzymes have been reported [4,5]. A non-enzymatic MCO system comprises of ascorbate, O_2 and Fe (III) or Fe (II). While an enzymatic MCO system has been shown to catalyze the O_2 and Fe (III) dependent oxidative modification of proteins. In the present paper by using non-enzymatic MCO system, we have proposed ASP^{352} as the co-ordination site for the metal binding in maize NADP-malic enzyme during catalysis.

2. Procedure

Purification of maize NADP-malic enzyme: Malic enzyme from maize leaves was purified to apparent homogeneity according to the method of Asami *et al,* [6]. The final specific activity of the enzyme was about 70 units per mg protein.

Gel Electrophoresis and isolation of Protein Fragments: The native and the inactivated proteins were run on SDS PAGE following the method of Laemmli [7]. The electroblotting was conducted in a blotting cassette (Banglore Genei) using a constant voltage of 20 V for 18 h at 6°C.

G. Garab (ed.), Photosynthesis: Mechanisms and Effects, Vol. V, 3591–3594.
© 1998 *Kluwer Academic Publishers. Printed in the Netherlands.*

N-Terminal Amino Acid Sequencing: The Transblot membrane containing the protein fragments was directly sequenced using Shimadzu PPSQ-10 Gas Phases Protein Sequencer.

3. Results and Discussion

Inactivation of NADP malic enzyme by ferrous sulfate: The incubation of maize malic enzyme with 20 µM Fe^{2+} in presence of 20 mM ascorbate resulted in an irreversible inactivation of the malic enzyme activity which followed pseudo-first-order reaction kinetics (Fig. 1). In contrast, the pigeon liver malic enzyme showed complex reaction kinetics. Similarly, the inactivation rates for maize malic enzyme were not temperature dependent and at room temperature reasonable rates were obtained. However, in case of pigeon liver malic enzyme the rates at room temperature were too fast, hence the studies were done at 0°C [2]. The presence of ascorbate, oxygen, and Fe^{2+} in the reaction mixture was essential for malic enzyme inactivation as suggested earlier for several other enzymes [1,2].

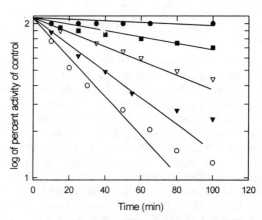

Figure 1. Time course of inactivation of malic enzyme at different Fe^{2+} concentration in presence of 20-mM ascorbate. The reaction mixture contained 1-2 µM malic enzyme in 50 mm Tris-HCl (pH 7.5). At the indicated time intervals aliquots were removed and assayed after termination of the reaction with 4-mM EDTA. Fe^{2+} concentrations were (●) 0.0 µM; (■) 5 µM ;(∇) ; 10 µM ; (▼) ;15 µM ; (○) ; 20 µM.

Protection by substrate and divalent cations: The preincubation of maize malic enzyme with divalent cations like $Mg^{2+,}$ Mn^{2+} and Ca^{2+} provided considerable protection against malic enzyme inactivation by Fe^{2+}. Mg^{2+} was most effective amongst the cations used in protection studies. NADP (1mM) along with 0.5 mM Mg^{2+} provided much better protection as compared to 0.5-mM Mg^{2+} alone (data not included). Since imidazole did not offer any protection against MCO inactivation, it seems that histidine residue may not be involved in metal binding.

Effect of catalase and n-propylgallate on Fe^{2+} ascorbate inactivation: Metal catalyzed oxidation systems generate active oxygen species H_2O_2, OH and O_2^- radicals through Fenton chemistry. These reactive species then react with the nearby susceptible amino acid residues mostly basic amino acids and proline. Such a mechanism has been proposed for oxidative modification of many enzymes. Ishida *et al.* showed that the cleavage by non-enzymatic MCO system of the large subunit of Rubisco into 37-kDa and 16-kDa polypeptide was due to the active oxygen species probably hydroxyl radicals [8]. We, therefore, studied the effect of the active oxygen scavengers on the loss of maize malic enzyme activity by Fe^{2+}-ascorbate. We had used catalase (for hydrogen peroxide) and n-propylgallate (for hydroxyl radicals) as scavengers. The results indicate that catalase totally prevented the inactivation of the enzyme. However, n-propylgallate did not offer any protection against inactivation by Fe^{2+}-ascorbate (data not included).

SDS-PAGE gel pattern of the Fe^{2+}-ascorbate inactivated malic enzyme: The inactivated enzyme (<10% residual activity) showed the presence of two major fragments of 31-kDa and 34-kDa which seem to have been derived by a single cleavage of the intact subunit of 65-kDa (Fig 2. Lanes c & d).

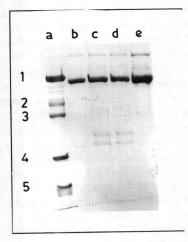

Figure 2. SDS-PAGE pattern of the MCO catalysed oxidised malic enzyme. Lane a, Protein standard. 1. Albumin bovine 66000. 2. Albumin egg 45000. 3.Trypsinogen 24000. 4. ß-Lactoglobulin 18400. Lane b, Native enzyme; Lanes c & .d, modified enzyme; Lane e, Mg^{2+} protected enzyme.

The presence of magnesium as mentioned earlier, not only protected the enzyme activity but also prevented the cleavage of the subunit (Fig. 2, lane e).

The N-terminal sequence analysis of the 31-kDa and 34-kDa fragments: The loss of malic enzyme activity to some extent also resulted from the protein cleavage (see Fig. 2 lane c and d). The sequence data shown in Fig. 3 imply that ASP^{352} was located in the 34-kDa fragment of maize NADP-malic enzyme, while in pigeon liver NADP-malic enzyme, the 31-kDa fragment was found to be N-terminally blocked and Asp^{258} was located in 31-kDa fragment [2]. The N-terminal analysis of the 31-kDa maize enzyme fragment for 14 cycles is shown in Fig 3. The results indicate that the bond between Cα of

▼

353
Maize (31-kDa)   I Q G T A S V V L A G L L-A

259
Pigeon (34-kDa) I Q G T A S V A V A G L-L-A

1
Maize (34-kDa)   V S N A Q T Q T

Fig. 3. N-terminal sequence analyses of maize malic enzyme 31 kDa and 34 kDa fragments. Comparison of the putative metal binding sites.

Asp^{352} and carbonyl carbon of Ile^{353} is cleaved and the Ile^{353} as the new N-terminal residue is indicated by an arrow. [2,9]. Our results suggest that aspartate is the metal binding ligand in maize malic enzyme in accordance with the general phenomena that Asp predominates in co-catalytic metal sites where binding frequency is Asp. > His >> Glu [10]. The N-terminal sequence of the 34-kDa fragment is also presented in Fig. 3. This sequence matches well with the first few N-terminal amino acids of the deduced protein sequence from c-DNA of maize malic enzyme [11].

Mechanism of Inactivation: Malic enzyme belongs to the most susceptible enzymes towards metal catalyzed oxidation and the system is highly specific for the metal binding site. Since the Fenton systems involve H_2O_2 and can be blocked by catalase, our data on protection by catalase clearly indicate the involvement of Fenton system in the inactivation mechanism. We do think that Fe^{2+} or Fe^{3+} occupies the metal binding site causing inactivation because in that case the inhibition should have been reversible by excess Mg^{2+}. The cleavage of the enzyme occurs subsequent to the process of inactivation. The correlation between enzyme inactivation and the peptide bond cleavage was not done, as the cleavage process was too slow. The inactivation proceeds in three steps: First, the enzyme forms a reversible complex with Fe^{2+} and divalent ions such as Mg^{2+} are able to offer protection at this step. Then, Fe^{2+}-ascorbate system generates some reactive species (H_2O_2, OH^-, O_2^-) which react with nearby Mg^{2+} ligand Asp^{352} causing inactivation. Finally, the modified enzyme is cleaved at the susceptible bond. This mechanism is consistent with our observations. The studies carried out by Soundar and Colman [5] using iron-ascorbate system on NADP specific isocitrate dehydrogenase indicated that affinity cleavage of the enzyme by iron alone (without substrate ie. isocitrate) produced peptides by specific cleavage between Tyr^{272} and Asp^{273}. Metal catalyzed oxidative modification of tyrosine has been reported by Kim et al [12]. However, it is also possible that the cleavage can occur subsequent to the oxidation of carbon either of Tyr^{272} or Asp^{273} [5,13]. Our results are in agreement with that of Wei et al. [2] that iron-ascorbate is an excellent affinity cleavage system for the identification of metal binding site. Asp^{352} may be the coordination site for binding of the metal ion to the maize malic enzyme during catalysis, which corresponds to the Mn^{2+} binding site Asp^{258} of pigeon liver malic enzyme. Our earlier study with Woodward's reagent 'K' has demonstrated presence of a carboxyl group at or near NADP binding site [14]. In both these studies, we have observed greater protection of maize malic enzyme activity by Me^{2+} + NADP. Thus, it may be possible that either a single carboxyl group may be involved in metal and NADP binding or there may be two different carboxyl groups located at the NADP and metal binding sites of maize malic enzyme.

4. References

1. Hsu, R.Y., Mildvan, A.S., Chang, G.G. and Fung, C.H. (1976) J. Biol. Chem. 251, 8053-8057.
2. Wei, C.H., Chou, W., Haung, S., Lin, C.C. and Chang, G. (1994) Biochemistry 33, 7931-7936.
3. Wei, C.H., Chou, W. And Chang, G.G. (1995) Biochemistry 34, 7949-7954.
4. Amici, A., Levin, R., Tsai, L. and Stadman, E.R. (1989) J. Biol. Chem. 264, 3341-3346.
5. Soundar, S. and Colman, R.F. (1993) J. Biol. Chem. 268, 5264-5271.
6. Asami, S., Inoue, K., Matsumoto, K., Murachi, A. and Akazawa, T. (1979) Arch. Biochem. Biophys. 194, 503-510.
7. Laemmli, U.K. (1970) Nature (Lond.) 277, 680-685.
8. Ishida, H., Nishimori, Y., Sugisawa, M., Makino, A. And Mae, T. (1997) Plant Cell Physiol. 38, 471-479.
9. Platis, I.E., Ermacora, M.R. and Fox, R.O. (1993) Biochemistry 32, 12761-12767.
10. Valle, B.L. and Auld, D.S. (1993) Biochemistry 32, 6493-6500.
11. Rothermal, B.A. and Nelson T. (1989) J. Biol. Chem. 264, 19587-19592.
12. Kim, K., Rhee, S.G. and stadman, E.R. (1985) J. Biol. Chem. 260, 15394-15397.
13. Hoyer, D., Cho, J. and Schultz, P.G. (1990) J. Am. Chem. Soc. 112, 3249-3250.
14. Rao, S.R., Kamath, B.G. and Bhagwat, A.S. (1997) Ind. J. Biochem. Biophys. 34, 253-258.

PYRUVATE, Pi DIKINASE OF CAM PLANTS: SPECIES VARIATION IN THE SUBCELLULAR LOCALIZATION AS REVEALED BY IMMUNOELECTRON MICROSCOPY

Ayumu Kondo[1,3], Akihiro Nose[1], Hiroshi Yuasa[2] and Osamu Ueno[3]
[1]Fac. Agric., Saga Univ., Saga 840-8502, Japan.
[2]Res. Inst. Evol. Biol., Setagaya, Tokyo 158-0097, Japan.
[3]Natl. Inst. Agrobiol. Resour., Tsukuba, Ibaraki 305-8602, Japan.

Key words: CAM, PPDK localization, immunogold electron microscopy.

1. Introduction

CAM plants can be divided into two groups, PCK (phosphoenolpyruvate carboxykinase)-CAM and ME (malic enzyme)-CAM plants, based on the difference in decarboxylation process. In ME-CAM plants, pyruvate, Pi dikinase (PPDK) activity is found, and it is involved in the conversion of pyruvate to PEP. PCK-CAM plants lack PPDK activities (1). Recently, we have revealed that in some ME-CAM species, PPDK is present in the cytosol as well as in the chloroplasts of the mesophyll cells (2), although it was conventionally thought to be localized in the chloroplasts (3,4). In *Mesembryanthemum crystallinum* PPDK was present only in the chloroplasts, whereas in *Kalanchoe pinnata* and *K. daigremontiana* it was found in both chloroplasts and cytosol (2). ME-CAM plants occur in various families, such as Aizoaceae, Cactaceae, and Crassulaceae. In this study, we investigated the subcellular localization of PPDK for ME-CAM species from various taxonomic groups.

2. Materials and Methods

2.1. *Plant materials*: Of the 12 CAM species examined here, *Kalanchoe* and *Alluaudia* species were collected in Madagascar, and other species were purchased at a local market. Plants were grown in a naturally illuminated green house maintained at 25℃ /20℃ (day / night temperature) for two months from November to December.

2.2. *Antiserum*: Antiserum raised against PPDK from maize leaves was generously provided by Dr. M. Matsuoka (Nagoya University, Nagoya, Japan).

2.3. *Immunoelectron microscopy*: Preparations and immunostainning of samples for immunoelectron microscopy were performed as described previously (5). Small segments of leaves were fixed in 3% gultaraldehyde in 50mM sodium phosphate (pH6.8) for 4 h on ice, and were then washed in phosphte buffer. They were dehydrated through an ethanol series at -20℃ and embedded in Lowicryl K4M resin at -20℃. Ultrathin sections were incubated in 0.5% (w/v) bovine serum albumin (BSA) in phosphate-buffered saline (PBS) that contained 10mM sodium phosphate (pH7.2), 150mM NaCl, and 0.1% (v/v) Tween 20 for 20 min, and then for 3 h in antiserum that had been diluted with 0.5% BSA in PBS. They were incubated in a suspension of 15nm protein A-gold particles (E.Y. Lab. Inc.) and stained with uranyl acetate and lead citrate.

2.4. *Quantitative analysis of immunogold particles*: The density of PPDK labeling was determined by counting the gold particles on electron micrographs and calculating the

G. Garab (ed.), Photosynthesis: Mechanisms and Effects, Vol. V, 3595–3598.

Table 1. Intracellular localization of PPDK and enzyme activities in photosynthetic tissues of various CAM plants

Taxa	Type	Immunogold labeling of PPDK		Enzyme activities (μmol mg Chl^{-1} hr^{-1})			
		chloroplasts	cytosol	PPDK	NADP-ME	NAD-ME	PCK
Dicotyledon							
Aizoaceae							
Lithops bromfieldii	Chlt-Cyt	++	+++	42.6	10.4	139.0	N.D.
*Mesembryanthemum crystallinum**	Chlt	++	−				
Cactaceae							
Nopalxochia ackermanni	Chlt-Cyt	+	+++				
Schlumbergera russelliana	Cyt	−	+++	47.3	16.1	60.5	N.D.
Crassulaceae							
Crassula argentea	Chlt	++	−	14.0	56.6	38.6	N.D.
Kalanchoe beharensis	Chlt-Cyt	++	+++	14.9	47.5	50.9	N.D.
*K. daigremontiana**	Chlt-Cyt	++	+++				
K. fedtschenkoi	Chlt-Cyt	++	+++	35.4	55.2	107.2	N.D.
K. hidebrandtii	Chlt-Cyt	++	+++	27.3	64.3	85.8	N.D.
K. miniata	Chlt-Cyt	++	+++	13.4	19.2	114.4	N.D.
K. orgyalis	Chlt-Cyt	++	+++	60.3	53.1	60.3	N.D.
*K. pinnata**	Chlt-Cyt	++	+++				
K. prorifera	Chlt-Cyt	+	+++	40.3	45.6	79.5	N.D.
Didiereaceae							
Alluaudia procera	Chlt-Cyt	+	+++	36.5	25.0	50.0	N.D.
Monocotyledon							
Dracaenaceae							
Sanseveria hahnii	Chlt-Cyt	++	+++	16.7	26.4	23.3	N.D.

(+) and (−) refer to the relative intensities of labeling, with (+++) indicating heavy labeling and (−) indicating no labeling. PPDK, pyruvate Pi dikinase; NADP-ME, NADP-malic enzyme; NAD-ME, NAD-malic enzyme; PCK, PEP-carboxykinase.
*Data from Kondo et al. (1998).

number per unit area (μm^2). For profiles of chloroplasts, the areas occupied by starch grains were excluded from estimation. Ten to 12 individual cells were examined in several immunolabeled sections from each species.

2.5. *Assay of enzymatic activities*: PPDK and three decarboxylating enzymes were assayed spectrophotometrically in 1-ml reaction mixtures at 25℃, as reported previously (5).

3. Results

Table 1 lists 12 species in 8 genera of 5 families examined in this study and 3 species examined previously (2). To determine the subtype of these CAM species, activities of PPDK and three decarboxylating enzymes, namely, NADP-ME, NAD-ME and PCK, were assayed (Table 1). PPDK activities were found in leaves of all CAM species examined. Although activities of NADP-ME and NAD-ME were also detected in the CAM species, PCK activities were undetectable. These data suggest that all CAM species examined are classified into ME-CAM type.

Western blot analysis confirmed that the antiserum against maize PPDK recognizes specifically PPDK from the ME-CAM species examined. The intracellular localization

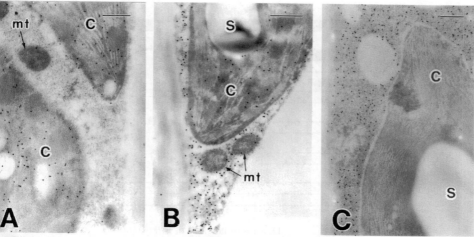

Fig. 1. Immunogold localization of PPDK in msesophyll cells of *Crassula argentea* (A), *Kalanchoe prorifera* (B), and *Schlumbergera russelliana* (C). Bars = 0.5µm. C, Chloroplast; mt, mitochondrion; S, starch grain.

of PPDK in the mesophyll cells of the ME-CAM species was investigated by immunogold electron microscopy using the antiserum. The pattern of labeling for PPDK was divided into three types, namely, Chlt type, Chlt-Cyt type, and Cyt type. In the Chlt type, labeling for PPDK is present only in the chloroplasts. In the Chlt-Cyt type, labeling is present in both the chloroplasts and cytosol, whereas in the Cyt type, it occurs only in the cytosol (Table 1). *M. crystallinum* (Aizoaceae) and *Crassula argentea* (Crassulaceae) represented the Chlt type (Fig.1A), and *Schlumbergera russelliana* (Cactaceae) did the Cyt type (Fig.1C). Other 12 species in 5 genera of 5 families were classified into the Chlt-Cyt type (Table 1), In this type, the density of labeling of PPDK was always higher in the cytosol than in the chloroplasts (Fig.1B). However, the ratios of the labeling density of the cytosol to that of the chloroplasts varied among species (Fig.2). In *Nopalxochia ackermanni* (Cactaceae), the labeling density in the chloroplasts was very low, and this pattern of labeling approached that in the Cyt type.

4. Discussion

This study revealed that the intracellular pattern of PPDK accumulation largely varies among the ME-CAM species. At present, the physiological significance of the species variation in PPDK localization is unclear. The results of our study prove only the accumulation of PPDK protein, but it is unknown whether the PPDK is an active form or not. Nevertheless, the existence of PPDK activities in *S. russelliana* with the Cyt type suggests that the PPDK in the cytosol is active and may play a functional role. Cytosolic PPDK has been found in seeds of maize and wheat (6), and it may be involved in the synthesis of seed storage protein (6). In the amphibious sedge *Eleocharis vivipara*, which expresses the C_4-like and C_3-like modes under terrestrial and submerged conditions, PPDK is also accumulated in both chloroplasts and cytosol of the photosynthetic cells (7). The genes encoding the chloroplastic and cytosolic PPDK have been isolated (8), and the expression patterns differed between the growth forms (7,8). Very recently, it has been reported that cytosolic PPDK is induced by

3598

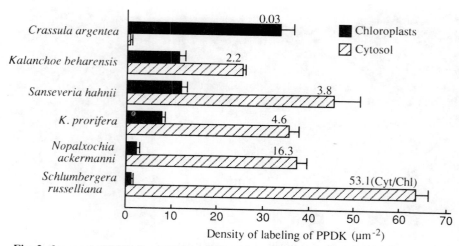

Fig. 2. Immunogold labeling of PPDK in mesophyll cells of ME-CAM species. Numbers show the ratio of the labeling density of PPDK in the cytosol to that in the chloroplasts. *C. argentea* and *S. russelliana* are the Chlt type and the Cyt type, respectively. Others are the Chlt-Cyt type.

low-oxygen stress and water deficit in roots of rice (9). Further research is necessary to clarify the functional significance of cytosolic PPDK in ME-CAM plants.

5. Acknowledgements

Part of this study was supported by a grant-in-aid from the Science and Technology Agency of Japan (Enhancement of Center-of-Excellence) to O.U.

6. References

1. Holtum, JAM. and Osmond, CB. (1981) Aust. J. Plant Physiol. 8, 31-44.
2. Kondo, A., Nose, A. and Ueno, O. (1998) J. Exp. Bot. 49 (in press).
3. Spalding, MH., Schmitt, MR., Ku, MSB. and Edwards, GE. (1979) Plant Physiol. 63, 738-743.
4. Winter, K., Foster, JG., Edwards, GE. and Holtum, JAM. (1982) Plant Physiol. 69, 300-307.
5. Ueno, O. (1992) Physiol. Plant. 85, 189-196.
6. Aoyagi, K. and Chua, N-H. (1988) Plant Physiol. 86, 364-368.
7. Ueno, O. (1998) Plant Cell 10, 571-583.
8. Agarie, S., Kai, M., Takatsuji, H. and Ueno, O. (1997) Plant Mol. Biol. 34, 363-369.
9. Moons, A., Valcke, R. and Montagu, M.V. (1998) Plant J. 15, 89-98.

Temperature and pH responses of phosphofructokinase from three CAM plants, *Ananas comosus, Kalanchoë pinnata*, and *K. daigremontiana*

Akihiro Nose, Ayako Miyata, Karin Kobayashi and Kikuo Wasano
Faculty of Agriculture, Saga University
1 Honjo-machi, Saga, 840-8502 JAPAN

Key words: ATP, diurnal change, glycolysis, enzymes, isoform, regulatory processes.

1. Introduction

CAM-type photosynthesis uses the glycolytic pathway in both directions to produce phosphoenolpyruvate (PEP) as the substrate of dark CO_2 fixation and to convert the C_3-residues of malate decarboxylation to sugars in the neoglycolitic direction.

The glycolysis is the metabolic pathway from glycogen to pyruvate in the living cell, and phosphofructokinase (PFK) has been recognized as a key enzyme to regulation of the metabolic flow(1). PFK catalyzed the reaction between fructose-6-phosphate (F-6-P) and fructose-1,6-bisphosphate (F-1,6-P_2). There were two isozymic variations of the PFK for the substrate of the reaction using pyrophosphate (PPi) and adenosine-triphosphate (ATP), called as PPi-PFK and ATP-PFK, respectively. The PPi-PFK catalyzed reversibly the reaction, F-6-P + PPi $\rightleftarrows$ F-1,6-P_2 + Pi, and located in the cytosol of the plant cell. The ATP-PFK proceeded the reaction irreversibly, F-6-P + ATP$\rightarrow$F-1,6-P_2 + ATP, and located in the cytosol or plastids.

Nose et al. (2) found that when a night temperature higher than 36℃ was maintained, *K. daigremontiana* and *K. pinnata* shifted from the CAM type photosynthesis to C_3 type photosynthesis, but *A. comosus* did not alter its CAM type photosynthesis. In addition, regarding malate sensitivity of PEP carboxylase (PEPCase), *K. pinnata* showed diurnal changes similar to *K. fedeschenkoi* (3) and *Mesembryanthemun crystallinum*(4), but in the case of both *K. daigremontiana* and *A. comosus* it was suggested that the diurnal changes in malate sensitivity of PEPCase did not relate directly to the diurnal regulation of CAM-type CO_2 fixation.

In this study, in order to determine the relation of glycolysis to CAM-type photosynthesis, the effects of extraction methods and substrates were first examined, then temperature and pH responses in the two forms of PFK were compared among three CAM

G. Garab (ed.), Photosynthesis: Mechanisms and Effects, Vol. V, 3599–3602.
© *1998 Kluwer Academic Publishers. Printed in the Netherlands.*

plant species.

2. Procedure

The plant materials were grown in a heated greenhouse at Saga University, Japan. Plants of *A. comosus* one to one-half year after planting and two species of *Kalanchoë* three to six months after planting were used in this study. Experimental leaf materials were collected after being kept for 6 days in a growth chamber (KG-50HLA, Koito Co.Ltd., Japan). Conditions in the chamber were as follows: day and night temperature was kept at 30 and 20℃, respectively; PAR on the sample leaves levels was ca. 330 μ mol/m²/s; and day length was 12hrs, from 7:00 to 19:00. Fully expanded leaves were used as experimental materials. The leaf disks were weighed just after collecting, fixed in liquid nitrogen, and kept at -80℃.

Enzymes were extracted with a buffer of 100mM Hepes-NaOH pH8.0, 150 mM CH_3COOK, 30mM β-mercaptoethanol, 5mM $MgCl_2$, 1mM EGTA, and 1%(W/V) PVP-40; 1% TritonX-100 or 0.5M NaCl was added appropriate to the extraction buffer.

The activity was measured spectrophotometically with a buffer of 100mM Hepes-NaOH pH 8.0, 2.5mM $MgCl_2$, 0.15mM NADH, 10mM F-6-P, 6 units Aldolase, 6 units triosephosphate isomerase (TPI), and 6 units α-glycerol-phosphodehydrogenase (GDH). The reaction was initiated with 1mM PPi-Na, ATP-Na, or GTP-Na.

3. Result and Discussion

3.1 Effects of the addition of TritonX-100 and NaCl at extraction, and differences of substrates on the PFK activity (Table 1):In pineapple plants, the activities of PPi-PFK were about 10-fold higher than those of ATP-PFK in three extracting conditions, and the ATP and GTP-PFK activities were almost similar to those of two *Kalanchoë* species. In the *Kalanchoë* species, ATP-PFK showed high activities in the extraction to which TritonX was added, and PFK activities using GTP as the substrate were also high in the TritonX block. These observations suggested that PPi-PFK was located in either the cytosol or chloroplast stroma in the pineapple leaf cells; that ATP-PFK was located in the chloroplast membrane of *Kalanchoë* leaf cells; and that GTP substituted for the ATP-PFK reaction as the phosphate donor.

3.2Temperature responses (Figs. 1 and 2):PPi-PFK activity in pineapple plants showed sigmoidal increasing in both day and night forms as temperature also increased from 10℃ to 40℃. ATP-PFK activity in pineapple plants increased gradually up to 40℃ in both day and night forms. In analyzing Arrhenius-plots, the breaking point of PPi-PFK was 25℃ and 27.5℃ for day and night forms, and that of ATP-PFK was 32.5℃ in both day and night forms, respectively.

ATP-PFK activity of *K. daigremontiana* showed maximum levels at 35°C for both day and night forms (Fig. 2). The breaking point of Arrhenius-plots was 32.5°C in both.
ATP-PFK activity of *K. pinnata* showed a maximum level at 35°C and 30°C for day and night, respectively. However the breaking point of Arrhenius-plots was 32.5°C for both.

As mentioned above, the two forms of PFK in pineapple leaves did not show an optimum temperature different from ATP-PFK of *Kalanchoë*. However, the breaking point

Table 1. Effects of application of TritonX and NaCl into the extraction buffer on PFK activities (μ mol/mg Chl./min.)

Plant species	Phosphate Donor	Control	TritonX-100 1%	NaCl 0.5M
Ananas comosus	PPi	9.583	10.933	11.021
	ATP	0.901	1.127	1.012
	GTP	0.769	0.928	0.904
Kalanchoe daigremontiana	PPi	0.045	0.051	0.061
	ATP	0.080	0.439	0.076
	GTP	0.059	0.431	0.060
K. pinnata	PPi	0.120	0.160	0.131
	ATP	0.111	0.778	0.143
	GTP	0.148	0.666	0.172

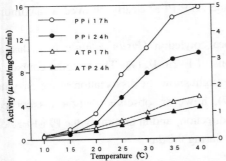

Fig. 1. Temperature responses of PFK from *A. comosus*.
Legends in the figure mean PFK forms and sampling times.

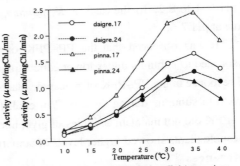

Fig.2. Temperature responses of ATP-PFK from *K. daigremontiana* and *K. pinnata*.
Legends in the figure mean plant species and sampling times, for example "daigre.17" shows K.daigremontiana and 17hr of sampling time.

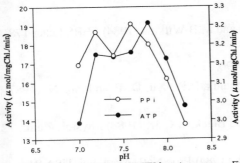

Fig.3. pH responses of PPi and ATP-PFK from *A. comosus*.
Sampling time is 17hr.

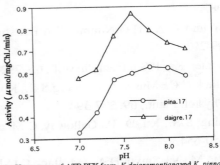

Fig. 4. pH responses of ATP-PFK from *K.daigremontiana* and *K. pinnata*.
Legendes in the figure see in Fig.2.

of ATP-PFK in Arrhenius plots was similar to that in three CAM plant species. As it was considered that ATP-PFK worked in maintenance glycolysis (1), the similarity of breaking points in Arrhenius-plots might show a characteristic of the maintenance pathway. In addition, in the three CAM plant species, PFK showed high levels of activity in day forms. Concerning the role of glycolysis in CAM type photosynthesis, higher levels of activity in day forms was inconsistent with predictions. It is well-known that PPi-PFK is regulated closely by $F-2,6-P_2$ content (5,6). The contradiction in this study might be due to in vitro conditions that did not examine the coexistence of $F-2,6-P_2$.

3.3pH responses (Figs.3 and 4):pH responses of the PFK activities were examined in the samples collected at 17:00. In pineapple plants, the PPi-PFK showed bimodal response, in which maximum values were observed at pH 7.2 and 7.6. The ATP-PFK of pineapple plants showed a complex response curve, which included a maximum value at pH 7.8 and a shoulder course from pH 7.2 to 7.6. The ATP-PFK of *K. daigremontiana* increased the level of activity from pH 7.0 to 7.4, and sustained high levels of activity under conditions of pH 7.4 to 8.2. In the case of *K. pinnata*, the ATP-PFK clearly showed a maximum value at pH 7.6.

It was observed that the PPi-PFK of pea cotyledon (*Pisun sativum* cv. Alaska) dissociated to subunits under lower concentrations of $F-2,6-P_2$ (7). The bimodal course of pH response in the PPi-PFK of pineapple plants might show the dissociation of the enzyme protein to subunits during the day. Wu et al. (7), however, observed that the dissociated PPi-PFK did not maintain activities of glycolitic direction, and in this study the PPi-PFK of pineapple plants showed higher levels of activity during the day.

References

1 Plaxton, W.C. (1996) Annu. Rev. Plant Physiol. Mol. Biol., 47,185-214.

2 Nose, A., Abe, S. and Kawamitsu, Y. (1994) Jpn. J. Crop Sci. 63 (Extra Issue 2), 107-108.

3 Nimmo, G.A., H.G.Nimmo, C.A.Fewson and M.B.Wilkins (1984) FEBS Letter 178, 199-203.

4 Winter, K. (1980) Plant Physiol., 65, 792-796.

5 Black,C.C., Mustardy,L., Sung,S.S., Kormanik, P.P., Xu, D.-P. and Paz, N. (1987) Physiol. Plant., 69, 387-394.

6 Sung, S.S., Xu, D.-P., Galloway, C.M. and Black, C.C.Jr. (1988) Physiol. Plant., 72, 650-654.

7 Wu, M.-X., Smyth,D.A. and Black,C.C.Jr. (1984) Proc.Natl. Acad. Sci. USA 81, 5041-5055.

FD-GOGAT IS NECESSARY FOR THE INTEGRATION OF PHOTOSYNTHETIC C- AND N-ASSIMILATION IN CYANOBACTERIUM

Hiroaki Okuhara, Tomohiro Matsumura, Yuichi Fujita and Toshiharu Hase
Institute for Protein Research, Osaka University, 3-2 Yamadaoka, Suita,
Osaka 565-0871, Japan

Key words: cyanobacteria, mutants, N metabolism, phycobiliproteins, glutamate synthase

1. Introduction

In the oxygenic photosynthetic organisms, ammonia is assimilated into glutamate through the combined actions of glutamine synthetase (GS) and glutamate synthase (glutamine 2-oxoglutarate amidotransferase or GOGAT). GS catalyzes the ATP-dependent amination of glutamate to yield glutamine. GOGAT catalyzes the reductive transfer of the amide group of glutamine to the keto position of 2-oxoglutarate to yield two molecules of glutamate. The GS/GOGAT pathway ultimately requires ATP and reducing power, which are generated by photosynthesis and oxidative catabolism of carbohydrates, and utilizes carbon skeletons. This pathway is thus involved in the integration of carbon and nitrogen assimilations.

In higher plants, GOGAT occurs as two distinct forms, Fd-dependent GOGAT (Fd-GOGAT) and NADH-dependent GOGAT (NADH-GOGAT), that differ in the specificity for electron donor and in the molecular architecture (Fig.1). In cyanobacteria, only the presence of Fd-GOGAT was reported in several species and no activity of pyridine nucleotide-dependent GOGAT was found. In our study of GOGATs from the cyanobacterium *Plectonema boryanum*, which is able to grow under both photoautotrophic and heterotrophic conditions, we detected two comparable activities of Fd-GOGAT and

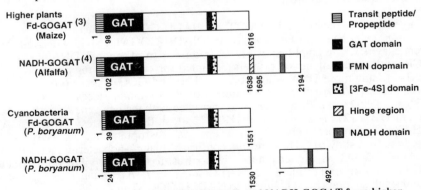

Figure 1. Schematic representation of Fd-GOGAT and NADH-GOGAT from higher plants and cyanobacteria, *P. boryanum*.

3603

G. Garab (ed.), Photosynthesis: Mechanisms and Effects, Vol. V, 3603–3606.
© 1998 *Kluwer Academic Publishers. Printed in the Netherlands.*

NADH-GOGAT in contrast to the reports for another species of cyanobacteria. We have cloned genes for both the GOGATs, and isolated mutants lacking each gene. In this report, we demonstrate that the Fd-GOGAT gene disruption causes the imbalance of the N- and C-assimilations in the cyanobacterium.

2. Procedures

2.1 Cyanobacterial strain and culture conditions

The cyanobacterium *Plectonema boryanum* IAM-M101 strain *dg5* and its derivative strains isolated in this study were cultivated in BG11 medium or its modified one containing 10 mM NH_4Cl instead of 17.6 mM $NaNO_3$. Growth was monitored by turbidity with the Klett-Summerson photoelectric colorimeter equipped with KS66 filter (640-700 nm). YFD1 carrying kanamycin resistant *neo* gene on a neutral site of the *P. boryanum* genome was used as a control strain (1).

2.2 Targeted mutagenesis of GOGATs genes

The *neo* gene cartridge was inserted into each GOGAT gene. A plasmid carrying the resulting interrupted or truncated GOGAT gene was linearized and introduced into *P. boryanum* cells by electroporation for homologous recombination. Mutants were selected on BG-11 with 15 µg/ml kanamycin. Gene disruption was confirmed by Southern analysis.

2.3 GOGAT assay

Glutamate synthase activity was determined by glutamate formation as described in (2). Ferredoxin (40 µM) from *P. boryanum* and NAD(P)H (1 mM) were used as electron donors for the GOGAT assay.

2.4 Measurements of O_2 evolution and ammonia uptake

Rate of oxygen evolution was determined with an oxygen electrode unit at 30°C at light intensity from 30 to 210 $\mu E\ m^{-2}s^{-1}$. Cyanobacterial cells were suspended in BG11 containing 20 mM Hepes (pH7.4) without $NaNO_3$, and total photosynthetic activity was measured after addition of 10 mM $NaHCO_3$ to the cell suspension. Ammonia uptake was determined as consumption of ammonia in the culture using an ammonia selected ione electrode.

3. Results and Discussion

3.1 Targeted mutagenesis of Fd-GOGAT and NADH-GOGAT genes

P. boryanum had both Fd-GOGAT and NADH-GOGAT activities derived from two different enzymes. We amplified two DNA fragments from genomic DNA of *P. boryanum* by PCR with a pair of mixed primers designed for conserved regions among various GOGATs. Using those DNA as a probe, we cloned Fd-GOGAT gene (*glsF*) and NADH-GOGAT genes (*gltBD*) from genomic library of *P. boryanum* (accession No. for DDBJ, D85735 and D85230), and each gene was disrupted with the *neo* gene cassette. The *glsF*-disruptant (HOF12), *gltB*-disruptant (HOB12) lost the activities of Fd-GOGAT and NADH-GOGAT, respectively (Fig.2). The *gltD*-disruptant had no NADH-GOGAT activity (data not shown). These results confirmed that cyanobacterium has Fd-GOGAT and NADH-GOGAT as higher plants. In contrast to a single polypeptide structure of NADH-GOGAT in higher plants, NADH-GOGAT in *P. boryanum* is composed of the large and small subunit (Fig.1), similar to NADPH-GOGAT in *E. coli*.

3.2 Growth experiments on NADH-GOGAT and Fd-GOGAT deficient mutant cells

Growth analysis was undertaken using Fd-GOGAT deficient mutant (HOF12) and NADH-GOGAT deficient mutant (HOB12), which were cultivated photoautotrophically in the medium containing 10 mM NH_4Cl as a nitrogen source under two light conditions of 25 and 165 $\mu E\ m^{-2}s^{-1}$. Under the higher light condition with aeration of 2% CO_2 in air, the growth curve of HOF12 showed a sigmoidal shape in contrast to hyperbolic shapes of HOB12 and the control strain, YFD1 (Fig.3). The color of HOF12 cells cultivated in this condition changed from normal green to yellow. The yellow coloration of HOF12 could not be attributed to a response to light intensity, such as chromatic adaptation. When cells were grown under the lower light intensity or when the supplement of 2% CO_2 to air was omitted even under the higher light intensity this strain grew as well as the control strain. Under the heterotrophic condition in the dark, all mutant strains grew in the same manner with the control strain.

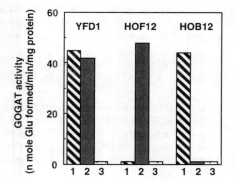

Figure 2. GOGAT activities of the control (YFD1), *glsF*-deficient (HOF12), and *gltB*-deficient (HOB12) strains of *P. boryanum*. Total cell extracts of YFD1, HOF12, and HOB12 were used for the assay of GOGAT activities with Fd (lane 1), NADH (lane 2), and NADPH (lane 3) as an electron donor.

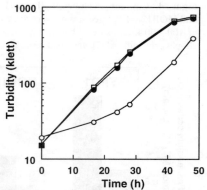

Figure 3. Growth curves of the control (YFD1, ●), *glsF*-deficient (HOF12, ○), and *gltB*-deficient (HOB12,□) strains of *P. boryanum*. The three strains were grown in a modified BG11 medium containing 10 mM NH_4Cl with 2% CO_2 aeration at 30°C under 165 $\mu E\ m^{-2}s^{-1}$ illumination.

3.3 Imbalance of N- and C-assimilations in the Fd-GOGAT deficient mutant

The yellow coloration of HOF12 could mean that a considerable changes in the contents of the cellular pigments and proteins were induced by the deficiency of Fd-GOGAT. In fact the contents of chlorophyll and phycobiliproteins in HOF12 were decreased by 2.5 to 3.3 folds compared with those in HOB12 and YFD1 (Fig.4). The capacities of carbon and nitrogen assimilations were compared between the HOF12 and YFD1 by measuring rates of O_2 evolution and ammonium consumption in the medium. The consumption rate of ammonia in the culture medium was increased with an increase of light intensity in the control strain, while Fd-GOGAT deficient mutant lacked such ability (Fig. 5). On the other hand, O_2 evolution were increased similarly in both strains in response to the light intensities (data not shown). The increased rate of the ammonium consumption under the higher light was dependent on the availability of CO_2, because the removal of CO_2 from aeration gas resulted in a decrease of the rate (data not shown).

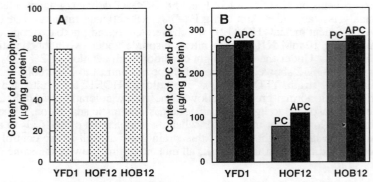

Figure 4. Contents of chlorophyll (panel A) and phycobiliproteins (panel B) of YFD1, HOF12, and HOB12. The three strains were grown under 165 μE m^{-2}s^{-1} illumination to reach the turbidity of 100–400 klett as described in Fig. 3, and the contents of chlorophyll, phycocyanin (PC) and allophycocyanin (APC) were analyzed.

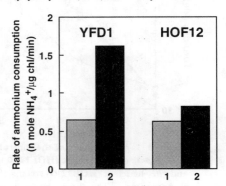

Figure 5. Ammonium consumption by YFD1 and HOF12. The rate of ammonium ion uptake from the modified BG11 medium were measured in the two strains grown under the two light conditions, 25 μE m^{-2}s^{-1}(lane 1) and 165 μE m^{-2}s^{-1}(lane 2).

These results suggested that the Fd-GOGAT deficient mutant was unable to assimilate ammonia as efficiently as the control strain under the sufficiency of light and CO_2, and that this inability induced an impairment in the integration of photosynthetic carbon and nitrogen assimilations, which resulted in the phenotype of nitrogen deficiency.

References

1 Fujita, Y., Matsumoto, H., Takahashi, Y. and Matsubara, H. (1993) Plant Cell Physiol. 34, 305-314
2 Martin, F., Suzuki, A. and Hirel, B. (1982) Anal. Biochem. 125, 24-29
3 Sakakibara, H., Watanabe, M., Hase, T. and Sugiyama, T. (1991) J. Biol. Chem. 266, 2028-2035
4 Gregerson, R.G., Miller, S.S., Twary, S.N., Gannt, J.S. and Vance, C.P. (1993) Plant Cell 5, 215-226

AMMONIUM ASSIMILATION IN CYANOBACTERIA. THE REGULATION OF THE GS-GOGAT PATHWAY.

Francisco J. Florencio, Mario García-Domínguez, Eugenio Martín-Figueroa, José L. Crespo, Francisco Navarro, M. Isabel Muro-Pastor and José C. Reyes.
Instituto de Bioquímica Vegetal y Fotosíntesis, Universidad de Sevilla-CSIC. Avda Américo Vespucio s/n, 41092-Sevilla, SPAIN

Keywords: Nitrogen metabolism, *Synechocystis* 6803, *Anabaena* 7120, light regulation, glutamine synthetase, glutamate synthase.

1. Introduction

The assimilation of ammonium by most of the photosynthetic organisms occurs through the sequential operation of glutamine synthetase (GS) and glutamate synthase (GOGAT), commonly know as the GS-GOGAT pathway [1,2]. The reaction catalyzed by GS involves the ATP-dependent amination of glutamate to yield glutamine. GOGAT then catalyzes the transfer of the amide group from glutamine to α-ketoglutarate to yield two molecules of glutamate. Carbon skeleton required for the ammonium assimilation is provided by the α-ketoglutarate which is synthesized by the isocitrate dehydrogenase (IDH). This metabolic route represents the connecting step between carbon and nitrogen metabolism and requires two typical photosynthetic products, ATP and reducing power. This central position in the metabolism make this pathway susceptible to be regulated by different environmental conditions, such as nitrogen and carbon availability, and by photosynthetic growth conditions.

Cyanobacteria are photosynthetic prokaryotes that carry out oxygenic photosynthesis like plants. Most cyanobacteria can use nitrate or ammonium ions as nitrogen sources and some strains are also able to fix dinitrogen. Although the existence of NADP-dependent glutamate dehydrogenase has been reported in cyanobacteria, ammonium is mainly assimilated through the GS-GOGAT pathway [1]. Recent biochemical, molecular and genetic studies are devoting to know the complexity of this assimilatory pathway in cyanobacteria. Our work has been carried out mainly in two cyanobacteria, *Synechocystis* 6803 and *Anabaena* 7120. The first one contains two different GS, GSI and GSIII and two GOGATs, one depending on ferredoxin (Fd-GOGAT) and another one on NADH (NADH-GOGAT) as electron donor, while *Anabaena* 7120 has only a GS, GSI and one Fd-GOGAT. In addition both strains only contain one NADP-IDH.

2. Results and discussion.
2.1 The enzymes and genes
2.1.1 Isocitrate Dehydrogenase
Cyanobacteria have an incomplete tricarboxilic acid cycle lacking α-ketoglutarate dehydrogenase activity. The isocitrate dehydrogenase reaction constitutes, in

G. Garab (ed.), Photosynthesis: Mechanisms and Effects, Vol. V, 3607–3612.
© *1998 Kluwer Academic Publishers. Printed in the Netherlands.*

cyanobacteria, a terminal step in carbon flow, and seems not to be involved in energy production as in other organisms. Although NADP⁺-dependent and NAD⁺-dependent have been described in prokaryotes, most bacteria have only the NADP⁺ linked enzyme. Cyanobacterial IDH is strictly dependent on NADP⁺ and no NAD⁺-IDH activity has been reported so far. We have purified and characterized the NADP⁺-IDH from the unicellular cyanobacterium *Synechocystis* sp. PCC 6803 and from the filamentous strain *Anabaena* sp. PCC 7120. The enzyme is composed of two identical subunits ($M_r = 57,000$) and shows kinetic and molecular parameters similar to the NADP⁺-IDH from *E. coli* . The *icd* gene from both cyanobacteria, gene encoding for IDH, was cloned by complementation of an *E. coli icd* mutant [3,4]. That suggest that even if cyanobacterial IDHs (involved only in the anabolic pathways instead of catabolic pathways) play a different role than the *E. coli* enzyme, both enzymes are functionally interchangeable. The deduced amino acid sequence of IDHs from both cyanobacteria is similar to those of other prokaryotic NADP-IDHs but presents an extra region which seems to be specific for the cyanobacteria [4].

2.1.2 Glutamine synthetase

Three different types of GS has been found so far. Most prokaryotes have a dodecameric GS (GS type I, GSI), composed of 12 identical subunits (M_r, about 50,000) arranged in two superimposed hexagonal rings . On the contrary, eukaryotic GS (GS type II, GSII) is an octameric enzyme with subunits of M_r, about 40,000. Members of the family Rhizobiaceae and certain actinomycetes harbor both a GSI and a GSII-like enzyme [5]. A third type of GS (GS type III, GSIII), composed of six identical subunits (M_r, about 75,000), was initially identified in *Bacteroides fragilis* [6]. These three types of GS are rather different in amino acid sequence. However, five domains of homology among all known GSs can be identified that are involved in the catalytic activity of the GSs [7].

GSI (encoded by the *glnA* gene) from cyanobacteria of the genera *Anabaena*, *Synechococcus*, *Synechocystis*, *Calothrix* and *Phormidium* have been purified to homogeneity and characterized. All cyanobacterial GSI were similar to each other in size and subunit composition and also similar to other prokaryotic GSI. *glnA* genes (gene encoding GSI) from several cyanobacteria have also been cloned and sequenced, showing a high amino acid sequence homology among them (more than 76 % identity).

The fact that a *glnA* mutant strain of *Synechocystis* 6803 was not glutamine auxotroph lead us to investigate and finally identified a second gene encoding for a glutamine synthetase in this organism. The sequencing of this gene, which was designed *glnN*, demonstrated that *glnN* product was a GSIII, homologs to GSs from *Bacteroides fragilis* (44 % identity) and *Butyrivibrio fibrisolvens* (41%) [7]. The GSIII from *Synechocystis* 6803 has been purified. Biosynthetic activity of GSIII requires the same substrates and cofactors than GSI and GSII enzymes. Apparent K_m values for ATP, glutamate and ammonium were also similar to those of *Synechocystis* GSI [8]. GSIII is present in many other non-nitrogen fixer cyanobacteria but not in nitrogen fixers[7, 8]. Recently we have also found the existence of at least one cyanobacterium, *Pseudanabaena*, 6903 that lack of GSI (and *glnA* gene) and harbor only GSIII. ☐Therefore, the putative three possibilities of diversity are present in cyanobacteria: only GSI, only GSIII and both, GSI and GSIII. A phylogenetic tree calculated from the alignment of different GSs sequences is shown in Figure 1.

Synechocystis glnA mutants are viable using nitrate as nitrogen source, but does not grow in ammonium containing medium. However, *glnN* mutants are viable under all the condition tested. Finally, a *glnA glnN* double mutant could not be obtained in any growth condition, suggesting that cyanobacteria require an active GS [9].

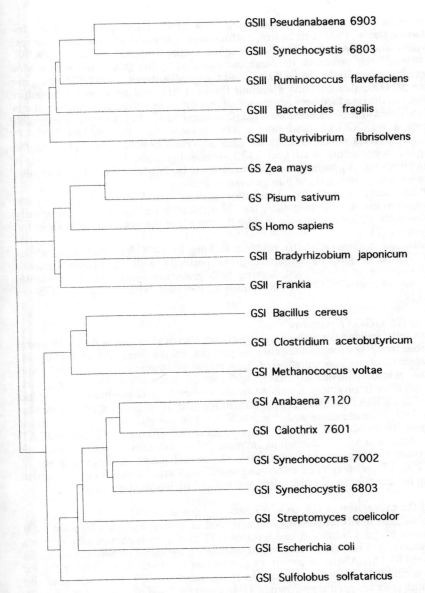

Figure 1. Phylogenetic relationships of the different types of glutamine synthetases. Sequences were obtained from the EMBL/GeneBank database, and were aligned using the Pileup program from the GCG software package.

2.1.3 Glutamate synthase

Two different types of glutamate synthase has been reported in photosynthetic organisms including cyanobacteria, one uses ferredoxin as electron donor, named Fd-GOGAT, encoded by the gene *gltS*, or *glsF* is composed of a single polypeptide, with a Mr≈ 170.000, contains a flavin, FMN and an iron-sulfur cluster (3Fe-4S) as prosthetic groups, the second type named NADH-GOGAT, is similar to the bacterial NADPH-GOGAT composed by two different subunits, the large one encoded by the gene *gltB*, with a Mr≈ 150.000 shows similarities to the Fd-GOGAT, and the small subunit encoded by the gene *gltD*, with Mr≈ 50.000, that contains a second flavin, FAD, and additional iron-sulfur clusters, probably in the interface between the small and the large subunits, in higher plants both subunits are fused and is encoded by a single nuclear gene [10]. In cyanobacteria all of them contain a Fd-GOGAT, and only a few present also the NADH-GOGAT, composed by two different subunits, as is the case of *Synechocystis* 6803 [11]. In the nitrogen fixing cyanobacterium *Anabaena* 7120, containing only Fd-GOGAT, this is absent of the heterocysts, suggesting that glutamate must be imported from the vegetative cells to produce glutamine inside these nitrogen-fixing cells.

Phylogenetic data clearly suggest that glutamate synthases found in higher plants come from the cyanobacterium-like that originates the chloroplast, since both genes are actually transferred to the nucleus, but in some algae, like the red algae the gene *gltS* remain in the chloroplast genome [12].

We have constructed in *Synechocystis* mutants lacking Fd-GOGAT (*gltS* mutant), or NADH-GOGAT (*gltB* or *gltD* mutants) that grow perfectly well in different growth conditions, however the double mutant, lacking both glutamate synthases is not viable, indicating that an operative GOGAT is required, together with an active GS in cyanoabacteria [11].

2.2 Regulation of GS-GOGAT pathway

Glutamine is not only utilized for protein synthesis but also acts as amide donor in many metabolic reactions that gives metabolites such as purines, pyrimidines and amino sugars. Therefore homeostasis of the concentration of intracellular glutamine is an important issue for most of the microorganisms. In fact, addition of ammonium to nitrate-grown *Synechocystis* 6803 cells produces a change in the intracellular concentration of amino acids, related to GS-GOGAT pathway [13]. Thus, in the first 30 seconds after ammonium addition, the pool of glutamate decreased dramatically, while the glutamine level increased reciprocally (about 30 to 60 fold change). This suggest that efficiency of GS-GOGAT pathway in the assimilation of ammonium increase 30 to 60 fold in the presence of ammonium. In other to maintain the homeostasis of internal amino acid pools levels of activity of GS-GOGAT pathway need to be readjusted. This regulatory process is carried out by regulating the expression of *glnA*, *glnN* and *icd* and by inactivating GSs.

2.2.1 Control of the expression of GS-GOGAT pathway genes

Expression of *glnA* gene has been partially characterized in several cyanobacteria. In most of the cases amount of *glnA* transcript and level of GSI protein is highest under nitrogen deficiency conditions, and lowest in ammonium containing medium, nitrate represents an intermedium level [14,15]. A similar pattern of expression has been described for the *icd* gene and IDH protein [3,4]. However, at least in *Synechocystis* 6803, the *glnN* gene and GSIII are only high after nitrogen step-down [14]. In contrast, level of expression of *gltB*, *gltD* and *gltS* seems not to be affected by the nitrogen availability.

Transcriptional control of *glnA*, *glnN* and *icd* seems to be positively regulated by the transcriptional activator NtcA. NtcA belongs to the CRP family of bacterial DNA-binding proteins and seems to be responsible for the induction of transcription of many genes in the absence of ammonium [16]. Structure of the NtcA-regulated promoters has been proposed to be constituted by a -10 *E. coli* consensus box, followed at 22 nucleotides by a NtcA-binding site (GTAN8TAC) that substituted for the canonical -35 box [17]. NtcA binds directly to the promoter region of *Anabaena* sp. PCC 7120, *Synechococcus* sp. PCC 7942, and *Synechocystis* 6803 *glnA* genes. In addition, NtcA mutants of *Synechococcus* sp. PCC 7942 and *Anabaena* sp. PCC 7120 are impaired in the expression of the *glnA* gene [17]. We have also demonstrated that NtcA is able to bind to a GTAN8TGC site present in the promoter of the *Synechocystis* 6803 *icd* gene[4]. Primer extension analysis also revealed the existence NtcA-dependent promoter upstream of the *Synechocystis glnN* gene. However, *E. coli*-expressed NtcA failed to bind to this site, suggesting that NtcA has low affinity for this site or that an additional modification of NtcA is required. It is worth noting that in *Synechocystis* 6803 expression of *glnN* is different respect to the pattern observed for *glnA* and *icd*. *glnN* is only transcribed in condition of nitrogen deficiency. In fact, in nitrate-growing *Synechocystis* cells, the *glnA* gene product (GSI) is responsible for 97% of the total GS activity, while the *glnN* product (GSIII) accounts for only about 3%. However, after 15 h of nitrogen deprivation, the activity corresponding to the GSIII represents about 20% of the total GS activity suggesting that *glnA* encodes for the housekeeping GS while *glnN* encodes for a nitrogen-starvation specific GS [7]. In nature, cyanobacteria are probably subjected to nutritional conditions that include long nitrogen starvation periods. The *glnN* expression would be induced under those conditions and GSIII would be involved to a significant extent in the assimilation of the little nitrogen available.

In addition to nitrogen conditions, other factors may also control the expression of the *glnA* gene. We have recently reported that photosynthetic and respiratory electron transport control transcription of the *glnA* gene from *Synechocystis* sp. PCC 6803 [18].

2.2.2 Regulation of GS activity
The activity of GS in enteric and other Gram-negative bacteria is modulated by a mechanism of adenylylation/deadenylylation of the enzyme, being active the deadenylylated form [5]. Studies from different group suggest that cyanobacterial GSI is not controlled by this system of modification. However, addition of ammonium to nitrate-growing cells of *Synechocystis* sp. PCC 6803, provokes a short-term inactivation of GSI (90-95% inactivation in 25 min). The inactive enzyme can be visualized because of its different mobility in non-denaturing PAGE but not in SDS-PAGE. The inactive GS can be reactivated *in vivo* by remotion of ammonium and *in vitro* by several treatment such as increasing the pH or the ionic strength of the crude extract or by incubation with alkaline phosphatase. Furthermore *in vivo* ^{32}P labeling of GS was never observed [13, 19, 20]. All these data suggest the existence of a non-covalent modification of the inactive enzyme.

Synechocystis GSI can also be inactivated by darkness or by the addition of DCMU in a process that do not depend of the light *per se* but rather to the redox state of the cell. Ammonium- and redox-mediated GS inactivation seems to occur through the same type of molecular modification. Dark- and DCMU-mediated inactivation of GS can be prevented by the presence of glucose in the culture medium suggesting that respiratory electron transport might also control GS activity [21].

In addition GSIII seems to be also inactivated after ammonium addition, but not nitrate, indicating that both Gss are subjected to regulatory processes that allow to reduce the GS activity level that ensuring an adequate intracellular glutamine level

3 Conclusions
In this review, we have presented different aspects of the function and regulation of the GS-GOGAT pathway in cyanobacteria that have being elucidated during the last 10 years. The complexity of the pathway has been evidenced by the existence of duplicated functions: two GSs, two GOGATs. We still do not understand well why this duplicity is required. Are GSI and GSIII or NAD-GOGAT and Fd-GOGAT just functionally

redundant or do they have specific roles non interchangeable in certain nutritional conditions? In the case of the GSs it seems that the clear regulatory differences and the fact that the *glnA* mutant is not viable in ammonium, suggest specific functions. Several aspects of the regulation have been addressed. However, to know what are the signals transduced from the sensors to effectors like NtcA remain unanswered.

References

1 Meeks, J.C. Wolk, C.P., Lockau, W., Schilling, N., Schaffer, P.W. and Chien, W.S. (1978) J. Bacteriol. 134, 125-130

2 Stewart, G.R., Mann, A.F. and Fentem, P.A. (1980) in The Biology of Plants (Stumpf K. and Conn, E.E., eds) 5, pp. 217-237, Academic Press, New York, USA

3 Muro-Pastor M.I., and Florencio, F.J. (1994) J. Bacteriol. 176, 2718-2726

4 Muro-Pastor, M.I., Reyes, J.C. and Florencio, F.J. (1996) J. Bacteriol. 178, 4070-4076

5 Merrick, M.J. and Edwards, R. A. (1995) Microbiol. Rev. 59, 604-622

6 Southern, J.A., Parker, J.R. and Woods, D.R. (1986) J. Gen. Microbiol. 132, 2827-2835

7 Reyes, J.C. and Florencio, F.J. (1994) J. Bacteriol. 176, 1260-1267

8 García-Dominguez, M., Reyes, J.C. and Florencio, F.J. (1997) Eur. J. Biochem. 244, 258-264

9 Reyes, J.C. and F.J. Florencio. (1994) J. Bacteriol. 176:7516-7523.

10 Temple, S.J., Vance, C.P. and Gantt, J.S. (1998) Trens Plant Sci. 3, 51-56

11 Navarro, F., Chavéz, S., Candau, P. and Florencio, F.J. (1995) Plant Mol. Biol. 27, 753-767

12 Valentin, K., Kostrzewa, M. and Zetsche, K. (1993) Plant Mol. Biol. 23, 77-85

13 Mérida, A., P. Candau, and Florencio, F.J. (1991) J. Bacteriol. 173, 4095-4100.

14 Reyes, J.C. Muro-Pastor, M.I. and Florencio, F.J. (1997) J. Bacteriol. 179, 2678-2689.

15 Mérida, A., Leurentop, L., Candau, P. and Florencio, F.J.(1990) J. Bacteriol. 172, 4732-4735.

16 Vega-Palas M. A., E. Flores and A. Herrero. 1992. Mol. Microbiol. **6**:1853-1859

17 Luque I., Flores, E. and Herrero, A.. 1994. EMBO J. 13, 2862-2869

18 Reyes, J.C. and Florencio, F.J. (1995) Plant Mol. Biol. 27, 789-799

19 Reyes, J.C. and Florencio, F.J. (1995) FEBS Letters, 367:45-48.

20 Mérida, A., Candau, P. and Florencio, F.J. (1991) Biochem. Biophys. Res. Commum. 181, 780-786.

21 Reyes, J.C., J.L. Crespo, M. García-Domínguez and F.J. Florencio. (1995) Plant Physiol. 109, 899-905

UNIDENTIFIED NITROGEN IN THE METABOLITES OF NITROGEN DIOXIDE IN PLANT LEAVES

Gen-ichiro Arimura, Misa Takahashi, Naoki Goshima, Hiromichi Morikawa
Grad. Dept. Gene Sci., Fac. Sci., Hiroshima Univ., Higashi-Hiroshima 739 Japan

Key words: ^{15}N labelling, environment, higher plants, N metabolism, phyotoremediation

1. Introduction

Nitrogen dioxide (NO_2) is a major atmospheric pollutant, and plants act as a sink for it (1, 2, 3, 4, 5, 6, 7). NO_2 molecules taken up into plant leaves through stomata are, at least in part, converted to organic nitrogens such as amino acids or proteins through the nitrate assimilation pathway. Formation of ^{15}N-labeled nitrate, nitrite, and amino acids in plants fumigated with ^{15}N-labeled NO_2 has been reported by the previous authors, and since then it has been assumed that NO_2 nitrogen is metabolized similarly to the nitrate in plant cells. Whether a stoichiometric relationship exists between the total nitrogen (N) derived from NO_2 and the sum of reduced and nitrate/nitrite N has not, however, been studied. We therefore used ^{15}N-labeled NO_2 ($^{15}NO_2$) and potassium nitrate ($K^{15}NO_3$) to clarify whether the total ^{15}N taken up in the leaves of *Arabidopsis thaliana* plants is equal to the amount of reduced ^{15}N plus nitrate/nitrite ^{15}N.

2. Procedure

Plant material *Arabidopsis thaliana* ecotype C24 plants were sown in vermiculite and perlite (1:1, v/v) in plastic pots, and the pots placed in a growth chamber, and seedlings were grown under continuous light (70 μE m^{-2} s^{-1}) at 22.0 ± 0.3 °C at a relative humidity of 70 ± 4% and irrigated at 4-day intervals with a half-strength solution of the inorganic salts of MS medium. Seedlings of 5- to 6-week-old plants were used.

Fumigation with $^{15}NO_2$ The *Arabidopsis thaliana* seedlings in the plastic pots were placed in a fumigation chamber and treated with $^{15}NO_2$ (51.6 atom%) at concentrations of 0.1 to 4.0 ppm (v/v) for up to 4 h under continuous light (70 μE m^{-2} s^{-1}), 340 ± 80 ppm CO_2 at 22.0 ± 0.3 °C, and a relative humidity of 70 ± 4% (6). Relative variation in the NO_2 concentrations among the experiments was less than 10%.

Treatment with $K^{15}NO_3$ The roots of *Arabidopsis thaliana* seedlings were treated with 1 mM $K^{15}NO_3$ (50.0 atom% ^{15}N), 20 or 50 mM $K^{15}NO_3$ (10.0 atom% ^{15}N) and kept in the fumigation chamber with no added NO_2 for up to 8 h under the conditions described above.

Preparation of leaf samples Leaves were harvested from *Arabidopsis thaliana* plants treated with $^{15}NO_2$ or $K^{15}NO_3$ were rinsed with distilled water. Unless otherwise stated, the leaves were dried at 80 °C for 12 to 24 h. The dried leaves of 6 to 10 plants (50 to 150 mg DW) were combined and ground into powder with a mortar and pestle, and stored in a desiccator until used.

G. Garab (ed.), Photosynthesis: Mechanisms and Effects, Vol. V, 3613–3616.

Determination of NO_2-derived (or nitrate-derived) nitrogen in the total, reduced, ammonium, and nitrate/nitrite N fractions

Total N fraction Samples of the powdered leaves were weighed in containers made of tin for use in an elemental analyzer (EA/NA; Fisons Instrument, Milano, Italy). The ^{15}N (atom% excess) content was determined with a mass spectrometer (Delta C; Finnigan MAT, Bremen, Germany) connected directly to the elemental analyzer (EA/NA; Fisons Instrument, Milano, Italy). The amount of NO_2-derived (or nitrate-derived) nitrogen, here defined as the total nitrogen (total N) expressed in nanograms of nitrogen per milligram of dry leaves [ng N (mg DW)$^{-1}$], was estimated from these values.

Reduced N fraction The reduced nitrogen (reduced N) fractions of the dried leaf samples were prepared by the Kjeldahl method, and the total amount of reduced N was determined by titration (6). To quantify the ^{15}N content in the distillates by mass spectrometry, the ammonia in the distillates was concentrated by the Conway method. Using the ^{15}N value and total amount of reduced N determined by titration, we estimated the quantity of NO_2-derived (or nitrate-derived) reduced N in each sample, expressed by [ng N (mg DW)$^{-1}$].

Ammonium N fraction A mortar and pestle made of agate was used to homogenize leaf powder in pure water from a Milli-Q system to prepare the ammonium nitrogen (ammonium N) fraction of the dried leaf samples was prepared. The homogenate was centrifuged, and total ammonia in the supernatant (sup) was determined by capillary electrophoresis (7). The ammonia in the sup was concentrated using Conway units, and the ^{15}N (atom% excess) content of each solution was determined by mass spectrometry.

Nitrate/nitrite N fraction Leaf powder in 0.1% SDS was homogenized with the agate mortar and pestle to prepare the dried-leaf fraction containing both nitrate and nitrite (nitrate/nitrite N fraction). The homogenate was centrifuged. The total nitrate and nitrite in the sup of each homogenate was determined by capillary electrophoresis (7). For quantification of the ^{15}N content in the fraction, the nitrate and nitrite in the homogenate were separated by ion exchange chromatography using Dowex 50W hydrogen form. The separated nitrate and nitrite were reduced by Devarda alloy and resulting ammonia was recovered by distillation, then concentrated in Conway diffusion units. The ^{15}N (atom% excess) content of each solution was determined by mass spectrometry.

Fractionation of leaf homogenate Leaf powder samples prepared from *Arabidopsis thaliana* plants fumigated with 4 ppm $^{15}NO_2$ for 4 h were homogenized in 0.1% SDS. The homogenate then was centrifuged and the contents of NO_2-derived total and reduced N in the sup and pellet fractions were determined by mass spectrometry.

Ion-exchange chromatography Leaf powder samples prepared from *Arabidopsis thaliana* plants fumigated with 4 ppm $^{15}NO_2$ for 4 h were homogenized in pure water, after which the sup was applied to a column of Dowex 50W hydrogen form, and cationic and non-cationic fraction fractions were separated. The contents of NO_2-derived total and reduced N in the cationic and non-cationic fractions then were determined by mass spectrometry.

Gel-permeation of chromatography Leaf powder samples prepared from *Arabidopsis thaliana* plants fumigated with 4 ppm NO_2 for 4 h were homogenized in Tris-HCl buffer. The homogenate was centrifuged. The sup was passed through a Superdex 75 HR 10/30 column and separated using FPLC. The NO_2-derived total N and reduced N contents of eluate fractions respectively were determined by mass spectrometry.

3. Results and Discussion

Fumigation with 0.1 ppm or 4 ppm NO_2 increased the total and reduced N derived from the NO_2 levels, and these remained relatively constant after transfer of the plants to an atmosphere without added NO_2. The nitrate/nitrite N derived from the NO_2 level, however, remained low [2 to 4% of the NO_2-derived total N, except at 1 h (13%) and 2 h (7%) after fumigation with 0.1 ppm NO_2] throughout the experiment. The value was too low to account for the difference between the NO_2-derived total N and reduced N. The total N derived from NO_2 therefore was not equal to the sum of the NO_2-derived reduced and nitrate/nitrite N, indicative of the presence of an unknown organic or inorganic nitrogen compound(s) in the leaves of plants fumigated with NO_2. This type of nitrogen, called unidentified nitrogen(UN) hereafter, was estimated to be approximately 20 to 30% of the total N derived from NO_2.   In the case of treatment with 1 mM or 50 mM $K^{15}NO_3$, the levels of the total, reduced, and nitrate/nitrite N derived from nitrate increased almost linearly with the period of treatment, and the nitrate-derived total N coincided, within the experimental errors, with the sum of the nitrate-derived reduced N and nitrate/nitrite N, much less or no UN being present in the leaves of plants fed $K^{15}NO_3$. In the case of treatment with $K^{15}NO_3$, substantial amounts of nitrate-derived total N (40 to 60%) were accounted for by the nitrate/nitrite N, in marked contrast to the finding for NO_2-derived nitrogen. These results are clear evidence that the metabolic fate of NO_2 nitrogen differs from that of nitrate nitrogen in leaves of Arabidopsis thaliana plants.

To further clarify the characteristics of the UN, leaf homogenates prepared from Arabidopsis thaliana plants fumigated with 4 ppm $^{15}NO_2$ for 4 h were fractionated into sup and pellet fractions, and their NO_2-derived nitrogen contents analyzed. The total N derived from NO_2 was distributed in the sup and pellet fractions in the ratio of 7 to 3. The NO_2-derived total N and reduced N contents were almost the same in the pellet but differed in the sup fraction. This suggests that the UN is present in the sup but not the pellet fraction. On the average, 18 percentage points of UN were estimated to be recovered in the sup fraction, indicative that the most of the UN is located in this fraction.

When the sup fraction was separated in a Dowex 50W column, about 90% of the NO_2-derived total N was recovered in the cationic fraction, and the rest in the non-cationic fraction. About 11 percentage points of the UN were estimated to be recovered in the cationic fraction, but only 1 percentage point of UN in the non-cationic fraction. This suggests that the most of the UN-bearing compound(s) is cationic.

The sup fraction of the leaf homogenate prepared from Arabidopsis thaliana plants fumigated with 4 ppm $^{15}NO_2$ for 4 h was separated by gel-permeation chromatography (GPC). Each of the GPC fractions was analyzed for NO_2-derived total N and reduced N. Approximately 8, 66, and 7% of the NO_2-derived total N of the sup fraction were recovered, respectively, in fractions 4, 5 and 6. Most of the UN appeared to be contained in fraction 5, with some in fraction 6. When nitrate (mol wt = 62), phenylalanine (165), and cystine (240) were co-chromatographed, fraction 5 contained 91% of the nitrate, 100% of the phenylalanine, and no cystine. The compound(s) bearing the UN appears to have a low molecular weight similar to the weights of nitrate and phenylalanine. Fraction 5 was estimated to contain 18 percentage points of the UN. Preliminary analysis of the fraction 5 by GC/MS detected ^{15}N-bearing N-nitrosodiphenylamine (mol wt = 198), which accounted for about 10% of the UN.

Compounds bearing oxidized forms of nitrogen, organic nitro, nitroso, and azo compounds are known not to be converted to ammonia under the Kjeldahl digestion conditions we used. In fact, no ammonia was detected after the digestion of acetaldehyde oxime, nitrotyrosine, KNO_3, $NaNO_2$, and N_2O under those conditions. This suggests that organic nitro, nitroso, and azo, or the inorganic nitrous oxide used are candidates for the UN-bearing compound(s). Therefore, the UN-bearing compound(s) can not be completely reduced to ammonia even under those digestion conditions.

Nitrogen dioxide has an unpaired electron which make it a strong one electron oxidant (8), and it efficiently initiates free radical oxidation of unsaturated lipids, thiols, and proteins to form nitroderivatives such as nitrotyrosine. The UN-bearing compound(s) is not likely to be a lipid because it appears to be hydrophilic. Nitration of proteins in the cells by NO_2, if any, also is unlikely because the UN-bearing compound(s) has a low molecular weight.

Fumigation of plants with NO_2 resulted in the accumulation of nitrite ion in the leaves of plants including *Arabidopsis thaliana* plants. At an acidic pH, nitrite ion reacts like nitrosonium ion (8) which also is highly reactive and forms nitroso compounds such as nitrosoamines with the amino acids and amines present in plant cells. Plant cells have substantial amounts of polyamines. The nitrosoamines formed between polyamines and nitrogen dioxide therefore are candidates for the UN-bearing compound(s). Consistent with this is the fact that preliminary analysis of the fraction 5 by GC/MS detected ^{15}N-bearing N-nitrosodiphenylamine (mol wt = 198) which accounted for about 10% of the UN (see above).

At a neutral pH, nitrite ion produces nitric oxide (NO) and nitrous oxide (N_2O). Formation of NO and N_2O and acetaldehyde oxime in the suspension of plant leaf tissues fed nitrate has been reported (9, 10, 11). Some of these compounds also are poorly digested by the Kjeldahl method or are poorly reduced by Devarda alloy. Therefore they are possible candidates for the UN-bearing compound(s). The characteristics of the UN must be studied further in order to identify the UN-bearing compound(s) and its physiological function. Such information is important if we are to clarify the effects of this air pollutant on plant growth and the global environment. We currently are working on this, and results will be published elsewhere.

We thank Professor Eiji Hirasawa, Osaka City University for his invaluable discussions of our work, also Mr. Yoshifumi Kawamura and Ms. Mihoko Kawahara of Hiroshima University for their technical help. This research was supported in part by the program Research for the Future, Japanese Society for the Promotion of Science (JSPS-RFTF96L00604).

References

1 Rogers, H.H., Campbell, J.C. and Volk, R.J. (1979) *Science* 206: 333-335.
2 Yoneyama, T. and Sasakawa, H. (1979) *Plant Cell Physiol.* 20, 263-266.
3 Lee, Y.-N. and Schwartz, S.E. (1981) *J. Phys. Chem.* 85: 840-848.
4 Okano K., Machida T. & Totsuka T. (1988) *New Phytologist* 109, 203-210.
5 Wellburn, A.R. (1990) *New Phytol.* 115: 395-429.
6 Morikawa, H., Higaki, A., Nohno, M., Takahashi, M., Kamada, M., Nakata, M., Toyohara, G., Okamura, Y., Matsui, K., Kitani, S., Fujita, K, Irifune, K. and Goshima, N. (1998) *Plant Cell Eviron* 21, 180-190..
7 Kawamura Y., Takahashi M., Arimura G., Isayama T., Irifune K., Goshima N. & Morikawa H. (1996). *Plant & Cell Physiology* 37, 878-880.
8 Beckman J.S. (1996) in *Nitric Oxide. Principles and Actions* (ed J. Lancaster, Jr.), pp. 1-82. Academic Press, San Diego.
9 Klepper L. (1979) *Atmospheric Environment* 13, 537-542.
10 Dean, J.V. & Harper, J.E. (1986) *Plant Physiology* 82, 718-723.
11 Mulvaney C.S. & Hageman R.H. (1984). *Plant Physiology* 76, 118-124.

IS NITRITE REDUCTASE ESSENTIAL IN THE METABOLISM OF NITROGEN IN PLANTS ?

Misa Takahashi, Michel Caboche[1], Hiromichi Morikawa
Graduate Department of Gene Science, Faculty Science, Hiroshima University,
Higashi-Hiroshima 739-8526, Japan
[1]Laboratoire de Biologie Cellulaire, INRA, Centre de Varsailles, Route de St
Cyr, 78026 Versailles Cedex, France

Key words: [15]N labelling, N metabolism, transgenic plants, particle bombardment

1. Introduction

Nitrite reductase (NiR) is the second enzyme in the primary assimilation pathway of nitrate to catalyze six-electron reduction of nitrite to ammonia, and thus this enzyme is essential in assimilatory nitrate reduction in plants. We have been studying the assimilation of nitrate and nitrogen dioxide in transgenic *Arabidopsis thaliana* plants that overexpress NiR gene. Interestingly, our data implied that plants of zero NiR activity could reduce nitrate and nitrogen dioxide. Therefore, we have studied nitrogen assimilation in transgenic plants that express antisense NiR RNA and hence lack or have much reduced NiR activity.

2. Procedure

Transgenic *Arabidopsis thaliana* plants
Transgenic plants of *Arabidopsis thaliana* ecotype C24 were obtained by particle bombardment as reported previously (1). Root explants were bombarded with pSNIRH, which contains spinach NiR cDNA (provided by Dr. S. Rothestein, University of Guelph, Canada) (2) and *hpt* (3) under the control of CaMV 35S promoter and NOS terminator, and transgenic plants were selected for hygromycin resistance. Transgenic nature of selected plants was confirmed by PCR using the primer which define a ca. 350 bp fragment at the 5' region of spinach NiR cDNA. Transgenic and nontransformed plants of *A. thaliana* that were grown on vermiculite and perlite (1:1, v/v) under feeding 1/2 MS medium (4) under continuous light at 22 C at a relative humidity of 70% for up to 5 to 6 weeks and used for the analysis of the ability to reduce nitrogen dioxide.
Transgenic tobacco
Transgenic tobacco plants that express the antisense RNA of tobacco NiR gene (5) were cultured aseptically for five weeks on solid B medium containing 10 mM di-ammonium succinate as the sole nitrogen source. These plants were used for the analysis of the ability to reduce nitrate. Alternatively, transgenic tobacco were grown for 11 weeks on B medium with 10 mM di-ammonium succinate as the sole nitrogen source, and used for the analysis of the ability to reduce nitrogen dioxide.
Analysis of the assimilation of nitrate and nitrogen dioxide
The ability of transgenic *A. thaliana* and tobacco plants, and respective nontransformed

G. Garab (ed.), Photosynthesis: Mechanisms and Effects, Vol. V, 3617–3620.

control plants to assimilate nitrogen dioxide was analyzed as reported (6). Briefly,plants were fumigated with ^{15}N-labeled NO_2 (4.0 $\pm$ 0.1 ppm) for 8 h in the light, after which the leaves of plants were harvested, washed and ground into fine powder. The reduced nitrogen in the leaf samples of these plants were determined by the Kjeldahl method. And the ^{15}N/^{14}N ratio in Kjeldahl nitrogen fraction was determined by an EA-MS mass spectrometer. From these two value we estimated the quantified of NO_2-derived reduced nitrogen.

Transgenic and nontransformed tobacco plants were fed ^{15}N-labeled 20 mM KNO_3 (10 atom%) for 1 weeks, after which the leaves of these plants were harvested, and subjected to Kjeldahl digestion. Nitrate derived reduced nitrogen was quantified in a similar manner to that for NO_2-derived reduced nitrogen as described above.

Analysis of NiR activity

NiR activity in the leaves of these transgenic and nontransformed plants were analyzed using methyl viologen as an electron donor as reported elsewhere (7).

3. Results and Discussion

Figure 1 shows the content of NO_2-derived reduced nitrogen plotted against NiR activity in twelve transgenic lines, where the value of the control sample being taken as 100. The NO_2-derived reduced nitrogen content increased almost linearly with the increase of NiR activity. There was a positive correlation (r=0.86) between the NO_2-N content and NiR activity. These results clearly indicate that the enrichment of NiR enzyme enhances the NO_2-assimilation ability of plants. The highest values for the

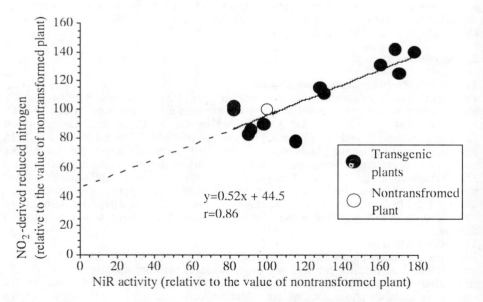

Figure 1. Content of NO_2-derived reduced nitrogen plotted against NiR activity in the leaves of transgenic *Arabidopsis thaliana* plants. Five-to-six-week old plants were

fumigated with 4 ppm ^{15}N-labeled NO_2 for 8 h, and the content of NO_2-derived reduced nitrogen and NiR activity were determined. See text for details.

content of NO_2-derived reduced nitrogen and the NiR activity in transgenic plants was respectively 1.4 and 1.8 times of the corresponding value of the control plants.

When the curve of Figure 1 is extrapolated to zero NiR activity, the content of NO_2-derived reduced nitrogen does not decrease to zero but it was approximately 50% of the value of the control nontransformed plants . This suggests, at least in theory, that even plants of zero NiR activity can reduce nitrate or nitrogen dioxide.

Vaucheret et al. (5) have reported transgenic tobacco plants that lack NiR activity thanks to the expression of NiR cDNA in antisense orientation. We then analyzed the ability of these plants (clone 271) to reduce nitrate and nitrogen dioxide. Results are shown in Figure 2 A and B. We also confirmed that NiR activity of these transgenic plants was close to the detection limit of this enzyme activity. The nitrate-derived reduced nitrogen in the leaves of transgenic plants was 16.70 ± 3.33 mg/g DW (average ± SD), and the respective value for control plants was 14.49 ± 2.27 mg/g DW (Figure 2 A). Clearly, the nitrate-reduced nitrogen of the transgenic plant was very close to that of nontransformed plants. Similarly, the NO_2-derived reduced nitrogen content in the transgenic tobacco plant was 0.37 ± 0.04 mg N/g DW, and the values for control plants was 0.52 ± 0.11 mg N/g DW (Figure 2 B). The average value of transgenic tobacco plants was ca. 70% of that of the control tobacco plant. Furthermore, formation of ^{15}N-labeled ammonia from clone 271 plants fumigated with $^{15}NO_2$ (Takahashi et al., unpublished results).

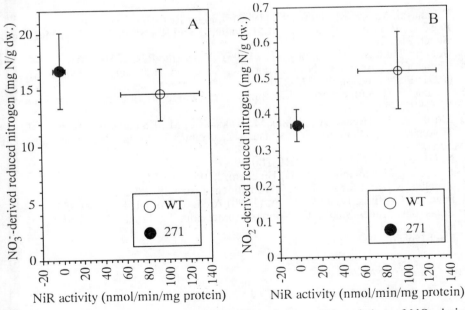

Figure 2. Content of nitrate-derived reduced nitrogen (A) and that of NO_2-derived reduced nitrogen (B) plotted against NiR activity in the leaves of transgenic tobacco plants (●) and nontransformed plants (○). Five-week old tobacco plants were fed ^{15}N-labeled KNO_3 for 1 week (A) or 11-week-old tobacco plants were fumigated with 4

ppm ^{15}N-labeled NO_2 for 8 h (B), after which they were analyzed for reduced nitrogen and NiR activity as described. WT; nontransformed control plants, 271; transgenic plants. See text for details.

Our present result that plants lacking NiR activity can reduce nitrate and nitrogen dioxide strongly suggests that there is an alternative pathway, in which NiR is not involved. Involvement of sulfite reductase (SiR) in the observed reduction of nitrate and nitrogen dioxide in clone 271 plants cannot be excluded, although SiRs are reported to have more than 250 times lower affinity to nitrite than that for sulfite in spinach (8). Isolated cytochromes such as cytochrome c from chlorella and cytochrome f from *Porphyra yezoensis* have nitrite reductase activity (T. Oku, personal communication). Therefore, involvement of cytochromes in our observed reduction of nitrate and nitrogen dioxide in NiR-lacking transgenic plants is also plausible. In *Escherichia coli*, in addition to NADH-NiR, NiRs which use formate, pyruvate or ethanol as an electron donor have been reported (e.g., 9). Formate dehydrogenase has been found in plant mitochondria (10). We are currently studying detailed mechanisms of the observed reduction of nitrate in NiR-lacking transgenic plants.

This research was supported in part by the program Research for the Huture, Japanese Society for the Promotion of Science (JSPS-RFTF96K00604).

References

1 Takahashi, M. and Morikawa, H. (1996) J. Plant Res. 109, 331-334.
2 Back, E., Burkhart, W., Moyer, M., Privalle, L. and Rothstein, S. (1988) Mol. Gen. Genet. 212, 20-26.
3 Goto, F. Toki, S. and Uchimiya, H. (1993) Transgenic Res. 2, 300-305.
4 Cheng, C.-L., Acedo, G. N., Dewdney, J., Goodman, H. M. and Conkling, M. A. (1991) Plant Physiol. 96, 275-279.
5 Vaucheret, H, Kronenberger, J., Lepingle, A., Vilaine, F., Boutin, J.-P. and Caboche, M. (1992) Plant J. 2, 559-569.
6 Morikawa, H., Higaki, A., Nohno, M., Takahashi, M., Kamada, M., Nakata,.M., Toyohara, G., Okamura, Y., Matsui, K., Kitani, S., Fujita, K., Irifune, K. and Goshima, N. (1998) Plant Cell Environ. 21, 180-190.
7 Joy, K and Hagemann, R (1966) Biochem. J. 100, 263-273.
8 Krueger, R.J. and Siegel, L.M. (1982) Biochemistry 21, 2892-2904
9 Page, L., Griffiths, L and Cole J.A. (1990) Arch. Microbiol. 154, 349-354.
10 Colas des Francs-Small, C., Ambard-Bretteville, F., Small, I.D. and Rémy, R. (1994) Plant Physiol. 102, 1171-1177.

NO$_2^-$ UPTAKE BY A NO$_2^-$ TRANSPORTER OF CHLOROPLAST ENVELOPE

Masaaki TAKAHASHI, Hiroshi HARUKI and Miwa SUGIURA*
Dep. Applied Biol. Chem., Osaka Pref. Univ., Sakai, 599-8531, Japan
*Inst. Phys. Chem. Res. (RIKEN), Wako, 351-0198, Japan

Key words: envelope, metabolite transport, nitrite transporter, N metabolism, proton transport

1. Introduction

Conversion of NO$_3^-$ to NH$_4^+$ in plants is comprised of two consecutive reduction steps catalyzed by reductases for NO$_3^-$ and NO$_2^-$. The reduction of NO$_3^-$ occurs in the cytoplasm while its product, NO$_2^-$, is reduced inside chloroplasts by plastid nitrite reductase [1, 2]. The uptake of NO$_2^-$ by intact chloroplasts is strictly light dependent [3, 4]. The dependence of NO$_2^-$ uptake on light was also proposed based on NO$_2^-$-induced phenomena such as swelling of chloroplasts [5], stromal pH shift [6-8], and inhibition of enzymes in CO$_2$ fixation [6-8]. Participation of a NO$_2^-$-specific transport system in the uptake of NO$_2^-$ by such photo-energized chloroplasts has been proposed based on the fact that the kinetics of NO$_2^-$ uptake was saturable with increasing NO$_2^-$ [3, 4], and inhibited by a thiol modifier [9]. Shingles et al. [10], on the other hand, proposed that the diffusional influx of HNO$_2$ is faster than the consumpsion of NO$_2^-$ by the reductive assimilation in chloroplast stroma if NO$_2^-$ is present at a sub-millimolar level with outside-acidic pH gradient.

To clarify the type of energy driving the influx of NO$_2^-$ from the cytosol to the stroma and to determine whether a transporter is involved, NO$_2^-$ transfer across chloroplast envelopes was directly determined by measuring NO$_2^-$ uptake by isolated chloroplasts in the dark at 5°C. In this paper, it is shown that NO$_2^-$ uptake is also driven by a H$^+$/NO$_2^-$ symport through chloroplast envelope. We cloned a cucumber homolog of H$^+$-coupled nitrate and peptide transporters of Arabidopsis thaliana. Its expression in cotyledons was uniquely induced by light illumination.

2. Procedure

2.1 Assay of NO$_2^-$ influx into chloroplasts

Intact chloroplasts were isolated from spinach leaves by Percoll density gradient centrifugation according to Perry et al. [11] and suspended in 50 mM Hepes-NaOH, pH 6.8 (or 8.1), 2 mM EDTA, 2 mM MgCl$_2$, 1 mM MnCl$_2$, 5 mM Na-pyrophosphate, and 0.33 M sorbitol (hereafter referred to as isolation buffer).

NO$_2^-$ influx into chloroplasts was determined by silicone layer filtering centrifugation [12]. Reaction mixture (425 μl) containing 50 mM Mes-NaOH, pH 5.5, 2 mM MgCl$_2$, 1 mM MnCl$_2$, 5 mM pyrophosphate, and 0.33 M sorbitol (hereafter

3621

G. Garab (ed.), Photosynthesis: Mechanisms and Effects, Vol. V, 3621–3624.
© 1998 Kluwer Academic Publishers. Printed in the Netherlands.

referred to as external medium), was layered onto 500 µl of silicone layer in an Eppendorf centrifugation tube. A chloroplast suspension (50 µl; 250 µg of Chl) was added to the reaction mixture. Five minutes later, NO_2^- influx was initiated by the addition of 25 µl of 10 mM $NaNO_2$ at 5°C under a dim light. After incubation for various times at 5°C, the reaction was stopped by a brief centrifugation at 10,000 x g . The tubes were frozen at -85°C and then the layer of the frozen external medium was removed by cutting the chilled tube at the silicone layer. The amount of NO_2^- entrapped in the precipitated chloroplasts was colourimetrically determined according to Hageman and Reed [13] .

2.2 *Cloning of a transporter of nitrogenous compound and RT-PCR analysis.*

mRNA was isolated from greening cucumber (*Cucumis sativus* L.) cotyledons and was used to prepare a cDNA library (λgt10). The library was screened by plaque hybridization with the oligonucleotide probe TTATA CCTCA CCGCT CTAGG AACGG GAGGC which corresponds to the CHL1 amino acid sequence Leu153-Gly162 of the most conserved region of *CHL1* [14] and the oligopeptide transporter, *NTR1* [15], of *A. thaliana*. A 1.6-kb insert from the *Not*I-digested DNA of the positive λgt10 clone was subcloned into *pBluescript* KS+ (Promega) and sequenced. To analyze the level of accumulated mRNA, cucumber cotyledons were grown for 72h in the dark at 28°C with or without 17 mM KNO_3 and then exposed to light for 0, 6, 18, and 36 h. Total RNA was extracted and reverse-transcribed with oligo(dT) primer according to Sambrook *et al.* [16]. Using resulting single stranded cDNAs, PCR-amplified 530-bp fragment in agarose gel was quantified from the intensity of ethidium bromide staining.

3. Results and Discussion

3.1 *NO_2^- is taken up with H^+ by spinach intact chloroplasts*

A small NO_2^- influx was observed when chloroplasts suspended at pH 6.8 (pH_i) were mixed with reaction medium at a neutral pH (pH_o) in the dark. However, when pHo was lowered to 5.5, chloroplasts took up NO_2^- from the medium even in the absence of light (*Figure 1*). The NO_2^- concentration in the stromal space after 10 - 15 s incubation was 4 to 5 times the equilibrium concentration, which was attained after 30 min of incubation. This transient increase implied an uphill transport against an inwardly directed concentration gradient of NO_2^- .

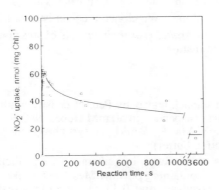

Figure 1 Kinetics of the NO_2^- influx into intact chloroplasts.

Intact chloroplasts (250 µg of Chl), suspended in 50 µl of 50 mM Hepes-NaOH, pH 6.8, 2 mM EDTA, 2 mM $MgCl_2$, 1 mM $MnCl_2$, 5 mM pyrophosphate, and 0.33 M sorbitol, were mixed with 425 µl of 2 mM EDTA, 2 mM $MgCl_2$, 1 mM $MnCl_2$, 5 mM pyrophosphate, 0.33 M sorbitol and 50 mM Mes-NaOH at pH 5.5, and incubated with 500 µM $NaNO_2$ for the indicated periods at 5°C.

3.2 Concentration dependence of NO_2^- influx into chloroplasts

The rate of NO_2^- influx into chloroplasts depended on the concentration of NO_2^- in the external medium, but was not linearly correlated with the NO_2^- concentration, showing at least two phases with different affinities to NO_2^- in the NO_2^- uptake by intact chloroplasts (*Figure* 2). Assuming that chloroplast envelopes have two independent uptake systems, one of which is a passive HNO_2 uptake as suggested by Shingles *et al.* [10], the NO_2^- influx rate was best approximated as a function of NO_2^- concentration. The K_m for a saturable NO_2^- uptake was estimated to be 1.83 mM and V_{max} was 12.5 nmol (mg Chl)$^{-1}$ s^{-1} at pH$_i$ 8.1/pH$_o$ 6.8 at 5°C.

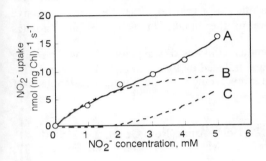

Figure 2 Concentration dependence of NO_2^- influx into intact chloroplasts.

NO_2^- uptake for the initial 10 s was determined with various concentrations of NaNO$_2$ at pH$_i$ 8.1/pH$_o$ 6.8 (o). The line (A) was drawn by a least squares approximation assuming the presence of single saturable uptake with a K_m of 1.83 mM and a V_{max} of 12.5 nmol (mg Chl)$^{-1}$ s^{-1} (B) and a passive diffusion (C). An equation for the passive diffusion of HNO_2 was derived from the data reported by Shingles *et al.* (10).

3.2 Molecular properties and light-induced expression of a transporter of nitrogenous compound.

When cucumber seedlings, dark-grown for 3 days after germination, were illuminated, the cotyledons became green 24 h after the onset of illumination, and oxygen evolution developed most rapidly at 36 h. Using a cDNA library of greening cucumber cotyledons after a 36-h illumination, we cloned a cDNA with a nucleotide sequence homologous to the *A. thaliana* NO_3^- (CHL1: ref. 14) and oligopeptide (NTR1: ref. 15) transporters. The cDNA is 1,587 bp long with a 1,452-bp open reading frame [EMBL accession number Z69370]. The molecular weight of the encoded protein of 484 amino acids is 54,182. The homology is 38.1% for CHL1 and 41.8% for NTR1 in its amino acid sequence, respectively. Encoded membrane protein have 10 transmembrane regions in which there is a 70-amino acid hydrophilic loop between the 4th and 5th hydrophobic domains. This structure with transmembrane domains and the interconnecting hydrophilic loop is characteristic of the transporter families [17].

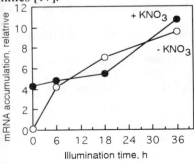

Figure 3. RT-PCR analysis of light- and NO_3^--induced expression of mRNA in the cucumber cotyledons. Etiolated seedlings of cucumber, grown for 72 h after germination under the dark, were placed under continuous light and the expression of the mRNA in cotyledons at 0-h, 6-h, 18-h, and 36-h illumination was determined by RT-PCR. Amplified 530-bp fragment of cDNA was stained with ethidium bromide and quantified from its intensity by using NIH image software.

In the illuminated seedlings, corresponding mRNA accumulated (*Figure 3*). When etiolated seedlings were supplemented with NO_3^-, about 40% of the amount at 36 h was present before the start of illumination. Light and NO_3^--induced expression of mRNA suggests that the function of the protein is related to the transport of nitrogenous compound in the nitrogen metabolism of photosynthesis.

4. Conclusion

With an outward acidic pH gradient, NO_2^- was taken up by intact chloroplasts during the first 10 - 15 s of incubation in the dark at 5°C and was transiently accumulated at a concentration 4 to 5 times the equilibrium concentration. There are two phases in the NO_2^- uptake by chloroplasts. One is a H^+/NO_2^- symport with a saturable NO_2^- concentration dependence and the other is a transfer of HNO_2 due to passive diffusion. The former participates more in the uptake of NO_2^- at physiological concentration.

A transporter cDNA of nitrogenous compound was cloned from greening cotyledons of cucumber. Its expression was induced by light and/or NO_3^- supply to the growth of etiolated seedlings indicating its involvement in the nitrogen metabolism of photosynthesis.

References

[1] Crawford, N. M. and Arst Jr., H. N. (1993) Annu. Rev. Genet. 27, 115-146.
[2] Hoff, T., Truong, H.-N. and Caboche, M. (19 h94) Plant Cell Environ. 17, 489-506.
[3] Brunswick, P. and Cresswell, C. F. (1988) Plant Physiol. 86, 378-383.
[4] Anderson, J. W. and Done, J. (1978) Plant Physiol. 61, 692-697.
[5] Heber, U. (1973) Biochim. Biophys. Acta 305, 140-152.
[6] Purczeld, P., Chon, C. J., Portis Jr., A. R., Heldt, H. W. and Heber, U. (1978) Biochim. Biophys. Acta 501, 488-498.
[7] Enser, U. and Heber, U. (1980) Biochim. Biophys. Acta 592, 577-591.
[8] Kaiser, G. and Heber, U. (1983) Planta 157, 462-470.
[9] Brunswick, P. and Cresswell, C. F. (1988) Plant Physiol. 86, 384-389.
[10] Shingles, R., Roh, M. H. and McCarty, R. E. (1996) Plant Physiol. 112, 1375-1381.
[11] Perry, S. E., Li, H.-M. and Keegstra, K. (1991) in: Methods in Cell Biology. pp. 327-344, Academic Press, New York.
[12] Werdan, K., Heldt, H. W. and Millovancev, M. (1975) Biochim. Biophys. Acta 396, 276-292.
[13] Hageman, R. H. and Reed, A. J. (1980) in: Methods in Enzymology (Pietro, A. S. ed.) pp. 270-275, Academic Press, New York.
[14] Tsay, Y.-F., Schroeder, J. I., Feldman, K. A. and Crawford, N. M. (1993) Cell 72, 705-713.
[15] Rentsch, D., Laloi, M., Rouhara, I., Schmelzer, E., Delrot, S. and Frommer, W. B. (1995) FEBS Lett. 370, 264-268.
[16] Sambrook, J., Fritsch, E. F. and Maniatis, T. (1989) *Molecular cloning. A laboratory Manual.* New York: Cold Spring Harbor Laboratory Press.
[17] Sadée, W., Drübbisch, V. and Amidon, G. L. (1995) Pharmaceut. Res. 12, 1823-1837.

ELECTRON FLOW FROM NADPH TO FERREDOXIN IN SUPPORT OF NO_2^- REDUCTION

Tie Jin, Heather C. Huppe, and David H. Turpin
Department of Biology, Queen's University,
Kingston, Ontario, Canada K7L 3N6

Keywords: electron transfer, FNR, N metabolism, oxidation/reduction, respiration, nitrite reduction, nitrite reductase

1.   Introduction

The assimilation of NO_3^- into amino acids involves three processes: nitrate reduction by nitrate reductase, nitrite reduction by nitrite reductase (NiR) and the assimilation of the resulting NH_4^+ into amino acids via the glutamine synthetase/GOGAT pathway (1). All three steps involve reduction reactions with the most significant being the transfer of six electrons to NO_2^- via NiR. The electron donor for all NiRs studied to-date is reduced ferredoxin (Fd). Fd is known to be reduced via the light reactions of photosynthesis (2) and subsequently used by Fd: $NADP^+$ oxidoreductase (FNR) to reduce $NADP^+$ to NADPH for use in CO_2 fixation. The reduction of NO_2^- is second only to CO_2 as a sink for photosynthetically generated electrons.

In many organisms and tissues NO_3^- assimilation also occurs in the dark (3,4). As nitrite reductase requires reduced ferredoxin to support dark NO_2^- reduction, there must be a method of producing reduced ferredoxin in the dark. Physiological studies have suggested that the onset of dark NO_3^- (NO_2^-) reduction results in the activation of the oxidative pentose phosphate (OPP) pathway in both higher plants (5,6,7,8) and algae (9,10) presumably producing NADPH which can be used for NO_2^- reduction. The question is, can the NADPH produced by the OPP pathway reduce ferredoxin, a reaction which is in the opposite direction of that occurring in photosynthesis, and thus provide a mechanism by which respiration can support NO_2^- reduction in the dark?

The purpose of this study was to test the hypothesis that electrons from NADPH may support NO_2^- reduction via a FNR- and Fd-mediated electron transfer pathway from NADPH to NO_2^-. We report the reconstitution of an *in vitro* electron-transfer system from NADPH to NO_2^- using FNR, Fd, and NiR purified from the green alga *Chlamydomonas reinhardtii*. Furthermore, we have isolated G6PDH from the same source and coupled G6PDH-dependent NADPH generation to NO_2^- reduction, providing direct evidence for the potential of G6PDH to support NO_2^- reduction in the dark.

G. Garab (ed.), Photosynthesis: Mechanisms and Effects, Vol. V, 3625–3628.
© 1998 *Kluwer Academic Publishers. Printed in the Netherlands.*

2. Procedures

Chlamydomonas reinhardtii CC-1183 was grown in NO_3^- sufficient chemostats as previously described (11) and cells were harvested daily, frozen in liquid nitrogen and stored at -80° C until use. The purification of Fd, NiR, FNR and G6PDH and the assay methods were as previously described (12).

NADPH-Dependent NO_2^- Reduction Assay: The reaction mixture (0.5 mL), containing 50 mM Tris-HCI (pH 7.8), 0.4 mM $NaNO_2$, 20 nM purified *C. reinhardtii* FNR, 20 μM purified *C. reinhardtii* Fd, and 0.1 unit of NiR (activity calculated by MV-NiR assay), was mixed in a rubber-capped glass chamber (3 mL). The reaction was started by injecting NADPH to a final concentration of 1 mM. The assay continued for 15 min. at 30°C.

G6PDH-Coupled NO_2^- Reduction Assay: The reaction mixture (0.5 mL) contained 50 mM Tris-HC1 (pH 7.9), 2 mM $NaNO_2$, 0.35 unit of G6PDH, 6 mM G6P, 0.1 mM $NADP^+$ (or NADPH), 20 nM purified *C. reinhardtii* FNR, 20 μM *C. reinhardtii* Fd, and 0.1 unit of NiR (calculated by MV-NiR activity). The reaction was performed at 30°C for 10 min.

For all assays, the disappearance of NO_2^- was used to calculate enzyme activity and 1 unit of NiR activity catalyzed 1 μmol NO_2^- reduction per min. NO_2^- concentration was measured (13) and a mean of at least three independent measurements was used to calculate NiR activity. The SE associated with the reported values was always less than 3%.

3. Results and Discussion

NO_2^- reduction can be supported *in vitro* by a reconstituted system including NADPH, FNR, Fd and NiR. The absence of any of the three protein components prevents NO_2^- reduction (Table 1). The rates of NO_2^- reduction were shown to be dependent on the concentrations of each of the proteins involved in the electron transfer pathway (Fig. 1).

Table 1. Examination of NADPH-dependent NO_2^- reduction reactions

Reaction	Specific Activity of NiR
Complete reaction	0.079
- FNR	0.000
- Fd	0.000
- NADPH	0.000

NADPH dependent nitrite reduction was strongly inhibited by $NADP^+$ (Table 2), but the activities of Fd and methyl viologen (MV) dependent reduction were not. Table 2 demonstrates that the inhibitory effect of $NADP^+$ was mediated via inhibition of FNR.

These results demonstrate that optimal electron transfer from NADPH to NO_2^- requires not only the provision of NADPH but also the consumption of $NADP^+$ to minimize the inhibitory effects on FNR.

By generating NADPH via G6PDH and G6P it is possible to consume the $NADP^+$ produced as a result of NADPH reduction of ferredoxin. Such a reaction system

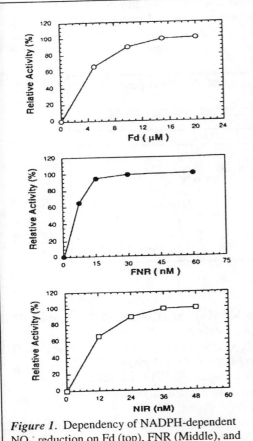

results in rates of NO_2^- reduction (Table 3) significantly greater than that supported by NADPH alone (Table 1), presumably because in the latter case NADP$^+$ produced builds up and inhibits FNR. This is confirmed by observing that removal of G6P from the reaction mixture in the presence of NADPH lowered rates of NO_2^- reduction to the NADPH dependent rates (Table 3).

The results of this study demonstrate that NADPH may reduce Fd via FNR and thereby support NO_2^- reduction. The reversible character of the reaction catalyzed by FNR is the key point of this reconstituted system. The equilibrium between reduced/oxidized Fd and NADPH/ NADP$^+$ drives the FNR reaction in the appropriate direction. Therefore, to have Fd serve as an efficient electron donor to NiR, the ratio of NADPH/ NADP$^+$ must be high.

Figure 1. Dependency of NADPH-dependent NO_2^- reduction on Fd (top), FNR (Middle), and NiR (bottom).

Table 2. Effect of NADPH/NADP$^+$ on NADPH-, Fd-, and MV- dependent NO_2^- reduction

	Relative Catalytic Ability %		
NADPH/NADP$^+$	NADPH-NiR	Fd-NiR	MV-NiR
mM			
1.0/0.0	100	100	100
1.0/0.1	73	96	105
1.0/1.0	37	95	94

When G6PDH is used to generate NADPH and to consume NADP$^+$, NiR activity increased dramatically, primarily due to the consumption of NADP$^+$, and release of

FNR from NADP$^+$ inhibition. Calculations show that the rates of NO$_2^-$ reduction supported in this way account for the observed rates of whole-cell NO$_2^-$ assimilation in the dark (12).

Table 3. Dependence of G6PDH-coupled NO$_2^-$ reduction on reaction components

Reaction	Specific Activity of NiR
	Units mg^{-1} protein
Complete reaction	0.49
-FNR	0.00
-Fd	0.00
-G6PDH	0.00
-G6P	0.00
-NADP$^+$	0.00
-NADP$^+$, +NADPH	0.57
-NADP$^+$, +NADPH, - G6P	0.076

References

1. Vance, C.P. (1997) in Plant Metabolism (Dennis, D.T., Turpin, D.H., Lefebvre, D.D. and Layzell, D.B., (eds) pp. 449-477.
2. Prézelin, B.B. and Nelson, N.B. (1997) in Plant Metabolism (Dennis, D.T., Turpin, D.H., Lefebvre, D.D. and Layzell, D.B., (eds) pp. 274-285.
3. Syrett, P.J. (1981) Can. Bull. Fish. Aquat. Sci. 219: 182-210.
4. Oaks, A. and Hirel, B. (1985) Ann. Rev. Plant Physiol. 36:345-365.
5. Emes, M.J. and Fowler, M.W. (1983) Planta 158: 97-102.
6. Oji, Y., Watanabe, M., Wakiuchi, N. and Okamoto, S. (1985) Planta 165: 85-90.
7. Bowsher, C.G., Dunber, B., and Emes, M.J. (1993) Protein Expr. Purif. 4:512-518.
8. Borchert, S., Harborth, J., Schünemann, D., Hoferichter, P., and Heldt, H.W. (1993) Plant Physiol. 101:303-312.
9. Vanlerberghe, G.C., Huppe, H.C., Vlossak, K.D.M. and Turpin, D.H. (1992). Plant Physiol. 99:495-500.
10. Huppe, H.C., Farr, T.J. and Turpin, D.H. (1994) Plant Physiol. 105: 1043-1048.
11. Huppe, H.C. and Turpin, D.H. (1996) Plant Physiol. 110:1431-1433.
12. Jin, T., Huppe, H.C. and Turpin, D.H. (1998) Plant Physiol. 117:303-309.
13. Hirasawa, M., Gray, K.A., Sung, J. and Knaff, D.B. (1989) Arch. Biochem. Biophys 275:1-10.

INFLUENCE OF NITROGEN AVAILABILITY ON SULFATE ASSIMILATION PATHWAY IN *ARABIDOPSIS THALIANA*.

Anna KOPRIVOVA, Christian BRUNOLD, Stanislav KOPRIVA
Institute of Plant Physiology, Altenbergrain 21, 3013 Bern, Switzerland

Key words: enzymes, gene expression, molecular biology, N metabolism, S metabolism

1. Introduction

Sulfate and nitrate, which contain sulfur and nitrogen in their most highly oxidized form, are the dominant species available to many plants for covering their needs for these elements (1-3). In amino acids, including sulfur containing amino acids cysteine and methionine, both sulfur and nitrogen are present in reduced form, therefore, sulfate and nitrate have to be reduced for their synthesis. The dominant portion of amino acids is used for protein synthesis hence the S/N ratio in plants remains quite stable. Plants appear to possess a mechanism to coordinate assimilatory sulfate and nitrate reduction so that the appropriate proportions of both sulfur containing and other amino acids are available for protein synthesis (4-7). These reports also indicate that the second enzyme of assimilatory sulfate reduction APS reductase is highly regulated by different nutritional conditions. In this study we were particularly interested in the regulation of APS reductase activity, mRNA expression, and protein accumulation under different conditions of nitrogen availability in *Arabidopsis thaliana* plants.

2. Procedure

2.1 Plant Material and Treatment
Arabidopsis thaliana var Columbia plants were grown in the pots filled with small clay balls held in trays with Hentschel nutrient solution (8). The plants were grown in day/night cycle of 10h/14h and light intensity of 115 to 160 umol sec^{-1}m^{-2}. All experiments were performed with 4-5 weeks old plants. Plants were incubated during 72h under nitrogen deficient conditions (all nitrogen components in Hentschel medium were substituted by corresponding chlorides). After 3 days standard nutrient solution was added and plants were treated for additional 24h.

2.2 APR Activity Measurement
For extractions whole shoot or root systems of 4-8 plants were used. Shoot and root material was extracted 1:10 and 1:20 (w/v), respectively, in 50 mM NaKPO$_4$ buffer (pH 8) supplemented with 30 mM Na$_2$SO$_3$, 0.5 mM 5'-AMP, and 10 mM DTE, using a glass homogenisator. APR activity was measured in extracts as the production of [^{35}S]sulfite, assayed as acid volatile radioactivity formed in the presence of [^{35}S]APS and DTE (9). The protein concentrations in the extracts were determined according to Bradford (10) with bovine serum albumin as a standard.

3629

G. Garab (ed.), Photosynthesis: Mechanisms and Effects, Vol. V, 3629–3632.

2.3 RNA Isolation and Analysis

Plant material was pulverized with mortar and pestle in liquid nitrogen and RNA was isolated by phenol extraction and selective precipitation with LiCl. Electrophoresis of RNA was performed on formaldehyde-agarose gels at 120 V. RNA was transferred onto Hybond-N nylon membranes (Amersham) and hybridized with 32P-labelled total cDNA probes. The membranes were washed four times at different concentrations of SSC in 0.1% SDS for 20 min, the final washing step being 0.5 x SSC, 0.1% SDS at 65°C. The membranes were exposed to a X-ray film (Fuji medical RX) at - 80°C for 3 to 8 days.

2.4 Western Blot Analysis.

Protein extracts were prepared as described by Zavgorodnyaya et al. (11). Aliquots representing 10 μg protein were subjected to SDS-PAGE and electrotransferred to nitrocellulose filter (0,2 μm pore size; Schleicher and Schüll, Dassel, Germany). The blots were analysed with antisera against recombinant APR2 and chloroplastic OAS-TL and developed with the SuperSignal Western Blotting System (Pierce).

3. Results and discussion

3.1. APR Activity.

The effects of the omission of nitrogen from the medium on extractable activity of APS reductase (APR) was measured during 96h in both shoots and roots extracts from *Arabidopsis* plants (Fig. 1). In both extracts a decrease of APR activity was evident after 48h. After 4 days of nitrogen starvation APR activity decreased to 50% in shoots. In roots during the first 24h activity was slightly elevated followed by a decrease to 30% of the original activity after 4 days. When medium with nitrate was applied to the plants after 3 days of nitrogen starvation a quick restoration of APR activity was observed in both shoots and roots. Already after 24h extractable APR activity was higher then the activity at the beginning of treatment. From these data we can conclude that APR activity is highly regulated during nitrogen starvation and restores quickly after addition of nitrogen source, what correlates well with data, obtained earlier by different authors (5, 6, 12). That would indicate also that this enzyme plays the key role in regulation of assimilatory sulfate reduction and its coordination with nitrate assimilation.

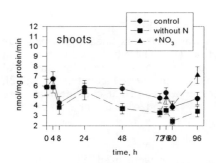 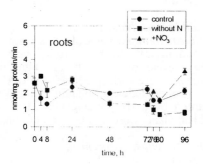

Figure 1. APR activity in shoots and roots of *A. thaliana* during 96h nitrogen deficiency and after addition of nitrate after 72h. Time point 0 corresponds to 9 a.m. Mean values and SD of 4-8 measurements are shown.

3.2. RNA Analysis

In order to understand the regulation of the nitrogen deficiency effects it was important also to investigate the mRNA level and protein accumulation during these conditions. Northern blot analysis with cDNAs coding for two isoforms of APR was performed to show regulation of mRNA synthesis. As shown in Fig. 2, the level of mRNA for both APR isoforms was decreased during nitrogen starvation. In addition, APR2 mRNA shows also strong diurnal fluctuations, after 4h from the beginning of the day the mRNA for APR2 is almost not detectable in shoots. In roots the fluctuations are not as strong as in leaves, but the APR2 mRNA level is lower then that of APR1 isoform. After addition of nitrate a quick induction of APR mRNA accumulation could be seen in roots after 4h, whereas in shoots only after 24h. Northern blot analysis of plastidic isoform of cysteine synthase revealed that mRNA coding for this enzyme is also reduced although not to such extent as APR mRNA.These results clearly show that APS reductase activity correlates well with the mRNA levels. Furthermore, in agreement with Takahashi et al. (1997) APR mRNA level was highly regulated in response to different nutritional status (13).

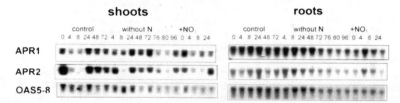

Figure 2. Northern blot analysis of APR1, APR2, and plastidic OAS-TL during 96h of nitrogen deficiency and during 24 h recovery with nitrate as nitrogen source.

3.3. Western Blot Analysis

Western blot analysis was performed with antibodies against APR2 and two isoforms of cysteine synthase. As shown in figure 3, protein accumulation under nitrogen deficiency decreased in shoots as well as in roots. In roots this effect was visible after 48h, whereas in shoots it started only after 72h. This decrease of protein accumulation corresponds well with activity measurements and mRNA expression under these conditions. This would mean that APS reductase is regulated on the transcriptional level. Similarly, accumulation of plastidic isoform of cysteine synthase decreased in both shoots and roots (Fig. 3). This effect was, however, not detected for cytoplasmic isoform. These results again correspond to cysteine synthase mRNA expression. But we have to mention here that decreasing of both mRNA level and protein accumulation did not correlate with the extractable activity for this enzyme that did not change during the experiment (data not shown).

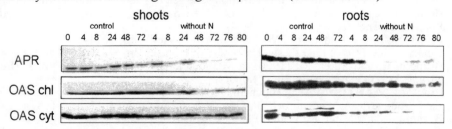

Figure 3. Western blot analysis of APR and plastidic and cytoplasmic cysteine synthase in *A. thaliana* during 4 days of nitrogen starvation.

3632

After addition of nitrate to the nutrient solution APR protein accumulation was immediately increased (Fig. 4). In shoots the effect was quicker then in roots, but in roots after 24h amount of protein was higher than in the control plants. For cysteine synthase protein the effect of nitrate addition was not as strong as for APR (Fig. 4). The plastidic isoform accumulation increased already after 4h in shoots while the increase of cytoplasmic isoform of CS protein amount was much slower.

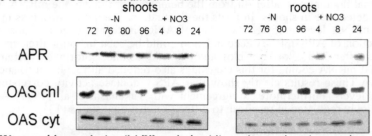

Figure 4. Western blot analysis of APR and plastidic and cytoplasmic cysteine synthase in *A. thaliana* during 24 h recovery with nitrate as nitrogen source.

4. Conclusions

Taken together, our data confirm that during nitrogen deficiency the key enzyme of sulfate assimilation, APS reductase, is highly regulated. We have demonstrated that the resulting decrease of APR activity is caused by a strong reduction of mRNA and protein accumulation. Further experiments will examine the effects of other N-containing compounds on the sulfate reduction and changes of flux through the sulfate assimilation pathway under different nitrogen status.

References

1   Schiff, J.A. (1983) in Encyclopedia Plant physiology (Läuchli,A. and Bielski,R.L. eds.) pp.402-421, Springer, Berlin
2   Cram, W.J. (1990) in Sulfur Nutrition and Sulfur assimilation in Higher Plants (Rennenberg, H., Brunold, C., De Kok, L.J., Stulen, I., eds.) pp. 3-11, Academic Publishing, The Hague, The Netherlands
3   Oaks, A. (1992) BioScience 42, 103-111
4   Saccomani, M., Cacco, G., Ferrari, G. (1984) J. Plant Nutr. 7, 1043-1057
5   Haller, E., Sutter, M., Brunold, C. (1986) J. Plant Physiol. 125, 275-283
6   Brunold, C. (1993) in Sulfur nutrition and sulfur assimilation in higher plants (Rennenberg, H., Brunold, C., De Kok, L.J., Stulen, I, eds) pp. 13-31, Academic Publishing, The Hague, The Netherlands
7   Anderson, S. and Anderson, J.W. (1997) Plant Physiol. 115, 1671-1680
8   Hentschel, G. (1970) PhD thesis, University of Hohenheim, Stuttgart, Germany
9   Brunold, C. and Suter, M. (1990) in Methods in Plant Biochemistry Vol.3, (Lea, P. ed.) pp 339-343, Academic Press, London
10 Bradford, M.M. (1976) Anal. Biochem. 72, 248-254
11 Zavgorodnyaya, A., Papenbrock, J., Grimm, B. (1997) Plant J. 12, 169-178
12 Brunold, C. and Suter, M. (1984) Plant Physiol. 76, 579-583
13 Takagashi H., Yamazaki, M., Sasakura, N., Watanabe, A., Leustek, T., de Almeida Engler, J., Engler, G., Van Montagu, M., Saito, K. (1997) Proc. Natl. Acad. Sci. USA 94, 11102-11107

RESPONSES OF THE MICROALGA *DUNALIELLA SALINA* TO SULFUR LIMITATION

Mario Giordano[1], Ruediger Hell[2] and Valerio Pezzoni[1]
[1]Laboratorio di Fisiologia Algale, Facoltà di Scienze, Università di Ancona, Italy
[2]Lehrstuhl für Pflanzenphysiologie, Ruhr Universität Bochum, Germany

Key words: algae, aquatic ecosystems, Dunaliella, metabolic processes, S-metabolism, tolerances.

1. Introduction

Sulfur is an essential element for plant and algal growth since it is a key component of amino acids, proteins, sulfolipids, vitamins and FE/S-clusters of the electron transport pathways (1). Traditionally, sulfur is thought never to limit productivity in aquatic ecosystems. This is likely to be true in most marine situations where sulfate ions, the primary source of sulfur, are present at concentrations of 20 –30 mmol L^{-1} (2). However, in many freshwaters, particularly oligotrophic lakes and rivers in upland areas where the sulfur content of the rocks in the catchment is low, the concentration of sulfate can be about 0.03 mmol L^{-1} (3). Furthermore, in such sites deposition from 'acid rain' can be the major source of sulfur (4) and the sulfur deposition is declining in Europe as air pollution is reduced (5) and may decline further as a result of the Oslo Protocol of 1994. Long-term records of water-chemistry in small lakes in the English Lake District, an upland area of the UK, have shown that sulfate concentrations have declined in response to reduced atmospheric deposition (3). The aim of this work is to find out whether sulfur may limit algal growth where sulfate concentrations are low and how algae respond to S-limitation.

2. Materials and methods

Dunaliella salina was selected as model organism. It was grown in batch cultures at [S] between 6 μmol L^{-1} and 100 mmol L^{-1}, in media and growth conditions previously described (6). Growth was followed with a Burker hemocytometer. The cell volume was estimated with a photonic microscope, assuming the cells were prolate ellipsoid (7). The organic C content of the cells was calculated according to Strathman (8). The protein cell content was determined according to Lowry (9). The intracellular SO_4^{2-} was measured on cell extracts by ion chromatography. Electrophoretic patterns were obtained by SDS-PAGE (10). ATP-S activity was measured determining spectrophotometrically the rate of change of $[NADP^+]$ resulting from the coupled

G. Garab (ed.), Photosynthesis: Mechanisms and Effects, Vol. V, 3633–3636.
© 1998 *Kluwer Academic Publishers. Printed in the Netherlands.*

reactions of hexokinase and glucose-6P dehydrogenase. Photosynthesis was determined with an O_2 electrode (Hansatech Instruments, Ltd.).

3. Results and Discussion

Dunaliella salina specific growth rate was not saturated by 50 µmol L^{-1} SO_4^{2-} in the medium (Table 1). However, the low-S treatment increased the volume and the estimated organic C content of the cells (Table 2).

$[SO_4^{2-}]$ (mM)	μ
0.006	**0.37**
0.025	**0.41**
0.050	**0.54**
100	**0.60**

Table 1. Specific growth rates (μ) of *Dunaliella salina* cultured in the presence of different SO_4^{2-} concentrations.

	$6\ \mu M\ SO_4^{2-}$	$100\ mM\ SO_4^{2-}$
Cell Volume (fL cell^{-1})	**513 (192)**	**179 (63)**
Cell organic C (pg cell^{-1})	**89 (31)**	**222 (83)**

Table 2. Volume and organic C of *Dunaliella salina* cells cultured in the presence of either 6 µM or 100 mM SO_4^{2-}. The numbers in parenthesis indicate the standard deviations (n = 10).

These data indicate that, while the division potential was reduced by S-limitation, the capacity of cells to produce biomass was not. The increased size of cells was not associated with an increase in cell protein (Table 3).

	$6\ \mu M\ SO_4^{2-}$	$100\ mM\ SO_4^{2-}$
Total protein (pg cell^{-1})	**44 (7)**	**41 (5)**
Methionine (pmol/10^6 cells)	**112 (5)**	**116 (6)**

Table 3. Total protein and methionine content of *Dunaliella salina* cells cultured in the presence of either 6 µmol L^{-1} or 100 mmol L^{-1} SO_4^{2-}. The numbers in parenthesis indicate the standard deviations (n = 3).

Also methionine did not vary as a function of [S] in the growth medium (Table 3). This could depend in part on the very high capacity of the S-limited cells to accumulate SO_4^{2-} internally: cells cultured in the presence of 6 µmol L^{-1} SO_4^{2-}, in fact, had intracellular $[SO_4^{2-}]$ over 1000 times higher than extracellular $[SO_4^{2-}]$ (data not shown).
In spite of the fact that the total amount of protein per cell was not appreciably affected by the S concentration used for growth, S-limitation induced a substantial

rearrangement of the protein pool. In particular, Rubisco protein decreased both in absolute (data not shown) and in relative abundance (Fig 1) with decreasing S, while a protein with electrophoretic characteristics corresponding to those of ATP-S (11) had a opposite behavior (Fig 1).

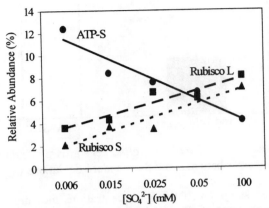

Figure 1. Changes of the relative abundance of L and S Rubisco subunits and ATP-S protein as a function of [SO_4^{2-}] in the growth medium. The relative abundance is expressed as % of total protein.

The changes in the relative abundance of Rubisco and ATP-S were statistically correlated (correlation coefficient = 0.92, degrees of freedom = 9). Photosynthesis (Fig 2) and ATP-S enzymatic activities (Fig. 3) varied accordingly with these observations.

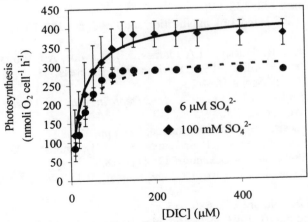

Figure 2. Response of photosynthesis to dissolved inorganic carbon (DIC). The cells were cultured in the presence of either 6 μM or 100 mM SO_4^{2-}. The error bars indicate the standard devations (n = 3).

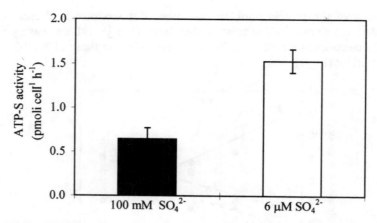

Figura 3. ATP-S activity of *Dunaliella salina* cells grown at either 100 mM or 6 μM SO_4^{2-}. The error bars indicate the standard deviations (n = 12).

The results reported here suggest that the S concentrations measured in some natural environments (2,3) could potentially limit the productivity of the ecosystems. *Dunaliella salina* responded to S-limitation activating very effective SO_4^{2-}-concentrating mechanisms that mitigated the effects of low S availability. These mechanisms, however, were not sufficient to prevent the cells from reallocating their resources: Rubisco, for instance, appeared to be degraded, possibly in order to recover amino acids for the construction of enzymes involved in S assimilation.

References cited

1. Hell, R. (1997), Planta **202**: 138-148
2. Carr, N.G. (1978), CRC Handbook in Nutrition and Food, pp. 283, CRC Press, Boca Raton, USA.
3. Tipping, E., Carrick, T.R., Hurley, M.A., James, J.B., Lawlor, A.J., Lofts, S., Rigg, E., Sutcliffe, D.W. and Woof, C (1998). Environ. Poll. (in press).
4. Whelpdale, D.M. (1992). *Scope* **48**: 4-26.
5. UK Review Group on Acid Rain (1997). *Acid Deposition in the United Kingdom*, AEA Technology, Abingdon, UK.
6. Norici, A.(1997). Thesis, University of Ancona, Italy, pp. 120
7. Bèrubè, K.A., Roessler, J., Jones, T.P. and Janes, S. (1994). Ann. Bot. **73**: 481-491.
8. Strathman, R.R. (1967). Limnol Oceanogr. **12**: 411-418.
9. Lowry, O.H.; Rosebrough, N.J., Farr, A.L., and Randall, R.J. (1951). J. Biol. Chem. **193**: 265-275.
10. Laemmli, U.K. (1970). Nature **277**: 680-685.
11. Li, J., Saidha, T. and Schiff, J.A.(1991), Biochim. Biophys. Acta **1078**: 68-76.

RIBOFLAVIN BIOSYNTHETIC ENZYMES

Douglas B. Jordan[1], Karen O. Bacot[2], Thomas J. Carlson[2], Michael P. Picollelli[1], Zdzislaw Wawrzak[2], Martin Kessel[3], and Paul V. Viitanen[2]

[1]E. I. DuPont de Nemours & Co., Stine-Haskell Research Center, Elkton Road, P.O. Box 30, Newark, DE 19714, [2]E. I. DuPont de Nemours & Co., Experimental Station, P.O. Box 80402, Wilmington, DE 19880-0402, [3]Laboratory of Structural Biology, National Institutes of Health, Bethesda, MD, 20892

1. Introduction

Riboflavin after conversion to flavin mononucleotide (FMN) and flavin adenine dinucleotide (FAD) serves as an essential cofactor for mainstream metabolic enzymes which mediate hydride, oxygen, and electron transfer catalytic functions (1). Enzymes of the riboflavin biosynthetic pathway are attractive targets for the rational design of antibiotics and crop protection chemicals since humans do not synthesize riboflavin (also known as vitamin B_2) and must obtain it through their diets (2). Indeed, there are design and synthesis reports regarding the penultimate and ultimate enzymes of the biosynthetic pathway, 6,7-dimethyl-8-(1'D-ribityl)-lumazine synthase (lumazine synthase) and riboflavin synthase, respectively (3-5).

Competitive Inhibitors of riboflavin synthase that were designed to mimic a proposed reaction intermediate (3), yield dissociation constants in the range of 10^{-8} M. Russupteridine, a natural product inhibitor of riboflavin synthase, also has a dissociation constant of 10^{-8} M (4), but its mode of inhibition is noncompetitive with respect to substrate , 6,7-dimethyl-8-(1'D-ribityl)-lumazine (4). Steady-state kinetic results on inhibitors of riboflavin synthase may be complicated by the fact that the enzyme has two binding pockets for lumazine (termed "donor" and "acceptor"), which are functionally nonequivalent having different substrate binding coefficients (6). The three-dimensional structure of riboflavin synthase has not been reported, but there are preliminary reports on crystals of the protein which are suitable for X-ray studies (7,8). The design and synthesis of lumazine synthase inhibitors is also reported in the literature (5), but to date all of these compounds are weaker binders than either of the enzyme's substrates. Recently, however, a 2.4 Å resolution crystal structure was reported for the *Bacillus subtilis* homolg, in which the enzyme is complexed with a

G. Garab (ed.), Photosynthesis: Mechanisms and Effects, Vol. V, 3637–3640.

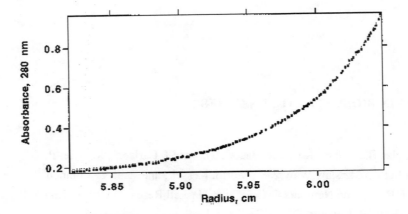

Figure 1. Analytical centrifuge study on spinach lumazine synthase. Data were collected after 72 h of centrifugation at 4500 rpm and 4° C using a XL-A Optima (Beckman) analytical centrifuge. The absorbance data were fit to a model for an ideal monodisperse system having a partial specific volume of 0.724 and the calculated mass was 991,000 Da.

ribitylpyrimidine analog of substrate 4-ribitylamino-5-amino-2,6-dihydroxypyrimidine (RAADP) and an active site orthophosphate which likely occupies the substrate binding site for dihydroxybutanone phosphate (DHBP; reference 9). Three-dimensional studies are crucial to the design of effective enzyme inhibitors using modern strategies, and there have already been some attempts to design and synthesize bi-substrate like, dual-affinity inhibitors of lumazine synthase based on its X-ray structure (5). The *B. subtilis* and *E. coli* lumazine synthases consist of 60 identical subunits that are organized as twelve pentameric units to form a hollow icosahedral capsid, while the native yeast protein would appear to be a pentamer (10).

Although plants are also capable of riboflavin biosynthesis, virtually nothing is known about the process. Indeed, according to the literature, the only pathway enzyme that has been cloned to date is GTP cyclohydrolase II which was obtained from arabidopsis (11). Below we summarize our results on the cloning, expression, purification, cellular localization, and three-dimensional homology modeling of spinach lumazine synthase.

2. Materials and Methods

Homology modeling of the three-dimensional structure of spinach lumazine synthase was through the computer program Biopolymer within the suite of programs of Sybyl (Tripos, St. Louis) Amino acids were mutated individually upon the X-ray structure of *B. subtilis* lumazine synthase (PDB accession code, 1RVV).

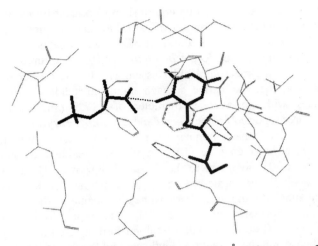

Figure 2. A homology model of the active site (4 Å radius around substrates) of spinach lumazine synthase suggesting the Michaelis complex. Substrates are in bold.

3. Results and Discussion

An *E. coli* lumazine synthase knockout mutant was created by insertional inactivation of the gene and P1 transduction, and the resulting auxotroph was used to screen a spinach cDNA expression library for growth in the absence of added riboflavin. One of the complementing plasmids, encoding the full length protein, was sequenced completely and used in further studies. The open reading frame encoded a protein of 223 amino acids, making it much larger than any of its microbial counterparts. From sequence alignments with yeast and bacterial lumazine synthases, it was predicted that the plant protein is synthesized as a larger molecular weight precursor with an N-terminal chloroplast targeting sequence of ~ 66 residues; the latter are rich in serine and threonine residues.

To test this notion, the full-length spinach lumazine synthase was labeled with [^{35}S] methionine and used for chloroplast import experiments. Classical protease protection experimental protocols for the uptake of the labeled protein into organelles were followed and they indicated that spinach lumazine synthase was imported into chloroplasts in an ATP-dependent fashion. The lumazine synthase protein imported into the chloroplast was found to have a molecular weight of about 16 kDA, more in line with the microbial enzymes.

In order to study mature spinach lumazine synthase, the chloroplast targeting sequence was removed by PCR and the resulting construct was cloned behind a T7 promoter for expression in *E. coli*. The recombinant protein (~15% of the total soluble protein) was purified using anion exchange and gel filtration chromatography. Purified spinach lumazine synthase was characterized by equilibrium ultra-centrifugation measurements as having a native molecular weight of 991,000 Da (Figure 1), which nicely corresponds to the predicted mass of a 60-mer of spinach lumazine synthase protomers (992,083 Da). Thus, in common to the *B. subtilis* and *E. coli* lumazine synthases, the spinach enzyme is a huge macromolecular assembly. It was further revealed through electron microscopy, that the native spinach lumazine synthase consists of hollow spherical particles with an average diameter of ~17.5 nm, consistent with the icosahedral structures reported for the *E. coli* and *B. subtilis* enzymes (10).

A homology model for spinach lumazine synthase was readily built from the X-ray structure of the *B. subtilis* enzyme (10). Active site residues of spinach lumazine synthase were about 80% identical to the *B. subtils* enzyme. Because the *B. subtils* lumazine synthase X-ray structure had a substrate analog mimicking the ribitylpyrimidine substrate and an orthophosphate molecule in the active site giving a position for the phosphorylated substrate, a Michaelis complex for the spinach lumazine synthase was modeled (Figure 2). The first bond-forming step in catalysis must be in forming a Schiff base and the formation is indicated by the dotted line in Figure 2.

References

1   Bacher, A. (1991) in Chemistry and Biochemistry of Flavoproteins, (Müller, F., ed.), vol. 1, pp. 215-259, CRC Press, Boca Raton.
2   Yagi, K. (1991) in Flavins and Flavoproteins 1990 (Curti, B, Ronchi, S. and Zanetti, G., eds.), pp. 3-16, Walter de Gruyter & Co., Berlin.
3   Al-Hassan, S. S., Kulick, R. J., Livingstone, D. B., Suckling, C. J., Wood, H. C. S., Wrigglesworth, R. and Ferone, R. (1980) J. Chem. Soc., Perkin Trans. 1: 2645-2656.
4   Smith, E. E., O'Kane, D. J. and Meighen, E. A.(1994) in Biolumin. Chemilumin., Proc. Int. Symp. (Campbell, A. K., Kricka, L. J.; Stanley, P. E., eds), pp. 568-571, Wiley, Chichester, UK.
5   Cushman, M., Mavandadi, F.; Kugelbrey, K.and Bacher, A. (1997) J. Org. Chem. 62, 8944-8947.
6   Harvey, R. A. and Plaut, G. W. E. (1966) J. Biol. Chem. 241, 2120-2135.
7   Jordan, D., Viitanen, P., Wawrzak, Z., Picollelli, M., Bacot, K., Schwartz, R. and Thompson, J. (1997) in Flavins and Flavoproteins 1996, pp. 945-948, University of Calgary Press, Calgary.
8   Meining W., Tibbelin G., Ladenstein R., Eberhardt S., Fischer M. and Bacher A. (1998) J. Struct. Biol. 121, 53-60.
9   Ritsert, K., Huber, R., Turk, D., Ladenstein, R., Schmidt-Bäse, K. and Bacher, A. (1995) J. Mol. Biol., 253, 151-167.
10 Mörtl, S., Fischer, M., Richter, G., Tack, J., Weinkauf, S. and Bacher, A. (1996) J. Biol. Chem. 271, 33201-33207
11 Kobayashi, M., Sugiyama, M., and Yamamoto, K. (1995) Gene 160, 303-304.

19. Molecular physiology

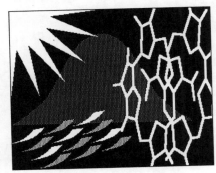

CONGRESS ON PHOTOSYNTHESIS

XIth INTERNATIONAL

BUDAPEST ✳ 1998

REGULATION OF CO_2 ASSIMILATION RATE BY THE CHLOROPLAST CYTOCHROME *BF* COMPLEX

Susanne von Caemmerer, Sari A. Ruuska, G. Dean Price John R. Evans, Jan M. Anderson, T. John Andrews, and Murray R. Badger. Research School of Biological Sciences, Institute of Advanced Studies, Australian National University, PO Box 475, Canberra, ACT 2612, Australia

Keywords: antisense suppression, ATP synthase, gas exchange, genetic manipulation, Rieske center, Rubisco

1. Introduction

The photosynthetic electron transport pathway is responsible for the provision of ATP and NADPH for photosynthetic carbon reduction (PCR) and as such it has the potential to exert strong regulation over photosynthesis and growth. Transduction of light energy by the photosynthetic apparatus of chloroplasts depends on the co-ordinated capture of light energy by the two photosystems, PSI and PSII, along with electron transfer between these two centers. Generation of a trans-thylakoid proton gradient is required for synthesis of ATP via the ATP synthase complex, and NADPH production is dependent on linear electron flow from PSII to PSI and through ferredoxin to ferredoxin:$NADP^+$ oxidoreductase. Functionally the cytochrome *bf* complex is located in a central position in the electron transport scheme between P680 and P700 because of its ability to operate as both a plastoquinol and plastocyanin oxidoreductase (1,2). The cytochrome *bf* complex can act in both linear electron transport (production of ATP and NADPH) and cyclic electron transport flow (ATP generation only). Price and coworkers have engineered transgenic tobacco plants where the nuclear-encoded Rieske FeS subunit of the cytochrome *bf* complex was specifically suppressed through antisense RNA technology (3). They produced plants with cytochrome *bf* content as low as 5% of wild type values. In this paper we have used these Rieske FeS antisense plants to examine the link between electron transport rate and PCR cycle activity. We conclude that the reduction in electron tranport rate reduces CO_2 assimilation rate both by reducing the RuBP concentration and by reducing Rubisco carbamylation.

2 Procedure

2.1 Plant material

Details on the construction and initial analysis of tobacco lines (*Nicotiana tabacum* L.cv W38) with an antisense construct directed against the transcript for the Rieske FeS protein of the chloroplast *bf* complex have been reported previously (3). T_1 seed from FeS antisense lines, B6F 2.2U, GUS2 23.3A and GUS2 23.1B (3) or selfed T_3 seed from

3643

line B6F-2.2-513-16, were germinated on soil and fertilised with a complete nutrient solution. Plants were grown in a growth cabinet at a photon flux of 100-150 µmol m^{-2} s^{-1} and a daily light period of 20 hours at a constant temperature of 25°C. Plants were used at 7 to 9 weeks post-germination. Transgenic tobaccos with an antisense construct to chloroplast glyceraldehyde 3-phosphate dehydrogenase (anti-GAPDH plants) were grown in the glasshouse as described (4).

2.2 Leaf gas exchange.
Gas-exchange measurements were made on attached leaves in a gas exchange chamber attached to a freeze clamp apparatus (4). Illumination was provided by a 150 W slide projector, the light being passed via a mirror to the leaf chamber. Assimilation rates were measured at an irradiance of 1.5 mmol quanta m^{-2} s^{-1}, 350 µbar CO_2 in air, a leaf temperature of 25 °C and a leaf to air vapour pressure difference of around 10 mbar.

2.3 Biochemical measurements
Western blotting and quantitation were performed as previously described using antibodies against tobacco Rieske FeS protein, the ATP synthase δ subunit and antibody for cytochrome f (3,5). For measurements of Rubisco and metabolites, leaves were allowed to equilibrate for 40 min under controlled gas exchange conditions and were then rapidly freeze-clamped in situ (4,5). The frozen leaf disk was divided into two halves, each 2.7 cm^2 in area. One half was used for measurement of Rubisco site concentration and activation state from CABP binding (6). The other half was used for the determination of RuBP and PGA levels as described in (4).

3. Results and Discussion

3.1 Effect of growth conditions on the phenotype of wild type plants
The tobacco transformants with an antisense construct directed against the Rieske FeS subunit of the chloroplast bf complex had to be grown in growth cabinets under low irradiance and long days to ensure a stable antisense phenotype (3, 5). Typically, shade leaves have distinctive features when compared to sun leaves (7,8). For example, light-saturated CO_2 assimilation is reduced compared to highlight grown plants and lower Rubisco and soluble protein content accompany this. Frequently the chlorophyll content per leaf area is greater in shade leaves and the chlorophyll a/b ratio is decreased. We found that a long day length could compensate for the low irradiance in the growth cabinet. This is shown in a comparison of wild type tobacco grown in the glasshouse with midday irradiances of 1000 µmol m^{-2} s^{-1} and wild type grown in growth cabinets with low light and long days (Table 1).

TABLE 1. Physiological characteristics of wild type tobacco grown in growth cabinets at an irradiance of 100 μmol quanta m^{-2} s^{-1} and 20 hour day length and wild type grown in a glasshouse with midday irradiances of 1000 μmol quanta m^{-2} s^{-1}. CO_2 assimilation rate, A and stomatal conductance, g, were measured at 25°C, 1000 μmol quanta m^{-2} s^{-1} and an ambient CO_2 of 350 μbar.

	Low light, long days	Glasshouse
A (μmol m^{-2} s^{-1})	12.8 ±1	13.7 ±1
g (mol m^{-2} s^{-1})	0.29 ±0.03	0.33 ±0.04
Sol protein(g m^{-2})	3.7 ±0.3	4 ±0.25
Rubisco (g m^{-2})	1 ±0.1	1.5 ±0.1
Chl (mg m^{-2})	365 ±14	343 ±31
Chl a/b	2.9 ±0.03	3.1 ±0

The CO_2 assimilation rate and stomatal conductance of the low-light and glasshouse-grown plants were surprisingly similar. Low light plants had slightly less soluble protein and Rubisco. This shows that with adequate carbon gain, ensured through extended day length, high photosynthetic capacities can also be obtained under growth conditions with low irradiance.

3.2 Cytochrome f, ATPδ and Rubisco
Western blot analysis showed a strong correlation between cytochrome f and Rieske FeS content confirming that the antisense resulted in a lowering of the of the cytochrome bf complex, and not just the Rieske FeS polypeptide (5)

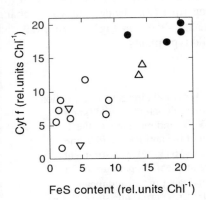

FIG. 1. Cytochrome f protein as a function of chloroplast Rieske FeS protein per unit chlorophyll in wild type (●), and Rieske FeS antisense plants (T1 progeny of line B6F2,2U (○), line Gus2.23.3A (Δ), line GUS2.23.1B (∇).

There was little change in either the δ subunit of the ATP synthase (ATPδ) or Rubisco content on a chlorophyll basis showing that the reduction in Rieske FeS had led to a specific reduction in the cytochrome *bf* complex (FIG. 2.).

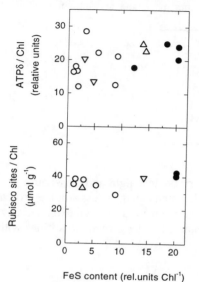

FIG. 2. ATPδ and Rubisco content on a chlorophyll basis as a function of chloroplast Rieske FeS protein in wild type and Rieske FeS antisense plants. Symbols are the same as in FIG. 1.

A lower transthylakoid ΔpH is evident as low non-photochemical quenching (qN) of chlorophyll fluorescence (3). Restriction of electron flow through the cytochrome b*f* complex causes the plastoquinone pool to be very reduced which leads to closure of PSII reaction centres and little photochemical quenching (qP) (3). Under our gas exchange conditions the mean qN and qP for the transformants shown in Fig. 1. Were 0.33 and 0.22, compared to values of 0.73 and 0.45 in the wild type plants.

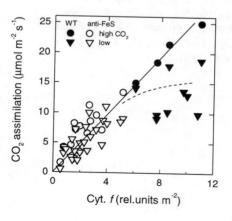

FIG. 3. CO_2 assimilation rate as a function of cytochrome *f* content in wild type tobacco (▼,●) and members of the T_3 population of the anti Rieske FeS line B6F-2.2-513-16 (∇,○). Measurements were made at a leaf temperature of 25 °C, 1000 μmol quanta m^{-2} s^{-1} and mean intercellular CO_2 partial pressures of 250 μbar (▼∇) or 550 μbar (●,○).

3.3 CO₂ assimilation rate is reduced in plants with low Rieske FeS content

The capacity of the chloroplast electron transport chain is expected to limit CO_2 assimilation rate at high irradiance and at high CO_2 (9). Since at the low irradiances of the growth conditions, light is likely to limit CO_2 assimilation rates in the wild type, we chose to compare the gas exchange characteristics of plants at an irradiance of 1000 μmol quanta m^{-2} s^{-1}. Fig.3 illustrates the close correlation between CO_2 assimilation rate and cytochrome f content that existed in the T_3 population of the anti Rieske FeS line B6F-2.2-513-16. We made measurements at ambient and elevated CO_2 partial pressures and found a close linear relationship between cytochrome f content and CO_2 assimilation rates at elevated CO_2. This supports the notion that electron transport limits CO_2 assimilation rate under these conditions (9). A more curvilinear relationship is evident at ambient CO_2 partial pressures, where presumably Rubisco content limits the CO_2 assimilation rate in some of the wild type plants. In the Rieske FeS antisense plants there was little difference in CO_2 assimilation rate between measurements made at ambient and elevated CO_2 partial pressures, showing that electron transport rate limited CO_2 assimilation rate under both conditions.

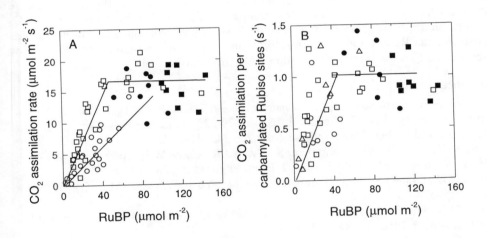

FIG. 4. CO_2 assimilation rate (A) and CO_2 assimilation rate per carbamylated Rubisco sites (B) as a function of leaf RuBP content in wild type (●■) and transgenic tobacco with reduced amount of GAPDH (□) or reduced amount of Rieske FeS content (○). Plants with the same symbol shape were grown together. Measurements were made under conditions described in Fig. 3. The range of Rubiso site concentration was 15 to 25 μmol m^{-2}. Lines were drawn by eye.

3.4 Are low RuBP pools the sole cause of low Rubisco carboxylation rates in anti Rieske FeS plants?

Lowering electron transport rate by reducing chloroplast Rieske FeS protein reduces the rate of RuBP regeneration evident in the strong correlation between Rieske FeS content and RuBP pools (5, 10). A reduction in the electron transport rate can reduce the rate of RuBP regeneration via a reduction in the rate of ATP and NADP production and via changes to the activation of light regulated enzymes of the photosynthetic carbon reduction cycle. We were interested to examine whether the lowered RuBP pools are the cause of low CO_2 assimilation rates.

We therefore compared the relationship between CO_2 assimilation rate and RuBP content in wild type tobacco and transgenic tobacco with antisense constructs to chloroplast glyceraldehyde 3-phosphate dehydrogenase (anti-GAPDH) or Rieske FeS plants (Fig. 4). Reduction in GAPDH activity also led to strong reduction in RuBP pools and RuBP regeneration rate (4). However the relationship between CO_2 assimilation rate and RuBP was clearly different between the two transgenic genotypes (Fig.4, A). In the anti GAPDH plants, Rubisco carbamylation was not affected by the reduction in RuBP content (4) and the relationship between CO_2 assimilation rate and RuBP is that expected for Rubisco given the high Rubisco site concentration in the chloroplast (11, 4). In the anti Rieske FeS plants Rubisco carbmylation was reduced (5, 10). When we examined the relationship between CO_2 assimilation rate per carbamylated Rubisco sites, we found no difference between the two antisense constructs. We therefore conclude that the changes in Rubisco carbamylation account for the difference seen in Fig.4A. Antisense FeS plants had reduced rates of electron transport which resulted in lower rates of CO_2 assimilation from the combined effects of lower RuBP content and lower Rubisco carbamylation. Ruuska and coworkers, this issue, address the intriguing question of how Rubiso carbamylation is reduced by reduced electron transport rate.

References

1 Cramer W.A., Furbacher P.N., Szczepaniak A., and Tae G.-S. (1991) in Current Topics in Bioenergetics, (Lee, C. P. ed.) vol. 16, pp. 179-222, Academic Press, San Diego

2 Anderson J.M. (1992) Photosynthesis Research 34, 341-357

3 Price G.D., Yu J-W., von Caemmerer S., Evans J.R., Chow W.S., Anderson J.M., Hurry V., Badger M.R. (1995) Aust. J. Plant. Physiol., 22, 285-297

4 Price G.D., Evans J.R., von Caemmerer S, Kell P., Yu J-W, Badger M.R. (1995) Planta 195, 369-378.

5 Price G.D., von Caemmerer S., Evans, J.R., Siebke K., Anderson J.M., Badger M.R. (1998) Aust. J. Plant. Physiol. 25, 445-452

6 Mate C.J, von Caemmerer S., Evans J.R., Hudson G.S, Andrews T.J. (1996) Planta 198, 604-613

7 Björkman O. (1981) in (Lange O.L., Nobel P.S., Osmond C.B., Ziegler H., eds), Encyclopedia of Plant Physiol., New Series, Vol 12A: pp57-107, Springer-Verlag, Berlin

8 Anderson J.M., Chow W.S., Goodchild D.J. (1988) Aust. J. Plant. Physiol. 15, 11-26

9 von Caemmerer S., Farquhar G.D. (1981) Planta 153, 376-387

10 Ruuska S.A., von Caemmerer S., Andrews T.J., Badger M.R., Price G.D, (1998) this issue.

11 Farquhar G.D. (1979) Arch Biochem Biophys 193, 456-468

The plastidic ATP/ADP-Transporter: Implications for starch metabolism in chloroplasts and storage plastids

Ekkehard Neuhaus, Joachim Tjaden, and Torsten Möhlmann
Universität Osnabrück, Pflanzenphysiologie, Barbarastr. 11,
49060 Osnabrück, Germany

Key words: Carbohydrate, cDNA cloning, higher plants, membrane protein, metabolite transport, transgenic plants

1. Introduction

Plastids are surrounded by an envelope consisting of two membranes. The outer membrane is due to the presence of porins permeable for compounds up to a molecular mass of about 10kDa. The inner envelope represents the barrier for most metabolites and gains permeability due to the presence of several types of specific transport proteins (1). Well characterised transporters are up to now the group of phosphate/sugar phosphate transporters (2,3,4), and a new type of the dicarbonic-acid exchanger (5).

In addition to these transporters, all types of plastids possess an ATP/ADP exchange system which differs in many respects significantly form the functional homologues in mitochondria. The mitochondrial ADP/ATP-transporter is under *in vivo* conditions solely able to mediate ATP export in counter exchange for cytosolic ADP (for review see Klingenberg (6)). This mechanism allows the provision of ATP consuming reactions outside the mitochondrium with energy. In contrast to this, the plastidic ATP/ADP transporter prefers to import ATP from the cytosol in counter exchange to stromal ADP and thus catalysing the opposite direction of transport (7). The mitochondrial ADP/ATP transporter is significantly inhibited by specific inhibitors like bongkrekic acid or atractylate (6) whereas the plastidic adenylate transporter system is hardly affected by these compounds (8). Finally, a polyclonal antiserum raised against the mitochondrial ADP/ATP transporter from *Neurospora crassa* recognises the homolog proteins from various types of tissues but does not cross react with any protein in highly enriched plastidic envelope membranes. In conclusion, these differences clearly indicated that the molecular nature of the plastidic ATP/ADP transporter differs substantially from the functional homologs in mitochondria.

2. Molecular identification of the plastidic ATP/ADP transporters in *Arabidopsis thaliana* and biochemical characterisation

By screening an *Arabidopsis thaliana* cDNA library for a copper transporter we identified a cDNA encoding for a protein exhibiting 66% similarity to the ATP/ADP transporter from the intracellular, human pathogenic bacterium *Rickettsia prowazekii* (named AATP1, At; ATP/ADP Transport Protein 1, *Arabidopsis thaliana*; 9). This bacterium represents a strictly intracellular living organism which exploits the host cell cytosol by uptake of various compounds necessary for metabolism, growing, and amplification. One of these compounds is ATP which is imported in counter exchange to bacterial ADP (10). The corresponding transport protein has been identified on the molecular level and comprises 497 amino acids named TlcRp (Translocase *Rickettsia prowazekii*; 11). According to a computeral hydrophocity analysis TlcRp exhibits twelve transmembrane domains and thus belongs to the large family of solute transporters exhibiting such configuration (12). As demonstrated for the plastidic ATP/ADP transporter the rickettsial transporter mediates uptake of ATP from the cytoplasm (the

3649

G. Garab (ed.), Photosynthesis: Mechanisms and Effects, Vol. V, 3649–3652.
© 1998 *Kluwer Academic Publishers. Printed in the Netherlands.*

host cell cytoplasm), is not inhibited by bongkrekic acid or atractylate, and does not cross react with polyclonal antisera raised against mitochondrial ADP/ATP transporters (13). These similarities between the rickettsial translocase and AATP1(At) indicated that plants contain a second type of eukaryotic adenylate transporter lacking substantial homology to the functional counterpart in mitochondria.

In contrast to the rickettsial ATP/ADP transporter the plastidic type of adenylate transporter possess a N-terminal amino acid extension of about 100 residues exhibiting similarity to the transit peptides of four other proteins residing in the inner plastidic envelope membrane (14). After *in vitro* translation of the *AATP1*(At) cDNA we incubated isolated chloroplasts with the radioactively labelled pre-protein and demonstrated that the pre-protein is targeted into the envelope membrane and that the pre-protein becomes proteolytically processed to the mature protein (14). In addition, this subcellular location of AATP1(At) has been demonstrated by using a peptide specific antiserum which strongly cross reacted with a protein of the expected molecular mass in enriched chloroplast envelopes, but not with proteins in enriched mitochondrial membranes (14).

The functional analysis of plant membrane proteins is relatively difficult and laborious. In most cases a newly identified cDNA has to be expressed heterologously in yeast, fission yeast or other eukaryotic expression systems, the proteins have afterwards to be reconstituted in proteoliposomes and transport activities monitored after addition of radioactivity. Therefore, we tried to express the AATP1 (At) cDNA functional in *Escherichia coli* to study nucleotide uptake directly on the intact bacterial cell. This approach was encouraged since it has been demonstrated that the rickettsial transporter (which shares a high degree of similarity to the plastidic ATP/ADP transporter, see above) can be functionally expressed in *E. coli* (13). After cloning of *AATP1*(At) into an *E. coli* expression vector a histidine tagged chimeric protein of AATP1(At) (His-AATP1,At) was synthesised in IPTG-induced cells. These cells were able to import radioactively labelled [α^{32}P]ATP whereas uninduced *E. coli* cells did not import radioactivity (15). In control experiments it was revealed that IPTG induction specifically stimulated adenylate transport, since import of other phosphorylated or unphosphorylated metabolites (e.g. glucose-6-phosphate or glucose) was not stimulated (15). Expression of *His-AATP1*(At) in *E.coli* represents a first example of a plant solute transporter functionally synthesised in this organism. Moreover, this system is powerful since the biochemical properties of the heterologously expressed *His-AATP1*(At) are very similar to the data raised on the authentic protein in enriched plastids (see Tjaden et al. (15) for a discussion). The high specificity of the plastidic ATP/ADP transporter was further demonstrated by inhibitor studies. ADP was a strong competitive inhibitor of [α^{32}P]ATP uptake ($K_{i\ ADP}$ 3.6 μM) whereas all other metabolites tested like AMP, ADPglucose, UTP, UDP, NAD, and NADP did not influence [α^{32}P]ATP import.

The presence of a second plastidic ATP/ADP transporter in *Arabidopsis thaliana* had been suggested by southern-blot analysis (9). Indeed, after rescreening the *A. thaliana* cDNA library with the entire *AATP1*(At) cDNA we identified an isoform, named *AATP2*(At) (16). As deduced from the amino acid sequence AATP2(At) exhibited 77.6% identity to AATP1(At) and 36% to the rickettsial protein. The biochemical properties of His-AATP2(At) are similar to the properties of His-AATP1(At) (16) and it remains to be analysed why *A. thaliana* possess two isoforms of this transporter.

3. Function of the ATP/ADP transporters in chloroplasts and heterotrophic plastids

One of the first reports on the plastidic ATP/ADP transporters was given by Heldt (7). Isolated chloroplasts from spinach leaves possesses an ATP/ADP exchange system preferentially importing ATP from the cytosol. As the Vmax of the chloroplastic ATP/ADP exchange system was only about 5μmol per mg of chlorophyll and hour a support of this transporter to photosynthesis appeared to be unlikely (7). However, isolated mesophyll chloroplasts from the C4 plant *Digitaria sanguinalis* exhibited ATP import rates in the rage from 50 to 70 µmol per mg of chlorophyll and hour (17) making it possible that photosynthetic metabolism in this species is positively influenced by uptake of ATP from the cytosol (17). It remains to be analysed whether such high ATP import rates resembles also a feature of mesophyll chloroplasts from other C4 species or not.

In conclusion, in case of leaf-chloroplasts it is still not clear how far ATP uptake influences photosynthetic processes. However, photosynthetic processes in fruit chloroplasts seem to be positively influenced by ATP uptake. It is well known that photosynthesis in fruit chloroplasts is extremely important for crop yield (for review see Blanke and Lenz (18). One feature of a wide range of fruit chloroplasts is that they contribute to starch-, lipid- and also amino-acid synthesis in these tissues (for review see Quick and Neuhaus (19). We demonstrated that isolated chloroplasts from sweet pepper fruits are able to synthesise starch (20). In analogy to plastids from totally heterotrophic tissues sweet pepper fruit chloroplasts import glucose-6-phosphate and convert the glucose moiety to starch. For such conversion ATP is required in order to energise the reaction of the ADPglucose pyrophosphorylase (AGPase). A part of this ATP is synthesised *via* photosynthetic electron transport coupled to ATPase activity in thylakoids. However, the additional presence of ATP in the incubation medium doubled the rate of starch biosynthesis (20) clearly indicating that ATP uptake into this type of chloroplast is required for maximal crop yield.

But even in leaf-chloroplasts there are situations were ATP uptake is required for metabolism. By using mutants from *Arabidopsis thaliana* lacking the ability for nocturnal starch degradation it was demonstrated that accumulation of starch induced a decrease of photosynthetic carbon fixation (21). This observation illustrates that a controlled degradation of transitory starch is highly important for regulation of photosynthesis. However, the metabolic events occurring in chloroplasts during nocturnal starch degradation are not very well analysed. By using high starch containing chloroplasts from *Mesembryanthemum crystallinum* (a facultative C3-CAM plant, called „ice plant") we demonstrated that degradation of internal starch was strongly promoted after addition of exogenous ATP (22). Such stimulatory effect holds true for both, chloroplasts isolated from C3 and from CAM-induced *Mesembyranthemum crystallinum* plants (22). This observation indicated that stromal ATP synthesis is not sufficient to drive starch mobilisation at highest rates and that the activity of the plastidic ATP/ADP transporter is required.

4. Analysis of transgenic potato plants with increased or decreased activities of the plastidic ATP/ADP transporter

To analyse the physiological function of the plastidic ATP/ADP transporter transgenic potato plants were created. Potato plants were chosen because of their heterotrophic storage organs (tubers) which accumulate most strongly *AATP1* mRNA as shown by northern blot analysis. For the transformation of potato plants an *Agrobacterium tumefaciens* based system was used in which the corresponding cDNA was controlled by the constitutive cauliflower 35S promotor. The activity of the plastidic ATP/ADP transporter was reduced by transforming potato plants with *AATP1* (*Solanum tuberosum*) cDNA cloned in antisense orientation after first identifying a cDNA encoding for the protein in potato (AATP1, St). To increase the activity of the plastidic ATP/ADP transporter *AATP1* (At) cDNA was cloned in sense orientation to avoid cosuppression. Regenerated potato plants were screened for mRNA levels using specific probes for either *AATP1* (At) or *AATP1* (St). Several sense lines contained significant amounts of *AATP1* (At) mRNA whereas the levels of endogenous *AATP1* (St) mRNA were reduced in several antisense lines. Altered mRNA levels correlated with changed activities of the transporter measured in a proteoliposome system. Four antisense- and three sense-lines were chosen for further investigations. Although no phenotype was detectable in the green components of the plants there were pronounced differences in tuber morphology between different lines. In antisense lines an increase in tuber number by a simultaneous decrease in size and budding of tubers was observed. These antisense lines exhibited reduced starch contents but substantially elevated levels of soluble sugars in tubers. In contrast, sense lines accumulated increased amounts of tuber starch. Iodine-stained tuber slices from antisense plants, but not from wildtype or sense plants, destained rapidly in water indicating that the amylose to amylopectin ratios were different. Starch isolated from wildtype tubers had an amylose content of 18.8%, starch from antisense plants contained 11.5 to 18.0% amylose whereas starch from sense plants reached levels of 22.7 to 27.0%.

From this data we conclude that the stromal ATP supply is important for the ability of starch synthesis in amyloplasts of potato tubers and that this supply is strongly mediated by the plastidic ATP/ADP transporter. The availability of ATP in the amyloplastic stroma is a prerequisite for the synthesis of ADPglucose which is the substrate for the different starch synthases. In general, starch synthases can be classified in granular bound- and soluble isoforms (23). In potato, the granular-bound isoforms are responsible for amylose synthesis (24,25) and exhibit lower affinities for ADPglucose than the soluble isoforms, which are mainly involved in amylopectin synthesis (23). Changes in the stromal ADPglucose concentration would therefore not only lead to altered starch contents but also to changed amylose to amylopectin ratios.

References

1 Emes M.J., and H.E. Neuhaus (1998) J. Exp. Bot. **48**: 1995-2005.
2 Flügge, U.I., Fischer, K., Gross, A., Sebald, W., Lottspeich, F., and Eckerskorn, C. (1989) *EMBO J.*, **8**, 39-46.
3 Fischer K., Kammerer B., Gutensohn M., Arbinger B., Weber A., Häusler R., Flügge U.I. (1997) Plant Cell **9**: 543-462
4 Kammerer B., K. Fischer, B. Hilpert, S. Schubert, M. Gutensohn, A. Weber, and U.I. Flügge (1998) Plant Cell **10**: 105-117.
5 Weber, A., Menzlaff, E., Arbinger, B., Gutensohn, M., Eckerskorn, C., and Flügge, U. I. (1995) *Biochem.* **34**, 2621-2627
6 Klingenberg M., (1989) Arch. Biochem. Biophys. **270**: 1-14.
7 Heldt H.W. (1969) FEBS Lett. **5**: 11-14.
8 Schünemann D., S. Borchert, U.I. Flügge, H.W. Heldt (1993) Plant Physiol. **103**: 131-137.
9 Kampfenkel K., T. Möhlmann, O. Batz, M. van Montagu, D. Inzé, and H.E. Neuhaus (1995). FEBS-Lett. **374**: 351-355.
10 Winkler, H. H. (1976) J. Biol. Chem. **251**: 389-396.
11 Williamson L.R., Plano G.V., Winkler H.H., Krause D.C., and D.O. Wood (1989) Gene **80**: 269-278.
12 Krämer R. (1994) Biochim. Biophys. Acta **1185**: 1-34.
13 Krause D. C., Winkler, H. H., and Wood, D. O. (1985) Proc. Natl. Acad. Sci. USA **82**: 3015-3019
14 Neuhaus H.E., E. Thom, T. Möhlmann, M. Steup, and K. Kampfenkel (1997) Plant J. **11**: 73-82.
15 Tjaden, J., C. Schwöppe, T. Möhlmann, and H.E. Neuhaus (1998) J. Biol. Chem. **273**: 9630-9636.
16 Möhlmann, T., J. Tjaden, C. Schwöppe, H.H. Winkler, K. Kampfenkel, and H.E. Neuhaus (1998) Eur. J. Biochem. **252**: 353-359.
17 Huber S.C., Edwards G.E. (1976) Biochim. Biophys. Acta **440**: 675-687.
18 Blanke M.M., F. Lenz (1989) Plant, Cell and Environ. **12**: 31-46.
19 Quick P. and H.E. Neuhaus *In:* A Molecular Approach to Primary Metabolism in Plants. C.H. Foyer ed., pp. 41-61, Taylor & Francis, London, UK.
20 Batz O., R. Scheibe and H.E. Neuhaus (1995). Planta **196**: 50-57.
21 Neuhaus H.E., M. Stitt (1990) Planta **182**: 445-454.
22 Neuhaus H.E. and N. Schulte (1996) Biochem. J. **318**: 945-953.
23 Smith, A.M., K. Denyer, and C. Martin (1997) Ann. Rev. Plant Physiol. Plant Mol. Biol. **48**: 67-87
24 Visser, R.G.F., Somhorst I., Kuipers, G.J., Ruys, N.J., Feenstra W.J., and Jacobson., E., (1991) Mol. Gen. Genet. **225**: 289-296.
25 Kuipers A.G.J., E. Jacobson, and R.G.F. Visser (1994) Plant Cell **6**: 43-52.

LIGHT-DEPENDENT REGULATORY MECHANISMS ACTING ON PHOTOSYNTHETIC PHOSPHOENOLPYRUVATE CARBOXYLASE IN C₄ PLANTS

J. Vidal, S. Rydz *, S. Coursol, J. Grisvard, J.-N. Pierre
Institut de Biotechnologie des Plantes, Université de Paris-Sud, Bâtiment 630, 91405 Orsay-Cedex, France.* Institute of Experimental Plant Biology, University of Warsaw, Pawinskiego 5a, 02-106 Warsaw, Poland.

Keywords: CO_2 uptake, C metabolisms, kinases, phosphorylation/dephosphorylation, gene regulation, signal transduction.

1. Introduction

Phosphoenolpyruvate carboxylase (EC 4.1.1.31; PEPC) catalyzes the β-carboxylation of PEP by HCO_3^- in the presence of a divalent cation. PEPC isozymes are widely distributed in plant tissues in which they are involved in a variety of physiological contexts. In C₄ plants, a specific isoform (C₄ PEPC) plays a key role in the primary CO_2 fixation in the photosynthesis pathway . C₄ PEPC is subject to a light-dependent transcriptional control resulting in the accumulation of high amounts of the corresponding protein in the mesophyll cell cytosol during greening of the etiolated C₄ leaf. In vitro studies have shown that the enzyme activity is regulated by photosynthesis-related metabolites, e.g., feedback inhibition by L-malate and allosteric activation by sugar-P. During the last decade, a posttranslational process, i.e., reversible phosphorylation, acting on C₄ PEPC has been established and shown to modulate both its functional and regulatory properties. The main purpose of the work reported here was to explore the light-dependent mechanisms which control the biosynthesis and covalent modification of C₄ PEPC.

2. Results and Discussion

2.1 Transcriptional control during leaf greening

In Sorghum, C₄ PEPC is nuclear encoded by a small multigene family (1). Three PEPC genes, namely SvC3, SvC3R and SvC4 have been cloned and sequenced, including 5' flanking sequences and 3' untranslated regions, providing specific probes for the study of individual gene expression. Each of them exhibited specific expression patterns: SvC3 encodes a housekeeping PEPC isoform and is constitutively transcribed at a low level in both roots and leaves; SvC3R is also transcribed in these organs, but its expression is stimulated in roots by NH_4^+ ions (2). In contrast, SvC4 is highly and specifically expressed in leaf in a light-dependent and tissue-specific (mesophyll cells) manner (3). "Run on" experiments showed that transcriptional activity of this gene was gained during the greening process of Sorghum leaves, via the photoreceptor phytochrome (3). This was correlated with the consistent accumulation of corresponding C₄ PEPC transcripts and protein in the mesophyll-cell cytoplasm (3). In an attempt to understand further the molecular mechanism governing SvC4 expression, nuclear proteins (PCC4-AT) binding to an AT-rich box of the SvC4 promoter were detected by gel shift experiments. The specificity of protein-DNA interactions was proved by a competition binding assay using

G. Garab (ed.), Photosynthesis: Mechanisms and Effects, Vol. V, 3653–3658.
© 1998 Kluwer Academic Publishers. Printed in the Netherlands.

an excess of cold, specific DNA probe. Binding activity of PCC4-AT was detected only with protein extracts from etiolated leaves, or corresponding nuclei, in which *SvC4* transcription is blocked. During leaf greening, loss of binding activity was found to be dependent upon phytochrome, as suggested by the classical photoreversibility test (R/FR) and correlated with the activation of *SvC4* transcription and accumulation of C_4 PEPC. The results support the view that PCC4-AT acts as a *SvC4* repressor in the etiolated leaf. *In vitro*, PCC4-AT binding activity was lost after incubation of nuclear protein extracts in the presence of ATP or GTP, but was maintained if K252a (a protein-kinase inhibitor) was added together with the nucleotide; it was subsequently restored when excess EDTA (chelating Mg^{2+} ions) was present in the assay, thereby suggesting that the protein factor was active under its non-phosphorylated form. The hypothesis that the dynamic equilibrium between phosphorylated and dephosphorylated forms of PCC4-AT could be responsible for the regulation of *SvC4* gene *in vivo* was supported by the observation that protein-phosphatase inhibitors okadaic acid and microcystine are able to trigger the accumulation of C_4 PEPC in etiolated *Sorghum* leaf maintained in the dark.

2.2 Posttranslational control

2.2.1 C_4 PEPC as a target for phosphorylation in mesophyll cells

C_4 PEPC is reversibly modulated by a light-dependent, regulatory phosphorylation process in the mesophyll cell cytosol. This modification changes the enzyme's functional and regulatory properties, increasing its Vm, affinity for glucose-6-P, and decreasing its malate sensitivity, when measured at suboptimal pH (7.3) and concentration of PEP (2.5 mM). The identification and sequence of the phosphorylation site have been obtained from both *in vivo* and *in vitro* (^{32}P) labelled C_4 PEPC. From trypsin-generated peptides, a single (^{32}P) labelled nonapeptide was isolated and sequenced by automated Edman degradation. Comparing this amino-acid sequence (His-His-**Ser(P)**-Ile-Asp-Ala-Gln-Leu-Arg) to the nucleotide sequence of the corresponding cDNA made it apparent that the phosphorylation domain is very close to the protein's N-terminus and contains a single seryl residue (serine 15 and 8 for maize and *Sorghum* enzyme, respectively) whose phosphorylation state undergoes changes in response to light/dark transitions (4,5). A survey of the deduced amino-acid sequences of C_4 PEPC subunits from diverse sources led to the conclusion that this phosphorylation site is conserved in all plant enzymes, but is not found in cyanobacterial and bacterial PEPCs (4,5). Structure/function relationships of *Sorghum* C_4 PEPC were further investigated by recombinant protein technology and site-directed mutagenesis. Changing serine 8 to aspartate (S8D), or cysteine (S8C) showed that phosphorylation can be functionally mimicked by the introduction of a negative charge in the N-terminal domain of the protein (4,5). These results also provided the unequivocal proof that there is cause-and-effect relationship between phosphorylation of a single serine in C_4 PEPC and modification of its enzymatic properties.

2.2.2 The phosphorylation cascade

2.2.2.1 PEPC kinase is a calcium-independent enzyme

In vitro, C_4 PEPC can be phosphorylated by calcium-dependent and independent protein kinases from leaf extracts and by the regulatory subunit of mammalian PKA as well. The calcium-independent enzyme (PEPC-PK) has been shown to be i, the main PEPC kinase of the mesophyll cell protoplast (6) and ii, upregulated by light *in vivo* (4,5), and thus should be physiologically relevant. This low molecular mass (30-39 kD) protein kinase is unique in that its activity is not modulated by second messengers or by phosphorylation/dephosphorylation processes, but rather through rapid changes in its turnover rate or that of a regulatory protein factor. Indeed, the cytosolic protein synthesis inhibitor cycloheximide (CHX) could block *in vivo* the light-induced increase in activity of this converter enzyme. Recent data have indicated that the light-regulation of maize PEPC-PK should be at the level of translatable mRNA (7). *In vitro*, the enzyme is highly sensitive to pH, in the same range/way as its target C_4 PEPC and the phosphorylation reaction is controlled by the ratio malate (inhibitor)/sugar-P (activator) via a presumably indirect target effect (4,5,8).

2.2.2.2 *An increase in cytosolic pH is an early event*

The light-transduction chain leading to C_4 PEPC phosphorylation by its requisite, Ca^{2+}-independent protein kinase has been studied by flow cytometry, confocal microscopy and cellular pharmacology techniques using mesophyll protoplasts from crabgrass (*Digitaria sanguinalis*) (9). It was found that addition of a weak base, such as methylamine or ammonium chloride, induces both alkalization of cytosolic pH (pHc) and the upregulation of PEPC-PK/phosphorylation of C_4 PEPC in illuminated protoplasts. Similar effects have been observed when 3-phosphoglyceric acid (3-PGA) was used instead of the weak bases. A consistent hypothesis is that the protonated form (2⁻) of this photosynthesis-related metabolite, while being taken up by the chloroplast, increases mesophyll protoplast pHc. These result provided the first physiological evidence that a rise in cytosolic levels of 3-PGA and pHc in the mesophyll is an early step of the light-transduction chain controlling the phosphorylation status of C_4 PEPC.

2.2.2.3 *Mobilization of vacuolar calcium*

Pharmacological investigations using calcium channels blockers (TMB-8, verapamil, diltiazem) revealed that calcium release from the protoplast vacuole was another key event of the light-dependent transduction pathway (9). Moreover, phosphoinositide-specific phospholipase C (PI-PLC) inhibitors (neomycin sulfate, U73122) were shown to markedly inhibit C_4 PEPC phosphorylation in light+weak base-treated protoplasts. Consistently, PI-PLC activity and $InsP_3$ (PI-PLC reaction product) levels were found to be enhanced early during the time course of protoplast induction (S. Coursol, unpublished). These results supported the view that a phosphoinositide pathway may be operating in the C_4 PEPC phosphorylation cascade. It is not yet clear whether changes in pHc can modulate PI-PLC activity in the mesophyll.

2.2.2.4 *Calcium-dependent enzymes are involved*

The calmodulin antagonists (W5, W7, C48/80) could also inhibit PEPC phosphorylation in illuminated protoplasts in the presence of the weak base, thus suggesting that a calcium/calmodulin protein kinase, or a CDPK, is involved in the transduction of light signals (9). This finding was consistent with the abovementioned light-dependent increase in cytosolic calcium. However, since PEPC-PK is a calcium-independent protein kinase, these pharmacological data suggested that the cascade is multicyclic, including a Ca^{2+}-dependent step upstream from the ultimate PEPC-PK. Unravelling the role played by calcium-dependent enzymes in the *de novo* synthesis of PEPC-PK (or related protein factor) is a prospect for future studies.

Our current model on the transcriptional and posttranslational regulation of C_4 PEPC in the C_4 leaf is summarized in Figure 1.

3. Physiological significance

An IRGA (InfraRed Gaz Analyser)-based experiment showed that uptake of CHX by an excised *Sorghum* or maize leaf performing steady state photosynthesis caused a progressive and well-correlated decrease in its CO_2 assimilation rate and C_4 PEPC phosphorylation status (4,5). It was demonstrated that this inhibitor-based observation was not due to secondary detrimental effects like a perturbation of the Calvin cycle,

3656

photosynthetic electron flow, photophosphorylation or stomatal functioning. These results favoured the view that phosphorylation of the carboxylase has a critical regulatory impact on the overall functioning of C_4 photosynthesis.

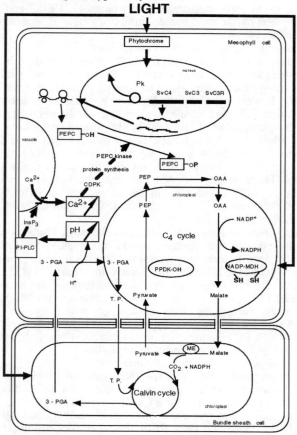

Figure 1. Transcriptional and posttranslational regulation of C_4 PEPC in the C_4 leaf. The transcription of nuclear *SvC4* gene is triggered by light, via phytochrome, during leaf greening. This process is correlated with phosphorylation-dependent unbinding of nuclear protein factors from the gene promoter. C_4 PEPC is subsequently accumulated in the mesophyll cell cytosol (PEPC-OH). The moving of 3-phosphoglyceric acid (3-PGA) from Calvin cycle in the bundle sheath to the chloroplasts of mesophyll cells results in the alkalization of pHc in the latter. Other components of the transduction pathway are: phosphoinositide-specific phospholipase C (PI-PLC), $InsP_3$, $InsP_3$-dependent tonoplast calcium channel, calcium, calcium-dependent protein kinase (CDPK) and finally calcium-independent PEPC kinase. Superimposed are the metabolic cross-talk between these two cell types and the corresponding enzymes of the C_4 photosynthesis pathway (PEPC, NADP-MDH (NADP-malate dehydrogenase), ME (Malic enzyme), PPDK (Pyruvate-Pi-dikinase), Calvin cycle).

In the illuminated leaf, C_4 PEPC is faced to mM concentrations of L-malate, high enough (10-20 mM) to severely block its catalytic activity (Ki for L-malate is close to 0.3 mM). *In vitro* experiments performed in the physiological range of pH and substrate/effector concentrations show that phosphorylation is regulatory, increasing substantially both the enzyme activity and its IC_{50} for L-malate. Interestingly, addition of the allosteric activator G-6-P to the phospho-enzyme antagonizes more efficiently the malate effect as it highly increases IC_{50} for this inhibitor (close to 15 mM). This finding may explain why C_4 PEPC phosphorylation i, allows the enzyme to cope with high, physiological concentrations of L-malate and ii, is critical for the functioning of C_4 photosynthesis *in planta*, as observed in the IRGA/CHX-based experiment. The collective data also point to a major role for pHc in this regulatory process. A light-induced shift in H^+ concentration in the mesophyll cytosol would be coupled, via 3-PGA, to the rate of the Calvin cycle/light intensity, in the bundle sheath. 3-PGA-induced increase in pHc is susceptible to alter PEPC activity and its sensitivity to effectors directly, and indirectly, via phosphorylation, following a pH/transduction cascade-dependent increase in PEPC-PK activity. Finally, the interaction between covalent and metabolite control of C_4 PEPC is reciprocal. Indeed, it has been found that L-malate can decrease the phosphorylation rate of C_4 PEPC, presumably via a target effect; conversely, sugar-P are able to antagonize this negative influence. This metabolite control has the potential for fine-tuning the phosphorylation status of the enzyme. It is, thus, apparent that the metabolite and enzymatic protagonists in this regulatory network are interdependent and ultimately connected to light. C_4 PEPC's covalent modification by this highly complex regulatory mechanism ensures the required coordination of the two metabolic cycles involved in C_4 photosynthesis. In the dark, dephosphorylation of PEPC by a 2A-type protein phosphatase (10) and a decrease in pHc to its original level would account for a more complete deactivation of the enzyme, as needed in a physiological context.

The posttranslational modification of C_4 PEPC is obviously slower than its catalytic rate and, consequently, a hysteretic behavior of the enzyme is expected. In the present case, a hysteresis of C_4 PEPC would be of critical importance during the build-up and decline of photosynthetic PEP in the mesophyll cytosol.

These complex mechanisms involving light-dependent control of C_4 PEPC at both transcriptional and posttranslational levels allow to adjust the intercellular carbon flow according to the demand by the Calvin cycle and ensure efficient functioning and homeostasis of C_4 photosynthesis.

References

1 Crétin, C., Santi, S., Keryer, E., Lepiniec, L., Tagu, D., Vidal, J and Gadal, P. (1991) Gene, 99, 87-94

2 Lepiniec, L., Vidal, J., Chollet, R., Gadal, P. and Crétin, C. (1994) Plant Sci. 99, 111-124

3 Thomas, M., Crétin, C., Vidal, J., Keryer, E., Gadal, P. and Mösinger, E. (1990) Plant Sci. 69, 65-78

4 Chollet, R., Vidal, J. and O'Leary, M.H. (1996) Annu. Rev. Plant Physiol. Plant Mol. Biol. 47, 273-298

5 Vidal, J and Chollet, R. (1997) TIPS. 2, 230-237

6 Nhiri, M., Bakrim, N., Pacquit, V., El Hachimi-Messouak, Z., Osuna, L. and Vidal, J. (1998) Plant Cell Physiol. 39, 241-246

7 Hartwell, J., Smith, L., Wilkins, M.B., Jenkins, G.I. and Nimmo, H. (1996) Plant J. 10, 1071-1078

8 Echevarria, C., Pacquit, V., Bakrim, N., Osuna, L., Delgado, B., Arrio-Dupont, M. and Vidal, J. (1994) Arch. Biochem. Biophys. 315, 425-430

9 Giglioli-Guivarc'h, N., Pierre, J-N., Brown, S., Chollet, R., Vidal, J. and Gadal, P. (1996) Plant Cell, 8, 573-586

10 Carter, P.J., Nimmo, H.G., Fewson, C.A. and Wilkins, M.B. (1990) FEBS Lett. 263, 233-236

MUTANTS OF C$_4$ PHOTOSYNTHESIS

R.C. Leegood[1], K.J.Bailey[1], R.J. Ireland[2], L.V. Dever[3] and P.J. Lea[3]

[1]Research Institute for Photosynthesis, Sheffield University Sheffield, S10 2TN, U.K. [2] Biology, Mount Allison University, Sackville, NB, E4L 1G7, Canada. [3] Biological Sciences, Lancaster University, Lancaster, LA1 4YQ, U.K.

Key words: amino acid, C metabolism, CO$_2$ concentration, N metabolism, PEPC, photorespiration,

1.Introduction

The method of screening for mutants was based on that originally described by Somerville and Ogren, who isolated a range of mutants of *Arabidopsis thaliana* (1) that lacked key enzymes of the photorespiratory carbon and nitrogen cycle (2). Somerville and Ogren argued that in C$_3$ plants grown at elevated concentrations of CO$_2$, glycollate-2-P formation would be prevented due to the inhibition of the oxygenase function of Rubisco. Thus there would be no flux of carbon and nitrogen through the photorespiratory cycle and any enzyme deficiencies would not be detrimental. Mutant plants would grow normally in elevated CO$_2$ but would exhibit severe stress symptoms when exposed to ambient air. Utilising this principle, a total of seven barley mutants (plus one double mutant) lacking enzymes of the photorespiratory cycle, have been isolated at Lancaster and Rothamsted (3). More recently heterozygous barley mutants containing enzyme activity (e.g. glutamine synthetase, glutamate synthase and glycine decarboxylase) varying from 40 to 100% have been studied in detail (4,5).

Applying a similar screen to that described above, we predicted that mutants of C$_4$ plants lacking an enzyme of the C$_4$ photosynthetic cycle would only be able to grow in an atmosphere of elevated CO$_2$. This prediction was based on the observations that bundle sheath cells have a low permeability to CO$_2$, but that this barrier can be overcome by high external CO$_2$ concentrations (6). In our initial experiments, batches of 100-200 azide mutagenised seed of *Amaranthus edulis* were grown in a glasshouse maintained at a CO$_2$ concentration of 7000 µmol mol^{-1} for two weeks and plants showing any abnormal features were discarded. The seedlings were then transferred to an identical glasshouse at ambient CO$_2$ and examined carefully every day. Plants that grew slowly or showed chlorosis were immediately transferred back to the glasshouse at 7000 µmol mol^{-1} CO$_2$, and allowed to grow for a further 2-3 weeks. Those that recovered and produced new healthy green leaves were screened for enzymes of the C$_4$ photosynthesis cycle and soluble amino acid content.

2. A mutant lacking PEP carboxylase

The first mutant that was positively identified was LaC$_4$ 2.16, which was shown to contain only 5% of the normal wild type PEP carboxylase activity. Western blot analysis indicated that the mutant lacked the major leaf PEP carboxylase protein, but that a minor

G. Garab (ed.), Photosynthesis: Mechanisms and Effects, Vol. V, 3659–3664.
© 1998 *Kluwer Academic Publishers. Printed in the Netherlands.*

polypeptide of higher molecular mass was still present. Light microscope analysis of immunogold labelled sections of leaves of the mutant LaC_4 2.16, confirmed that the PEP carboxylase protein was absent from the mesophyll cells of the leaf. Two PEP carboxylase proteins were detected in the roots, stems, petioles and flowers of the wild type A. edulis, which were also present at the same concentration in the mutant LaC_4 2.16 (7,8).

In order to establish the molecular basis of the mutation, Grisvard and colleagues (9) isolated poly (A^+) RNA from wild type and mutant leaves and the $[^{35}S]$-methionine labelled translation products were subjected to immunoprecipitation. Polyclonal antisera raised against full length sorghum PEP carboxylase recognised a predominant in vitro translation product that was 9kDa shorter than the normal PEPC protein. This 100kDa PEP carboxylase polypeptide was recognised by N-terminal-specific antisera, but not by C-terminal-specific antisera, suggesting that the mutant protein lacked the C-terminal domain. As the truncated form of the PEP carboxylase polypeptide was not detected in the mutant leaves, it must be assumed that it is subject to proteolysis immediately after synthesis. Sequence analysis showed that there was a point mutation at the 3' end of intron 9 of the C_4 PEP carboxylase gene, where the G of the AG splice site had been changed to A. Such a mutation prevented the normal splicing of intron 9 and induced a shift in the intron/exon boundary to the next AG sequence. At one such AG sequence, the splicing introduced a frame shift that lead to the production of a truncated polypeptide of 100kDa, due to the presence of a stop codon in the new reading frame. From the in vitro translation results described above, this latter inaccurate splicing system would appear to be operating in the mutant (9)

The original LaC_4 2.16 plant was grown to maturity in elevated CO_2 and both self-fertilised homozygous seed and backcrosed F_1 heterozygous seed were obtained. Analysis of the self-fertilised F_1 plants indicated that the mutation segregated in a normal Mendelian fashion in the F_2 generation. Homozygous mutant F_2 seedlings containing less than 10% of the wild type PEP carboxylase, activity grew very slowly in air and never reached a hight greater than 10 cm even after 4 months, although the leaves were pale, they did survive and were not totally chlorotic. Heterozygous F_1 plants containing approximately 50% of the wild type PEP carboxylase activity appeared to grow normally in air, however a comparison of the dry weights of the root, shoot and seed indicated that there was a reduction in the biomass of approximately 15% (7).The reduction of PEP carboxylase activity had no effect on the chlorophyll and protein content of the leaf nor on the activities of a number of enzymes involved in C_4 photosynthesis (10).

The rate of photosynthetic CO_2 assimilation by the homozygous mutant LaC_4 2.16 at ambient C_i was close to zero, which is in agreement with the data obtained by Brown (11) using an inhibitor of PEP carboxylase. However the rate displayed a linear response to increasing C_i, up to the maximum concentration tested. This response suggests that photosynthetic CO_2 assimilation in the mutant is directly dependent on the rate of CO_2 diffusion into the bundle sheath cells, and explains how the mutant is able to grow at the elevated atmospheric CO_2 concentration of 7000 $\mu mol\ mol^{-1}$. The heterozygous plants containing approximately 50% of PEP carboxylase activity exhibited a similar A/C_i response curve to that shown by the wild type, but the maximum rate of CO_2 assimilation was approximately 10% lower.

Photosynthetic CO_2 assimilation by wild type A. edulis has an optimum O_2 concentration of approximately 5kPa, below this optimum the decrease in rate is associated with lower PSII activity, whereas above the optimum, photorespiration accounts for the inhibition of photosynthesis (12). In the homozygous mutant, the optimum O_2 concentration was reduced to 1-2 kPa, a value normally found for C_3 plants. Maroco *et al.* (13) concluded that the high O_2 optimum of C_4 photosynthesis is linked to the O_2-dependent production of ATP by pseudocyclic/cyclic phosphorylation required for the synthesis of PEP and this requirement is absent in the LaC_4 2.16 mutant.

The degree of control exerted by PEP carboxylase on photosynthetic CO_2 assimilation was estimated using the principles of Metabolic Control Analysis, as originally proposed by Kacser and Burns (13). At ambient C_i (130μmol mol^{-1}), the flux-control coefficient (C^J) for PEP carboxylase increased from 0.35 in the wild type to 0.49 at 55% PEP carboxylase. At moderate C_i (60 μmol mol^{-1}), C^J increased from 0.65 in the wild type to 0.77 at 55% PEP carboxylase, whilst at low C_i (30 μmol mol^{-1}) C^J was 0.70 in the wild type increasing to 0.81 at 55% PEP carboxylase. The control exerted by PEPC on photosynthetic flux in *A. edulis* is therefore relatively high and comparable to the values obtained by Furbank *et al.* (14) for pyruvate P_i dikinase and Rubisco using antisense constructs in transgenic *Flaveria bidentis*

For plants containing PEPC activities below 55%, there was an upturn in the rate of CO_2 assimilation (10). This upturn was observed in each of the three years of study, using independently generated heterozygous plants. These findings suggest that a compensation mechanism begins to operate once PEPC activity falls below a critical amount. A similar response has been seen in the release of ammonia in barley plants which contain decreased amounts of glutamine synthetase activity (4).

The malate sensitivity of PEP carboxylase from the wild type and heterozygous plants was compared over a physiological concentration range. PEP carboxylase isolated from the leaf of a fully illuminated heterozygous plant was shown to be less sensitive to malate inhibition than the corresponding wild type enzyme over the concentration range 0.3-1mM. The apparent increased in phosphorylation state of the enzyme isolated from the heterozygous plant was confirmed by ^{32}P-labelling and subsequent autoradiography of the immunoprecipitated PEP carboxylasae protein (15). In plants containing less than 55% of the wild type PEP carboxylase activity, there was evidence of an upturn in the concentration of compounds that would activate PEP carboxylase, these included triose phosphates, glucose-6-phosphate, fructose-6-phosphate, glycine and serine. It has been suggested that as the photorespiratory rates increase, due to reduced decarboxylation in the bundle sheath cells, the glycine and serine are able to move into the mesophyll cells and activate PEP carboxylase (10).

A mutant lacking NAD-malic enzyme

The mutant LaC_4 73, was shown to contain approximately 5% of the NAD-ME activity normally found in wild type *A. edulis* plants (16). When the leaf proteins of the mutant and wild type were subject to Western blot analysis following SDS-PAGE, using antisera raised against both the α and β subunits (17), both subunits were clearly visible in LaC_4 73 . To investigate the aggregation state of NAD-ME, Western blot analysis was also carried out following native gradient PAGE, utilising the antisera described above and that raised against the α-subunit only (18). Both antisera recognised proteins in the octameric (480 kDa), tetrameric (240 kDa) and dimeric (120kDa) regions of the blot. NAD-ME was present mostly in the octameric form in the wild type leaf tissue, whether isolated from illuminated or darkened leaves. The leaf extracts of the homozygous mutant LaC_4 73, although still containing predominantly octamers, exhibited a higher proportion of the dimeric NAD-ME protein. It is possible that this higher proportion is due to the low rate of photosynthetic CO_2 fixation. It is also possible that the low activity of NAD-ME in LaC_4 73 is due to a direct amino acid change in the active site of the enzyme protein or the site(s) involved in the binding of metabolite activators, e.g. CoA, fumarate or fructose 1,6-bisphosphate.

As described previously for the mutant lacking PEP carboxylase activity, LaC_4 73 was maintained in an atmosphere of elevated CO_2 and back-crossed to the wild type to produce F_1 seed. Analysis of plants from the F_2 generation indicated that the mutation segregated in a normal Mendelian fashion and that leaves from heterozygous plants contained approximately 50% of the wild-type NAD-ME activity. There was no evidence of any pleiotropic effect of the deficiency on all the key enzymes of the C_4 pathway tested. Biomass measurements indicated that there was no difference in the growth rate in air between the wild type and hetrozygous plants, whether based on total plant dry weight or leaf area. However the homozygous mutant grew very slowly in air and the total dry weight of the plants was only 1% and the leaf area 4% of the wild type values (15).

The A/C_i curves obtained for the homozygous LaC_4 73 mutant confirmed the biomass growth data, in that the rate of photosynthetic CO_2 assimilation at ambient C_i was very low (8). However, as was seen for the PEP carboxylase mutant, the rate increased in a linear fashion with elevated C_i. The heterozygous plants exhibited CO_2 assimilation rates that were 10% lower than the wild type at high C_i, however at ambient C_i values the rates were the same even up to light intensities of 1700 μmol m^{-2} s^{-1}. These data for the heterozygous plants indicate that NAD-ME has little control over the rate of C_4 photosynthsis and that in the wild type, the enzyme will have a very low flux-control coefficient (13).

Bundle sheath strands isolated from the wild type *A. edulis* were able to oxidise malate in the dark and convert the pyruvate formed to alanine by the action of glutamate:pyruvate aminotransferase, as would be predicted from the data for other NAD-ME type C_4 plants (19)). Although bundle sheath strands isolated from the homozygous mutant $LaC_4$73 were able to oxidise malate at rates only slightly lower the wild type, the rate of alanine synthesis was less than 10% of the wild type, confirming that the NAD-ME protein was unable to convert malate to pyruvate *in vivo*. The changes in concentration of metabolites in the leaves of the mutant following transfer from elevated CO_2 to air are consistent with a lack of NAD-ME activity . First the substrate malate was elevated in the mutant , whereas the products pyruvate (and alanine) were much lower, whether in high CO_2 or air. Secondly these differences became much more pronounced upon transfer to air with a considerable accumulation of malate in the homozygous mutant and a marked deficiency in pyruvate and alanine. The accumulation of glycine and serine in the homozygous mutant are consistent with the view that the rate of photorespiration is enhanced, due to the lack of CO_2 production by NAD-ME in the bundle sheath cells.

Mutants that accumulate glycine.

Five different mutants of *A. edulis* have been isolated that accumulate glycine following exposure to air. Of these LaC_4 2.11, LaC_4 25, and LaC_4 30 have been the most extensively characterised. Self fertilised seed of the original mutants can be germinated in an atmosphere of 7000 μmol mol^{-1} CO_2, where the plants grow at a slower rate than the wild type, but they are not viable in air. Unlike the two mutants described previously deficient in PEP carboxylase and NAD-ME, the glycine accumulating mutants exhibit severe symptoms of stress with a rapid loss of chlorphyll and extensive bleaching after transfer to air from elevated CO_2, the onset of which is dependent upon light intensity. The concentration of glycine in the leaves of the mutants can reach over 30 μmol g^{-1} fresh weight after 24 hours and may constitute over 80% of the total soluble nitrogen, which correlates with a dramatic fall in the concentration of all other soluble amino acids. The rates of photosynthetic CO_2 assimilation of the mutants decreased dramatically in the first hour of exposure to air and fell to zero after 6 hours.

The properties of the mutants are similar to those reported for photorespiratory mutants of barley that lack proteins of the glycine decarboxylase complex (5). Western blot analysis of the four proteins of the glycine decarboxylase complex (P, H. T and L), indicated that none of the mutants exhibited a severe loss of any of the specific proteins. The activity and protein content of serine hydroxymethyl transferase were also similar in the wild type and mutant plants. When bundle sheath strands were isolated in the same manner as described previously for the NAD-ME mutant (19), wild type bundle sheath strands had the capacity to oxidise both glycine and malate. However the bundle sheath strands isolated from all the mutants were unable to oxidise glycine, whilst still maintaining the capacity to oxidise malate at similar rates to the wild type. The results confirm that the accumulation of glycine is due to a loss of glycine metabolising capacity in the bundle sheath cells, probably due to a lesion in the glycine decarboxylase complex.

The magnitude of the rate of photorespiratory CO_2 release in C_4 photosynthesis has been the subject of considerable debate, due to possible reassimilation before it is lost to the atmosphere. Working on the hypothesis that the glycine accumulating in the mutants is derived from the photorespiratory carbon and nitrogen cycle, we attempted to use the mutants to establish the rate of photorespiration in *A. edulis*. Ammonia and CO_2 are released at the same rate in the conversion of glycine to serine during photorespiration. If leaves are treated with an inhibitor of glutamine synthetase, such as phosphinothricin (PPT), there is a rapid accumulation of ammonia in a wide range of C_3 plants and C_4 plants , which may be derived from a number of metabolic reactions (20). The rate of PPT dependent ammonia accumulation in the leaves of wild type *A. thaliana* was reduced by 60%, either by incubation in 7000 μmol mol^{-1} CO_2 or treatment with the glycine decarboxylase inhibitor aminoacetonitrile (AAN), indicating that this fraction was derived from photorespiration. The addition of glycine to wild type leaves in the presence of PPT stimulated the accumulation of ammonia, indicating that there is spare glycine decarboxylase capacity within the bundle sheath mitochondria. However when the leaves of the glycine accumulating mutants were treated with PPT, ammonia accumulation was much lower than detected in the wild type and was not stimulated by glycine or inhibited by AAN . By comparing the rates of PPT dependent ammonia accumulation in the wild type and mutants in the presence and absence of AAN, it proved possible to calculate a minimum value of 6% for the rate of photorespiration as expressed as a proportion of net photosynthesis (21). This value is somewhat higher than the photorespiratory rate of 3%, determined by Jenkins (6).

1.Somerville, C. R. (1986) Ann. Rev. Plant Physiol. 37, 467-507

2. Keys, A. J., Bird, I.F., Cornelius, M.J., Lea, P.J., Wallsgrove, R.M. and Miflin, B.J. (1978) Nature 275, 741-743.

3. Leegood, R. C., Lea, P.J., Adcock, M.D. and Häusler, R.E. (1995) J. Exp. Bot. 46, 1397-1414.

3664

4. Häusler, R.E., Bailey, K.J., Lea, P.J. and Leegood,R.C. (1996) Planta 200, 388-396.

5. Wingler, A., Lea, P.J. and Leegood, R.C. (1997) Planta 202,171-178

6. Jenkins, C.L.D. (1997) Aust. J. Plant Physiol. 24, 543-547

7. Dever, L.V., Blackwell, R.D., Fullwood, N.J., Lacuesta, M., Leegood, R.C., Onek, L.A., Pearson, M.A. and Lea, P.J. (1995) J. Exp. Bot. 46, 1363-1376

8. Dever, L.V., Bailey, K.J., Lacuesta, M., Leegood, R.C. and Lea,P.J. (1996) Comptes Rendus Acad. Sci. III. La Vie. 319, 951-959

9. Grisvard, J., Keryer, E., Takvorian, A., Dever, L.V., Lea, P.J. and Vidal, J. (1998) Gene 213, 31-35

10. Bailey, K.J., Battistelli, A., Dever, L.V., Lea, P.J. and Leegood, R.C. (1999) (submitted)

11. Brown, R.H. (1997) Aust. J. Plant Physiol. 24, 549-554

12. Maroco, J.P., Ku, M.S.B. and Edwards, G.E. (1997)Plant Cell Environ. 20, 1525-1533

13. Maroco, J.P., Ku, M.S.B., Lea, P.J., Dever, L.V., Leegood, R.C., Furbank, R.T. and Edwards, G.E. (1998) Plant Physiology 116, 823-832

14. Furbank, R.T., Chitty, J.A., Jenkins, C.L.D., Taylor, W.C., Trevanion, S.J., von Caemmerer, S. and Ashton, A.R. (1997) Aust. J. Plant Physiol. 24, 477-485

15. Dever, L.V., Lea, P.J., Bailey, K.J. and Leegood, R.C. (1997) Aust. J. Plant Physiol. 24, 469-476

16. Dever, L.V., Pearson, M., Ireland, R.J., Leegood,R.C. and Lea,P.J. (1998) Planta (in press)

17. Murata, T., Ikeda, J.L. and Ohsugi, R. (1989). Plant Cell Physiol. 30, 429-437

18. Long, J. J., Wang, J-L. and Berry, J.O. (1994) J. Biol. Chem. 269, 2827-2833

19. Agostino, A., Heldt, H.W. and Hatch, M.D. (1996) Aust. J.Plant Physiol.23, 1-7

20. Lea, P.J. (1991) in Topics in Photosynthesis (Baker N.R. and Percival M., eds.) pp. 267-298, Elsevier, Amsterdam, The Netherlands

21. Lacuesta, M., Dever, L.V., Muñoz-Rueda, A. and Lea, P.J. (1997) Physiol. Plant. 99, 447-455

OPERATION OF COUPLED AND NON-COUPLED PATHWAYS OF MITOCHONDRIAL ELECTRON TRANSPORT IN PHOTOSYNTHETIC PLANT CELL

Abir U. Igamberdiev[1,2], Natalia V. Bykova[1,2] and Per Gardeström[1]
[1] Department of Plant Physiology, University of Umeå, S-901 87 Umeå, Sweden
[2] Permanent address: Department of Plant Physiology and Biochemistry, Voronezh University, Voronezh 394693, Russia

Key words: alternative electron transport, glycine, NAD(P)H dehydrogenase, photorespiration, protoplasts.

1. Introduction

Mitochondrial electron transport plays a central role in regulation of the redox state and energetics of photosynthetic plant cells. During photosynthesis in C_3 plants, in addition to oxidation of the tricarboxylic acid cycle substrates, which commonly decreases to more or less extent in the light [1-3], an intensive flux of glycine is provided by photorespiratory reactions. Glycine oxidation is important to be maintained at very high level to provide continuous operation of photorespiratory cycle and prevent accumulation of toxic photorespiratory products, such as glyoxylate and its derivatives.

The electron transport chain of plant mitochondria contains, in addition to the complexes common to all organisms, a number of pathways non-coupled to energy conservation. These are the alternative oxidase bypassing during electron flow from ubiquinone to oxygen [4] and the four NAD(P)H dehydrogenases oxidizing external and internal NADH and NADPH correspondingly bypassing complex I [5,6]. They all serve to maintain the redox level of the plant cell and to prevent the formation of reactive oxygen species. They may provide oxidation of photosynthetically formed substrates in an 'overflow' manner, however being highly regulated. The alternative oxidase is activated by reduction and ketoacids, mechanisms of regulation of alternative NAD(P)H dehydrogenases have to be elucidated. There are indications that these non-coupled pathways play an important role during photosynthesis, but information in this field is not sufficient.

The goal of this paper is to investigate operation of coupled and non-coupled pathways of mitochondrial electron transport in connection with photosynthetic function of plant cells. In particular we wanted to examine if photorespiration had a specific effect on the partitioning of electrons to non-phosphorylating pathways of mitochondrial electron transport.

G. Garab (ed.), Photosynthesis: Mechanisms and Effects, Vol. V, 3665–3670.
© 1998 *Kluwer Academic Publishers. Printed in the Netherlands.*

2. Material and Methods

Objects of investigation were mitochondria and protoplasts of barley (*Hordeum vulgare* L. var. Gunilla, Svalöf).

Mitochondria were isolated from green and etiolated leaves of 8-day-old plants according to Lernmark et al. [7] and incubated in 0.4 M mannitol, 10 mM KCl, 10 mM KH_2PO_4, 0.5 mM $MgCl_2$, 0.1% BSA, 10 mM MOPS-KOH (pH 7.2) in the presence of 0.5 mM NAD^+ and 2 mM ADP. Exogenous $[1-^{14}C]$glycine (at concentration 10 mM) was applied after pre-incubation with inhibitors for 5 min. The reaction was followed in 1 ml volume for 10 min and stopped by addition of 0.1 ml 2 N H_2SO_4. $^{14}CO_2$ evolved was absorbed by 20% NaOH. The inhibitors were supplied in the following concentrations: 20 μM rotenone (the inhibitor of complex I), 1 mM salicylhydroxamic acid (SHAM, the inhibitor of alternative oxidase), 5 μM antimycin A (the inhibitor of complex III). Malate was supplied at concentration 10 mM.

Protoplasts were isolated from green leaves as described earlier [8] and incubated in assay medium (0.25 M sucrose, 0.25 M sorbitol, 10 mM KCl, 0.5 mM $MgCl_2$, 0.06% BSA, 0.2% PVP, 10 mM HEPES, pH 7.2) in darkness or upon illumination at 500 μmol quanta m^{-2} s^{-1}. For measurements in 'no CO_2' a CO_2-free medium was prepared and for measurements at high CO_2 10 mM $NaHCO_3$ was added. The inhibitors [20 μM rotenone for complex I, 1.5 mM SHAM for the alternative oxidase, 25 μM antimycin A for complex III, 10 mM aminoacetonitrile (AAN) for glycine decarboxylase complex, 10 mM hydroxypyridinemethane sulfonate (HPMS) for glycolate oxidase] were pre-incubated with protoplasts in darkness at room temperature for 4-7 min prior to illumination.

Separation to subcellular compartments was performed using the apparatus of rapid fractionation based on the principle of membrane filtration [9]. For providing moderate photorespiratory conditions at 'low CO_2' (50% of maximal CO_2 fixation) 0.2 mM $NaHCO_3$ was added to the medium. For strong photorespiratory conditions CO_2-free assay medium was prepared. Protoplasts were pre-incubated with inhibitors in the dark during 7 min and then illuminated at 500 μmol quanta m^{-2} s^{-1} for 5 min for obtaining conditions of steady-state photosynthesis.

Chlorophyll concentration was measured according to [10]. ATP was determined by the firefly luciferase method [9]. ADP was converted to ATP by pyruvate kinase.

3. Results

3.1. Glycine decarboxylation by isolated mitochondria

In mitochondria isolated from green leaves both SHAM and antimycin A inhibited glycine decarboxylation about 50% (Table 1). Rotenone revealed only about 20% inhibition. In mitochondria from etiolated leaves the suppression by antimycin A was stronger (only about 20% of initial activity remained present), whereas inhibition by SHAM was not observed. Rotenone revealed strong inhibitory effect (more than 50%). Malate stimulated glycine decarboxylation strongly in mitochondria from green leaves and slightly decreased its rate in mitochondria from etiolated leaves.

Table 1. Decarboxylation rates (nmol CO_2 min^{-1} mg^{-1} protein) of [1-^{14}C]glycine by mitochondria isolated from green and etiolated barley leaves

	Green	Etiolated
Control	182 ± 11	66 ± 4
Rotenone	141 ± 14	31 ± 3
SHAM	98 ± 12	67 ± 7
Antimycin A	94 ± 5	14 ± 3
Malate	342 ± 16	55 ± 7

3.2. O_2 consumption by protoplasts at high and very low CO_2

We examined the effect of inhibitors on protoplast respiration in darkness at high CO_2 (10 mM $NaHCO_3$) and in the absence of added bicarbonate (Table 2). An increase in respiration after illumination was observed. At high CO_2 it was more than 40%, and in the absence of bicarbonate applied 20%. During the initial 30 s it was 1.6-1.8 higher than in darkness which corresponds to the effect of light-enhanced dark respiration.

Rotenone had only a small effect on respiration after illumination at high CO_2 (10%). In the absence of bicarbonate added (CO_2-free medium) its effect was greater (50%). Before illumination it inhibited respiration by about 30% both in the presence and in the absence of bicarbonate. SHAM revealed greater inhibition of O_2 uptake after illumination compared to darkened protoplasts, especially when bicarbonate was absent. Upon illumination antimycin A had a smaller effect than SHAM on respiration (less than 10%) and its inhibition was even less than that before illumination.

The effect of AAN was small even after illumination (about 10%) and it was similar at high and very low CO_2. An effect of HPMS after illumination in photorespiratory conditions (35% inhibition) was observed which was stronger than the effect of AAN.

Table 2. The effect of inhibitors on the respiratory rate (μmol O_2 mg^{-1} Chl h^{-1}) of barley protoplasts in darkness before and after illumination (rates measured 1 min after light turned off)

	Before illumination		After illumination	
	High CO_2	No CO_2	High CO_2	No CO_2
Control	5.0 ± 0.5	5.4 ± 0.6	7.2 ± 0.7	6.5 ± 0.7
Rotenone	3.5 ± 0.4	4.1 ± 0.4	6.5 ± 0.8	3.2 ± 0.3
SHAM	4.0 ± 0.4	4.1 ± 0.2	5.0 ± 0.5	3.7 ± 0.5
Antimycin A	3.9 ± 0.4	4.2 ± 0.4	6.6 ± 0.2	5.9 ± 0.3
AAN	4.5 ± 0.3	4.9 ± 0.4	6.5 ± 0.6	5.8 ± 0.5
HPMS	4.4 ± 0.4	4.7 ± 0.3	6.6 ± 0.3	4.1 ± 0.3

3.3. ATP/ADP ratios in cellular compartments in photorespiratory conditions

Protoplasts were incubated in steady-state photosynthesis in 'strong' photorespiratory conditions (0% of photosynthetic O_2 evolution [CO_2-free buffer, 'no CO_2')] and 'moderate' photorespiratory conditions [50% of photosynthetic O_2 evolution (0.2 mM $NaHCO_3$, 'low CO_2')]. Rotenone had no effect on chloroplastic and cytosolic ATP/ADP ratios. In mitochondria it strongly decreased this ratio in the conditions of 'no CO_2' and 'low CO_2'. AAN lowered ATP/ADP ratio in the cytosol and mitochondria only in the conditions of low CO_2, but not in the absence of photosynthetic CO_2 evolution (Table 3).

Table 3. ATP/ADP ratios in chloroplasts, cytosol and mitochondria of barley protoplasts incubated in light in conditions of the absence of photosynthetic CO_2 evolution (no CO_2) and of 50% of saturating CO_2 evolution (low CO_2) in the presence of rotenone and aminoacetonitrile (AAN)

| | Chloroplasts | | Cytosol | | Mitochondria | |
	Low CO_2	No CO_2	Low CO_2	No CO_2	Low CO_2	No CO_2
Control	2.6 ± 0.2	4.7 ± 0.7	4.7 ± 0.5	7.3 ± 2.1	2.7 ± 0.4	6.3 ± 1.4
Rotenone	3.1 ± 0.1	3.7 ± 0.3	4.6 ± 0.2	6.4 ± 1.7	1.8 ± 0.9	2.2 ± 0.8
AAN	3.7 ± 0.3	5.1 ± 1.0	3.0 ± 0.3	6.6 ± 1.0	1.3 ± 0.3	7.0 ± 1.3
AAN+ Rotenone	4.0 ± 0.4	4.7 ± 0.6	3.3 ± 0.7	5.2 ± 0.8	2.1 ± 1.0	2.7 ± 0.6

4. Discussion

Glycine oxidation in mitochondria of photosynthetic tissue is coupled in more degree with cyanide-resistant and rotenone-resistant paths of electron transport contrary to etiolated leaves where these pathways are involved in much less extent (Table 1). This confirms previous data on the effect of inhibitors on O_2 consumption by mitochondria from green and etiolated barley leaves oxidizing glycine [11]. The pathways of electron transport non-coupled to energy conservation acquire importance in photosynthetic tissue particularly for rapid oxidation of photorespiratory glycine and preventing over-reduction of the photosynthetic cell. Stimulation of glycine decarboxylation by malate is also a feature of mitochondria from photosynthetic tissue. It may be connected with synergistic operation of GDC and malate dehydrogenase [12].

During photosynthesis under photorespiratory conditions glycine is believed to be the most important substrate for mitochondrial respiration [3]. The inhibitor of glycolate oxidase HPMS inhibited respiration of protoplasts after illumination in photorespiratory conditions ('no CO_2') more than the inhibitor of glycine decarboxylase AAN (35% versus 10%, Table 2). This may indicate that photorespiratory intermediates before glycine in the glycolate pathway contribute to respiration to some extent. The suppression of glycine oxidation by AAN resulted in very little inhibition of O_2 uptake even under photorespiratory conditions. This can be explained by the possible participation of alternative pathways of glycolate metabolism during inhibition of glycine decarboxylation: the direct non-enzymatic oxidation of glyoxylate by H_2O_2 is one possibility, another possibility is condensation of glyoxylate with succinate [13].

Electron transport via the alternative oxidase increases strongly after illumination particularly in photorespiratory conditions (Table 2) which reflects its role during oxidation of photorespiratory glycine. The pathway via rotenone insensitive dehydrogenase(s) reveals a more complex picture: in the experiments on protoplasts it was more active in non-photorespiratory conditions, although glycine decarboxylation revealed more rotenone-insensitivity in mitochondria isolated from green leaves. Experiments with rapid fractionation of protoplasts showed that rotenone in photorespiratory conditions had a significant effect on ATP/ADP ratios in mitochondria but not in the cytosol. This is probably explained by alterations of nucleotide transport across mitochondrial membrane in these conditions. For further elucidation of the role of rotenone-insensitive dehydrogenases during photosynthesis the information is needed about their regulatory properties.

Experiments on rapid fractionation of protoplasts (Table 3) reveal that in the conditions of steady-state photosynthesis at high light intensities the role of mitochondria may consist mainly in the prevention of the 'over-reduction' of the plant cell. In steady-state photosynthesis the role of rotenone-resistant pathway becomes more important, especially for glycine oxidation in photorespiratory conditions. When CO_2 is depleted in the medium, this oxidation becomes non-coupled with ATP production possibly passing via alternative oxidase. In these strong photorespiratory conditions ATP/ADP ratio is very high and the paths of mitochondrial electron transport non-coupled with energy conservation are most active. Therefore strong photorespiratory conditions, in which glycine flux is prevailing in mitochondria, lead to interruption of supply of cytosol with ATP. In moderate photorespiratory conditions in which CO_2 fixation is present, glycine oxidation supplies cytosol with ATP which is evident from the decrease of cytosolic ATP/ADP ratio under inhibition of glycine decarboxylase by AAN.

The involvement of paths non-coupled with ATP synthesis in strong photorespiratory conditions may provide the defence from formation of active oxygen species which are intensively produced when ubiquinone pool is in a highly reduced state [14]. Mechanisms providing connection of glycine oxidation with paths non-coupled with ATP synthesis may be important in the prevention of over-reduction in mitochondria and therefore of oxygen damage.

5. Conclusion

The results presented here highlight the flexibility of mitochondrial electron transport and photorespiratory carbon metabolism. The ability of plant mitochondria to switch between rotenone-sensitive and rotenone-resistant pathways and between the alternative oxidase and cytochrome oxidase makes up a very flexible system for coupling between electron transport and ATP production. Mitochondria of photosynthetic tissues acquired the mechanism of involvement of cyanide-resistant and rotenone-insensitive pathways during glycine oxidation in higher extent than during oxidation of other substrates which may play a role in redox regulation during photosynthetic metabolism and in ensuring of the continuous operation of the photorespiratory cycle.

References

1 Pärnik, T. and Keerberg, O. (1995) J. Exp. Bot. 46, 1439-1447
2 Gardeström, P. and Lernmark, U. (1995) J. Bioenerg. Biomembr. 27, 379-385
3 Krömer, S. (1995) Annu. Rev. Plant Physiol. Plant Molec. Biol. 46, 45-70
4 Vanlerberghe, G.C. and McIntosh, L. (1997) Ann. Rev. Plant Physiol. Plant Molec. Biol. 48, 703-734
5 Melo, A.M.P., Roberts, T.H. and Møller, I.M. (1996) Biochim. Biophys. Acta 1276, 133-139
6 Agius, S.C., Bykova, N.V., Igamberdiev, A.U. and Møller, I.M. (1998) Physiol. Plant. (in press)
7 Lernmark, U., Henricson, D., Wigge, B. and Gardeström, P. (1991) Physiol. Plant. 82, 339-344
8 Igamberdiev, A.U., Zhou, G., Malmberg, G. and Gardeström, P. (1997) Physiol. Plant. 99, 15-22
9 Gardeström, P. and Wigge, B. (1988) Plant Physiol. 88, 69-76
10 Bruinsma (1961) Biochim. Biophys. Acta 52, 576-578
11 Igamberdiev, A.U., Bykova, N.V. and Gardeström, P. (1997) FEBS Lett. 412, 265-269
12 Wiskich, J.T., Bryce, J.H., Day, D.A. and Dry, I.B. (1990) Plant Physiol. 93, 611-616.
13 Igamberdiev, A.U. and Kleczkowski, L.A. (1997) In Handbook of Photosynthesis (M.Pessarakli, ed.) pp. 269-279, Marcel Dekker, New York
14 Skulachev, V.P. (1996) Quart. Rev. Biophys. 29, 169-202

RELATIONSHIP BETWEEN PHOTOSYNTHETIC RATE AND TURNOVER OF THE D1 PROTEIN OF PSII

Belinda C. Morrison and Christa Critchley
Department of Botany, The University of Queensland
Brisbane QLD 4072, Australia

Key words : D1 turnover, gas exchange, electron transport, O_2 evolution, wheat, N metabolism

1. Introduction

The D1 protein is one of the photosystem II reaction centre proteins and plays an integral role in its structure and function. The D1 protein has been the focus of much attention because its rate of turnover, relative to its abundance, is extremely high (2). This turnover has been linked to high light photodamage of the protein (3). However, it has been demonstrated that high turnover rates occur at light intensities far below those which cause photoinhibition and a maximum turnover rate is seen at a threshold light intensity (4).

In a recent study by Russell et. al. (personnal communication) a positive correlation was observed between D1 turnover rates and CO_2 fixation rates. In this work CO_2 levels were altered to change photosynthetic capacities of leaves and compared with D1 turnover. Russell suggests that D1 turnover responds to positive feedback from the Calvin cycle.

The aim of our experiments was to provide further evidence for the link between D1 turnover and photosynthetic rate. *Triticum aestivum* plants were grown under identical, non-photoinhibitory light intensities and the nitrogen status of the plants was altered to yield plants with intrinsically different photosynthetic rates. The photosynthetic capacity of the + and -N plants were tested and their D1 synthesis rate was assessed.

2. Materials and Methods

2.1 *Growth conditions*

Triticum aestivum plants were grown in a temperature controlled (18/15°C day/night) glasshouse at ambient light intensities (max.1500 μmol m^{-2} s^{-1}). Five plants were grown per 10L pot in alluvial soil. After germination all plants were watered daily. Leaves were sampled in the 6th growth week and the youngest fully developed leaves were used for all measurements.

G. Garab (ed.), Photosynthesis: Mechanisms and Effects, Vol. V, 3671–3674.
© 1998 Kluwer Academic Publishers. Printed in the Netherlands.

2.2 *Experimental treatments*
Following germination control plants were fertilised with solution that contained nitrates and other necessary nutrients. The nutrient solution for the -N plants was identical except that it lacked nitrates (5). During the experimental week all leaves were sampled and various parameters measured.

2.3 *Labelling experiment*
Leaf segments were cut, the lower epidermis slightly punctured and segments floated on the labelling solution (3 mM bicarbonate, 0.2% Tween, 1 mCi ^{35}S-methionine [Trans ^{35}S label; ICN Biomedicals Inc., Irvine, Cal., USA; specific activity 42-48 TBq mmol^{-1}). Labelling period was three hours and samples were taken every 30 min. Uptake was conducted at 800 μmol m^{-2} s^{-1} because maximum photosynthetic rates were achieved at this light intensity.

2.4 *Chlorophyll a fluorescence*
Dark-adapted (10 min) chlorophyll fluorescence measurements of unlabelled leaf segments were made every 30 min prior to and during the labelling period, using a Plant Efficiency Analyser (Hansatech, King's Lynn, UK). The MINI-PAM fluorometer (Walz GmbH, Effeltrich, Germany) was used to calculate electron transport rates (ETR), also every 30 min during labelling.

2.5 *Thylakoid extractions and electrophoresis*
Eight and sixteen leaf segments from the control and -N treatments, respectively, were ground in frozen glass homogenisers in ice-cold extraction buffer (2.5 mM Tris/glycine, pH 8.5). The thylakoid membranes were spun down in an Eppendorf centrifuge at 14,000 rpm for 5 min, washed twice in washing buffer (0.15 M NaCl, 2.5 mM Tris/glycine, pH 8.5) and resuspended in 200 μL of extraction buffer. Chlorophyll concentrations were measured in 80% buffered acetone (6).

Thylakoid samples were solubilised (0.1 M Tris.HCL, pH 8.8, 5% glycerol, 3% LDS and 0.3% DTT), heated for 5 min at 70°C and loaded on an equal chlorophyll basis on a LiDS/15-23% polyacrylamide/4M urea gradient gel. Electrophoresis was carried out at 4°C. One gel was stained and the others were transferred to nitrocellulose membrane. The membrane was exposed to phosphorimager plates (PhosphorImager SI, Molecular Dynamics) for 3 days and scanned to detect radioactive protein bands. Radiolabelled D1 protein was quantified using ImageQuaNT, Molecular Dynamics software.

2.6 *Gas exchange measurements*
Photosynthetic O_2 evolution was measured using a Hansatech Leaf Disc Electrode (Hansatech, Kings Lynn, UK.). Replicate leaf samples were taken 3 hours into their photoperiod. Photosynthetic rates were measured using light intensities from 0 to 1800 μmol m^{-2} s^{-1}. CO_2 levels in the chamber were saturating and the temperature was kept constant at 23°C by water circulation.

3. Results and Discussion

After 6 weeks growth there were large differences between the -N and the control plants. Nitrogen deficient leaves were visibly chlorotic and smaller and contained approximately 50% less N than the controls (Table 1). The total amount of chlorophyll was decreased by half. To understand fully the relationship between D1 turnover and photosynthesis it was important to know the relative numbers of photosystems present in the two types of leaves. The chlorophyll a/b ratio is indicative of the relative proportions of photosystems to absorbing pigments. Interestingly, the chl a/b ratios were the same for the control and -N leaves, thus indicating that the nitrogen deficient leaves had less total chlorophyll but similar proportions of PSI and PSII.

Table 1: Nutritional and physiological parameters of control and -N plants

Parameter	+N	-N
% N in leaf	7.5	3.5
Chlorophyll/area (μg Chl/mm^2)	0.25	0.12
Chl a/b ratio	3.4	3.5
Photosynthetic rate (μmol O_2 m^{-2} s^{-1})	25	16
Photosynthetic rate (μmol O_2 mg Chl^{-1} hr^{-1})	352	484
D1 synthesis rate (D1 mg Chl^{-1} hr^{-1})	31	46
ETR (MINI-PAM)	184	129

Photosynthetic capacities were determined for all leaves. When expressed on a leaf area basis the -N leaves had significantly lower rates of O_2 evolution and lower ETRs measured with the MINI-PAM fluorometer. These reduced rates clearly reflect the reduced chlorophyll content of the -N leaves. However, when the photosynthetic rates were examined on a unit chlorophyll basis, the -N leaves had significantly higher rates than the controls. The -N leaves contained only half the number of photosystems, but performed 38% more photosynthesis. Consistent with this were 46% higher D1 turnover rates, also measured per unit chlorophyll (Table 1). Apparently the -N plants are more efficient than the control plants. As photosynthetic rate is determined by the number of active photosystems, it may be inferred that the control leaves contain PSII that is not active.

D1 turnover and photosynthetic rates expressed per unit chlorophyll when plotted against each other show a positive correlation (Fig. 1), similar to that observed by Russell *et al.* (personal communication), although not as close. In their experiments photosynthetic rate was measured as CO_2 fixation. It is reasonable to suggest that the observed variation may arise from the fact that electrons released during the oxygen evolving process may be channelled into other metabolic pathways in the chloroplast. When measuring photosynthetic rate as O_2 evolution, these rates may be overestimated.

3674

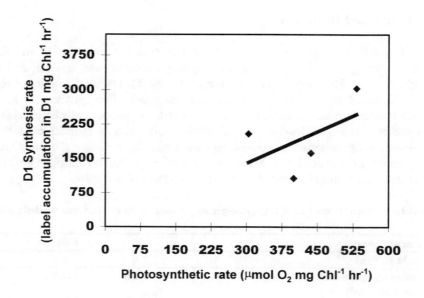

Photosynthetic rate (μmol O_2 mg Chl^{-1} hr^{-1})

In conclusion, nitrogen deficiency provides a system in which the link between photosynthetic rate and D1 turnover can be investigated because the deficiency results in leaves that have intrinsically different photosynthetic rates. The research presented here shows that D1 turnover rates can be altered by factors other than light and high maximum rates can be achieved at light intensities where photoinhibition does not occur. Increasing evidence points towards the conclusion that D1 turnover is a process that responds to the metabolic demands of the leaf.

Acknowledgements

BCM was supported by an Australian Postgraduate Award.

References

1 Hansson, O. and Wydrzynski, T. (1990) Photosyn. Res. 23, 131-162
2 Weinbaum, S.A., Gressel, J., Reisfeld, A. and Edelman, M. (1979) Plant Physiol. 64, 28-832
3 Arntzen, C.J., Kyle, D.J., Wettern, M. and Ohad, I. (1984) in Biosynthesis of the Photosynthetic Apparatus: Molecular Biology, Development and Regulation. Alan R Liss, Inc, New York. pp 313-324
4 Flanigan, Y.S and Critchley, C. (1996) Planta. 198, 319-323
5 Hewitt and Smith (1975) in Plant Mineral Nutrition. English Universities Press. London. pp 32-34
6 Porra, R.J., Thompson, W.A., Kriedemann, P.E. (1989) Biochim. Biophys. Acta. 975, 384-394

CEMA HOMOLOGUE IN CYANOBACTERIA (PXCA) INVOLVED IN PROTON EXCHANGE

M. Sonoda[1], H. Katoh[1], W. Vermaas[2] and T. Ogawa[1]

[1]Bioscience Center, Nagoya University, Nagoya 464-8601, Japan, and
[2]Center for the Study of Early Events in Photosynthesis, Arizona State University, Tempe, AZ 85287-1601, USA

Key words: chloroplast genes, CO_2 uptake, electron transport, mutants, photosystem 2, proton transport

1. Introduction

When suspension of cyanobacterial cells are illuminated, there is an acidification of the medium followed by an alkalization [1-4]. Both acidification and alkalization are specifically stimulated by Na^+. The acidification was assumed to be due to a light-dependent uptake of CO_2 that is converted to HCO_3^- [2,3] and the alkalization due to extrusion of OH^- produced as a result of conversion of HCO_3^- to CO_2 that is fixed by photosynthesis [1]. Ambiguity, however, remains on the source of protons or hydroxyl ions extruded in the light and electron transport involved in the proton extrusion is not yet known.

pxcA (formerly known as *cotA*) is a cyanobacterial gene homologous to *cemA* or *ycf10* in chloroplast genomes and has been cloned from *Synechocystis* PCC6803 and *Synechococcus* PCC7942 [5-7]. Mutants with inactivated *pxcA* were unable to grow in low Na^+ medium or in acidic medium and did not show light-dependent proton extrusion [4,7]. Western analysis indicated that PxcA is located in the cytoplasmic membrane [8]. These results indicated that PxcA is involved in light-dependent proton extrusion and is essential to cell growth under acidic or low salt conditions. The present study aims to clarify which mode of electron transport is involved in the light-dependent proton extrusion and to see the effect of *pxcA* inactivation on the uptake of CO_2, HCO_3^- and NO_3^- .

2. Materials and Methods

Cells of wild-type (WT) and mutants (*pxcA*[4], *psaAB*[9] and *psbDIC*/*psbDII*[10]) of *Synechocystis* PCC6803 were grown at 30°C in BG-11 medium [11] buffered at pH 8.0 during aeration with 3% (vol/vol) CO_2 in air. Continuous illumination was provided at 40 μmol photosynthetically active radiation/m²s (400-700 nm) for *psaAB* cells and at 100 μmol /m²s for the other strains. Glucose (5 mM) was added to the above medium for the growth of *psaAB* and *psbDIC*/*psbDII*.

Changes in pH of the cell suspension (3 ml) kept at 30 °C were monitored by a pH electrode with a meter (Inlar 423 and Delta 350; Mettler Toledo, Halstead Essex, UK) [4].

G. Garab (ed.), Photosynthesis: Mechanisms and Effects, Vol. V, 3675–3678.
© *1998 Kluwer Academic Publishers. Printed in the Netherlands.*

3. Results and Discussion

3.1. *Proton exchange by the WT and pxcA⁻ mutant.*

When cell suspension of WT *Synechocystis* was illuminated, there was an acidification followed by an alkalization of the medium (Fig. 1, curve A). In contrast, only alkalization was observed with the *pxcA⁻* mutant in the light (D). The result indicates that *pxcA* is involved in light-dependent proton extrusion. In the presence of glyceraldehyde (GA; an inhibitor of CO_2 fixation) the alkalization was inhibited both in the WT and mutant (B and E), which supports the view that the alkalization is due to extrusion of OH⁻ produced as a result of HCO_3^- to CO_2 conversion [1]. The inhibition was not complete, indicating the presence of alkalization independent of this reaction. When WT cells were illuminated in the presence of 2,5-dimethyl-p-benzoquinone (DMBQ), an electron acceptor from photosystem(PS) II, the net proton extrusion ceased after a minute of illumination and a post-illumination influx of protons was observed (C). This suggests that in the light both extrusion and influx of protons occur, reaching a stationary level where there is no net proton exchange. After the light is turned off (causing proton extrusion to cease) proton influx continues for a short time until new steady-state level is attained. In the presence of DMBQ, inorganic carbon uptake does not occur [12] and the observed proton fluxes are independent of this reaction. Both light-dependent proton extrusion and postillumination proton influx were very small in the *pxcA⁻* mutant in the presence of DMBQ (F).

3.2. *Net proton exchange in mutants defective in PS I or PS II.*

To clarify with which part(s) of photosynthetic electron transport proton extrusion may be associated, mutants lacking either PS I or PS II were investigated. The *psaAB⁻* (PSI-less) strain showed light-dependent proton extrusion (curve A in Fig. 2). On a per-chlorophyll basis, the amplitude of proton extrusion was 2-3 fold larger than that in WT (compare with curve A in Fig. 1). The result indicates that PS II-mediated electron transfer can drive a significant amount of proton extrusion. No proton uptake was observed in the PS I-less mutant in the light, consistent with the lack of CO_2 fixation in this strain. In the presence of DMBQ, a more extensive acidification was observed with this mutant followed by proton uptake (Fig. 2 B). Thus, PS II-driven electron transport from water to DMBQ or, to a lesser extent, to oxygen can lead to proton extrusion. A small amount of light-induced

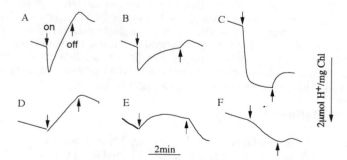

Figure 1. Net proton movements in suspensions of WT (curves A-C) and *pxcA⁻* (D-F) cells of *Synechocystis* PCC6803 upon switching the light on and off. Cells were suspended in 0.2 mM Tes-KOH buffer (pH 8.0) containing 15 mM NaCl in the absence (A, D) and presence of 20 mM GA (B, E) or 1 mM DMBQ (C, F).

proton extrusion was observed when a cell suspension of the *psbDIC⁻/psbDII⁻* strain was illuminated in the absence of DMBQ (C) but not in its presence (D).

The above results indicate that the activity of proton extrusion is correlated with the activity of photosynthetic water splitting and electron transport through the cyt.b_6/f complex, both of which produce a proton gradient across the thylakoid membrane and thereby can lead to the generation of ATP. PxcA does not have an ATP-binding motif and is unable to hydrolyze ATP by itself. PxcA could be a regulator of an ATP-dependent proton extrusion pump, of which activity may be very low in the absence of PxcA.

3.3. Effect of Na^+ and pH on the uptake of CO_2, HCO_3^- and NO_3^- in WT and pxcA⁻.

Cells have a mechanism to maintain a homeostasis with respect to the intracellular pH and electroneutrality during transport of nutrients. To test whether the PxcA-dependent proton exchange is involved in maintaining this homeostasis, the uptake of CO_2, HCO_3^- and NO_3^- was monitored in the WT and *pxcA⁻* strain at pH 8.0 and 6.5 in the presence of normal concentration (15 mM) of NaCl (N-Na⁺) or KCl with a contaminating concentration of Na⁺ (L-Na⁺; ~100μM Na⁺). There was no difference between the WT and mutant in their HCO_3^- uptake at least at pH 8; the activity was high at N-Na⁺ and low at L-Na⁺ (Fig. 3, middle column).   At L-Na⁺, the activity of NO_3^- uptake was very low in *pxcA⁻* at pH 8.0

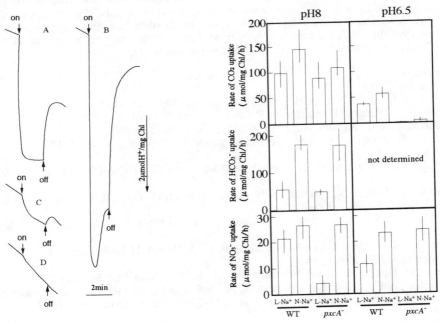

Figure 2. (left) Net proton movement in the suspensions of *psaAB⁻* (A, B) and *psbDIC⁻/psbDII⁻* (C, D,) cells upon switching the light on and off. DMBQ was added prior to illumination for curves B and D.

Figure 3. (right) The rates of CO_2, HCO_3^- and NO_3^- uptake in WT and *pxcA⁻* cells of *Synechocystis* at pH 8.0 and 6.5 at normal (N-) and low (L-) Na⁺ concentrations.

and was zero at pH 6.5 (bottom columns). At N-Na$^+$, no significant effect of $pxcA$ inactivation was observed on CO_2 and NO_3^- uptake at pH 8.0 but CO_2-uptake activity was reduced significantly at pH 6.5 (upper and bottom columns). No CO_2 uptake was observed in the mutant at pH 6.5 and L-Na$^+$. It is evident that the inactivation of $pxcA$ strongly affected the CO_2 uptake under acidic conditions and the NO_3^- uptake at low Na$^+$ concentrations.

We propose a working hypothesis involving two complementary proton exchange systems, $i.e.$ PxcA-dependent and PxcA-independent proton exchange systems, in maintaining homeostasis with respect to the intracellular pH and electroneutrality. The PxcA-independent system could be Na$^+$/H$^+$ antiport. The proton exchange catalyzed by these systems is stimulated by Na$^+$. Both systems are essential to CO_2 transport at pH 6.5 but PxcA-independent system alone is sufficient at pH 8 at N-Na$^+$. However, both systems are required at L-Na$^+$ even at this alkaline pH. At L-Na$^+$, NO_3^- uptake requires the PxcA-dependent system. When PxcA-independent proton exchange is active at N-Na$^+$, PxcA-dependent system is not required for NO_3^- uptake. Uptake of HCO_3^- requires high activity of PxcA-independent proton exchange at N-Na$^+$ either in the WT or in $pxcA^-$.

Proton exchange catalyzed by the PxcA-independent system should be observed as the pH changes of the suspension medium of $pxcA^-$ cells. The slow alkalization observed with $pxcA^-$ in the light and the slow acidification in the dark may be due to proton influx and efflux by the PxcA-independent system; it is also possible that rapid influx and efflux of protons via the PxcA-independent system system with a small net proton movement that can not be measured by the pH electrode used in this study.

References

1 Miller, A.G. and Colman, B. (1980) J. Bacteriol 143, 1253-1259.
2 Scherer, S., Riege, H. and Boger, P. (1988) Plant Physiol. 86, 769-772.
3 Ogawa, T. and Kaplan, A. (1987) Plant Physiol. 83: 888-891.
4 Katoh, A., Sonoda, M., Katoh, H. and Ogawa, T. (1996) J. Bacteriol. 178, 5452-5455.
5 Sasaki, Y., Sekiguchi, K., Nagano, Y. and Matsuno, R. (1993) FEBS Lett. 316, 93-98.
6 Katoh, A., Lee, K.S., Fukuzawa, H. Ohyama, K. and Ogawa, T. (1996) Proc. Natl. Acad. Sci. USA 93, 4006-4010.
7 Sonoda, M., Katoh, H. Ohkawa, H. and Ogawa, T. (1997) Photosynth. Res. 54, 99-105.
8 Sonoda, M., Kitano, K., Katoh, A., Katoh, H., Ohkawa, H. and Ogawa, T. (1997) J. Bacteriol. 179, 3845-3850.
9 Shen, G., Boussiba, S. and Vermaas, W.F.J. (1993) Plant Cell 5: 1853-1863.
10 Vermaas, W.F.J., Charit, J. and Eggers, B. (1990) in Current Research in Photosynthesis (M. Baltscheffsky, ed.), Vol. 1, pp. 231-238, Kluwer, Dordrecht, The Netherlands.
11 Stanier, R.Y., Kunisawa, R., Mandel, M. and Cohen-Bazire, G. (1971) Bacteriol. Rev. 35, 171-205.
12 Ogawa, T., Miyano, A,. and Inoue, Y. (1985) Biochim Biophys Acta 808, 77-84.

EVIDENCE FOR THE WIDESPREAD DISTRIBUTION OF MEMBRANE-BOUND INORGANIC PYROPHOSPHATASES AMONG PHOTOSYNTHETIC PROKARYOTES

I. García-Donas[1], J.R. Pérez-Castiñeira[2], P.A. Rea[2], Y.M. Drozdowicz[2], M. Baltscheffsky[3], M. Losada[1] and A. Serrano[1]
[1]Instituto de Bioquímica Vegetal y Fotosíntesis, CSIC-Universidad de Sevilla, CICIC, 41092-Sevilla, Spain; [2]Plant Science Institute, University of Pennsylvania, Philadelphia, PA 19104, USA; [3]Department of Biochemistry, Arrhenius Laboratories, Stockholm University, S-10691, Stockholm, Sweden

Key words: Cyanobacteria, green sulfur bacteria, inorganic pyrophosphate, membrane protein, purple bacteria

1. Introduction

Two types of proton-pumping, membrane-bound inorganic pyrophosphatases (PPases), both from photosynthetic organisms, have been characterized to date and their genes cloned: the vacuolar PPase (V-PPase) of higher plants [1,2] and the PPi synthase/PPase of chromatophores of the non-sulfur purple bacterium *Rhodospirillum rubrum* [3,4]. Despite their occurrence in phylogenetically very distant organisms, these proteins exhibit remarkable structural similarities, a property we have exploited to survey the distribution of membrane-bound PPases and their genes in different groups of photosynthetic prokaryotes, namely, non-sulfur purple bacteria, Chromatiaceae, Chlorobiaceae and cyanobacteria. Evidence obtained both at protein and DNA levels indicates that membrane-bound PPases are present in a broad range of phototrophic prokaryotes, being therefore more widely distributed than was previously thought.

2. Procedure

2.1 Bacterial strains and growth conditions

All microorganisms were grown under phototrophic conditions at 30°C. Photosynthetic bacteria were kindly provided by A. Verméglio (CEA, St. Paul Lez Durance, France), M. Baltscheffsky (Arrhenius Laboratory, Stockholm University, Sweden) and J. Mas (Autonomous University of Barcelona, Spain). The cyanobacterial strains were obtained from the Collection of microorganisms of the Institute Pasteur (PCC, Paris, France). *Escherichia coli* K-12 and DH5α strains were cultured in Luria Broth medium supplemented with ampicillin (100 µg/ml) when necessary at 37°C.

2.2 Protein techniques

Bacterial cell membrane preparations were obtained in the cold by sonication (70W, 2 min) followed by centrifugation at 40,000 *g* for 25 min. Vacuolar (tonoplast) membranes were isolated from leaves of *Antirrhinum majus* as described in [5]. Two monospecific

G. Garab (ed.), Photosynthesis: Mechanisms and Effects, Vol. V, 3679–3682.
© 1998 Kluwer Academic Publishers. Printed in the Netherlands.

polyclonal antibodies were raised in rabbits against KLH-conjugated synthetic peptides corresponding to conserved regions involved in substrate-binding of plant V-PPase [2]: the so-called antibody 324 (Ab.324), against the peptide TKAADVGADLVGKIE (putative hydrophilic loop V), and antibody 326 (Ab.326), against the sequence HKAAVIGDTIGDPAK (hydrophilic loop XII). The antisera were used (1:1,000 dilution) for Western blot analysis after SDS-PAGE (10%) and transfer to nitrocellulose of cell membrane preparations from cyanobacteria and anoxygenic photobacteria or purified plant tonoplast membranes that were used as internal controls. Protein was determined using a modification of the method of Lowry carried out in the presence of SDS.

2.3 DNA techniques

Two 20-mer oligonucleotides corresponding to the 5'- (sense) and 3'- (antisense) ends of the ORF encoding the putative PPi-synthase of *R. rubrum* [4] were used for PCR amplification of a single 2-kb long DNA fragment using genomic DNA of this bacterium as template and the *Taq* Plus-Precision System (Stratagene). After electrophoresis on agarose gel the amplified band was excised and cloned in pGEMt plasmid (Promega). A cDNA clone containing the *Arabidopsis thaliana* V-PPase gene (cloned in the *Sma*I site of pBS SKII) and a partial cDNA clone of the V-PPase gene of *A. majus* strongly hybridized with this band. Two 18-mer degenerate primers were constructed from amino acid sequences conserved in both V-PPase and PPi-synthase proteins (GDNVGD, for sense primer; GGIAEM, for antisense primer) and used to amplify an internal 0.7-kb DNA fragment in PCR experiments with genomic DNAs as templates.

Southern blots were performed using completely cleaved genomic DNAs from different bacteria run on 0.7% agarose gels and blotted to nylon membranes. The PCR amplified PPi-synthase gene of *R. rubrum* labelled with ^{32}P dCTP was used as a probe under heterologous (55°C) hybridization conditions. Screenings of genomic DNA libraries (partially *Sau*3A1-restricted DNA cloned in the *Bam*H1 site of pBS SK(+)) of different cyanobacteria and anoxygenic bacteria were performed by standard procedures using the same probe after alkaline lysis of *E. coli* colonies and transfer to nitrocellulose.

3. Results and Discussion

Two peptide-specific polyclonal antibodies raised against sequences conserved in all known plant V-PPases [2] detect polypeptides of the sizes expected for membrane-bound PPases (60-75 kDa) on Western blots of SDS-PAGE-separated membrane proteins not only from the non sulfur purple bacterium *R. rubrum*, as was expected from the virtually identical sequences found in its PPi-synthase gene [4], but also in membranes from a broad range of anoxygenic photosynthetic bacteria, namely, other non sulfur purple bacteria (*Rhodopseudomonas palustris*, *Rhodopseudomonas viridis*, *Rhodobacter capsulatus*, *Rhodobacter sphaeroides*), Chromatiaceae (sulfur purple bacteria: *Amoebobacter roseus*, *Chromatium vinosum*), and Chlorobiaceae (green sulfur bacteria: *Chlorobium limicola*) (Fig. 1). Interestingly, protein bands of similar molecular masses were also immunodetected in membrane preparations of cyanobacteria (oxygenic photobacteria) of different taxonomic groups, either unicellular (*Synechocystis* sp. PCC6803), filamentous (*Pseudanabaena* sp. PCC6903) or N2-fixing heterocystous (*Anabaena* sp. PCC7120, *Calothrix* sp. PCC7601, *Fischerella* sp. UTEX1829) strains (Fig. 2). It should be noted that no bands were detected in the soluble protein fractions of all these microorganisms. Overall, these results indicate a general expression of membrane-bound PPase proteins among photosynthetic prokaryotes.

Figure 1. Western blot analysis of cell membrane preparations (100 μg of protein per lane) of different anoxygenic photosynthetic bacteria probed with two polyclonal antibodies, namely Ab.324 *(left)* and Ab.326 *(right)*, raised against conserved regions of plant V-PPase. Immunodetected bands of the expected size for membrane-bound PPases are indicated by the arrow. A.r.:*Amoebobacter roseus*; C.l.:*Chlorobium limicola*; C.v.:*Chromatium vinosum*; R.c.:*Rhodobacter capsulatus*; R.p.:*Rhodopseudomonas palustris*; R.r.:*Rhodospirillum rubrum*; R.s.:*Rhodobacter sphaeroides*; R.v.:*Rhodopseudomonas viridis*.

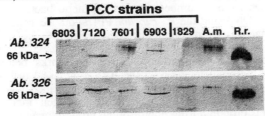

Figure 2. Western blot analysis of cell membrane preparations (100 μg of protein per lane) of cyanobacteria of different taxonomic groups probed with the same polyclonal antibodies used in Fig. 1. PCC numbers are, in general, provided for cyanobacteria. Immunodetected bands of the expected size for membrane-bound PPases are indicated by the arrow. 6803: *Synechocystis* sp. PCC6803; 7120: *Anabaena* sp. PCC7120; 7601: *Calothrix* sp. PCC7601; 6903: *Pseudanabaena* sp. PCC6903; 1829: *Fischerella* sp. UTEX1829; A.m.: *A. majus* tonoplast membranes; R.r.: *R. rubrum* cell membranes.

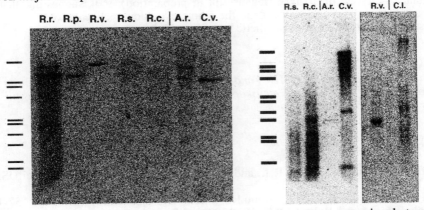

Figure 3. Southern blots of genomic DNAs from diverse anoxygenic photosynthetic bacteria probed with the *R. rubrum* PPi-synthase gene. *(Left)* BamH1-digested DNAs; *(right)* SacII-digested DNAs. Either 2 μg (*R. rubrum*) or 10 μg of DNA were used per lane. EcoRI-HindIII-restricted lambda DNA was used as standard. Strain abbreviations used were the same that in Fig. 1.

Figure 4. Western blot probed with antibody 326 of membrane preparations (100 μg of protein per lane) from *R. rubrum* cells taken at different times of culture under phototrophic conditions (Hutner medium, anaerobiosis, light).

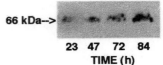

Southern blot experiments performed in heterologous conditions of restricted genomic DNAs from diverse anoxygenic photosynthetic bacteria using the PCR-generated PPi-synthase gene of *R. rubrum* as a probe, showed clear hybridization bands in most cases (Fig. 3), indicating again a wide distribution of genes encoding membrane-bound PPases among the microorganisms tested.

Degenerate primers designed on the basis of sequences conserved in the plant vacuolar and *R. rubrum* chromatophore PPases, yielded PCR amplification products of the expected size (ca. 0.7 kb) for the coding sequences of membrane-bound PPase when genomic DNA from photosynthetic bacteria were used as template (data not shown). On the other hand, screens of genomic DNA libraries from a number of photosynthetic bacteria and cyanobacteria, using the plant V-PPase cDNA and the *R. rubrum* PPi synthase gene as probes, yielded several putative clones that are currently under characterization.

It has been reported that the PPi-synthase of *R. rubrum* works better under low light conditions, thus working with advantage over the ATP-synthase —that needs a membrane potential for activation— under these conditions [3]. In accordance with this, time course analysis by Western blots of the PPi-synthase protein in cell membrane preparations revealed an increase of this protein during phototrophic growth, reaching a maximum in the stationary phase in which light should be limitant (Fig. 4). Studies are in progress to clarify if it is the case also under other stressing conditions.

On the basis of all these results, membrane-bound PPases are concluded to be far more widely distributed among photosynthetic prokaryotes than was previously thought. Preliminary data indicate that similar membrane-bound proteins appear also in different unicellular eukaryotic microalgae (manuscript in preparation), thus suggesting that these ionic pumps have been conserved along the whole evolutionary lineage of photosynthetic organisms, from primitive photosynthetic bacteria to higher plants.

Supported by grants PB 94-33 (MEC) and PAI-CVI198 (Junta de Andalucía), Spain, and the Carl Tryggers Stiftelse för Vetenskaplig Forskning and the Magnus Bergvall and Wenner Gren Foundations, Sweden.

References

1 Sarafian V., Kim Y., Poole R.J. and Rea P.A. (1992) Proc.Natl.Acad.Sci.USA 89: 1775-1779

2 Rea P.A. and Poole R.J. (1993) Annu.Rev.Plant Physiol. Plant.Mol.Biol. 44:157-180

3 Baltscheffsky M. and Baltscheffsky H. (1992) in Molecular Mechanisms in Bioenergetics (Ernster L., ed.) pp. 331-348, Elsevier, Amsterdam

4 Baltscheffsky M., Nadanaciva N., and Schultz A. (1998) Biochim. Biophys. Acta 1364:301-306

5 Warren M., Smith J.A.C. and Apps D.K. (1992) Biochim. Biophys. Acta 1106:117-125

STRUCTURAL DIVERSITY AND FUNCTIONAL CONSERVATION OF SOLUBLE INORGANIC PYROPHOSPHATASES FROM PHOTOSYNTHETIC PROKARYOTES AND PLASTIDS

R. Gómez[1], W. Löffelhardt[2], M. Losada[1] and A. Serrano[1]
[1]Instituto de Bioquímica Vegetal y Fotosíntesis, CSIC-Universidad de Sevilla, CICIC, 41092-Sevilla, Spain; [2]Institut für Biochemie und Molekulare Zellbiologie der Universität Wien und Ludwig-Boltzmann-Forschungsstelle für Biochemie, A-1030 Vienna, Austria

Key words: Chloroplast, cyanobacteria, microalgae, P metabolism, purple bacteria

1. Introduction

Inorganic pyrophosphatases (sPPase; pyrophosphate phosphohydrolase, EC 3.6.1.1) are essential enzymes found in the cytosol and cellular organelles that drive anabolism through the hydrolysis of the energy-rich pyrophosphate (PPi) produced in the synthesis of biological polymers, making these processes thermodinamically irreversible. Since this hydrolysis replenishes Pi to the energy-converting systems, sPPases should play an important role in the intracellular "phosphate cycle". In plants and some bacteria, sPPase also participates in sulfur metabolism by coupling PPi hydrolysis to sulfate activation steps. Moreover, PPi (like ATP) can be synthesized in photosynthetic bacteria as a result of photophosphorylation by a PPi-synthase [1].
The cytoplasmic sPPases of *Saccharomyces cerevisiae* (homodimer, 32 kDa subunit) and *E. coli* (homohexamer, 19 kDa subunit), although quite different in aminoacid sequence and molecular mass, exhibit a well conserved active site and similar protein core structures [2]. They are the best studied eukaryotic and prokaryotic sPPases and have been postulated as examples of divergent structural evolution. Searching for other possible different sPPases should therefore be useful to validate this proposal.
Our aim is to clarify the relationships between the sPPases of a wide range of photosynthetic organisms, from photosynthetic bacteria to higher plants, a comparative study with phylogenetically diverse microorganisms and organelles being presented here. These sPPases exhibit clear structural differences that could be relevant in the context of the molecular phylogeny of plastids.

2. Materials and Methods

2.1 Organisms and growth conditions
All organisms were grown under phototrophic conditions. Photosynthetic bacteria were kindly provided by A. Verméglio (CEA, St. Paul Lez Durance, France), M. Baltscheffsky (Arrhenius Laboratory, Univ. of Stockholm, Sweden) and J. Mas (Aut. Univ. of Barcelona, Spain). The cyanobacterial and algal strains were obtained from the collections of microorganisms of the Institute Pasteur (PCC, Paris, France) and the Univ. of Göttingen (SAG, RFA), respectively. The endocyanome *Cyanophora paradoxa* was grown at the Institut für Biochemie und Molekulare Zellbiologie (Univ. of Vienna, Austria). Spinach chloroplasts, isolated as described in [3], *C. paradoxa* cyanelles, obtained after host-cell breakage by gentle stirring, and *Chlamydomonas reihardtii* chloroplasts, isolated after freeze-thawing cell disruption, were eventually disrupted by sonication.

2.2 Protein techniques
Cell-free extracts were obtained in the cold by sonication followed by centrifugation at 40,000 g for 25 min. An optimized purification method (sonication, $(NH_4)_2SO_4$ fractiona-

G. Garab (ed.), Photosynthesis: Mechanisms and Effects, Vol. V, 3683–3686.
© 1998 *Kluwer Academic Publishers. Printed in the Netherlands.*

tion of soluble proteins, and anionic exchange(DEAE), Phenyl-Sepharose and hydroxyapatite column chromatographies) was used to purify to electrophoretic homogeneity most sPPases studied in this work. The following protein characterization techniques were used: a) SDS-PAGE (apparent subunit molecular mass); b) FPLC gel filtration on Superose12 (Pharmacia) under native conditions (pH 7.5, 2 mM Mg_2Cl) (apparent native molecular mass); c) column chromatofocusing (isoelectric point); and d) MALDI-TOF mass spectroscopy (absolute subunit molecular mass). N-terminal sequences were obtained by Edman degradation using automatic aminoacid sequencers. Monospecific polyclonal antibodies raised in rabbits against the sPPases of *Synechocystis* sp. PCC6803 and *C. reinhardtii* chloroplasts were used (1:500) in Western blots and cross-reaction tests after SDS-PAGE of both purified sPPases and cell-free extracts. Pi generated by enzymatic PPi hydrolysis was determined by a modified Fiske-Subbarow method (Sigma). Protein was estimated with the Coomassie Blue method (BioRad).

2.3 DNA techniques

The *ppa* gene of *Synechocystis* sp. PCC6803 was cloned in the pBS SK(+) vector using a PCR strategy (*Taq*Plus-Precision, Stratagene) and the encoded sPPase overexpressed in *E.coli* DH5α. Southern blots were performed using completely restricted genomic DNA of different strains run on 0.7% agarose gels and blotted to nylon membranes. The *ppa* gene of *Synechocystis* labelled with ^{32}P dCTP was used as a probe in heterologous (55°C) hybridization conditions. Analysis of biological sequences was performed using programs of the GCG package.

3. Results and Discussion

The sPPases of a variety of strains representative of phylogenetically diverse groups of photosynthetic organisms have been purified to electrophoretic homogeneity and their basic structural parameters have been determined: a remarkable structural diversity was found (Table 1) (R. Gómez *et al.*, in preparation). The enzymes of photosynthetic prokaryotes –anoxygenic bacteria and cyanobacteria– have subunits of virtually identical mass (ca. 23 kDa, SDS-PAGE) except those of some sulfur bacteria, which in Western blots appear larger (Fig. 1). However, whereas cyanobacterial sPPases are homohexameric (120 kDa), diverse oligomeric states (tetramer, dodecamer) were found for the enzymes of the purple bacteria analyzed (Table 1). Western blots using antibodies against the sPPase of the cyanobacterium *Synechocystis* sp. PCC6803 indicate cross reaction with the enzymes of other cyanobacteria and, unexpectedly, of most anoxygenic bacteria tested (Fig. 1, right). Overall, these sPPases exhibit a prokaryotic-type structure. In contrast, the sPPases of eukaryotic photosynthetic organisms have larger subunits (ca. 35kDa, SDS-PAGE, Fig. 1) and, in general, are monomers (Table 1). Subcellular localization experiments showed that in photosynthetic eukaryotes virtually all sPPase activity is located in their plastids (Table 2). Antibodies raised against the plastidic *C. reinhardtii* sPPase show cross reaction with the enzymes from other green algae, euglenoids and, weekly, with *C. paradoxa*, but not with sPPases of red and brown microalgae (Fig. 1, right) and plant chloroplasts. The sPPases from *Euglena gracilis* and *Ochromonas danica* are being characterized in detail.

Cyanelles are considered to represent a different and very early diverging branch of plastid evolution [4]. The unexpected finding of different sPPases in cyanobacteria –that resemble the ancestor of chloroplasts– and cyanelles –very similar in structure to free-living unicellular cyanobacteria but with a cellular integration level comparable to chloroplasts– prompted us to carry out a detailed comparative study of these enzymes. The N-terminal regions of four cyanobacterial sPPases, as well as the whole protein sequence deduced from the *Synechocystis* PCC 6803 *ppa* gene, exhibit a marked similarity with thermophilic bacteria sPPases. MALDI-TOF analysis of the *Synechocystis* sp. PCC6803 sPPase subunit showed an absolute molecular mass of 19,187 Da, in fine agreement with the value deduced from the *ppa* gene. In fact, the recombinant *Synechocystis* sPPase overexpressed in *E.coli* was purified and is indistinguishable from the natural one. The cloned *Synechocystis* sp. PCC6803 *ppa* gene able us to demonstrate by Southern hybridization

Table 1. Some structural parameters of sPPases from phylogenetically diverse photosynthetic microorganisms and organelles.

Source	Subunit molecular mass (kDa)[a]	Native molecular mass (kDa) and oligomeric structure[b]	Isoelectric point[c]
PROKARYOTES			
Anoxygenic bacteria			
Non-sulfur purple bacteria			
Rhodopseudomonas viridis	23	240 (dodecamer?)	<7.0
Rhodopseudomonas palustris	23	85 (tetramer)	<7.0
Rhosdospirillum rubrum	23	90 (tetramer)	<7.0
Rhodobacter sphaeroides	23	-- [d]	--
Sulfur purple bacteria			
Amoebobacter roseous	21	--	--
Chromatium vinosum	60	--	--
Sulfur green bacteria			
Chlorobium limicola	42	--	--
Cyanobacteria			
Synechococcus sp. PCC7942	23	120 (hexamer)	<7.0
Synechocystis sp. PCC6803	23 [19,187[e]]	120 (hexamer)	4.70
Dermocarpa sp. PCC7437	23	120 (hexamer)	<7.0
Pseudanabaena sp. PCC6903	24	120 (hexamer)	<7.0
Anabaena sp. PCC7120	23	120 (hexamer)	<7.0
Fischerella sp. UTEX1829	23	120 (hexamer)	<7.0
EUKARYOTES			
Cyanidium caldarium	40	160 (tetramer)	5.20
Cyanophora paradoxa	34	34 (monomer)	5.95
Ochromonas danica	38	--	ca. 7.0
Chlamydomonas reinhardtii	36	35 (monomer)	ca. 7.0
Chlorella fusca	36	--	--
Monoraphidium braunii	37	--	--
Euglena gracilis	37	--	ca. 7.0
Astasia longa[f]	63	--	--
Spinacea oleracea (leaves)	37	40 (monomer)	>7.0

[a] Except otherwise specified, values were estimated by SDS-PAGE and/or Western blot analysis (means of at least three independent determinations).
[b] Values estimated by FPLC gel filtration under native conditions in the presence of 2 mM $MgCl_2$ (means of three independent determinations).
[c] Estimated by column chromatofocusing (exact values) or from the protein behaviour under anion-exchange chromatography.
[d] --, not determined.
[e] Absolute subunit molecular mass determined by MALDI-TOF.
[f] Non-photosynthetic euglenophycean strain.

Table 2. Subcellular distribution of sPPase activity in several photosynthetic organisms.

Source		Enzyme activity (U/mg of protein)
Cyanophora paradoxa	cytosol	<0.001
	cyanelles	5.4
Chlamydomonas reinhardtii	cytosol	<0.001
	chloroplasts	4.1
Spinacea oleracea (leaves)	cytosol	<0.001
	chloroplasts	5.3

homologous genes in all cyanobacteria tested (Fig. 1, left), those of *Anabaena* PCC7120 and *Pseudanabaena* PCC6903 were cloned by screenings of genomic DNA libraries.

3686

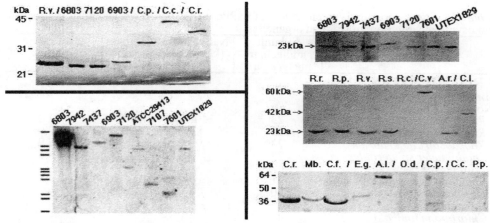

Figure 1. *(Left up)*. SDS-PAGE of purified sPPases from different photosynthetic microorganisms (5 μg of protein per lane); *(left bottom)* Southern blot of *Hind*III-digested genomic DNAs from diverse cyanobacteria probed with the *Synechocystis ppa* gene. *Eco*RI-*Hind*III-restricted λ DNA was used as standard; *(right)* Western blots of cell-free extracts (80 μg of protein per lane, twice for C.l., C.v. and A.r.): from cyanobacteria *(up)* and anoxygenic bacteria *(center)*, tested with antibodies against the *Synechocystis* sPPase, and from different photosynthetic protists *(bottom)* tested with antibodies against the *C. reinhardtii* sPPase. PCC numbers are, in general, provided for cyanobacteria. Immunodetected bands are indicated by arrows. A.l.:*Astasia longa*; A.r.:*Amoebobacter roseus*; C.c.:*Cyanidium caldarium*; C.f.:*Chorella fusca*; C.l.:*Chlorobium limicola*; C.p.:*Cyanophora paradoxa*; C.r.:*Chlamydomonas reinhardtii*; C.v.:*Chromatium vinosum*; E.g.:*Euglena gracilis*; M.b.: *Monoraphidium braunii*; O.d.: *Ochromonas danica*; P.p.:*Porphyridium purpureum*; R.c.:*Rhodobacter capsulatus*; R.p.:*Rhodopseudomonas palustris*; R.r.:*Rhodospirillum rubrum*; R.s.:*Rhodobacter sphaeroides*; R.v.:*Rhodopseudomonas viridis*.

In contrast with the cyanobacterial proteins, the sPPase of the cyanelle of *C. paradoxa* is a monomeric enzyme of 34 kDa (Table 1). Although the presence of mitochondrial sPPses cannot be probed, all the activity should be located in the photosynthetic organelle (see Table 2). Comparative analysis of the N-terminal sequence of the *C. paradoxa* sPPase clustered it clearly with eukaryotic sPPases, in agreement with its physico-chemical parameters. The cyanelle sPPase should be therefore an eukaryotic sPPase, encoded by the host cell genome, that during the integrative process that gave rise to this organelle substituted its homologous enzyme of the cyanobacterial-like endosymbiont (Fig.2). Oligonucleotides have been constructed from the N-terminal sequence of the cyanelle sPPase and are being used to clone the corresponding *ppa* gene by RT- PCR.

Figure 2. The proposed evolutionary relationships between the sPPases of cyanobacteria, cyanelles and chloroplasts.

Supported by grants PB94-0033 of DGICYT, PAI CVI-198 (Junta de Andalucía) and Acciones Integradas Hispano-Austriacas.

References

1 Baltscheffsky M. and Baltscheffsky H. (1995) Photosynth.Res. 46:87-91
2 Cooperman B.S., Baykov A. and Lahti R. (1992) TIBS 17:262-266
3 Kuwavara T. and Murata N. (1992) Plant Cell Physiol. 23:533-539
4 Löffelhardt W., Bohnert H.J. and Bryant D.A. (1997) Crit.Rev.Plant Sci. 16:393-413

CONSTRUCTION AND CHARACTERISTICS OF TWO HYBRIDS BETWEEN PEA THIOREDOXINS *m* and *f*

J. López Jaramillo, A. Chueca, <u>M. Sahrawy</u>, O. Wangensteen, R. Cazalis and J. López Gorgé. Department of Plant Biochemistry, Estación Experimental del Zaidín (CSIC), Granada, Spain

Key words: Calvin-cycle, cDNA cloning, design of proteins, light activation, metabolic processes, overexpression.

1. Introduction

Fructose-1,6-bisphosphatase (FBPase) is a key enzyme of the Calvin cycle. It is active under light by a light-induced rise of stromal pH and $[Mg^{2+}]$, and by reduction of a S-S group via the ferredoxin-thioredoxin system (1). The electron transfer from ferredoxin to thioredoxin (Trx) is catalyzed by ferredoxin-thioredoxin reductase, whereas the Trx-mediated reduction of FBPase is a thiol-disulfide exchange between both molecules. Two Trx isoforms, similar in size and folding but with different primary structures, are in the chloroplast: Trx *f*, especially operative in the activation of FBPase, and Trx *m*, functional in $NADP^+$-malate dehydrogenase (NADP-MDH) modulation (2). However, spinach FBPase is also activated by Trx *m* in the presence of fructose-1,6-bisphosphate and Ca^{2+} (3). Similar results were found by ourselves with pea FBPase and Trx *m*, even in the absence of modulators (4). It is noteworthy the strong interaction between pea photosynthetic FBPase and chloroplast Trxs (5), weaker in spinach and other species. Looking for molecular differences that can explain this behaviour, we have built up two hybrid Trxs between pea Trxs *f* and *m*, interchanging the NH_2- and COOH-halves respect to the common modulatory cluster.

2. Procedure

2.1 *Construction of the Trx f-Trx m hybrid recombinants*
Using 6 synthetic oligonucleotides corresponding to the coding sequences of the NH_2- and COOH-terminal ends of pea Trxs *f* and *m*, and to the NH_2- and COOH-sides of the *Ava*II restriction site, 4 different cDNA clones encoding for the two halves of pea Trxs *f* and *m* were built up, taking as splitting point the *Ava*II site betwen the 2 cysteines of the regulatory cluster. Four PCR reactions were carried out with the couples Trx *f* (N)-Ava(C), Ava(N)-Trx *f* (C), Trx *m* (N)-Ava(C), and Ava(N)-Trx *m*(C), using as templates pET-Trx *f* (6) for the first two, and pET-Trx *m* (4) for the second two, obtaining the recombinants TfN, TfC, TmN and TmC, respectively. Hybrid Trx *f/m* was generated by ligation TfN and TmC with T4DNA- ligase, and subcloning in the *Nco*I and *Bam*HI sites of the vector pET-3d. The Trx *m/f* hybrid was constructed through a new PCR with the oligonucleotides Trx *m*(N) and Trx *f*(C) , using TmN and TfC as templates. The fused DNA was then digested with *Nde*I and *Bam*HI, and subcloned at the same sites of the vector pET-12a similarly digested.

2.2 *Expression and purification of hybrids Trx m/f and Trx f/m*
After expression of recombinants pET-Trx *f/m* and pET-Trx *m/f* in *E. coli* BL21(DE3),

G. Garab (ed.), Photosynthesis: Mechanisms and Effects, Vol. V, 3687–3690.

cells were broken in a French-press, and the Trx m/f hybrid purified from the supernatant by 10 min 70^0 heating, precipitation with 90% sat. $(NH_4)_2SO_4$, and filtration through Sephadex G-50. On the contrary, Trx f/m appeared as inclussion bodies in the insoluble material of the cell lysate, and because of its low stability could only be partially purified. After 1M urea treatment of the pellet, the inclusion bodies were solubilized in 50 mM K_2HPO_4-K_3PO_4 buffer pH 10.5, 1 mM EDTA-Na_2 by 5 hours gentle stirring at room temperature.

2.3 *Other techniques*

The Trx-dependent FBPase and NADP-MDH activities were analyzed by the two step procedures (7,8). Protein determination, SDS-PAGE, western- and northern-blots, and ELISA assays were performed by conventional methods. Isoelectric points were determined by electrofocusing in a Multiphor II (Pharmacia) equipment. The RASMOL program allowed the final arrangement of the predicted tertiary structures for pea Trx m, Trx f and the hybrid Trx m/f, on the basis of the a.a. sequence of these proteins, and of the primary and tertiary structures of *E. coli* and human Trxs.

3. Results and Discussion

As outlined in Procedure, we have built up the recombinants pET-Trx m/f and pET-Trx f/m, with the coding sequence for pea hybrids Trx m/f and Trx f/m, and the splitting point in the *Ava*II site between both cysteines of the regulatory cluster. pET-Trx m/f expression in *E. coli* provided the hybrid Trx m/f in a soluble form; it was purified to homogeneity as a protein of 12 kDa, showing by western blot positive reaction with pea Trx m and Trx f antibodies. On the contrary, pET-Trx f/m expression did not provide Trx f/m signals in cell supernatants. However, northern blots against a pET-Trx f/m probe showed the presence of the related mRNA, whose translation product appeared as inclussion bodies. After solubilization we obtained a preparation which showed by western blotting a 40 kDa band positive to pea Trx f antibodies, but negative against pea Trx m ones. This fraction, an oligomerization state of Trx f/m, was unstable and becomes inactive quickly.

Trx m/f has a thermal stability similar to that of pea Trx m. It loses 10% capability for FBPase activation by 5 min heating at 90°C, a stability higher than that of pea Trx f, which loses 50% of FBPase activation ability by 5 min heating at 90^0. Trx m remains fully active under this treatment. It was not possible to study the thermal stability of Trx f/m, since it became inactive few min after partial purification. It appears evident that the NH_2-side of Trx active center is essential in maintaining the functional structure for FBPase activation.

```
                          ↓ ↓↓
Trx f VGKVTEVNKDTFWPIVNAAGDKTVVLDMFTKWCGPCKVIAPLYEELSQKYLD.
Trx m AVNEVQVVNDSSWDELVIGSETPVLVDFWAPWCGPCRMIAPIIDELAKEYAG.
. VVFLKLDCNQDNKSLAKELGIKVVPTFKILKDNKIVKEVTGAKFDDLVAAIDTVRSS
.KIKCYKLNT DESPNTATKYGIRSIPTVLFFKNGERKDSVIGAVPKATLSEKVEKYI
```

Figure 1. Alignment of pea Trxs f and m. Homologous amino acids appear underlined, and the conservative ones in bold types. The regulatory cysteines are marked with arrows, and the splitting site with a double one.

Pea Trx m shows 15 acid residues in the whole molecule, 6 of them on the NH_2-side of the regulatory cluster, and 9 on its COOH-side (Fig. 1) (Table 1); the basic residues are 13, all of them on the COOH-side. The net charge of pea Trx m is then -2, with a shift of 6 negative charges on the NH_2-side, to 4 positives on the COOH-side. The situation is similar in Trx m/f, with a net charge of -5, and -6 and +1 charged residues in the NH_2- and COOH-sides of the regulatory cysteines, respectively. An opposite situation exists in Trx f and Trx f/m,

Table 1.- Charged residues of pea Trx f, Trx m, Trx m/f and Trx f/m.

	a.a.(n°)		acid a.a.(n°)		basic a.a.(n°)		net charge		pI	
	N-side	C-side	N-side	C-side	N-side	C-side	N-side	C-side	pred.	determ.
Trx f	34	75	4	11	4	12	0	+1	7.9	8.0
Trx m	34	74	6	9	0	13	-6	+4	5.0	5.4
Trx m/f	34	75	6	11	0	12	-6	+1	4.5	5.0
Trx f/m	34	74	4	9	4	13	0	+4	9.2	-

which show 4 acid and 4 basic residues on the NH_2-side of the modulatory cluster, with 0 net charge in both cases. The number of acid residues on the COOH-side of Trx f and Trx f/m is 11 and 9, respectively, as oppose to 12 and 13 basic residues. This means that the net charge for the COOH-segments of Trx f and Trx f/m is +1 and +4, which are also the net charges for the whole molecules. These values are in agreement with the pIs determined by isoelectric focusing (5.4 and 5.0, respectively), and with those predicted with the PeptideStates program (5.0 and 4.5, respectively). On the contrary, Trx f and Trx f/m showed computer-deduced pIs of 7.9 and 9.2, respectively. A value of 8.0 was determined for Trx f by electrofocusing, a technique which could not be applied to Trx f/m because of its low stability. We then conclude that acid Trxs are more stable than basic ones, and that their stability can be correlated with a net negative charge on the NH_2-side of the modulatory cysteines.

The same occurs in the hybrids built up by similar fusion of *E. coli* and *Anabaena* Trxs (9). The wild type proteins have pIs of 4.5 and 6.5, respectively; those of the hybrids are between these values. Both chimeric Trxs show a net negative charge of -3, focused on the NH_2-side of the regulatory cluster, whereas the COOH-domain supports a net basic nature. Accordingly both hybrids are stable, with a higher stability of the A-E form than the E-A one.

Trx m/f showed with antibodies against pea Trx m (ELISA) only 15% reactivity of that with pure Trx m, whereas with pea Trx f antibodies the reactivity was 21% of that with Trx f. If we consider the lack of cross reaction between pea Trxs f and m (4), the 15% reactivity between pea Trx m/f and Trx m must be related to their Trx m 34 common a.a., and the 21% cross reactivity between Trx m/f and Trx f to their Trx f 75 common ones. We can then deduce a higher antigenicity of the m-side than the f-side of the Trx m/f hybrid.

Fig. 2 shows the kinetic behaviour of pea Trx m/f in FBPase activation. Its affinity for FBPase is higher than that of Trx f, and its saturation ratio surprisingly low, with a Trx/FBPase quotient of about 0.4. Nevertheless, the catalytic rate is lower than those of both wild type Trxs. As occurs with pea Trx f, the saturation curve is sygmoid. The same has been described with the two hybrids between *E. coli* and *Anabaena* Trxs. Because of a lower K_m value, the A-E form shows a higher catalytic efficiency than both wild types when coupled to Trx-reductase and ribonucleotide-reductase (9). On the contrary, the opposite hybrid E-A has a much lower efficiency than the wild type Trxs. Moreover, as occurs with *E. coli* Trx, the hybrid Trx A/E promotes the growth of T7 bacteriophage, not supported by the hybrid Trx E/A. Trx m/f appears unable in activating pea NADP-MDH, at least up to a Trx/enzyme ratio of 60. This is unexpected if we consider that the pea enzyme becomes saturated at a Trx m/enzyme ratio of 6, and at a Trx f/NADP-MDH one about 40.

The tridimensional structure of Trx m/f (RASMOL program) shows some differences with that of *E. coli* Trx (Fig. 3). The β_1-strand is absent, and the axis of the α_4-helix shifts its

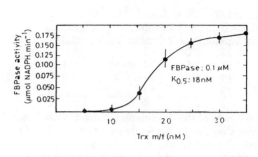

Figure 2.- Saturation curve of pea chloroplast FBPase by pea Trx m/f. Each value is the mean ± SE (bars) of four determinations. The indicated amounts of FBPase and Trxs are the concentrations in the final incubation mixture. $K_{0.5}$ values are also indicated in the figure.

Figure 3.- Three dimensional modelling of pea Trx m/f (RASMOL). The N and C inserts means the amino- and carboxi-terminus.

parallel position to that of helix α_2, to a new one perpendicular to the axis of the latter. In addition, the β_5-strand seems poorly developed. However, its high thermal stability seems to minimize these differences with E. coli and m-type Trxs. From these results we inferre parallel conclusions to those of E. coli-Anabaena Trx hybrids (9). In spite of the changes introduced on the NH_2-side of Trx f in the Trx m/f hybrid, the f-side of the latter keeps a structure for an efficient FBPase binding. Moreover, the introduction of the NH_2-side Trx m features into Trx f improves the stability of the whole molecule; on the contrary, these changes seem to disturb the catalytic rate of FBPase activation. The situation is different in Trx f/m. The stability provided by the NH_2-domain of Trx m to the Trx m/f form disappears. From these alternatives nature has selected the high stable structure of Trx m, and the catalytically efficient of Trx f; the hybrid Trx m/f could be a compromise between them.

References

1 Scheibe, R. (1990) Bot. Acta 103,323-334

2 Buchanan, B.B. (1992) Photosynth. Res., 33,147-162

3 Schürmann, P., Roux, J. and Salvi, L. (1985) Physiol. Vég. 23,813-818

4 López Jaramillo, J., Chueca, A., Jacquot, J.P., Hermoso, R., Lázaro, J.J., Sahrawy, M. and López Gorgé, J. (1997) Plant Physiol. 114,1169-1175

5 Hermoso, R., Castillo, M., Chueca, A., Lázaro, J.J., Sahrawy, M. and López Gorgé, J. (1996) Plant Mol.Biol. 30,455-465

6 Hodges, M., Miginiac-Maslow, M., Decottignies, P., Jacquot, J.P., Stein, M., Lepiniec, L., Cretin, C. and Gadal, P. (1994) Plant Mol. Biol. 24,225-234

7 Hertig, C. and Wolosiuk, R.A. (1980) Biochem. Biophys. Res. Commun. 97,325-333

8 Fickenscher, K. and Scheibe, R. (1983) Biochim. Biophys. Acta 749,249-254

9 Lim, Ch.J., Gleason, F.K., Jacobson, B.A. and Fuchs, J.A. (1988) Biochemistry 27,1401-1408

ANTISENSE DOWNREGULATION OF SPS IN *ARABIDOPSIS THALIANA*

Åsa Strand[1], Rita Zrenner[2], Mark Stitt[2], Petter Gustafsson[1] and Per Gardeström[1]
[1]Department of Plant Physiology, Umeå University, S-901 87 Umeå, Sweden
[2]Botanisches Institut, Universität Heidelberg, D-69120 Heidelberg, Germany

Key words: C metabolism, C partitioning, carbohydrate, sucrose

1. Introduction

Sucrose-phosphate synthase (SPS) is the major regulatory enzyme in sucrose biosynthesis catalyzing the synthesis of sucrose from Fru6P and UDPGlu. There are two distinct mechanisms to control the enzymatic activity of the SPS protein: allosteric control by Glu6P and Pi, and covalent modification by protein phosphorylation[1]. The amount of SPS transcript has also been reported to respond positively to glucose feeding[2]. At low temperatures *Arabidopsis* has been shown to increase SPS activity and the amount of transcript for SPS[3]. Further, the flux of carbon into sucrose also increases after growth at low temperatures indicating an important role for SPS in cold acclimation. Our aim is to elucidate the role of SPS in the cold acclimation process. *Arabidopsis* plants with SPS downregulated using antisense technique were constructed to test the performance at low temperatures, although data is only presented in this report from warm temperature grown plants.

2. Material and Methods

Plant material
Arabidopsis plants, ecotype Columbia, wt and 3 transgenic lines were grown under controlled environment conditions: 150 μmol photons m^{-2}, sec^{-1}; day/night regime 23/18°C; photoperiod 8 h.

Preparation of transgenic plants
The *Arabidopsis* EST clone 109I5T7 encoding SPS was cloned in an antisense orientation into the plant transformation vector pBI120 Km between the CaMV 35S promotor and the *nos* terminator. The vector was transformed into the *Agrobacteria* strain C58 rif^{R}/EHA 105 and *Arabidopsis* plants, ecotype Columbia, were transformed using vacuum infiltration. 20 independent lines with various degrees of down regulation of SPS were obtained from the transformation.

Carbohydrates and enzymes
Carbohydrates were determined according to Stitt et al., (1989). Samples for SPS, cytFBPase and AGPase were collected 4 h into the photoperiod and the enzyme activities were measured according to Hill et al., (1996) and Strand et al., (1997). For imunoblotting SPS polyclonal antibodies from potato[6], AGPase polyclonal antibodies from tomato[7], and cytFBPase polyclonal antibodies from *Arabidopsis* (Strand et al., upublished) were used.

G. Garab (ed.), Photosynthesis: Mechanisms and Effects, Vol. V, 3691–3694.

CO₂ exchange

In situ photosynthetic CO_2 exchange was measured under ambient conditions usi   a open flow gas exchange system.

3. Results and Discussion

The antisense construct

The *Arabidopsis* EST clone 109I5T7 encoding SPS was cloned in an antisense orientation into the plant transformation vector pBI120 Km between the CaMV 35S promotor and the *nos* terminator (Fig 1). The EST clone 109I5T7 is a partial clone which is 1800 bp long and covering the 3′end starting ~1700 bp into the spinach and potato sequences.

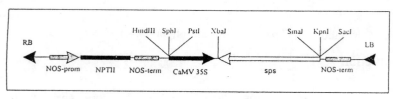

Fig. 1. The EST clone 109I5T7 was cloned in an antisense orientation into the plant transformation vector pBI120 Km.

Transgenic lines

Three independent transgenic lines (T1, T5 and T6), with various degrees of repression of SPS were analyzed and compared to wt. The amounts of SPS protein was determined by Western blot analysis and a clear reduction in the amount of SPS protein was observed in all 3 transgenic lines (Fig. 2). SPS activity was also measured in those lines and the activity correlated with the different levels of SPS protein. T1 showed 18%, T5, 32%, and T6, 22%, of wt SPS activity respectively. The activation state increased from approximately 30% in wt to around 40% in the 3 transgenic lines. However, this increase is not sufficient to compensate for the lower amounts of SPS protein in the transgenic lines.

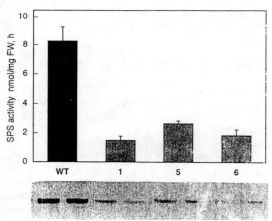

Fig. 2. Total enzyme activity and Western blot for SPS for wt and the 3 transgenic lines. The activity represent the mean (±SD) of at least 4 different plants. Samples from two different plants of each transgenic line are shown on the Western.

CO$_2$ exchange

CO$_2$ exchange was determined at ambient CO$_2$ and at growth light. The CO$_2$ exchange rate observed for wt was 8.8 µmol CO$_2$ m^{-2}, s^{-1}. All the transgenic lines were impaired in their CO$_2$ exchange rate: the T1 plants 6.3, T5, 7.1 and T6, 5.3 µmol CO$_2$ m^{-2}, s^{-1}. The CO$_2$ exchange rates were mean values from at least 6 different leaves and the SD was less then 10%. Clearly, the reduction of SPS activity affected the carbon assimilation.

Carbohydrates

Samples for carbohydrates were collected at various intervals throughout the photo period. All 3 transgenic lines accumulated less sucrose compared to wt (Fig. 3).The level of sucrose accumulation also coincided with the amount of SPS remaining in the transgenic lines. T1 and T6 contained only 30 and 20 %, respectively, of the wt levels of sucrose at the end of the light period, T5 which has higher SPS activity compared to T1 and T6, contained about 50 % of the wt sucrose. Surprisingly the starch pool was also effected in T1 and T6. T5 accumulated wt levels of starch during the light period but T1 and T6 contained less then 20 % of the wt starch. Consequently, the largest difference in carbohydrates was found in starch content, and it appears as if the transgenic lines do not compensate for the reduced activity of SPS by increasing the production of starch. Differences were also found in the pools of free hexoses. Fructose increased in wt during the light period whereas it remained low throughout the day in T1 and T6. The T5 plants showed a small increase in the amount of fructose during the day, which correlated with the higher sucrose levels observed in those plants. The amount of glucose was almost constant throughout the day in wt. The T5 plants also showed the same glucose levels as wt. T1 and T6 showed almost undetectable levels of free glucose. The level of SPS protein clearly affects the accumulation of carbohydrates, which coincides with the reduced rate of CO$_2$ exchange observed in the transgenic lines.

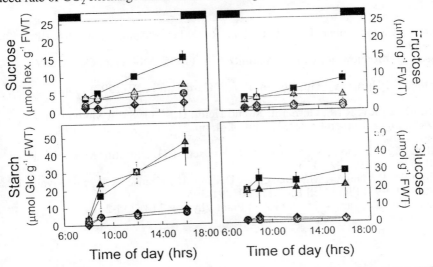

Fig. 3. Carbohydrate content for wt (■) and the 3 transgenic lines, T1 (◉), T5 (▲) and T6 (♦). Each point represent the mean (±SD) of at least 4 different plants. The open bars on the top of the figure represents the photoperiod.

Enzyme activity and protein levels for cytFBPase and AGPase

Activity for another important regulatory enzyme in the pathway for sucrose synthesis, cytosolic fructose-1,6-bisphosphatase (cytFBPase) was determined for the 3 transgenic lines and wt. cytFBPase activity was similar to wt (3 nmol mg FW^{-1}, h^{-1}) in all 3 transgenic lines. This was further confirmed by Western blot analysis, which showed the same protein levels for cytFBPase in wt and the 3 transgenic lines suggesting that cytFBPase is not coregulated with SPS. The activity for the key enzyme in starch biosynthesis, ADP-glucose pyrophosphorylase (AGPase) was also determined. The wt activity for AGPase was 20 nmol mg FW^{-1}, h^{-1}. In T1 slightly less activity for AGPase was found, 14 nmol mg FW^{-1}, h^{-1}, whereas the other two transgenic lines, T5 and T6, showed wt activity. Western blot showed no reduction in the amount of AGPase protein in any of the transgenic lines. In summary no major changes were observed in the transgenic plants for these two key enzymes in carbon metabolism and carbon allocation.

4. Conclusion

We have shown that reduced activity of SPS decreased the rate of CO_2 exchange, indicating that the activity of SPS is closely linked to the CO_2 assimilation. Further, our data suggest that the transgenic plants do not compensate for the reduced activity of SPS in the cytosol by increasing the accumulation of starch. Clearly, the level of SPS protein affects the accumulation of carbohydrates. Further, if the accumulation of sucrose in the cytosol does not exceed a threshold level it appears as though no carbon will be allocated for starch synthesis. The plant cell may somehow have the ability to sense the amount of sucrose in the cytosol and consequently the activity of SPS might play an important role in this regulation of carbon allocation.

5. References

1 Huber S.C. and Huber, J.L. (1996) Ann. Rev. Plant Physiol. Plant Mol. Biol. 47, 431-444
2 Hesse, H., Sonnewald, U. and Willmitzer, L. (1995) Mol. Gen. Genet. 247, 515-520
3 Strand, Å., Hurry, V., Gustafsson, P. and Gardeström, P. (1997) Plant J.,12(3), 605-614
4 Stitt, M., Lilley, R.M., Gerhardt, R. and Heldt, H.W. (1989)in Methods in Enzymology: Biomembranes, Volume 174, pp. 518-552, Academic press, Amsterdam
5 Hill, L.M., Reimholz, R., Schröder, R., Nielsen, T.H. and Stitt, M. (1996) Plant Cell Environ. 19, 1223-1237
6 Reimholz, R., Geiger, M., Haake, V., Detting, U., Krause, K.P., Sonnevald, U. and Stitt, M. (1997) Plant Cell Environ. 20, 291-305
7 Chen, B.Y. and James, H.W. (1997) Plant Physiol. 113, 235-241

MOLECULAR CROWDING AND CYTOSKELETAL PROTEINS AFFECT THE ALLOSTERIC REGULATORY PROPERTIES OF ADP GLUCOSE PYROPHOSPHORYLASE.

Diego F. Gómez Casati, Miguel A. Aon, Alberto A. Iglesias. Instituto Tecnológico de Chascomús (INTECH, SECyT-CONICET), Camino Circunv. Laguna km 6, CC 164, Chascomús, 7130, Bs. As., Argentina.

Key words: carbohydrate, cyanobacteria, enzymes, polysaccharides, protein-protein interaction.

Introduction.

Starch is the major product of photosynthesis in plants (1). Starch synthesis occurs in chloroplasts (in green tissues) or amyloplasts (reserve tissues) through a pathway which uses ADPGlc as the glucosyl donor, in a similar form as that taking place during glycogen synthesis by bacteria (for reviews see 1-4). The key regulatory step for this biosynthetic route is the production of the sugar-nucleotide, a reaction catalyzed by ADPGlc pyrophosphorylase (AGP; EC 2.7.7.27):

$$Glc1P + ATP \Leftrightarrow ADPGlc + PPi$$

In plants, AGP is allosterically regulated by 3P-glycerate (3PGA, activator) and Pi (inhibitor) and the fine regulation of the enzyme by these metabolites has been clearly demonstrated to directly determine levels of starch accumulated in plant tissues (2,3,5). AGP from cyanobacteria constitutes a good experimental system, since it possesses properties intermediate between bacterial and plant enzymes (3).

Studies on the characterization of enzymes playing key roles in plant metabolism were performed with pure enzymes in aqueous media. A few reports deal with the behavior in environments which could be closer to physiological or stress conditions occurring *in vivo* (6-9). In the present work we study the effect of polyethyleneglycol (PEG)-induced molecular crowding and microtubular protein (MTP) on the activity and regulation of AGP from cyanobacteria as an approach to analyze the possible kinetic properties of the enzyme *in vivo*.

Materials and Methods.

AGP from *Anabaena* PCC 7120 used in this study is the recombinant enzyme obtained from *Escherichia coli* B, strain AC70R1-504 transformed with pAnaE3a (10). The enzyme was purified after (11).

Enzyme activity was assayed in the ADPGlc synthesis direction as previously described (11). The standard reaction mixture (200 µl final volume) contained 100 mM MOPS-KOH buffer pH 7.5, 6.25 mM $MgCl_2$, 2.5 mM ATP, 0.5 mM [^{14}C]Glc1P (about 1000 cpm/nmol) and 0.15 units of inorganic pyrophosphatase. Assays were initiated by the addition of enzyme and, after 10 min at 37°C, they were stopped by heating at 100°C during 45 s. Activity values are expressed as relative to the activity of the enzyme (about

3695

G. Garab (ed.), Photosynthesis: Mechanisms and Effects, Vol. V, 3695–3698.
© 1998 *Kluwer Academic Publishers. Printed in the Netherlands.*

3 units/mg) assayed in the absence of additions. Modifications to the assay medium by additions of 3PGA, Pi, PEG, MTP, or others are specified in each case.

MTP was purified from *Saccharomyces cerevisiae* by two cycles of polymerization-depolymerization (G. Sosa, D. Gomez Casati, A. Iglesias, M. Aon, unpublished data). AGP activity was assayed in the presence of 0.5 mg/ml of polymerized or depolymerized MTP. The polymerization of MTP was started by the addition of 1 mM GTP and performed at 37°C during 10 min. After MTP polymerization, the AGP reaction was started by the addition of 10 mM [^{14}C]Glc1P to the assay medium. ADPGlc synthesis was measured as described above.

Results and Discussion.

Studies were carried out using a highly purified recombinant AGP from the cyanobacterium *Anabaena* PCC 7120 (10), and the regulatory properties of the enzyme were analyzed in diluted aqueous conditions, in the presence of a molecular crowding agent such as PEG 8000, and in the presence of proteins which compose the cytoskeleton, such as MTP.

Table I shows the effects of PEG 8000 and polymerized MTP on the activity of AGP. As previously described (11), 3PGA activated the enzyme by 15-fold when assayed in a buffered aqueous medium (Table I). It can be seen that both compounds affect the enzyme activity in a distinctive manner and they are dependent on the absence or the presence of the allosteric activator of AGP in the medium. Thus, as shown in Table I, PEG inhibited the enzyme activity by 80% or activated the enzyme by near 2-fold depending on the absence or the presence of 2.5 mM 3PGA, respectively. On the other hand, polymerized MTP (the non-polymerized protein exhibited no effect, see Table I) activated AGP by 3-fold when assayed in the absence of 3PGA, whereas exhibited no effect in its presence (Table I).

Table I. Effect of a molecular crowding agent (PEG 8000) and the extent of polymerization of MTP on the activity of AGP from *Anabaena* PCC 7120.

Compound	Relative activity	
	- 3PGA	+ 3PGA[a]
None	1.0	15.0
PEG 8000 9 % W/W	0.2	29.0
non polymerized MTP 0.5 mg/ml	1.1	14.9
polymerized MTP 0.5 mg/ml	3.0	13.5

[a] 3PGA was added at 2.5 mM

The effects of PEG and MTP on AGP were dependent upon their respective concentrations in the assay medium, as illustrated in Figure 1. In the absence of 3PGA, the enzyme was 90% inhibited at PEG concentrations of 12% (W/W), with 50% inhibition reached at 5% PEG (Fig. 1A). The inhibitory effect of PEG could be partially overcome by the addition of Pi. In the presence of 3PGA (with or without Pi) a quite different result was obtained, since a 2-fold maximal activation of AGP (which represents 30-fold when compared with assays in the absence of additions, see Table I) was observed at concentrations of PEG of 7% or higher, with half-activation obtained at 2.5% PEG (Fig. 1A).

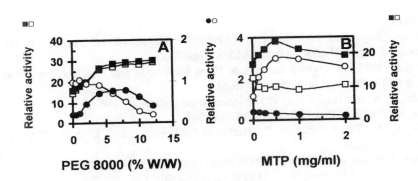

Figure 1. Effect of microtubular protein and PEG-induced molecular crowding on AGP activity. The enzyme was assayed as described under Materials and Methods in the presence of the indicated concentrations of PEG 8000 (A) or MTP (B), and without further additions (open circles), plus 2.5 mM 3PGA (open squares), plus 0.1 mM Pi (filled circles) or in the presence of 3PGA and Pi (filled squares).

Concerning the effect of different concentrations of polymerized MTP on AGP activity, data are shown in Figure 1B. As can be seen, MTP (in the range 0-2 mg/ml) had little effect on the enzyme activity assayed in the presence of 3PGA or Pi. However, when assayed in the absence of further additions, AGP was activated by 3-fold at a concentration of 0.5 mg/ml of polymerized MTP (Fig. 1B).

Several lines of evidence support a structured view of the cytoplasm, giving rise to the question of the consequences for the dynamics of processes occurring in such a highly crowded but organized enviroment (8). Crowding may alter the average degree of self-association of an enzyme, in turn changing its average catalytic activity (8,14). Besides, MTP was shown to specifically induce flux activation at steps catalyzed by carbon metabolism-related enzymes (8).

At present it is unclear the actual conditions of the intracellular milieu of bacteria (like cyanobacteria) or in the chloroplast stroma (where AGP localizes) of higher plants. It is a subject of some speculation whether bacteria contain microtubules (12) although a molecular crowding condition at least exist (8,13). According to this scenario, the results showed above suggest that *in vivo*, the allosteric regulation of an enzyme involved in starch biosynthesis may be effected in a more effective and subtle way than that deduced from *in vitro* studies.

Acknowledgements. This work was supported, in part, by grants from Fundación Antorchas and Comisión de Investigaciones Científicas (CIC). MAA and AAI are research careers from CONICET, Argentina and DFGC is a fellow from the same Institution.

References
1. Iglesias, A.A. and Podestá, F.E. (1996) In: Handbook of Photosynthesis (M. Pessarakli, ed.) pp. 681-698. Marcel Dekker, New York.
2. Preiss, J. (1991) In: Oxford Surveys of Plant Molecular and Cell Biology (J. Miflin, ed.) pp. 59-114. Oxford Univ. Press, Oxford.
3. Iglesias, A.A. and Preiss, J. (1992) Biochem. Education 20, 196-203.
4. Okita, T.W. (1992) Plant Physiol. 100, 560-564.
5. Stark, D.M., Timmerman, K.P., Barry, G.F., Preiss, J.and Kishore, G.M. (1992) Science 258, 287-292.
6. Podestá, F.E. and Andreo, C.S. (1989) Plant Physiol. 90, 427-433.
7. Podestá, F.E. and Plaxton, W.C. (1993) Plant Physiol. 103, 285-288.
8. Aon, M.A. and Cortassa, S. (1997) Dynamic Biological Organization. Fundamentals as Applied to Cellular Systems. Chapman & Hall, London., 549 pp.
9. Aon, M.A., Cortassa, S., Gomez Casati, D. and Iglesias, A.A. (1998). (submitted).
10. Charng, Y-y., Iglesias, A.A. and Preiss, J. (1994) J. Biol. Chem. 269, 24107-24113.
11. Iglesias, A.A., Kakefuda, G. and Preiss, J. (1991) Plant Physiol. 97, 1187-1195.
12. Bermudes, D., Hinkle, G. and Margulis, L. (1994) Microbiol. Rev. 58, 387-400.
13. Garner, M.M., and Burg, M.B. (1994) Am. J. Physiol. 266 (Cell Physiol. 35) C877-C892.
14. Minton, A.P. and Wilf, J. (1981) Biochemistry 20, 4821-4826.

THE ROLE OF THE C-TERMINAL EXTENSION OF SORGHUM NADP-MALATE DEHYDROGENASE IN THE INHIBITION OF ITS REDUCTIVE ACTIVATION BY NADP.

Miginiac-Maslow M., Ruelland E., Schepens I., Issakidis-Bourguet E. and Decottignies P.
ERS CNRS 569. Institut de Biotechnologie des Plantes. Bât. 630.
Université de Paris-Sud 91405 Orsay Cedex France

Key words: C metabolism, chloroplast, disulphide bond, light activation, site-directed mutagenesis, thioredoxin.

1. Introduction

In addition to being submitted to thiol-regulation by reduced thioredoxin, the chloroplastic NADP-dependent malate dehydrogenase (NADP-MDH: EC. 1.1.1.82) is regulated by the redox state of its cofactor. Indeed, the oxidized cofactor NADP inhibits the reductive activation of the enzyme. This feature, first reported 15 years ago [1, 2] is still poorly understood. The activity of the fully activated enzyme is not inhibited by NADP, thus the effect on activation must be linked to a subtle difference between the conformation of the acitve vs. inactive enzyme. NAD does not inhibit significantly: the specificity of the inhibition parallels the specificity of the enzymatic reaction. Modeling of the 3-D structure of NADP-MDH based on the coordinates of the permanently active NAD-dependent porcine heart enzyme [3] showed only one coenzyme binding site, accessible from the outside. The accessibility of the coenzyme binding site in the oxidized enzyme is also substantiated by the fact that triazine-dye affinity columns mimicking the structure of the cofactor (matrex red A) can be used to purify the enzyme. The hypothesis put forward in relation with the model was that the oxidized cofactor would bind in a wrong orientation [3]. In the course of our investigation on the mechanism of the reductive activation process of the enzyme by thiol/disulfide interchange with reduced thioredoxin, we created a number of partially deregulated mutants where selected disulfides were suppressed by site-directed mutagenesis by exchanging cysteines for serines or alanines [3,4]. The proteins mutated at the C-terminal cysteines 365 and/or 377, quite unexpectedly, lost their sensitivity to inhibition by NADP. This result suggested that in some way, the C-terminal end of the enzyme was implicated in the inhibition effect. In the present paper, the effect of NADP on NADP-MDH activation has been further investigated by additional site-directed mutagenesis experiments targeting either the last two C-terminal residues, or residues supposed to be implicated in the cofactor specificity. A tentative explanation is proposed for the observed effects: the positively charged oxidized cofactor would intaract with the negatively charged C-terminus of the enzyme, thus enhancing the locking of the active site.

3699

G. Garab (ed.), Photosynthesis: Mechanisms and Effects, Vol. V, 3699–3702.
© *1998 Kluwer Academic Publishers. Printed in the Netherlands.*

2. Procedures

Recombinant sorghum leaf NADP-MDH either wild-type or modified by site-directed mutagenesis [3-5] was used in these studies. Additional mutations were performed either by the method of Kunkel [6] (C-terminal mutants E387Q and ΔEV) or by PCR (cofactor specificity mutant S87I/R89N/S90A). The proteins were obtained by expressing the corresponding cDNAs in *E.coli* BL21 strain using the very efficient pET system [7]. The overproduced proteins were purified to homogeneity by already described procedures [4]. The activation tests were performed with *E. coli* thioredoxin reduced by DTT. Enzyme activity measurements were done spectrophotometrically at 340 nm.

3. Results and discussion

3.1. *Effects of the suppression of the C-terminal disulfide bridge*
When the cysteines of the C-terminal disulfide bridge were substituted for alanines, either individually or together, the activation of the mutant proteins was not inhibited by NADP (Fig.1), showing that the effect was due to the removal of the C-terminal disulfide, mimicking the situation where the enzyme is reduced (activated).

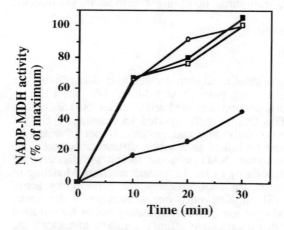

Fig. 1: Effect of NADP on reductive activation of wild type (WT) and mutant NADP-MDHs. 1 mM NADP was added to the activation medium which contained 10mM DTT and 15 uM thioredoxin m.

—○— WT Control

—●— WT NADP

—□— C365/377A Control

—■— C365/377A NADP

3.2. *Effects of the suppression of the C-terminal negative charges*
In relation with the hypothesis that the C-terminal end of the enzyme inhibits its activity in the oxidized form because the C-terminal bridge maintains the C-terminus within the active site, moving upon reduction (see Ruelland et al., ref.8), mutants E387Q and ΔEV (deleted by the 2 most C-terminal residues) were created in order to investigate the possible electrostatic interactions between the negatively charged C-terminus, and positive charges within the active site of the enzyme. These mutants, which had no spontaneous activity, but could be considered as having a partially accessible active site, showed no susceptibility to inhibition by NADP (Fig.2).

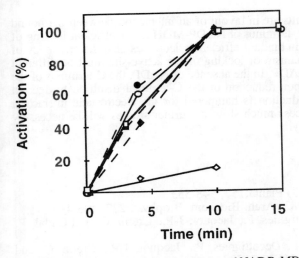

Fig. 2: **NADP effect on the activation of NADP-MDH mutated on the C-terminal negatively charged residuess**
The purified enzymes were activated by reduced thioredoxin in presence (open symbols) or absence (closed symbols) of 1 mM NADP.
Circles: E387Q MDH; Squares: ΔEV MDH; Diamonds: Wild-type MDH.

3.3. *Changing the cofactor specificity of NADP-MDH*

The thiol-regulated MDH is NADP-dependent, whereas the permanently acive forms are all NAD-dependent [9]. To investigate the relationship between cofactor specificity and sensitivity to its oxidized form, a mutant enzyme has been engineered with the aim to render it able to use NADH. Based on examples in the literature where specificity of NAD-dependent forms was changed into NADP-dependent forms [10], and on sequence comparisons, a S87I/R89N/S90A mutant has been created by PCR (Ser and Arg are supposed to coordinate the phosphate group). The specificity of the mutant protein was not completely reversed, but it became able to use both cofactors with similar efficiencies.

The mutation markedly changed the cofactor specificity of the inhibition by the oxidized form (Fig.3): inhibition by NADP was strongly decreased and inhibition by NAD dramatically increased.

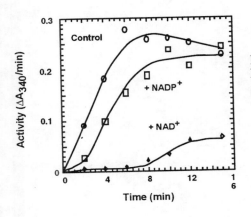

Fig. 3: Inhibition of the activation of a mutant MDH (S87I/R89Q/S90A) with changed cofactor specificity by NADP$^+$ or NAD$^+$

4. Conclusions

In conclusion, the present results are in favour of an interaction between the bound cofactor and the negatively charged C -terminus of NADP-MDH for an efficient locking of the access to the active site. Decrease in binding efficiency decreases also the efficiency of inhibition of the activation. The mechanism of locking of the active-site must be different in the presence or in the absence of NADP. In the absence of NADP, the C-terminus of the enzyme should be easily displaced upon reduction of the C-terminal disulfide bridge. In the presence of NADP, either the reduction is hampered, or the electrostatic interaction stronger, rendering the activation process much slower. Further studies will be necessary to elucidate this mechanism completely.

5. References

1. Scheibe, R. and Jacquot, J-P. (1983) Planta,157, 548-553
2. Ashton, A.R. and Hatch, M.D. (1983) Arch. Biochem. Biophys., 227, 406-415
3. Issakidis, E., Saarinen, M., Decottignies, P., Jacquot, J-P., Crétin, C. and Gadal, P. (1994) J. Biol. Chem. 269, 3511-3517
4. Issakidis, E., Miginiac-Maslow, M., Decottignies, P., Jacquot, J-P., Crétin, C. and Gadal, P. (1992) J. Biol. Chem. 267, 21577-21583
5. Ruelland, E., Lemaire-Chamley, M., Le Maréchal, P., Issakidis-Bourguet, E., Djukic, N. and Miginiac-Maslow, M. (1997) J. Biol. Chem. 272, 19851-19857
6. Kunkel, T.A. (1985) Proc. Natl. Acad. Sci. USA, 82, 488-492
7. Studier, F.W., Rosenberg, A.H., Dunn, J.J. and Dubendorff, J.W. (1990) Methods in Enzymol., 185, 60-89
8. Ruelland, E., Decottignies, P., Djukic, N. and Miginiac-Maslow, M. (1998) These proceedings.
9. Goward, C.R. and Nicholls, D.J. (1994) Protein Science, 3, 1883-1888
10. Nishiyama, M., Birktoft, J.J. and Beppu, T. (1993) J. Biol. Chem. 268, 4656-4660

MECHANISM OF AUTO-INHIBITION OF NADP-MALATE DEHYDROGENASE BY ITS C-TERMINAL EXTENSION.

Ruelland E., Decottignies P., Djukic N. & Miginiac-Maslow M.
ERS CNRS 569. Institut de Biotechnologie des Plantes. Bât. 630.
Université de Paris-Sud 91405 Orsay Cedex France

Key words: C metabolism, chloroplast, disulphide bond, light activation, site-directed mutagenesis, thioredoxin.

1. Introduction

NADP-malate dehydrogenase (NADP-MDH: EC. 1.1.1.82) is a chloroplastic enzyme activated in the light by the ferredoxin/thioredoxin system. Its totally inactive oxidized form contains two disulfide bridges per subunit, located in specific sequence extensions, one at the N-terminus [1, 2] and the other at the C-terminus [3]. The C-terminal extension shields the access to the active site [4]. Upon reduction, the N-terminal bridge is isomerized and the newly created disulfide is reduced. During this reduction process, the active site undergoes a conformational change towards a high catalytic efficiency conformation [5]. The reduction of the C-terminal bridge leads to a displacement of the C-terminal extension, uncovering the access to the active site. Up to now, the molecular mechanism of the shielding of the active site by the oxidized C-terminal extension is poorly understood. The extension might act as a lid, as suggested by proteolysis experiments [6], or it might enter the active site and bind to specific active-site residues, as suggested by molecular modeling [7]. In this regard, it can be noted that the C-terminal end of the protein bears two negative charges: the C-terminal carboxyl group and the lateral carboxyl group of the penultimate glutamate residue and thus can mimick the dicarboxylic substrate of the enzyme.

In the present study, the biochemical basis of the interaction of the C-terminal extension with the core part of the protein has been examined by substitution of selected amino-acid residues by site-directed mutagenesis techniques. The results highlight the role of the two most C-terminal residues in the locking of the access to the active site.

2. Procedures

Recombinant sorghum leaf NADP-MDH either wild-type or modified by site-directed mutagenesis [8] was used in these studies. Combined mutations were performed by cloning previously mutated fragments of the enzyme into appropriate restriction sites of the newly created mutants. The proteins were obtained by expressing the corresponding cDNAs in *E.coli* using the very efficient pET system [9]. The overproduced proteins were purified to homogeneity by already described procedures [2,3]. The activation tests were performed with *E. coli* thioredoxin reduced by DTT. Enzyme activity measurements were

G. Garab (ed.), Photosynthesis: Mechanisms and Effects, Vol. V, 3703–3706.

done spectrophotometrically at 340 nm. Chemical derivatization with the histidine-specific reagent diethyl pyrocarbonate (DEPC) was done as described previously [10].

3. Results and discussion

3.1. Effects of the suppression of the C-terminal negative charges

To address the problem of the locking of the active site through electrostatic interaction between the C-terminal negative charges and the core part of the protein, two mutants were created: a deletion mutant where the two most C-terminal residues have been suppressed (ΔEV mutant) and a mutant where the penultimte Glu has been replaced by a Gln (E387Q mutant). When the activity of the proteins was measured, no basal activity was observed: the mutant proteins still required preactivation by reduced thioredoxin and their activation rates were similar to those of the WT protein (Fig. 1). However, the mutations conferred on them some of the characteristics of the proteins mutated on the cysteines of the C-terminal disulfide bridge. In particular, treatment of the unactivated mutants with diethyl pyrocarbonate (DEPC), a reagent specifically derivatizing histidine residues, partially inhibited the activity of the activated protein, suggesting that the active-site histidine became partially accessible in the mutants. This observation was substantiated by the observation that NADPH protected against inactivation. In these conditions, the unactivated WT enzyme is insensitive to DEPC.

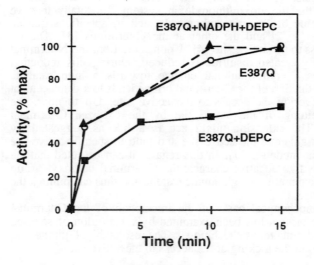

Fig.1. Inhibition of the E387Q NADP-MDH mutant by pretreatment with DEPC of the unactivated enzyme. Protection by NADPH. The unactivated enzyme was pretreated with 1 mM DEPC in the presence or in the absence of 1 mM NADPH. Then 20 μM thioredoxin + 10 mM DTT in 100 mM Tris buffer 7.9 were added and the appearance of activity measured as a function of time.

3.2. *Combined effcts of suppression of the C-terminal negative charges and of both regulatory disulfides at the N-terminus.*

Our previous studies demonstrated that the reduction/isomerization of the N-terminal diuslfides induced a conformational change of the active site, shaping it in a highly active conformation. In order to investigate the role of the C-terminal negative charges in an enzyme having reached its fully active conformation, the mutations of the last C-terminal residues have been combined with the mutation of the N-terminal cysteines (mutants C29S/C207A/E387Q or C29S/C207A/ΔEV and control mutant C29S/C207A).

The control mutant, where the N-terminal disulfides were removed, exhibited the fast activation kinetics of all of the N-terminal cysteine mutants [2,5] and had no spontaneous activity in the oxidized form (data not shown). When the mutations E387Q or ΔEV were superimposed on the mutation of the N-terminal cysteines, the unactivated enzymes acquired 50% of the activity of the activated forms and this spontaneous activity was fully inhibited by DEPC (Fig.2). Upon increasing oxaloacetate concentration in the reaction medium, the spontaneous activity could reach 95%, the Km oxaloacetate of the oxidized enzyme being much higher than the Km of the activated enzyme (data not shown).

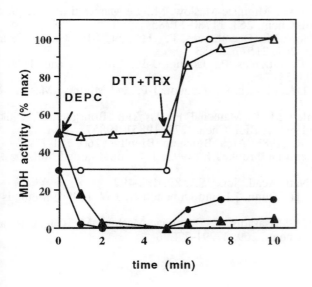

Fig. 2 Activity of the combined C29S/S207A/E387Q or C29S/S207A/DEV mutants before and after activation with reduced thioredoxin. Sensitivity to pretreatment with DEPC. Circles: E387Q mutation. Triangles: ΔEV mutation. Open symbols: controls. Closed symbols: Pretreated with 1mM DEPC.

This result clearly shows that the C-terminal negative charge is involved in the locking of the active site, but demonstrates also that there is an interaction between the conformational change triggered by the reduction of the N-terminal disulfides and the release of the C-terminus from the active site.

When only one of the two possible N-terminal disulfides was suppressed, the results were much less dramatic: only a tiny spontaneous activity (2% to 4% of the activity of the fully activated enzyme) could be observed upon removal of the C-terminal negative charge (data not shown). Thus a full conformational change of the active site is necessary to release the C-terminus.

4. Conclusions

The present results clearly demonstrate that the negative charge of the penultimate C-terminal residue of NADP-MDH is involved in the locking of its active site but show also that there is a cooperation between the C-terminal end and the conformational changes linked to the reduction of the N-terminal disulfides to uncover the access to the active site. They are consistent with the hypothesis [7] that the C-terminal end of the enzyme would act as an internal inhibitor mimicking the substrate oxaloacetate.

5. References

1. Decottignies, P., Schmitter, J-M., Miginiac-Maslow, M., Le Maréchal, P., Jacquot J-P., and Gadal, P. (1988) J. Biol. Chem. 263, 11780-11785
2. Issakidis, E., Miginiac-Maslow, M., Decottignies, P., Jacquot, J-P., Crétin, C. and Gadal, P. (1992) J. Biol. Chem. 267, 21577-21583
3. Issakidis, E., Saarinen, M., Decottignies, P., Jacquot, J-P., Crétin, C. and Gadal, P. (1994) J. Biol. Chem. 269, 3511-3517
4. Issakidis, E., Lemaire, M., Decottignies, P., Jacquot, J-P. and Miginiac-Maslow, M. (1996) FEBS Lett. 392, 121-124
5. Ruelland, E., Lemaire-Chamley, M., Le Maréchal, P., Issakidis-Bourguet, E., Djukic, N. and Miginiac-Maslow, M. (1997) J. Biol. Chem. 272, 19851-19857.
6. Fickensher, K. and Scheibe, R. (1988) Arch. Biochem. Biophys. 260, 771-779
7. Jackson, R., Sessions, R.B. and Holbrook, J.J. (1992) J. Comput.-aided Mol. Des. 6, 1-18
8. Kunkel, T.A. (1985) Proc. Natl. Acad. Sci. USA, 82, 488-492
9. Studier, F.W., Rosenberg, A.H., Dunn, J.J. and Dubendorff, J.W. (1990) Methods in Enzymol., 185, 60-89
10. Lemaire, M., Schmitter, J-M., Issakidis, E., Miginiac-Maslow, M., Gadal, P. and Decottignies, P. (1994) J. Biol. Chem., 269, 27291-27296

DIFFERENCES BETWEEN *CHROMATIUM MINUTISSIMUM* AND *RHODOSPIRILLUM RUBRUM* IN PHYTOHORMONAL COMPOSITION

Olga P. Serdyuk, Lidiya D. Smolygina, Elena. P. Ivanova
Institute of Basic Biological Problems, Russian Academy of Sciences,
142 292 Pushchino, Moscow Region, Russia.

Key-words: purple bacteria, phylogeny, hormones, cytokinins, auxins

1. Introduction

Plant hormones belong to compounds widespread in nature. Their structure and functions have been well studied not only in higher plants but also in non-phototrophic microorganisms entering into various kinds of interactions with plants. However, a rather large group of phototrophs, namely the phototrophic purple bacteria, have fallen out of the field of vision of phytohormone researchers and these substances have not been studied in these bacteria up to the present (1). The phototrophic bacteria in physiological and morphological properties represent a heterogenous group of microoganisms. In their phenotypical features they are divided into several large families consisting of numerous genera. But phylogenetic analysis of 16S rRNA nucleotide sequences permitted to divide phototrophic bacteria into just a few classes, namely the α –, β – and γ – subdivisions; some representatives of purple bacteria proved to be in closer relationship to non-phototrophic microorganisms than between themselves (2). As mentioned above some non-phototrophic bacteria have the ability to synthesize phytohormones. Cytokinins or auxins play a key role under parasitic or symbiotic interections of non- phototrophic bacteria with plants. Purple bacteria are free-living microorganisms. But sometimes these bacteria can be found in the rhizosphere of such aqueous cultures as rice or *Azolla*, but their importance for plants has not been clearly determined. Thus, the question arose of the presence and content of phytohormones as well as of their function in these bacteria. Besides this, the formation and selection of structures concerning photosynthesis and its regulation, ocurring namely in these bacteria, is the result both of similarity and the difference of these features between the bacteria themselves as well as between the bacteria and plants. Phytohormones apparently are not the exeption. Thus, purple bacteria can be an interesting object not only in studies of structure and function of phytohormones but of their evolution.

G. Garab (ed.), Photosynthesis: Mechanisms and Effects, Vol. V, 3707–3710.
© 1998 *Kluwer Academic Publishers. Printed in the Netherlands.*

2. Materials and methods

2.1 Object and conditions of cultivation

The *Rhodospirillum rubrum* 1R and *Chromatium minutissimum* were kindly supplied by the Department of Microbiology, Moscow State University (Moscow, Russia). The cells of *Rs. rubrum* were grown in 1 litre flasks on Ormerod's medium, pH 6.8 and *Cm. minutissimum* on Larsen's medium, pH 7.3 and harvested at the stationary growth phase.

2.2 Extraction of auxins and cytokinins

Sonicated *Rs. rubrum* and *Cm. minutissimum* cells were extracted by methanol to a final concentration of 80%. After incubation of these suspensions for 16 h at 4 ^{0}C, the methanol was evaporated and the aqueous phases were adjusted to pH 2.8. After extraction of acidic aqueous phase by ethylacetate, the crude fraction of auxins was obtained. After that the aqueous phase was adjusted to pH 7.8. Extraction of the alkaline aqueous phase by *n*-butanol gave the crude fraction of cytokinins. The crude fractions of auxins and cytokinins from the media of *Rs. rubrum* and *Cm. minutissimum* were obtained by the analogous method as for biomass without preliminary extraction by methanol.

2.3 Thin-layer chromatography

The crude fractions of auxins and cytokinins were separated on Silufol plates in a solvent system of isopropanol:benzene:ammonia (4:1:1). Individual bands were scraped and then extracted with 96.6 % ethanol.

2.4 Spectral measurements

The UV-spectra were recorded with a Specord M-40 (Germany) and a Shimadsu UV-160 (Japan) instruments. The electron-impact mass spectra (MS) were measured with a high resolution Finnigan MAT 8430 mass spectrometer.

2.5 Cytokinin bioassays

The cytokinin activity of individual substances was tested in specific bioassays according to the procedure of Letham (3).

3. Results and Discussion

3.1 Phytohormones of purple non-sulfur bacteria Rs. rubrum of the α - subdivision

More early studies have shown that the main phytohormone of *Rs. rubrum* is cytokinin of the purine row, namely zeatiribozide, which was found both in the biomass (1) and in the medium of this bacteria (4). One more compound exhibiting high cytokinin activity in *Amaranthus caudatus* bioassay has been also found in the *Rs. rubrum* cultural medium (5). It was identified as 2-(p-hydroxyphenyl)-ethanol by MS and ^{1}H NMR spectra (Fig. 1). This compound exibited cytokinin activity in other specific for this class of phytohormone bioassays: on cuttings of radish leaves (an increase in their weight), on wheat seedlings (an increase in chlorophyll content), on tobacco callus (a stimulation of cell division), on leaf explants of strawberry, actinidia and cherry (a stimulation of organogenesis). Futhermore, 2-(p-hydroxyphenyl)-ethanol, just as the purine line cytokinins, have the capability to supress proliferation of transformed sarcoma human cells in a tissue culture (6). All the above said allows to assume that this compound can apparently be attributed, along with diphenylurea and some other natural compounds (7), to the scarce class of non-purine cytokinins.

Another interesting property of the 2-(p-hydroxyphenyl)-ethanol compound is the high level of fluorescence which is of an order higher than in purine line cytokinins (5). It is known that phenol compounds in plants serve both as a mechanical and a chemical barriers against bacterial infection. Thus, at the moment of infection of any part of plant tissue a zone with a high concentration of simple phenolic compounds exibiting a high level of fluorescence is detected (8). It is dificult to say at the pesent whether 2-(p-hydroxyphenyl)-ethanol plays a protective role in purple bacteria.

3.2 Phytohormones of purple sulfur bacteria Cm. minutissimum of the γ- subdivision

Analysis of phytohormonal composition of the *Cm. minutissimum* cultural medium showed the absence of classic cytokinins of the purine line. However, four compounds in it were detected and identified by MS as follows: in the ethylacetate extract - para-hydroxybenzoic acid and 2-(p-hydroxyphenyl)-ethanol; in the butanol extract - phenylacetic acid and indolyl-3-propionic acid (Fig. 1). Both compounds from the

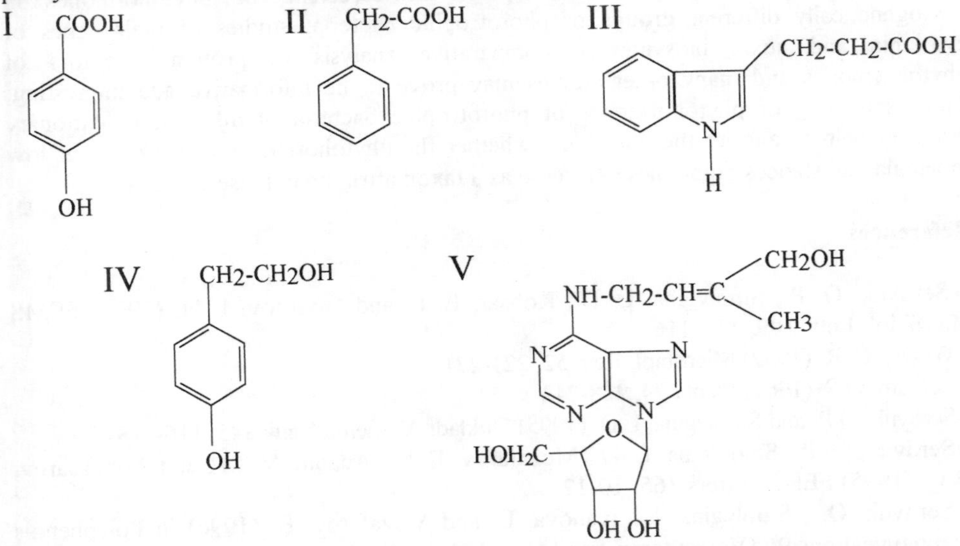

Figure 1. Structures of the substances with phytohormonal activity isolated from *Cm. minutissimum and Rs. rubrum.*

I - p-Hydroxybenzoic acid; II -Phenylacetic acid; III - Indolyl - 3 - propionic acid; IV - 2-(p-Hydroxyphenyl)-ethanol; V - Zeatinriboside. The compounds I - IV - in *Cm. minutissimum* and IV - V - in *Rs. rubrum* were detected.

ethylacetate extract exhibited cytokinin activity in bioassay. Biological activity in phenylacetic acid and indolyl-3-propionic acid isolated from butanolic extract were not tested, but the presence of auxin activity in them according to the literature data is expected.

In the *Cm. minutissimum* biomass no compounds with phytohormonal activity were detected.

3.3 Comparison of data on the phytohormonal composition of Rs. rubrum and Cm.minutissimum

As follows from our data, essential differences are observed in the content of phytohormones in the phototrophic purple bacteria related to the phylogenetically various subdivisions. Thus, the main biological activity of *Rs. rubrum* of the α - subdivision bacterium is connected with compounds of a purine and phenolic nature, exhibiting cytokinin activiy, while the biological activity of the γ - subdivision bacterium Cm. minutissimum is correlated with compounds of an indole and phenolic nature exibiting both auxin and cytokinin actvities. At this step of studies it is difficult to explain this phenomen and also to determine the functions of each substance with phytohormonal activity. It can be assumed that phytohormone diversity is connected with the formation and selection of some functions and mechanism of action of phytohormones namely in these bacteria. From this point of view the screening of phytohormones in phylogenetically different groups of phototrophic bacteria, studies of their genes of cytokinin and auxin biosynthesis, comparative analysis of protein receptors of phytohormones and many other studies may prove to be informative and interesting. Thus, screening of phytohormones of phototrophic bacteria of different evolutionary branches help to answer the question - whether the phytohormones can, like other low molecular substances , quinones (9), serve as a taxon attribute in these bacteria?

References

1 Serdyuk, O. P., Smolygina, L. D., Kobsar, E. F. and Gogotov, I. N. (1993) FEMS Microbiol. Lett. 109, 113-116

2 Woese, C. R. (1987) Microbiol. Rev. 52, 221-271

3 Letham, D.S. (1967) Planta 74, 228-242

4 Serdyuk, O.P. and Smolygina, L.D. (1995) Dokladi Akademii Nauk 345, 116-118

5 Serdyuk, O. P., Smolygina, L. D., Muzafarov, E. N., Adanin, V. M., and Arinbasarov, M.U. (1995) FEBS Letters 365, 10-12

6 Serdyuk, O. , Smolygina, L., Ivanova, E. and Muzafarov, E. (1996) in Polyphenols Communications 96 (Vercauteren J., ed.) pp. 161 - 162, Universite Bordeaux , Bordeaux

7 Moore, T.G. (1989) in Biochem. and Physiol. of Plant Hormones (Moore T.G., ed.) pp. 158-195, Springer-Verlag, NY, Heidelberg, Berlin

8 Harborne, J.B. (1988) in Introduction to Ecological Biochemistry (Harborn J.B., ed.) pp. 302-340, Academic Press, London, N.Y., Boston, Tokio

9 Kato, S. and Komagata K. (1985) J. Gen. Appl. Microbiol. 31, 381-398

REGULATION OF PHOTOSYNTHESIS AND TRANSLOCATION OF PHOTOSYNTHATES BY ENDOGENIOUS PHYTOHORMONS IN LEAVES DURING ONTOGENESIS

Kiselyova I.S., Borzenkova R.A. The Urals State University, 51 Lenin Av., P.O. Box 620083, Ekaterinburg, Russia

Key words: leaf physiology, source-sink, cytokinin, IAA, ABA, barley, potato

1. Introduction

The role of different leaves in sink-source relations and plant productivity is defined by their photosynthetic activity, growth rate and duration, ability to export or to retain photosynthates. Are ontogenetic changes in photosynthesis and export function of different leaves connected with changes in endogenous phytohormone content? The role of exogenous hormones in regulation of photosynthesis and assimilate translocation is well known (1-6). Simultaneous study of the changes in endogenous phytohormones, growth and formation of photosynthetic apparatus and export function during leaf ontogenesis will allow to maintain the correlation between the hormonal status of native leaves and these processes.

2. Procedure

Different leaves of potato, *Solanum tuberosum L.*, cv. Malachite (7-th, 11-th and 15-th) and barley, *Hordeum vulgare L.*, cv. Varde (preflag and flag) consequently emerging on a shoot, grown in field conditions, were studied during plant ontogenesis. Growth rate, photosynthetic activity per unit leaf area and chloroplast, assimilate translocation and the content of endogenous cytokinin, IAA and ABA were studied in each leaf during it's development.

The rate of CO_2 assimilation was determined by the uptake of labeled $^{14}CO_2$ (350 ppm), and assimilate translocation - after 24 h as the percentage of "label" in leaf in comparison with the total radioactivity of a plant (2).

Concentration of endogenous IAA, ABA and zeatin-like cytokinin were determined by ELIZA method ("Pharmcheminvest" kits of reagents, Russia).

3. Results and Discussion

Leaves consequently appeared during plant ontogenesis differed in absolute and relative growth rates, determined by changes both in area and in dry weight (Table 1). In potato the largest area and dry mass were revealed in the 7-th, lower leaf formed earlier compared with upper 11-th and 15-th, that had smaller growth rate and maximal area. The same changes were revealed in preflag and flag leaves of barley (Table 1, Fig.1).

G. Garab (ed.), Photosynthesis: Mechanisms and Effects, Vol. V, 3711–3714.
© 1998 *Kluwer Academic Publishers. Printed in the Netherlands.*

Table 1. Absolute (AGR) and relative (RGR) growth rates of potato and barley leaves during the phase of linear growth.

Leaf type	AGR				RGR			
	cm^2/day		mg/day		cm^2/day		mg/day	
	mean	max	mean	max	mean	max	mean	max
Potato								
7-th	2.6	3.2	6.8	10.2	0.24	0.88	0.50	0.72
11-th	1.9	2.4	5.2	8.0	0.24	0.60	0.37	0.60
15-th	1.5	1.6	5.0	6.9	0.17	0.53	0.24	0.42
Barley								
Preflag	0.35	0.44	-	-	0.044	0.067	-	-
Flag	0.25	0.40	-	-	0.032	0.050	-	-

Ontogenetic changes of photosynthetic rate in each leaf of both species were of typical 1-peak curve (Fig.1). The largest photosynthetic activity in potato was revealed in the 7-th (lower) leaf when it reached 80% of a maximal area. The 11-th and 15-th leaves showed maximal rate of photosynthesis on the earlier stages of their development, when they were 50% and 40% of area. The decrease of photosynthetic rate in late ontogenetic stages was slower in upper leaf. The rate of photosynthesis per each chloroplast in lower leaf was also higher in comparison with upper ones. Mean values of this parameter were 5.6; 3.3 and 2.7 mg $CO_2/10^9$chloroplasts*h in 7-th; 11-th and 15-th leaves respectively. In barley flag leaf had smaller assimilating rate than preflag during all stages of their development, mainly by low activity of each chloroplast.

The enhancement of photosynthesis during leaf development was accompanied by the increasing ability to export assimilates. It is known, that leaves begin to export photosynthates when their area reaches 15-30 % from maximal (2,7). Fig.2 shows that the development of translocation function is more rapid in upper leaves (11-th and 15-th in potato and flag in barley) than in lower. In the age of 7 days the 7-th leaf of potato exported 16% of assimilates, whereas 11-th - 24% and 5-th - 40% in the age of 5 days. In the upper flag barley leaf assimilate export exceeded the same in preflag about 7-10%. Both in potato and barley upper leaves were characterized not only by more rapid development of the export function, but by its maximal level also. So lower leaves retained for their growth more assimilates than upper ones, that correlates with higher growth rate in lower leaves. The enhancement of export in upper leaves during late stages of plant ontogenesis was stipulated, probably, by the intensive formation of sink at that period: tuber in potato and ear in barley.

Ontogenetic changes of growth rate, photosynthetic activity and export function in each leaf was closely connected with the content of endogenous hormones (Fig.3). Maximal level of cytokinin in lower leaves was about 2-times higher than in upper. Irrespective of concentration, the period of cytokinin accumulation and its maximal level coincided with the period of the most intensive leaf growth. Later cytokinin level decreased. The same tendencies were revealed for IAA. Though concentrations of cytokinin and IAA in upper leaves were smaller than in lower, its later decrease was more slow. Evidently, this promoted longer duration of upper leaves compared to lower, that senesced earlier.

In potato the high correlation between absolute growth rate and cytokinin concentration was observed (r=0.90; 0.95; 0.74 for the 7-th,11-th,15-th leaves respectively). Changes in cytokinin concentration correlated also with photosynthetic activity per unit leaf area and chloroplast both in potato and barley ("r" varied from 0.65 to 0.90).

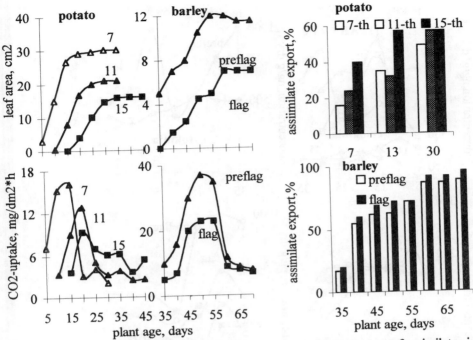

Fig.1 Ontogenetic changes in leaf area and photosynthetic activity of different leaves

Fig.2. Export of assimilates in different leaves

Maximal ABA level was reached to the moment when leaf finished its growth and then decreased probably due to ABA-conjugates synthesis. This tendency characterized all studied leaves, but in upper ones the decrease was more slow. This coincide with their smaller growth rate and area compared with lower leaves (7-th in potato and preflag in barley). Thus, in upper leaves, that developed later, the more high level both of ABA and cytokinin lasted longer, that defined smaller growth rate, area and duration in these leaves.

The development of export function in each leaf was accompanied by the increase of cytokinin and IAA concentration first, and then by decrease of photosynthetic activity, IAA and cytokinin content but slow accumulation of ABA. At that time negative correlation between export ability and cytokinin/ABA ratio (r=-0.86) and IAA/ABA ratio (r=- 0.87) was shown. This was also marked for different leaves: the increase of export function in upper leaves in comparison with lower coincided with the decrease of cytokinin/ABA and IAA/ABA ratio. The stimulation of assimilate export from leaves was shown earlier in experiments with exogenous hormone treatment of both sink and source (leaves) organs (4,6).

In summary the above observations confirm the data of other authors, obtained by using exogenous hormones, that IAA, cytokinin and ABA influence on ontogenetic changes in leaf photosynthesis, assimilate translocation and source-sink relations in plant.

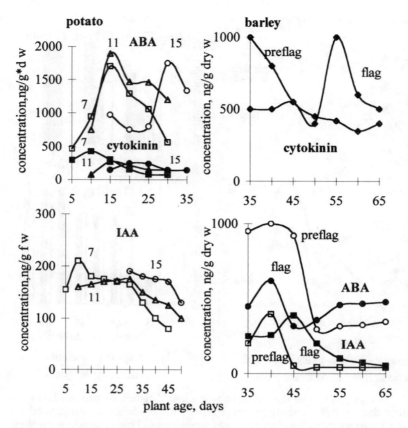

Fig.3. Ontogenetic changes of phytohormone concentration in different leaves

This study was supported by RBRF grant #97-04-48374.

References

1. Feierabend J. (1970) Planta 94, 1-15
2. Mokronosov, A.T. (1981) Ontogenetic aspect of photosynthesis Nauka, Moscow, Russia
3. Pavar, S.S., Klyachko, N.L., Romanko, E.G., Zimmermann, K.-C., Kulaeva, O.N. (1983) Rus. Plant Physiol. 30, 459-466
4. Dorflling, K., Tietz, A., Fenner, R., Naumann, R., Dingkuhn, M. (1984) Ber. Dtsch. Bot. Ges. 97, 87-99
5. Makeev, A.V., Krendelyova,T.E., Mokronosov, A.T. (1992) Rus. Plant Physiol. 39,170-182
6. Mauk, C.S., Baucher, M.G., Yelenosky, G. (1986) J. Plant Growth Regul. 5, 11-120
7. Kursanov, A.L. (1976) Transport of assimilates in plants Nauka, Moscow, Russia

PHOTORESPIRATORY METABOLISM OF *WOLFFIA ARRHIZA*

Natalia V. Bykova, Inna V. Popova and Abir U. Igamberdiev
Department of Plant Physiology and Biochemistry, Voronezh University,
Voronezh 394693, Russia

Key words: alternative electron transport, glycine, mitochondria, photorespiration

1. Introduction

The genus *Wolffia* of the family of duckweeds (*Lemnaceae*) contains the most simplified aquatic higher plants characterised by dedifferentiation of their morphological structure (absence of roots and of the vascular system, infrequent flowering) being possibly a result of evolutionary adaptation to photoheterotrophic growth conditions. The growth rate of *Wolffia* plants in a medium supplemented with an organic substrate (e.g. sucrose) is several fold higher then on bicarbonate as a sole carbon source [1]. This means that *Wolffia* plants can simultaneously utilise exogenous carbohydrates and perform photosynthesis. The incorporation of carbohydrates in metabolism via glycolysis and/or the oxidative penthose phosphate pathway is not suppressed by operation of photosynthetic pathways [2].

Wolffia is the C_3 plant [1], therefore in low CO_2 the photorespiratory flux may be high enough and the significant amounts of photorespiratory intermediates are formed. Photorespiratory metabolism in aquatic angiosperms has some peculiarities similar to that in green algae. Thus, some aquatic higher plants excrete glycolate to the environment [3]. It was also demonstrated that the two marine flowering species contain glycolate dehydrogenase, the enzyme abundant mostly in unicellular algae [4]. This means that the transition to aquatic environment is connected with acquirement of some features characteristic for algae.

Since *Wolffia* is the oversimplified aquatic higher plant, we investigated its photorespiratory metabolism in order to demonstrate what evolutionary metabolic changes (accompanied by morphological simplification) appeared during its adaptation to life in aquatic environment.

2. Material and Methods

Growth conditions. Sterile culture of *Wolffia arrhiza* (L.) Hork. ex Wimmer was a generous gift of Prof. R. Ehwald (Humboldt University of Berlin, Germany). The plants were grown as described earlier [2]. For providing heterotrophic conditions, 1% sucrose was added. To eliminate CO_2, a CO_2-free medium was prepared by introducing a vial containing 10% KOH solution into the flask with *Wolffia* plants. The CO_2 concentration of 1 and 10% was obtained by introducing a vial with the Warburg buffer [5].

G. Garab (ed.), Photosynthesis: Mechanisms and Effects, Vol. V, 3715–3718.
© 1998 *Kluwer Academic Publishers. Printed in the Netherlands.*

Experiments with labelled substrates. *Wolffia* plants (100 mg) were put into vials and covered by 1 ml of cultivation medium (without sucrose) containing [1-^{14}C]glycine, [1,4-^{14}C]succinate or [1,4-^{14}C]malate (10 mM, 0.1 MBq ml^{-1}) and corresponding inhibitors or metabolites. ^{14}CO$_2$ evolved was trapped by 20% NaOH during incubation and for 1 h after reaction was stopped by adding 0.1 ml 2 N H$_2$SO$_4$. Radioactivity was measured by liquid scintillation counter. The inhibitors were supplied in following concentrations: 50 µM rotenone (the inhibitor of complex I), 3 mM salicylhydroxamic acid (SHAM, the inhibitor of alternative oxidase), 1 mM KCN (the inhibitor of complex IV). Malate and succinate were supplied at concentration 10 mM.

Determination of glycolate oxidase and glycolate dehydrogenase. The enzymes were measured after homogenisation of plants in 50 mM HEPES-KOH buffer, pH 7.5, containing 1% PVP, 0.1% Triton X-100, 5 mM MgCl$_2$ and 5 mM DTT. Both enzymes were measured spectrophotometrically by two methods: by monitoring the formation of glyoxylate phenylhydrazone at 324 nm [6] or by monitoring the dichlorophenolindophenol reduction at 600 nm [7]. For inhibition of glycolate dehydrogenase, 2 mM KCN was added in incubation medium, since glycolate dehydrogenase is inhibited by KCN completely but not glycolate oxidase [4]. Protein was measured by the modified method of Lowry et al. [8].

Determination of phenolics, organic acids, and amino acids. The total content of phenolics in the cultivation medium was measured by azo-coupling with sulfanilic acid [9]. Amino acids and organic acids were determined by paper chromatography after separation by ion exchange techniques [9]. Glycolate content was estimated with 2,7-dihydroxynaphtalene by absorption at 530 nm [10]. The values presented are average means of 3-4 independent measurements and their standard deviations.

3. Results

Heterotrophic growth significantly enhanced biomass accumulation of *Wolffia*. The initial inoculum fresh weight was 200 mg. After three-week growth in 1% sucrose it was 7.3 ± 0.6 g which exceeded the fresh weight of *Wolffia* plants grown in 10% CO$_2$ and in 1% CO$_2$ by a factor of two and five, respectively. When sucrose was present in the medium, inorganic carbon practically did not affect biomass accumulation. However, protein content was higher in heterotrophic plants grown in the absence of CO$_2$ (4.4 ± 0.1 mg g^{-1} fresh weight) compared to both heterotrophic (3.4 ± 0.1) and autotrophic (3.5 ± 0.1) plants grown in 10% CO$_2$. We present below (if not indicated) the data obtained for *Wolffia* plants grown heterotrophically without CO$_2$ which reflect photorespiratory conditions.

Glycine decarboxylation by heterotrophically grown *Wolffia* plants (12.2 ± 1.1 nmol CO$_2$ min^{-1} g^{-1} fresh weight) was 4-5 times lower than decarboxylation of tricarboxylic acid cycle substrates succinate (63 ± 7 nmol CO$_2$ min^{-1} g^{-1} fresh weight) and malate (56 ± 9 nmol CO$_2$ min^{-1} g^{-1} fresh weight). It was strongly inhibited by rotenone, and it was coupled to both cyanide sensitive and cyanide resistant pathways (Table 1). Malate and succinate when added with glycine suppressed its decarboxylation. The addition of glycine did not suppress the rate of ^{14}CO$_2$ evolution from [1,4-^{14}C]succinate, it was 71 ± 8 nmol CO$_2$ min^{-1} g^{-1} fresh weight.

Glycolate oxidase activity was low, the rate measured by the method with phenylhydrazine was 11 ± 3 nmol min^{-1} g^{-1} fresh weight and measured by the method with dichlorophenolindophenol 10 ± 4 nmol min^{-1} g^{-1} fresh weight. There was no any detectable effect on the activity in the presence of 2 mM KCN in the incubation medium. In autotrophically grown *Wolffia* (1% CO_2) the glycolate oxidase activity was 15 ± 2 nmol min^{-1} g^{-1} fresh weight measured by the method with phenylhydrazine.

Table 1. Decarboxylation rates of [1-^{14}C]glycine by Wolffia plants in the presence of inhibitors and metabolites

Variant	nmol CO_2 min^{-1} g^{-1} fresh weight
Glycine alone	12.2 ± 1.1
+Rotenone	3.6 ± 0.6
+SHAM	7.4 ± 0.8
+KCN	6.4 ± 0.6
+Malate	7.8 ± 0.7
+Succinate	9.4 ± 0.8

Wolffia plants actively excreted organic compounds in the surrounding medium. After 3 weeks of heterotrophic growth, the medium contained phenolics, organic acids, and amino acids at detectable concentrations (Table 2).

Table 2. Concentration of phenolics, organic acids, and amino acids after 3-week growth of Wolffia plants under photoheterotrophic conditions

Excreted compound	Concentration, mM
Total phenolics	0.45 ± 0.15
Malate	0.03 ± 0.01
Glycolate	0.05 ± 0.02
Citrate	0.01 ± 0.005
Glycine	0.04 ± 0.01
Lysine	0.01 ± 0.005
γ-Aminobutyrate	0.02 ± 0.01

4. Discussion

The data indicate essential peculiarities of metabolism of *Wolffia arrhiza* reflecting its evolutionary simplification during adaptation to aquatic environment. *Wolffia* is dependent on organic substances in the environment since its growth is retarded when only inorganic carbon is present.

Photorespiratory metabolism in *Wolffia* did lost significant peculiar characteristics common to photosynthetic higher plants. Glycine conversion in this plant proceeded with low rate (Table 1) which may be connected with low level of glycine decarboxylase complex in mitochondria. In other C_3 plants (wheat, barley) the rate of glycine oxidative decarboxylation was shown to be at the same level that of succinate and of other tricarboxylic acid cycle substrates [11, 12]. Even C_4 plants (e.g. maize) revealed much higher rates of glycine decarboxylation [11, 12]. Glycine oxidation in C_3 plants is important to be maintained at very high level to prevent accumulation of toxic photorespiratory products, such as glyoxylate and its derivatives, and the special mechanisms of preferential oxidation of glycine in mitochondria have been developed allowing its oxidation at appreciable rates in the presence of other mitochondrial substrates. These mechanisms include involvement of rotenone insensitive bypass and cyanide insensitive path of mitochondrial electron transport both non-coupled to energy conservation, as well as an increase of glycine oxidation in the presence of malate and succinate [11-13]. As it is evident from the data obtained, glycine decarboxylation in *Wolffia* lacks these essential characteristics: it was strongly suppressed by rotenone and decreased in the presence of malate and succinate, whereas decarboxylation of succinate was not affected in the presence of glycine. However, the cyanide insensitive pathway was highly active in *Wolffia*.

Glycolate oxidase activity in *Wolffia* was low (terrestrial higher plants reveal several times higher activity [6]). We could not detect any glycolate dehydrogenase activity. Thus, similarly to other higher plants, and contrary to most algae, *Wolffia* possesses common glycolate pathway which includes glycolate oxidase and glycine decarboxylase, but activities of these enzymes are relatively low and the glycine metabolism lost characteristic features of its preferential (to other mitochondrial substrates) oxidation.

The reduction of photorespiratory pathway is consistent with the observation that some amounts of glycolate and glycine were excreted into surrounding medium (Table 2). This property, common for algae, was revealed earlier in some aquatic angiosperms [3]. It was shown that *Wolffia* had lower capacity to refixation of photorespiratory ammonium compared to other higher plants including the plant of the same family, *Lemna gibba* L. [14]. Excretion of other compounds, especially phenolics, reflects a significant reduction of secondary metabolism characteristic for all *Lemnaceae* [1].

The photorespiratory metabolism of *Wolffia* has undergone a significant reduction consisting in the excretion of glycolate and glycine into surrounding medium, low activities of glycolate oxidase and glycine decarboxylase, strong rotenone sensitivity of glycine decarboxylation and its suppression in the presence of succinate and malate.

References

1 Landolt, E. and Kandeler, R. (1987) The Family of Lemnaceae - a Monographic Study. Geobot. Inst. Veröffentlichungen, Zürich
2 Igamberdiev, A.U. and Zabrovskaya, I.V. (1994) Russ. J. Plant Physiol. 41, 208-214
3 Kolesnikov, P.A., Zore, S.V., Mutuskin, A.A. and Einor, L.O. (1985) Sov. Plant Physiol. 32, 282-288
4 Tolbert, N.E. (1976) Aust. J. Plant Physiol. 3, 129-132
5 Semikhatova, O.A., Chulanovskaya, M.V. and Metzner, H. (1971) In Plant Photosynthetic Production (Z.Šesták, ed.), pp. 238-256, Junk, Hague
6 Frederick, S.E., Gruber, P.J. and Tolbert, N.E. (1973) Plant Physiol. 52, 318-323
7 Hess, J.L., Tolbert, N.E. (1967) Plant Physiol. 42, 371-379
8 Lowry, O.H., Rosebrough, N.J., Farr, A.L. and Randall, R.J. (1951) J. Biol. Chem. 193, 265-275
9 Harborne, J.B. (1973) Phytochemical Methods: A Guide to Modern Techniques of Plant Analysis, Chapman and Hall, London
10 Calkins, V.P. (1943) Anal. Chem. 15, 762-765
11 Igamberdiev, A.U., Bykova, N.V. and Gardeström, P. (1997) FEBS Lett. 412, 265-269
12 Igamberdiev, A.U. and Bykova, N.V. (1994) Russ. J. Plant Physiol. 41, 344-347
13 Wiskich, J.T., Bryce, J.H., Day, D.A. and Dry, I.B. (1990) Plant Physiol. 93, 611-616
14 Monselise, E.B.I. and Kost, D. (1993) Planta 189, 167-173

NON-ENZYMIC CLEAVAGE OF KETOSES AS A SOURCE OF GLYCOLATE

Vuk Maksimović & Željko Vučinić[1], Center for Multidisciplinary Studies, Belgrade University, 29.Novembra 142, 11060 Beograd, F.R.Yugoslavia

Key words: carbohydrate, free radicals, H_2O_2, photorespiration, sugar, temperature dependence

1. Introduction

Two experimentally verified mechanisms of synthesis of glycolate are the oxidation of a transketolase reaction intermediate (1) and the oxygenase reaction of Rubisco (2). Although the current opinion is that the second enzymatic reaction is the primary source of glycolate in photosynthetic tissue, neither of the two reactions can account, with their stechiometries of CO_2 release and measured rates of production, for the dramatic increase of photorespiratory activity with temperature at high light intensities (3). The aim of this work was to determine whether glycolate can be produced non-enzymatically by H_2O_2 breakdown of sugars, and if so, whether free radicals participate in the reaction.

2. Procedure

Assays were performed in 20 ml aliquots held in test-tubes kept in a thermostated water bath. Reactions were allowed to run 20 min, following which the samples were adjusted to pH 4 and passed through strongly basic Dowex 1 x 8 anion exchanger with 200-400 mesh particle size in glass columns ($\varnothing$ =10mm), sealed with glass fiber cotton in the capillary bottom end and kept at 25°C. Glycolate was eluted with 4M acetic acid, the 5-15 ml fraction being collected. The eluent was derivatized with 2,7-dihydroxynaphtalene according to Calkins (4) and measured at 534 nm using Shimadzu 101 spectrophotometer. Following elution, columns were regenerated with NH_4Cl and washed until no chloride could be detected. In each set of experiments (50 columns) a calibration curve was performed on randomly selected columns. The level of detection based on such calibration was 20 µg of glycolate. In the experiments standard conditions were: sugar concentration: 30mM; [H_2O_2]: 10mM; temperature 40°C and pH 8, except where explicitly stated otherwise. The data

---

[1]To whom correspondence should be addressed to. E-mail: *evucinic@ubbg.etf.bg.ac.yu*

G. Garab (ed.), Photosynthesis: Mechanisms and Effects, Vol. V, 3719–3722.
© 1998 *Kluwer Academic Publishers. Printed in the Netherlands.*

of three independnent experiments, two samples of each being passed through columns. All chemicals were of analytical grade.

3. Results and Discussion

Initial experiments to test the feasibility of the reaction were performed on 4 aldoses (glucose, arabinose, manose and glyceraldehyde) and 2 ketoses (fructose and dihyroxyacetone) of varying lenght of carbon skeleton. Of the six monosaccharides tested only the two ketoses had the capacity to produce glycolate, the rate of formation under the above mentioned standard conditions being 6.5 and 22.2 mmol·glycolate·hr⁻¹ for fructose and dihydroxyacetone, respectively. Thus, in one hour 8.2% of fructose and 26.4% of dihyroxyacetone initial concentration were cleaved.

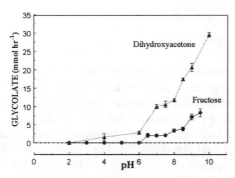

Figure 1. The effect of pH on the rate of release of glycolate from dihyroxyacetone and fructose

Concentration dependence of the reactions exhibited saturation kinetics, and analysis in the sugar concentration range from 1 to 40 mM (at $[H_2O_2]$ = 10mM) demonstrated that the reaction approached saturation at concentrations > 20 mM of either of the two sugars analysed. The rate of glycolate synthesis increased very rapidly up to 5 mM H_2O_2, approaching saturation when $[H_2O_2]$ > 10 mM, in the experiments where H_2O_2 concentration varied from 0.5 to 20 mM ([sugar]=30mM). DHA was more susceptible to breakdown, 7.95 mmol hr⁻¹ of glycolate being released at 0,5 mM H_2O_2, in comparison to 2.45 mmol hr⁻¹ yielded by fructose.

In Fig. 1 we show the effect of varying pH on the reaction (pH 2-10), demonstrating the very prounounced effect of higher pH values on the yield of the reaction. In the case of fructose no glycolate synthesis could be detected below pH < 7, while dihydroacteone released detectable amounts of glycolate even at pH 4. At alkaline pH values a dramatic increase in glycolate production occured for both of the sugars analysed, a three-fold rise being observed when going from pH 7.5 to 9.5.

The reaction exhibited a strong temperature dependence, assays in the temperature range from 25-55°C having activation energies of 52.2 and 14.4 kJ mol⁻¹ for fructose and dihydroxyacetone, respectively, when approximated by a linear function in the Arrhenius plot (Fig. 2). The coefficient Q_{10} in the range from 25-45°C was 3.8 and 1.5 for fructose and dihydroxyacetone, respectively.

Time course experiments revealed that the reactions of fructose and dihydroxyacetone with H_2O_2 cannot be considered as simple first or second order reactions. Derivation diagram shows a more linear dependence in the case of dihyroxyacetone compared to fructose. The rate of glycolate synthesis from fructose decreased more gradually than that of dihyroxyacetone, the second reaction having higher initial rates but not exhibiting the increase in glycolate synthesis with time of almost three-fold demonstrated by fructose.

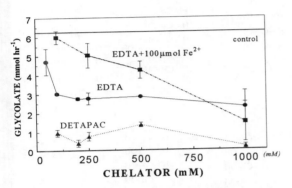

Figure 2. The effect of the presence of chelators EDTA and DETAPAC on the rate of glycolate release from frucose. Experimental conditions as in Procedure.

Addition of Fe^{2+} or Co^{2+} to the assay medium increased the yield of reaction between fructose and H_2O_2 more than 30 and 50%, respectively, when the metal concentration was 100μM. Saturation was reached with concentrations above 75μM, but the stimulation was clearly visible at concentration < 50μM.

When chelators were added to the assay, rates of glycolate synthesis decreased 30 (EDTA) and 70% (DETAPAC). Maximal relative inhibition of the reaction was achieved with 100 μM concentration of chelators, but when Fe^{++} was simultaneously added in equimolar concentrations with EDTA, the maximal relative inhibition was observed at concentrations of EDTA above 1mM (Fig. 2).

We propose that the mechanism of peroxide oxidation of polyhydroxy-ketones is a relevant model for describing the synthesis of glycolate from ketose monosacharides and other keto metabolic intermediates, in which peroxide performs a nucleophylic attack on the carbonyl C atom, forming a very unstable intermediate, which further undergoes intramolecular changes that can lead to the breakdown of a C-C bond (5). The cleavage of the bond in the two 2-ketoses would always occur between the C_2 and C_3 atom, forming glycolate as the product. Pronounced effect of alkaline pH on the reaction stabilizes the reaction intermediates providing easier breakdown of the C-C bond. Such a reaction would be catalyzed by metal ions that can facilitate a peroxide attack and/or produce a more potent oxidant such as the ˙OH radical (6). Our results with chelating agents demonstrate that chelated iron (with EDTA) can stimulate glycolate synthesis, while DETAPAC inhibits the reaction even at very small concentrations, such results being in very good agreement with previous reports (7). This proves that free radicals influence the reaction, althought, if the proposed mechanism is valid, their participation is not obligatory. The reason for the observed effect of the fall in the rate of glycolate synthesis with increasing tempeatures (observed in the case of (observed in the case of dihydroxyacteone) could be sought in the formation of very reactive intermediates, which in turn alter the mechanism of reaction. It is obvious that peroxide oxidation of dihydroxyacetone leads to the formation of a C_1 residue, whose fate is not clear.If the C_1 residue were a radical, its reaction with dihydroxyacetone could compete with the H_2O_2 induced reaction increasingly with temperature (8).

fructose

erithrose

Many of the keto intermediates in plant and animal biochemical cycles are 2 ketoses, and via the above mentioned mechanism glycolate could be formed if there is a potential oxidant available. The yield and efficiency of non-enzymatic synthesis of glycolate *in vivo* is a matter of further experimental analysis, but the results presented provide an ubiquitous mechanism for oxidative degradation of keto intermediates, where glycolate is a characteristic and perhaps an inevitable product.

Acknoledgement: The authors wish to thank Prof. U.Heber for his helpful coments. This work was financed through a grant in biophysics (0E322) from the local Ministry of Science and Technology.

4. References

1. Wilson, A.T. and Calvin, M.J. (1955) J.Am.Chem.Soc. 77, 5948-5953
2. Ogren, W.L. and Bowes, G. (1971) Nature, 230, 159-160
3. Vučinić, Ž. and Grodzinski, B. (1986) Fiziol.Biokhim.Kulturnih Rastenii (Kiev), 18, 6-24
4. Calkins, V.P. (1943) Anal.Chem.,15,762-763
5. De la Mare, H. at al. (1963) J.Chem.Soc. 85, 1437-1444
6. McCord, J. and Day, D.E. (1978). FEBS 86, 139-142
7. Haliwell, B. (1978) FEBS, 92 321-327
8. Grodzinski, B. (1984) Plant Pysiol. 74, 781-786

THE RELEASE OF CO_2 FROM TALL FESCUE LEAVES FOLLOWING PHOTOSYNTHESIS UNDER PHOTORESPIRATORY AND NON-PHOTORESPIRATORY CONDITIONS

Eugeniusz Parys and Elzbieta Romanowska
Department of Plant Physiology, University of Warsaw, Poland

Key words: CO_2 burst, gas exchange, respiration, temperature dependence.

1. Introduction

The two main transient phenomena in the course of CO_2 evolution can be recorded when leaf of C_3 plant photosynthesizing in air is suddenly darkened. The first represents the postillumination burst of CO_2 (PIB), a phenomenon originally described by Decker (1) from that the existence of photorespiration was first deduced. After that follows a period of increased respiration, a phenomenon termed light-enhanced dark respiration (LEDR) (2). Generally similar sensitivity of PIB and enhanced respiration to oxygen observed in tall fescue leaves (3) suggests the contribution of photorespiratory substrates. However, the solution of bicarbonate introduced into these leaves greatly enhanced both phenomena under photorespiratory conditions. The promotion of LEDR by bicarbonate have also been observed in mesophyll protoplasts of pea (4). Information on the effect of temperature on the transient CO_2 exchange phenomena under various O_2 concentrations is limited. Only the response of PIB to temperature at 21% O_2 has been studied (5). The purpose of the present study was to examine of how a temperature affects PIB and enhanced respiration in tall fescue leaves under photorespiratory (21 and 50% O_2) and non-photorespiratory (1% O_2) conditions.

2. Procedure

2.1 The middle segments of the leaves of tall fescue (*Festuca arundinacea* Schreb.) detached from plants grown in the garden of Warsaw University were placed into photosynthesis chamber in temperature controlled, and it was connected to an infrared CO_2 analyzer arranged to a close circuit system as described in (3).
2.2 Photosynthetically active radiation (PAR, 400-700 nm) at the leaf surface was 300 Wm^{-2}. Leaf temperature was maintained within $\pm 0.5°C$ of the desired temperature over the range from 15 to 35°C during the light and dark periods.
2.3 Postillumiantion burst of CO_2 (PIB) was measured at CO_2 concentration near the CO_2 compensation point (Γ) in the closed circuit system filled with either 1, 21 or 50% O_2. The amounts of free CO_2 evolving during the PIB were estimated as described in (3). The rates of dark respiration following photosynthesis (R) were calculated after 3-4 minutes of darkness. The volume of PIB and rate of R were expressed in terms of leaf area.

G. Garab (ed.), Photosynthesis: Mechanisms and Effects, Vol. V, 3723–3726.
© 1998 Kluwer Academic Publishers. Printed in the Netherlands.

3724

2.4 The rates of net photosynthesis (Pn) were calculated from the change in CO_2 concentrations in the range 380-340 µl/l. The CO_2 compensation point (Γ) of leaves treated with various temperature and O_2 concentrations were determined. The result shown are mean values ± SE.

3. Results and Discussion

The recorder in figure 1 shows the course of CO_2 evolution after darkening of tall fescue leaves photosynthesizing at CO_2 compensation point (Γ) at air (21% O_2). The generally accepted view is that a burst a CO_2 (PIB) represents remnant of photorespiration (7). An alternative interpretation suggests that PIB reflect release of CO_2 from a pool of inorganic carbon accumulated within leaf cells in response to rapid decrease of intracellular pH during light-dark transition (8). Figure 2 illustrates that the rate of respiratory CO_2

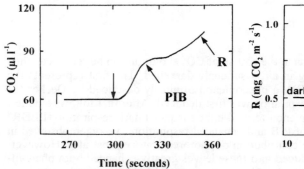

Figure 1. Typical tracings showing the change of CO_2 concentration at 25°C in an atmosphere of air (21% O_2) after rapid darkening of tall fescue leaves at CO_2 compensation point as indicated by arrow. PIB, postillumination CO_2 burst; R, enhanced respiration.

Figure 2. Respiration rate of tall fescue leaves before and after illumination at 25°C in relation to oxygen concentration. a, 1% O_2; b, 21% O_2.

evolution from tall fescue leaves after illumination was enhanced by about 80% as compared to its rate before illumination. No PIB and enhanced respiration (R) were observed at 1% O_2. The phenomenon of light-enhanced dark respiration (R) has also been demonstrated in mesopyll protoplasts of pea (4), however, their origin is not established. The suppression of CO_2 evolution during both the PIB and R following photosynthesis at 1% O_2 and the stimulation at higher (21 and 50%) O_2 concentrations as we noted in tall fescue leaves (3) suggest the contribution of photorespiratory substrates. Besides oxygen the temperature is also powerfull factor affecting CO_2 exchange in plants. Its combination with oxygen concentration caused in tall fescue leaves a shift in the temperature optima of the net photosynthesis (Pn) and of the PIB in the opposite directions with increasing of both the O_2 concentration from 1 to 50% and the temperature from 15 to 35°C (Fig. 3 and 4). The different temperature optima of PIB and Pn might result from its effect on the internal O_2/CO_2 solubility ratio, which increases with increasing leaf temperature (9) and from the higher Arrhenius activation energy (Ea) of oxygenase reaction than carboxylase

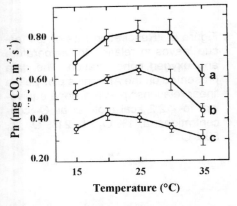

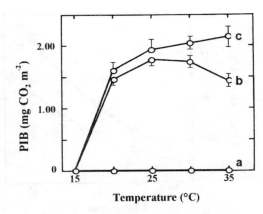

Figure 3. Net photosynthesis rate of tall fescue leaves in relation to temperature and oxygen concentration. a, 1% O_2 ; b, 21% O_2 ; c, 50% O_2 .

Figure 4. Postillumination CO_2 burst of tall fescue leaves in relation to temperature and oxygen concentration. a, 1% O_2 ; b, 21% O_2 ; c, 50% O_2 .

reaction of Rubisco (6). The lack of PIB in tall fescue leaves under none of the temperatures used (15-35°C) at 1% O_2 suggests that either the temperature did not affect the rate of photorespiration or the PIB had a non-photorespiratory origin. The CO_2 compensation point (Γ) in these leaves (Fig. 5) was affected by temperature identically with the Arrhenius activation energy (Ea) of -8.1 kcal mol^{-1} and Q_{10} = 1.5 for 15 to 35°C only at both 21 and 50% O_2 . A quite similar Ea of Γ (-7.6 kcal mol^{-1}) has been noted in other C_3 plant (10). The activation energy of -29.4 kcal mol^{-1} (Q_{10} = 5.1 for 20 to 35°C) as found for Γ in tall fescue leaves at 1% O_2 is much higher than reported Ea (7-16 kcal mol^{-1}) for most enzyme catalized reaction over the physiological temperature range (11). Such a high value of Ea and Q_{10} as we found for Γ at 1% O_2 was noted for photorespiration based on the PIB in tobacco leaf tissue photosynthesizing at 21% O_2 (Ea = 30.1 kcal mol^{-1}, Q_{10} = 5.2) (5). One might speculate that the increase of Γ in tall fescue leaves with increasing of temperature at 1% O_2 have resulted from the respiration.

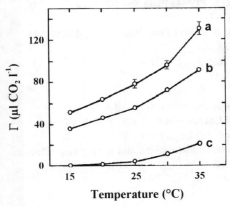

Figure 5. CO_2 compensation point of tall fescue leaves in relation to temperature and oxygen concentration. The data presented on the Arrhenius plot yield a linear relationships with a slopes equivalent to -29.4 kcal mol^{-1} and -8.1 kcal mol^{-1} at 1% O_2 and at 21% or 50% O_2 respectively. a, 1% O_2 ; b, 21% O_2 ; c, 50% O_2 .

3726

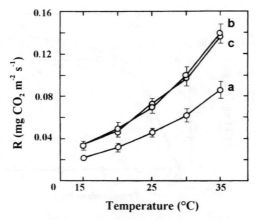

Figure 6. Respiration rate of tall fescue leaves in relation to temperature and oxygen concentration. The data presented on the Arrhenius plot yield a linear relationship with a slope equivalent to -12.2 kcal mol^{-1} at all O_2 concentrations. a, 1% O_2 ; b, 21% O_2 ; c, 50% O_2 .

However, the respiration rate in these leaves following photosynthesis, which was significantly lower (by about 40-50%) at 1% than that at 21 and 50% O_2 (Fig. 6), increased with temperature according to $Q_{10} = 2.0$ and with Ea = -12.2 kcal mol^{-1} in all O_2 concentrations applied (1-50%). These values of Ea and Q_{10} are typical reported for mitochondrial respiration in various plant species (11). Then, the lack of PIB in tall fesuce leaves under none of the temperatures used (15-35°C) at 1% O_2 can be caused by efficient refixation of CO_2 by phosphoenolopyruvate carboxylase (12) in mesophyll cells before it escaped form the leaf. The different dependence of PIB and R on temperature and oxygen concentration in tall fescue leaves demonstrates that enhanced respiration is distinct from a PIB and represents mitochondrial activity.

Acknowledgment. This work was supported by the Polish Committee for Scientific Research Grant 6 PO4C 001 14.

References

1   Decker, J. P. (1955) Plant Physiol. 30, 82-84
2   Raghavendra, A. S., Padmasree, K., Saradadevi, K. (1994) Plant Sci. 97, 1-14
3   Parys, E. (1990) Plant Physiol. Biochem. 28, 711-717
4   Malla Reddy, M., Vani, T. and Raghavendra, A. S. (1991) Plant Physiol. 96, 1368-1371
5   Peterson, R. B. (1983) Plant Physiol. 73, 983-988
6   Chen, Z. and Speitzer, R. J. (1992) Photosynth. Res. 31, 151-164
7   Zelitch, I. (1992) BioScience 42, 510-516
8   Bown, A. W. (1982) Plant Physiol. 70, 803-810
9   Ku, S. B. and Edwards, G. E. (1977) Plant Physiol. 59, 986-990
10 Björkman, O. Gauhl, E. and Nobs, M. A. (1969) Carnegie Inst. Wash. Year Book 68, 620-633
11 Raison, J. K. (1980) in The Biochemistry of Plants, 2, Metabolism and Respiration (Davies, D. D., ed.) pp. 613-626, Academic Press, New York
12 Bryce, J. H. and ap Ress, T. (1985) Phytochemistry 24, 1635-1638

MITOCHONDRIAL OXIDATIVE ELECTRON TRANSPORT OPTIMIZES PHOTOSYNTHESIS IN MESOPHYLL PROTOPLASTS OF PEA

K. Padmasree and A.S. Raghavendra
Department of Plant Sciences, School of Life Sciences,
University of Hyderabad, Hyderabad 500 046, INDIA

Key Words: Alternative pathway, ATP, bicarbonate, cytochrome pathway, mitochondria, protoplast

1. Introduction

Recent reports suggest that photosynthetic carbon assimilation is dependent strongly on mitochondrial metabolism (1-5). Mitochondria of higher plants, possess two routes of oxidative electron transport: the cyanide sensitive cytochrome pathway and cyanide-resistant alternative pathway. Although the role of AOX pathway is known for the rapid oxidation of substrates and generation of heat in some aroid tissues, its role in plant metabolism is not completely understood (6). In the present study, we tried to investigate the implication of the AOX pathway by employing SHAM or propyl gallate, as the typical inhibitors of AOX pathway (7). Further, we compared these results with those of oligomycin (inhibitor of mitochondrial oxidative phosphorylation) or antimycin A (inhibitor of cytochrome pathway).

Our results demonstrate that alternative pathway of mitochondrial electron transport plays a significant role in optimizing photosynthesis at both optimal and limiting CO_2.

2. Materials and Methods

Pea (*Pisum sativum* L. cv. Arkel) plants were grown outdoor (natural photoperiod of approximately 12h, average daily temperatures of 30 °C day/20 °C night). The mesophyll protoplasts were isolated from the first and second fully expanded leaves of 8-10 day old plants as already described (8). Photosynthetic activities of protoplasts were determined by monitoring their oxygen evolution at limiting (0.1 mM $NaHCO_3$) or optimal (1.0 mM $NaHCO_3$) CO_2. Test compounds were included in the reaction medium so as to give the required final concentration, and protoplasts were pre-incubated in darkness at 25 °C for 5 min, before switching on the light. The levels of ATP and ADP were measured using enzymatic assays coupled to NADPH formation and NADH utilization, respectively by a modified procedure of Stitt et al. (9).

G. Garab (ed.), Photosynthesis: Mechanisms and Effects, Vol. V, 3727–3730.
© *1998 Kluwer Academic Publishers. Printed in the Netherlands.*

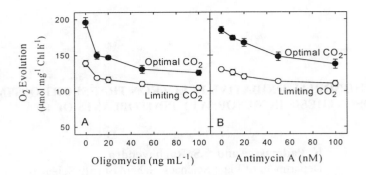

Figure 1. Effect of oligomycin or antimycin A on photosynthetic O_2 evolution in pea protoplasts at optimal (1.0 mM $NaHCO_3$) or limiting (0.1 mM $NaHCO_3$) CO_2.

3. Results and Discussion

The effect of mitochondrial inhibitors on photosynthetic O_2 evolution was assessed at limiting (0.1 mM $NaHCO_3$) or optimal (1.0 mM $NaHCO_3$) CO_2. The extent of inhibition of photosynthesis by oligomycin (inhibitor of oxidative phosphorylation) or antimycin A (inhibitor of cytochrome pathway) was more at optimal CO_2 (25 to 35% decrease) than that at limiting CO_2 (15 to 25% decrease) (Fig. 1). SHAM and propyl gallate also restricted photosynthetic activity, the suppression being stronger at optimal CO_2 than that at limiting CO_2. However, the inhibition due to SHAM in presence of antimycin A was similar at limiting or optimal CO_2 (Fig. 2).

The significant decrease in the rate of photosynthesis under both limiting and optimal CO_2 in presence of these mitochondrial inhibitors suggests that mitochondrial metabolism is essential for maximal photosynthesis under limiting CO_2 (photorespiratory conditions) as well as optimal CO_2. As most of the studies on mitochondrial inhibitors indicate that AOX pathway is engaged only when the electron flow through cytochrome pathway is either saturated or inhibited (10). We have therefore studied the effect of SHAM, both in the absence or presence of 100 nM antimycin A. In the absence of antimycin A, the addition of SHAM alone is likely to increase the flow of electrons through cytochrome pathway, while the presence of antimycin A would ensure the effective restriction of alternative pathway electron transport by SHAM. For e.g. there was 28% inhibition in the rate of photosynthesis at optimal CO_2 in presence of SHAM alone, while there was 54% inhibition with SHAM in presence of 100 nM antimycin A.

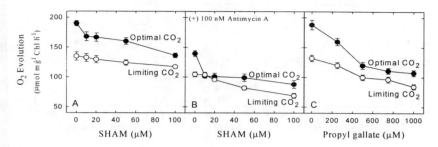

Figure 2. Effect of SHAM ($\pm$ antimycin A) or propyl gallate on photosynthetic O_2 evolution in pea protoplasts at optimal (1.0 mM $NaHCO_3$) or limiting (0.1 mM $NaHCO_3$) CO_2.

Since ATP is generated during both photosynthetic and respiratory processes, changes in adenine nucleotides could be one of the reasons for the biochemical interaction between chloroplast and mitochondria. Therefore, the effects of these mitochondrial inhibitors on intracellular ATP/ADP ratio was assessed at both limiting and optimal CO_2. The decrease in ATP/ADP ratio was more pronounced in presence of oligomycin than that of antimycin A, particularly at limiting CO_2 (Fig. 3). However, there was a small, but distinct decrease in ATP/ADP ratio with SHAM in presence of antimycin A compared to that of SHAM and propyl gallate (Fig. 4).

In contrast to photosynthetic activity, the intracellular ATP/ADP ratio was more sensitive to mitochondrial inhibitors at limiting CO_2. Although the mitochondrial contribution to cellular ATP is high, photosynthesis was less sensitive to oligomycin

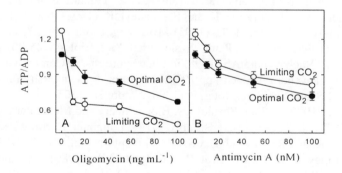

Figure 3. Effect of oligomycin or antimycin A on ATP/ADP in pea protoplasts at optimal (1.0 mM $NaHCO_3$) or limiting (0.1 mM $NaHCO_3$) CO_2.

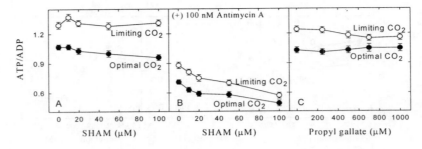

Figure 4. Effect of SHAM (± 100 nM antimycin A) or propyl gallate on ATP/ADP in pea protoplasts at optimal (1.0 mM NaHCO₃) or limiting (0.1 mM NaHCO₃) CO_2.

or antimycin A at limiting CO_2 than that at optimal CO_2. These results suggest that under photorespiratory conditions, mitochondrial oxidative electron transport is more crucial than oxidative phosphorylation in optimizing photosynthesis. We conclude that both cytochrome and alternative pathways of mitochondrial metabolism are essential for optimal performance of photosynthesis at both optimal and limiting CO_2. Further, alternative pathway does contribute significantly to the cellular ATP demands in a photosynthesizing cell, but to a limited extent.

Acknowledgements

This work was supported by Career award grant (F. 1/5-90 SR IV/SA II) from the University Grants Commission, New Delhi and KP was a recepient of UGC-SRF.

References

1. Krömer, S., Malmberg, G. and Gardeström, P. (1993) Plant Physiol. 102, 947-955
2. Raghavendra, A.S., Padmasree, K. and Saradadevi, K. (1994) Plant Sci. 97, 1-14
3. Gardeström, P. and Lernmark, U. (1995) J Bioenerg. Biomembr. 27, 415- 421
4. Krömer, S. (1995) Ann. Rev. Plant Physiol. Plant Mol. Biol. 46, 45-70
5. Padmasree K, Raghavendra AS (1998) in Photosynthesis: A Comprehensive Treatise (Raghavendra A.S., ed.) pp.197-211, Cambridge University Press, UK
6. Seymour, R.S. and Schultze-Motel, P. (1996) Nature 383, 305
7. Lambers, H. (1990) in Plant Physiology, Biochemistry and Molecular Biology (Dennis D.T. and Turpin D.H., eds.) pp.124-143, Longman, Harlow, UK
8. Saradadevi, K. and Raghavendra, A.S. (1992) Plant Physiol. 99, 1232-1237
9. Stitt, M., Lilley, R.Mc.C., Gerhardt, R. and Heldt, H.W. (1989) in Methods in Enzymology (Fleischer S. and Fleischer B., eds.) Vol 174, pp.518-552, Academic Press Inc, New York
10. Møller, I.M., Bérczi, A., van der Plas, L.H.W. and Lambers, H. (1988) Physiol Plant. 72, 642-649

REGULATION OF THE PHOTOSYNTHETIC AND RESPIRATORY CO2 EXCHANGE IN LEAVES BY EXTERNAL FACTORS IN THE LIGHT

Tiit Pärnik[1], Per Gardeström[2], Hiie Ivanova[1], Olav Keerberg[1]
[1]Institute of Experimental Biology at the Estonian Agricultural University, 76902, Harku, Estonia
[2]Department of Plant Physiology, University of Umeå, S-901 87 Umeå, Sweden

Key words: C3, CO_2 uptake, higher plants, light regulation, photorespiration, respiration

1. Introduction

The problems of interrelation of photosynthesis and respiration are actual up today. It has been shown that respiratory processes are functioning in parallel with photosynthesis in the light (1, 2, 3, 4). Respiratory activity of mitochondria may be suppressed or enhanced by light depending on availability of substrates and the adenylate status of the cell (5). Both the primary and stored photosynthates may be the substrates of the Krebs cycle and the rate of mitochondrial decarboxylations to great extent is determined by their availability in different environmental conditions. The substrate level may be regulated by the mitochondrial pyruvate dehydrogenase complex (mPDC) which activity is reduced under photorespiratory conditions (2). Photorespiratory decarboxylation of glycine is the main CO_2-producing reaction in mitochondria of C_3 plants under normal conditions. Respiratory reactions made 3-15% the total rate of the decarboxylation of primary photosynthates. Photorespiratory decarboxylation of stored photosynthates 3 to 5 times exceeded the rate of respiratory decarboxylation of these compounds (6). At present the mechanisms of the main reactions of CO_2 exchange in photosynthesizing leaves are known but a question of their environmental control needs the further examination. In this work by the means of a radio-gasometric method the basic components of photosynthetic and respiratory CO_2 fluxes under different environmental conditions were measured and possible regulatory mechanisms discussed.

2. Methods

Seedlings of winter rye (*Secale cereale* L., cv. Musketeer) and wheat (*Triticum aestivum* L., cv. Saratovskaya-29) were grown 3 weeks in vermiculite under fluorescent tubes (PPFD 250 µmol m^{-2} s^{-1}) at 25°C , photoperiod 16 h..All measurements were carried out on fully expanded third or fourth attached leaves. Photosynthetic and respiratory CO_2 fluxes were determined by radio-gasometric method (6). Four components of respiration in the light were distinguished according to the substrates and oxygen dependence of

G. Garab (ed.), Photosynthesis: Mechanisms and Effects, Vol. V, 3731–3734.
© 1998 *Kluwer Academic Publishers. Printed in the Netherlands.*

decarboxylation reaction: (1) photorespiratory decarboxylation of primary photosynthates; (2) photorespiratory decarboxylation of stored photosynthates; (3) respiratory decarboxylation of primary photosynthates; and (4) respiratory decarboxylation of stored photosynthates. Photorespiratory decarboxylation is linearly dependent on oxygen concentration at least up to 21% , respiratory decarboxylation becomes saturated with oxygen at about 1.5%. Dark respiration was measured 10 minutes after switching off the light when substrates of photorespiration were depleted.

3. Results and Discussion

In normal conditions (see control in Table 1) the total rate of decarboxylation (R_P+R_R) in winter rye leaves was 30 % the rate of carboxylation. It is somewhat more than in other cereals, wheat and barley, measured earlier (6). In winter rye leaves primary photosynthates were the main substrates of decarboxylations, the contribution of stored photosynthates was 19% (Fig.1). In previous experiments with C_3 species studied (barley, wheat and tobacco) the proportion of primary photosynthates was on an average 56% and that of stored photosynthates 44% (6).

Table 1

Rates of photosynthetic and respiratory CO_2 fluxes in leaves of winter rye at different environmental conditions

CO_2 Fluxes $\mu mol\ CO_2$ $m^2\ s$		Control	Low O₂	High CO₂	Low light	Low temp.
	[O_2] $mL\ L^{-1}$	210	15	210	210	210
	[CO_2] $\mu L\ L^{-1}$	300	300	2300	300	300
	PPFD - $\mu mol\ m^{-2}s^{-1}$	750	750	750	80	750
	T - $°C$	26	26	26	26	5.2
Carboxylation P_C :		12.7	14.6	18.3	2.32	6.5
Photorespiratory decarboxylation						
primary photosynthates		2.16	0.14	0.50	0.25	0.36
stored photosynthates		0.48	0.04	0.04	0.11	0.33
Total R_P :		2.64	0.18	0.54	0.36	0.69
Respiratory decarboxylation						
primary photosynthates		0.94	0.94	0.60	0.14	0.21
stored photosynthates		0.24	0.24	0.45	0.40	0.15
Total R_R :		1.18	1.18	1.05	0.54	0.36
Respiration in the dark						
R_D :		0.45	0.37	0.45	0.45	0.33

In control plants the rate of photorespiration (R_P) was 20.8% and that of respiration (R_R) 9.3% the rate of carboxylation. As respiration is saturated at low O_2 concentration (at. about 1.5 mL L^{-1}) the rates of the respiratory decarboxylation at 210 and 15 mL L^{-1} were assumed to be equal (Table 1).

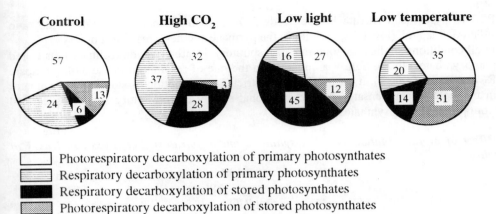

Control   **High CO$_2$**   **Low light**   **Low temperature**

☐ Photorespiratory decarboxylation of primary photosynthates
▨ Respiratory decarboxylation of primary photosynthates
■ Respiratory decarboxylation of stored photosynthates
▦ Photorespiratory decarboxylation of stored photosynthates

Fig. 1. Per cent distribution of respiratory components in leaves of winter rye
in the light (experimental conditions as in Table 1).

Under high CO$_2$ carboxylation was enhanced from 12.7 to 18.3 µmol CO$_2$ m^{-2}s^{-1} (44 %)
and photorespiration decreased 5 times (Table 1). The total respiration was almost the
same as in control plants, but the proportion of stored photosynthates increased to 31%
(Fig. 1). The contribution of stored photosynthates was the highest in low light - 57% the
rate of respiration (Fig. 1). In the limiting light the total photorespiration made15.5% the
rate of decarboxylation compared to 20.8% in control. The reason of this decrease is not
clear because the CO$_2$ concentration in mesophyll was at low light even somewhat lower
(6.9µM) than in normal conditions (7.9µM). Temperature shift from 26 °C to 5.2°C
resulted in a 2-fold decrease of carboxylation (Table 1). This was accompanied by an
increase of stomatal resistance (from 220 to 1050 s m^{-1}) and of the CO$_2$ concentration in
mesophyll (from 7.9 to 12.0µM). Under low temperature the relative rate of
photorespiratory decarboxylation of primary photosynthates decreased from 57% to 35%
but that of stored photosynthates increased from 13% to 31 % (Fig. 1).
Of four respiratory components measured the respiratory decarboxylation of stored
photosynthates is the only component operating in the dark. Earlier measurements on
other C$_3$ species have shown that in the light its rate was between 2% and 5% the rate of
carboxylation (6). In this work with rye leaves the respiratory decarboxylation of stored
photosynthates made only 1.9% of carboxylation in normal conditions but enhanced up
to 17.2% in the limiting light. This component was suppressed by light in control
conditions, under low O$_2$ and low temperature (compare respiratory decarboxylation of
stored photosynthates in the light and respiration in the dark, Table 1). No inhibition was
detected in nonphotorespiratory conditions, at high CO$_2$ and low light. The inhibition can
not be explained only by the functioning of photorespiration with concomitant
inactivation of mPDC complex (2) as the same phenomena was detected also at low O$_2$.
Another possible explanation is the substrate limitation in the level of reaction converting

PEP to pyruvate. Pyruvate kinase needs ADP and in the conditions where the ratio ATP/ADP is high (low CO_2, high light) the pyruvate synthesis may be suppressed (7). In all conditions studied the total rate of respiratory decarboxylation (primary plus stored photosynthates) in the light was higher than the rate of respiration in the dark (compare R_R and R_D in Table 1). It means that the restricted consumption of stored photosynthates in the light is compensated by the additional supply of respiratory substrates derived from primary photosynthates.

Table 2

Effect of preillumination on photosynthetic and respiratory CO_2 fluxes in leaves of wheat

	Preilluminated 28h		Darkened 34h	
	$\mu molCO_2$ $m^2 s$	% respiration	$\mu molCO_2$ $m^2 s$	% respiration
Carboxylation P_C :	9.54		8.18	
Photorespiratory decarboxylation				
primary photosynthates	1.14	58.9	0.91	68.4
stored photosynthates	0.49	25.1	0.23	17.3
Respiratory decarboxylation				
primary photosynthates	0.22	11.3	0.12	9.0
stored photosynthates	0.09	4.7	0.07	5.3
Respiration in the dark R_D :	0.63		0.43	

The effect of long-term illumination and darkening of plants on photosynthetic and respiratory CO_2 fluxes was also studied (Table 2). Photosynthesis of preilluminated wheat leaves was 15 % higher than that of darkened leaves. Preillumination enhanced photorespiratory decarboxylation of stored photosynthates and respiratory decarboxylation of primary photosynthates in the light. The rate of respiration in the dark was in preilluminated leaves about 50% higher than in darkened ones. The light pretreatment did not affect the relative rates of photorespiration and respiration. Respiratory decarboxylation of stored pohotosynthates was almost completely inhibited by the light in both preilluminated and darkened leaves, the rate of this component in the light was less than 1% the rate of carboxylation.

References

1 Raghavendra, A.S., Padmasree, K. and Saradadevi, K. (1994) Plant Science 97, 1-14

2 Krömer, S (1995) Annu. Rev. Plant Physiol. Plant Mol.Biol. 46, 45-70

3 Laisk, A. and Loreto, F. (1996) Plant Physiol. 110, 903-912

4 Atkin, O.K., Evans, J.R. and Siebke, K. (1998) Aust.J.Plant Physiol. 25, 437-443

5 Mamushina, N.S., Zubkova,E.K. and Voitsekhovskaya, O.V. (1997) Russian Journal of Plant Physiology 44, 390-400

6 Pärnik, T. and Keerberg, O. (1995) Journal of Experimental Botany 46, 1493- 1447

7 Gardeström, P. and Wigge, B. (1988) Plant Physiol. 88, 69-76

δ^{13}C OF CO_2 RESPIRED IN THE DARK AND LEAF CARBOHYDRATES IN BEAN PLANTS (*Phaseolus vulgaris* L.) UNDER PROGRESSIVE DROUGHT

Muriel Duranceau[1], Jaleh Ghashghaie[1], Franz Badeck[2], Eliane Deléens[3] and Gabriel Cornic[1]
[1] Laboratoire d'Ecophysiologie Végétale, CNRS, URA 2154, Bât.362, Université de Paris XI, F-91405 Orsay Cedex, France
(E-mail: jaleh.ghashghaie@eco.u-psud.fr)
[2] Potsdam Institute for Climate Impact Research (PIK), P.O. Box 601203, 14412 Potsdam, Germany
[3] Laboratoire de la Structure et Métabolisme des Plantes, IBP, Université de Paris XI, F-91405 Orsay Cedex, France

Key words : carbon isotope, C3, Gas exchange, leaf physiology, respiration, sucrose.

1. Introduction

Plants discriminate against ^{13}C during photosynthetic CO_2 fixation.
A theoretical model has been developed (1) which predicts a linear relationship between the carbon isotope composition (Δ) and the ratio of intercellular to atmospheric partial pressures of CO_2 (p_i/p_a) in C_3 plants:
$$\Delta = a + (b - a) \cdot p_i/p_a \quad \textbf{(Eq.1)}$$
where a (= 4.4‰) is the fractionation due to diffusion of CO_2 in air and b (about 27‰) is the net fractionation during carboxylation by both Rubisco and PEPc (2). According to this model, all factors decreasing p_i/p_a should also decrease Δ. This model does not take into account the fractionation due to CO_2 dissolution, liquid phase diffusion, photorespiration and respiration. The magnitudes of the discrimination during both photorespiration and dark respiration are uncertain. Nevertheless, they have been considered to be very small and to not significantly modify the net discrimination during on-line measurements (3). Therefore, the simple model, which has been validated for many species is usually used. However, it has been recently shown (4) that the difference between Δ observed on-line and Δ predicted by Eq.1 increased when photorespiration or respiration makes a larger contribution to net CO_2 exchange i.e. at low light intensities or at high oxygen concentration. There is only a few contradictory data in the literature concerning δ^{13}C of respired CO_2 in the dark. It has been reported to be ^{13}C-enriched (1-8‰) or ^{13}C-depleted (2-4‰) compared to plant or leaf material varying within species (2). It may be suggested that substrates with different δ^{13}C are used for respiration and/or a discrimination during dark respiration occurs. It has also been recently shown that no fractionation occurred during dark respiration by protoplasts isolated from bean and maize leaves and incubated with sucrose, fructose or glucose (5)
Moreover, there is no data in the literature about the carbon isotope composition (δ^{13}C) of leaf material in advanced stages of dehydration cycles where stomata are closed and leaf photosynthesis very low. Indeed, under such conditions, a higher refixation of

G. Garab (ed.), Photosynthesis: Mechanisms and Effects, Vol. V, 3735–3738.
© 1998 *Kluwer Academic Publishers. Printed in the Netherlands.*

respired CO_2 compared to assimilation of ambient CO_2 could modify the ^{13}C content of carbohydrates formed under these conditions.

The main objectives of the present work were (i) to examine whether any discrimination occurs during dark respiration in intact leaves, (ii) to compare the carbon isotope composition of respired CO_2 to that of leaf carbohydrates and (iii) to determine the carbon isotope composition of leaf carbohydrates synthesised under water deficit.

2. Materials and Methods

French bean were grown in pots in a controlled-environment cabinet under well-watered conditions for a 10-15 d period. Subsequently, two treatments were imposed : drougth stress and well-watered conditions. Plants were subjected to drought by withholding water from the pots. All the measurements were made on cotyledonary leaves. Leaf gas exchanges were measured using an open gas-exchange system as previously described (6). Leaf respired CO_2 was collected, at the beginning of the night, in a CO_2-free closed system. Carbon isotope composition of purified CO_2 was then determined. Leaf discs were harvested at mid-afternoon and at the beginning of the night (after respiration) in order to follow the variations of carbon isotope composition ($\delta^{13}C$) in leaf carbohydrates. Carbon isotope composition were determined with a stable isotope ratio mass spectrometer (Optima, Fison, UK) with a high precision (±0.2‰).

3. Results and Discussion

During the dehydration cycle, two phases were observed:

During the first phase of the dehydration cycle, leaf net photosynthesis (fig.1b) rapidly declined in dehydrated plants, reaching zero. In control plants, it remained constant and then decreased, reaching 3 µmol $CO_2.m^{-2}.s^{-1}$ at the end of the experiment. In both control and dehydrated plants, the decline in net CO_2 uptake was accompanied by a parallel decline in leaf conductance (Fig.1c). The decrease in leaf net photosynthesis and leaf conductance in control plants in spite of RWC maintain may indicate leaf ageing in these plants. As expected, sucrose became heavier (enriched in ^{13}C) during the first phase of the cycle (fig.1d), the discrimination by Rubisco decreasing with the stomatal closure (fig.1c). This is in agreement with Farquhar's model.
But during the second phase, when stomata remained closed (Fig.1c) and leaf net photosynthesis was very low (Fig.1b), the tendency reversed and sucrose became lighter (depleted in ^{13}C) (fig.1d). The expected variations of discrimination, calculated for each value of pi/pa relative to a reference value of $pi/pa=0.7$, (using Farquhar's model, Fig.2) roughly correspond to the variations of sucrose $\delta^{13}C$ (Fig.1d) suggesting that the variations of sucrose $\delta^{13}C$ could be globally explained by the variations of pi/pa during both phases of the experiment. However, the values of pi/pa under severe water deficit (when leaf net CO_2 uptake is very low) can not be estimated with precision and should be taken with caution (7)
Another possible explanation for this depletion in ^{13}C in sucrose pool may be, as hypothesised before, a higher refixation of respired CO_2 compared to assimilation of ambient CO_2. Indeed, respired CO_2 being depleted in ^{13}C compared to ambient CO_2, the carbohydrates formed under such conditions should also be progressively depleted in [13]C. Interestingly, the variations in $\delta^{13}C$ of CO_2 respired in the dark were similar (Fig.1e), with an offset, to those of sucrose for both dehydrated and well-watered plants, respired CO_2 always being ^{13}C-enriched compared to sucrose.

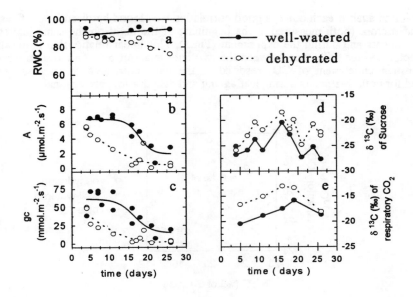

Figure 1- Changes in a) leaf relative water content (RWC%), b) leaf net CO_2 assimilation rate (A), c) leaf conductance to CO_2 (gc), d) carbon isotope composition ($\delta^{13}C$) of leaf sucrose in the afternoon and e) carbon isotope composition ($\delta^{13}C$) of CO_2 respired in the dark as a function of time for both well-watered (closed symbols, solid line) and dehydrated plants (open symbols, dotted line). Day zero corresponds to the last watering for dehydrated plants. Each data point represents a measurement done on an individual plant.

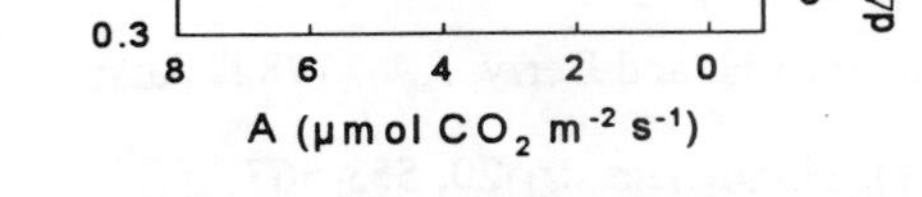

Figure 2 - Changes in p_i/p_a during the experiment as a function of leaf net CO_2 assimilation (A) for both control (closed symbols) and dehydrated plants (open symbols). According to Farquhar's simple model (1), for each observed value of p_i/p_a an expected variation of the discrimination (dΔ) relative to a reference value when $p_i/p_a = 0.7$ was determined (an increased dΔ indicates a lower discrimination by the carboxylating enzymes than in the reference case).

When plotted against each other, a good correlation was found between $\delta^{13}C$ of respired CO_2 and sucrose (both measured at the beginning of the night on the same leaves) for both treatments and during the experiment (Fig.3). This linear relationship is parallel to the bisector and the difference between the two lines is about 6‰. Our results, showing this constant enrichment of CO_2 respired in the dark compared to sucrose (Fig.3), obtained for both treatments during leaf ageing and dehydration, are original.

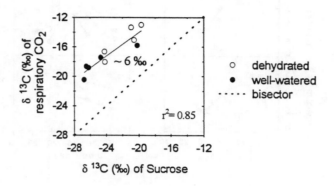

Figure 3 - Relationship between carbon isotope composition ($\delta^{13}C$) of CO_2 respired in the dark and sucrose measured on the same leaves at the beginning of the night (using a tight closed system) for both well-watered plants (close symbols) and dehydrated plants (open symbols). The values of $\delta^{13}C$ of CO_2 respired in the dark from Fig.1 are used for the relationship.

The most simple interpretation of this correlation between carbon isotope composition of sucrose and CO_2 respired in the dark for all the plants used is that sucrose or a closely linked substance is the main substrate for dark respiration. It is concluded that a discrimination by about -6 ‰ occurs during dark respiration processes in intact cotyledonary leaves of bean plants during both leaf ageing and plant dehydration.

References

1 Farquhar G.D., O'Leary M.H. and Berry, J.A. (1982) Aust. J. of Plant Physiol. 9, 121-137.
2 O'Leary, M.H. (1981). Phytochemistry 20, 553-567
3 Evans J.R., Sharkey T.D. and Farquhar G.D. (1986) Aust. J. of Plant Physiol. 13, 281-292.
4 Gillon J.S. and Griffiths H. (1997) Plant Cell Environ. 20, 1217-1230
5 Lin G. and Ehleringer J. R. (1997) Plant Physiol 114, 391-394.
6 Cornic G. and Ghashghaie J. (1991) Planta 185, 255-260
7 Bro E., Meyer S. & Genty B. (1996) Plant Cell Environ. 19, 1349-1358.

TRANSPIRATION HAS A FUNCTION OF ASSIMILATE REDISTRIBUTION IN THE WHOLE PLANT.

Vladimir Chikov, Golsoyar Bakirova, Nadejda Avvakumova
Kazan Institute of Biology of Russian Academy of Sciences, P.O. Box 30, Kazan 420503, Russia   E-mail : chikov@sci.kcn.ru

Key words: metabolite transport, photosynthesis, transpiration, apoplast.

There is some reason to suppose that assimilates flow out of the cells to apoplast not only during the loading of phloem ends but also in course of their transport along the shoot phloem [Minchin,Mc Naughton,1987]. This suggests the involving of apoplast in the regulation of photosynthetic function at the level of whole plant. This investigation is a verification of the given idea.

Methods.
The experiments were carried out on the long-fibred flax plants (Linum usitatissimum L.). The choice of the object was explained by intensive growth and big height of flax plant shoot. Plants were cultivated in green-house and were of 40 cm high. Part of plants had the increased level of nitrogen nutrition($CaNO_3$).   The extraction of labeled assimilates from apoplast was performed as previously described [Chikov,1987]. The shoot section, assimilating $^{14}CO_2$ included 2,5 cm length in the middle of plant. The assimilation time of labeled $^{14}CO_2$ was equal to 2 min. In 5 minuts after the beginning of $^{14}CO_2$ assimilation experimental plants were cut into three parts(Fig1). The section, assimilating $^{14}CO_2$ was located in the middle part after cutting the plant into sections. AII 3 sections were infiltrated by water. After infiltration each section was placed into "pressure chamber" and gaseous nitrogen was simultaneously supplied to all there chambers to extract the liquid from intercellular space. After collecting the apoplast liquid, each shoot section was divided in parts (see Fig.1 and Tabl. 1,2). All plant parts were fixed separately in the boiling ethanol and the radioactivity of each part was determined.

Results and discussion.
5 min just after $^{14}CO_2$ assimilation the labeled assimilates reached quite distant parts of plant, the rate of thier transport to top(more 4m/h) being considerably larger than the linear rate of transport along phloem. During first minutes after the $^{14}CO_2$ assimilation, the labeled products of photosynthesis were transfered in the ascending direction more intensively than in the descending one (Tabl 1). However, the study of ^{14}C distribution along the plant on the next day after the start of $^{14}CO_2$ assimilation showed that the most part of labeled assimilates occured in the lower part of shoot (Table

G. Garab (ed.), Photosynthesis: Mechanisms and Effects, Vol. V, 3739–3742.

2). As a result, the ratio of radioactivity of the plant parts located above the donor section and below it (in the tables it was designated as top/bottom of shoot) was decreased during 1 day chase to 18-30-fold.

The intensive influx of the labeled assimilates to the upper part of shoot (in the Table1. The ^{14}C distribution along the plant and influx of labeled assimilates into apoplast in 5 min after $^{14}CO_2$ pulse in the middle part of shoot (in % of total organ radioactivity).

Part of plant	non-fertilized plant	N-fertilized plant
The part which located above the ^{14}C-donor part including: top	$0,9 \pm 0,04$	$0,3 \pm 0,04$
leaves	$3,8 \pm 0,60$	$4,0 \pm 0,25$
bust	$1,8 \pm 0,08$	$2,3 \pm 0,30$
wood	$1,2 \pm 0,06$	$0,4 \pm 0,05$
apoplast	$0,7 \pm 0,04$	$0,3 \pm 0,02$
The ^{14}C-donor part (in all)	$89,4 \pm 0,4$	$92,5 \pm 0,6$
including: leaves	$71,7 \pm 0,4$	$73,4 \pm 0,6$
bust	$14,5 \pm 0,3$	$13,3 \pm 0,2$
wood	$1,4 \pm 0,2$	$2,1 \pm 0,1$
apoplast	$1,7 \pm 0,1$	$1,9 \pm 0,1$
The lower part (in all)	$1,5$	$1,9$
including: leaves	$0,9 \pm 0,1$	$1,1 \pm 0,1$
bust	$0,3 \pm 0,05$	$0,4 \pm 0,05$
wood	$0,2 \pm 0,02$	$0,3 \pm 0,02$
apoplast	$0,1 \pm 0,07$	$0,1 \pm 0,03$
top/bottom of shoot	$5,6$	$3,8$

first minutes after $^{14}CO_2$ assimilation) was shown to correlate with high ^{14}C amount in the apoplast. As follows from this, the labeled products of photosynthesis, entering the extracellular space from the phloem vessels, were transported upwards with the transpiration flux of water.The assimilates, leaving the apoplast in lower part of shoot were washed away by water, flowing out from the roots and having no labeled carbon and, thereby, fill up the ascending ^{14}C-flux. It should be noted that in apoplast itself the concentration of assimilates is very low. This fact can be explained by high dilution of substances dissolving in apoplast. It should be kept in mind the value of the intensity of water transpiration flux along the plant per unit of leaf surface is 1000-fold more than the value of carbon flux via photosynthesis. The concentration of assimilates which are translocated within apoplast, can become pronounced only in their "ultimate destination", i.e. in the top or in the leaf mesophyll due to their accumulation. However during the next day, the major part of labeled assimilates, which were translocated from donor part of plant to the upper leaves, flux to the ofher parts of plant (Tabl. 1,2).The data on the absolute ^{14}C content in upper leaves also support this idea (Fig.1). After 1 day of chase only about 25% of total amount of labeled assimilates translocated primary to the upper leaves, have been left there. It means that assimilates transfering within apoplast with water flux in the ascending direction are repeatly loaded in the phloem ends without entering the cells reaching the free space of upper leaves. When translocating down along the shoot phloem, they could been washed away from the phloem by the contrary water flow. Thus, they are circulating and redistributed along the plant with the help of transpiration flux. As a result, after 1 day the main amount of labeled assimilates are in the lower part of plant, basically in the wood. In the bust, only part of ^{14}C

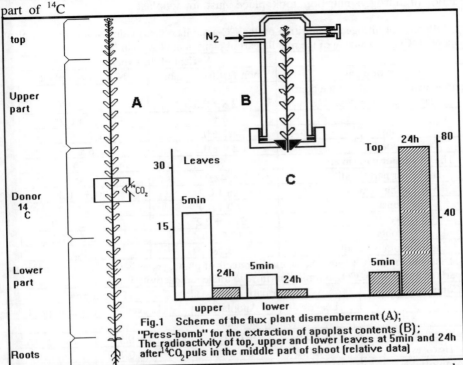

Fig.1   Scheme of the flux plant dismemberment (A);
"Press-bomb" for the extraction of apoplast contents (B);
The radioactivity of top, upper and lower leaves at 5min and 24h
after $^{14}CO_2$ puls in the middle part of shoot (relative data)

flowing within apoplast seem to be used for the formation of phloem structure elements, while the others are washed away by the transpiration flux and translocated to apex where they are utilized. Unlike the leaves or bust, the assimilates which entered the top are used on the spot and therefore are accumulated(Tabl.1,2, Fig.1). The cultivation of the high level of nitrogen nutrition led to the known effect — to a relatively decrease of assimilate export from leaves. The conditions of nitrogen nutrition affect not only the processes occuring in mesophyll celles but also the state of apoplast and systems of distant transport, of which the loading and permeability support the specific mode of the assimilate distribution along the plant(Tabl.1,2). The ^{14}C distribution among photosynthetic products showed that sucrose had the greater part with respect to labeled products of photosynthesis which were detected in apoplast, compared to mesophyll cells (Table 3). Taking into account these all things considered, we can suggest that the noncarbonhydrate part of compounds translocated within apoplast might be more greater than that one seems at first glance, since sucrose is retransported, while the non-carbohydrate products are accumulated. The invertase takes part in the regulation of the disfant transport of assimilates. Invertase action in free space of leaf might be one of the mechanisms affecting the inhibition of assimilate efflux out of the leaves under action of enhanced nitrogen nutrition. The increase of sucrose hydrolysis just in the apoplast of fertilized plants is effectively illustrated in Table 3.

The most pronounced changes of the sucrose to hexoses ratio (in table sucrose/hexoses) under action of nitrogen nutrition took place just in apoplast

Table 2. Influence of nitrogen of nutrition on the ^{14}C-assimilate distribution along the flax plant at 1 day after 14CO 2 fertilizing of the middle part of plant in the light (in %).

The part of plant	Variant of the experiment	
	non-fertilized plant	N-fertilized plant
The part which located above the donor part, including: top	$2,8 \pm 0,5$	$2,3 \pm 0,3$
leaves	$1,0 \pm 0,2$	$1,0 \pm 0.2$
bust	$1,1 \pm 0,3$	$1,7 \pm 0,3$
wood	$2,9 \pm 0,5$	$2,8 \pm 0,3$
The ^{14}C-donor part (in all)	50,3	56,7
The lower part (in all)	42,0	35,5
including: leaves	$0,6 \pm 0,2$	$0,5 \pm 0,1$
bust	$7,0 \pm 0,9$	$7,3 \pm 0,7$
wood	$21,8 \pm 2,2$	$19,2 \pm 2,0$
roots	$12,5 \pm 0,4$	$8,5 \pm 1,2$
top/bottom of shoot	0,18	0,22

Table 3. Distribution of ^{14}C among the labeled products of photosynthesis in leaves or in apoplast of flax parts at 5 min after ^{14}CO$_2$ assimilation(in % to totale radioactivity fraction)

Compounds	Upper part		^{14}C-donor part	
	Leaves	Apoplast	Leaves	Apoplast
Control plant				
Sucrose	$73,5 \pm 0.6$	$89,7 \pm 0.2$	$60,9 \pm 0,5$	$89,6 \pm 0,5$
Hexoses	$4,9 \pm 0.2$	$0,6 \pm 0,2$	$3,6 \pm 0,1$	$0,7 \pm 0,1$
Amino acids	$10,0 \pm 0,6$	$4,1 \pm 0,2$	$15,1 \pm 0,3$	$3,9 \pm 0,4$
Sucrose/hexoses	15,0	149,5	16,9	128,0
Nitrogen-fertilized				
Sucrose	$61,5 \pm 0,3$	$77,1 \pm 0,5$	$57,1 \pm 1,0$	$76,2 \pm 1,1$
Hexoses	$5,3 \pm 0,2$	$2,1 \pm 0,2$	$4.2 \pm 0,1$	$2,0 \pm 0,2$
Amino acids	$15,7 \pm 0,5$	$11,7 \pm 0,5$	$17,2 \pm 0,6$	$11,1 \pm 0,7$
Sucrose/hexoses	11,6	36,7	13,6	38,1

space. In leaves of fertilized plants this ratio was found to be decreased only by 20%, whereas in apoplast this index was reduced by 70%. Monosugars formed as a result, returned back to the cells where they were used in intracellular metabolism.

In conclusion, the present data confirm the important role of processes occuring within apoplast, for the regulation of nitrogen and carbon interaction at photosynthesis and assimilate transport.

ROBUSTNESS AND TIME-SCALE HIERARCHY IN BIOLOGICAL SYSTEMS

I. Rojdestvenski*, M. Cottam[+], Y.-I. Park*, G.Samuelsson*, G. Oquist*,
* Department of Plant Physiology, Umea University, Umea, Sweden
[+] Dept. of Physics, University of Western Ontario, London, Canada

Key words: model systems, stability, CO_2 uptake, carbonic anhydrase, adaptation

Introduction

The phenomenon of plant life is associated with systems that efficiently utilize solar energy to sustain plant structure, growth and development. This high level of efficiency is achieved through considerable complexity of the plant organization involving a multitude of metabolic pathways and associated regulatory mechanisms. It is interesting how the system can, in principle, combine robustness and stability with the elaborate regulation allowing high efficiency. The metabolic pathways are seemingly so interwoven that a change in the parameters of one regulatory mechanism should inevitably influence other regulatory activities, swinging the whole system away from equilibrium. To provide for necessary stability, or "robustness" against such changes, a certain mechanism for "decoupling" different regulatory mechanisms must exist. A prominent property of biological systems is that the metabolic reactions comprise processes with significantly different time scales. The aim of this contribution is to discuss possible roles of time-scale hierarchy in decoupling regulatory mechanisms. A model example of the recently suggested CO_2 pumping mechanism in *Chlamidomonas reinhardtii* is presented to illustrate the concepts discussed.

Robustness and time-scale hierarchy

Suppose there are n processes in the system, and their characteristic times, τ_i, form a time-scale hierarchy such that $\tau_1 \ll \tau_2 \ll \tau_3 \ldots \ll \tau_n$. Let us choose one such process (or group of processes), k, its characteristic time being τ_k. With respect to τ_k the other characteristic times can be divided in two groups. For the "slow" processes, with $\tau_i \gg \tau_k$ we have effectively $\tau_i = \infty$, while for the "fast processes", $\tau_i \ll \tau_k$, we may safely assume $\tau_i = 0$. The term "robustness" in the given context means the time evolution is independent of certain parameters in the wide range of their variation [1]. A formal mathematical discussion of time-scale decoupling, as applied to metabolic pathways control equations, was given in [2]. Here we present only qualitative arguments. Suppose that the "fast" processes are described by the following formal set of dynamic equations:

$$L(y_1,\ldots,y_s,x_1,\ldots,x_r,y_{s+1},\ldots,y_k,t) = 0 \tag{1}$$

Here, as previously, the y variables stand for the internal parameters of the system (metabolite concentrations, rates of processes, etc.) and the x variables represent the parameters of the environment, while L is the time evolution operator. We attribute the

G. Garab (ed.), Photosynthesis: Mechanisms and Effects, Vol. V, 3743–3746.
© 1998 Kluwer Academic Publishers. Printed in the Netherlands.

first s of the y variables to the parameters of the fast processes, whilst the s+1 to k of the y variables describe the parameters of the slow processes and, hence, enter into (1) as static parameters. Suppose also that the equation (1) has an *equilibrium* solution corresponding to the values $\{y^{(0)}_1,....,y^{(0)}_s\}$ at given values of parameters $x_1,...,x_r,y_{s+1},...,y_k$. In general, kinetics of all the processes slower than τ_{k-1} depend on the above *equilibrium* solution $\{y^{(0)}_1,....,y^{(0)}_s\}$, rather than on the complete solution of (1). Hence, the effective number of parameters describing the slower processes may be reduced to only those that determine the equilibrium solution. The slower kinetics are insensitive to the other parameters, or, in other words, are *robust against changes in them*. We further illustrate how robustness manifests itself due to the time-scale hierarchy by taking a concrete model system.

Carbonic anhydrase driven CO_2 pump in *Chlamydomonas reinhardtii*.

In (Park et al., this issue) and (Rojdestvenski et al., unpublished) we studied the bicarbonate (BC) depletion effects on photosynthesis in high-CO_2 grown *Chlamydomonas reinhardtii* cell wall-less mutants with (cw92) or without (cia3/cw15) active carbonic anhydrase (CA) associated with photosystem II (PSII)[3]. We observed the shutdown of oxygen evolution in cw92 that we attributed to CO_2 depletion of the Calvin cycle. We suggested the role of CA being to pump CO_2 from the thylakoid lumen into the thylakoid stroma, as a result of the low pH value in the lumen due to the PSII-ATPase regulation. By admitting a (possibly PSII-activated) BC diffusion into the lumen, a quick CA-facilitated CO_2 production from the incoming BC and a CO_2 diffusion into the stroma, we achieve a sort of "engine", which converts stromal BC into stromal CO_2. The backward conversion of CO_2 into BC in the stroma in the absence of CA is slow and the Calvin cycle efficiently competes for CO_2, leading to inorganic carbon depletion in the system. In the case of the mutant cia3, which has no active CA in the lumen, the depletion happens much slower and its effects are less pronounced.

We suggested a simple model including five processes (BC diffusion from stroma to lumen and back; CO_2 diffusion from stroma to lumen and back; BC/CO_2 interconversion in the stroma; BC/CO_2 interconversion in the lumen, facilitated by CA; CO_2 consumption by the Calvin cycle) and four pools (stromal and lumenal BC and CO_2). We took the following rates for the above processes (square brackets stand for concentrations, superscripts denote lumenal and stromal compartments, C_i stands for the total inorganic carbon content (i.e. quantity of added bicarbonate, and subscripts h and -h refer to hydration (CO_2 to BC) and dehydration (BC to CO_2) processes) (see also [4,5]).

Process	Rate function	Rate constants values
CO_2 diffusion from/to stroma	$v_{dCO2} = K_D([CO_2]^{lum}/C_i - [CO_2]^{str}/C_i)$	$K_D = 3 \cdot 10^{-14}$ mol/cell/s
BC diffusion from/to stroma	$v_{dBC} = K_D([BC]^{lum}/C_i - [BC]^{str}/C_i)$	same as above
BC/CO_2 conversion in stroma (No CA)	$v_{-h} = ([BC]^{str}/C_i)K_{-h}^{str}$ $v_h = ([CO_2]^{str}/C_i)K_{-h}^{str} \cdot 10^{pHL - 6.25}$	$K_{-h}^{str} = 1.5 \cdot 10^{-17}$ mol/cell/s
BC/CO_2 conversion in lumen (CA present)	$v_{-h} = ([BC]^{lum}/C_i)K_{-h}^{lum}$ $v_h = ([CO_2]^{lum}/C_i)K_{-h}^{lum} \cdot 10^{pHL - 6.25}$	$K_{-h}^{lum} = 1.5 \cdot 10^{-12}$ mol/cell/s
CO_2 assimilation by CC	$V_{max}[CO_2]^{str}/([CO_2]^{str} + K_m)$	$V_{max} = 0.9 \cdot 10^{-16}$ mol/s/cell $K_m = 1 \cdot 10^{-14}$ mol/cell.

In Fig. 1 we present our modeling results for the time dependences of the BC and CO_2 concentrations in the lumen and in the stroma. Initial inorganic carbon concentration (in the form of BC in stroma) is denoted by C_0.

(a)   (b)

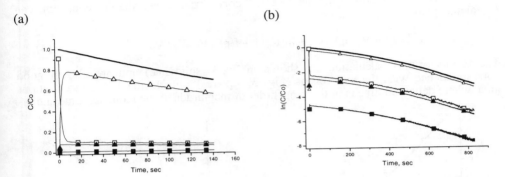

Figure 1. Time dependences of the concentrations of bicarbonate (squares) and CO_2 (triangles) in stroma (open symbols) and lumen (solid symbols), as well as total inroganic carbon content (bold solid line). $pH_S=7.8$, $pH_L = 5.3$. (a) induction period, linear scale; (b) longer times, logarithmic scale.

A characteristic feature of the presented data is that after a short induction period (Fig.1a) the concentrations at all times remain proportional to each other (Fig.1b). The above result hints at the onset of a certain partial equilibrium reached during the induction period. We also checked our results against the variations in the diffusion and conversion rates. Our data showed (Rojdestvenski et al., unpublished) that even an order-of-magnitude change in the diffusion and conversion rates affected only the duration of the induction period, having no influence on the results after the induction was over. On the other hand, changes in the Calvin cycle parameters V_m and K_m had almost no influence on the induction period, significantly changing the further evolution of the system (Rojdestvenski et al, unpublished). Thus, the overall time evolution of the model proved robust against substantial variations in four parameters out of six, and the induction period proved robust against variations in the remaining two parameters. Mathematically this robustness is a consequence of the fact that the diffusion rates, conversion rates and Calvin cycle rates represent three different time scales, each differing by 1 or 2 orders of magnitude.Then, with respect to "slow" steady state kinetics we can simplify our model, stating $K_D = K_h^{lum} = \infty$; $K_h^{str} = 0$. This led to formulation of a "quasi-equilibrium" theory with only two parameters – V_{max} and K_m of the Calvin cycle (Rojdestvenski et al, unpublished). The theory predicts that the time dependence of the inorganic carbon concentration in the experiment with cw92 obeys Michaelis-Menthen kinetics, but with the K_m constant replaced by $K_m(1 + f(pH_L))$ where $f(pH_L) = 10^{pHL - 6.25}$,

$$dC_i/dt = - V_{max}C_i/(C_i + K_m(1 + f(pH_L))) \tag{2}$$

with the solution:

$$(C_i - C_0) + K_m(1 + f(pH_L))\ln[C_i/C_0] = -V_{max}t; \tag{3}$$

The results of this theory showed good correspondence with computer modeling results (see Fig.2). To calculate the depletion time we first noticed that at $C_i \sim K_m(1+f(pH_L))$ Calvin cycle rate slows down considerably (Fig.2). Then, if the PSII quenching by Calvin cycle retardation is responsible for the rapid fluorescence drop observed *in vivo*, the value $K_m(1 + f(pH_L))$ can be taken for the "cut-off" concentration, and the depletion time can be calculated as

$$T_D = -\{2K_m(1 + f(pH_L) - C_0 + \ln[K_m(1 + f(pH_L)/C_0] \}/ V_{max} \qquad (4)$$

The results of this calculation are shown in Fig.3 (solid line) and are also in good correspondence with the experiment (Fig. 3, squares) outside the low concentrations area, where depletion time is of the same order of magnitude as the induction time and

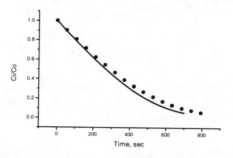

Figure 2. C_i concentrations as function of time: computer experiment (circles) and quasi-equilibrium theory (solid line).

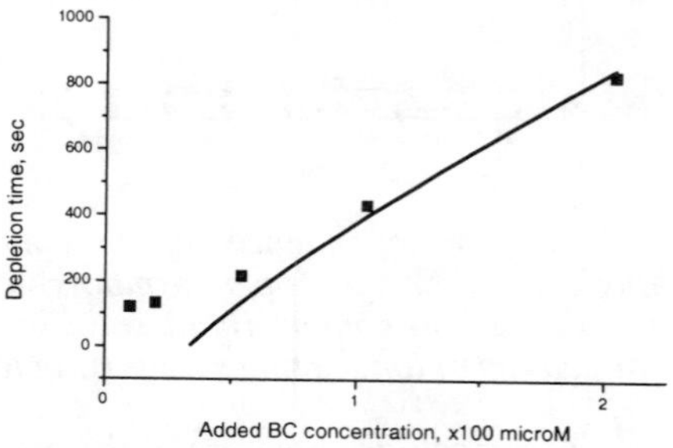

Figure 3. Depletion times vs added BC concentration: theory (solid line) compared to experimental data (squares).

the theory becomes transcendental. That the theory adequately describes experimental and modeling results with only two parameters means that the modeled system is robust against changes in the other four rate constants provided the time scale hierarchy is not violated.

Conclusions

Time-scale hierarchy leads to the decoupling of regulatory mechanisms and the emergence of robustness against variations in certain parameters of the discussed system.
This work has been supported by the grant No. I-AG/WB 04830-334 by NFR, Sweden.

References
1. Barkai, N. and S.Leibler, Nature, 387, 913-7 (1997)
2. Delgado, J.C.Liao, Biosystems, 36, 55-70 (1995)
3. Karlsson, J. *et al. EMBO J.* submitted (1997)
4. Raven, J. Plant Cell Environ. 20, 147-154 (1997)
5. Spalding, M.H. and A.R.Portis Jr., Planta, 164, 308-320 (1985)

20. Photosynthesis and agricultural production; herbicides

CONGRESS ON PHOTOSYNTHESIS

XIth INTERNATIONAL

BUDAPEST * 1998

THE CIRCADIAN RHYTHM IN THE EXPRESSION OF NITRATE REDUCTASE IN TOMATO IS DRIVEN BY CHANGES IN PROTEIN LEVEL AND NOT BY PROTEIN PHOSPHORYLATION

Dawn E. Tucker[1] and Donald R. Ort[1,2] , [1]Department of Plant Biology and [2]Photosynthesis Research Unit, USDA/ARS, 1201 W. Gregory Drive University of Illinois, Urbana, IL 61801

Key words : chilling, gene regulation, N metabolism, temperature stresses, phosphorylation/dephosphorylation, regulatory processes

Introduction

The presence of clock-controlled processes in virtually all living organisms is suggestive of a fundamental biological relevance. Circadian rhythms are involved in diverse processes ranging from daily sleep/wake cycles in humans, to gene transcription in prokaryotes, as well as leaf movements, stomatal, and photosynthesis in higher plants. Attempts to understand both the mechanism and purposes of circadian oscillations in higher plants have revealed numerous genes that exhibit circadian regulation of transcription. By comparison, there are relatively few examples of circadian rhythms in enzyme activity. One reason is that the protein products are often abundant and quite stable, so that a circadian oscillation in gene transcription and subsequent translation often does not result in any detectable oscillation at the protein level. This, for example, is the case for the chlorophyll *a/b* binding protein and rubisco activase (1). However, because circadian rhythms are observed in numberous whole plant processes it is apparent that many enzymes must have circadian rhythms in activity. Circadian oscillations in enzyme activity could be conferred in various ways including cycling protein levels, changing substrate concentrations, cycling levels of allosteric effectors, or oscillations in protein phosphorylation or redox state.

Our interest in circadian rhythms in enzyme activity has focused on reactions that might underlie the chilling sensitivity of photosynthesis in chilling sensitive plant species. An overnight low-temperature treatment (4°C) impairs next day photosynthesis by as much as 60% in chilling sensitive plant species such as tomato (2). The mechanism for this inhibition of photosynthesis is unknown, however one striking effect of the chilling treatment is to delay the progress of the circadian clock regulating the transcription of certain nuclear encoded genes (1). Our interest in the mechanism of the chilling sensitvity of photosynthesis led us to investigate ways that uncoordinated gene expression or enzyme activity, as a result of low temerature induced mistiming of transcriptional circadian rhythms, might lead to feedback inhibition in the metabolic reactions of photosynthesis. Our results caused us to focus attention on nitrate reductase, a key enzyme in plant cell nitrogen metabolism that has been shown to have an endogenous rhythms in activity in several plant species.

G. Garab (ed.), Photosynthesis: Mechanisms and Effects, Vol. V, 3749–3754.
© 1998 *Kluwer Academic Publishers. Printed in the Netherlands.*

NR catalyzes the first and rate-limiting step in nitrogen metabolism reducing nitrate to nitrate in the cytoplasm. NR is subject to complex transcriptional and post-transcriptional regulation and displays a robust circadian rhythm in activity levels in many higher plants (3,4,5,6) including *Lycopersicon esculentum* (1). NR protein level can be regulated by synthesis and degradation but light/dark regulation of enzyme activity is dominated by post-translational modification (7,8,9) involving phosphorylation of specific seryl residues (10,11,12), followed by the divalent cation dependent association of 14-3-3 inhibitor proteins that inactivate the enzyme (13,14,15,16,17). Identifying the factors responsible for NR rhythmic behavior is important to understanding the chilling response of circadian rhythms in cold sensitive species.

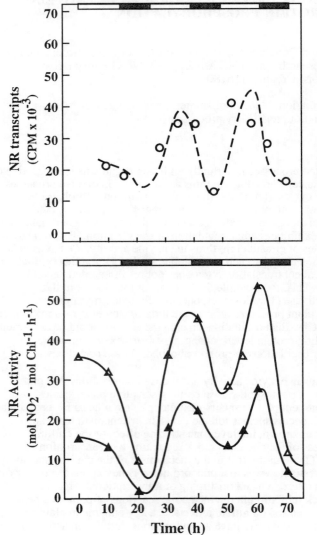

Figure 1. NR has a circadian rhythm in transcript and protein levels which is responsible for an endogenous rhythm in NR activity in tomato. Top, total leaf RNA was isolated from tomato leaves under 3 d, constant-light (450 µmol quanta m^{-2} s^{-1}) and constant-temperature (26°C) circadian conditions and probed with a dCTP^{32} labeled tobacco nia -2 cDNA (O). The radioactivity was quantified on a phosphor imager. Bottom, NR activity was assayed under in vitro conditions in the absence (△) and presence (▲) of Mg^{2+}. The lack of Mg^{2+} prevents the binding of 14-3-3 type inhibitor proteins to phosphorylated NR allowing all of the enzyme present to be in its active state. Modified from Jones et al (18).

Materials and Methods

Tomato (*Lycopersicon esculentum* Mill. cv Floramerica) plants were grown from seed in growth chambers under a 14 h (26°C) light/10 h (21°C) dark cycle at 75% RH, as described by Jones and Ort (19). *In vitro* NR activity assays conducted using crude tomato tissue extracts followed a protocol modified from Kaiser and Brendle-Benisch (20). Desalted samples were measured either in the presence of 10 mM $MgCl_2$ or 5 mM EDTA. The reactions were incubated for 3 min at 30°C and the reaction stopped by the addition of zinc acetate. NO_2^- produced was quantified by standard colorimetric reagent addition and the measurement of A_{540}. NR activity was calculated on a chlorophyll basis. Transcript rhythms were determined from total RNA extracts collected over a 3 d constant condition time course. RNA was probed with a P^{32} labeled *nia*-2 cDNA probe. Ethidium Bromide staining ensured equal RNA loaded per lane.

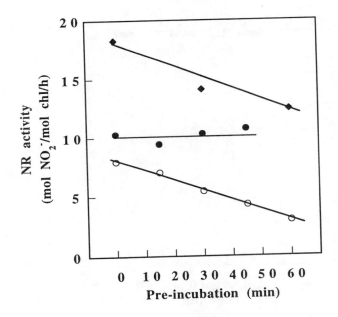

Figure 2. Continuous-light circadian samples do not show slow activation of NR in the absence of divalent cations. Pre-incubation with EDTA ensured the NR remained unbound by 14-3-3 inhibitor proteins for up to 60 min. Leaf samples were taken at the peak (32 h, ◆) or trough (24 h, O) of continuous-light circadian NR activity rhythms and after 20 h of dark adaptation (●). The data shown are the average of 3 samples.

Results and Discussion

There was a robust circadian rhythm in NR activity in tomato under constant light and temperature conditions (fig. 1, bottom). Light activation of NR activity can involve both changes in NR enzyme level and changes in enzyme phosphorylation state. The *in vitro* assay allowed resolution of activity changes due to phosphorylation state and subsequent Mg^{+2} dependent binding of 14-3-3 inhibitor proteins to phospho-NR (13,14,15,16). Under continuous light conditions in the absence of Mg^{2+} (Δ) NR cannot be inactivated by inhibitor protein thus indicating that NR protein level is subject to endogenous rhythmic behavior (fig. 1). Kaiser and Huber (21) showed that, when phosphorylated, full activation of spinach NR by the removal of Mg^{+2} can require up to 30 min. This effect could obscure regulation of activity by changes in NR phosphorylation states. In tomato, leaf samples from both high and low activity points, incubated in EDTA for up to 60 min, show no slow activation in the absence of Mg^{2+} (fig. 2, $\blacklozenge$, O). This slow activation can, however, be induced in tomato by prolonged dark adaptation (fig. 2, $\bullet$). Artifactual

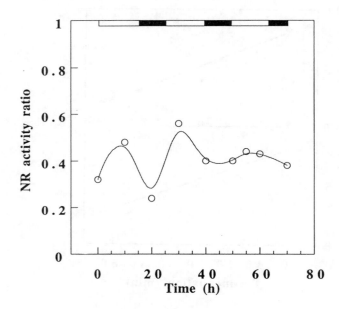

Figure 3. Data represent the difference between fully active NR (Mg^{2+} absent) and NR active under limiting conditions such as are present in the plant cell (Mg^{2+} present). This graph illustrates the absolute amount of NR activity over a circadian time course that is due solely to changes in the phosphorylation state of the enzyme. Less enzyme is phosphorylated when less total protein is present as in low activity periods suggesting that higher phosphorylation states of the enzyme do not correspond to the subjective night periods that would be necessary to drive such a rhythm.

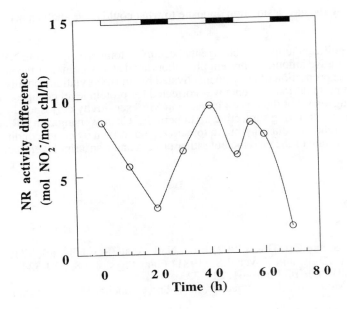

Figure 4. The ratio of activity due to phosphorylation changes of NR is illustrated in order to correct for protein changes that occur over the 3 d constant condition time course. Data clearly shows the proportion of enzyme inactivation due to post translational modification during subjective night periods does not increase as would be expected if phosphorylation state played a significant role in the NR circadian rhythm.

protein degradation occurring during these assays was minimized by protease inhibitors and low temperature. The circadian rhythm occurring in the presence of Mg^{2+} reflects the actual activity of NR in plant cells because Mg^{2+} normally present in the cytoplasm is sufficient to allow inactivation of phospho-NR. The clear rhythm in enzyme level prompted investigation of NR transcript levels that also show a distinct endogenous rhythm directly shadowing protein level oscillations (fig. 1, top). The NR circadian rhythm in tomato appears to be driven by oscillations in transcript level which in turn result in oscillations of the enzyme level.

Previous work has indicated that phosphorylation and subsequent inactivation by inhibitor protein binding plays the predominant role in light/dark activity of NR. The robust rhythm in activity in the absence of Mg^{2+} exhibited during this 3 d constant-light time course suggests that the NR circadian rhythm activity in tomato is likely driven by protein and transcript changes rather than covalent modification. To determine the magnitude of activity driven by phosphorylation changes rather than protein level oscillations, the difference between the activity of NR with and without Mg^{2+} present was calculated and graphed (fig. 3). This data illustrates that the absolute difference in NR activity due to phosphorylation at any given time during the circadian rhythm could not be driving the oscillations in activity observed. Clearly less NR is phosphorylated during subjective night periods than during the day, a situation opposite to what could drive such an activity rhythm.

Because of the significant amplitude change that occurs in total protein during NR activity oscillations, the relative amount of protein phosphorylated at any point in time is obscured. To determine if the proportion of enzyme inactivated by phosphorylation actually increases during subjective night periods the data was corrected for protein level changes. The resulting graph suggests that there is no increase in Mg^{2+} sensitivity corresponding to the low activities that occur during subjective night periods. Clearly changes in NR activity during a constant light circadian rhythm in tomato are not driven by corresponding changes in phsophorylation state of the enzyme but rather protein and transcript level fluctuations.

References

1. Martino-Catt, S. and Ort, D.R. (1992) Proc. Natl. Acad. Sci. USA 89, 3731-3735
2. Martin, B., Ort, D. R. and Boyer, J.S. (1981) Plant Physiol. 68, 329-334
3. Lillo, C. (1984) Physiol. Plant. 61, 219-223
4. Deng, M-D., Moureaux, T., Leydecker, M-T. and Caboche, M. (1990) Planta 180, 257-261
5. Cheng, C-L., Acedo, G.N., Dewdney, J., Goodman, H.M. and Conking, M.A. (1991) Plant Physiol. 96, 275-279
6. Pilgrim, M.L., Caspar, T., Quail, P.H. and McClung, C.R. (1993) Plant Mol. Bio. 23, 349-364
7. Kaiser, W.M. and Spill, D. (1991) Plant Physiol. 96, 368-375
8. MacKintosh, C. (1992) Biochim. Biophys. Acta. 1137, 121-126
9. Kaiser, W.M. and Huber, S.C. (1994) Plant Physiol. 106, 817-821
10. Douglas, P., Morrice, N. and MacKintosh, C. (1995) FEBS Lett 377, 113-117
11. Bachmann, M., Shiraishi, N., Campbell, W.H., Yoo, B.C., Harmon, A.C. and Huber, S.C. (1996) Plant Cell 8, 127-131
12. Su, W., Huber, S.C. and Crawford, N.M. (1996) Plant Cell 8, 519-527
13. Spill, D. and Kaiser, W.M. (1994) Planta 192, 183-188
14. Bachmann, M., McMichael, R.W., Huber, J.L. Kaiser, W.M. and Huber, S.C. (1995) Plant Physiol. 108, 1083-1092
15. Glaab, J. and Kaiser, W.M. (1995) Planta 195, 514-518
16. MacKintosh, C., Douglas, P. and Lillo, C. (1995) Plant Physiol. 107, 451-457
17. Lillo, C., Kazazaic, S., Ruoff, P. and Meyer, C. (1997) Plant Physiol. 114, 1377-1383
18. Jones, T.L., Tucker, D.E. and Ort, D.R. (1998) Plant Physiol. 118, in press
19. Jones, T.L. and Ort, D.R. (1997) Plant Physiol. 113, 1167-1175
20. Kaiser, W.M. and Brendle-Benisch, E.(1991) Plant Physiol. 96, 363-367
21. Kaiser, W.M. and Huber, S.C. (1997) J. Exp. Bot. 48, 1367-1374

PHOTOSYNTHETIC RESPONSES TO DROUGHT OF TWO GRAPEVINE VARIETIES OF CONTRASTING ORIGIN

Hans R. Schultz[1,2], [1]UFR Viticulture, INRA/ENSA, 34060 Montpellier, France, [2]present adress: Institut für Weinbau und Rebenzüchtung, Forschungsanstalt, D-65366 Geisenheim, Germany

Key words: grapevine, gas exchange, fluorescence, stomata, water stress

1. Introduction

Water shortage is among the most dominant environmental constraints for grape production. Even in moderate temperate climates, grapevines often face some degree of drought stress during the growing season. The large genetic diversity of available grapevine species and cultivars and their characteristic geographical distribution may indicate different strategies to cope with the regional climate and its limitations for production. It is therefore likely that grapevine cultivars from different origines may have evolved different ways to adapt to dry soil conditions.

Several studies suggest differences in stomatal sensitivity to water deficit between grape species and cultivars judging from reported differences in water use and water use efficiency (WUE) (e.g. 1, 2, 3). Nevertheless, comparative results from gas exchange studies under field conditions with different grape varieties are rare, yet suggest that differences in WUE may arise from cultivar differences in their sensitivity to high temperature and/or low humidity (2, 4, 5), frequent environmental co-factors of water deficit and dominant meteorological characteristics of different climates.

In the Mediterranean bassin, the question of drought tolerance is of economic importance since most vineyards are non-irrigated. In this respect, the tendency in recent years to replace the typical spectrum of cultivars with varieties from other climatic zones for reasons of wine quality is not without risks, since little is known about their respective response to summer drought. It seemed therefore necessary to conduct a detailed evaluation of different components of the genetic variability of 'resistance or tolerance to drought' using a traditional variety, Grenache, and an introduced variety of mesic origin, Syrah.

One objective of this study was to analyse stomatal behaviour of these 2 varieties under naturally fluctuating environmental conditions and soil water status, and to quantify the effects of stomatal control on single leaf and canopy gas-exchange, and on viticultural performance, i.e. yield components and fruit quality.

G. Garab (ed.), Photosynthesis: Mechanisms and Effects, Vol. V, 3755–3760.
© 1998 *Kluwer Academic Publishers. Printed in the Netherlands.*

2. Material and methods

2.1 *Location*

The larger part of this study was conducted in 1994 and 1995 on 8-year old Grenache and Syrah vines grown side by side in a commercial vineyard near Montpellier, France (43°45'N, 3°40'E). The site is on calcareous soil and has a planting density of 3860 vines ha^{-1}. Of each variety, 7 vines were irrigated weekly at 30l/vine after bloom (15. June) for the first time in 1994. In 1995, about 30 vines were irrigated at maximum capacity (up to about 40l per vine/week), the rest of the vineyard received only natural precipitation.

2.2 *Gas exchange and stomatal sensitivity*

Leaf gas exchange measurements were conducted on mature, well exposed leaves with an ADC-LCA 3-type gas exchange system equipped with a Parkinson leaf cuvette. Leaf carbon balance was estimated from diurnal and nocturnal measurements of gas exchange. Stomatal sensitivity was determined from the relationship of stomatal conductance (g) to the product of photosynthetic rate (A), measured under varying light intensities, vapour pressure deficits and temperatures, and a factor consisting of the relative humidity (hs) divided by the external CO_2 concentration (Ca) (6, here termed BWB-model) ($g = g_0 + k \cdot A \cdot [hs/Ca]$). Where g_0 is a residual stomatal conductance (as A—> 0 when PFD—> 0), and k is termed the stomatal sensitivity factor (7).

2.3 *Chlorophyll Fluorescence*

Chlorophyll Fluorescence (photosystem II) in the light adapted state was measured with a portable pulse amplitude modulation fluorometer (PAM 2000, Walz, Germany). Using diurnal changes in the efficiency of open PSII centers (Fv'/Fm'), the fraction of light absorbed in PSII antennae and dissipated thermally (D) (D=1-Fv'/Fm') and the fraction utilized in PSII photochemistry (P) (P=Fv'/Fm' qp) (qp=coefficient of photochemical quenching), as well as the rates of D and P (multiplied by PFD, photon flux density) were estimated (8).

2.4 *Water relations, leaf area and fruit characteristics*

Water potential was determined with a pressure chamber (Soilmoisture Corp. Santa Barbara, USA). Leaf area was determined on 15 vines per treatment (8). Fruit was harvested at the beginning of September. The juice of representative subsamples (250 berries) was analysed for pH, and soluble solids by refractometry. Phenolics were determined by absorption at 280 nm (A 280). Colour was analysed at 520nm (A 520).

3. Results and Discussion

3.1 *Gas-exchange*

The course of maximum daily photosynthesis (A_{max}) and maximum daily stomatal conductance (g_{max}) during the season was similar for Syrah and Grenache for any given day (Fig 1A, B), yet the degree of water deficit (expressed as pre-dawn leaf water potential, ψ_{PD}), and midday water potential, ψ_M, at which these values occurred evolved different for the two varieties during dry-down (Fig. 1C, D). The A_{max} and g_{max} were more sensitive to water deficit in Grenache, where ψ_{PD} decreased only to a low of -0.85 MPa at the end of August (DOY 240), as compared to Syrah, where ψ_{PD} at the same date had reached -1.4 MPa (Fig. 1C), yet with similar photosynthetic rates and stomatal conductances (Fig. 1A, B, C). This implies very different strategies in water use (9). Whereas ψ_{PD} and ψ_M were always lower for stressed as compared to control vines for Syrah, ψ_M of stressed Grenache was not different from the control (Fig. 1D), despite large differences in ψ_{PD} (Fig. 1C). This behaviour is not strictly concurrent with the typical anisohydric or isohydric grouping of species (10), suggesting that within species, some intermediate forms may exist.

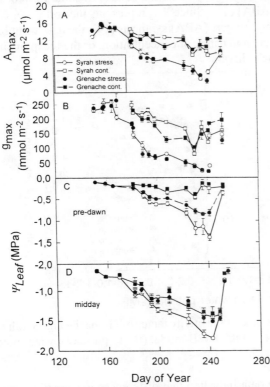

Figure 1. Change in A_{max}, (A), g_{max}, (B), pre-dawn - (C), and midday leaf water potential (D) for irrigated and water-stressed Grenache and Syrah vines during 1994 (5).

3.2. *Stomatal sensitivity*

The data from Fig. 1 suggested differences in stomatal control between the grape varieties during water deficit, so the BWB-model was used to quantify stomatal sensitivity. Although there is concern whether there is any mechanistic ground to the BWB model (11), some have used it as a tool to simulate water stress effects on photosynthesis by expressing k as a function of ψ_{PD} (7, 12). Applying this method to the data set on Grenache and Syrah, k was significantly smaller, thus stomatal sensitivity higher for Grenache than for Syrah, below a ψ_{PD} of about -0.10 - 0.15 MPa (Fig. 2A). The coefficients of determination (R^2), which can be regarded as a measure of the degree of coupling between g and the BWB-model, were usually higher for Grenache than Syrah and rapidly increased for both varieties with increasing water deficit (Fig. 2B). The degree of linearity or coupling is remarkable, considering the variation in environmental data at any date (30% to near 70% hs, 100-1700 PFD, leaf temperatures from 18°C to 44°C). Maximum stomatal coupling to the BWB-index was already achieved at a Ψ_{PD} of between -0.3 and -0.4 MPa, indicating that moderate stress improved this relationship (Fig. 2B).

Stomatal closure in grape may be under the control of some form of root to shoot communication, probably [ABA] (13). Since Ψ_{PD} represents the soil water content in the rooting zone which presumably triggers [ABA] synthesis (14), the data would indicate that either the signals may be different in nature or concentration, or the receptor site sensitivity may be different between the cultivars.

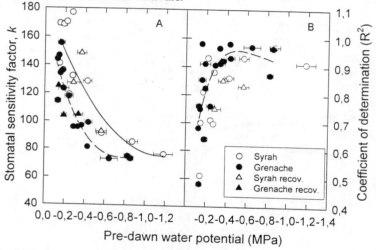

Figure 2. Relationship of the stomatal sensitivity factor of Grenache and Syrah (A), and the degree of coupling between g and the BWB-model (B), to pre-dawn water potential (5).

3.3 *Chlorophyll Fluorescence*

Diurnal gas exchange data indicated that differences in stomatal control between varieties were pronounced at mid-afternoon (13-16hrs). Estimating the rates of photochemistry (P) and heat dissipation (D) (11) for this period by probing Chlorophyll Fluorescence confirmed the gas exchange data and showed, that Syrah was capable of maintaining higher rates of photochemistry at lower soil and leaf (data not shown) water potentials (Fig. 3A). The capacity to dissipate energy as heat was not different between the varieties (Fig. 3B). There were only small differences in the efficiency of PSII photochemistry, Fv'/Fm' and qp in mid-afternoon and thus the rate of photochemistry might have been more affected by possible differences in leaf absorptance (8).

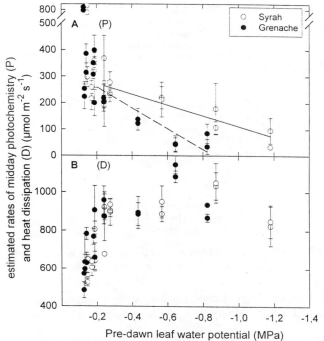

Figure 3. Estimated rates of mid-afternoon photochemistry (A) and heat dissipation (B) as a function of ψ_{PD} for the 2 grapevine varieties.

3.4 *Leaf Carbon Balance*

Figure 4 demonstrates the apparent trade-off between high sensitivity or large degree of stomatal control and production capacity. The leaf carbon balance of stressed Syrah vines was much more positive than that of Grenache vines. This was also reflected in its Viticultural performance. In terms of total amount of sugar produced at harvest, Syrah had 16.3% more carbohydrates invested in yield and total sugar concentration in the fruit, and had higher levels of Phenolics and Anthocyanins, whose construction costs in terms of glucose equivalents are very high.

In natural ecosystems, plant survival during drought will be closely coupled to efficient strategies for water use. In the most "pessimistic" of all strategies, the annual growth cycle will be completed before the onset of drought, or water use will be extremely conservative, as will be, as a consequence, the production of dry matter. These types of strategies are of no interest for agricultural or viticultural production, since they minimize yield. So for any environment in dry (no irrigation) Viticulture, the ideal adapted cultivar would be that whose behaviour tends to maximize assimilation in relation to the amount of water available (9). In this sense the variety with high WUE, Grenache, exhibited a 'pessimistic' strategy with negative consequences for production. Syrah demonstrated an 'optimistic' approach, which in this particular case, proved to be superior.

3760

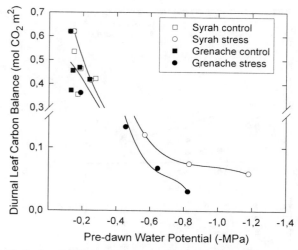

Figure 4. Daily leaf carbon balance for Grenache and Syrah during a water deficit.

Acknowledgements

I thank Gérard Bruno and Eric Lebon for help with irrigation, harvest, and fruit analysis, and Serge Rambal, CEFE, Montpellier, for loaning the PAM fluorometer.

References

1   Füri, J. and Kozma, F. (1977) Wein-Wiss. 21: 103-121
2   Chaves, M.M., Harley, P.C., Tenhunen, J.D. and Lange, O.L. (1987) Physiol. Plant. 70: 639-647
3   Düring, H. (1988) Vitis 27: 199-208
4   Loveys, B.R. and Düring, H. (1984) New Phytologist 97: 37-47
5   Schultz, H.R. (1996) Acta Hort. 427: 251-266
6   Ball, J.T., Woodrow, I.E. and Berry, J.A. (1987) In: Progress in Photosynthesis Research. (Biggins J., ed.) pp. 221-224,. Martinus Nijhoff, Dordrecht, The Netherlands
7   Harley, P.C. and Tenhunen, J.D. (1991) In: Modeling crop photosynthesis - from biochemistry to canopy (Boote, K.J., Loomis, R.S. eds.) CSSA Pub. 19: 17-39
8   Demmig-Adams , B., William W. Adams III., Barker, D.H., Logan, B.A., Bowling, D.R.and Verhoeven, A.S. (1996) Physiol. Plant. 98: 253-264
9   Jones, H.G. (1992) Cambridge University Press, Cambridge, U.K. 428p
10   Tardieu, F. and Simonneau, T. (1998) J. Exp. Bot. 49: 419-432
11   Aphalo, P.J. and Jarvis, P.G. (1993) Ann. Bot. 72: 321-327
12   Lloyd, J. (1991) Aust. J. Plant Physiol. 18: 649-660
13   Correia, M.J., Pereira, J.S., Chaves, M.M., Rodrigues, M.L. and Pacheco, C.A. (1995) Plant, Cell Environ. 18: 511-521
14   Davies, W.J., and Zhang, J. (1991) Ann. Rev. Plant Physiol. 42:55-76

FLURIDONE- AND LIGHT- AFFECTED CHLOROPLAST ULTRASTRUCTURE AND ABA ACCUMULATION IN DROUGHT-STRESSED BARLEY

Losanka P. Popova
"M. Popov" Institute of Plant Physiology, Bulgarian Academy of Sciences,
Acad. G. Bonchev str., bl. 21, 1113 Sofia, Bulgaria

Key words: chloroplast development, electron microscopy, etioplasts, guard cells, herbicides, leaf anatomy.

Introduction

Fluridone and nonflurazon are selective inhibitors of carotenoid synthesis in cells. They interfere with desaturation steps of carotenoid biogenesis and thus blocks the accumulation of carotenoids (1,2,3). Carotenoids and chlorophylls are essential constituents of photosynthetic membranes. The lack of carotenoids results in a block in membrane formation and the concurrent bleaching of the organism. Carotenoid-deficient mutants produce photobleached seedlings owing to the photooxidation of chlorophyll in the absence of carotenoids. Non-mutant seeds treated with fluridone or norflurazon also produce albino seedlings due to inhibition of carotenoid synthesis and subsequent photooxidation of chlorophyll (4). In all these treatments plastidogenesis is destroyed and plastid ribosomes are completely absent. The degree of destruction of the internal structure of chloroplasts varies widely depending on the intensity of irradiance under which the plants are grown. In strong white light the cotyledones of norflurazon-treated mustard seedlings did not contain normal chloroplasts, but only small chlorophyll-free rudiments with completely destroyed internal structure (5). Treatment of etiolated bean leaves with norflurazon had little effect on the formation of normal prothylakoids and prolamellar bodies (6). This is in marked contrast to *Euglena* where inhibition of carotenoid synthesis with norflurazon resulted in proplastids with undeveloped thylakoid system (7). By growing barley seedlings in darkness, it was showed that fluridone inhibited carotenoid accumulation but did not alter plastid biogenesis and protein composition. However, data on the ultrastructure of proplastids were not reported (8). In addition, it was found that carotenoid-deficient plants are ABA-deficient (4), thus leading investigators to suggest a link betwen chloroplast function and ABA synthesis.

In this study we used fluridone as a tool to study to what extent the integrity of chloroplast is important for the synthesis of ABA. And second, is the observed inhibition of stress-induced accumulation of ABA after fluridone treatment due to

G. Garab (ed.), Photosynthesis: Mechanisms and Effects, Vol. V, 3761–3766.
© *1998 Kluwer Academic Publishers. Printed in the Netherlands.*

blocking of carotenogenesis or to plastid dysfunction, which would results in the inability to produce ATP and reducing power. An attempt has been made to separate the photobleaching effect of fluridone on plastid development and drought-induced accumulation of ABA from the same effect under non-photooxidative conditions.

Methods

Plant material and fluridone treatment. Seeds of *Hordeum vulgare* L.,*var. Alfa.* were imbibed for 24 h on water containing 0 or 10 μM fluridone. The hydrated seeds were planted in a Fafard M 2 soiless medium and fertilized twice weekly. Plants were cultured in a Percival growth cabinet at day / night temperatures 25°C/ 20°C, RH 60%, and a 16 h photoperiod at a fluence of 600 μmol m^{-2} s^{-1}. Young, fully expanded second leaves of 10-d old plants were used in all experiments.

In another set of experiments the fluence was reduced to 40 μmol m^{-2} s^{-1} by shading with several layers of "Miracloth".

For dark treatment experiments, plants were grown for 10 days under the standard growth conditions and in continuous darkness

Microscopic observations. For scanning electron microscopy (SEM) , freeze- dried samples were sputter coated with Au/Pd and examined with a JEOL JSM-840 operated at 10 kV.

Ttransmission electron microscopy (EM) was follwed after the procedure descibed in (9). Monitor sections (1 μm) from the same material were used for the light microscopy (LM).

ABA analysis. ABA was estimated by enzyme-amplified ELISA as earlier described (10).

Results

Leaf anatomy. Treatment of barley plants that had been grown at a PPF of 600 μmol m^{-2} s^{-1} with 10 μM fluridone caused a strong reduction in the rate of growth. The leaves, although fully developed, were small, and completely photobleached. Herbicide treatment led to some stimulation of growth in darkness. Leaves of treated plants were more elongated but less expanded than those of untreated plants. Despite non-photooxidative conditions in darkness, fluridone treatment caused albino appearance of foliar tissue (data not shown).

Control plants had a normal leaf anatomy, with chloroplasts appressed to the plasmalemma of mesophyll cells (*fig*.1A) Fluridone-treated plants differed in their leaf anatomy. Most of the mesophyll cells either did not have well developed chloroplasts or lacked them. In the latter case the intracellular space was filled with granullar substance (*fig*.1B). Etiolated, dark-grown plants showed a typical leaf anatomy with clear differentiation between cell types and well arranged etioplastids against the cell walls (*fig*.1C). Dark-grown and fluridone-treated barley plants showed similar leaf anatomy to the controls. Some of the mesophyll cells contained only a few etioplastids or completely lacked them and their intracellular space was granullar (*fig*.1D).

The number of etioplasts per cell counted from TEM cross sections was twice less for the fluridone-treated barley leaves (*fig*.2).

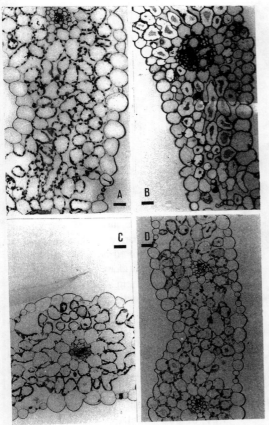

Figure 1. Effect of fluridone (10 µM) on leaf anatomy of barley plants. **A** and **B**, light micrographs of cross sections of barley leaves of control (**A**) and fluridone-treated (**B**) plants, grown at an irradiance of 600 µmol m^{-2} s^{-1}; **C** and **D**., control (**C**) and fluridone-treated (**D**) plants grown in darkness. Bars = 100 µm.

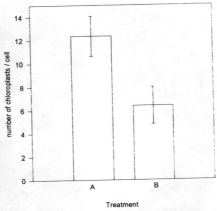

Figure 2. Effect of fluridone on plastid biogenesis of dark-grown barley plants. The number of ethioplastids were counted from 320 cells of 12 cross sections for the control (**A**) and from 500 cells of 12 cross sections for the fluridone-treated plants (**B**).

Ultrastructure of plastids. Chloroplasts of green leaves grown under 600 µmol m^{-2} s^{-1} PPFD were elongated and contained grana consisting of several normaly arranged thylakoids (*fig*.3A). Treatment of plants with fluridone led to the internal destruction of the plastids. The thylakoids had disappeared and mostly vesicles were present. The changes were mainly restricted to the internal structure but also ruptures and discontinuities of the chloroplast's envelope were observed (*fig*. 3B, arrowheads). Starch was always absent from these plastids

The ultrastructure of chloroplasts of control plants grown at a PPF of 40 µmol m^{-2} s^{-1} was not affected (*fig*.3C). Plastids of fluridone-treated plants grown under the same fluence were characterized with a completely destroyed internal structure (*fig*.3D).
Plants that had been grown in darkness (etiolated plants) contained well developed etioplasts showing thylakoids and big crystalline prolamellar body area. Darkly stained plastoglobules were numerous and usually arranged into groups (*fig*. 3E). The etioplasts from fluridone-treated leaves were smaller in size, with less prolamellar body area and thylakoid length. The degree of grana staking was lower and many vesicles were present (*fig*.3F).

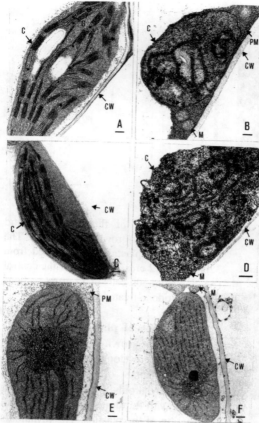

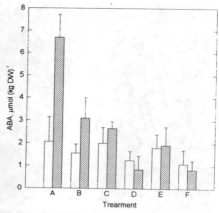

Figure 4. ABA contents of control and fluridone-treated barley plants and grown at different levels of irradiance. **A**, control plants, and **B**, fluridone treated plants and grown at 600 μmol m^{-2} s^{-1} PPFD; **C**, control plants, and **D**, plants treated with fluridone and grown at 40 μmol m^{-2} s^{-1} PPFD; **E**, control plants, and **F**, plants treated with fluridone and grown in darkness.

Figure 3. Electron micrographs of chloroplasts of barley plants grown at different levels of irradiance. **A**, chloroplast of control plants and **B**, chloroplast of fluridone-treated plants and grown at 600 μmol m^{-2} s^{-1} PPFD; **C**, chloroplast of control plants, **D**, chloroplast of fluridone-treated plants and grown at 40 μmol m^{-2} s^{-1} PPFD; **E**, etioplast of control plants, **F**, etioplast of fluridone-treated plants and grown in darkness.

Effect of photosynthetic photon flux density on stress-induced ABA.

The ABA level of control plants grown at a PPF of 600 μmol m^{-2} s^{-1} was low but increased more than 15-fold after dehydration of the leaves. (*fig.*4A). Plants grown in the presence of fluridone and high fluence had diminished levels of ABA in both unstressed and dehydrated leaves (*fig.*4B). Growing plants at low irradiance (40 μmol m^{-2} s^{-1}) diminished the level of stress-induced ABA accumulation. Fluridone treatment reduced the level of ABA in dehydrated leaves (*fig.*4C and 4D). Dark-grown control plants had a detectible level of ABA but twice less as compared to the light-gtown plants. Plants which had been dehydrated followed by incubation for 4 h accumulated 40-fold more ABA than the turgid leaves (*fig.*4E). Growth of plants in darkness in the presence of 10 μM fluridone reduced the levels of ABA in both unstressed and dehydrated leaves (*fig.* 4F).

Discussion

The results in this study showed that treatment of barley seedlings with fluridone caused albino appearance of foliar tissue. The degree of bleaching did not depend on the intensity of irradiance under which the plants were grown. Treatment of barley with fluridone in darkness also led to appearance of albino seedlings (data not shown). The most obvious effect of fluridone was not limited only on the growth, but the leaf's anatomy was also affected (fig.1).The alterations concerned mainly the structure of different cell types, including guard cells. The leaves of light-grown and fluridone-treated plants did not contain chloroplasts but only small chlorophyll-free rudiments whose internal structure had almost disappeared. Treatment of barley seedlings with fluridone in darkness produced effects on plastid development (fig.2) and ultrastructure (fig.3F). These results partially would be explained with the inhibitory effect of fluridone on carotenoid synthesis. The lack of carotenoids results in a block in membrane formation, the formation of membrane constituents and assembly (5). A reduction in chlorophyll levels also occurs in higher plants grown with norflurazon in dim light (11, 12).

These results allowed us to test the accumulation of stress-induced ABA in the presence of photooxidative damage for the chloroplasts (light-grown plants), and in the absence of plastid photodestruction (dark-treated plants). ABA level of control light-grown plants increased more than 15-fold after dehydration of plants (fig.4A). Plants grown in the presence of 10 μM fluridone under high irradiance had diminished levels of ABA in both unstressed and dehydrated leaves. They had no capacity to accumulate ABA after dehydration (fig. 4 B).

When barley seedlings were grown under low irradiance (40 μmol m^{-2} s^{-1}) the levels of stress-induced ABA decreased. Fluridone-treated plants grown under the same fluence were not capable to accumulate ABA after imposition of dehydration (fig. 4C and 4D). It is possible that plants grown under low light had sufficient carotenoids to support ABA accumulation after leaf dehydration, while in fluridone-treated plants either reduced carotenoid content or damage to plastids could have limited ABA accumulation. Fluridone-treated and dark-grown plants failed to accumulate ABA after imposition to dehydration (fig. 4E and 4F). Thus, in etiolated barley plants, fluridone had effects on stress-induced ABA accumulation and on the organization of the thylakoid membrane system that are independent of the photochemical damage. The bleaching effect of fluridone on barley plants grown in darkness points that treated plants are carotenoid-deficient. This suggestion is consistent with the observation that norflurazon at concentration of 10 μM caused inhibition of carotenoid synthesis in the dark (1).

Several reports on the effect of light on endogenous levels of ABA in vegetative tissues have appeared. It was found that the levels of ABA in intact unstressed plants, and the amounts induced by water stress were greater in light than in dark-grown plants (13), and the accumulation of stress-induced ABA may be favoured by higher than lower light intensities (14). This is in agreement with our data presented in figure 4 and also with our previous experiments showing that light strongly affected the capacity of *Vicia* leaves to accumulate ABA in response to water stress (9). Our data presented here showed that etiolated barley plants produced higher level of stress-induced ABA compared with the light-grown plants (*fig.* 4A and 4E). In summary, we have shown that treatment of barley seedlings with fluridone caused alterations in leaf anatomy and led to internal destruction of the plastids. These effects did not depend on the irradiance under which the plants are grown. Fluridone treatment reduced the levels of ABA in both unstressed and dehydrated leaves. The capacity of leaves to accumulate ABA in response to water stress is affected by light irradiance. One explanation of these results is that only leaves that lack precursor carotenoids are unable to make ABA. Another explanation is that lack or low levels of carotenoids cause variety of anomalies, and that the lack of ABA accumulation is an indirect result of carotenoid deficiency. As the lack of carotenoids destroyed the chloroplast ultrastructure it is very difficult to determine which of the two factors is more important for ABA accumulation. This would imply that structurally intact and /or functionally active plastids are required for drought stress to elicit a rise in ABA

References

1 Bartels, P.G. and Watson, C.W. (1978) Weed Sci., 26, 198-203.
2 Eder, F.A. (1979) Z. Naturforsch., 34c, 1052-1054.
3 Fong, F. and Schiff J.A. (1979) *Planta,* 146, 119-127.
4 Henson, I.E. (1984) Z. Phlanzenphysiol., Bd. 114, 35-43.
5 Reib, T., Bergfeld, R., Link, G., Thien, W. and Mohr, H. (1983) Planta, 159,518-528.
6 Pardo, A.D. and Schiff, J.A. (1980). *Can. J. Bot.,* 58, 25-31.
7 Vaisberg, A.G. and Schiff, J.A. (1976) Plant Physiol., 57, 260-269.
8 Gamble, P.R. and Mullet, J.E. (1986) Eur. J. Biochem., 160, 117-121.
9 Popova, L.P. and Riddle, K.N. (1996) Physiologia Plantarum, 98, 791-797.
10 Zhang, S., Hite, D.R.C. and Outlaw, W.H., 1991 Physiol. Plant., 83, 304-306.
11 Bartels, P.G. and McCullough, C. (1972) Biochem. Biophys. Res. Commun., 48, 16-22.
12 Kunert, K.J. and Boger, P. (1979) Z. Naturforsch., 34c, 1047-1051.
13 Simpson, G.M. and Saunders, P.F. (1972) *Planta,* 102, 272-276.
14 Moore, R. and Smith, J.D. (1985) Planta, 164, 126-128.

INORGANIC NITROGEN SIGNAL TRANSDUCTION FOR EXPRESSION OF MAIZE *C4PPC1*.

Tatsuo Sugiyama, Hitoshi Sakakibara, and Mitsutaka Taniguchi
Dept. Biological Mechanisms and Functions, Graduate Sch. of
Bioagricultural Sciences, Nagoya Univeristy, Nagoya, Japan 464-8601

Key words: C4, cytokinins, photosynthesis genes, maize, N-responsive regulation, whole plants

1. Introduction

Plants live under various environmental conditions, including nutrients, light, acidity, temperature, and humidity, that can change unexpectedly and rapidly. Among nutrients, nitrogen (N) is of particular importance since this nutrient is needed in greatest abundance by plants and most crucially limits the rate determining step of photosynthesis and thereby vegetative growth rate. Inorganic N-sources act as substrates for assimilation and signals for modification in metabolism and gene expression. Inorganic N-sources taken up by plants through the roots are assimilated utilizing energy and C skeletons that are produced primarily in the leaves where C is assimilated. This fact envisages that regulation of gene expression for photosynthetic C assimilation by N in plants must require a complex network of communication at cellular, intercellular, and organ-to-organ levels.

Inorganic N-responsive gene expression for C4 enzymes *in vivo* has been studied most extensively with maize phospho*enol*pyruvate carboxylase (C4Ppc1) and pyruvate-Pi dikinase (Ppdk) of which levels are of primary importance in terms of photosynthetic productivity as well as N economy in C4 plants (see a review, 1). Allocation of N into C4Ppc1 and Ppdk is selectively up-regulated by inorganic N-sources such as nitrate and ammonium ions, both transcriptionally and posttranscriptionally (2-4). A similar mode of regulation has also been demonstrated for carbonic anhydrase in maize (5, 6) and alanine aminotransferase (7) and asparate aminotransferase (8) in *Panicum miliaceum*, which function in the C4 pathway.

We describe here the isolation and characterization of ZmCip1 in maize, a homologue of bacterial response-regulators, and its possible involvement in the inorganic N-signal transduction for *C4Ppc1* gene expression which is mediated by cytokinin as a root-to-leaf signal. We also discuss our recent findings in relation to the question of how plants recognize inorganic N-sources to allocate N into C4 proteins in response to its avaiilability.

2. Procedure

Plant Materials and Growth Conditions

G. Garab (ed.), Photosynthesis: Mechanisms and Effects, Vol. V, 3767–3772.
© 1998 *Kluwer Academic Publishers. Printed in the Netherlands.*

Maize (*Zea mays* L. cv. Golden Cross Bantam T51) plants were grown for about 18 days in vermiculite or for about 3 weeks in an aerated hydroponic system with limited N (0.8 mM NaNO3 for vermiculite, 0.08 mM NaNO3 for hydroponic system) as descried previously (4). The youngest, fully developed leaves of plants were cut and used as detached leaves. The basal ends where photosynthetic cells are maturing were used as materials for extraction of nuclei, RNA, and proteins.

Methods for Aanalysis

Nuclear run-off transcription was measured by a method described previously (9). Differential display, screening for cDNA clone, DNA sequencing, Northern analysis, preparation of antibody, Western blot analysis, and identification and quantification of cytokinin species were conducted according to methods described previously (10).

3. Result and Discussion

3.1 Nature and Mode of Inorganic N-Responsive Gene Expresion for C4Ppc1

We have utilized the basipetal gradients of C4Ppc1 in the youngest developed leaves of maize plants to look at the inorganic N-responsive gene expression in the whole-plant. The inorganic N-responsive accumulation of C4Ppc1 occurs most conspicuously in the basal region where photosynthetic cells are maturing (4). The results obtained have demonstrated the following:
1. Allocation of N into C4Ppc1 is selectively up-regulated by inorganic N-sources such as nitrate and ammonium ions and the accumulation of the protein is primarily regulated by the rate of protein synthesis accompanying a concomitant increase in level of mRNA (2-4).
2. The inorganic N-responsive gene expression is regulated at least both transcriptionally and posttranscriptionally specific to the C4 type of form (9).

To identify a possible messenger(s) in the inorganic N-responsive gene expression of *C4Ppc1*, which mediates communication between leaf and root tissues, we have chosen detached leaves as the experimental material. The results obtained have led to the following conclusions:
1. Glutamine and/or its metabolite(s) is a metabolic signal for the inorganic N-responsive accumulation of *C4Ppc1* mRNA(6).
2. Cytokinin can be another internal signal of inorganic N-sources and up-regulates the transcription of *C4Ppc1* (9, 11).
3. The cytokinin-mediated transcription may require *de novo* synthesis of protein(s) (9).

Based on these results, we propose a working hypothesis for the inorganic N-responsive regulation of *C4Ppc1* in the whole-plant, postulating : (i) an external inorganic N-source transiently induces the accumulation of cytokinin(s) in roots and (ii) an unknown protein(s) is an integral link for the cytokinin-mediated transcription of *C4Ppc1* according to the suggestion described previously (9).

3.2 Accumulation of Cytokinins in Roots in Response to External Nitrate

There is a certain body of evidence that the accumulation of cytokinins in roots, shoots, and xylem sap is up-regulated by N-availability (12-15). To test the possibility of N-responsive accumulation of cytokinin(s) in roots, we measured the levels of cytokinins in roots during recovery from N-starvation by supplying nitrate to N-starved plants. In support of postulation (i), several species of cytokinins increased in roots during N recovery (K.Takei et al. unpublished). Among the cytokinins, N^6-(3-Methyl-2-butenyl)-adenosine (i.e., iPA) has been identified by mass-spectrometry in the phosphatase-treated nucleotide fractions of cytokinins, increasing from 0.89 to 11.0 pmoles per g fresh weight in roots 2 hour after supply of nitrate to N-starved plants (10). The real molecular species may be iPMP, which is postulated to be the precursor to most other cytokinins in plants (16)

3.3 Identification of cDNA Encoding Cytokinin-Inducible Protein 1 (ZmCip1) by Differential Display

We attempted to identify the gene(s) for protein(s) that is possibly induced by cytokinin, using differential display technique. A cDNA clone, pZmCip1, encoding maize cytokinin-inducible protein 1 was isolated by visualizing and comparing the abundance of individual RNAs that were obtained from total RNA prepared from detached leaves treated with or without *t*-zeatin for 45 min and 90 min (10). The deduced open reading frame encodes 157 amino acids with a predicted mass of 16.7 kDa and exhibits significant homology to bacterial response regulators such as *Escherichia coli* CheY, which is one of the most extensively characterized response-regulators among the bacterial receivers and functions as a communication module of chemotaxis by mediating changes in locomotion in response to signals sensored (17). The best alignment between the two proteins shows strong similarity (30% identity, 47% similarity) as illustrated in Fig.1, but ZmCip1 has an extra 30 amino acids in the amino-terminal region. All of the three conserved residues in the response regulators that form an acidic pocket, consisting of the putative phosphate binding site at Asp90 and two other active sites, Asp44 and Lys142, are also conserved in ZmCip1.

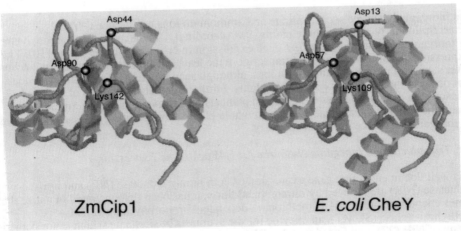

Figure 1. Comparison of ZmCip1 with *E. coli* CheY. Models are constructed according to Swiss Model (18). Asp44, Asp90, and Lys142 in ZmCip1 are amino acid residues forming a putative acidic pocket.

The simplest bacterial two component systems have two proteins, sensor and response regulator, that mediate input signals into output elements by relaying phosphate (17).

Proteins similar to bacterial sensors have been identified in eucaryotes including protozoa, fungi, and plants. These signaling proteins found in unicellular organisms handle a wide variety of signaling tasks, including metabolic adaptation to changes in nutrients such as C, N, and phosphate. In plants, receptor-type histidine kinases have been identified in *Arabidopsis thaliana* and tomato (19). ETR1 and its cognates have been demonstrated to be an ethylene receptor in ethylene signal transduction (20-22) and CKI1 has been shown to be involved in cytokinin signal transduction possibly as a cytokinin receptor (23). Quite recently, a series of homologs of the orthodox type of bacterial response regulators like ZmCip1 have been identified also in *Arabidopsis thaliana* (24-26).

3.4 *Expression of* ZmCip1 *in Detached Leaves*

We have studied the gene expression of *ZmCip1* in the detached leaves of N-starved plants (10). The transcript accumulates within 30 min after incubation with 5 μM*t*-zeatin. This accumulation is transient and precedes the transcription of *C4Ppc1*, which is also a cytokinin-inducible gene (9). It is noteworthy that the accumulation of *ZmCip1* transcript by nitrate and ammonium ions is very low. The accumulation of *ZmCip1* transcript by *t*-zeatin is dose-dependent, being induced at physiological concentrations ranging from 10^{-9} M to 10^{-7} M and the effect of *t*-zeatin is replaceable by isopentenyladenosine (iPA) and isopentenyladenosine-5'-monophosphate (iPMP) that accumulate in roots in response to nitrate. The induction pathway does not require protein synthesis and therefore may depend on signal transduction element(s) that is already present when necessary, since the cytokinin-responsive expression of *ZmCip1* is insensitive to cycloheximide.

3.5 *Expression of* ZmCip1 *in Whole-Plants*

An inorganic N-source such as nitrate and ammonium ions can induce *C4Ppc1* transcription in N-starved whole plants (9). Accordingly, we have studied responsiveness of inorganic N-sources of *ZmCip1* gene expression in N-starved whole plants (10). The expression is leaf specific and the transcript in the leaves accumulates transiently 2-6 hours after supply of nitrate or ammonium ions, although accumulation of the transcript does not occur in the leaves of N-sufficient plants. Furthermore, accumulation of *ZmCip1* transcript in detached leaves of N-starved plants is insensitive to nitrate or ammonium ions. The results indicate that *ZmCip1* in whole plants is transiently expressed in repsonse to inorganic N-source during N recovery.

3.6 *Transduction of Inorganic N-Source for* C4Ppc1 *Gene Expression*

It is well documented that gene expression of both nitrate reductase (NR) and nitrite reductase (NiR) are induced by nitrate. In addition, it has been demonstrated in maize that genes encoding S-adenosyl-L-methionine-dependent uroporphyrinogen III C-methyltransferase (SUMT), an enzyme for the synthesis of siroheme which is a prosthetic group of NiR (27), and ferredoxin-NADP$^+$ oxidoreductase (FNR) (28), an enzyme for electron supply to NR, are nitrate-responsive. Furtheremore, in the mesophyll cells of maize, plastidic glutamine synthetase is also nitrate-responsive, whereas ferredoxin-dependent glutamate synthase (Fd-GOGAT) is nitrate-insensitive at the level of protein (29). The up-regulation by nitrate of these genes implies that glutamine may preferentially

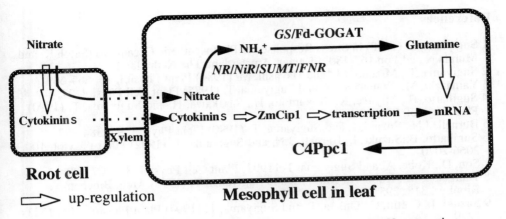

Figure 2. *A Scheme for cytokinin as a root-to-leaf signal in inorganic N-responsive expression for C4Ppc1 in maize.* Italics represent nitrate-inducible genes. See text for abbreviations.

accumulate in the mesophyll cells of maize in response to nitrate resulting in a posttranscriptional up-regulation of *C4PPc1* to lead the inorganic N-responsive expression of its protein.

The findings described here provide four lines of evidence to support a hypothetical scheme as depicted in Fig. 2: that *ZmCip1* may be involved in the inorganic N signal transduction for *C4Ppc1* transcription mediated by cytokinin in maize. The reasons for this are as follows: 1) *ZmCip1* transcription is induced by cytkinins in detached leaves, whereas its transcription is induced by inorganic N-sources in whole-plants. 2) An active cytokinin for the induction of *ZmCip1* in detached leaves, iPMP, and/or its dephosphorylated form iPA, accumulates in roots in response to nitrate. 3) In terms of speed of appearance and disappearance of gene expression without the requirement of protein synthesis, *ZmCip1* seems to be an early responsive gene to its signals, either cytokinins or inorganic N-sources. 4) *C4Ppc1* transcription requires cytokinins and inorganic N-sources as signals as does *ZmCip1* transcription, and induction of *ZmCip1* by the signals precedes that of *C4Ppc1*. The possibility of *C4Ppc1* as one of the ZmCip-targeting genes needs further study because such a correlation does not necessarily show cause-and-effect. Our next goals will be to determine the cellular and sub-cellular localization of ZmCip1 in leaves, and to elucidate other communication component(s) in the transduction pathway. In addition, it is interesting to understand validity of the inorganic N-signal transduction mediated by cytokinin in other plant species since ARRs 3 to 7, response regulator homologs identified in *Arabidopsis thaliana* (24), are also nitrate- and cytokinin-responsive as in the case of ZmCip1in maize (30).

Acknowledgements: This work was supported by grants from the Ministry of Education, Sports, Science, and Culture, Japan (Scientific Research on Priority Areas Grants 09274101 and 09274102) and by Japan Tobacco Inc., Plant Breeding and Genetics Labolatory to TS.

3772

References

1 Sugiyama, T. (1998) in Stress Responses of Photosynthetic Organisms (Satoh K. and Murata N., ed.) pp.167-180, Elsevier, Amsterdam, The Netherlands

2 Sugiyama, T., Mizuno, M., and Hayashi, M. (1984) Plant Physiol. 75, 665-669

3 Yamazaki, M., Watanabe, A. and Sugiyama, T. (1985) Plant Cell Physiol. 27, 443-452

4 Sugiharto, B., Miyake, K., Nakamoto, H., Sasakawa, H. and Sugiyama, T. (1990) Plant Physiol. 92, 963-969

5 Burnell, J.N., Suzuki, I. and Sugiyama, T. (1990) Plant Physiol. 94, 384-387

6 Sugiharto, B.,Suzuki, I., Burnell, J.N. and Sugiyama, T. (1992) Plant Phsyiol. 100, 2066-2070

7 Son, D., Kobe, A. and Sugiyama, T. (1992) Plant Cell Physiol. 33, 507-509

8 Taniguchi, M., Kobe, A., Kato, M. and Sugiyama, T. (1995) Arch. Biochem. Biophys. 318, 295-306

9 Suzuki, I., Crétin, C., Omata, T. and Sugiyama, T. (1994) Plant Physiol. 105, 1223-1229

10 Sakakibara, H., Suzuki, M., Takei, K., Deji, A., Taniguchi, M. and Sugiyama, T. (1998) Plant J. 14, 337-344

11 Sugiharto, B., Burnell, J.N. and Sugiyama, T. (1990) Plant Physiol. 100, 153-156

12 Salama, A.M.S.E.A. and Waering, P.F. (1979) J. Exp. Bot. 30, 971-981

13 Horgan, J.M. and Waering, P.F. (1980) J. Exp. Bot. 31, 525-532

14 Kuiper, D., Kuiper, P.J.C., Lambers, H., Schuit, J. and Staal, M. (1989) Physiol. Plant. 75, 511-517

15 Thorsteinsson, B and Eliasson, L. (1990) Plant Growth Regul. 9, 171-181

16 Binns, A.N. (1994) Annu. Rev. Plant Physiol. Plant Mol. Biol. 45, 173-196

17 Parkinson, J.S. and Kofoid, E.C. (1992) Annu, Rev. Genet. 26, 71-112

18 Peitsch, M. C. (1996) Biochem. Soc. Trans. 24, 274-279

19 Bleeker, A.B. and Schaller, G.E. (1996) Plant Physiol. 111, 653-660

20 Chang, C., Kwok, S.F., Bleeker, A.B. and Meyerowitz, E.M. (1993) Science 262, 539-544

21 Schaller, G.E. and Bleeker, A.B. (1995) Science 270, 1809-1811

22 Bleeker, A.B. and Schaller, G.E. (1996) Plant Physiol. 111, 653-660

23 Kakimoto, T. (1996) Science 274, 982-985

24 Imamura, A., Hanaki, N., Umeda, H., Nakamura, A., Suzuki, T., Ueguchi, C. and Mizuno, T. (1998) Proc. Natl. Acad. Sci. USA 95, 2691-2696

25 Urao, T., Yakubov, B., Yamaguchi-Shinozaki, K. and Shinozaki, K. (1998) FEBS Lett. 427, 175-178

26 Brandstatter, I. and Kieber, J.J. (1998) Plant Cell 10, 1009-1019

27 Sakakibara, H., Takei, K. and Sugiyama, T. (1996) Plant J. 10, 883-892

28 Ritchie, S.W., Redinbaugh, M.G., Shiraishi, N. Vrba, J.M. and Campbell, W.H. (1994) Plant Mol. Biol. 25, 679-690

29 Sakakibara, H., Kawabata, S., Hase, T. and Sugiyama, T. (1992) Plant Cell Physiol. 33, 1193-1198

30 Taniguchi, M., Kiba, T., Sakakibara, H., Ueguchi, C., Mizuno, T. and Sugiyama, T. (1998) FEBS Lett. 429, 259-262

IMPROVED GROWTH OF PEPPER UNDER REDUCED LIGHT INTENSITY

Mauro Centritto[1], Sebastiano Delfine[2], Maria C. Villani[1], Annalisa Occhionero[2], Francesco Loreto[1], Arturo Alvino[2] and Giorgio Di Marco[1].

[1] CNR - Istituto di Biochimica ed Ecofisiologia Vegetali, Via Salaria km. 29,300 - 00016 Monterotondo Scalo (Roma), Italy
[2] Dipartimento SAVA, Università del Molise, 86100 Campobasso, Italy

Key words: gas exchange, productivity, Rubisco, shading

1. Introduction

Mediterranean summers are typically hot and dry. Under these conditions, environmental stresses such as high temperatures and drought are frequent and constitute the main limitation to plant growth and production. It is well known that crop evapotranspiration can be decreased by screen shelters, leading to increased water use efficiency and lower water use (1, 2). However, reduced incident light intensity may lower air temperature but may also limit carbon fixation by photosynthesis and, consequently, depress plant growth. If the amount of incident light is reduced to an extent to which its effect on the air temperature and vapor pressure difference between leaf and air (VPD) above the canopy is greater than that on plant photosynthetic rate, a higher rate of growth could be induced. We tested whether manipulation of light regimes could improve growth of annual crop.

2. Procedure

Sweet pepper (*Capsicum annum*) plants, similar in height and number of leaves, were transplanted on June 3, 1997, and grown under two light regimes in the field near Rome, Italy. Six plots were made with rows spaced 0.8 m apart. Three plots (control) were maintained under natural sunlight conditions, and three plots (shaded) were permanently covered by a white fiberglass net that reduced incident radiation up to 30%, depending on the incident light intensity and the hour of the day. To avoid any water and nutrient limitation, the saplings were watered regularly to soil water capacity and supplied with mineral nutrients at free access rates, following Ingestad principles (3).

Seven harvests were made to determine growth on 10 plants per plot-treatment during the crop cycle, each plant was oven dried at 70 °C to obtain the dry mass of the plots. Leaf area, measured using a leaf area meter (LI 3100, LI-COR Inc., Lincoln, NE, USA), was used to calculate the leaf area index (LAI) of the plots.

G. Garab (ed.), Photosynthesis: Mechanisms and Effects, Vol. V, 3773–3776.
© 1998 *Kluwer Academic Publishers. Printed in the Netherlands.*

Diurnal trends of photosynthesis (*A*) and stomatal conductance (*g_s*) were measured in the field with the gas-exchange portable system LiCor 6400 (LI-COR Inc., Lincoln, NE, USA) on fully expanded leaves. Leaf samples (3.8 cm^2) were frozen in liquid nitrogen immediately after gas-exchange measurements. Frozen leaves were ground in a chilled mortar with 30 mg polyvinylpolypyrrolidone (PVPP), quartz sand, and 2 cm^3 of extraction buffer (100 mol m^{-3} bicine pH 8, 10 mol m^{-3} MgCl$_2$, 5 mol m^{-3} dithiothreitol (DTT), 1 mol m^{-3} ethylendiaminetetraceticacid (EDTA), and 0.02% (weight/volume) bovine serum albumin, BSA). The solution was centrifuged at 10,000 g for 10 s. A fraction of the supernatant was used to determine radiometrically the total carboxylase activity of Rubisco (4). Another fraction of the supernatant (400 μl) was added to 100 μl of the denaturing solution (20% sodium dodecyl sulfate (SDS), 20% β-mercaptoethanol and 200 mol m^{-3} Tris HCl pH 6.8) at 95 °C for 5 min. Rubisco content was determined on the denatured solution by SDS-PAGE, using 14% acrylamide gel. Gels were stained with Coomassie brilliant blue R-250, de-stained and scanned at 550 nm using a Dual - Wavelength Flying Spot Scanner (CS-9000, Shimadzu, Tokyo, Japan).

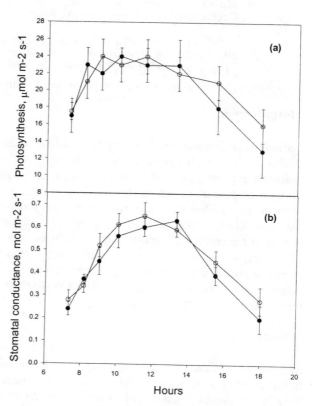

Figure 1. Diurnal trend of (a) photosynthesis, and (b) stomatal conductance of control (light circle) and shaded (dark circle) pepper crops 78 dat. Data are means of 10 plants per treatment ± SEM.

3. Result and Discussion

The overall effect of shading on gas exchange parameters, measured at midday at the field conditions during the whole crop cycle, was not significant (data not shown). Also

Table 1. Time course of Rubisco activity *in vitro* and Rubisco content (on a leaf area basis) of control and shaded pepper crop. Data are means of 8 leaves per treatment 1 ± SEM measured at different days after transplanting (DAT).

DAT	Rubisco activity (μmol m^{-2} s^{-1})		Rubisco content (g m^{-2})	
	Control	Shaded	Control	Shaded
47	116 ± 6	115 ± 7	4.02 ± 0.18	4.18 ± 0.21
83	112 ± 7	111 ± 6	3.98 ± 0.19	4.01 ± 0.20
91	101 ± 6	98 ± 6	4.11 ± 0.21	4.21 ± 0.19
99	103 ± 7	98 ± 7	4.20 ± 0.20	3.95 ± 0.21

the diurnal trend of A measured in the growth conditions was not affected by shading (Figure 1a).

There were no significant differences also in Rubisco activity and amount (on a leaf area basis) between control and shaded plants (Table 1). Thus, shading did not cause any down-regulation of both photosynthetic rate and capacity (maximum Rubisco activity).

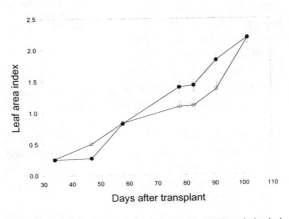

Figure 2. Leaf area index of control (light circle) and shaded (dark circle) pepper crops. Data are means of 10 plants per treatment ± SEM.

However, the diurnal trend of stomatal conductance showed that g_s was slightly smaller, although not significantly, in shaded plants (Figure 1b), leading to improved water use efficiency (data not shown).

Figure 2 shows LAI and fruit dry mass production over the crop cycle. Both at the beginning and at the end of the cycle, shading did not improve LAI. We attribute this effect to the low temperature and to the light conditions of late spring and end of summer (for example daily maximum temperatures were ~25°C).

However, as temperature, light intensity and light duration increased during summer (i.e. when daily maximum temperatures were > 25°C), the leaf area index of shaded plants became significantly larger than that of controls (Figure 2). As temperatures increased, the beneficial effect of shading also led to a significant increase in pepper dry mass production (Figure 3) and quality (data not shown). However, at the end of summer, because of

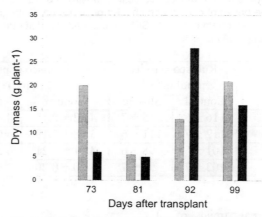

Figure 3. Pepper dry mass per plant of control (light bars) and shaded (dark bars) crops. Data are means of 10 plants per treatment ± SEM.

the decreasing temperatures and light duration, there were no longer significant differences in either LAI or pepper dry mass production. However, the fruit quality was always improved by shading (not shown).

This study confirms results obtained with clonal cherry saplings grown under reduced light intensity (M.Centritto *et al.*, unpublished), and shows that a moderate reduction of light intensity, which does not affect photosynthesis, can be a useful practice for improving growth and dry mass production when plants are exposed to high light intensity and temperatures.

References

1 Allen, L.H. Jr. (1975) Agronomy Journal 67, 175-181
2 Loomis, R.S. (1983) in Limitations to Efficient Water Use in Crop Production (Taylor, H.M., Jordan, W.R. and Sinclair, T., eds.) pp. 345-374, ASA-CSSA-SSSA, Madison, USA.
3 Ingestad, T. and Ågren, G.I. (1992) Physiologia Plantarum 84, 177-184
4 Di Marco, G. and Tricoli, D. (1983) Photosynthesis Research 4, 145-149.

MICROELEMENTS AS PHOTOSYNTHESIS REGULATORS IN PEACH TREES

Nina Titova, George Shishkanu
Plant Physiology Institute, Academy of Sciences of Moldova,
Padurii str., 22, Kishinev 2002, Republic of Moldova

Key words: net photosynthesis, environmental stress, grown, pigments, peroxidase activity, source-sink relations.

1. Introduction

In research of photosynthesis and productivity process a special attention is paid to exogenous tegulation of plant photosynthesis activity. Regulation of photosynthesis by means of microelements is one of the most important way to achieve high productivity and yealds of plants. Very much information about microelement influence on plant growth, metabolism and photosynthesis are collected in literature, but these issnes are studied in fruit trees to a lesses degree (1,2,3). Regulation of peculiarities of peach trees photosynthesis and overground part and root system activities, specific relation of the scoin-stock combinations with different compatibility remains less studied. These resarches are very important for the Republic of Moldova, where soils, as a rule, have a low content microelements /4/.

2. Procedure

2.1 *Plant material*
During 1989-1995 rootstock seedlingls with various requirements for water supply and temperature of peach (Persica vulgaris Mill. var. Hibrid №2, relatively drought-resistant , less frostproof, exigent to light and water) and apricot (Armeniaca vulgaris Lam. var. №10-11-12, with frostproof root system, slightly compatible with peach) and annual peach trees (Persica vulgaris Mill. s. Zolotoi Jubilei) grafted on them have been studied.

2.2
Plants were grown in the fruit tree nursery of the Plant Physiology Institute using conventional growing techniques.

2.3
The foliar treatment with the microelements of Zn and Mn (as 0.05% solution $ZnSO_4$ and $MnSO_4$) was carried out during the intensive growth period at the end of June.

2.4
Specific leaf mass (SLM), leaf area index (L1), photosynthetic potential (PhP), net photosynthetic productivity (Phn) are estimated according to A.A. Nichiporovich (5).

3777

G. Garab (ed.), Photosynthesis: Mechanisms and Effects, Vol. V, 3777–3780.

2.5
Chlorophyll and carotenoid content was measured in 80% acetone according to A.A.Shlyk (6).

2.6
Peroxidase activity in the leaf, shoot and root homogenates is determined by the reaction rate with 3% H_2O_2 and 0.1% benzydin on SF-46 LOMO.

3. Results and Discussion

As it was described earlier (7), the zinc and manganese positive action on the CO_2 assimilation is displayed 2 days and afterwards 3,7,10 days after treatment. The photosynthetic capacity of treated peach plants are predetermined by rootstock peculiarities and then compatibility with the scion, Significantly morphological differences were discovered in the control and treated plants (Tab. 1)

Table 1. Some morphological and photosynthetical characteristics of peach rootstock seedlings treaded with microelements. Juli 1994, 21 days after treatment. The means of 10 plants, significantly different from 0 at $P \leq 0.05$ (t test).

	Biomass, g dry matter					Leaf area	SLM	LAI	PhP	Ph$_n$
	Leaf	Shoot	Trunk	Root	Plant	dm^2	mg/dm^2	m^2/m^2	m^2/ha day	g/ m^2 day
Control	6.22	1.88	3.78	1.30	13.50	8.92	6.94	0.45	459.00	2.84
Zn	7.13	2.12	4.78	1.65	15.68	9.03	7.89	0.45	460.02	3.41
Mn	10.31	4.80	3.25	3.26	26.52	16.56	6.28	9.83	844.56	3.16

Under conditions of drought the vegetative period of 1994 all the biometric parameters and, as a result, net photosynthetic productivity Zn and Mn treated peach plants are significantly increased. It is known, that the root/leaf mass ratio and root/overground organs mass ratio show an ecological stability and source-sink relations in the plants (8). The grown basic attractive center root in the treated plants is more active than that of the basic source center-leaf.The root/leaf mass ratio exceeds the control by 1.5-2.0 times, which is evidenced by the source-sink relations change in plant. The Zn and Mn effect on chlorophyll and carotenoid content in the leaves of the rootstock and peach trees grafted on them is also expressed more significantly after 10-15 days (Tab.2).

Table 2. Pigment content in intact leaves of the peach and apricot rootstocks and annual peach/peach plants 12 days after the treatment. All the values are expressed in mg/dm^2 n=4.

		Chlorophyll a	Chlorophyll b	Carotenoids
Peach rootstock	Control	1.19 ± 0.03	0.37 ± 0.05	0.72 ± 0.01
	Zn	2.18 ± 0.04	0.50 ± 0.02	0.89 ± 0.01
	Mn	2.54 ± 0.04	0.62 ± 0.01	1.01 ± 0.03
Apricot rootstock	Control	1.56 ± 0.08	0.40 ± 0.02	0.66 ± 0.03
	Zn	2.03 ± 0.14	0.48 ± 0.03	0.78 ± 0.04
	Mn	2.15 ± 0.09	0.60 ± 0.04	0.83 ±0.02
Peach/peach plants	Control	2.45 ± 0.08	0.66 ± 0.02	1.09 ± 0.01
	Zn	2.84 ± 0.01	0.79 ± 0.01	1.24 ± 0.02
	Mn	3.36 ± 0.08	0.94 ± 0.03	1.54 ± 0.05

The chlorophyll a, chlorophyll b and carotenoid contents in the treated plant are by 20% higher than in the control. At the same time the chlorophyll a/ chlorophyll b ratio is practicaly unchanged, which indicates a good status of pigment fund and a similar photosynthetic capacity in the plants studied. The results show a marked stimulation of the key metabolism enzyme peroxidase activity by the microelements in all the organs of the plant studied /Tab. 3/

Table 3. Peroxidase activity in the organs of the peach and apricot rootstock seedlings and annual peach/peach plants 12 days after treatment. The values are presented as conditional units, n=4.

		Leaf	Shoot	Root
Peach rootstock	Control	3.92 ± 0.62	0.88 ± 0.06	1.65 ± 0.05
	Zn	11.01 ± 0.62	120.70 ± 9.75	2.63 ± 0.15
	Mn	5.33 ± 0.67	54.61 ± 2.51	4.62 ± 0.25
Apricot rootstock	Control	0.56 ± 0.05	2.02 ± 0.18	8.69 ± 0.66
	Zn	0.65 ± 0.10	34.18 ± 0.38	16.26 ± 1.12
	Mn	1.03 ± 0.08	63.46 ± 2.12	13.81 ± 1.78
Peach/peach plants	Control	11.81 ± 1.05	135.20 ± 11.10	1.53 ± 0.01
	Zn	17.91 ± 1.11	249.99 ± 14.50	2.24 ± 0.21
	Mn	23.27 ± 3.01	172.61 ± 12.05	2.12 ± 0.04

To a most degree this is observed in intensively growing shoots, which proves the metabolic process activation in the treated plants. It has been established that the variability of the physiological reaction of different peach grafts to the envoronmental factors, as we previously reported (9), remains specific upon the application of the microelements. The water supply and temperature conditions in the first half of the vegetative period are the determining factors upon the microelements action after 10-20 days. This action especially affects the parameters of qwick response: water content in organs, peroxidase activity and pigment fund. In the second half of the vegetative period the constituents of the productivity process (SLM, LAI, PhP, sourse-sink relations, Ph$_m$) are less dependent on the environmental factors and more on the sensitivity and individual respond of the plant under the exogenous microelement action, on the requirment for each element. The Zn and Mn influence jn the mineral element content is expressed most considerably in quantitative increase of iron and phosphorus in assimilative polar ograns-leaf and root. These elements play a special role in plant energetics, chloroplast electron transport, photochemical reaction intensity (10).

4. Conclusions

These studies have demonstrated the high sensibility of the rootstock seedlings and peach plants grafted on them to the foliar microelement treatment. This response Involves the specific regulations of the source-sink relations in plant and the change of the constituents of the productivity process parameters. The main factors of this adjustment regulation include the increase of the pigment content and functional activity of peroxidase, the change of the mineral element content, especially Fe and P, the growth of the attractive role of plant roots under conditions of the negative action soil and air drought, which results in the intensification of photosynthesis in the treated plants.

References

1. Bergman V. (1979)   Erkennen and Vermelden von Micronährsttoffmangel im Obstbau. Kurzdokumentation, 178 p. Expert: Jena.

2. Physiologie der Obstgeholze (1978), 416 p., Akademie-Verlag, Berlin.

3. Shuruba G.A. (1985) Necornevoe pitanie plodovyh i jagodnyh cultur microelementami, 176 p. Vyshcha shcola, Lvov, Ukraine.

4. Toma S.I. (1977) In. Microelementy v selscom hoziastve Moldavii, p.3-11, Shtiintsa Press, Kishinev, Moldova.

5. Nichiporovich A.A. (1956) Fotosintez i teoria poluchenia vysokih urojaev. 56 p., Acad. Nauc SSSR Press, Moscow.

6. Chlorophyll (1974) A.A. Shlyk red., 400 p., Nauka i tehnika Press, Minsk.

7. Shishkanu G.V., Titova N.V. (1992) Izvestia Acad. Nauc Moldovy, s. biol. chim. nauk, 6, 21-25.

8. Latiffe H., Travis R. (1984) Crop. Sci., 24, 447-452.

9. Titova N., Shishkanu G. (1991) In Proc. International Symposium Biochem. Mecanisms involved in Growth Regulation. Milan, 11-13 Sept, p. 113.

10. Brecht A.E. (1990) Biol. Rundshaau, 28, 73 –82.

COMPARATIVE EFFECTS OF IRON, MANGANESE AND BORON DEFICIENCIES IN VINE-LEAF (VITIS VINIFERA) BASED ON POLYPHASIC CHLOROPHYLL A FLUORESCENCE TRANSIENT.

[1.]Eyletters M., [2.]Bitaud C. and [3.]Lannoye R.
[1] Université Libre de Bruxelles, Laboratoire de Physiologie et d'Agrotechnologies
Végétales, Av F. Roosevelt 50 CP169, 1050 Bruxelles Belgium
[2] Pole Agroenvironnemental Aspach-Le-Bas, 68700 Cernay, France

Key words: chl fluorescence induction, photochemistry, minor element, deficiency, early diagnosis, grapevine

1. Introduction

Micronutrients such as B, Fe and Mn are essential elements directly involved in plants nutrition and in physiological and biochemical mechanisms.

Many photosynthetic performance parameters are influenced by Fe-deficiency: ultrastructure of chloroplasts (1) and protein and lipid composition of thylacoid membrane (2).This deficiency is accompanied by the decrease in photosynthetic pigments but not uniformly: xanthophylls being less affected than chlorophyll and β caroten (3). Iron deficiency reduces the light harvesting complexe (LHC) and core complex (CC) of PSII and PSI (4).

Manganese is also a micronutrient transition metals and activates several enzymes in vitro, particularly decarboxylases and dehydrogenases (5). The most well known function of Mn in green plants is its involvement in photosynthetic O_2 evolution (6): the Hill reaction- the water-splitting and O_2 evolving system in photosynthesis. Photosystem II (PSII) contains a manganoprotein which catalyses the early stages of O_2 evolution. The Mn deficiency impairs the first step of the electron transport chain of the light reaction and has negative effects on subsequent reactions such as photophosphorylation and the reduction of CO_2.

B deficiency causes many anatomical, physiological and biochemical changes. B is involved in the membrane structure which interferes with numerous processes including enzyme function and transport ions, metabolites and hormones (7).

It is well known that *Vitis vinifera* is very sensitive to the B, Fe and Mn deficiency (8). The factors influencing these micronutrients concentration in grape bush above - ground tissues are soil type (chemical composition and pH value), water regime, temperature, variety and vegetation stage. Deficiency frequently results in chlorosis which greatly damages viticulture (9).

G. Garab (ed.), Photosynthesis: Mechanisms and Effects, Vol. V, 3781–3784.
© 1998 Kluwer Academic Publishers. Printed in the Netherlands.

The purpose of the present research was to use photochemical parameters derived of chl a fluorescence kinetic to investigate the possibility to diagnose early these element deficiencies.

2. Material and methods

2.1. *Plant material*
Greenhouse experiments were carried out on vine-shoot (*Vitis vinifera* Cep. Sylvaner) grown hydroponically in nutritive solution (10) with or without B or Fe or Mn. The photoperiod was 16h. After 3-4 weeks, plants had developed about 5 leaves. These young, rapidly expanding leaves were used for all measurements.

2.2. *Chlorophyll a fluorescence measurement- fast fluorescence kinetics*
Chl a fluorescence induction curve was monitored by a Plant Efficiency Analyser (PEA, manufactured by Hansatech Ltd King's Lynn, Norfolk, England) with 600 W.m-2 of ligh intensity as described in detail by (11).

2.3. *Chlorophyll a fluorescence measurement- slow fluorescence kinetics in steady state*
The slow phase of chl a fluorescence emission was measured by using modulated chl a fluorescence system (PAM 101-103 Waltz, Effeltrich, Germany)(12).

2.4. *Treatments*
These chl a fluorescence parameters were measured 30 min after dark or after irradiation of the sample with actinic light (for PAM measurement) or with sunlight (for PEA measurement). During 1 sec pulse of saturating light (300 $W.m^{-2}$), the maximum fluorescence emission (Fm) was measured.

3. Results and discussion

Chlorophyll a fluorescence transients provides informations on the filling up of the plastoquinone pool and the associated heterogeneity, but they are affected by both the electron donor and acceptor sides of PSII of oxygen evolving organisms (13).

For each deficiency , the percentage of control of different chl a fluorescence parameters: the photochemical k_p and non photochemical k_N deexcitation rate constants and the maximum quantum efficiency (Opo = 1-(Fo/Fm)) were deduced from the equations in figure 1 based on the measurment of Fo and Fm (minimal and maximal fluorescence); These early observations were made for the same pigment content and no visual symptom on leaf is detected at this stress intensity.

The main characteristic of the Fe deficiency was the rise of Fo compared to its value in control plants, whereas Fm exhibited no significant variation. The relative increase in Fo may suggest an increased proportion of PSII are of the non-Qb reducing type (or of the Qa reducing type Qa-) (14) inducing a slowing down of electron transport due to the disturbed activity of electron carrier having lost their Fe atom. This perturbation is reflected in the decrease in k_p but also in the increase in the rate constant of non photochemical deexcitation events (expressed by k_N).

Mn deficiency induced changes in the maximum quantum yield of primary photochemistry (Opo) due to the decrease in k_p (decrease in 1/Fo-1/Fm) and to the increase in k_N (increase of 1/Fm). Because of the well-known involving of Mn in the early processes of photosynthesis.

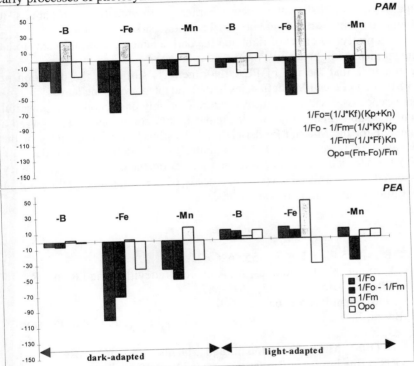

$$1/Fo=(1/J*Kf)(Kp+Kn)$$
$$1/Fo - 1/Fm=(1/J*Kf)Kp$$
$$1/Fm=(1/J*Ff)Kn$$
$$Opo=(Fm-Fo)/Fm$$

Figure1: Percentage of control for different chl a fluorescence parameters in grape leaves submitted to boron (-B), iron (-Fe) and manganese (-Mn) deficiency. Leaf total chlorophyll (in µg.mm^{-2})was respectively 1.38, 1.50, 1.40 and 1.50 for control, B, Mn and Fe deficient plant.

B deficiency showed global perturbations of chl a fluorescence parameters especially in dark-adapted sample studied with PAM fluorimeter.

Except for B, Fe and Mn deficiencies are well detected by both apparatus

Globally, PSII primary photochemistry in dark-adapted sample was almost perturbed in B-deficient plants. However in light adapted sample, the improvement of the photosynthetic apparatus is due to light adaptation which protects and reduces the impair caused by deficiency.

Based on the analysis of the maximum quantum yield in dark-adapted leaves, we can concluded that the primary photochemistry of grapevine is damaged in the presence of B, Mn and Fe deficiencies.In state «light adapted», we observed a protective effect of the light on PSII.However, this state change also showed the damages of the photosynthetic apparatus due to the deficiencies but with smaller variations.The high increase in the kN demonstrates a high energy state (15).

For foliar diagnostic establishment to detect earlier photosynthetic perturbations, it seems that bigger percents variation are obtained after the dark adaptation.

We can observe a protective effect of light and the dark adapted condition seems to be more convenient than the light adapted one.

Overall, we conclude that the use of PEA fluorimeter is sufficient the detection of damages in PSII due to Fe and Mn defiencies in dark-adapted grapevine leaves.

The fast fluorescence rise, within the measuring time of only one second is a powerful tool to screen in a short time many samples under in situ conditions. We observed a specific behaviour for the Mn and Fe deficiency consisting in typical changes of the fluorescence signal: increase in Fo (+102% of control) for Fe deficiency and decrease in Fm (-15%) for Mn deficiency. For micronutrient deficiency, the Plant Efficiency Analyser appeared to be well adapted of our needs of screening and comparing of many samples directly in the field in a short time.

References

1   Stocking, C.R., (1975 Plant Physiol. 55,. 626-631.
2   Abadia, A., Lemoine, Y., Tremolieres, A., Ambard-Bretteville, F.and Remy, R., (1989) Plant Physiol. Biochem. 27(5),. 679-687.
3   Terry, N., (1980). Plant Physiol. 65, 114-120.
4   Fodor, F., Boddi, B., Sarvari, E., Zaray, G., Cseh, E.and Lang, F. (1995) Physiol. Plantarum 93,.750-756.
5   Marschner, H. (1986).Mineral nutrition of higher plants, Academic Press.
6   Amesz, J. (1983) Biochimica et Biophysica Acta, 726,.1-12.
7   Kastori, R., Plesnicar, M., Pankovic, D.and Sakac, Z. (1995) J.Plant Nutr. 18(9), 1751-1763
8   Loué, A.(1993) Oligoéléments en agriculture (SCPA-Nathan, ed.).
9   Keller, M., Hess, B, Schwager, H, Scharer, H.and Koblet, W(1995)Vitis 34(1), 19-26.
10 Terry, N. and Abadia, J., 1986. Journal of Plant nutrition 9,.609-646.
11 Strasser, R.J., Srivastava A. and Govindjee (1995. Photochem.Photobiol.61,.32-42.
12 Schreiber, U., Schliwa, U. and Bilger, W. (1986) Photosynth. Res. 10, 51-62.
13 Krause, G.H. and Weis, E. (1991). Ann. Rev. Plant. Physiol. 42, 313-349.
14 Morales, F., Abadia, A. and Abadia, J.(1991). Plant Physiol. 97,.886-893.
15 Ouzounidou,G., Moustaka,M. and Strasser,R.(1997)Aust.J.Plant Physiol. 24, 81-90

THE PHOTOSYNTHETIC ACTIVITY OF THE VINE BUSH WITH THE NEW ARCHITECTONICS

R. Malina, G. Scurtu, H. Roshca
Plant Physiology Institute, Academy of Sciences of Moldova,
Padurii str., 22, Kishinev 2002, Republic of Moldova

Key words: biomass, pigments, net photosynthesis, bioproductivity, grape vine, source-sink.

1. Introduction

Viticulture is one of the main branches of the Republic of Moldova agriculture, and the urgent problem is to develop high producing and profitable plantations which would maximally utilise PhAR at the expense of the plantation structure, efficient regimes of nutrition and water consumption. Solid plantinqs annual crops contribute to the production of maximum leaf surface and appropriate biological yields under optimal conditions. This task being complicated for perennial plantings and grape plants. The problem may be solved in two ways-to increase the planting density or to increase the loading of the number of shoots per plant. These possiblities are restricted as the interrow width should be sufficient for the cultivating machines and the overloading of a plant with shoots leads to high density within the crown, decrease of fruit bearing capacity and a shift towards, the development of vegetative mass. Therefore the PhAR coefficient is extremly low in grape plantings (1). It has been established that the shoot overloading contributes to weak vine maturing or bud death during cold winters on the high trunk grape culture (2,3). The frost-resistance decrease is due to the reduction of the total carbohydrate content in shoots. The long winter growth results in the exhaustion of shaded shoots, they have no time to prepare for winter (4). For grape like liane, spatial orientation of the grown and the shape of its man-made attachment type are of a great importance. A multitude of the buch architectonics types currently used have a number of short comings, the main of which are different qualities of shoots, uneven illumination of leaves, labour-intensive operations to maintain the biomass development equilibrium towards the berry growth and a high quality of yields. All these hinder the creation a dequate models to Programme the vine-yard producing capacity.

In 1991 the reseacher G.Scurtu developed and patented a new shape of the grape bush-Elliptic Horizontal Cordon (EHC)(5). This shape provides for a good acces of lights and air to all the shoots, their development is uniform, this creating a large surface of intesively working leaves. The major part of the shoots are fruit bearing contributes to yield increments by 2-2.5 times. Since the new shape increases the loading per bush significantly, a problem has been set to comparatively study the photosynthetic activity and producing of the grape bush and the new architectonics.

G. Garab (ed.), Photosynthesis: Mechanisms and Effects, Vol. V, 3785–3788.

2. Material and methods

Four-year-old grape bushes (cv Chardonnay) were grown in ferro-concrete lyzimeters (1 x 1 x 2.5) m and shaped according to the method of the Elliptic Horizontal Cordon. The skelet shoot was fixed on a support at height of 0.8 in the shape of spirally fixed rings of a diameter up to 1m. The number of the spiral turn was from one to five. The The shoots inside the spiral were removed and the construction center was accesible to the air circulation and light energy flow. The shoots of current growth hung down outside the cordon, and the bunches of grapes were situated near the axial ring. The other fundamental difference in the experiment was the utilization of a fruit bearing Almond trees (Amigdalus communis) shaped as a bowl rather than a conventional support. Such a natural symbiosis creates conditions for the development of both plants. The grape shoots supported this way do not shade each other as the Almond tree branches are oriented upwards, and the grape shoots hang lowsely down. This tiered position of shoots rules out their competition in soil. The man receives double yield of nuts and grape berries and a significant reduction of expenses linked with the fixation of supports. The control plants had a conventional high trunk shape on a support. During the vegetation period from May to September the following photosynthetic parametres were estimated: the content of chlorophyll and carotenoids, the balance of dry biomass and organs the number and area leaves, specific leaf area, the number, length and diameter of shoots PhP, the content of sugars and titratable acids were measured in berry sap after havesting.

3. Results and duscussions

The reduction in crop level generally accelerates maturation and improves grape and vin quality but reduces final production. Therefore it is necessary to identify anual define the cropping of grapevin which determines the equilibrium between the vegetative and productive apparatus. In fact the source-sink status of a plant is dynamic, so organs may change from being a consumers to a applier or vice-versa. The specific light regime inside the crown shaped according to EHC has created preconditions for equilibrated sourse-sink relations during the accumulation of dry biomass and development of an original leaf cannopy. The author of the new architectonics G. Scurtu has succeeded in the overcoming of the problem associated with the different quality of shoots which is of a paramount importance at the modelling of progarammed yields. The shoots in the EHC are uniform for length, diameter, leaf area sizes and weight of bunches of grapes. At the flowering stage the number of shoots per bush with the new architectonics was by 2-3 times higher and consequently the number of leaves increased(Fig. 1). In the group of control plants the length, the number of leaves varied considerably (from 0.3m to 3m, from 4 to 30 respectively). The main difference in the shapes of the grape bush consists in the direction of the shoot growth. Thus with the conventional architectonics the shoots growth upwards or horizontally along the support forming a spatially oriented wall, and then possibilities to develop without reciprocal shading are restricted. The leaf canopy of the new architectonics resembles a column or a truncated cone with an open center through which light penetrates uniformly to all the leaves (Tab. 1). At the flowering stage when the average shoot length did not exceed 1m the number shoots In EHC was almost by three times higher with a similar number of leaves. In total, the number of leaves and flowering bunches was by three times higher on the grape with the new architectonics.

Figure 1
Histogram of shoot elongation rate with the two types of the vine bush

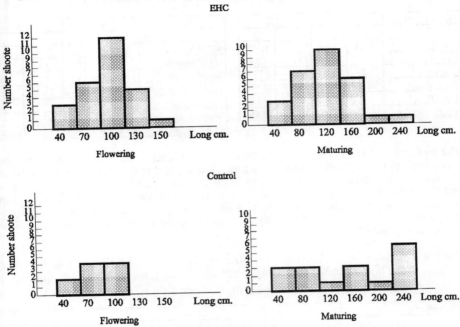

The increase loading with shoots did not deteriorated the light regime of the plant. At the stage of berry maturing, new shoots appeared on the control plants resulting in a 30% decreax of the fruit-bearing coefficient, and the bushes with the new architectonics maintained their equilibrium. The uncontrollable shoot growth in the control plant led to a three-fold increase in leaf number, to a considerable density of the crown and a shift of the yields/biomass balance towards the vegetative part. Under the similar conditions of nutrition and water supply the EHC bushes increased their leaf number only by 60%. Itence, in the plant with the new architectonics, the photosyntheticaly active leaf apparatus had developed earlier and its viability was longer. A comparative analysis of the parameters of one leaf showed that the leaves in the new architectonics were inferior for weight, area and leaf specific area by 10%-20%, but their total weigth and area per bush higher. It means that the architectonics pattern affects the leaf structure, the differences retaining during the whole vegetation period. In the studies on the pigment content in leaves, we set a problem to identify, which leaf in the new architectonics accumulated more of them. It turned out, that in the EHC shoots the 7-8-th leaf was the most active one in comparison with the conventional shape where the best was the 5-6-th leaf. The total content of chlorophyll and carotenoids in the EHC leaves was larger, especialy chlorophyll "a", therefore the a/b ratio was higher. The EHC leaves were more resistant to the import of environmental factors, especialy to diseases.

Table 1 Biomass measurements for the bush of the control and Elliptical Horizontal Cordon plants during the flowering period (1) and berry maturing (2)

Dry weigth (g.)	C		EHC	
	1	2	1	2
One leaf	0.84	0.65	0.77	0.63
Total leaves	83.27	296.98	217.91	312.80
One shoot	7.83	24.41	7.85	17.22
Total shoots	78.27	341.77	242.73	482.07
Total bunches	-	172.74	-	483.67
Veget. biomass	161.53	635.05	460.64	794.92

	C		EHC	
	1	2	1	2
No of leaves	100	316	285	484
No of bunches	10	10	28	28
No of shoots	10	14	27	28
Leaves/shoots	1.06	0.85	0.89	0.65
Area of leaf	2.36	1.66	1.74	1.2
Specific area leaf	0.39	-	0.36	-
Prod/veget/ratio	-	0.27	-	0.68

The leaf index at the flowering stage was twice as higher in the EHC leaves. The photosynthetic potential then was by 2-2.5 times higher, and at the stage of berry maturing the difference between the treatments made 25% Pn index was more conservative and the difference between the architectonics types did not exceed 10%-20%. During the vegetation period the bush biomass with the new architectonics was by 40%, and the yields by three times higher. Thus, the yield biomass balance was twice as better in the EHC plants (Tab.2)

Table 2. Leaf area index (LAI) m^2 m^{-2}, Potential Photosynthetic [Pp] mln m^2 ha^{-1} day^{-1} Net photosynthetic rate (Pn) g m^{-2} day^{-1} ,for bush grape vines differing in architectonics

Architectonic type	LAI		Pp		Pn	
	Flowering	Berry maturing	Flowering	Berry maturing	Flowering	Berry maturing
Control	2.04	5.43	0.510	1.457	3.17	4.41
EHC	4.93	6.55	1.233	2.139	3.74	3.65

At the same time, the berry quality, the sugar content were higher.

4. Conclusion

Thus, the Elliptic Horizontal Cordon architectonics studied is most promising in obtaining fundamentally new plantings will the increased utilisation of PhAR, iproved light regime inside the crown. This type of the grape vine architectonics will allow the 2-3-fold increase in the economic producing capacity of plants grown on the same area and the improvement of berry quality.

5. References

1. Bondarenco S.G. (1990), Sistemny analis v vinogradarstve, 231p.,Kishinev, (Shtiintsa)
2. Mozer L. (1971), Vinogradarstvo po novomu, Moscow, 276
3. Litvinov P.J. (1978), Vinogradarstvo, Kiev, 197-204
4. Sherbatuc N.A. (1974), Vliianie nagruzki na uglevodny obmen v vinogradnom rastenii. Sadovodstvo , vinogradarstvo i vinodelie Moldavii, 7, 26-28
5. Scurtu G. (1995), Brevet de inventie al Republicii Moldova, №25, Kishinau.

CHANGES IN PHOTOSYNTHETIC PARAMETERS OF EVOLUTIONARY DIFFERENT SPECIES OF *TRITICUM* AND *AEGILOPS* GROWN AT HIGH AND LOW NITROGEN AVAILABILITY

Gloser, J., Barták, M., Gloser, V.
Dept. of Plant Physiology, Faculty of Science, Masaryk University
Kotlářská 2, 611 37 Brno, Czech Republic

Key words: gas exchange, genotypes, Chl fluorescence induction, leaf physiology, nutrients, wheat

1. Introduction

Modern highly productive cultivars of hexaploid crop wheat (*Triticum aestivum*) developed from wild diploid and tetraploid wheat species and from two species of the genus *Aegilops*. Several attempts have been undertaken recently to explain the physiological background of differences in growth rate and yield among species representing certain stages of the evolution of crop wheat. It was soon recognized, that no simple relation could be defined between productivity and specific rate of net photosynthetic CO_2 uptake (P) by leaves. Evans and Dunstone (1) reported surprising finding that maximum P in flag leaves of modern wheat cultivars is lower than in the wild progenitors. As found in some other studies, the decrease is usually connected with lower nitrogen content in leaves (2), lower rate of Hill reaction (3), and lower content of chlorophyll per leaf mass unit (4). However, the amount of Rubisco was not different among genotypes with different ploidy and its specific activity was found to be higher in hexaploid species (5). It was shown in several other experiments that the potential rate of electron transport reached highest value in tetraploid species (4), and that P may be in some cases considerably higher in hexaploid cultivars than in primitive diploid species. The contrasting results found in the literature could be caused, to some extent, by different cultivation conditions, including variation in nutrient supply. To test this hypothesis we have done our experiments at two levels of mineral nitrogen in substrate.

2. Procedure

Comparative measurements of photosynthetic and growth parameters were done in seven species of the genera *Triticum* and *Aegilops,* each of them representing different stages of the evolution of crop wheat: *T. aestivum* cv. Sparta (modern hexaploid cultivar, genome ABD), *T. spelta* (primitive hexaploid culture wheat, genome ABD), *T. durum* (culture tetraploid wheat, genome AB), *T. dicoccoides* (wild tetraploid wheat, genome AB), *T. boeoticum* (wild diploid wheat, genome A), *Aegilops speltoides*

G. Garab (ed.), Photosynthesis: Mechanisms and Effects, Vol. V, 3789–3792.
© 1998 Kluwer Academic Publishers. Printed in the Netherlands.

(diploid, genome B), *A. tauschii* (diploid, genome D). The seeds were obtained from the National Gene Bank in the Research Institute for Crop Production (Prague, Czech Rep.). After germination, the seedlings were transferred to hydroponic cultivation vessels, and placed in a growth chamber (photosynthetic photon flux 400 μmol m^{-2} s^{-1}, temperature 20/15 °C, photoperiod 16 h). The plants were cultivated in aerated modified Hoagland solution with two levels of nitrate nitrogen: 2 mM (denoted as high N treatment), and 0.2 mM (low N treatment). The physiological characteristics were measured in six weeks old plants (after germination) using gasometric and fluorometric apparatus (CIRAS-1 and PAM-2000, respectively). The first and the second fully developed leaves (from apex) were used for measurements.

3. Results and Discussion

The basic photosynthetic characteristics derived from gasometrically estimated light curves (*P max* and Φ *app*, see Table 1) of all measured species grown at high N availability were surprisingly similar.

Table 1. Basic photosynthetic and morphological characteristics of leaves of some *Triticum* and *Aegilops* plants grown at different N availability: maximum net photosynthetic CO_2 uptake by leaves at light saturation (*P max*, μmol m^{-2} s^{-1}), apparent quantum yield of photosynthesis (Φ *app*, mol mol^{-1}), specific leaf area (*SLA*, projected leaf area per leaf dry mass unit, m^2 kg^{-1}), leaf weight ratio (*LWR*, leaf dry mass to whole plant dry mass). Mean values from four plants, standard errors are given in italics.

	High N				Low N			
	P max	Φ app	SLA	LWR	P max	Φ app	SLA	LWR
T. aestivum	24.0	0.062	30.2	0.44	13.5	0.049	29.1	0.35
	2.2	*0.002*	*1.6*	*0.02*	*1.6*	*0.003*	*1.8*	*0.01*
T. spelta	24.8	0.058	30.1	0.43	13.0	0.044	26.0	0.35
	2.0	*0.003*	*2.8*	*0.02*	*1.4*	*0.003*	*1.4*	*0.01*
T. durum	25.4	0.061	24.3	0.40	12.5	0.046	24.0	0.37
	1.4	*0.001*	*2.1*	*0.03*	*1.8*	*0.004*	*1.8*	*0.01*
T. dicoccoides	26.3	0.062	27.5	0.38	14.6	0.050	26.0	0.33
	2.7	*0.003*	*1.5*	*0.02*	*1.4*	*0.001*	*1.6*	*0.02*
T. boeoticum	24.9	0.061	25.0	0.43	14.5	0.051	26.2	0.35
	2.1	*0.001*	*2.2*	*0.02*	*1.2*	*0.001*	*2.0*	*0.03*
A. speltoides	26.6	0.063	28.2	0.45	13.5	0.051	29.1	0.38
	2.6	*0.002*	*1.6*	*0.04*	*1.2*	*0.002*	*2.0*	*0.03*
A. tauschii	24.1	0.059	30.0	0.38	12.9	0.050	28.5	0.34
	2.7	*0.002*	*1.8*	*0.06*	*1.5*	*0.002*	*1.9*	*0.03*

At low N, the *P max* decreased by about 50% and *Φ app* by 20 to 25%. No significant interspecific differences were found in the response to low N availability. From morphological parameters, the values of LWR were much more influenced by low N than SLA.

Induced chlorophyll fluorescence parameters showed very slight variability among the species studied, and the differences were in most cases not significant. At low N, the values of *Fv/Fm* ratio, *Fo* and *qP* were very similar as in plants grown at high N availability. The most remarkable response to N deficiency was found only in *qN*.

Table 2. Basic parameters of induced chlorophyll fluorescence in leaves of some *Triticum* and *Aegilops* plants grown at different N availability: Variable fluorescence to maximum fluorescence ratio *(Fv/Fm)*, background fluorescence *(Fo)*, photochemical quenching *(qP)* and non-photochemical quenching *(qN)*. Mean values from four plants, standard errors are given in italics.

	High N				Low N			
	Fv/Fm	Fo	qP	qN	Fv/Fm	Fo	qP	qN
T. aestivum	0.787	0.389	0.897	0.580	0.777	0.341	0.915	0.341
	0.002	*0.016*	*0.023*	*0.033*	*0.015*	*0.016*	*0.012*	*0.101*
T. spelta	0.788	0.400	0.862	0.579	0.792	0.388	0.930	0.286
	0.005	*0.006*	*0.042*	*0.028*	*0.006*	*0.027*	*0.013*	*0.051*
T. durum	0.799	0.340	0.888	0.523	0.768	0.331	0.918	0.282
	0.005	*0.044*	*0.011*	*0.015*	*0.022*	*0.017*	*0.005*	*0.087*
T. dicoccoides	0.788	0.371	0.886	0.492	0.768	0.367	0.909	0.337
	0.005	*0.033*	*0.011*	*0.028*	*0.007*	*0.016*	*0.019*	*0.114*
T. boeoticum	0.777	0.334	0.873	0.572	0.789	0.306	0.899	0.251
	0.010	*0.033*	*0.033*	*0.071*	*0.006*	*0.010*	*0.006*	*0.020*
A. speltoides	0.780	0.307	0.889	0.563	0.774	0.225	0.916	0.342
	0.014	*0.012*	*0.037*	*0.053*	*0.009*	*0.018*	*0.035*	*0.082*
A. tauschii	0.795	0.354	0.835	0.504	0.761	0.329	0.928	0.249
	0.005	*0.020*	*0.029*	*0.041*	*0.019*	*0.016*	*0.012*	*0.044*

Our results have not confirmed the previously reported data on higher specific P in wild and more primitive culture wheat species (1), and also have not indicated any advantageous photosynthetic characteristics of polyploidic species. Very similar responses of all tested species to changes in N availability have also not supported our initial hypothesis that nutrition conditions could change considerably interspecific differences in photosynthetic characteristics of leaves. Most probably, the more efficient allocation and use of the photosynthate produced, together with higher morphogenetic potential is associated with higher productivity and grain yield of the modern wheat genotypes, as discussed in detail by Evans (6).

It should be stressed that our results are of preliminary character. The research was done in context of a more complex project focused mainly on nitrogen uptake regulation and use in plants. The results presented in this contribution were derived from single-term measurements done in rather young plants. Interspecific differences in photosynthetic characteristics could be more pronounced later during ontogenesis. Faster senescence of the wild species has already been reported by several authors (1, 3). In addition, the flag leaf of adult plants is probably the most important and sensitive organ reflecting potential interspecific differences (7), and, therefore, more attention should be paid to its performance in our next work.

4. Addendum

This work was supported by Grant 521/97/0284 from the Grant Agency of the Czech Republic.

References

1 Evans, J.R. and Dunstone, R.L. (1970) Aust. J. Biol. Sci. 23, 725-741
2 Khan, M.A. and Tsunoda, S. (1970) Japan J. Breed. 20, 133-140
3 Zelenskii, M.I., Mogileva, G., Shitova, I.and Fattakhova, F. (1978) Photosynthetica 12 428-435
4 Hieke, B. (1983) Photosynthetica 17, 578-589
5 Evans, J.R. and Seemann, J.R. (1984) Plant Physiol. 74, 759-765
6 Evans, L.T. (1996) Crop Evolution, Adaptation and Yield, Cambridge University Press, Cambridge, UK
7 Dunstone, R.L., Gifford, R.M. and Evans, L.T. (1973) Aust. J. Biol. Sci. 26, 295-307

REGULATION OF SUCROSE METABOLISM IN SPRING AND WINTER WHEAT.

Leonid V. Savitch, Tracy Harney, and Norman P.A. Huner
Department of Plant Sciences, The University of Western Ontario, London
Ontario, N6A 5B7, Canada. . . . .

Key words: Acclimation, environmental stress, C metabolism, C partitioning, enzymes, net photosynthesis.

1. INTRODUCTION

We have shown previously that exposure of wheat plants to the low temperature or high irradiance is accompanied by increase in sucrose/starch ratio, decrease in Glc 6-P/Fru 6-P ratio, triose-P accumulation, decreased ATP/ADP ratio and concomitant limitations of electron transport (1). It has been shown previously that restriction in phosphate recycling and photophosphorylation and an occurrence of feedback limited photosynthesis under low temperature stress and the high irradiance results from inhibition of starch synthesis (2). However, cold acclimated Monopol exhibited a 1.6-fold stimulation of CO_2 assimilation rates compared to controls due to the increased capacity of Calvin cycle enzymes and differential capacity to adjust sucrose phosphate synthase activity. In contrast, cold acclimated Katepwa exhibited a reduction in CO_2 assimilation rates compared to controls (1). Alternatively, growth at high irradiance resulted in a limited adjustment of Calvin cycle enzymes and SPS activity, but comparable photosynthetic capacity to that of controls and cold acclimated winter wheat plants (3). We have suggested that differential cultivar-specific adjustment of CO_2 assimilation rates is related to the differential cultivar-dependent adjustment of carbon partitioning to sucrose and fructan biosynthesis (1).

2. MATERIALS AND METHODS

Two cultivars of wheat (*Triticum aestivum L.*), cv Katepwa (spring wheat) and cv Monopol (winter wheat), were grown with a 16-h photoperiod under controlled environmental conditions with a temperature regime of 20/16°C (day/night) at a PFD of 250 and 800 μmol/m²s and 5°C/5°C at 250 PFD. In addition, to study the effect of low temperature stress some of the 20°C/250 PFD plants were transferred to the 5°C (20→5°C/250 PFD) for 12 h when the fourth leaf was fully developed. All measurements were made 4 h after the start of photoperiod on fully expanded fourth leaves.

Sucrose phosphate synthase (SPS), sucrose synthase in the synthetic direction (SS(s)) and in the breakdown direction (SS(d)), and neutral invertase were assayed according to Huber (4). Sucrose:sucrose fructosyl transferase (SST) and acid invertase were assayed according to Jeong and Housley (5) and the abundance of carbohydrates was measured according to Savitch et al. (1). ATP-dependent phosphofructokinase (ATP-PFK) were assayed according to Podesta and Plaxton (6), and NADP-dependent glucose-6-phosphate dehydrogenase (G6PDH) according to Srivastava and Anderson (7).

G. Garab (ed.), Photosynthesis: Mechanisms and Effects, Vol. V, 3793–3796.
© 1998 *Kluwer Academic Publishers. Printed in the Netherlands.*

3. RESULTS and DISCUSSION

In response to either low temperature stress (20→5°C), cold acclimation (5°C) or growth at high irradiance (20°C/800), leaves of both cultivars similarly increased sucrose/starch ratio by 10 to 12-fold in the spring wheat and by 4 to 5-fold in the winter wheat (Table I). No major cultivar differences were observed at the level of starch abundance or glucose, fructose and sucrose levels, while fructans were accumulated only in a leaves of cold acclimated winter wheat (data not shown). Low temperature stress, cold or high light acclimation resulted in 5 to 10-fold increase in the level of glucose and fructose in both wheat cultivars (data not shown). However, only cold acclimated wheat plants increased Glc+Fru/Sucrose ratio by 2-fold (Table I).

It is generally recognized that sucrose synthase is located in the cytosol of leaf tissue, while soluble acid invertase as well as SST are considered to be a vacuolar enzymes (4,8). As a result of low temperature stress the maximum extractable activities of most of the enzymes tested were not effected. However winter wheat increased SST activity by 60% (Table I). Cold acclimation of spring wheat did not result in any adjustment at the level of sucrose or fructan biosynthesis, as indicated by minor changes in the activities of SPS, SS(s) and SST. However, cleavege of sucrose associated with an induction of SS(d) and 2-fold increase in neutral invertase activities in spring wheat was induced during cold acclimation. In contrast, cold acclimation of winter wheat resulted in 3-fold higher SPS, 1.6-fold higher SS(s), 3.9-fold higher SST activities, when compared to controls (20°C/250). Moreover, neutral invertase activity increased by 2.5-fold and acid invertase activity increased by 6.6-fold, while SS(d) activity was not effected upon cold acclimation of winter wheat (Table I). Growth at high irradiance did not result in any stimulation of SPS, SS(s), SST, and acid invertase activities in both wheat cultivars. Moreover, high irradiance grown winter wheat plants showed limited capacity for fructan biosynthesis as was seen in 50% lower SST activity when compared to the controls. However, growth at high irradiance resulted in 85% increase in neutral invertase activity in winter wheat, and an induction of SS(d) and 30 % increase in neutral invertase activity in spring wheat.

We have shown that cold acclimation of winter wheat, but not the spring wheat resulted in a adjustment to the control values of the CO_2 assimilation rates associated with increased capacities of the SST, SPS, SS(s) and overall the capacity for fructan accumulation in the leaves. Our data are consistent with previously reported results that cold acclimation results in the stimulation of SPS, SS, and SST (9). Moreover, the increased carbon flux through SPS and SS(s) relative to that through cytosolic FBPase (3) might be a result of reutilization of hexose by phosphorylation and resynthesis into sucrose (10) associated with an induction of fructan biosynthesis and facilitation of phosphate recycling at low temperature (11). The data presented support the contention of a strong positive correlation between CO_2 assimilation rate and SPS activity (12) and fructan biosynthesis (13).

Finally, despite the different mechanisms of sucrose breakdown increased hexose levels resulted in a 5 to 7-fold increase in ATP-PFK activity, and a 3.5 to 7.3-fold increase in G6PDH activity upon cold acclimation and 3-fold increase in ATP-PFK and 1.3 to 1.8-fold increase in G6PDH activities upon high light acclimation of both wheat cultivars. These data are consistent with an increased flux of metabolites through the glycolytic and pentose phosphate pathways.

Crowns of wheat are an important sink organ and are essential for winter survival. Due to the fact, that sucrose is considered to be a most abundant export form of carbohydrates, enzymes of sucrose metabolism should reflect the organ sink capacity. During cold acclimation most of the monosacharides and sucrose accumulate in a extravacuolar space, where they can contribute to the cryoprotection (14).

Table 1. Effects of low temperature stress, cold or high light acclimation on accumulation of carbohydrates and regulatory enzymes of fructan metabolism. Enzyme activities were expressed in $\mu mol\ m^{-2}\ s^{-1}$ (for leaves) and in $\mu mol\ g^{-1}\ fr\ wt\ h^{-1}$ (for crowns). Fructan sucrose and fructan metabolism. Enzyme activities were expressed in $\mu mol\ m^{-2}\ s^{-1}$ (for leaves) and in $\mu mol\ g^{-1}\ fr\ wt\ h^{-1}$ (for crowns). Fructan abundance was expressed in $\mu mol\ g^{-1}$ dry wt. Ratios of carbohydrates were calculated on a w/w basis. Data represent the mean ±SE, n=4.

	Katepwa				Monopol			
	20°C/250 PFD	20→5°C	5°C	20°C/800 PFD	20°C/250 PFD	20→5°C	5°C	20°C/800 PFD
Leaves								
SPS	1.6 ± 0.2	1.3 ± 0.2	1.6 ± 0.1	2.0 ± 0.1	1.6 ± 0.1	1.2 ± 0.2	5.0 ± 1.2	2.0 ± 0.2
SS(s)	1.2 ± 0.2	1.5 ± 0.2	0.8 ± 0.1	0.9 ± 0.1	1.5 ± 0.2	1.8 ± 0.2	2.5 ± 0.2	0.8 ± 0.1
SS(d)	ND	ND	1.4 ± 0.1	1.4 ± 0.1	ND	ND	ND	ND
Acid invertase	0.3 ± 0.0	0.3 ± 0.1	0.3 ± 0.1	0.5 ± 0.1	0.5 ± 0.1	0.6 ± 0.1	3.1 ± 0.2	0.5 ± 0.1
Neutral invertase	0.6 ± 0.1	0.5 ± 0.0	1.1 ± 0.1	0.7 ± 0.1	0.7 ± 0.1	0.7 ± 0.1	1.8 ± 0.1	1.3 ± 0.1
SST	0.2 ± 0.0	0.3 ± 0.1	0.3 ± 0.1	0.4 ± 0.0	0.5 ± 0.1	0.8 ± 0.0	1.9 ± 0.2	0.2 ± 0.0
ATP-PFK	1.0 ± 0.1	0.6 ± 0.1	5.2 ± 1.3	3.4 ± 0.3	1.0 ± 0.1	0.9 ± 0.1	7.4 ± 0.1	3.0 ± 0.2
G6PDH	1.0 ± 0.1	0.9 ± 0.1	3.5 ± 0.3	1.3 ± 0.1	1.2 ± 0.2	1.1 ± 0.2	8.7 ± 0.3	2.2 ± 0.3
Suc/starch	0.2	2.4	3.3	2.2	0.6	2.2	3.0	3.0
Glc+Fru/suc	0.14	0.12	0.31	0.18	0.26	0.15	0.56	0.21
Crowns								
SST	ND	ND	ND	ND	98 ± 11	164 ± 10	688 ± 32	360 ± 18
SS(d)	ND	ND	ND	174 ± 4	ND	ND	175 ± 24	87 ± 7
Acid invertase	14 ± 2	32 ± 3	40 ± 3	73 ± 10	9 ± 2	66 ± 10	214 ± 27	69 ± 8
Neutral invertase	10 ± 1	11 ± 2	79 ± 8	64 ± 5	13 ± 1	48 ± 5	66 ± 7	66 ± 5
Glc+Fru/suc	0.94	0.50	1.38	2.07	1.19	0.39	2.44	3.26
Fructans (DP>4)	187 ± 20	325 ± 12	270 ± 10	119 ± 10	437 ± 32	516 ± 28	833 ± 10	675 ± 68
Starch/fructans	0.75	0.34	0.18	1.17	0.17	0.20	0.04	0.49
Suc/starch+fructans	0.13	0.27	0.14	0.68	0.09	0.27	0.08	0.22
TNC(% from dry wt.)	8.8	13.8	9.5	23.1	12.7	19.1	23.7	50.3

Due to the fact that hexoses generally do not accumulate in a vacuoles (4), increased activities of SST and acid invertase might help to regulate the osmotic potential of the vacuole and an induction of SS(d) and neutral invertase might be an important factor in increased cryoprotection. As a result of cold or high light acclimation glc+fru/sucrose ratio increased in both wheat cultivars by 1.5 to 2.7-fold. However, despite the similar 5-fold increase in neutral invertase activity for both cold or high light acclimated wheat cultivars, the increase in the glc+fru/sucrose ratio was associated with cultivar and stress-factor dependent adjustment of sucrose metabolizing enzymes. In addition to the 4-fold higher SST activity in winter wheat, both high light acclimated wheat cultivars stimulated SS(d) activity and increased acid invertase activity by 6 to7-fold. In contrast, cold acclimation of spring wheat resulted in 3-fold increase in acid invertase activity, when compared to the controls, while cold acclimated winter wheat stimulated SS(d) and increased acid invertase activity by 24-fold in addition to the 7-fold higher than controls SST activity. Despite the stimulation of sucrose metabolizing enzymes upon cold acclimation, both wheat cultivars maintained a similar sucrose/starch+fructans ratio to controls (20°C/250). In contrast, acclimation to high light resulted in a 2 to 5-fold increase in sucrose/starch+fructans ratio, when compared to controls maintained at 20°C. We have shown that a decrease in the starch/fructan ratio upon cold acclimation and an increase in this ratio upon high light acclimation of wheat plants reflect different temperature dependent enzyme sensitivity of starch and fructan biosynthesis. Moreover, spring wheat appears to exhibit a lower capacity for fructan biosynthesis upon cold acclimation, and lower capacity for fructan and starch biosynthesis upon high light acclimation, than winter wheat.

4. CONCLUSIONS

We suggest that the higher capacity for CO_2 assimilation in the cold acclimated winter wheat, but not the spring wheat was associated with an increased sucrose-fructan biosynthesis in leaves and a concomitant increased fructan biosynthesis in a crowns. This is consistent with an overall increased crown sink capacity as indicated by increased SST, SS(d) and acid invertase activity. Higher crown sink capacity and higher CO_2 assimilation rates in high light acclimated winter wheat were associated with an increased capacity for fructan and starch biosynthesis. As a result, the accumulation of total non-structural carbohydrates in crowns was always higher in winter wheat during cold as well as high light acclimation.

5. REFERENCES

1. Savitch LV, Gray GR, Huner NPA (1997) Planta. 201: 18-26
2. Sharkey TD, Vassey TL (1989) Plant Physiol 90: 385-387
3. Gray GR, Savitch LV, Ivanov AG, Huner NPA (1996) Plant Physiol 110: 61-71
4. Huber SC (1989) Plant Physiol. 91: 656-662
5. Jeong BR, Housley TL (1990) Plant Physiol. 93: 902-906
6. Podesta FE, Plaxton WC (1994) Planta 195: 891-896
7. Srivastava DK, Anderson LE (1983) Biochim. Biophys. Acta 724: 359-369
8. Wagner W, Keller F, Wiemken A (1983) Z. Pflanzenphisiol. 112: 359-372
9. Tognetti JA, Calderon PL, Pontis HG (1989) J. Plant Physiol. 134: 232-236
10. Collis BE, Pollock CJ (1992) J.Plant Physiol. 140: 124-126
11. Pollock CJ, Lloyd EJ (1987) Ann Bot 60: 231-235
12. Galtier N, Foyer CH, Huber J, Voelker TA, Huber SC (1993) Plant Physiol 101: 535-543
13. Chatterton NJ, Harrison PA, Bennett JH, Asay KH (1989) J.Plant Physiol. 134: 169-179
14. Koster KL, Lynch DV (1992) Plant Physiol 98: 108-113

OPTIMAL PHOTOSYNTHETIC TYPE OF WHEAT PLANT AND BIOTECHNOLOGICAL APPROACHES TO CROP IMPROVEMENT.
Olga I. Kershanskaya

Institute of Plant Physiology, Genetics and Bioengineering, Ministry - Academy of Sciences, Timiryasev Str., 45, Almaty, 480090, Kazakstan

Key words: chloroplast, electron transport, gas exchange, biomass, optimal photosynthetic models, genetic determination.

1. Introduction

Agriculture needs new sorts with optimal combination of economical - valuable properties and resistance to biotic and abiotic limiting factors of environments. For creation of such sorts crop breeding needs new ideas, approaches and methods - new ways of biotechnology of adjacent biological sciences: plant physiology, biochemistry, genetic. The crop breeding enters into a new stage, which may be called «synthetic crop breeding» based on physiology-genetic management of photosynthesis, growth, development processes and is characterized by passage from extensive to intensive way of crop production improvement (1,2). A real approach to plant production process' regulation is the creation of physiology - genetic active photosynthetic apparatus by crop breeding. Increasing of photosynthetic apparatus activity can be created by a way of photosynthetic tests' elaboration and application and its genetic determination in favorable and limiting environmental conditions. In these connection its necessary new ways' stages: a) systematic study of photosynthetic indices on different levels of photosynthetic apparatus organization on natural gene - controlled models of wheat, b) differentiation of selective collections by photosynthetic tests based on its genetic determination, c) revealing of forms - donors of high photosynthetic activity, d) creation of a model of Optimal photosynthetic type wheat plant (OPTWP) with maximum productivity and adaptability to environments, e) creation of new productive and adaptive sorts with optimal combination of economic-valuable properties.

The mechanism of relation between primary photosynthetic reaction, CO_2 - gas - changes activities and high crop production (3) in wheat and it importance for creation of the (OPTWP) Concepts as the biotechnological model is not known yet. But namely such concept may be used as the base for physiology - genetic increasing of photosynthetic activity on different levels of photosynthetic apparatus organization: from photosystem - to plant; determination of mechanism of relation between primary energetic processes of photosynthesis in chloroplast, high CO_2 - fixation in leaf and high crop production; understanding of possible mechanisms of photosynthetic apparatus adaptation to different environments and it importance for creation of maximum productivity.

2. Procedure

G. Garab (ed.), Photosynthesis: Mechanisms and Effects, Vol. V, 3797–3800.
© *1998 Kluwer Academic Publishers. Printed in the Netherlands.*

2.1 *Materials*
It have been used more 100 contrast productive and environmental adaptive wheat objects during 1980 - 1996 years.

2.2 *Methods*
Spectrophotometric, polarographic and chlorophyll fluorescence methods were used for the determination of primary photosynthetic processes in photosystems II and chloroplasts on vies, species, lines and heteroseous hybrids of winter wheat as objects with high yield potential. Gas - exchanges measurement were conducted for determination of photosynthetic CO_2 - assimilation in leaves. The measurements of leaf area and biomass distribution and redistribution by organs for study of source - sink regulation in whole plant as well as the determination of productivity elements were conducted. Photosynthetic parameters were calculated by dispersion, polygene, combinative ability and correlative analysis methods.

Morpho-physiological indices have been determined by measuring of leaf area, distribution and redistribution of biomass by organs in ontogenesis, elements of yield structure. CO_2-gas-changes have been measured in open and closed system with using of IRGA Infralyt, Junkalor in special chambers. Polygene, combine ability and correlation analysis were conducted by Hayman, Griffing and Dospehov methods (4).

3. Result and Discussion

3.1 *The main thesis of OPTWP concept*
We have shown the intensification of the reaction center of photosystem II activity, Hill reaction, non cyclic and separate electron transport activity, total photophosphorylation, light coupling of energetic processes in chloroplast; the increasing of the chlorophyll a contents for OPTWP with maximum productivity in comparison with photosynthetic apparatus of wheat standard or parents forms in concrete environments.

This increasing was not due to enlargement of the sizes of light - harvesting chlorophyll a/b protein complexes of photosystem II or number of chloroplasts in cell and suspension. These structural parameters can be inherent to some genotypes of OPTWP in wheat. The important factor in regulation of OPTWP forming was optimal functional - structural interactions between activity of primary bioenergetic reaction and structural organization of photosynthetic apparatus of photosystem and chloroplast. Namely photosystem II reaction center activity, rate of electron flow, intensification of photophosphorylation and tight coupling of primary bioenergetic processes in chloroplast were the reasons for net photosynthesis rate increasing in leaf of OPTWP.

The main thesis of OPTWP concept in wheat are: have observed united chain of compatible photosynthetic processes from photosystem to whole plant and crop production; is determined the reproductive type of production process and the role of organs of high part of shoot («flag» leaf, steam after ear, ear's elements) in accumulation and redistribution of assimilates; is ascertained the optimal combination of extensive (structure) and intensive (function) indices of photosynthetic activity; is confirms the relatively independent (discrete) character of revealing and inheritance of photosynthetic indices that determined their different interactions and relations; in total high photosynthetic productivity is determined both by high autonomy of photosynthetic structures of low levels of photosynthetic apparatus organization (photosystem, chloroplast, leaf) and by complicated system of integration of photosynthetic processes in the whole plant (5-7).

3.2 Denetic determination study of photosynthetic processes

The mechanism of endogenous regulation and the role of integrated approach in OPWPT can be formulated in genetic determination study. It confirms the significance of photosynthetic activity integration on all of the levels of photosynthetic apparatus organization for maximum productivity and adaptability creation. The mobile system of physiology - genetic correlations between photosynthetic indices and productivity elements in limiting factors of environmental conditions are suggested.

It have been shown photochemical activity of chloroplasts is controlled by gene-systems with additive and non-additive effects. Effects of sub-dominance were typical, ones stipulated for positive and negative heterosis in wheat hybrids and could be inherit it. High photochemical activity predominates over low ones in these group of forms, low photochemical activity is inherited by recessive type, that facilitated selection by this index from F_2. Many of heteroseous hybrids were characterized by low number of chloroplasts in cell and high photosynthetic activity of one chloroplast. It correspondes to idea about possibility of existence of genetics «strong» chloroplasts and their using as a test in crop breeding on productivity. Genetic variation of Net photosynthesis was operated by polygene nuclear system with non-additive and less additive effects. High rate of Net photosynthesis was controlled by recessive genes as rule, independently on gene-type. Leaf area was characterized by polygene type of inheritance with genetics polymorphism. For nuclear genes which controlled the development of assimilation apparatus, inherited to sub-domination effects. Revealing forms were sufficiently contrast and hybrids showed intermediate inheritance of leaf area (8, 9).

3.3 Photosynthetic indices of high productivity

It has been shown wheat crop production based on functional activity and structure on all of the levels of photosynthetic apparatus organization: from photosystems - to whole plant. The more suitable indices for crop breeding in field conditions were the morpho-physiological ones and CO_2 -gas - components in leaf and whole plant. We have characterized 30 photosynthetic indices for tests on productivity and drought adaptability elaboration. More informative among them were: the shoot biomass, ear and leaf biomass interaction in flowering, organ biomass distribution from flowering to wax ripeness, net photosynthesis, leaf area. Biological and economic-valuable meanings of photosynthetic tests permit to unite them into photosynthetic test-system: I - extensive indices of photosynthetic apparatus' sizes, II - intensive indices of functional activity of photosynthetic apparatus, III - ^{Inq} - «inquiry» of ear in assimilates, IY - ^{Real} - realization of ear's inquiry in assimilates in ontogenesis, Y - ^{Incr} - increase of ear mass, YI - ^{LInq} - leaves' function in grain mass forming, YII - rate of biomass accumulation in reproductive and vegetative period, YIII - distribution and redistribution of assimilates by source and sink organs in ontogenesis, IX - part of harvesting-valuable grain mass in common plant biomass, X - real organ's biomass creation, XI - real crop yield mass creation (10).

3.4 Types of wheat production processes in Kazakstan

Based on these indices with using of polygene, combine ability and correlation analysis, we have elaborated the ways of wheat productivity and ecological adaptability determination. As a result of wheat differentiation by photosynthetic tests, four types of winter wheat production processes in watered conditions of South - East of Kazakstan were determined. For each type inhered in concrete photosynthetic indices interaction for production forming during vegetation. I Intensive - adaptive and productive type of production process is characterized by high photosynthetic activity and not so large size of leaf area. It is a perspective type for new future of wheat crop breeding with grain yield more 100 c/ha. II - Semi - intensive, but relatively adaptive type of production process is produced relatively more high demands to environments because of higher leaf are. It is a good type for modern wheat agriculture. III - Semi -extensive - relatively productive in watered conditions but don't adaptive type of production process, which needs favorable environments and intensive agrotechnology for growth of large leaf area. IY - Extensive type - crop and biomass production in which is possible only on the base of high extensive parameters - sizes of photosynthetic apparatus, large leaf area. It is a traditional morpho-physiological type of crop breeding, result of green revolution with productivity not more 55-60 c/ha.

3.5 New productive wheat lines and forms creation

With using of revealing ways was conducted search of wheat genotypes with high photosynthetic activity. The new lines of wheat - donors of high photochemical activity, high net photosynthesis and low leaf area were revealed and gathered in own core collection. Forms-donors of high photochemical activity of chloroplasts were: sort Omskaya, heteroseous hybrids Omskaya x Progress with heterosis and epystaz phenomena. The highest heterosis on number chloroplasts in cell was observed on hybrids Progress x Omskaya. Rate of net photosynthesis were highest in hybrids of Gostianum-88, Progress and Omskaya. Eritrospermum-350 had the best combine ability on photosynthetic CO_2 gas-changes parameters. Recessive genes of leaf area were prevailed in Gostianum-88 and Eritrospermum-350. Their hybrids had low value of these indices. The contrast productive and drought resistant wheat lines were created. The first productive sort of winter wheat with high photosynthetic activity was produced and given to State examination

Results of these researches can be used for development of new crop biotechnology as the model of high photosynthetic production and adaptation to different environments.

References:

1 Nichiporovich, A.A. (1982) in Physiology of Photosynthesis (Nichiporovich A. A., ed.) pp. 7-33, «Science» Publishers, M., Russia

2 Mokronosov, A.T. (1983) Photosynthetic Function and Integration of Plant Organism, 42nd Timiryasev Reading, «Science» Publishers, M, Russia

3 Krendeleva, T.E. (1985) Structural-Functional Heterogeneity of Primary Photosynthetic Processes in Plant, 44 p. Thes. Doc. diss., Moscow University Publishers, M, Russia

4 Sidorenko, O.I., Bedenko, V.P. and Urazaliev R.A. (1990) Photosynthesis of Heteroseous Hybrids in Winter Wheat, 156p., «Gilim» Publishers, Almaty, Kazakstan

5 Sidorenko, O. (1996) Plant Phys. and Bioch., Special Issue, 10th FESPP Congress, Florence, Italy, Sept. 9-13, 87.

6 Sidorenko, O. (1997) in XYIII Congress SPPS, pp.170-171, The Scandinavian Society for Plant Physiology Publishers, Uppsala, Sweden

7 Kershanskaya, O.I. J. Exp. Bot. 49, 29-30

8 Sidorenko, O.I. and Gaziyantz, S.M. (1994) Proc. Nat. Ac. Sci. Kaz. 3, 9-18.

9 Kershanskaya, O. I. and Shpak, E. V. (1998) Proc. Ninth Int. Wheat Gen. Symp., pp. 4, Univ. Saskatchevan Publishers, Sascatoon, Canada

10 Kershanskaya, O.I. and Urazaliev, R.A. (1998) J. Exp. Bot. 49, 31-32

CONCOMITANT MEASUREMENTS OF BIOMASS INCREASE AND PHOTO-SYNTHETIC ACTIVITY OF MAIZE DURING THE VEGETATION PERIOD

Lj.Milovanović[1], S.Veljović-Jovanović[1], Lj.Jovanović[2] & Ž. Vučinić[1*],
[1]Center for Multidisciplinary Studies,Belgrade University, 29.Novembra 142;
[2]Center for Pesticides, Institute for Agricult.Research; Belgrade, F.R.Yugoslavia

Key words: bioproductivity, Chl fluorescence, CO_2 uptake, electron transport, net photosynthesis, transpiration, *Zea mays* L.

1. Introduction

An apparent paradox exists in literature, in that numerous short-term measurements of photosynthetic rates and bioproductivity could not be correlated (1). Only when photosynthetic measurements were performed on very young plants, in the exponential phase of growth, could some form of a relationship be established between photosynthetic rates and biomass accumulation. Since plants exhibit a typical sigmoid form of biomass accumulation during the vegetation period, we were interested in determining whether there occurs a diminishment of the rates of CO_2 assimilation and the capacity of the plants to absorb light and perform photosynthesis, during the period of vegetation when the plants rapidly decrease their rate of dry weight accumulation, while environmental conditions (light and temperature) are not limiting factors.

2. Procedure

2.1 Material
Maize hybid seed (ZP SC 599) were planted in twenty-liter plastic containers in high quality soil and grown to full maturity. Planting was performed on 19th May 1997 and final harvesting on 15th October (D151 - 151th day from planting) of the same year. Plants were supplied with unlimiting quantities of water, and supplemented with NPK fertilizer every two weeks from D25 till D85, the final concentration of nitrogen, P_2O_5 and K_2O provided to the plants being 24, 12 and 8 g·plant^{-1}, respectively.

2.2 Biomass analysis
In the period from D65 till the end of vegetation 3-5 plants were harvested every week.

---

*To whom correspondence should be addressed to. E-mail: *evucinic@ubbg.etf.bg.ac.yu*

G. Garab (ed.), Photosynthesis: Mechanisms and Effects, Vol. V, 3801–3804.
© 1998 *Kluwer Academic Publishers. Printed in the Netherlands.*

Samples of leaf were taken for chlorophyll analysis, and the rest placed in a ventilated drying oven at 105°C for 1 hr, followed by 60°C till no further change in weight could be observed. Roots were previously carefully washed of soil and dried with laboratory filter paper.

2.3   Photosynthetic activity

Photosynthetic activity measurements were performed on bright, sunny days. Gas exchange (CO_2 uptake and transpiration), together with light, temperature and humidity were recorded with a portable infra-red gas analyzer (Walz GmbH model CQP130). Leaf fluorescence measurements were performed simultaneously on the same leaf as the gas exchange measurements, using a modulated photosynthetic fluorimeter (Walz GmbH model PAM101), the F_m' level being determined by using saturating pulses (4000 $\mu mol \cdot m^{-2} \cdot s^{-1}$) from a Schott lamp. Measurements were performed on the third or fourth well developed leaf above the ear, by placing the leaf chamber on a part of a leaf and light guide for fluorescence measurements on the adjoining part of the same leaf at equal distance from the vein. Care was taken to ensure even exposure to light of both areas of the leaf. The data presented are averaged results obtained in the temperature range of 25-28°C and light intensities of 1550-1800 μmol photons$\cdot m^{-2} \cdot s^{-1}$. Steady state values of CO_2 assimilation and transpiration were obtained after approximately 15 minutes of air flushing through the leaf chamber, and such values were used for averaging and presentation. Quantum efficiency of electron transport through photosystem II (Φ_{PSII}) was calculated according to Genty et al. (2). Three to four different plants were rotated for measurements every day.

3. Results and Discussion

In Fig. 1 we show the rate of accumulation of dry weight of the whole plant, and its main components (stem with leaf, root and cob). Maximal rates of dry weight accumulation were observed in the period from D75 to D90, with an increase in weight of 10 g·plant^{-1}·day^{-1}, falling to 8.3 g·plant^{-1}·day^{-1} in the period D90-D100, and 5.9 g·plant^{-1}·day^{-1} in the period D100 to D110. From the data presented it is obvious that the stem and leaf, as well as the root, upon achieving maximal dry weights (D90) lose a part of their accumulated weight, stem with leaf going from a maximum of 220 g·plant^{-1} to 150 g·plant^{-1} at the end of the season (32% loss), the roots decreasing from 110 g·plant^{-1} to 40 g·plant^{-1} at the end

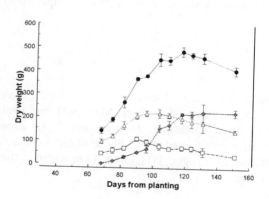

Figure 1. Dry weight increase in experimental plants during the vegetation period: (filled circles) whole plant; (open triangles) stem and leaf; (open squares) root; (shaded rhomb) cob.

of the season (64% loss). The ratio of the above ground part of the plant (excluding the cob) to that in the soil was approximately 2:1 till the plants began realocating their carbon skeleton. The observed rates of dry weight accumulation are approximately 30% greater than those published for maize (3), our data comprising root dry weight. At the time the stem and root began losing a part of their weight, the weight of the cob was approximately 100 g, the final cob weight at the end of the season being 220 g·plant[-1]. Thus, one could hypothesize that the later part of the growth period of

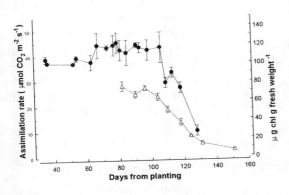

Figure 2. Assimilation rates (●) and chlorophyll content of plants (▽) during the vegetation period.

plants is utilized by the system to realocate a part of the already fixed carbon and total dry weight in the stem and the root to the reproductive organ.

Figures 2. and 3. show the pattern of midday values of CO_2 assimilation rates, transpiration, quantum efficiency of PS_{II}, as well as the chlorophyll content during the whole vegetation period from the D30 in June to the D130 in late September. The rate of CO_2 assimilation increased to maximal values (43 μmol $CO_2 \cdot m^{-2} \cdot s^{-1}$) in July and stayed at the same level till after D100, that is in our experiment the end of August. The decrease in the assimilation and transpiration rates observed in the first half of September was accompanied by decreases in the quantum efficiency of electron transport through PS_{II}. The somewhat earlier decline in the transpiration rates that could be observed can be correlated with the decrease in the total dry weight of the root system after D90. A similar gradual decrease in chlorophyll content could be also noted. The relationship between electron transport and CO_2 fixation was not affected during whole vegetation period.

We have also monitored the change in the photosynthetic parameters during the day, and in Fig.4 we show

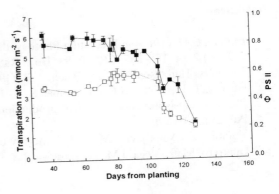

Figure 3. Transpiration rates (■) and quantum efficiency of PS_{II} - Φ_{PSII} (□) of plants during the vegetation period.

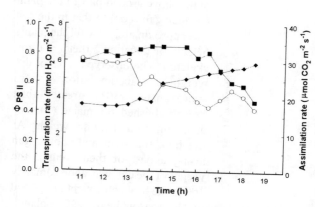

Figure 4. Diurnal variation of CO_2 assimilation rate (○), transpiration rate (■) and PS_{II} efficiency Φ_{PSII} (◆) measured on the 54th day from planting.

an example of the change in the assimilation, transpiration and quantum efficiency on D53. In some instances we observed a decline in the assimilation rate around and after noon without the closure of stomata. Such a decrease in CO_2 assimilation rate was not accompanied by a decrease in quantum efficiency of PS_{II} as shown in Fig.4. This phenomenon was especially noticed in the later stages of growth (second half of August), even in the late morning hours, although the light and temperature were not limiting. These results indicate that under such conditions a stimulation of electron transport to alternative acceptors occurs.

Thus, the answer to the aforementioned apparent paradox in photosynthetic literature could be explained either by diurnal fluctuations in carbon fixation and release, where large quantities of CO_2 fixed during the day are released at night (4), and/or by the occurrence of alternative photosynthetic reactions (5).

Acknowledgment: This work was financed through a grant in biophysics (03E22) from the local Ministry of Science and Technology. The authors are grateful to the Serbian Orthodox Church Students Association that enabled this presentation to Lj.M. by covering the cost of the registration fee.

4. References

1. Elmore, C.D. (1980) in Predicting Photosynthesis for Ecosystem Models (Hesketh, J.D. ed) 2, pp.155-167, CRC Press, Boca Raton, USA
2. Genty, B., Harbinson, N.R. and Baker, N.R. (1989) Plant Physiol.Biochem., 28, 1-10
3. Beadle, C.L., Long, S.P., Imbamba, S.K., Hall, D.O. and Olembo, R.J. (1985) Photosynthesis in relation to plant production in terrestrial environments, Tycooly Pub.Ltd., Oxford, UK
4. Vučinić, Ž., Vilotijević, M. and Stanojlović, R. (1997) Arch.Biol.Sci. 49, 25-29
5. Guerrero, M.G., Vega, J.M. and Losada, M. (1981) Ann.Rev.Plant Physiol. 32, 109-122

QTL MAPPING ASSOCIATED TO PHOTOSYNTHESIS IN BEAN
(*Phaseolus vulgaris* L.)

Ricardo Bressan-Smith and Messias G. Pereira. Plant Breeding Lab., State University of the North Fluminense, Campos, RJ, 28.015-620 BRAZIL

Key words: net photosynthesis, photosynthesis genes, productivity, crop improvement, genetic manipulation, molecular biology

1. Introduction

The genetic control of agronomic traits associated with yield, including complex metabolic processes such as photosynthesis, is an important goal for plant breeders because photosynthesis is the major determinant of biomass production in plants (1). Research has shown the existence of genetic variability in net photosynthesis among species and among cultivars (2), however there has been no utilization of these differences in plant breeding programs.

The inheritance of photosynthesis is better understood if the numbers of factors involved in its expression is considered. Net photosynthesis is governed by a large number of genes with controlled expression in the nucleus and in the chloroplasts, therefore it is possible to utilize a strategy which increases the number of favourable alleles. In order to obtain such desirable traits, it is possible to utilize classical and advanced breeding methods associated with DNA markers.

DNA markers facilitate the construction of genetic linkage maps which permit the localization and evaluation of quantitative trait loci (QTL) in many crops (3). The search for QTL which control metabolic pathways, such as photosynthesis, could provide a new evidence for the understanding of plant physiology.

The aim of this work was to develop a genetic linkage map and the evaluation of genetic variation and QTL mapping of several traits associated with photosynthesis in *Phaseolus vulgaris*.

2. Procedures

Segregating population was obtained by crosses between snap bean (cv. HAB-52) and common bean (BAC-6). The F_2 population was cultivated in the field. CO_2 assimilation and stomatal conductance were determined on light saturation. Photosynthetic pigments and leaf soluble protein were evaluated by biochemical procedures.

G. Garab (ed.), Photosynthesis: Mechanisms and Effects, Vol. V, 3805–3808.
© 1998 *Kluwer Academic Publishers. Printed in the Netherlands.*

Genomic DNA from the parentals was isolated and tested in terms of polymorphism using several random oligodecameres (primers). The Random Amplified Polymorphic DNA (RAPD) procedure was utilized (4). Selection of polymorphic primers was based on the presence/absence of amplified bands. These selected primers were then utilized for evaluation of polymorphism on the F_2 population.

The construction of a genetic linkage map was achieved through segregation of DNA amplified bands of the F_2 population. The software MAPMAKER/EXP 3.0 (5) was utilized to accomplish the linkage analysis from 114 RAPD loci and three morphological markers. Linkage groups were declared with a minimum LOD score of 2.0 and a maximum distance of 50 cM.

Interval mapping (6) was performed by software MAPMAKER/QTL 1.1 (7) for identification of QTL on genetic linkage map which had been previously constructed. Interval mapping calculated and composed the LOD scores at each 2 cM on linkage map, and the results plotted graphically at each linkage group.

3. Results and discussion

3.1. Analysis and distribution of phenotypic data

The parents were contrasting for most of the traits (Table 1). Quantitative inheritance of CO_2 assimilation rate was demonstrated by Shapiro-Wilk test in F_2 population, which generated an approximately normal distribution curve (*Figure 1*). In bean, the inheritance of net photosynthesis (*Pn*) is quantitative as revealed in cross between cultivars Red Kidney and Michelite-62 (8).

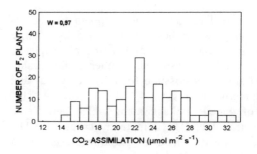

Figure 1 - CO_2 assimilation rate histogram from 193 F_2 plants. W represents the normality test of Shapiro-Wilk.

CO_2 assimilation rate was very high in some plants of the F_2 population reaching 32.34 $\mu mol\ m^{-2}\ s^{-1}$ when compared to the F_2 mean of 22.51$\mu mol\ m^{-2}\ s^{-1}$. The occurrence of transgressive segregation in the F_2 population is the result of substantial genetic variation between the parents and/or a combination of positive or negative complementary alleles from both parents, resulting in distinct performance. These high values of CO_2 assimilation in some plants can be used as a breeeding tool for generating superior inbred lines, with high photosynthetic rates.

With exception of CO_2 assimilation ($h^2_b = 0.29$) and stomatal resistance ($h^2_b = 0.40$), the other traits demonstrated heritabilities close to or higher than 0.50. These relatively high values of heritability indicate the possibility of satisfactory genetic gain in a selection program.

Table 1 – Summary of phenotypic trait means from parents, F_1 and F_2 population, showing the amplitude of variation in F_2, and the broad sense heritability (h^2_b)

Traits	Mean Values				Amplitude of variation in F_2	h^2_b
	HAB-52	BAC-6	F_1	F_2		
A (μmol CO_2 m^{-2} s^{-1})	19.55	22.93	21.92	22.51	14.46-32.24	0.29
g_s (mol H_2O m^{-2} s^{-1})	0.95	1.37	1.31	1.49	1.1-4.15	0.47
C_i (μmol mol^{-1})	257.78	262.48	253.85	262.57	194-378	0.46
r_s (s cm^{-1})	0.55	0.46	0.57	0.35	0.1-0.92	0.40
Chlorophyll a content (μM)	831.08	1035.91	757.04	899.87	272-1307	0.68
Chlorophyll b content (μM)	182.58	277.90	218.46	249.39	99-400	0.61
Total Chlorophyll (μM)	1013.00	1313.00	989.00	1187.00	340-2711	0.66
Carotenoid content (μM)	289.65	401.95	296.50	354.30	103-496	0.67

A: CO_2 assimilation g_s: Stomatal conductance
C_i: CO_2 internal concentration r_s: Stomatal resistance

3.2. QTL analysis

From the linkage analysis, the genetic linkage map was constructed and provided 13 linkage groups. Although the linkage map did not attained 11 linkage groups (number of haploid chromosomes of bean) it provided strong evidence for the identification of QTL associated with photosynthetic traits. A graphical representation of QTL of stomatal conductance, identified in group E, is shown in *Figure 2*.

For CO_2 assimilation only one QTL was detected, however, this does not imply that the CO_2 assimilation is necessarily monogenic, even though the QTL explained 55,1% of phenotypic variation. CO_2 assimilation is a consequence of multiple genes participating in photosystems I and II, the Calvin cycle and the biochemical pathways of chlorophylls and carotenoids. These results indicate that genomic regions affecting CO_2 assimilation of the genome from both parents are very similar and therefore did not exhibit polymorphism detected by DNA markers.

Three QTL controlling chl a, total chl and carotenoid content were identified closely linked in linkage group B. This indicates the possibility of linkage or pleiotropy, in which a gene is responsible for more than one characteristic (9). The fact that these traits show high correlation (data not shown) indicates that a pleiotropic effect is highly probable because chl a and carotenoid have similar function on photo absorption at antenna system, located in thylacoid membranes of the chloroplast. As chlorophylls and carotenoids are highly associated through linking proteins in antenna system, it may be that the QTL identified are related to them. This may be confirmed by the utilization of specific primers for some of these proteins (10).

Table 2 - Localization and QTL effects affecting photosynthetic traits

Characteristic[a]	Linkage group	R^2 (%)	LOD[b]	Gene effects a	d	d/a	Gene[c] action	DIR[d]
A	F	55.1	2.21	2.28	-5.05	2.21	SD	BAC
g_s	B	68.3	2.79	-0.59	-0.99	0.17	DP	HAB
	C	58.4	2.29	0.70	-0.85	-1.21	SD	BAC
	E	53.5	3.19	0.73	-0.83	-1.14	DC	BAC
	G	62.8	3.48	0.44	-1.20	-2.73	SD	BAC
	L	58.5	2.34	0.48	-1.11	-2.31	SD	BAC
Multiple QTL model		75,2	5.21	0.75	-1.76	-2.34	SD	BAC
C_i	B	35.6	3.42	59.16	-61.16	-1.03	DC	BAC
r_s	L	57.2	2.50	-0.0087	0.2750	-31.61	SD	HAB
	M	50.6	2.35	0.0871	-0.2225	-2.55	SD	BAC
Multiple QTL model		88,0	8.39	0.1306	0.1087	0.83	DC	BAC
Chl a	C	50.6	2.37	-129.13	210.24	1.63	SD	HAB
Chl b	B	24.1	2.58	-19.18	51.93	-2.71	SD	HAB
Chl ab	B	24.2	2.26	-102.03	259.72	-2.55	SD	HAB
Crt	B	25.7	2.44	-27.03	74.52	-2.75	SD	HAB

[a]A: CO_2 assimilation g_s: Stomatal conductance Chl a: Chlorophyll a content
Ci:. Internal conc. of CO2 r_s: Stomatal resistance Chl b: Chlorophyll b content
Chl ab: Chlorophyll a and b content Crt: Carotenoid content
[b] a: additive; d: dominance
[c]Additive effects are associated with BAC-6 allele. Thus, a negative value means this allele decreases the value of the trait;
[d]Direction of response is the parent whose additive value of a allele increased the value of the trait

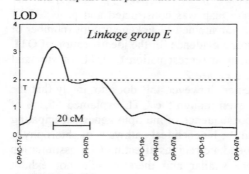

Figure 2 - Log-likelihood plots of linkage group E for stomatal conductance. The vertical axis represents LOD score (LOD 2.0 was used as significant threshold - T). The RAPD markers are identified in the horizontal axis.

4. References

1 Jeuffroy, M.-H. and Ney, B. (1997). Field Crop Res., 53, 3-16
2 González, A., Lynch, J., Tohme, J.M., Beebe, S.E. (1995) Crop Sci., 35, 1468-1476
3 Tanksley, S.D. (1993) Annu. Rev. Gen., 27: 205-233
4 Williams, J.G.K., Kubelick, A.R., Livak, K.J. (1990) Nucleic Acids Res., 18, 6531-6535
5 Lander, E.S., Green, P., Abrahanson, J., Barlow, A. (1987) Genomics, 1: 174-181.
6 Lander, E.S. e Botstein, D. (1989) Genetics, 121, 185-199
7 Paterson, A.H. et al (1988) Nature, 335: 721-726.
8 Izhar, S. e Wallace, D.H. (1967) Crop Sci., 7, 457-460
9 Falconer, D.S. (1989) in Introduction to quantitative genetics. 3rd ed., Longman, 438p.
10 Oberschmidt, O. et al (1995) in Photosynthesis: from light to biosphere (Mathis, P., ed) Vol. 3, Kluwer Academic Publishers, Dordrecht, The Netherlands

THE EFFECT OF ELEVATED [CO₂] ON PHOTOSYNTHESIS, CARBON AND NITROGEN ALLOCATION IN CHERRY LEAVES

Mauro Centritto[1], Helen Lee[2] and Paul Jarvis[2]

[1] CNR - Istituto di Biochimica ed Ecofisiologia Vegetali, Via Salaria km. 29,300 - 00016 Monterotondo Scalo (Roma), Italy
[2] Institute of Ecology & Resource Management, University of Edinburgh, Darwin Building, Mayfield road, Edinburgh, EH9 3JU, UK.

Key words: allocation, cherry, elevated [CO₂], gas exchange, Rubisco

1. Introduction

It is well established that in C_3 species there are short term increases in the photosynthetic rate (A) of plants grown in elevated [CO₂] (1). However, in the long term feedback loops may operate resulting in downward acclimation of the photosynthesis process (2). This down-acclimation in photosynthetic capacity may be the result of feedback mechanisms which arise because increased rates of CO_2 uptake disrupt the carbon-nitrogen functional balance. These feedback mechanisms can operate even when plants are supplied with free access to nutrients (3), and may arise from source-sink imbalance caused by inadequate rooting volume (4). However, a new perspective on down-acclimation of photosynthetic capacity in elevated [CO₂] has been recently put forward (5): downregulation of photosynthesis in elevated [CO₂] was found to be the result of earlier onset of the timing of natural ontogenetic decline of photosynthesis, normally associated with progressive leaf senescence. We tested this model of ontogenetic decline of photosynthesis on cherry seedlings grown for two growing seasons in elevated [CO₂]. The experiment was done at the height of the second growing when the seedlings had already developed maximum leaf area (i.e. after 115 days into the second growing season).

2. Procedure

Cherry (*Prunus avium* L.) seedlings were grown in ambient (~350 µmol mol⁻¹) or elevated (ambient + ~350 µmol mol⁻¹) CO_2 concentrations in six OTCs located inside a glasshouse at the University of Edinburgh for two growing seasons. A further set of trees was maintained without chambers as an absolute outside control. In the first growing season the seedlings were grown in 7.5 dm³ pots; in the second growing season the seedlings were transplanted before budburst into 15 dm³ pots. The seedlings were

G. Garab (ed.), Photosynthesis: Mechanisms and Effects, Vol. V, 3809–3812.

3810

fertilised once a week, in order to supply mineral nutrients at free access rates in both growing seasons, and kept constantly at soil water capacity.

To study whether ontogenetic decline of photosynthesis occurred, three different leaf positions, corresponding to three different leaf age groups (i.e. three different ontogenetic stages), along the plants were selected: top (newly-expanded leaves), middle (intermediate age leaves), and bottom leaves (oldest leaves). Gas exchange measurements were made inside the glasshouse at the University of Edinburgh, using a portable gas exchange system (ADC-LCA-3, Analytical Development Co. Ltd., Hoddesdon, Herts, UK). To enable measurements of PPFD-saturated photosynthetic rates, illumination of the leaf cuvette by natural sunlight was supplemented with artificial light. Leaf A was measured in ambient $[CO_2]$ conditions on three different leaves per leaf positions per plant, three plants per $[CO_2]$ treatment.

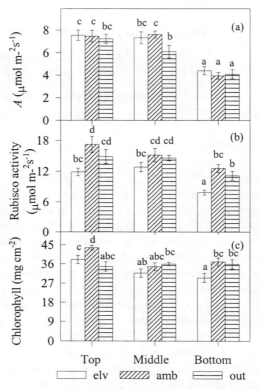

Figure 1 The effect of leaf position on a) A, b) Rubisco activity, and c) total chlorophyll in cherry seedlings grown in ambient $[CO_2]$ (amb), elevated $[CO_2]$ (elv), or outside (out). Letters (a,b,c,d) indicate significant differences at $P < 0.05$; n = 9 ± 1SEM.

Immediately after gas exchange measurements, samples were taken from each top, middle and bottom leaf (i.e., three leaves per position per plant, three plants per $[CO_2]$ treatment) to determine Rubisco activity (6) and concentration of total chlorophyll (7), starch, sugar, and nitrogen (8). Data were tested using a simple factorial ANOVA (one-way interaction) to determine the main effects of $[CO_2]$ treatment on all dependent variables. Where appropriate, the treatment means were compared using Duncan's multiple range test.

3. Result and Discussion

There were no significant differences in A measured at 350 µmol mol^{-1} of CO_2, between ambient and elevated $[CO_2]$ in any leaf position. Thus, there was no downregulation of A in response to elevated $[CO_2]$ (Figure 1a). However, Rubisco activity (Figure 1b) and total chlorophyll concentration (Figure 1c) of both top and bottom leaves (i.e. the newly-expanded and the oldest leaves, respectively) were significantly reduced in plants grown in elevated $[CO_2]$, although no differences were shown by middle leaves (i.e. intermediate age leaves).

It has been proposed that increase in mass of sugars would regulate gene expression of the photosynthetic apparatus, and hence Rubisco activity and amount (9). However, significant differences in soluble sugar concentrations between elevated and ambient $[CO_2]$ were found only in bottom leaves (Figure 2a). Starch concentration did increase ($P < 0.001$) in response to

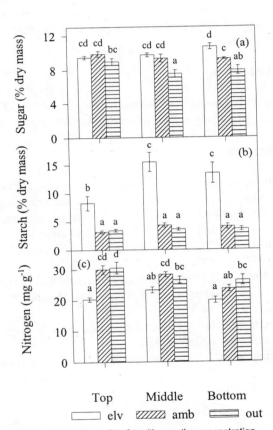

Figure 2 The effect of leaf position on the concentration of a) sugar, b) starch, and c) nitrogen in cherry seedlings grown in ambient $[CO_2]$ (amb), elevated $[CO_2]$ (elv), or outside (out). Letters (a,b,c,d) indicate significant differences at P < 0.05; n = 9 ± 1SEM.

elevated $[CO_2]$ in all leaf age groups (Figure 2b). Starch accumulation in leaves, by maintaining stroma P_i cycling, allows A to continue, and by lowering the amount of soluble sugar in the cytosol reduces the source of the regulatory signal that, according to Van Oosten and Besford (1996), may effect coarse control of the photosynthetic genes. Nitrogen concentration was also affected ($P < 0.001$) by elevated $[CO_2]$, although it was significantly reduced in top and middle leaves but not in bottom leaves (Figure 2c).

Our result do not support the ontogenetic decline of photosynthesis model (5), since A was not downragulated by elevated $[CO_2]$ in any leaf age group. Moreover, the increased leaf carbon/nitrogen ratio (i.e. equal soluble sugar concentration and lower nitrogen concentration in top leaves, and increased soluble sugar concentration and equal nitrogen concentration in bottom leaves), may account for the loss in Rubisco activity and chlorophyll concentration found in top and bottom leaves of the cherry seedlings grown in elevated $[CO_2]$.

4. Addendum

This research was done within the EU Project ECOCRAFT (Contract no. ENV4-CT95-0077) "The likely impact of rising CO_2 and temperature on European forests".

References

1 Von Caemmerer, S. and Farquhar, G. D. (1981) Planta 159, 376-387
2 Eamus, D. and Jarvis, P. (1989) Advances in Ecological Research 19, 1-55
3 Centritto, M. (1997) PhD thesis, University of Edinburgh.
4 Curtis, P.S. and Wang, X. (1998) Oecologia 113, 299-313.
5 Miller, A., Tsai, C-H., Hemphill, D., Endres, M., Rodermel, S. and Spalding, M. (1997) Plant Physiology 115, 1195-1200
6 Besford, R.T. (1984) Journal of Experimental Botany 35, 495-504
7 Porra, R.J., Thompson, W.A. and Kriedemann, P.E. (1989) Biochimica et Biophysica Acta 975, 384-394
8 Allen, S.E. (1989) Chemical Analysis of Ecological Materials. Blackwell Scientific Publications, Oxford, UK.
9 Van Oosten, J-J. and Besford, R.T. (1996) Photosynthesis Research 48, 353-365

FRUIT PHOTOSYNTHESIS

Michael M. Blanke, Institut für Obstbau und Gemüsebau
Auf dem Hügel 6, D-53121 Bonn, Germany

Key words: Chl, CO_2 uptake, CO_2 concentration, PEPC, respiration, stomata.

1. Introduction

European fruit orchards produce 10 - 50 t of apple fruit ha^{-1} $year^{-1}$, equivalent to 0.4-2 t of carbon, originating from photosynthesis in both leaf and fruit. Fruits like apple, grape, avocado, grape and orange grow on plants with C_3 photosynthesis leaves. However, recent results show that fruit photosynthesis bears little resemblance with that of their respective leaves (4).

Photosynthesis and carbon metabolism of fruit seem to differ from that of leaves in that organic acids contribute to the carbon storage pool despite starch and sugar (4). Fruit possess a system which refixes CO_2 from mitochondrial respiration to produce malic acid involving phosphoenolpyruvate carboxylase (PEPC, EC 4.1.1.31). The objective of this paper is to provide a synthesis of morphological, physiological and biochemical fruit features to categorise its photosynthesis. Fruits were hence examined for the presence and frequency of stomata, transpiration, CO_2 assimilation, intercellular CO_2 concentration and chlorophyll as well as activity and kinetics of PEPC.

2. Material and methods

Stomatal frequencies were determined using a SEM and counts on 0.25 x 0.25 mm squares. Chlorophyll was extracted by direct immersion of fresh fruit sections in 5 % (w/v) DMF and reading the OD at 647 nm and 664.5 nm after eight hours (1). Intercellular CO_2 was measured by gc and fruit respiration and transpiration were determined by fruit porometry (2,3,5).
PEPC was purified from apple and avocado fruit. They were homogenised with a pestle and mortar in a chloride-free 50 mM Tricine-NaOH buffer (3:1 w/v) pH 7.8 containing 2.5 mM $MgSO_4$, 5 mM $NaHCO_3$ and 5 mM DTT. Sand aided grinding and 5% (w/w) PVP prevented browning. After 30 mins centrifugation at 20,000 g, the supernatant was 65% saturated with ammonium sulfate, centrifuged as before and the dissolved pellet desalted on a Sephadex G-25 column. The eluate in MOPS-NaOH buffer pH 7.0 contained 2.5 mM $MgSO_4$, 5 mM $NaHCO_3$ and 5 mM DTT and was assayed for enzyme activity. All extracts were made at 4°C. The activity of PEPC was determined spectrophotometrically (6,7) at 340 nm by coupling the reaction to the oxidation of NADH in the presence of MDH. Kinetics were determined using a

G. Garab (ed.), Photosynthesis: Mechanisms and Effects, Vol. V, 3813–3816.
© *1998 Kluwer Academic Publishers. Printed in the Netherlands.*

dedicated commercial enzyme kinetics program ('Enzkin', D. Hucklesby, 4, Royal York Crescent, Clifton, Bristol, UK) and initial rates. Kinetic parameters were calculated based on Hanes' transformation of s/v against s.

3. Results and Discussion

3.1 Stomata

All fruits examined exhibited stomata which were of similar size as on their respective leaves and regulatory when the fruit were young. However, their frequency was considerably lower on fruits than on their respective leaves and declined with fruit development, whereas the number of stomata per fruit remained constant.

Table 1: Stomata on fruits compared with those of respective leaves

Species	Stomatal size [μm]	Stomatal **frequency** per mm^2	Stomata **per fruit**
Apple **fruit**	13-21 x 21-33	< 1 mm^{-2}	2,000-4,000
Avocado **fruit**	9-12 x 16-19	< 5 mm^{-2}	20,000-30,000
Grape **berry**	13-15 x 16-18	< 2 mm^{-2}	18-30 per berry
Orange **fruit**	12-16 x 19-22	26 - 64	190,000-240,000
L e a f	14-16 x 19-22	300- 400	n / a

3.2 Chlorophyll

Fruits contained less chlorophyll than their respective leaves with a closer chlorophyll a:b ratio. Fruits contained from 0.2 to 3 mg chlorophyll dm^{-2} compared with 5 mg in the respective leaves (Table 2), with the chlorophyll content of fruits declining with the onset of maturation.

Table 2: Chlorophyll content and chlorophyll a:b ratio

Species	Chlorophyll [mg dm^{-2}]	Chlorophyll a:b ratio
Apple **fruit**	0.6-1.8	1-2 :1
Avocado **fruit**	2 - 3	2.8 :1
Grape **berry**	0.2-0.6	2-3 :1
Orange **fruit**	1 - 3	1-2 :1
L e a f	5	**3-4** :1

3.3 Respiration and intercellular CO_2 concentration

Table 3: Respiration and intercellular CO_2 concentration

Species	Respiration per fruit [mg $CO_2 h^{-1}$]	Respiration per fresh matter	Intercellular CO_2 concentration [%]
Apple **fruit**	1.7-3.5	9 - 19	2 - 5 %
Avocado **fruit**	40 - 60	150 - 200	1 - 15 %
Orange **fruit**	3.5 - 4.5	30 - 50	0.6 - 2 %
L e a f	n / a	n / a	**0.034 %**

3.4 CO_2 refixation and kinetics of fruit PEPC

The developing fruit is largely heterotrophic and imports most of its carbon from the adjacent leaves mostly in the form of sorbitol and sucrose. Within the fruit, these two transport forms are converted to storage forms such as glucose, fructose and starch. Malate metabolism in fruits is coupled to glycolysis in which hexoses are metabolised to provide substrates including PEP and energy for metabolic conversions within the fruit. Energy is ultimately provided in the form of ATP in the TCA cycle.

PEPC uses HCO_3 rather than CO_2 as one and PEP as the other substrate. Kinetic analysis using 'Enzkin' showed a non-competitive relationship between these two substrates. With a low Km of ca. 0.17 for HCO_3 and of ca. 0.09 mM for PEP, apple and avocado fruit PEPC appear very efficient isoforms relative to those in CAM or C_4 photosynthesis. Regulation of PEPC activity in these fruits is by glucose-6-phosphate stimulation and end product inhibition by malic acid in a strongly pH-dependent manner. PEPC has a pH optimum at pH 7.8 and is salt-sensitive at concentrations above 160 mM salt; hence chloride-free buffer and zwitter-ionic acidic buffers were used in all extractions.

Fruit PEPC exhibited a Ki of between 10 μM at pH 6.8 and 46 **mM** L-malate at pH 8. This strong pH-dependency of the malate inhibition of PEPC enzyme activity recycles CO_2, but also fine-regulates the cytoplasmic pH in the apple fruit. An increase in cytoplasmic pH would be countered by an increase in PEPC activity to produce more malic acid and a decrease in pH would activate malic enzyme to convert malate to the less acidic pyruvate.

Fruit PEPC exhibited a large affinity for its substrates. The fruit isoenzyme is very responsive to its substrates and effectors and a very effective form of PEPC, i.e. more effcient than its CAM or C_4 counterparts. It provides fruits with an elegant fine regulation of cellular pH while recycling respiratory CO_2.

3.5 Fruit Photosynthesis Concept

Table 4: Photosynthetic features of fruits

Fruit characteristics	Photosynthetic feature
Low stomatal frequency	CAM
Low transpiration	CAM
Net CO_2 efflux	N.N.
Large intercellular CO_2 conc.	CAM
Organic acid accumulation	C_4, CAM
CO_2 concentrating mechanism	C_4, CAM
PEPC kinetics	C_3, CAM

Photosynthetic features of fruits such as apple, avocado, grape and orange differs from that of the respective leaves, which exhibit C_3 photosynthesis (Table 4). Fruits possess **fewer stomata** (Table 1) than their respective leaves, associated with **smaller transpiration rates**. Fruits also contain **less chlorophyll** with a **closer chlorophyll a:b ratio** (Table 2). Fruits accumulate **large amounts of intercellular CO_2 (1-15%)**, compared with 0.034 % in the respective leaves, resulting in net CO_2 efflux from the fruit (Table 3). Phosphoenolpyruvate carboxylase, **PEPC refixes this respiratory CO_2**. These findings, the CO_2 concentrating mechanism by PEPC and the kinetics of fruit PEPC imply the definition of a new type of 'fruit photosynthesis' rather than its categorisation within an existing type of photosynthesis.

4. Acknowledgements

I gratefully acknowledge a Visiting fellowship in Plant Physiology from Bristol university where most of the work was carried out and financial support from grants Le 281/21, Le 281/25 and Bl 263/2/3 from Deutsche Forschungsgemeinschaft (DFG).

References

1 Blanke, M.M. (1990a) Wein-Wissenschaft 45, 76-78.
2 Blanke, M.M. (1990b) Gartenbauwissenschaft 55, 282
3 Blanke, M.M. (1991) Postharvest, News and Information 2, 429-433
4 Blanke, M.M. and Lenz, F. (1989) Plant Cell & Environment 12, 31-46
5 Blanke, M.M. and Lovatt, C.J. (1993) Annals of Botany 71, 543-547
6 Blanke, M.M., Notton, B.A. and Hucklesby, D.P. (1986) Phytochemistry 25, 601-606
7 Notton, B.A. and Blanke, M.M. (1993). Phytochemistry 33, 1333-1337

THE ULTRASTRUCTURE OF THE CHLOROPLASTS IN KIWIFRUIT PULP *

Gui-Xi Wang[1] Yi-Chen Zong[1] Hong-Jie Wang[2]
Li-Song Liang[1] Ze-Pu Zhong[3] Ding-Ji Shi[3]

([1]The Research Institute of Forestry, Chinese Academy of Forestry, Beijing 100091,
[2]Capital Normal University, Beijing 100037, [3]Institute of Botany, Chinese Academy of
Sciences, Beijing 100093, China)

Key words: Actinidia chinensis Planch, chloroplast, ultrastructrue, granum, starch
granule, thylakoid membranes

Introduction

Chloroplasts as photosynthetic apparatus, usually exist in leaves, but it has been found
that a few young fruits also contain chloroplasts, such as apple[1], orange[2], avocado[3],
grape[4], pod wall of pea[5], etc. The size of the chloroplasts is smaller than those in
leaves, with lamella structrue, large starch granules and few thylakoids per granum[6].
These chloroplasts are gradually disintegrated with the fruit development. And they can
be hardly found when the fruits are in full maturity[7]. From young to ripen, the pulp
tissues of kiwifruit always contain chlorophyll[8]. The chloroplasts having primary
structure had been found in pericarp of the mature kiwifruit by Possingham, etc [9]. The
chloroplasts of kiwifruit exist in a microenvironment where there are thick pulp, light
scarcity and low pH value. It has not been found any reports about the structure and
photosynthetic activity of the kiwifruit chloroplasts. In this work the structure of
kiwifruit chloroplasts which are in pericarp, mesocarp and core have been viewed by
transmission electron microscope, and compared with kiwifruit leaves.

1. Materials and Methods

1.1 *Fruit material*
Kiwifruit (*Actinidia chinensis* Planch. Var. 36) was taken from Botanical Garden of
Institute of Botany, Chinese Academy of Sciences. The leaves with or without fruit
were also picked.

1.2 *Microscopy*
The fruits and leaves were cut into slices (approx. 1mm³). They were fixed with 2.5%
glutaraldehyde in phosphate buffer (0.1M at pH 7.4) and post-fixed with 1% osmium
tetroxide in phosphate buffer. Then the samples were dehydrated in an alcohol series

---

* This project is financed by National Nature Sciences Foundation of China

G. Garab (ed.), Photosynthesis: Mechanisms and Effects, Vol. V, 3817–3820.
© 1998 *Kluwer Academic Publishers. Printed in the Netherlands.*

and embedded in Spurr's resin. Sections were cut with ultramicrotome (LKBⅣ), stained with urangyl acetate and lead citrate. The samples were viewed in a HITACHI –600 microscope.

2. Result and Discussion

2.1 *The size and shape of kiwifruit chloroplasts*
Mature kiwifruits have a rough brown epidermis covered with brown hairs, an emerald green pericarp about 10~15 mm thick, mesocarp containing brown seeds and a central semi-fibrous core[9] (Fig 1). The electron microscopy of kiwifruit showed that the shapes of pulp chloroplasts are short ellipses and the leaf chloroplasts are long ellipse. The sizes of chloroplasts tend smaller from pericarp to mesocarp to core (Table 1). They are all smaller than that of the leaf.

Table 1 The comparison of the chloroplasts exist in pericarp, mesocarp and core of the pulp, and leaf of kiwifruit*

Chloroplast	Pericarp	mesocarp	core	Leaf with fruit	Leaf without
Length(um)	4.0±0.5	3.4±0.4	2.8±0.5	5.7±0.3	4.1±0.2
Breadth(um)	2.8±0.3	2.0±0.5	1.5±0.3	1.6±0.4	2.2±0.6
L/B	1.4	1.7	1.9	3.6	1.9
Granum length(um)	0.55± 0.10	0.67± 0.10	0.70±0.10	0.38±0.10	0.36±0.05
Stack thickness(um)	0.15± 0.10	0.13± 0.10	0.17±0.05	0.39±0.10	0.25±0.10
Cross area of a granum	0.083	0.087	0.119	0.148	0.090

*means value of 3~10 samples.

Fig 1　The character of kiwifruit(Actinidia chinensis Planch. Var.36). A: intact fruit B: cross section　C: Longitudinal section
Fig 2　The pericarp chloroplast of kiwifruit pulp.(×17,000)
Fig 3　The core chloroplast of kiwifruit.(×25,000)
Fig 4　The mesocarp chloroplast of kiwifruit pulp. (×30,000)
Fig 5　The chloroplast of leaves without fruit.(×10,000)
Fig 6　The chloroplast of leaves with fruit.(×20,000)

Gui-Xi Wang etc. — Ultrastructure of the Chloroplasts in Kiwifruit Pulp

2.2 *The structure of the kiwifruit chloroplasts*

Intact lamellae exist in the chloroplasts of pericarp, mesocarp and core. Starch granules and osmophilic lipid droplets are distributed among the lamellae. The numbers of grana and thylakoids of the pulp are much less than those of leaves. Although the thickness of grana lamellae is thinner than those of leaves, the lengths of thylakoid membranes become longer and the lengths of stroma membranes become shorter from pericarp, mesocarp to core of the pulp (See Table 1). It is suggested that the photosynthetic capacity of chloroplasts drop gradually from pericarp to core. From lamellae arrangement it can be seen that the pericarp chloroplasts consist of parallel lamella membrane structure (Fig 2), however, in the mesocarp and core, the lamella membrane is distorted (Fig 3 and 4). It is possible attributed to the light scarcity and low pH value. The chloroplasts in pericarp, mesocarp and core all contain starch granules which may be from kiwifruit photosynthesis, but the size and numbers are less than those of leaves. The numbers of starch granules in leaves with fruits are much more than those in leaves without fruits, but it are contrary in the volume (Fig 5 and 6).

After mature and storage chloroplasts kept intact lamella membranes and starch granules in the three parts of the kiwifruit pulp. This result turned out contrary to H. Clijsters' conclusion[10]. Our data shown that the situation of kiwifruit chloroplasts is not as same as in other fruits.It provides a cytology basis for further studying fruit photosynthesis of kiwifruit.

References

1 Kidd, F., West C.(1947) New Phytol.46, 274
2 Bean, R.C. and Todd, G.W.(1960) Plant Physiol.35, 425
3 Todd, G.W. etc.(1961) Plant Physiol.36, 69
4 Kriedemann, P.E.(1968) Aust.J.Agric.Res.19, 775
5 Craig, A. ATKINS, John KUO and John S. PATE(1977) Plant Physiol.60,779-786
6 Blanke, M.M. and Lenz F.(1989) Plant Cell and Environment 12, 31-46
7 Abdul, W., Ejaz R., Altaf R.R.and Rana M.I.
8 Yoko, F., Kenji, S. and Hiroatsu, M. (1985) Journal of Food Science. 50,1220-1223
9 Possingham, J.V., Coote, M. and Hawker, J.S.(1980) Ann. Bot. 45,529-533
10 Clijsters, H. (1969) Plant Mater Veget. 19, 129

INFLUENCE OF ROOT SYSTEM RESTRICTION ON PHYSIOLOGICAL CHARACTERISTICS OF FOUR PAPAYA GENOTYPES

E. Campostrini[1,2]; Yamanishi, O.K.[1,2] [1]UENF/CCTA, 28015-620, Campos-RJ, Brazil
[1,2]Current address: UnB/FAV, CP 04508, 70910-970, Brasília-DF, Brasil

Key words: Abiotic stress, environmental stress, fluorescence, net photosynthesis, quantum yield, pigments

1. Introduction

Rooting volume restriction (RVR) is a physical stress imposed on roots when plants are grown in small containers. It occurs mostly with greenhouse-grown horticultural crops. Although it has been well established that root confinement induced by RVR severely suppresses shoot growth (1, 2, 3, 4, 5), the physiological and biochemical processes involved are largely unknown (6). A probable but controversial explanation for the poor root growth is a physiological drought stress in shoots as a result of the higher hydraulic resistance of roots and shoots (7), insufficient nutrients to the shoot (8) and hormones, such as the gibberellins and cytokinins produced in roots, might be involved, as suggested by (2), ethylene (9) and abscisic acid (6). (10) suggest that imbalanced root/shoot ratios caused the development of internal water stress and the consequent reduction in stomatal aperture, culminating in leaf and whole plant senescence. Thus, the present work was carried out to study the influence of root system restriction on some physiological characteristics of four papaya (*Carica papaya* L.) genotypes.

2. Materials and Methods

Four papaya genotypes were used: three from the 'Solo' group (Sunrise Solo TJ, Sunrise Solo 72/12 and Baixinho de Santa Amália) and one from the 'Formosa' group (Know-You). The plants were grown in 6 L plastic pots under greenhouse environmental condition on a clay-sandy soil, limed, fertilized and without either water restraint or excess throughout the experiment. Two evaluation times were established for both the physiological characteristics stomatal conductance (g_s), maximum net photosynthetic rate (A_{max}), soluble leaf proteins, *Chl* **a** (Chlorophyll **a**), *Chl* b (Chlorophyll b), *Chl* **total** (Chlorophyll **total**) and *Chl* **total**:*Car* ratio, F_v/F_m ratio at 14:00 and quantum yield (ϕ). The first evaluation was at 100 days after planting (DAP) which represented a period of growth without restriction of the root system. The second was made at 200 DAP, at which time the plants were considered to have restricted root system. This information was based in morphological and physiological external aspects (weekly measure of the trunk diameter and tree height) (11). The maximum net photosynthesis and quantum yield was obtained by drawing a curve [net photosynthesis (**A**) *versus* quantum flux of photons (**Q**)]. All leaf measurements were made using the seventh leaf from the apex. The stomatal

G. Garab (ed.), Photosynthesis: Mechanisms and Effects, Vol. V, 3821–3824.

conductance and maximum net photosynthesis were determined using the LI-6200 Portable Photosynthesis System, LI-COR, USA. The concentration of leaf photosynthetic pigments was made according to (12). The total soluble leaf proteins was determined multiplying the nitrogen values obtained by 6.25. Nitrogen content was determined using sulfuric digestion and the final extract was analyzed at spectrophotometric visible region (SPEKOL, Germany). The fluorescence emission was monitored with a Plant Efficiency Analyzer (PEA, Hansatech, England).

2.1 Data Analysis

Results according to genotype and periods were statistically analyzed as a randomized design with eight treatments [genotype (four) x periods (two)] with four replications per treatment by analysis of variance with the statistical package of Statistica (Statsoft, Inc., USA). The values of **A** were adjusted according to the model:

$$A = [a + bQ + c\sqrt{(Q)} + d\exp(-Q)]$$

where:

A net photosynthesis, **Q** quantum flux of photons and **a**, **b**, **c** and **d** are coefficients.

3. Results and Discussion

All genotypes were significantly affected by RVR, but no difference was observed among genotypes. Breakdown of chlorophylls may be one of the earliest symptoms of senescence (13). The ratio between chlorophyll and carotenoids has been much less widely used diagnostically, but (12) have found this ratio to be a sensitive marker distinguishing natural full-term senescence and senescence due to environmental stress. Changes in *Chl total:Car* ratio are then potentially sensitive indicators of stress (14). The concentration of leaf photosynthetic pigments and its relations in four genotypes at two periods of are shown in evaluation (100 and 200 DAP) are shown in *Figure 1*. All papaya genotypes showed a significant reduction in photosynthetic pigments and *Chl total/Car* ratio when exposed to RVR condition (*Figure 1*). Similar results were observed by (15) in spreading euonymus and by (10) in alder seedlings for *Chl a* and *Chl total* in plants growing under root restriction condition. (16) has suggested that *Chl a* is destroyed more rapidly during leaf senescence than *Chl b*. (17) and (18) related that senescence process reduce the concentration of chlorophyll and (19) mentioned that ethylene is the principal hormone involved in this process. Many reports (9, 20) indicate that nonwounding physical stress like soil impedance also increases ethylene production, which in turn acts as an endogenous growth regulator. Thus, we supposed that ethylene production by all genotypes would be higher at 200 DAP than 100 DAP. Our hypothesis about senescence induction by RVR may be confirmed by reduced values of photosynthetic characteristics observed in papaya trees grown under RVR (*Table 1 and Figure 2*).The values of ϕ, F_v/F_m, A_{max} and g_s were significantly reduced by RVR. It led suppose that senescence process caused stomatal closure, chlorophylls breakdown, progressive loss of membrane integrity (21)

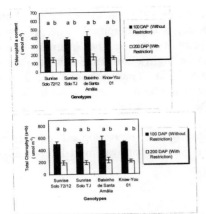

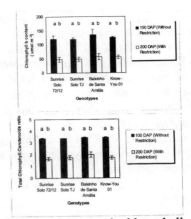

Figure 1. Chlorophyll *a* content, chlorophyll *b* content, total chlorophyll, total chlorophyll:carotenoids ratio in leaves at 100 Days After Planting (Without Restriction) and 200 Days After Planting (With Restriction) of four papaya genotypes grown in 6 L pots under greenhouse condition. Vertical bars indicate standard error (n=4). Different letters show significant difference at 5% level (Duncan's multiple range test)

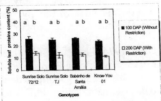

Figure 2. Soluble leaf proteins content in leaves at 100 Days After Planting (Without Restriction) and 200 Days After Planting (With Restriction) of four papaya genotypes grown in 6 L pots under greenhouse condition. Vertical bars indicate standard error (n=4). Different letters show significant difference at 5% level (Duncan's multiple range test)

Table 1. Quantum yield (ϕ), F_v/F_m ratio at 14:00, maximum net photosynthetic rate (A_{max}) and stomatal conductance (g_s) in leaves at 100 DAP and 200 DAP of four papaya genotypes grown in 6 L pots under greenhouse condition

	ϕ		F_v/F_m		A_{max}		g_s	
	(μmol CO_2 μmol fótons^{-1})		ratio		(μmol m^{-2} s^{-1})		(mol m^{-2} s^{-1})	
Genotypes	100	200	100	200	100	200	100	200
Sunrise Solo TJ	0.0349 Bb^z	0.0204 Aa^z	0.826 Aa^y	0.749 Ba^y	22.76	10.81	0.113 Aa^y	0.044 Ba^y
Sunrise Solo 72/12	0.0338 Bb	0.0195 Aa	0.825 Aa	0.718 Ba	20.45	11.10	0.093 Aa	0.043 Ba
Baixinho de S. Amália	0.0357 Bb	0.0226 Aa	0.822 Aa	0.761 Ba	23.29	12.18	0.127 Aa	0.047 Ba
Know -You 01	0.0332 Bb	0.0173 Aa	0.827 Aa	0.793 Ba	20.51	10.69	0.132 Aa	0.039 Ba

^{z} Horizontally, average followed by the same capital letters for each analyzed characteristic did not differ at level of 5% of probability using the "t" test. Vertically average followed by the same small letters for each analyzed characteristic are not significantly different at level of 5% of probability using the "t" test

^{y} Horizontally, average followed by the same capital letters for each analyzed characteristic did not differ at level of 5% of probability using the Duncan's multiple range test. Vertically, average followed by the same small letters for each analyzed characteristic did not differ at level of 5% of probability using the Duncan's multiple range test.

and protein (22), decline in photosynthetic capacity (23) and the lowering of protein content, since the CO_2-fixing enzyme, Rubisco, is a preferential target of proteases during senescence (13). The F_v/F_m is proportional to the quantum yield of photochemistry (24) and has been shown to be highly correlated with the quantum yield of net photosynthesis of intact leaves (25). Thus, reduced values of the chlorophyll fluorescence at 200 DAP show that photosynthetic capacity was very affected by RVR. Another factor that may have contributed to senescence process is the hormonal imbalance.

We concluded that RVR reduced significantly the physiological characteristics evaluated in the four papaya genotypes and, it is reduction was most likely caused by senescence induction.

References

1 Carmi, A. and Heuer B. (1981) Annals of Botany 48: 519

2 Carmi, A. Hesket, J.D. Enos, W.T. Peters, D.B. (1983) Photosynthetica 17: 240

3 Carmi, A. and Shalhevet, J. (1983) Crop Science 23: 875

4 Dubik, S.P. Krizek, D.T. Stimart, D.P. (1989) Journal of Plant Nutrition 12: 1021

5 Choi, J.H. Chung, G.C. Suh, S.R. Yu, J.A. Sung, J.H. Ju, K. (1997) Plant Cell Physiology 38: 495

6 Liu, A. and Latimer, J.G. (1995) Journal of Experimental Botany 45: 1011

7 Hameed, M.A. Reid, J.B. Rowe, R.N. (1987) Annals of Botany 59: 685

8 Rieger, M. and Marra, F. (1994) Journal of the American Society for Horticultural Science 119: 223

9 Kays, S.J. Nicklow, C.W. Simons, D.H. (1974) Plant and Soil 40: 565

10 Tschaplinski, T.J. and Blake, T.J. (1985) Physiologia Plantarum 64: 167

11 Campostrini, E. (1997) Comportamento de quatro genótipos de mamoeiro (Carica papaya L.) sob restrição mecânica ao crescimento do sistema radicular. PhD Thesis, Campos dos Goytacazes

12 Hendry, G.A.F. and Price, A.H. (1993) in Methods in comparative plant ecology (Hendry, G.A.F. Grime, J.P., eds) 148, Chapman and Hall, London

13 Penarrubia, L. and Moreno, J. (1995) in Handbook of plant and crop physiology (Pessarakli, M., ed) 461, Marcell Dekker, New York

14 Buckland, S.M. Price, A.H. Hendry, G.A.F. (1991) New Phytologist 119: 155-160

15 Dubik, S.P. Krizek, D.T. Stimart, D.P. (1990) Journal of Plant Nutrition 13: 677

16 Wolf, F.T. (1956) American Journal Botany 43: 714

17 Brown, S.B. Houghton, J.D. Hendry, G.A.F. (1991) in Chlorophyll's (Scheer H., ed) 465, CRC Press, Boca Raton

18 Hendry, G.A.F. Houghton, J.D. Brown, S.B. (1987) New Phytologist 107: 255

19 Matoo, A.K. and Suttle, J.C. (1991) The plant hormone ethylene CRC Press, Boca Raton

20 Sarquis, J.I. Jordan, W.R. Morgan, P.W. (1991) Plant Physiology 96: 1171

21 Penarrubia, L. Moreno, J. García-Martínez (1988) Physiologia Plantarum 73: 1

22 Brady, C.J. (1988) in: Senescence and aging in plants (Noodén, L.D. Leopold, A.C. eds) 147, Academic Press, London

23 Gepstein, S. (1988) in: Senescence and aging in plants (Noodén, L.D. Leopold, A.C. eds) 85, Academic Press, London

24 Kitajima, M. and Butler, W.L. (1975) Biochemistry Biophysic Acta 723: 169

25 Demmig, B. and Björkman, O. (1987) Planta 171: 171

PHOTOSYNTHETIC PIGMENT CONTENT AND ESSENTIAL OIL YIELD OF OCIMUM BASILICUM L. DURING DIFFERENT STAGES OF GROWTH IN THE FIELD.

Meena Misra
Aromatic and Medicinal Plants Division,
Regional Research Laboratory (CSIR), Bhubaneswar-751013, India.

Key words: basil oil, biomass, chlorophyll, *Ocimum*, productivity, sweet basil

1. Introduction

Ocimum basilicum L. commonly known as sweet basil or french basil is one of the members of family Lamiacea. Plants belonging to this family have natural oil of commercial importance. Sweet basil oil is used in perfume, cosmetic and flavour industry (1). The oil is conventionally extracted by steam distillation from leaves and flowering tops. Basil oil has sweet test and has a typical aniside like odour due the presence of methyle chavicol.

Sweet basil is a native to India and it has been naturalised in different parts of the world. This is commercially cultivated in France, Italy, Bulgaria, Hungary, Egypt, South Africa and USA. Average global production of sweet basil oil is 2 tonnes per annum. Because of its maximum production in Europe, sweet basil oil is classified under European or French basil category. Commercially this is also grown in Australia as a replacement crop in traditional cash cropping areas (2).

Agronomic practices and climatic factors (3) as well as on the phenological stages of the crop (4) influence the quality of basil oil, which is correlated with its composition. The commercial viability of this crop depends on the appropriate phenological stage at which this crop should be harvested. So in the present study, sweet basil oil yield at different growth stages is investigated.

G. Garab (ed.), Photosynthesis: Mechanisms and Effects, Vol. V, 3825–3828.
© 1998 *Kluwer Academic Publishers. Printed in the Netherlands.*

2. Materials and methods

Ocimum basilicum L. seeds were collected from a commercial farm at Aligarh, UP and shown in the raised nursery beds during rain season (July). Seedlings were transplanted after 2 weeks and plants were grown for two months for establishment. Plants with uniform growth were selected for experiment. The experiment was conducted in a RBD with 10 replicates. Growth and oil yield measurements were done at weekly intervals. Photosynthetic pigment content was measured as described by Misra (5,6). Leaves and flowers were collected and shade dried. Ocimum oil was extracted by steam distillation of these shade dried materials.

3. Results and Discussion

Sweet basil is a moncarpic crop. Flowering of the crop leads to leaf yellowing and shading in the field condition. Fig.1 shows the changes in plant height, number of brances and fresh weight of total herbage of the plant at different growth periods. Growth was slow till 52 days and from 52-70 days ocimum plants grew rapidly, followed by a slow phase of growth. The later phase is accompanied with the yellowing leaves and flowcring of plants. This is characterised by maturation and senescence phase of the plants. Number of branches increased from 7 at 48 day to 18 at 84 days. Branching increased as the vegetative stage progressed. However, branching ceased at post anthesis period in *Ocimum* and so the contribution of branching to total herbage yield at this stage is minimum.

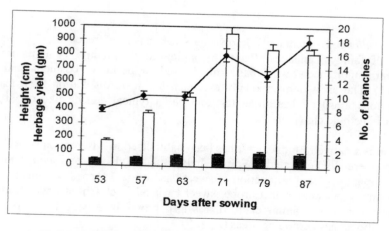

Figure 1. Changes in plant height (hatched histograms), number of branches and herbage yield (gm per plant) of sweet basil. The mean values of five different harvests are shown. The vertical lines indicate standard error.

The changes in photosynthetic pigment content of a plant characterises its developmental phase. During plant development, photosynthetic pigment content increased and at the onset of aging, pigment contents start to decrease (7). Photosynthetic pigment content of the plant increased gradually till 96 days and started to decline there after (Fig.2). So in the present study, growth of *Ocimum* took place till 94 days and senescence occurs later.

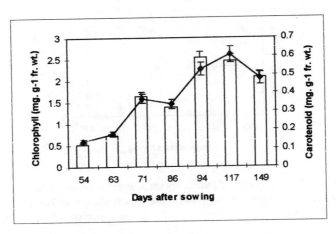

Figure 2. Changes in chlorophyll (graph) and carotenoid (histograms) content of sweet basil leaves. The mean values of five different harvests are shown. The vertical lines indicate standard error.

However, growth in height and increase in fresh herbage yield ceases much before this period. Misra (5,6) suggested that an increase in photosynthetic pigment content reflects the increased photosynthetic ability of patchouli plants. However, there could be a translocation barrier and/or increase in the translocation of photosynthates to the developing sink in the floral parts after the anthesis stage, resulting in a gradual decrease in the growth of *Ocimum* plants (8). An increase in the photosynthetic efficiency in *Ocimum* suggests the potential for increase in production even after anthesis stage.

Changes in the *Ocimum* oil yield is shown in Fig.3. Basil oil yield increased with an increase in plant growth. Component analysis showed that leaf is the principal contributor for oil yield, with an average yield of about 0.9% yield at different growth stages. However, the oil yield from floral parts increased gradually and reached maximum 0.55% at 87 days (Fig. 3). Although, this is a minor contributor for total yield of *Ocimum* oil, reports are available that sweet basil oil from the floral parts have better quality compared to that from the leaves. However, looking at the labour intensive nature of *Ocimum* cropping and harvesting, major contribution to *Ocimum* oil yield is done by the foliage, as fresh leaf and herbage yield is more than 800gram per plant at the post-

3828

anthesis stage. From this study it is concluded that increasing the foliage yield, ocimum oil yield can be improved.

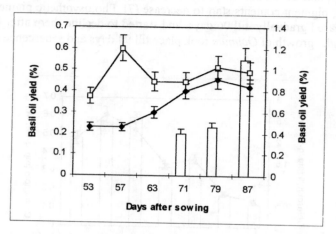

Figure 3. Basil oil yield of leaves (open squares), total herbage (closed blocks) and flowers (histograms). The mean values of five different harvests are shown. The vertical lines indicate standard error.

Acknowledgements

The authors are grateful to Prof. H.S.Ray, Director, RRL for providing necessary facilities and Dr. H.O.Saxena, Head, AMPD for encouragement. MM is recepient of a Research Associateship of CSIR and Scientist (Fellow) of RRL during this study.

References

1 Guenther, E. (1952) The Essential Oils. p.403, Van Nostrand, New York.
2 Lachowicz, K.J., Jones, G.P., Briggs, D.R., Brienvenu, F.E., Palmer, M.V., Ting, S.Y. and Hunter, M. (1996) J. Agric. Food Chem. 44, 877-881.
3 Randhawa, G.S., Gill, B.S. and Saini, S.S. (1994) Indian Perfumer 38, 123-128.
4 Gupta, S.C. (1996) Indian Perfumer 40, 17-22.
5 Misra, M. (1995 a) Biol. Plant. 37, 219-223.
6 Misra, M. (1995 b) Biol. Plant. 37, 635-639.
7 Misra, A.N. and Biswal, U.C. (1980) Protoplasma 105, 1-8.
8 Misra, A.N. (1994) Acta Physiol. Plant. 17, 41-46.

IMPORTANCE OF PHOTOSYNTHESIS OF VARIOUS ORGANS IN PROTEIN SYNTHESIS IN GRAIN OF WHEAT GENOTYPES UNDER WATER STRESS

Jalal ALIEV
Institute of Botany, Patamdar Shosse 40, 370073 Baku, AZERBAIJAN

Key words: storage proteins, CO_2 assimilation, enzyme, photorespiration, electron transport, source-sink relation

1. Introduction

The investigation of the reaction of the organism to water supply have a great interest due to the fact that a moisture accessibility defines the possibility and the bounds of cultivation of agricultural crops and the areal of distribution of species. In this respect, the study of the nature of water stress, the mechanism of its influence on organism, the search of ways and means of safeguard from this unfavorable environmental factor and the increase of plant resistivity are very pressing problems presenting wide field for scientific investigation. There are known in nature species, forms, and genotypes with high tolerance to water stress with good crop capacity. At the same time in many arid zones of the world the qualitative value of agricultural crops undergoes the danger of destructive influence of water deficit (1,2).

2. Procedure

There were investigated ancient evolutionary formed forms – aboriginal wheat, widely cultivated well known varieties and drought-resistant genotypes (**Pic. 1**). The samples were chosen out of the genotypes numbering tens of thousands, including received from ICARDA, CIMMYT and other regional centers of the world. Contribution of photosynthesis of ear and other organs in protein synthesis of grain was defined by means of blocking photosynthesis of these organs and by exposition of $^{14}CO_2$ in atmosphere of all plants in sowing (3).

3. Results and Discussion

The percentage content of storage proteins in the grain of all genotypes appropriately increases in average by 5% of absolute value in water stress. At the same time total protein content decreases in all genotypes variously. On the whole the protein content loss fluctuates within 30-45% of their total content in *Triticum durum L*. genotypes differing with their great tolerance to water stress than those of *Triticum aestivum L*. which protein loss reaches up to 70% (**Fig. 1**).

In the given direction the results of combined research, starting from primary photochemical processes in an ear and including photosynthetic carbon metabolism, activity of key enzymes fixing CO_2, interconnected with the intensity of photosynthesis on an ear and its elements (awn, glume, caryopsis) are of interest. The important role of

3829

G. Garab (ed.), Photosynthesis: Mechanisms and Effects, Vol. V, 3829–3832.
© 1998 *Kluwer Academic Publishers. Printed in the Netherlands.*

3830

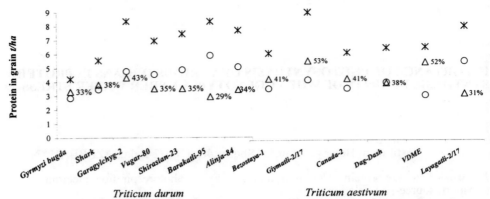

Triticum durum   *Triticum aestivum*

Figure 1. Protein accumulation in the grain of wheat varieties grown in normal (✻) and deficit (○) water supply. Losses under drought (△) is also given. Average for 1989-97.

this organ in general photosynthetic activity of the whole plant is defined by the fact, that photosynthetic apparatus of an ear has powerful system of transformation energy and restoration potentials, necessary for photosynthetic assimilation of CO_2 and formation of organic substances.

Ascertaining the contribution of an ear in photosynthetic function of plant, it is necessary to notice, that synthesized products of photosynthesis already at an early stage of reproductive period are used more effectively in formation of grain.

On the other hand the process of intensified protein synthesis by ears of high-productive genotypes is realized synchronously with intensive photosynthetic assimilation of CO_2. In compact sowings of this genotypes with optimum assimilating surface and sufficient donor ability the value of photosynthesis intensity is always high. Together with high photosynthetic activity and attractive force of ear it constitutes the basis of high crop. Supplementing data on ear yield and its protein content by results of intensity of CO_2 assimilation by leaves we obtain a more clear picture of work of this organ.

The activation of attractive forces blocking an ear appreciable stimulates intensity of carbon dioxide assimilation with marked differences among genotypes. Non-photosynthesized acceptor with intensified attractive force creates condition for increase of intensity of CO_2 assimilation by leaves to 25-35 $mgCO_2/dm^2hour$. At the same time if an ear doesn't photosynthesize, then we observed the changes in photosynthetic metabolism of carbon and the reduction of quantity of assimilators in an organ. Especially these variations are considerably displayed in high productive forms with high intensity of CO_2 assimilation, with increased activity of key enzymes of photosynthesis as well as distinguishing by great speed of biosynthesis and total value of glycine + serine fund. It is necessary to pay attention on the fact, that the certain share of protein in non-photosynthesized ears can be synthesized at the expense of increase of intensity of CO_2 assimilation by leaves about 10-15%, due to great attractive force of blocked ear.

It is interesting that of non-protein nitrogen increases more than twice in grain as well as in the straw of non-photosynthesized ears with prevalence increase in low productive genotypes. Therefore, it is possible to speak about a great amount of nitrogen which didn't used for ear synthesis. By other words the low protein percentage is connected not with the lack of nitrogen content, but with shortage of photosynthesis products (carbohydrates) necessary for protein synthesis. Rather low content of free nitrogen in grain and in the straw of non-photosynthesized ears of high productive varieties ascertaining significant increase of intensity of CO_2 assimilation by leaves at the expense of high attractive forces of non-photosynthesizing ear is also another proof to given conclusion. The presence of sufficient amount of free nitrogen both in a grain and in other parts of an ear, deprived carbon dioxide assimilation on the one hand, and reduction of crop of all elements and protein content in a grain on the other hand – is obvious proof on a photosynthetic role of this organ in formation of protein in a grain. An established situation in a complex with considering parameters of photosynthetic activity allows to assert that 30-45% of grain protein of various wheat genotype is synthesized at the expense of photosynthesis of an ear.

The results of the study of the state of source-sink relations between assimilating and consuming organs of a plant also testify a great contribution of ear photosynthesis in synthesis of protein. At each of these genotype pairs the contribution of organs in total assimilation of $^{14}CO_2$ has close values (in percentages) and can be characterized by its average value.

The analysis of results shows that both in intensive and in extensive varieties in tillering stage, in providing an ear with assimilators the basic share is made by two upper leaves (the eight and the seventh) and by an ear. At the same time in extensive varieties the seventh and the sixth leaves with sheaths as well as stalks being above and below the flag leaf also gives an essential contribution similar to an ear (**Table 1**). While passing to

Table 1. Contribution of separate organs in production process of intensive (1) and extensive (2) wheat genotypes on account of current photosynthesis. *In % of total $^{14}CO_2$ assimilation.*

Organs	Stem		Sheath of the 7th and 6th leave		Sheath of the 8th leaf		The 7th leaf		The 8th leaf		Ear	
Stages	1	2	1	2	1	2	1	2	1	2	1	2
Earing	5,6	8,3	5,1	8,85	5,6	3,25	18,7	18,75	29,2	26,6	31,8	16,5
Flowering	5,15	7,7	3,85	3,85	6,8	8,85	15,7	10,25	23,2	25,25	37,5	60,1
Milky ripeness	5,25	9,76	1,31	2,25	8,6	11,05	3,85	7,35	18,1	19,8	60,1	33,9

following stage of vegetation (flowering, milk stage) contribution of flag leaf in provision with assimilators filling grain remains on enough high level at the general tendency to reduction. All other organs of plant, with the exception of sheath of the eight leaf, the stalk above flag and an ear regularly reduce the intensity of photosynthesis and accumulation of assimilators. The share of the sheath of the eighth leaf in general fund of assimilators increases more than 3 times and the share of stalk above flag leaf – 1.5 time, that is especially typical for intensive varieties. As to ear, in intensive genotypes its contribution in production process starting from heading stage is the highest and by the end of vegetation constitutes 60%. Extensive genotypes, in comparison with intensive ones has almost twice less share of an ear in total $^{14}CO_2$ assimilation.

Thus while in extensive genotypes in initial stage of development each of vegetative organs constitutes a greater contribution in formation of grain at the expense of current photosynthesis, in intensive ones preponderance is found in an ear (31.8% against 16.2% in extensive) and in two upper leaves. In subsequent development (flowering and milk stages) with appreciable reduction of the share of participation of leaves and the stalk below flag leaf in intensive varieties, the share of ear and stalk is going on to increase during all vegetation in intensive ones.

The latter, apparently, has a greater CO_2 assimilation ability of an ear, which by that is explained an essential contribution of an ear in providing assimilators for own needs at the expense of photosynthesis. At the reutilization the powerful acceptor ability of ear significantly increases upon blocking photosynthesis of the ear.

3832

As a result, leaves and other photosynthesizing organs are quickly released from assimilators, and due to that intensifies the photosynthesis of these organs. While blocking the ear the formation of grain is completed by accumulation of protein in it up to 45% protein. Other action occurs in photosynthesis of intact plants in atmosphere of $^{14}CO_2$, when all organs function normally in natural condition and is revealed true potential contribution of an ear and its elements in formation of grain and synthesis of protein in it (up to 60%). Some advantages of intensive genotypes, such as great assimilating ability of an ear in current photosynthesis and the best acceptor activity in utilization and reutilization of reserve products of photosynthesis are decisive in creation of both high productive and tolerant to water stress genotypes. It follows from the fact, that when leaves lose photosynthetic function in drought condition, the ear is the basic contributor to photosynthesis in period of ear formation and grain-filling. In tolerant genotypes $^{14}CO_2$-assimilators are transported more vigorously and carbon dioxide assimilation by ears is 2 times intensive. For this reason there is significant protein crop in a grain of high productive intensive genotypes, with high photosynthetic function of an ear in extreme water supply condition. And that confirms a decisive role of an ear under water stress.

A certain role in adaptation to stress play C_4-cycle enzymes (PEP-carboxylase, NAD- and NADP-malatedehydrogenase, Aspartateaminotransferase) in C_3-plants. In genotypes tolerant to water insufficiency enzyme activity shifts toward to C_4-plants. In these conditions the activity of C_4-cycle enzymes in leaves and in the ear elements considerably increases – appears some activity of piruvate-ortophosphate dikinase enzyme too, whereas it is absent in normal water supply condition.

Pic. 1. Genotypes under drought

By comparative investigations of photosynthetic metabolism of carbon it was established that the values of relation of net photosynthesis to photorespiration in leaves and in the elements of ear in genotypes with various rate of CO_2 assimilation are close and is equal on the average 3:1 with the increase of given relations in intensive genotypes. Thus it is possible to speak about parallel increase of rate, of net photosynthesis and photorespiration in ontogenesis of leaves. Therefore the conception of photorespiration wastefulness aspiration to reduce its intensity with the purpose of increase of productivity is insolvent. The rate of photorespiration is in some degree in inverse correlation with water supply. By increasing tolerance of genotypes to water stress or with strengthening drought it takes place a fall of the rate of photorespiration in a great degree in the elements of ear. Investigations of primary photosynthesis processes allowed to specify that chloroplasts of high productive genotypes are characterized by high rate of electron transport and photophosphorylation, by great value of ΔpH in thylakoid membrane, and also to approve the presence of dependence among photosynthetic electron transport, CO_2 assimilation and productivity (4).

References

1 Massacci, A. et al., (1995) Inter Drought-95 in Proceedings of Int. Cong. of Integrated Studies on Drought Tolerance of Higher Plants. Montpellier, France. P. VIII A
2 Cornic, G. And Massacci, A. (1996) in Advances in Photosynthesis (Baker N.R., ed.), Vol. 5 pp. 347-366, Kluwer Academic Publishers, Dordrecht, The Netherlands
3 Voznesensky, V.L. (1977) Fotosintez pustynnyx rasteniy, L. Nauka, 256 p.
4 Aliev, J.A. (1996) Progress in Biophysics & Molecular Biology, Vol. 65, S. 1, p.154

PHOTOSYNTHETIC CHARACTERISTICS OF ROLLING LEAF WHEAT LINES IN RESPONSE TO DROUGHT STRESS

H. G. Jones[1], O. I. Kershanskaya[1,2,] E. D. Bogdanova[2]

Department of Biological Sciences, University of Dundee, Dundee, DD1 4HN, UK
Institute of Plant Physiology, Genetics and Bioengineering, Ministry Sciences - Academy Sciences of Kazakstan, 45 Timiryazev Str., Almaty, 480090, Kazakstan

Key words: rolling leaf genes, genotypes, abiotic stress, gas exchange, fluorescence, crop improvement

1. Introduction

Drought is the most important factor limiting photosynthesis and crop productivity world-wide (1). The problem of drought resistance is very important for wheat production in Kazakstan, where spring drought very often suppresses wheat production through effects on the main physiological phases: vegetative growth, earing, flowering and grain forming . An important objective of the wheat crop improvement programme is to use physiological information (2, 3) on the genetic basis of differences in reaction to stress to improve production under drought conditions. A possible character that may be important in drought tolerance is the capacity of leaves to roll under stress. Specific genes relating to this have been identified in wheat (RL_1 and RL_2) and they have been located on chromosomes 6A and 4D, respectively (4). It is not clear, however, whether possession of these genes only acts through water conservation, or whether they have consequences for photosynthetic performance. This paper attempts to address this question.

2. Procedure

2.1 Material
Five new Kazakstan lines of spring wheat containing different expression of the RL_1 and RL_2 (rolling leaf) genes: [1] Grecum 476 (the donor of RL_1 and RL_2); [2] BC6 (Omskaya 9 x Grecum 476) x Omskaya 9; [3] Omskaya 9; [4] KG-1 (a monogene dihaploid: mono-6A Kazakstanskaya 126 x Grecum 476) [4]; [5] a disomic F_2 and BC3 (Omskaya 9 x Grecum 476) x Omskaya 9 (crop breeder E.D. Bogdanova; 4). Line [3], the drought resistant but non-rolling cultivar Omskaya 9 was used as a reference.

2.2 Methods
The photosynthetic responses of these five lines to drought were evaluated in a pot experiment in a glasshouse at Dundee University, UK. Two cycles of water deficit were imposed by witholding water during the vegetative and grain - forming stages. Measurements were made of soil moisture content (Theta probe, Delta-T Devices, UK), leaf water potential (pressure chamber), leaf gas-exchange (CIRAS-1, PP-Systems, UK)

3833

and chlorophyll fluorescence (FMS-1, Hansatech, UK). In addition, total photosynthetic area and biomass per main stem, per plant and per pot were estimated. Gas exchange measurements of net CO_2 exchange (P) and stomatal conductance (G) were made under two conditions: 'real' (R; natural PAR and 350 vpm CO_2), and 'potential' (P; 1500 mol m^{-2} s^{-1} PAR and 650 vpm CO_2). The efficiency of PS II was estimated from F_v/F_m, measured after a dark preincubation of at least 20 minutes.

3. Results and Discussion

3.1 *Soil moisture content*
The soil moisture contents achieved by the two sequential drying periods are shown for the different genotypes in Figure 1. In general the control plants were maintained near field capacity which was c. 50% soil moisture by volume for the peat used in this experiment, while the stressed treatments dried the soil down to approximately 10%. There were significant differences between the genotypes in the amount of soil drying, with the reference line [3] drying the soil the most in the first drought period.

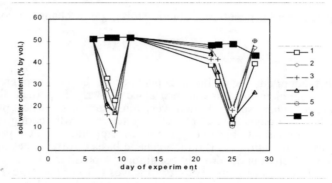

Figure 1. Sequence of soil moisture content (% by volume) for the different genotypes (numbered 1 to 5) in the two stress treatments (open symbols). The average moisture content of the control treatments is indicated by the closed symbols.

3.2 *Photosynthetic area*
It is apparent from the data in Figure 2 that the differences in soil drying by the various genotypes were not related to differences in leaf area. The corresponding leaf area per pot of the control (continuously watered plants) at the end of the second drought period was largest in cultivars [1] and [2], followed by [4] and [5] with Omskaya [3], having the smallest leaf area at this stage.

3.3 *Photosynthesis*
As expected, the water deficits depressed both photosynthesis (whether measured under natural conditions or at high light and with high CO_2) and stomatal conductance (Figure 3). There was no clear evidence from either the first or second stress periods that photosynthetic rate was affected independently of stomatal conductance, thus providing no evidence for any differential non-stomatal control of the assimilation rate. Similarly, there was little difference in the response of stomata or photosynthesis to drought whether measured under natural conditions or at high light and CO_2. There was, however, a clear indication that the different genotypes responded differently to the first

and the second drought periods, with [1] being least sensitive in the first stress period and most sensitive in the second, while the reference line [3] showed the opposite response, being most sensitive to drought early and less sensitive later. These differences were not related to differences in soil moisture content attained by these two genotypes. In contrast to the results during the two stress periods, there was evidence from data collected two days after the final rewatering of all plants, that photosynthesis continued to be inhibited in genotype [4] even after the stomata had reopened nearly to the control values (data not shown).

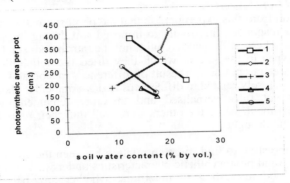

Fig. 2. The relationship between photosynthetic area per pot and soil moisture content at the peak of each drought cycle. The smaller leaf area of each pair is the leaf area at the peak of the first drought period.

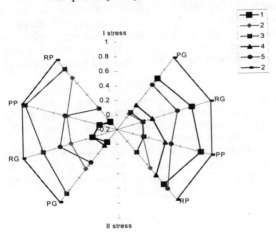

Figure 3. Summary of the percentage inhibition of photosynthesis and stomatal conductance by water stress in the first (Right hand half of figure) and second (Left hand half of figure) drought periods. RP and RG refer to 'real' estimates of net assimilation and stomatal conductance, respectively, and PP and PG are the corresponding 'potential' values.

As might be expected from previous results (e.g. [5]), the chlorophyll fluorescence data indicated that the efficiency of PSII was relatively little affected by the stress treatment imposed in this experiment. Detailed analysis of the quenching curves following illumination of the leaves (data not shown), however, showed some differences in the development of non-photochemical quenching (qNP). Interestingly, however, the qNP increase in response to stress was greatest in [3] in the first stress period and greatest in [1], [2] and [4] in the second stress.

3.4 *Discussion and Conclusions*

The main conclusion from this experiment is that the photosynthetic performance of the different genotypes responded differently to imposed soil drying in the two successive cycles of drought. These differences could not be attributed solely to differences between the genotypes in the rate at which they dried the soil (either as a result of differences in leaf area per pot or as a result of differences in stomatal conductance) and were probably more closely related to differences in developmental phases. For example genotype [1] was not fully vernalised and therefore remained mostly vegetative throughout the experiment, while the others were well into their reproductive phase by the time of the second drought.

It would be very interesting to find some correlation between the content of rolling leaf genes (RL_1 and RL_2) and photosynthetic characteristics under water stress conditions. In particular it would be interesting to determine whether any benefits for drought tolerance are related solely to their effects on water conservation, or whether they have any more direct effect on photosynthesis. The small differences between photosynthesis of the different genotypes when grown with ample water in this experiment suggests that the any effects are most likely to be related to water conservation. It is possible that leaf rolling might have a greater effect on photosynthetic assimilation than it does on water loss, hence actually reducing the potential water use efficiency of these lines. This is because leaf rolling would be expected to reduce light absorbtion by the photosynthetic system rather more than it is likely to reduce transpiration (because transpiration is dependent on both light absorption and on the vapour pressure deficit of the ambient air).

It is, of course, important to recognize that responses of leaf area development to drought are even more important than differences in photosynthetic or stomatal characters in determining the drought tolerance of crops. Though leaf area development is best analyzed in field experiments it is worth noting that in this experiment, genotypes [1], [2] and [3] tended to have greatest stability of photosynthetic area in response to drought. Genotypes [4] and [5] were relatively unstable in terms of photosynthetic area and biomass development. These latter genotypes have a smaller expression of RL_1 and RL_2 genes than do [2] or [1].

References

1 Jones, H.G. (1995) . J. Exp. Bot. 46, 1415-1422
2 Sidorenko, O.I. and Gaziyanz, S.M. (1994) Proc. Kaz. Ac. Sci. 4, 9-18.
3 Kershanskaya, O.I. (1998) J. Exp. Bot. 49, 29-31
4 Omarova, E.I. and Bogdanova, E.D. (1996) R. J. Plant Phys. 6, 817-822.
5 Massacci, A. and Jones, H.G. (1990) Trees 4, 1-8.

HEAT STRESS ON RICE LEAVES PROBED BY THE FLUORESCENCE TRANSIENT OKJIP

[1]B. Vani, [1]P. Mohanty, [2]P. Eggenberg, [2]B.J. Strasser and [2]R.J. Strasser
[1]Dept. Photobiochemistry and Bioenergetics, JNU, Delhi-110067, India
[2]Bioenergetics Lab., Geneva Univiversity, CH-1254 Jussy, Switzerland

Key words: Chl fluorescence induction, crop improvement, heat stress, OKJIP

1. Introduction

Our goal is to find a practical way to characterize rice plants in respect to their behavior upon different heat treatments. Screening of cultivars is planed (1). This way we hope to be able to improve the selection of seeds for a certain agricultural area and to improve the prediction of crop yield of cultures under heat stress by fast fluorescence techniques.

2. Procedure

Rice leaves were heat stressed either at 40°C for 3 h (Fig. 1) or heat stress from 25°C to 50°C for 10 hours (Fig 2) and were analyzed at tip, middle and base of the leaf. The polyphasic fluorescence rise (OKJIP) was deconvoluted according to the JIP-test (2-5) which allows the determination of specific (per active reaction center II) and phenomenological (per active cross section) fluxes such as light absorption (ABS), trapping (TR), electron transport beyond Q_A^- (ET). Flux ratios (yields) for primary photochemistry (Φ_{Po}) energy conversion of Trapping to Electron Transport (Ψ_0) and energy trapped per absorption (Φ_{Eo}). Average antenna size, ABS/RC, and density of RCs per cross section (RC/CS) are estimated as well. Fluorescence induction was measured by using a shutterless fluorometer (Plant Efficiency Analyzer, PEA, built by Hansatech Instruments Ltd. King's Lynn, Norfolk, PE30 4NE, GB). Figure. 1 (see legend on next page).

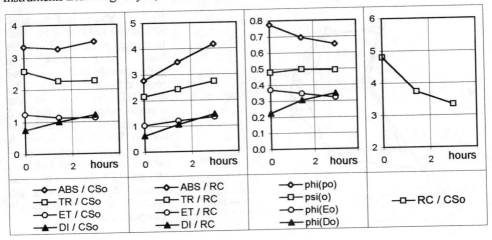

3837

G. Garab (ed.), *Photosynthesis: Mechanisms and Effects, Vol. V,* 3837–3840.
© 1998 *Kluwer Academic Publishers. Printed in the Netherlands.*

3. Results and Discussion

Figure 1 shows the effect of 3 hours of mild heat treatment at 40°C. The phenomenological fluxes (per active cross section, CS) change very little, while all specific activities (per reaction center, RC) increase due to the fact that some active RCs are converted into heat sinks. This is seen as a decrease in the density of active RCs per leaf cross section (RC/CS).

Figure 2 shows the temperature dependence of heat after 10 hours. The fluorescence rise at the origin, dV/dt_0 increases due to the appearance of the K-step at 300 μs in the polyphasic fluorescence rise, OKJIP, measured as $V_K / V_J = (F_{300 μs} - F_{50μs}) / (F_{2ms} - F_{50μs})$. Once the K-step appears the ET/CS decreases sharply due to the inactivation of the oxygen evolving system (6-7) in parallel to the decrease of the yield for primary photochemistry $\Phi_{Po} = TR_0/ABS$.

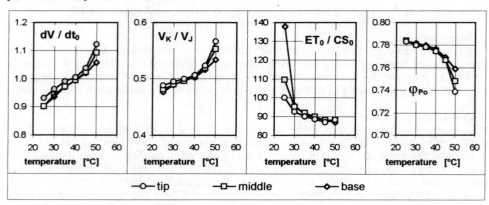

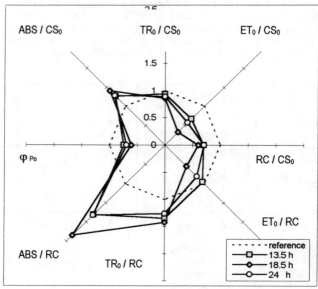

Figure 3 shows in a so called SPIDER-PLOT a long heat treatment from 13 to 24 hours in the dark. The phenomenological and specific fluxes for ABS, TR_0, and ET_0 as well as the corresponding yield Φ_{Po}, and the density of reaction centers RS/CS are shown. A typical acclimation phenomenon can be seen between 18 and 24 hours of heat stress. Apart of the strain reached at 18 hours is reversed at 24 hours.

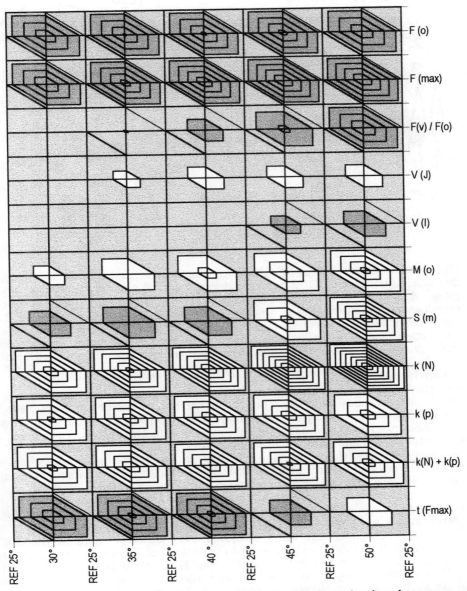

Labels (vertical axis, top to bottom): F (o), F (max), F(v) / F(o), V (J), V (I), M (o), S (m), k (N), k (p), k(N) + k(p), t (Fmax)

Labels (horizontal axis): REF 25°, 30°, REF 25°, 35°, REF 25°, 40°, REF 25°, 45°, REF 25°, 50°, REF 25°

Figure 4 shows a so called CARPET-PLOT. This plot has been developed to compare many experimental parameters (vertical axis; here original experimental fluorescence data e.g., F_0, F_M etc. and derived expressions) versus environmental parameters (horizontal axis; temperature, or soil composition, or cultivars etc.). The contour lines are spaced in this graph with an equidistance of +5% (white areas) or -5% shaded area. Carpet plots allow to detect the optimal combinations and constellations for future experiments.

3840

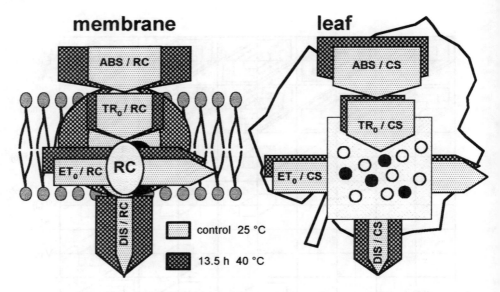

Fig. 5 shows the two PIPELINE MODELS which can be drawn of each experimental fluorescence transient O(K)JIP measured during 1 second only. The <u>membrane pipeline model</u> shows the specific (per RC) values and the <u>leaf pipeline model</u> shows the phenomenological fluxes after 13.5 hours of heat treatment at 40°C compared to 25°C. The width of each arrow indicates the value of the flux and the size of the oval indicates the average antenna size. The small circles correspond to the active reaction centers (open circles), and the heat sinks or inactive reaction center complexes (closed circles).

4. Conclusions

Rice plants create heat sink centers to cope with heat stress. We concludes that the big number of independent expressions one can derive from the polyphasic fluorescence rise using the JIP-test as an analytical procedure, allows to present the data in many different ways. Each plot has its advantages for a given set of data. Different plots may clarify our understanding for the dynamic behavior of our samples. The visualization of the Structure-Function-Relationship is a challenge in quantitative plant physiology and ecology. (Supported by Gov. of India, DST B6.9 to PM and Swiss National Foundation and Sociéte Academic de Genève to RJS).

References.

1 Srivastava, A. and Strasser, R.J. (1996) J. Plant Physiol. 148, 445-455

2 Strasser, R.J., Srivastava, A. and Govindjee (1995) Photochem. Photobiol. 61, 32-42

3 Strasser, R.J., Eggenberg, P. and Strasser, B.J. (1996) Bull. Soc. Royal Sci. Liège 65, 330-349

4 Strasser, R.J., Tsimilli-Michael, M. and Greppin, H. (1997) in Travelling Shot on Plant Development. (Greppin, H, Penel, C. and Minon, P., eds.) pp. 99-129, Rochat-Baumann, Imprimerie Nationale, Genève

5 Strasser, B.J. and Strasser, R.J. (1995) in Photosynthesis: From Light to Biosphere (Methis, P., ed.) pp. 977-980, Kluwer Academic Publisher, The Netherlands

6 Srivastava, A. Guissé, B. and Strasser, R.J. (1997) Biochim. Biophys. Acta 1320, 95-106

7 Strasser, B. (1997) Photosynth Res. 52, 147-155

DIFFERENTIAL SENSITIVITIES OF PHOTOFUNCTIONS OF THYLAKOID MEMBRANES TO INVIVO ELEVATED TEMPERATURE STRESS IN ORYZA SATIVA (Var. Kalinga III) SEEDLINGS

B.Vani[1], P. Pardha Saradhi[1] and Prasanna Mohanty[2]
1 Centre for Biosciences, Jamia Millia Islamia, New Delhi-110025.
2 School of Life Sciences, Jawaharlal Nehru University, New Delhi-110067.

Key Words : Chl fluorescence induction, D1 protein, D2 protein, electron transport, energy transfer, heat stress.

1. Introduction

Heat treatment (HT) of leaves, isolated chloroplasts or intact plants, causes inhibition of electron transport activities of both PSII and PSI (1). The order of thermal sensitivity of various photosynthetic light reactions is oxygen evolution, photochemical activity of PSII, Photophosphorylation, Cyt b_6/f complex and photochemical activity of PSI (2). *In vitro* studies with isolated thylakoids suggest that loss of extrinsic protein namely 33 KDa Mn-stabilizing protein at the donor side of PSII and subsequent loss of Mn are the primary causes of inhibition of electron transport by heat stress (3). However, unlike the donor side, very few reports exist on the heat induced changes in the PSII acceptor side. Further, Michalski and Wettern,(4) observed that incubation of algal cells at 41.5 °C for 90 min resulted in membrane bound break down products of D1-Polypeptide of 27 and 25 KDa. They suggest that heat shock might have led to these intermediates as a result of a heat-dependent inactivation of a protesase that degrades damaged D1. They also reported that, D1 digestion during heat shock incubation in darkness resulted in an additional membrane bound intermediate of 15 KDa. Heat shock incubation in the presence of light did not induce any degradation products of D2 while treatment in dark led to the formation of two degradation products of 26 KDa. These results prompted us to study the degradation pattern of D1 and D2 upon *in vivo* HT in higher plants and also Q_A to Q_B electron transfer to ascertain changes if any. Most of the studies relating to heat stress deal with *in vitro* experiments using isolated chloroplasts or PSII/PSI sub-membrane fractions. Therefore, it was felt necessary to study stress response of PSII in *in vivo*.

2. Procedure

Paddy seeds were grown on mineral nutrient medium as given in ref (5). Seven day old rice seedlings were subjected to 40-42 °C for 24 h, in dark. Simultaneously, seedlings were kept at 25±2 °C served as control. Leaves were harvested immediately after the stress period for experimental analysis. The thylakoid membranes were isolated by the procedure of (6). Chl was estimated according to (7). Various photochemical activities of thylakoids were measured at 25 °C according to (8). Radiolabelled (^{14}C) atrazine binding experiment was carried out according to (9) and western blot analysis was done according

G. Garab (ed.), Photosynthesis: Mechanisms and Effects, Vol. V, 3841–3844.
© 1998 *Kluwer Academic Publishers. Printed in the Netherlands.*

3842

to procedure described in ref (10). Fast fluorescence transient was measured with 10 μs resolution in a time span of 40 μs to 1 s according to (11).

3. Results and Discussion

3.1 Partial Photochemical Activities

Loss in the uncoupled PSII activity using different quinone acceptors was about 60-65% in HT samples indicated significant reduction in the rate of primary photochemistry upon elevated temperature treatment. Heat induced loss in PSII electron transport has been reported earlier by several groups (1). Unlike earlier reports, a loss in PSI activity was observed in the present study. Whole chain (WC) electron transport from H_2O to MV showed 50-65% loss in HT seedlings as compared to the control samples (Table1).

Assay	Control 25 °C	Treated 40°C	%Change
PS II			
$H_2O \rightarrow pBQ$	241	87	64
$H_2O \rightarrow DCBQ$	276	100	63
$H_2O \rightarrow PD \rightarrow FeCN$	196	64	67
$H_2O \rightarrow DCPIP$	173	109	37
$H_2O \rightarrow FeCN$	104	52	50
PS I			
$DCPIPH_2 \rightarrow MV$	409	360	12.2
$TMPDH_2 \rightarrow MV$	679	539	20
$DAD \rightarrow MV$	1200	1108	8
WC			
$H_2O \rightarrow MV$	186	62.5	66

Table 1 Rate of PSII, PSI and WC catalysed electron transport measured as μmol O_2 evolution or consumption/mg Chl//h using different electron acceptors and donors in the thyakoids isolated from control and heat-treated rice plants.

3.2 Relative Quantum yield Of PSII

Uncoupled pBQ supported PSII activity was measured as a function of white light intensity (Fig. 1A and B).

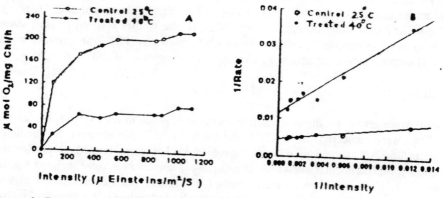

Figure 1: Rate of PSII induced O_2 evolution vs light intensity. A direct plot, B. Double reciprocal plot

Slope of the double reciprocal plot of rate vs light intensity increased from 0.335 in control to 1.96 in HT sample suggesting significant decrease in the relative quantum yield of photochemistry and/or change in the functional antenna size of PSII. Concomitant decrease in maximal (Vmax) PSII activity was observed from 220 µmol O_2 mg Chl^{-1} h^{-1} in control to 88 µmol mg Chl^{-1} h^{-1} in HT seedlings which is nearly 60% less than the control. Vmax reflects the number of active PSII centers. Decrease in the number of active PSII centers could be due to uncoupling of reaction centers from Chl antenna.

3.3 The Acceptor Side Of PSII

Atrazine binding properties were studied with ^{14}C labeled inhibitor. The number of electron carrier molecules Q_B correspond to the number of specific binding sites. Double reciprocal plot of bound vs free inhibitor for control and HT samples showed only a small change (i.e. 5.5-5.6 nmol/mg Chl) in the number of specific binding sites due to heat stress. However, the slope of the plot which refers to inhibitor dissociation constant (9,12) increased significantly from 0.04 µM in control to 0.06µM in HT samples (Fig. 2). The higher dissociation constant of atrazine for HT samples implies significantly reduced affinity of binding site (Q_B site on D1 protein) for the inhibitor. It is well known that the semiquinone form of Q_B^- has very high affinity for the binding site on D1, and consequently lowers the affinity for atrazine to bind to the same site (9). Our results indicate that mild treatment of rice seedlings induced decrease in the binding affinity of atrazine, possibly heat treatment *in vivo* caused accumulation of high concentration of Q_B^- and resulted in obstruction in electron transfer due to non-protonation of Q_B. This indicates that upon elevated temperature in vivo induces heterogeinity in the PSII reaction centers (active i.e. Q_B reducing and inactive i.e. Q_B non-reducing centers). Thus we suggest a retardation in electron transfer occurs at the acceptor site upon heat treatment.

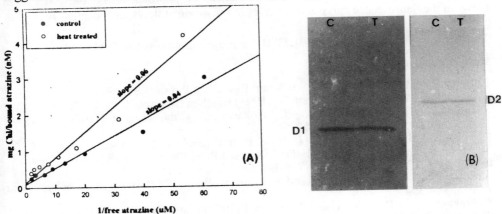

Figure 2: (A) Double reciprocal plot of bound vs free radio-labeled atrazine, (B) Western blot analysis of D1 and D2 polypeptides, in the thaylakoids isolated from control (●) and heat treated (O) rice seedlings

Such an inhibition in electron transfer from Q_A to Q_B has been reported earlier by chlorophyll fluorescence transient analysis (13,14). The authors suggest that HT probably convert some of the active PSII centers into inactive ones. This could be true in our experimental conditions also. However, immunoblot analysis of D1 and D2 did not show any change in steady state levels of the two proteins (Fig. 3). Therefore, our results suggest significant modification at binding site of D1 upon mild HT of the rice plants.

3.4 Chl a fluorescence studies

Polyphasic fluorescence rise has been designated as OJIP according to Strasser et al. 1995 (11). This transient provides signature for heat induced alteration *in vivo* in leaf. Fo, minimal fluorescence level arising from the antenna complex increased by 58% under the influence of mild heat accompanied by 63% decrease in variable fluorescence Fv (Table. 2) This may be due to uncoupling of the light harvesting antenna from the reaction center. The area above the fluorescence curve decreased upon heat treatment suggesting reduced quinone pool size and reduction in the number of active PSII centers. Our results thus shows that *in vivo* mild heat stress alters PSII structure and function.

Temp.	F_o	%Change	F_m	%Change	F_v	%Change	F_v/F_m	%Change	Area
25 °C(Control)	310	100	1786	100	1476	100	0.83	100	26496
40°C(Treated)	490	158	1030	57.7	540	36.6	0.524	62.7	16134

Table 2: Values taken from Fast fluorescence transient measured in inatact leaves of control and heat treated rice plants. F0: minimal fluorescence, Fm: maximal fluorescence, Fv: variable fluorescence, Fv/Fm: Quantum yield of photochemistry and Area: area between the fluorescence curve and the level of the maximal fluorescence intensity.

This work was supported by DST/ILTP B-6.9 project grant given to P.Mohanty

References

1. Berry, J. and Bjorkman, O. (1980) Annu. Rev. Plant Physiol. 31, 491-543.
2. Mamedov,M.,Hayashi,H., Murata,N. (1993) Biochim.Biophys. Acta. 1142,1-5.
3. Enami,I., Kitamura, M. Tomo, T., Isokawa, Y. Ohta, H. and Katoh, S. (1994) Biochim. Biophys. Acta 1186, 52-58.
4. Michalski, B. and Wettern, M. (1994) J. Plant Physiol. 144, 424-426.
5. Arora, S., and Pardha saradhi, P. (1995) Aust.J.Plant Physiol. 22, 383-386.
6. Nakatani,H.Y.and Barber,,J (1977) Biochim.Biophys.Acta 461,510-512.
7. Porra, R.J., Thempson, W.A. and Kriedmann, P.E. (1989) Biochim. Biophys. Acta 975, 384-394.
8. Prakash,J.S.S., Baig, M.A. and Mohanty, P(1997) Photosynthetica (in press)
9. Jursnic, P. and Stemler, A. (1983) Plant Physiol. 73, 703-708.
10.Towbin, H., Staehelin, T. and Gordon, J. (1979) Proc. Natl. Acad. Sci. USA 76, 4350-4354.
11.Strasser, R.J., Srivastava,A. and Govindjee (1995) Photochem.Photobiol.61,32-42.
12.Tischer,W. and Strotmann,H(1977) Biochim.Biophys.Acta 460,113-125.
13.Cao, J. and Govindjee (1990) Biochim. Biophys. Acta 1015, 180-188.
14.Lavergne,J. (1982) Photobiochem.Photobiophys. 3, 272-28

EFFECTS OF CHILLING ON PHOTOSYNTHESIS IN THE C$_4$ GRASS *MISCANTHUS X GIGANTEUS*

P.K. Farage, N.R. Baker, J.I.L.Morison & S.P. Long, Department of Biological Sciences, University of Essex, Colchester, Essex, CO4 3SQ, UK

Key words: biomass, CO$_2$ uptake, fluorescence, gas exchange, temperature stresses.

Introduction

Miscanthus x *giganteus* is a C$_4$, rhizomatous, grass that is currently being considered by the European Union as a biomass energy crop on account of its high annual productivity. Most C$_4$ plants have evolved in tropical or sub tropical climates and are therefore sensitive to low temperature. However, *M.* x *giganteus* originates from temperate regions in SE Asia and can achieve high rates of productivity in the cool climate of southern England where yields in excess of 20 t ha^{-1} have been obtained (1). However, *M.* x *giganteus* is still more sensitive to chilling than native C$_3$ species of NW Europe and in particular this can delay early season establishment and canopy closure.

Photosynthetic production of a biomass crop is more closely related to economic yield than for most other crops because all above ground material is harvested and consequently partitioning has less significance. Photosynthetic efficiency is therefore fundamental in the ability of *M.* x *giganteus* to realise high rates of productivity. Photosynthetic performance of *Zea mays*, the only C$_4$ crop grown in the UK, suffers seriously during periods of low temperature (2). To investigate which aspects of the photosynthetic apparatus provide *M.* x *giganteus* with a superior performance at low temperature, and whether further improvement of chilling tolerance is feasible, the effect of growth temperature on photosynthesis is being investigated *in vivo* using the combined approach of leaf gas exchange and chlorophyll fluorescence.

Methods

Plant Material and Growing Conditions
Rhizomes of *Miscanthus x giganteus* originating from a single clone (Picoplant, Oldenburg, Germany) were planted in peat-based compost (F2, Levington Horticultural Ltd., Ipswich, UK.) and grown in controlled environment chambers (Fitotron SGC066.CHX, Sanyo Gallenkamp PLC, Leicester, UK.). Fertilisation was provided by irrigating with Hoagland's nutrient solution (3). Day/night temperatures were either

G. Garab (ed.), Photosynthesis: Mechanisms and Effects, Vol. V, 3845–3848.
© 1998 *Kluwer Academic Publishers. Printed in the Netherlands.*

25°C/20°C or 10°C/8°C and the vapour pressure deficit (VPD) was kept below 1 kPa. Photon flux density at leaf height was 700 μmol m^{-2} s^{-1} and the photo period 14 hours.

Leaf Gas Exchange and Chlorophyll Fluorescence Measurements
Fluxes of CO_2 and water vapour to and from the most recently expanded leaf were measured with an infrared gas analyser (LI-6400, LI-COR Inc, Lincoln, USA). Carbon dioxide response curves were made over the range 50-500 μmol mol^{-1} CO_2 using a photon flux density of 1000 μmol m^{-2} s^{-1}, a VPD < 1 kPa and were analysed according to the model of Collatz *et al* (4). Steady-state chlorophyll fluorescence was measured simultaneously with gas exchange using a modulated fluorimeter (FMS1, Hansatech Ltd, Kings Lynn, UK). The chlorophyll fluorescence signals were analysed as described previously (5).

Results and Discussion

The effect of decreasing leaf temperature on the light saturated rate of photosynthesis (A_{sat}) for plants grown at 25°C and 10°C is illustrated in Fig. 1. Below 20°C CO_2 uptake decreases directly with temperature. Growth at chilling temperature confers no acclimatory advantage because rates of the 25°C grown plants are higher at all measurement temperatures. However, although photosynthesis is impaired in leaves that have developed under chilling conditions, *M. x giganteus* still achieves rates of CO_2 uptake that are in excess of most other C$_4$ species e.g. *Zea mays*, Fig. 1.

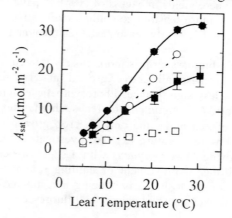

Figure 1. Rates of light saturated photosynthesis (A_{sat}) of *M. x giganteus* grown at 25°C ● and 10°C ■; and of *Z. mays* grown at 22°C ○ and 14°C □.

Limitation to photosynthesis by low temperature is generally associated with a reduction in the rate of enzymically controlled reactions. This was investigated *in vivo* by analysing the response of CO_2 uptake to changes in CO_2 concentration, A/c_i analysis. Plants grown at 10°C and measured at 20°C have similar rates of photosynthesis at CO_2

saturation, i.e. the plateau of the A/c_i curve (A_{max}), as those grown at 25°C and measured at 20°C, Fig. 2. The plateau of the A/c_i curve relates to the level of *in vivo* Rubisco activity (4). In contrast, the initial slope of the A/c_i response, i.e. carboxylation efficiency, is reduced by 50% in the 10°C grown plants and therefore suggests a decrease in the *in vivo* activity of phosphoenolpyruvate carboxylase (PEPc). Lowering the measurement temperature to 10°C produces a 50% reduction in A_{max} of the plants grown at 25°C. Those grown at 10°C do not exhibit any further reduction in carboxylation efficiency but A_{max} is reduced by over 60%, Fig. 2.

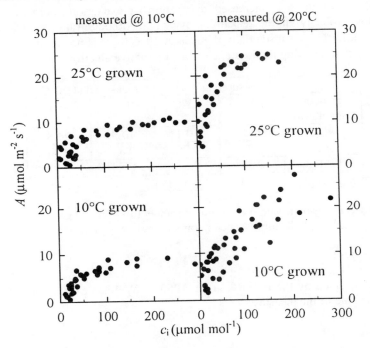

Figure 2. Effect of growth and measurement temperature on the response of CO_2 uptake (A) to change in intercellular CO_2 concentration (c_i) for *M.* x *giganteus*.

Examination of the A/c_i curves indicates that the operating points, i.e. the c_i that occurs at the normal atmospheric CO_2 concentration, are on the plateau of the A/c_i curves, even for those plants stressed by chilling. Consequently, any decrease in PEPc activity will not affect A_{sat}. However, at those times when A_{max} is decreased, which is when the measurement temperature is reduced to 10°C, Rubisco activity or reactions down stream of this enzyme (4) will limit photosynthesis. This would include pyruvate, orthophosphate dikinase which is a particularly cold-sensitive C_4 enzyme (6).

Low temperature induced decrease in CO_2 fixation also has implications for photosynthesis through the increased possibility of photoinhibition. In *Z. mays* this results in a down regulation of photosynthesis and eventually photodamage to the reaction centres (2). At 20°C the 10°C grown plants have much lower rates of A_{sat} than those grown at 25°C and will therefore be absorbing more light energy than would normally be dissipated in carbon fixation. However there is no decrease in the quantum efficiency of electron transport through PSII, (ϕ_{PSII}), suggesting that there must be an electron sink acting as an alternative to CO_2 (Table 1). When the measuring temperature of the 10°C grown plants is reduced to 10°C both ϕ_{PSII} and the efficiency of excitation energy capture by open PSII reaction centres, (F_v'/F_m'), are reduced. The decrease in F_v'/F_m' implies that there is increased dissipation of excitation energy in the PSII antennae, most probably via the zeaxanthin cycle (7). Although rates of CO_2 uptake by plants grown at 25°C are substantially reduced at 10°C there is no significant reduction in ϕ_{PSII}, once again implying the existence of an alternative electron sink.

	measured at 10°C		measured at 20°C	
growth temperature	10°C	25°C	10°C	25°C
A_{sat}	6.0	10.1	10.9	24.8
ϕ_{PSII}	0.22	0.29	0.31	0.33
F_v'/F_m'	0.37	0.55	0.54	0.57
q_P	0.59	0.55	0.6	0.58

Table 1. Effect of growth and measurement temperature on leaf gas exchange and chlorophyll fluorescence parameters of *M.* x *giganteus*.

Acknowledgements

Funding was provided by the BBSRC grant A05060. We thank Courtenay Brown for assistant with some of the measurements.

References

1 Beale, C.V. and long, S.P. (1995) Plant Cell Environ. 18, 641-650
2 Baker, N.R., Farage, P.K., Stirling, C.M. and Long, S.P. (1994) in Photoinhibition of Photosynthesis (Baker N.R and Bowyer J.R. eds.) pp.349-363, BIOS Scientific Publishers Ltd., Oxford, UK
3 Arnon, D. and Hoagland, D.R. (1940) Soil Sci. 50, 463-484
4 Collatz, G.J., Ribas-Carbo, M. and Berry, J.A. (1992) Aust. J. Plant Physiol. 19, 519-538
5 Andrews, J.R. Bredenkamp, G.J. Baker, N.R. (1993) Photosynth. Res. 38, 15-26
6 Usami, S., Ohta, S., Komari, T. and Burnell J.N. (1995) Plant Mol Biol. 27, 969-980
7 Demmig-Adams, B. and Adams, W.W. (1992) Annu. Rev. Plant Physiol. Plant Mol. Biol. 43, 599-626

EFFECTS OF GROWTH TEMPERATURE ON *LUPINUS ALBUS* PHOTOSYNTHETIC RESPONSES TO PHOSPHORUS AVAILABILITY

Osório, M.L.[1,2], Ricardo, C.P.P.[2] and Chaves, M.M.[1]
[1]Instituto Superior de Agronomia, Tapada da Ajuda,1399, Lisboa, Portugal
[2]ITQB, Quinta do Marquês 2780, Oeiras, Portugal

Key words: abiotic stress, gas exchange, fluorescence, photochemistry, down regulation

1. Introduction

It is well known that phosphate deficiency influence growth and metabolism of plants, CO_2 assimilation being one of the major physiological processes affected. Photosynthesis can be phosphate limited *in vivo* when a sudden or large inhibition of end-product synthesis (due, for example, to low temperature) leads to an imbalance between CO_2 fixation and sucrose synthesis. Previous data (1) indicate that the cytosolic concentration of P for maximal photosynthesis is dependent on actual temperature, being larger at lower than at higher temperatures. However, plants can acclimate to low temperature and "escape" from this limitation (2). On the other hand, PSII is very sensitive to the physical environment of the plant (3), including the nutrition status. Inadequate supply of P increased non-photochemical dissipation of energy and down-regulated PSII activity in intact leaves of sunflower and maize (4).

The objectives of this study were to evaluate how growth temperature affected: (i) long-term acclimation of lupin plants photosynthesis to low P and (ii) the efficiency of PSII photochemistry when P deficiency limited the rate of photosynthesis.

2. Material and methods

Lupin (*Lupinus albus* L.) plants were grown in plastic pots filled with washed sand under controlled conditions of temperature (25/20°C or 15/10°C, day/night), PPFD (300 μmol m^{-2} s^{-1}) and photoperiod (14 h). Plants in half the pots were irrigated with a modified Hoagland solution containing 0.2 mM K_2HPO_4 (P-sufficient treatment, HP) and in the other treatment plants were given only a 0.01mM K_2HPO_4 solution (P-deficient treatment, LP). All measurements were carried out on recently expanded leaves of plants in a similar developmental stage. Gas exchange was measured using a Minicuvette System (H.Walz, Effeltrich, Germany) attached to a BINOS CO_2/H_2O

G. Garab (ed.), Photosynthesis: Mechanisms and Effects, Vol. V, 3849–3852.
© 1998 *Kluwer Academic Publishers. Printed in the Netherlands.*

differential infrared gas analyser. Values reported were obtained under constant environmental conditions in the measuring cuvette: 650.5 ± 1.5 μmol m^{-2} s^{-1} of PPFD, $24.9\pm0.0°C$ of leaf temperature and a leaf to air vapour pressure deficit of 17.0 ± 0.43 Pa KPa^{-1}. Leaf conductance to water vapour (g_S) and net photosynthetic rate (A) were calculated according to von Caemmerer and Farquhar (5). Photosynthetic capacity (A_{max}) was measured as oxygen evolution at saturating CO_2 (10%) and PPFD (700 μmol m^{-2} s^{-1}) and 25°C, with an oxygen electrode (Hansatech, Norfolk, UK). Chlorophyll a fluorescence was measured with a pulse amplitude modulation fluorometer (PAM-101/103 system, Walz, Germany) on five leaf discs placed in the chamber of an oxygen electrode transformed to receive the optical fibres of the fluorometer. Maximal photochemical efficiency of PSII was estimated by the fluorescence ratio (F_v/F_m) in pre-darkened leaves for 30 minutes. The efficiency of excitation capture by open PSII reaction centres (F'_v/F'_m), as well as photochemical quenching (q_p) and non-photochemical quenching (q_N) were calculated according to Bilger and Schreiber (6). The quantum yield of non-cyclic electron transport (ϕ_e) in the steady state was calculated as ($F'_m-F_s)/F'_m$ ratio after Genty *et al.* (7). The concentration of non-structural carbohydrates was enzymatically determined as in Stitt *et al.* (8). The activity of RuBisCo was assayed by the coupled enzyme method of Liley and Walker (9).

3. Results

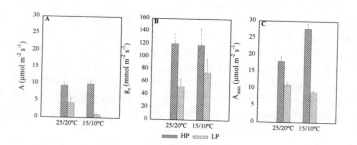

Figure 1. Leaf photosynthetic rate (A), leaf stomatal conductance (g_s) and photosynthetic capacity in P-sufficient (HP) and P-deficient (LP) lupins grown at 25/20°C or at 15/10°C (day/night). Values shown are mean ± standard error of four replicates.

Leaf photosynthetic rate (A) was reduced by phosphate deficiency, with plants grown at 25/20°C and 15/10°C exhibiting, on average, decreases of 54% and 90%, respectively, relative to the phosphate-sufficient plants (Fig.1A). However, at low temperature, the inhibitory effect of P-deficiency on g_s (37%) was less pronounced than on A (Fig.1B). A_{max} (expressed on a leaf area basis) was highest at low temperature and normal P, but the inhibitory effect of low P was stronger under this temperature (67%) than under high temperature (39%) (Fig.1C).

The inhibitory effect of low P on photosystem II (PSII) was pronounced, particularly at the thermal regime 15/10°C. Phosphate deficiency decreased ϕ_e by 54% and 29%, q_p by

26% and 9%, F'_v/F'_m by 37% and 34% in plants grown at 15/10°C and 25/20°C, respectively (Fig.2).

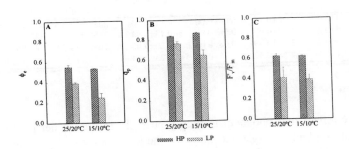

Figure 2. Quantum yield of non-cyclic electron transport (ϕ_e), photochemical quenching (q_p) and efficiency of excitation capture by open PSII reaction centres (F'_v/F'_m) in P-sufficient (HP) and P-deficient (LP) lupins grown at 25/20°C or at 15/10°C (day/night). Values shown are mean ± standard error of four replicates.

On the other hand, P deficiency increased the thermal dissipation of excitation by PSII (q_N), but did not affect significantly F_v/F_m (Fig.3).

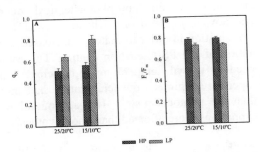

Figure 3. Non-photochemical quenching (q_N) and maximal photochemical efficiency of PSII (F_v/F_m) in P-sufficient (HP) and P-deficient (LP) lupins grown at 25/20°C or at 15/10°C (day/night). Values shown are mean ± standard error of four replicates.

As is observed in Table I, both leaf sucrose and starch concentrations are particularly sensitive to low growth temperature, the accumulation of these carbohydrates being higher in plants grown at low P. Total and initial carboxylase activities of RuBisCo were stimulated by low growth temperature and were significantly decreased by low-P at this thermal regime. However, P-deficiency had no significant effect in activities of this enzyme in plants grown at 25/20°C, as in RuBisCo activation state in both temperatures (Table I).

Table I. Starch and sucrose concentrations, initial and total activity and activation state of RuBisCo in P-sufficient (HP) and P-deficient (LP) lupins grown at 25/20°C (HT) or at 15/10°C (LT) (day/night). Values shown are mean ± standard error of five replicates.

| | Starch (μmol cm^{-2}) | | Sucrose (μmol cm^{-2}) | | RuBisCo (μmol mg chl^{-1}h^{-1}) | | |
	Predawn	Midday	Predawn	Midday	Initial Activity	Total Activity	Activation (%)
HPxHT	0.67±0.09	1.23±0.15	0.06±0.00	0.11±0.00	253.2±52.3	407.1±74.3	63.9±10.3
LPxHT	0.65±0.09	1.24±0.09	0.06±0.00	0.13±0.04	240.4±54.2	338.8±70.4	70.6±8.5
HPxLT	2.86±0.23	2.12±0.49	0.08±0.00	0.20±0.02	705.7±39.3	744.8±45.6	94.9±2.1
LPxLT	3.81±0.66	2.58±0.42	0.11±0.01	0.23±0.04	397.8±118	454.4±130	85.8±3.9

4. Discussion

Our results confirm that phosphorus supply strongly influence net photosynthetic rate (A), stomatal conductance (g_s) and photosynthetic capacity (A_{max}) of lupin leaves and that these responses are modulated by growth temperature. In phosphate deficient plants both A and g_s decreased, but the former was much more affected than the latter, especially at the lower temperature regime. This suggests that photosynthesis was more limited by the metabolism of the mesophyll than by stomata in phosphate deficient plants and confirms the findings in other species (10). Our data also support the findings by Furbank *et al.* (11) and Plesnicar *et al.* (12), showing that photosynthesis under P-limiting conditions is regulated both at the thylakoid and the carbon metabolism level. Low-P treatment decreased the efficiency of excitation capture by open PSII reaction centres, the photochemical quenching of PSII and the quantum yield of non-cyclic electron transport and increased the non-photochemical quenching. Chlorophyll fluorescence parameters showed that PSII photochemistry was down-regulated by non-photochemical energy dissipation in order to match a decreased carbon metabolism and hence avoid overexcitation of the PSII reaction centres. The capacity to cope with P starvation was influenced by growth temperature, P deficiency having more dramatic effects at the low temperature regime. Low temperature had more drastic effects on carbon metabolism than on photochemistry. The accumulation of starch observed indicates that carbon export was more affected than photosynthesis.

Acknowledgement: This research was supported by the project Praxis/2/2.1/BIA/227/94.

References

1 Labate, C. A. and Leegood, R. C. (1988) Planta, 173, 519-527.
2 Labate, C. A. and Leegood, R. C. (1989) Plant Physiol., 91, 905-910.
3 Baker, (1991) Physiol. Plant. 81, 563-570.
4 Jacob, J. and Lawlor, D. W. (1993) Plant Cell and Environm., 16, 785-795.
5 von Caemmerer, S. and Farquhar, G. D. (1981) Planta, 153, 376-387.
6 Bilger, W. and Schreiber, U. (1986) Photosynth. Res., 10, 303-308.
7 Genty, B., Briantais, J. M. and Baker, N. R. (1989) Biochem. Biophys. Acta, 990, 87-92.
8 Stitt, M., Liley, R., McC., Gerhard, R. and Heldt, H.W. (1989) Methods of Enzymol., 174, 518-552.
9 Liley, R., McC. and Walker, D. A. (1974) Biochem. Biophys. Acta, 358, 226-229.
10 Jacob, J. and Lawlor, D. W. (1991) J. Exp. Bot. 42, 1003-1011
11 Furbank, R. T., Foyer, C. H. and Walker, D. A. (1987) Biochem. Biophys. Acta, 894, 552-551.
12 Plesnicar, M., Kastori, R., Petrovi, N, and Pankovic, D, (1994) J. Exp. Bot. 45, 919-924

ACCLIMATIZATION OF MICROPROPAGATED TOBACCO PLANTLETS

P. Kadleček[1], I. Tichá[1], V. Čapková[2], C. Schäfer[3]
[1]Department of Plant Physiology, Faculty of Science, Charles University, Viničná 5, CZ-12844 Praha 2, Czech Republic
[2]Institute of Experimental Botany, Academy of Science of the Czech Republic, Rozvojová 135, CZ-16502 Praha 6 – Lysolaje, Czech Republic
[3]Lehrstuhl für Pflanzenphysiologie, Universität Bayreuth, Universitätsstraße 30, D-95440 Bayreuth, Germany

Key words: biotechnology, fluorescence, *in vitro/ex vitro* transfer, net photosynthesis

1. Introduction

Micropropagated plantlets are grown conventionally under conditions of low irradiance (30-70 μmol photons m^{-2} s^{-1}), high relative air humidity, variable CO_2 concentration (from 2% to compensation concentration) and presence of carbohydrates in the medium (1, 2). The consequences of these conditions may be changes in plant anatomy (e. g., stomata, cuticle – for review see 3) and physiology (e. g., low photosynthetic ability – for review see 2). However, photosynthetic ability and starch accumulation of plants *in vitro* may favour acclimatization to *ex vitro* environment (1, 4). Acclimatization of plants after transfer to *ex vitro* environment is considered for the final, but often critical stage of micropropagation of many plant species. Thus, the success of micropropagation is considerably dependent on *in vitro* growth. The aim of this study was to clarify whether the presence of sucrose in the medium during *in vitro* culture affects the long-term development under *ex vitro* conditions.

2. Procedure

2.1 Material
Tobacco (*Nicotiana tabacum* L. cv. Samsun) plantlets were cultured *in vitro* on solidified Murashige-Skoog medium under 200 μmol photons m^{-2} s^{-1} and CO_2 saturation photoautotrophically (PA; without sucrose in the medium) or photomixotrophically (PM; with 3% sucrose in the medium) at 25/18 °C day/night temperatures, 16-h photoperiod. The 35-days-old plantlets were transplanted into pots with soil and transferred to the greenhouse. For all the plants the mean insolation in the greenhouse varied from 30 to 90 μmol photons m^{-2} s^{-1} and the mean temperature from 18 to 26 °C. After further 20 days the plants were transferred to open air. There the mean insolation varied between 200 and 1400 μmol photons m^{-2} s^{-1} and the mean temperature between 24 and 31 °C.

G. Garab (ed.), Photosynthesis: Mechanisms and Effects, Vol. V, 3853–3856.
© *1998 Kluwer Academic Publishers. Printed in the Netherlands.*

2.2 Methods

Photosynthetic oxygen evolution was measured with a Clark type gas-phase leaf-disc oxygen electrode (LD2/2, Hansatech, King's Lynn, UK) at 25 °C in CO_2-enriched air (2%). Chlorophyll a fluorescence emission from the upper surface of the leaves was measured in modulated light with a pulse amplitude modulation fluorometer (PAM, Walz, Effeltrich, Germany) simultaneously with O_2 evolution. Chlorophyll $a+b$ contents were determined spectrophotometrically after 24 h or 48 h of elution in N,N'-dimethylformamide. Starch was extracted and measured using enzymatic and spectrophotometric methods.

3. Results and Discussion

After 35 days of *in vitro* growth significant higher chlorophyll $a+b$ content (Fig. 1A), photosynthetic capacity (Fig. 1B), maximum photochemical efficiency (Fig. 1C) and starch content (Fig. 1D) in PM grown plantlets than in PA grown plantlets were found

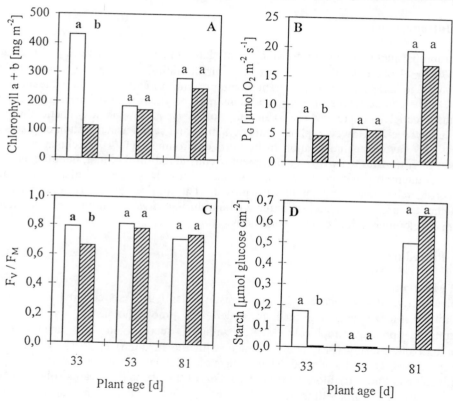

Figure 1.

Chlorophyll $a+b$ content (A), photosynthetic capacity, i. e., gross photosynthetic rate (P_G) at light and carbon dioxide saturation (B), maximum photochemical efficiency F_V/F_M in dark adapted leaves (C) and starch content (D) during transfer of tobacco plants from *in vitro* to *ex vitro* conditions. Data are means of three plants. On the 35th day of culture the plants were transferred from *in vitro* to the greenhouse, on the 55th days from the greenhouse to open air. Different letters above the columns indicate significant differences at P=0.05 and P=0.01 (bold). Open columns – originally photomixotrophically grown plants, hatched columns – originally photoautotrophically grown plants.

(cf. also paper 5). The positive influence of exogenous carbohydrates in the medium on chlorophyll *a+b* content and photosynthetic capacity was shown (6), but contradictory results were described, too (4). From the decrease in F_V/F_M, in photosynthetic capacity and from the breakdown of chlorophylls in PA grown plantlets, it was deduced that in these plantlets the irradiance 200 µmol photons m^{-2} s^{-1} led to photoinhibition. Similar results showed (6). However, in PM grown plantlets photoinhibition was not observed, sugar feeding probably prevented the occurrence of photoinhibition. Plant growth was also significantly enhanced by the sucrose pretreatment *in vitro* (higher plants, more leaves, larger leaf area and higher dry matter accumulation – Tab. 1) in PM grown plantlets. The PM plantlets used for biomass production not only their photosynthetic abilities, but also exogenous sucrose as carbon and energy sources (7).

Table 1.
Growth parameters of tobacco plants during transfer from *in vitro* to *ex vitro* conditions. For further explanations see Fig. 1.

	In vitro 32 d		*Ex vitro* (greenhouse) 52 d		*Ex vitro* (open air) 80 d	
	PM	PA	PM	PA	PM	PA
Plant height [cm]	2.2 a	1.1 b	14.8 a	4.9 b	46.3 a	25.6 b
Number of leaves per plant	7.3 a	5.3 a	11.3 a	8.7 b	24.3 a	17.0 b
Total leaf area [cm²]	52.8 a	34.9 b	250.8 a	112.0 b	4683.7 a	1797.6 b
Total dry matter [g]	0.15 a	0.06 b	0.51 a	0.20 b	17.80 a	5.96 b

During the 20 days of growth under identical conditions in the greenhouse, the differences described above in *in vitro* plantlets step by step disappeared. At the end of the first stage of acclimatization in the greenhouse, no significant differences between originally PA and PM grown plants were found (Fig. 1A-D). The originally PA *in vitro* grown plants recovered from the influence of excessive irradiance (lower irradiance in the greenhouse – see 2.1). As the consequence, a slight increase in F_V/F_M (Fig. 1C), in photosynthetic capacity (Fig. 1B) and in chlorophyll *a+b* content (Fig. 1A) were found. Starch accumulated during *in vitro* growth in PM plants was used for plants growth in the greenhouse. Starch accumulation during *in vitro* increased the success of the transfer from *in vitro* to *ex vitro* environment (4). Four weeks after the transfer to open air, chlorophyll *a+b* content (Fig. 1A), photosynthetic capacity (Fig. 1B) and starch content (Fig. 1D) were increased. Nevertheless, F_V/F_M (Fig. 1C) was slightly decreased both in the originally PA and PM grown plants. These changes were probably the consequence of the expressive increase in irradiance in open air (see 2.1). The significant differences in growth parameters found at the end of the *in vitro* period between PM and PA grown plantlets, were present also during the whole acclimatization period (in the greenhouse and in open air).

Conclusions:
1. All the plants survived the transfer from *in vitro* to *ex vitro* environment.
2. Photosynthesis, fluorescence, chlorophyll *a+b* content and starch content during *ex vitro* acclimatization were not significantly affected by the different culture conditions *in vitro*. On the other hand, growth of plants was significantly decreased during acclimatization in the plants grown originally *in vitro* without sucrose.

4. Acknowledgements

We gratefully acknowledge financial support by the European Commission (Copernicus-Project ERBCIPACT 930115), the Ministry of Agriculture and the Ministry of Education and Culture of Hungary and the Ministry of Education of the Czech Republic (Projects VS 96145 and PG 98234).

References

1   Kozai, T. (1991) in Micropropagation. Technology and Application (Debergh P.C., Zimmerman R.H., eds.) pp.447-469, Kluwer Academic Publishers, Dordrecht-Boston-London
2   Pospíšilová, J., Čatský, J. and Šesták, Z. (1997) in Handbook of Photosynthesis (Pessarakli M., ed.) pp.525-540, Marcel Dekker, New York-Basel-Hong Kong
3   Preece, J.E. and Sutter, E.G. (1991) in Micropropagation. Technology and Application (Debergh P.C., Zimmerman R.H., eds.) pp.71-93, Kluwer Academic Publishers, Dordrecht-Boston-London
4   Capellades, M., Lemeur, R. and Debergh, P. (1991) Plant Cell Tiss. Organ Cult. 25, 21-26
5   Tichá, I., Čáp, F., Pacovská, D., Haisel, D., Čapková, V. and Schäfer, C. (1998) this Volume
6   Tichá, I., Čáp, F., Pacovská, D., Hofman, P., Haisel, D., Čapková, V. and Schäfer, C. (1998) Physiol. Plant. 102, 155-162
7   Tichá, I. (1996) Photosynthetica 32, 475-479

CHLOROPHYLL FLUORESCENCE CAN BE USED TO IDENTIFY PLANT SPECIES AUTOMATICALLY

Esa Tyystjärvi*,Mika Keränen*, Antti Koski[†], Olli Nevalainen[†] and Eva-Mari Aro*
*Dept. of Biology, Lab Plant Physiology, BioCity A and
[†]Dept. of Computer Science, DataCity;
*,[†]University of Turku, FIN-20014 Turku, Finland

Key words: Chl fluorescence induction, herbicides, optical properties, precision farming, remote sensing, weeds.

1. Introduction

New agricultural techniques, so called precision farming, aim at cost efficient production with a smaller environmental load. The key idea is to continuously analyse the field and to deliver agrochemicals only at locations that need them. For example, fertilizers can be distributed according to measurements of the grain yield per each square metre of the field. An important way to reduce the amounts of chemicals is to spray leaf herbicides only on such field locations that contain weeds. Agricultural machines that can detect plants and spray herbicides on plant-containing spots only, have been under development for several years. The detection was earlier based on the reflectance properties of plants (1), but the reflection spectrum of certain earth types can be near to the reflection spectra of plants. Chlorophyll a fluorescence is a straightforward answer to the selectivity problem, and a report on an experimental machine using fluorescence as a plant detection method was recently published (2).

The present weed detection methods cannot distinguish between weeds and cultivated plants and are therefore limited to mainly to applications on fallow fields. We studied the possibility to use chlorophyll fluorescence as a method for automatic identification of plant species. We show that the species-specific features of fluorescence induction curves can be used for identification. The identification is carried out with a computer running a pattern recognition software. The method is called 'fluorescence fingerprinting'.

In our first experiments (Tyystjärvi, Koski & Nevalainen, unpublished) we showed that by measuring 8-s long curves from dark adapted leaves we could identify the species, and the leaf side (adaxial or abaxial) where appropriate, from a set of seven species with 97 % accuracy. However, a shorter measurement time is required for practical agricultural purposes, and here we report on the use of 3-s curves to identify five species.

G. Garab (ed.), Photosynthesis: Mechanisms and Effects, Vol. V, 3857–3860.
© 1998 Kluwer Academic Publishers. Printed in the Netherlands.

2. Materials and Methods

We used a regular PAM-101 fluorometer (Walz GmbH, Effeltrich, Germany) to measure fluorescence induction curves from five different species. The fluorometer was remote controlled and the curves were recorded with the FIP fluorescence system (Q_A-Data Oy, Turku, Finland). In order to reveal as many species-specific features as possible, each curve was composed of several short periods of illumination with different types of light (red, far red and white saturating light). We used dark adapted leaves in the first experiments and routinely recorded 199 curves from each species. The upper and lower sides of the leaves were treated as different classes.

Maize (*Zea mays*) and tobacco (*Nicotiana tabacum*) were grown in a growth chamber and pine (*Pinus sylvestris*), birch (*Betula pendula*) and *Polytrichum commune* (a moss) were collected from the nature. The leaves were dark adapted for 2 hours before measuring the fluorescence curves. We used the 'magic' trigger of the FIP program to allow the software to detect automatically the presence of a leaf at the fluorometer's probe. The 3-s curves consisted of 0.2 s of measuring beam alone, 0.6 s of moderate red light, 0.2 s of far red light, 0.2 s of darkness, 0.4 s saturating light, 0.7 s darkness and 0.6 s far red light.

To be able to treat the data with a pattern recognition software, we first extracted 16 parameters from each curve by calculating nine partially overlapping regression lines along the curve. After normalization, these parameters were used as inputs for pattern recognition procedures. Pattern recognition requires a training set on which the features of a certain class (in our case, a species) are based. We picked 100 curves randomly from each class to the training set and used 99 remaining curves to test the reliability of the identification.

Three different pattern recognition methods were tested. *The distance-to-mean method* starts with calculating mean parameter values for each class of the training set. An unknown

Input layer   Hidden layer   Output layer **Fig. 1.** The idea of a neural network. The input parameters are multiplied by a_{ij} to calculate the hidden layer, and the classification is completed by calculating the output layer using b_{ij}.

signal is classified by determining which group mean is nearest to it. In *the k-nearest-neighbour method*, one determines the distance of an unknown signal to all signals of the training sets. The parameter k is a small integer. The class to which the majority of k smallest distances belong to, is considered the correct one. The third method is *a neural network* (Fig. 1) in which the parameter values are used as an input layer and the classes form the output layer. A so called hidden layer between the input and output layers helps in defining a set of complicated border lines between the classes. The factors a_{ij} (Fig. 1) define the contributions of the input cells to the hidden layer and the factors b_{ji} define how much each cell of the hidden layer affects each cell of the output layer. Using the backprogation algorithm, the factor sets a_{ij} and b_j are optimised so that the maximum amount of correct results is obtained with the training set. When the network has been trained, the identification of an unknown signal is therefore a very simple calculation.

In the application reported here, we had 16 input cells, 20 cells in the hidden layer and 5 or 8 output cells, as indicated.

3. Results

All three pattern recognition methods can be used to identify the plant species. The distance-from-mean method and the 9-nearest-neighbour method recognized the correct class (species and leaf side) with a 80 and 78 % accuracy, respectively, and the neural network yielded 76 % correct recognitions (Table 1).

Table 1. Percentage of correct recognitions obtained from five species with the three pattern recognition procedures, described in Materials and Methods. Data in the two first columns were obtained by including measurements from both leaf sides of maize, tobacco and birch whereas for pine and *Polytrichum* only one side measurement was used.

Method	Correct classes (8 output cells)	Correct species (8 output cells)	Correct species* (Upper side only; 5 output cells)
	%	%	%
Distance-from-mean	69.7	81.0	89.3
9-nearest-neighbour	77.8	91.4	97.0
Neural network	75.6	95.8	98.0

The highest recognition accuracy was always reached if only curves measured from the upper side were included. However, even if curves from both sides were taken into the analysis, a very high accuracy was obtained for species recognition with the neural network algorithm (Table 1).

A closer look on the recognition matrix obtained with the neural network (Table 2) shows that the curves recorded from pine needles and moss leaves can be easily distinguished both from each other and from the curves of higher plants. Interestingly, the adaxial leaf side was recognized as the abaxial side more frequently than vice versa with the neural network (Table 2) and the 9-nearest neighbour method while the distance-from-mean method did not show this behavior (data not shown).

Table 2. Recognition matrix. The identification was done with a neural network algorithm. The percentage of correct recognitions is 75.6 and the percentage of correct species is 95.8 The letters u and l refer to the upper (adaxial) and lower (abaxial) side of leaf, respectively.

Species				Recognized as				
	Pinus	Betula (u)	Betula (l)	Nicotiana (u)	Nicotiana (l)	Zea (u)	Zea (l)	Polytrichum
Pinus	99	0	0	0	0	0	0	0
Betula (u)	0	88	6	5	0	0	0	0
Betula (l)	0	0	88	5	5	0	1	0
Nicotiana (u)	0	1	1	36	60	0	1	0
Nicotiana (l)	0	0	0	0	93	0	5	1
Zea (u)	0	0	0	0	4	2	93	0
Zea (l)	0	3	0	0	1	1	94	0
Polytrichum	0	0	0	0	0	0	0	99

4. Discussion

The fluorescence induction curves are known to have species-specific features. For example, conifer needles show a fast increase in the fluorescence yield to the P-peak and a rapid decrease after the peak. However, the differences usually consist of features and combinations of features at several phases of the induction curve. The use of an efficient pattern recognition procedure is therefore essential in using fluorescence curves for species identification. The number of features of a fluorescence curve can be made high either by using a long measuring time (which is undesirable for practical purposes) or by applying a series of short illumination phases of different types of light. The present data shows that a three-second illumination is long enough for an efficient identification tool. We assume that the time required for the identification can still be substantially reduced.

The present experiments show that chlorophyll fluorescence is a potent candidate for the development of an automatic method for identification of plant species in field conditions. However, several problems remain to be solved before a field-compatible method can be developed: how short can the dark adaptation be and how dark should it be; how short curves can identify the species with a high enough accuracy for agricultural purposes; what are the optimal features of the series of light treatments. The best results will probably be obtained by using several methods, e.g. machine vision together with fluorescence.

References
1 Thompson JF., Stafford JV & Miller PCH. (1991) Crop Protection 10, 254-259
2 Visser R & Timmermans AJ (1996) In Meyer GE and DeShazer JA (eds) Optics in Agriculture, Forestry and Biological Processing 2, paper # 2907-13.
3 Schalkoff RJ (1994) Pattern Recognition: Statistical, Structural and Neural Approaches, John Wiley & Sons, Singapore

PLANT *P*-HYDROXYPHENYLPYRUVATE DIOXYGENASE: A TARGET FOR NEW BLEACHING HERBICIDES

I. Garcia[1], M. Rodgers[2], R. Pépin[2], Tzung-Fu Hsieh[3] and M. Matringe[1]
[1]Unité Mixte CNRS/Rhône-Poulenc (UMR 41); [2]Département des biotechnologies, Rhône-Poulenc Agrochimie, 14-20 rue Pierre Baizet, 69263 Lyon cedex 09, France; [3]Department of Biology, Texas A&M University, College Station, TX 77843 USA

Key words: herbicide, plastoquinone pool, isoforms, transgenic plants, overexpression

INTRODUCTION

p-Hydroxyphenylpyruvate dioxygenase (HPPD, EC 1.13.11.27) catalyses the formation of homogentisate from *p*-hydroxyphenylpyruvate (HPP) and molecular oxygen. In plants, this enzyme plays a specific and crucial role. Indeed, homogentisate is the aromatic precursor of all plastoquinones and tocopherols which are respectively essential components of the photosynthetic electron- transfer chain and antioxydative elements. Also plant HPPD is involved, as in other organisms, in the catabolism of the aromatic amino acid tyrosine. Interest in the plant enzyme was raised recently by the demonstration that HPPD is the target of new bleaching herbicide families (sulcotrione-1; isoxasoles-2). It is believed that this effect results from an indirect inhibition of phytoene desaturase activity, as a consequence of the depletion of the plastoquinone-cofactor pool. The involvement of plant HPPD activity in metabolic processes as different as prenylquinone biosynthesis and tyrosine catabolism raises the question of the presence of different isoforms of HPPD in plants, associated with different subcellular compartments. Recently, we have purified to near homogeneity the HPPD of cultured carrot cells and isolated the correspondant cDNA (3). Carrot HPPD resembles the mammalian enzyme since it behaves as a homodimer with 48 kDa subunits. In this previous study we demonstrated that this enzyme activity was associated with the cytosol. We report here the biochemical characterisation of carrot recombinant HPPD, and the molecular characterisation and the sub-cellular localization of the corresponding enzyme from *A. thaliana* .

RESULTS AND DISCUSSION

Biochemical characterisation of carrot recombinant HPPD

The complete open reading frame present in the carrot cDNA was cloned in the pTRc 99A plasmid and the recombinant protein was produced in *E. coli* JM 105 cells. The recombinant HPPD was purified by a sequence of chromatographic steps including an EMD DEAE 650(M) chromatography, an hydroxyapatite column and a source Q chromatography. The activity of the purified carrot HPPD was assayed by monitoring the oxygen consumed during the catalytic reaction. The specific activity of recombinant HPPD, defined under optimized conditions to restore the activity, was 0.4-0.8 µmol of homogentisate formed per min per mg of protein, which was in the range of specific

G. Garab (ed.), Photosynthesis: Mechanisms and Effects, Vol. V, 3861–3864.

activity reported for the purified mammalian HPPD (4).The Km for HPP was found to be around 5µM. The recombinant HPPD, like the native enzyme, was inhibited by isoxazoles. The diketonitrile of isoxaflutole behaves as a slowly reversible tight binding inhibitor of HPPD with a rate constant of association of 0.07 $\mu M^{-1} s^{-1}$. This result is in agreement with association rate constant found with crude native carrot HPPD (5) and with that determined for sulcotrione binding to rat liver HPPD (6). Presumably, inhibitor binding is limited by a chemical step and/or a conformational change of the enzyme. Moreover, this binding is competitive with respect to the HPP substrate.

Subcellular localization of plant HPPD

Plant HPPD is involved in two different processes, the catabolism of tyrosine and the biosynthesis of prenylquinones. This dual involvement of plant HPPD raises the question of the presence of HPPD isoforms associated with different subcellular compartments. In our previous study, we demonstrated that this activity was cytosolic in cultured carrot cells. Such a subcellular localization is in apparent contradiction with the situation previously described in spinach by *Fiedler et al.* (7).This study suggested the existence of a chloroplastic form of HPPD, responsible of the biosynthesis of prenylquinones and a peroxisomal form, implicated in the degradation of tyrosine. We have thus verifyied that the cytosolic localization in non-chlorophyllous carrot cells was also the situation encountered in chlorophyllous tissues.

For that purpose, we have cloned *A. thaliana* HPPD. The largest *A. thaliana* HPPD cDNA we isolated was 1614-bp long. This cDNA includes one large open reading frame of 1338-bp which encodes a predicted polypeptide of 445 amino acids with a molecular mass of 48761 Da. This sequence presents 75% identity with that of the carrot HPPD polypeptide. Southern analysis was carried out to examine the copy number of the gene encoding this *A.t haliana* HPPD mRNA. The result was consistent with the existence of a single gene copy. This was corroborated by northern blot analysis which revealed a single transcript of 1.6- 1.7 kb. The *A. thaliana* recombinant protein presents the same biochemical characteristics than carrot HPPD and was specifically recognized by the polyclonal antibody raised against the purified carrot HPPD (3).

We have overexpressed the complete coding sequence in tobacco using *Agrobacterium tumefaciens* transformation and examined the subcellular localization of the recombinant protein. The rationale of these experiments is as follows. If the complete coding sequence encodes an amino-terminal or carboxy-terminal transit peptide, tobacco cells will target the recombinant HPPD to the corresponding organelle. Conversely, if this coding sequence does not contain a signal sequence, the recombinant HPPD will remain in the cytosol.

The presence of *A. thaliana* cDNA HPPD in the genome of transformed tobacco was verified by PCR amplification using specific primers corresponding to the 5' and 3' ends of the coding sequence. The overproduction of recombinant *A. thaliana* HPPD in transformed tobacco plants was further confirmed by immunodetection, using a polyclonal antibody raised against carrot HPPD. The subcellular localization of the recombinant HPPD in transformed tobacco plants was examined by immunocytochemistry. In sections obtained from tobacco transformed with *A. thaliana* HPPD no reaction over the level of the background was observed inside chloroplasts, mitochondria, or peroxisomes. A specific reaction occurred exclusively in the cytosolic compartment, indicating that the recombinant HPPD remained in the cytosol (Fig. 1). This result demonstrates that *A. thaliana* HPPD does not contain any targeting signal.

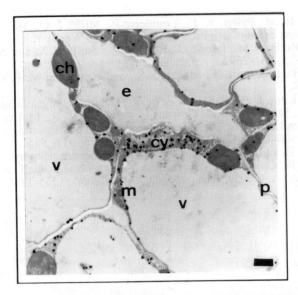

Figure1: Subcellular localization of the recombinant *A. thaliana* HPPD overproduced in parenchyme cells of transgenic tobacco leaves by immunocytochemistry.

Transgenic tobacco transformed by the complete coding sequence of *A. thaliana* HPPD (x5700); spongy mesophyll (AB1 1/25; AB2--gold φ 10 nm, 1/50, amplification 2 min). The recombinante HPPD was specifically revealed in the cytosol, the others sub-cellular compartments were at background level.
Abbreviations: ch: chloroplast; cy: cytosol; m: mitochondrion; v: vacuol. Primary antibody: immunopurified rabbit IgG raised against carrot HPPD (AB 1). Secondary antibody: goat anti rabbit conjugated with colloidal gold, particles of φ 10 nm.
Scale bar = 1 μm.

CONCLUSION

The cytosolic localization of recombinant *A. thaliana* HPPD overexpressed in tobacco is in agreement with our previous study on cultured carrot cells. Moreover, this result is also consistent with the situation encountered in mammalian liver cells. Thus, it has been known for many years that α-ketoisocaproate dioxygenase activity is associated with the cytosolic compartment in rat liver cells (8) and recently, it has been demonstrated that this enzyme activity is in fact catalysed by HPPD (9). The involvment of the plant HPPD in metabolic processes as different as tyrosine catabolism and prenylquinone biosynthesis raises the question of whether the HPPD presently studied is involved only in tyrosine degradation or in both processes. In our different studies we always detect only the cytosolic isoform of HPPD. However, these results do not rule out the possible existence of a chloroplastic HPPD isoform involved in prenylquinone biosynthesis. We could imagine that the chloroplastic activity is too weak to be detected or does not crossreact with the polyclonal antibody raised against the cytosolic HPPD.

REFERENCES
1- Shulz A., Oswald O., Beyer P. and Kleinig H. (1993) FEBS Lett. 318, 162-166.

2- Luscombe B. M., Pallett K. E., Loubierre P., Millet J. C., Melgareja J. and Vrabel T. E. (1995) Proceedings Brighton Crop Prot. Conf. Weeds 1, 35-42.

3- Garcia I., Rodgers M., Lenne C., Rolland A., Sailland A. and Matringe M. (1997) Biochem. J. 325, 761-769.

4- Lindblad B., Lindstedt G., Lindstedt S., and Rundgren M. (1977) J. Biol. Chem. 25, 5073-5084.

5- Viviani F., Little J. and Pallett K. E. Pestic. Biochem. Physiol. (submitted)

6- Ellis M. K., Whitfield A. C., Gowans L. A., Auton T. R., McLean Provan W., Lock E. A. and Smith L. L. (1995) Toxicology and applied pharmacology 133, 12-19.

7- Fiedler E., Soll J. and Shultz G. (1982) Planta 155, 511- 515.

8- Sabourin J. E. and Bieber L. L. (1981) Arch. Biochem. Biophys. 206, 132-144.

9- Baldwin J. E., Crouch N. P., Fujishima Y., Lee M. H., MacKinnon C. H., Pitt J. P. N. and Willis A. A. (1995) Bioorg. Med. Chem. Lett. 5, 1255-1260.

EFFECTS OF AMITROLE AND NORFLURAZON ON CAROTENOGENESIS IN BARLEY PLANTS GROWN AT DIFFERENT TEMPERATURES

LA ROCCA N.*, BONORA A.°, DALLA VECCHIA F.*, BARBATO R.* AND RASCIO N.*
*Dipartimento di Biologia, via U. Bassi 58/B, I-35131 Padova (Italy)
°Dipartimento di Biologia, via Porta a Mare 2, I-44100 Ferrara (Italy)

Key words: Bleaching, carotenoids, chloroplast, ultrastructure, herbicide, T°dependence

1. Introduction

A carotenoid deficiency can account for serious alterations of thylakoids, due to the essential roles played by these pigments in membrane protection against photo-oxidative damage (1,2). In spite of carotenoid importance in preserving the photosynthetic apparatus, information on their biosynthesis in green tissues is rather scarce, and the knowledge of carotenogenesis and its regulation mainly comes from studies on chromoplasts of ripening fruits (3,.4). Previous research on the lycopinic *tigrina o³⁴* mutant of barley (5) and on barley plants treated with amitrole (6), a herbicide inhibiting lycopene cyclization (7), showed that plant growth temperature could affect carotenoid biosynthesis in leaf chloroplasts, suggesting the existence of thermo-modulated steps bypassing the block of mutation or herbicide. Recently, in tomato (4) and *Arabidopsis* (8) alternative reactions, which allow ß-carotene and xanthophylls to be synthesized without involving lycopene as intermediate, have been proposed. In order to better define the relationship between temperature and carotenoid synthesis in chloroplasts, barley plants have been grown at two rather close temperatures. They have been treated with two herbicides, to analyze whether interruption of the carotenogenic pathway at distinct points would cause differentiated responses to the change in plant growth temperature. The two chemicals used were norflurazon (NF) and amitrole (AM), which inhibit phytoene desaturation and lycopene cyclization, respectively (7). In leaves of plants grown in different experimental conditions, the total and relative quantities of carotenoids synthesized and the ratios between the pigment forms were analyzed. The ultrastructural organization, chlorophyll contents and photosynthetic activity of chloroplasts were also investigated, to verify the photooxidative damage caused by the disturbed carotenoid biosynthesis.

2. Materials and methods

Grains of barley (*Hordeum vulgare* L.) were germinated and seedlings were grown at 20°C and 25°C on water (control), 125 µM AM (2-amino-1,2,4-triazole) or 100 µM NF (4-chloro-5-(methylamino)-2-(α,α,α,-trifluoro-m-tolyl)-3-(2h)-pyridazinone) in a

G. Garab (ed.), Photosynthesis: Mechanisms and Effects, Vol. V, 3865–3868.
© 1998 Kluwer Academic Publishers. Printed in the Netherlands.

growth chamber with a 12h photoperiod and light of 200 $\mu mol\cdot m^{-2}\cdot s^{-1}$. All the analyses were carried out on the first leaf of 7-day-old plants. Carotenoids and chlorophylls were analyzed by reverse fase HPLC according to (9,10). The *in vivo* O_2 evolution from leaf tissues was measured as described in (11). Samples of leaf tissues were processed for electron microscopy and observed as described in (6).

3. Results and Discussion

At the growth temperature of 20°C ß-carotene and xanthophylls were present in the control leaf chloroplasts (Tab.1). The organelles contained large amounts of chloro-phylls and a well organized thylakoid system (Fig.1). In NF-treated leaves (Tab.1), on the contrary, phytoene, with its oxidation product OH-phytoene (12), was the only carotenoid precursor accumulated in chloroplasts. The total inhibition of the photo-protective pigment biosynthesis led to the absence of chlorophylls and to the complete dismantling of thylakoids.(Fig.2). Finally, in AM-treated leaves (Tab.1) chloroplasts contained carotenoid precursors up to lycopene, whose amount was about 10% of the compounds synthesized. Besides the persistence of precursors, however, a production of very limited quantities of ß-carotene and xanthophylls could also be detected in the organelles, and this accounted for the maintainance of some chlorophylls (Tab.1) and few, although altered, thylakoids (Fig.3). The rise in plant growth temperature to 25°C did not significantly affect either the pigment composition (Tab.2) or the ultrastructural organization (not shown) of the control chloroplasts. Also in the NF-treated organelles the herbicide effects were unchanged (Tab.2), thus showing that the temperature did not interfere with the first step of the carotenogenic pathway. In AM-treated plants, on the contrary, the rise in growth temperature caused noticeable changes in the responses to herbicide. In the leaf cells well organized chloroplasts (Fig.4) were present, whose chlorophyll contents and photosynthetic activity (Fig.5) were comparable with those of the control organelles. As far as the carotenoids were concerned, both the quantities and the component spectrum showed a remarkable recovery. The disturbing action of AM was still evidenced by the persistence of precursors up to lycopene. However the concentrations of these compounds were drastically reduced, while the ß-carotene and xanthophyll levels were sharply increased (Tab.2). The recovery of the carotenoid amounts in AM-treated leaves was connected with the normalization of several parameters pertinent to the pigments present in chloroplasts (Tab.3), such as the xanthophyll/carotene and chlorophyll *a/b* ratios. Also the chlorophyll/carotenoid ratio, indicative of the photoprotective ability of these latter pigments, approached that of the control organelles. Furthermore, the analysis of percentage of the different carotenoid forms, referred to the total quantities of pigments synthesized (Tab.4), clearly evidenced that in all the experimental conditions the proportions between ß,ε-carotenoids and ß,ß-carotenoids remained almost constant, around values of about 40% and 60%, respectively. This suggested the existence of a tight regulation of the ß and ε cyclization ways, which was unaffected by the AM treatment with the consequent situation of photooxidative stress. The different experimental conditions, on the contrary, showed an evident effect on the relative proportions of the ß,ß-carotenoids and, in particular, on the ratios between ß-carotene and xanthophylls of the violaxanthin cycle (Tab.4). Thus, while in the control leaves the (violaxanthin +antheraxanthin)/ß-carotene ratio was about 0.6, in the photooxidative conditions caused by AM at 20°C the value of this ratio rose to 4.9, with a quantity of antheraxanthin reaching 23% of the synthesized carotenoids, against the 0.8% of the control. This situation, although very subdued, persisted in chloroplasts of AM-treated leaves at 25°C, where the (violaxanthin+antheraxanthin)/ß-carotene ratio was about 0.9 and the antheraxanthin proportion (4.2%) was still higher than in the control organelles (1.2%).

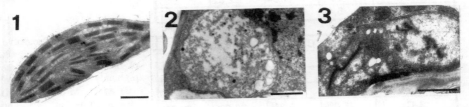

"Figures 1-3." Plant grown at 20°C. *"Figure 2."* Well organized chloroplast of a control plant. Fig.2. Damaged chloroplast of a NF-treated plant. *"Figure3."* Damaged chloroplast of an AM-treated plant. (bars = 1 µm).

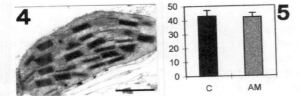

Figure 4." Well organized chloroplast of an AM-treated plant grown at 25°C (bar = 1 µm)
"Figure 5." Oxygen emission (µmol O_2 /h/mg chl (a+b)) from leaves of control (C) and AM-treated (AM) plants grown at 25°C.

Tab.1	20°C			25°C		
	C	AM	NF	C	AM	NF
Car. precursors						
Phytoene	/	33,38	83,1	/	3,91	72,2
OH-phytoene	/	5,51	9,0	/	0,73	12,1
Phytofluene	/	1,55	/	/	0,17	/
ξ-carotene	/	1,93	/	/	/	/
Lycopene	/	5,94	/	/	0,22	/
Carotenes						
ß-carotene	62,9	0,89	/	71,3	47,4	/
Xanthophylls						
Antheraxanthin	1,8	2,2	/	3,0	8,8	/
Violaxanthin	39	2,13	/	41,9	37,1	/
Neoxanthin	26,6	0,69	/	30,1	29,1	/
Lutein	93,6	3,67	/	104,7	87,0	/
Chlorophylls						
Chlorophyll a	891,5	15,99	/	1067,5	775,1	/
Chlorophyll b	326,5	11,93	/	365,6	308,8	/

"Table 1." Precursor and pigment quantities (µg/g f.w.) in leaves of control (C), AM-treated (AM) and NF-treated (NF) plants grown at 20°C and 25°C.

Tab.2	20°C		25°C	
	C	AM	C	AM
C+X	223,90	9,60	251,00	209,60
X	161,00	8,70	179,70	162,20
C	62,90	0,90	71,30	47,40
X/C	2,56	9,76	2,52	3,42
Chl (a+b)	1218,00	27,90	1433,10	1083,90
Chl a/b	2,73	1,34	2,92	2,51
Chl(a+b)/X+C	5,44	2,91	5,71	5,17

"Table 2." Total quantities (µg/g f.w.) and ratios of the different pigments in leaves of control (C) and AM-treated (AM) plants grown at 20°C and 25°C. (C = carotenes, Chl = chlorophyll, X = xanthophylls)

Tab.3	20°C		25°C	
	C	AM	C	AM
ß,ε-car (L)	42,0%	38,3%	42,0%	41,5%
ß,ß-car	58,0%	61,7%	58,0%	58,5%
ß-C	28,0%	9,3%	28,0%	22,5%
A+V+N	30,0%	52,4%	30,0%	36,0%
A+V/ß-C	0,65	4,90	0,63	0,97

"Table 3" Photoprotective carotenoid composition as % of the total amount of these pigments in leaves of control (C) and AM-treated (AM) plants grown at 20°C and 25°C . (A = antheraxanthin, C = carotene, car = carotenoids, L = lutein, N = neoxanthin, V = violaxanthin)

4. References

1 Young, A.J. (1991) Physiol. Plant. 83, 702-708

2 Havaux, M. (1988) Trends Plant Sci. 3, 147-151

3 Fraser, P.D., Truesdale, M.R., Bird, C.R., Schuck, W. and Bramley, P.M. (1994) Plant Physiol. 105, 405-413

4 Pecker, I., Gabbay, R., Cunningham, F.X. Jr. and Hirschberg, J. (1996) Plant Mol. Biol. 30, 807-819

5 Casadoro, G., Høyer-Hansen, G., Kannangara, C.G. and Goug, S.P. (1983) Carlsberg Res. Commun. 48, 95-129

6 Agnolucci, L., Dalla Vecchia, F., Barbato, R., Tassani, V., Casadoro, G. and Rascio N. (1996) J. Plant Physiol. 147, 493-502

7 Barry, P. and Pallett, K.E. (1990) Z. Naturforsch, 45, 492-497

8 Cunningham, F.X. Jr., Pogson, B., Sun, Z., McDonald, K.A., Della Penna, D. and Gantt, E. (1996) Plant Cell 8, 1613-1626

9 Bonora, A., Pancaldi, S., Gualandri, R. and Fasulo, M.P. (1996) 10th FESPP Congress, Florence, Italy (Sept. 9-13), Plant Physiol. Biochem. (special issue) S09- 05, 106

10 Pancaldi, S., Bonora, A., Gualandri, R., Gerdol, R., Manservigi, R. and Fasulo, M. P. (1998) Bot. Acta 111, 1-11

11 Ishii, R., Yamagishi, T.Y. and Murata, Y. (1977) Jap. J. Crop Sci. 46, 53-57

12 Sandmann, G. and Albrecht, M. (1990) Z. Naturforsch. 45, 487-491

EFFECTS OF SALICYLIC ACID AND RELATED COMPOUNDS ON PHOTOSYNTHETIC PARAMETERS IN YOUNG MAIZE (*ZEA MAYS* L.) PLANTS

Janda, T.[1], Szalai, G.[1], Antunovics, Zs.[1] Ducruet, J.-M.[2] and Páldi, E.[1]

[1]Agricultural Research Institute of the Hungarian Academy of Sciences, H-2462, Martonvásár, POB 19, Hungary
[2]Section de Bioenergetique, DBCM, CEA Saclay, 91191, Gif-sur-Yvette CEDEX, France

Key words: Chl fluorescence induction, CO_2 uptake, cold stress, gas exchange, non-photochemical quenching, stomata.

1. Introduction

Salicylic acid is known to be a signal molecule in acquired resistance to pathogens in several plant species (1). It was also reported that SA accumulates during exposure to ozone or UV light (2,3), and SA treatment improved the chilling tolerance of maize (4) and the heat-shock tolerance of mustard plants (5). Recently we observed that some compounds similar to salicylic acid (benzoic acid, acetyl-salicylic acid, but not sulpho-salicylic acid) also reduced the symptoms of chilling injury in young maize plants. The effects on the photosynthetic parameters of these compounds will be discussed.

2. Procedure

2.1 Plant material

Sterilized seeds of maize (*Zea mays* L., hybrid Norma) were germinated for 4 days at 26 °C, then grown for 2 weeks in Hoagland solution at 22/20 °C with a 16/8-h light/dark periodicity in a Conviron PGV-36 plant growth chamber. The PPFD at leaf level was 200 $\mu mol\ m^{-2}\ s^{-1}$ provided by metal halide lamps, with a relative humidity of 75 %. The cold treatment was carried out in the same chamber at 2 °C.

2.2 Chlorophyll fluorescence induction measurements

The chlorophyll fluorescence induction parameters of the youngest fully expanded leaves were determined after 30 min dark adaptation at room temperature using a pulse

3869

G. Garab (ed.), Photosynthesis: Mechanisms and Effects, Vol. V, 3869–3872.

amplitude modulated fluorometer (PAM-2000, Walz, Effeltrich, Germany) with PPFD = 120 μmol m^{-2} s^{-1} red actinic light at room temperature. Quenching parameters were calculated after 15 min according to Schreiber *et al.* (6).

2.3 Gas exchange measurements

The youngest fully developed leaves were chosen for net photosynthesis measurements, which were made on attached leaves using a LI-6400 infrared gas analyser (LICOR, Lincoln, Nebraska, USA) operated with a 6400-02 LED light source. The measurements were carried out at room temperature. Gas exchange parameters were calculated according to von Caemmerer and Farquhar (7). Pigment contents were determined from 100 % acetone extracts.

2.4 Statistical analysis

The results are the means of at least 3 measurements and were statistically evaluated using the standard deviation and T-test methods.

3. Results and Discussion

Young maize plants grown at 22/20 °C in a hydroponic solution were treated with 0.5 mM salicylic acid, benzoic acid, aspirin (acetyl-salicylic acid) or sulpho-salicylic acid for 1 day under normal growth conditions (22/20 °C) and then chilled at 2 °C.

After a few days of cold treatment severe damage symptoms could be observed visually in untreated plants, while those which were pretreated with salicylic acid, benzoic acid or aspirin were hardly affected by low temperature: these plants looked much as they did before the cold treatment. Sulpho-salicylic acid had no effect on the chilling symptoms. However, it was also observed that SA, benzoic acid and acetyl-salicylic acid may cause severe damage to the plants under normal growth conditions. The damage symptoms are more pronounced in the light than in the dark (data not shown).

The effects of pre-treating of young maize plants with 0.5 mM salicylic acid, benzoic acid and acetyl-salicylic acid at growth temperature (22/20 °C) on the gas exchange parameters are shown in Table 1.

After 1 day a significant decrease occurred in net photosynthesis. The dark CO_2 consumption does not change significantly after 1 day of SA treatment (4). Similarly, there was a strong decrease in stomatal conductance (g_s), transpiration (E) and intercellular CO_2 content (C_i). These results suggest that the compounds investigated may cause a decrease in the photosynthetic capacity under normal growth conditions. As was the case for protection against chilling injury, sulpho-salicylic acid had no significant effect on gas exchange parameters either (data not shown).

Table 1. Effect of 1 day of 0.5 mM salicylic acid, aspirin or benzoic acid treatment at 22/20 °C on net photosynthesis rate (P_N; $\mu mol\ CO_2\ m^{-2}\ s^{-1}$), stomatal conductance (g_s; $mol\ H_2O\ m^{-2}\ s^{-1}$), transpiration (E; $mmol\ H_2O\ m^{-2}\ s^{-1}$) and intercellular CO_2 concentration (C_i; $\mu mol\ CO_2\ mol^{-1}$ air) in young maize plants at 200 $\mu mol\ m^{-2}\ s^{-1}$. *, **, ***: significant at the 0.05, 0.01 and 0.001 levels, respectively, compared to the untreated plants.

Treatment	P_N	g_s	E	C_i
No addition	10.6 ±0.6	0.108 ±0.023	1.229 ±0.252	194 ±26
Salicylic acid	7.9 ±1.4***	0.058 ±0.012***	0.759 ±0.127***	139 ±12***
Aspirin	9.1 ±1.8*	0.065 ±0.019***	0.871 ±0.240***	124 ±23***
Benzoic acid	9.4 ±0.6***	0.072 ±0.017**	0.986 ±0.268**	138 ±24***

There was no significant change in the chlorophyll and carotenoid contents after 1 day of treatment (Table 2).

Table 2. Effect of 1 day of 0.5 mM salicylic acid, aspirin or benzoic acid treatment at 22/20 °C on the chlorophyll and carotenoid contents (μg pigment g^{-1} FW) and chlorophyll(a/b) ratio in young maize plants.

Treatment	Chl-a	Chl-b	Chl(a/b)	Carotenoid
No addition	1688 ±286	459 ±98	3.698 ±0.149	426 ±83
Salicylic acid	1941 ± 90	547 ±33	3.552 ±0.064	539 ±36
Aspirin	1922 ± 61	540 ±18	3.558 ±0.019	517 ±36
Benzoic acid	1636 ±228	463 ±69	3.536 ±0.093	454 ±58

The effects of pre-treating of young maize plants with 0.5 mM salicylic acid, benzoic acid and acetyl-salicylic acid for 1 day at 22/20 °C on the chlorophyll fluorescence induction parameters are shown in Table 3.

Table 3. Effect of 1 day of 0.5 mM salicylic acid, aspirin or benzoic acid treatment at 22/20 °C on the chlorophyll fluorescence induction parameters in young maize plants. Measurements were carried out after at least 30 min dark adaptation. *, ***: significant at the 0.05 and 0.001 levels, respectively, compared to the untreated plants.

Treatment	F_v/F_m	$(F_m'-F_s)/F_m'$	q_P	q_N
No addition	0.769 ±0.005	0.652 ±0.006	0.917 ±0.011	0.154 ±0.009
Salicylic acid	0.763 ±0.013	0.596 ±0.091	0.863 ±0.106	0.261 ±0.056*
Aspirin	0.781 ±0.005	0.595 ±0.056	0.906 ±0.012	0.428 ±0.132*
Benzoic acid	0.775 ±0.012	0.618 ±0.020*	0.903 ±0.009	0.328 ±0.010***

It can be seen from Table 3 that there was no significant difference in the F_v/F_m and photochemical quenching (q_P) parameters after 1 day of treatment under these experimental conditions. There was a slight decrease in the quantum yield of PS II [(F_m'-F_s)/F_m'], which was statistically significant in the benzoic acid-treated plants. These results suggest that the decrease in net photosynthesis is not due to the depression of PS II, but is possibly due to the decrease in stomatal opening or to a decrease in the activity of the enzymes responsible for CO_2 fixation, as was suggested for barley plants treated with salicylic acid (8). However, it should be mentioned that using higher concentrations and irradiance, a significant decrease in F_v/F_m and the quantum yield of PS II can be observed (data not shown). A significant increase occurred in the non-photochemical quenching (q_N) parameter after 1 day of salicylic acid, benzoic acid or acetyl-salicylic acid treatment even under the low irradiance used in these experiments.

From these results it can be concluded that exogenous salicylic acid and certain related compounds (for example benzoic acid or acetyl-salicylic acid) may cause severe damage in young maize plants. This stress can be observed as a decrease in the photosynthetic capacity. However, after this stress effect the plants become more resistant to other stresses, such as low temperature.

4. Addendum

This work was supported by OTKA grants Nos. F16001, F26236 and T21115.

References

1 Raskin, I. (1992) Ann. Rev. Plant Physiol. Mol. Biol. 43, 439-463
2 Yalpani, N., Enyedi, A.J., León, Y. and Raskin, I. (1994) Planta 193, 372-376
3 Sharma, Y.J., Leon, J., Raskin, I. and Davis, K.R. (1996) Proc. Natl. Acad. Sci. USA 93, 5099-5104
4 Janda, T., Szalai, G., Tari, I. and Páldi, E. (1997) in Crop Development for the Cool and Wet Regions in Europe (Sowinski P., Zagdanska B., Aniol A. and Pithan K. eds) pp.179-187, European Communities, Belgium
5 Dat, J.F., Lopez-Delgado, H., Foyer, C.H. and Scott, I.M. (1998) Plant Physiol. 116, 1351-1357
6 Schreiber, U., Schliwa, U. and Bilger, W. (1986) Photosynth. Res. 10, 51-62
7 von Caemmerer, S. and Farquhar, G.D. (1981) Planta 153, 376-387
8 Pancheva, T.V., Popova, L.P. and Uzunova, A.N. (1996) J. Plant Physiol. 149, 57-63

PHOTOSYNTHETIC APPARATUS OF MAIZE PLANTS AS INFLUENCED BY SPRAYING WITH UREA

Moursi, M.A., Nour El-Din, N.A.; El-Bagoury,O.H.,Saad,A.O.M. and Ashour,N.I.

Fac.of Agric.Ain Shams Univ. and National Res. Center,Dokki- Giza Egypt.

key words: Foliar, pigments,carotenoids,chloroplasts,porphyrin, stability, enzyme.

Abstract:

An increase in chlorophyll and carotenoids content , stability of chlorophyll- protein-lipid complex, photosynthetic activity and protein in isolated chloroplasts were found with increasing urea concentration up to 4 % in foliar sprays. Increasing urea in the aqueous solution from zero to 6 % significantly increased the ratio of chlorophyll a /b, whereas it depressed chlorophyllase enzyme activity and protein chlorophyll ratio in isolated chloroplasts.

Introduction

The photosynthetic apparatus (chloroplast) contains nitrogen as atoms in the chlorophyll molecule. Nitrogen is added to the plant generally in two forms , i.e,oxidated form (NO_3) and reduced form (NH_4).

It is well known that nitrate is reduced in the leaves to NH_4. It seems that the availability in the leaves of the reduced form of nitrogen such as urea as well as the product of its hydrolysis may spare the photosynthetic reducing agent , and this in turn may increase plant productivity. Part of the photosynthetic energy is consumed in this reduction (Voskresenskaya and Greshina, 1962).

A positive effect of foliar spraying with urea on photosynthetic pigments content of maize plants was reported by several investigators (Yakout *et al*, 1980, Saad *et al*. , 1981 and Ashour *et al* 1983)

Increasing the tonnage of maize can be achieved by horizontal expansion and / or by vertical expansion. Vertical expansion could be realized in many ways , the most important of which is nitrogen fertilization. The yield capacity of maize is influenced by nitrogen fertilization through many factors among which photosynthetic efficiency plays a significant role. Therefore this study was designed to investigate the effect of different concentrations of urea solution on the photosynthetic apparatus of maize.

Procedure

Pot experiment was conducted to study the effect of different concentrations of urea solution on photosynthetic apparatus of maize plant.

Earthenware pots 40 cm. in diameter and 40 cm. in depth were used and each pot was filled with 28 kg of sandy loam soil . Grains of Giza -2 zea maize cultivar were sown in each pot. :plants were thinned to one plant per pot 15 days after sowing . Calcium super phosphate (16 % P_2O_5) was applied at the rate of 4.49 g/plant. The experiment included four treatments and nine replicate for each treatment . Seedling were sprayed 30 days after sowing with aqueous solution of urea at a concentration of 0.0, 2.0 , 4.0 and 6.0 % . The treatments were arranged in a complete randomized system. After 50 days from sowing, samples were taken from every treatment for the photosynthetic apparatus studies. The fourth upper leaf was chosen for all studies.

G. Garab (ed.), Photosynthesis: Mechanisms and Effects, Vol. V, 3873–3876.

Photosynthetic pigments content on the basis of blade area (mg/ dm2) (chl. a, chl. b, chl. a+b and carotenoids) was calculated by means of Wettstein 's formula (von Wettstein, 1957). The stability of chlorophyll - protein lipid complex was estimated by using Sapozhnikov and Chernomorsky (1960) method. The activity of chlorophylase was determined according to the method of Vorobyova et al. (1963). The photosynthetic activity in isolated chloroplast was determined by the method of Holt et al. (1951). Chlorophyll₇ protein content and protein/ chlorophyll ratio in isolated chloroplasts were also determined.

The analysis of variance of the completely randomized design was used according to the method described by Snedcor and Cochran (1967). Means followed by the same alphabetical letters did not differ significantly according to Duncan's multiple range test (Duncan , 1955).

Results and Discussion
1. Effect on photosynthetic pigments content:

The greatest photosynthetic pigments content (chlorophyll a, chlorophyll b, chlorophyll a+b and cartoenoids) in blades of maize plant were obtained by spraying with 4.0 % of urea in aqeous solution. The increase in photosynthetic pigments content by increasing urea concentration (nitrogen level) in the spray solution might be attributed to the incorporation of nitrogen in the porphyrin ring which constitutes the main skeleton of the chlorophyll molecule.

Table 1: Effect of spraying with different concentration of urea solution on photosynthetic pigments in leaf bladesof maize after 50 days from sowing (mg/dm

Urea conc.	Chl. a	Chl. b	Chl. a+b	Carot.	Chl. a/b	Chl.a +b/ carot.
0	0.68 a	0.20 a	0.88 a	0.39 a	3.48 a	2.24 a
2 %	0.97 a	2.4 b	1.21 b	0.50 b	4.02 b	2.46 a
4%	1.26 c	0.28 c	1.54 c	0.69 d	4.53 b	2.24 a
6 %	1.03 b	0.18 a	1.21 b	0.55 c	5.73 c	2.19 a

Moreover, carotenoids protect chlorophyll a from destruction (Ashour, 1964). Carotenoids absorb the high -energy waves , i.e short waves and this in turn protects the enzymes of the cell from destruction.On the other hand , the reduction in photosynthetic pigments content with increase in spraying solution from 4.0 % to 6.0 % might be attributed to an increase in chlorophyllase activity.

Increasing urea concentration from 0 to 6.0 % enlarged the ratio of chlorophyll a/ chlorophyll b significantly in maize leaves, and did not affect the ratio of chlorophyll a+b / carotenoids (Table 1).

2. Effect on the stability of chlorophyll- protein- lipid complex and the activity of chlorophylase enzyme.

Spraying maize plants with increasing concentration of urea solution up to 4.0 % caused an increase in the stability of chlorophyll complex in tissues of maize leaves . The difference between 4.0 % and 6.0 % was not great enough to reach the 5 % level of significance. Apparently a large portion of chlorophyll in the leaves of maize plants sprayed with high concentration of urea become well bound to the lipo- protein in lamellae in chloroplasts. Increasing urea concentration from 0 to 4.0 % increased stability of chlorophyll in maize leaves by 19.7 %.

Table 2: Effect of spraying with different concentrations of urea solution on photosynthetic apparatus efficiency in leaf blades of maize and its isolated chloroplasts after 50 days from sowing

Urea conc.	Stability of chl.-protein-lipid complex*	chloro-phyllase activity *	Photosynthetic activity in isolated chloroplast ***	Protein content in isolated chloroplast mg/g fresh leaves	Chl. content in isolated chloroplasts mg/g fresh leaves	Protein chlorophyll ratio in isolated chloroplasts wt: wt.
0	50.8 a	9.56 a	342.9 a	6.06 a	.074 a	81.9
2 %	54.2 a	8.01 a	7119.2 b	6.23 a	.097 b	64.5
4 %	70.5 b	6.75 a	9427.1 c	8.57 c	1.49 c	57.4
6 %	69.1 b	7.43 a	7327.0 b	6.97 b	.145 c	48.0

* Percentage of chlorophyll not extracted with 1 % ethanol in petroleum ether.
** Percentage of chlorophyll split during 1 hr incubation in the dark room temperature.
*** μ moles of DIP which was reduced by 1 mg chlorophyll hr.

There was no relationship between chlorophyllase activity and the concentration of urea in the foliar spray . However , the treatments that caused no statistically significant reduction in chlorophyllase activity , tended to increase the chlorophyll content in the leaves. In this connection , Sivetsev *et al (1973)*. reported that an inverse correlation exists between activity of chlorophyllase and the content of chlorophyll in tomato leaves . It was observed also that the leaves in which the chlorophyll was more stable had a lower chlorophyllase activity. Ashour and Thalooth (1971) observed that the leaves of sugar beet plants in which the chlorophyll was more stable had lower chlorophyllase activity.

3. Effect on photosynthetic activity in isolated chloroplasts :

The maximum photosynthetic activity in the isolated chloroplasts was obtained with spraying maize plants with 4 % urea solution . Increasing urea solution concentration from 0 to 4 % increased micro moles of DIP which was reduced by 1 mg chlorophyll / hr by 5943.2 μ moles , Confirming results obtained by Tanaka and Hara (1970). They found that increasing nitrogen concentration in the nutrient solution from 0 to 50 ppm , augmented chlorophyll content and photosynthetic rate of maize plants whereas opposite results were obtained by increasing nitrogen concentration from 150- 400 ppm where chlorophyll content remained the same , but photosynthetic rate decreased. The greatest photosynthetic activity was obtained by spraying maize plants by 4 % urea..

However, many invistigators, Kirichenko (1964) ' Saric et al. (1965) and Nevins and Loomis (1970) found an increase in photosynthetic activity with adding nitrogen.

It was observed that the plants which possesed higher photosynthetic activity (sprayed with 4 % urea) had a more stable chlorophyll- protein- lipid- complex than the others , and their chlorophyll was less destructible by chlorophyllase . In this regard Osipova and Ashour (1964)' Ashour and Osipova(1965) found a correlation between the chlorophyll stability and the rate of photosynthesis in maize plants.

The increase in photosynthetic activity in maize leaves with increase in concentration of urea might be attributed to the existence of reduced form of nitrogen (NH_3) in urea. In this respect , Voskresenskaya and Greshina (1962)reported that the more oxidized form of nitrogen such as NO_3 is known to use part of the photosynthetic energy for its reduction to NH_4.

4. Effect on protein content in isolated chloroplasts:

The greatest protein content in isolated chloroplasts was obtained with spraying maize plants with 4 % urea in urea solution (Table 2).

There was a consistent depression in protein / chlorophyll ratio in isolated chloroplasts by increasing the concentration of the urea solution up to the greatest level ,i.e. 6% . Bazynski et al.,1972 found that maize plants suffering from nitrogen deficiency had a very pronounced increase in the ratio of chloroplast protein to chlorophyll.

The results in Table 2 show an inverse relationship between the protein / chlorophyll ratio in the isolated chloroplast and the stability of chlorophyll -protein- lipid complex. These results indicate that most of the additional protein is located in the stroma rather than in the grana, i.e. it adds to the enzymatic proteins. In other word , the protein responsible for improving the stability of chlorophyll is only a small portion of the treatment- induced protein increase.

References

1- Voskresenskaya, N.P. and Greshina, G.S. (1962).Fiziol. Rast, 9, 7-15.

2- Yakout, G.M.,Saad, A.O.M., El- Moursy, A.,and Ashour, N.I. (1980). Egypt. J. Agron. 5,No.1, 35-44.

3- Saad, A.O.M., El- Moursi, A., and Yakout, G.M. (1981). Res.Bull. of Fac. of Agric. Ain Shams Univ. Jan. 1991 (1441).

5- Ashour, N.I., Saad, A.O.M. and Thalooth, A.T. (1983). Proc. of the 1st conf. of Agron. Egypt. Soc. of Crop Sci.,1 (A), 12, 257- 272.

6- von Wettestein(1975). Exptl. Cell Res., 12, 427- 433.

7- Sapozhnikov, D.L. and Chernomorky, S.A. (1960). Sov. Plant Physiol., 7, 664.

8- Vorobyova, L.M., Plstrova, M.E. and Krasonovsky, A.A. (1963). Sov. Biochem.,28, 524-529.

9- Snedecor, G.W., and Cochran, W.G. (1967). Statistical Methods: Iowa State College Press, 6 th Ed. USA.

10- Duncan,D.B. (1968). Biometries, 11: 1-42.

11- Ashour, N.I. (1964). Ph.D. Thesis, Acad. Sci.,USSR, Moscow.

12-Sivetsev, M.V., Ponomareva, S.A. and Kuznetzova, E.A. (1973). Fiziol. Rast, 20, 62- 65.

13- Tanaka, A., Yamaguchi, J. and Imoi, M.(1974). Nippon Dojohiryagoaku Zasshi, 42, 33-36.

14 Kirchenko, E.B. (1964). Kishinev, 2, 36-38.

15 Saric, M. Cupina, T. and Geric, I. (1965). Opzockstns Gent, 30, 1045-53.

16 Nevins, D.S. and Loomis, R.S. (1970). Crop Sci., 10, 21-25.

17 Ashour, N.I. and Osipova, O.P. (1965). Dokl. Akad. Nauk. USSR, Moscow, 163, 11- 514.

18 Bazynski, T. Br. and J. Barr, R., Krogmann, D.W. and Crane, F.L.(1972). Plant Physiol., 50, 410- 411.

IMPAIRMENT OF PHOTOSYSTEM II ACCEPTOR SIDE OF SPINACH CHLOROPLASTS INDUCED BY TRICOLORIN A

L. Achnine[1], R. Pereda-Miranda[2], R. Iglesias-Prieto[3] and B. Lotina-Hennsen[1]

Departamento de Bioquímica; [2]Departamento de Farmacia. Facultad de Química. UNAM. México, D.F. 04510. [3]Estación de Investigaciones Marinas "Puerto Morelos", ICMyL-UNAM. Apartado Postal 1152, Cancún, Quintana Roo 77500, México.

Key words: photosystem II, inhibitors, electron transport, Q_B, fluorescence, spinach.

Introduction

Farmers in the southern intertropical Mexican state of Morelos use *Ipomoea tricolor* Cav. (Convolvulaceae) during the fallow period in sugar-cane fields. This plant has the useful property of suppressing the growth of other plants, including invasive weeds. Tricolorin A, the major phytogrowth inhibitor present in the active mixture of resin glycosides, was characterized as a tetrasaccharide of jalapinolic acid with a macrocyclic lactone-type structure (Fig. 1). Bioassays showed that radicle growth of *Amaranthus leucocarpus* Watts. and *Echinochloa crus-galli* (L.) Beauv was inhibited by this allelopathic agent with IC_{50} values ranging from 12 to 37 μM (1). However, the primary mode of action of tricolorin A has not been established, nor any biochemical insights as to how this oligosaccharide might contribute to the chemical ecology of *Ipomoea tricolor*, including its possible interference with the photosynthetic metabolism of other plant species. In this context, this work reports the results of further studies on the effects of tricolorin A on several photosynthetic activities in isolated spinach chloroplasts.

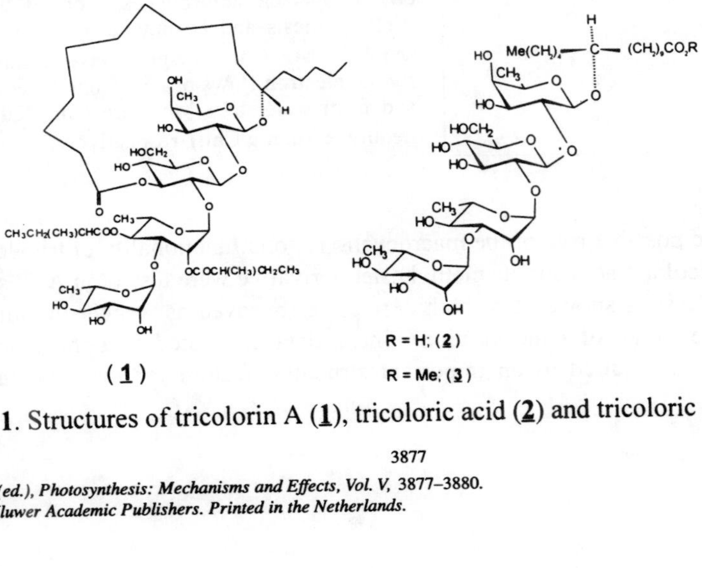

Figure 1. Structures of tricolorin A (**1**), tricoloric acid (**2**) and tricoloric acid methyl ester (**3**).

G. Garab (ed.), Photosynthesis: Mechanisms and Effects, Vol. V, 3877–3880.
© 1998 Kluwer Academic Publishers. Printed in the Netherlands.

2. Procedures

Tricolorin A was obtained by HPLC purification from the $CHCl_3$-soluble glycoresins of *Ipoma tricolor* Cav. (Convolvulaceae) according to the procedures previously described (1). On alkali hydrolysis, tricolorin A liberated an H_2O-soluble glycosidic acid, designated as tricoloric a (Fig. 1). This compound was alkylated by CH_2N_2 treatment to obtain tricoloric acid methyl es (Fig. 1). Chloroplasts were obtained from spinach leaves (*Spinacea oleraceae* L.) as descri (2). Chlorophyll levels were determined as reported (3). Bioassays including proton uptake, A synthesis, electron transports in spinach chloroplasts were performed according to establish procedures (2, 4-6). KCN (0.1 mM) was added to inhibit catalase activity. Chloroph fluorescence induction curves were captured with a portable shutter less apparatus (PE Hansatech UK) as previously described (7). Kinetic analyses of the relative variable fluorescer were performed by deconvolution of data collected during the first 2 ms with 10 μs resoluti Deconvolution analyses were performed with a non linear fitting procedure.

3. Results and Discussion

3.1 Polarographic studies. Tricolorin A inhibited the ATP formation and H^+-uptake with I_{50} of and 12 μM, respectively (Fig. 2). Photosystem I activity, measured from reduced dichlorophe indophenol ($DCPIPH_2$) to methylviologen (MV), was insensitive to tricolorin A (Fig. 2). T allelochemical did not affect the electron flow from water to silicomolibdate (SiMo), inhibited photosystem II electron transport from water to DCPIP (Fig. 2). Consequently, t natural product (20 μM) inhibited the redox chain of spinach chloroplasts at the level of Q_B. site of inhibition and the concentration that totally inhibited electron transport were similar those exhibited by DCMU.

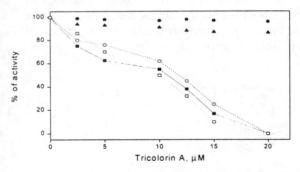

Figure 2. Effects of increasing concentrations tricolorin A on ATP synthesis ($\bar{}$), proton upt (O) and partial reactions of photosystem I (from water to DCPIP (■), and from water to Si (●) in spinach chloroplasts. Control values ATP synthesis and H^+-uptake were 350 and $\mu mol/(h.$ Mg Chl.), respectively. Control val for the electron flows of PSI, from water to DC and from water to SiMo were 650, 720 and $\mu equiv. e^-/(h.$ mg Chl.), respectively.

In order to investigate the possible role of the macrocyclic lactone functionality of tricolorin A the inhibitory activity, tricoloric acid and its methyl ester derivative were tested on ATP synthe and electron transports. Results showed that both derivatives behaved as weaker inhibitors t tricolorin A at the same range of concentration. These data suggested that the macrocy structure of tricolorin A is required as an important structural feature for maximal inhibit activity.

2. Chlorophyll a *fluorescence studies.* Isolated spinach chloroplasts showed a polyphasic ﬂorescence curve with regular *O-J-I-P* sequence of transients (Fig. 3A) similar to those eviously described for several intact organisms (7,8). Addition of DCMU resulted in the loss of e regular *OJIP* sequence and the formation of an *OJ* sequence (Fig. 3A). This is due to the cumulation of Q_A^- during the first 2 ms of the induction curve, in contrast to control cloroplasts ich required approximately 900 ms to close all PSII reaction centers. Comparison of the orescence induction curves of chloroplasts infiltrated with DCMU with those obtained from loroplasts exposed to tricolorin A clearly showed remarkable similarities (Fig. 3B). Consistent th the loss of electron transport capacity, chloroplasts exposed to different concentrations of colorin A showed a significant concentration-dependent reduction in their relative Q_A^- re-idation capacity (Fig. 4).

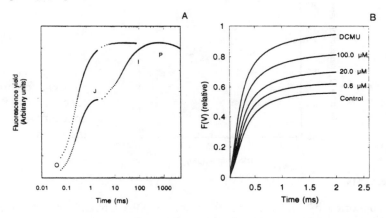

igure 3. A) Fluorescence induction curves of control chloroplasts and samples with thenticated acceptor side damage (+DCMU). B) Chlorophyll fluorescence induction curves of ntrols and samples exposed to increasing concentrations of tricolorin A.

he rise of variable fluorescence ($V=(F_t-F_o)/(F_m-F_o)$) during the first 2 ms of the induction curve control chloroplasts required two components to be accurately described: a fast sigmoidal mponent with time constant of around 0.1 ms^{-1} and a slow exponential one with time constant ound 0.4 ms^{-1}. Kinetic analysis of induction curves revealed that acceptor side inactivation of SII, produced by infiltration with DCMU, resulted in increases in the amplitude of both xponential and sigmoidal components without any significant changes in their rate constants lative to control chloroplasts. Comparison of the kinetic characteristics of samples with thenticated acceptor side damage, by tricolorin A, and those obtained from control chloroplasts dicated that the increase in fluorescence yield resulted from an increment in the amplitude of e sigmoidal components. The results showed no detectable variation in the rate constants of oth sigmoidal and exponential components. However, connectivity of the sigmoidal components as been lost with the increase of tricolorin A concentrations.

From the results presented, it is concluded that tricolorin A, the major constituent of
allelopathic glycoresins of *Ipomoea tricolor* (Convolvulaceae), is a potent natural inhibitor of
electron transport between Q_A^- and Q_B at the acceptor side of PSII. Evidence of this inhibiti
based on Chl. *a* fluorescence analysis was consistent with polarographic data. Moreover, t
target and the concentration of tricolorin A needed for total inhibition of the electron transp
were similar to those exhibited by the commercial herbicide DCMU.

Thus, tricolorin A is suitable for further development as an effective "green" herbicide and/or
the design of new biodegradable agents with herbicide activity.

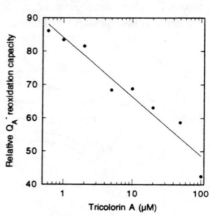

Figure 4. Q_A^- re-oxidati
capacity (relative)
spinach chloroplasts
function of increas
concentrations
tricolorin A.

4. Acknowledgements

This work was supported by grants IN205197 from DGAPA-UNAM; 005341 from PADE
UNAM and 400313-5-2358PN and 25118N from CONACyT. L. Achnine acknowledges
fellowship awarded by DGEP-UNAM to carry out his Ph.D. studies.

References

1. Bah, M. and Pereda-Miranda, R. (1996) Tetrahedron, 52, 13063-13080.
2. Peña-Valdivia, C.B.; Rodríguez-Flores, L.; Tuena de Gómez Puyou, M. and Lotina-Henns
   B. (1991). Biophys. Chem. 41. 169-174.
3. Strain, H.H.; Coppe, B.T. and Svec, W.A. (1971) In Methods in Enzymology (San Pietro,
   ed.) Vol. 3, pp. 452-466, Academic Press, New York.
4. Dilley, R.A. (1972). In Methods in Enzymology (San Pietro, A. ed.) Vol. 24. pp 68-
   Academic Press, New York.
5. Lotina-Hennsen, B.; Roque-Reséndiz, J.L.; Jiménez, M. and Aguilar. M. (1991)
   Naturforsch. 46c, 777-780.
6. Allen, J.F. and Holmes. N.G. (1986) In Photosynthesis Energy Transduction: A Practi
   Approach. (Hipkins. MF & Backer, NR (eds.). pp. 103-141, IRL Press. Oxford. U. K.
7. Strasser. R.J.: Srivastava. A. and Govindjee (1995) Photochem. and Photobiol. 61. 32-42.
8. Iglesias-Prieto. R. (1995) In Photosynthesis: From Light to Biosphere. (Mathis. P., ed.) V
   IV. pp. 793-796. Kluwer Academic Publishers, Dordrecht, The Netherlands.

[3]H-LEUCINE INCORPORATION IN SHORT AND LONG-TERM HERBICIDE TREATMENTS IN VARIOUSLY TOLERANT HIGHER PLANTS

E Pace[1], D Esposito[1], M Rizzuto[1], L Pompili[1] and MT Giardi[2]
1) Institute of Nuclear Chemistry. 2) Ibev-CNR. Via Salaria km 29.3- 00016 Monterotondo Scalo, Italy.

Key words: photosystem 2, D1 turnover, herbicides, tolerance

1. Introduction

Higher plants have evolved several mechanisms of resistance to Photosystem II-directed herbicides. Such mechanisms include a slower translocation into the chloroplast, immobilisation and detoxification by endogenous enzymes through conjugation and/or degradation and mutations in the target protein which preclude the binding of herbicides. Often, more than one mechanism acts at the same time.

Photosystem II (PSII) is essential to the regulation of Photosynthesis since it catalyses the oxidation of water into oxygen and supports electron transport. PS II consists of a core, light-harvesting antennae and an oxygen-evolving system. The core is comprised of the reaction centre proteins, D1 and D2, cytochrome b-559, the internal antennae chlorophyll-proteins CP43 and CP47, PsbH and several other proteins of unknown function. All pigments and prosthetic groups necessary for charge separation and stabilisation are bound to the D1 and D2 proteins. The action site of the PSII-directed herbicides (triazine, diazine, urea and phenol-compounds) is in/close to the Q-B pocket of the D1 protein. These compounds act by competitive displacement of the secondary quinone acceptor PQ from its binding site Q-B. Under optimal physiological conditions, the D1 protein exhibits a light-dependent turnover several times higher than that of other chloroplast proteins (1-3).

In the present work we study the response of the main PSII proteins to short and long-term treatment with atrazine herbicide by fluorescence, ^{3}H-Leucine incorporation and immunoblot analyses. This study was performed in higher plants with varying tolerances to the herbicide. It is observed that herbicide strongly acts at two different levels on the D1 protein: i) displacement of PQ from its Q-B site; ii) alteration of its turnover. The results also suggest that D1 protein turnover, due to its compartmentalisation in the thylakoid membrane, is probably involved in the resistance of plants to herbicide action.

2. Procedure

For short-term experiments on the combination effects of inhibitors and herbicide, the seedlings (5 cm height) were treated with the inhibitors for 10 hours. D1 turnover was then

G. Garab (ed.), Photosynthesis: Mechanisms and Effects, Vol. V, 3881–3884.

followed by pulse-chase experiments with ^{3}H-Leucine incorporation in vivo, extraction of thylakoids, SDS-PAGE analysis, treatment with Amplify and fluorography (4-5) Immunoblottings were performed with polyclonal antibodies against the main PSII proteins. For long-term experiments, the seedlings were treated after germination (1 cm height) with both inhibitors and herbicide (twice 200 ml for 30x30 cm vermiculite pots, about 40 seedlings) seven days before measurements. At, 33 mg/L; CAP, 200 mg/L; Cycloheximide, 2mg/L; SE was about 11%. The experiments were replicated ten times.

3. Results and Discussion

We treated various plant species with atrazine herbicide in combination with synthesis inhibitors. The protein synthesis was blocked with chloramphenicol (CAP), a translation inhibitor on 70-S plastidic ribosomes, or with cycloheximide, an inhibitor of protein synthesis at the nucleus instead of the chloroplast. The effect of herbicide action under the different conditions reported above was determined in sensitive, moderately sensitive and resistant plants in a long-term experiment. A synergistic effect between herbicide action and the inhibitor of protein synthesis CAP was previously observed (6). It is important to emphasise that no synergism was present with cycloheximide, indicating that the observed effect should be attributed to a primary action on proteins synthesised in the chloroplasts. The synergism was weak in *Avena sativa L., Senecio vulgaris, Vicia faba* greater in plants which in our conditions were moderately resistant (*Pisum sativum, Triticum durum*), and absent in herbicide resistant plants, *Zea mays* and *Senecio vulgaris* mutant. It is known that in resistant plants the herbicide is immobilised by glutathione in the former and does not bind to the Q-B site in the latter.

We treated various plant species with atrazine herbicide in combination with CAP in short-term experiments as well. A great reduction of fluorescence was observed in all tested cases which was of similar extent indicating that the inhibitors equally reach the PSII apparatus; however, the synergistic effect was seen primarily in pea.

With the purpose to understand the mechanism of the herbicide action, we analysed the PSII proteins by incorporation of ^{3}H-Leucine and immunoblot analyses. The seedlings were treated with the inhibitors for 10 hours. D1 turnover was then followed by pulse-chase experiments with radioactive incorporation *in vivo* in the light. The figure shows the course-time distribution of radioactivity on the D1 protein and the protein content. The experiments were repeated three times and the figure represents typical results.

CAP and herbicide affected both synthesis and degradation of D1 protein in all tested species. However, CAP and herbicide showed an opposite effect on D1 content, the former

Pulse-chase experiments with ³H-Leucine radioactivity into D1 protein in relative units

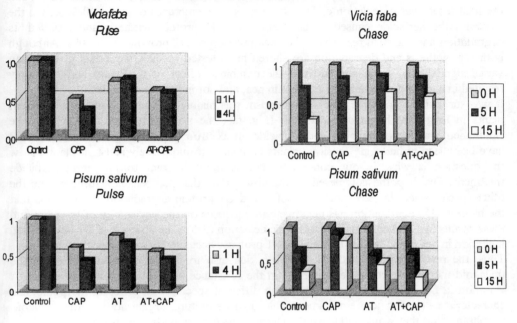

D1 content per unit of chlorophyll

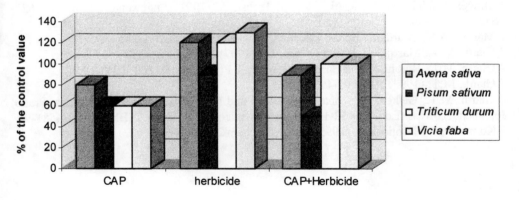

caused a great decrease while the latter an accumulation. The D1 synthesis was inhibited and the degradation retarded by CAP, in accordance with previous observations (6-7), and the final result was a depletion of D1 from thylakoid membranes of about 40-50%. On the contrary, the herbicide caused a slight effect on D1 protein synthesis and retarded its degradation and the final result was the accumulation of D1 protein of 10-30%. Although both D1 synthesis and degradation are altered in all tested species, the extent of inhibition varies significantly. As a consequence the combined effect of CAP and herbicide only resulted in a great decrease in D1 content in pea, and not in the other species.

It was surprising to observe that the synergism was mainly present in moderately resistant plants in long-term herbicide treatment. If in fact the synergism were only due to the simultaneous action of inhibitor and herbicide on electron transfer, a greater effect would have been seen in herbicide sensitive plants compared to moderately resistant plants. Now the question is why, in our conditions, pea is more tolerant to long-term herbicide treatment? One hypothesis is based on the observation that pea behaviour differs from the other species since the herbicide does not retard D1 protein degradation. That means that the block of D1 degradation has negative consequences on the physiology of the plant and these results in a higher sensitivity to herbicide action (3-5). The absence of D1 degradation inhibition in pea could be due to a higher D1 protease accessibility in this species.

Thus, the results suggest that D1 turnover, due also to its compartmentalisation in the thylakoid membrane, could be important in the resistance of plants to herbicide action. A variation in D1 turnover is expected among different species with different morphological characteristics (leaf structure, thickness, chloroplast structure, organisation of the thylakoid membrane etc.) that would influence the interaction between herbicide and D1 protein.

References

1 Draber, W., Tietjen, K., Kluth, J.F. and Trebst, A. (1991) Angewand. Chem. 30, 1621-1633

2 Mattoo, A.K., Marder, J.B. and Edelman, M. (1989) Cell 56, 241-246

3 Giardi, M.T., Masojidek, J., and Godde, D. (1997) Physiol Plant 101, 635-642

4 Giardi, M.T., Rigoni, F. and Barbato, R. (1992) Plant Physiol. 100, 1948-1954

5 Giardi, M.T. (1993) 190, 107-113

6 Giardi, M.T., Geiken, B., Cona, A. In:Weed and Crop Resistance to Herbicides. Prado, Jorrin and Torress (Eds) pp 132-138. Kluwer Academic Publishers. ISBN 0-7923-4581-9

7 Komenda, J, Masojidek, J. (1995) Eur J. Biochem 233, 677-682

EFFECTS OF EXOGENOUS ELECTRON ACCEPTORS ON KINETIC CHARACTERISTICS OF PROMPT AND DELAYED FLUORESCENCE IN ATRAZINE INHIBITED THYLAKOID MEMBRANES

V.Goltsev , L.Traikov and V.Hristov
Dept. Biophysics and Radiobiology, Faculty of Biology, University of Sofia,
8, Dr. Tzankov Blvd., 1421 Bulgaria

Key words: electron transport, luminescence, chlorophyll fluorescence, photosystem 2, herbicide

1. Introduction

The emission of light quanta by preilluminated photosynthetic objects occurs over two mechanisms - photophysical, by direct radiation of excited chlorophyll molecules (prompt fluorescence - PF), and photochemical, by means of secondary, chemical excitation of chlorophylls (delayed fluorescence - DF) (1). During the both luminescence species the light quanta origin from almost the same population of antennae pigment molecules (2), predominantly in Photosystem II (PS II). The changes of PF and DF quantum efficiency during dark-light state transition of photosynthetic apparatus, denoted as induction kinetics (IK), correlate with: a) electron transport rate in both donor and acceptor sides of PS II; b) thylakoid membrane (TM) energization; c) secondary ion transfer; d) excitation energy distribution between both Photosystems; e) enzymatic dark processes of Photosynthesis (1, 3-6), etc. Comparison of kinetic characteristics of simultaneously radiated prompt and delayed luminescence can contribute to better understanding of both luminescent processes. In this study the effects of PS II electron acceptors and herbicide atrazine on kinetic characteristics of simultaneously registered PF and DF in pea thylakoids were analyzed. The contribution of sub-millisecond and millisecond decay kinetics components of DF in the different induction kinetics phases was estimated.

2. Procedure

Chloroplasts from leaves of 14-day-old pea seedlings (*Pisum sativum* L., var. Ran 1) were isolated according to (7). The thylakoids were stored in liquid nitrogen as described in (8). Prior to use, they were thawed and diluted with 25 mM Tricine-NaOH buffer (pH 7.8), 5 mM $MgCl_2$, 0.33 M Sorbitol to a chlorophyll concentration of 1 mg.ml^{-1} and stored in the dark at 0°C. The luminescence measurements of 3 min dark adapted chloroplast suspensions were carried out at 22°C. Every probe contained the same buffer and thylakoids at chlorophyll concentration of 20 μg.ml^{-1}. When neded the electron acceptors 2,5-dichloro-*p*-benzoquinone (DCBQ), 2,6-dimethyl-*p*-benzoquinone (DMBQ), $K_3Fe(CN)_6$ (FeCy), 2,6-dichlorophenol indophenol (DCPIP) to final concentration 10^{-5} M and/or 2-chloro-4(ethylamino)-6-(isopropylamino)-*s*-triazine (atrazine) to concentration 10^{-6} M were added 3 min before measuring. The simultaneous registrations of both prompt (PF) and delayed (DF) chlorophyll fluorescence as well as kinetics of DF

G. Garab (ed.), Photosynthesis: Mechanisms and Effects, Vol. V, 3885–3888.
© *1998 Kluwer Academic Publishers. Printed in the Netherlands.*

dark relaxation were made using the apparatus *Fl-2006* (*Test*, Russia). Each work cycle of the phosphoroscope had a duration of 11.33 ms and included interval of object illumination (5585 µs), period of registration of DF (5585 µs), and two dark periods (75 µs each) between them (5). The maximum modulated irradiance at the level of object was 1200 µmol.m^{-2}.s^{-1}.

3. Results and Discussion

Using simultaneous registration of PF and DF induction kinetics (PFIK and DFIK), the correlation between the both luminescence species can be established (5). The DFIK in control suspension of TM (Fig. 1 A) shows two phases - in the millisecond and in the second time range. The fast phase is splitted in 2 maximums (I_1, ~20 ms and I_2, ~100 - 150 ms) witch are well pronounced in the native objects - leaves, alga cells (5) and weakly - in TM. It have been supposed that the shape of DFIK in fast phase reflects the transitions between different redox states of PS II acceptor complex. The slow IK phase (I_3), is due to photoinduced transmembrane ΔpH as well as secondary ion transport (9,6). Both DF intensity and DFIK shape are changed by addition of PS II electron acceptors (see Table 1 and Fig. 1, respectively).

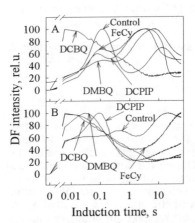

Figure 1. Effect of electron acceptors on the induction kinetics of delayed fluorescence in control (A) or treated with 10^{-5} M atrazine (B) pea thylakoids.

The DF intensity is powerfully decreased by DCBQ and fastest IK phase (I_1) only is appeared. Such IK is a typical for diuron poisoned photosynthetic objects (5), which are presented only by Q_B non-reducing reaction centers. We can assume that accepting of electrons from Q_A^- DCBQ (10) greatly competes with a charges recombination. Thus the light quanta emitting may occur only in Q_B non-reduced centers of PS2. The other three acceptors quite not make significant changes of DF amplitude in the I_2 maximum, but eliminate slope after I_2 maximum. This slope is due to "closing" of the PS2 reaction centers as result of PQ pool reduction during illumination. It supposes that above mentioned acceptors keep a high stationary oxidizing level of Q_B by means of its direct or indirect oxidizing by PQ. The herbicide atrazine in 1 µM concentration suppress predominantly I_2 and following decline by decreasing of PQ pool reduction rate. The atrazine does not change DFIK in presence of DCBQ, which confirms direct oxidizing of Q_A^- by this acceptor (10). It is interesting that FeCy and DCPIP multiply the inhibition activity of atrazine (see Fig. 1 and Table 1). In opposite, DMBQ significantly gets over the atrazine blockage through competition for Q_B pocket (10).

Table 1. Inhibition coefficients for maximal DF intensity in TM treated with atrazine, electron acceptors, or their combination. n = 6

Acceptor	Atrazine	
10^{-5} M	None	10^{-6} M
None	1.00	5.68±0.71
DCBQ	16.37±0.96	16.44±1.42
DMBQ	0.35±0.06	3.72±0.27
FeCy	1.56±0.22	19.65±2.53
DCPIP	1.08±0.04	13.04±1.06

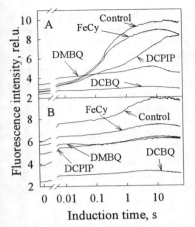

Figure 2. Effect of electron acceptors on the induction kinetics of prompt fluorescence in control (A) or treated with 10^{-5} M atrazine (B) pea thylakoids.

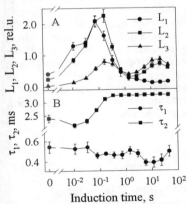

Figure 3. Dynamics of kinetics parameters of DF dark relaxation in pea thylakoids. A - amplitudes, and B - relaxation times.

PFIKs are modified by above mentioned compounds in good correlation with DF changes (Fig. 2). It must be emphasized that in millisecond time domain of IK the position of I_1 coincides with O-I fluorescence transition, and of I_2 position - with the moment of maximal fluorescence growth of D-P transition. For example in control and treated with FeCy and DCPIP thylakoid membranes where the I_2 maximum is well pronounced and I_1 is relatively weak, the variable part of the fluorescence is high and O-I transition is negligible. In oppose, for the samples with pronounced I_1 (DMBQ and DCBQ respectively), the part of O-I transition in variable fluorescence dominates (it especially is valid for all atrazine treated samples).

In our study the DF kinetics of dark relaxation in sub-milliseconds and milliseconds time interval measured at different moments of the induction curve were also analyzed. The kinetics of dark relaxation were described as exponential decay, and the fitting procedure was applied according the equation:

$$I_{DF} = L_1 \cdot e^{-t/\tau_1} + L_2 \cdot e^{-t/\tau_2} + L_3 \,,$$

where Li are amplitudes, and τ_i are characteristic times of decay components. It was established that the relaxation kinetic parameters were changing during induction period (Fig. 3). DF dark relaxation kinetics is depicted by three kinetic components, as follows: a) fast component with characteristic time τ_1 about 400-600 μs; b) intermediate component with τ_2 varying between 1.5 up to 3.5 ms; c) slow component with characteristics time τ_3, > 10 ms (the data for τ_3 are not presented). We suppose that the emission, decayed in the sub-milliseconds domain was connected with the recombination of $Z^+P_{680}Q_A^-$, and the lifetime (τ_1) was determined by the possibilities of direct and back electron transport in this state. The millisecond characteristic time (τ_2) represent emission of PS II reaction centers in $Z^+P_{680}Q_B^=$ state, and L_2 reflects the moment concentration of the PS II reaction centers in $Q_AQ_B^-$ state. In the time course of the amplitudes the maximums similar to those of DFIK recorded in integral regime are founded. But for definite kinetic components this maximums differ by the amplitude and their position in the time scale. The changes of sub-millisecond and millisecond decay component amplitudes are connected with the dynamics of different "light emitting" states of PS II reaction centers in initial part of IK (up to 1 s). During slow peak (1 - 5 s) this possibly is consequence of transmembrane proton and electric gradients (11). The τ_1 is diminished after 5 s of illumination (Fig. 3B). We suppose that the transmembrane proton gradient accelerates the separated charges recombination in PS II reaction centers. τ_2 grows within the first 700 ms of IK from 2.11 ± 0.08 up to 3.29 ± 0.01 ms, that probably correlates with PQ pool reduction, and the closure one of the ways leading to disappearance of "light emitting" states.

The time course of amplitudes of the three components are strongly dependent on presence of electron acceptors and herbicide atrazine (Fig. 4). At the presence of electron acceptors used the L_1 forms the first induction maximum (Fig. 4A,D), and the second one is well pronounced in DMBQ treated TM only. Atrazine suppresses slower maximums and the first one is relatively saved. At the slow induction time the maximum at about 3 s is formed after addition of DMBQ, DCPIP and particularly - FeCy. In other samples the amplitude of sub-millisecond component is not ΔpH dependent. The L_2 practically does not form first IK maximum (Fig. 4B,E). The second one is presented between 77 and 168 ms. The slowest maximum appears at about 3 s for DCPIP and FeCy and between 6 and 12 s - for control and DMBQ treated TM. DCBQ as well as atrazine induce only the first maximum formation. DCPIP, DMBQ and FeCy induce development of slow phase (~6 s) for the third decay component (Fig. 4F). In other variants (excluding control) the first maximum ~11 ms is followed by DF intensity growth up to end of induction kinetic (Fig. 4C).

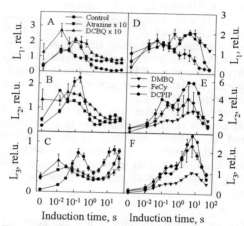

Figure 4. Time course of DF decay components amplitudes

We can conclude that in the first DF induction maximum dominates sub-millisecond decay component, in the second maximum - the millisecond one, and in the I_3 induction phase formation participate millisecond and slower decay component.

References

1 Lavorel, J. (1975) In Bioenergetics of Photosynthesis (Govindjee, ed.) pp. 223-317, Academic Press, New York, San Francisco, London
2 Amesz, J. and Van Gorkom, H.J. (1978) Annu. Rev. Plant Physiol. 29, 47-66
3 Krause, G.H. and Weis, E. (1991) Annu. Rev. Plant Physiol. 42, 313-349
4 Radenovich, C., Markovic, D. and Jeremic, M. (1994) Photosynthetica, 30, 1-24
5 Goltsev, V. and Yordanov, I. (1997) Photosynthetica 33, 571-586
6 Gaevskii,N.A. and Morgun,V.N.(1993) Fiziologia Rastenii, 40,136-145 (in Russian)
7 Whatley, F.R. and Arnon, D.I. (1963) In Methods in Enzimology (Colowick S.P. and Kaplan N.O., eds.) v. 6, pp. 308-313, Academic Press, New York
8 Goldfeld, M.G., Mikojan, V.D., Suskina, V.I., Timofeev, V.P. and Shapiro, A.B. (1980) Fiziologia Rastenii, 27, 1143-1153 (in Russian)
9 Wraight, C.A. and Crofts, A.R. (1971) Eur. J. Biochem. 19, 386-397
10 Kashino, Y. (1996) Plant Cell Physiol. 37, 976-982
11 Schmidt, W. and Schneckenburger, H. (1995) Photochem. Photobiol. 62, 745-750

EVIDENCE THAT PHENOLIC INHIBITORS OF Q_B HAVE LONG-RANGE EFFECTS ON THE S-STATE TRANSITIONS.

Arthur Roberts, J. Scott Townsend and David M. Kramer

Institute of Biological Chemistry, Washington State University. Pullman, WA 99164-6340

Key Words: photosystem 2, structural changes, thermoluminescence, Q_A, fluorescence, inhibitors

Introduction

In this paper, we describe the effects of phenolic Q_B site inhibitors (trinitrophenol (TNP), bromoxynil and dinoseb) on the energetics of photosystem II (PS II). Our main interest stemmed from two unusual aspects of these inhibitors. Firstly, because of the acidic pK_a values ($pK_a < 5.00$) on their hydroxyl groups, these inhibitors most likely bind to the Q_B site in their anionic forms (see [1]). This makes them potentially useful analogs of the Q_B semiquinone, and thus may help us to understand certain biophysical aspects of the two-electron gate. Secondly, these inhibitors have been shown to have large effects on the thermoluminescence (TL) bands associated with $S_2Q_A^-$ recombination [2] . At first glance, our data suggested that the phenolic inhibitors may influence the midpoint potential of Q_A^- by electrostatic interaction. We show, however, that it is more likely that they influence the S1-S2 state transition.

Materials and Methods

Thylakoids were prepared as previously described (see [3]) and resuspended in "resuspension buffer" containing 50mM HEPES (pH 7.6), 400mM sucrose, 10mM NaCl and 5mM $MgCl_2$. Flash-induced changes in fluorescence yield were measured as described in [4,5] at a chlorophyll concentration of 5 $\mu g \ ml^{-1}$ with 5 $\mu g \ ml^{-1}$ gramicidin and 10 μM p-benzoquinone as an electron acceptor. Samples were dark adapted for a minimum of 5 minutes, followed by a series of single turn-over xenon flashes. After each actinic flash the fluorescence yield was probed by a series of weak flashes from light emitting diodes. Thermoluminescence experiments were performed as in [6], with further details given in the figure legends. Redox titrations were performed by incubation of samples in the presence of variable ratios of ferricyanide/ferrocyanide [7]. The ambient redox potential (E_h) was measured by Pt vs Ag/AgCl electrodes calibrated with saturating quinhydrone.

Results and Discussion

Figure 1 shows that TNP blocks $Q_A \rightarrow Q_B$ electron transfer with an I_{50} of about 175-225 nM (Figs. 1A, B). The onset of blockage is unusually slow (Fig. 1C), with a second order rate constant of about 25 $mM^{-1}s^{-1}$. Such slow on rates appear to be a general feature of the phenolic inhibitors (see [8] and [9]) and we suggest that this is due to a slow transfer of the anion into the lipid bilayer. Recombination from $S_2Q_A^-$ in the presence of TNP was strikingly rapid ($t_{1/2}$ of ~ 0.4 s, Fig. 1B), compared to that from DCMU (3-(3,4-dichlorophenyl)-1,1-dimethylurea) -blocked centers ($t_{1/2}$ of 2-3 s, Fig. 1B) (see review [10]), indicating that TNP affects the energetics of the $S_2Q_A^-$ state. The

G. Garab (ed.), *Photosynthesis: Mechanisms and Effects, Vol. V,* 3889–3892.

acceleration of $S_2Q_A^-$ recombination was independent of any ADRY (acceleration of the deactivation reactions of the water-splitting enzyme Y) effects of TNP (see [11]) since these only appeared at much higher concentrations of TNP (not shown). Upon further addition of high concentrations (10 μM) of DCMU, the recombination time gradually increased to about 3 s, indicating that DCMU was able to displace TNP. Thus the binding sites of these two inhibitors strongly overlap (see [12] and [13]). The time required for DCMU to displace TNP was used to estimate the off-rate for TNP binding to be ~0.013 s^{-1} (Fig. 1D).

Figure 1: k_{on}, k_{off}, and I_{50} determined by flash fluorescence spectroscopy. Fig. 1A: Normalized fluorescence yields at ~10ms after flash excitation as a function of [TNP] Fig. 1B: Flash-induced fluorescence decay kinetics at various times after the addition of 1μM TNP. Fig. 1C: Onset of TNP-blockage measured as a rise in fluorescence yield at 10 ms after flash excitation. Fig. 1D: Displacement of TNP by DCMU (10 μM) measured by the slowing of $S_2Q_A^-$ recombination.

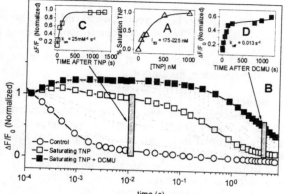

The binding constant estimated by k_{on} and k_{off} was within a factor of 2 of the I_{50}, indicating a straightforward binding mechanism. We note that, even at saturating concentrations, a small fraction (~10%) of centers appeared to be unaffected by TNP (Fig. 1B). These centers were nevertheless rapidly blocked by DCMU, indicating that they possessed relatively normal Q_B sites. Figure 2 shows that the effects of other phenolic inhibitors on the energetics of $S_2Q_A^-$ are similar to those of TNP. Since bound phenolics are charged, it is reasonable to suggest that they alter the redox properties of Q_A by electrostatic interactions, thus decreasing the energetic barrier to recombination. These findings were largely substantiated by TL experiments (Fig. 3 A) which showed that the Q-bands from PS II, blocked with phenolic inhibitors generally had lower temperatures than that in the presence of DCMU (see [2]).

We next compared the effects of DCMU and phenolic inhibitors on recombination from $Yz^•Q_A^-$ (see also [2] and [14]). Figure 4 shows that, in hydroxylamine (HA)

Figures 2, 3 and 4: The effects of phenolics on $S_2Q_A^-$ recombination (Fig. 2), the Q-band (Fig. 3A), the A_T band (Fig. 3B) and $Yz^•Q_A^-$ recombination (Fig. 4). For TL curves, a heating rate of 1°C/s was used. B- and Q-bands were charged with a single-turnover flash at 0°C, whereas A_T bands were charged at −30°C with 30 s continuous red light (50 μmole photons m^{-2} s^{-1})

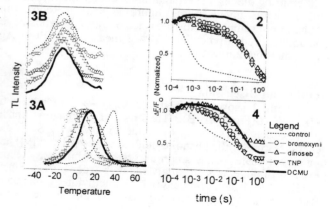

washed thylakoids, the phenolics bound to PS II, and effectively blocked $Q_A \rightarrow Q_B$ electron transfer. However, the rates of $Yz^\bullet Q_A^-$ recombination in the presence of phenolics (90-160 ms) were virtually indistinguishable from that in the presence of DCMU (140-230 ms). Furthermore, the TL curves from hydroxylamine (HA) washed thylakoids in the presence of phenolics were virtually identical to those in the presence of DCMU (Fig 3B). This data calls into question the hypothesis that phenolic inhibitors electrostatically influence Q_A redox potential.

Direct measurements of Q_A redox potential are fraught with difficulty [15]. However, any electrostatic effects of phenolics on Q_A should also influence the redox properties of the non-heme iron (Fe, also known as Q_{400}), which is located between Q_A and the Q_B pocket (see review in [16]). We therefore performed redox titrations of the Fe using flash fluorescence yields as an assay of Fe redox state, in general as described in [7]. After incubation in ferricyanide/ferrocyanide in complete darkness for 10-60 min, a series of closely-spaced single turnover actinic flashes were given and fluorescence yields were determined. In the presence of Fe^{3+}, oxidation of Q_A occurs within 7 μs and low fluorescence yield is observed in the 100-microseconds kinetic domain. However, after the first flash, all Fe^{3+} is reduced, leading to normal fluorescence kinetics. Thus a ratio of the first flash to subsequent flash data can be used as a measure of Fe redox state. Figures 5A-D show that blockage of Q_B with inhibitors had a small effect (~30 mV) on the E_m of Fe, but no differences were observed among DCMU, atrazine, or the phenolic inhibited samples.

We conclude, therefore, that it is unlikely that the phenolic inhibitors influence the redox properties of Q_A. Since significant acceleration of recombination in the presence of phenolics is only observed from the $S_2Q_A^-$ state, we suggest that these inhibitors alter the redox properties of the S1-S2 state transition. There is precedence for apparent long-range effects between the donor- and acceptor sides of PS II. For examples, HA or Tris washing [10], or donor side mutations (e.g. [17] and [18]) appear to have significant effects on the $Q_A \leftrightarrow Q_B$ kinetics and equilibrium. The present data is, perhaps, the clearest example of an S1-S2 state effect in a wild type Photosystem II.

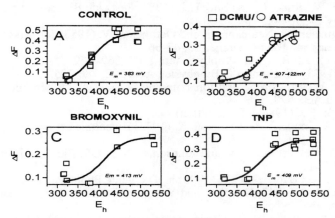

Figure 5: Redox titration on PS II non-heme iron in a control and DCMU (Fig. 5B), atrazine (Fig. 5B), bromoxynil (Fig. 5C), and TNP (Fig. 5D) treated thylakoids.

Conclusions

Phenolic inhibitors represent a class of compounds that are known to be good Q_B site inhibitors and ADRY agents. Using various biophysical techniques, it was possible to elucidate the donor and acceptor side kinetic behavior of phenolic inhibitors on Photosystem II. First, the low pK_a and anionic form of these inhibitors results in a slow partitioning into the membrane and a subsequent, slow k_{on}. Second, they appear to accelerate the $S_2Q_A^-$ recombination rate without having an effect on the $Y_Z^{\bullet}Q_A^-$ recombination. We also found that the effects on the S1-S2 state transition is independent of the ADRY effect. Therefore, we attribute the effects on the donor side to conformational changes which originate on the acceptor side from inhibitor binding.

References

[1]   Renger, G. (1972) FEBS Letters 23, 321-324.
[2]   Demeter, S., Droppa, M., Vass, I. and Horvath, G. (1982) Photobiochemistry and Photobiophysics 4, 163-168.
[3]   Kramer, D.M. and Crofts, A.R. (1990) in: Current Research in Photosynthesis, Vol. III, pp. 283-286 (Balscheffsky, M., Ed.) Kluwer Academic Publishers, Dordrecht, The Netherlands.
[4]   Kramer, D.M., Robinson, H. R. & Crofts, A. R. (1990) Photosyn. Res. 26, 181-193.
[5]   Kramer, D.M., DiMarco, G. and Loreto, F. (1995) in: Photosynthesis: From Light to Biosphere, Vol. I, pp. 147-150 (Mathis, P., Ed.) Kluwer Academic Publishers, The Netherlands.
[6]   Townsend, J.S., Kanazawa, A. and Kramer, D.M. (1997) Phytochemistry 4, 641-649.
[7]   Bowes, J.M., Crofts, A.R. and Itoh, S. (1979) Biochim. Biophys. Acta 547, 336-346.
[8]   Oettmeier, W., Kude, C. and Soll, H.J. (1987) Pesticide Biochemistry and Physiology 27, 50-60.
[9]   Johanningmeier, U., Neumann, E. and Oettmeier, W. (1983) Journal of Bioenergetics and Biomembranes 15, 43-66.
[10]   Debus, R. (1992) Biochim. Biophys. Acta 1102, 269- 352.
[11]   Vater, J. (1973) Biochimica et Biophysica Acta 292, 786-795.
[12]   Oettmeier, W., Masson, K. and Johanningmeier, U. (1982) Biochimica et Biophysica Acta 679, 376-383.
[13]   Vermaas, W.F.J. (1984). (Dissertation)
[14]   Ichikawa, T., Inoue, Y. and Shibata, K. (1975) Biochimica et Biophysica Acta 408, 228.
[15]   Krieger, A., Rutherford, A.W. and Johnson, G.N. (1995) Biochim. Biophys. Acta 1229, 193-201.
[16]   Diner, B.A. and Babcock, G.T. (1996) in: Oxygenic Photosynthesis: The Light Reactions, pp. 213-247 (Ort, D.R. and Yocum, C.F., Eds.) Kluwer Academic Publishers, The Netherlands.
[17]   Roffey, R.A., Kramer, D.M., Govindjee and Sayre, R.T. (1994) BBA 1185, 257-270.
[18]   Kanazawa, A. (1997) U. of Illinois, Urbana-Champaign.

RESPONSE OF MAIZE PLANTS TO PRE-EMERGENCE APPLICATION OF HERBICIDES MONITORED BY FAST CHLOROPHYLL FLUORESCENCE INDUCTION

Martina Matoušková[1], Jan Nauš[1] and Marie Flašarová[2]
[1]Department of Physics, Palacky University, tř. Svobody 26, 77146 Olomouc, Czech Republic
[2]Agricultural Research Institute, Havlíčkova 2787, 76701 Kroměříž, Czech Republic

Key words: abiotic stress; plant topography

1. Introduction

A large group of herbicides causes alterations of function of photosynthetic apparatus. Curves of fluorescence induction with a typical O-J-I-P transient reflect basically the state of electron transport through photosystem II (PSII) [1,2] so that allow to study action of herbicides affecting this transport (PSII herbicides). It has been found that the most sensitive parameter of the O-J-I-P curve to PSII herbicides action is a relative height of the J step [3,4]. Two phases of damage to photosynthetic apparatus due to herbicide can be discerned. The first phase, characterized by increasing relative height of the J step without changes of Fv/Fp and Fo parameters, reflects accumulation of Q_B-nonreducing centres of PSII [4]. In the second phase, a decrease of Fv/Fp and an increase of Fo indicate a more pronounced injury consisting in formation of inactive PSII centres. In this work, an effect of three pre-emergently applied herbicides on photosynthetic apparatus of maize plants was studied using the O-J-I-P curve. It is shown that parameters of this curve allow also a detection of the effect of herbicides which do not influence directly the photosynthetic apparatus. The observed changes of fluorescence parameters seem to be related with an acceleration of senescence due to herbicide action.

2. Procedure

Seeds of maize (*Zea mays* L. cv. Falcet FAO 200) were sown in soil in pots on June 17, 1997. Herbicides <u>Gesagard 80</u> (Ciba-Geigy AG, CH), <u>Bladex 50 SC</u> (American Cyanamid Co., USA) and <u>Eradicane 6E</u> (Zeneca Agrochemicals Ltd., GB) (for active substances see Tab. 1.) were applied immediately after sowing in amounts of 0.3 mmol per m^2 of the projected surface area. The maize plants emerged on June 24, 1997. The chlorophyll fluorescence induction was measured using the portable fluorometer PEA (Hansatech Instruments Ltd., England). The measurements were carried out non-destructively on gradually accruing leaves of maize plants from the first (L1) to the sixth one (L6) within one month after emergence (see Fig. 1.). The measured leaf region, a central part of the adaxial side of leaf blades, was dark-adapted for 30 min in the leaf clips (Hansatech). A fluorescence detection time of 2 s and an excitation intensity of about 3000 μmol m^{-2} s^{-1} were used. Fluorescence parameters Fo, Fv/Fp = (Fp-Fo)/Fp, rFj = (Fj-Fo)/(Fp-Fo) and rFi = (Fi-Fo)/(Fp-Fo) were determined. The values of fluorescence Fj and Fi at the J and I steps were evaluated from the experimental curves

3893

G. Garab (ed.), Photosynthesis: Mechanisms and Effects, Vol. V, 3893–3896.

3894

at times 2 and 30 ms, respectively. The fluorescence parameters are presented using a general statistical description (medians) as they may not obey the normal distribution [5,6]. A method of fluorescence topography was applied. At the same time of fluorescence measurements, length, assimilation area of consistent leaves and dry weight of the above-ground part of plants (shoot) were estimated.

Table 1. Active substances of herbicides pre-emergently applied on maize plants.

herbicide	active substance
Gesagard 80	prometryn: 2,4-bis(isopropylamino)-6-methylthio-1,3,5-triazine)
Bladex 50 SC	cyanazine: 2-[[4-chloro-6-(ethylamino)-1,3,5-triazin-2-yl]amino]-2-methylpropane-nitrile
Eradicane 6E	S-ethyldipropylthiocarbamate and dichrormide

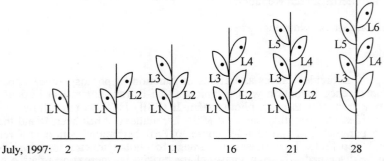

July, 1997:   2   7   11   16   21   28

Figure 1. Leaf numbering and leaf regions used for fluorescence measurement of gradually accruing leaves of maize plants.

3. Results and Discussion

3.1 *Effect of Gesagard 80*
The Gesagard 80 had the most pronounced effect on fluorescence parameters in comparison with the other used herbicides (Fig. 2 a,b). The leaves in younger phases of their ontogeny were characterized by an increased rFj which corresponded to accumulation of Q_B-nonreducing centres of PSII. As the leaves grew old, their Fv/Fp decreased which can be understood as the second phase of damage to photosynthetic apparatus consisting in an increase of the amount of inactive PSII centres. Changes of fluorescence parameters (especially rFj) signalizing a worse state of treated plants appeared already 14 days before the visually observed changes.

3.2 *Effect of Bladex 50 SC*
Bladex 50 SC did not influence Fv/Fp of the older leaves (L1 and L2) in the first five terms of measurement (Fig. 2c). In the last term (July 28), fluorescence induction of the L1 and L2 leaves was unmeasurable so that values of Fv/Fp (and rFj) are missing. However, a visual observation revealed an accelerated senescence of these leaves as compared with the control plants. The Fv/Fp parameter in younger leaves (L3-L6) of treated plants was unchanged within the whole time-range of measurement (Fig. 2c). The pre-emergently applied Bladex 50 SC caused a mild increase of rFj in L1, L2 and L3 leaves (Fig. 2 d). The most pronounced rFj increase was observed in the L1 leaves (of about 10-20% compared with the control plants). The increase of rFj of L3 leaves in the last term (July 28) seems to reflect an incipient senescence of these leaves.

The above mentioned results shown that the pre-emergently applied Bladex 50 SC caused only the first phase of damage of photosynthetic apparatus, i. e. an increased amount of Q_B-nonreducing centres of PSII. This effect was most pronounced in the oldest leaves (L1), in the younger leaves was weaker. It can be supposed that in gradually accruing maize leaves an amount of Bladex 50 SC decreased. In the final phases of measurement an accelerated senescence of the older leaves was apparent.

3.3 *Effect of Eradicane 6E*
The pre-emergence application of Eradicane 6E caused no changes in parameters of fluorescence induction (Fig. 2e,f) except a mild rFj increase of L3 leaves because of an accelerated senescence of these leaves. The earlier start of senescence in the treated plants was confirmed by a visual observation when the extent of yellowing of L1, L2 and L3 leaves was much higher than in control plants. The results of fluorescence measurements indicate that photosynthetic apparatus was not primarily affected by Eradicane.

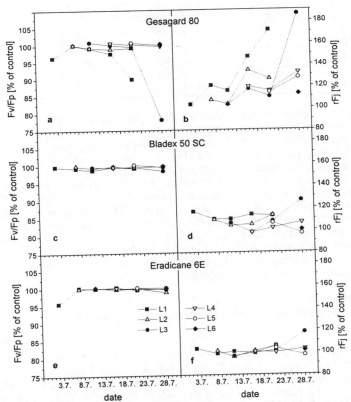

Figure 2. Fv/Fp and rFj parameters of individual leaves of maize plants pre-emergently treated by Gesagard 80, Bladex 50 SC and Eradicane 6E. Leaves numbered from bottom (see Fig. 1.). Expressed in % of values of control plants, calculated from medians, n=10.

Interestingly, the dry weight of the above-ground part of plants was not affected by the used herbicides. The assimilation leaf area was found to be lower in all treated variants for the first (L1) and second (L2) leaves. The assimilation area of other younger leaves (L3-L6) was within 10 % the same as that of the control plants. The same conclusion can be drawn from measurements of the leaf length. Generally, the pre-emergently applied herbicides did not influence the dimensions and area of the leaves in spite of the mentioned accelerated senescence of the older leaves. This means that the matter of the senescing leaves was transported into the younger parts or the growth of the younger parts compensated for the matter loss in older leaves.

Table 2. Fluorescence topography of control plants and pre-emergently affected plants by three herbicides 34 days after emergence (July 28, 1997). CE - characteristic extreme, TG - topographic gradient.

		control	Gesagard 80	Bladex 50 SC	Eradicane 6E
Fv/Fp	CE	0.820	0.822	0.818	0.820
	TG	0.029	0.244	0.043	0.041
Fo	CE	295	293	296	302
	TG	0.059	0.597	0.057	0.012
rFj	CE	0.465	0.524	0.465	0.468
	TG	0.086	0.792	0.371	0.271
rFi	CE	0.732	0.769	0.728	0.753
	TG	0.103	0.260	0.135	0.134

To evaluate the effect of herbicides on the whole green part of plant in a more quantitative and condensed way, a method of fluorescence topography was applied [6]. It enables a presentation of more fluorescence parameters (e.g. also Fo and rFi). A series of four topographic points was chosen (Fig. 1- since July 16). Two quantities of the topography for each fluorescence parameter are present in Table 2 (for the date July 28): a characteristic extreme (CE - the best value in the series) and a topographic gradient (TG - a difference between the maximal and minimal value devided by the CE). The CE characterizes the state of the whole plant, whereas TG represents the gradient of the quantity within the plant. Since the CE of Fv/Fp and Fo were not changed the whole plants were not under a strong stress. A higher amount of the Q_B-nonreducing centres in the whole plant was found only for Gesagard (a higher CE of rFj). TGs of rFj and Fv/Fp document that all the used herbicides caused an accelerated senescence of older leaves, the largest effect was caused by Gesagard.

Acknowledgement
Supported by the grant No. 521/96/0713 from Grant Agency of Czech Republic.

References
1 Strasser, R.J. and Govindjee (1991) in Regulation of Chloroplast Biogenesis (Argyroudi-Akoyunoglou J.H., ed.) pp. 423-426, Plenum Press, New York, USA
2 Strasser, R.J. and Govindjee (1992) in Research in Photosynthesis, Vol.2, (Murata N., ed.) pp. 29-32, Kluwer Academic Publishers, Dordrecht, The Netherlands
3 Lazár, D., Brokeš, M., Nauš, J. and Dvořák, L. (1998) J. Theor. Biol. 191, 79-86
4 Lazár, D., Nauš, J., Matoušková, M. and Flašarová, M. (1997) Pestic. Biochem. Physiol. 57, 200-210
5 Lazár, D., Nauš, J. (1998) Photosynthetica 35, 121-127
6 Matoušková, M., Nauš, J., Flašarová, M. and Fiala, J. (1996) Acta Univ. Palacki. Olomuc., Fac. Rer. Nat., Physica 35, 195-208

MEASUREMENTS OF HERBIZIDE ACTION ON THE PHOTOSYNTHETIC COMPONENTS OF WHOLE LEAVES WITH THE LEAF - FLASH SPECTROMETER

H. Kretschmann, Ö. Saygin

Inst. of Environmental Sciences, Bogazici University, 80815 Bebeck, Istanbui, Türkey

1. Introduction

Flash spectroscopic technics are applied to leaves, to obtain the performance of photosynthesis in leaves. The method is based on measuring absorption or flourescence changes of a leaf using flashes. At the moment five signals are being measured at specific wavelengths partly from enzymes: - water-splitting enzyme, -Qa (the primary acceptor of Photosystem II), P700 (the primary donor of Photosystem I), -electro-chromy (indicator of membrane electric potential, decay by ATP -Synthase), -flourescence induction and Kautsky -effect. This overall -analysis on the basic of determination of the most essential molecular parameters in the photosynthetic process, led to diagnostic criteria for plant injury and optimisation of the resources water, compost and herbicides. We have investigate the influence of the herbicides DCMU, BASTA and ROUNDUP on the S-state signals, the charge on Qa, P700 and the Kautsky -effect. We get also information about a different dark-adapted S-state distribution for higher plants and blue-green algea.

2. Material and Methods

The measurements were done with a new commercial avaible Leaf-Flash spectrometer (LFS, for more information call one of the author and N.Aksoy, http//:www.photosynthesis.org/Abyss/) This was a PC -controlled unit with LEDs and laserdiodes as light sources. Flourescence oscillation are measured 800ms after each flashes (600nm). In the 1.Flash measurement we excite with 571nm. The Kautsky -effect are measured at ≽ 670nm with continous light of 650nm.P700 signals are inverted ! DCMU: Diuron, ROUNDUP: 41% Glyphosat, Polyoxiethylammonium, BASTA: Glufosinateammonium.

G. Garab (ed.), Photosynthesis: Mechanisms and Effects, Vol. V, 3897–3900.
© 1998 Kluwer Academic Publishers. Printed in the Netherlands.

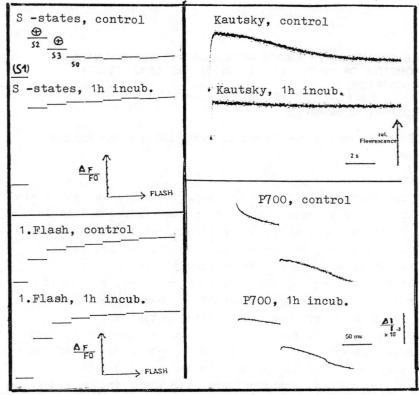

Fig.1

left : Flourescence change after single flashes at λ > 695 nm, shows surplus charges
upper : net- charge oscillation of the OEC; lower: 1. Flash shows charge on Q_A ?
right: upper : Kautsky- effect; lower : P700 oxidation at 800nm, exc. with 735 nm
Control measurements and incubation with 2 mM DCMU, 1 h .

3. Results and Discussion

In Fig. 1 the Signals after blocking the Qa - Qb transition
through the Diuron DCMU can been seen. The S- state oscillation
disappear, wheras a greater 1. Flash appear, because of the now
stored charge on Qa (Fig.1, upper part, left). For the 1.Flash
measurements, we can observe also a greater 1.Flash change (Fig.1
lower part, left). Because of the Diuron effect, the flourescence decline in the Kautsky measurment get very slowly and the
rise -time very fast.(Fig.1, upper part, right).For the P700
measurement, we cannot seen a significant effect of DCMU (Fig.1
lower part, right). For treatment with ROUNDUP, we get the
result shown in fig.2. No significant changing of all signals
can be observed, after incubation for 1 hour. A small relax -
ation of the flourescence oscillation signal occur (Fig.2, left)
it can be flash induced. Glyphosat is a known ihibitor of the

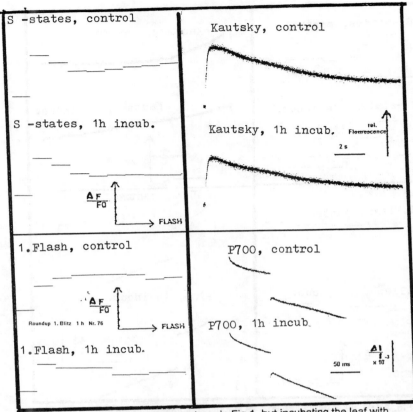

Fig.2 The same measurements as in Fig.1, but incubating the leaf with
ROUNDUP.

EPSPS Enzym, a enzym in the aromatic amino biosynthetic pathway and therefore should have no direct influence on the photosynthetic components. In Fig.3 the result with BASTA shown. In
the flourescence oscillation measurements, the 1.Flash get be
smaller and the S_2-S_3 transition also (Fig.3,left). It's possible
that Glufosinateammonium react like hydroxylamin, which led to a
shifting of the S -states (1). For the Kautsky effect we see a
slower relaxation and for P700 no change of the signal after incubation (Fig.3,right). The measurements shows, how explicit the
LFS can get results from the photosyntnetic components. In Fig.4
we have compare the flourescence oscillation pattern from whole
leaves and whole cells from a blue-green algea. We find that the
blue-green algea starts in the dark with S_0, whereas higher plants
starts witn mainly S_1. This may be a result of a evolutionary
different structural environment of the Mn-complex, because we
have snown, that a different lenght of a CP43 sequence led to a
different dark adapted S -state distribution (2).

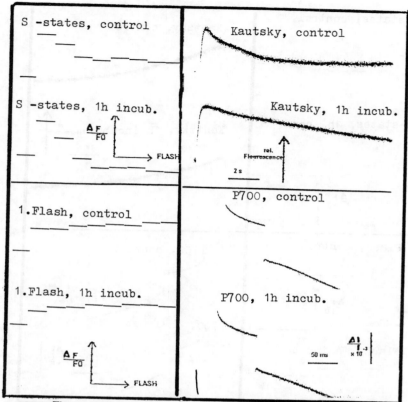

Fig. 3 The same measurements as in Fig.1, but incubating the leaf with BASTA .

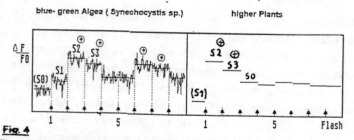

blue- green Algea (Synechocystis sp.)   higher Plants

Fig. 4

Flourescence Oscillation measurments of whole plants (right side) and blue-green Algea (left side), shows the surplus charges of the S- states.

Conclusion : Dark- adapted whole blue- green Algea starts with S_0, whereas whole higher Plants starts with S_1 . This can be consider under Evolution aspects.

(1) Saygin,O and Witt,H.T. (1987) Biochim.Biophys. Acta 893,p.452-469

(2) Kretschmann,H., Proceedings of the Xth International Photosynthesis Congress,1995, Kluwer Academic Publisher

ALTERED LIGHT RESPONSE OF XANTHOPHYLL CYCLE IN HERBICIDE-RESISTANT *ERIGERON CANADENSIS* BIOTYPES IN THE PRESENCE OF PARAQUAT

[1]Váradi, Gy., [2]Darkó, É. and [2]Lehoczki, E.
[1]Research Institute for Viticulture and Enology, Kecskemét,
[2]Department of Botany, József Attila University, Szeged, Hungary

Keywords: Chl fluorescence induction, quenching, pH-gradients, whole plants, paraquat resistance

1. Introduction

The excess of the absorbed energy must be sufficiently dissipated to maintain the integrity of the photosynthetic apparatus. One of these protective energy dissipating mechanisms the xanthophyll epoxidation cycle working on the lumenal side of grana thylakoids recently has been thoroughly investigated in many species and under a wide range of environmental and physiological conditions [1,2,3,4].

Paraquat (methyl-viologen) has been widely used for a long time as an artificial electron acceptor at PS1 in *in vitro* systems and sometimes it is also used in experiments on xanthophyll cycle [5,6]. The first step in paraquat effect is considered to be an overexpressed electron capture catalysed by the paraquat molecule resulting in (1) the formation of a highly reactive free radical which readily react with molecular oxygen generating superide radical and (2) it enhances the ΔpH formation accross thylakoid membranes providing favourable conditions for violaxanthin deepoxidation as well. Since paraquat-resistant weed biotypes have emerged many efforts were done to explore the key steps in the mechanism of resistance to paraquat. However, no generally accepted explanation has been found [7,8,9]. The most characteristic phenomenon related to paraquat resistance in *E. canadensis* is the recovery of resistant plants after some hours of a serious transient inhibition of their photosynthetic process [10,11]. However, we have only indirect evidences that paraquat can penetrate into the chloroplast of the paraquat-resistant biotypes of *E. canadensis*. Earlier experiments revealed that the light was essential to initiate and maintain this recovery process [11,12].

In the present paper the *in vivo* effect of paraquat on the light response of the xanthophyll cycle was studied in the wild (S) and paraquat-resistant (PQR), atrazine-

G. Garab (ed.), Photosynthesis: Mechanisms and Effects, Vol. V, 3901–3904.
© *1998 Kluwer Academic Publishers. Printed in the Netherlands.*

resistant (AR) and paraquat-atrazine co-resistant (PQAR) biotypes of *Erigeron canadensis* (horseweed) in order to reveal further details of the mode of action of paraquat *in vivo* and of the resistance to paraquat in horseweed as well. Some chlorophyll fluorescence quenching measurements also have been done in the case of the S and PQR biotypes to learn a small part of the background of altered xanthophyll cycle responses.

2. Materials and methods

Seeds of different *E. canadensis* biotypes were collected from vineyards near to Kecskemét (Hungary) and the plants were grown under natural environmental conditions. Treatments and measurements were carried out on 5 to 6 months old plants having 12 to 15 fully expanded leaves in the rosette stage.

Plants were sprayed under dim light conditions with diluted commercial Gramoxone containing 3×10^{-4} M paraquat and allowed to desiccate (5 min). Light intensity (PPFD) during treatments was 1400 μmol m^{-2} s^{-1}. Sampling for xanthophyll analysis and timing of fluorescence measurements are indicated in the figures. HPLC determination of xanthophyll cycle pigments were carried out according to [14] and the fluorescence quenching measurement method was the same as in [15].

3. Results and discussion

The xanthophyll cycle in S biotype of *E. canadensis* strongly responded to light plus paraquat treatments (Fig. 1) and the leaves nearly lost their photosynthetic functions at the 4th hour of the treatment (Fig. 2), as was expected. There was only, however, a transient enhancement of violaxanthin deepoxidation during the first hour in PQR and PQAR biotypes, suggesting that paraquat can reach its site of action in the chloroplast and exerts its primary electron diverting effect in all biotype. In the first 15 minutes of paraquat treatment PQR biotype responded similarly to S biotype. The AR biotype was so affected as S plants.

The relatively deep transient effect on violaxanthin deepoxidation in PQR plants seemed to be similar to the well known transient inhibition reflected in Fv/Fm values [11] and the time course of the qP, qN and EQY fluorescence quenching parameters (Fig. 2) highly support our hypothesis that paraquat is present in the chloroplast.

One of the most surprising phenomena in our experiment was the extremely pronounced response of the xanthophyll cycle in the AR biotype, since upon our earlier observations on the moderate reaction of the xanthophyll cycle in AR plants to high PPFD [14,15] we would predict a rather dampened reply of that biotype to paraquat plus light treatments.

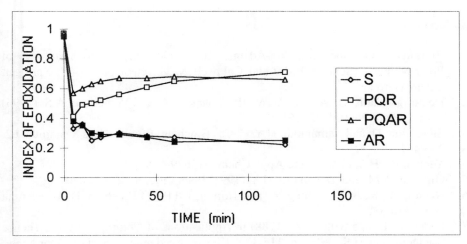

Figure 1
Time course of epoxidation index of the xanthophyll cycle in leaves of the wild (S) and paraquat-resistant (PQR), atrazine-resistant (AR) and paraquat-atrazine co-resistant (PQAR) biotypes of *Erigeron canadensis* in the presence of paraquat (intact plants sprayed with 3×10^{-4} M paraquat) under high-light (1400 μmol m^{-2} s^{-1}) conditions.

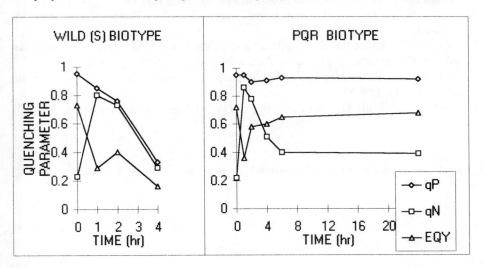

Figure 2
Time course of the fluorescence quenching parameters qP, qN and EQY in response to moderate light conditions (500 μmol m^{-2} s^{-1} PPFD) in leaves of the wild (S) and paraquat-resistant (PQR) biotypes of *Erigeron canadensis* in the presence of paraquat (intact plants sprayed with 3×10^{-4} M paraquat).

4. References

1 Björkman, O. and Demmig-Adams, B. (1994) in Ecophysiology of Photosynthesis (Schulze E.-D. and Caldwell M.M., eds.) pp. 17-47, Springer, Berlin, Germany

2 Demmig-Adams, B., Adams, W.W., III., Logan, B.A. and Verhoeven, A.S. (1995) Aust. J. Plant Physiol. 22, 261-276

3 Sieferman, D. and Yamamoto, H.Y. (1975) Biochem. Biophys. Res. Commun. 62, 456-461

4 Yamamoto, H.Y. (1979) Pure Appl. Chem. 51, 639-648

5 Gilmore, A.M. and Yamamoto, H.Y. (1991) Plant Physiol. 96, 635-643

6 Noctor, G., Rees, D., Young, A. and Horton, P. (1991) Biochem. Biophys. Acta 1057, 320-330

7 Lehoczki, E. and Szigeti, Z. (1988) in Applications of Chlorophyll Fluorescence (Lichtenthaler H.K., ed.) pp. 115-120, Kluwer Academic Publishers, Dordrecht, The Netherlands

8 Shaalthiel, Y. and Gressel, J. (1986) Pesticide Biochem. Physiol. 26, 22-28

9 Vaughn, K.C., Vaughan, M.A. and Camilleri, P. (1989) Weed Sci. 37, 5-11

10 Pölös, E., Mikulás, J., Szigeti, Z., Matkovics, B., Hai, D.Q., Párducz, A. and Lehoczki, E. (1988) Pesticide Biochem Physiol. 30, 142-154

11 Lehoczki, E., Laskay, G., Gaál, I. and Szigeti, Z. (1992) Plant Cell. Environ. 15, 531-539

12 Váradi, Gy., Lehoczki, E., Szigeti, Z. and Pölös, E. (1990) in Proc. 9th Australian Weeds Conference, 6-10 August, Adelaide, South Australia, pp. 257-259

13 Váradi, Gy., Darkó, É., Pölös, E., Szigeti, Z. and Lehoczki, E. (1994) J. Plant Physiol. 144, 669-674

14 Darkó, É., Váradi, Gy. and Lehoczki, E. (1996) Plant Physiol. Biochem. 34, 843-852

This work was supported by the Hungarian Research Fund, OTKA T-16445.

EFFECT OF PARAQUAT MEASURED VIA IN VIVO P-700 OXIDATION AT 820 nm ON PARAQUAT-SUSCEPTIBLE AND RESISTANT *ERIGERON CANADENSIS* (CRONQ.) BIOTYPES

R. Cseh, L. Almási and E. Lehoczki

Department of Botany, József Attila University, H-6701 Szeged, POB 657 Hungary

1. Introduction

The absorption of paraquat (methylviologen: 1,1'-dimethyl-4,4'-bipyridinium) through the leaf cuticle after spraying, is very rapid. The immediate electron donor of paraquat is the iron sulphur cluster $Fe-S_A/Fe-S_B$ (also known as P-430) (Hiyama and Ke 1971[1]), its action is to divert electrons from NADPH and transfer them to molecular oxygen, producing very reactive toxic oxygen species (Halliwell 1984[2]).

Light-induced absorbance changes around 820 nm, first described by Harbinson and Woodward (1987)[3] used in our experiments as a new approach of paraquat action and resistance. Considering the changes in P-700[+] accumulation and other photosynthetic properties of paraquat-resistant (**PQR**) and susceptible (**S**) *E. canadensis* biotypes, could lead to a better understanding of the mechanism(s) of paraquat action and resistance.

2. Procedure

Two *E. canadensis* biotypes were investigated in the present experiments: S and PQR; these were grown under field conditions from May to middle-August, and 5-6-month-old plants having 12-15 fully-expanded leaves in the rosette stage were used in all experiments.

3905

G. Garab (ed.), Photosynthesis: Mechanisms and Effects, Vol. V, 3905–3908.
© 1998 *Kluwer Academic Publishers. Printed in the Netherlands.*

Intact, individual leaves of the two bitypes of E. *canadensis* were sprayed with a 2 µmol ml^{-1} aqueous solution of commercially formulated paraquat (Gramoxone, 25%) from a hand atomizer. After spraying, the leaves were irradiated with actinic light at 150 µmol m^{-2} s^{-1} for 5 min to promote the appropriate penetration of paraquat into the leaves. All treated leaves of the S plants underwent desiccation and died within 24 h after paraquat treatment, but all the treated leaves of the PQR plants survived the herbicide stress.

In vivo PSII chlorophyll fluorescence was measured at room temperature by means of a Hansatech dual-channel modulated fluorescence measuring system (MFMS-2; Hansatech, Kings Lynn, Norfolk, UK).

Changes in the oxidation state of P-700 in leaves under herbicide stress relative to untreated (control) leaves were measured at 820 nm with a Hansatech P-700 measuring system. From ΔA_{820} recorded during the period of actinic irradiation and produced by far-red irradiation, the quantum efficiency for electron transport by PSI was calculated as described previously (Weiss, Ball and Berry 1987[3]).

The effect of paraquat action on CO_2 fixation was measured at room temperature and 150 µmol m^{-2} s^{-1} by an infrared gas analyser (LCA-III ADC Ltd., Hoddeston, UK) with calculations via the formula presented by Caemmerer and Farquar (1981)[4].

3. Results and Discussion

The same kinetic changes were observed after paraquat treatment in both S and PQR biotypes as reported by Harbinson and Hedley (1993[6]) after methylviologen treatment.

Once paraquat reached PSI and started to exert its primary electron-diverting effect, strong changes occurred in the characteristic photosynthetic parameters. The S biotype displayed a continuous deterioration in photosynthetic parameters after paraquat treatment (except for qP), whereas the PQR biotype gave a minimum 20 min after treatment and thereafter exhibited a recovery.

Figure 1. reveals that the nonphotochemical quenching (NPQ) of the S biotype rose considerably higher ($\approx$600% of the untreated control level) than that of PQR biotype. The changes were observed in NPQ were mainly due to the energy-dependent quenching qE (data not shown), indicating a decrease in the intrathylakoidal pH, because the rate of reaction between PQH_2 and cyt f via the Reiske Fe-S centre is determined by the intrathylakoidal pH and the concentration of PQH_2 in the membrane (Bendall 1982[8]).

ΔA_{820} obtained by using the modulated technique is a linear function of $t_{1/2}$ of the reduction of P-700$^+$, which is a measure of the 'resistance' between the PQH_2 pool and P-700$^+$ (Harbinson and Hedley 1989[7]). ΔA_{820} underwent a slight increase in the PQR biotye, in contrast with a strong increase in the S biotype in the first 20 min of paraquat treatment; thereafter, it started to decrease in the PQR biotype, but increased further in the S biotype (Fig. 2). These changes were consistent with the changes in the quantum efficiency of PSI (Φ_{PSI}) and the rate of CO_2 fixation (Fig. 2). The increase in ΔA_{820} resulted in decreases in Φ_{PSI} and CO_2 assimilation. A significantly higher ΔA_{820} was observed in the untreated PQR than the S leaves (p>1%, n=18). This slower electron

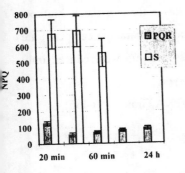

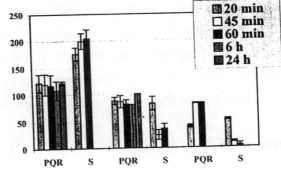

Figure 1. The effect of paraquat tretment on NPQ (at the % of untreated control).

Figure 2. The effect ofparaquat treatment on $\Delta A820$ (left) ΦPSI(mid.) and CO2 assimilation (right) (at the % of control).

transport could participate in a protection mechanism against paraquat inside the PQR chloroplast.

Evidence was obtained that paraquat can reach its site of action near PSI and exerts its primary electron-diverting effect in both S and PQR *E. canadensis* biotypes. Our results suggested a possible process of paraquat action. When a dark-adapted leaf was irradiated after paraquat treatment, the action of paraquat caused a decrease in the intrathylakoidal pH (measured via the NPQ increase). The low intrathylakoidal pH slowed the electron transfer between PQH_2 and the cytb/f complex (measured via the increase in ΔA_{820}) which decreases the quantum efficiency of PSI and the rate of CO_2 assimilation; in parallel with these processes, the toxic oxygen species formed disorganized the photosynthetic membranes in the S biotype, whereas in the PQR biotype the membrane remained intact.

4. Conclusions

The main differences between PQR and S *E. canadensis* biotypes under the above-mentioned measuring conditions are as follows: **1.** At low paraquat concentrations, significantly (P>0.1%, n=18) higher NPQ (qE) was observed in the S biotype than in the PQR biotype. **2.** In untreated control leaves of the PQR biotype, ΔA_{820} was significantly higher (P>1%, n=18), and the electron transfer between the PQH_2 and the cytb/f complex is therefore significantly lower than in the S biotype. These results suggested that significantly lower concentrations of paraquat can penetrate into the PQR than into the S chloroplast. **3.** The CO_2 assimilation was reduced only transiently in the PQR biotype, but irreversibly in the S biotype.

Supported by the Hungarian National Research Fund (OTKA Grant T016455).

References

1. Hiyama T., Ke B. Arch Biochem Biophys 47: 99-108 (1971)
2. Halliwell B. What's New in Plant Physiology 15: 21-24 (1984)
3. Harbinson J, Woodward F. I. Plant, Cell and Environment 9: 131-137 (1987)
4. Weiss E., Ball J. T., Berry J. in Photosynthesis Research Vol II pp 553-556 (1987)
5. von Caemmerer S., Farquhar G. D. Planta 153: 376-387 (1981)
6. Harbinson J., Hedley C. Plant Physiology 103: 649-660 (1993)
7. Harbinson J., Hedley C. Plant Cell and Environment 12: 352-369 (1989)
8. D. S. Bendall Biochimica and Biophysica Acta 683: 119-151 (1982)

LOSS OF H_2O_2-SCAVENGING CAPACITY DUE TO INACTIVATION OF ASCORBATE PEROXIDASE IN METHYLVIOLOGEN-FED LEAVES, AN ESTIMATION BY ELECTRON SPIN RESONANCE SPECTROMETRY.

Jun'ichi Mano, Chiaki Ohno, and Kozi Asada[1]
Res. Inst. Food Sci., Kyoto Univ., Uji, 611-0011, Japan, [1]Dept. Biotechnol., Fukuyama Univ., Gakuen-cho 1, Fukuyama, 729-0292 Japan

Key words: abiotic stress, EPR, oxidative stress, monodehydroascorbate radical, H_2O_2.

1. Introduction

Ascorbate (AsA) plays a critical role in the protection of chloroplasts and plant cells against photooxidative stress by scavenging H_2O_2 as the electron donor to ascorbate peroxidase (APX) or by scavenging O_2^- and •OH nonenzymically [1]. In these reactions AsA is univalently oxidized to monodehydroascorbate radical (MDA), which then is reduced to AsA by the reduced ferredoxin (Fd) or by NAD(P)H via MDA reductase (Fig. 1). We have shown that MDA, which is detectable by electron spin resonance spectrometry (ESR) in vivo, is a sensitive endogenous probe of the status of oxidative stress in leaves [2, 3]. In this study we have established a method to evaluate H_2O_2-scavenging capacity in leaves from MDA measurement. In MV-fed and illuminated leaves, APX is the most labile enzyme in the chloroplastic system for scavenging reactive oxygen species. Loss of APX by MV-phototreatment limited the H_2O_2-scavenging capacity in leaves.

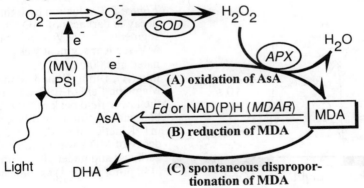

Figure 1. A scheme of production and scavenging of MDA in chloroplasts. Steady-state concentration of MDA is determined by the balance between the oxidation rate of AsA (A) and the rate of MDA scavenging by reduction (B) and by spontaneous disporportiation (C). In healthy leaves (A < B + C) MDA is undetectalble by ESR even under illumination. Under stress conditions, production of O_2^- is enhanced or MDA-reducing system is inhibited (A > B + C) and the MDA concetration is increased to a detectable level.

G. Garab (ed.), Photosynthesis: Mechanisms and Effects, Vol. V, 3909–3912.

2. Materials and Methods

Spinach leaves from local markets were cut into segments (5 × 30 mm), infiltrated with 0.1 mM methylviologen (MV), and incubated in darkness for 3 h at 25°C. MV-fed leaf segment was pasted with silicone grease on a quartz tissue cell of ESR (JEOL, Tokyo, Japan) and illuminated with a tungsten lamp at 1,000 μmol photon m^{-2} s^{-1} for various periods (MV-phototreatment). After an incubation in darkness for 5 min, MDA in the treated leaf section was determined at room temperature with an X-band ESR spectrometer (JES-RE2X, JEOL, Tokyo, Japan)[2]. MDA signals in suspensions of chloroplasts and thylakoids were measured in a quartz flat cell (thickness, 0.4 mm). Superoxide dismutase (SOD) was determined from the inhibition of ferricytochrome c reduction by O_2^-, which was produced using xanthine and xanthine oxidase. Electron transport activities of whole chain ($H_2O \rightarrow NADP^+$) and PSI (AsA/DCIP $\rightarrow$ MV/O_2) were determined by $NADP^+$-photoreduction at 340 nm [4] and by O_2-consumption, respectively.

3. Results and Discussion

3.1 *Changes in the MDA production and decay due to the MV-phototreatment*

As shown in Fig. 2, production upon illumination, and decay after turning off the light, of MDA in MV-fed leaf segment were affected by preillumination in two major points; (i) the photoinduced MDA signal level was decreased, and (ii) the lifetime of the MDA signal after turning off the actinic light was elongated. For isolated chloroplasts, illuminated in 0.1 mM MV and 5 mM AsA, the elongation of the post illumination lifetime of MDA was also observed (not shown). These indicated that in MV-phototreated leaves MDA is produced even after illumination, by a stable, photoaccumulated oxidant, namely, H_2O_2. Thus, H_2O_2-scavenging capacity in chloroplasts is decreased by MV-phototreatment.

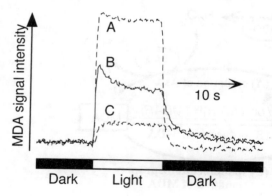

Figure 2. Production by light and decay in darkness of MDA in leaf segments after MV-phototreatment. MV-fed leaf segment was preilluminated at 1,000 μmol m^{-2} s^{-1} for 0 min (A), 1 min (B) and 20 min (C). Magnetic field of ESR was fixed at the lower-field peak of MDA, a doublet with a splitting width of 0.23 mT. Kinetics of production and decay of MDA was monitored in darkness and under 10-s actinic light (440 μmol m^{-2} s^{-1}) as indicated.

3.2 *Ascorbate peroxidase is the most labile enzyme to MV-phototreatment*

APX activity was decreased by the MV-phototreatment (Fig. 3). Both thylakoid-bound and stromal APXs were inactivated to the same extent. MDA reductase (20 μmol mg Chl^{-1} h^{-1}) and SOD (5850 units mg Chl^{-1}), which could affect the scavenging and production rate of MDA, were unchanged by MV-phototreatment when APX was inactivated by 75%. Activity of PSII was not affected. Electron transport activity of PSI (135 μmol O_2 mg Chl^{-1} h^{-1}), and whole chain (192 μmol NADPH mg Chl^{-1} h^{-1}) were decreased only by 10% and 18%, respectively, when APX was inactivated by 75%. Glutathione reductase (GR) was not inhibited in spinach chloroplasts by the MV-phototreatment to inacitvate 75% of APX. DHA reductase (DHAR) was decreased by 23% (Table 1), as reported for pea leaves treated for longer period [5]. Thus, APX is the most sensitive to MV-phototreatment among the chloroplastic enzymes for scavenging reactive oxygen species. The extension of the post-illumination lifetime of MDA was therefore most likely caused by the inactivation of chloroplastic APX.

The inactivation mechanism of APX has been proposed [6]: APX is oxidized first by H_2O_2 to form compound I, which is reduced back to the initial state by two AsA molecules to produce two MDA molecules. Alternatively, the compound I is further oxidized by H_2O_2 to be an inactive form, to result in the heme-decomposition. Therefore, under a high $[H_2O_2]/[AsA]$ condition, the compound I is more likely oxidized by H_2O_2 than reduced by AsA, leading to the irreversible loss of APX activity. Although AsA concentration in chloroplasts was unchanged by MV-phototreatment (data not shown), the ratio $[H_2O_2]/[AsA]$ in the microenvironment around APX molecules must have been raised by MV due to the higher O_2^- production and lower MDA reduction.

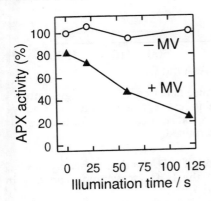

Figure 3. Time course of the activity changes of APX in chloroplasts after MV-phototreatment. Chloroplasts in 0.1 mM MV and 5 mM AsA in an isotonic medium were illuminated and collected by centrifugation. APX activities of thylakoids and stroma were determined separately. The activity ratio thylakoidal APX: stromal APX was always 4:1, therefore only total activities are shown. MV at this concentration did not inactivate APX in darkness. Activity at 100% was 1,400 μmol mg Chl^{-1} h^{-1}.

Table 1. Activity changes in APX, DHAR and GR in intact chloroplasts by illumination in 0.1 mM MV. Chloroplasts at 30 μg Chl ml^{-1} in 5 mM AsA in an isotonic medium were illuminated at 1,000 μmol m^{-2} s^{-1} at 25°C. Chloroplasts were collected by centrifugation and the enzyme activities in them determined. Values are averages of 3 runs. No inactivation of APX and DHA reductase was observed in the absence of MV.

Illumination	Relative activity (%)		
(min)	APX	DHAR	GR
0	100	100	100
10	24	78	118

3912

3.3 Half-life of the post-illumination production of MDA negatively correlates with APX activity in chloroplasts and in leaves

We quantitated the post-illumination production of MDA by the half-life of the decay of post illumination MDA signal. A negative correlation between APX activity and the half-life was found for MV-phototreated chloroplasts (Fig. 4, left). The correlation was not colinear; APX activity was decreased as the chloroplasts were treated for longer time, but the half-life of MDA was unchanged until APX activity went down to 700 μmol mg Chl^{-1} h^{-1}. Below that point the half-life is increased depending on the loss of APX activity. A similar negative correlation between the activity of thylakoid APX, namely of the chloroplastic APX, and the half-life of MDA signal was observed in leaves as well (Fig. 4, right).

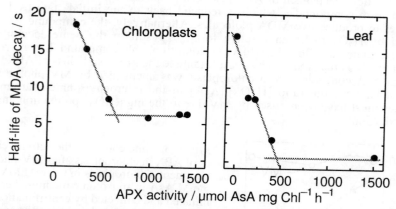

Figure 4. Relationship between APX activity and thee half-time of MDA signal after actinic illumination in intact chloroplasts (left) and in leaf segments (right). APX activity in chloroplasts represents the sum of thylakoidal and stromal APXs. For leaf segments only thylakoid-bound APX activity was determined.

Thus, the elongation of post illumination production of MDA as determined by in vivo ESR measurement is a quantitative measure of the H_2O_2-scavenging capacity in chloroplasts, which is primarily limited by APX activity.

References

1 Asada, K., Endo, T., Mano, J. and Miyake, C. (1998) in Stress Responses of Photosynthetic Organisms (Satoh K. and Murata N., ed.) pp. 37-52, Elsevier, Tokyo, Japan

2 Heber, U., Miyake, C., Mano, J., Ohno, C. and Asada, K. (1996) Plant Cell Physiol. 37, 1066-1072

3 Hideg, É., Mano, J., Ohno, C. and Asada, K. (1997) Plant Cell Physiol. 38, 684-690

4 Mano, J., Ushimaru, T. and Asada, K. (1997) Photosynth. Res. 53: 197-204

5 Iturbe-Ormaetxe, I., Escuredo, P.R., Arrese-Igor, C. and Becana, M. (1998) Plant Physiol. 116, 173-181

6 Miyake, C. and Asada, K. (1996) Plant Cell Physiol. 37, 423-430

DAMAGE CAUSED TO THE PHOTOSYNTHETIC APPARATUS DUE TO UNCONTROLLED APPLICATION OF THE PESTICIDE, ENDOSULFAN: THERMOLUMINESCENCE AND FLUORESCENCE STUDIES

Jyoti U. Gaikwad, Sarah Thomas, S.D. Aghav and P.B. Vidyasagar
Biophysics Laboratory, Department of Physics, University of Pune,
Pune 411 007 INDIA

Key words: Environmental stress, grapevine, metabolic processes, source-sink, electron transport, photosystem II

1. Introduction

Grape production in India has been showing an upward trend in recent years due to its high export potential in the international market. Subsequently there has been an impressive increase in vineyards. However, proportional increase in production and quality is not observed. This may be attributed to the changes in the climatic conditions and increased severity of general infections(1). Among the various grape varieties, mainly Thompson seedless and its successor varieties are being cultivated in India. Though these varieties are sweet in taste they are more susceptible to infections and diseases. Hence, in order to increase yield, uncontrolled and unscientific application of pesticides is a common practice among the grape growers. However, it has been observed that the use of pesticides for causal organisms have proved to be more serious on the plants Their application may also disturb plant metabolism and invite diseases like cracks, rots or possibly other physiological disorders like pink berries. Moreover, regular and heavy applications of most pesticides initiated resistance development and severe phytotoxicity. There are also evidences of pesticide residues in plants exposed to pesticides. It has been recently reported that most of the pesticides that may enter a human body are usually from plants(2).

Hence it is high time to concentrate our efforts on studying possible damage caused due to uncontrolled pesticide application and to detect symptoms of the disorders at an early stage. Such studies would help in formulating effective and safer dose levels of pesticides that would ward off insects and other predators without affecting the plant and also the quality of grapes.

Work related to the analysis of pesticide residues in the grape berry is being carried out to some extent. But, since the ripening berry is a strong sink for solute from photosynthesis and reserve organs it would also be interesting to study the effect of uncontrolled pesticide application on the photosynthetic apparatus of the grape leaves. In the present work the biophysical techniques namely absorption, thermoluminescence (TL) and fluorescence have been adopted.

G. Garab (ed.), Photosynthesis: Mechanisms and Effects, Vol. V, 3913–3916.
© 1998 *Kluwer Academic Publishers. Printed in the Netherlands.*

2. Materials and Methods

Mature leaves from about the 7th nodal position of the grape plant (*Vitis Vinifera* L. cv. Thompson Seedless) were taken. Among the various pesticides that are generally applied on the grapevine plant, a widely used one, namely Endosulfan (6, 7, 8, 9, 10, 10-hexachloro-1, 5, 5a, 6, 9, 9a-hexahydro-6, 9-methano-2, 4, 3-benzodioxathiepin 3-oxide) was selected.

The effects of endosulfan were observed on isolated photosynthetic pigments and also on the photosynthetic apparatus in intact leaf discs, as a function of incubation period and pesticide concentration. For the former study, samples were incubated with two different doses of the pesticide viz. 1.4 ppm (prescribed dose) and 4.2 ppm (equivalent to three times above dose) for different time periods ranging from 1 hr to 120 hr. while for the latter one, samples were treated with different concentrations of endosulfan ranging from 4.2 ppm to 63 ppm for a fixed incubation period of 3 hrs.

Chloroplasts were isolated using Demeter's method. Extraction was done using phosphate buffer of 0.2M and pH 7. The isolated chloroplast suspension was treated with the selected dose of the pesticide and this mixture kept for a pre-decided incubation period, in the refridgerator alongwith untreated controlled chloroplasts. Thereafter, the pesticide residues were removed by centifuging the treated samples using the suspension medium. Chlorophyll was then extracted with 80% chilled acetone and the absorption spectra were recorded using Hitachi Spectrophotometer Model 330 in the range from 360 nm to 720 nm.

For fluorescence and TL studies, leaf discs cut out from mature leaves and soaked for 10 min in the pesticide solution. Thereafter they were kept in the refridgerator for the fixed incubation period in clean petriplates devoid of the pesticide solution. After pesticide treatment, leaves were washed with water to remove any pesticide residue.

Steady state fluorescence was measured from control and treated leaf discs of diameter 1.5 cm, using Perkin Elmer LS 50 spectrofluorometer. Fluorescence emission spectra were measured by keeping the excitation wavelength fixed at 440 nm.

TL glow curves were recorded from intact leaf discs of 2 cm diameter, using a PC-based TL apparatus built in our laboratory(3). The leaf disc was illuminated at room temperature for 15 s and again at $-80\ ^\circ$C for 5 s with a laser diode (wavelength: 670 nm and power: 3 mW) A linear heating rate of 15 °C/min was obtained using a microprocessor based temperature controller. TL was recorded in the temperature range of -80 to $+50°$C. The fitting and analysis of the glow curve was done using a model based on the general order kinetics theory, to obtain various thermodynamical parameters, viz. activation energy (E), frequency factor (s) entropy (ΔS), and free energy (F.E.) (4).

3. Results and Discussion

The absorption spectra of chlorophyll extracted from chloroplasts treated with 1.4 and 4.2 ppm of endosulfan after 1hr. of incubation is shown in Fig. 1. No definite trend was observed in these spectra on varying the incubation period. Also, there was no change in peak position in all cases. Similar observations were obtained from the studies related to the effect of different concentration of pesticides on pigments. These results indicated that the pesticide endosulfan might not be having a direct effect on the pigments.

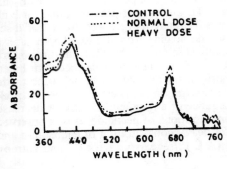

Fig. 1. Absorption spectra of chlorophyll

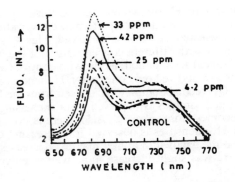

Fig. 2. Fluorescence emission spectra

Fig. 2 shows the effect of increasing concentrations of endosulfan, (ranging from 4.2 ppm to 42 ppm) for three hours incubation, on the steady state fluorescence emission spectra recorded from the leaf disc. No shift in either the 684 nm peak attributed to PSII components or in the 730 nm peak attributed to PSI can be observed. An increase in fluorescence intensity with increasing concentration of pesticide was observed upto 32 ppm while for the 42 ppm concentration a decrease in fluorescence intensity is observed. As seen from Fig. 3, a plot of the F730/F684 ratios calculated from the fluorescence records shows that with increasing pesticide concentration. The ratio shows decreasing values. This is because, the variation in fluorescence intensity at 684 nm was more than that at 730 nm. That is there is a more rapid increase in the PSII fluorescence as compared to PSI fluorescence as evident from Fig. 2. This may indicate that for increasing concentrations of the pesticide PSII acceptor side is affected to a larger extent than PSI. The decrease in fluorescence at 52 ppm concentration may be attributed to overall degradation of the photosynthetic apparatus itself. Varying the incubation periods from 1hr to 120 hrs for 1.4 ppm and 4.2 ppm concentration, did not result in any significant changes in the fluorescence observations.

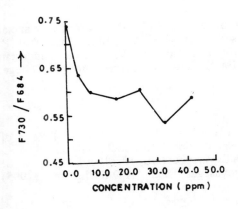

Fig. 3. F730/F684 ratios

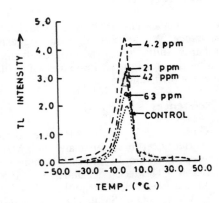

Fig.4. TL glow curves

In case of TL studies related to different pesticide incubation periods varying from 1hr. to 120 hrs., no definite pattern in the variation of TL intensity at peak maxima, for both the selected doses, was observed. Fig. 4 shows TL patterns recorded from leaf discs treated with different concentrations of the pesticide ranging from 4.2 to 63 ppm. after a 3hr incubation period, along with control (untreated) sample. The TL from the control sample showed only one major TL band at around −5 °C. This peak has been attributed to Q_a recombination (5). As seen from the figure no shift in peak position for pesticide treated samples is observed. For increasing concentration of pesticide the TL peak becomes more narrow. Also, there is no emergence of any other TL peak. A decrease in TL intensities is observed for increasing concentrations of pesticide. It was observed that the bandwidth at half maxima decreases for heavy concentration of endosulfan and there was slight increase in activation energy (A.E.), frequency factor (s) and entropy (ΔS).

The decrease in TL intensity for heavier doses indicates that electron chain is getting affected prior to the primary electron acceptor (Q_a) site since the peak is originating from Q_a itself. Therefore, the effect of the pesticide may be prior to the primary electron acceptor (Q_a) site or at the core RC / LHC.

Summary:

The present study was planned to see the possible effects of the pesticide, Endosulfan, on photosynthetic activity of *Vitis vinifera* L. (cv. Thompson seedless). Absorption studies of the chlorophyll pigments indicated that there was no direct effect of pesticide on pigments. Hence studies were planned to see the effects on the photochemical activity of grape leaf disc using fluorescence and TL techniques. Fluorescence observations showed that there is a definite effect of pesticide on the acceptor side of electron transfer chain. TL studies on the other hand, indicated that the site of action of the pesticide was prior to Q_a or at the core RC/LHC. However, since absorption studies indicated that the pigments are not affected, hence it can be said that the target of pesticide action is on the PSII acceptor site prior to Q_a.

Acknowledgements:

Jyoti U. Gaikwad and Dr Sarah Thomas, acknowledge C.S.I.R., INDIA for the financial assistance. We acknowledge the suggestions by Prof. Govindjee, Illinois and Prof. P. Mohanty, India during the progress of this work and also Mr U.V. Gaikwad, Hindustan Antibiotic Ltd, Pimpri, Pune and Dr. Milan Banerjee, Armed Forces' Medical College, Pune for their inputs

4. References

1   Gaikwad U.V., Kale A.R., and Shelke S.N. (1994) The Nat. Symp. of Pests of Agricultural Importance and their Management, Bioved Res Soc, Allahabad, 38
2   Ames B N, Gold L S, (1997) Environ Health Perspect (US) 105(4) 865-73
3   Sarah Thomas, M.Banerjee, P.B.Vidyasagar and A.D.Shaligram (1995) Measurement Sci. Technol., 6, 554-559
4   P.B.Vidyasagar, Sarah Thomas, M.Banerjee, U.Hegde and A.D.Shaligram, (1993) J. Photobio. B: Biol., 19, 125-123
5   Jyoti Gaikwad, Sarah Thomas, S.D.Aghav and P.B.Vidyasagar, (1998) Vitis 37 (1), 11-14

DICOTYLEDONS SENSITIVE TO HERBICIDES THAT INHIBIT ACETYL-COENZYMEA CARBOXYLASE.

John T. Christopher, Joseph A. M. Holtum

Department of Tropical Plant Sciences, James Cook University of North Queensland, Douglas, Townsville, Queensland 4811, Australia

Key words: herbicides, lipid synthesis, weeds, tolerances, toxicity.

1. Introduction

The key regulatory enzyme in fatty acid biosynthesis, acetyl-coenzymeA carboxylase (ACCase), is known to occur in two structural forms. In prokaryotes three types of subunit make up ACCase while in eukaryotes a single multifunctional peptide carries out all three functions (1). In most plants both forms of ACCase occur. A multisubunit "prokaryotic" form occurs in the chloroplasts of dicots and most monocot species (2, 3). A multifunctional "eukaryotic" form of ACCase occurs in the plant cytosol and has recently been reported to occur in the chloroplasts of *Brassica napus* (4). The prokaryotic form of ACCase is relatively insensitive to inhibition by ACCase-inhibiting herbicides compared to the eukaryotic form (2) rendering most plants insensitive to these herbicides. In contrast, the Gramineae have only the eukaryotic form of ACCase but not the prokaryotic (2, 3) and are susceptible to ACCase-inhibiting herbicides which have become known as "graminicides". Until recently, sensitivity to ACCase-inhibiting herbicides was only known from the Gramineae suggesting that the phenomenon had evolved only once. However, the ACCase-inhibiting herbicide haloxyfop has recently been registered for the control of two dicots *Erodium botrys* and *E. moschatum* in Australia.

Aim

To determine why the dicotyledon *E. moschatum* is susceptible to haloxyfop herbicide preciously only known to kill Gramineae and whether closely related dicots are also susceptible.

2. Procedure

2.1 *Whole plant response to haloxyfop herbicide of E. moschatum compared to known resistant and susceptible plant types.*
Seedlings were grown outdoors in 5 L pots containing sand- and peat-based potting soil at James Cook University ($19^{\circ}19'43"$S, $146^{\circ}45'36"$E). The commercial formulation of haloxyfopethoxyethyl was applied at 0, 2, 7, 26 and 104 grams active ingredient ha^{-1} for

G. Garab (ed.), *Photosynthesis: Mechanisms and Effects, Vol. V,* 3917–3920.

susceptible monocotyledons *D. ciliaris* and wild type *L. rigidum* and for dicotyledon *E. moschatum*. Resistant dicotyledon *N. tabacum* and a resistant *L. rigidum* known to have mutant haloxyfop-insensitive ACCase (5) were treated at 26, 104, 416 and 832 grams active ingredient ha^{-1}. Mortality was recorded 21 days after treatment and LD$_{50}$ values calculated by probit analysis.

2.2 *Haloxyfop-inhibition of ACCase activity*
ACCase activity in leaf tissue was assayed by following acid-stable incorporation of ^{14}C from NaH$^{14}CO_3$ as described previously (5).

2.3 *Western blot analysis*
Leaf proteins were extracted and separated on a 7.5 % polyacrylamide gel before blotting on to PVDF membranes. Biotin-containing proteins were hybridised with streptavidin-conjugated alkaline phosphatase and visualised using Fast Red (Pierce, Rockford, Illinois). Biotinylated peptides of certain sizes are known to accumulate in correlation with increased ACCase activity (6).

2.4 *Response to haloxyfop of dicotyledons from taxa closely related to Erodium*
Plants were cultured and treated with haloxyfopethoxyethyl herbicide at 104 grams active ingredient ha^{-1} as described in section 2.1. Twenty one days after treatment mortality was recorded.

3 Results

3.1 *Whole plant response to herbicides*
The amount of haloxyfop herbicide required to control *E. moschatum* is similar to that required to control known susceptible graminaceous species *D. ciliaris* and *L. rigidum* (Table 1). In contrast, the dicot *N. tabacum* and the resistant *L. rigidum* with mutant haloxyfop-insensitive ACCase exhibited no mortality at doses of up to 832 g ai ha^{-1} (Table 1). Thus, the response of the dicot *E. moschatum* is very different to that of the dicot *N. tabacum* and a mutant monocot known to have haloxyfop-insensitive ACCase (5) but is similar to that of susceptible monocots.

3.2 *Inhibition of ACCase.*
Susceptiblity to haloxyfop in *E. moschatum* is due to herbicide-sensitive ACCase. The I_{50} for haloxyfop activity in the leaves of *E. moschatum* is similar to the I_{50}s of the two susceptible monocot species *D. ciliaris* and susceptible wild type *L. rigidum* (Table 1) but two orders of magnitude lower than the I_{50}s of the dicots *N. tabacum* and *P. sativum* and the haloxyfop-resistant *L. rigidum* with mutant ACCase (Table 1). Thus, there is good agreement between herbicide susceptibility of whole plants and herbicide-sensitivity of ACCase.

Table 1. Dose of haloxyfopethoxyethyl herbicide (grams active ingredient ha^{-1}) required to kill 50% of pot-cultured plants (LD$_{50}$) and concentration of haloxyfop acid (μM) required to inhibit ACCase activity in leaf extracts by 50% (I$_{50}$). Manufacturer's recommended rates of application are 78, 156 or 104 grams active ingredient ha^{-1} for control of *Erodium, Digitaria* and *Lolium* respectively.

	Herbicide LD$_{50}$	ACCase I$_{50}$
Monocotyledons		
Digitaria ciliaris	9	2
Lolium rigidum susceptible wild type	8	0.4
L. rigidum resistant with mutant ACCase	>832	350
Dicotyledons		
Nicotiana tabacum	>832	180
Erodium moschatum	10	1

3.3 *Western blot analysis.*

Leaf extracts from the herbicide-insensitive dicots *P. sativum* and *N. tabacum* contained a biotinylated protein at approximately 220 kD and another at less than 43 kD corresponding to the expected size ranges for the eukaryotic ACCase and prokaryotic biotinylated subunits of ACCase respectively (Figure 2; 6). The herbicide-sensitive monocot *D. ciliaris* exhibited only the 220 kD eukaryotic subunit as has previously been reported for a number of species of Gramineae (Figure 2; 3, 4). Dicot *E. moschatum* exhibited only the 220 kD subunit as does *D. ciliaris* indicating that herbicide sensitivity in these two widely separate taxa has evolved by the same mechanism, loss of the herbicide-insensitive prokaryotic form of ACCase.

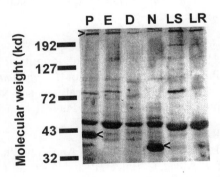

Figure 2. Western blot of leaf extracts from *Pisum sativum* (P), *E. moschatum* (E), *D. ciliaris* (D), *N. tabacum* (N), suceptible wild type *L. rigidum* (LS) and resistant *L. rigidum* with mutant ACCase (LR). Biotinylated proteins at 220 kd (>) and less than 43 kd (<) are in the size ranges expected for the biotinylated proteins of eukaryotic and prokaryotic ACCase respectively (6).

3.4 *Response to haloxyfop herbicide of dicotyledons from taxa closely related Erodium.*

In the family *Geranaceae* plants from the genera *Erodium* and *Pelargonium* were susceptible to haloxyfop herbicide while two species in the genus *Geranium* were not (Table 2). Species in the families *Oxalidaceae* and *Tropaeolaceae* were insensitive to haloxyfop herbicide (Table 2).

Table 2. Mortality of pot-cultured plants 21 days after treatment with haloxyfop herbicide (104 grams active ingredient ha^{-1}) classified as susceptible (S) or resistant (R)

		Plants tested	Mortality (%)	
Geraniaceae				
	Erodium moschatum	11	100	S
	Geranium canariense	3	0	R
	G. rubescens	4	0	R
	Pelargonium spp. (hybrid)	2	100	S
	P. molleconium	4	100	
Oxalidaceae				
	Oxalis corniculata	12	0	R
Tropaeolaceae				
	Tropaeolum majus	15	0	R

4. Discussion

Our observations provide evidence that a dicotyledon *E. moschatum* is susceptible to an ACCase-inhibiting herbicide. The mechanism is a lack of herbicide-insensitive prokaryotic form of ACCase, a form found in the chloroplasts of most plants. This situation was previously known only for Gramineae. In Gramineae it has been shown that ACCase in the chloroplasts is a multi-functional eukaryotic form (3). It has recently been demonstrated that the multi-functional eukaryotic ACCase is encoded by a multi-gene family including at least two chloroplastic isozymes in *Brassica napus* indicating that dicot species may also have eukaryotic ACCase isozymes in the chloroplasts (4). Although the structural form of ACCase in *E. moschatum* chloroplasts has not been elucidated, it seems likely that multifunctional eukaryotic ACCase will be found.

Acknowledgements
Thanks to Dow Elanco Ltd. for gifts of haloxyfop herbicide and haloxyfop acid, Drs. C. Preston and S. Powles for *L. rigidum* seed, Drs J. Burnell and J. Woodward for assistance with protein analyses, and Mr. M. Barber and Mr A. Rice who carried out preliminary experiments. Funded by a James Cook University, Merit Research Grant.

References
1   Harwood, J.L. (1988) Ann. Rev. Plant Physiol. l 39, 101-138
2   Alban, C. Baldet, P. and Douce R. (1994) Biochem J 300,557-565
3   Konishi, K. Shinohara, K. Yamada, K. and Sasaki, Y. (1996) Plant Cell Physiol. 37,117-122
4   Schulte, W. Topper, R. Stracke, R. Schell, J. and Martini, N. (1997) Proc Natl Acad Sci 94,3465-3470
5   Tardif, F. Preston, C. Holtum, J.A.M. and Powles, S.B. (1996) Aust. J. Plant Physiol. 23,15-23
6   Wurtele, E.S. and Nikolau, B.J, (1992) Plant Physiol. 99,1699-1703

21. Ecological aspects

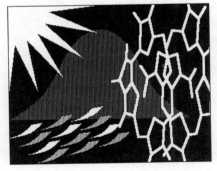

CONGRESS ON PHOTOSYNTHESIS

XIth INTERNATIONAL

BUDAPEST ✳ 1998

Elevated CO_2 and Plant Productivity in the 21st Century: Can we feed billions and preserve biological diversity?

F.A. Bazzaz
Harvard University
Department of Organismic and Evolutionary Biology
Cambridge, Massachusetts

Sometime near the middle of the next century the human population will have doubled. It will , according to projections, increase from about six billion to almost twelve billion people. We must feed and shelter these additional humans. Additionally, it is expected that, this increase in human population will result in a major increase in the amount of fossil fuel consumption above and beyond the current per capita consumption. Therefore, the emission of carbon dioxide will increase dramatically and may become a much more critical environmental issue than at present (Fig. 1). Currently, a person in the United States of America, on average, uses 22 tons of carbon per year, whereas a person in India, for example, uses only 0.7 tons of carbon per year (Fig. 2). Thus, an individual in the United States consumes almost 40 times as much as an individual in India. Progress, which is the aim of many developing nations will be accomplished only by increased per capita energy consumption. As a result, the carbon output per person, in these developing countries will dramatically increase. It may even double. For example, if we take current emission in Russia (currently between the United States and India) as a standard and assume that all developing countries at the moment, aspire for development, we reach an astonishing fact. The amount of energy used in the amount of carbon dioxide emitted in the atmosphere, would show that the emission of carbon dioxide could go up by a factor of 4 or there about. Apparently, there is no escape from the fact that the concentration of carbon dioxide in the atmosphere will significantly rise and that, in addition to direct effects on plant photosynthesis, growth allocation and phenology, there will be changed in weather patterns. CO_2 is predicted to influence primary productivity on the land and the ocean (Fig.

G. Garab (ed.), Photosynthesis: Mechanisms and Effects, Vol. V, 3923–3940.
© 1998 Kluwer Academic Publishers. Printed in the Netherlands.

3). Thus, the need to better understand the mechanisms of photosynthesis and photosynthate allocation.

Models show that when you increase a carbon dioxide concentration (or the concentration of other greenhouse gases) in the atmosphere temperature will increase globally. Currently there are several models which purport to predict this rise. These are called General Circulation Models (GCM). More recently, ocean models are coupled with these terrestrial models to produce what is called Coupled General Circulation Models (CGCM). These models differ in detail but agree on two things. The first is an increase in temperature, which is expected to be about 3 to 5°F, with a doubling of CO_2 will not be equitably distributed in that polar regions will receive much higher temperatures than equatorial regions. Some models predict, for example, that in the Arctic a temperature can go up by 16° (Houghton et al. 1990). The second critical issue that models agree upon is that there will be a change in the pattern of global precipitation. For example, some models predict that there is going to be an intensification of the Indian monsoon resulting in an increase in precipitation in the Indian subcontinent. These models also predict a decline in the rainfall in the Amazon Basin and a decline in Central and Tropical Africa as well. Precipitation models are not currently exact and there is a great deal of variations among models in this respect. However, most of these models show that there will be globally, an increase in precipitation of about 15%. This change may be regionally undesirable but together with increased water use efficiency may result in an increase in yield globally (Idso et al. 1987). Some scientists argue (Kimball 1993) that aerosol in the atmosphere may modify both the global mean temperature and pattern of precipitation.

Regions which already suffer greatly from excess water (flooded), such as Bangladesh, could suffer more severely and regions which need more water can end up with less water. The Intergovernmental Panel of Climate Change (IPCC) has published a second report in

1995 in which several scenarios for the increase in carbon dioxide in the atmosphere are discussed. And their "business as usual scenario" the doubling of carbon dioxide in the atmosphere could occur by the middle of the next century. In fact, some scientists speculate that even if we hold global CO_2 emission at the 1994 level, the CO_2 concentration in the atmosphere will continue to rise can exceed 1000 ppm.

Being an important substrate for photosynthesis, elevation of carbon dioxide in the atmosphere could have significant impacts on this process. Agricultural scientists deserve much of the credit for pointing to us as far back as the '50's and '60's the importance of the elevation of carbon dioxide to plants productivity. Greenhouse growers in countries like the Netherlands have been using CO_2 enrichment for a very long time to increase the productivity and the yield of their crops. A survey by Kimball (1983) who looked at hundreds of reports concludes that, on average, crop productivity in a high CO_2 environment could increase by about 34%. Experiments with crops in many parts of the world have shown that crops do differ in their responsiveness to elevated CO_2 (Allen et al. 1996). For example, cotton is very responsive to elevated CO_2 with an increase of 80%, whereas wheat is less responsive with an increase somewhere in the neighborhood of 10 to 50% (Gifford 1989). These sorts of results were obtained under controlled environments in glasshouses, in open top chambers, and in the FACE experiments (which is the release of carbon dioxide around the crop without having any chambers). There are now several FACE studies in crops and in simple ecosystems in Arizona, North Carolina, Switzerland and other parts of the world.

In 1990 I wrote a review on this the impact of elevated CO_2 on plants. For example, based on available information at that time, there will be an asymptotic increase in photosynthesis in response to elevated CO_2 with concomitant decline in transpiration because it is well established that elevated CO_2 causes stomatal closing and therefore the instantaneous water use efficiency (WUE) will improve. Plants can photosynthesize at a higher rate while expending less water (Fig. 4). I also speculated that this enhancement of photosynthesis does not necessarily remain high for all species. They experience what is called "down-regulation". If there are sinks for the manufactured photosynthate will go to these sinks and the enhancement of photosynthesis by high CO_2 remains high. If there are no sinks, the accumulation of sucrose (and perhaps glucose) can lead to a down regulation of photosynthesis, i.e., a reduction of photosynthesis, irrespective of the concentration of CO_2 around the plant. E.G. Reekie did an experiment with 8 species of *Brassica*, one pair has a large sink in flowers, a second has a large sink in stems, a third has a large sink in roots, and the fourth has a no particular sinks. He grew these plants in ambient CO_2 (about 350 ppm) and in elevated CO_2 (700 ppm) and measured the rate of photosynthesis. He found that during the early stages of the experiment, all four species responded positively to elevated CO_2. In every case high CO_2-grown plants had a significantly higher photosynthetic rate than ambient CO_2 grown plants. However, during the last harvest, he found that the plants with sinks, the first, second and third pair of species of *Brassica* all still have higher photosynthetic rates in a high CO_2 environment than in a low CO_2 environment but the fourth pair did not, apparently because it did not have any available sink (Fig. 5). The mechanisms of decline in photosynthesis after long-term exposure of plants to CO_2, have been elucidated by several investigators, especially Besford et al. (1993), and J. Sheen (1990). There is regulation at the biochemical, physiological and the molecular level of photosynthesis. When a leaf is subjected to elevated CO_2 surrounds a leaf of the there is an increase in photosynthesis. That increase in CO_2 around a leaf generates a large quantity of sucrose and the sucrose is transported to other parts of the plant through the usual channels of sucrose transport. However, if there are no sinks sucrose accumulates in plants and is broken down into glucose and fructose. And, there hexoses the increase in the activity and concentration of the enzyme hexokinase, which sends a signal to the nucleus, which turns off or reduces the expression of the genes in the nucleus encoding for the small subunit of *Rubisco*. The activity of *Rubisco* declines and the photosynthesis decline as well. However, we still do not know what the role of the enzyme activase which is responsible for the activation carbamilation of *Rubisco*. For Rubisco to be an active form, it has to have both carbon dioxide and magnesium attached to it. And this attachment, called carbonation, is apparently under the control of the enzyme activates. Activases also capable of deactivating *Rubisco* of taking the magnesium and carbon dioxide of *Rubisco* (Fig. 6). There is now evidence that different plant species differ in their ability to down-regulate photosynthesis even within the same ecosystem. For example in a 3-year experiment we found no down-regulation in red oak while there was great down-regulation in yellow birch. Modelling these responses lead to the dominance of the stand by red oak. Norby found down-regulation in tulip poplar but not in white oak.

We must keep in mind, when discussing the impact of elevated CO_2 on plants, particularly in the northern hemisphere, that there is a large additional deposition of nitrogen on land and in the ocean. This nitrogen comes from two sources 1) from vehicles and other machines that use fossil fuel and 2) from fertilizer application in agriculture. The amount of total nitrogen deposition could be as high as 150 kilograms per hector per year as the hot spots in Central Europe, such as parts of the Netherlands, down to small quantities of, 10 to 15 kilograms per hector per year such as in parts of the eastern United States (Holland et al. 1997) (Fig. 7). Some scientists have suggested that the co-presence in carbon dioxide and nitrogen predicted to happen under global change conditions will be of great dividends in terms of the increase productivity and yield in crops (Allen et al. 1996). This may be the case particularly if the breeders and the molecular biologists prepare varieties of these crops that can perform best under the changing carbon dioxide and nitrogen conditions. While this may be a free fertilizer evidence shows that this enormous deposition is having negative feedbacks on biological diversity.

Thus, with global change, we are simultaneously adding more carbon and more nitrogen. both of which could enhance productivity as long as nitrogen is limiting. It is also critical to realize that in order to gain nitrogen from this soil the plant must expend carbon to grow more roots and forage widely in the soil for essential nutrients. And in order to gain carbon, the plant needs nitrogen which is the major component in the enzyme *Rubisco*, the critical enzyme for dark reactions photosynthesis and also a component of the chlorophyll molecule involved in light reactions. The nitrogen and carbon are linked to each other and any understanding and assessment of the impact of global change must consider both together.

When nitrate (NO_3^-) is absorbed from the soil it has to be reduced to ammonia (NH_4) before it can be used in biosynthesis. Once it produced NH_4 it has to be attached on carbon skeletons which come from photosynthesis to make amino acid and proteins. It stands to reason then, that the uptake of carbon and nitrogen and co-regulated. Recent evidence suggests that when nitrogen is present in the soil (in the form of nitrate or ammonia) in large quantities, the roots manufacture cytokinen which is transported up to the leaves and enhances the ability of the leaves that fix carbon (Fig. 8).

Models predict that productivity in developed countries (like western Europe, the United States and Japan), could increase up to 15 to 20% because of global change. This may be true without considering the possibility of temperature rise and the expansion of crops in the northern latitudes which, at the moment, are not suitable. The increase in agricultural productivity must be tempered by other events.

We must keep in mind, is that the same models that predict an increase in the productivity in the developed countries predict a decrease in productivity in developing countries (Fig. 9). Developing countries, because they do not have the resources to develop, or purchase, the new crop varieties, and other needed materials such as biosites to increase production, can suffer greatly. It is very likely that their agricultural production can not keep up with their rapidly increasing populations. This is a potentially dangerous situation because the difference between developed countries and developing countries will actually be larger (an increase of 15 to 20% in developed countries and a decrease of about 10 to 15% in developing countries). This major difference between may cause many social, economic, political and security upheavals. Of course, there are other implications to the situation. In the developing countries with forests there is also going to be an increase in deforestation. As the human population in these countries increases, humans are going to encroach on natural vegetation and, therefore, as the cities expands, because of the increased human population, and migration to cities, the area under agriculture must expand and natural vegetation will be diminished. There will be a smaller area for natural vegetation (Fig. 10). We must also remember that with the increased contact between people and the forest, particularly with the animals of the forest, there may be an increase in communicable diseases in particular those that can move from animal to humans. The recent incidence for example the hemoragic fever or the Ebola virus attests to the danger in increased contact between humans and animals in the forest (Table 1).

The other fact which needs attention is that studies have shown that the quality of the product under high CO_2 environment may actually be less than desired. For example, there is a tendency for the carbon to nitrogen ratio to actually increase which may mean that, even if there is an increase in yield, the protein or nitrogen content of that yield will be lower. For example, we grew wheat in an ambient and high CO_2 environment and compared the suitability of the grain for bread-making. The analysis done by a USDA laboratory show that the wheat from the high CO_2 plants are not suitable to bread-making.

Increased C/N ratio may have influences on the growth of insects and other herbivores (Kimball and Lincoln et al. 1993, and Lindroth et al. 1993). We have shown in one case that larvae ate more and grew less when fed on plants grown in a high CO_2 environment (Fig. 11) (Lincoln et al. 1993). Furthermore, we observed in another experiment that the female or the voracious gypsy moth were much more impacted than the male (Traw et al. 1996). There is also an increase in consumption by phloem feeding aphids (Hughes and Bazzaz 1997).

The impact on natural ecosystems can be very significant in two areas 1) the reduction in the area of these vegetation types and 2) the decline of biological diversity. For example, we are all aware of the fact that much of the tropical rainforest is being cut, the vegetation burned, and grasses are planted for the cattle feeding and people involved in agriculture. We also learn that, in some areas, the productivity of these lands declines quickly with time, because of severe erosion. Models suggest that because of the cutting of the forest there is going to be a loss in nutrients, an increase in the soil temperature and a general decline in the amount of evaporation from these surfaces which may lead to a decline in the rainfall over a region which, in turn, could impact what remains of the forest.

There may be feedbacks on ecosystem function caused by global change. For example, a change in litter chemistry may have a negative feedback on the systems ability to take extra carbon (Fig. 12). It is known for example that plants grown in elevated CO_2 environment and high nitrogen may have a higher carbon content relative to nitrogen content in their leaves at least under certain conditions. And, when these leaves fall to the ground they may decay slowly releasing nutrients slowly and this slow decay could be a negative feedback on the system in the sense that nutrients do not become available through the decay process for plants to take them. In other words, there could be a decline in the rate of the chemical cycling in ecosystems. Recent evidence, however, seems to cast doubts about the generally this scenario. Another possibility is that because of enhanced root exudates there can be an increase in microbial biomass and microbial activity in the soil, which could mean two things. 1) Either the microbes will compete with the plants for ammonia (NH_4) and therefore reducing available nitrogen and resulting in a negative feedback on the carbon cycle for the uptake of nutrients (Diaz et al. 1993) or they could potentially increase the decomposition rate and therefore make the nutrients more available (Zak et al. 1993). These are two opposing activities and we not know under what circumstances or in what system does one override the other. And these feedbacks, (in addition to regard to the down regulation of photosynthesis) could be critical factors in assessing the future impact of global change on photosynthesis and productivity. Another critical issue that needs to be investigated further is the fact that when the increase in the concentration of other gases can impact negatively the ability of the plant to photosynthesize at the expected rate. An example of that is the increase in O_3 and SO_2 levels.

In our work on photosynthesis and allocation, we must pay attention to what is happening in developing countries. It is conceivable that the four horsemen of the apocalypse are on the horizon in these countries and that the major responsibility of the people in the field of photosynthesis is to better understand the process. The advances in this field have been most gratifying but we need to carry on research for the future which aims at the increased photosynthetic, increased production and crop yield and that would allow us to feed the billions of people that we expect in the next century. This is a great and challenges opportunity for researchers interested in photosynthesis.

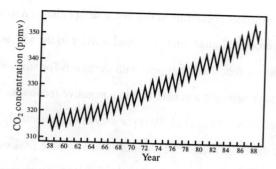

Fig. 1 Changes in the global CO_2 concentration as measured in Mauna Loa, Hawaii (from Keeling et al. 1989a).

Table 1.
Tropical Diseases Likely to Spread with Global Warming

Disease	Vector	Change with global warming
Malaria	mosquito	+++
Schistosomiasis	water snail	++
Dengue	mosquito	++
Filariasis	mosquito	+

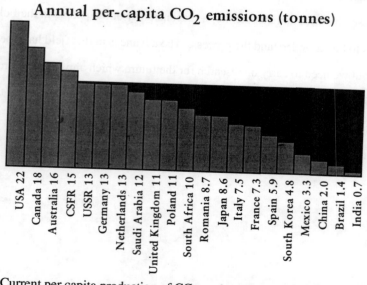

Fig. 2 Current per capita production of CO_2, and expected increase in the future as developing countries improve their standard of living (current trends based on data from Enquete Commission "Protecting the Earth's Atmosphere of German Bundestag (ed.) 1992).

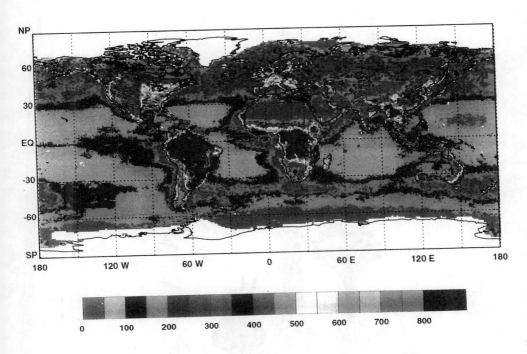

Fig. 3   Global terrestrial and oceanic productivity (from Field et al 1998).

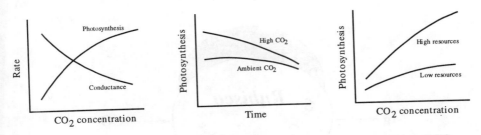

Fig. 4   General trend in photosynthesis and stomatal conductance in elevated CO_2 concentrations (modified from Bazzaz 1990).

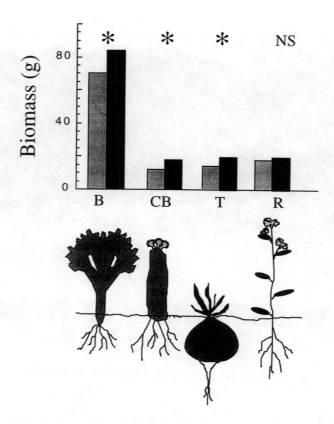

Fig. 5   Down regulation of photosynthesis occurs when there are no sinks for
photosynthates in *Brassica* (from Reekie 1996).

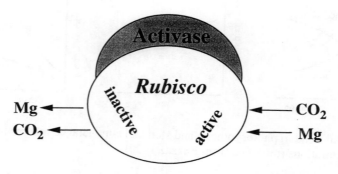

Fig. 6   The enzyme activase may play in the process of activation and deactivation of the
enzyme *Rubisco* by adding or deleting CO_2 and Mg to *Rubisco*.

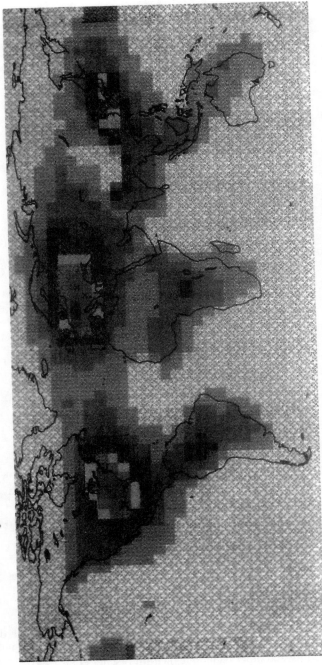

IMAGES NOy deposition

Fig. 7 Global distribution of nitrogen deposition estimated using IMAGES model (from Holland et al. 1997).

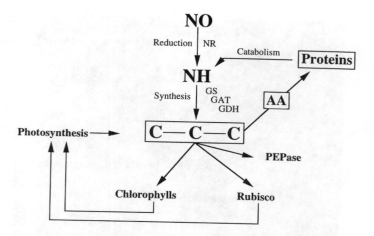

Fig. 8 The dependance of nitrogen assimilation on the supply of carbon (and vice versa) from photosynthesis in plants.

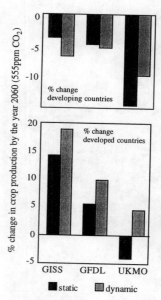

Fig. 9 Model prediction of the response of agriculture in developed and in developing countries using both static (no adaptation) and dynamic (with adaptation through use of breeding and other techniques) (from Fischer et al. 1996).

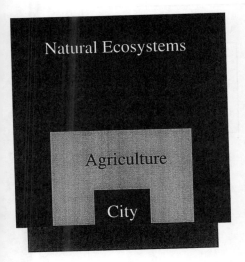 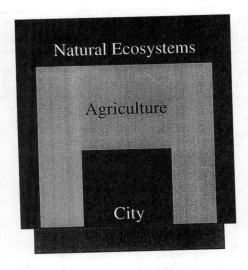

Fig. 10 Representation of the possible impact of the growth of the human population on agricultural land use and the extent of natural ecosystems.

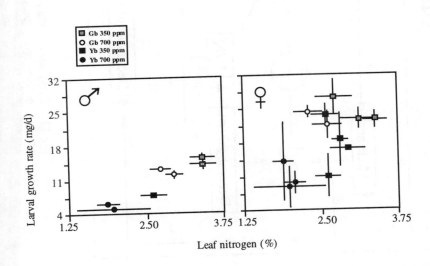

Fig. 11 The relationship between nitrogen concentration and the response of male and female gypsy moth, an important generalist herbivore (from Traw et al. 1996).

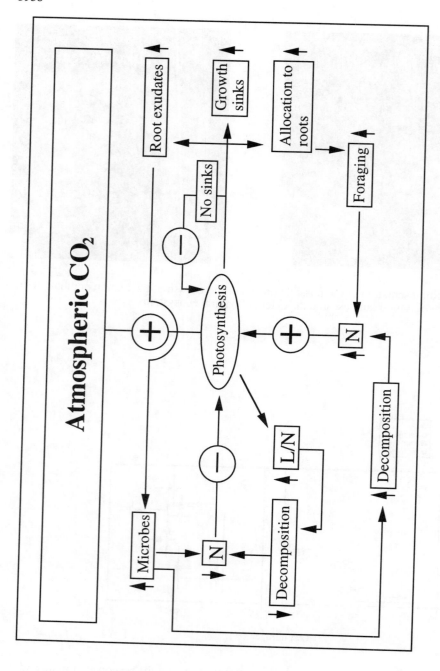

Fig 12   Diagram showing the impact of elevated CO_2 on growth and allocation and the possible positive and negative feedback on photosynthesis and primary production in ecosystems.

References

Allen, L.H. Jr., Baker, J.T. and Boote, K.J. 1996. The CO_2 Fertilization Effects Higher Carbohydrate Production and Retention as Biomass and Seed Yield. Pages 61-100 in F.A. Bazzaz and W. Sombroek (editors). Global Climate Change and Agricultural Production. John Wiley & Sons, New York.

Bazzaz, F.A. 1990. The response of natural ecosystems to the rising global CO_2 levels. Ann. Rev. Ecol. Syst. 21:167-196.

Diaz, S., Grime, J.P., Harris, J., McPherson, E. 1993. Evidence of a feedback mechanism limiting plant response to elevated carbon dioxide. Nature, 364:616-617.

Enquete Commission "Protecting the Earth's Atmosphere" of the German Bundestag (ed.) 1992. Climate Change-A Threat to Global Development. Economica Verlag, Bonn. Verlag C.F. Müller, Karlsruhe.

Field, C.B., Behrenfeld, M.J., Randerson, J.T., Falkowski, P. 1998. Primary production of the biosphere: Integrating terrestrial and oceanic components. Science 281:237-240.

Fischer, G., Frohberg, K., Parry, M.L., Rosenzweig, C. 1996. The potential effects of climate change on world food production and security. Pages 199-235 in F.A. Bazzaz and W. Sombroek (eds.). Global Climate Change and Agricultural Production. John Wiley & Sons, New York.

Gifford, R.M. 1989. The effect of the build-up of carbon dioxide in the atmosphere on crop productivity. Proceedings of the Fifth Australian Agronomy Conference, Perth, WA. pp. 312-322.

Holland, E.A., Braswell, B.H., Lamarque, J.-F., Townsend, A., Sulzman, J., Müller, J.-F., Dentener, F., Brasseur, G., Levy, H. II, Penner, J.E., and Roelofs, G.-J. 1997. Variations in the predicted spatial distribution of atmospheric nitrogen deposition and their impact on carbon uptake by terrestrial ecosystems. Journal of Geophysical Research 102:15,849-15,866.

Houghton, J.T., Jenkins, G.J., J.J. Ephraums (Editors) 1990. Climate Change: The IPCC Scientific Assessment (362 pages). Cambridge University Press, Cambridge, UK.

Hughes, L. and Bazzaz, F.A. 1997. Effect of elevated CO_2 on interactions between the western flower thrips, Frankliniella occidentalis (Thysanoptera: Thripidae) and the common milkweed, Asclepias syriaca. Oecologia 109:286-290.

Idso, S.B., Kimball, B.A., Anderson, M.G. and Mauney, J.R. 1987. Effects of atmospheric CO_2 enrichment on plant growth: The interactive role of air temperature. Agric. Ecosystems Environ. 20:1-10.

Keeling, C.D., Bacastow, R.B., Carter, A.F., Piper, S.C., Whorf, T.P., Heimann, M., Mook, W.G., and Roeloffzen, H. 1998a. A three dimensional model of atmospheric CO_2 transport based on observed winds: 1. Analysis of observational data in: Aspects of climate variability in the Pacific and the Western Americas, D.H. Peterson (ed.), Geophysical Monograph, 55, AGU, Washington (USA) 165-236.

Kimball, B.A., Mauney, J.R., Nakayama, F.S. and Idso, S.B. 1993. Effects of increasing atmospheric CO_2 on vegetation. Vegetatio 104/105:65-75. Also: pp. 65-75. In: CO_2 and Biosphere. (Advances in Vegetation Science 14). J. Rozema, H. Lambers, S.C. van de Geijn and M.L. Cambridge (eds.) Kluwer Academic Publishers, Dordrecht.

Kimball, B.A. 1983. Carbon dioxide and agricultural yield: An assemblage and analysis of 430 prior observation. Agronomy Journal 75:779-788.

Lincoln, D.E., Fajer, E.D.. and Johnson, R.H. 1993. Plant-insect herbivores interactions in elevated CO_2 environments. Trends Ecol. Evol. 8:64-68.

Lindroth, R.L., Kinney, K.K., Platz, C.L. 1993. Responses of deciduous trees to elevated atmospheric CO_2: Productivity, phytochemistry, and insect performance. Ecology 74:763-777.

Reekie, E.G. 1996. The effects of elevated CO_2 on developmental processes and its implications for plant-plant interactions. Pages 333-346 in Ch. Körner and F.A. Bazzaz (eds.) Carbon Dioxide, Populations, and Communities. Academic Press, London.

Sheen, J. 1990. Metabolic repression of transcription in higher plants. Plant Cell 2:1027-1147.

Traw, M.B., Lindroth, R.L. and Bazzaz, F.A. 1996. Decline in gypsy moth (Lymantria dispar) performance in an elevated CO_2 atmosphere depends upon host plant species. Oecologia 108:113-120.

Van Oosten, J.-J. and Besford, R.T. 1996. Acclimation of photosynthesis to elevated CO_2 through feedback regulation of gene expression: Climate of opinion. Photosynthesis Research 48:353-365.

Zak, D.R., Pregitzer, K.S., Curtis, P.S., Teeri, J.A., Fogel, R. and Randlett, D.L. 1993. Elevated atmospheric CO_2 and feedback between carbon and nitrogen cycles. Plant and Soil 151:105-117.

THE OCEANIC PHOTOSYNTHETIC ENGINE: ORIGINS, EVOLUTION, AND ROLE IN GLOBAL BIOGEOCHEMICAL CYCLES

Paul G. Falkowski
Environmental Biophysics and Molecular Biology Program
Institute of Marine and Coastal Science, Rutgers University
New Brunswick, New Jersey 08901-8521

1. Introduction

The combination of vast areas of liquid water on its surface together with a high concentration of free molecular oxygen in its atmosphere, is unique to Earth in this solar system. Proxy geochemical evidence suggests that the concentration of oxygen in Earth's atmosphere reached approximately 1% of the present value between 2.2 and 2.0 billion years ago as a consequence of oxygenic photosynthesis by marine unicellular algae, the phytoplankton [1]. The oxidation of Earth's atmposphere necessarily was accompanied by the production of massive quantities of organic carbon that are preserved in the lithosphere. A small fraction of this fossil phytoplanton carbon literally fuels the industrial world in the present geological epoch. These biologically driven changes reflect a key transition in the geochemical evolution of Earth that profoundly influenced the subsequent course of evolution of life. Here I briefly examine the controls and feedbacks between oceanic phytoplankton photosynthesis and global geochemical processes. A more extensive review can be found in Falkowski et al. [2]

2. Phytoplankton diversity and biogeochemical stability

Unlike higher plants, from a phylogenetic standpoint phytoplankton are a highly diverse group of organisms. These single-celled photoautotrophic organisms comprise 12 taxonomic divisions spanning three Kingdoms [3]. Fossils with structural similarities to extant cyanobacteria extend at least to 3.45 Ga before present [4]. This group numerically dominates many phytoplankton assemblages in the contemporary ocean. Molecular phylogenetic trees suggest that these procaryotes gave rise, via a series of endosymbiotic associations, to all eucaryotic algae and, ultimately to higher plants [5]. Although phylogenetic diversity within the phytoplankton is deeply branching, many fundamental metabolic pathways are remarkably conserved and can be grouped into a handful of biogeochemical functions. Sequence homology of reaction center proteins that comprise the core of PSII and PSI are prime examples of this conservation, whilst the

3941

G. Garab (ed.), Photosynthesis: Mechanisms and Effects, Vol. V, 3941–3947.
© 1998 Kluwer Academic Publishers. Printed in the Netherlands.

relative diversity of light harvesting pigment protein complexes are indicative of the deeply branching evolutionary diversity [6]. The combination of deep genetic diversity and high functional redundancy has helped to assure the continuity of oxygenic photosynthesis in the oceans from the Archean epoch to present despite considerable changes in Earth's climate and the ocean environment.

3. Phytoplankton photosynthesis and productivity in the contemporary ocean

The global carbon cycle is critically dependent on the fraction of photosynthetically fixed carbon that is not respired by the photoautotrophs themselves [7]. Consequently, this fraction, called net primary production (NPP), is available to other trophic levels for growth and respiration, or can be sequestered as stored plant biomass. In both marine and terrestrial environments NPP can be estimated from remotely sensed information [7]. In the ocean, the selective absorption of blue and blue-green wavelengths by photosynthetic pigments, especially chlorophyll *a,* allows quantification of photosynthetic biomass based on satellite-derived measurements of ocean color [8]. Chlorophyll concentrations and the taxonomic composition of phytoplankton communities are qualitatively correlated with the physical circulation and dynamics of the ocean, especially as these processes influence the fluxes of essential nutrients from the subsurface nutrient reservoir into the euphotic zone. Based on thousands of measurements of NPP from radiocarbon tracer studies conducted in the ocean since the mid 1950's, mathematical models are developed and parameterized that relate irradiance profiles through the upper portion of the euphotic zone to daily photosynthetic rates in specific regions of the ocean [9]. Together with knowledge of sea surface temperature, incident solar irradiance and mixed layer depths, chlorophyll data can be used in conjunction with the radiocarbon data to estimate NPP for any region of the ocean from remotely sensed information [10]. The results of such calculations suggest that global oceanic NPP is ca. 45 to 50 Pg C per annum [7, 10]. This carbon flux, which amounts to approximately 46% of the total global NPP, is driven by a phytoplankton biomass of approximately 1 Pg C, which is amounts to only 0.2% of the photosynthetically active carbon biomass on Earth. Consequently, on average, phytoplankton biomass in the oceans turns over on the order of once per week. This turnover time sharply contrasts with terrestrial plant ecosystems, where the turnover time of plant carbon averages a decade or more [3].

4. Supplies of nutrients

To sustain a net carbon flux through marine ecosystems, essential nutrients must be supplied. The mean elemental ratio of marine organic particles is 106C/16N/1P by atoms and is highly conserved. This specific elemental stoichiometry, called the "Redfield ratio", is unique to marine ecosystems and reflects the average biochemical composition of marine phytoplankton and their early degradation products [11]. When incident solar radiation is sufficient and the depth of the upper mixed layer constrains cells within the euphotic zone, transient physical processes, such as eddies, coastal upwelling, and wind or convective mixing, can promote the transient blooms of phytoplankton , by increasing the flux of nutrients from deeper, nutrient-rich waters. As such blooms are supported by the import of nutrients into the euphotic zone, in the steady-state, phytoplankton and their degradation products must be exported to the deep ocean, or the concentration of nutrients in the deep, nutrient-rich waters would be slowly depleted [12]. From a perspective of oceanic ecology, the production, accumulation and export of phytoplankton sustains the world's fisheries and sets the upper bound for sustainable yields [13].

Based on the conservation of elemental ratios (i.e. the Redfield ratios), it follows that in the steady state the sinking flux of particulate organic matter from the euphotic zone represents a stoichiometric loss of associated nutrients that must be balanced by "new" nutrient inputs. Because the physical processes that promote new production are transient, there is ample observational data demonstrating that steady-state export of organic matter is seldom reached on time scales of days, weeks, or interannually. Storms, mesoscale eddies, and El Nino conditions are but a few examples where nutrient delivery processes are affected by transient physical processes.

5. The exchange of CO_2 between the ocean and atmosphere

Biological processes in the ocean affect atmospheric gas composition through phase state transitions of specific elements. For example, photosynthetic sequestration of inorganic carbon converts CO_2 into organic solutes and particles with a corresponding decrease in the partial pressure in the surface layer [14]. The resulting change in the partial pressure gradient at the air-sea interface potentially provides a driving force for the invasion of CO_2 from the atmosphere into the ocean. Conversely, heterotrophic oxidation of organic solutes and particles leads to the formation of CO_2, and the potential efflux of CO_2 from the ocean to the atmosphere. The export of organic material from the euphotic zone to the ocean interior, followed by biological oxidation of the organic carbon, leads to an inverse concentration gradient in inorganic carbon. In the contemporary ocean, this flux is ca. 16 Pg C per annum, or approximately one-third of the total ocean production [2]. Once the organic carbon sinks beneath the main ocean thermocline, it is effectively sequestered from the atmosphere for centuries to millenia. It is frequently assumed in biogeochemical models that this "biological carbon pump" is in a steady state on time scales of decades to centuries [15]. Such an assumption leads to the unsubstantiated conclusion that the biologically-mediated *net* exchange of CO_2 between the ocean and atmosphere is virtually nil .

Changes in the *net* biologically mediated flux of CO_2 between the atmosphere and ocean requires changing (a) ocean nutrient inventories, (b) the utilization of unused nutrients in enriched areas, and/or (c) the average elemental composition of the organic material [16]. Based on the analysis of the proximate elemental composition of marine organisms, Redfield concluded that phosphorus limits primary production in the oceans [11]. It should be stressed that in this context, "limiting" denotes a factor that constrains the formation of total phytoplankton *biomass* in the euphotic zone, not necessarily the *rate* of formation of the biomass (i.e. the photosynthetic rate). The distinction between these two processes is critical. Light, for example, always limits the average instantaneous rate of photosynthesis in the euphotic zone, but as long as nutrients brought into the euphotic zone are eventually consumed in the production of biomass, the vertical fluxes of the nutrients (including inorganic carbon) will be sustained.

Redfield's tenet of phosphorus limitation was based on three fundamental concepts. First, because nitrogen is the most abundant gas in the atmosphere and can, in principle, be biologically reduced to the level of NH_3, Redfield assumed that nitrogen fixation should keep apace with demands of the photoautotrophs. Second, phosphorus has no biological or atmospheric source, but rather is supplied to the ocean largely from fluvial sources. As the rate of supply of phosphorus is lower than that for nitrogen, the residence time of the former in the oceans is significantly longer than that of the latter element. The longer residence time implies that the source of the element limits its availablity to consumers (i.e. phytoplankton). Finally, the ratio between biological fixation of nitrogen and losses of fixed nitrogen resulting from denitrification determines the concentration of fixed nitrogen in the ocean interior. Redfield proposed that the maximum concentration of fixed nitrogen in the oceans is ultimately determined by the availability of phosphorus, such that the fixed N/P ratio will adjust to the ratio of the two elements in the sinking flux of particulate organic matter (i.e. about 16/1 by atoms).

In the contemporary ocean, vertical profiles of dissolved fixed inorganic nitrogen (primarily nitrate) and phosphorus (as phosphate) invariably reveal a deficiency of fixed nitrogen relative to phosphorus in the ocean interior of the three major oceans. Below the upper 500 m, the average N/P ratio for the world oceans is ca. 14.7, corresponding to a deficit of ca. 2.9 µmoles N/kg of seawater [17-19]. The deficiency in fixed inorganic nitrogen reflects losses mediated by anaerobic denitrifying bacteria, which are primarily found on continental margins and in hypoxic regions of the open ocean It should be stressed that nitrogen fixation and denitrification are not coupled; that is, imbalances between these processes can readily occur on geological time scales, thereby leading to a change in the inventory of fixed nitrogen in the oceans.

6. The role of iron

Over the past several years, it has become increasingly apparent that, for the ocean as a whole, nitrogen fixation is itself limited by some factor, most probably iron [20]. The vast majority of nitrogen fixation in the open ocean appears to be accomplished by a handful of species of cyanobacteria, most notably *Trichodesmium* spp [21]. The paucity of marine N_2 fixing species (i.e. functional singularity) suggests that the net flux of fixed nitrogen to the ocean *via* this critical biogeochemical pathway is especially sensitive to environmental perturbations. In *Trichodesmium* spp., nitrogen fixation occurs within the same cells that evolve O_2 (i.e. the genus does not form heterocysts), and the ratio of photosystem (PS) I to PSII is extremely high, on the order of ca. 25. PSI contains 12 iron atoms, and is responsible for both protecting the oxygen sensitive nitrogenase from molecular oxygen generated by PSII through the Mehler reaction [22], as well as generating ATP for the energy-intensive nitrogen fixation pathway. Moreover, nitrogenase itself has a strict iron requirement and the iron use efficiency (i.e.

enzymatic catalysis rate per unit of iron) is amongst the lowest of any iron-containing enzyme known [23]. The limitation of nitrogen fixation by iron is especially apparent in the South Pacific, where the aeolian flux of iron-rich dust is exceptionally low [24].

As the fourth most abundant element in Earth's crust, iron would appear to be unlikely to limit biochemical processes, however, the bioavailability of iron is critically dependent upon its redox state. Reduced (ferrous) iron is highly soluble in seawater while the oxidized (ferric) form is virtually insoluble. Hence, the oxidation state of the ocean, which, on geological time scales is determined by the ratio of photosynthetic oxygen evolution to the sum of biological respiration and chemical reducing equivalents, is a critical determinant of trace element selection and supply. It should be noted, in this regard, that other key transition metals used in redox reactions for water, nitrogen, or carbon, including Mo, Mn, and Cu are relatively abundant in the ocean, and are not thought to limit phytoplankton photosynthesis or growth.

Prior to the evolution of oxygenic photosynthesis, the ocean was anoxic and mildly reducing; iron was present in very high concentrations (estimated at about 25 mM). Oxygenic photosynthesis ultimately led to the oxidation and subsequent precipitation of iron. In the contemporary ocean, soluble iron concentrations rarely exceed a few nanomolar. The massive precipitation of iron by oxygenic photoautotrophs is a clear example of a negative feedback in the evolution of biogechemical cycles on Earth. Evidence of this biologically induced change in Earth's chemistry is etched in iron-stress induced gene sequences of extant cyanobacteria that have been preserved from the Proterozoic epoch. Similarly, the selective use of iron rather than copper to ferry electrons to PSI in diatoms suggests that the evolution of Heterokonts in the early Triassic corresponded with reducing conditions in the oceans that mark the extinction at the Permian-Triassic boundary [25].

A major source of iron for the oceans is wind-blown continental dust [26]. This aeolian flux limits phytoplankton photosynthesis throughout much of the contemporary Pacific, but not Atlantic oceans. By "limitation" in this case, I am referring to a true physiological limitation of the quantum yield of photosynthesis that can be assessed from changes in variable fluorescence [27, 28]. Iron limitation appears to dramatically diminish the capacity of phytoplankton to synthesize cyt b_{559} and other heme complexes, thereby effectively inactivating a fraction of PSII reaction centers [Greene, 1992 #2596;[29]. The effect of iron limitation on PSI in the ocean is not known.

Examination of aeolian dust particles obtained from ice-cores suggests that during glacial periods, the supply of iron to the world oceans was at least approximately an order of magnitude higher than during interglacial periods. As glaciation led to a reduction in sea level of ca. 120 m, denitrification on continental shelves essentially ceased while the supply of iron would have stimulated biological nitrogen fixation, thereby leading to an increase in the global pool of fixed nitrogen in the oceans. The iron-based nitrogen enrichment would enhance phytoplankton carbon fixation until ultimately primary production became limited by phosphate [19, 30]. Assuming that the average proximate elemental composition of phytoplankton during glacial epochs conformed to that of the contemporary ocean (i.e. the ratio of inorganic N/P in the ocean interior increased from 14.7 to 16), simple box model calculations suggest that the enhanced production would have lowered atmospheric CO_2 from ca. 275 ppm to 245 ppm, accounting for 30% of the interglacial-glacial difference in CO_2. If strong deviations in the average elemental composition of phytoplankton occurred, this effect could be much greater.

7. Global feedbacks

To a first order, the aeolian flux of iron to the ocean is regulated by the hydrological cycle, which is, in turn, strongly influenced by atmospheric radiative forcing. Soil moisture variability determines the episodic aridity, which is the primary regulation on dust supply. The transport of aeolian iron is related to wind speed and direction, which are related to the temperature contrast between the continents and ocean. Silica supply, essential for diatom blooms, is primarily dependent upon riverine fluxes and upwelling from the ocean interior. The distribution of nitrogen fixing cyanobacteria is not only dependent upon iron fluxes but also a warm, relatively quiescent euphotic zone. Foodweb structure, which is a strong determinant of export flux, is critically dependent upon mesoscale physical processes that promote nutrient pulses. The rapid increase in CO_2 at the end-terminal glacial periods is consistent with ocean biological control of the termination process yet the data do not conclusively indicate which process(es) specifically dominated. Given the lack of mechanistic understanding of many of the feedbacks, and the paucity of observational data on decadal to century time scales, it is presently beyond credible capability to quantitatively represent these processes within coupled time-dependent, 3-dimensional ocean-atmosphere models [31]. Hence our ability to quantitatively predict ecosystem structure and biogeochemical feedbacks based on forecasted climate change scenarios is limited.

Given the complexity of and feedbacks within biogeochemical cycles, it is clear that our understanding of the roles of phytoplankton in the Earth's global environment is far from complete yet is critically important. The photosynthetic processes in the oceans are particularly amenable to techniques such as fluorescence spectroscopy and optical reflectance analysis, and the application of such techniques has been extremely rewarding over the past decade. However, detailed analyses regarding the role of photosynthetic processes in the selection of key taxonomic groups remains at the forefront of research [3]. This issue critically determines the role that phytoplankton play in the structure of marine ecosystems. It is my hope that future generations of researchers in photosynthesis will cross disciplinary boundaries and examine how the reductionist experimental approach that traditionally characterizes laboratory research can be integrated into truly synoptic, global studies of photosynthetic processes, where aquatic photoautotrophs play a profound role.

References

1 Holland, H.D., (1984) The Chemical Evolution of the Atmosphere and Oceans, Princeton, N.J.: Princeton University Press
2 Falkowski, P., R. Barber, and V. Smetacek, (1998) Science. 281: p. 200-206.

3 Falkowski, P.G. and J.A. Raven, (1997)Aquatic Photosynthesis, Oxford: Blackwell Scientific Publishers. 375 pp.
4 Schopf, J., (1993) Science. 260: p. 640-646.
5 Bhattacharya, D. and L. Medlin, (1998) Plant Physiol. 116: p. 9-15.
6 Bhattacharya, D., et al., (1992) Evolution. 46: p. 1801-1817.
7 Field, C., et al., (1998) Science. 281: p. 237-240.
8 Morel, A., (1991) Prog. Oceanogr. 26: p. 263-3
9 Behrenfeld, M. and P. Falkowski, (1997) Limnol Oceanogr. 42: p. 1479-1491.
10 Behrenfeld, M.J. and P.G. Falkowski, (1997) Limnol Oceanogr. 42: p. 1-20.
11 Redfield, A.C., (1958) Am Sci. 46: p. 205-221.
12 Eppley, R.W. and B.J. Peterson, (1979) Nature. 282: p. 677-680.
13 Ryther, J.H., (1969) Science. 166: p. 72-77.
14 Sarmiento, J.L. and U. Siegenthaler, New production and the global carbon cycle. in Primary Productivity and Biogeochemical Cycles in the Sea, P.G. Falkowski, Editor. 1992, Plenum Press: New York and London. p. 316-317.
15 Sarmiento, J.L. and M. Bender, (1994) Photosyn. Res. 39: p. 209-234.
16 Falkowski, P.G. and C. Wilson, (1992) Nature. 358: p. 741-743.
17 Fanning, K., (1992) J. geophys Res. 97C: p. 5693-5712.
18 Gruber, N. and J. Sarmiento, (1997) Glob. Biogeochem. Cycles. 11: p. 235-266.
19 Falkowski, P., (1997) Nature. 387: p. 272-275.
20 Martin, J.H., Iron as a limiting factor in oceanic productivity., in Primary Productivity and Biogeochemical Cycles in the Sea, P. Falkowski and A. Woodhead, Editors. 1992, Plenum, in press: New York. p. 123-137.
21 Capone, D., et al., (1997) Science. 276: p. 1221-1229.
22 Kana, T.M., (1993) Limnol. Oceanog. 38: p. 18-24.
23 Raven, J.A., (1988) New Phytol. 109: p. 279-287.
24 Martin, J.H., R.M. Gordon, and S.E. Fitzwater, (1990) Nature. 345: p. 156-158.
25 Knoll, A., et al., (1996) Science. 273: p. 452-457.
26 Duce, R.A. and N.W. Tindale, (1991) Limnol. Oceanogr. 36: p. 1715-1726.
27 Kolber, Z.S., et al., (1994) Nature. 371: p. 145-149.
28 Behrenfeld, M., et al., (1996) Nature. 383: p. 508-511.
29 Vassiliev, I.R., et al., (1995) Plant Physiol. 109: p. 963-972.
30 Codispoti, L., (1995) Nature. 376: p. 724.
31 Sarmiento, J., et al., (1998) Nature. 393: p. 245-249.

The author is grateful for support from NASA, the Office of Naval Research and the U.S. Dept. of Energy.

PHOTOSYNTHESIS OF RAIN FORESTS AT A LARGE SCALE

John Grace, Institute of Ecology & Resource Management, The
University of Edinburgh, Edinburgh EH9 3JU, UK

Key words: global climate changes, carbon cycle, water stress

1. Introduction

Tropical rain forests occupy about 17.5×10^{12} m^2 of land, corresponding to 12 % of the terrestrial surface[1,2]. They constitute a large store of carbon in the form of biomass, although there is considerable uncertainty in the quantity. A recent compilation of data[3] suggests they hold, on average, about 150 tC ha^{-1} above-ground and 35-50 below-ground, making a total stock of 185-200 tC ha^{-1}. Assuming the mid-range figure of 192 tC ha^{-1} to apply over the entire 17.5×10^{12} m^2, this implies a global total carbon stock of 336 Gt C which is more than previously thought and half of the estimated[4] global biomass stock of 610 Gt C.

The current deforestation rate, averaged over the biome, is about 1 % per year[5,6]. Taking account of the incomplete combustion, it is estimated that the carbon flux to the atmosphere globally is at least 1.7 Gt each year[4]. This is equal to about one-quarter of the current (1990's) emissions from the combustion of fossil fuels. The pastures and other land surface covers which replace rain forests have relatively small stores of carbon above-ground, although they may have surprisingly deep roots and a high below-ground biomass[7]. Most importantly, they have different biophysical and physiological properties than the forest they replace[8,9], and, given the large area that they occupy, it is often hypothesized that deforestation may influence regional or even global climates.

Although the deforestation rate is rapid, and does constitute a substantial flux of carbon to the atmosphere, there remains a large area of forest that is undisturbed. Recently it has been suggested that these undisturbed forests are not in a steady state with respect to their carbon balance, and thus they are *not merely stocks of carbon but also sinks*. It has been speculated that their sink strength may be increasing as a result of anthropogenic emissions of carbon dioxide and an enhanced deposition rate of nitrogen and other minerals[10,11,12,13]. If this is true, then the sink strength may offset the emissions caused by deforestation. Can the sink strength of such forests be measured? Can we depend on the sink to continue over decades, as the CO_2 continues to rise?

G. Garab (ed.), Photosynthesis: Mechanisms and Effects, Vol. V, 3949–3954.
© 1998 *Kluwer Academic Publishers. Printed in the Netherlands.*

2. Methodologies

2.1 *Micrometeorological approaches*

The introduction of micrometeorological techniques has enabled ecophysiological approaches to be used at the scale of the whole ecosystem[14,15,16,17]. Eddy covariance provides more-or-less continuous, all-weather measurements of the fluxes of energy, carbon dioxide and water vapour over the land surface, with a spatial sample size in the order of 1 km². In a forest, there are tens of thousands of trees in a km². The flux data gained in this way may be used to fit models, which may then be run in the predictive mode to estimate the net ecosystem CO_2 exchange rate over longer periods than the measurements, and over larger spatial scales.

2.2 *Permanent forestry plots*

There are many observation plots established by foresters and ecologists throughout the world for the purpose of examining growth and mortality of the trees. Such plots, usually one hectare each, are remeasured every few years, using a tape measure to assess the girth of every tree in the plot. It is possible to estimate the biomass of a tree from its girth, and as biomass is about 48 % carbon, the biomass-carbon can be estimated. From time to time, plots are created and abandoned or destroyed, and so a statistical procedure is required to avoid discontinuities. By tracking the biomass changes in these plots over decades, it is possible to infer any net absorption or loss of carbon over the period. It must be kept in mind that this approach ignores any carbon changes in soil organic matter, which may be appreciable but not so important in tropical forests as in forests in colder regions where the soil organic matter is much higher.

3. Results

The sites from which data are available in the Brazilian Amazon are at Ducke and Cuieras, both near Manaus (central Amazonia), and at Reserva Jaru in Rondonia (SW Amazonia). The Manaus sites are dense humid rain forests with a leaf area index of 5-6, whilst the Rondonia site is more open, with less biomass and a leaf area index of about 4. The climate at Rondonia is somewhat more seasonal. Eddy covariance measurements at these three sites show a high degree of consistency when the mean diurnal curves are compared (Fig. 1).

They show a respiratory loss of 5-10 μmol CO_2 m^{-2} s^{-1} at night, and a photosynthetic gain by day of 15-20 μmol CO_2 m^{-2} s^{-1} with a maximum uptake before noon (Fig. 1). All three data sets indicate a net uptake of carbon dioxide. If the data are fitted to models[18,19] to scale up to 365 days, we have an estimated uptake of 1.0, 2.2 and 5.9 tC ha^{-1} year^{-1} at Rondonia, Ducke and Cuieras respectively. We have examined the data for 'nocturnal losses' of CO_2, and we conclude that losses are unlikely to be high enough to alter the conclusion[15,17].

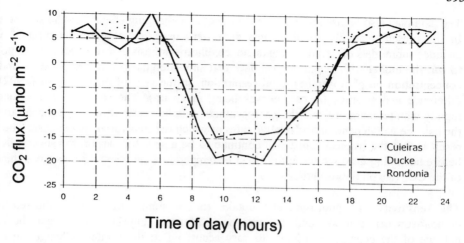

Fig. 1 Uptake of CO_2 by three rain forests in the Brazilian Amazon[9,12,17,25]. Fluxes from the atmosphere to the canopy are negative by convention. The four lines are for Ducke, near Manaus (solid line), Jaru in Rondonia (pecked line) and Cuieiras near Manaus in the dry season (upper-dots) and wet season (lower-dots).

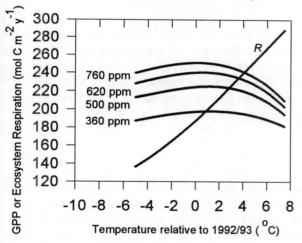

Fig. 2 The sensitivity of the Rondonia rain forest to changes in temperature and CO_2 concentration, estimated from the model[18] fitted to the flux data[9,15], assuming no acclimation. The model has been re-run several times after subtracting or adding a temperature change to the real climatological data from 1992. Gross primary productivity (labelled according to CO_2 concentration) and total ecosystem respiration (labelled R).

The data may also be used to examine the effect of a changing climate on the sink strength. In the present example the Rondonia data were used to calibrate the model[18] which includes submodels of stomatal conductance, photosynthesis and respiration. It is possible to run the model with climatological data that simulate changes in temperature and CO_2 concentration on photosynthesis and respiration. In the year of measurement (1992), this difference was positive (ie carbon was being taken up). The respiratory flux increases rapidly with temperature, and the photosynthetic flux increases with CO_2 concentration, so the balance will depend on the relative rates at which the two variables increase. The likely result is that the tropical forest will continue to be a sink for carbon for several decades before becoming a source when the effect of temperature on respiration exceeds the effect of the CO_2 rise on photosynthesis.

The sensitivity of photosynthetic uptake to the supply of water, indicated in the comparison between wet and dry seasons at Manaus (Fig 1), implies that the carbon balance of the ecosystem is likely to vary according to the rainfall. Changes in rainfall occur from year to year, as a result of climatological phenomena such as the droughts induced by El Niño, and also there may be trends over decades. The importance of drought is currently being investigated using an experimental set-up in which water is witheld.

Turning now to the permanent forestry plots, we found 151 from the humid tropics, of which 94 were from Amazonia. The entire data set represented more than 500,000 individual tree measurements[20]. The data show an increase in biomass since records began of 1.13 ± 0.55 t biomass ha^{-1} year^{-1} . For Amazonian inventories there have been biomass gains since the mid-1970s, with an average accumulation rate of 0.99 ± 0.54 t biomass ha^{-1} year^{-1}. However, the paleotropics showed no significant gain at all.

3. Discussion

The forests discussed in this paper are said to be 'undisturbed' , and contain large and old trees, and a range of species which are suggestive of an undisturbed status[21]. However, most Amazonian forests are likely to have been disturbed at some time in the distant past by traditional slash-and-burn agriculture, and many of them may be recovering from flooding or wind-damage, so we cannot preclude the possibility and we may be seeing a slow accumulation of biomass over a few centuries as the steady state is approached. Inspection of biomass data from young secondary forests suggest that most of the biomass increase during regrowth is likely to occur over a shorter time, perhaps a century[22], and so we think it unlikely that the biomass increases reported here are connected with natural agradation during succession.

It has long been held that undisturbed forests are more-or-less at a steady state, with gains of carbon from photosynthesis being balanced by losses from autotrophic and heterotrophic respiration. Hitherto, there has been no way to test the hypothesis. The two approaches to measuring the ecosystem carbon balance, outlined above and described in detail in the cited papers, contradict the steady-state hypothesis. However, the steady state hypothesis is unlikely to apply if the environment is changing, as we know it is at the present time. In addition to the anthropogenic changes in concentrations of CO_2 and deposition rates of nitrogen, there are also changes in rainfall which occur in response to large-scale climatological phenomena such as El Niño Southern Oscillation.

Several other groups of workers have constructed mathematical models that represent the basic process of photosynthesis and respiration. Currently, there are several published models that suggest a current modest increase in carbon uptake in reponse to elevated CO_2 and deposition rates of nitrogen[13,23]. However, as these anthropogenic effects are global, we would expect such enhancement to be present in all tropical forests, whereas the sample plot data suggest they occur only in the neotropics. One possibility, not considered by these models, is that trends in rainfall, which differ in different parts of the equatorial region, may have a strong effect upon the sink strength. Indeed, there are suggestions from a time series of satellite remote sensing that the net primary productivity of the forest may be strongly influenced by precipitation[24]; although the Amazonian permanent plots[20] do not show an obvious response to the drying which characterised the strong El Niño in 1982-3. We will continue to analyse these data to assess the impact of the recent El Niño.

Acknowledgements

The work described in this paper was funded by the Natural Environmental Research Council, UK. Thanks are due to my co-workers Y Malhi, P Meir and B Kruijt, and many collaborators elsewhere: A & H Miranda (Universidade de Brasilia), A Nobre & N Higuchi (Instituto Nacional Pesquisas da Amazonia), J Lloyd (formerly of the Research School of Biological Sciences, Institute of Advanced Studies, Australian National University), C Nobre (Centro de Previsao de Tempo e Estudos Climaticos, Cachoeira Paulista, SP), J Gash (Institute of Hydrology, Wallingford) and O Phillips (University of Leeds).

References

1 Whittaker, R.H. & Likens, G.E. (1975) The biosphere and man In "Primary Productivity of the Biosphere" (Eds. R.H. Whittaker & G.E. Likens). Pp. 305-328. Springer, Berlin.
2 Taylor JA & Lloyd J (1992) Australian Journal of Botany 40, 407-418.
3 Grace, J., Malhi, Y., Higuchi, N.& Meir, P (1998) Productivity and carbon fluxes of tropical rain forests. In press. Springer, Berlin.
4 Schimel, D.S. (1995) Global Change Biology 1, 77-91.
5 FAO (1991). Forest Resources Assessment 1990, Global Synthesis, FAO Forestry Paper 124, FAO, Rome, Italy.
6 Dixon RK, Brown S, Houghton RA, Solomon AM, Trexler MC & Wisniewski J (1994) Science 263, 185-190.

7 Nepstad, D.C., de Carvalho, C.R., Davidson, E.A., Jipp, P.H., Lefebvre, P.A., Negreiros, G.H., da Silva, E.D., Stone, T.A., Trumbore, S.E. and Vieira, S. (1994). Nature 372, 666-669.

8 Gash, C.A. Nobre, J.M. Roberts & R.L. Victoria (1996). Amazonian Deforestation and Climate. Wiley, Chichester.

9 Grace, J., Lloyd, J., McIntyre, J., Miranda, A.C., Meir, P., Miranda, H., Nobre, C., Moncrieff, J.M., Massheder, J., Malhi, Y., Wright, I.R. & Gash, J. (1995b). Science 270, 778-780.

10 Körner C (1993) CO_2 fertilization: the great uncertainty in future vegetation development. Pp 53-70. In AM Solomon, HH Shugart (eds) Vegetation Dynamics and Global Change. Chapman & Hall, London.

11 Gifford RM (1994). Australian Journal of Plant Physiology, 21, 1-15.

12 Grace, J., Lloyd, J., McIntyre, J., Miranda, A.C., Meir, P., Miranda, H., Moncrieff, J.M., Massheder, J., Wright, I.R. & Gash, J. (1995a). Global Change Biology 1, 1-12.

13 Wang, Y.P. & Polglase, P.J. (1995). Plant, Cell & Environment 18, 1226-1244.

14 Baldocchi, D., Valentini, R., Running, S., Oechel, W. & Dahlman, R (1996). Global Change Biology 3, 159-168.

15 Grace, J., Malhi, Y., Lloyd, J., McIntyre, J., Miranda, A.C., Meir, P. & Miranda, H.S. (1996). Global Change Biology 2, 209-218.

16 Moncrieff, J.B., Massheder, J.M., de Bruin, H., Elbers, J., Friborg, T., Heusinkveld, B., Kabat, P., Scott, S., Soegaard, H. and Verhof, A. (1997). Journal of Hydrology 189, 589-611.

17 Malhi, Y., Nobre, A., Grace, J., Kruijt B, Pereira, M., Culf A. & Scott, S. (1998) in press, Journal of Geophysical Research.

18 Lloyd, J., Grace, J., Miranda, A.C., Meir, P., Wong, S.C., Miranda, H., Wright, I., Gash, J.H.C., & McIntyre, J. (1995). Plant, Cell & Environment 18, 1129-1145.

19 Williams, M., Malhi, Y., Nobre, A.D., Rastetter, E.B., Grace, J. & Pereira, M.G.P. (1998). Plant, Cell & Environment (in press).

20 Phillips, O.L, Malhi, Y., Higuchi, N., Laurance, W.F., Nuñez V., P., Vásquez M., R., Laurance, S.G., Ferreira, L.V., Stern, M., Brown, S. and Grace, J. (1998). Accepted by Science.

21 Chambers, J.Q., Higuchi, N., & Schimel, J.P. (1998). Ancient Trees in Amazonia, Nature, 391, 135-136.

22 Alves, D.S., Soares, J.V., Amaral, S., Mello, E.M.K., Almeida, A.S., da Silva, O.F., Silveira, A.M. (1997). Global Change Biology, 3, 451-462.

23 Lloyd, J. & Farquhar, G.D. (1996). Functional Ecology 10, 4-32.

24 Lui, WTH, Massambani O. & Nobre C (1994). International Journal of Climatology, 14, 343-354.

25 Fan, S-M, Wofsy, S.C., Bakwin, P.S. & Jacob, D.J. (1990). *Journal of Geophysical Research*, 95(D10), 16851-16864.

PHOTOSYNTHETIC CHARACTERISTICS OF CAM SUCCULENTS WITH HIGH PRODUCTIVITY

Park S. Nobel
Department of Biology—Organismic Biology, Ecology, and Evolution
University of California, Los Angeles, California 90095-1606
E-mail: psnobel@biology.ucla.edu

Abstract

Responses of net CO_2 uptake over 24-h periods to total daily photosynthetic photon flux (PPF), day/night air temperatures, and soil water status under controlled conditions in environmental chambers have been determined for 9 species exhibiting Crassulacean Acid Metabolism (CAM) as well as one C_3 and one C_4 species. For the CAM species, this has enabled accurate predictions of net CO_2 uptake under field conditions and the PPF responses have led to the design of field plant spacings that maximize annual net CO_2 uptake per unit ground area. Two *Agave* species and two *Opuntia* species have an average measured dry weight productivity of 43 metric tons hectare^{-1} year^{-1}, which exceeds that of all C_3 agronomic species and C_3 trees and is only slightly less than the most highly productive C_4 species. The high productivity of these CAM species can be explained based on daily net CO_2 exchange and cellular photosynthetic properties. However, highly productive agaves and cacti are not tolerant of low temperatures, a problem for their cultivation in regions of the United States and other countries where air temperatures annually fall below about $-5\,°C$.

Key words: biomass, bioproductivity, CO_2 uptake, environmental stress, global climate changes, whole plants

Introduction

A net uptake of CO_2 is necessary for growth and biomass productivity by plants. The uptake of CO_2 is influenced by environmental factors such as light, temperature, and soil water status. To quantify the effect of these factors on net CO_2 uptake as it relates to productivity, the instantaneous rate of net CO_2 uptake must be integrated over 24 h so that periods of net CO_2 loss, when respiration predominates, are included. Because various environmental factors can simultaneously affect net CO_2 uptake, a first-order environmental productivity index (EPI) has been proposed and tested [1, 2]:

$$\text{EPI} = \text{Fraction of maximal net } CO_2 \text{ uptake over 24 h}$$
$$= \text{Light Index} \times \text{Temperature Index} \times \text{Water Index} \qquad (1)$$

Each component index of EPI ranges from 0.00, when limitations by that environmental factor abolish net CO_2 uptake, to 1.00, when that factor is optimal for net CO_2 uptake. For instance, if drought leads to stomatal closure and the Water Index (Eq. 1) becomes 0.00, then EPI equals 0.00 no matter what the values are for the Light Index or the Temperature

G. Garab (ed.), Photosynthesis: Mechanisms and Effects, Vol. V, 3955–3960.
© *1998 Kluwer Academic Publishers. Printed in the Netherlands.*

Index, and no net CO_2 uptake occurs over 24 h.

Component indices are determined while holding the other factors at optimal levels. For example, the Light Index is determined by varying the photosynthetic photon flux (PPF, 400 to 700 nm) for plants at optimal day/night air temperatures for net CO_2 uptake (Temperature Index = 1.00) and under conditions of wet soil (Water Index = 1.00). For the Crassulacean acid metabolism (CAM) species studied using the EPI approach, the main focus of this review, most CO_2 uptake occurs at night in the absence of light. Hence, the instantaneous PPF (μmol m^{-2} s^{-1}) is integrated over 24-h periods to obtain the total daily PPF (mol m^{-2} day^{-1}), which is then related to the total daily net CO_2 uptake (mol m^{-2} day^{-1}), expressed as a fraction of maximal daily net CO_2 uptake. For non-optimal day/night temperatures or during drought, the dependence of the Light Index on PPF differs somewhat from the dependence under ideal conditions [2]; such secondary influences occur when EPI is considerably below 1.00 and generally have little impact on the mean EPI predicted on a seasonal or annual basis.

Values for Component Indices of EPI

CAM species can take up CO_2 during both the daytime and at night, so measurements of net CO_2 uptake for such plants re conventionally made over 24-h periods. In particular, the dependencies of net CO_2 uptake over 24 h on total daily PPF, day/night air temperatures, and soil water potential have been determined under controlled conditions for *Agave deserti* [1,3], *A. fourcroydes* [4, 5], *A. lechuguilla* [5-7], *A. salmiana* [8], *A. tequilana* [9], *Ferocactus acanthodes* [3, 10], *Hylocereus undatus* [11], *Opuntia ficus-indica* [3, 12, 13], and *Stenocereus queretaroensis* [14]. Recently, the EPI approach has been successfully extended to the C_3 species *Encelia farinosa* and the C_4 species *Pleuraphis rigida* [15]. In all cases, net CO_2 uptake over 24-h periods is determined for plants in environmental chambers for which only one factor (light, temperature, or soil water potential) is varied while the other two are maintained near optimal values.

The response of net CO_2 uptake to light level, as quantified by the PPF index, is remarkably similar for *Agave deserti*, the common desert agave of the southwestern United States and the original species for which EPI was determined, *Opuntia ficus-indica*, a platyopuntia currently cultivated in over 20 countries for its fruits and cladodes, and *Stenocereus queretaroensis*, a columnar cactus cultivated for its fruit in Jalisco, Mexico (Fig. 1A). For all three species, the light compensation point is at a total daily PPF of about 2 mol m^{-2} day^{-1}, 30% of maximal net CO_2 uptake occurs at 10 mol m^{-2} day^{-1}, and 90% of PPF saturation occurs at 21 mol m^{-2} day^{-1} (Fig. 1A).

The temperature responses are also remarkably similar among these three species, with an optimal nighttime temperature of about 15°C in all cases (Fig. 1B). Because most net CO_2 uptake occurs at night for CAM plants, the mean nighttime temperature is more critical than the daytime temperatures in determining net CO_2 uptake over 24-h periods [2]. The optimal temperatures are relatively low (Fig. 1B); tropical epiphytic CAM species, which constitute most of the 7% of vascular plant species that exhibit CAM [18, 19], may have higher optimal temperatures for net CO_2 uptake in keeping with higher nocturnal temperatures in their native habitats.

In contrast to their similar PPF and temperature responses, the three species considered have different responses of net CO_2 uptake to drought duration (Fig. 1C). For the first few days of drought, little change in daily net CO_2 uptake occurs relative to the well-watered condition, as considerable amounts of water are stored in the succulent shoots. Thereafter, effects of the mean tissue depth for water storage (shoot volume for water storage divided by shoot surface area for water loss) become apparent. In particular, the Water Index becomes 0.5 for the leaves of *A. deserti* (mean water storage depth of 7 mm) at 11 days, for the flat stem segments of *O. ficus-indica* at 22 days (water storage depth of

23 mm), and for the cylindrical stems of *S. queretaroensis* at 36 days (water storage depth of 48 mm; Fig. 1C). Thus the Water Index differs among the three species, reflecting the ability of the shoot to store water that can support stomatal opening and consequent net CO_2 uptake.

Nutrients/Atmospheric CO_2 Concentration

Besides water uptake, plants also take up nutrients from the soil, which in turn affects their rates of net CO_2 uptake. Indeed, a multiplicative Nutrient Index can be incorporated into EPI [20]. Responses to soil elements for four agave species and eleven cactus species indicate that five elements (nitrogen, phosphorus, potassium, boron, and sodium) have major influences on daily net CO_2 uptake, growth or biomass accumulation; other macronutrients and micronutrients can be limiting in specific soils. Net CO_2 uptake, growth, and biomass productivity for these CAM plants are optimal at an average of about 3 mg N g^{-1} soil (0.3% by soil mass), 60 μg P g^{-1}, 250 μg K g^{-1}, and 1.0 μg B g^{-1} [2, 20, 21]. Soil nutrient levels leading to half maximal values of net CO_2 uptake, growth, or biomass productivity are about 0.7 mg N g^{-1}, 5 μg P g^{-1}, 3 μg K g^{-1}, and 0.04 μg B g^{-1}. Salinity inhibits growth for the agaves and cacti examined, 20% inhibition occurring at about 60 μg Na g^{-1} soil and 50% inhibition at about 150 μg g^{-1} [2, 20]. Instead of determining the levels of individual nutrients in the soil

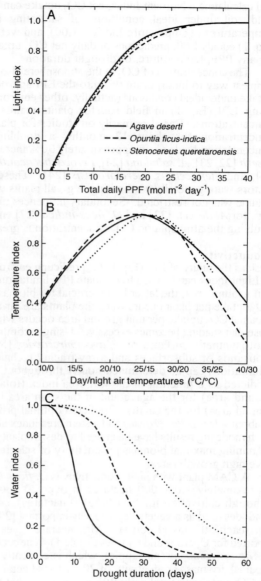

Figure 1. Component indices of EPI (Eq. 1) for the CAM species *Agave deserti* [1-3], *Opuntia ficus-indica* [16, 17], and *Stenocereus queretaroensis* [14; P. S. Nobel, unpublished]: (A) dependence of the Light Index on the total daily PPF; (B) dependence of the Temperature Index on the day/night air temperatures; and (C) dependence of the Water Index on the duration of drought (defined as when the shoot water potential is less than the bulk soil water potential).

and calculating a Nutrient Index, net CO_2 uptake can be measured over a 24-h period in the field soil under ideal conditions of saturating PPF (Light Index = 1.00), optimal temperatures (Temperature Index = 1.00), and wet soil (Water Index = 1.00) so that EPI (Eq. 1) equals 1.00; predictions of daily net CO_2 uptake can then be made for that field site for any PPF, temperature, or drought duration.

The concentration of CO_2 in the atmosphere also affects net CO_2 uptake by plants. The simplest way to incorporate this into the EPI approach is to determine total daily net CO_2 uptake under ideal conditions (actually, other environmental conditions can be adjusted for using EPI, Eq. 1) in field soil for plants maintained under various atmospheric CO_2 concentrations. Based primarily on studies for plants maintained under a particular CO_2 concentration for about 12 months, a doubling of the current atmospheric CO_2 concentration leads to an approximately 30% increase in biomass productivity for *Agave deserti* [22, 23], *A. salmiana* [24], *Ferocactus acanthodes* [22], *Opuntia ficus-indica*, [25-28], and *Stenocereus queretaroensis* [24]. These studies were performed when other factors were generally not limiting, e.g., all plants were regularly watered and the rooting volume was not restricted. Secondary influences on the component indices of EPI occur; e.g., for *A. deserti* [23] and *O. ficus-indica* [27] enhancement of daily net CO_2 uptake by doubling the atmospheric CO_2 concentration is greater at lower PPF and during drought.

Productivity
Besides the utility of EPI (Eq. 1) for predicting environmental influences on net CO_2 uptake, its Light Index can be used to evaluate PPF limitations on net CO_2 uptake per plant and per unit ground area, the latter being crucial for biomass productivity. In particular, maximal CO_2 uptake per plant occurs when the plants are far apart and do not shade each other, but maximal CO_2 uptake per unit ground area occurs when plants are close together yet not so close that shading becomes excessive. Using a computer model that subdivides the opaque photosynthetic surfaces of *Agave fourcroydes* [5] and *Opuntia ficus-indica* [12] into thousands of subsurfaces and a ray-tracing technique so that PPF on subsurfaces at all orientations can be determined hourly throughout a year, biomass productivity has been predicted as a function of the leaf area index (total area of both sides of leaves per unit ground area) for the agave and of the stem area index (total stem surface area per unit ground area) for the cactus. Predicted optimal productivity occurred at a leaf area index of about 4 for *A. fourcroydes* and a stem area index also of about 4 for *O. ficus-indica* [18]. Such modeling results have been used to design plant spacing intervals for field experiments examining maximal biomass productivity of selected CAM species that were observed to have high growth rates.

A CAM plant with high productivity is *O. ficus-indica*, which is currently cultivated on approximately 1,050,000 hectares in over 20 countries, mainly for fodder and forage, although cultivation for fruit (95,000 hectares), cochineal dye, chemicals, and nopalitos (cladodes used as a vegetable) is locally important [29-31]. Its annual dry-mass productivity can reach 50 tons (1 ton = 1 Mg) hectare^{-1} year^{-1}, which exceeds the productivities reported for C_3 crops and C_3 trees (Table 1). The second most widely cultivated CAM plant is *Ananas comosus* (pineapple), which is currently grown on 720,000 hectares and has a maximal dry-mass productivity of 20-31 tons hectare^{-1} year^{-1} [31, 35, 36]; its response of net CO_2 uptake to total daily PPF is similar to that for other CAM species (Fig. 1A), as is its optimal nighttime temperature for net CO_2 uptake of 15°C. Other opuntias and agaves with high biomass productivities include *Agave mapisaga* and *A. salmiana* (Table 1), whose leaves can achieve a fresh mass exceeding 40 kg each, and *O. amyclea*, a Mexican horticultural variety of prickly pear that is tolerant of low temperatures. The high annual biomass productivity of these four CAM species (Table 1) is surpassed by that of only a few C_4 species, in particular by that of *Saccharum officinarum* (sugar cane), whose high

Table 1. Annual aboveground-weight productivities under near optimal conditions. Data are from 18, 32-34.

Species	Productivity (tons hectare^{-1} year^{-1})
CAM crops	
Agave mapisaga	38
A. salmiana	42
Opuntia amyclea	45
O. ficus-indica	47-50
C$_3$ crops (*Beta vulgaris, Elaeis guineensis, Lolium perenne, Manihot esculenta, Medicago sativa, Triticum aestivum*)	29-45, ave 36
C$_3$ trees (*Cryptomeria japonica, Eucalyptus globulus, E. grandis, Pinus radiata, Salix purpures, Tsuga heterphylla*)	36-44, ave 39
C$_4$ crops (*Cynodon dactylon, C. plectostachyas, Pennisetum purpureum, Saccharum officinarum, Sorghum bicolor, Zea mays*)	32-70, ave 49

productivity of up to 67 tons hectare^{-1} year^{-1} has been documented in various tropical areas [18, 32].

The exceptionally high biomass productivities of certain CAM plants is consistent with measurements of net CO_2 uptake per unit photosynthetic surface area. Specifically, CAM plants take up most CO_2 at night but can take up some during the daytime, whereas C$_3$ and C$_4$ plants take up CO_2 only during the daytime, leading to slightly higher daily net CO_2 uptake for certain CAM species compared with the C$_3$ species with the highest productivities [18]. Moreover, photorespiration greatly increases the energetic cost per CO_2 fixed for C$_3$ species compared with C$_4$ or CAM species. In particular, the energetic cost per net mole of CO_2 fixed (assuming that the oxygenase activity is 25% of the carboxylase activity of Rubisco for C$_3$ plants but is negligible for C$_4$ and CAM plants) is 867 kJ mol^{-1} CO_2 for C$_3$ plants, 640 or 690 kJ mol^{-1} CO_2 for C$_4$ plants, and 753 or 790 kJ mol^{-1} CO_2 for CAM plants [19]. Thus the higher productivity for certain CAM species than C$_3$ species can be explained at both a leaf and a cellular level.

What about the future productivity of currently highly productive CAM species ? With respect to the increasing atmospheric CO_2 concentration involved in global climate change, each 10 μmol mol^{-1} (10 ppm by volume) increase in CO_2 concentration should increase CO_2 uptake and biomass productivity by about 1% [22-28]. The increasing air temperatures associated with increasing atmospheric CO_2 concentrations should have an even larger influence, as highly productive agaves and cacti are not very tolerant of subzero temperatures. For instance, commercially important agaves and cacti such as *Agave fourcroydes, A. mapisaga, A. salmiana, A. tequilana, O. ficus-indica,* and *Stenocereus queretaroensis* can be killed by tissue temperatures of -6°C to -8°C [2, 24]. The approximately 3°C increase in air temperature predicted to accompany a doubling of the atmospheric CO_2 concentration should increase the area where *O. ficus-indica* can be cultivated by about 20% worldwide [13, 37] and by about 60% in the United States [13, 25]. Indeed, selection for greater low-temperature tolerance may be the most important factor in the increased usage of agaves and cacti in temperate regions.

3960

References

1   Nobel, P.S. (1984) Oecologia 64,1-7
2   Nobel, P.S. (1988) Environmental Biology of Agaves and Cacti, Cambridge University Press, New York
3   Nobel, P.S. and Hartsock, T.L. (1986) Plant Cell Environ. 9, 741-749
4   Nobel, P.S. (1985) J. Appl. Ecol. 22, 157-173
5   Nobel, P.S. and Garcia de Cortázar, V. (1987) Photosynthetica 21, 261-272
6   Nobel, P.S. and Quero, E. (1986) Ecology 67, 1-11
7   Quero, E. and Nobel, P.S. (1987) J. Appl. Ecol. 24, 1053-1062
8   Nobel, P.S. and Meyer, S.E. (1985) Physiol. Plant. 65, 397-404
9   Nobel, P.S. and Valenzuela, A. (1987) Agric. Forest. Meteorol. 39, 319-314
10 Nobel, P.S. (1986) Amer. J. Bot. 73, 541-547
11 Raveh, E., Gersani, M. and Nobel, P.S. (1995) Physiol. Plant. 93, 505-511
12 Garcia de Cortázar, V., Acevedo, E. and Nobel, P.S. (1985) Agric. Forest. Meteorol. 34, 145-162
13 Nobel, P.S. (1991) Plant Cell Environ. 14, 637-646
14 Nobel, P.S. and Pimienta-Barrios, E. (1995) Environ. Exp. Bot. 35, 17-24
15 Nobel, P.S. and Zhang, H. (1997) Aust. J. Plant. Physiol. 24, 787-796
16 Nobel, P.S. and Hartsock, T.L. (1983) Plant Physiol. 71, 71-75
17 Nobel, P.S. and Hartsock, T.L. (1984) Physiol. Plant. 60, 98-105
18 Nobel, P.S. (1991) New Phytol. 119, 183-205
19 Nobel, P.S. (1996) in Crassulacean Acid Metabolism. Biochemistry, Ecophysiology and Evolution (Winter K. and Smith J.A.C., eds.) pp.255-265, Springer-Verlag, Berlin
20 Nobel, P.S. (1989) J. Appl. Ecol. 26, 635-645
21 Nobel, P.S., Quero, E. and Linares, H. (1988) J. Plant Nutrition 11, 1683-1700
22 Nobel, P.S. and Hartsock, T.L. (1986) Plant Physiol. 82, 604-606
23 Graham, E.A. and Nobel, P.S. (1996) J. Exp. Bot. 47, 61-69
24 Nobel, P.S. (1996) J. Arid Environ. 34, 187-196
25 Nobel, P.S. and Garcia de Cortázar, V. (1991) Agron. J. 83, 224-230
26 Cui, M., Miller, P.M. and Nobel, P.S. (1993) Plant Physiol. 103, 519-524
27 Nobel, P.S. and. Israel, A.A. (1994) J. Exp. Bot. 45, 295-303
28 Cui, M. and Nobel, P.S. (1994) Plant Cell Environ. 17, 935-944
29 Barbera, G., Inglese, P. and Pimienta-Barrios, E., eds. (1995) Agro-Ecology, Cultivation and Uses of Cactus Pear, FAO, Rome
30 Nobel, P.S. (1994) Remarkable Agaves and Cacti, Oxford University Press, New York
31 Nobel, P.S. (1999) Climate Change and Global Crop Productivity (Reddy K.R. and Hodges H.F., eds.) in press, CAB International, Wallingford, England
32 Loomis, R.S. and Ferakis, P.A. (1975) Photosynthesis and Productivity in Different Environments (Cooper J.P., ed.) pp.145-172, Cambridge University Press, Cambridge, England
33 Nobel, P.S., Garcia-Moya, E. and Quero, E. (1992) Plant Cell Environ. 15, 329-335
34 Garcia de Cortázar, V. and Nobel, P.S. (1992) J. Amer. Soc. Hort. Sci. 117, 558-562
35 Bartholomew, D. and Rohrbach, K.G., eds. (1993) First International Pineapple Symposium, Acta Hort. 334, 1-471
36 Zhu, J., Bartholomew, D.P. and Goldstein, G. (1997) Acta Hort. 425, 297-308
37 Garcia de Cortázar, V. and Nobel, P.S. (1990) Agric. Forest. Meteorol. 49, 261-279

CO$_2$ DIFFUSION INSIDE LEAF MESOPHYLL OF LIGNEOUS PLANTS

Bernard Genty, Sylvie Meyer, Clément Piel, Franz Badeck and Rodolphe Liozon. Groupe Photosynthèse et Environnement, Laboratoire d'Ecophysiologie Végétale, Bât. 362, CNRS URA 2154, Université Paris Sud, Orsay 91405, France. Fax (33 1) 01 69 15 72 38

Key words : fluorescence imaging, gas diffusion, gas exchange, leaf physiology, ligneous plants.

1. Introduction

CO$_2$ diffusion into the leaf to the chloroplast not only depends on stomatal conductance but also on the conductance along the path CO$_2$ follows into the leaf mesophyll from the substomatal cavities to the catalytic sites of Rubisco. Recent methodological advances showed that a significant drop in the CO$_2$ partial pressure may occur along this path [1-5]. Such a drop reveals a low conductance to CO$_2$ diffusion in the mesophyll. A lower conductance in the mesophyll has been shown in leaves of woody plants than in herbaceous plants [4-7].

Two components determine the internal conductance in the mesophyll, g$_i$: i) the gaseous phase diffusion through the net of intercellular air spaces inside the leaf, ii) the liquid phase diffusion through the cell wall and into the cell to the carboxylation site [5]. However their relative contribution in determining g$_i$ remains poorly documented. Comparison of leaf gas exchange in air with that in helox (air with nitrogen replaced by helium) in which the diffusivity of CO$_2$ is largely enhanced (gas diffusivities are about 3.6 times higher in He than in N$_2$) has been used to estimate the conductance to CO$_2$ transfer through intercellular air spaces (g$_{ias}$) [8]. This approach provides a strong evidence for low g$_{ias}$ in hypostomatous leaves [8, 9]. In contrast, comparative analysis of leaf anatomy, cell ultrastructure and g$_i$ have emphasised the role of the conductance through the liquid phase (g$_{liq}$) in determining g$_i$ [10, 11]. A modelling work supported this view for leaves of various ligneous species [11].

In this study we introduce an approach combining measurements of leaf gas exchange and chlorophyll fluorescence in N$_2$ and He based atmospheres to experimentally estimate the relative contribution of g$_{ias}$ and g$_{liq}$ in determining low g$_i$ in leaves of two ligneous plants.

2. Procedure

CO$_2$ and H$_2$O gas exchange measurements and chlorophyll fluorescence imaging were simultaneously performed on detached leaves of *Rosa rubiginosa* L. and the hybrid *Populus koreana.trichocarpa* cv. "Peace" as described in [12, 13]. Measurements were done under a subsaturating photon flux density of 400-450 µmol m^{-2} s^{-1} at various CO$_2$ mole fractions in air, helox and in O$_2$ depleted (0 or 0.6%) atmosphere with N$_2$ and He background. Leaf temperature was monitored by infrared thermometry. Gas exchange

G. Garab (ed.), Photosynthesis: Mechanisms and Effects, Vol. V, 3961–3966.
© 1998 *Kluwer Academic Publishers. Printed in the Netherlands.*

parameters in various atmospheres were calculated according to [8]. Boundary layer conductances for water vapor in the different gas mixtures were estimated for each leaf from leaf temperature using an energy balance method corrected for physical characteristics of each gas mixture [13].

The partial pressure of CO_2 at the carboxylation sites p_c and g_i were estimated in air and in helox as described in [3] as :

$$p_c = \frac{(J_t + 8(A_n + R))\Gamma^*}{J_t - 4(A_n + R)} \quad \text{and} \quad g_i = \frac{A_n P}{P_{es} - p_c}$$

where A_n and R are net CO_2 assimilation and dark respiration; J_t is the rate of electron transport estimated from the empirical relationship between the photochemical yield of photosystem II and the quantum yield of gross CO_2 assimilation under non-photorespiratory conditions [7,14] (see Fig. 2); P is the atmospheric pressure; p_{es}, is the partial pressure of CO_2 at the evaporating sites and Γ^* is the CO_2 compensation point in the absence of R.

In a rather simplistic diffusion model in one dimension (see [9]), g_i can be expressed as :

$$g_i^{-1} = g_{ias}^{-1} + g_{liq}^{-1} \tag{1}$$

where g_i, g_{ias} and g_{liq} have same dimension.

Then, assuming that g_{liq} remains unchanged in various background atmospheres, from g_i estimated in air and in helox and the reported ratio of diffusivities of CO_2 for helox/air [8], g_{ias} was calculated as :

$$g_{ias} = \frac{(1 - K^{-1})}{g_i^{-1} - g_{i\,helox}^{-1}} \tag{2}$$

where K is the diffusion ratio for CO_2 (2.33 for helox/air); $g_{i\,helox}$ is the internal conductance estimated in helox and g_i, g_{ias} are conductances estimated in air. Finally, g_{liq} was calculated from g_i and g_{ias} by rearranging equation (1).

3. Results and Discussion

Ratio of diffusivities in N_2 and He background atmospheres. Populus "Peace" has been reported to have low g_i [15] and almost no leaf stomatal response to ABA, light and CO_2 [16,17]. "Peace" leaf stomatal conductance was also unresponsive to leaf water pressure deficit (VPD) (Fig. 1). Therefore, we used this material as calibration leaf for checking the relative variation of the diffusivity of water vapor when the leaf background atmosphere was changed from N_2 to He. Figure 1 shows the stomatal conductance for H_2O in N_2 and He. Mean ratios of conductances for He/N_2 and helox/air were 3.12 and 1.99 respectively which is lower than those calculated from reported diffusivities, 3.49 and 2.33 respectively [8]. A similar difference has been reported for helox by [8]. This may result from "Knudsen diffusion" instead of free diffusion in the stomatal pores and the intercellular paths which participate to water diffusion or from contribution of different evaporating surfaces in the mesophyll for N_2 and He background [8,9]. However, the relatively small difference between experimental and theoretical values should not largely affect the p_{es} estimated here (the conductances for CO_2 were calculated from the reported difffusivities for water vapor and CO_2 in given gas background as described by [8]).

Figure 1. Relationship between stomatal conductance for H_2O and transpiration obtained in pure N_2 (o) and pure He (◊) when VPD of a *Populus* "Peace" leaf was varied in the dark. Boundary layer conductance for H_2O were 2.53 and 2.57 mol m^{-2} s^{-1} in N_2 and He respectively. Leaf temperature was 24 °C.

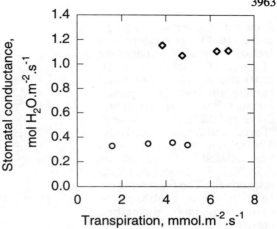

Estimates of electron transport rate using chlorophyll fluorescence in N_2 and He background. In non photorespiratory conditions (O_2 depleted atmosphere), when switching from N_2 to He background, the relationships between the photochemical yield of photosystem 2 estimated by fluorescence and the quantum yield of linear electron transport estimated by CO_2 exchange were similar (Fig. 2). Similar results were obtained in *Rosa* (data not shown). These data suggest that the activity profile probed by fluorescence and the one probed by gas exchange remained not significantly changed in the two gaseous backgrounds. Thus, estimates of p_c and g_i could be compared in N_2 and He based atmosphere.

Figure 2. Relationship between the photochemical yield of PS 2, $1-\Phi/\Phi_m$, and the quantum yield of linear electron transport, Φ_e, in leaves of *Populus* "Peace" in N_2 (o) and He (◊) O_2 depleted background. Φ_e was calculated on the basis of incident irradiance as $\Phi_e=4\Phi_{CO_2}$ where Φ_{CO_2} is the quantum yield of gross CO_2 assimilation estimated as in [12, 14]. Measurements were made at photon flux densities of 250 and 420 μmol m^{-2} s^{-1} under varying CO_2 ambient partial pressure.

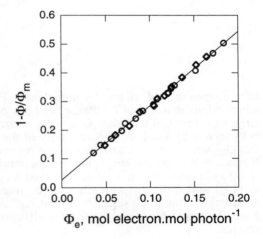

Figure 3. A_n versus p_{es} (o, ◊) and p_c (●,◆) relationships in a leaf of *P*. "Peace" (A) and one of *Rosa* (B) obtained in air (o,●) and in helox (◊,◆) under varying ambient CO_2 partial pressure. P was 101 kPa. g_i was 0.119 and 0.231 mol m⁻² s⁻¹ in *P*. "Peace" and *Rosa* respectively. The minimum variance of g_i was obtained for Γ^* of 3.66 and 3.58 Pa in *P*. "Peace" and *Rosa* respectively. The fits were done for ribulose-biphosphate saturation as described previously [7].

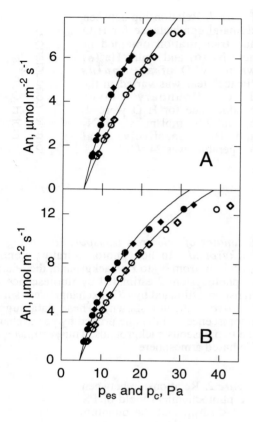

CO_2 response and internal conductance. Figure 3 shows A_n versus p_{es} and p_c relationships in a leaf of *P*. "Peace" and one of *Rosa* obtained in air and in helox. The large shift between the relationships A_n versus p_{es} and A_n versus p_c indicates that in both leaves, g_i was low. The similarity of the relationships in air and helox illustrates that g_{ias} was large. Similar relationships A_n versus p_{es} were also obtained in 0.6% O_2 in N_2 and He background atmospheres (data not shown). A mean g_i of 0.181±0.067 (±SD) and 0.212±0.020 mol m⁻² s⁻¹ were obtained for 4 leaves of *P*. "Peace" (higher value than in [15]) and *Rosa* respectively and the intercellular diffusion limitation was never measurable. This result suggests that intercellular gaseous diffusion should contribute to a small proportion of the large CO_2 diffusion limitation inside leaves of these two species.

Conclusion. We have shown that low g_i of leaves from two ligneous plants corresponded to low g_{liq} with no detectable contribution of g_{ias}. It is worth noting that if in these leaves evaporative area were not only located in the substomatal cavities but also deeper in the mesophyll, then a contribution of intercellular gaseous diffusion limitation would have been included in the stomatal conductance estimates and would have not been taken into account by g_i and g_{ias}. In this context, a likely small contribution of intercellular gaseous diffusion limitation cannot be discarded but in any case it would not alter g_i as it has been calculated here. This problem has not been taken into account in the modelling approach used by [11].

Low g_i and g_{liq} were observed for two rather different leaf anatomies. *Rosa* has hypostomatous heterobaric leaves [13, 18]. Leaflets were 190 µm thick on average with a fairly packed mesophyll (volumic fraction of intercellular air spaces in the leaf: 0.32). "Peace" leaves are amphistomatous with a stomatal ratio of 0.09 for adaxial/abaxial sides. Since many bundle sheath extensions associated with the reticulate venation only span across palisade and spongy mesophylls (ie see Fig 4C in [19]), and do not reach a loosely arranged non assimilatory spongy mesophyll which allows lateral gas diffusion over several reticules, the leaf should be described as "semi-heterobaric". According to previous data using helox [8], a significant intercellular CO_2 diffusion limitation would have been more likely in *Rosa* leaves but this was not the case. Further experiments with thicker hypostomatous leaves in which g_{ias} is expected to be low are needed to confirm that g_{liq} remains the major determinant of g_i.

Thus, our data provide first experimental evidence that it is a lower g_{liq} that can explain the lower g_i in sclerophyllous leaves than in mesophytic leaves as previously suggested [5, 11] . This may indicate that a low g_{liq}, and therefore an important limitation of CO_2 diffusion through the cell wall to the carboxylation sites are traits of leaves of ligneous plants. Understanding where the limitations reside in sclerophyllous leaves would need further insights into structure, composition, and properties of the determinants of CO_2 diffusion in the liquid phase (cell wall, plasmalemma, cytosol, chloroplast membrane, stroma, see [5, 10]). It should be possible to examine the role of carbonic anhydrase using ethoxyzolamide to inhibit carbonic anhydrase activity [20].

Aknowledgments
We thank E. Dreyer (INRA, Nancy) for providing cuttings of *P.* "Peace".

References
[1] Evans J.R., Sharkey T.D., Berry J.A. and Farquhar G.D. (1986) Aust. J. Plant Physiol. 13, 281-292
[2] von Caemmerer S. and Evans J. R. (1991) Aust. J. Plant Physiol. 18, 287-305
[3] Harley P. C., Loreto F., Di Marco G. and Sharkey T. D. (1992) Plant Physiol. 98, 1429-1436
[4] Loreto F., Harley P. C., Di Marco G. and Sharkey T. D. (1992) Plant Physiol. 98, 1437-1443
[5] Evans J. R. and von Caemmerer S. (1996) Plant Physiol. 110, 339-346.
[6] Lloyd J., Syvertsen J.P., Kriedemann P.E. and Farquhar G.D. (1992) Plant Cell Environment 15, 873-899
[7] Epron D., Godard D., Cornic G. and Genty B. (1995) Plant Cell Environment 18, 43-51.
[8] Parkhurst D.F. and Mott K.A. (1990) Plant Physiol. 94, 1024-1032
[9] Parkhurst D.F. (1994) New Phytol. 126, 446-479
[10] Evans J.R., von Caemmerer S., Setchell B.A. and Hudson G.S. (1994) Aust. J. Plant Physiol. 21, 475-495
[11] Syvertsen J.P., Lloyd J., McConchie C., Kriedemann P.E. and Farquhar G.D. (1995) Plant Cell Environment 18, 149-157
[12] Genty B. and Meyer S. (1995) Aus. J. Plant Physiol. 22, 277-284
[13] Meyer S. and Genty B. (1998) Plant Physiol. 116: 947-957

[14] Genty B., Briantais J-M. and Baker N.R. (1989) Biochim. Biophys. Acta 990, 87-92

[15] Ridolfi M. and Dreyer E. (1997) New Phytol. 135, 31-40

[16] Furukawa A., Park S.Y. and Fujinuma Y (1990) Trees 4, 191-197

[17] Ridolfi M., Fauveau M.L., Label P., Garrec J.P. and Dreyer E. (1996) New Phytol. 134, 445-454

[18] Terashima I. (1992) Photosynth. Res. 31 : 195-212

[19] Hinckley T.M., Ceuleman R., Dunlap J.M., Figiola A., Heilman P.E., Isebrands J.G., Scaracia-mugnozza G., Schulte P.J., Smit B., Stettler R.F., Van Volkenburgh E. and Wiard B.M. (1989)in Structural and functional responses to environmental stresses. ed. Kreeb K.H., Richter H., Hinckley T.M.. SPB Academic Publishing, The Hague, The Netherlands. pp. 199-217

[20] Peltier G., Cournac L., Despaux V., Dimon B., Genty B. and Rumeau D. (1995) Planta 196, 732-739

DRY MATTER PARTITIONING IN *PHASEOLUS VULGARIS VAR. NANUS* CULTIVATED UNDER ELEVATED CO_2 MODIFIED BY LIGHT INTENSITY AND THE MODE OF NITROGEN SUPPLY.

H.R.Bolhàr-Nordenkampf, M.Th.Unger and M.H.Meister, Institute of Plant Physiology, Division of Horticultural Plant Physiology and Primary Production, University of Vienna, P.O.Box 285, A1091 Vienna, Austria

Key words: C partitioning, elevated CO_2, N fixation, nodules, *Phaseolus vulgaris*.

1. Introduction

Carbon dioxide is a factor which contributes to climatic change and will continue to increase for at least fifty years. Since carbon dioxide is a gaseous fertiliser for all green plants, modifications of carbon partitioning and also of CO_2 sequestration in biomass will induce changes in species composition and ecosystem structure (1,2) Photosynthesis in C_3 plants responds linearly to a doubling in CO_2 concentration under high light (> 700 $\mu mol.m^{-2}.s^{-1}$) and high temperatures (> 20°C) (3). In addition, the response to elevated CO_2 is strongly controlled by the sink capacity for assimilates. Many trials have found that the elevated CO_2 concentrations gave only a small increase in above ground biomass whereas a large proportion of the additional carbon was transferred to below ground biomass (4). This partitioning promoted root growth and resulted in a better carbon supply for mycorrhizza, nodules of legumes and for microorganisms in the rhizosphere. Biomass accumulated under elevated CO_2 showed higher C:N ratios which resulted in a better growth of nodules lower content of RubisCO protein (5,6).

This study was designed to clarify i. does additionally fixed carbon promote the development of nodules and ii. in case of with higher activity for nitrogen fixation, does this compensate for the decrease in nitrogen contend under elevated CO_2 ? (7, 8, 9, 10)

1. Material and Methods

Bean plants (*Phaseolus vulgaris* L. ssp. *vulgaris* var. *nanus* (L.) Aschers. Cv. `Forum`) were cultivated in indoor open top chambers. High light and low light treatment was combined with ambient and elevated CO_2 concentrations to study the effect of inoculation by *Rhizobium leguminosarum* bv. *phaseoli* (tab.1). Plants were studied in the frame of PHOTOGROW, a protocol to study photosynthesis, carbon partitioning,

G. Garab (ed.), Photosynthesis: Mechanisms and Effects, Vol. V, 3967–3972.
© 1998 *Kluwer Academic Publishers. Printed in the Netherlands.*

pigments and functional leaf anatomy. Data from growth analysis, nodule development and C/N determinations will be discussed.

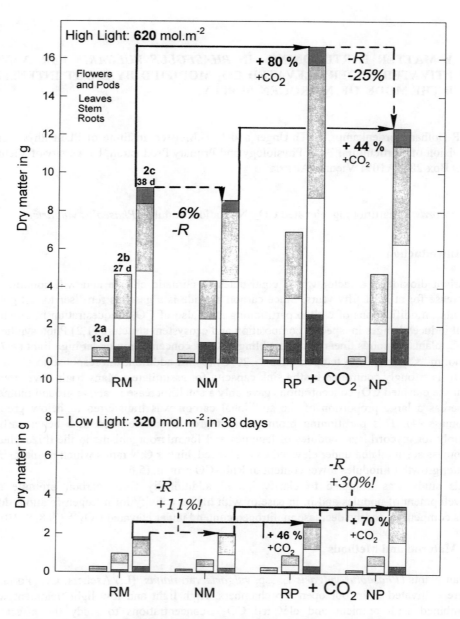

Figure 1: Dry matter partitioning during ontogenesis in *Phaseolus vulgaris*: Harvests 13, 27 and 38 days after sowing (see tab. 1)

Table 1: Growing conditions, indoor open top chambers
    air-conditioned glasshouse; day: 28±5°C; night: 22±3.5°C; r.H. 60±15%,
L low light: day length 14±0.5h, 320 mol photons.m^{-2};
H high light: day length 16h, 620 mol photons.m^{-2}; +1.8°C, -10% r.H.
M ambient CO_2 : 320-420 $cm^3.m^{-3}$; **P** elevated CO_2 : 670-770 $cm^3.m^{-3}$
R inoculated with *Rhizobia*, mineral fertiliser -N added
N fertilised with mineral fertiliser +N

3. Results and Discussion

Dry matter accumulation was strongly controlled by light intensity (fig. 1). In high light, plants inoculated with *Rhizobia* showed an significant increase in biomass under elevated CO_2 after three weeks of ontogenesis. Low light grown plants respond with a decrease in biomass due to *Rhizobium* inoculation, whereas the CO_2 induced stimulation of biomass accumulation was maintained. Dry matter accumulation of stems was similar to that of the whole plant, but leaves and roots developed differently (11, 12, 13). Under conditions mentioned above, leaf growth was stimulated much earlier during ontogenesis and differences between inoculated and N fertilised plants were more pronounced. Specific leaf area ($cm^2 .g^{-1}$ d.wt. plant) was strongly reduced by elevated CO_2 and stimulated by the activity of *Rhizobia* only during late stages of ontogenesis. Leaves were shown to be thicker, but nevertheless it could be concluded that less leaf area was needed to maintain a given amount of dry matter under elevated CO_2(data not shown) (14). Surprisingly, in elevated CO_2 partitioning of carbon to below ground biomass was strongly increased during ale stages of ontogenesis. Calculations of root shoot ratios revealed an increase in high light with *Rhizobium* inoculation during early ontogenesis and at the and of ontogenesis for low light grown plants (data not shown) (15, 16). After 38 days of growth, dry matter accumulation was significantly enhanced by elevated CO_2. Under high light conditions, *Rhizobia* inoculation nearly doubled the positive effect of CO_2, whereas in low light less biomass was found, revealing a parasitic feature of *Rhizobium* infection under these conditions (fig. 1). At the final harvest after 38 days of growth, roots of all plants displayed nodule development regardless of whether they were inoculated or not, although showing remarkable variance in numbers and sizes of nodules (tab. 2). In contrast, significant differences were seen during the first 4 weeks of ontogenesis and between high light and low light grown plants. In both groups, elevated CO_2 stimulated nodule development of inoculated plants significantly, but in high light grown plants nodule density was 5 to 10 times higher (tab. 2).
Dry matter accumulation is clearly controlled by the sink source relationship (17,18, 19). This can be demonstrated by calculating the enhancement in biomass production as a percentage of the control (ambient CO_2). Data from low and high light treatment can be normalised by plotting them versus the sum of incident light (fig. 2). Plants grown in low light and inoculated by *Rhizobia* were responsible for negative values at the beginning of ontogenesis (parasitic features of *Rhizobia*). The steep increase up to 300 moles of photons was caused by two sinks: leaf growth and nodule development (see fig. 1 and tab. 2). The plateau in the graph (fig. 2) marks the change from vegetative to generative

growth. Finally, the commencement of pod filling processes caused a second increase of CO_2 stimulated growth.

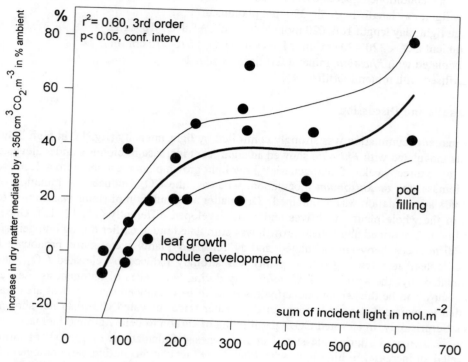

Figure 2: Changes in the response to elevated CO_2 during ontogenesis of *Phaseolus vulgaris*. Relative increase in dry matter accumulation versus the sum of incident light since emergence. Mean values, $n = 6$

Table 2: Nodules attached to the root system of 1 plant:

d.a.s.	RML	RPL	NML	NPL	RMH	RPH	NMH	NRH
15	3	8	0	0	67	133	0	9
26	18	42	0	3	189	311	210	147
38	76	77	28	17	260	284	325	296

means values $n = 3$; d.a.s days after sowing; R inoculated with *Rhizobium*; N no inoculation; M ambient, P elevated CO_2; L low light; H high light;

After 50 days of growth, the nitrogen content of harvested green pods was significantly reduced by CO_2 enrichment in spite of inoculation with *Rhizobium* (tab. 3). Higher yields were achieved with inoculated plants, and this effect was strongly enhanced by elevated CO_2 concentrations (fig. 3). In low light, the dry weight of green pods per plant was strongly reduced by the parasitic action of *Rhizobia*, and in addition, CO_2 enrichment has no effect. High light grown pants also revealed a positive effect of *Rhizobium* inoculation and CO_2 enrichment with respect to the amount of harvested nitrogen equal to protein content in green pods (fig. 3).

Table 3: Green pods harvested after 50 days of growth (1 plant)

	RML	NML	RPL	NPL	RMH	NMH	RPH	NPH
% nitrogen in d.m.	3,77b	3,63ab	3,21a	3,09a	3,34a	3,79b	3,11a	3,09a

Duncan-test: different letters stand for significant differences, $p < 0,1$; $n = 9$,
d.m. dry matter of green pods

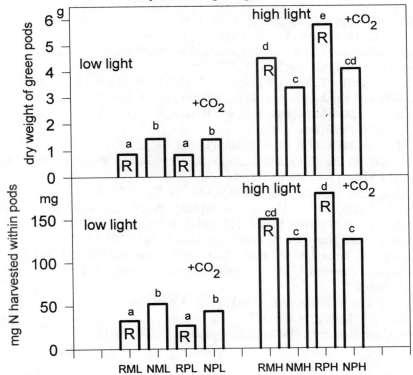

Figure 3: *Phaseolus vulgaris*, green pods after 50 days of ontogenesis, yield in dry matter and nitrogen

5. Conclusions

The amount of assimilates available determined on one hand the point of time for root infection by *Rhizobia* and on the other hand parasitic or symbiotic effects of nodule development. This was clearly shown by the comparison of low light and high light grown plants. Neither the supply of nitrogen rich fertiliser nor CO_2 stimulated nitrogen fixation by *Rhizobia* were able to offset the increase in C/N ratios of plant material. Higher availability of nitrogen clearly promoted plant growth in general. During vegetative growth, additional nitrogen from *Rhizobia* was predominately invested in leave development. This trend was in turn enhanced by CO_2 enrichment in high light, which gives rise to a higher source capacity, feeding *Rhizobia* better with assimilates.

5. Acknowledgements
This research project was supported by a grant from the Ministry for Science, Research and traffic: GZ.30.545/1-IV/8b/95

6. References

1. Lawlor, D.W. (1995) J.Exp.Bot. 46, 1449- 1456
2. Bazzaz, C.L. (1990) Annu.Rev.Ecol.Syst. 21, 167-196
3. Bolhàr-Nordenkampf, H.R. (1976) Biochem.Physiol.Pflanzen 169, 121-161
4. Curtis, P.S. (1996) PCE 19 (2), 127-137
5. Curtis, P.S., Drake, B.G. and Whifham, D.F. (1989) Oecologia 78, 297-301
6. Akey, D.H. and Kimball, B.A. (1989) SW Entomol. 14, 255-260
7. Coleman. J.S., McConnaughay, K.D. and Bazzaz, R.A. (1993) Oecologia 93, 195-200
8. Finn, G.A. and Brun, W.A. (1982) Plant Physiol. 69, 327-331
9. Mjwara, J.M., Botha, C.E.J. and Randolff, S.E. (1996) Phys. Plantarum 97, 754-763
10. Noby, R.J. (1987) New Phytol. 111 491-500
11. Radoglou, K.M. and Jarvis. P.G. (1992) Ann.Bot. 70, 245-256
12. Cure, J.D., Rufty, R.W. and Israel, D.W. (1988) Crop.Sci. 28, 671-677
13. Wardlaw, I.F. (1990) New Phytol. 116, 341-381
14. Thomas, J.F. and Harvey, C.N. (1983) Bot.Gaz. 144 (3), 303-309
15. Gregory, P.J., Palta, J.A. and Battis, G.R. (1997) Plant and Soil 187, 221-228
16. Rogers, H.H., Peterson, C.M., McCrimmon, J.N. and Cure, J.D. (1992) PCE 15, 749-752
17. Farrar, J.F. and Williams, M.L. (1991) PCE 14, 819-830
18. Arp, W.J. (1991) PCE 14 869-875
19. Carmi,A. and Koller, D. (1978) Photosynthetica 12 (2) 178-184

PROTECTIVE STRATEGIES IN PLANTS EXPOSED TO HIGH VISIBLE AND UV-B (280-315 nm) RADIATION

J.F. Bornman
Lund Univ., Plant Physiology, Box 117, S-221 00 Lund, Sweden

Key words: abiotic stress, acclimation, environmental stress, flavonoids, light penetration, UV radiation

1. Introduction

Regulatory, acclimatory and long-term adaptive mechanisms of plants may be affected by an increased amount of ultraviolet-B radiation (UV-B, 280-315 nm) as a consequence of a decreased stratospheric ozone layer. A pertinent question in this respect is whether there is indeed cause for concern with regard to the stratospheric ozone layer. Apart from more than 60% seasonal decreases in Antarctic ozone, relatively large decreases have been measured over the Arctic. Ozone-deficient air around the polar regions moves towards lower latitudes, causing transient decreases (1,2). Globally, analyses show a decline by ca 4-5% since 1979 (3) indicating that the ozone loss is a global problem. Although the Montreal Protocol and its Amendments should result in repair of the ozone layer within the next few decades, greenhouse gases and consequent global warming may delay the recovery, and may actually be responsible for the ozone loss recorded over the Arctic (4), since warming of the earth leads to a cooler stratosphere which in turn provides favourable conditions for ozone depletion (5).

2. Protection of photosynthesis by flavonoid compounds

2.1 *UV filters*
Enhanced UV-B radiation often results in several acclimative and adaptive plant responses, one of which is particularly common, namely, the induction of UV-screening phenolic compounds such as flavonoids. These compounds usually provide an effective, protective shield, reducing the amount and internal distribution of UV radiation within leaves. It has been shown *in situ*, using quartz optical fibres to probe the light environment within the leaf, that there is good correlation between UV penetration and UV-absorbing pigments (6). These screening compounds also seem to provide fairly good protection to the photosynthetic apparatus, as has been shown in a few studies using mutants deficient in these compounds (7,8). For example, in a mutant soybean (*Glycine max*) deficient in flavonoids, CO_2 fixation was reduced compared with that of a cultivar containing higher amounts of pigments (7). In barley (*Hordeum vulgare*)

3973

G. Garab (ed.), *Photosynthesis: Mechanisms and Effects, Vol. V,* 3973–3978.
© 1998 *Kluwer Academic Publishers. Printed in the Netherlands.*

leaves, good correlations were found between phenolic compounds and photosynthetic efficiency (measured as quantum yield).

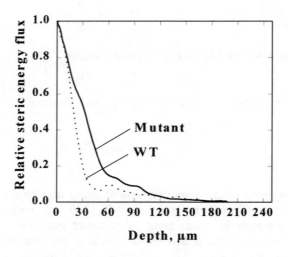

Figure 1. Penetration of UV-B (310 nm) in wild type (WT) and flavonoid-deficient mutant barley leaves from plants grown under visible and UV-B radiation. Values were divided by the incident light (xenon arc) and normalised to 1. (Redrawn after 8).

The penetration and distribution of UV-B radiation (310 nm) shows the importance of the phenylpropanoid compounds as effective UV-B shields when comparing the flavonoid-deficient species with the wild type (Figure 1). However, Figure 1 also shows that the mutants are not transparent to UV-B radiation, despite the fact that leaves only contained 7% of total extractable flavonoids relative to the wild type (8). This is indicative of a constitutive shield by other UV-absorbing compounds throughout the leaf, which are fairly unresponsive to additional UV radiation. The photochemical quantum yield (Figure 2) of the mutant was lower after 7 days exposure to relatively high light (HL, 720 μmol m^{-2} s^{-1} photosynthetically active radiation, 400-700 nm) and enhanced UV-B radiation (13 kJ m^{-2} day^{-1} weighted irradiance; 9) compared to that of the wild type.

2.2 Selective increases in flavonoid compounds
Apart from actual increased amounts of UV-absorbing pigments in response to enhanced UV-B there may also be large increases in specific, major UV-absorbing pigments such as certain flavonoids in some plants (8,10-13). These differential shifts in content have been suggested to favour antioxidant activity, and would thus be of

strategic importance for many plant processes, including photosynthesis, since UV-B radiation has been shown to induce active oxygen species (14).

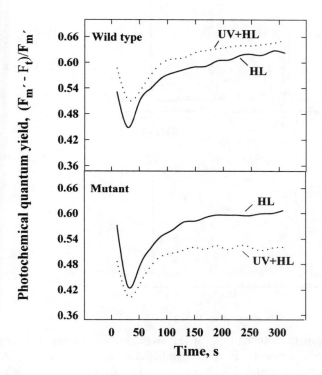

Figure 2. Photochemical quantum yield of photosystem II in leaves of WT and flavonoid-deficient barley plants (mutant) grown under visible (HL) and UV-B radiation (UV) for 1 week (Redrawn after 8).

In barley, for example, the HPLC profiles of the soluble UV-absorbing compounds showed a 500% increase in the major compound lutonarin (with additional hydroxyl group in ring B of the flavonoid skeleton, favouring antioxidant activity, 15) after growth with supplemental UV-B radiation. The other major compound, saponarin, increased only by 30% (8). This same potential, protective response to enhanced levels of UV-B radiation also occurs in *Brassica napus* (13, Figure 3), where quercetin glycosides relative to the other major compounds, kaempferol glycosides, increased 36-fold after exposure for 16 days to 13 kJ m^{-2} day^{-1} weighted UV-B irradiance and visible light (800 μmol m^{-2} s^{-1} photosynthetically active radiation, 400-700 nm).

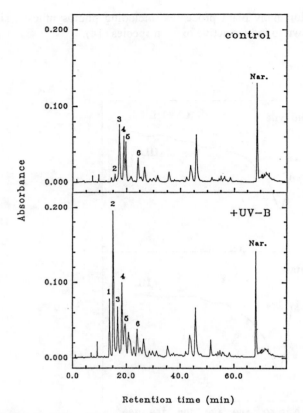

Figure 3. HPLC profiles of methanolic extracts of leaves of *Brassica napus,* cv. Paroll, grown for 16 days under 800 μmol m^{-2} s^{-1} photosynthetically active radiation (PAR, 400-700 nm; control) or supplemental UV-B radiation (13 kJ m^{-2} day^{-1} weighted irradiance; 9; +UV-B). Nar., naringenin, internal standard; 1, quercetin glycoside; 2, quercetin-3-sophoroside-7-glucoside; 3, kaempferol glycoside; 4, kaempferol-3-sophoroside-7-glucoside; 5, quercetin-3-2‴-sinapoylsophoroside-7-glucoside; 6, kaempferol 3-2‴-sinapoylsophoroside-7-glucoside. (Modified from 13 with permission).

2.3 *Reverse acclimation*

Despite examples of the protective role of flavonoids, plant response may not necessarily show a positive effect. This was seen in the case where UV-B induced a shade-type response at the level of the chloroplast in *Brassica napus* (16). Among several other changes induced by supplementary UV-B radiation which mimicked a shade type response, there were increases in the surface area of the appressed and non-appressed thylakoid membranes in the palisade tissue (16). Figure 4 shows the effect of supplemental UV-B radiation on the surface area of the appressed thylakoid membranes after only 30 min of supplemental UV-B radiation. Further analyses at the end of one day and one week exposure gave similar results, although the 30 min irradiation treatment induced the most marked response. This response could conceivably increase the statistical probability of photon interception and at the same time lead to an increased energy cost that would be required to sustain the membrane surface area during exposure to UV-B radiation. This may not necessarily be seen as a direct damaging effect to the plant but it does suggest an increased regulatory influence by UV-B radiation, which does not appear to be a positive acclimation response in all respects. It is possible that this response is triggered by the redox state of the plastoquinone pool which is thought to influence thylakoid membrane dynamics (17).

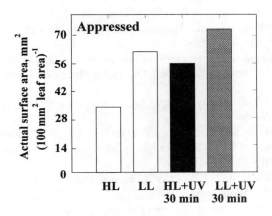

Figure 4. Thylakoid surface areas in palisade tissue per leaf volume for appressed or granal thylakoids after one week of radiation treatment. HL, 700 μmol m^{-2} s^{-1} photosynthetically active radiation (PAR, 400-700 nm); LL, 200 μmol m^{-2} s^{-1} PAR; HL+UV and LL+UV with 30 min enhanced UV-B radiation (2.34 kJ m^{-2} day^{-1} weighted irradiance; 9). (Redrawn after 16).

3. *In conclusion*
Apart from phenolics, and their multiple role, e.g. as UV-screening compounds and potential antioxidants among others, morphological features such as leaf thickening, decreased leaf area (18), and UV-absorbing trichomes (19) to mention a few, also afford protection from UV-B radiation. In addition, physiological and biochemical mechanisms within the leaf span a wide range of processes including induction of specific enzymes and proteins upon irradiation of plant material with supplemental UV-B radiation (18). Dissipation of excess excitation energy may also increase after exposure to UV-B radiation, partly reflecting the activity of the xanthophyll cycle (20).

Although not a perfect screen, the induction of UV-absorbing pigments is a key process that aids, together with other protective mechanisms, in decreasing potential harmful effects of UV-B radiation in plants. The amount, type of compound and its localisation within leaves are characteristics that can be used as indicators of plant tolerance.

References

1. Roy, C., Gies, H. and Elliott, G. (1990) Nature 347, 235.
2. WMO (1991) World Meteorological Organization, Scientific Assessment of Ozone Depletion: 1991. Global Ozone Research and Monitoring Project - Report #25.
3. SORG Stratospheric ozone (1996) United Kingdom Stratospheric Ozone Review Group. Fifth Report. HMSO, London.
4. Shindell, D.T., Rind, D. and Lonergan, P. (1998) Nature 392, 589-592.
5. Austin, J., Butchart, N. and Shine, K.P. (1992) Nature 360, 221-225.
6. Ålenius, C.M., Vogelmann, T.C. and Bornman, J.F. (1995) New Phytol. 131, 297-302.
7. Middleton, E.M. and Teramura, A.H. (1993) Plant Physiol. 103, 741-752.
8. Reuber, S., Bornman, J.F. and Weissenböck, G. (1996) Plant Cell Environ. 19: 593-601.
9. Caldwell, M.M. (1971) In Photophysiology (A.C. Giese, ed.), Vol. 6, pp. 131-177. Academic Press, New York.
10. Cen Y-P., Weissenböck, G. and Bornman, J.F. (1993) In Physical, Biochemical and Physiological Effects of Ultraviolet Radiation on Brassica napus and Phaseolus vulgaris. ISBN 91-628-1051-0.
11 Liu, L., Gitz, D.C. and McClure, J.W. (1995) Physiol. Plant. 93, 725-733.
12. Bornman, J.F., Reuber S., Weissenböck, G. and Cen, Y-P. (1997) In Plants and UV-B: Responses to Environmental Change (P. Lumsden, ed.), pp. 157-168. Cambridge Univ. Press, UK.
13. Olsson, L.C., Veit, M., Weissenböck, G. and Bornman, J.F. (1998) Phytochemistry (In press).
14. Hideg, É. and Vass, I. (1996) Plant Sci. 115, 251.
15. Rice-Evans, C., Miller, N.J. and Paganga, G. (1997) Trends Plant Sci. 2, 152.
16. Fagerberg, W.R. and Bornman, J.F. (1997) Physiol. Plant. 101, 833-844.
17. Allen, J.F., Alexciev, K. and Håkansson, G. (1995) Curr. Biol. 5, 869-872.
18. Bornman, J.F. and Teramura, A.H. (1993). In Environmental UV Photobiology (A.R. Young, L.O. Björn, J. Moan and W. Nultsch, eds), Plenum Publ. Co., New York. pp. 427-471. ISBN 0-306-44443-7.
19. Karabourniotis, G., Kyparissis, A. and Manetas, Y. (1993) Environ. Exp. Bot. 33, 341-345.
20. Bornman, J.F. and Sundby-Emanuelsson, C. (1995) In Environment and Plant Metabolism: Flexibility and Acclimation. Environmental Plant Biology Series (N. Smirnoff, ed.), pp. 245-262. BIOS Sci. Publ., Oxford, UK. ISBN 1-872748-93-7.

FLAVODOXIN AS AN IN SITU MARKER FOR THE DETECTION OF IRON STRESS IN MARINE PHYTOPLANKTON FROM HIGH NUTRIENT LOW CHLOROPHYLL WATERS

[1]La Roche J., [2]McKay R.M., [3]Boyd P. , [1]Brookhaven National Lab, Upton NY 11973, U.S.A; [2]Dept. of Biological Sciences, Bowling Green State Univ., Bowling Green, OH 43403, USA; [3]NIWA, Dept. of Chemistry, Univ. of Otago, Dunedin, NZ

Key words: diatoms, environmental stress, ferredoxin, global climate changes, marine algae

1. Introduction

Although iron is the second most abundant metal in the earth crust, the equilibrium concentration of free ferric ion (FeIII) in an oxygenic, aqueous environment, is approximately 10^{-17} M. The specific chemistry of Fe in the current oxygenic atmosphere therefore makes this element among the least biologically available (1). Yet, the chemical properties of iron together with the broad redox potential observed in Fe-containing enzymes make Fe an invaluable component of diverse, but fundamental metabolic processes such as the electron transport reactions of photosynthesis and respiration.

The pioneering work of Martin and colleagues (2-4) indicated that iron could be a limiting factor constraining phytoplankton productivity and standing crop in vast regions of the oceans, and in particular, those regions characterized as "high nutrient and low chlorophyll" (HNLC). Notably, results from a recent mesoscale iron fertilization effort in the HNLC equatorial Pacific have been compelling in their support for the "iron hypothesis" (5, 6). Iron may also be a limiting factor in some coastal regions, particularly in those regions lacking an extensive continental shelf (7).

Microorganisms confronted with Fe deficiency commonly respond by replacing ferredoxin, a non-heme iron-sulfur protein with flavodoxin, a protein that contains flavin-mononucleotide as a prosthetic group (8). Flavodoxin can replace ferredoxin in many enzymatic reactions including the downstream side of Photosystem I (9), and as an electron donor to both nitrite reductase (10) and glutamate synthase (11). Because of its mode of regulation by Fe, flavodoxin in marine diatoms is strongly induced by iron

3979

deficiency whereas it is usually absent under iron-replete conditions (12, 13). All species of diatoms we examined to date have showed a similar pattern of repression of flavodoxin by high iron and de-repression in low iron culture medium. Here we present additional results demonstrating the inverse pattern of flavodoxin and ferredoxin expression in laboratory cultures. In addition, we describe results showing the *in situ* flavodoxin detection in Northeastern subarctic Pacific and the Subantarctic, two HNLC regions with contrasting physical characteristics (14).

2. Material and methods

The abundances of ferredoxin and flavodoxin were assessed for the marine diatom *Phaeodactylum tricornutum* grown in batch culture with different additions of Fe. *P. tricornutum* was grown in an artificial seawater medium as previously described (12). Cultures were provided continuous illumination with cool-white fluorescent lamps at a photon fluence rate of 200 μmol quanta$\cdot$m$^{-2}\cdot$s^{-1} and maintained at 20°C. Following 48 h of growth, aliquots were removed for analysis of *in vivo* chl fluorescence using fast repetition-rate fluorometry (15). Total protein was extracted and analyzed as described (13).

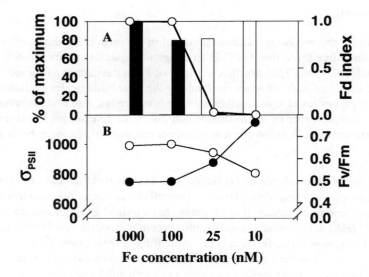

Figure 1. A) Accumulation of flavodoxin (Flv; white bars) and ferredoxin (Fd; black bars) as a function of initial [dFe] in batch cultures of *P. tricornutum*. A ferredoxin (Fd) index (O) was calculated as Fd/(Fd+Flv) using Flv and Fd values expressed as a percent of the maximum values measured over the range of [dFe] additions. B) Measurements of Fv/Fm (O) and σ_{PSII} (●) as a function of [dFe] in the same cultures.

In situ flavodoxin was measured from discrete 50L water samples collected underway. Water samples were filtered and the particulate material concentrated into pellets by centrifugation. Flavodoxin was detected by immunological methods as previously described (16). Dissolved Fe ([dFe]) measurements were made by graphite furnace atomic absorption spectrophotometry (GFAAS) on samples collected using trace metal clean techniques as previously described (17)

3. Results and discussion

We have recently prepared polyclonal antisera directed against diatom ferredoxin. Consistent with prior observations of flavodoxin accumulation (13), there is an inverse relationship between flavodoxin and ferredoxin abundance in batch cultures of the diatom *P. tricornutum*, with only ferredoxin detected above 100 nM Fe, and only flavodoxin detectable below 25 nM Fe (Fig. 1a). The switch from ferredoxin to flavodoxin between 100 and 25 nM is coincident with an initial decline in photochemical quantum efficiency (Fv/Fm) and increase in absorption cross-section (σ_{PSII}) of PSII (Fig. 1b).

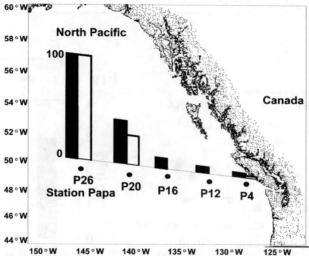

Figure 2. Relative abundance of Flv in bulk protein samples collected along an E-W transect during May (black bars) and Sept (white bars) 95. Amounts of Flv for each date are expressed as a percentage of the maximum value detected at station P26.

The immunological probe raised against flavodoxin purified from *P. tricornutum* (12) has recently been used to assess the iron status of natural populations of diatoms in the Northeastern subarctic Pacific (16). Along an east to west transect, flavodoxin was mainly undetectable in coastal areas and increased westward (16), where it reached

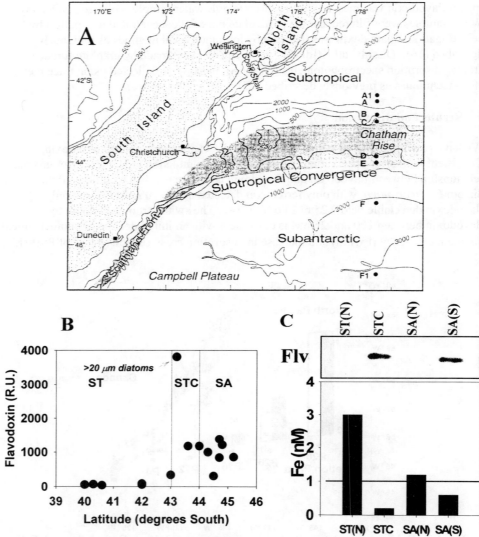

Figure 3. A) Map of the Subantarctic water study area in relation to New Zealand. Subantarctic (SA) and Subtropical (ST) waters are separated by the Subantarctic convergence (STC; shaded area). B) Relative abundance of flavodoxin were collected at stations A1 to F1 along the 178° 30' E meridian in May 1997. Vertical lines indicate arbitrary boundaries between the ST, STC and SA water masses. C) Western blot showing the accumulation of flavodoxin (top panel) at selected stations where [dFe] measurements are also available (lower panel) in the three different water masses. (N) and (S) represent the northern and southern boundaries of water masses, respectively.

maximal levels at the former weather station Papa, the archetypal HNLC locale (2). Field data demonstrated that diatom assemblages from estuarine and coastal regions did not contain flavodoxin *in situ* yet the populations were easily induced to accumulate flavodoxin by increasing the ratio of macronutrients relative to dissolved iron in nutrient enrichment bioassays (12, 16, 17). The patterns of flavodoxin distribution across the oceanic basin (Fig. 2) corroborate previous work supporting that iron is the major factor limiting phytoplankton growth in the Northeastern subarctic Pacific (17-19). Since 1995, additional data along the same E-W transect have demonstrated a consistent pattern of increasing flavodoxin abundance westward, as dissolved iron is increasingly depleted from the surface waters. (R.M. L. McKay, P. Boyd and J. La Roche, unpublished data). In addition, a single-cell immunofluorescence assay was used to confirm that diatoms are targeted (16).

The northeastern subarctic Pacific has a sluggish circulation and well stratified water column allowing only limited transport of nutrients from the deep waters into the surface mixed layer. In contrast to this relatively simple system, the subantarctic (SA) waters northeast of New Zealand are part of a much more turbulent and dynamic system (Fig. 3a). It has been hypothesized that low [dFe] may also limit phytoplankton growth in SA waters. However, the low silicate levels and the relatively deep mixed layers characteristic of summer conditions suggest that the factors controlling the magnitude of algal biomass are probably complex in this region (20). We have measured the relative abundance of flavodoxin on a N-S transect along the 178° 30' E meridian from 42.30°S to 47°S. Flavodoxin levels were near the detection limit throughout the subtropical waters (ST), where [dFe] are relatively high, while significantly higher levels were observed in the SA waters where [dFe] are low (Fig. 3b). Discrete [dFe] measurements taken concurrently with bulk protein samples suggest that the relative abundance of flavodoxin rises sharply below 1nM [dFe] (Fig. 3c).

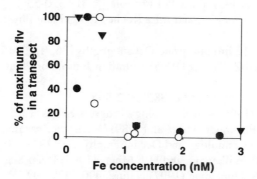

Figure 4. Scatter plot of flavodoxin as a function of [dFe]. Data are from May (●) and Sept (○) 95 cruises to the Northeastern subarctic Pacific and from the transect along the 178° meridian in Subantarctic waters (▲)

Similar results were obtained in the Northeastern subarctic Pacific (Fig. 4). The relationship between [dFe] and flavodoxin presented here is significant in light of the observation that the growth rate of natural phytoplankton populations saturates at 1 nM [dFe] in the HNLC waters of the Equatorial Pacific (6). However, there may be some exception to this tight regulation of flavodoxin and ferredoxin by iron, especially among some large species of diatoms for which iron acquisition may be ultimately limited by diffusion (21). In support of this hypothesis, preliminary data for *Rhizosolenia formosa*, an isolate from the Sargasso Sea, suggests that flavodoxin may be constitutively expressed in this large diatom (R. L. M. McKay and T. A. Villareal, unpublished).

4. References

1   da Silva, F.J.J.R. and R.J.P. Williams, *The biological chemistry of the elements*. 1 ed. 1991, Oxford: Clarendon Press. 561p.
2   Martin, J.H. and S.E. Fitzwater (1988) Nature 331, 341-343
3   Martin, J.H., R.M. Gordon, and S.E. Fitzwater (1990) Nature 345, 156-158
4   Martin, J.H., R.M. Gordon, and S.E. Fitzwater (1991) Limnology and Oceanography 36, 1793-1802
5   Behrenfeld, M., *et al.* (1996) Nature 383, 508-511
6   Coale, K.H., *et al.* (1996a) Nature 379, 621-624
7   Hutchins, D.A. and K.W. Bruland (1998) Eos, Transactions, American Geophysical Union 79, OS21
8   Geider, R.J. and J. La Roche (1994) Photosynthesis Research 39, 275-301
9   Smillie, R.M. (1965) Plant Physiology 40, 1124-1128
10   Zumft, W.G. (1972) Biochimica et Biophysica Acta 276, 363-375
11   Vigara, A.J., C. Gomez-Moreno, and J.M. Vega (1995) Bioelectrochemistry and Bioenergetics 38, 21-24
12   La Roche, J., *et al.* (1995) Journal of Phycology 31, 520-530
13   McKay, R.M.L., R.J. Geider, and J. La Roche (1997) Plant Physiology 114, 615-622
14   Gargett, A.E. (1991) Limnology and Oceanography 36, 1527-1545
15   Falkowski, P.G. and Z. Kolber (1995) Autralian Journal of Plant Physiol. 22, 341-355
16   La Roche, J., *et al.* (1996) Nature 382, 802-805
17   Boyd, P.W., *et al.* (1996) Marine Ecology Progress Series 136, 179-193
18   Martin, J.H., S.E. Fitzwater, and e. al. (1989) Deep-Sea Research 36, 649-680
19   Coale, K.H. (1991) Limnology and Oceanography 36, 1851-1864
20   Dugdale, R.C., F.P. Wilkerson, and H.J. Minas (1995) Deep-Sea Research I 42
21   Sunda, W.G. and S. Huntsman (1997) Nature 390, 389-392

THE ANALYSIS OF THE PHOTOSYNTHETIC APPARATUS OF FRESH-WATER PHYTOPLANKTON: PROBLEMS - PROGRESS - PERSPECTIVES

C. Wilhelm, A. Becker, A. Domin, M. Gilbert, T. Jakob, C. Lohmann,
University of Leipzig, Institute of Botany - Plant Physiology -,
Johannisallee 19-23, D-4103 Leipzig, Germany

Key words: chlorophyll, electron transport, pigments, phytoplankton, primary production, xanthophyll-cycle

1. Introduction

Freshwater habitats have to cope with dynamic changes in nutrient availability, temperature and light. Most freshwater habitats become increasingly eutrophicated with the result of high primary production. High „standing crop" of pigmented phytoplankton cells absorb PAR efficiently in the upper surface layer leading to a steep light gradient characterised by dynamic changes in light quality and quantitiy. In addition, freshwaters are charged with organic carbon from waste waters or from natural sources. In many cases these compounds themselves absorb visible light („gilvin") leading to strong attenuation of light, especially when the waters are shallow and turbid [1]. The light climate together with changes in nutrient availabily have selective influence on the growth rates of the different phytoplankton taxa leading to very fast changes in phytoplankton population structure during the year.

In the last few years many eutrophicated lakes have been redeveloped by a wide variety of techniques to prevent or to reduce blooming of phototrophs. Mass developments of toxic cyanobacteria or algae have to be controlled due to the risk for public health not only in the context of drinking water, but also for recreation activities [2]. Many restoration technologies induce permanent destratification especially in summer, either to prevent anaerobiosis at the sediment/water interface or to reduce light availability for the phototrophs. [3]. The success of these measures has to be verified by permanent surveillance. Therefore, environmental protection agencies ask for new scientific instruments not only for monitoring, but also for predicting water quality. The goal is to develop easy to use procedures to analyse the composition of the phytoplankton community and to measure the actual growth potential. The functional analysis of the photosynthetic apparatus of the phytoplankton can help to develop such tools. The main targets in the photosynthetic apparatus are the pigments and the parameters of the P-I curve.

2. Procedure

We studied lakes with different trophic state in Germany: the reservoirs of Neunzehnhain (oligotrophic) of Dröda (eutrophic) and of Bleiloch (hypertrophic). Neunzehnhain (Saxonia) and Dröda (Saxonia) are dimictic lakes with strong stratification in summer with clear water stages. The reservoir of Bleiloch (Thüringen) is

G. Garab (ed.), Photosynthesis: Mechanisms and Effects, Vol. V, 3985–3990.

characterised by a high charge of gilvin which attenuates penetrating blue light totally at 1m depth. In these lakes we have taken samples at different depths every two weeks by filtering the phytoplankton cells on Whatman-filters GF and frozing them in liquid nitrogen. The samples have been freeze dried and the pigments were extracted and chromatographed according to [4]. One has to emphasize that this procedure does not measure „free" porphyrins dissolved in the water but only membrane-bound pigments. The chromatographic procedure allows to quantify not only all chlorophylls (a,b, c_{1-3}) but also all taxonomically relevant carotenoids as well as the major chlorophyll derivatives as chlorophyllide a and b, pheophorbide a and b and pheophytin a and b. Comparative measurement of pheophytin a with determination of the numbers of PSII reaction centers allowed the assessment of artificially produced pheophytin (see results). Pigment standards of xanthophylls were prepared from unialgal cultures, the standards for chlorophyll-derivatives were prepared by chemical standard procedures. The percentage contribution of the major algal groups were calculated on the basis of xanthophyll/chlorophyll ratios from unialgal cultures or via statistic methods as mentioned in the results [5]. All samples were comparatively analysed by means of the Utermöhl-technique [6], which determines the cell volume in the inverse-microscope and calculates the percentage contribution of the major taxa to total biomass on the basis of cell volumes.

Electron flow measurements in the field were done by a Xenon-PAM [7]. The in-situ light field was measured with an spectral resolution of 1nm. The light saturation curves were obtained from fluorescence quenching analysis according to Genty et al. [8]. A spectrally resolved bio-optical model [9] was applied to assess primary production rates (for details see[10]). To assess physiological response of algae exposed to prolonged dark periods in a dynamic light climate, cells of *Phaeodactylum tricornutum* were grown in growth chambers according to [11]. Fluorescence and pigment analysis was carried as described earlier [12].

3. Results and Discussion

Figure 1 shows that changes in community structure can be documented with high accuracy by means of pigment-analysis giving evidence that its result is in good agreement with cell countings. Cell countings refer the biomass to the biovolume, whereas the pigment analysis uses chlorophyll a as biomass indicator. Because the ratio chlorophyll/biovolume depends on the physiological status of the cell and on its morphology, it can not be expected that both methods yield the same results. Because Chl a is a parameter of routine surveillance, in practice Chl a is a much better biomass indicator than biovolume which is difficult to estimate. The calculation of the percentage contributions of the major taxonomic groups is based on xantho-phyll/chlorophyll a ratios as shown earlier [4]. The ratio of the taxonomically relevant xanthophyll per chlorophyll can be assessed either on laboratory cultures or on statistical correlation analysis in the data matrix of the pigment measurements. This can be done either by multi-variate regression analysis or simply by data fitting.

Measuring the absolute amount of marker xanthophyll, it is possible to calculate the theoretical amount of Chl a (Chl_{calc}) which should be present in the sample based on the Xan/Chl a ratios defined. The ratio of Chl a_{calc} to the amount of Chl a found in the sample (Chl a_{meas}) is a critical parameter which can help to identify artifacts of the method or to collect additional physiological information of the sample analysed. Ideally, this ratio accounts for 1 and plotting Chl_{meas} against Chl_{calc} leads to a linear relationship with a very high correlation coefficient as shown in figure 2. Here Chl_{calc} was determined on the basis of iterative data analysis.

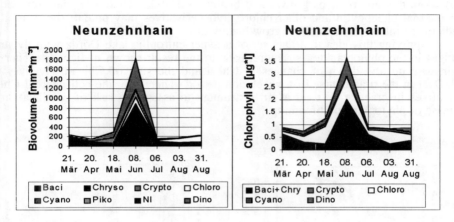

Figure 1:

Changes in the community structure in the barrage of Neunzehnhain at 5 m depth in 1995. Comparison of cell countings (left hand side) with HPLC-Fingerprinting (right hand side). Chlorophyll was calculated on Xan/Chl ratios derived from mulit-variate regression analysis.

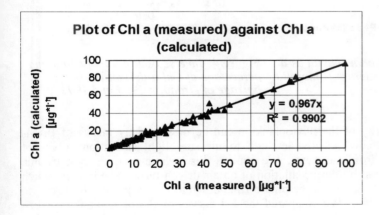

Figure 2.
Correlation analysis of measured and calculated Chl a (see text). The analysis includes all samples taken at different depths and sites in 1997 from the barrage of Bleiloch.

However, deviations from this relationship occur very often and lead us to the analysis of chlorophyll derivates in the samples. The ratio of Chla/PS II is not universal in oxygenic phytoplankter. It was reported to be 550 in green algae, 350 in cyanobacteria [13], and 450 in diatoms and dinoflagellates [unpublished data]]. If the taxonomic composition of a sample is known, one can calculate the minimal content of pheophytin a. Applying our extraction procedure to logarithmically growing cultures of green algae we detect a ratio of Chla/Phaeo about 265 and a Chl/PSII ratio of 530. In Fig. 3 we have plotted the ratio $Pheo_{calc}$ /$Pheo_{meas}$ as a function of time during culture. It is obvious that during the logarithmic growth phase this ratio is constant and rises when the culture stops growth and begins to decline. Therefore, the presence of cell internal pheophytin and other chlorophyll derivatives may potentially be used as predictor of future phytoplankton growth activities.

Applying this idea to a data set of a hyper-eutrophic lake Dörda with dynamic changes in the phytoplankton structure leads to the result shown in figure 4. There is an inverse up and down of changes in Chl a and the sum of Chl a-derivatives (S-ChlaDer) in the time course of a vegetation period. Therefore, the cell-internal Chl a-derivatives in practice be used as an instrument to assess the risk of continuation or increase of phytoplankton blooming.

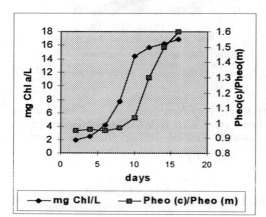 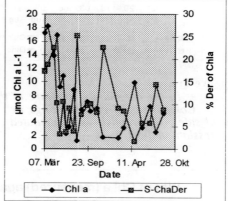

Figure 3: *Cell internal phaeophytin as a function of time during culture (Chlorella fusca)*

Figure 4: *Inverse oscillation of Chl a and percentage of Chl a-derivatives in the barrage of Dröda in 1994 and 1995*

The actual activity of the phytoplankton can be measured by means of PAM-fluorometry, which is now sensitive enough to measure light saturation curves of samples directly taken from the water. We applied this technique in an hypertrophic lake which is redeveloped by the installation of an air-driven turbulence system which drifts the phytoplankton cells from the photic zone down to the dark with the intention to limit primary production. We compared the P-I curves of the phytoplankton taken from deeply mixed parts („Mauer") with that of a well stratified area („Piere") (Fig. 5). Obviously, deep mixing induced in the cells a reduction of photosynthetic efficiency and capacity. If one knows (1) the light quality and quantity from different depth, (2) the absolute pigment concentrations from the phytoplankton community and (3) the photosynthetic quantum yield as a function of light, one can calculate the absolute primary production on the basis of a spectrally resolved model [9]. This has been done by Gilbert et al [10] showing that this procedure yields realistic results.

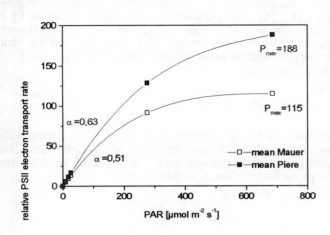

P-I curve measured with the Xenon-PAM according to [] at two different sites in the barrage of Bleiloch. At the site „Mauer" the water column is totally mixed up to a depth of 25m whereas at „Piere" the water column is stratified.

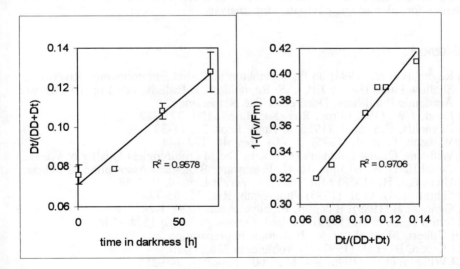

Figure 6:
The de-epoxidation state of Phaeodactylum as a function of prolonged dark periods in the dynamic light regime (left hand side) and the correlation between increased de-epoxidation state and non-photochemical quenching according to Stern-Vollmer.

3990

To understand the reasons why the phytoplankton at the „Mauer"-Site showed lower efficiency, we studied the effect of prolonged dark periods in a dynamic light regime in the laboratory. Surprisingly, cells of *Phaeodactylum* increase after several light/dark cycles the de-epoxidation state of the xanthophyll cycle pigments during the dark with an enhanced diatoxanthin-dependent non-photochemical quenching the longer the cells are kept in the dark (Fig. 6). This paradox response can be explained by a persistant pH-gradient in the dark due to chlororespiration. At the moment we do not know if the dynamic light regime really mimikes the light climate in the reservoir, but the results clearly show that simple measurements of xanthophyll cycle pigments do not allow the characterisation of light stress.

Conclusions

The simultanuous application of HPLC-pigment analysis and PAM fluorometry yields a data set which allows a rather complete characterisation of phytoplankton. The data show that isolated fluorescence and/or pigment data sets can lead to important misinterpretations of the photophysiological fitness of the phototrophs.

Acknowledgements:

The authors acknowledge finanacial support from the German Ministery of Education and Research 02 WU9607/6 and Drs. Horn from the Saxonian Academy of Sciences for microscopic phytoplankton analysis.

References

[1] Reynolds et al. (1994) ub Phytoplankton in Turbid Environments: Rivers and Shallow Lakes (Descy J.P. C.S. Reynolds & J. Padisák, eds.) pp.1-9, Kluwer Academic Publishers, Dordrecht, The Netherlands.
[2] Lund, J.W. (1972) Proc. Roy. Soc. Lond. 180, 371-382
[3] Reynolds, C.S. et al. (1986) J. Appl. Ecol. 21, 11-39
[4] Wilhelm, C. et al. (1995) J. Wat. Res. 44, 132-144
[5] Wilhelm, C. and Lohmann (1997), in The Microbiological Quality of Water (Sutcliffe D.W. ed.) pp. 81-91, Freshwater Biological Association, Cumbria, U.K.
[6] Utermöhl, H. (1958) Mitt. Internat. Verein Limnol. 9, 1-38
[7] Schreiber, U. et al. (1993) Photosynth. Res. 36, 65-72
[8] Genty et al (1989), Biochim. Biophys. Acta. 1020, 1-24
[9] Smith R. C. et al. (1989): Limnol. Oceanogr. 34 (8) 1524-1544
[10] Gilbert, M. et al. Arch. Hydrobiol. in preparation
[11] Kroon, B. et al. (1992) Hydrobiologia 238, 63-70
[12] Wilhelm et al. (1996) Sci. Mar. 60, (Suppl 1), 249-255
[13] Melis, A. (1991) Biochim. Biophys. Acta. 1058, 87-106

DEVELOPMENT OF THE PHOTOSYNTHETIC APPARATUS IN AUSTRALIAN RAINFOREST LEAVES

Sharon A. Robinson and A. Wendy Russell
Plant Molecular Ecophysiology Research Group,
University of Wollongong, Northfields Avenue, NSW 2522, Australia

Keywords: carotenoids, electron transport, fluorescence, leaf development, photoprotection, pigments

Introduction

Although light is often considered a limiting factor in rainforests, one of the major stresses rainforest seedlings may face, particularly in disturbed sites, is overabundance of light leading to photoinhibition (1). In reality, rainforest plants must be able to withstand considerable variation in light from the very low levels in the understorey to the high intensities imposed by sunflecks, gap formation or at canopy level (2).

In leaves of rainforest species the photosynthetic apparatus often develops in an asynchronous fashion, a factor which may increase the susceptibility of young leaves to photoinhibition (3). Leaves of rainforest species thus show variations in cyanic pigmentation and rates of greening, as well as in leaf longevity. In particular, red juvenile leaves have been noted in many shade-tolerant species. Cyanic pigmentation could confer photoprotection via screening of UV radiation or the prevention of photooxidative reactions due to the antioxidant properties of anthocyanins (4,5).

This study examined the relationship between light environment and the pattern of photosynthetic development in a gap-specialist, *Omalanthus populifolius,* and a shade tolerant species, *Syzygium luehmannii.* The photoprotective strategies employed during the developmental process were examined in each species.

Materials and Methods

Plant material: Two Australian rainforest species were chosen for this study. *O. populifolius* is a gap specialist whilst *S. luehmannii* is shade tolerant and produces flushes of pink new leaves. Plants of both species were established in full sun and shade (10% full sun) conditions and leaf growth rate and longevity were determined on tagged leaves.

Light harvesting and electron transport function: Chlorophyll fluorescence techniques were used to study the development of the light reactions of photosynthesis. Modulated chlorophyll fluorescence was measured on individual leaves at various stages of development using a PAM 2000 chlorophyll fluorometer (Walz, Germany). Light response curves of electron transport rate (ETR) and nonphotochemical quenching (NPQ) were determined as described in Barker *et al.* (6). Mean values of leaf reflectance for each leaf developmental category were obtained using an integrating sphere and used in calculations of ETR as described by Barker *et al.* (6). ETR and NPQ measured at a standard photon flux density (PFD) of 500 μmol.m^{-2}.s^{-1} are described here to enable comparison between species and growth conditions.

Pigment analysis: Chlorophyll and carotenoid pigments were extracted in acetone and were quantified by HPLC using the method of Gilmore and Yamamoto (7).

G. Garab (ed.), Photosynthesis: Mechanisms and Effects, Vol. V, 3991–3994.
© *1998 Kluwer Academic Publishers. Printed in the Netherlands.*

Results and discussion

Leaf development in *O. populifolius* and *S. luehmannii*

O. populifolius has fast-growing, short-lived leaves. Leaves grown in full sun develop significantly faster than those in the shade (mean time to full expansion 35.5 d (sun) versus 43.9 d (shade), but leaf lifespan is identical (abscission occurring after 126 d for sun and shade grown plants). In contrast *S. luehmannii* produces long-lived leaves which are considerably smaller. Like the *O. populifolius* leaves, development is faster in the sun (28.3 d) than in the shade (38.9 d). All *S. luehmannii* leaves tagged in 1997 are still present on the plants (Winter 98) so mean leaf longevity is at least one year.

Photosynthetic development in *O. populifolius* and *S. luehmannii*

Distinct differences between the 2 species are apparent in the establishment of photosynthetic capacity during leaf development In *O. populifolius*, ETR developed linearly with increasing leaf area in both the sun and shade grown plants (Fig. 1A).

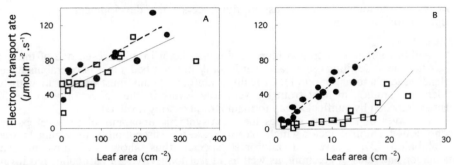

Figure 1. Electron transport rate during development of sun (●) and shade (□) leaves of *O. populifolius* (**A**) and *S. luehmannii* (**B**).

However in *S. luehmannii*, the development of ETR was linearly related to leaf area only in the sun grown plants, whilst the shade plants showed a definite lag in development (Fig. 1B). In *S. luehmannii* shade plants ETR did not fully develop until after full leaf expansion.

Maximum photosynthetic efficiency (measured as dark adapted F_v/F_m) was high throughout development in both sun and shade leaves of *O. populifolius* whereas in *S. luehmannii* it was very low in young leaves from both sun and shade plants and increased slowly during leaf development (data not shown).

Photosynthetic pigments in developing leaves of *O. populifolius*

Chlorophyll concentration was similar in the youngest leaves of sun and shade grown *O. populifolius* and increased steadily with leaf development. Mature shade leaves contained more chlorophyll than sun leaves (Fig. 2A). The chlorophyll a/b ratio was constant (3.0) throughout development in shade leaves but increased from 2.65 to 3.38 in sun leaves.

The concentration of xanthophyll cycle pigments (V+A+Z expressed on a chlorophyll basis) declined with leaf development in shade leaves of *O. populifolius*, but increased in sun leaves (Fig. 3A). Levels of lutein declined slightly with increasing age and ß-carotene levels were similar at all leaf stages (data not shown).

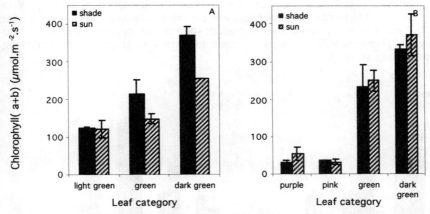

Figure 2. Chlorophyll content during development of sun and shade leaves of *O. populifolius* (**A**) and *S. luehmannii* (**B**).

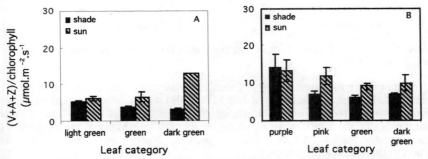

Figure 3. Xanthophyll cycle (V+A+Z) pigment content during development of sun and shade leaves of *O. populifolius* (**A**) and *S. luehmannii* (**B**).

NPQ (measured at 500 μmol.m^{-2}.s^{-1}) was highest in young leaves and declined in both sun and shade leaves with increasing maturity (Fig. 4A). NPQ declined more rapidly in sun leaves than in the shade leaves. Young light green leaves had lower photosynthetic capacity indicating that they dissipate more energy as heat than mature leaves which have greater photochemical capacity.

Photosynthetic pigments in developing leaves of *S. luehmannii*

Chlorophyll concentration was very low in young leaves of both sun and shade grown *S. luehmannii* and did not increase until leaves were fully expanded (Fig. 2B). However mature sun and shade leaves contained similar concentrations of chlorophyll. The chlorophyll a/b ratio increased throughout development in both sun (2.1 up to 3.0) and shade leaves (2.3 up to 2.8).

Xanthophyll cycle pigments were highest in purple leaves of both sun and shade grown leaves of *S. luehmannii* (Fig. 3B). V+A+Z pigments made up 49% and 40% of the total carotenoid pool in purple shade and sun leaves, respectively. The relative concentration of these pigments declined to 21% and 27% of the total carotenoid pool in the dark green leaves. The level of xanthophyll cycle pigments was generally higher in sun than shade leaves. The other carotenoid pigments showed an opposite pattern to the xanthophyll cycle pigments with lutein and ß-carotene increasing slightly with increasing leaf maturity (data not shown).

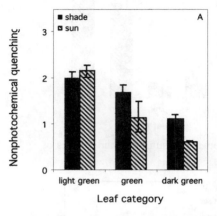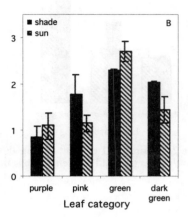

Figure 4. NPQ during development of sun and shade leaves of *O. populifolius* (**A**) and *S. luehmannii* (**B**).

Despite containing the highest levels of xanthophyll cycle pigments per chlorophyll, young purple leaves had the lowest levels of NPQ (Fig. 4B) which then increased as leaves developed from pink to green. This is surprising given that the ETR is also low in these plants and suggests that the xanthophyll cycle pigments play an additional photoprotective role in these younger leaves. Other pigments such as anthocyanins may also offer photoprotection to these developing leaves.

Conclusions

Our results show that there are distinct differences in photosynthetic development between the gap specialist *O. populifolius* and the shade tolerant *S. luehmannii*. Development of leaves in high light appears to follow a similar pattern in both species but *S. luehmannii* shows a more plastic developmental response to variations in light environment and this may reflect its ecological niche as an understorey seedling that develops into a canopy tree

Acknowledgments: This work was supported by the Australian Research Council and the Australian Flora and Fauna Research Centre, University of Wollongong.

References
1. Lovelock, C.E., Jebb, M. and Osmond, C.B. (1994) Oecologia 97, 297-307
2. Watling, J.R., Robinson, S.A., Woodrow, I.E. and Osmond, C.B. (1997) Aust. J.Plant Physiol. 24, 17-25
3. Krause, G., Virgo, A., Winter, K. (1995) Planta 197, 583-591
4. Gould, K.S., Kuhn, D.N., Lee, D.W. and Oberbauer, S.F. (1995) Nature 378, 241-2
5. Woodall, G.S., Dodd, I.C. and Stewart, G.R. (1998) J. Ex. Bot. 49,79-87
6. Barker, D.H., Seaton, G.G.R. and Robinson, S.A. (1997) Plant Cell Environ. 20, 617-624
7. Gilmore, A.M. and Yamamoto, H.Y. (1991) J. Chromat. 543, 137-145

SLOW LEAF DEVELOPMENT IN EVERGREEN TREES

S.-I. Miyazawa and I. Terashima
Department of Biology, Graduate School of Science, Osaka University, 1-16
Machikaneyama-Cho, Toyonaka, Osaka, 560-0043 JAPAN

Key words: full leaf expansion, leaf dry mass per area (LMA), steady-state LMA

1. INTRODUCTION
It is widely recognized that the rate of light-saturated net photosynthesis on leaf area basis ($P_N$max) peaks at or before the completion of leaf area expansion. This is true for most herbs and deciduous trees. However, in other species such as cocoa (*Theobroma cacao* L.;[1]) and many tropical rainforest trees, $P_N$max is very low at their full leaf expansion and attains its maximum more than 10 days afterwards. These species are distinguished as 'delayed greening' species from 'normal greening' species [5].

In delayed greening species from tropical rainforests, valuable resources such as nitrogen are not transported into lamina until the leaf is fully expanded and becomes tough. This would minimize loss of the resources by herbivores. Moreover, the cost for forfeited photosynthesis at full leaf expansion (low chlorophyll and low Rubisco) is low under shady conditions [5]. Therefore, Coley and Kursar [3] speculated that the delayed greening species are adapted to under shady conditions with high herbivory pressure.

One typical characteristic of the delayed greening species is that their young leaves are not green, but red, brown or white. Many Japanese evergreen broad-leaved trees also have non-green young leaves. However, they grow on sunny sites. In this study, we followed changes in photosynthetic characteristics and leaf anatomy during leaf development in these Japanese species to clarify whether leaf development of these species are different from that of the delayed greening species from tropical rainforests [5].

2. MATERIALS AND METHODS
Experiment I: We followed changes in $P_N$max, dark respiration rate (Rd), leaf dry mass per area (LMA) and leaf nitrogen content on area basis (LNa) during leaf development of six typical canopy species of warm-temperate forests in Japan; *Castanopsis sieboldii, Quercus myrsinaefolia, Quercus glauca, Machilus thunbergii, Cinnamomum japonicum* and *Neolitsea sericea* [6]. Mature trees were selected. Development of sun leaves was followed for all the trees except for *N. sericea*. For the *C. sieboldii* tree, we also examined development of shade leaves.

Experiment II: We planted 2-year-old seedlings of *Castanopsis sieboldii* in soil in 5L pot. They were grown in an open site. Changes in the relationship between the rate of photosynthesis and the intercellular partial pressure of CO_2 (A-Ci curves) were followed during leaf development. Leaf sections of about 1 mm x 3 mm were fixed in 2.5% glutaraldehyde. After post-fixation in 2.5% osmium tetraoxide, the sections were

G. Garab (ed.), Photosynthesis: Mechanisms and Effects, Vol. V, 3995–3998.
© 1998 *Kluwer Academic Publishers. Printed in the Netherlands.*

dehydrated in acetone series and embedded in Spurr's resin. Changes in height and width of mesophyll cells and the volume ratio of the intercellular air spaces were examined.

3. RESULTS AND DISCUSSION

Changes in photosynthetic characteristics during leaf development

In *C. sieboldii* and *Q. glauca*, $P_N$max levels at full leaf expansion were far below their maxima and another 43 and 21 days, respectively, were required for the attainment of the maximum levels (Fig. 1B). LNa and LMA changed in a similar manner (Fig. 1D, E). The Rd levels were highest in unfolded leaves and decreased gradually (Fig. 1C). The Rd levels at full leaf expansion were still high and decreased for more than one month to attain the steady-state levels. All these results indicated that the construction of the leaves was still underway at the completion of leaf area expansion in these two species.

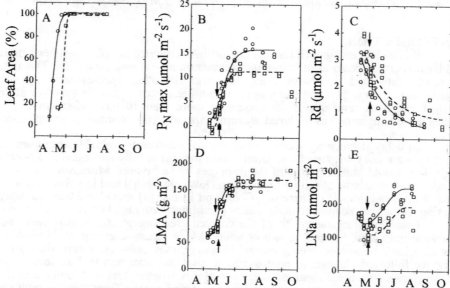

Figure 1. Seasonal changes in leaf area (A), $P_N$max (B), Rd (C), LMA (D) and LNa (E) during leaf development of *Castanopsis sieboldii* and *Quercus glauca*. ○ and solid lines, *C. sieboldii*; □ and dashed lines, *Quercus glauca*. Downward arrows and upward arrows indicate full leaf expansion of *C. sieboldii* and *Q. glauca*, respectively.

Changes in photosynthetic and anatomical characteristics of C. sieboldii during leaf development

In Experiment II with seedlings of *C. sieboldii*, we also observed the increase in $P_N$max well after the full leaf expansion (Fig. 2A). The slope of A-Ci curve also increased after the end of leaf area expansion (Fig. 2D). An extension of mesophyll cells, in particular palisade tissue, occurred after full leaf expansion (Fig. 2B,C). The extension of mesophyll cells after full leaf expansion contributes to an increase of mesophyll surface area per unit leaf area, which increases an area available for CO_2 diffusion. Since large proportion of LMA is composed of cell wall materials; cellulose, hemicellulose and lignin [2], the large increase of LMA after full leaf expansion would be due to the construction of cell walls which was contributed by the extension/expansion of mesophyll cells and cell wall thickening. These results indicate that photosynthetic competence of the chloroplast develops in concert with the construction of cell walls after the completion of leaf area expansion in *C. sieboldii*.

The volume ratio of intercellular air spaces increased sharply and reached almost steady-state level at full leaf expansion (Fig. 2E). Small proportion of intercellular air spaces in young leaves probably interfere with CO_2 gas diffusion.

C_i in young developing leaves was higher than that of mature leaves (Fig. 2F). Low stomatal conductance and high dark respiration rate in young developing leaves would contribute to high C_i.

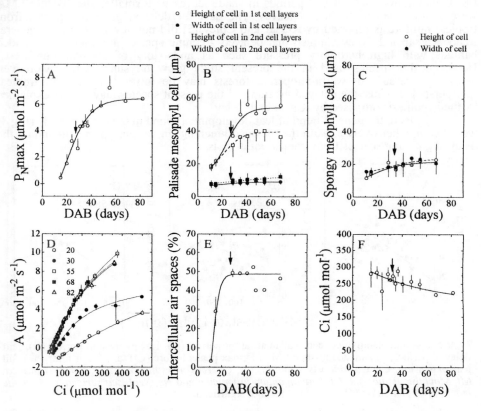

Figure 2. Changes in $P_N$max (A), height and width of mesophyll cells (B, C), intercellular air spaces (E) and C_i (F) during leaf development. C_i was obtained in the same conditions of $P_N$max measurement. A-C_i curves were obtained during leaf development (D). The numbers in figure D indicate DAB (days after bud break). The arrows indicate full leaf expansion.

The cause of the difference in leaf developmental patterns

The steady-state LMA varies greatly among the species. Although the phenomenon is not recognized widely, the data indicating synchronous increases of $P_N$max and LMA during leaf development have been published for many species. If this is generally the case, we expected that the leaves with larger steady-state LMA would require longer periods for photosynthetic maturation (leaf maturation period) because the leaves with larger steady-state LMA require a larger amount of resources for constructing cell walls. We calculated leaf maturation period from the data so far published for various species and plotted the maturation periods of these species as well as the present species against the steady-state LMA. We found a significant correlation between them (Fig. 3A). Moreover, since leaf expansion periods were less variable than leaf maturation periods (Fig 3B), there was also a strong correlation between the delayed period, which is the difference in maturation period and expansion period, and the steady-state LMA (Fig. 3C). In the shade leaves of *C. sieboldii* with small steady-state LMA, maturation and delayed periods in shade leaves with smaller steady-state LMA were shorter than those in sun leaves. The data for delayed greening species from tropical rainforests reported by Kursar and Coley [5] did not deviate from the general trend. Kursar and Coley claimed that delayed greening species are adapted to shaded habitats with high herbivory pressure such as understory of tropical rainforests. However, the species used in the present study grow on sunny sites. Moreover, herbivory pressure in warm-temperate forests may be lower than that in tropical rainfoersts [4]. Thus, the delayed greening of the present species may not be explained by the hypothesis proposed by Kursar and Coley.

    Instead, the general trend of leaf development found in the present study (Fig.3) would comprehensively explain the 'delayed greening' in various species including the delayed greening species from tropical rainforests.

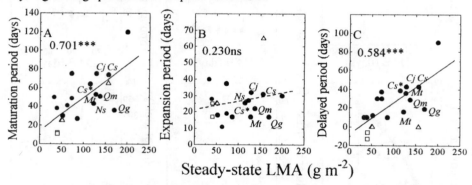

Figure 3. Relationships between leaf maturation period (A), leaf expansion period (B), leaf delayed period (C) and steady-state LMA. $P_N$max peaks before (O), at ($\triangle$) or after (●) full leaf expansion. *Cs, Castanopsis sieboldii; Qs, Quercus myrsinaefolia; Qg, Quercus glauca; Mt, Machilus thunbergii; Cj, Cinnamomum japonicum* and *Ns, Neolitsea sericea. Cs**, shade leaves of *C. sieboldii*. *** ; $P < 0.01$. ns ; not significant

References

1 Baker, N.R. and Hardwick, K. (1973) New Phytologist 72, 1315-1324
2 Chapin, F. S. (1989) American Naturalist 133 (1), 1-19.
3 Coley, P.D. and Kursar, T.A. (1997) *Tropical Forest Plant Ecophysiology*. New York: Chapman and Hall, 305-336.
4 Coley, P.D. and Aide, T.M. (1991) *Plan-animal interactions: evolutionary ecology in tropical and temperate regions*. New York: Wiley & Sons, 25-49.
5 Kursar, T.A and Coley, P.D. (1992) Functional Ecology 6, 411-422.
6 Miyazawa, S.-I., Satomi, S. and Terashima, I. Annals of Botany (in press) .

FAST INDUCTION OF PHOTOSYNTHESIS IN LIGHT DEPENDENT AND SHADE TOLERANT FOREST TREES.

U. Hansen, M. Schulte, Humboldt-Universität zu Berlin, Institut für Biologie, Pflanzenphysiologie, Unter den Linden 6, D-10099 Berlin, Germany

Key words: light acclimation, Chl fluorescence induction, sunfleck, sun and shade leaves

1. Introduction

Natural habitats of forest trees differ in respect to light availability (Tab.1). Pine and birch trees are known to be less shade tolerant than beeches. The aim of the study was to investigate if the lower shade tolerance of birch and pine is coupled to specific characteristics of photosynthesis. The fast induction of photosynthesis in response to lightflecks and the acclimation of the photosynthetic apparatus along the vertical light gradient in the canopy was compared between the species listed in Tab.1.

pine	*Pinus sylvestris* (L.) KARST.	7	high light, exceptionally in shade
birch	*Betula pendula* (ROTH.)	7	high light, exceptionally in shade
oak	*Quercus petraea* (MATT.)LIEBL.	6	between 5 and 7
spruce	*Picea abies* (L.) KARST.	5	half shade, >10% rel. PPFD
beech	*Fagus sylvatica* (L.)	3	shade, <5% rel. PPFD

Table 1. Light indicator values on a scale from 1-9 (1) and light regime of the natural habitat (1) of the investigated species.

2. Material and Methods

Photosynthesis of mature trees of a beech/oak, a pine/birch and a spruce measuring site was investigated from June to August 1997. The sites which are located in Northern Germany, were desribed in detail in (2), the spruce site in (3). At different height positions in the canopy CO_2-gas exchange measurements were carried out using a LI6400 gas exchange system (LI-COR, NE, USA) and samples were taken for HPLC analysis of leaf/needle pigments. With a PAM 103 system (Walz, Effeltrich, D) fluorescence measurements were carried out as reported in (4). Electron transport rates were calculated from the incident photon flux density (PPFD), the measured proportion of absorbed quanta and the apparent quantum yield of PSII determined according to (5). The photon flux density at the sampling positions relative to the PPFD above the canopy was recorded using LI-190S sensors (LI-COR, NE, USA).

G. Garab (ed.), Photosynthesis: Mechanisms and Effects, Vol. V, 3999–4002.
© *1998 Kluwer Academic Publishers. Printed in the Netherlands.*

3. Result and Discussion

3.1 *Fast Induction of Photosynthesis*

Fig. 1 shows the variation in the rate of electron transport (ETR) during the first 3 min of light exposure applied after a dark adaptation of 30 min. Upon onset of irradiation ETR was high and decreased rapidly during the first 10-30 s. The fast decrease in electron transport is in accordance with results gained from O_2 gas exchange measurements (6,7) and model calculations (8). Subsequent to the fast decrease an increase in ETR was found for all species except pine. This induction of ETR can be due to the increasing ATP and NADPH demand of dark reactions which results from the activation of RuBP regenerating calvin cycle enzymes. This enzyme activation is known to limit carbon uptake during the first phase of photosynthetic induction (see below). Other processes can also cause a variation in ETR with time, since the proportion of electrons, ATP and NADPH fed into alternative pathways may vary. The temperature dependence of the ETR induction state reached within 3 min in the light (Fig.2) suggests that enzymatic reactions are involved in ETR induction. At the same temperature oak leaves of the upper sun exposed crown part reached a higher ETR within 3 min than shade leaves. This can be explained by a higher energy demand of dark reactions in sun leaves compared to shade adapted leaves. **In respect to the comparison between tree species differing in shade tolerance the data indicate that the photosynthetic response to a steep increase in PPFD found with pine differed from the response recorded on the other species.**

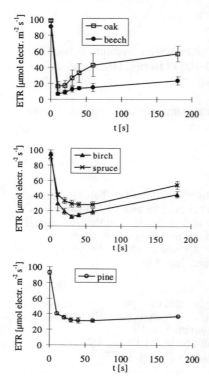

Figure 1. Electron transport rates plotted against the duration of light exposure (white light, 272 µmol photons m^{-2} s^{-1}) applied after a 30 min dark adaptation. Mean values and standard deviation of 3 recordings taken on different days are given, the variation is a result of differing temperature conditions (see Fig.2); ETR was calculated from fluorescence data sampled on leaf/needle material taken in the upper sun exposed crown part.

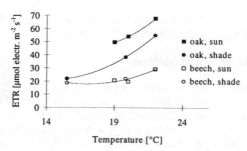

Figure 2. Electron transport rate reached within 3 min in the light (see. Fig.2) as a function of temperature.

In order to further investigate the limitation of photosynthesis during the first minutes of high light exposition the CO_2 gas exchange rate was recorded under saturating light given after a dark adaptation of 5 min (Fig.3). The induction pattern of CO_2 gas exchange shows the typical phases reported in literature for beech leaves (9): an instantaneous CO_2 uptake and a slower increase in CO_2 uptake rate referred to as the fast phase of induction of photosynthesis. During the fast induction the CO_2 uptake rate is limited by the activation state of RuBP regenerating enzymes of the calvin cycle (10). During a slower third phase of induction (up to about 30 min) the CO_2 uptake can be limited by the CO_2 supply and the carbon uptake parallels stomatal conductance. The data shown in Fig. 3 indicate that during the first minutes of irradiation the CO_2 uptake rate was not limited by a low CO_2 concentration in the leaf. Carbon uptake increased at a nearly constant stomatal conductivity (Fig.3b). The CO_2 supply inside the leaf was sufficient as the internal CO_2 concentration did not fall below 230 $\mu mol\ mol^{-1}$ (Fig.3c).

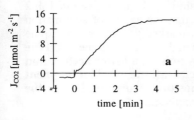

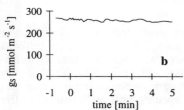

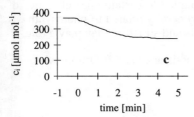

3.2 Photosynthetic acclimation along the vertical light gradient in the canopy

In order to compare the ability of the photosynthetic apparatus to adapt to different light regimes light response curves of CO_2 gas exchange were recorded at different height positions in the canopy. Sun leaves show a higher CO_2 uptake rate at saturating light compared to shade leaves (Fig.4). A similar relationship between J_{CO2m} and the relative PPFD (% of PPFD above canopy) was found for oak, birch and beech. At the same relative PPFD the maximum CO_2 uptake rate of birch leaves was slightly higher than

Figure 3. Time course of CO_2 exchange rate (a), stomatal conductance (b) and internal CO_2 concentration of an attached leaf in the upper sun exposed crown part of a mature beech tree. After 5 min in the dark exposure to saturating red light (1000 $\mu mol\ phot.\ m^{-2}\ s^{-1}$) started at time zero. The leaf temperature was 23°C and the water vapour pressure deficit at the leaf surface 1.1kPa.

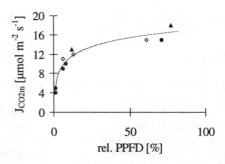

Figure 4. Maximum CO_2 exchange rate as a function of rel. PPFD (PPFD at the height position relative to the PPFD above the canopy) recorded on beech (dots), birch (triangles) and oak (diamonds); measurements were carried out at saturating light, ambient CO_2 concentration, optimum leaf temperature and optimum relative humidity of the air.

the rates detected on oak and beech. Anyway, **the J_{CO2m} acclimation pattern of birch did not differ markedly from the pattern found with the more shade tolerant tree species**. Acclimation of the photosynthetic apparatus to the prevailing light regime occurs also at the level of leaf pigmentation. Tab.2 gives the chlorophyll a/b ratio at height positions receiving 20% and 70% of the PPFD above the canopy respectively. In coincidence with studies reported in literature the chl a/b ratio was generally higher at crown positions with high light availability. The data given in Tab.2 show that the extent of acclimation was markedly lower in pine compared to the other species. **Pine seems to have a lower acclimation capacity at the level of pigment composition compared to the other species.**

	chl a/b 20% rel. PPFD	chl a/b 70% rel. PPFD
pine	3.22	3.27
birch	3.07	3.32
oak	3.27	3.76
spruce	3.10	3.36
beech	3.18	3.44

Table 2. Ratio of chlorophyll a to chlorophyll b at 20% and 70% rel. PPFD.

3.3 *Discussion*

Both birch and pine trees are known to exhibit a low shade tolerance. Whereas the photosynthetic response of pine needles to a sudden increase in light intensity differed from that found with the other species, the response of birch was similar to that of spruce, beech and oak. Pine differed from the other species also with respect to the ability to adapt the pigment composition to the prevailing light regime, whereas birch showed an acclimation pattern similar to that of the shade tolerant species, both with respect to the acclimation of maximum CO_2 uptake rate and leaf pigmentation. The data suggest that, in contrast to birch, the low shade tolerance of pine is coupled to certain characteristics of photosynthesis. The light dependency is therefore not generally related to photosynthesis in any case, other factors like carbon allocation, nutrients and structure may play a role.

Acknowledgements: We thank A. Fischer for technical assistance and B.Grimm who made available the LI6400.

References

(1) Ellenberg, H. (1986) Vegetation Mitteleuropas mit den Alpen, Ulmer Verlag,Stuttgart, D
(2) Rode, M. W. (1995), Plant and Soil 168-169:337-343.
(3) Ellenberg, H., Meyer, R., Schauermann, J. ed. (1986) Ergebnisse des Sollingprojektes, Ulmer Verlag, D
(4) Hansen, U. and Höflich, V. (1995) in Photosynthesis from Light to Biosphere (Mathis P., ed.) Vol V, 853-856, Kluwer Academic Publ., Dordrecht, NL
(5) Genty, B., Briantais, J.-M., Baker, N.R. (1989) BBA990,87-92
(6) Kirschbaum, M.U.F. and Pearcy, R.W. (1988) Planta 174,527-533
(7) Krall, J.P. and Pearcy, R.W. (1993) Plant Physiol. 103,823-828
(8) Kirschbaum, M.U.F. et al. (1998) Planta 204,16-26
(9) Küppers, M. and Schneider, H. (1993) Trees 7, 160-168
(10) Sassenrath-Cole, G. F. and Pearcy, R.W. (1994) Plant Physiol 105,1115-1123

IS THERE SHADE ACCLIMATION OF PHOTOSYNTHESIS IN A SCOTS PINE CANOPY?

Sari Palmroth, Department of Forest Ecology, P.O.Box 24, FI-00014 University of Helsinki, Finland

Key words: gas exchange, light gradient, sun and shade leaves, respiration

1. Introduction

The efficiency of the energy conversion, from photosynthetically active radiation (PAR) into biomass, depends on the balance between the availability of different resources (light, water, nutrients) at the individual leaf level. When the stand level production is estimated, the differences in the efficiency in utilising radiation energy within the canopy should be known. As a response to the variation in the radiation regime, both physiological changes, e.g. changes in the light saturated photosynthetic rates (1), as well as morphological adjustments, like differences in specific leaf area (SLA) (2, 3), have been found in several species. It is, therefore, important to specify the basis on which physiological processes are reported and compared (4, 5).

Scots pine (*Pinus sylvestris* L.) is considered as a light demanding tree species. The canopies are characterised by narrow needles and deep crowns. This kind of canopy structure is giving rise to penumbras and also allows substantial penetration of PAR into the lower levels of the canopy (6). This paper focuses on the characteristics of the photosynthetic light response curve of Scots pine shoots from different crown positions. The data presented here is part of a study aiming to develop methodology for characterisation of the radiation regime and for calculating canopy photosynthesis in coniferous canopies.

2. Materials and methods

The light response curve measurements were performed at Hyytiälä forest field station in central Finland (61° 51'N, 24° 17'E). Age, LAI (all-sided), mean height and stocking density of the experimental stand were 34 years, 9 m^2 m^{-2}, 13 m and 1500 trees ha^{-1}. Sampled trees were divided into three size classes and the crowns into three zones. Measurements were performed three times during the summer 1997 (in June, July and August). Each time 25 - 30 shoots from different size classes and crown zones were measured.

The gas exchange measurements were performed in the lab with infra-red gas analysers (URAS 3G, Hartmann & Braun AG). A daylight lamp (Metallogen HMI, 1200 W/GS, OSRAM) was used as a light source and irradiance was measured with a PAR-sensor

G. Garab (ed.), Photosynthesis: Mechanisms and Effects, Vol. V, 4003–4006.
© 1998 Kluwer Academic Publishers. Printed in the Netherlands.

(Delta-T Devices). CO_2 and H_2O exchange were measured at 11 PAR levels from 2000 to 0 μmol m^{-2} s^{-1}. Temperature was 19 ± 0.5 °C and water vapour deficit between 11 and 15 mmol mol^{-1} during the measurements. Photosynthetic rate, transpiration and dark respiration rates and stomatal conductance to water vapour were calculated. Needle dry weight and all-sided needle area of the shoots were measured. PAR was measured in the sample plot at different heights (10.7 - 8.3 m) with 168 sensors (EG&G Vactec Optoelectronics) and also above the canopy (15 m) with a quantumsensor (Li-190 SZ, Li-Cor Inc.).

3. Results and discussion

The average PAR values at the highest (10.7 m) and lowest (8.3m) levels were 90% and 20% of the maximum value above the canopy. These averages were obtained using data from two measurement periods (11.00-11.30 and 14.00-14.30 local summer time) on a clear day, 16.7.1998. Hence, they do not represent all sun angles. Distributions of PAR at two levels within the canopy (Figures 1a, b) are typical for a Scots pine stand, where the shape of the distributions is not clearly bimodal (observation points in either total shade or total sun) due to penumbra (6).

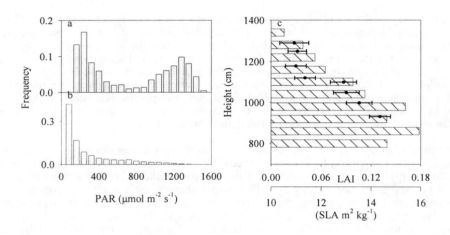

Figure 1 Distributions of PAR at 10.7m (a) and 8.3m (b) on 16.7.1997; 11.00-1130 and 14.00-14.30 local summer time. Leaf area index (LAI, bars, in relative units) and specific leaf area (SLA, m^2 kg^{-1}) ± standard error of the mean in the experimental stand (c).

An increasing trend was found in SLA as a function of depth in the canopy (Figure 1c). SLA varied between 10.4 and 14.2 m^2 kg^{-1}. However, there was considerable between tree variation, which could be genetic or due to differences in the local radiation regime or availability of other resources. Suppressed trees, in general, had higher SLA. These results are in agreement with the previous findings in morphological differences within a Scots pine canopy.

During the first measurement period (early June) photosynthesis seemed not to have completely recovered from the winter depression. The dark respiration rate was considerably higher and photosynthetic rate at saturating light lower than in June and August. Despite these level differences, similar features were distinguishable in all the datasets. Irrespective of the decrease in SLA, no clear trends in the photosynthetic rate at saturating light or in the initial slope of the response curve (on a mass or an area basis) with depth were found (Figure 2a). The whorl level averages (over all the measurements) in the photosynthetic rate at saturating light varied no more than 20%. The dark respiration rate (on a mass and an area basis) and the light compensation point increased with increasing average PAR (Figures 2b, c). No clear trend in the stomatal conductance to water vapour (PAR = 1600 μmol m^{-2} s^{-1}) was observed (July data in Figure 2d).

Although the differences in the photosynthetic rates were small, there were distinct changes in the dark respiration rates suggesting that acclimation occurs. Scots pine seems to utilise lower values of PAR more efficiently in photosynthesis (decreased respiration and light compensation point) in the lower parts of the crown.

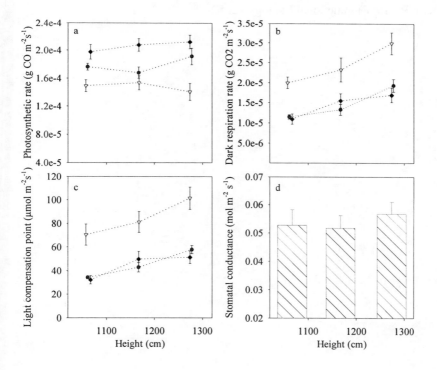

Figure 2. Photosynthetic rate (PAR= 1600 μmol m^{-2} s^{-1}) (a) and dark respiration rate (b) (in g CO_2 m^{-2}s^{-1}), light compensation point (μmol m^{-2} s^{-1}) (c) and stomatal conductance (only July data, mol m^{-2} s^{-1}) (d) against mean heights of the crown zones. June (triangles), July (diamonds) and August (circles).

Acknowledgements

This study was financed by APFE (reasearch group on Atmospheric Physics and Forest Ecology, University of Helsinki)

References

1 Björkman, O. (1981) in Physiological Plant Ecology, Vol. 1, Encyclopedia of Plant Physiology 12A (Lange, O.L. Nobel, P.S. Osmond C.B. and Zeigler H., eds) pp. 57-107. Springer-Verlag, Berlin.
2 Kull, O. and Koppel, A. (1987) Scand. J. of For. Res. 2:157-166.
3 Leverenz, J. W. (1996)Tree Physiology 16, 109-114.
4 Stenberg, P., DeLucia, E.H., Schoettle, A.W. and Smolander, H. (1995) in Resource Physiology of Conifers (Smith, W.K. and Hinkley T.M., eds) Physiological Ecology Series. pp. 3-38. Academic Press, London.
5 Niinemets, Ü. And Tenhunen, J.D. (1997) Plant, Cell and Enviroment 20, 845-866.
6 Stenberg, P. (1998) Functional Ecology 12, 82-91.

LIGHT AND STRESS TOLERANCE IN HIGH MOUNTAIN PLANTS IS ACHIEVED BY MULTIPLE STRATEGIES

Peter Streb, William Shang & Jürgen Feierabend
Botanisches Institut, J.W. Goethe Universität, Postfach 111932, D-60054 Frankfurt, Germany

Introduction

The growing season at high altitude in the Alps is very short and accompanied by adverse environmental conditions, such as high light, low and high temperature, high wind speed and low CO_2-concentration (1). Whereas these conditions are very unfavourable for photosynthesis, alpine plants are well adapted to low temperature and high light intensity (2). Early and wide-spread symptoms of photodamage are the photoinactivation of the photosystem II (PSII) reaction centre protein D1 and of the enzyme catalase. Both proteins usually have a rapid turnover in light and require permanent de novo synthesis (3, 4). The two alpine plants Soldanella alpina and Ranunculus glacialis were found to be very resistant to photoinactivation of PSII and catalase, even when protein synthesis was suppressed by inhibitors or low temperature (1), indicating that the turnover of the D1 protein and of catalase was slow in these plants. Since PSII of isolated thylakoid membranes and isolated catalase obtained from these alpine plants were as sensitive to photoinactivation as from lowland plants, very efficient mechanisms of photoprotection exist in these plants in vivo (1). The strategies of protection appeared to be, however, very divergent in the alpine plants S. alpina and R. glacialis, as indicated by striking differences in their contents of both enzymic and non-enzymic antioxidants and of xanthophyll cycle pigments (1, 5). The relevance of different protective strategies for light stress tolerance was now further evaluated in leaves of S. alpina and R. glacialis.

Material and Methods

Leaves of S. alpina and R. glacialis were collected during the months June and July at sites between 2400 - 2700m altitude in the western parts of the French Alps and investigated in the laboratory at the Col du Lautaret as described (1). The non-alpine plants Taraxacum officinale and Ranunculus acris were collected from unshaded sites of the Botanical Garden of the university of Frankfurt. For experimental treatments leaves were exposed to white light of 500 to 2000µmol $m^{-2}s^{-1}$ PAR or to sunlight of an average light intensity of 1800µmol $m^{-2}s^{-1}$ PAR. Incubations were either performed in an ice bath or at ambient temperature between 18 to 25°C. For experiments in the presence of phosphinothricin (PPT) and dithiothreitol (DTT) leaves were preincubated for 2h at weak room light or for 14h in darkness, respectively. Extraction procedures and analytical methods were described previously (1). Radioactive labeling, pulse chase experiments and gel electrophoresis were performed as described (6). Chlorophyll fluorescence was analysed with a portable photosynthesis yield analyser Mini-Pam (H. Walz GmbH, Effeltrich).

Results and Discussion

The synthesis and turnover of the D1 protein of PSII was compared by 2h labeling with L-[^{35}S]-methionine and subsequent chase experiments in segments from leaves of the alpine species R. glacialis and S. alpina and of the lowland species T. officinale. In leaves of T. officinale the D1 protein was more rapidly labeled than in the alpine plants. During chase experiments in light in the presence of unlabeled methionine the radioactivity incorporated into the D1 protein was rapidly lost in leaves of T. officinale but largely retained in leaves of the alpine plants. After a 3h chase period radioactivity in the D1 protein of T. officinale had declined to 14% of its maximum but was still very high in leaves of S. alpina and R. glacialis (Fig. 1). The results demonstrate that the turnover of the D1 protein in light was much slower in leaves of the alpine plants than usually observed in lowland plants (3, 6), such as T. officinale. This confirms previous observations (1) that in leaves of the alpine plants S. alpina

4007

G. Garab (ed.), Photosynthesis: Mechanisms and Effects, Vol. V, 4007–4010.
© 1998 Kluwer Academic Publishers. Printed in the Netherlands.

and R. glacialis PSII was only to a minor extent inactivated in light, while rapid inactivation occurred under these conditions in lowland plants. Similarly, catalase which is usually also distinguished by a high turnover in light, appeared to be much more stable in leaves of alpine plants (1). The general high resistance of alpine plants to photoinhibitory damage raised the question of which protective mechanisms might be involved. Since repair processes appeared to be only of minor significance, effective scavenger systems for the removal of reactive oxygen species must be present or the production of the latter must be largely avoided.

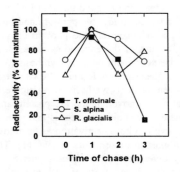

Fig. 1 Change of the radioactivity incorporated into the D1 protein of PSII during a 3h chase period in the absence of radioactivity, following 2h labelling with L-[^{35}S] methionine. Labeling and chase experiments were performed with leaf sections of the alpine plants R. glacialis and S. alpina and of the lowland plant T. officinale in white light of 1000 μmol m^{-2} s^{-1} PAR. Quantitative estimations were performed by scanning fluorographs obtained from electrophoretic separations of total cellular membranes (6). Per unit chlorophyll the initial radioactivity incorporated into the D1 protein of the alpine plants was for R. glacialis only 10% and for S. alpina only 18% of that determined for T. officinale.

In order to assess the role of antioxidative protection, leaves of S. alpina and R. glacialis were exposed to different paraquat concentrations in light and compared with two lowland plants, T. officinale and R. acris. Photoinactivation of catalase and the bleaching of chlorophyll were assayed as symptoms of photodamage. Paraquat is known to enhance the production of superoxide and H_2O_2 in light and thus to increase oxidative stress (9). At 10μM paraquat catalase activity was totally lost and chlorophyll greatly degraded in the lowland plants R. acris and T. officinale., while leaves of S. alpina were much more resistant to paraquat-induced photodamage (Table 1). Even at 100μM paraquat 54% of the chlorophyll content were retained in S. alpina leaves. Leaves of R. glacialis were, however, as sensitive to paraquat-induced photodamage as the lowland plants (Table 1). These observations suggest that the high capacities of antioxidative scavenger systems that were found in S. alpina (1) may essentially contribute to the efficient photoprotection in leaves of this plant. However, in leaves of R. glacialis the production of reactive oxygen species must be avoided, because the antioxidative scavenger system would be insufficient.

The formation of the xanthophyll-cycle pigment zeaxanthin has been found to correlate with the appearance of non-photochemical fluorescence quenching and is believed to reduce the utilization of light energy by photosynthesis in favour of a harmless dissipation as heat (10). In order to investigate the potential role of the xanthophyll cycle for photoprotection in the alpine

Table 1
Catalase activity and chlorophyll content (as % of control without paraquat) in leaves of the alpine plants S. alpina and R. glacialis and the lowland plants T. officinale and R. acris after 24h exposure to light of 1000μmol m^{-2}s^{-1} PAR at 25°C in the presence of different paraquat concentrations.

	S. alpina		R. glacialis	T. officinale		R. acris
Paraquat:	10μM	100μM	10μM	10μM	100μM	10μM
Catalase (%)	49.8±77	13.8±3.4	9.8±4.0	0±0	0±0	0±0
Chlorophyll (%)	91.7±4.8	53.6±11.9	23.4±10.4	25.2±7.9	0±0	57.1±1.0

plants, leaves were incubated in the presence of DTT, that was shown to block zeaxanthin formation (11). The contents of xanthophyll cycle pigments were much higher in S. alpina than in R. glacialis (1). During 14h dark-incubation, both in the presence and absence of DTT, zeaxanthin was totally depleted in leaves of R. glacialis, while a considerable amount (13% of total xanthophyll cycle carotenoids) was retained in leaves of S. alpina. During light exposures the contents of zeaxanthin increased rapidly in both plants and accounted for about 30% of total xanthophyll cycle carotenoids in R. glacialis and 50% in S. alpina after 3h. DTT prevented the light-induced formation of zeaxanthin in leaves of both plants, but in leaves of S. alpina some zeaxanthin was always present (data not shown). When leaves of both plants were irradiated for 3h in sunlight in the presence of DTT at either ambient temperature or on ice, the chlorophyll fluorescence ratio Fv/Fm, which was determined as an indicator of photoinhibition of PSII, decreased in both plants, relative to controls without DTT. The decline of Fv/Fm was, however, much stronger in S. alpina, accounting for about 50% at ambient temperature and 40% on ice. In leaves of R. glacialis the decline of Fv/Fm was only about 10 % at ambient temperature and 20% on ice (Table 2). Whereas zeaxanthin was still present in leaves of S. alpina, the suppression of additional zeaxanthin formation by DTT greatly enhanced photoinhibition of PSII in S. alpina, but hardly affected R. glacialis. Xanthophyll cycle-related protection from photoinhibition of PSII appeared to be considerably more important for S. alpina than for R. glacialis.

Photorespiration is thought to protect the photosynthetic electron transport chain from overreduction and thus to prevent acceptor side-induced photoinhibition (12, 13). The operation of the photorespiratory cycle depends on the recycling of ammonia by glutamine synthetase (13). The herbicide PPT specifically inhibits glutamine synthetase (14) and was used as a tool to assess the role of photorespiration in the alpine plants. When leaves of S. alpina and R. glacialis were incubated for 2h in the presence of PPT in weak light, the activity of

Table 2
Changes of the Fv/Fm ratio (as % of control without inhibitor) in leaves of S. alpina and R. glacialis during a 3h exposure to sunlight at ambient temperature or on ice in the presence of either 30 mM dithiothreitol (DTT) or 1 mM phosphinothricin (PPT)

	Fv/Fm (%)			
	30 mM DTT	30 mM DTT + Ice	1 mM PPT	1 mM PPT + Ice
S. alpina	53.7±7.4	61.3±9.1	63.0±7.4	75.0±11.4
R. glacialis	90.0±2.6	80.0±4.3	55.1±3.9	58.5±8.6

glutamine synthetase decreased by 90%, whereas the maximum photosynthetic oxygen evolution at saturating CO_2 was only slightly affected. This demonstrates that PPT did not exert major direct effects on photosynthesis under non-photorespiratory conditions (data not shown). During exposure to sunlight at ambient temperature in the presence of PPT the Fv/Fm ratio decreased more strongly in leaves of R. glacialis (45%), than in leaves of S. alpina (37%), relative to the controls without inhibitor. The difference between the two alpine species in their sensitivity to PPT in light was, however, much higher at low temperature. At low temperature the decline of Fv/Fm ratio was about 41% in leaves of R. glacialis but only 25% in S. alpina. These results suggest that photorespiration provided a major protection from photoinhibition in leaves of R. glacialis but was less important in S. alpina. While photorespiration was usually not expected to play a role at low temperature (15), our present results suggest that it may serve as a most essential photoprotective mechanism for R. glacialis, even at low temperature.

Conclusions

The slow turnover of the D1-protein of PSII in S. alpina and R. glacialis demonstrates that these alpine plants were less dependent on de novo protein synthesis than lowland plants. Since protein synthesis might be severely restricted by the prevailing low temperature conditions, alpine plants have apparently developed efficient mechanisms for the avoidance of, or protection against, photooxidative damage. Protective mechanisms appeared to be, however, greatly divergent in the two species S. alpina and R. glacialis. In S. alpina antioxidative scavenger systems and zeaxanthin-related energy dissipation appeared to be most important, but to play only minor roles for R. glacialis. In R. glacialis acclimation to high light and environmental stress appeared to strongly depend on photorespiration and carbon metabolism.

Acknowledgements

We thank the Deutsche Forschungsgemeinschaft for financial support. We are very grateful to Dr. Bligny for the great hospitality and support at the Station Alpine du Lautaret, Université Joseph Fourier. Phosphinothricin was kindly provided by the AgrEvo GmbH, Frankfurt am Main.

References:

1. Streb P., Feierabend J., Bligny R. (1997) Plant Cell Environ, 20, 1030-1040
2. Körner Ch., Larcher W. (1988) In Plants and Temperature. (Long S.P.Woodward F.J. eds.) Society of Experimental Biology, Cambridge, 42, 25-57
3. Hertwig B., Streb P., Feierabend J. (1992) Plant Physiol, 100, 1547-1553
4. Aro E.M., Virgin I., Andersson B. (1993) Biochim Biophys Acta, 1143, 113-134
5. Wildi B., Lütz C. (1996) Plant Cell Environ, 19, 138-146
6. Shang W., Feierabend J. (1998) FEBS Lett, 425, 97-100
7. Streb P., Michael-Knauf A., Feierabend J. (1993) Physiol Plant, 88, 590-598
8. Foyer C.H., Lelandais M., Kunert K.J. (1994) Physiol Plant, 92, 696-717
9. Dodge A.D. (1994) In: Causes of Photooxidative Stress and Amelioration of Defence Systems in Plants (Foyer C.H., Mullineaux P.M. eds.) CRC Press, Boca Raton, , pp 220-236
10. Demmig-Adams B., Adams III W.W. (1996) Trends Plant Sci, 1, 21-26
11. Demmig-Adams B., Adams III W.W., Heber U., Neimanis S., Winter K., Krüger A., Czygan F.C., Bilger W., Björkman O. (1990) Plant Physiol, 92, 293-301
12. Heber U., Bligny R., Streb P., Douce R. (1996) Bot Acta, 109, 307-315
13. Kozaki A., Takeba G. (1996) Nature, 384, 557-560
14. Wendler C., Barniske M., Wild A. (1990) Photosynth Res, 24, 55-61
15. Hurry V., Keerberg O., Pärnik T., Öquist G., Gardeström P. (1996) Plant Physiol, 111, 713-719

EVALUATION OF PHOTOINHIBITION ON SINGLE-LEAF AND WHOLE-PLANT PHOTOSYNTHESIS OF MEDITERRANEAN MACCHIA SPECIES

C. Werner[1], O. Correia[2], R.J. Ryel[3] & W. Beyschlag[1]
[1]Lehrstuhl f. exp. Ökologie u. Ökosystembiologie, Univ.-Bielefeld, W4-114, 33615 Bielefeld, Germany; [2]DBV FCUL, Univ. Lisboa, Portugal; [3]Dep. Rangeland Res., Utah State Univ., Logan, UT, 84322-5230 USA

Key words: canopy, fluorescence, light stress, modelling, non-photochemical quenching, leaf orientation

1. Introduction

Mediterranean macchia species have developed several protective mechanisms against high radiation levels during summer. These involve changes in plant structure and function that prevent irreversible damage of the photosynthetic apparatus, such as a reduction of light interception (e.g. adjustment of leaf angle) or controlled dissipation of excess excitation energy within the photochemical system.

While it is well recognised that the processes of non-radiative dissipation of excess energy are important mechanisms of photoprotection, little is known about its costs for whole-plant carbon gain (1). Canopy photosynthesis models are an increasingly important means to evaluate whole plant primary production, but few attempts have been made to incorporate photoinhibitory effects into these models.

The objectives of this work were: (i) to evaluate the species specific structural and physiological adaptations to the high light environment under natural conditions; (ii) to integrate photoinhibitory effects into a whole-plant photosynthesis model through a quantitative relationship between the extent of photoinhibition and the light micro-environment (iii) to evaluate the effect of photoinhibition on whole-plant carbon gain during summer.

2. Material and methods

2.1 Study site
The study was performed in the *Parque Natural da Serra da Arrábida*, in southwest Portugal. The study site is a ca. 30 year old macchia stand, situated on a southeast facing slope at the coast (38° 27' 34" N, 9° 0' 20" W, elevation 270 m).

2.3 Chlorophyll a fluorescence
In *situ* chlorophyll fluorescence was measured with a portable, pulse-modulated fluorometer using the leaf-clip holder 2030-B (PAM-2000, Walz, Effeltrich, Germany).

2.4 Canopy Structure
Leaf area index (LAI) was measured with a LAI-2000 Plant Canopy Analyzer (PCA, LiCor, Inc. Lincoln, Nebraska, USA) and verified by stratified clipping. Leaf angles were measured using a compass-protractor.

2.5 Model simulations
Model simulations were performed using the three-dimensional canopy model of RYEL et al. (2), with a mechanistic photosynthesis submodel, based on equations of FARQUHAR et al. (3).

G. Garab (ed.), Photosynthesis: Mechanisms and Effects, Vol. V, 4011–4014.

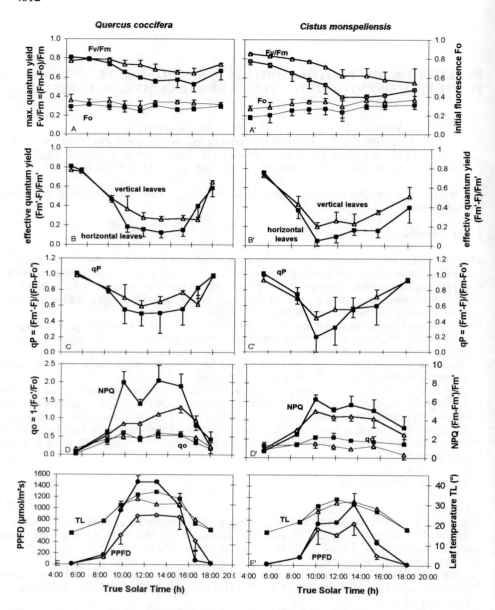

Figure 1. Diurnal course of dark (A, A'), and light adapted (B,B') quantum-use efficiency of PSII, photochemical (C,C') and non-photochemical (D,D') quenching coefficients, incident PPFD and leaf temperature (TL) (E,E') of vertically (open symbols) and horizontally (closed symbols) orientated leaves of *Q. coccifera* (A-E) and *C. monspeliensis* (A'-E') during August. Leaves were dark-adapted for 15 min. prior to Fv/Fm measurements (n=6-12,+SD).

3. Results and Discussion

Plant performance under high radiation environments was studied during a two-year period on eleven Mediterranean macchia species. Data from an evergreen sclerophyllous species (*Quercus coccifera*) and a semi-deciduous species (*Cistus monspeliensis*), which represent the opposing ends of the spectrum of potential responses observed in this ecosystem, are discussed. A diurnal course of fluorescence parameters is shown for August (Fig.1), when the combined environmental stresses lead to the strongest decrease of maximum photochemical efficiency (Fv/Fm). Measurements were recorded separately in vertically and horizontally orientated sun leaves, which differed in the amount of intercepted light and leaf temperature during the midday hours, when radiation load was maximal (Fig.1E,E'). Over the course of the day, the increase and subsequent decrease in PPFD was mirrored by a pronounced decrease and subsequent increase of maximum (Fv/Fm, Fig.1A,A') as well as effective quantum yield of PSII (ΔF/F, Fig.1B,B'). In both species a reduction in quantum yield was less accentuated in vertical leaves which maintained higher photochemical quenching rates throughout the day (Fig.1C,C'), whereas non-photochemical quenching was enhanced in horizontal leaves (Fig.1D,D'). Initial fluorescence (Fo) always remained higher in steeply oriented leaves (Fig.1A,A').

In *Q. coccifera* no significant variation in Fo was observed during the day (Fig.1A). In general this species showed a high capacity of non-radiative dissipation of the excess radiation energy, preventing irreversible photodamage to the leaves. A pronounced initial fluorescence quenching (qo, Fig.1D) suggests that nonradiative energy dissipation in the antennae complex dominated (4), e.g. xanthophyll cycle-dependent energy dissipation, which might contribute to a faster recovery in the afternoon.

The semi-deciduous *C. monspeliensis* was more susceptible to photoinhibition and recovered only slowly in the late afternoon (one vertical leaf did not recover, Fig. 1A'). Non-photochemical quenching processes might have not been sufficient in dissipating all excess energy, leading to a rise in Fo during the day.

Table 1. - Canopy structure of *Quercus coccifera* and *Cistus monspeliensis* during summer: crown height and diameter, leaf area index (LAI) and leaf angles (° to horizon).

Species	height (m)	diameter (m)	LAI	leaf angle
Q. coccifera	1.70	2.50	4.35	53.1 ± 20
C. monspeliensis	1.12	1.10	0.84	71.7 ± 12 (n=497)

Semi-deciduous *C. monspeliensis* showed pronounced structural adaptations to summer drought, e.g. a reduction in leaf area and changes from a preferential horizontal leaf orientation during the favourable seasons (spring and autumn, data not shown) to a predominant vertical leaf orientation in summer (Tab.1). *Q. coccifera* did not exhibit remarkable structural changes in during summer.

To evaluate photoinhibitory effects on whole-plant photosynthesis, model simulations were performed. Photoinhibition (Fv/Fm) was linked to the model via its effect on light-use-efficiency (initial slope of the light response curve) (5). Furthermore, diurnal changes of maximum quantum yield of PSII were strongly related to the combined effect of incident light intensity on the leaf surface and the duration of exposure (see also 6). Diurnal changes in Fv/Fm were linearly related to cumulative incident light integrated over a defined time period. An 8 hour period best explained the diurnal variation in Fv/Fm for *C. monspeliensis* (r^2=0.78), whereas the best fit was found for a shorter time period of 4 hours in *Q. coccifera* (r^2 =0.74). This is probably due to the effective dissipation

4014

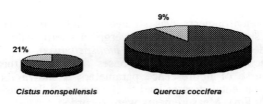

Figure 2. Reduction in potential whole-plant daily carbon gain due to photoinhibitory effects during summer.

mechanisms or an enhanced repair cycle in *Q. coccifera*, resulting in a faster recovery in the afternoon (Fig.1).

These species-specific differences reflected their susceptibility to photoinhibitory decline in whole-plant photosynthesis. The loss in potential canopy carbon gain on a summer day was estimated as 9 % and 21% for *Q. coccifera* and *C. monspeliensis*, respectively (Fig. 2). This reduction of whole-plant photosynthesis of *C. monspeliensis* was more pronounced in the afternoon (Fig. 3) which is consistent with the diurnal pattern of fluorescence measurements (Fig. 1). It was possible that irreversible photodamage may have occurred in this semi-deciduous species, which can result in shedding of some leaves during summer.

These species exhibit different ecological characteristics. *C. monspeliensis* has a lower capacity to protect leaves from excessive radiation during summer, resulting in a loss of one-fifth of potential carbon gain due to photoinhibition. This species uses changes in structure to regulate light interception by reducing leaf area and having steeply inclined leaves that may prevent further photoinhibitory damage.

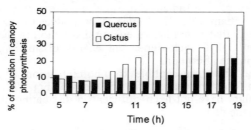

Figure 3. Diurnal pattern of reduction in carbon gain due to photoinhibition relative to potential carbon yield of each species simulated without photoinhibition.

Q. coccifera showed a higher capacity to dissipate the excess light energy resulting in a lower percentage of carbon loss due to photoinhibitory effects. The evergreen sclerophyllous leaves are costly, and avoidance of irreversible damage through protective processes might be critical, even at the expense of some loss in carbon gain.

This might be compensated by a longer leaf life span. It might also be that the diurnal photoinhibition represents a smaller cost than the hypothetical cost of avoiding photoinhibition (1).

These results indicate that photoinhibition can have a marked effect on whole-plant photosynthesis under high light environments during periods of additional environmental stresses, and highlights the importance of integrating photoinhibition into simulation models of plant primary production in these ecosystems.

References

1 Ögren, E. 1994. in Photoinhibition of photosynthesis from molecular mechanism to the field (Baker N.R. and J.R. Bowyer, eds.), pp. 433-447. BIOS Scientific Publishers
2 Ryel R.J., Beyschlag W. and Caldwell M.M. 1993. Funct. Ecol. 7, 115-124
3 Farquhar, G-D., von Caemmerer S. and Berry, J.A. 1980. Planta 149, 78-90
4 Spunda V., Kalina,J. Marek, M.V. and Naus, J. 1997. Photosynthetica 33, 91-102
5 Björkman, O. & Demmig, B. 1987. Planta 170, 489-504
6 Ögren, E. & Sjöström, M. 1990. Planta 181, 560-567.

DAILY PATTERNS OF PHOTOSYNTHESIS OF TWO MEDITERRANEAN SHRUBS IN RESPONSE TO WATER DEFICIT.

Sergi Munné-Bosch, Salvador Nogués and Leonor Alegre

Department of Plant Biology, University of Barcelona, Av. Diagonal 645, E-08028 Barcelona, Spain.

Key words: diurnal changes, dew, drought stress, gas exchange, non-photochemical quenching, photoinhibition

Introduction

The effect of water deficit on gas exchange in Mediterranean vegetation has been widely reviewed by a number of authors [1,2]. Drought affects not only the rate of gas exchange, but also results in diurnal changes in activity [3,4].

The effect of water shortage on photosynthesis depends on the rate and magnitude of dehydration [5,6]. A comparison between species might help to understand the photosynthetic responses of plants to stress [7].

The aim of this study was to determine the photosynthetic response of *R. officinalis* and *L. stoechas*, grown in Mediterranean field conditions, to summer drought, dew and recovery after the first autumn rains. Daily time courses of gas exchange and chlorophyll fluorescence were used to determine the photosynthetic response of these xerophyte shrubs to stress.

Materials and Methods

One-year-old plants of rosemary (*Rosmarinus officinalis* L.) and lavender (*Lavandula stoechas* L.) were grown in the Experimental Fields of the University of Barcelona (NE Spain). Two water treatments were imposed: half of the plots were watered with *ca.* 80 mm per month (irrigated [IR] plants), whereas the remainder were not irrigated (water-stressed [WS] plants). Plant and soil water potential, and relative water content were determined. A LI-6200 portable measuring system (LICOR Inc., Lincoln, Nebraska) was used to estimate net CO_2 assimilation (A) and stomatal conductance (g_s) rates. Steady-state modulated chlorophyll fluorescence of single attached leaves of each apical non-woody shoot was measured using a portable fluorimeter (mini-PAM, Walz, Effelrich, Germany).

G. Garab (ed.), Photosynthesis: Mechanisms and Effects, Vol. V, 4015–4018.
© 1998 *Kluwer Academic Publishers. Printed in the Netherlands.*

Results and discussion

The effects of summer drought (from 16 August to 26 September), dew (11 October) and rainfall (16 and 31 October) on RWC and Ψ is shown in Table 1. RWC decreased by ca. 26% from 16 August to 11 October in both IR and WS R. officinalis plants. In L. stoechas RWC decreased by ca. 35 and 63% in IR and WS plants respectively from 16 August to 26 September, but then it increased by ca. 47 % in IR and more than two fold in WS plants from 26 September to 11 October. A rapid recovery of RWC was observed in both species after rain. The recovery observed in the RWC of L. stoechas between 26 September and 11 October might be due to water absorption from dew by the adaxial leaf surface.

Table 1. Relative leaf water content (RWC, %) and leaf water potential (ψ) of irrigated (IR) and water-stressed (WS) R. officinalis and L. stoechas shrubs from 16 August to 31 October. All the measurements were made early in the morning. * indicates statistical significance at p<0.05 probability level comparing IR and WS treatments. Dashed line indicated water recuperation.

| | RWC (%) | | | | ψ (-MPa) | | | |
| | R. officinalis | | L. stoechas | | R. officinalis | | L. stoechas | |
	IR	WS	IR	WS	IR	WS	IR	WS
16Aug	77.58±2.74	77.07±1.35	59.22±4.24	60.16±4.81	0.50±0.02	1.05±0.09*	0.36±0.01	0.44±0.06
26Sep	62.51±2.88	55.49±5.50	38.62±3.04	22.06±5.52*	0.68±0.14	1.17±0.13*	0.69±0.03	1.23±0.24*
11Oct	57.37±1.38	56.63±2.73	56.73±4.42	57.83±4.93	0.45±0.04	1.86±0.10*	0.50±0.06	0.88±0.09
31Oct	77.79±2.61	78.61±4.45	62.28±5.29	63.76±1.11	0.39±0.03	0.30±0.03	0.38±0.03	0.41±0.03

The effects of the summer drought and the first autumn rain on the net CO_2 assimilation rate (A), stomatal conductance (g_s) and relative quantum efficiency of photosystem II photochemistry (ϕ_{PSII}) at natural incident PPFD in R. officinalis and L. stoechas shrubs are shown in Figure 1. At the beginning of the measuring period. R. officinalis and L. stoechas had a typical one-peaked daily course of photosynthesis, with a maximum peak of CO_2 assimilation and stomatal conductance rates in the early morning. As drought progressed, maximal CO_2 assimilation and stomatal conductance rates decreased by ca. 12% and 70% in IR and WS R. officinalis plants and ca. 45 and 78 % for IR and WS L. stoechas plants respectively. Moreover, the daily pattern of photosynthesis of IR L. stoechas changed to two daily peaks courses of A with a midday depression. After rain, A and g_s increased to a level similar to that early in the summer and one-peaked daily course of A and g_s, but with a maximum wide plateau over several hours (from 10 to 14 h) was observed. ϕ_{PSII} decreased by ca. 50% during the summer drought in WS and was not affected in IR plants of both R. officinalis and L. stoechas. After the first autumn heavy rains, ϕ_{PSII} increased to a level higher to that early in the summer. In R. officinalis F_v/F_m remained constant at ca. 0.75 and 0.74 on IR and WS plants respectively. In L. stoechas there was a slight decrease (from 0.76 to 0.72) in F_v/F_m at midday on 26 September in both IR and WS plants. No variations in F_v/F_m were observed in the rest of

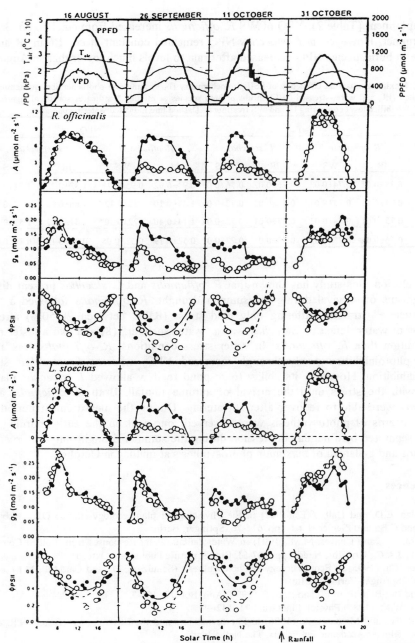

Fig 1 Daily courses of photon flux density (PPFD, μmolm-2s-1), temperature (Tair, °C), vapour pressure deficit (VPD, KPa), net CO_2 assimilation rate (A), stomatal conductance (gs) and PSII efficiency (φPSII) of irrigated (solid symbols) and water-stressed (open) *R. officinalis* and *L. stoechas*

the experiment (Table 2). NPQ of WS *R. officinalis* increased from *ca.* 1 to *ca.* 3 during the summer. However, in *L. stoechas* NPQ remained constant at *ca.* 1.6 in IR and WS plants. In both species NPQ decreased after rain to levels much lower than in August.

Table 2. Maximum quantum efficiency of PSII photochemistry (F_v/F_m) and non-photochemical quenching (NPQ) of irrigated (IR) and water-stressed (WS) shrubs of both species at midday.* indicates statistical significance at $p<0.05$ probability level comparing IR and WS treatments. Dashed line indicated water recuperation.

| | F_v/F_m | | | | NPQ | | | |
| | *R. officinalis* | | *L. stoechas* | | *R. officinalis* | | *L. stoechas* | |
	IR	WS	IR	WS	IR	WS	IR	WS
16Aug	0.75±0.01	0.74±0.02	0.76±0.02	0.76±0.03	0.93±0.02	0.95±0.21	1.33±0.23	1.51±0.13
26Sep	0.74±0.02	0.74±0.01	0.72±0.02	0.72±0.01	1.24±0.10	3.32±0.12*	1.66±0.05	1.77±0.07
11Oct	0.75±0.01	0.74±0.03	0.77±0.01	0.78±0.03	1.14±0.40	2.72±0.30*	1.41±0.12	1.51±0.30
31Oct	0.76±0.02	0.75±0.02	0.75±0.01	0.77±0.01	0.52±0.19	0.52±0.05	0.82±0.13	0.88±0.16

In conclusion, this study has shown that *R. officinalis* and *L. stoechas* present different mechanisms of survival during the summer drought. *R. officinalis* followed a "water conservative" strategy suffering only mild stress (RWC between 70-50%) after two months of water deficit. *L. stoechas* was a "water spender" (8) and had a faster rate of dehydration than *R. officinalis*. In water stress conditions, *R. officinalis* was able to avoid photoinhibition partly through increased NPQ, whereas *L. stoechas* showed photoinhibition. However, its ability to respond to dew allowed *L. stoechas* plants to cope with the stress until the arrival of autumn rainfall. Both *R. officinalis* and *L. stoechas* were able to recover after rewatering. During the measurement period three daily patterns of photosynthesis were observed: one-peak in the early morning, two-peaks separated by a midday depression during the summer drought only observed in *L. stoechas* and a plateau of maximum photosynthesis at midday in October.

References

1 Schulze, E.D. and Hall, A.E. (1982) in the Physiological plant ecology (Lange O.L., Nobel P.S., Osmond C.B., and Ziegler H. eds) pp. 615-76, Springer, Berlin
2 Pereira, J.S., and Chaves, M.M. (1993) in Water deficits, Plant Responses from Cell to Community (Smith, J.A.C., Griffiths, H. eds) pp. 237-51. βios Scientific Publishers, Oxford
3 Körner, Ch. (1995) in Ecophysiology of Photosynthesis (Schulze, E.D., and Caldwell, M.M. eds) pp. 463-90, Springer-Verlag, Berlin
4 Mäkëla, A., Berninger, F., and Hari, P. (1996) Ann. Bot. 77, 461-7
5 Kaiser, W.M. (1987) Physiol. Plantarum 71, 142-49
6 Cornic, G., and Massacci, A. (1996) in Photosynethesis and the Environment (Baker, N.R. ed) pp. 347-66, Kluwer Academic Publishers, The Netherlands.
7 Cornic, G., Papageorgiou, I., and Louason, G. (1987) J. Plant Physiol 126, 399-18.
8 Levitt, J. (1980) Responses of plants to environmental stresses. Vol. 1. Academic Press, N.Y.

PHOTOSYNTHETIC ACTIVITY DURING AUTUMNAL BREAKDOWN OF CHLOROPHYLL IN TREE SPECIES

Fatbardha Babani[1], Maria Balota[2] and Hartmut K. Lichtenthaler[3]
[1] Institute of Biological Research, Academy of Sciences, Tirana, Albania
[2] Cereals and Industrial Crops Research Institute, 8264 Fundulea, Romania
[3] Botanical Institute II, University of Karlsruhe, 76128 Karlsruhe, Germany

Key words: Carotenoids, Chl fluorescence induction, Chl fluorescence spectra, pigment ratios, Rfd-values.

1. Introduction

The chlorophyll fluorescence analysis has been applied to describe and investigate the photosynthetic light processes and quantum conversion as well as to detect damage to the photosynthetic apparatus and its function due to the inverse relationship between photosynthetic performance and Chl emitted fluorescence (1-6). The Chl fluorescence ratio F690/F735 has been established as a non-invasive indicator of the *in vivo* Chl content of leaves (1, 2-5, 7). In a pre-darkened leaf the fast Chl fluorescence rise (from ground fluorescence Fo to a maximum fluorescence level Fm) and the slow fluorescence decline (from Fm to steady-state fluorescence Fs), known as Kautsky effect, reflect the functioning of the photosynthetic apparatus (4, 6). The variable Chl fluorescence decrease ratio Rfd, defined as (Fm-Fs)/Fs, measured at 690 and 735 nm is a measure of the potential photosynthetic capacity of a leaf and is correlated to the photosynthetic net CO_2 assimilation (4, 5, 7). The aim of this work was to determine the decline of photosynthetic activity and chlorophyll levels during the autumnal breakdown of the photosynthetic pigments via the variation of Chl fluorescence parameters.

2. Procedure

2.1 *Plants*
Leaves , in the long-lasting summer period 1997, with different pigment content (green to greenish-yellow) of 6 tree species *Platanus hybrida, Prunus spec., Ulmus spec., Cornus mas* L., *Fagus sylvatica* L. and *Philadelphus coronarius* L. (Karlsruhe University campus) were collected, starting from early October until the end of November.

2.2 *Pigment determination*
Leaf pigments were extracted with 100% acetone and determined spectrophoto-metrically according to Lichtenthaler (8). Mean of 4 determinations.

G. Garab (ed.), Photosynthesis: Mechanisms and Effects, Vol. V, 4019–4022.
© 1998 *Kluwer Academic Publishers. Printed in the Netherlands.*

2.3 *Chlorophyll fluorescence induction kinetics*

Chl fluorescence induction kinetics were measured by the portable two-wavelength fluorometer LITWaF (λ_{exc} = 632.8 nm, PFD 500 µmol m^{-2}s^{-1}) (4). The ratio F690/F735 at Fm and Fs and the Rfd-values (Rfd690 and Rfd735) were calculated.

2.4 *Chlorophyll fluorescence emission spectra*

Chl fluorescence emission spectra were recorded during the fast rise and slow decline of the induction kinetics using the Karlsruhe CCD-OMA spectrofluorometer (λ_{exc} = 632.8 nm, PFD 40000 µmol m^{-2} s^{-1}) (7). The fluorescence ratio F690/F735 and the Rfd-values (Rfd690 and Rfd735) were determined from the spectra (in a 3 nm range).

3. Results and Discussion

3.1 *Pigment content and pigments ratios*

The decrease of the total chlorophyll level (from 40 to 1 µg cm^{-2} leaf area) was recorded during autumnal Chl breakdown. The rate of decline of Chl a and Chl b was nearly the same in all tree species analyzed down to a Chl level of 5-7 µg cm^{-2} leaf area, as seen from the values of about 3 for the pigment ratio Chl a/b (Fig. 1). The latter indicate that

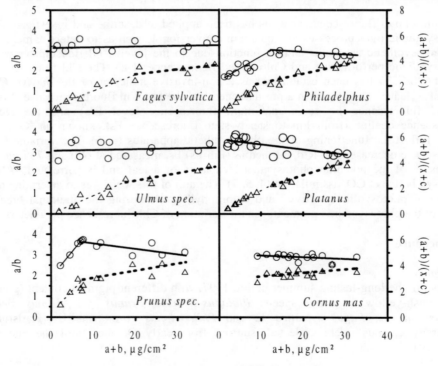

Figure 1. Pigment ratios Chl a/b (O—O) and chlorophylls to carotenoids (a+b)/(x+c) (△ −△) during autumnal chlorophyll breakdown. Mean of 4 determinations. SD< 5%.

despite the progressing Chl breakdown, the remaining Chl is functionally organized. This is also emphasized by the ratio of chlorophylls to carotenoids $(a+b)/(x+c)$ which remained in the range of 3 to 4 in the initial stage of Chl breakdown. Only at a Chl level below ca. 15 µg cm^{-2} the decline of chlorophylls was faster than that of carotenoids.

3.2. Chlorophyll fluorescence measurements

The CCD laser-induced Chl fluorescence emission spectra showed a fast rise to Fm within 200-240 ms and then a slow decline to Fs, measured 300 s after onset of illumination. The spectra showed characteristic changes during Chl breakdown (Fig. 2). At the initial stage with a Chl content of 40-30 µg cm^{-2} the far-red fluorescence band near 735 nm (F735) was much higher than the red band near 690 nm (F690). During the decline of Chl content (from 20 to 10 µg cm^{-2}) F690 considerably increased with respect to F735 yielding two separate maxima. At very low Chl content only the red band near 690 nm was detectable, whereas the second one near 735 nm was reduced to a shoulder.

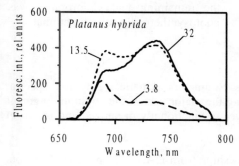 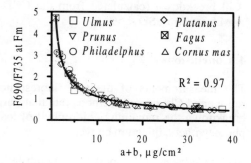

Figure 2. Shape of Chl fluorescence emission spectra at Fm during autumnal Chl breakdown. The total chlorophyll levels are indicated (µg cm^{-2} leaf area). Mean of 3 spectra. SD < 5%.

Figure 3. Inverse curvilinear relationship between the Chl fluorescence ratios F690/F735 and Chl content during autumnal Chl breakdown measured at Fm (power function $y = ax^{-b}$). SD < 5%. (CCD-OMA)

The values of the Chl fluorescence ratio F690/F735 increased with decreasing Chl content of leaves beeing very sensitive at a Chl level less than 20 µg cm^{-2} (Fig. 3). This can be explained by the decreasing reabsorption of emitted red Chl fluorescence by the declining *in vivo* Chl content and the saturation of Chl absorption near 680 nm at a high Chl content (1, 2, 4, 5). The correlation between the ratio F690/F735 and Chl content can be expressed by a curvilinear function ($y = ax^{-b}$, $R^2 = 0.97$) which is valid for all tree species analyzed (a: 4.39, 3.34 and and b: 0.59, 0.56 at Fm and Fs, respectively). This indicates that the ratio F690/F735 mainly depends on the Chl content and only very little on the structure of leaves or the plant species proving that it is an excellent indicator of the *in vivo* Chl content of leaves.

At the initial stage of the autumnal Chl breakdown (30 to 40 µg cm^{-2}) the Rfd-values, Rfd690 and Rfd735, showed high values of 4 - 6 and these remained fairly constant

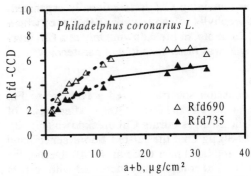

Figure 4. Variable fluorescence decrease ratios Rfd690 and Rfd735 during autumnal Chl breakdown (calculated from the CCD-OMA spectra). SD < 5%.

up to Chl content of 12-15 μg cm^{-2} leaf area (*Platanus, Prunus, Ulmus, Fagus, Philadelphus*). In *Cornus mas* the Rfd-values were lower (2 - 3) and also remained at the same level. Only at an advanced stage of senescence at a Chl level less than 10 μg cm^{-2} the Rfd-values decreased very fast (Fig. 4). Similar results were obtained with the LITWaF instrument. The high values of the plant vitality indices, the fluorescence ratios Rfd690 and Rfd735, indicate that during a major part of the autumnal Chl breakdown the remaining leaf Chlorophyll was photosynthetically fully functional.

4. Conclusions

The measurements of the Chl fluorescence parameters (kinetics, emission spectra, ratios) and Chl content demonstrate that during the major part of the autumnal pigment breakdown the remaining chlorophyll of leaves is functionally organized and active in photosynthetic quantum conversion.

Acknowledgement: Support by 3-month fellowships from DAAD, Bonn, for M. Balota and F. Babani for a stay in Karlsruhe, is gratefully acknowledged.

References

1 D'Ambrosio, N., Szabo K. and Lichtenthaler, H.K. (1992) Radiat. Environ. Biophys. 31, 51-62

2 Gitelson, A.A., Buschmann C. and Lichtenthaler, H.K. (1998) J. Plant Physiol. 152, 281-296

3 Krause, G.H. and Weis, E. (1991) Ann. Rev. Plant Physiol. Plant Mol. Biol. 43, 313-349

4 Lichtenthaler, H.K. and Rinderle, U. (1988) Crit. Rev. Anal. Chem. 19, Supl. 1, 29-85

5 Lichtenthaler, H.K. and Miehe, J.A. (1997) Trends in Plant Sciences 2, 316-320

6 Schreiber, U. and Neubauer, C. (1990) Photosynth. Res. 25, 279-293

7 Babani, F. and Lichtenthaler, H.K. (1996) J. Plant Physiol. 148, 555-566

8 Lichtenthaler, H. K. (1987) Methods in Enzymol. 148, 350-382

9 Szabo, K., Lichtenthaler, H.K., Kocsanyi, L. and Richter, P. (1992) Radiat. Environ. Biophys. 31, 153-160

EFFECTS OF ELEVATED CO_2 AND TEMPERATURE ON RUBISCO IN SCOTS PINE AFTER ONE YEAR OF EXPOSURE IN CLOSED TOP CHAMBERS

Tuhkanen E-M.[1], Laitinen K.[2], Kellomäki S.[2] & Vapaavuori E.[1]
[1]Finnish Forest Research Institute, Suonenjoki Research Station,
FIN-77600 Suonenjoki, Finland. [2]Faculty of Forestry,University of Joensuu,
P.O.Box 111, FIN-80101 Joensuu, Finland

Key words: C metabolisms, Chl, down regulation, global climate changes, Pinus

1. Introduction

The objective of the present experiment is to study the effect of elevated CO_2 (700 ppm) and temperature (ambient +4 °C) on biochemical characteristics of photosynthesis in field-grown Scots pine (*Pinus sylvestris* L.) before and after one year of exposure.

2. Material and methods

The experiment was conducted in a naturally regenerated Scots pine (*Pinus sylvestris* L.) stand close to the Mekrijärvi Research Station (62°47'N, 30°58'E, 145 a.s.l.) of the University of Joensuu in eastern Finland. Trees of the stand were 25 - 30 years old and the forest type on the sandy iron-podsole soil represents the Vaccinium type (1). 16 closed top chambers (CTC) were constructed around individual trees, and following treatments were applied:
- 4 chambers with ambient CO_2 and ambient T (ambient chamber, AmbC)
- 4 chambers with ambient CO_2 and elevated T (+T)
- 4 chambers with elevated CO_2 and ambient T (+CO_2)
- 4 chambers with elevated CO_2 and elevated T (+CO_2+T)
- four trees were selected for controls of the chamber effect (Ctrl)
The target of the elevated CO_2 concentration was 700 μmol mol^{-1} and the mean elevation of the ambient temperature 4 °C. Current and 1-yr-old needles of third whorl from the stem apex were collected before the beginning of the exposure in August 1996 and after one year of exposure in August 1997. For biochemical determinations two replicates of current and previous year needles collected under saturating light were frozen in liquid-N and stored at -80 °C. The frozen needles were homogenized in ice

G. Garab (ed.), Photosynthesis: Mechanisms and Effects, Vol. V, 4023–4026.
© *1998 Kluwer Academic Publishers. Printed in the Netherlands.*

cold extraction buffer containing 50 mM MES, pH 6.8, 20 mM $MgCl_2$, 50 mM β-mercaptoethanol and 1 % TWEEN (2). Aliquots of the crude extract were analyzed for the chlorophyll content (3), and activity of Rubisco was determined as incorporation of ^{14}C into acid-stable products (4). The assay was performed at 25 °C in a reaction mixture containing 50 mM Epps-NaOH, pH 8.2, 20 mM $MgCl_2$, 0.26 mM EDTA, 10 mM $NaH^{14}CO_3$ and 0.3 mM RuBP. The amount of Rubisco protein was determined by PAGE (5) and the soluble protein content by the method of Bradford (1976) (6). From each shoot 20 needles were collected for determination of needle fresh weight and needle area. The projected area of needles was measured by leaf area meter (LI-3050A, LI-COR Inc., Lincoln, USA) and the total surface area was obtained by multiplying the projected needle area by π (7). Area-based biochemical results presented are calculated on the total surface area of needles.

3. Results

3.1 Total activity and amount of Rubisco

The total activity of Rubisco in Ctrl trees was decreased by 56%, when 1-yr old needles in Aug 1996 are compared to 1-yr old needles in Aug 1997 (Fig. 1A). In AmbC and in $+CO_2$ the decrease was 44% and 49% respectively, but in +T only 2% and in $+CO_2+T$ 8%. In current yr needles of Ctrl trees the total activity was, contradictory, increased by 12% from Aug 1996 to Aug 1997. In AmbC the total activity of current yr needles increased by 55%, in $+CO_2+T$ by 191% and in +T by 161%. In $+CO_2$ the increase was smallest, only 2%. In summary, one yr of growth in +T or in $+CO_2+T$ increased the total activity of Rubisco relative to the change observed in AmbC in both needle age classes, whereas growth in $+CO_2$ tended to decrease. These trends are obvious also when considering the development of current yr needles of 1996 into 1-yr old needles of 1997 (Table 1). The changes in total activity reflect the amount of Rubisco protein (Fig. 1B) on an needle area basis. Also here it is obvious, when the development of current yr needles of 1996 into 1-yr old needles of 1997 is studied, that +T and $+CO_2+T$ increased the amount of Rubisco relative to the change occurring in AmbC, while $+CO_2$ decreased (Table 1).

Table 1. The development of current yr needles of Aug 1996 into 1-yr old needles of Aug 1997. Change (%) of total activity of Rubisco ($\mu mol\ m^{-2}\ s^{-1}$), amount of Rubisco protein ($g\ m^{-2}$), amount of Rubisco relative to chlorophyll ($mg\ mg^{-1}$) and soluble protein (%) in control trees (Ctrl), ambient chambers (AmbC), elevated temperature (+T), elevated CO_2 ($+CO_2$) and elevated CO_2 and temperature ($+CO_2+T$).

	Total activity	Rubisco protein	Rubisco Chlorophyll	Rubisco Soluble protein
Ctrl	+63.5	+113.1	+65.8	+67.0
AmbC	+60.2	+133.3	+22.7	+49.9
+T	+118.0	+239.5	+60.4	+122.6
$+CO_2$	+11.3	+81.3	+10.2	+44.4
$+CO_2+T$	+183.4	+163.9	+1.7	+66.0

3.2 *Amount of Rubisco relative to soluble protein and chlorophyll contents*

In Ctrl trees, the proportion of Rubisco of needle soluble proteins (R/SP) was at the same level in 1-yr old needles in Aug 1997 as in Aug 1996 (Fig. 1C). The chamber caused a slight increase (7%), whereas +T increased R/SP most (72%), +CO_2+T also considerably (40%) and +CO_2 decreased (35%). In current yr needles of Ctrl trees R/SP was increased by 13% from Aug 1996 to Aug 1997. As in 1-yr old needles, the chamber, +T and +CO_2+T increased R/SP (36%, 42% and 100%, respectively), but here also +CO_2 had a positive effect (53%). When considering the growth of current yr needles of 1996 into 1-yr old needles of 1997, it is again observed that, compared to AmbC, +T increased R/SP, as also +CO_2+T did, but +CO_2 decreased (Table 1). The amount of Rubisco relative to chlorophyll (R/Chl) (Fig. 1D) behaved somewhat similarly as R/SP, with one distinction: in current yr needles R/Chl is decreased in +T relative to the change in AmbC (increase 0% and 8%, respectively). But during the development of current yr needles of 1996 to 1-yr old needles of 1997, +T increased R/Chl relative to AmbC, while +CO_2+T and +CO_2 decreased (Table 1).

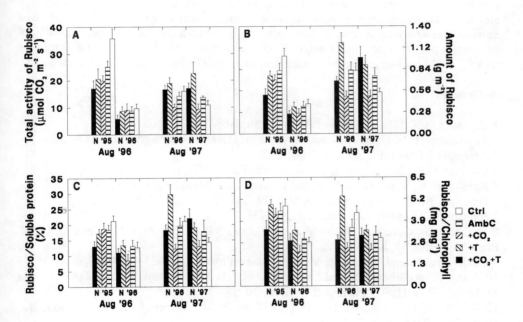

Figure 1. A) Total activity of Rubisco (μmol m^{-2} s^{-1}) B) amount of Rubisco protein (g m^{-2}) C) amount of Rubisco relative to chlorophyll (mg mg^{-1}) and D) soluble protein (%) in control trees (Ctrl), ambient chambers (AmbC), elevated temperature (+T), elevated CO_2 (+CO_2) and elevated CO_2 and temperature (+CO_2+T). N '95 = needles formed in 1995; N '96 = needles formed in 1996; N '97 = needles formed in 1997.

4. Discussion

Our previous results of a branch bag experiment with unfertilized Scots pine trees at the same site indicated down-regulation of the amount and activity of Rubisco in elevated CO_2 grown branches (Tuhkanen et al., unpublished). Also in this study elevated CO_2 down-regulated the amount and activity of Rubisco, when 1-yr old and current yr needles in Aug 1996 are compared to the respective age classes in Aug 1997 (Fig. 1A, 1B). This down-regulation is apparent either as a smaller relative increase or a greater decrease of Rubisco in $+CO_2$ compared to the change observed in AmbC between these two seasons. Also the amount of Rubisco relative to soluble proteins or Chl 1997 (Fig. 1C, 1D). was decreased in $+CO_2$ with one exception (R/SP in current yr needles). Elevation of mean T had an opposing effect: Rubisco was up-regulated with one exception (R/Chl in current yr needles). Also elevated CO_2 and T together up-regulated the amount and activity of Rubisco. These trends are maintained, when the development of current yr needles of 1996 into 1-yr old needles of 1997 is considered (Table 1). In general, $+CO_2$ suppressed, while $+T$ and $+CO_2+T$ enhanced the increase of Rubisco compared to the change in AmbC. Decline in the amount and/or activity of Rubisco at elevated CO_2 has frequently been reported (8,9). Growth at elevated T has decreased maximum photosynthetic rate in Scots pine (10), or had little effect on net assimilation of loblolly pine (11) or white birch (12). Our preliminary results show that elevation of mean T could counteract the downward acclimation of Rubisco caused by elevated CO_2.

References

1 Cajander, A.K. (1949) Acta Forest. Fenn. 56, 1-69
2 Vapaavuori, E.M., Rikala, R. and Ryyppö, A. (1992). Tree Physiol. 10, 217-230
3 Arnon, D.I. (1949) Plant Physiol. 24, 1-15
4 Lorimer, G.H., Badger, M.R. and Andrews, T.J. (1977) Anal. Biochem. 78, 66-75
5 Ruuska, S.A., Vapaavuori, E.M. and Laisk., A. (1994) J. Exp. Bot. 45, 343-353
6 Bradford, M.M. (1976) Anal. Biochem. 72, 248-254
7 Johnson, I.D. (1984) Forest Sci. 30, 913-921
8 Drake, B.G., Gonzàles-Meler, M.A. and Long, S.P. (1997) Ann. Rev. Plant Physiol. Plant Mol. Biol. 48, 609-639
9 Ceulemans, R. and Mousseau, M. (1994) New Phytol. 127, 425-446
10 Wang, K., Kellomäki, S. and Laitinen, K. (1995) Tree Physiol. 15, 211-218
11 Teskey, R.O. (1997) Plant Cell Environ. 20, 373-380
12 Koike, T., Lei, T.T., Maximov, T.C., Tabuchi, R., Takahashi K. and Ivanov, B.I. (1996) Tree Physiol. 16, 381-385

PROTEIN ADJUSTMENTS OF WHEAT FLAG LEAVES IN RESPONSE TO ATMOSPHERIC CARBON DIOXIDE ENRICHMENT.

Richard C. Sicher & James A. Bunce
U.S.D.A., Agricultural Research Service,
Climate Stress Laboratory, Bldg. 046-A,
Beltsville Agricultural Research Center,
10300 Baltimore Avenue,
Beltsville, MD 20705 USA

Key words: Acclimation, Rubisco, Chl, Senescence, Down Regulation, Elevated CO_2

1. Introduction

Rubisco is an extremely abundant enzyme and it typically comprises about 30% of the total N invested in mature leaves. Rubisco protein concentrations in leaves are almost certainly in excess during plant growth in shade or in atmospheres enriched with carbon dioxide. A down regulation of Rubisco activity and of Rubisco protein concentrations occurs during photosynthetic decline in response to long-term growth in elevated carbon dioxide . Changes of leaf protein concentrations during carbon dioxide enrichment potentially could serve to optimize the photosynthetic apparatus and to achieve a better balance between carboxylation and other reactions in the chloroplast (1, 2). According to this hypothesis specific photosynthetic proteins are up or down regulated in a coordinated manner. Alternatively, protein adjustments in response to carbon dioxide enrichment could represent an acceleration of development and this would result in early maturation and premature leaf senescence (3). In this report, we have compared changes of soluble protein, Rubisco protein and total Chl in flag leaves of wheat plants grown in the field with either ambient or elevated carbon dioxide.

2. Procedure

2.1 *Plant materials*
Winter wheat (*Triticum aestivum* L. cv. Coker) was grown in open-topped field chambers (1.2 m^2) at Beltsville, Maryland, USA, during the 1995, 1996 and 1997 growing seasons as previously described (4, 5). Plots were seeded in October and measurements were performed during the following April and May. Plants were grown in a silt-loam soil using standard agronomic practices for this area (4). Carbon dioxide treatments were either ambient (35 ± 5 Pa in the daytime) or elevated (ambient plus 35 ± 5 Pa) and were continuous from seeding to harvest. Three chambers of each carbon dioxide treatment were planted annually.

2.2 *Leaf measurements*
One or two flag leaves from each chamber were sampled at about solar noon on bright, sunny days. Chambers were harvested at 2 to 7 d intervals beginning just before anthesis and sampling ended when the leaves had completely senesced. Leaf segments of 3 to 4

G. Garab (ed.), Photosynthesis: Mechanisms and Effects, Vol. V, 4027–4030.
© 1998 *Kluwer Academic Publishers. Printed in the Netherlands.*

cm length were collected from the lamina, transferred to Al bags and frozen with liquid N_2. One-half of the frozen leaf segments were extracted with 2 mL buffer containing 50 mM Hepes-NaOH (pH 8.1), 5 mM $MgCl_2$ and 1 mM dithiothreitol. The resultant homogenates were spun in a microfuge for three minutes at full line voltage and the supernatants were assayed for total soluble protein and for Rubisco protein (4, 5). Total soluble protein was measured with Coomassie brilliant blue R using a standard dye-binding method. To measure Rubisco protein, aliquots of the buffer extracts were mixed with an equal volume of gel loading buffer containing 100 mM Tris-HCl (pH 6.8), 33% glycerol and 3.3% SDS. The mixtures were boiled for 2 min and loaded directly onto a 12% SDS-PAGE gel with a 5% stacking gel. After electrophoresis for 1 h at 200 V, the gels were stained with 1% Coomassie blue and destained overnight with 20% methanol and 7% acetic acid. Bands corresponding to the large subunit of Rubisco were excised and dye was eluted at 50°C with 1 mL of formamide (6). Absorbance was measured at 595 nm and standard curves were prepared with purified tobacco Rubisco. Additional leaf segments were extracted with methanol:chloroform:water and Chl was measured in 80% acetone based on absorbance at 665 and 649 nm.

3.   Results

3.1 *Soluble Protein and Rubisco Protein*
Previously published findings (see 4, 5) documented changes of various components in

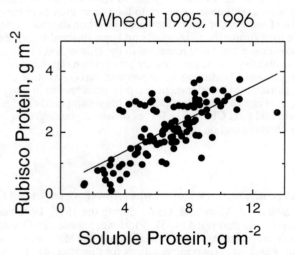

Figure 1. Ratio of Rubisco protein to total soluble protein in wheat flag leaves during growth in ambient or elevated carbon dioxide.

wheat flag leaves grown in open-topped chambers in the field from 1995 to 1997. Protein and Chl levels decreased with leaf age and this occurred 5 to 7 days earlier in the elevated compared to ambient carbon dioxide plots.The correlation between changes of Rubiso protein and total soluble protein in wheat flag leaves in response to carbon dioxide enrichment is shown in Figure 1. Data were combined for all individual leaf measurements performed on wheat flag leaves in the ambient and elevated carbon dioxide plots during 1995 and 1996. The correlation was linear (y = 0.33 + 0.27x, r² = 0.70) and was significant at P ≤ 0.001.

3.2 *Soluble Protein and Total Chl*
The ratio of changes in total Chl to changes in total soluble protein in wheat flag leaves is shown in Figure 2. Comparisons were for plot means for wheat flag leaves in the ambient and elevated carbon dioxide treatments during 1996 and 1997. This correlation was also linear (y = 0.29 + 18 42x), r² = 0.79) and was significant at P ≤ 0.001).

4. Discussion

4.1 *Mechanistic Considerations*
Photosynthesis is probably not limited by Rubisco activity during plant growth at

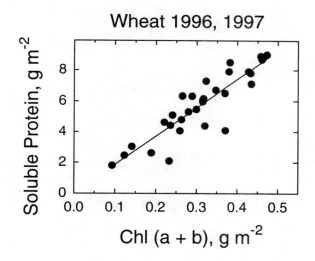

Figure 2. Ratio of total soluble protein to total Chl in wheat flag leaves during growth in ambient or elevated carbon dioxide.

elevated carbon dioxide partial pressures. In order to optimize N use efficiency under these conditions, leaf N would have to be re-allocated to other proteins that might limit photosynthesis, particularly those involved in light capture or in carbohydrate synthesis. In the current study, the relationship of total soluble protein to Rubisco protein and to total Chl, which is primarily associated with membrane proteins, was the same in both the ambient and elevated plots. Although correlation is not causation, these results suggested that a re-allocation of N from saturating to limiting processes did not occur in wheat flag leaves in response to carbon dioxide enrichment. Similar results have been reported for rice plants (*Oryza sativa* L) grown in a glasshouse study (1). These observations are best supported by the suggestion that carbon dioxide enrichment accelerated leaf development and hastened maturation (3, 5).

5. References

1. Nakano, H. Makino, A. and Mae, T. (1997) Plant Physiol. 115: 191-198.
2. Long, S.P. (1991) Plant, Cell Environ. 14: 729-740.
3. Miller, A., Tsai, C.-H., Hemphill, D., Endres, M., Rodermel, S. and Spalding, M. (1997) Plant Physiol 115: 1195-1200.
4. Sicher, R.C. and Bunce, J.A. (1998) Int. J. Plant Sci. 159: 798-804.
5. Sicher, R.C. and Bunce, J.A. (1997) Photosyn. Res. 52: 27-38.
6. Makino, A., Mae, T. and Ohira, K. (1986) Agric. Biol. Chem. 50: 1911-1912.

THE EFFECTS OF INCREASED ATMOSPHERIC CO$_2$ ON GROWTH, CARBOHYDRATES AND PHOTOSYNTHESIS IN RADISH, *Raphanus sativus*.

Usuda, H. and Shimogawara, K.
Lab. of Chem., Teikyo Univ. , Hachioji, Japan

Key words: CO$_2$ uptake, elevated CO$_2$, source-sink, radish.

1. Introduction

The effects of increased atmospheric carbon dioxide on plant growth and photosynthesis are very important issue because the atmospheric concentration of CO$_2$ will likely rise to about 700 μmol mol^{-1} by the end of the next century. Enhanced levels of CO$_2$ generally stimulates the net photosynthetic fixation of CO$_2$ in C$_3$ plants. Many studies have shown, however, that short-term gain may be offset, in the longer term, by a negative acclimation in photosynthetic capacity. The negative acclimation might be related to an imbalance between source capacity and sink capacity (1). In this study, we first addressed the question of the effect of sink capacity on the acclimation of photosynthesis in leaves grown in an atmosphere with an elevated CO$_2$, investigating the rate of photosynthesis and the growth rate during the developmental stages in 15- to 46-d radish seedlings. In radish, a major sink is the storage root, and its thickening growth is initiated early. Therefore, radish seemed to be a very suitable plant to address this question. Secondly we initiated the research to identify the critical component of sink capacity.

2. Procedure

For the first series of the experiments, seeds of radish, *Raphanus sativus* L. (cv. White Cherish), were grown in a controlled growth chamber with a 14-h light (400 ± 50 μmol m^{-2} s^{-1} [400-700 nm], 25°C, 7:00-21:00)/10-h dark (20°C) cycle. The concentration of CO$_2$ in the chamber was maintained at 350 (ambient CO$_2$) or 750 μmol mol^{-1}(elevated CO$_2$). See reference 2 for the details.

For the second series, seeds were germinated on the wet paper in the dark. On the third day hydroponic culture was started with 14-h light/10-h dark cycle in the presence of ambient CO$_2$. Other conditions were the same as mentioned above. Samples taken at 3 DAP (days after planting) were from continuous dark. The rest of the samples were harvested around 17:00.

3. Results and Discussion

We investigated the effects of elevated (750 μmol mol^{-1}) and ambient (350 μmol mol^{-1}) atmospheric CO$_2$ on growth, carbohydrate levels, and photosynthesis in

4031

G. Garab (ed.), Photosynthesis: Mechanisms and Effects, Vol. V, 4031–4034.

radish seedlings from 15 to 46 DAP. In radish, a major sink is the storage root, and its thickening is initiated early. Table 1 shows the summary of the effects of increased carbon dioxide on growth, carbohydrates, and photosynthesis in radish, *Raphanaus sativus*. Elevated CO_2 increased the accumulation of dry matter but had no effect on the acclimation of the rate of photosynthesis or on the levels of carbohydrates in leaves at dawn. The rate of photosynthesis with the leaves grown under elevated CO_2 decreased relatively rapidly during growth, perhaps because of earlier initiation of leaf senescence. Actually the ontogenesis is accelerated under elevated CO_2. The elevated CO_2 had an effect of leaf ontogeny in tobacco (3). The effect of an elevated CO_2 on the contribution of the storage root to "total dry weight" was compared in terms of "total dry weight" to correct for the accelerated ontogenesis. The patterns of changes in the contribution for plants grown under elevated and ambient CO_2 were similar in terms of "total dry weight" (line 4 of Table 1 and see Fig 4 of reference 2). Sink activity (= $(\ln W_2 - \ln W_1)/T$) in terms of "total dry weight" was, however, significantly increased under elevated CO_2. This enhanced capacity seemed to be responsible for absorption of elevated levels of photosynthate and to result in the absence of any over-accumulation of carbohydrates in source leaves and the absence of negative acclimation of photosynthetic capacity at the elevated level of CO_2.

Table 1 Summary of the effects of increased carbon dioxide on growth, carbohydrates, and photosynthesis in radish, *Raphanaus sativus*.

	Elevated/Control
DW[1] of Leaf after 46 DAP[2]	140%
DW of Shoot after 46 DAP	147%
DW of Storage Root after 46 DAP	205%
DW of Storage Root/DW of "Total DW[3]" vs "Total DW" from 15 DAP to 46 DAP	almost similar
The rate of photosynthesis at 350 µmol/mol CO_2	almost similar[4]
The rate of photosynthesis at 750 µmol/mol CO_2	almost similar[4]
Photosynthesis-C_i[5] curve	almost similar
Carbohydrates content	slightly higher during the day but almost similar at dawn
Sink activity (g/g/day) vs "Total DW"	significantly increased

[1] DW=dry weight
[2] DAP=days after planting
[3] Total DW=DW of shoots plus storage root without fibrous roots
[4] See the text.
[5] Ci=partial pressure of CO_2 in the intercellular spaces of leaves.

The underlying mechanism for the stimulated sink activity of the storage root under elevated CO_2 is interesting issue to be examined. We have to consider at least two possibilities: accelerated sink growth might be simply due to an elevated supply of sugars, and/or it might be due to the up-regulation of the expression of genes for proteins responsible for sink capacity by an enhanced supply of sugars. A specific set of genes for proteins that function in the storage, utilization, and import of carbon is known to be positively regulated by elevated levels of sugars (4). For the first step to understand the sink activity in radish storage root (hypocotyl), we investigate the changes in the growth of whole plant, and sink activity and levels of carbohydrates, enzyme activities, and protein profile in the hypocotyl during early thickening growth under ambient CO_2 of 350 μmol mol^{-1} from 3 to 21 DAP.

The dry weight of hypocotyl (shoot)/plant were 1.55 (6.51), 1.62 (7.12), 2.13 (14.7), 2.97 (34.0), 6.30 (70.0), 21.6 (272), 290 (1017) and 1813 (1787) mg at 3, 4, 6, 8, 10, 13, 17 and 21 DAP, respectively (n=4-24). The leaf area/plant were 1.24, 1.72, 4.01, 10.5, 18.9, 70.5, 245.8, and 348.9 cm^2 at similar DAPs mentioned above, respectively. The carbohydrate (glucose + fructose + sucrose + starch) content in the hypocotyl were 13.5, 10.1, 6.0, 3.1, 13.1, 66.5, 2600, and 22175 μmol carbon atom/hypocotyl at similar DAPs mentioned above, respectively. Glucose and fructose were major sugars and starch content was very low. All these result indicate that soon after photosynthetic carbon gain initiated in the leaves, thickening growth of hypocotyl is initiated and the increase of sucrose synthase activity, sugar contents and sink activity seemed to be highly related (Fig. 1). The level of protein having molecular weight of ca. 90 kD increased during thickening growth. The amount of sucrose synthase in the hypocotyl also increased during thickening growth (Fig. 2) . All these results suggest that initiation of photosynthesis induced translocation of photosynthate into the hypocotyl and thickening growth of the hypocotyl was initiated. Sugar-induced up-regulation of sucrose synthase might be responsible for the increased sink activity. Further studies are needed to clarify the mechanism, timing, and localization of expression of sucrose synthase in the hypocotyl considering structural development of the hypocotyl.

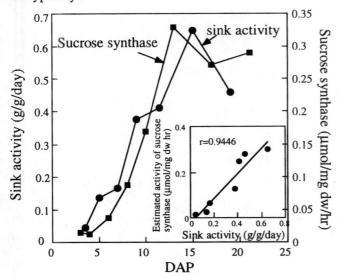

Figure 1. Changes in the sink activity and sucrose synthase activity in the hypocotyl of radish. Insert is the relationship between sink activity and calculated sucrose synthase activity in the hypocotyl. Sucrose synthase activity is the mean value of that of the samples from t1 and t2 shown in the figure.

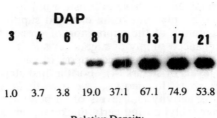

DAP

3 4 6 8 10 13 17 21

1.0 3.7 3.8 19.0 37.1 67.1 74.9 53.8

Relative Density

Figure 2. Changes in the level of sucrose synthase in the hypocotyl. Each sample (1 µg protein) was separated on SDS-PAGE of 8.8% (w/v) acrylamide and transferred to a nitrocellulose membrane. Sucrose synthase was detected using antibody against sucrose synthase of mung bean kindly provided from Dr. H. Mori, Nagoya University and ^{125}I-protein A using BAS 2000 Bio-Imaging Analyzer (Fuji Film, Tokyo, Japan)

Acknowledgments
    This work was supported in part by a grant-in-aid for "Research for the Future" Program (JSPS-RFTF97R16001) from the Japan Society for the Promotion of Science and "Bio-design Plan" from the Ministry of Agriculture, Forestry and Fisheries of Japan.

References

1. Bowes, G. (1996) in Photosynthesis and the Environment (Baker, N.R., ed.) pp.387-407, Kluwer Academic Publishers, Derdrecht, The Netherlands
2. Usuda, H. and Shimogawara, K. (1998) Plant Cell Physiol.39, 1-7.
3. Miller, A. et al. (1977) Plant Physiol. 115, 1195-1200.
4. Koch, K. E. (1996) Carbohydrate-modulated gene expression in plants. *Annu. Rev. Plant Physiol. Plant Mol. Biol.* 47: 23-48.

ACCLIMATION OF *LOLIUM TEMULENTUM* TO GROWTH AT ELEVATED CO_2

Charlotte E. Lewis[1], David R. Causton[2], Giovanni Peratoner[1], Andrew J. Cairns[1] and Christine H. Foyer[1], [1]IGER, Aberystwyth, SY23 3EB, UK; [2]University of Wales Aberystwyth, Aberystwyth SY23 3DD, UK.

Key words: Rubisco, respiration, carbohydrate, CO_2 uptake, ryegrass, biomass

1. Introduction

Stimulation of photosynthesis and growth at elevated CO_2 has been demonstrated in a variety of plant species, at least in the short-term (1,2,3). Long-term growth with CO_2 enrichment frequently results in marked increases in foliar carbohydrate accumulation. Many studies on CO_2 enrichment have examined species that accumulate sucrose and starch in their leaves. Relatively few have concentrated on plants that partition a large proportion of their assimilate into other sugars or sugar alcohols. We have therefore examined the responses of the fructan-forming monocotyledonous ryegrass *Lolium temulentum* to continuous growth in elevated CO_2 (700 μmol mol^{-1}) at two irradiances, 500 μmol m^{-2} s^{-1} (HL) and 150 μmol m^{-2} s^{-1} (LL).

2. Procedure

Lolium temulentum Ba3081 (summer annual) was grown from seed in controlled environment chambers at air (350 μmol mol^{-1}) and elevated CO_2 (700 μmol mol^{-1}) with free access to nutrient solution (4). Light was provided at either 150 or 500 μmol m^{-2} s^{-1}. At intervals plants were removed from the chambers for growth analysis. All other measurements were performed on the fourth leaf at 35 days after sowing, a point at which the fourth leaf had reached full expansion in all growth conditions. Photosynthesis was measured under saturating light using a combined infra-red gas analyser (CIRAS, Version 2.7, PP Systems, Hitchin, UK). The response of photosynthetic CO_2 uptake (A) to varying intercellular CO_2 concentrations (c_i) was used to calculate the *in vivo* capacity for primary carboxylation ($V_{c,max}$) of ribulose-1,5-bisphosphate carboxylase/oxygenase (Rubisco) and the maximum capacity for regeneration of the primary CO_2 acceptor, ribulose-1,5- bisphosphate (RuBP) (J_{max}), by using the equations of Evans and Farquar (5). Leaf respiration was measured in the dark by a "Clarke-type" oxygen electrode

G. Garab (ed.), Photosynthesis: Mechanisms and Effects, Vol. V, 4035–4038.

(Hansatech Ltd., King's Lynn, UK). Rubisco activity was determined according to the method of Parry et al. (6). The amount of Rubisco protein was determined using antibodies specific to the small and large subunits following Western blotting and the number of RuBP binding sites determined by ^{14}C[2-carboxy-D-arabinitol-1,5-bisphosphate] (CABP) binding according to Hall et al. (7). Total carbon and nitrogen contents of leaves were determined by isotopic analysis (Europa Scientific 20-20, Crewe, UK). Foliar carbohydrate composition and content were determined enzymatically (8,9,10), and the amount of fructan present measured by high performance liquid chromatography (11).

3. Results and Discussion

Light-saturated rates of photosynthesis (A_{sat}) in the youngest fully expanded leaves were stimulated by growth at high CO_2 in both LL and HL plants (Table 1). Increases in A_{sat} at elevated CO_2 were accompanied by increases in the dark respiration rates of the leaves (Table 1), particularly in plants grown at low irradiance. The A/c_i response curves indicated that while A_{sat} was stimulated at elevated CO_2, the carboxylation rates, determined by $V_{c,max}$, were decreased (Table 1). Similarly, while maximal Rubisco activities were comparable in plants grown in air and high CO_2, a pronounced decrease in the activation state of the enzyme was found in plants exposed to CO_2- enrichment (Table 1).

Growth Treatment		A_{sat} (μmol m^{-2}s^{-1})	$V_{c,max}$ (μmol m^{-2}s^{-1})	J_{max} (μmol m^{-2}s^{-1})	Rubisco Activation State %	Dark Respiration (μmol O$_2$ m^{-2}s^{-1})
AIR	LL	13.9 ± 1	61.5 ± 4	127.5 ± 7	64.5	0.63 ± 0.1
	HL	14.4 ± 1	67.0 ± 7	162.2 ± 12	63.4	1.34 ± 0.2
CO_2	LL	20.0 ± 1	56.9 ± 4	150.1 ± 5	58.1	1.06 ± 0.1
	HL	19.4 ± 1	43.4 ± 5	170.0 ± 16	46.5	1.56 ± 0.2

Table 1. Estimates of $V_{c,max}$, J_{max}, and measurements of A_{sat}, Rubisco activation states and dark respiration rates in the fourth leaf of *Lolium temulentum*. Values are means ± SE for four replicates per treatment.

Total foliar Rubisco contents, determined by immunodetection, and the number of available RuBP binding sites, determined by binding of radio-labelled CABP, were increased by growth at elevated CO_2 (Fig. 1). The carbon to nitrogen ratios of the leaves of plants grown with CO_2 enrichment were lower than those of plants grown in air (Table 2). Despite the marked increase in A_{sat} in plants grown continuously with elevated CO_2, relative growth rates and total plant biomass were comparable to those of plants grown in air (Table 2). CO_2-induced increases in leaf carbohydrate accumulation were observed only in plants grown at the low irradiance (Fig. 2). Fructan biosynthesis was not stimulated by growth with CO_2 enrichment. Fructan accounted for less than 10% of the total carbohydrate pool and leaf carbohydrate composition was similar in all growth conditions.

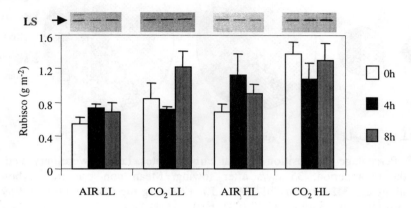

Figure 1. Determination of Rubisco contents by immunodetection and by binding of $^{14}C[CABP]$ throughout the photoperiod. LS (large subunit) of Rubisco.

Growth Treatment		Biomass (mg DW)	RGR (g g^{-1} d^{-1})	Foliar C:N
AIR	LL	105.3 ± 11	0.065 ± 0.014	5.15 ± 0.08
	HL	233.6 ± 29	0.082 ± 0.013	7.73 ± 0.20
CO_2	LL	108.4 ± 7	0.058 ± 0.014	4.58 ± 0.06
	HL	232.7 ± 20	0.056 ± 0.008	6.69 ± 0.18

Table 2. Total plant dry weight (DW) and foliar C:N ratios, 35 days after sowing. Relative growth rates (RGR) were calculated between 35 and 45 days after sowing. Values are means ± SE for six replicates per treatment.

Light-saturated photosynthesis in *L. temulentum* was stimulated by elevated CO_2, despite decreases in the activation state of Rubisco. In contrast to many plant species, growth at high CO_2 did not cause repression of photosynthetic capacity, Rubisco activity or Rubisco protein content in *L. temulentum*. Foliar N contents were similar or even enhanced compared to plants grown in air. Since total plant biomass was similar in plants grown in air and with CO_2 enrichment, we conclude that allocation of carbon for growth and dry matter production in *L. temulentum* is modified at elevated CO_2 such that increases in the use of carbon by other processes such as respiration or root exudation offset increased carbon gain.

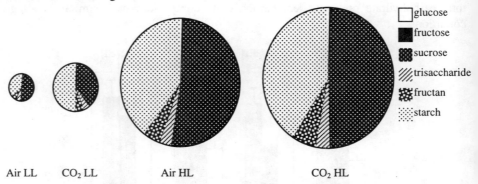

Air LL   CO₂ LL   Air HL   CO₂ HL

Figure 2. Percentage composition of total carbohydrates. Leaves were harvested at the end of the photoperiod, 35 days after sowing. Mean non-structural carbohydrate concentrations are 6420 µg mgChl⁻¹ (Air LL), 12000 µg mg Chl⁻¹ (CO₂ LL), 25954 µg mgChl⁻¹ (Air HL), 26545 µg mg Chl⁻¹ (CO₂ HL).

Acknowledgements. We are indebted to Martin Parry for the generous gift of ¹⁴C-labelled CABP, and to Steven Long for collaboration regarding A/c$_i$ analysis.

References

1 Cambell W.J., Allen L.H. & Bowes G. (1988) Plant Physiol. 88, 1310-1316

2 Lawlor D.W., Delgado E., Habash S.P., Driscoll S.P., Mitchell R.A.C & Parry M.A.J. (1995) in Photosynthesis: from light to biosphere (P. Mathis, ed.) pp.989-992, Kluwer Academic Publishers, Dordrecht, The Netherlands

3 Fordham M.C., Barnes J.D., Bettarini I., Griffiths H.G., Miglietta F. & Raschi A. (1997) in Plant responses to elevated CO₂: evidence from natural springs (A. Raschi, F. Miglietta, R. Tognetti & P.R. Gardingen, eds.) pp.174-196, Cambridge University Press, UK

4 Hoagland D.R & Arnon D.I. (1938). University of California Ag. Exp. St. Circular, 347

5 Evans J.R. & Farquhar G.D. (1991) in Modelling Crop Photosynthesis: from Biochemistry to canopy (K.J. Boote & R.S. Loomis, eds) pp.1-16, Crop Soc. America Inc

6 Parry M.A.J., Keys A.J., Foyer C.H., Furbank R.T. & Walker D.A. (1988) Plant Physiol. 87, 558-561

7 Hall N.P., Pierce J. & Tolbert N.E. (1981) Arch. Biochem. Biophys. 212, 115-119

8 Stitt M., Wirtz W. & Heldt H.W. (1980) Biochim. Biophys. Acta 593, 85-102

9 Lunn J.E. & Hatch M.D. (1995) Planta 197, 385-391

10 Bergmeyer H.V., Bernt E., Schmidt F. & Stauk H. (1974) in Methods of enzymatic analysis (H.V. Bergmeyer, ed.) pp.1196-1202, Verlag Chemie, Weinheim

11 Cairns A.J. & Pollock C.J. (1988) New Phytol. 109, 399-405

PHOTOSYNTHETIC CHARACTERISTICS OF LEAVES OF A DIOECIOUS SPECIES (*PISTACIA LENTISCUS L.*) UNDER SUMMER STRESS CONDITIONS

Correia O., Diaz Barradas M.C.[1]
Departamento de Biologia Vegetal. Faculdade de Ciências da Universidade de Lisboa. Campo Grande. C2.1700 Lisboa. Portugal. [1]Departamento de Biología Vegetal y Ecología. Universidad de Sevilla. Spain

Key Words: ecology, dioicy, drought stress, gas exchange, fluorescence.

1. Introduction
Pistacia lentiscus is a common dioecious evergreen shrub occurring in the mediterranean ecosystems. Previous studies of spatial distribution of male and female shrubs demonstrated that less disturbed areas, with a well developed vegetation cover, had male-biased sex ratios, whereas in old abandoned agricultural areas there were no significant differences between the number of male and female plants (1). Periodic droughts, are characteristic of mediterranean ecosystems and may be largely responsible for the dominance of male plants in more advanced stages of sucession when competition and stress are enhanced. The male-biased sex ratios in low resource environments have been supported by other studies (2,3). What are the characteristics that would confer a competitive advantage to these male individuals on dry sites and poor soils? Females usually show higher reproductive effort than males because they produce fruits in addition to flowers, and thus allocate more biomass to reproduction than males do (4,5,6). In this study we investigated if differences in ecologically important photosynthetic characteristics between male and female plants, could play a role in spatial sex segregation.

2. Procedure
2.1 Species and site characteristics
The field study was conducted in central Portugal, Serra da Arrábida (38° 27' 34" N; 9° 0' 20" W, elevation 270 m), in a mixed sclerophyll scrub, 20 to 30 years old, on a south facing slope. *Pistacia lentiscus* L.(Anacardiaceae) is a dioiceous sclerophyllous evergreen species that forms up to 2 m high shrubs, sometimes attaining a tree pattern in more humid and protected sites.

2.2 Gas exchange measurements
Measurements of net photosynthetic rates (A) stomatal conductance (g), and transpiration rates (Tr) in the field were made using a portable compact CO_2/H_2O porometer (Walz, Effeltrich, Germany). Ten measurements were made on four well exposed leaves from current years twigs in mature male and female shrubs growing adjacent to each other, during two summer days of 1996. Parallel measurements of leaf water potential (Ψ) was also conducted using a pressure chamber. Total N

G. Garab (ed.), Photosynthesis: Mechanisms and Effects, Vol. V, 4039–4042.

contents of leaves where photosynthetic measurements were done were determined with a C/N Elemental Analyzer EA 1108 (Fisons).

2.3 Measurements of chlorophyll fluorescence
Measurements of optimal quantum yield of PSII of dark-adapted leaves (F_v/F_m), efective quantum yield (Φ_{II}) and the apparent photosynthetic electron transport rate (ETR) were measured with the pulse-amplitude-modulation technique with a PAM-2000 portable fluorometer (Walz, Effeltrich, Germany) (7).

3. Results and Discussion

3.1 Gas exchange
Maximum photosynthetic rates and stomatal conductance were found only in midmorning in both sexes and declined throughout the day. These restrictions of main activity agree with the behaviour exhibited by several mediterranean sclerophyllous species during summer with one peak in the morning and a long shoulder in the afternoon (8). Light saturation of net photosynthesis occurred before noon at a photon flux density of 1500 μmol m^{-2} s^{-1} for both sexes. Female plants presented lower A and g than male. Daily variation in water potential (Ψ) was similar in both sexes, with a minimum around midday, but with lower values for female plants (Table I).

Table I. Values of maximum photosynthetic rates (A_{max}), maximum stomatal conductance (g_{max}), water use efficiency (WUE) expressed as A/Tr, photosynthetic nitrogen use efficiency (PNUE) and minimum water potential (Ψ_{min}) for male and female plants observed during the diurnal courses of the days 24 July and 1 August. Mean values of 6 leaves $\pm$ standard deviation.

Sex	A_{max} μmol m^{-2} s^{-1}	G_{max} mmol m^{-2} s^{-1}	WUE mmol mol^{-1}	N mmol m^{-2}	PNUE μmol mol^{-1} s^{-1}	Ψ_{min} MPa
Male	9.10±1.18	98.63 ± 21.52	4.87 ± 0.95	217.48 ±37.83	43.46 ± 12.46	-2.95± 0.07
Female	4.32±1.60	29.53±10.59	8.11± 3.87	182.38 ±22.05	23.85 ± 8.38	-3.05±0.21
t-test	p<0.001	p<0.001	NS	NS	P<0.05	NS

Male plants displayed significantly higher values of A_{max} and g_{max} during the study period. A greater water use efficiency was found in female plants than in male plants. The higher leaf nitrogen concentration in male plants was associated with a proportionally higher A. In female plants a high WUE was associated with a relatively low photosynthetic nitrogen use efficiency (Table I). A trade-off between the efficiency of the use of water and the use of nitrogen in photosynthesis has been found in others evergreen species (9).

3.2 Chlorophyll fluorescence
The F_v/F_m around midday declines relatively to the maximal values (around 0.80) probably reflecting the influence of high radiation and summer water stress.
On the second day, as water stress was progressively increasing (Ψ_{min}= -3.2 MPa) a moderate decline in F_v/F_m was observed, with the female plants showing lower values (Table II). A diurnal change in the F_v/F_m ratio on sunny days has been

described for other species reaching minimum values at noon when radiation is maximal, with a sustained recovery in late afternoon (10).

Table II. Maximum photochemical efficiency of PS II (F_v/F_m) of dark-adapted leaves, measured at midday (12:00 -15:00, solar time) on male and female plants. Values are means of 10 measurements per shrub ± standard deviation (SD).

Day	Male $F_v/F_m \pm SD$	Female $F_v/F_m \pm SD$
24 July	0.78±0.013	0.78±0.021
1 August	0.69±0.047	0.66±0.066

In Fig.1 the PFD dependence of the efficiency of energy conversion in PS II (Φ_{II}) at steady state for the two sexes is shown. The courses of this efficiency displays the same shape for both plants.

The Φ_{II} determined in the absence of excess light was high for both plants, and showed a strong decline at very moderate PFDs. The decline in efficiency was more pronounced in male plants, although the slopes of the linear regression were not significantly different between sexes.

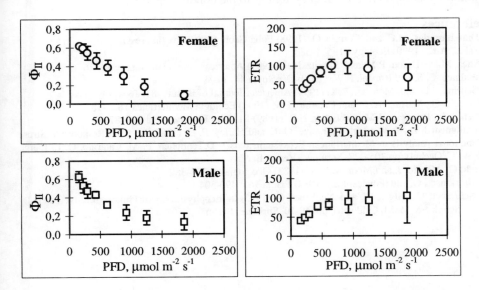

Figure 1. Efficiency of photosystem II (Φ_{II}) and the electron transport rate of PS II (ETR) at steady state in relation to the PFD incident on the leaf.

The electron transport rate of PS II photochemistry (ETR) increased with PFDs. Both plants exhibited the same pattern with a maximum of ETR around a photon flux density of 1800 μmol m^{-2} s^{-1} for male plants and 800 μmol m^{-2} s^{-1} for female plants with a decline for higher values of photon flux density. The differences obtained in light response curves of electron transport, with female plants being photosynthetically more efficient at low radiation and male the most active at higher radiations, are consistent with the values of A_{max} (Table I). Probably the limitation on photosynthesis during water stress is more pronounced in female plants due only to a stomatal restriction of CO_2 availability.

A reduction of the photosynthetic reactions under summer drought stress and high photon flux density is a widespread phenomenon, which can be a consequence of both stomatal closure causing increased constraint on CO_2 diffusion, and non-stomatal limitation such as decreased chloroplast activity. The extent to which a given light intensity is excessive depends on many factors, including light and temperature acclimation, water status and stomatal opening, the developmental state and the previous treatments (11).

These data suggest that male plants are relatively more tolerant to conditions of severe soil and atmospheric stress than female plants and are able to maintain a substantial positive carbon balance. From this study we suggest that the differences between male and female are due exclusivelly to a stomatal control and not to differences in photosynthetic apparatus. In fact, although female plants substantially decrease g during summer days, their water potential decreases to the same values as in male plants. Despite the higher control of water loss by female plants they still have to reduce considerably their water potential, possibly because of a lower water availability associated to the root system or to the conducting xylem.

References

1. Diaz Barradas, M.C. and Correia, O. (1998) Folia Geob. et Phitotax (in press)
2. Sakai, A.K. (1990) Ecology 71, 571-580
3. Shea, M.M., Dixon, P.M. and Sharitz, R.R. (1993) American Journal of Botany 80, 26-30
4. Wallace, C. S. and Rundel, P.W. (1979) Oecologia 44, 34-39
5. Gehring, J.L. and Monson, R.K. (1994) American Journal of Botany 81, 166-174
6. Correia, O., Martins, A. C. and Catarino, F.M.(1992) Ecologia Mediterranea 18, 7-18
7. Genty, B., Briantais, J-M. and Baker, N.R. (1989) Biochim Biophys Acta 990, 87-92
8. Tenhunen, J. D., Beyschlag, W., Lange, O. L. and Harley, P. C. (1987) In: Plant Response to Stress Functional Analysis in Mediterranean Ecosystems.(eds J. D.Tenhunen, F. M. Catarino, O. L. Lange & W. C. Oechel), pp 305-327. Springer -Verlag.Berlin
9. Field,C., Merino, J. and Mooney, H.A. (1983) Oecologia 60,384-389
10. Björkman , O. and Powles, S.B. (1984) Planta 161: 490-504
11. Schreiber, U., Bilger, W. and Neubauer, C. (1995) In: Ecophysiology of Photosynthesis. (eds E-D. Schulze & M. M. Caldwell), pp. 49-70. Springer, Berlin

ADAPTIVE RESPONSES OF LEAVES TO DROUGHT IN PRODUCTION PROCESS: WHICH OF THEM ARE ADVANTAGEOUS FOR MAINTAINING PHOTOSYNTHESIS AND PRODUCTIVITY?

M. Brestic - K. Olsovska - J. Mika - A. Kostrej
Dept. of Plant Physiology, The Slovak Agricultural University,
A. Hlinku 2, 949 76 Nitra, Slovakia, brestic@afnet.uniag.sk

Key words: drought stress, adaptation, net photosynthesis, stomata, osmotic pressure, productivity

1. Introduction

The plant production process depends primarily on photosynthetic activity, which is affected frequently by limiting ecological interactions. Plants evolved high level of adaptation to various ecological constraints that is usually more complex, including major and minor factors influenced crop productivity. The plant seedlings with intensively growing leaves dispose of a potential to adaptation, which is different from the potential of mature plants with fully expanded leaves. In cereals, the drought after the developmental stage of anthesis, when only the upper leaves remain functional acts considerably in a tendency towards limitation of production process. In this period, the penultimate, below-flag leaf of barley plays a crucial role in the yield formation. Therefore, the study of plant adaptive responses and drought tolerance seems to be highly actual where the photosynthetic activity and production stability under the shortage of available soil water are advantageous.

This experiment was performed to describe the problems as related to gas exchange limitations on the whole plant level as well as connected with the selection of barley genotypes tolerant to water stress.

2. Materials and methods

The 3 ecologically distant genotypes of barley (Hordeum vulgare L.), such as Kompakt (Slovakia), Agnette (Sweden) and Tagide (Portugal) were cultivated in the natural climatic conditions in 25 l plastic pots with a soil substrate supplemented by mineral nutrients up to the level of 1:0,88:2,01 (N:P:K). 60 pots (40 plants per one pot) were placed to simulate a compact canopy. Additional pots eliminated the marginal effect of canopy. The plants were watered regularly to maintain the 70 % of total available water.

7 days after anthesis the pots were replaced into the greenhouse ($27\pm2/20\pm1$°C day/ night temperature, 50 ± 5 % relative air humidity) for the plant acclimation. In the period of

G. Garab (ed.), Photosynthesis: Mechanisms and Effects, Vol. V, 4043–4046.
© 1998 *Kluwer Academic Publishers. Printed in the Netherlands.*

intensive grain formation, after simulation of maximal soil water hydration the plants were subjected to 9-day dehydration by withholding water.

We measured the net CO_2 assimilation, transpiration and stomatal conductance of the penultimate, below-flag leaf closed in the 250 cm^3 leaf chamber of the portable photosynthetic system LI-COR 6200 (LI-COR, Lincoln, Nebraska, USA) at 40% of relative air humidity, 650 $\mu mol.m^{-2}.s^{-1}$ of irradiation (PAR), 350 ± 10 $cm^3.m^{-3}$ of CO_2 and 28 °C of air temperature. Water (ψ_w) and osmotic (ψ_Π) potentials (Wescor, Wescor, Logan, Utah, USA), leaf diffusion resistance (Rs, porometer Delta-T-Devices, Cambridge, UK) and relative water content (RWC, from fresh, saturated and dry mass weights) were measured immediately after the photosynthesis measurements.

In the period of full grain maturity the current agronomic characteristics of whole plant, main stem and tiller productivity were quantified and the detailed analyses of ear productivity, such as spikelet sterility and dry mass accumulation into the individual grains in the ear structure were done.

3. Results and discussion

In cereals, drought evoked during the period of grain formation decreases the net CO_2 assimilation, shortens the leaf life durability and fastens the grain filling [1]. We supposed to study the drought effect on barley genotypes rather in this period.

In our experiment, drought was simulated in order to realise the mechanisms of plant adaptation to developing water deficit. It is known for years that the photosynthetic

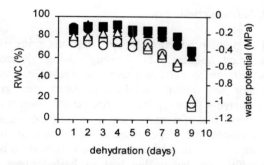

Figure 1: Water relations: RWC in % (black symbols) and water potential in MPa (empty symbols) during dehydration of intact barley plants (square - Kompakt, circle - Agnette, triangel - Tagide).

apparatus is resistant to water stress up to the level of 25-30 % water deficit [2, 3, 4]. The slowly developing water stress induces the ABA-dependent stomata closure [5] and stomatal inhibition of photosynthesis [6].

As seen from Figure 1 the RWC and ψ_w on the 7th day of dehydration remained on the level of 87-80 % and -0,422 to -0,479 MPa, respectively. In this period, measured CO_2 assimilation varied from 12-15 $\mu mol.m^{-2}.s^{-1}$ (Figure 2). Decrease in photosynthesis of dehydrated plants compared to well watered plants was caused by lowered total leaf conductance as related to heterogenous

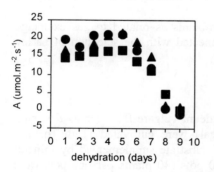

Figure 2: The net CO_2 assimilation of penultimate, below-flag leaf during barley dehydration (square - Kompakt, circle - Agnette, triangel - Tagide).

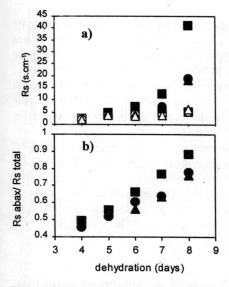

Figure 3: a) Below-flag leaf diffusion resistance in abaxial (black symbols) and adaxial (empty symbols) sides b) The abaxial to total leaf diffusion resistance ratio as related to dehydration (square - Kompakt, circle - Agnette, triangel - Tagide).

stomata closure in the abaxial and adaxial leaf sides. Our data (Figure 3) present the 2-4 day delaying the conductance decrease in the abaxial leaf side leading to modified Rs_{abax}/Rs_{total} ratio and to amplified the differences among studied genotypes. The phase of heterogenous decrease of leaf conductance accompanied the leaf osmotic adjustment (OA). Kompakt with higher heterogeneity in stomata closure maintained the lowest osmotic potential from the 4th day of dehydration. The ψ_Π variability among barley genotypes became larger as stomata started to close (data not shown). However, there is already a considerable flow of assimilates into the grains during the grain filling and we found a less significant decrease of ψ_Π than was already observed during the intensive plant growth [7, 8]. The morphological responses, such as leaf twisting, spiralling and leaf to stem angle and green to dry leaves ratio preserved high water content. Complexity and speciality of the leaf morphological symptoms are more difficult to relate to the level of water deficit than OA presented by lowered osmotic potential (ψ_Π). The leaf twisting and reduced leaf to stem angle (Tagide) seem to be more effective for maintaining the photosynthetic activity than interactions of other symptoms. This trait did not correlate with higher OA (Agnette).

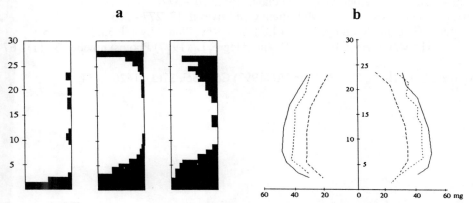

Figure 4: Possible limitations of barley grain fertility (a), accumulative effect (b) and genotype variability after drought (black spaces - degree of reduction of grain number: 0 - 30)

The question is whether the plant adaptive responses typical for different genotype origins and regulating the photosynthetic activity are advantageous for final plant productivity. As we have observed, the genotypes with higher MA during the grain filling period disposed of a higher capacity for adaptation to drought. Resistance to water stress includes more components improving plant ecostability, but the morphological attributes, heterogeneity of stomata closure and osmotic adjustment play a decisive role in the maintaining photosynthetic potential and in the screening barley genotypes for selection under drought.

In our present studies of plant adaptive responses attention has been paid to $\delta^{13}C$ variations, ear sterility and accumulative effect of individual grains (Figure 4) in the ear structure. From the agronomic point of view their correlations are convenient to be included in the models of ear structure formation. To understand the work of photosynthetic apparatus in conditions limiting the crop production as predicted by models of global climatic changes, the multimethodical approach is discussed.

Acknowledgment

This work was supported by a national research project of the Ministry of Agriculture. We thank to Livia Remzova for her technical assistance at the plant analyses.

References:

1 Brestic, M., Olsovska, K. and Kalaji, H.M. (1996) in Proceed. „Progress in Plant Sciences: from Plant Breeding to Growth Regulation" (Ordog V., Szigetti eds.), PAU Mosonmagyaróvár, Hungary, 47 - 52

2 Chaves, M. M. (1991) J. Exp. Bot. 42, 1 - 16

3 Cornic, G., Ghashghaie, J., Genty, B. and Briantais, J.M. (1992) Photosynthetica 27(3), 295 - 309

4 Tourneaux, C.and Peltier, G. (1995) Planta 195, 570 – 577

5 Zhang, J. and Davies, W.J. (1990) Plant Cell Environ. 13, 277 - 285

6 Farquhar, G.D. and Sharkey, T.D. (1982) Ann. Rev. Plant Physiol. 33, 317 - 345

7 Arneau, G., Monneveux, P., This, D. and Alegre, L. (1997) Photosynthetica 34 (1), 67 - 76

8 Nguyen, H.T., Babu, C. and Blum, A. (1997) Crop Sci. 37(3), 1426-1434

SPECIFIC FEATURES OF PHOTOSYNTHETIC APPARATUS OF PLANTS IN HIGHT MOUNTAIN DESERTS OF THE PAMIRS UNDER ECOLOGICAL STRESS

Kondratchouk A., Pyankov V., Urals State University,
51 Lenin Av., 620083, Ekaterinburg, Russia

Key words: alpine plants, cell structure, environmental stress, leaf anatomy, adaptation, microscopy.

1. Introduction

Alpine plants are the interesting objects for investigation of adaptation mechanisms, because they are influenced on by the complex of unfavourable treatments as low temperature, high insolation, increased UV irradiation, and low partial CO_2 pressure. The possibility for habitation of plants in alpine conditions are connected with adaptation of basic physiological functions, and photosynthesis in particular to extreme environment.

Obviously, plant adaptation to the extreme climatic conditions touches on the structure of assimilating organs that supply assimilation of carbon dioxide and transpiration, i.e. control carbon metabolism and water relations at the leaf level.

The investigations of the structure of assimilating organs in alpine plants which inhabited Central Alps, New Zealand, Himalayas (1) defined same common directions of adaptation of plants on the level of assimilating apparatus to the conditions in high altitudes. On the other hand difference between mountain areas are unlike and characterised by particular environmental conditions and specific boundaries of altitude belts. This causes specific features of adaptive mechanism in plants, inhabiting different mountains.

Eastern Pamirs distinguishes from other mountain system by an extremely dry continental climate that lead to formation of a specific type of xerophilic communities, the so-called mountain deserts. Thus, in addition to the extreme factors listed above, plants inhabiting Eastern Pamirs are influenced on by moisture deficit, both in soil and air, which is comparable in its severity to desert areas. The combined effects of environmental factors in Eastern Pamirs are resulted in a specific type of plant community that combines features of both alpine and deserts species.

The studies were focused on the changes of quantitative mesophyll structure in the same species, inhabiting different elevations and moisture conditions of the Eastern Pamirs.

2. Procedure

This study was conducted in the Eastern Pamirs, region of biological station of Academy of Sciences of Tadjikistan. 12 species of dicotyledonous from 12 families were collected in altitudinal range from 3800 m to 4750 m (Tabl.1). The same species growing at the sub-alpine belt 3600 - 4100 m and alpine belt 4100 - 4700 m were compared.

Plant samples were collected in Chechecty river valley (3860m), near the salt lake Rang-Kul (3800m), in the closed humid circus Mukor-Chechecty (4200-4300m), in glacial circus Zor-Chechecty (4750m).

G. Garab (ed.), Photosynthesis: Mechanisms and Effects, Vol. V, 4047–4050.
© *1998 Kluwer Academic Publishers. Printed in the Netherlands.*

Table 1. Ecological characteristics and mesophyll structure of 12 species of the Eastern Pamirs inhabiting different altitudinal belts.

Species	Life form	Ecomorph	Area type	Type of mesophyll structure	Leaf structure sub-alpine belt	alpine belt
Ajania tibetica	ssh	x	ca	ip	2p/2p	2p/2p
Taraxacum dissectum	ph	m	ea	dv	3p/2-3s	3p/3-4s
Taraxacum leucanthum	ph	m	ca	dv	3p/3-4s	3p/3-4s
Smelowskia pectinata	cp	x	ma	ip	5-6p/5p	4p/4-3p
Ceratoides papposa	ssh	x	pa	ip	2p/1-2r+2p	2p/1-2s+2p
Hedysarum minjanense	ph	xm	ma	ip	2-3p/2p	2-3p/2p
Corydalis stricta	ph	xm	ca	dv	2p/3-4s	3p/4-5s
Swertia marginata	ph	m	ca	hom	8s	6-7s
Polygonum pamiricum	ph	xm	ca	ip	3p/3-4p * 4p/4p**	
Polygonum viviparum	ph	m	ga	dv	3p/3-4s	4p/4-5s
Ranunculus rufocepalus	ph	m	ca	dv		3p/3-4s*** 3p/2-3s****
Potentilla moorcroftii	ph	xm	ca	dv	2p/4-5s	3p/4-5s

Symbols: ssh - semishrab, ph - perennial herb, cp - cushion plant; x - xerophyte, m - mesophyte, xm- xeromesophyte; ca - central asian, ma - mid-asian, ea - eurasian, pa - palearctic; ip - isopalisade, dv - dorsoventral, hom - homogenous, p - palisade mesophyll layer, s - spongy mesophyll layer; * - 3800m, ** - 3860 m, *** - 4750m, **** - 4300m.

Plant samples were collected in Chechecty river valley (3860m), near the salt lake Rang-Kul (3800m), in the closed humid circus Mukor-Chechecty (4200-4300m), in glacial circus Zor-Chechecty (4750m). Vertical climatic belts were distinguished by average monthly temperature and fluctuations of daily temperatures, the amount of precipitation, presence of summer frost period and the length of vegetation period (Tabl.2.). Sampling period were mid June and July characterised by maximum of vegetative development and peak flowering at each altitude.

Table2. Description of the same climatic parameters of Eastern Pamirs

Belts	Elevation	Temperature, ^{0}C		Amount of precipitation	Morning frost-free period	Vegetation period
	m	June	Annual	mm	days	days
alpine	4100 - 4700	0-5	n.d.	150-250	not	60 - 80
sub-alpine	3600 - 4100	5-10	-1-2	60-125	25-72	120 - 180

The quantitative anatomy of the mesophyll was examined by «mesostructure method» according to Mokronosov (2,3), Pyankov et al. (1998).

It is based on a series of quantitative determinations of photosynthetic tissues: leaf thickness, leaf mass per unit leaf area, size and number of mesophyll cells per unit leaf area and per unit leaf volume, size and number of chloroplasts per cell. These data are then used to calculate derived parameters such as surface area and volume of cells and chloroplasts and total surface area of mesophyll cells (A_{mes}/A) and chloroplasts (A_{chl}/A) per unit leaf area.

3. Result and Discussion

12 species growing in different altitudinal areas include a wide variety of life forms, ecological groups, types of mesophyll structure.
We have distinguished three structural groups of plants, differs in mesophyll organization (homogenous, dorsoventral and isopalisade), and have shown their relation to different ecological conditions of habitation. Plants with isopalisade type of mesophyll grew in dry sites. This group consisted of edificators and dominants of deserts communities and cushion plants (Ajania tibetica, Ceratoides papposa, Smelowskia pectinata) and species which inhabited in the dry slopes (Hedysarum minjanense, Polygonum pamiricum). Mesophyll of such species consisted of 4-10 layers of palisade cells. The plants, which inhabited in more humid places had dorsoventral structure of leaf blade (Taraxacum dissectum, Polygonum viviparum). This species had 2-4 layers of palisade and 2-5 layers of spongy cells. One species - Swertia marginata was characterized by homogenous mesophyll type, which consisted of 8-10 layers of asymmetric cells.

Table 3. The basic and derivative characteristics of photosynthetic apparatus in 12 alpine species inhabiting different altitudinal and moisture conditions of the Eastern Pamirs. Parameters: Tlf - leaf thickness; V_{cell} - cell volume; N_{cell}/S - number of cells (P+S) per unit leaf area; N_{chl}/S - number of chloroplasts (P+S) per unit leaf are; A_{chl}/A - surface of chloroplasts (P+S) per unit leaf area; A_{mes}/A - surface of cells (P+S) per unit leaf area; P+S - palisade and spongy mesophyll. The mean value of 30 determinations are presented.

Species	Site	Leaf area, 10^{-4} m^2	Tlf, µm	V_{cell}, 10^3µm^3	N_{cell}/S, 10^7/m^2	N_{chl}/S, 10^{10}/m^2	A_{chl}/A m^2/m^2	A_{mes}/A, m^2/m^2
Ajania tibetica	bs	0,24	267	10,0	1092,9	47,8	24,1	37,4
A. tibetica	zch	0,16	274	6,2	1041,9	48,7	22,8	26,5
Taraxacum dissectum	bs	4,96	378	32,8	295,4	15,2	7,9	7,6
T. dissectum	zch	0,87	418	16,0	354,3	12,7	7,2	17,4
T. leucanthum	bs	2,18	490	30,3	555,2	19,7	15,4	15,6
T. leucanthum	p	0,45	398	15,3	1104,4	44,4	29,6	9,9
Smelowskia pectinata	bs	0,14	586	7,9	595,3	19,7	8,9	11,4
S. pectinata	cm	1,27	237	3,3	1273,4	27,9	12,5	13,7
Ceratoides papposa	bs	0,93	349	14,0	706,5	35,9	23,3	17,8
C. papposa	zch	0,10	271	7,0	1983,7	70,3	47,5	31,7
Hedysarum minjanense	bs	2,90	238	2,9	1189,7	41,6	17,4	18,9
H. minjanense	zch	2,56	262	3,0	1501,1	46,1	22,8	23,6
Corydalis stricta	p	8,73	350	10,6	1813,2	56,3	44,6	18,6
C.stricta	ran	9,52	299	8,8	750,7	30,0	14,8	5,8
Swertia marginata	bs	4,45	393	29,7	259,3	22,9	11,1	12,0
S. marginata	cm	6,43	435	46,8	192,9	15,2	9,7	12,1
Polygonum pamiricum	bs	1,27	662	13,1	1049,2	66,3	36,4	41,7
P. pamiricum	ran	1,04	590	29,0	369,7	11,1	7,3	26,3
P. viviparum	bs	6,86	295	6,4	1120,1	33,7	17,4	13,6
P.viviparum	cm	2,75	443	11,4	611,6	18,8	10,6	7,7
Ranunculus rufocepalus	cm	4,22	446	27,4	428,0	29,5	16,2	12,0
R. rufosepalus	zch	1,68	413	36,2	309,4	28,6	17,6	11,7
Potentilla moorcroftii	bs	1,60	237	3,0	1482,8	72,5	29,5	22,5
P. moorcroftii	zch	1,51	325	7,6	2880,5	61,6	31,4	46,9

The quantitative analysis permits us to determine the specific traits in adaptation of plants with elevation on the level of assimilating apparatus, which are expressed in microphyllia of plants, increasing of leaf blade thickness and LMA (leaf mass per unit leaf area), development of palisade mesophyll, larger number of chlorenchyma cells and chloroplasts per unit leaf area.

Probably these peculiarities stipulate for the development of internal assimilation surface (A_{mes}/A and A_{chl}/A) and decrease of mesophyll resistance to CO_2 diffusion in stress conditions of Pamir mountains.

The comparative analysis of quantitative parameters of photosynthetic apparatus indicated the changes of leaf characteristics with elevation. Leaf area is continuing to reduce and thickness of blade to enlarge with elevation. However, we could not find general trend of changes in cell volume, number of cells and number of chloroplasts per unit leaf area along altitudinal gradient for the Pamirs plants. Nevertheless all this changes are regular and in xerophytes they are mainly connected with increase of the amount of mesophyll layers, decrease of cell volume and therefore enlarging of the number of cells per leaf area; in mesophytes - with increase of cell volume and decrease of cells amount per leaf area.

The revealed changes of quantitative parameters of photosynthetic apparatus firstly can be connected with their adaptation to the lack of moisture and low temperatures, because the level of solar radiation and CO_2 concentration insignificantly changes in the studied altitudinal range.

The study was supported in part by grant PBRF # 97-04-49900 and grant # 454 of the program «Universities of Russia» of the Ministry of High Education.

References

1. Korner, C., Neumayer, M., Pelaez Menendez-Riedl, S. and Smeets-Scheel, A. (1989) Flora 182, 353-383
2. Mokronosov, A.T. (1978) in the Mesostructure and functional activity of photosynthetic apparatus (Mokronosov, A.T., ed.) pp. 5-34, Urals State University, Sverdlovsk (in Russian)
3. Mokronosov, A.T. (1981) Ontogenetic aspect of photosynthesis. Nauka, Moscow (in Russian) 196 p.
4. Pyankov, V.I., Ivanova, L.A. and Lambers, H. (1998) in Inherent variation in plant growth. Physiological mechanisms and ecological consequences.(Lambers H. Et al., ed.) pp.71-87, Backhuys Publishers, Leiden, The Netherlands

EXPLORING THE ROLE OF EXTRACELLULAR ASCORBATE IN MEDIATING OZONE TOLERANCE

Enikö Turcsányi[1], Matthias Plöchl[2] and Jeremy Barnes[1]

[1]The Air Pollution Laboratory, Dept. Agricultural and Environmental Science, Ridley Building,The University, Newcastle Upon Tyne, NE1 7RU UK [2]Institute of Agricultural Engineering - Bornim (ATB), Max-Eyth-Allee 100, 14469 Potsdam-Bornim, Germany

Key words: apoplast, ozone, ascorbate, abiotic stress

Introduction

Tropospheric concentrations of ozone (O_3) are known to pose a growing threat to the vitality of natural and managed ecosystems in many parts of the industrialized world (1,2). However, the mechanisms underlying the phytotoxicity of this ubiquitous air pollutant are just beginning to be unravelled (3,4,5).

The flux of O_3 to the leaf interior is known to be predominantly controlled by stomatal conductance (6). Once inside the leaf, the pollutant is assumed to dissolve into (and react with constituents of) the aqueous matrix of the cell wall - the apoplast - resulting in an intercellular space O_3 concentration that is demonstrably close to zero (7).

Two main factors are considered to govern the resistance of plants to O_3; the physical exclusion of the pollutant from sensitive intracellular targets (8,9) and the capacity to intercept and detoxify O_3, and/or its reactive dissolution products - ideally, before reaction with components of the plasmalemma and cytosol (10). Over the past decade, growing attention has been paid to the role of certain constituents of the apoplast in attenuating the flux of O_3 impinging on the plasmalemma. In terms of potential detoxification capacity, attention has focused chiefly on the reaction of O_3 with ASC - the compound believed to top the hierarchical series of cell wall reaction-targets (11,12,13).

The aim of the present study was to employ broad bean (*Vicia faba* L.) as a model to examine the impacts of exposure to environmentally-relevant O_3 concentrations on ASC/DHA levels in the leaf apoplast/symplast *in vivo*, probe the relationships between ASC/DHA status and O_3 resistance, and estimate the theoretical extent of O_3 protection afforded by the reaction of O_3 with apoplastic ASC.

Results and conclusions

1. Exposure to acute levels of O_3 (150 nl L^{-1} O_3) caused a significant decline in the concentration of apoplastic ASC in both CFA- and O_3- grown plants, without affecting the content/redox state of ASC in the symplast. There was also evidence of a recovery

G. Garab (ed.), Photosynthesis: Mechanisms and Effects, Vol. V, 4051–4052.

in apoplastic ASC content in O_3-grown plants following acute O_3 exposure. These findings suggest that ASC in the leaf apoplast is consumed *in vivo* upon exposure to O_3

2. Chronic O_3 exposure results in lower stomatal conductance, and increased apoplastic ASC content.

3. A linear relationship exists between the concentration of ASC in the apoplast and symplast, consistent with suggestions that diffusion-limited processes constitute the predominant mechanism controlling the replenishment of apoplastic ASC.

4. Model calculations suggested that the concentration and rate of turn-over of ASC in the leaf apoplast is *potentially* sufficient to intercept 60-70% of the incoming O_3 at under external fumigation with 150 nl L^{-1} O_3.

Further work is urgently needed to establish whether (i) the potential scavenging of O_3 by apoplastic ASC is fully realized *in vivo,* and (ii) the role played by other constituents of the leaf apoplast in mediating the flux of O_3 to the plasmalemma.

Acknowledgments
The authors sincerely thank Dr. T. Lyons (AES, University of Newcastle) for his invaluable support, and also Phil Green, Alan White and Keith Taylor for technical assistance. The work was financed by through the Royal Society's East-West Postdoctoral Exchange Scheme with additional support to cover the day-to-day running costs of the fumigation facility provided by The Royal Society, Newcastle University Equipment Fund, The Swales Foundation and the Overseas Development Agency. The work was performed in JBs laboratory during his tenure of a Royal Society Research Fellowship.

References

1 UNECE (1996) Critical Levels for ozone in Europe: testing and finalizing the concepts (Kärenlämpi, L. and Skärby, L. eds.), University of Kuopio, Dept Ecol & Environ Sci, Kuopio, Finland.

2 Davison, A.W., Barnes, J.D. (1992) in Managing the human impact on the natural environment: patterns and processes (Newson, M.D. ed.), pp. 109-129, Bellhaven, New York

3 Kangasjärvi, J., Talvinen, J., Utriainen, M., Karjalainen, R. (1994) Plant Cell Environ 17, 783-794

4 Sharma, Y.K., Davis, R. (1997) Free Rad Biol Med 23, 480-488

5 Lyons, T.M., Plöchl, M., Barnes, J.D. (1998) in Environmental pollution and plant responses (Agrawal, S., Agrawal, M., Krizek, D.T., eds.), CRC Press, Boca Raton. In the press

6 Kersteins, G., Lendzian, K.J. (1989) New Phytol 112, 13-19

7 Laisk, A., Kull, O., Moldau, H. (1989) Plant Physiol 90, 1163-1167.

8 Runeckles, V.C. (1992) in Surface-level ozone exposures and their effects on vegetation (Lefohn, A.S. ed.), pp. 157-188, Lewis, Chelsea

9 Barnes, J.D., Bender, J., Borland, A.M. (1998) in Air pollution and biotechnology in plants (Omasa, K. ed.) Springer-Verlag, Tokyo Berlin. In the press

10 Chameides, W.L. (1989) Environ Sci Technol 23, 595-600

11 Kanofsky, J.R., Sima, S. (1995a) Arch Biochem Biophys 316, 52-62

12 Kanofsky, J.R., Sima, S. (1995b) J Biol Chem 270, 7850-7852

13 Mudway, I.S., Kelly, F.J. (1998) Toxicol Appl Pharmacol 148, 91-100

ECOPHYSIOLOGY OF PERENNIAL GRASSES UNDER WATER DEFICITS AND COMPETITION.

Pedrol N., Ramos P., Reigosa M.J.
Plant Ecophysiology. Dept. Bioloxía Vexetal e Ciencias do Solo.
Universidade de Vigo. Apd. 874, 36200 Vigo, Spain.

Key words: acclimation, biotic stress, drought stress, ecophysiology, osmoregulators, polyamines, salt stress.

1. Introduction

Galician peninsulas are characterised by the existence of strong bioclimatic gradients with regards to soil water availability, specially during summer drought, providing very different life conditions to plants from near places. Coastal populations of perennial grasses in the extreme of the gradient suffer the most severe water deficits, aggravated by salinity and the presence of sandy and thin soils. Some of these grasses occupy close ecological niches, competing closely for limited resources. Previous studies on grass species populations distributed along a summer drought gradient, in a maximum distance of 20 Km (O Morrazo peninsula and Cíes Islands. Galicia, NW Spain) have not shown genetic distances or physiological and morphological differences leading to consider different genotypes or ecotypes; thus, acclimation mechanisms and plasticity of metabolism are expected. The main objective of this work is the ecophysiological characterisation of three wild grasses from this area: *Holcus lanatus* L., *Koeleria glauca* (Schrader) DC., *Dactylis glomerata* L., compared to a pair of commercial cultivars widely used in this area: *D.glomerata* cv. Micol, and *Lolium perenne* cv. Sun., with regards to their capacities of acclimation to water stress: salinity and drought, besides to their competitive abilities.

2. Materials and methods

Pure stands of mentioned species were stablished with plants obtained from seeds collected from field and from commercial grain (20 seedlings per pot). The following treatments were applied, under greenhouse at controlled conditions, during 18 weeks: (i) two levels of salinity, watering pots with NaCl solutions 0.06 and 0.12 M, respectively, and (ii) two levels of water availability (stablished from the maximum water retention capacity of the soil=100%): well watered (100%) and water deficit (40%). Effects of treatments on all species were measured through relevant ecophysiological parameters: net photosynthetic rate (using a LI-COR 6200), yield (above ground biomass, *DW*), leaf size (leaf surface index=leaf length·leaf width), specific leaf area (SLA=dry weigh/unit of leaf area), water use efficiency (WUE index=yield/water used), leaf relative water content (RWC), leaf carbon/nitrogen rate (C/N), photosynthetic nitrogen use efficiency (PNUE), leaf protein content and leaf concentrations of stress metabolites: free proline, polyamines (by TLC) and phenolic acids (by HPLC).
Simultaneously, competition among wild grasses was studied under different culture

G. Garab (ed.), Photosynthesis: Mechanisms and Effects, Vol. V, 4053–4056.

conditions: (i) varying plant densities under well watered conditions, following an *additive* design: pure stands (20 plants per pot) and mixed stands (20+20 plants per pot) and (ii) varying contribution of each species to a total fixed density, following a *replacement* design: pure stands (20 plants per pot) and mixed stands (15+5, 10+10 and 5+15 plants per pot). Effects of competition were measured on productivity (yield, relative yield, leaf size, net photosynthetic rate and leaf protein contents) and stress metabolites (leaf free proline and polyamines).

3. Results and discussion

3.1 *Water deficit and salinity*

H.lanatus suffered the least reduction of leaf size due to stress. Under water deficit, it produced much more yield, showing a high WUE (Fig.1). Its SLA was the biggest when well watered, showing more plasticity in response to stress for this character than the other grasses (Fig.2). All these characteristics of a fast-growing species, will confer *H.lanatus* some competitive advantages in its natural environment, where water is a limited resource. A fast early growth can be an escape strategy to avoid water stress, and allows a greater soil cover, diminishing evaporation of water from the soil (1, 2).

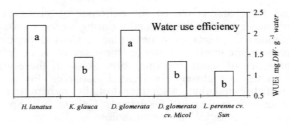

Figure 1. Water use efficiency in five perennial grasses. Bars labelled with distinct letters are significantly different (P≤0.05, n≥3).

Leaves of this grass accumulated big quantities of free proline, specially under salt stress and had big RWCs in all treatments (Fig.2). Proline accumulation contributes to osmotic adjustment, allowing plants to maintain water uptake under water deficit and salinity. *H.lanatus* was the unique species increasing quantities of leaf proteins significantly due to water deficit (Fig.2). It could be an adaptative response involving *de novo* synthesis of proteins with protective function against water deficit effects (3). Water deficit also induced an increase in leaf polyamine (spermidine and spermine) concentration (data not shown) with respect to diamines (putrescine).This response has been described as a symptom of stress tolerance, because induction of polyamine synthesis from diamines is related with the inhibition of senescence and the protection of cellular structures and macromolecules (4). Qualitative phenolic composition of *H.lanatus* changed with stress, increasing the diversity of phenolic acids both in high salinity and water deficit (with occurrence of *p*-hydroxibenzoic, syringic, and *p*-coumaric acids, all absent when well watered; data not shown).*K.glauca*, in spite of the stress induced reduction of yield and leaf size, did not decrease net photosynthetic rates significantly with drought and salinity (data not shown). Although the contribution of free proline to osmoregulation was minimum, leaf RWCs were high under water stress (Fig.2). This wild grass showed little WUE (Fig.1), but a very low SLA and a C/N ratio (Fig.2). This species must better invest photosynthate in storage of carbon-based compounds than in leaf surface, showing a scarce growth. The wild *D.glomerata* resulted very efficient in water use (Fig.1) and, besides to *H.lanatus*, produced the biggest yield under water deficit (data not shown). This grass accumulated higher quantities of free proline under water deficit than under salinity (Fig.2), possibly contributing to the maintenance of a high RWC. Under salt stress, the decrease in RWC was not an impediment to photosynthesis, showing higher values than those of *H.lanatus*. Other authors have reported big decreases in RWCs of *D.glomerata* from dry habitats, without significant effects on photosynthetic rates (5). Nevertheless, both photosynthesis and PNUE decreased a lot under adverse conditions (Fig.). *D.glomerata* cv. Micol was, as a whole, the most affected by stress. It showed a little plasticity in response to water deficit and salinity, mainly in characters related to growth, with a little WUE (Fig.1). Although *L.perenne* was very productive when well watered, it had low WUE values (Fig.1). Nevertheless, it showed symptoms of metabolism accommodation to adverse conditions.

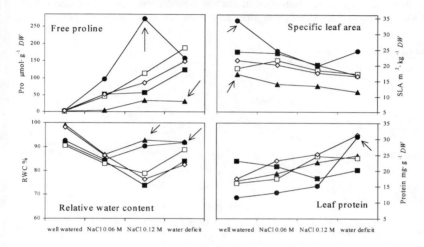

Figure 2. Effects of salinity and water deficit on *H.lanatus* (circles), *K.glauca* (triangles), wild *D.glomerata* (black squares), and two commercial cultivars of *D.glomerata* (white squares) and *L.perenne* (diamonds) measured on four ecophysiologycal parameters. Relevant results of statistical treatments are marked with arrows.

3.2. *Competition*

Results of well watered plants under *additive* competition (Fig.3) showed no significant effects of biotic stress on neither free proline nor polyamine contents (data not shown). Competitive pressure of *D.glomerata* caused a significant decrease in *H.lanatus* yield, but an increase in leaf size that could be due to light competition. No effect was observed on net photosynthetic rate of *H.lanatus* (data not shown), but its leaves accumulated lot of proteins. We argue a possible adaptative response to indirect water stress, caused by a fast water shortage at high plant density, maybe related to the improvement and protection of photosynthetic machinery. *K.glauca* produced no significant competitive effects on the development and metabolism of *H.lanatus*. Similar effects were found in *D.glomerata* under competition in additive mixtures (data not shown). Both leaf size and yield of *K.glauca* were drastically reduced in additive mixtures. Competition with *D.glomerata* reduced significantly its net photosynthetic rate, thus showing the importance of photosynthesis assessment (and other ecophysiological parameters) in

4056

order to discuss the outcome of competition (6). With respect to plants grown under *replacement* series (data not shown) relative yields in mixtures *H.lanatus-K.glauca* at different planting ratios, separated significantly from "null model of no significative competitive interactions" (7). When well watered, the absolute winner is *H. lanatus*. This superiority decreases under water deficit, where *H.lanatus* is maybe negatively affected by intraspecific competition for limited water. Mixtures of *H.lanatus-D.glomerata* were close to null model in most of the planting ratios (data not shown), showing similar competitive abilities and intensities. Possible resource complementarity (8) was observed in *D.glomerata-K.glauca* mixtures under high salinity, where *K.glauca* was closer to expected values and *D.glomerata* showed the highest relative yields.

H.lanatus and *D.glomerata* from the studied area conjugate high competitive abilities, a big WUE and fast growth with wide acclimation capacity. *K.glauca*, typical grass from arid zones, must storage large quantities of carbon based compounds that contribute to its sclerophyllous character. Its strategy against stress seems to be based on persistence.

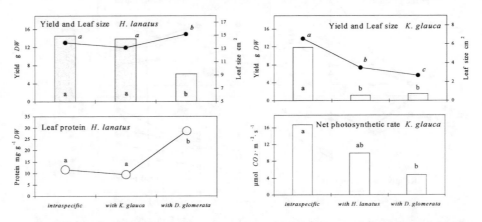

Figure 3. Effects of competition on different parameters measured on pure and mixed stands of three perennial grasses in an additive design. Bars labelled with distinct letters are significantly different ($P \leq 0.05$, $n \geq 3$).

References

1 Ehleringer J.R. (1993) in Water deficits-Plant responses from cell to community (Smith J.A.C. and Griffiths H., eds) pp.265-284, Bios Scientific Publishers, Oxford, UK.
2 Fischer R.A. (1980) in Adaptation of plants to water and high temperature stress (Turner N.C. and Kramer P.J., eds) pp.323-339, Wiley-Liss Inc. Publishers, New York, USA.
3 Ingram J. and Barlets D. (1996) in Annu. Rev. Plant Physiol. Plant Mol. Biol. 47, 377-403.
4 Tiburcio A.F., Campos J.L., Figueras X. and Besford R.T. (1993) in Plant Growth Regul. 12, 331-340.
5 Roy J. and Lumaret R. (1987) in Evol. Trend. Plant. 1, 9-19.
6 Szente K., Tuba Z., Nagy Z. and Csintalan Zs. (1993) in Weed Res. 33, 121-129.
7 Harper J.L. (1977) The population biology of plants, Academic Press, London, UK.
8 Snaydon R.W. and Satorre E.H. (1989) J. Appl. Ecol. 26: 1043-1057.

THE ROLE OF PHOTOSYNTHETIC ACTIVITY IN THE VULNERABILITY OF AN INSULAR BIOME TO INVASION BY ALIEN SPECIES.

Sándor Dulai[1], István Molnár[1], Endre Lehoczki[1], Tamás Pócs[2]
[1]Departments of Plant Physiology and [2]Botany, Eszterházy Teachers' College, H-3301 Eger POB, 43, Hungary

Key words: Chl fluorescence induction, gas exchange, environment, evolution, productiviy, plant invasion.

1. Introduction

Biological invasion by alien plants has been documented to have destructive effects on the native vegetation of insular biomes. This phenomenon is widespread throughout our investigation area: the islands in the Indian Ocean. The question is why the native plant communities of small islands are so vulnerable when faced with the invasion of alien, continental species. We hypothesise that these species came from a continental environment where, over a much larger area and usually during a much longer time, competition between a higher number of species with a much smaller realised niche space and a greater degree of niche overlap led to the evolution of a higher degree of competitiveness (1). On the small islands, on the other hand, the limited number of indigenous species, living in a broad niche space with little overlap, provided little selection pressure for competitive ability. How can we define and measure competitive ability? To test our hypothesis, we have compared ecophysiological parameters of indigenous, often endemic species on the island of Réunion with those of invasive species of continental origin. There are appreciable numbers of both types of species (2), and it was easy to select five typical species from each of the groups (Table 1). Any life process taken into account must be of fundamental significance, must be closely interconnected with other life processes, must reflect well the physiological state and vitality of the whole plant individual and must have an influence on the competitive ability. These requirements are well met by photosynthetic activity: it has an essential energy-transforming function, photosynthetic metabolites and chemically bound photoenergy influence other metabolic and physiological processes, and consequently the photosynthetic productivity is reflected in some form in the competitive ability. The functioning of photosystem II (PS II) is the most sensitive indicator of environmental and interspecific effects in the photosynthetic apparatus (3-5). Measurements of chlorophyll fluorescence associated with PS II activity, and of CO_2 gas-exchange, provide a means of rapidly and non-destructively probing the biophysical and biochemical bases of photosynthetic characteristics in the field (6-9). These methods can also be used to estimate plant growth, dry-matter production (10-12) and hence competitive ability.

G. Garab (ed.), Photosynthesis: Mechanisms and Effects, Vol. V, 4057–4060.
© 1998 *Kluwer Academic Publishers. Printed in the Netherlands.*

2. Procedure

The introduced and indigenous plant species to be investigated were selected from an *Acacia heterophylla*-dominated mountain rain forest on the Beluve plateau (1400 m above sea level) on the island of Réunion. All experiments were performed on intact leaves of these introduced and indigenous plants (Table 1).

The *in vivo* chlorophyll fluorescence measurements on intact leaves of investigated plants were carried out with a computerised portable chlorophyll fluorometer after a 30-min dark adaptation. The fluorescence was excited by a light-emitting diode (Stanley KR5004X) of 200 μmol m^{-2} s^{-1} photon flux density, and detected by a BPX-60 (Siemens) photodiode.

The light saturation of CO_2 assimilation was measured in normal air with an infrared gas analyser in an open gas-exchange system at 20 °C. The rates of net CO_2 fixation were calculated by using the equations of von Caemmerer and Farquhar (9).

3. Resuts and discussion

In forest ecology, one of the most frequently used fluorescence parameters is the ratio F_v/F_m measured in dark-adapted leaves, which indicates the maximum possible quantum yield of PS II and, in a general understanding, correlates well with the CO_2 fixation under certain conditions. The significant difference observed between the F_v/F_m values of the introduced and the indigenous species (Table 1) clearly shows that primary charge separation is more efficient in the introduced plants. Further, the higher F_i values (data not shown) of the indigenous plants indicate a deceleration of the $Q_A \rightarrow Q_B$ electron transfer within the photosynthetic electron transport.

Table 1. F_v/F_m and F_s values of introduced and indigenous species.
*Endemic taxa of Réunion; **Mascarene endemics (Réunion, Mauritius).

Introduced species group	F_v/F_m	F_s	Indigenous species group	F_v/F_m	F_s
Psidium cattleyanum	0.74	32,38	Cyathea glauca*	0.67	39,59
Hedychium flavescens	0.76	33,03	Acacia heterophylla*	0.72	37,49
Duchesnea indica	0.78	31,85	Cordyline mauritiana**	0.71	33,85
Erigeron karwinskyanus	0.79	29,2	Aphloia theaeformis	0.73	35,91
Fuchsia x exoniensis	0.78	28,23	Forgesia borbonica*	0.74	36,06
Average values for introduced group	0.77	30,9	Average values for indigenous group	0.71	36,59

Consequently, the elevated proportion of Q_A^- (higher F_i) in the indigenous group results in a larger fraction of closed reaction centres incapable of stable charge separation, and thus in a decrease in the light-limited quantum efficiency of the PS II photochemistry. Although the significance of these primary processes cannot be questioned, they do not in themselves explain the strong competitive ability of the introduced plant species, a precondition of invasivity, since the dry-matter production and vigorous growth at optimum quantum yield are influenced by other photosynthetic processes too.

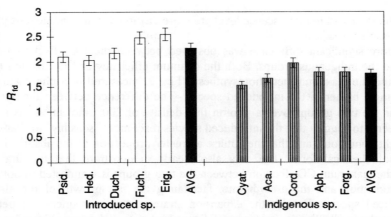

Fig. 1. Fluorescence decrease ratio (R_{fd}) values for the introduced (open bars) and indigenous (shaded bars) plats. AVG: average R_{fd} values of two groups (closed bars).

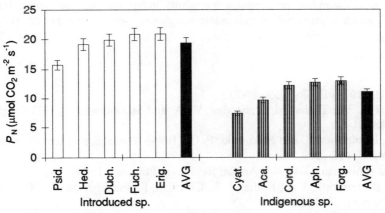

Fig. 2. Maximum net CO_2 fixation (P_N) values for the introduced (open bars) and indigenous (shaded bars) plats. AVG: average P_N values of two groups (closed bars).

At the steady-state level of photosynthesis, fluorescence and photosynthesis are complementary processes. The steady-state fluorescence level (Table 1) is determined jointly by photochemical and non-photochemical quenching, the latter being linked to the protecting and regulating mechanisms of the photosynthetic apparatus. The value of the fluorescence decrease ratio (R_{fd}) values characterises the potential photosynthetic activity of the leaves (22, 23); similarly to F_s, this was found to differ significantly in the invasive and indigenous plants (Fig. 1). The average R_{fd} value of 2.3 for the introduced plants reveals a very promising potential photosynthetic activity level, whereas the indigenous group possesses only a weaker level (R_{fd} = 1.7). This may imply that the processes determining the steady-state fluorescence level are more disturbed in the indigenous plants than in the invasive plants.

An even more significant difference was observed between the CO_2 fixation of the introduced and the indigenous groups. Both the quantum efficiency of CO_2 fixation (data not shown), and the maximum net photosynthesis (Fig. 2) measured in a photosaturated state are strikingly higher in the introduced species. The difference between the average parameters for the two groups over a photon flux density of 200 μmol m^{-2} s^{-1} is nearly 100%. This seems to suggest that the introduced species react more "sensibly" to rhapsodic changes of light conditions, and that they utilise a greater proportion of the absorbed light energy in photochemical form. The data also clearly demonstrate that the dramatic difference in the maximum CO_2 fixation between the two groups is manifested in both the dry-matter production and the growth rate, facilitating the fast growth of the already settled introduced species. Besides the expansion strategy, the consequent competitive advantage is undoubtedly the most important factor and a partial cause of the invasive behaviour of the continental species introduced into the insular environment with a long evolutionary past behind them. Taking advantage of their more efficient photosynthetic activity, these plants supplant or suppress the plants indigenous to the island, thereby depleting the original populations of vulnerable indigenous species and greatly reducing the biodiversity.

References

1. Hulbert, S. H. (1978) Ecology 59, 67-77 (
2. Macdonald, I. A. W., Thébaoud, C., Strahm, W. A. and Strasberg, D. (1991) Environ. Conservation 18, 51-61
3. Long, S. P., Humphries, S. and Falkowski, P. G. (1994) Annual Rev. Plant Physiol. Plant Mol. Biol. 45, 633--662
4. Berry, J. and Björkman, O. (1980) Annual Rev. Plant Physiol. 31, 491-543
5. Ball, M. C., Butterworth, J. A., Roden, J. S., Christian, R. and Egerton, J. J. G. (1995) Aust. J. Plant Physiol. 22, 311-319
6. Krause, G. H. and Weiss, E. (1991) Annual Rev. Plant Physiol. Plant Mol. Biol. 42, 313-349
7. Schreiber, U. and Bilger, W. (1993) Progress in Bot. 54, 151-173
8. Schreiber, U., Bilger, W. and Neubauer C. (1994) in Ecophysiology of Photosynthesis (Schulze, E. D. and Caldwell, M. M., eds.) pp. 49-70, Springer, Berlin
9. von Caemmerer, S. and Farquhar, G. D. (1981) Planta 153, 376-387
10. Farage, P. K.and Long, S. P. (1991) Planta 185, 279-286
11. Holly, C., Laughlin, G. P. and Ball, M. C. (1994) Aust. J. Bot. 42, 139-147
12. Königer, M. and Winter, K. (1991) Oecologia 87, 349-356
13. Lichtentaler, H. K., Buschmann, C., Rindle, U. and Schmuck, G. (1986) Rad. Environ. Biophys. 25, 297-308
14. Strasser, R. J., Schwarz, B. and Bucher, J. (1987) Europ. J. Forest Pathology 17, 149-157

CHLOROPHYLL FLUORESCENCE INDUCTION IN ANNUAL AND PERENNIAL LEAVES

Lesleigh E. Force, Christa Critchley, and Jack J.S. Van Rensen*
Department of Botany, The University of Queensland, Brisbane, Australia
*Department of Plant Physiology, Agricultural University, Wageningen, The Netherlands

Key words: electron transport, higher plants, photochemistry, photosystem 2, Q_a, Q_b

1. Introduction

Empirical parameters derived from the Chlorophyll-*a* fluorescence transient have long been used to investigate the photosynthetic performance of plants. This transient, when plotted on a logarithmic time scale, yields a polyphasic signal, termed the OJIP fluorescence transient [1] - (see Fig. 1). However, a full interpretation of the photosynthetic reactions underlying this transient based on experimental data or theoretical analyses, is unavailable. In order to fully utilise this sensitive, non-invasive probe into *in vivo* photosynthesis, the inherent complexity of the fluorescence transient must be unravelled.

We have performed chlorophyll-*a* fluorescence measurements on leaves of perennial and annual species. The experimental data were interpreted within the context of a theoretical analysis of the OJIP transient published by Stirbet and Strasser, 1995 [2]. From that analysis it was suggested that the OJIP fluorescence transient reflected not only the reduction of Q_A but also a specific distribution in time of reduced Q_B forms: $Q_A^- Q_B$ was thought to predominate at J, while mixtures of $Q_A^- Q_B^{2-}$ and $Q_A^- Q_B H_2$ predominated at I and P. We demonstrate that the transient reflects the time dependent reduction of Q_B, and that the speed of reduction increases after saturating light exposure.

2. Experimental Approach

Chlorophyll-*a* fluorescence transients were recorded from the leaves of three perennial species: *Monstera deliciosa*, *Philodendron selloum*, and *Schefflera arboricola*, and two annual species: *Brassica napus* and *Pisum sativum*, using a Plant Efficiency Analyser (PEA, Hansatech Instruments Ltd., King's Lynn, UK). A typical transient for a high light grown pea leaf is shown in Fig. 1. Fo=O, 'L' designates the first inflection point occurring at 50 µs (sometimes used as O in other studies), with J, I, and P, measured at 1.9, 30 and 350 ms respectively, thereby yielding an OLJIP

G. Garab (ed.), Photosynthesis: Mechanisms and Effects, Vol. V, 4061–4064.

fluorescence transient. Data analysis was performed by quantifying the change in fluorescence from one level to the next, then dividing by the total variable fluorescence (Fv=Fm-Fo) to allow comparison between species and leaf types. The four resulting fluorescence components FoL, LJ, JI, and IP for Fig. 1 are shown in Fig. 2. Measurements were made on leaves exposed to low light intensities ($<50\mu$moles photons.m^{-2}.s^{-1}) (controls) and those exposed to saturating (550-650μmoles photons.m^{-2}.s^{-1}) or higher light intensities (up to 1700μmoles photons.m^{-2}.s^{-1}). Dark adaptation periods were minimum times required for a stable Fv/Fm ratio following light exposure.

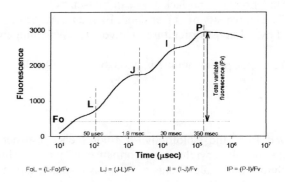

Figure 1 - Typical chlorophyll-a fluorescence transient for a control high light grown Pisum leaf. Dark adaptation period 15 min.

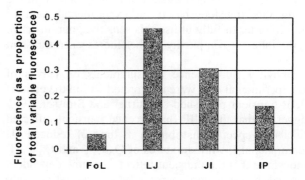

Figure 2 - Fluorescence changes from O to L, L to J, J to I and I to P (expressed as a proportion of total variable fluorescence [Fv Fm-Fo]) for the transient shown in Fig. 1.

3. Results and Discussion

The LJ component was the dominant contributor to total variable fluorescence in both annual and perennial species (Figs. 2&3). The rise in fluorescence to J has been reported as representing the net reduction of Q_A to Q_A^- [3], while in the interpretation of Stirbet and Strasser, (1995) [2], was thought to represent the state $Q_A^- Q_B$. Our study shows that in control leaves, fluorescence emanated predominantly from PSIIs in the state $Q_A^- Q_B$ with minor contributions from PSII populations in the states $Q_A^- Q_B^{2-}$ (JI) and $Q_A^- Q_B H_2$ (IP).

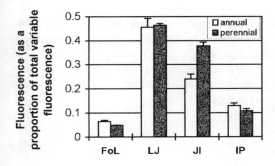

Figure 3 - Components of the OLJIP transient (FoL, LJ, JI, and IP) expressed as a proportion of total variable fluorescence (Fm-Fo) for <u>control</u> - averaged annual species (Brassica + Pisum) and <u>control</u> - averaged perennial species (Monstera + Philodendron + Schefflera).

Annual and perennial species markedly differed in their JI components, with annual species having a lower JI and a higher IP component than seen in the perennial species (Fig. 3). The rise in fluorescence from J through I and up to P, is thought to reflect the movement of electrons from Q_A to Q_B, yielding the states $Q_A^- Q_B^{2-}$ (JI) and $Q_A^- Q_B H_2$ (IP) [2]. Hence, over the two seconds of the fluorescence recording, more PSIIs in the annual species progressed to a later state of Q_B reduction (i.e. $Q_A^- Q_B H_2$) than did perennials. This suggests a faster rate of electron transport through PSII in the annual species compared to the perennial species. This is consistent with the higher photosynthetic rates reported for annual species [4].

Of the four fluorescence components, only JI fluorescence was found to be significantly correlated with the total variable fluorescence (p values of 0.0007 and 0.0002 for perennials and annuals, respectively) and the Fv/Fm ratio (p values of <0.0001 and 0.0024 for perennials and annuals, respectively). This suggests that the quantum yield of primary photochemistry of PSII, as indicated by the Fv/Fm ratio, is mainly defined by the concentration of closed PSIIs in the state $Q_A^- Q_B^{2-}$. The higher the concentration of $Q_A^- Q_B^{2-}$, the higher the Fv/Fm ratio.

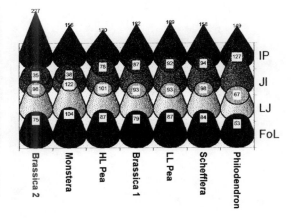

Figure 4 — Changes in FoL, LJ, JI, and IP upon light exposure expressed as a percentage of the individual control values for each listed species.

After light exposure, the FoL, LJ, and JI components generally decreased while the IP component increased, relative to the controls (Fig. 4). This implies that after light exposure, more PSIIs progressed to a later state of Q_B reduction (i.e. $Q_A^-Q_BH_2$) than seen in the controls. We believe that the light activation of photosynthetic processes allows the time dependent reduction of Q_B, to proceed more quickly after light exposure compared with control (low light, non-activated) conditions.

Acknowledgments

This work was supported by an Australian Research Council Large Grant to C. Critchley and a University of Queensland Travelling Fellowship to J.J.S. Van Rensen

References
[1] Strasser R.J., and Govindjee (1991), In 'Regulation of Chloroplast Biogenesis'. (Ed. J.H. Argyroudi-Akoyunoglou.) pp. 423-426 (Plenum Press: New York)
[2] Stirbet A.D. and Strasser R.J. (1995) Archs Sci. Genève, 48, 41-60
[3] Srivastava A., Guissé B., Greppin H., Strasser R.J. (1997) Biochim. Biophys. Acta, 1320, 95-106
[4] Salisbury F.B. (1992), In 'Plant Physiology'. (Eds. F.B. Salisbury and C.W. Ross.) pp. 529 (Wadsworth Publishing Company: California)

Fv/Fm AS A STRESS INDICATOR FOR WOODY PLANTS FROM URBAN-ECOSYSTEM.

Makarova V., Kazimirko Yu., Krendeleva T., Kukarskikh G., Lavrukhina O., Pogosyan S., and Yakovleva O.
Biophysics Department, Biological Faculty,
M.V. Lomonosov State University, Moscow, 119899, Russia

Key words: ecosystem, environmental stress, fluorescence, higher plants, whole plant

1. Introduction

Plants grown under urban environmental conditions are exposed to different anthropogenic stress factors: air pollutants e.g. SO_2, NO, NO_2, photochemical smog, photooxidants, acid rains, mineral soil deficiency induced by acid rains, heavy metal load, enhanced UV-B radiation. Photosynthetic apparatus of plants is sensitive to stress conditions caused by anthropogenic and natural stress factors [1,2]. Chlorophyll fluorescence measurements provide a powerful non-destructive method for fast outdoor screening of trees. Some parameters of chlorophyll fluorescence can be widely used to characterize photosynthetic activity of a whole plant *in vivo* [3]. Phelloderm cells of bark of woody plants containing a lot of chlorophyll demonstrate photosynthetic activity. Application of chlorophyll fluorescence measured on intact shoots of trees make it possible to monitor urban woody plantations since early spring till late autumn in many east and west European parts of Russia. The goal of this work was to conduct reliable measurements of chlorophyll fluorescence parameters on one-year-old shoots of trees for stress detection on plants under urban conditions.

2. Procedure

Fluorescence parameter Fv/Fm can characterize PS2 photosynthetic activity related to light energy utilization in plants [4,5]. Fv/Fm parameter was measured by the portable chlorophyll fluorometer (Model PAM-2000, WALZ, Germany). Chlorophyll fluorescence emitted from bark samples of one-year-old shoots of lower tree-tier was determined. Sample was fixed in a specially made device where it could be put in the dark with fiberoptics exit plane oriented. Standard site for our measurements was chosen after many preliminary experiments. It was located under crown bud on the ventral surface of a shoot. In experiments we measured fluorescence emitted from 5-8 shoots located on each of the 5 branches of one tree. Fluorescence determinations have been made during summer-autumn seasons on leaves and bark of one-year-old shoots of trees (*Betula verrucosa, Populus pyramidalis, Tilia vulgaris, Syringa vulgaris, Sorbus aucuparia, Fraxinus excelsior, Ulmus campestris, Acer rubrum, Quercus acuminata*).

G. Garab (ed.), Photosynthesis: Mechanisms and Effects, Vol. V, 4065–4068.
© *1998 Kluwer Academic Publishers. Printed in the Netherlands.*

3. Result and Discussion

It was shown that Fv/Fm values measured on bark of shoots of different kinds of trees grown under favorable environmental conditions were approximately similar during the same vegetation period. Stress factors decrease the efficiency of photosynthesis and suppress the variable fluorescence of dark-adapted chlorophyll-containing objects correspondingly. The ratio Fv/Fm also decreases with the increasing effect of stress factors.

Our results showed that the decrease in Fv/Fm values was higher in plants located near the main traffic roads. The Fv/Fm ratio for resting plants also declined in autumn when air temperature was lower as compared to summer time. The Fv/Fm values were determined on leaves and bark of trees grown at different distances from main roads of Moscow-City. These data were obtained on one-year-old shoots of lime-trees in October at the air temperature of 7-8 C° (*Figure 1.*) The decrease in Fv/Fm values reflected destructive changes in photosynthetic apparatus caused by pollutants generated on the main roads.

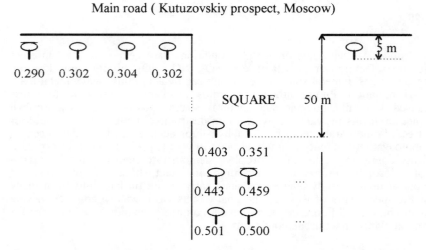

Figure 1. Fv/Fm values of lime-trees grown at different distances from the main traffic road.

The decrease in Fv/Fm values for bark samples may be observed under any ecological conditions in winter-rest period. In plants grown under favorable ecological conditions at the distances more than 25 m from the main road the efficiency of primary processes of photosynthesis decreased gradually as plants reached their winter resting state. The Fv/Fm changes during autumn for two groups of trees growing at 25 m and 5 m from the main road are shown in *Figure 2*. The difference in Fv/Fm values between these two groups was about 25% at any temperature (*Figure 2*). The bending point on curves in *Fig. 2* of Fv/Fm values appeared at temperatures below 0 C°. Apparently, photosynthetic processes in chloroplasts of bark cells is strongly suppressed at these temperatures.

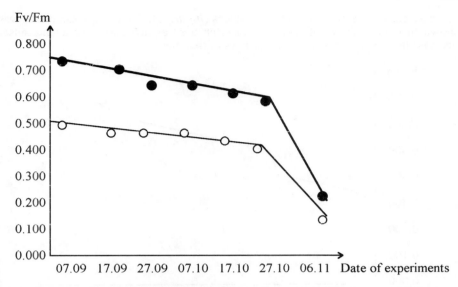

Figure 2. Seasons changes in the Fv/Fm values of one-year-old shoots of lime-trees grown at the different distances (dark circle - 25 m; white circle - 5 m) from the main road in Moscow.

To conduct the idea that changes of Fv/Fm values actually reflect the physiological status of trees, special laboratory experiments were carried out on branches detached from trees. Diuron which is known as an inhibitor of primary processes of photosynthesis slowly penetrates into an isolated shoot through the xylem stream. Registration of Fv/Fm permits to observe the dynamics of diuron penetration (*Table 1*). Diuron penetration was accompanied by the decrease in Fv/Fm values. This effect was mostly pronounced in bark as compared to petioles and leaves.

Table 1. Effects of diuron treatment of lilac shoot on Fv/Fm value.

Site of measurements	Fv/Fm	
	before treatment	after 24 hours of treatment
top leaf	0.745	0.727
top leaf petiole	0.740	0.602
bark under top leaf petiole	0.731	0.498
next leaf	0.710	0.667
next leaf petiole	0.714	0.483
bark under next leaf petiole	0.714	0.058

Investigations of fluorescence allow to obtain important information concerning physiological status of trees during cold seasons of the year. When the resting shoot was put in water vessel at room-temperature, it left the rest state. It was accompanied by the Fv/Fm increase up to 0.500 and 0.650 in 1 or 2 weeks correspondingly. Drying of branches caused the decrease in Fv/Fm values (*Figure 3*). On the bark samples it was also shown that the addition of NaCl (1%) often used as anti-glassy-ice agent caused the decrease in Fv/Fm values twice after 5 days of treatment.

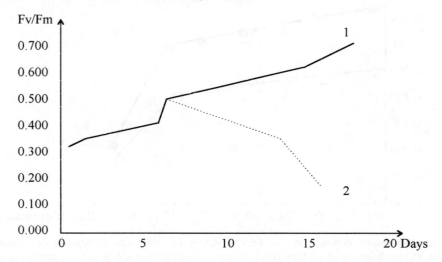

Figure 3. Changes in Fv/Fm values for detached birch shoots in water leaving rest period (1) and under drying in air (2) at room-temperature. Experiment was carried out in November.

Fv/Fm values does not depend on the total chlorophyll fluorescence intensity. Therefore this method can be used to test plant samples with different chlorophyll content. In summary, proposed express method based on Fv/Fm determinations is a powerful to evaluate the physiological state of plant in urban ecosystems.

References

1 Ranger, G. and Schreiber, U. (1986) in Light Emission by Green Plants and Bacteria (Govindjee, Amesz J., Fork D.C., eds.) pp. 587-619, Academic press, Orlando
2 Rubin, A.B., Venediktov, P.S., Krendeleva, T.E. and Paschenko, V.Z. (1986) Photochem. Photobiol. 12, 185-189
3 Lichtenthaler, H.K. (1988) in Applications of Chlorophyll Fluorescence in Photosynthesis Research, Stress Physiology, Hydrobiology and Remote Sensing (Lichtenthaler H.K., ed.) pp. 129-143, Kluwer Academic Publishers, The Netherlands
4 Schreiber, U. (1986) Photosynth. Res. 9, 261-272
5 Genty, B., Briantais, J.M. and Baker, N.R. (1989) Biochim. Biophys. Acta 990, 87-92

ESTIMATION OF THE PHOTOSYNTHETIC ELECTRON FLOW IN INTACT LEAVES: EFFECT OF MODULATED LIGHT INTENSITY ON THE FLUORESCENCE PARAMETERS Fv/Fm and ΔF/Fm'

Michito Tsuyama and Yoshichika Kobayashi

Faculty of Agriculture, Kyushu University, Hakozaki, Higashi-ku, Fukuoka 812, Japan

Key words: quantum yield, photosystem 2, CO_2 uptake, photorespiration, Mehler reaction, leaf physiology

1. Introduction

Previous studies have confirmed the validity of the measurement of quantum yield in PS II for estimating the rate of linear electron transport in leaves (Je) by simultaneous measurements of chlorophyll fluorescence and gas exchange [1], because the product of the quantum yield of PS II (ΦPSII) and the photon incidence absorbed by PS II (IPSII) permits the calculation of Je from the equation Je=(ΦPSII)(IPSII). Genty et al.[2] showed that a linear relationship is obtained between the quantum yield of PS II (ΔF/Fm') and the apparent quantum yield of CO_2 assimilation in non-photorespiratory conditions. However, the actual values of ΦPSII and IPSII are difficult to determine in intact leaves. At present, two different methods for estimating Je have been proposed. Edward and Baker [3] calculated Je from the equation, Je=1/2(ΔF/Fm')·(PFDabs), under the assumption that the photon flux density absorbed by a leaf (PFDabs) is equally distributed between PS I and PS II. Another method is based on the linear relationship between the relative measurement of quantum yield of PS II (ΔF/Fm') and the apparent quantum yield of gross CO_2 assimilation under non-photorespiratory conditions [2, 4, 5, 6]. They used the relationship for estimating the electron flow under photorespiratory conditions (Je) in leaves.

We found that the value of ΔF/Fm' is greatly affected by the intensity of measuring beam operating at the frequency of 100 kHz in the PAM system. In this study, we analyzed the effect of the measuring-pulse intensity on the fluorescence parameter ΔF/Fm'. Characteristics of the leaf coefficient α, which is the slope of linear regression line fitted through relationship between (ΔF/Fm')(PFD) and the flux of electrons consumed by gross CO_2 assimilation under non-photorespiratory conditions (Jco_2), are described.

2. Materials and Methods

Cucumber plants grown for 4 to 6 weeks in a phytotron at 25°C were used for the experiments. Chlorophyll fluorescence and CO_2 exchange were simultaneously measured with the PAM chlorophyll fluorometer (Walz) and an infra-red gas analyzer (Li-COR) in attached leaves. Measurements of net CO_2 uptake and transpiration were performed using an open system with the infra-red gas analyzer. The CO_2-free air and compressed air containing 100% CO_2 were mixed at different rates according to the desired CO_2 concentration. The flow rate of

G. Garab (ed.), Photosynthesis: Mechanisms and Effects, Vol. V, 4069–4072.
© *1998 Kluwer Academic Publishers. Printed in the Netherlands.*

the flushing air passing through a leaf chamber was 670ml/min. The partial pressure of water was measured with a dew-point hygrometer (Li-COR). For the chlorophyll fluorescence measurement, the sample leaf was illuminated with the three beams, modulated measuring light, actinic white light and saturating pulse, through the five-armed fiber optic system. The frequency of modulated measuring light was 1.6 or 100 kHz, and the intensity was changed between 0.4 and 12.3 μmol photons/m^{2}sec^1 to examine the effect of modulated measuring light intensity on the fluorescence parameters Fv/Fm and ΔF/Fm'. Saturating pulses (1 sec) of white light (7,500 μmol photons/m^{2}sec^1) were provided by a KL 1500 Schott light source at intervals of 100 sec.

3. Results and Discussion

Effect of the measuring light intensity on Fv/Fm and ΔF/Fm' in non-photorespiratory conditions.

Figure 1 shows the relationships between Fv/Fm, ΔF/Fm' and the intensity of modulated light used for the measurement of fluorescence emission.

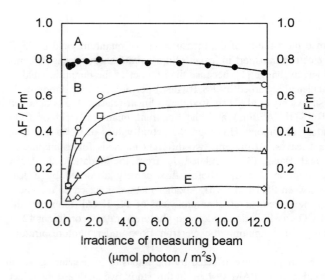

Figure 1. The relationships between Fv/Fm, ΔF/Fm' and intensity of fluorescnece-measuring beam. The intensities of actinic light were 0 (A), 20 (B), 260 (C), 480 (D) and 1750 (E) μmol photons/m^{2}s^1.

In the experiments, two fluorescence yields were measured: Fo and Fm where Fo is the minimal fluorescence determined under the illumination with the indicated intensities of measuring beam and Fm, the maximal fluorescence induced by the 1 s saturating pulse. The value of Fv/Fm=(Fm - Fo)/Fm which is a measure of quantum yield of photochemistry by open PS II trap was not largely affected by the measuring beam intensity, and the values determined at different intensities of measuring beam were near to 0.8 which coincidences with the maximal quantum yield of PS II in most green plants (Curve A). In contrast, the values of ΔF/Fm'=(Fm' - Fs)/Fm' which were determined under actinic illumination at different PFDs decreased strikingly when the irradiance of measuring beam was below 2 μmol photons/m^{2}sec^1.

Relationship between ΔF/Fm' and the rate of photosynthetic electron transport (JCO₂) in non-photorespiratory conditions.

Figure 2a shows the extent of (ΔF/Fm')(PFD) as a function of JCO_2 which was calculated from the rate of gross photosynthesis multiplied by four (4 e- per CO_2). The values of ΔF/Fm'

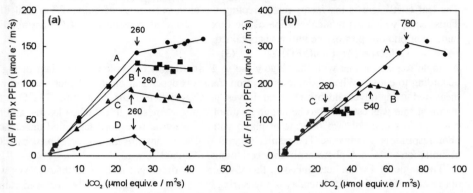

Figure 2. The relationships between (ΔF/Fm')(PFD) and JCO_2.

were measured using the same leaf under illumination of measuring beam at the irradiance of 0.4 (D), 1.0 (C), 2.3 (B) or 12.3 μmol photons/m²sec[1] (A). The activity of CO_2 assimilation was varied by changing PFD of actinic light between 23 and 1,764 μmol photons/m²sec[1] in air containing 1% O_2 and 1,000 ppm CO_2. As shown by Curves A-D, the relationship between (ΔF/Fm')(PFD) and JCO_2 was linear within the range under JCO_2=25.9 μmol equiv.e⁻/m²s[1], and the slope of the regression curves became larger with increasing of the irradiance of measuring beam. A break point was observed when JCO_2 and (ΔF/Fm')(PFD) approached toward saturation under the actinic illumination of 260 μmol photons/m²sec[1]. Then, the value of (ΔF/Fm')(PFD) decreased linearly with increasing of JCO_2. The negative slope became steeper with the decrease in the irradiance of measuring beam.

Similar measurements were performed using three types of cucumber leaves which possessed very different activities of net CO_2 uptake (Fig. 2b). The purpose of the experiments was to examine the effect of leaf photosynthetic activity on the correlation curve between (ΔF/Fm')(PFD) and JCO_2. In the leaves, ΔF/Fm' was determined at the same irradiance of measuring beam (2.3 μmol photons/m²sec[1]). There was no significant difference in the slope of the regression curves, but the break points varied considerably. The break point was observed at different rates of JCO_2 at the actinic light intensities of 260, 540 and 780 μmol photons/m²sec[1]. These results suggest that the slope of the correlation curve of (ΔF/Fm')(PFD) versus JCO_2 is primarily dependent on the measuring beam intensity, and the break point is associated with the redox state of electron carriers.

Characteristics of the leaf coefficient α

From the definition of quantum yield, the rate of linear electron transport (Je) is given as the product of the quantum yield of PS II (ΦPSII) and the photon incidence absorbed by PS II (IPSII):

$$Je = (\Phi PSII)(IPSII) \qquad (1)$$

On the basis of the results in Fig. 2, the value of leaf coefficient (α) was determined from the slope of linear regression line fitted through the plot of ($\Delta F/Fm'$)(PFD) versus J_{CO_2} according to eq(2):

$$(\Delta F/Fm')(PFD) = \alpha \cdot J_{CO_2} \qquad (2)$$

The leaf coefficient is the factor for correcting the observed values of both $\Delta F/Fm'$ and PFD. Thus, eq(3) which permits calculation of the flux of electrons consumed by CO_2 assimilation under non-photorespiratory conditions is obtained.

$$J_{CO_2} = (\Delta F/Fm')(PFD) /\alpha \qquad (3)$$

A single regression line with the slope close to 1 was drawn by replotting the data in Fig. 2 according to eq(3). When the rate of electron transport dependent on the Mehler reaction was measured using glyceraldehyde-treated leaves in CO_2-deficient air containing 1% O_2, the maximal rate was estimated to be about 15 μmol equiv. e^-/m^2sec^1 in detached leaves, and the rate was less than 10% of the rate of photosynthetic electron transport determind under non-photorespiratory conditions. The results suggest that electrons are mainly consumed by CO_2 assimilation under non-photorespiratory conditions. (data not shown).

The value of Fv/Fm determined in the absence of actinic light in dark-adapted leaves was not largely affected by the intensity of measuring beam (Fig. 1, Curve A). The observed values of Fv/Fm were close to 0.8, which is comparable with the maximal quantum yield of photosynthesis. In the dark, there is no formation of fluorescence quenchers, and the primary quinone acceptor Q_A is fully oxidized. In such leaves, if all PS II complexes have similar efficiency of photochemistry, an identical value of Fv/Fm is available irrespective of the difference in the intensity of measuring beam.

When a leaf was illuminated from the upper side, actinic light is absorbed during traversing leaf tissue, and only a small amount of photons reach to chloroplasts on the lower side. Under the light conditions, chlorophyll fluorescence quenching is induced more strongly in chloroplasts on the upper side. Similarly, upon illumination of low modulated light (measuring beam), fluorescence is dissipated by scattering and re-absorption during traversing the leaf tissue from the lower side to the upper side, and fluorescence signal from the lower side is less detected. At high measuring light, however, fluorescence signal from the lower side is detected effectively. There is a possibility that the above complex events affect the measurement of $\Delta F/Fm'$. Our results indicate that $\Delta F/Fm'$ is solely a measure of the relative value of quantum yield. The leaf coefficient (α) is a factor for correcting the value of ($\Delta F/Fm'$)(PFD) by using the reliable value of J_{CO_2}. Factors relating to leaf absorptance and distribution of light energy between PS I and PS II are also involved in α. The measurement of $\Delta F/Fm'$ has an advantage when the rate of electron flow under photorespiratory conditions is estimated by using the leaf coefficient, α, determined in non-photorespiratory conditions.

References

[1] Tsuyama M., Kobayashi H., Shinya M., Yahata H. and Kobayashi Y. (1996) J. For. Res 1, 79-85
[2] Genty B., Briantais J.M., Baker N.R. (1989) Biochim. Biophys. Acta 990, 87-92
[3] Edwards G.E. and Baker N.R. (1993) Photosynth. Res. 37, 89-102
[4] Cornic G. and Briantais J.M. (1991) Planta 183, 178-184
[5] Ghashghaie J. and Cornic G. (1994) J.Plant Physiol. 143, 643-650
[6] Heber U., Bligny R., Streb P. and Douce R. (1996) Bot. Acta 109, 307-315

SCREENING GLYPHOSATE-INDUCED CHANGES IN THE PHOTO-SYNTHESIS OF DUCKWEED *LEMNA GIBBA L.* BY USING CHLOROPHYLL FLUORESCENCE INDUCTION METHOD

Ilona Mészáros[1], Gyula Lakatos[2], Szilvia Veres,[1] Anikó Papp[1] Imre Lánszki [3]
[1]Department of Botany, Kossuth University, Debrecen, Hungary;
[2]Department of Ecology, Kossuth University, Debrecen, Hungary;
[3]ZENACA, Hungary Co., Budapest, Hungary

Key words: chlorophyll fluorescence, duckweed, glyphosate, ecotoxicology

1. Introduction

Herbicides that exert no direct effect on photosynthesis eventually induce severe secondary impairment of chloroplast functioning. Glyphosate primarily inhibits the shikimate pathway but it may also reduce the carbon fixation rate through an indirect effect on the RuBP levels and acts as an inhibitor of chlorophyll synthesis (1). Beside revealing the precise physiological action of herbicides in target and non-taget species there is an increasing need to introduce sensitive methods for laboratory testing and in situ monitoring of their effects in the recent ecotoxicological research and toxicity assessment (2, 3). Being widely used to detect plant stress, the screening of chlorophyll fluorescence induction kinetics has the advantage of such methods, and moreover, it offers a rapid non-destructive technique for detecting the herbicide induced changes in photosynthetic processes and physiological state of the plants. (4). In the present study, the effects of glyphosate on the photochemical activity of aquatic macrophyte *Lemna gibba* L. (duckweed) were investigated by *in vivo* chlorophyll fluorescence measurements comparing two commercial forms of glyphosate, Medallon (N-(phosphono-methyl-glycine)-trimethylsulfonium) and Glialka 480 (N-(phosphono-methyl-glycine)-isopropylamine).

2. Materials and Methods

For this study axenic test cultures of duckweed (*Lemna gibba* L.) were grown in Hutner medium. Each culture was derived from the same stock culture and containing 10-10 three-frond plants. Stocks and experimental cultures of *Lemna gibba* were grown axenically in Hunter medium in growth chambers (16/8 h light/dark period, 250 µmol m^{-2} s^{-1}, 25±2 °C). Glyphosate was introduced into the test cultures of duckweed in two

G. Garab (ed.), Photosynthesis: Mechanisms and Effects, Vol. V, 4073–4076.
© 1998 *Kluwer Academic Publishers. Printed in the Netherlands.*

commercial formalizations, Medallon (N-(phosphono-methyl-glycine)-trimethyl-sulfonium) and Glialka 480 (N-(phosphono-methyl-glycine)-isopropylamine). The chlorophyll fluorescence screening of the effects of glyphosate on the photosynthetic activity was performed in 48 hour treatments under continuous light conditions. In these experiments, glyphosate was applied to the test cultures at 5-15 mM for Glialka 480 and 15-50 μmol for Medallon. The effects of Glialka 480 and Medallon on the growth of duckweed were also studied by recording the temporal change in the number of fronds in the tradional two-week growth tests (4). In the growth tests, glyphosate was applied at 100-1000 μmol for Glialka 480 and 1-15 μmol for Medallon. Chlorophyll fluorescence induction kinetics were detected with PAM 2000 fluorometer (WALZ, Germany) and chlorophyll fluorescence parameters were calculated after Scheiber et al.(5) and Lichtenthaler (6).

3. Results and Discussion

Different dose responses of duckweed to glyphosate were obtained for Medallon and Glialka 480 considering both the growth rate and chlorophyll fluorescence parameters. Medallon proved to be a more toxic form of glyphosate for duckweed at every comparable concentration.

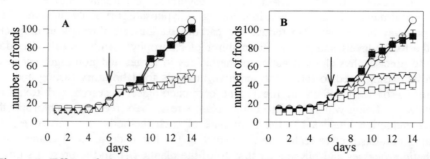

Figure 1. Effect of glyphosate on the growth rate of duckweed estimated by the number of fronds. A) Time course of growth when glyphosate was applied in the form of Glialka 480. *Symbols* ○,■, ▽, *and* □ *show control and treatments by 100 μmol, 500 μmol and 1000 μmol glyphosate, respectively* B) Time course of growth when glyphosate was applied in the form of Medallon. *Symbols* ○,■, ▽, *and* □ *show control, 1 μmol, 10 μmol and 15 μmol (glyphosate) concentrations.* Arrays indicate the start of treatments on the 6th day of test cultures containing 10-10 three-leaved fronds.

Glyphosate in both forms induced significant reduction in the growth rate of duckweed (Figure 1). However, in the presence of Medallon, not only the growth was supressed (Figure 1B) with the increasing concentration, but a break-up of three-frond colonies and drastic dying of roots were also found. Similarly to the results on growth, Glialka 480 was also less toxic to the functioning of the photosynthetic apparatus, at least with an order of magnitude (Figure 2), and induced chlorophyll loss at higher concentration than Medallon did. While Medallon caused a significant decrease in the potential photochemical activity (Fv/Fm) (Figure 3A) with a parallel decline of Fm and an increase of ground flourescence Fo, Glialka 480 had smaller effects on Fv/Fm even at 500-1000 times higher concentrations (Figure 2A).

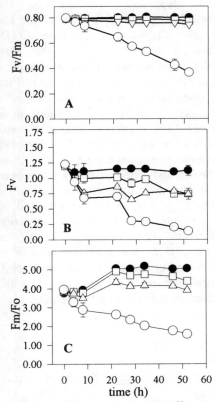

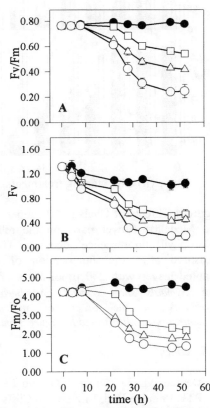

Figure 2. Time course of effects of glyphosate applied in the form of Glialka 480 on chlorophyll fluorescence parameters: A) Maximal photochemical efficiency of PSII (Fv/Fm) B) Variable fluorescence C) Ratio of maximal to ground fluorescence. *Symbols* ●,□, Δ *and* ○ *show control and treatments by 5, 10 and 15 mmol glyphosate, respectively.*

Figure 3. Time course of effects of glyphosate applied in the form of Medallon on chlorophyll fluorescence parameters A) maximal photochemical efficiency of PSII (Fv/Fm) B) Variable fluorescence C) Ratio of maximal to ground fluorescence *Symbols* ●,□, Δ *and* ○ *show control and treatments by 15, 30 and 50 μmol glyphosate, respectively.*

4076

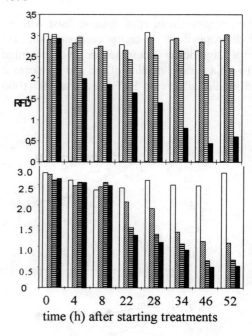

time (h) after starting treatments

Figure 4. Effects of Glialka 480 (upper part) and Medallon (lower part) on the relative chlorophyll fluorescence decrease (RFD) measured after the illumunation of dark adapted leaves with 200 μmol m^{-2} s^{-1} for 5 min. *Symbols show control and herbicide treatments in the order of increasing concentrations as illustrated in Figure 3 and 4*

Within 48 hours Medallon has resulted a 50% inhibition of the potential quantum yield of PSII (Fv/Fm) at 30 μmol dose, while Glialka 480 did it at 15 mmol. Parameters obtained from the slow fluorescence kinetics also showed differences in the inhibitory effects of the two glyphosate herbicides on the physiological activity of duckweed. The relative fluorescence decrease (RFD) obtained at the end of the slow chlorophyll induction phase were around 3 in the untreated control plants and the glyphosate treatments resulted in a decrease until below 1 at higher dosis within 48 h. Paralell to this, an increase of the non-photochemical chlorophyll fluorescence quenching occurred.

The methods used in this work proved that glyphosate-induced changes in the physiological state can be screened by chlorophyll a fluorescence induction technics. The combination of the traditional *Lemna* growth test with *in vivo* mesurement of the chlorophyll fluorescence are suggested to test the effects of environmental chemicals.

References

1. Devine, M., D., Duke, S.O. and Fedtke, C. (1993) Physiology of herbicide action. PTR Prentice Hall, Englewood Cliffs, New Jersey.
2. Lewis, M. A. (1995) Environ Pollut., 87, 319-336
3. Wang, W. (19986a) Environ Pollut., 11, 1-14
4. Scheiber, U., Bilger, W. and Neubauer, C. (1994) In Ecophysiology of Photosynthesis (Schulze and Caldwell eds) pp.49-70. Springer Verlag, Berlin.
5. Lichtenthaler H.K. (1996) J. Plant Physiol., 148:4-14
6. Lakatos, Gy., Mészáros, I., Bohátka, S., Szabó, S., Makádi, M., Csatlós, M. and Langer G. (1993) Sci. Total Environ 44, 773-778.

PHOTOSYNTHETIC PROPERTIES OF *PHRAGMITES AUSTRALIS* IN THE HEALTHY AND DIE-BACK SITES OF LAKE FERTŐ/NEUSIEDLERSEE

Gyula Lakatos[1], Ilona Mészáros.[2], Mária Dinka, M.[3], Szilvia Veres[1]
[1]Department of Ecology, Kossuth University, Debrecen, Hungary; [2]Department of Botany, Kossuth University, Debrecen, Hungary; [3]Institute of Ecology and Botany of the Hungarian Academy of Sciences, Vácrátót, Hungary

Key words: abiotic stress, chl fluorescence induction, gas exchange, macrophytes

1. Introduction

The common reed (*Phragmites australis* /Cav./ Trin. ex Steudel), the dominant macrophyte species in the littoral zones of shallow lakes has experienced a large-scale die-back in Europe since the early 80's (1, 2). The strong reduction in the vegetative growth and production of declining reeds has raised the problem that photosynthesis may be a target process to the factors affecting the vitality of *Phragmites*. The reed is a perenial plant and its survival and successful spread depend largely on the balance and allocation of carbon primarily assimilated in the photosynthesizing tissues of the rhizoma. In the recent studies of EUREED Project in Lake Fertő (3), comparative measurements have been performed in healthy and die-back reed stands to reveal the alteration in the leaf photosynthetic rate, photochemical activity and pigment composition of *Phragmites*.

2. Materials and Methods

In the summer of 1997, investigations were carried out in two different phases of the reed development: in June, when the common reed showed intense vegetative growth, and in August when the development of plants arrived at the reproductive phase. *In situ,* ecophysiological measurements and the collection of plant samples were performed in five permanent EUREED research sites selected in the Hungarian part of Lake Fertő. In three sites (Eu2, Eu5 and Eu9), symptoms of reed decline cannot be visually observed and here the reed forms dense and homogenous stands. In two sites (Eu3 and Eu7), the die-back of reeds occur in large extents and the reed stands are seriouly damaged. In site Eu3, there were separate measurements in largely degraded (Eu3A) and in visually healthy (Eu3B) parts of the reed stand. The leaf gas exchange was measured by the application of a portable infrared gas analyser (LCA 2, ADC, UK), the chlorophyll fluorescence induction

G. Garab (ed.), Photosynthesis: Mechanisms and Effects, Vol. V, 4077–4080.
© 1998 Kluwer Academic Publishers. Printed in the Netherlands.

stimulating the accumulation of photoprotecting pigment components (especially that of the β-carotene) in the chloroplast tylakoids.

In June, the values of Fv/Fm describing the potential photochemical activity (the potential photochemical efficiency of PSII) were very similar (0.79-0.81) in all the sites (Figure 3A). But in August, the Fv/Fm considerably decreased in the die-back sites (Eu3, Eu 7) and reached lower values (0.74-77) than that of the vigorous stands (Figure 3B).

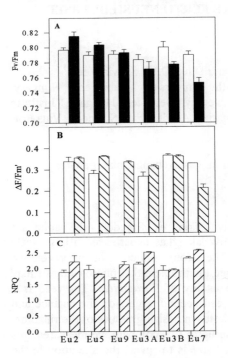

Figure 3. Chlorophyll fluorescence parameters measured in the 3rd and 4rth detached leaves of reeds on 9-10 June (□) and 5-6 August 1997 (■ ,▨ ▧) in vigorous (Eu2, Eu5, Eu9) and die-back sites (Eu3,Eu7). A) Maximal photochemical efficiency of PSII (Fv/Fm) measured after 30 min dark adaptation period (n=14) B) Actual photochemical efficiency of PSII (ΔF/Fm') measured after the illumination of dark adapted leaves by 1200 μmol m^{-2} s^{-1} actinic light for 5 min by saturation pulse method B) Non-photochemical chlorophyll fluorescence quenching (NPQ) measured after the illumination of dark adapted leaves by 1200 μmol m^{-2} s^{-1} actinic light for 5 min by saturation pulse method (n=6)

The actual photochemical efficiency of PSII and non-photochemical quenching measured after the dark adapted leaves was illuminated by actinic light (1200 μmol m^{-2} s^{-1}) also indicate the stress state of leaves in the declining sites by August (Figure 3B). The development of non-photochemical quenching (Figure 3C) was characteristic in the leaves of declining reed plants, which was in proper correlation with the increase of the VAZ pool of leaves (Figure 2).

The measurements of gas exchange performed on leaves at various canopy positions revealed a reduction in CO_2 assimilation capacity in die-back stands (Figure 4). By August the decrease of CO_2 assimilation became more apparent in all declining sites (Figure 5) and reached a much larger extent than that of the maximal photochemical activity (Figure 3A).

parameters of dark and light adapted leaves were determined with PAM 2000 fluorometer (WALZ, Germany) and was calculated as recommended by Scheiber (4). The leaf photosynthetic pigments were extracted with 80 % acetone and were determined by means of spectrophotometric and HPLC methods.

3. Results and Discussion

In the vigorous reed stands (Eu2, Eu5 és Eu9), the chlorophyll content in the leaves was high and ranged from 4.5 to 5.5 mg (chla+chl b) g^{-1} (d.w.), which reflects the proper functioning of the photosynthetic apparatus (Figure 1). In the Eu3 die-back site, where the stand community was highly degraded, however, the chlorophyll content of leaves was found to be strongly reduced and reached only 2.5 -3 mg (chla+chlb) g^{-1} (d.w.). In June, the concentration of chlorophylls was suprisingly the highest (5.5.-6.5 mg g^{-1}) in the die-back site Eu7, but a 50 % reduction occurred by August.

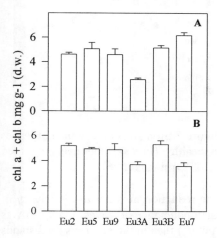

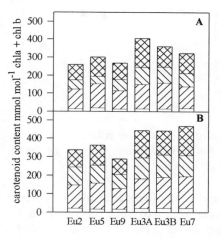

Figure 1. Leaf chlorophyll content of reed plants in the healthy (Eu2, Eu5,Eu9) and the declining stands (Eu3, Eu7) on 9-10 June (A) and 5-6 August 1997 (B). Mean ± S.E. (n=4).

Figure 2 . Carotenoid composition of reed leaves in healthy (Eu2, Eu5,Eu9) and declining stands (Eu3, Eu7) measured on 9-10 June (A) and 5-6 August 1997 (B).
☐ neoxanthin   ▨ lutein
◩ Viol+Ant+Zea   ▩ β-carotene

In the declining reed stands, the total carotenoid content, size of violaxanthin cycle pool and the β-carotene content significantly increased as compared to the vigorous stands (Figure 2). Changes at the level of carotenoid composition suggest that the photosynthetic apparatus of plants at the die-back sites are subjected to photoinhibiton

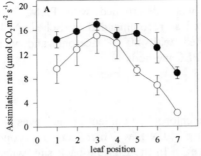

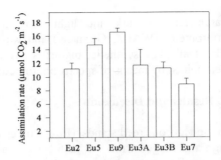

Figure 4 CO_2 assimilation of leaves of reed in healthy (Eu9:●) and declining (Eu7: ○) sites. June 1997 (n=6)

Figure 5. CO_2 assimilation rate in healthy (Eu2, Eu5, Eu9) and declining sites (Eu3, Eu7). August 1997 (n=6)

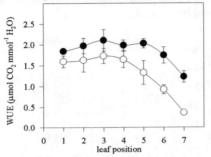

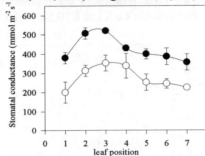

Figure 6. Water use efficiency of leaves in the healthy (Eu9: ●) and declining (Eu7: ○) sites. (n=6)

Figure 7 Stomatal conductance of leaves in healthy (Eu9:●) and declining reed (Eu7: ○) stands (n=6)

The decline of reeds can also be described with the reduction in water use efficiency of leaves (Figure 6). As it was observed, the stomatal conductance of leaves (Figure 7) was much lower in the die-back sites (200-350 mmol m^{-2} s^{-1}) than in the vigorous reed stands (380-550 mmol m^{-2} s^{-1}). This alteration can induce a further functional impairment of the lacunar, gas ventilation system of declining reeds and later it may result in oxigen deficiency and damages of the rhizoma.

Acknowledgement
This work was supported by the EUREED Project No. EVSV-CT92-0083/CIPDCT925064.

References
1. Armstrong, W., Afreen-Zobayed, Armstrong, W. (1996) New Phytol., 134, 601-614
2. Lakatos, Gy. (1993) Studies in Environmental Science 55:365-372
3. Dinka, M., Szeglet, P., and Szabó, I. (1995). In Reednews, the Newsletter of the EC-EUREED Project (van der Putten, W.H. ed.), 4, 53-58, Heteren

PHOTOSYNTHETIC RESPONSES TO SPATIAL AND DIURNAL VARIATIONS OF LIGHT CONDITIONS IN SEEDLINGS OF THREE DECIDUOUS TREE SPECIES

Ilona Mészáros, R. Viktor Tóth, Szilvia Veres
Department of Botany, Kossuth University, Debrecen, Hungary

Key words: chlorophyll fluorescence, gas exchange, photoinhibition, sun/shade acclimation, xanthophyll cycle

1. Introduction

In natural environment plants are continuosly subjected to the fluctuation in light conditions and their growth and survival basically depend on their capacity to acclimate their physiology, biochemistry and morphology (1). The function of photosynthetic apparatus in sun and shade adapted leaves and plants show considerable differences in terms of photosynthetic energy conversion, CO_2 assimilation rate and in the efficiency of protective mechanisms against the effects of high light or excess light under stress conditions (2, 3). This paper presents results of a comparative study on the relationship between photosynthetic activity and susceptibility to photoinhibition of seedlings of canopy trees *(Fagus sylvatica, Fraxinus excelsior, Carpinus betulus)* of a beech forest grown under different light conditions.

2. Materials and Methods

Field measurements were performed in the understory of a closed beech forest, in the forest edge of southern exposure and in the adjacent clear-cut area at an permanent experimental site in the Bukk Mountains (NE Hungary) at 500 m above the sea level (4) in summer of 1996 and 1997. Gas exchange and chlorophyll fluorescence parameters of attached leaves weres measured by the means of a portable infrared gas analyser (LCA 2, ADC, UK) and PAM 2000 fluorometer (WALZ, Germany), respectively. The light response of photochemical efficiency of PSII was also studied on detached branches illuminating the leaves with gradually increasing intensity of actinic light for 5 min. Pigment composition of leaves was measured in 80 % acetone extract by reversed phase HPLC method with application of zeaxanthin as standard.

G. Garab (ed.), Photosynthesis: Mechanisms and Effects, Vol. V, 4081–4084.
© 1998 *Kluwer Academic Publishers. Printed in the Netherlands.*

3. Results and Discussion

The three habitats represent significantly different light climate both concerning the intensity and diurnal variation of PFD (Figure 1). Along the forest edge the penetration of light toward the forest interior and the illumination of plants are primarily dependent of the solar movement resulting in a transitional morphological and physiological character of the tree seedlings. It is reflected well in the specific leaf weight (Figure 2) which reaches the highest values in the sunny clear-cut area.

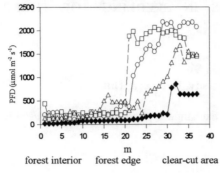 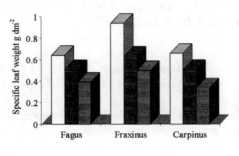

Figure 1. Changes in photosynthetic active radiation along transects perpendicular to the forest edge. Symbols O, □, Δ and ◆ represents means of data collected in the morning, at noon, in the afternoon and in early evening hours in July 1996.

Figure 2. Effects of light conditions on the specific leaf weight of tree seedlings. (July 1997) Symbols □ ▨ and ▤ indicate clear-clear-cut area, forest edge and forest interior.

Figure 3 and 4 show the daily course of Pn in the different habitats on the measurement period in July and August 1996. The CO_2 assimilation rate of species changed significantly with the habitat light conditions and even in the same habitat with the environmental stress (especially drought). The photochemical efficiency of PSII of light adapted leaves in the sunny habitats decreased (from 0.74-0.80) with increasing light during the day while in the forest interior the $\Delta F/Fm'$ values remained almost as high as at the dawn Fv/Fm values (0.79-0.82) (Figure 5). When the light dependence of photochemical efficiency of PSII in detached leaves was studied it was found that the shade adapted seedlings of each species tended to reach a smaller values at lower light intensities (Figure 6). However, in the case of seedlings grown in the clear cut area the $\Delta F/Fm'$ remains relatively high at saturation light intensity. The $\Delta F/Fm'$ values of seedlings grown in the forest edge reflect transitional sun adaptation features. Comparisons of the slope of $\Delta F/Fm'$ light response curve reveal that Carpinus is the mostly susceptible species to the photoinhibition.

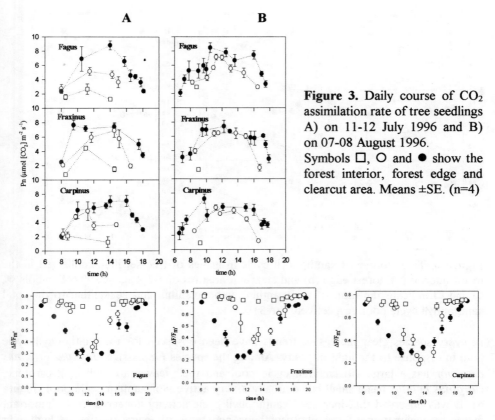

Figure 3. Daily course of CO_2 assimilation rate of tree seedlings A) on 11-12 July 1996 and B) on 07-08 August 1996.
Symbols □, O and ● show the forest interior, forest edge and clearcut area. Means ±SE. (n=4)

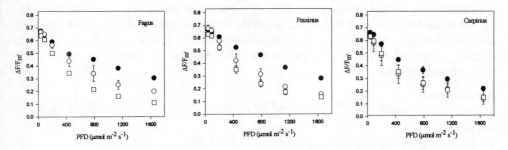

Figure 4. Changes in PSII photochemical efficiency measured in field conditions on 07 August 1996. Symbols ●, O and □ indicate clear-cut area, forest edge and forest interior (n=4)

Figure 5. Responses of PSII qauntum yield to PFD measured in laboratory conditions Symbols ●, O and □ indicate clear-cut area, forest edge and forest interior (n=4)

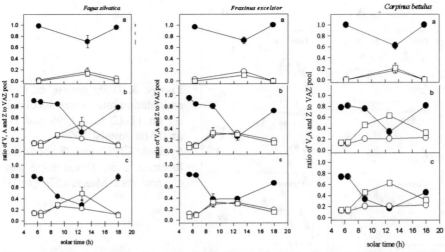

Figure 6. Time course of xanthophyll cycle in leaves of tree seedlings collected in the forest interior (a), forest edge (b) and clear-cut area (c) on 07 August 1997. Symbols ●, ○ and □ indicate the ratio of violaxanthin, antheraxanthin and zeaxanthin to the total xanthophyll cycle pool, respectively. (n=3)

In every species clear increasing trends have been observed for the xanthophyll cycle pool in response to the light intensity. Among the species *Fagus sylvatica and Fraxinus excelsior* has a large violaxanthin cycle pool and their leaves are rich in β-carotene, *Carpinus betulus* has a small but diurnally a very active violaxanthin cycle pool (Figure 6). It was observed that increased xanthophyll cycle activity appeared as an important factor in acclimation of the photosynthetic apparatus of every species to high-light conditions after the canopy openings. For each species a considerable retardation of the epoxidation process was characteristic during the night hours in the clear-cut area at dawn in the summer period and a considerable amount of de-epoxidized forms were peresent in the leavesConcernig the role of carotenoids in the protection against high light it is important to note that Fagus sylvatica and Fraxinus excelsior has a large VAZ pool and the leaves of both species and especially the latter one is rich in β-carotene. But among the species in leaves of Fraxinus the lutein and the neoxanthin occurred in the lowest concentrations.

Acknowledgements
This study was supported by COST Project ERP CIPA CT 93 0202 and OTKA No 25582.

References
1. Powles, S.B. (1984) Annu. Rev. Plant Physiol., 35:15-44
2. Demmig-Adams, B., Adams, W.W. (1992) Plant Cell. Environ., 15:411-419
3. Hager, A., Holocher, K. 1994: Planta, 192:581-589
4. Mészáros, I. (1990). In 1990: Spatial processes in plant communities.(Krahulec, F., Agnew, A.D.Q., Agnew, S., Willems, H.J. (eds.) Academia Prague, 59-71.

INVESTIGATION OF THE PHOTOSYNTHETICAL PROPERTIES OF *HYPNUM CUPRESSIFORME* MOSS TYPES ADAPTED TO DIFFERENT STRESS FACTORS

Zoltán Marschall[1], Evelin Ramóna Péli[1],
Roland Cseh[2]

[1] Department of Botany, Eszterházy Károly Teacher Training Collage, Eger H-3301, P.O. BOX 43, Hungary
[2] Department of Botany, József Attila Unyversity, Szeged H-6702, P.O. BOX 657, Hungary

Key words: adaptation, Chl fluorescence induction, environment, fluorescence, gas exchange, net photosynthesis.

1. Introduction

Mosses as the sensitive indicators of their environments (or microhabitats), dominant elements of several ecosystemes i.e. zonal biom therefore understanding of their given physiologycal response reactions for certain environmental effects are important for the mapping of complex relations of individual symbiosis. *Hypnum cupressiforme* can be found in many different habitats all over the world - in four microhabitats of our sampling area, with different morphological features and photosynthetical properties .On the basis of our investigations, the Hungarian taxa of *Hypnum cupressiforme* can be divided into more or less well-separable biotypes and ecotypes. After transplantation the behavioural types, collected from different microhabitats, were kept under identical conditions, which enabled us to acquire a better knowledge of the plasticity of their physiological properties. The temperature-dependent F_0-T curves proved the most informative in separating the photosynthetic properties of the behavioural types adapted to a low and a relatively high temperature of 10^0C and 25^0C, respectively. The nine-month adaptation of the types from various habitats under identical environmental conditions resulted in different morphological changes and photosynthetical responses. By the help of fluorescence induction methods applied we managed, for the first time, to show significant differences in the photosynthetical properties of the various types of *Hypnum cupressiforme*.

4085

G. Garab (ed.), Photosynthesis: Mechanisms and Effects, Vol. V, 4085–4088.

2. Procedure

Ecotypes of *Hypnum cupressiforme* were collected in four different microhabitats close to Felsôtárkány (HUN): shady beech forest (I.), semi opend xerophilous oak forest (II.), semi opened mesophilous oak forest (III.), half shady grassland (IV.). The moss samples were adapted in the growth cabinet at 10^0C and 25^0C, respectively. The water content of the shoots was kept at optimal level for photosynthesis (approx. 400% of dry weight in samples of *Hypnum cupressiforme*).

Samples collecting:1996.02.28. Adaptation was nine-month in the instrument place of the meteorological office of Eger. Measuring: 1996.11.05-16. Control samples collecting: 1996.11.12. Measuring: 1996.11.20-26.

A quantitative analysis of chlorophylls was undertaken by extraction in 80% ammoniacal aceton, as described by Arnon, using a Cary 3E UV-Visible Spectrophotometer.

The changes of CO_2 fixation between the transplanted and the control samples were determined by an IR gas analyzer (LAC-II, ADC Ltd., Hoddesdon, UK).

In vivo chlorophyll a fluorescence was measured in dark adapted leaves with a pulse amplitude modulation fluorometer (PAM 101-102-103, Walz, Effeltrich, FRG). 1^0C m^{-1} temperature elevation was used in heat stress measurements in the range of 20^0C -50^0C.

3. Results and discussion

Mosses have the same chl-protein complexes as higher plants (1). Similar pattern of LHCP-s can be observed in mosses as in higher plants (3). Therefore, changes in the F_0 are response to increasing temperature, can be considered as the way as higher plants (2). The temperature dependence of F_0 in higher plants proved to be a convenient tool in characterising the heat sensitivity of chloroplast thylakoids membranes. There are three critical points in a typical F_0-T curve, T1, T2 and Tp. T1 is the point where the heat start to be a stress factor for the plant and T2 in the case of *Hypnum cupressiforme* is the point where the CO_2 assimilation starting to decrease from expressed heat adaptation was observed in T1 and T2 points of the IV. and III. biotypes, after a month of heat adaptation from 10^0C to 25^0C, which was more expressed in IV. biotype. Because of its morfological properities the IV. biotype perhaps couldn`t dessicate quick enough to protect the photosynthetic membranes and it could be necessary for this biotype to the expressed heat adaptation. Different temperature values were obtained in mosses originated from different microhabitats before the temperature adaptation treatment (Table 1, 2).

IV.	T1	T2
10°C	26.498 (+1.881)	31.39 (+1.424)
25 °C	31.5 (+1.693)	34.36 (+1.135)
III.		
10°C	27.662 (+0.811)	32.09 (+0.389)
25°C	30.528 (+0.965)	32.78 (+0.330)

Table 1. Changing of T1, T2 point of F_0-T curves after a month
(T1: n=10; P<1% and T2: n=10; P<5%)

	T1	T2	Tp
I.	26.984 (±1.321)	33.054(±0.876)	41.143 (±1.124)
II.	32.805 (±0.585)	35.666 (±0.768)	42.803 (±1.112)
III.	27.750 (±2.296)	33.831 (±1.810)	41.001 (±2.353)
IV.	28.830 (±2.102)	33.666(±1.297)	41.083 (±1.289)

Table 2. The values of F_0-T curves of H. cupressiforme from the different microhabitats
(T1: n=10; P<1% and T2: n=10; P<5%)

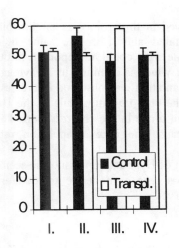

Fig. 1. Chl a/b ratio in the in the biotypes measured in autumn.

Differences were observed in the chl. *a/b* content of the mosses of different microhabitats the highest value were obtained from the I. sample and the lowest from the IV. biotype.Transplantating all the biotypes to a shady microhabitat increased the chlorophyll a+b content in all biotypes. However, changes in the chl. *a/b* ratio only in the II. and in the III. biotypes were occured.Chl. *a/b* ratio decreased in the II. biotype which is chracteristic in shade adapted mosses. Contrary to this an increase was observed in the chl. *a/b* ratio of III. biotype of *Hypnum cupressiforme* after nine-months of transplantation (Fig. 1). This is consistent with the observation of Valanne et al.(3). They reported that in the dark adapted *Ceratodon purpureus* chl. *a/b*-protein increased at the expense of antenna clorophyll *a*. The difference between the two ways of light adaptation was expressed also in the light saturation curves of CO_2 assimilation (Fig. 2).

In the case of both biotypes the slope of initial part of light saturation curve was increased, which is a typical shade-adapted property. In the II. biotype the max. assimilation rate declined while in the III. biotype the max. assimilation rate remained almost constant.

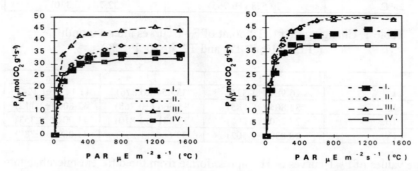

Fig. 2. Light response curves of net CO2 fixation rate in transplanted biotypes (on left) and control biotypes (on right).

4. Conclusions

Many interesting adaptation mechanisms were observed in three biotypes of the four examined. However, the I. biotype was transplantated almost in the same light conditions. Heat adaptation was observed in IV. biotype what could be important in the protection of this type of moss against the excess heat which is common in an average summer day in its microhabitat. Two types of light adaptation was observed in the case of II. biotype an increase chl. *b* content decreased the chl. *a/b* ratio while in the III. biotype an increase in chlorophyll a content resulted in a rise in the chl. *a/b* ratio. The latter mechanisms seems to be more efficient, because the max. assimilation rate remained approx. at the same value.

Considering the adaptation mechanisms the III. biotype of Hypnum cupressiforme seems to be the most plastic biotype, because in this moss type a moderate temperature adaptation and an efficient light adaptation can be observed.

Refrences

1 Aro, E.M. and Valanne (1981) In G. Akoyunoglou (ed.) Photosynthesis III. 327-335
2 Schreiber, U. and Berry, J. (1977) Planta, 136, 233-238
3 Valanne N.-E.M. Aro-H. Niemi (1982) Journ. Hattori Bot. Lab. 53, 171-179.

ECOPHYSIOLOGICAL RESPONSES OF THE DESICCATION TOLERANT MOSS, *TORTULA RURALIS* TO ELEVATED CO$_2$ LEVEL AND HEAVY METAL STRESS.

Takács, Z.[1], Csintalan, Zs.[1], Lichtenthaler, H.K.[2], Tuba, Z.[1]
[1] Department of Botany and Plant Physiology, Agricultural University of Gödöllő, 2103 Gödöllő, Hungary
[2] Botanisches Institut der Universität Karlsruhe, Lehrstuhl II, D-76128 Karlsruhe, Germany

Key words: cadmium, lead, fluorescence, respiration, CO$_2$ uptake, photosystem 2

1. Introduction

Interaction of elevated CO$_2$ concentration with different abiotic stress factors have been widely investigated. It can help plants to cope with drought (1) mostly by decreasing stomatal conductance, with heat stress (2), with salinity (3) and even with air pollutants like SO$_2$ (4) or O$_3$ (5). There was research providing evidence that increased CO$_2$ concentration may act at the antioxidant system. It decreases intrinsic oxidative stress and strengthens acclimation at the onset of stress (6).

Although bryophytes are used to biomonitor aerial dispersion of heavy metals using their ability to tolerate elevated quantities of heavy metals (7), heavy metals are by no means harmless to them. These ions denature enzymes and block their functional groups as direct toxic effects. Besides this they also induce free radical toxicity (8). We chose two metals with different toxicity (9) but having a similarly high importance in environment quality.

Fluorescence emission spectra excited by UV-A light and recorded in the visible range make it possible to calculate four characteristic fluorescence ratios (F450/F535, F450/F690, F450/F735, F690/F735) used for detecting various stresses (10). Blue and green fluorescence signals originate from secondary metabolites, mostly from phenolics (11), some of them inducible by several stress factors (12). Production of phenolics and terpenoids can be accelerated even in response to elevated CO$_2$ level (13).

In this work we aimed to determine whether elevated CO$_2$ level ameliorates heavy-metal induced damage at the already highly stress-tolerant moss, *Tortula ruralis*. We also investigated if the induced effects are identifiable with the novel plant stress detecting method, the fluorescence emission spectrum measurement.

G. Garab (ed.), Photosynthesis: Mechanisms and Effects, Vol. V, 4089–4092.
© *1998 Kluwer Academic Publishers. Printed in the Netherlands.*

2. Materials and Methods

2.1 *Plant material*
The moss *Tortula ruralis* (Hedw.) Gaertn et al. ssp. ruralis was used in this experiment.

2.2 *Treatments*
Mosses with their original soil substrates were transplanted into boxes and put in Plexiglas chambers situated in the botanical garden. High-CO_2 (700 µl l^{-1}) treatment was maintained as described in previous studies (14). Since chambers were closed on the top, plants had to be sprayed every morning with [1] deionized water (control); [2] 350 µmol l^{-1} Pb^{2+}; [3] 350 µmol l^{-1} Cd^{2+} or [4] 175 µmol l^{-1} Pb^{2+} + 175 µmol l^{-1} Cd^{2+}. Metals were given as nitrate salts. Exposure lasted for one month.

2.3 *Measurements*
The measurements were made on the upper green parts of the moss. To ensure full photosynthetic activity the plants were kept close to the optimum water contents (220-240%) for 24 h at a PPFR of 400 µmol m^{-2} s^{-1} at 20 °C before all gas-exchange and chlorophyll-fluorescence measurements on fully hydrated material. Chlorophyll fluorescence was recorded with a modulated one-channel Hansatech fluorometer. Dark respiration, light-saturated net photosynthesis rates were measured using an IRGA system (Type LCA2, ADC Co. Ltd., Hoddesdon, UK.), at a PPFR of 400 µmol m^{-2} s^{-1}. Fluorescence emission spectra were performed on the Perkin-Elmer LS-50 Luminescence Spectrometer (Perkin-Elmer, Übelingen, Germany). The fluorescence yield between the wavelengths 400 and 800 nm was recorded at an excitation wavelength of 340 nm. A 390 nm cut-off filter was applied in order to exclude scattered light.

3. Results and Discussion

3.1 *Chlorophyll fluorescence*
Photosynthetic quantum yield of PSII (Fv/Fm) (Fig 1A) decreases in response to heavy metals. At the visibly bleached Cd-treated plants it falls to 17-28 % of control level. At the mixed treatment inhibition is less, although the difference from the Cd treatment is significant only at high CO_2-level. Although the uptake efficiency for Cd is 40-65 % of that for Pb (15), its toxicity is much higher. Pb has a strong inhibitory effect on isolated chloroplasts *in vitro* but much of the Pb is sequestered as phosphate precipitation (16) or bound to the cell wall (17). "Yield" values ((F_m'-F_t)/F_m') (Fig 1B) decrease in response to Pb and mixed treatment, reach zero at Cd-treatment. Rfd values (Fp/(Fp-Fs)) (Fig 1B) that represent the functionality of the thylakoid membrane also decline at Pb-treated plants. The presence of Cd decreases Rfd to the noise level of the fluorometer even at its lower concentration (175 µmol l^{-1}) in the mixed treatment. These results indicate that these metals intervene in photosynthetic electron transport drastically, but PSII itself is less damaged. Rfd, reflecting also the state of carbon fixation decreased most among the fluorescence parameters due to the heavy metal treatments. There are no significant differences between CO_2-treatments in any of the fluorescence parameters.

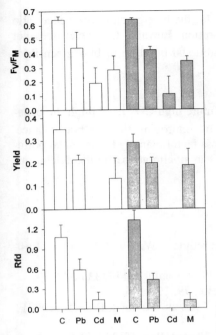

Fig 1. Flurescence values of
*T.r.*grown at 350 ppm (open bars)
and 700 ppm (solid bars). C –
control, Cd – 350 μmol l⁻¹, Pb – 350
μmol l⁻¹, M – 175 μmol l⁻¹ of both
elements.

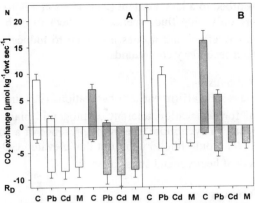

Fig 2. CO_2 gas exchange of *T.r.*grown at 350
ppm (open bars) and 700 ppm (solid bars)
measured at 350 ppm (A) and 700 ppm (B)
CO_2.

3.2 *CO_2 gas exchange*

CO_2 gas exchange measurements (Fig 2.) reveal the complete elimination of photosynthetic activity (P_N) due to the applied Cd doses independent of the CO_2 concentration. Pb cuts assimilation severely, too. Considering the net photosynthesis values of the mosses measured on their own CO_2 exposure level, CO_2 seems to enhance the assimilation of the Pb-treated samples more (by a factor of 3) than that of control ones (by a factor of 2). This also confirms that the primary attack site for heavy metals is at the Calvin cycle (18). Even though photosynthesis of the high-CO_2 mosses shows a slight downward regulation, but still they do better than those grown and measured at present CO_2 level. Dark respiration increases by approximately 300 % due to all heavy metal treatments. This phenomenon was attributed to the uncoupling of the mitochondrial membrane (19). Elevated CO_2 suppresses dark respiration because high intercellular CO_2 might affect indirectly respiratory enzymes, intercellular pH, dark CO_2 fixation (20). Dark respiration of control plants responded less to high CO_2 than heavy-metal stressed ones.

3.3 *Fluorescence spectra*

Fluorescence spectral ratios (data not shown) only increase only in response to the drastic Cd treatment. All the changes can be explained by the visible loss of chlorophyll. Loss of chlorophyll increases red/far-red fluorescence ratio because red fluorescence is reabsorbed to a lesser extent (21). The same is true for the blue/green fluorescence ratio where only blue fluorescence is subject to reabsorption. Elevated CO_2 level does not have any effect and it does not seem to induce additional build-up of blue/green fluorescing secondary compounds.

3.4 Conclusions

These results affirm earlier observations (22) that future high CO_2 level might be beneficial for desiccation tolerant cryptogamic plants by improving their carbon balance. This can be even more true in the presence of a stress factor, namely heavy metal pollution. We would like to point out that elevated CO_2 can ameliorate partly the deleterious effects of heavy metal stress as well.

References

1 Arp, W. J., Van Mierlo, J. E. M., Berendse, F. and Snijders, W. (1998) Plant Cell Environ. 21, 1-11

2 Wayne, P. M., Reekie, E. G. and Bazzaz, F. A. (1998) Oecologia 114, 335-342

3 Reuveni, J, Gale, J and Zeroni, M (1997) Annals Bot. 79, 191-196

4 Lee, E. H., Pausch, R. C, Rowland, R. A., Mulchi, C. L. and Rudorff, B. F. T. (1997) Environ. and Exp. Bot. 37, 2-3

5 McKee, I. F., Farage, P. K. and Long, S. P. (1995) Photos. Res. 45, 111-119

6 Polle, A., Eiblmeier, M., Sheppard, L. and Murray, M. (1997) Plant Cell Environ. 20, 1317-1321

7 Wells, J. M., Brown, D. H. and Beckett, R.P. (1995) New Phytol. 129, 477-486

8 Gadd, G. M. (1993) New. Phytol 124, 25-60

9 Wells, J. M. and Brown, D. H. (1995) Annals of Bot. 75, 21-29

10 Schweiger, J., M. Lang, and H.K. Lichtenthaler (1996) J. Plant Physiol. 148, 536-547

11 Lichtenthaler, H.K., and J. Schweiger (1998) J. Plant. Physiol. 152, 272-282

12 Lichtenthaler, H. K., Lang, M. and Stober, F. (1991) In: Internat. Geoscience and Remote Sensing Symposium, Helsinki University of Technology

13 Penuelas, J., Estiarte, M. and Llusia, J. (1997) Photosynthetica 33, 313-316

14 Tuba, Z., Szente, K., Nagy, Z., Csintalan, Zs and Koch, J. (1996) J. Plant Physiol. 148, 356-361

15 Berg, T., RØyset, O. and Steinnes, E. (1995) Atmosph. Environ. 29, 353-360.

16 Koeppe, D.E. (1981) In: Effect of heavy metal pollution on plants (N. W. Lepp, ed.) Applied Science Publishers, London and New Jersey.

17 Basile, A., Giordano, S., Gafiero, G., Spagnuolo, V. and Castaldo-Cobianchi, R. (1994) J. of Bryol. 18, 69-81

18 Krupa, Z. and Baszynski, T. (1995) Acta Physiol. Plant. 17, 177-190

19 Brown, D. H. and Wells, J. M. (1990) Annals of Bot. 66, 641-647

20 Ryan, M. G. (1991) Ecol. Applications 1, 157-167

21 Lichtenthaler, H.K., Hák, R. and Rinderle, U. (1990) Photos. Res. 25, 295-298

22 Tuba, Z., Csintalan, Zs. and Proctor, M. C. F. (1996) New Phytol. 133, 353-361

LICHEN'S ADAPTATION TO ALTITUDE.

E.Fernández, W. Quilhot, C. Rubio and E. Barre.
School of Chemistry and Pharmacy, Faculty of Medicine, University of Valparaíso, Chile.

ABSTRACT

Lichens from alpine regions and from areas of high altitudes are subject to the effects of high UV radiation levels, which have increased in the last decades due to the decrease of ozone, and they have developed chemical strategies to diminish the stress of the environment.The phenolic products absorbing properties of UV-A, UV-B, or both, agree with their photoprotector capacity determined by-*in vivo* and *in vitro* methods. *Rhizoplaca chrysoleuca.*, *Xanthoparmelia farinosa* and *Xanthoparmelia oleosa* accumulate usnic acid, appart from other secondary products. In transects performed in the Chilean Altiplano, between 3100 and 4500m a.s.l., a growing increment of usnic acid, with altitude and total UV radiation, was observed. The usnic acid increments are evidenced in intervals of some kilometers, in zones with high solar irradiance.The lichen adaptative capacity to radiation is expressed in the spatial and temporal variations of secondary products in habitats having different degrees of insolation.

INTRODUCTION

Chile has probably one of the richest and more varied flora in the world. The extraordinary diversity of habitats, from the hot desert and dry altiplano in the north, to the southern temperate rainforests and the cold desert of the Antarctic Territory, offers microhabitats appropiate for the development of lichens. Ultraviolet radiation levels (UV) have increased in the last years due to the decrease of stratospheric ozone. This phenomenon was first observed in the Antarctic in 1979; since then, the hole has been increasing both in size and depth . By means of satellite measurements, it has been possible to learn that this global change depends on the geographical location and the season of the year .

UV radiation (UV-A and UV-B) induce significant damaging effects. Living organisms have developed strategies to reduce this damage, one of them is the synthesis of UV radiation absorbing compounds, and which they may dissipate as fluorescence or heat. Because of the possible effects over ecosystems, the increase of UV radiation levels has led to the search of biological models to evaluate the magnitude and direction of this global change. It has been suggested that in these models lies the key to the understanding of what is happening in specific sites, in any part of the world (1).

Lichens accumulate ultraviolet radiation photoprotector compounds whose spectroscopic, photochemical and photophysical properties have been broadly studied (2,3). The photoprotector capacity of these compounds has been determined by means of *in vivo* and *in vitro* methods (4).

Usnic acid - the most frequent compound in lichen species - absorbs UV-B radiation (5). This compound presents chemical and photochemical stability, these properties permit the analysis of herbarium specimens, of any age in order to know the evolution of some environmental changes.

G. Garab (ed.), Photosynthesis: Mechanisms and Effects, Vol. V, 4093–4096.
© 1998 *Kluwer Academic Publishers. Printed in the Netherlands.*

It has recently been demonstrated that, in two antarctic lichen species, in a time scale of 30 years, usnic acid accumulation rates correlate with ozone levels measured during the same period of time (6).

A study was conducted in order to know if usnic acid concentration rates increase in species of *Xanthoparmelia* and *Rhyzoplaca* from the Chilean Altiplano. This zone is characterized by higher UV radiation levels (7).

EXPERIMENTAL

Lichen Material. *Xanthoparmelia farinosa* (Vainio) Nash, Elix & Johnston, *Xanthoparmelia oleosa* (Elix & P.Armstrong) Elix & Johnston, and *Rhyzoplaca chrysoleuca* were collected at the Chilean Altiplano. Voucher specimen are deposited in the Herbarium of the Escuela de Química y Farmacia, Universidad de Valparaíso. The location and characteristics of collection sites are included in Table 1.

Extraction and quantification of usnic acid. Lichen was extracted with acetone during 72h. The extracts were dissolved in acetone and spotted over HF 254 Merck, and were eluted in toluene-ethyl acetate-formic acid (35:5:1 v/v). The plates were scanned in a HPTLC photodensitometer at $\lambda=313$nm.

Table 1.-

Characteristics of *X. farinosa* , *X. oleosa* and *R. chrysoleuca* collection sites

Species	Site	Location	Altitud (m a.s.l.)
X. farinosa			
	Zapahuira-Putre	18°17'S;69°35'W	3350
	Enquelga	19°15'S;68°43'W	3690
	Chusmiza	19°41'S;69°10'W	3650
	Isluga	19°14'S;68°47'W	3740
X. oleosa			
	Zapahuira	18°20'S;69°37'W	3150
	Zapahuira-Putre	18°17'S;69°35'W	3530
	Colchane	19°23'S;68°45'W	3292
	Enquelga	19°15'S;68°43'W	3690
R. chrysoleuca			
	Putre	18°14′S;69°32′W	3650
	Chusmiza	19°41'S;69°10'W	3688
	Cerca de Isluga	19°14'S;68°47'W	3700
	Quebe	19°18'S;68°41'W	3700
	Isluga	19°13'S;68°45'W	3840
	Chucuyo	18°13'S;69°18'W	3530

RESULTS

The variation in usnic acid accumulation rates, and its relation with altitud, for *X. Farinosa, X. oleosa ,R. chrysoleuca* are presented in Table 2.

Table 2.-

Species	Site	Altitude (m.a.s.l.)	Usnic acid g/100	SD	cv
X.farinosa					
	Zapahuira-Putre	3350	1.694	0.313	0.188
	Chusmiza	3297	1.723	0.033	0.019
	Enquelga	3690	1.711	0.045	0.026
	Isluga	3740	3.404	0.006	0.002
X.oleosa					
	Zapahuira	3150	1.436	0.112	0.078
	Colchane	3292	1.708	0.113	0.067
	Zapahuira-Putre	3530	1.539	0.261	0.075
	Enquelga	3690	6.043	0.456	0.075
R. chrysoleuca					
	Putre	3530	0.670	0.098	0.146
	Chusmiza	3650	0.770	0.099	0.128
	Cerca de Isluga	3688	0.770	0.090	0.117
	Quebe	3700	0.930	0.090	0.097
	Isluga	3840	1.100	0.087	0.079
	Chucuyo	4500	2.300	0.344	0.150

DISCUSSION

There are evidences which allow to postulate that the increase of UV radiation induces an increment in lichen usnic acid concentration (8). The thinning of the ozone layer has produced an increase of 25% in the UV radiation levels in the Antarctica, if compared to the values considered normal. *In Neuropogon aurantiaco-ater and Ramalina terebrata* a significant increase was observed in the usnic acid synthesis in those years in which the ozone decreased to its minimum values (6).

The species of *Xanthoparmelia* and *Rhizoplaca chrysoleuca* were collected from an area with the highest radiation levels in high locations in Chile (18° - 19°S) (7), in a gradient between 321.1 and 324.7 Wh/cm^2. With UV radiation levels lower than 327.0 Wh/cm^2, no significant differences were observed in the accumulation of usnic acid, which allows to postulate the existence of physiological tolerance thresholds, once these are overcome, an increment in the biosynthesis of usnic acid would be produced, which would increase up to 3.8g/% in the *X.farinosa*, 6.06 g/% in *X.oleosa* and 2.30 g% in *Rhizoplaca chrysoleuca*.

The significant increase in usnic acid concentration, in the species collected from the locations with highest altitudes, might be due to a response of the species to the stress induced by the increase in UV radiation. *X. farinosa X. oleosa* and **Rhizoplaca chrysoleuca** show a similar trend in front of UV radiation, but they are different in the magnitude of the response which seems to be characteristic of each lichen species. In thalli from *X. oleosa* collected at 4690 m, no usnic acid was detected.

Quilhot *et al.* (6) formulated a biological model to evaluate changes in UV radiation levels, justifying the use of lichens as indicators of environmental changes. The validity and probable universality of the model would be tested by similar studies with other lichen species. The model is based in the variability study of accumulation rates for photoprotective metabolites of UV-A, UV-B radiation, or both. The low quantum yield of photoprotective metabollites (5) allows the use of herbarium specimens with the certainty that compounds do not experiment significant photodegrading changes with time.

REFERENCES

1. Galloway,D.J.(1993) Biblioth. Lichenol. 53, 87-95.
2. Hidalgo, M.E., Fernández, E., Quilhot, W. & Lissi, E.A. (1992) J. Photochem. Photobiol. A: Chem. 67, 245-254.
3. Quilhot, W., Fernández, E. & Hidalgo, M.E. (1994a) II Simposio Internacional Química de Productos Naturales y sus Aplicaciones (Silva M. & Bittner M. eds) Concepción, Chile, pp 148-154.
4. Fernández,E., Quilhot, W., González, I., Hidalgo, M.E., Molina, X. & Meneses, I. (1996) Cosmetic & Toiletries 111, 69-74.
5. Quilhot, W., Fernández, E. & Hidalgo, M.E. (1994b) British Lichen Soc. Bull. 75, 1-5.
6. Quilhot, W., Fernández, E., Rubio, C., Cavieres, M.F, Hidalgo, M.E., Goddard, M. & Galloway, D.J. (1996) Ser. Cient. INACH 461, 105- 111.
7. Cabrera, S., Bozzo, S. & Fuenzalida, H. (1995) J. Photochem. Photobiol. B: Biol. 28, 137-142.
8. Fahselt, D. (1994) Symbiosis 16, 117-165.

Acknowledgements. The research was supported by the Fondo Nacional de Ciencia y Tecnología (FONDECYT grant N° 1960353).

DIVERSITY OF KRANZ-ANATOMY AND BIOCHEMICAL TYPES OF CO$_2$ FIXATION IN LEAVES AND COTYLEDONS AMONG *CHENOPODIACEAE* FAMILY.

V. Pyankov[1], A. Kuzmin[2], M. Ku[3], C. Black[4], E. Artyusheva[1], G. Edwards[3].
[1]Dept. Plant Physiology, Urals State University, Ekaterinburg, 620083, Russia. [2]Institute Soil Sciences and Photosynthesis RAS, Pushchino, 142292, Russia. [3]Dept. Botany, Washington State University, Pullman, WA 99164-4238, USA. [4]Dept. Biochemistry and Molecular Biology, University of Georgia, Athens, 30602, USA.
Key words: adaptation, C$_3$, C$_4$, cotyledons, gene expression, ontogenesis.

1. Introduction

The *Chenopodiaceae* contains species with C$_3$, C$_4$ and possibly C$_3$-C$_4$ intermediate photosynthesis. This family has a second place in world flora in abundance of plants with C$_4$ photosynthesis (about 350-400 species) and it displays a large diversity of biochemical mechanisms of CO$_2$ fixation and Kranz-anatomy in mature leaves. C$_4$ *Chenopodiaceae* species dominate in arid desert regions of Africa, Asia and Europe, and are adapted to extreme environment. Previous studies were concentrated on investigation of C$_4$ photosynthesis in basic assimilative organs (leaves and assimilatory shoots) and demonstrated a large structural and biochemical diversity of C$_4$ photosynthesis (1). At the same time, leaf-like organs of plants such as cotyledons were, studied poor, but give evidence of occurrence both C$_3$ and C$_4$ photosynthetic types in these organs (2-3). The purpose of this study was a comparative structural and biochemical investigation of photosynthetic types in leaves and leaf-like organs in *Chenopodiaceae* species differed in taxonomy, evolution position, structural and biochemical features of C$_4$ photosynthesis.

2. Procedure

Types of photosynthesis were studied in fully expanded leaves and cotyledons of about 30 C$_4$ chenopods from Central Asia (Uzbekistan, Turkmenistan and Mongolia) belonging to different subfamilies (*Chenopodioideae* and *Salsoloideae*), tribes (*Salsoleae, Camphorosmeae* and *Suaedeae*) and genera (*Aellenia, Climacoptera, Camphorosma, Gamanthus, Halimocnemis, Haloxylon, Iljinia, Kochia, Salsola, Suaeda*). Both types of assimilative organs were tested for determination of the specific features of C$_4$ syndrome. Plants were grown in a greenhouse at the Washington State University (USA) at 25/20^0 C day/night, 1500 μmol quanta m^{-2} s^{-2} and in natural conditions at the field station of the Urals State University, Ekaterinburg, Russia.

Light microscopy. Samples of shoots and cotyledons were fixed in a 2% paraformaldehyde-2% glutaraldehyde solution in 0.1 M phosphate buffer (pH 7.2) and embedded in a mixture of Epon and Araldite. Cross sections obtained on ultramicrotome were stained with 1% Toluidine blue and 1% Na$_2$B$_4$O$_7$.

Enzyme assay. Enzymes were extracted from about 0.1 g of illuminated tissue with ml of chilled grinding medium (50 mM Tris HCl (pH 7.0), 20 mM MgCl$_2$, 2 mM MnCl$_2$, 1

G. Garab (ed.), Photosynthesis: Mechanisms and Effects, Vol. V, 4097–4100.

mM EDTA-Na$_2$, 5 mM DTT, 10% glycerol, 5% insoluble PVP and 1% soluble PVP). Activity of RUBP carboxylase (RUBPC) were assayed by the rate of ^{14}C-bicarbonate incorporation into acid-stable products. PEP carboxylase (PEPC), NAD-ME (malic enzyme), NADP-ME and aspartate aminotransferase (AAT) activities were measured spectrophotometrically following to the oxidation of NADH in case of PEPC, NAD-ME, AAT and reduction of NADP in case of NADP-ME.

Immunoblotting. For westerm immunoblot of PEPC and both ME, control species with different photosynthetic subtypes were used: *Flavesria robusta* (C$_3$), *Zea mays* (C$_4$, NADP-ME subtype), *Amaranthus cruenthus* (C$_4$, NAD-ME subtype) and *Urochloa panicoides* (C4, PEP-CK subtype). Protein was extracted using a phenol extraction procedure. Proteins were resolved in acrylamide gel. Antibodies used were anti-maize NADP-ME IgG, anti-maize PEPC IgG, anti-*Amaranthus hypochondriacus* mitochondrial NAD-ME IgG.

Primary ^{14}C-labeled products. Leaves and cotyledons were exposed in photosynthetic chamber in ^{14}CO$_2$ (400ppm) with specific activity about 5 mCi L^{-1} during 10 sec with further plunging into boiling alcohol. Labeled products were identified after two dimensional paper chromatography.

δ^{13}C carbon isotope determination. Carbon isotope fractionation values were determined on dry material using standard procedure relative to PDB limestone carbon isotope standard.

3. Result and Discussion

C$_4$ photosynthesis was found in cotyledons among all C$_4$ species studied belonging to the *Chenopodioideae* subfamily (genera *Bassia*, *Camphorosma*, and *Kochia*), and tribe *Anabaseae*. True leaves and cotyledons of these species have the same NADP-ME biochemical subtype. The main assimilatory organs (leaves or young green shoots) of plants of *Salsoleae* tribe had C$_4$ type of CO$_2$ fixation with Salsoloid type of Kranz-anatomy, but structural and biochemical diversity of photosynthetic types in cotyledons were demonstrated. The C$_3$ type of photosynthesis was found among the representatives of Salsola genus belonging to some primary sections, i.e. *Cardiandra* and *Belanthera*, and their derivative genera such as *Gamanthus*, *Halocharis*, *Halimocnemis*. Both C$_3$ and C$_4$ types of photosynthesis have been found in the genus *Climacoptera*. The C$_3$ mechanisms of CO$_2$ fixation was demonstrated in large shrubs, the derived *Coccosalsola* section (*Aellenia*, *Haloxylon*). C$_4$ cotyledons in the Salsola genus differ in Kranz-anatomy structure and NAD-/NADP-ME biochemical subtypes. More ancient Atriplecoid type of Kranz-anatomy and primary NAD-ME pathway were found in *S. laricina*, belonging to primitive *Caroxylon* section. All species studied of the youngest section Salsola (*S. kali*, *S. praecox*, *S. paulsenii*, and derived genera) and some species of *Coccosalsola* section (*S. arbuscula*, *S. richteri*) have a NADP-ME biochemical subtype , as in true leaves as in cotyledons. Interestingly, the cotyledons of C$_4$ Salsola species demonstrate more structural-biochemical diversity in photosynthetic types than true leaves do. Our investigation revealed the occurence of both C$_3$ cotyledons and C$_4$ cotyledons occuring in the following forms: 1) C$_3$ type without Kranz-anatomy; 2) C$_4$ type of photosynthesis with the same C$_4$ biochemistry and Kranz-anatomy as in mature leaves; and 3) the same biochemical type of CO$_2$ fixation, but a different types of Kranz-anatomy compared to mature leaves. Our finding shows the independent evolution of photosynthetic pathways of CO$_2$ fixation in cotyledons and leaves of C$_4$ *Chenopodiaceae* species. At the same time, the evolutionary lineages in both heteroblastic organs evolved in a similar directions:

Table. Structural and biochemical characteristics of photosynthetic pathways in leaves and cotyledons of *Salsola* and derivative species belonging to different systematic and ecological groups.

Section, species[1] Characteristics	Caroxylon		Malpigipila	Belanthera (derivative)	Cocco-salsola
	Salsola laricina	*Salsola orientalis*	*Salsola gemmascens*	*Halocharis hispida*	*S.richteri*
Life form	semishrub	semishrub	semishrub	annual	shrub
$^{13}C/^{12}C$ isotope discrimination					
L	-14.48	-12.78	-13.23	-13.73	-12.8
C	-14.78	-20.89	-23.21	-20.92	-15.16
Mesophyll structure[1]					
L	Kranz SALS	Kranz SALS	Kranz SALS	Kranz SALS	Kranz SALS
C	Kranz ATR	non Kranz IP	non Kranz DV	non Kranz DV	Kranz SALS
Activity of enzymes[2], μmol mg chl^{-1} h^{-1}					
RUBPC L	90.5	265.0	180.0	186.5	80.3
C	65.0	173.0	320.6	378.3	139.9
PEPC L	2190.4	1671.8	840.8	1377.4	1884.1
C	102.3	224.0	48.8	59.2	1290.6
NADP-ME L	19.8	398.9	9.0	0.0	210.6
C	0.0	11.8	42.7	7.4	143.4
NAD-ME L	105.4	17.1	105.4	34.8	76.2
C	67.5	31.5	24.4	33.5	43.0
AAT L	1373.0	1480.4	1430.5	1890.6	205.1
C	1607.5	318.8	414.8	259.6	444.5
Immunobloting analysis of enzymes in protein extracts[2]					
PEPC L	+++	+++	nd	+++	+++
C	++	+	nd	-	+++
NADP-ME L	-	+++	nd	-	+++
C	-	-	nd	-	+++
NAD-ME L	+++	+++	nd	+++	+++
C	+++	++	nd	+++	+++
Biochemical subtype					
L	NAD-ME	NADP-ME	NAD-ME	NAD-ME	NADP-ME
C	NAD-ME	C3	C3	C3	NADP-ME

[1]Mesophyll types: ATR- atriplicoid, SALS-salsoloid, IP-isopalisade, DV-dorsoventral.
[2]Enzyme amount: "+++" - high content, "++" - middle content, "+" - tracks, "-" - absent.

C_3 photosynthesis→ C_4 NAD-ME→ C_4 NADP-ME and were induced by environmental

4100

factors. The evolutionary level of the development of photosynthetic apparatus in leaves was usually higher than that in cotyledons, which can be reflected in more progressive biochemical/anatomy subtypes or specific features of quantitative mesophyll anatomy.

Photosynthetic types in leaves and cotyledons were identified using the protein extracted from both cotyledons and leaves, western blots were probed with antibodies raised against C_4-specific PEPC, NAD-ME, and NADP-ME to determine the presence or absence of the three key photosynthetic proteins. Determining protein content by western blot avoids problems that can occur due to inactivation of catalysis or unfavorable enzyme assay conditions. Twelve C_4 species of Chenopodiaceae were examined for the presence of three key C_4 enzymes, PEPC, NADP- and NAD-ME (Table). The results showed that six species possess C_3 cotyledons with C_4 leaves (Aellenia subaphylla, Climacoptera lanata, Haloxylon aphyllum, H. persicum, Salsola orientalis and S. sclerantha), while the other species have C_4 cotyledons and C_4 leaves (*Anabasis salsa, Salsola arbuscula, S. laricina, S. richteri, Climacoptera aralensis, Kochia prostrata*). As such, the results clearly determined the occurrence of C_4 photosynthesis in the cotyledons and the C_4 subtype for the twelve species examined, and demonstrated a possible evolutionary trend of C_4 photosynthesis in *Chenopodiaceae*. Interestingly, that molecular size NADP-ME in Chenopodiaceaea is larger in comparison with NADP-ME grass (Zea mays).

Identification of structural and biochemical traits of the C_4 syndrome in leaves and cotyledons may be important marker for phylogenetic reconstruction of the *Chenopodiaceae* family.

The study was supported in part by a Civilian Research and Development Foundation Grant (RB1-264) and grant #454 of the program «Universities of Russia» of the Russian Ministry of High Education

References
1. Pyankov, V.I., Voznesenskaja, E.V., Kondratchuk, A.V. and Black C.C. (1997) Amer. J.Bot. 84, 597-606.
2. Butnik, A.A., Nigmanova, R.N., Paizieva, S.A. and Saidov D.K. (1991) Ecological Anatomy of Desert Plants of the Central Asia. FAN, Tashkent. (In Russian).
3. Pyankov, V.I., Black, C.C., Artyusheva, E.G., Voznesenskaja, E.V., Ku, M.S.B. and Edwards, G.E. (1998) Plant Cell Physiol. (in press).

GENOME CHANGES IN PINE SEEDS AFTER IRRADIATION.

Stanislav V. ISAENKOV, Nikolay V. SOKOLOV and Boris V. SOROCHINSKY
Institute of Cell Biology and Genetic Engineering, 148 Zabolotnogo str., Kiev-252022, Ukraine

Key words: DNA, protoplasts, repair mechanism, Chernobyl

1. Introduction.

The problem of exposure of biological objects to low-intensity γ-irradiation has lately recieved a lot of interest. The data on genetic damages induced by low dose of irradiation did not show clear dependence of DNA damage on the irradiation dose [1,2]. Often, the exposure to low-intensity ionizing radiation induced more DNA damage than irradiation at much larger dose [3]. These effects can be explained by presence of two different repair systems, constitutive and inducible, in plant cells. For induction of the inducible repair system a certain level of DNA damage in cell might be required. When the number of DNA breaks is lower than the threshold, the inducible repair systems are not switched on and the constitutive repair system might not be able to repair all damages. Therefore, the biological objects irradiated by low-intensity irradiation might be strongly permanently impaired [4].

The coniferous plants that grow in constant low dose irradiation in Chernobyl zone often have different morphological abnormalities, e. g., pines with diminutive leaves. These can be already caused by permanent impairments in pine genome. Since the pine tree is the main forestry species at the Chernobyl region it is necessary to check the quality of seeds used for planting new forests. The aim of this study was to develop methods for detection of DNA damages in irradiated pine seeds and to investigate the level of DNA damages in seeds and seedlings exposed to irradiation.

2. Procedure

2.1 Plant material

Seeds of pine trees (*Pinus silvestris*) were collected from three parts of the Chernobyl zone with different level of radioactive pollution. Gamma-irradiation dose absorbed by the trees in the different regions has been measured by Staropetrovsks Forest Experimental Station as 0.03 Gy (location Dityatki), 3 Gy (Kopachi), and 6 Gy (Chistogalovka).

2.2 Irradiation

Gamma-irradiation of seeds and protoplasts was carried out in apparatus "Issledovatel" (Russia) with intensity of 0.063 Gy/sec. X-irradiation with different dose intensities was performed using devise RUM-17 (Russia). In order to compare the influence of chronic and acute irradiation seeds have been additionally irradiated with 3 or 6 Gy γ-irradiation.

G. Garab (ed.), Photosynthesis: Mechanisms and Effects, Vol. V, 4101–4104.
© 1998 Kluwer Academic Publishers. Printed in the Netherlands.

The seeds have been exposed to X-rays at doses 0.25 Gy and 0.5 Gy with intensities 4.2 R/min, 8.3 R/min, 26 R/min, 48 R/min, 87.5 R/min, and 123 R/min. Protoplasts isolated from seedlings were supplementary irradiated with 10, 20, 40, and 60 Gy γ-radiation.

2.3 Isolation of Protoplasts

Seeds have been germinated for 14 days on wet filter paper in Petri dishes in the dark at 25°C. 0.5 g of the shoot material were incubated in 3 ml of enzyme solution [1% Onozuka R-10 (Sigma, USA); 0.85% Macerase (Calbiochem, USA); 0.75% Cellulase (Calbiochem, USA); 0.75% Pectinase (Ferak, Germany) in 50 mM K-phosphate pH 5.8, 0.5 M sucrose, 50 mM $CaCl_2$]. The mixture was incubated 18 h at 25 C. Protoplasts were collected by centrifugation at 100g and washed twice with 10 ml of W5 solution [4].

2.4 DNA Isolation and Analysis

DNA was isolated from protoplasts according to [4] and from shoots and seeds according to [5]. The DNA damage has been monitored after 30 min, 2 hours, and 4 hours intervals between irradiation and cell lysis. Number of DNA breaks in DNA isolated from seeds and shoots which have been irradiated by X-rays was determined by the alkaline unwinding method followed by hydroxylapatite chromatography according to [6]. In order to increase the level of double-strand breaks of DNA the samples have been treated by S1 nuclease according to [7]. DNA from protoplasts exposed to additional external irradiation was analyzed by pulsed-field gel electrophoresis and hydroxylapatite chromatography as well. Pulsed-field gel electrophoresis has been performed in horizontal gel box (Diapuls, Russia) at 85 V, 199 mA at 8 C, pulse interval 45 min for 14 hours, after that pulse-interval has been changed to 20 min for 7 hours and then to 30 sec for 3 hours. The electrode buffer was 0.5xTBE [8].

3. Results and Discussion

The estimation of germination efficiency and viability of protoplasts derived from seedlings of pines from different Chernobyl zones revealed a non-linear response to the irradiation [4]. For further analysis, the appearance of smear in DNA samples in pulsed-field gel electrophoresis was used as the criteria for DNA damage. As shown in table 1, the intensity of DNA degradation significantly differs at different doses of γ-radiation.

Table 1. **Influence of supplementary γ-irradiation on degradation of DNA isolated from protoplasts absorbing different doses of chronic irradiation.**

Dose of supplementary irradiation	Time between irradiation and DNA isolation		Dose of chronic irradiation			Dose of acute irradiation	
		control	0.03 Gy	3 Gy	6 Gy	3 Gy	6 Gy
0	0	+	+	+	+	+	-
10 Gy	0	++	++	++	++		
	4 h	++	++	++	++		
20 Gy	0	++	++	++	++		
	4 h	++	+	-	-		
40 Gy	0	++	++	++	++		
	4 h	+++	+++	-	+		
60 Gy	0	+++		+++	+++	++	-
	2 h	+++		+++	+++		
	4 h	+++		+++	+++	+	-

DNA degradation kinetics varied among protoplasts having absorbed different doses of chronic irradiation. DNA from protoplasts supplementary irradiated with 10 Gy contained small amount of smear from degraded DNA that has identical density within the whole track. Character of this smear did not change during 4 hours after irradiation. At 20 Gy the appearance of DNA smear changes to clear DNA fragments. When protoplasts were incubated for 4 hours after irradiation the DNA smear decreased in dimension, except for control samples. These results revealed that chronical irradiation induces DNA reparation systems, which are able to correct DNA degradation caused by further irradiation. After supplementary acute irradiation at 40 Gy the repair system worked more effectively in protoplasts from seeds from 3 Gy zone than 0.03 Gy zone. When the protoplasts were irradiated with supplementary dose of 60 Gy the DNA degradation was very intensive and could not be repaired any more.

In next experiment biological effectiveness of the chronic and acute irradiation to induce the DNA repair systems was compared. Control seeds that were acutely gamma-irradiated at 3 and 6 Gy and seeds from 3 and 6 Gy zone were germinated and protoplasts were isolated from the seedlings. The protoplasts were then γ-irradiated with a 60 Gy dose. The DNA analysis has demonstrated that the effect of chronic irradiation is much higher than that of acute irradiation. DNA from seeds that were irradiated acutely at 3 Gy was almost completely repaired within 2 hours of incubation and also the DNA smear from 6 Gy seedlings was decreased within this time interval (Table 1). The same 60 Gy dose led to irreparable DNA degradation in seedlings from chronically irradiated seeds.

In order to confirm the results obtained by pulse-field electrophoresis we carried out similar experiments using alkaline unwinding method with subsequent chromatography on hydroxylapatite [8]. They resulted in quantitative indexes of DNA breaks in protoplasts supplementary irradiated at doses of 20 and 40 Gy (Fig.1). The chromatography basically confirmed the results obtained by pulse-field electrophoresis. DNA degradation process after supplementary irradiation with 40 Gy has been prevented in protoplasts from seeds from the 3 Gy zone but has increased in seeds from the 0.03 Gy zone.

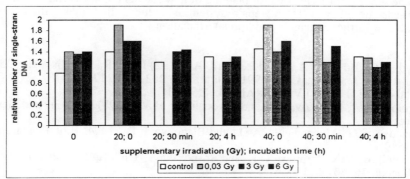

Figure 1. **Dependence of DNA damage in pine protoplasts on dose of chronic irradiation and time of recovery.**

In order to estimate the level of DNA damage in irradiated seeds and seedlings we have exposed them to X-rays of different intensities. DNA was analyzed by alkaline unwinding method with subsequent hydroxylapatite chromatography. The DNA damage in irradiated seeds was significantly higher than in control seeds (Fig. 2). Also the DNA from irradiated seedlings contained more single-strand DNA than the control (Fig. 3). In both seeds and seedlings the 0.5 Gy dose caused more severe DNA damage than the 0.25 Gy one. However, at the dose of 0.5 Gy and intensity 87,5 R/min the DNA damages level in both tissues was only as low as in control samples. The 0.25 Gy dose caused DNA degradation in seedlings only when applied with low intensity, thus confirming that an inducible system is responsible for the DNA repair. DNA from seeds was relatively more impaired than DNA from seedlings which could be explained by small level of metabolic processes and repair abilities of dry seeds.

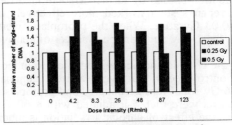

Figure 2 **Dependence of DNA damage in seeds on dose intensity of X-ray irradiation**

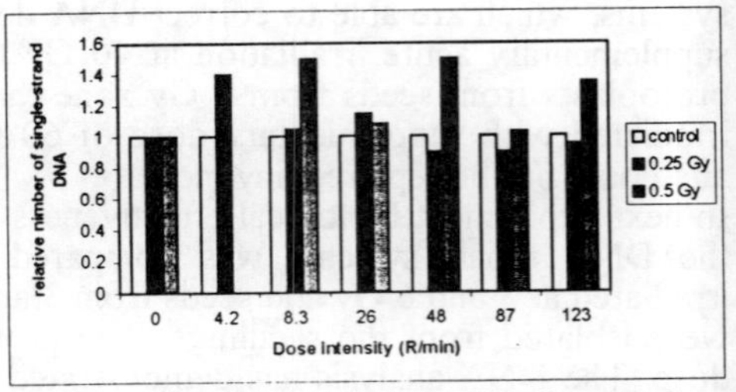

Figure 3 **Dependence of DNA damage in seedlings on dose intensity of X-ray irradiation**

In conclusion, we have successfully applied two different methods to estimate the level of DNA damage in irradiated biological material. We have observed that in pine seeds that were exposed to low dose irradiation the DNA is harmed, but during germination and seedling development part of this damage is eliminated by endogenous repair systems. We could also demonstrate the presence of an inducible DNA repair system that prevents further DNA degradation due to additional irradiation. Exposition of seeds to low intensity chronic irradiation causes more DNA damage than an acute exposition to the same radiation dose. Finally, as a practical result from our experiments, we could recommend using pine seeds from 3 Gy zone for planting of new forests in the Chernobyl zone, since in these seeds an efficient DNA repair system is induced that make these seeds less susceptible to further damage by additional irradiation.

References
1 Lloyd, D.C., Edvards, A.A. and Leonard, A. (1992) Int. J. Radiat. Biol. 61, 335-343
2 Pohl-Ruling, J. and Haas, O. (1991) Mutat. Res. 262, 209-217
3 Burlakova, E.B (1994) Vestn. RAN 64, 425-451
4 Sokolov, N.V., Isaenkov, S.V. and Sorochinsky, B.V. (1998) Cytology and Genetics 5 in press
5 Chancova, S.G., Mehandjeiev, A.D., and Blagoeva, E.D. (1994) Biol. Plantarum 36, 583-589
6 Reshotnikov, V.N., Lapteva, O.K. and Sosnovskaya, T.F. (1996) Radiat. Biol. Radioecol. 36, 567-571
7 Maniatis T., Fritisch E. and Sambrook J. (1982) Molecular Cloning. A laboratory Manual. E.L.Press, Oxford, UK
8 Davies K.E. ed., (1988) Genome analysis. A practical Approach. IRL Press, Oxford. Washington DC. 246p.

MODELING NET PHOTOSYNTHESIS BASED ON TEMPERATURE AND LIGHT IN COLONIAL *PHAEOCYSTIS ANTARCTICA* KARSTEN

Tiffany A. Moisan and B. Greg Mitchell, Scripps Institution of
Oceanography, University of California, La Jolla, CA 92093-0218

Key words: aquatic ecosystems, light acclimation, marine algae, optical properties, prymnesiophyte, temperature acclimation

1. INTRODUCTION

The genus *Phaeocystis* plays an important role in polar ecosystems and biogeochemical cycles. It blooms frequently and when abundant often contributes more than 50% of the total phytoplankton biomass (1). When blooms of this organism do occur, they are often monospecific in nature or either codominate with very few other species. By nature of the size difference between the single cells (3-8 μm) and millimeter-sized colonies of its lifecycle, it may preferentially offer a food source to the microbial food web while in its single cell stage or alternatively to the traditional food web while in its colonial life stage. It may serve as a significant dissolved organic matter source through exudation or colonial lysis to the microbial food web.

A goal of phytoplankton ecologists is to estimate primary productivity based on limited knowledge of light, temperature, and nutrient status. Oceanographers have approached this goal by semi-analytical models and regression methodology which correlates phytoplankton pigments to primary production estimates. Many contemporary models (2, 3, 4) utilize the chl-specific absorption coefficient ($a^*_{ph}(\lambda)$) and quantum yield for growth (ϕ_μ) as photophysiological parameters in models of net primary production,

$$P_{net} = chla \int_{350nm}^{700nm} \phi_\mu a^*_{ph}(\lambda) E_o(\lambda) d\lambda. \qquad [1]$$

The potential for estimating primary production from satellites, autonomous drifters or mooring data has stimulated a renewed interest in the inherent optical properties of phytoplankton including the spectral absorption coefficients which are the link between pigment concentrations (e.g. chlorophyll *a*) and ocean color or the reflectance spectrum of the ocean,

$$R(\lambda) \propto b_b(\lambda)/a(\lambda) \qquad [2]$$

where $b_b(\lambda)$ is the total backscatter coefficient and $a(\lambda)$ is the total absorption coefficient. The absorption coefficient is related to phytoplankton by the following equation

$$a_{ph}(\lambda) = a^*_{ph}[chl] \qquad [3]$$

where [chl] denotes the phytoplankton chlorophyll concentration.

In this paper, we present a model to estimate net primary production of *Phaeocystis* using optical variables which can be utilized with remote sensing instrumentation. We

G. Garab (ed.), Photosynthesis: Mechanisms and Effects, Vol. V, 4105–4108.

have determined the magnitude to which temperature and light affect the variables in the model. We also present a model of the quantum yield for growth which is a key parameter that links absorption to net photosynthesis, allowing for an estimate of net primary production through remote sensing.

2. PROCEDURE

Cultures of *Phaeocystis antarctica* (CCMP 1374) were grown semi-continuously for 5-8 generations in f/2 medium (5) at −1.5, 0, 2, and 4°C. Cultures were grown under continuous blue light at 5 levels ranging from 11 to 214 µmol quanta m^{-2} s^{-1}. The treatments at 4 °C were grown at 7 levels ranging from 14 to 542 µmol quanta m^{-2} s^{-1}. The culturing system simulated the spectral irradiance of ocean surface waters and is described in Moisan and Mitchell (6). The specific growth rate (µ) was calculated from the slope of daily estimates of the natural log of *in vivo* fluorescence of cell suspensions measured in a Turner Designs Model 10 fluorometer. Spectral absorption coefficients were estimated in culture suspension in an integrating sphere within a Perkin Elmer Lambda 6 dual beam spectrophotometer. The quantum yield for growth was estimated by rearranging Equation 1 and was based on the carbon-specific growth rate, whole cell *in vivo* spectral absorption, and the spectral irradiance in each treatment.

3. RESULTS AND DISCUSSION

To generate a model to predict net primary production, we assessed how the variables in the model behaved with respect to light and temperature. The chlorophyll-specific absorption coefficient, a^{*}_{ph} (λ), varied with irradiance (Figure 1A) and the magnitude was related to pigment packaging effects which were induced by the individual cell diameter, pigment content, and thylakoid stacking (Moisan and Mitchell 1998). Values of a^{*}_{ph} (λ) varied little if any over the range of temperatures. An example is shown in Figure 1B.

The specific growth rate, µ, was limited by both light and temperature. In a compilation of data from this study and the literature, values of µ increased with irradiance and saturated above 400 µmol quanta m^{-2} s^{-1} (Figure 1C). An overall comparison of our data compared to other data for *Phaeocystis* studies resulted in a saturated value for growth, $E_{k\mu}$, of 114 µmol quanta m^{-2} s^{-1} (Figure 1C). A strong temperature dependence was observed for growth rate; values of µ were low at −1.5 °C and peaked between 0 and 2 °C and were depressed at 4 °C. Relative growth as a function of temperature was similar for all light treatments (Figure 1D).

Quantum yields for growth (ϕ_{μ}) ranged from 0.0006 to 0.09 mol C fixed per mol quanta absorbed. Values of ϕ_{μ} increased with decreasing irradiance and were highest at 4 °C and decreased with decreasing temperature (Figure 1E). It should be noted that the relationship between ϕ_{μ} and irradiance were similar at 0 and -1.5 °C.

We modeled ϕ_{μ} as the product of the maximal quantum yield for growth and a Poisson probability function that an open photosynthetic unit will be hit (7, 8),

$$\phi_{\mu} = \phi_{mEo} \frac{1 - \exp^{-E_o / E_{k\mu}}}{E_o / E_{k\mu}}, \qquad [4]$$

where ϕ_{mEo} is the maximal quantum yield dependent on light for a specific temperature,

E_0 is the photosynthetically available irradiance, and $E_{k\mu}$ is the irradiance at which growth rate saturates.

Given our values of ϕ_μ, we estimated a light-dependent ϕ_{mEo} at each discrete temperature (Table 1). To find a model which would be applicable to all *Phaeocystis* populations, we used our composite $E_{k\mu}$ value of 114 µmol quanta m^{-2} s^{-1} which was statistically similar to the value estimated from our data. We then predicted the quantum yield for growth based on the light and temperature dependent values of ϕ_{mEo}. The equations given in Table 1 and the final functions of ϕ_μ are plotted in Figure 1E.

Using our model for quantum yield and measured values of a^*_{ph} (λ) for the corresponding irradiance values, our model for quantum yield does a reasonable job at estimating growth rates. The predicted growth rates underestimate the observed growth rates by about 10% (Figure 1F).

It is our hope in the future to find a continuous function for ϕ_μ so that it can be applied to estimate the growth rate of monospecific blooms of *Phaeocystis* in Polar Regions. A temperature and light- limited matrix of quantum yield is ideal for estimating the growth rate of *Phaeocystis*. During spring and early summer, nutrients are generally in excess in coastal or ice edge regions of the Antarctic. Thus, temperature and light are the relevant controlling factors that limit the onset of *Phaeocystis* blooms. Understanding the ecology of *Phaeocystis* in polar ecosystems is important because it can dominate blooms and make a significant contribution to the carbon and sulfur cycle.

Table 1. Model for maximal quantum yield based on irradiance at discrete temperatures.

Temperature (C)	Model for Maximal Quantum yield
-1.5	$0.0303 \exp^{-0.0061*E_0}$
0	$0.0132 \exp^{-0.0113*E_0}$
2	$0.0730 \exp^{-0.0067*E_0}$
4	$0.1231 \exp^{-0.0044*E_0}$

4. REFERENCES

1 Stoecker, D.K., Putt, M. and Moisan, T.A. (1996) J. Mar. Biol. Ass. U.K. 75: 815-832.
2 Kiefer, D.A. and Mitchell, B.G. (1983) Limnol. Oceanogr. 28: 770-76.
3 Sakshaug, E., Andresen, A. and Kiefer, D.A. (1989) Limnol. Oceanogr. 34: 198-205.
4 Cullen, J.J. (1990) Deep-Sea Res. 37: 667-83.
5 Guillard, R.R.L. and Ryther, J.H. (1962) Can. J. Microbiol. 8: 229-39.
6 Moisan, T.A. and Mitchell, B.G. Submitted to Limnol. and Oceanogr.
7 Mauzerall, D. (1978) Photochem. Photobiol. 28: 991-998.
8 Dubinsky, Z., Falkowski, P.G. and Wyman, K. (1986) Plant Cell Physiol. 27: 1335-1349.

5. ACKNOWLEDGEMENTS

We thank Scott Cheng and JL Swan for excellent technical assistance. This work was supported by Office of Naval Research grant N0014-91-J-1186 to BGM and NASA graduate fellowship NGT5-30036 to TAM. Travel support from NASA's Ocean Biogeochemistry Program supported travel by TAM to participate in the congress. We thank PG Falkowski for the opportunity to participate in the congress.

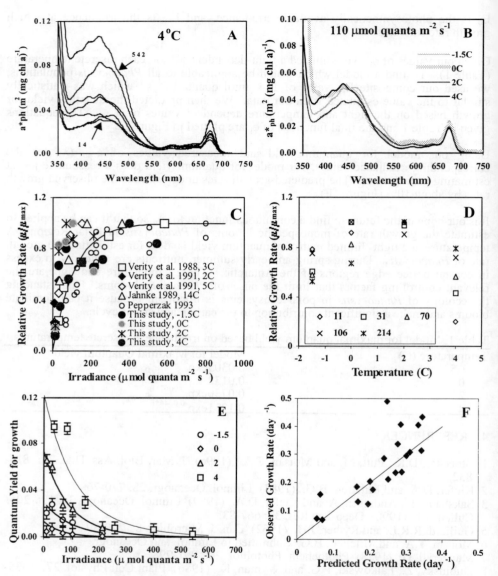

Figure 1. The relationship of chlorophyll-specific absorption (Panel A and B) and relative growth rate (Panel C and D) with light and temperature, respectively. Treatments shown in Panel A and D are in units of μmol quanta m^{-2} s^{-1}. The quantum yield for growth (mol C fixed per mol quanta absorbed) as a function of temperature and light (Panel E). The relationship between the observed growth rate and the predicted growth rate for our model of net photosynthesis which is given in Equation 4 and Table 1 compared to the 1:1 line (Panel F).

RELATIONSHIP BETWEEN CHLOROPHYLL FLUORESCENCE QUENCHING AND O₂ EVOLUTION IN MICROALGAE

João Serôdio, Jorge Marques da Silva* and Fernando Catarino
Instituto de Oceanografia and Centro de Engenharia Biológica*, Departamento de Biologia Vegetal, Faculdade de Ciências da Universidade de Lisboa, 1700 Lisboa, Portugal.

Key words: Chl fluorescence induction, cyanobacteria, diatoms, green algae, productivity, quantum yield

1. Introduction

The verification of a virtually identical quantitative relationship between the chlorophyll (Chl) *a* fluorescence parameter ΔF/Fm' and the quantum efficiency of photosynthesis in a large number of different plant species [1], enabled the establishment of a rapid and noninvasive way to estimate gross photosynthetic rates from simple Chl *a* fluorescence measurements. Although this technique has been applied to microalgal species and communities ([2],[3],[4]), little is known about the relationship between the fluorescence parameter ΔF/Fm' and the quantum yield of photosynthesis in organisms other than higher plants. Only very recently the first studies were published addressing this question [5]. The present study aims the characterisation of the relationship between the PSII photochemical quantum yield, determined by the Chl *a* fluorescence quenching analysis (ΔF/Fm'), and photosynthetic oxygen production in three species of microalgae representative of the main taxonomic groups found in the phytoplankton and microphytobenthos, with the purpose of evaluating the applicability of the technique to natural microalgal communities.

2. Material and Methods

2.1 Microalgal cultivation

Chl *a* fluorescence and O₂ evolution were measured in species of microalgae representative of the main groups found in the phytoplankton and microphytobenthos. The chlorophyte *Chlorella vulgaris* Beijerinck, the cyanobacterium *Spirulina maxima* (Setch. et Gard.), and the diatom *Phaeodactylum tricornutum* (Böhlin), were grown in semicontinuous unialgal batch cultures at 20 °C and 50 µmol m^{-2} s^{-1} (PAR, photosynthetically active radiation) in a 14/10 h light/dark cycle. *S. maxima* was grown in Zarrouk medium [6] and *P. tricornutum* was grown in natural seawater enriched with f/2 nutrients [7]. The medium for *C. vulgaris* was prepared according to [8]. Cultures were gently shaken daily. Cells were harvested by centrifugation (1000 × g, 5 min) during the exponential growth phase, at room temperature, and were resuspended in fresh growth medium supplemented with NaHCO₃ (10 mM final concentration). Self-shading was avoided by using diluted algal suspensions.

G. Garab (ed.), Photosynthesis: Mechanisms and Effects, Vol. V, 4109–4112.
© 1998 Kluwer Academic Publishers. Printed in the Netherlands.

2.2 *Measurement of oxygen and Chl* a *fluorescence*

Oxygen evolution and Chl *a* fluorescence were measured in a Hansatech DW2 oxygen and fluorescence chamber (Hansatech, Norfolk, UK) coupled to a magnetic stirrer. Stirring intensity was kept as low as possible to minimise cell damage. To avoid anaerobiosis, the chamber was left open to allow the dissolution of oxygen. Temperature of the sample was controlled by a waterbath (Haake NK22, Haake, Berlin, Germany). Oxygen evolution was measured using a Hansatech oxygen electrode (CBD1 control box), and Chl *a* fluorescence was measured with a pulse amplitude modulation fluorometer (PAM 101 Chlorophyll Fluorometer, Heinz Walz, Effeltrich, Germany). Chl *a* fluorescence and oxygen evolution were measured under different irradiance levels, ranging from ca. 50 to 1600 μmol m^{-2} s^{-1}, at 20 °C. White

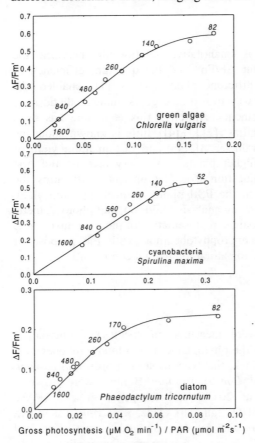

light was provided by a FL-102 fibre illuminator (Schott KL1500 halogen lamp; Schott, Mainz, Germany) together with neutral density filters. Incident irradiance was measured at the end of the fiberoptics. Saturation pulses of 0.8 s duration and ca. 7000 μmol m^{-2} s^{-1} were applied using a FL-103 fibre illuminator (Schott KL1500 halogen lamp). Samples were allowed to adapt to each irradiance level until a steady-state in fluorescence signal and a linear increase in O$_2$ concentration were observed. The gross rate of photosynthesis was calculated from the rate of increase of O$_2$ concentration in the light plus the measured dark respiration. ΔF/Fm' was measured according to [1], immediately after the determination of the photosynthetic rate.

Figure 1. Relationship between ΔF/Fm' and gross oxygen production per incident quanta in three species of microalgae. Numbers represent incident irradiance (PAR, μmol m^{-2}s^{-1}) level.

3. Results and Discussion

The same general pattern was observed in all the tested species (Figure 1). ΔF/Fm' and oxygen production per incident quanta vary linearly for most of the range of irradiance applied above 140 μmol m^{-2} s^{-1} (*C. vulgaris*, R=0.997, P<0.001; and *S. maxima*, R=0.971, P<0.001) or 170 μmol m^{-2} s^{-1} (*P. tricornutum*, R=0.994, P<0.001). Under

Table 1. Slope of the linear part of the relationship between $\Delta F/Fm'$ and gross O_2 production, expressed per incident PAR, and further normalized to F_o.

Species	Incident PAR	Incident PAR/F_o
Chlorella vulgaris	4.39	1.36
Spirulina maxima	2.11	1.03
Phaeodactylum tricornutum	4.08	1.02
c.v. (%)	35.1	17.0

lower irradiances, the trend departs from linearity, as observed for higher plants ([9],[10]). However, important differences are observed when the different curves are compared (Figure 2a), with the slope of the linear portion of the curve being found to differ by a factor of two among species (Table 1). Most of this variability is caused by differences in the fraction of incident irradiance that is actually absorbed for photosynthesis, due to differences in cell concentration and pigmentation. In contrast with plant leaves, in which the variations in this fraction represent a minor source of error in the applicability of the technique, the wide range of concentrations found in natural microalgal suspensions prevents the verification of a unique relationship between $\Delta F/Fm' \times$ PAR and photosynthetic rate if calculated on the basis of incident irradiance alone. Due to the low variability of its quantum yield of fluorescence emission, the dark-level fluorescence, F_o, may be expected to be proportionally related to the rate of light absorption [11], therefore providing a way to trace variations in the fraction of incident irradiance absorbed for photosynthesis. By using F_o as a normalisation factor (using the same instrument settings with all species), the variability among slopes was reduced by more than 50% (c.v. decreased from 35.1% to 17.0%, Table 1) significantly increasing the similarity between the curves obtained for different species (Figure 2b). However, still some significant variability remains in the shape of the curves. While $\Delta F/Fm'$ varies within a comparable range of values in *C. vulgaris* and *S. maxima* (maximum values between 0.55 and

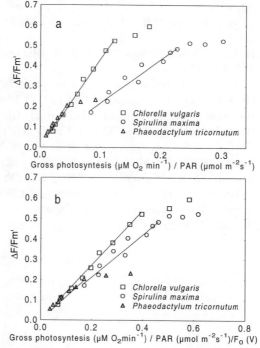

Figure 2. Relationship between $\Delta F/Fm'$ and oxygen production per incident quanta (a) and normalized to F_o (b).

0.60), for *P. tricornutum* it reaches much lower values, which do not exceed 0.30. Such differences cannot be expected to be caused by variations in cell concentration, and are likely to be due to basic differences in the organisation and pigmentary composition of the photosynthetic membranes among different taxonomic groups [12]. Nevertheless, the results indicate that in a situation representative of a natural microalgal community, formed by species from groups with different photophysiological characteristics, mixed in different and variable concentrations, photosynthetic production may be predicted from Chl *a* fluorescence quenching analysis.

Acknowledgements

This work was supported by Junta Nacional de Investigação Científica e Tecnológica, grant PRAXIS XXI BD/5045/95 to João Serôdio, and by project NICE, MAS3-CT96-0048.

References

1 Genty, B., Briantais, J.-M., Baker, N.R. (1989). Biochem. Biophys. Acta 990: 87-92.
2 Bilger, W., Büdel, B., Mollenhauer, R., Mollenhauer, D. (1994). J. Phycol. 30: 225-230.
3 Hofstraat, J.W., Peeters, J.C.H., Snel, J.F.H., Geel, C. (1994). Mar. Ecol. Prog. Ser. 103: 187-196.
4 Kromkamp, J., Barranguet, C., Peene, J. (1998). Mar. Ecol. Prog. Ser. 162: 45-55.
5 Geel, C., Versluis, W., Snel, J.F.H. (1997). Photosyn. Res. 51: 61-70.
6 Richmond, A. (1988). in Micro-algal Biotechnology (Borowitzka M.A. and Borowitzka L.J., eds.) pp.85-121, Cambridge University Press, Cambridge, U.K.
7 Guillard, R.R.L., Ryther, J. H. (1962). Can. J. Microbiol. 8: 229-239.
8 Veloso, V. (1995). MSc. Thesis. Faculty of Sciences, University of Lisbon.
9 Seaton, G.G.R., Walker, D.A. (1990). Proc. R. Soc. London B 242: 29-35.
10 Hormann, H., Neubauer, C., Schreiber, U. (1994). Photosyn. Res. 40: 93-106.
11 Serôdio, J., Marques da Silva, J., Catarino, F. (1997). J. Phycol. 33: 542-553.
12 Büchel, C., Wilhelm, C. (1993). Photochem. Photobiol. 58: 137-148.

BIOMONITORING OF CORAL REEF AND TEMPERATE FORAMINIFERS BY THE CHL *a* FLUORESCENCE RISE O-J-I-P OF THEIR SYMBIONTS

Merope Tsimilli-Michael[1,2], Martin Pêcheux[1] and Reto J. Strasser[1]
[1]Bioenergetics Laboratory, University of Geneva, CH-1254, Jussy-Geneva, Switzerland, [2]Cyprus Ministry of Education and Culture.

Key words: adaptation, bleaching, dinoflagellates, endosymbiosis, light thermoprotection, structural changes

1. Introduction

Massive bleaching of reef corals and foraminifers, involving the loss of their photosynthetic symbionts and/or their pigments, affects since the early 80's the reef ecosystem. The phenomenon is still poorly understood. However, temperature, irradiation and CO_2 are assumed to be primary factors, though it is not yet clear whether the host or the symbionts are more susceptible to them and thus responsible for the symbiosis rupture. Chl *a* fluorescence measurements were recently introduced to follow the behaviour of photosystem II (PSII) in coral associations or their isolated symbionts (1,2).

In order to establish a procedure for monitoring the vitality of the symbiotic associations, we here investigated in three genera of large foraminifers the PSII behaviour of their symbionts *in hospite* by means of the fast polyphasic Chl *a* fluorescence transients O-J-I-P (3) they exhibit upon illumination. The tested species were: the coral reef foraminifers *Amphistegina lobifera* (harbouring the diatom *Fragilaria* sp.) and *Amphisorus heimprichii* (harbouring the dinoflagellate *Symbiodinium* sp.), and the temperate foraminifer *Sorites variabilis* (harbouring *Symbiodinium* sp.). The measurements, carried out by the PEA fluorimeter (Plant Efficiency Analyser), can be conducted continuously even on a single cell in a test tube, as well as on the reef. The PSII behaviour of the photosynthetic apparatus at different physiological states established by different culture temperature and light conditions, was quantified through a constellation of parameters provided by the analysis of the fluorescence transients according to the JIP-test (4-6). By this screening test that we widely use for studies of land plant stress, many samples can be analysed in a short time as it needs a measuring time of only 1 to 5 seconds. The calculated parameters, all referring to time zero (onset of fluorescence induction), are (a) the specific (per reaction centre) energy fluxes for absorption (ABS/RC), trapping (TR_0/RC) and electron transport (ET_0/RC) and (b) the flux ratios or yields, i.e. the maximum quantum yield of primary photochemistry (φ_{Po}), the efficiency with which a trapped exciton can move an electron into the electron transport chain (ψ_0), and the quantum yield of electron transport (φ_{Eo}).

The aim is to investigate whether and how adaptive processes are revealed as regulating the different parameters, which can further permit the correlation of adaptability and resistance to bleaching, and to establish a rapid and easy test for the biomonitoring of symbiotic associations *in situ*. This would offer an access to the understanding of the causes of bleaching and possibly serve as well in foreseeing the future of the reefs.

G. Garab (ed.), Photosynthesis: Mechanisms and Effects, Vol. V, 4113–4116.
© 1998 *Kluwer Academic Publishers. Printed in the Netherlands.*

2. Procedure

The foraminifer cells were cleaned and distributed in glass tubes with 5.5 ml of daily exchanged Mediterranean sea-water (CO_2/HCO_3^--controlled pH at about 8.2), kept in a thermostated water bath and exposed to light-dark cycles of 12 h light (L) at 70 $\mu E/m^2 s$ - 12 h dark (D). The temperature was gradually elevated avoiding any unnatural conditions. Chl a fluorescence transients, induced by a red light (peak at 650 nm) of 600 W m^{-2} (~6000 $\mu E/m^2 s$), were recorded every 6 hours by a Plant Efficiency Analyser (PEA, built by Hansatech Instruments Ltd. King's Lynn Norfolk, PE 30 4NE, GB) with a 12-bit resolution (3) and analysed according to the JIP-test (4-6). The experimental protocol in Figure 1 shows the light conditions (LC) and the temperature of the culture (CT) for the 6 h preceding each measuring period (MP). The cultures were exposed to high light (HL) of 550 $\mu E/m^2 s$ for 6 h (between the 19th and 20th MP).

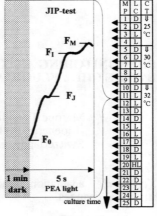

Figure 1.
The experimental protocol.

3. Results and Discussion

The shape of the fast polyphasic Chl a fluorescence transient depends strongly on the cultivation light and temperature conditions, as shown in Figure 2. The temperature elevation resulted in a pronounced down-regulation of F_t/F_0 when the cultures were in the

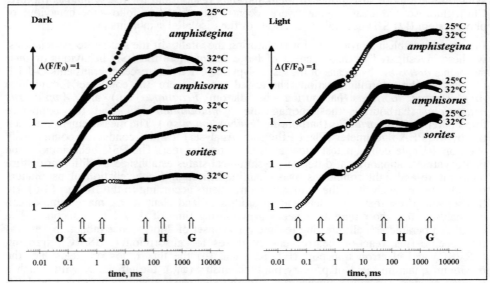

Figure 2. The fast Chl a fluorescence transients of the three organisms being for 12 h in the dark or in the light of 70 $\mu E/m^2 s$, at 25°C or 32°C. The transients, vertically displaced, are presented on a logarithmic time scale, revealing a sequence of more steps after the J-step, i.e. O-(K)-J-I-H-G. Depending on the experimental conditions any step can be the highest (P-step). The K-step (7) appears and becomes dominant if the cells suffer from strong heat stress (data not shown).

dark phase, while in the light phase the transients are almost identical at both temperatures, indicating that *the warm water during the night may be a severe stressor triggering coral reef bleaching*. The expression of the low-light thermoprotection is of different extent in the three studied organisms and, for each of them, in the several parameters probing the PSII behaviour, as revealed by the response patterns of the yields (Figure 3a) and the specific fluxes (Figure 3b). The protection is realised both by preventing the deformations and by reversing those occurring in the dark. It is also demonstrated that the deformations that the various parameters undergo in the dark are

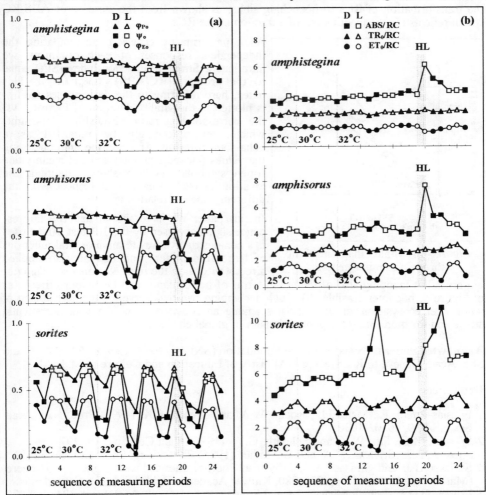

Figure 3. The response of the three species to the sequence of light-dark cycles, under a gradually elevated cultivation temperature and for a temporary (6 hours) increase of the light intensity to 550 $\mu E/m^2 s$ (HL), expressed (a) by the yields (or ratios of fluxes) φ_{Po}, ψ_o and φ_{Eo} and (b) by the specific fluxes ABS/RC, TR_0/RC and ET_0/RC. Closed symbols stand for the dark phase and open symbols for the light phase (70 $\mu E/m^2 s$).

4116

of different extent, thus indicating that different survival strategies are employed in response to stress. ψ_o appears more sensitive to the cultivation conditions than φ_{Po}. Though they are completely independent, their changes are in phase and, thus, the changes of φ_{Eo} are even more pronounced ($\varphi_{Eo} = \varphi_{Po} * \psi_0$). The homeostatic behaviour of TR_0/RC upon HL shows that regulation mechanisms maintain a constant excitation rate of the open RCs, thus avoiding their over-excitation and photo-destruction. Indeed, no loss of photosynthetic pigments, neither any impairment of PSII was detected, as judged from the stability of the F_0 level. It can well be speculated that these mechanisms involve the transformation of RCs to quenching sinks, as shown from the increase of ABS/RC which reflects (5) the inactivation of a fraction of the RCs.

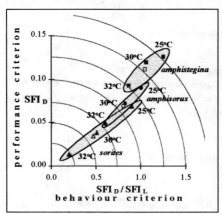

Figure 4. A vitality diagram of the three foraminifer genera.

As a performance criterion, accumulating the independent responses of each organism, we introduce a Structure-Function-Index (SFI) equal to the product (RC/ABS) $*$ (φ_{Po}) $*$ (ψ_0). This index, here referring to the dark phases (SFI_D), is plotted in the vitality diagram of Figure 4 vs. the corresponding ratio SFI_D/SFI_L (dark/light) which can accordingly be regarded as a behaviour criterion. An envelope line, enclosing the values (closed symbols) and the mean value (open symbols), is also shown, defining the location of each organism in respect with the different zones of relative vitality.

In this paper we have shown that the fluorescence signal analysed by the JIP-test can be used in a non-destructive way to monitor vital functions of large foraminifers and can be also applied directly *in situ* on the reef. Not only screening of big areas but mapping also of vitality in terms of performance and behaviour criteria, as in Figure 4, becomes feasible (8). Such a mapping can thus serve for the comparison of whole reef ecosystems, or of organisms among an ecosystem, upon any environmental stress and for monitoring the general impact of global changes.

Acknowledgements: Swiss National Foundation (SNF 31-46.860.96/ 31-52.541.97) and Soc. Acad. Genève to R.J.S. Cyprus Ministry of Education and Culture to M.T.-M.

References

1 Iglesias-Prieto, R., Mata, J.L., Robins W.A. and Trench, R.K. (1992) Proc. Natl. Acad. Sci. USA 89, 10302-10305
2 Warner, M.E., Fitt, W.K. and Schmidt, G.W. (1996) Plant Cell Env. 19, 291-299
3 Strasser, R.J., Srivastava, A. and Govindjee (1995) Photochem. Photobiol. 61, 32-42
4 Strasser B.J. and Strasser, R.J. (1995) in Photosynthesis: From Light to Biosphere, (Mathis P., ed.) Vol. V, pp.977-980, Kluwer Academic Publishers, The Netherlands
5 Krüger, G.H.J., Tsimilli-Michael, M. and Strasser, R.J. (1997) Physiol. Plant. 101, 265-277
6 Strasser, R.J., Srivastava. A. and Tsimilli-Michael, M. (1999) in Probing Photosynthesis: Mechanism, Regulation and Adaptation (Yunus, M., Pathre, U. and Mohanty, P., eds.) (in press)
7 Srivastava, A., Guisse, B. Greppin H. and Strasser, R.J. (1997) Biochim. Biophys. Acta 1320, 95-106
8 Tsimilli-Michael, M., Pêcheux, M. and Strasser, R.J. (1998) Archs. Sci. Genève (in press)

A SIMPLE VOLUMETRIC METHOD FOR PRIMARY PRODUCTION MEASUREMENTS IN SEDIMENTS

Adam Latała
Institute of Oceanography, University of Gdańsk
Al. Piłsudskiego 46, PL 81-378 Gdynia, Poland

Key words: *aquatic ecosystems, gas exchange, marine algae, microalgae, net photosynthesis, respiration*

1. Introduction

Microphytobenthos assemblages abundant in shallow sea water, often play an important role in organic matter production and regulation of fluxes of compound at the sediment-water interface. In investigations of photosynthetic activity of microbenthic algae the following methods are commonly applied: ^{14}C fixation, oxygen exchange e.g. incubation techniques with transparent and dark chambers connected with Clark's oxygen electrodes, and recently oxygen microprofile technique which requires the application of oxygen microelectrodes with a sensing tip of a few microns (1). Each method has its advantages and disadvantages (2, 3). In the present studies a simple microvolumetric method was applied. Since the number of studies concerning sublittoral microphytobenthic production remains more scarce than the intertidal, this work concerns a spatial study of the microphytobenthic production in the Gulf of Fos down to a depth of 12 m, which represents more than 50% of the Gulf area.

2. Material and Methods

Samples of sediments were collected by diving at eleven stations with sediments composed of fine sands (0.1; 0.5 and 1 m depth) and muddy sands (1.5; 2; 3; 4; 5; 6; 8 and 12 m depth) along a transect in the Carteau Bay, a shallow cove (average depth 4 m) located in the Gulf of Fos (43°25' N; 4°56' E, French Mediterranean Coast). The top layer of sediments (0-1 cm) was cut out from the core of 2.1 cm diameter and freeze-dried. Pigments (chlorophyll *a* and pheopigments) were extracted for 24 h with 90% acetone and measured according to (4). Organic carbon, total nitrogen and sulphur contents were determined on a CHN analyzer. The measurements of dark respiration and photosynthesis at saturated irradiance of PAR (650 μmol photons·m^{-2}·s^{-1}) were performed in September 1995. It was found that the optimum irradiance for the highest net and gross production in sediments form the Gulf of Fos is around 200 μmol photons·m^{-2}·s^{-1} (5) and many literature data indicate that photosynthesis of microbenthic algae becomes light-saturated at irradiance above 600 μmol photons·m^{-2}·s^{-1} (6). The gas exchange rates were determined by the microvolumetric method described in detail by

G. Garab (ed.), Photosynthesis: Mechanisms and Effects, Vol. V, 4117–4120.
© 1998 *Kluwer Academic Publishers. Printed in the Netherlands.*

(7). This method enables evaluation of gas exchange (photosynthesis and respiration) in small biological objects, and allows the estimation of oxygen production or consumption at rates varying from 10^{-1} to 10^{-4} µl $O_2 \cdot h^{-1}$ with an accuracy of ±5% (7). The small subsamples taken from the top layer of sediment (0-2 mm) were placed on the bottom of measuring chambers whereas carbonate buffer was settled as a suspending drop (Fig. 1). All measurements were made at 18°C, near the same, as it was in situ at the time of collection and at constant and relatively high CO_2 level (0.03% in the gas phase of the measuring chambers maintained by Warburg no. 10 carbonate buffer). The measurements were made immediately after collecting sediments from the field, at least in 4 replicates.

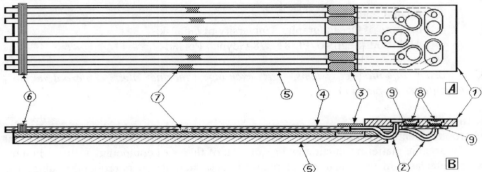

Fig. 1. Microrespirometer chamber systems: A - view from above; B - cross-section; 1 - brass plate with five microchambers; 2 - capillaries; 3 - plastic tubes; 4 - measurement capillaries; 5 - mounting base; 6 - comb; 7 - small column of coloured kerosene; 8 - carbonate buffer as a suspending drop; 9 - samples of sediment on the bottom of measuring microchambers.

3. Results and Discussion

The most important features of the superficial (first cm) sediments at the stations studied in the Gulf of Fos and microphytobenthic rates of gas exchange determined in surface layer (0-2 mm) were presented in the Table.

Depth [m]	NPP [dw]	NPP [chla]	R [dw]	R [chla]	GPP/R [dw]	Chla	Pheop	Chla/ Pheop	430/ 665	Corg	S
0.1	0.0164	2.509	-0.0094	-1.637	2.74	5.76	0.85	6.79	2.26	0.248	0.349
0.5	0.0171	2.829	-0.0032	-0.274	6.34	6.04	1.73	3.48	2.32		
1.0	0.0057	1.468	-0.0019	-0.654	4.11	3.87	1.14	2.89	2.54	0.092	0.098
1.5	-0.0240	-5.317	-0.0287	-7.473	0.16	3.84	6.44	0.54	3.65	0.674	0.339
2.0	-0.0646	-18.436	-0.0763	-21.750	0.15	3.51	9.71	0.36	4.19	0.655	0.827
3.0	-0.0421	-2.590	-0.0690	-4.240	0.39	16.28	35.70	0.46	3.43	1.587	1.058
4.0	-0.0263	-9.980	-0.0334	-12.680	0.21	2.63	12.70	0.21	3.45	0.737	0.464
5.0	-0.0044	-0.058	-0.0134	-0.325	0.67	41.24	91.05	0.46	3.18	1.098	0.468
6.0	-0.0123	-0.480	-0.0285	-2.103	0.57	13.54	28.05	0.48	2.95	0.756	0.346
8.0	-0.0083	-1.081	-0.0078	-1.033	0	7.63	28.90	0.26	3.56	1.773	0.782
12.0	0	0	-0.0054	-8.205		0.66	4.04	0.16	4.61	0.887	0.293

NPP[dw] - rates of net primary production expressed in µl $O_2 \cdot mg^{-1}dw \cdot h^{-1}$; NPP[chla] - rates of net primary production normalized to chlorophyll a [µl $O_2 \cdot µg^{-1}chla \cdot h^{-1}$]; R[dw] - dark respiration rates expressed in $\mu l\ O_2 \cdot mg^{-1}dw \cdot h^{-1}$; R[chl$a$] - dark respiration rates normalized to chlorophyll a [$\mu l\ O_2 \cdot \mu g^{-1}chla \cdot h^{-1}$]; GPP/R[dw] - ratio of gross primary production to dark respiration [$\mu l\ O_2 \cdot mg^{-1}dw \cdot h^{-1}$]; Chl$a$ - chlorophyll a contents in the top layer (0-1 cm) of sediments [mg chl$a \cdot g^{-1}dw$]; Pheop - concentration of pheopigments in the first cm of sediments [mg pheop $\cdot g^{-1}dw$]; Chl$a$/Pheop - ratio of chlorophyll a to pheopigments [mg $\cdot g^{-1}dw$]; 430/665 - ratio of OD at 430 nm to OD at 665 nm; C_{org} - organic carbon expressed in % sediment dry weight; S - total sulphur contents in the top layer (0-1 cm) of sediments [mg S $\cdot g^{-1}dw$]

At the stations located along the transect in Carteau Bay with an increase in depth characteristic changes in the gas exchange rate in the top layer of sediments could be observed. Only the most shallow 3 stations (0.1; 0.5 and 1 m) show positive values of net primary production and relatively low dark respiration rate. Consequently, the ratio of gross primary production to dark respiration is above 2. It was noted (8) that in the shallowest part of the Gulf of Fos (<1 m) the oxygen production of microphytobenthos exceeds that of phytoplankton. In sediments chlorophyll a prevails pheopigments and their ratio attains values of about 3 or more and organic carbon content is below 3 mg $C_{org} \cdot g^{-1}dw$. The stations of intermediate depths (1.5-4 m) show in the top layer negative, high values of net primary production and the highest dark respiration values. It could indicate intensive development of microorganisms. In sediments of these station organic carbon content is from above 6 to a dozen or so mg $C_{org} \cdot g^{-1}dw$. The above features observed at the stations of intermediate depths could be explained by intensive farming in the Carteau Bay covering an area of 0.053 km^2 at the depth 4-5 m. Mussel cultures on suspended ropes were shown to have a strong impact on benthic community by increasing organic matter and pigment contents in the sediments. Cumulative effects of biodeposition induce higher respiration rates (9). At deeper stations (5-8 m) in spite of the highest chlorophyll a and pheopigment concentrations, dark respiration rate and negative values of net primary production are low. At the deepest station (12 m) there was no net primary production. It is in accordance with the results of (8). Dark respiration (-0.0054 $\mu l\ O_2 \cdot mg^{-1}dw \cdot h^{-1}$) and pigment content in the sediment top layer were relatively low.

In the earlier investigations carried out in the Gulf of Fos no significant correlation was found between benthic primary production and chlorophyll a contents at spatial scale in all sublittoral sediment samples (5, 8, 9). Other authors have found significant correlations between chlorophyll a and gross primary production in sublittoral sediments. In the present studies such a correlation was found in the case of changes in gross primary production at different depths, normalized to chlorophyll a [$\mu l\ O_2 \cdot \mu g^{-1}$ Chl$a \cdot h^{-1}$]. The linear model of gross primary production was: depth = 6.54 - 1.63·GPP and R = -0.815. Also the relationship between net primary production and 430/665 ratio (linear model: NPP = 0.10237 - 0.037 · ratio 430/665) and between gross primary production normalized to organic carbon [$\mu l\ O_2 \cdot \mu g^{-1}C_{org} \cdot h^{-1}$] and the ratio of chlorophyll a to chlorophyll a + peopigments (linear model: GPP = -0.0029 + 0.01469 · ratio chl$a$/pigments) in sediments are well correlated R = -0.901 and 0.9775, respectively. Gross primary production normalized to dry weight of sediments or chlorophyll a does not show a correlation at p<0.05 with the pigment content in sediment. Moreover, it was found a relationship between net primary production and dark respiration normalized to dry weight (linear model: NPP = 0.01061 + 0.91664 · Respiration) or to chlorophyll a (linear model: NPP = 1.7255 + 0.98329 · Respiration). These parameters showed a strong correlation, R = 0.939 and 0.983, respectively. Linear relationship between respiration and gross photosynthesis in the cyanobacterial mat was described by (10), R value was above 0.96.

Until now microvolumetric methods were not commonly applied in the investigations of primary production in benthic microalgae assemblages. The employed microvolumetric method, till now generally used in the measurements of gas exchange rate in small biological objects, enabled successfully to determine precisely photosynthetic and respiration rates in benthic microalgae. The measured values of primary production and respiration rates are in accordance with the literature data, also with the values noted in the Fos Bay. The measurements performed in that Bay by incubation techniques with transparent and dark chambers and oxymeter application (5, 8) showed that in autumn the net primary production rate at the station of 0.5 m depth is ca. 1.3 μl $O_2 \cdot \mu g^{-1} chla \cdot h^{-1}$. Similar values were found in spring (1.29) and winter (1.1), and only in summer net primary production was lower, about 0.4 μl $O_2 \cdot \mu g^{-1} chla \cdot h^{-1}$. In my experiments the data from sediment samples collected in autumn at the same depth (0.5 m) and place were twice as large, i.e. about 2.8 μl $O_2 \cdot \mu g^{-1} chla \cdot h^{-1}$. However, it should be stressed that microrespirometer measurements enable saturation by PAR, whereas the production in situ at 0.5 m depth could not be measured in saturated conditions. The applied microvolumetric method appeared to be very effective. It is relatively simple, cheap, sensitive and makes it possible to test the same sediment sample at different temperatures or irradiances in the short time. Owing to the data obtained photosynthetic light curves commonly used in primary production modelling could be determined.

4. References

1 Revsbech, N.P. (1983) in Polarographic Oxygen Sensors (Gnaiger/Forstner ed.) pp. 265-273, Springer-Verlag, Berlin Heidelberg

2 Revsbech, N.P. Jørgensen, B.B. & Brix, O. (1981) Limnol. Oceanogr. 26, 717-730

3 Jönsson, B (1991) Limnol. Oceanogr. 36, 1485-1492

4 Lorenzen, C. (1967) Limnol. Oceanogr. 12, 343-346

5 Barranguet, C Cervetto, G & Fontaine M.-F. (1996) C. R. Acad. Sci. Paris, Sciences de la vie 319, 51-56

6 Blanchard, G.F. Guarini, J.-M. Gros, P. & Richard, P. (1997) J. Phycol. 33, 723-728

7 Zurzycki, J. & Starzecki, W. (1971) in Plant Photosynthetic Production. Manual of Methods (Šesták Z. Čatský J. & Jarvis P.G., eds) pp.257-275, Dr W. Junk N.V.Pub., The Hague

8 Barranguet, C. Plante-Cuny, M.R. & Alivon, E. (1996) Hydrobiologia 333, 181-193

9 Barranguet, C. Alliot, E. Plante-Cuny M.-R. (1994) Oceanol. Acta 17, 211-218

10 Epping, E.H. & Jørgensen, B.B. (1996) Mar. Ecol. Prog. Ser. 139, 193-203

5. Acknowledgements

The author is grateful to R. Plante from Station Marine d'Endoume, Marseille for providing laboratory facilities. I would like to thank M.-R. Plante-Cuny and R. Plante for their help and very useful suggestions.

This research was partly supported by ATP EO1 and BW

22. Biotechnological applications

CONGRESS ON PHOTOSYNTHESIS

XIth INTERNATIONAL

BUDAPEST * 1998

REDUCTION OF TELLURITE AND SELENITE BY PHOTOSYNTHETIC BACTERIA

Monique SABATY, Magali BÉBIEN, Cécile AVAZÉRI, Vladimir YURKOV*, Pierre RICHAUD and André VERMÉGLIO
CEA/ DEVM Laboratoire de Bioénergétique Cellulaire CEA Cadarache 13108 Saint Paul-lez-Durance, Cedex France
* present address : The University of British Columbia, Department of Microbiology and Immunology, 300-6174, University Boulevard, Vancouver, BC, Canada, V6T1Z3

Key words : heavy metals, nitrate reductase, purple bacteria, R. sphaeroides, resistance, toxicity

1. Introduction

Oxyanions of tellurium and selenium in the form of potassium or sodium salts are toxic for most microorganisms and animals (1, 2). However high level of resistance has been reported for purple (3) and aerobic anoxygenic photosynthetic bacteria (4) of the alpha subclass of *Proteobacteria*, and for extreme thermophiles of the genus *Thermus* (5). This constitutive high level of resistance confers to some of these bacteria the ability to grow at concentrations higher than 2 mg/ml of tellurite (4). This is two to three orders of magnitude higher than the minimal inhibition concentration (MIC) usually determined for most gram-negative bacteria. The toxicity of tellurite and selenite for microorganisms is probably linked to the strong oxidant property of these compounds. Although the mechanisms of resistance to tellurite and selenite have not been characterized in details, several hypotheses have been made (see ref. 5 for a detailed discussion). This includes the repair of cellular damage, reduced uptake and increased efflux or sequestration of the oxyanions. Recently it has been shown that two loci are involved in tellurite resistance in the case of *Rhodobacter sphaeroides* (6) but the exact role of their products is still unclear. One possible mechanism of protection is the reduction by the bacteria of the oxyanions to the lesser oxidant metallic form. Indeed, formation of black or red colonies due to the intracellular deposition of metallic tellurium or selenium respectively has been observed (7, 8) for bacteria grown in the presence of their corresponding oxyanions. The reduction of the soluble forms (TeO_3^{2-}, SeO_3^{2-}) to the solid metallic forms (Te^0, Se^0) leads to the intracellular accumulation of metal by the bacteria (see Figure 1 as example). In addition to these reduction processes, *Rhodobacter sphaeroides* exhibits high resistance to a large number of toxic heavy metals (3). These properties open potential applications in detoxification and bioremediation of polluted waters and soils by photosynthetic bacteria. The aim of our work is to describe the molecular mechanisms involved in these reduction processes. Identification of the genes encoding the different enzymes involved may allow the construction of recombinant plants with new bioremediation capabilities.

G. Garab (ed.), Photosynthesis: Mechanisms and Effects, Vol. V, 4123–4128.

2. Procedures

2. 1 *Bacterial strains*
Rhodobacter sphaeroides strains 2.4.1 and forma sp. *denitrificans* IL 106 were grown under photosynthetic conditions at 30° C in Hutner medium as described in (9). Obligate aerobic photosynthetic bacteria (personal collection of Dr. Vladimir YURKOV) were grown as previously described (10-13).

2. 2 *Polyacrylamide gels*
Electrophoresis under non-denaturing conditions was carried out as previously described (9). To visualize the enzymatic activities, the non-denaturing gels incubated with reduced methyl or benzyl viologen were placed in the presence of different substrates (KNO_3, K_2TeO_3, Na_2SeO_4 and Na_2SeO_3).

2. 3 *Electron microscopy*
Transition microscopy pictures were obtained as previously described (14).

2. 4 *Spectrophotometry*
Light-induced absorption changes in whole cells were detected with an apparatus similar to that described in (14). Excitation was provided by xenon flashes (3 μs half time). The light-induced carotenoid band-shift was measured as the difference $\Delta A_{522nm} - \Delta A_{508nm}$.

3. Results and Discussion

3. 1 *Reduction of tellurite by aerobic photosynthetic bacteria*

Aerobic anoxygenic photosynthetic bacteria have been discovered recently in a variety of habitats (15). They belong to the alpha subgroup of *Proteobacteria* but present specific features which distinguish them from typical anoxygenic phototrophs. These features are inability to grow photosynthetically under anaerobic conditions and strong inhibition of bacteriochlorophyll synthesis in the light (15). We have recently observed that several species of obligate aerobic photosynthetic bacteria present high level of resistance to tellurite and intracellular accumulation of metallic tellurium crystals (4). For some species the tellurium crystals are so large that they are running between daughter cells (see Figure 1 as example). The level of resistance is highly dependent upon the carbon sources and the considered species. The highest values for MIC, between 2.3 and 2.7 mg of TeO_3^{2-} per ml, were observed for cells of *Erythromicrobium hydrolyticum*, *Erythromonas ursincola* (formerly *Erythromicrobium ursincola*) and *Erythromicrobium ramosum* grown in the presence of acetate (4). Although in most cases, growth in the presence of tellurite was concomitant with the appearance of a black coloration of the suspension due to the formation of metallic Te^0, high level of resistance was observed without tellurite reduction for *Roseococcus thiosulfatophilus* and *Erythromicrobium ezovicum* grown in the presence of acetate for example (4). This implies that tellurite reduction is not the only factor conferring tellurite resistance. This conclusion is in agreement with previous studies. In *Enterobacteriaceae* for example, five different determinants, not related to tellurite reduction, have been shown to confer tellurite resistance (16-19).

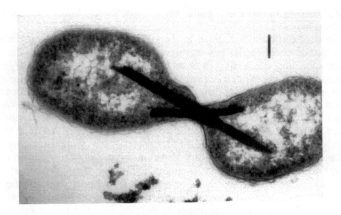

Figure 1 : Intracellular accumulation of metallic tellurium in cells of *Erythromicrobium ramosum*. The cells have been grown in the presence of 1 mg tellurite per ml. The bar represents 100 nm.

3. 2 *Nitrate reductase is able to reduce tellurite, selenate and selenite*

Although reduction of tellurite and selenite by bacteria is a very efficient process, the enzymes responsible for this transformation have not been purified and characterized in details. Early report of Chiong et al. (19) describes the purification of a protein fraction from *T. thermophilus* possessing an NADH/NADPH tellurite reductase activity. More recently and since tellurium and selenium are within the same group in the periodic table as sulfur, the possibility that the sulfate reducing pathway is also involved in their oxyanions reduction has been investigated (20). The results show that only trace amounts of selenate can be reduced by this pathway. A periplasmic selenate reductase has been purified to homogeneity from *Thauera selenatis* (21), a species which is able to grow at the expense of selenate reduction under anaerobic conditions. This enzyme is not able to reduce nitrate, nitrite, sulfate or chlorate. In *Rhodobacter sphaeroides*, a membrane-localized flavin-dependent tellurite reductase activity has been reported (3). We confirmed such activity for the membrane fraction of *Rhodobacter sphaeroides denitrificans* but found a much higher tellurite reductase activity in the soluble fraction when supplied with NADH, benzyl or methyl viologens as electron donors. To further characterize this enzymatic activity we monitored tellurite and selenite reductase activities using non-denaturing gels electrophoresis of the soluble fraction (data not shown). A single clear band linked to the oxidation of the exogenous electron donor benzyl viologen appears after addition of tellurite, selenate or selenite. The enzymes responsible for the reduction of tellurite, selenate and selenite present a R_F identical to the R_F observed for the nitrate reductase activity (data not shown). The capability of the periplasmic nitrate reductase to reduce tellurite, selenate or selenite is not specific of *Rhodobacter sphaeroides*. Similar activities have been also observed for periplasmic or membrane-bound nitrate reductases of different bacteria species (22).

To confirm the role of the nitrate reductase in the reduction of tellurite and selenite, we have constructed a mutant deficient in this enzyme by insertion of an antibiotic resistance cassette in the corresponding gene (nap⁻ mutant). This mutant is unable to grow under denitrification condition and, as expected, present no detectable nitrate, tellurite or selenite reductase activities in the soluble fraction. However this mutant still reduces in large amount tellurite and selenite as clearly shown by the appearance of black and red color of their respective metallic forms. We therefore conclude that, although the soluble nitrate reductase possesses a high tellurite and selenite reductase activity at least *in vitro*, other reducing pathways are present *in vivo*. Such reducing pathways are probably related to the tellurite reductase activity identified and localized in the membrane fraction (3).

The resistance level to tellurite or selenite is not affected by the deletion of the nitrate reductase gene in *Rhodobacter sphaeroides*. For both WT and nap⁻ mutant, the measured MIC was equal to 200 μg of tellurite or selenite per ml. This situation is however different from the one encountered in *E. coli* where mutants deleted into the two membrane-bound nitrate reductases are hypersensitive to tellurite with MICs of 0.03 μg of TeO_3^{2-} per ml compared to 2 μg per ml for the wild type (22). This implies that a minimal threshold level of nitrate reductase is required to confer tellurite resistance in *E. coli* contrary to what we observed for *Rhodobacter sphaeroides*.

3 .3 Why a high capacity for oxyanions reduction in photosynthetic bacteria?

Moore and Kaplan have postulated that the high capacity for oxyanions reduction, present in photosynthetic bacteria, is to maintain an optimal intracellular redox state for photosynthetic activity (23). This was based on the observation that the extent of TeO_3^{2-} reduction *in vivo* was inversely related to the oxidation state of the available carbon source (23). Under light anaerobic conditions, the photosynthetic bacteria generate excess of reducing equivalents. This reducing power can be removed by tellurite or selenite acting as electron acceptors. A similar mechanism has been proposed for DMSO, TMAO or NO_3^- acting as electron sinks (24). In the absence of auxiliary electron acceptors, a large fraction of the primary electron acceptor is reduced and therefore the photosynthetic reactions are partly blocked. This type of hypothesis will be consistent with the high values of the midpoint of the primary electron acceptor of anoxygenic photosynthetic bacteria which are 65 to 120 mV more positive than those determined for purple bacteria (25). In order to test this hypothesis we have measured, by flash-spectroscopy on intact cells, the light-induced carotenoid band-shift under anaerobic conditions in the presence or absence of tellurite or selenite. The results are presented in Figure 2 in the case of *Rhodobacter sphaeroides*.

Under anaerobic conditions, a slow phase (halftime of 1-2 ms) in the carotenoid band-shift is observed. This slow phase corresponds to the electron transfer occurring at the cytochrome bc_1 complex level. The rate of this phase is determined by the redox state of the quinone pool phase. A fast rate corresponds to a quinone pool fully reduced. In the presence of tellurite or selenite the slow phase is not observed (Fig. 2) demonstrating that these compounds have oxidized the quinone pool in the dark under anaerobic conditions in agreement with the hypothesis developed above.

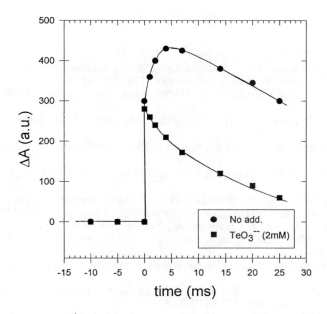

Figure 2 : Light-induced absorption changes linked to the carotenoid band-shift for cells of *Rhodobacter sphaeroides* placed under anaerobic conditions in the presence (■) or the absence (●) of 2 mM TeO_3^{2-}.

4. Conclusions

Since the industrial revolution, human activities have resulted in the release of toxic chemicals into the environment. Bacterial bioremediation is a potentially attractive, ecological and economical method for removing these toxic chemicals. The reduction of soluble form of different metal oxyanions to solid form with intracellular accumulation by photosynthetic bacteria could be an important mechanism for their removal from polluted environment.

Using a biochemical approach we have shown that the nitrate reductase can reduce *in vitro* tellurite, selenate and selenite. However reduction of these compounds are still occurring in a mutant deleted in nitrate reductase implying that other reducing pathways are involved in this reduction process. We have recently isolated mutants selected for their inability to reduce tellurite or selenite. Their characterization will provide new insights on the molecular mechanisms of metals oxyanions reduction by photosynthetic bacteria.

4128

References

1 Summers, A.O. (1985) Trends Biotechnol. 32, 122-125
2 Schroeder, H.A., Buckman J. and Balassa J.J. (1967) J. Chron. Dis. 20, 147-161
3 Moore, M.D. and Kaplan, S. (1992) J. Bacteriol. 174, 1505-1514
4 Yurkov, V., Jappé, J. and Verméglio, A. (1996) Appl. Environ. Microbiol. 62, 4195-4198
5 Turner, R.J., Hou, Y., Weiner, J.H. and Taylor, D.E. (1992) J. Bacteriol. 174, 3092-3094
6 O'Gara, J.P., Gomelsky, M. and Kaplan, S. (1997) Apll. Environ. Microbiol. 63, 4713-4720
7 Turkey, F.L., Walper, J.F., Appleman, M.D. and Donohue, J. (1962) J. Bacteriol. 83, 1313-1314
8 Gerrard, T.L., Telford, J.N. and Williams, H.H. (1974) J. Bacteriol. 119, 1057-1060
9 Sabaty, M., Gagnon, J. and Verméglio, A. (1994) Arch. Microbiol. 162, 335-343
10 Yurkov, V., Stackebrandt, E., Holmes, A., Fuerst, J.A., Hugenholtz, P., Golecki, J., Gad'on, N., Gorlenko, V.M., Kompantseva, E.I. and Drews, G. (1994) Int. J. Syst. Bacteriol. 44, 427-434
11 Yurkov, V. Gorlenko, V.M. and Kompantseva, E.I. (1992) Microbiology 61, 169-172
12 Yurkov, V., Stackebrandt, E., Buss, O., Verméglio, A., Gorlenko, V. and Beatty, J.T. (1997) Int. J. Syst. Bacteriol. 47, 1172-1178
13 Yurkov, V and Gorlenko, V.M. (1993) Microbiology 61, 163-168
14 Sabaty M., Jappé, J., Olive, J. and Verméglio, A. (1994) Biochim. Biophys. Acta 1187, 313-323
15 Shimada, K. (1995) in Anoxygenic Photosynthetic Bacteria (Blankenship, R.E., Madigan, M.T. and Bauer, C.E., eds) pp. 105-122, Kluwer Academic Publishers, Dordrecht, The Netherlands
16 Whelan, K.F., Colleran, E. and Taylor, D.E. (1995) J. Bacteriol. 177, 5016-5027
17 Walter, E.G., Weiner, J.H. and Taylor, D.E. (1991) Gene 101, 1-7
18 Turner, R.J., Hou, Y., Weiner, J.H. and Taylor, D.E. (1992) J. Bacteriol. 174, 3092-3094
19 Chiong, M., Gonzalez, E., Barra, R. and Vasquez, C. (1988) J. Bacteriol. 170, 3269-3273
20 Zehr, J.P. and Oremland, R.S. (1987) Appl. Environ. Microbiol. 53, 1365-1369
21 Schröder, I., Rech, S., Krafft, T. and Macy, J.M. (1997) J. Biol. Chem. 272, 23765-23768
22 Avazéri C., Turner R.J., Pommier, J., Weiner J.H., Giordano, G. and Verméglio, A. (1997) Microbiology 143, 1181-1189
23 Moore, M.D. and Kaplan, S. (1994) A.S.M. News 60, 17-23
24 Ferguson, S.J., Jackson, J.B., and McEwan, A.G. (1987) F.E.M.S. Microbiol. Rev. 46, 117-143
25 Yurkov, V., Menin, L., Schoepp, B. and Verméglio, A. (1998) Photosynth. Res. (in press)

APPLICATION OF LIGHT-EMITTING DIODES (LED'S) IN ALGAL CULTURE: IN CULTURE LIGHT HARVESTING EFFICIENCY OF THE GREEN ALGA CHLORELLA AND THE CYANOBACTERIUM CALOTHRIX.

Hans C.P. Matthijs, Hans Balke, Udo M. van Hes and Luuc.R. Mur
ARISE/Microbiology, University of Amsterdam, Nieuwe Achtergracht 127,
1018 WS Amsterdam, the Netherlands

Key words: bioproductivity, bioreactors, light emitting diodes, light harvesting complexes, micro-algae, phycobiliproteins, state transitions

1. Introduction.

Future exhaustion of fossil reserves may require extended use of renewable resources. Photosynthesis is the only effective long term solution for recycling of CO_2 and other mineral waste products. Climate conditions and otherwise permitting sunlight obviously is the ideal energy pump. In selected cases special product formation from algal or cyanobacterial biomass or in horticulture may call for indoor cultivation and artificial light supply. The cost of the electrical power needed in those conditions accounts to a large extent for the unfavourable competitive position of indoor cultivation. The present work is a spin off from a research program launched by the European Space Agency which was meant to study means for bio-regeneration of waste from human metabolism during manned space flight of longer duration. In that program the idea to use light emitting diodes (LED's) as light sources for algal culture was tested (1). The LED lamps are energy efficient, they can be selected with narrow chromatic windows, such that photons emitted can be optimally colour tuned to power photosynthesis, and last but not least LED lamps can be modulated. In the latter mode short light pulses (pulses of 5 µs duration were used in the present work) can be exchanged by periods of darkness in which LED's are off and consume no energy. A research goal was to observe whether the growth rates would decline proportionally with the length of the dark period in between flashes or by preference would decrease less, and if so to understand this actually arrived at observation. In this report we show application of LED light sources in culture of the green alga *Chlorella pyrenoidosa* and the cyanobacterium *Calothrix* sp. The green alga and the cyanobacterium largely differ in the spectral window that can be used for photosynthesis. Red light is best for the chl Qy excitation in the reaction centers and the LHC antenna of the green alga. In contrast to *Chlorella*, *Calothrix* demonstrates quite distinct differences between the spectral properties of its chl a containing reaction centers and its phycobilisome antenna. In addition, the latter also show complementary chromatic adaptation. That process enables cells to inducibly adapt the phycobilisome antenna to contain more phycocyanin (absorbance at 628 nm) or phycoerythrine (absorbance at 580 nm) after growth in orange or green light respectively. The results acquired by the use of LED light sources have engaged research directed towards understanding of the dynamic nature of light in actually growing algal cultures. Attractive energy economy in algal culture is feasible.

2. Procedure

Both *Chlorella* and *Calothrix* were grown in BG11 mineral medium as in (1). In continuous culture steady states were equilibrated at OD750 = 0.15 by adjustment of the dilution rate (growth medium pump speed) to equal the growth rate µ reached at the particular rate limiting condition, the supply of light. Each steady state lasted at least 4 days. On average a total of two weeks was needed to install equilibrium for each light supply condition and intensity. The light output from flashing LED's at the lowest frequencies is up to 3 times higher than from continuously on LED's, above 50 kHz, the LED's become warm and less efficient. Light flux measurements were made at each actual setting at the back of a culture vessel filled with medium instead of the algal culture. Growth in culture was monitored directly by measurements of the transmission of LED light and from samples taken to adjust the flow rate according to the OD750 reading, to record absorbance spectra, and to perform HPLC analysis (2), oxygen evolution assays and 77K fluorescence emission spectroscopy (3).

3. Results and Discussion

Fig. 1 shows the design of the airlift fermentator and LED light sources used in the present work. The electronic circuits are shown in fig. 2. The flat front and/or back side of a culture vessel were covered with the LED mounted screens. For *Chlorella* one sided illumination with red LED's (emission 665 nm, half height bandwith 15 nm) was used for excitation of the chlorophyll a and b Qy bands. Illumination from one side only was also used for culture of the filamentous cyanobacterium *Calothrix* in orange light. The light output of the green LED light sources was relatively low. In that case culture of *Calothrix* proceeded with illumination provided from both front and back sides. In order to control the temperature water jackets were mounted in between the light sources and the culture vessel. Air was bubbled continuously to provide mixing and CO_2 supply. The light path (thickness of the culture vessel) was varied from 3 to 10 cm.

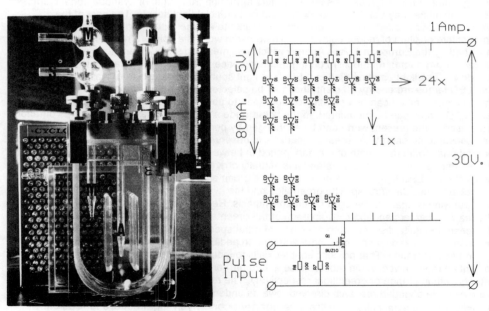

Fig. 1 (left). A flat glas airlift fermentor vessel, flat cooling water jacket, culture medium in and outlet and a LED panel. Sizes of culture vessels ranged from 300 to 2000 ml.
Fig. 2 (right). Technical drawing of the LED light sources and the steering electronics.

With LED's mounted on just one side of the culture vessel, a 5 cm light path rendered the highest number of cells produced by the culture in steady state growth conditions. These observations were made both in light modulated cultures and in cultures with continuously supplied light. Spectral and HPLC analysis indicated that access light conditions were sensed by the *Chlorella pyrenoidosa* cells from a reduced LHC content and a higher carotenoid content per cell after growth in the 3 cm light path culture vessels (data not shown).

The effects of light modulation are depicted in Fig. 3. At very low frequencies the light output was insufficient to support net growth. Above 2.5 kHz which corresponds to light 'on' during 1.25% of the time (Table I) and at higher frequencies net growth occured. The dilution rate (D) was equilibrated to equal µ in steady state conditions at the given light dose. Interestingly, the same number of photons provided as flashes yielded a higher growth yield, as can be seen by comparision of the upper -flashed and lower continuous light growth yield curves in fig. 3. At light intensities near to saturation of the requirements for the maximum growth rate in our conditions of cultivation the effects of modulation disappeared. Comparable results were also obtained in measurements of the quantum yield for oxygen production (data not shown). Initial studies with *Calothrix* in orange light indicated similar energetically profitable effects of modulated light supply. The results obtained indicate that a substantial part of the photons provided in continuous light are not effectively used for cell growth. Modulated light likely lowers the number of reaction centres that on average are found closed while a photon is trapped by the antennae. Reduction of the number of unusable photons is also likely helpful to minimize maintenance energy that is otherwise lost in repair processes. The modulation frequency of 10 kHz corresponds to dark intervals of 95 µs (see Table 1), which on average in culture may correspond to the time required to complete charge separation and Q_B reduction.

Table 1.

kHz	% of time on	luminiscence
2.5	1.25	12
5	2.5	20
10	5	28
15	7.5	39
25	12.5	58

Table 2.

light source	power usage (W·h^{-1})	
luminescent	106	
LED's on continuously	56	(Lum. 58)
LED's flashing (25 kHz)	22	(Lum. 58)

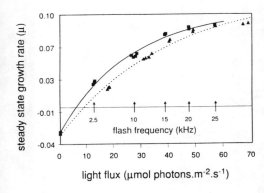

Fig. 3. Effect of the photon flux density from red (665 nm max) LED light sources on the growth rate of Chlorella pyrenoidosa in continuous culture. The right hand x-axis represents the growth rate in continuously supplied light, the left Y-axis indicates the growth rate in flash-wise provided light. It is clear that flash light gives rise to a higher growth rate per photon than continuously supplied light.

Table 1 (top). Duration of dark intervals at various flash frequencies, each flash had a duration of 5 µs. The luminiscence (Lum.) is expressed in µmol photons·m^{-2}·s^{-1}.

Table 2 (bottom). Energy consumption in indoor algal culture with luminiscent light sources, with continously on LED's and with modulated LED's. The energy needed for the pulse generator was taken into account

A further gain in growth yield on light may be realized in experiments in which PS2 and PS1 light are separately modulated. That way the PQ pool may be emptied by PS1 light just before excitation of PS2. Such an approach likely improves the quantum yield of PS2. In green algae with chl a (and b) the differences in wavelength optima for the LHC antenna and the reaction centers of PS1 and PS2 are not very outspoken. This makes the spectral resolution in time in these experiments less well feasible.In cyanobacteria, the phycobilisome antenna has spectral properties that are quite distinct from the reaction centers of PS1 and PS2. In case of *Calothrix* the blue phycocyanin pigment absorbs at 628 nm, and the reddish phycoerythrin pigment absorbs in the green around 580 nm. In green light culture *Calothrix* extends the rods of its phycobilisome antenna and incorporates phycoerythrin. In orange light culture the phycobilisome pigments are limited to allophycocyanin and phycocyanin only. Studies in the cyanobacterium *Calothrix* sp. with monochromatic LED light sources are ongoing, using green (570 nm), orange (625 nm) or red (665 nm) emitting LED light sources. For these various different colours of LED's supplier prices have gone down substantially over the last few years and the light output of the LED's has increased several fold. The spectral properties of the LED's used and the complementary chromatic adaptation effects observed are shown in Fig. 4. In continuous culture at the same net photon flux the growth rate was less with green LED's, than when growth proceeded in orange LED light. The μ max in orange light at 20 °C was estimated at 0.018 and in green light at to 0.011 per hour. These differences have never become apparent with light from luminescent light sources tuned to green or orange with 'monochromatic' plastic filters. An explanation for this observation followed from the rise in oxygen production which was observed in *Calothrix* cells from green LED light cultures tested in green LED light after the addition of farred light (Fig. 5). The earlier used green plastic filters actually transmitted light > 690 nm quite well, which explains the difference with the current results, the green LED's were not emitting in this spectral range. Orange light grown *Calothrix* tested with orange light in oxygen evolution assays showed no increase with added farred light.

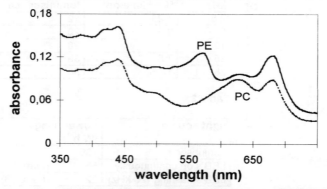

Fig. 4. Complementary chromatic adaptation in the cyanobacterium *Calothrix* as a function of the spectral output of green (top trace) and orange (bottom trace) LED's.

The observations in culture and in oxygen production gave rise to interesting questions about light energy distribution in *Calothrix*. Why would green light be transferred less efficiently to PS1 than orange light, or is the excitation of the chlorophyll Qx band by orange light sufficient for excitation of PS1? In 77 K fluorescence emission spectroscopy a substantial dynamic response of phycobilisome excitational coupling to PS1, PS2 or to neither one of both was observed. Currently, the mechanisms at play during state transitions in cyanobacteria are not yet fully resolved. Models exist that favour a linkage of the phycobilisome (PBS) antenna with a doublet of PS2 reaction centers. The PS1 centers are surrounding these PBS/PS2 centers. In case of desequilibrium of excitation energy distribution towards PS1 and PS2, suggestions have been made that various mechanisms may come into play. Spill over from PS2 to PS1 may occur, the PS1 centers may get nearer to the PBS/PS2 assemblies to share in light harvesting via the phycobilisome and newly suggested has been that the PBS itself may move rapidly over the surface of the thylakoids to balance the light distribution towards the reaction centers of PS1 and PS2 (4, 5).

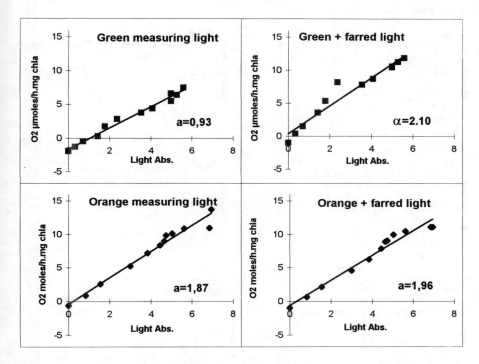

Fig. 5. Oxygen evolution as a function of LED delivered photon flux, effects of farred light. Cells from green LED supplied cultures (upper 4 frames) and from orange Led light grown cultures were used (lower 2 frames).

The data presented in Fig. 6 show fluorescence emission from *Calothrix* cells out of green or orange light LED cultures. The tested samples were taken from the culture and directly subjected to incubation in one of the following conditions: darkness, green (570 nm) LED light, orange (625 nm) LED light, red (665 nm) LED light or in farred (720 nm) light. The photon flux in these incubations was made equivalent to the flux in the culture. Samples from the incubation chamber (the oxygen electrode chamber served the purpose) were quickly taken with a capilairy tube and immediately frozen in liquid nitrogen. Fluorescence emission spectra from the samples kept at 77K were recorded from to 640 to 740 nm after excitation of chl at 436 nm, of phycoerythrine at 570 nm, and of phycocyanin at 625 nm. The peaks revealed in these assays have according to (6) been attributed to phycocyanin (650 nm), to allophycocyanin (665 nm), to the terminal emitter of the PBS (680 and 685 nm), to the PS2 reaction center (685 and 695 nm) and to PS1 (736 nm). Clear differences according to the preincubation and culture conditions were observed in emission after excitation with the phycobilisome wavelengths of 570 and 625 nm. In 438 nm chlorophyll light the emission spectra remained almost unchanged indifferent of the nature of the

pretreatment. The largest changes were rendered in the comparison between farred
pretreatment in which case a clear shift of the PBS away from PS1 was observed. On the other
hand, conditions giving rise to a reduced PQ pool (darkness, PBS light, PS2 light) showed PBS
movement away from PS2. Just uncoupling of the PBS from both PS1 and PS2 can be
discriminated from tighter coupling to either one of the reaction centers from the balance
between the 685 nm and the 695 nm peaks and fluorescence rendered from aloophycocyanin
and phycocyanin in uncuopled antennae.

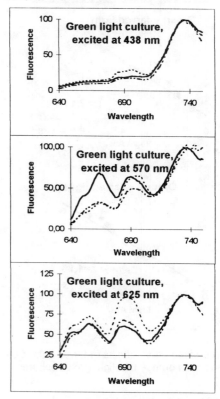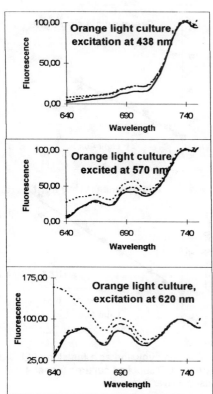

Fig. 6. 77K fluorescence emission spectra of *Calothrix* from green LED culture (left) and from
orange LED culture (right). Samples were excited at 438 nm (top), at 570 nm (middle) and
at 625 nm (bottom). Samples were pretreated differently (see text), ————; dark pretreatment,
············ farred pretreatment, -------- phycobilisome light pretreatment, according to
the applied growth condition i.e. green for the green and orange for the orange LED cultures

1 Matthijs, H.C.P., Balke, J., Van Hes, U.M., Kroon, B.M.A., Mur, L.R. and Binot, R.A. (1996)
   Biotechnol. Bioeng. 50, 98-107
2 Schubert, H., Kroon, B.M.A. and Matthijs, H.C.P. (1994) J. Biol. Chem. 269, 7267-7272
3 Garcia-Mendoza, E. et al., these proceedings
4 Mullineaux, C.W., Tobib, M.J. and Jones, G.R. (1997) Nature 390, 421-424
5 Van Thor, J.J., Mullineaux, C.W., Matthijs, H.C.P. and Hellingwerf, K.J., Acta Botanica
   (review), accepted for publication

EFFECT OF LIGHT-DARK CYCLE ON PHOTO-HYDROGEN PRODUCTION BY PHOTOSYNTHETIC BACTERIUM *RHODOBACTER SPHAEROIDES* RV.

Tatsuki Wakayama, Akio Toriyama,Tadaaki Kawasugi, Yasuo Asada and Jun Miyake , Research Institute of Innovative Technology for the Earth,1-1 Higashi, Tsukuba, Ibaraki 305-0046,Japan[1], Kubota Corp.,5-6 Koyodai, Ryugasaki,Ibaraki,301-0852,Japan[2], National institute of Bioscience and Human-Thechnology[3] , National Institute for Advanced Interdisciplinary Research[4], AIST, MITI, 1-1 Higashi, Tsukuba, Ibaraki 305-0046, Japan

Keywords; Photosynthetic bacteria, photo-hydrogen production,*Rhodobacter sphaeroides*, Light-dark cycle, Light distribution,Light penetration,

1. Introduction

We are studying to clarify the relationship between light irradiation and photo-hydrogen production by photosynthetic bacterium *Rhodobacter sphaeroides* strain RV(1-5). In particular, we will perform various investigations of conditions for utilizing sunlight.

Sunlight exhibits Gaussian distribution-like changes with a maximum light intensity of about 1 kw/m^2 in a fine weather day.

From indoor/outdoor hydrogen production experiments conducted thus far, we have found, however, that photo-hydrogen production by photosynthetic bacteria becomes saturated at a sunlight intensity around 1 kw/m^2 at the meridian transit. Efficiency of light energy conversion to hydrogen decreases under strong illumination at 1 kw/m^2. We are now examining various methods based on the premise of effective utilization of excessive sunlight not accompanied by a decrease in the amount of produced hydrogen. Then we chopping the light reduces the energy influx into a reactor(*Figure 1*).

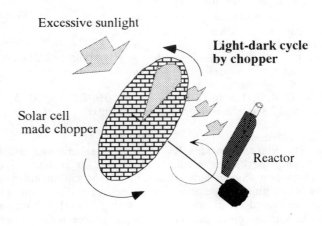

Figure 1 . Conceptual illustration of light-dark cycle.

G. Garab (ed.), Photosynthesis: Mechanisms and Effects, Vol. V, 4135–4138.
© 1998 *Kluwer Academic Publishers. Printed in the Netherlands.*

We propose a combination with solar batteries to utilized the chopped light for the increase of the total efficiency.

We examined effect of pulsed light at the interval of hour/min/second with keeping constant the total energy supply. We reported the influence of irradiation in light/dark cycles at intervals of about 12, 8, 6, 4, and 1 hour on hydrogen production at Biohydrogen'97 in Hawaii last year.

In this study, we considered the time dispersion of excessive sunlight for hydrogen production and obtained findings on the influence of light in light/dark cycles on the region of hour/minute/second considered to be caused by a chopper on the surface of a reactor.

2. Procedure

2. 1 *Bacterial strain*

The *Rhodobacter sphaeroides* strain RV, which products hydrogen with high stability was used(6).

2. 2 *Light condition*

The Kondo-Sylvania M-26 halogen lamp was used as light source.

The M-26 halogen lamp does not contain the infrared light by combining with a wavelength-selective dichroic mirror, and provides spectra comparatively similar to the sunlight (*Figure 2.*)(7).

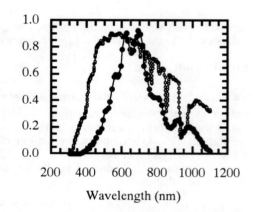

Figure 2. Spectrum of ERDA AM 1.5 and M-26. Symbols; ○:ERDA AM 1.5, ●:M-26

We measured light intensity change of sunlight in one day with fine weather as a reference, the average light intensity was set to 0.58kW/m^2, and total light energy in one day to 7 kWh/m^2, and the irradiation time to 12 hours. Light energy was measured with a radiometer(SJI ,model-4090). The Each of light/dark cycles was set at 12, 8, 6, 4, 3, 1.3 and 1 hours and 40, 30, 20, 15, 10, 5, 2 and minutes and 45,30, 15, 10, 5 and 2 seconds using a digital stop timer.

2. 2 *Cultivation*

The *Rba.sphaeroides* RV strain was subcultured using the aSy broth obtained by modifying the culture medium of Miyake *et al.*(6). Preculture of the bacteria was conducted on aSy medium containing 36mM sodium succinate as a carbon source and 10 mM ammonium sulfate as a nitrogen source. The main culture for hydrogen production was performed on gL medium containing 50mM sodium lactate as a carbon source and 10mM sodium glutamate as a nitrogen source under anaerobic light conditions (30°C, 15000 Lux tungsten lump).

A 200-ml Roux flask with an irradiation area of 70cm^2 was immersed in a water-bath at 30°C. The Roux flask was covered with black sheeting, except for the irradiation front area, in order to eliminate the influence of external light. Producted hydrogen was captured in a measuring cylinder through a Tygon tube.

2. 3 *Measurement*

Samples were taken every hour and the volume of produced hydrogen, production rate, optical density(660nm), and organic acid concentration in the medium(HPLC) were measured.

2. 4 *Estimation*

Energy conversion efficiency from light to hydrogen was calculated by the following equation: Efficiency (%) = (combustion enthalpy of hydrogen)/(absorbed light energy)*100

3. Results and discussion

The time-course of H_2 production in irradiation of light/dark cycles of 12,6,3, and 1 hour and 30,15, 5, and 1 minute intervals was examined. Among the all intervals tested, the 12-hour light/dark exhibited the highest H_2 production (70 $l/m^2/72h$, 2.0 $l/m^2/h$) ,as in the results obtained from two different size of Roux flasks. At shorter intervals of light/dark cycle, H_2 production was decreased. The shortest intervals of 1-minute produced only 40 l/m^2 of hydrogen. However, total volume of produced hydrogen was a sub-optimum around 20-minute interval which was comparable to that at the 12-hour light/dark cycle(*Figure 3.*). Similar results were obtained for the maximun efficiency of light to hydrogen conversion. This results was considered to show the effect of using a chopper.

In addition, maximum hydrogen production rates with intervals of light/dark cycle was similar as total volume of hydrogen production. There was a also sub-optimum at around 20-minute interval that was about half of 12-hour intervals(about 1.1 $l/m^2/h$).

In order to confirm this phenomena, the intervals of light/dark cycles were changed to 12 and 1.3 hours, 40, 20, and 10 minutes, and 30 and 15 seconds centering on minute/second units.

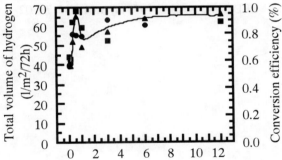

Figure 3. Effect of light-dark cycle on photo-hydrogen production. Symbols;●:1st.,■:2nd.,▲:3rd.

Similar experiments were conducted. It was also confirmed that there was a sub-optimum point in the hydrogen production rates around of 20 minutes intervals.

It is not clear whether this penomena were under the influenence of irradiation. Therefore, now we are conducting a study using monochromatic cycle light on the region of seconds/milliseconds to know the relationship between photo-hydrogen production and photosynthetic function, etc.

However, these results are considered to indicate that the method of dispersing light using a chopper is effective in avoiding excessive sunlight. Using chopper-equipped solar cells on the upper part, etc., the utilization efficiency of light energy will be improved.

We are now examining various methods based on the premise of effective utilization of excessive sunlight not accompanied by a reduction in the amount of produced hydrogen.

Acknowledgements

This work was done under the management of RITE as a part of the R&D Project on Environmentally Friendly Technology for the Production of Hydrogen, supported by NEDO.

References

1. Miyake, J., Kawamura, S.,(1987)*Int.J.Hydrogen Energy*, 39,147-149
2. Nagamine, Y.,Kawasugi, T.,Miyake, M.,Asada, Y.,Miyake, J., (1996)
    J.Mar.Biotechnol. 4,34-37
3. Nakada, E., Nishikata, S., Asada, Y.,Miyake, J.(1996),
    *J.Mar.Biotechnol.*4,38-42
4. Tsygankov, A.,Hirata, Y.,Miyake ,M.,Asada ,Y.,Miyake, J.(1994)
    Journal of Fermentation Bioengineering, 77,575-578.
5. Nakada, E., Asada, Y.,ARAI, T.,Miyake, J.(1995)
    Journal of Fermentation Bioengineering, 80,53-57.
6. Miyake, J., Mao, X.,Kawamura, S. (1984)
    Journal of Fermentation Technology, 62, 531-535.
7. ERDA and NASA (1977) *NASA TM* 73702

HYDROGEN PRODUCTION BY PHOTOSYNTHETIC BACTERIA : THE RELATIONSHIP BETWEEN LIGHT WAVELENGTH AND HYDROGEN PRODUCTION

Eiju Nakada[1], Satoshi Nishikata[1], Yasuo Asada[2], Jun Miyake[3]
Research Institute of Innovative Technology for the Earth (RITE) / Fuji
Electric Corporate R&D, Ltd., 2-2-1 Nagasaka, Yokosuka, 240-0194
Japan[1] Nat. Inst. of Bioscience and Human-Technology, 1-1 Higashi,
Tsukuba 305-0046, Japan[2] Nat. Inst. of Advanced Interdisciplinary
Research, 1-1-4 Higashi, Tsukuba 305-8562, Japan[3]

Key words: bacterial photosynthesis, bioreactors, environment, hydrogen production, light
penetration, photoconversion

1. Introduction

Solar energy technology is desirable to prevent environmental problems (e. g., the
greenhouse effect, acid rain) caused by fossil energy sources. Hydrogen is a well-known,
high weight density energy carrier and its combustion does not pollute the environment.
Among various photosynthetic microorganisms, photosynthetic bacteria have high
hydrogen production rates [1]. Photoproduction of hydrogen by photosynthetic bacteria
has been extensively studied from the viewpoint of the development of environmentally
friendly technology [2]. For the realization of hydrogen production systems based on
photosynthetic bacteria, light energy conversion efficiency to hydrogen is the most
important factor. The highest conversion efficiency of light to hydrogen yet recorded in
this class of organism was ca. 7% using light from a solar simulator [2,3].
The improvement of the photoreactor is needed to enhance the efficiency. We have
previously analyzed the light energy utilization in a photoreactor [4]. In the present work,
we examined the relationship between light wavelength and hydrogen production in
photoreactor. To clarify the wavelength effect on hydrogen production, the light source
(tungsten lamp) was used with wavelength-selective optical filters.

2. Procedure

2.1 Organisms and culture conditions

Rhodobacter sphaeroides RV was used [5]. The growth medium (aSy) was
composed of a basal medium (inorganic salts and vitamins), 0.1 % yeast extract, 10 mM
ammonium sulfate, and 75 mM sodium succinate [5,6]. Cells were grown for 3 days with
the tungsten light (30 W/m^2) at 30°C.

2.2 Light source

A tungsten lamp of 150 W (color temperature of 2,900 K) was used as the light
source. A water vessel was placed in front of the photoreactor to filter out infrared light
from light source. The light energy at the surface of the photoreactor (only light passed

G. Garab (ed.), Photosynthesis: Mechanisms and Effects, Vol. V, 4139–4142.
© 1998 Kluwer Academic Publishers. Printed in the Netherlands.

through the water vessel) was adjusted to 400 W/m². A radiometer (SJI, Model 4090, USA) was used to measure the total light energy.

The eight wavelength-selective optical filters (F420, F480, F560, F640, F760, F820, F860, F960) were used with the tungsten lamp. For example, a F560 filter is transparent for light above 560 nm but reflects light below 560 nm. The relationship between the specific wavelength of the filter and the absorption spectrum of *Rb. sphaeroides* RV was shown in Fig. 1.

2.3 *Hydrogen production*

A photoreactor (70-ml flat glass bottle) was used. It had an irradiation area of 27 cm² and thickness of 2.5 cm. The hydrogen production medium (gL) was prepared with the basal medium, 75 mM sodium D, L-lactate, 10 mM sodium glutamate, and 18 mM sodium bicarbonate [2]. Cell suspensions (100 ml) were mixed with freshly prepared gL medium (300 ml). The cells were next cultured in a 400-ml flat glass bottle with the tungsten lamp (100 W/m²) at 30°C for 24 to 48 h until the hydrogen production rate became constant. The cell suspension (13 mg dry wt ml-1) was then transferred to the photoreactor. The cell density of the photosynthetic bacteria was measured using a spectrophotometer (Hitachi, Model 110, Japan).

The photoreactor was illuminated by the tungsten lamp for 5 to 8 h until the hydrogen evolution rate became constant. And then a wavelength-selective optical filter was put on the irradiation area of the photoreacter, hydrogen evolution rate was measured furthermore 5 to 8 h. Evolved hydrogen gas was collected in glass syringes [7]. The light energy conversion efficiency to hydrogen was calculated according to the equation below:

Efficiency (%) = (combustion enthalpy of hydrogen) / (absorbed light energy) x 100   (1)

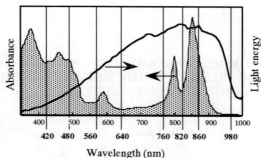

Figure 1. The relationship between the spectrum of the light source and absorption spectrum of the cell. Bold numbers show the specific wavelength of the wavelength-selective optical filters.

—— : spectrum of the tungsten lamp (at the surface of the photoreactor).

▦ : absorption spectrum of *Rhodobacter sphaeroides* RV.

3. Results and discussion

Hydrogen evolution rate under illumination with light passed through the wavelength-selective optical filter was shown in Table 1. The hydrogen evolution rate under illumination with tungsten lamp without the wavelength-selective optical filter was 6.9 L/m²/h and it decreased stepwise by using the filters. The tungsten lamp have its peak energy at around 800 nm (Fig. 1). Therefore the hydrogen evolution rate under illumination with light around 640-960 nm reached to 77% of the hydrogen evolution rate under illumination with tungsten lamp. (77% = (6.1-0.8) / 6.9 x 100).

Light energy and hydrogen evolution rate under illumination with light that has specific wavelength range were calculated by using the data of Table 1 (see Table 2). For example, light energy under illumination with light around 560-640 nm (E [560-640]) and hydrogen evolution rate under illumination with light around 560-640 nm ($H_{2[560-640]}$) were calculated according to the equation below, respectively:

$$E_{[560-640]} = E_{[560-1,100]} - E_{[640-1,100]} \tag{2}$$
$$H_{2[560-640]} = H_{2[560-1,100]} - H_{2[640-1,100]} \tag{3}$$

Light energy conversion efficiencies to hydrogen (Eff.) were calculated from light energy and hydrogen evolution rate according to the equation (1) (see Table 2).

Table 1 Hydrogen evolution rate and light energy under illumination with light passed through the wavelength-selective optical filters.

Filter	Wavelength(nm)	Light energy (W/m²)	H_2(L/m²/h)
no	ca. 350 - 1,100	400	6.9
F420	420 - 1,100	348	6.8
F480	480 - 1,100	344	6.6
F560	560 - 1,100	328	6.3
F640	640 - 1,100	292	6.1
F760	760 - 1,100	204	5.0
F820	820 - 1,100	160	4.1
F860	860 - 1,100	132	3.5
F960	960 - 1,100	88	0.8

Table 2 The relationship between wavelength and light energy conversion efficiency.

Wavelength(nm)	Light energy (W/m²)	H_2(L/m²/h)	Eff. (%)
ca. 350 - 1,100	400	6.9	5.2
350 - 420	52	0.1	0.6
420 - 480	4	0.2	15.0
480 - 560	16	0.3	5.6
560 - 640	36	0.2	1.7
640 - 760	88	1.0	3.4
760 - 820	44	1.0	6.8
820 - 860	28	0.6	6.4
860 - 960	44	2.7	18.4
960 - 1,100	88	0.8	2.7

Light energy conversion efficiencies to hydrogen under illumination with tungsten lamp without the filter was 5.2%. It was considerably high in comparison with the past experiments. We consider that narrow irradiation area of photoreactor was contributed to the high efficiency. More higher efficiencies were observed at light around 420-480 nm and 860-960 nm. Wavelength around 420-480 nm corresponds to the absorption wavelength of the carotenoid, and 860-960 nm corresponds to the absorption wavelength of the bacteriochlorophylls, respectively.

The results showed that cells in the photoreactor could effectively utilize light around 420-480 nm and 860-960 nm to hydrogen production. Also light around 640-760 nm, which does not corresponds to the absorption maxim of the photosynthetic bacteria, was found to be effective for hydrogen production. This is inconsistent with the results of Nogi et al. They reported that light around 600-780 nm was not effective for hydrogen evolution [8]. We consider that more detail analysis will be necessary to clarify the wavelength dependence of the light energy conversion efficiency to hydrogen. Because we neglected the light energy conditions on the discussion of the relationship between wavelength and efficiency. The effect of light energy on light energy conversion efficiency has been investigated by Miyake et al. and they confirmed that the efficiency became high at low light intensity [2]. Further work is in progress to investigate the relationship with light wavelength and hydrogen evolution by using monochromatic light or band passed light.

Acknowledgments

This work was performed under the management of RITE as a part of the Research & Development Project on Environmentally Friendly Technology for the Production of Hydrogen supported by New Energy and Industrial Technology Development Organization (NEDO).

References

1 Miyamoto, K. (1994) in Recombinant microbes for industrial and agricultural applications (Murooka Y. and Imanaka T., ed.) pp.771-786, Marcel Dekker, New York.

2 Miyake, J. and Kawamura, S. (1987) Int. J. Hydrogen Energy 39, 147-149

3 Miyake, J., Asada, Y. and Kawamura, S. (1989) in Biomass Handbook (Hall C.W. and Kitani O., ed.) pp.362-370, Gordon and Breech Scientific Publishers, New York.

4 Nakada, E., Asada, Y., Arai, T. and Miyake, J. (1995) J. Ferment. Bioeng. 80, 53-57

5 Miyake, J., Mao, X.-Y. and Kawamura, S. (1984) J. Ferment. Technol. 62, 531-535

6 Mao, X.-Y., Miyake, J. and Kawamura, S. (1986) J. Ferment. Technol. 64, 245-249

7 Miyake, J., Tomizuka, N. And Kamibayashi, A. (1982) J. Ferment. Technol. 60, 199-203

8 Nogi, Y., Akiba, T. and Horikoshi, K. (1985) Agric. Biol. Chem., 49, 35-38

SOME CHARACTERISTICS OF REGULATION OF NITROGENASE AND HYDROGEN PRODUCING ACTIVITIES IN CYANOBACTERIA

Masukawa, H., Mochimaru, M. and Sakurai, H.
Dept. of Biology, Sch. of Education, Waseda University, Shinjuku, Tokyo
169-8050, Japan

key words: biotechnology, cyanate, heterocyst

1. Introduction

In recent years, there is growing expectation for development of photobiological production of hydrogen as a supplement to and a substitute for fossil fuels due to growing concern about rapidly rising anthropogenic emission of global warming gases. Cyanobacteria have been suggested as the ideal candidate among H_2-producing phototrophs, because they can produce H_2 under aerobic conditions using water as an electron donor [1, 2]. Hydrogen production by cyanobacteria can be mediated either by hydrogenase or nitrogenase. Once hydrogenase was favored over nitrogenase as the hydrogen evolving system from a view point of energy efficiency, subsequent studies indicated that continued production of H_2 in air was mediated by nitrogenase. Several types of nitrogenase were found in N_2-fixing organisms. Most nitrogen fixing organisms contain molybdenum-containing nitrogenase (Mo-nitrogenase), but some of them contain vanadium containing nitrogenases as well (V-nitrogenase) [3]. Some cyanobacteria contain V-nitrogenase [4] and the gene has been sequenced from one of them [5]. V-nitrogenase was found to possess a higher ratio of H_2 production to C_2H_2 reduction activity than Mo-nitrogenase, but it is more labile to oxygen than the latter.

We report some factors which affect the activity levels of nitrogenase and hydrogen production in heterocystous cyanobacteria.

2. Procedures

Anabaena sp. PCC 7120, *A. cylindrica* (strain M1, IAM Center, Univ. of Tokyo) and *A. flos-aquae* (UTEX 1444) were cultured in a modified Detmer's medium (MDM) containing 1 mM NH_3 in place of 10 mM nitrate. These cells were washed with a medium containing no combined nitrogen (MDM(minus N)). For induction of nitrogenase, the standard medium contained 1 μ M molybdenum (Mo-MDM (minus

G. Garab (ed.), Photosynthesis: Mechanisms and Effects, Vol. V, 4143–4146.
© *1998 Kluwer Academic Publishers. Printed in the Netherlands.*

These results indicate that V-nitrogenase is also repressed by cyanate. In Mo-grown cells, nitrogenase activity declined at 8 hr after addition of cyanate, and increased again after that. The hydrogen production activity dropped almost to zero at 8 hr after the addition of cyanate, and very slowly rose again after that. Cyanate spontaneously and slowly decomposes into NH_3 and CO_2, and cyanobacteria have cyanase activities. These results may suggest that cyanate was decomposed more quickly in Mo-grown cells than in V-grown cells.

Effects of Mo and V on pigment contents

It is generally observed that V-grown cells are denser in blue-green color than Mo-grown cells (Fig. 2). The reason for this is not certain at present. When 0.1 mM cyanate was added to V-MDM (minus N), cells became much paler than the control which contained no cyanate. In Mo-grown cells, the effects of cyanate was not pronounced in harmony with the results in Figure 1.

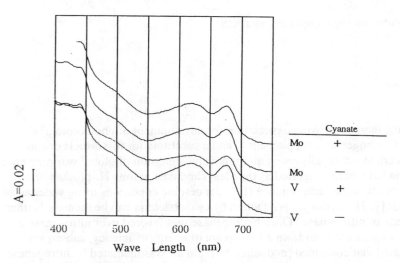

Figure 2 Effects of molybdenum and vanadium on pigment contents *Anabaena* PCC 7120 cells were transferred to either Mo-MDM or V-MDM which contained either 0.1 mM (+) or no cyanate (-). Absorption spectra were recorded after 76 hr.

Effects of MSX

MSX is an inhibitor of glutamine synthetase.. Addition of 1 mM NH_3 to MDM (minus N) medium, decreased nitrogenase activity to almost undetectable level in 18-24 hr. When 1 μ M MSX was added to the medium containing 1 mM NH_3, nitrogenase activity did not significantly decrease compared with the control cells which were grown in MDM(minus N) medium. These results are compatible with the notion that MSX is an inhibitor of glutamine synthetase.. Suzuki et al. [7] stated that cyanate may act as a regulator of the NH_3-repressive genes involved in carbon and nitrogen assimilation in the cyanobacterium. Vega-Palas et al. [8] identified a regulatory gene

N)) (Mo-grown cells). In some media, molybdenum was omitted, and 1 μ M vanadium was added (V-MDM (minus N)) (V-grown cells).

Nitrogenase activity was determined from C_2H_2 reduction. The reduction was assayed in an air-tight flask with 12% (v/v) C_2H_2 in air as the gas phase. H_2 evolution was assayed in an air-tight flask under Ar. H_2 and C_2H_2 were determined by gas chromatography, with a thermal conductivity and a hydrogen flame-ionizing detector, respectively.

The light intensities for cell culture and assays were 20 and 100 μ mol m^{-2} s^{-1} PAR, respectively.

l-Methionine-D,L-sulfoximine (MSX) was purchased from Sigma.

3. Results and Discussion

Time course of appearance of nitrogenase activity
When stationary phase cells of *Anabaena* PCC 7120 and *A. cylindrica* were transferred to fresh MDM(minus N), nitrogenase activity rose to 10-40 μ mol mg Chl a^{-1} hr^{-1} in 10 hr. In *A. flos-aquae* (UTEX), however, the activity were lower than 2 μ mol mg Chl a^{-1} hr^{-1}.

Effects of cyanate on the levels of nitrogenase
Ammonia inhibits nitrogenase synthesis. However, ammonia itself is not the direct regulator of nitrogenase synthesis, because it does not inhibit the synthesis when methylsulfoximine (MSX), an inhibitor of glutamine synthetase, was added to the growth medium. Lawrie (6) found that carbamoyl phosphate and cyanate inhibit the synthesis of nitrogenase. We found that nitrogenase and hydrogen production activities were induced in cells grown in either Mo-MDM (minus N) or V-MDM (minus N) medium. In V-grown cells, these activities steadily decreased when 0.1 mM cyanate was added to the medium 26 hr after transfer to V-MDM (minus N) medium (Fig. 1).

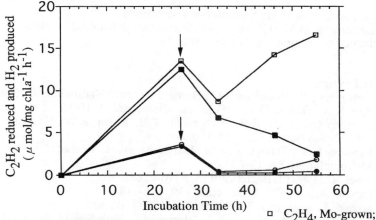

Washed cells were transfered to media, at zero time, containing either 1 μ M Mo or 1 μ M V. After 26 hr, 0.1mM cyanate was added ($\downarrow$).

□ C_2H_4, Mo-grown;
o H_2, Mo-grown;
■ C_2H_4, V-grown;
● H_2, V-grown cells.

(*ntcA*) in *Synechococcus* sp. PCC 7942 for full expression of proteins subject to ammonium repression.

When MSX concentrations were raised to 100 μM or 1 mM in a medium containing 1 mM NH_3, nitrogenase activities declined to almost undetectable levels in 18 hr. MSX also inhibits active uptake of some amino acids including methionine, and may inhibit intercellular transport of the amino acids (9). It may be that high concentrations of MSX decreased nitrogenase by severely inhibiting glutamine synthetase, uptake or intercellular transport of amino acids.

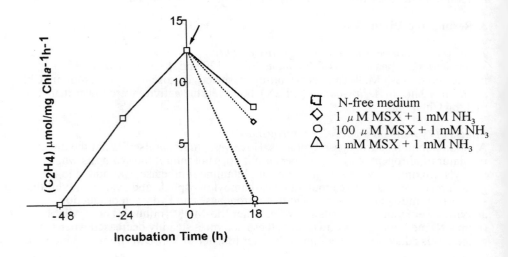

Figure 3 Effects of MSX on nitrogenase activity
Anabaena sp. PCC 7120 cells were transferred to Mo-MDM (N-free), and after 48, indicated concentrations of MSX were added.

4. References

1. Kentemich, T., Haverkamp, G., Bothe, H. (1990) Naturwissenshaften 77, 12-18
2. Rao, K.K., Hall, D.O. (1996) J. Mar. Biotechnol. 4, 10-15
3. Robson, R.L., Eady, R.R., Richardson, T.H., Miller, R.W., Hawkins, M., Postgate, J.R. (1986) Nature 322, 388-390
4. Kentemich, T., Dannenberg, G., Hundeshagen, B., Bothe, H. (1988) FEMS Microbiol. Lett. 51, 19-24
5. Thiel, T. (1993) J. Bacteriol. 175, 6276-6286
6. Lawrie, A.C. (1979) J. Bacteriol. 139, 115-119
7. Suzuki, I., Sugiyama, T., Omata, T. (1996) J. Bacteriol. 178, 2688-2694
8. Vega-Palas, M.A., Madueno, F., Herrero, A., Flores, E. (1990) J. Bacteriol. 172, 643-647
9. Montesinos, M.L., Herrero, A., Flores, E. (1995) J. Bacteriol. 177, 3150-3157

EFFECT OF pH ON POLY-β-HYDROXYBUTYRATE ACCUMULATION BY *RHODOBACTER SPHAEROIDES*

Emir Khatipov,[1,2] Anatoly Tsygankov,[3] Masato Miyake,[2], Nikolay Zorin,[3] Jun Miyake,[4,2] and Yasuo Asada [2]

[1] Research Institute of Innovative Technology for the Earth; [2] National Institute of Bioscience and Human Technology, AIST/MITI, Tsukuba, Ibaraki, Japan; [3] Institute of Basic Biological Problems, Pushchino, Russia; [4] National Institute for Advanced Interdisciplinary Research, AIST/MITI, Tsukuba, Ibaraki, Japan

Key words: NADH, NADPH, ATP, pH-regulation, bioreactors, PHA

1. Introduction

Phototrophic bacterium *Rhodobacter sphaeroides* strain RV is an effective biotechnological producer of hydrogen gas (1, 2). Recent works also indicate high potential of the strain to accumulate poly-β-hydroxybutyrate (PHB) (3, 4). Both PHB accumulation and H_2 evolution represent alternative ways for discharging excess reducing power of the cell (5, 6). Thus, the knowledge on regulation of both processes would be advantageous in biotechnological practice for applying conditions favoring for elevated yields of either H_2, or PHB. On the other hand, it is reasonably practical to use bacterial biomass derived from H_2 production as a source of PHB, and conditions to combine both production processes should be determined.

PHB accumulation is normally occurring under unbalanced growth conditions, that imply changes in substrate availability, unfavorable physical factors of the medium (5). In our previous work, we have shown that acetate, unlike lactate, is the best substrate for accumulation of PHB by *R. sphaeroides* RV grown in the batch culture (4). It was also shown that accumulation of PHB by the bacterium on lactate as a sole nitrogen source can be controlled by the pH of the medium, provided the conditions of growth are produced to build up final increase in pH of the medium in pH-uncontrolled batch cultures. In this study we monitored PHB accumulation by a pH-stat culture of *R. sphaeroides* RV grown in a photobioreactor, and elucidated the effect of quick alkalization of the medium on the final yield of the polymer. Intracellular contents of pyridine nucleotides and adenylates were also measured, as were determined some inhibitory patterns by these compounds of PB synthase and citrate synthase, and data obtained were analyzed from the point of view of cell energy/reductant balance.

4147

G. Garab (ed.), Photosynthesis: Mechanisms and Effects, Vol. V, 4147–4150.
© 1998 *Kluwer Academic Publishers. Printed in the Netherlands.*

2. Procedure

2.1 *Bacteria*

The culture of *R. sphaeroides* strain RV was grown at 150 W/m^2 and 30°C photo-heterotrophically under anaerobic conditions as described earlier (4) on the mineral medium supplemented with 40 mM sodium salts of lactate, or acetate as alternative carbon sources, 10 mM ammonium sulfate, 0.1% yeast extract.

2.2 *Photobioreactor*

Experiments on the effect of pH on intracellular concentrations of nucleotides and PHB were performed in a 3-liter cylindrical stirred photobioreactor (3) with internal illumination (halogen lamps, 450-500 W per m^2). Cultures were grown under pH-controlled conditions (pH 7.0, if not otherwise stated). Experiments were made with cells grown in logarithmic phase until early stationary phase. When cells reached first-third of logarithmic growth phase, pH of the medium was quickly changed (within 10 min) to 9.0 by addition of 0.5 N NaOH. Samples of cell suspensions were withdrawn from the bioreactor for further analysis of cellular constituents.

2.3 *Analytical methods*

PHB content and dry cell weight were determined gas chromatographically and gravi-metrically, respectively, as previously described (4). Reduced pyridine nucleotides (NADH, NADPH) or oxidized pyridine nucleotides (NAD, NADP) and adenylates (ATP, ADP, AMP) were extracted from the cells quenched by 0.5 N NaOH or HCl, respectively (7). The obtained extracts were neutralized and clarified by centrifugation prior to analysis. Nucleotides were determined by HPLC on Inertsil ODS 5 x 250 mm column equilibrated with 100 mM K-phosphate buffer (flow rate 1 ml/min, 22°C), pH 6.0, supplemented with 3.75% methanol, on a Shimadzu 10A system using UV (254 nm) and fluorescence (420 nm, excited by 350 nm light) detectors. Cell-free extracts were prepared by sonication (20,000 Hz, 3 min, 4°C) of cells washed and resuspended in 50 mM Tris-HCl buffer, pH 7.5, and subsequent removal of debris. Citrate synthase was assayed spectrophotometrically at pH 7.5 and 30°C using the conventional method (8). Polyhydroxybutyrate synthase was assayed (30°C) spectrophotometrically in 50 mM Tris-HCl buffer pH 7.5, in the presence of 0.3 mM DTNB, 1 mM DL(-)-3-hydroxybutyryl-CoA (9). Protein was quantified using BioRad protein assay kit with Coomassie brilliant blue reagent. Data shown are mean values of representative results of independent experiments that were at least duplicated.

3. Results and Discussion

3.1 *Intracellular content of PHB and nucleotides*

Cells of *R. sphaeroides* strain RV grown in a photobioreactor in a pH-stat mode at pH 7.0 on lactate as a sole carbon source accumulated PHB to about 7-9% of dry cell weight (dcw) (Table 1.). Cells grown under the same conditions on acetate maintained almost constant PHB content of about 30-32% of dcw. Rapid alkalization of the medium to pH 9.0 resulted in increase of final PHB yield up to 12-14% and 48-52% of dcw in lactate and acetate cultures, respectively.

Table 1.
Changes in the intracellular content of PHB, pyridine nucleotides and adenylates in R. sphaeroides RV cells in response to alkalization of the growth medium

Cell constituents[*]	Control growth, pH-stat 7.0		Rapid pH-shift 7.0 → 9.0	
	Early phase	Late phase	Before	After
Acetate culture				
PHB	31	32	31	50
Adenylate charge[**]	0.8	0.9	0.9	0.9
NADH	0.03	0.03	0.04	0.03
NADPH	~0	~0	~0	~0
NAD	1	1	1	0.1
NADP	0.5	0.5	0.5	0.6
Lactate culture				
PHB	4	9	5	13
Adenylate charge	0.7	0.8	0.7	0.4
NADH	0.5	0.3	0.5	0.3
NADPH	0.3	0.2	0.4	0.2
NAD	3.2	1.8	3.7	2.2
NADP	0.5	0.2	0.4	0.3

[*]PHB content is expressed in as percentage of dcw, content of all other constituents expressed in micrograms per mg dcw.
[**]Adenylate charge (E) was calculated from data on intracellular content of ATP, ADP, and AMP using equation $E = (ATP + 0.5 ADP) / (ATP + ADP + AMP)$.

Determination of pyridine nucleotides and adenylates in pH-stat cultures and in cultures subjected to alkalization of the medium have revealed a complex pattern of changes of intracellular content of these compounds in response to pH change (Table 1). In both acetate and lactate cultures cell adenylate charge was high, indicating that cells were grown under energy/light non-limited conditions. Adenylate charge in lactate cells was decreased substantially upon pH shift from 7 to 9, whereas in acetate cells that value remained constant after pH change. However, after several subsequent hours of growth energy charge in lactate cells was restored (data not shown).

Content of NADH and NADPH in acetate cells was considerably lower than in lactate-grown cells. Content of NAD in lactate cells was more than 3 times higher than in acetate cells. At the same time, content of NADP was similar in both cultures and was not changed considerably in response to pH increase.

3.2 *Citrate synthase and PHB synthase*

Effect of nucleotides (NADH, NADP, NADPH, NADP, ATP, and ADP) on the activity of citrate synthase and PHB synthase in cell-free extracts of acetate and lactate cells has been determined. Citrate synthase, as in other Gram-negative bacteria (8), was totally inhibited by 300 mM NADH. On the other hand, PHB synthase was inhibited by oxidized pyridine nucleotides (45 and 40% inhibition by 1 mM NAD and NADP, respectively).

The data presented indicate that, as it was proposed earlier (5), balanced cell growth depends on NADH. It is probable, that NADH↔NAD redox control system is involved in the regulation of PHB synthesis (5, 10). Cell synthesis and energy production are in general competing with the PHB accumulating pathway. However, transition from balanced to unbalanced growth conditions, as shown in this work, is apparently triggered by decrease in cell energy in response to pH shift (Table 1). Despite energy charge in lactate grown cells is gradually restored after fall-down upon pH shift, cells destabilized by pH shift cannot gain the normal rate of carbon substrate utilization on synthetic needs, and PHB synthesis pathway prevails as a consumer of carbon and reducing equivalents. The latter phenomenon can be useful in biotechnological practice to apply programmed changes in pH of the growth medium at a predetermined growth phase to obtain high yields of PHB.

References

1 Miyake J. and Asada Y. (1993) in Proceedings of the 1st Intl. Conference New Energy Systems and Conversions (Ohta, T. and Homma, T., Eds.). pp. 219-223. Universal Academy Press, Inc., USA

2 Nakada, E., Y. Asada, T. Arai, and J. Miyake (1995) J. Ferment. Bioeng. 80, 53-57

3 Suzuki, T., Tsygankov, A., Miyake, J., Tokiwa, Y. and Asada, Y. (1995) Biotechnol. Lett., 17/4, 395-400

4 Khatipov, E., Miyake, M., Miyake, J., and Asada, Y. (1998) FEMS Microbiol. Lett. 162/1, 39-45

5 Fuller, R.C. (1995) in Anoxygenic Photosynthetic Bacteria (Blankenship, R., Madigan, M., and Bauer, C., Eds.) pp. 973-990, Kluwer Academic Publishers, Dordrecht, The Netherlands

6 Hustede, E., Steinbüchel, A. and Schlegel, H.G. (1993) Appl. Microbiol. Biotechnol., 39, 87-93

7 Litt, M.R., Potter, J.J., Mezey, E., and Mitchell, M.C. (1989) Anal. Biochem., 179, 34-36

8 Weitzman P.D.J. and Danson, M.J. (1976) Curr. Top. Cell. Regul., 10, 161-204

9 Valentin, H.E. and Steinbüchel, A. (1994) Appl. Microbiol. Biotechnol., 40, 710-716

10 Henderson R. A., and Jones, C. W. (1997) Arch. Microbiol., 168, 486-492

This work was performed under the management of Research Institute of Innovative Technology for the Earth (RITE) as a part of the R&D Project on Environmentally Friendly Technology for the Production of Hydrogen supported by New Energy Development Organization of Japan (NEDO).

IMMOBILIZATION OF CHLOROPLASTS : PHOTOBIOREACTOR WITH P450 MONOOXYGENASE

Masayuki Hara,[1,3] Svetlana Iazvovskaia[1] Hideo Ohkawa,[2] Yasuo Asada[1] and Jun Miyake[1,4]

[1]NIBH, AIST, MITI, 1-1 Higashi, Tsukuba, Ibaraki 305-8566, Japan, [2]Department of Biological and Enviromental Science, Faculty of Agriculture, Kobe University, Rokkodai-cho 1, Nada-ku, Kobe, Hyogo 657-0013, Japan, [3] NAIR, AIST, MITI, 1-1-4 Higashi, Tsukuba, Ibaraki 305-8562,Japan.

Key words: NADPH, NADP, bioreactors, chloroplasts, spinach, photoconversion

1. Introduction

We have interest in application of redox enzymes for bioreactos. Methodololy for constructing bioreactor of the redox enzyme reactions have not been established well due to the difficulty in the supply of redox compounds such as NADPH or NADH. Photosynthetic reaction can supply such these redox compounds under illumination. In chloroplasts, photosystem II has the ability to split water molecules to evolve molecular oxygen. Photosystem I reduces $NADP^+$ to form NADPH. These activity can be used as a light-driven NADPH-regenarating system in bioreactors if the chloroplasts were immobilized in a suitable gel matrix.

P450 monooxygenase is an example of the redox enzyme which needs NADPH. Microsomal P450 monooxygenases and NADPH-P450 oxidoreductase in eucaryotic cells, involves the oxidative metabolism of xenobiotics, including drugs, food additives, and environmental chemicals and therefore attracted increasing attention because of their potential biotechnological applications. A number of P450 species with broad, overlapping substrate specificity catalyse different oxidative reactions toward lipophilic substrates in the presence of O_2 and NADPH (1).

2. Procedure

Spinach and cactus chloroplats were prepared as described previously, quickly frozen in liquid nitrogen and stored at -80°C before use (2).

The expression plasmid pAFCR1 for the fusion enzyme between rat CYP1A1P450 and yeast NADPH-cytochrome P450 oxidoreductase was transformed into *Saccharomyces cerevisiae* strain AH22 (3). Microsomes containing the fusion enzyme were prepared, and stored at -80°C before use (2).

G. Garab (ed.), Photosynthesis: Mechanisms and Effects, Vol. V, 4151–4154.

Immobilization of microsome was done in buffer A containing 20 mM Tris-HCl (pH 7.8) and 20%(w/v) glycerol except otherwise stated. Immobilization of chloroplasts was done in buffer B containing 50 mM Tricine-NaOH (pH 7.8), 10 mM NaCl, 3 mM $MgCl_2$ and 1 mM NH_4Cl except otherwise stated. Buffer B was used when both microsomes and chloropolats were immobilized simultaneosly in the same agarose gel. Microsome-immobilized gel and chloroplast-immobilized gel were mechnically disrupted into particles by passing through a stainless mesh and washed with the buffer by gentle filtration. Amount of immobilized protein was measured by the method of Bradford (4. Amount of immobilized chlorophyll (Chl) was measured by the method of Arnon (5).

Measurement of NADPH-Hill reaction (reduction of $NADP^+$ to NADPH) in chloroplasts was measured at 30 °C under illumination using a reaction mixture containing 50 mM Tricine-NaOH (pH 7.8), 10 mM NaCl, 3 mM $MgCl_2$, 1 mM NH_4Cl, 160 μg/ml ferredoxin (Fd), 0.046 unit/ml ferredoxon-NADP oxidoreductase (FNR) and an aliquot of immobilized chloroplasts. A tungsten lamp positioned at 40 cm above the test tube. An amount of NADPH was determined

o-Deethylation of 7-ethoxycoumarin (7-EC) to 7-hydroxycoumarin (7-HC) in gel containing immobilized chloroplasts and microsomes was measured 30°C under 6200 lux illumination. The reaction medium contained 50 mM Tricine-NaOH (pH 7.8), 10 mM NaCl, 3 mM $MgCl_2$, 1 mM NH_4Cl, 0.4 mM 7-EC, 0.6 mM $NADP^+$, 160 μ g/ml Fd, 0.046 unit/ml FNR, an aliquot of chloroplasts, and an aliquot of microsomes.

3. Results and Discussion

We tried various method to immobilize chloroplasts and found the entrapment in low temperature-melting agarose gels gave the best results for retaining the activity of NADPH-Hill reaction. Chloroplast-immobilized gel produced NADPH under illumination although the efficiency was lower than the suspension of unimmobilized chloroplasts. The reason was partly because some portion of chloroplats were not immobilized and washed out during the experimental procedure for immobilization. Another reason was probably that diffusion of electron carriers like Fd or FNR was restricted in gels. We choosed spinach chloroplasts for further reaerch.

We prepared the column reactor containing the concentrated spinach chloroplasts immobilized in 4%(W/V) agarose Nu Sieve. The column showed the high activity of NADPH production as shown in Fig 1(A). Concentration of NADPH increased within 20 min and reached a plateau at 70 μM. This concentration is much higher than the Km value of the fusion enzyme for NADPH. We thought that the immobilized chloroplasts could supply sufficicient NADPH for the fusion enzyme.

We immobilized both spinach chloroplasts and microsome membranes containing the fusion enzyme in agarose Nu Sieve gels and measeured the light-driven conversion of substrate (7-ethoxycoumarin) to product (7-hydroxycoumrain). The coupling reaction between the fusion enzyme and chloroplasts is schematically illustrated in Fig 2(A). We made a one-phase reactor in which mixture of both chloroplasts and microsomes were mmobilized as shown in Fig 2(B) and a two-phase reactor in which both components were immobilized separately as shown in Fig 2(C).

Figure 1(B) shows the results with those column reactors. Total volume of the reaction medium was 7.5-8 ml in the coumn-type reactor in this experiment. We took 1 ml of the reaction medium for determination of 7-HC and added back 1 ml of the new medium at each measurement during the time course experiment. Concentration of the substrate and that of the product could change not only by enzymatic conversion but also by removal and addition of the medium. We corrected the measured 7-HC concentration to be "predectable concentration" as if we measured 7-HC using a flow cell cuvette without loss nor addition of the medium when we did the measurement.

The two-phase column reactor as shown in Fig2(C) showed better conversion capability than one-phase reactor in Fig 2(B) after 180 min run although there was initial lag phase in the range between 0 and 20 min. The reason of this difference seemed to be that too much supply of oxygen formed by the chloroplasts might denature the fusion enzyme when the mixture of both component was immobilized in the same gel particles. Actcually we observed the light-induced bubble formation in the chloroplast-immobilized gel by oxygen evolving activity of photosystem II. The linitial lag phase was derived from the delay time that NADPH produced in the chloroplasts-immobiloized gel reached the microsomes-immobilized gel during the circulation of medium. The flow rate (16 ml/h) was relatively low in our column-type reactors.

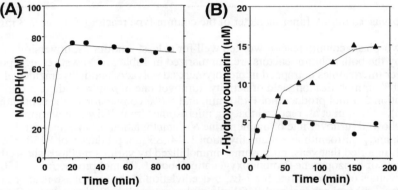

Fig. 1 NADPH Production by the spinach chloroplasts-immobilized column reactor (A) and conversion ot 7-EC to 7-HC by the one-phase reactor (closed circles) and the two-phase reactor (closed triangles) (B). The column in (A) contained 10.5 ml reaction medium, 2.5 ml gel and 150 mg Chl.

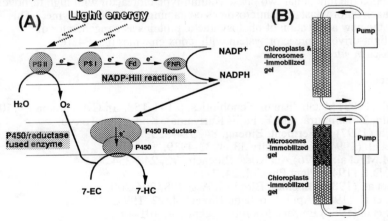

Fig 2 Schematic representation of the coupling reaction between the fusion enzyme and chloroplasts (A), that of one-phase reactor (B) and two-phase reactor (C).

Table 1. Conversion of 7-EC to 7-HC after 180 min run.

Type of bioreactor (volume of the reaction medium, volume of the gel in column, P450 concentration, & Chl concentration)	7-HC (μM)*	Yield of conversion (%)**	Turnover rate (7-HC/P450 /min)***
One-phase column reactor (7.5 ml medium, 4.0 ml gel, 80 nM P450, 21.1 mg/ml Chl)	4.6	0.80	0.32
Two-phase column reactor (8.0 ml, 1.2 ml microsome gel, 1.7 ml chloroplast gel, 33 nM P450, 13.2 mg/ml Chl)	14.8	3.1	2.49

Flow rate was 16 ml/h. Inner diameter of the column-type reactor (ϕ) was 7.8 mm.

This two-phase column reactor worked well for at least 3 hours. Conversion of 7-EC to 7-HC by the both column reactors are summarized in Table 1. Yawetz et al. reported that rat liver microsomes entrapped in prepolymerized polyacrylamide hydrazide gel showed 88% immobilization yield of activity, turnover rate of p-nitroanithole o-demethylation 2.3 mol product/mol P450/min, and 5.6% conversion of the substrate after 5 h run (6). Ibrahim et al. reported that pig microsomes immobilized by various methods showed turnover rate of aminopyrine N-demethylation 3.4-4.3 mol product/mol P450/min, and p-nitroanithole o-demethylation 1.0-0.85 mol product/mol P450/min (7). Azari et al. reported that yeast microsomes immobilized by various methods showed turnover rate of benzopyrene-3-monooxygenation about 0.01 mol product/mol P450/min (8). Our values of turnover rate for 7-EC o-deethylation in the two-phase reactor in Table 1 was 2.49 mol product/mol P450/min and comparable with those data. Our reactor worked well about the turnover rate although we can not compare the results exactly because other groups used different substrates and different molecular species of P450.

We developped the functional two phase column-type bioreactor with high turnover rate using spinach chloroplasts and microsomes containing the fusion enzyme. This results shows the new application of photosynthetic proteins as a regenerator of NADPH in a light-driven bioreactor system with redox enzymes like P450 monooxygenases *in vitro*.

4. References

1 Kamataki, T. (1993) Metabolism of Xenobiotics, p.141-158. In Cytochrome P-450, (ed. T.Omura, Y. Ishimura, and Y. Fujii- Kuriyama), Kodansha, Tokyo.
2 Hara, M. et al (1997) J. Ferment. Bioeng. 84, 324-329.
3 Sakaki, T. et al. (1994) Biochemistry 33, 4933-4939.
4 Bradford, M.M. et al. (1976) Analytical Biochem. 72, 248-254.
5 D.I. Arnon, D.I. (1949) Plant Physiol. 24, 1-15.
6 Yawetz, A. et al. (1984) Biochim. Biophys. Acta 798, 204-209.
7 Ibrahim, M. et al. (1984) Appl. Biochem. Biotechol. 12, 199-213.
8 Azari, M.R. et al (1982) Enzyme Microb. Technol. 4, 401-404.

ISOLATION AND ANALYSIS OF POLY-ß-HYDROXY BUTYRATE GRANULES FROM CYANOBACTERIA

Masato Miyake, Emir Khatipov, Kiminori Kataoka, Makoto Sirai, Ryuichiro Kurane, and Yasuo Asada
National Institute of Bioscience and Human-Technology, Higashi 1-1, Tsukuba, Ibaraki 305, and Division of Biotechnology, Ibaraki University, Ami, Inashiki, Ibaraki 300-03, Japan.

1. Introduction

Poly-beta-hydroxybutyrate (PHB) has become an important material in industrial point of view. Accumulation of PHB in cyanobacteria is one of interesting research fields because of the following reasons: i) biological role of PHB in cyanobacteria may be different from the other bacteria, since PHB in cyanobacteria is a minor storage compound otherwise they accumulate a large amount of glycogen as an energy storage compound. ii) Only cyanobacteria accumulate PHB photoautotrophically.

PHB is found in several cyanobacteria [1, 2, 3, 4, 5, 6, 7, 8]. PHB in cyanobacteria forms insoluble granules as well as in other bacteria [5, 8, 9]. The biological role of PHB in cyanobacteria is considered as a carbon storage material formed under nitrogen deprived conditions [7, 10]. PHB accumulation in cyanobacteria occurrs under certain nutrient deprived conditions [5, 7, 8, 9].

Recently, metabolic regulation of PHB accumulation has been studied in cyanobacteria. A nitrogen-dependent PHB accumulation in a cyanobacterium, *Synechococcus* sp. MA19, is controlled by a regulation of PHB synthase activity [11]. PHB synthase in this strain is a membrane-bound protein, unlike in other bacteria whose the enzyme is soluble [11, 12, 13, 14, 15].

Bacterial PHB granules are covered with lipids and some specific proteins including PHB synthase [14, 16, 17, 18]. Structure of PHB granules synthesized by membrane-bound PHB synthase in cyanobacteria is mostly unclear.

In this report we focus on isolation of native PHB granules from cyanobacteria and analysis of PHB-granule-associated proteins, to study the structure of cyanobacterial PHB granules.

G. Garab (ed.), Photosynthesis: Mechanisms and Effects, Vol. V, 4155–4158.
© 1998 *Kluwer Academic Publishers. Printed in the Netherlands.*

2. Procedure

2.1 Cyanobacteria and culture conditions
PHB granules were isolated from a PHB accumulating cyanobacterium, *Synechococcus* sp. MA19 grown under conditions described earlier [8].

2.2 Separation of PHB granules
Cells of 1 g (wet) were harvested by centrifugation at 10,000 x g for 10 min at 4 °C and washed twice and resuspended (up to OD_{660} = 1.3) with 25 mM Tris·HCl (pH 7.5). The cells were disrupted by sonication at 70 W. Crude extract solution (5 ml) was applied onto 20 ml 90% Percoll (Pharmacia, in 0.15 M NaCl) in a centrifuge tube (ϕ 25.5 mm, 30 ml), and centrifuged at 13,000 x g for 55 min at 4 °C. The band containing PHB was recovered and diluted 10 times by the same buffer. PHB granules were concentrated by centrifugation (10,000 x g, 20 min).

2.3 Preparation of PHB-granule-associated proteins
PHB granules were suspended into 1% sodium dodecylsulfate (SDS) and incubated for 10 min at 37 °C. Then the supernatant was mixed with four volumes of acetone and centrifugated (10,000 x g, 10 min, 4 °C) to obtain a white pellet of granule-associated proteins.

2.4 Quantitation of PHB and chlorophyll a
PHB was quantitated by gas chromatography as described by Miyake et al. [8]. Extraction and quantitation of chlorophyll a was carried out as described by Tandeau de Marsac and Houmard [19].

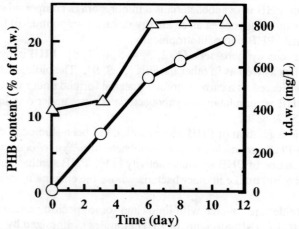

Figure 1. PHB accumulation by Synechococcus sp. MA19 under nitrogen-deprived conditions. Time indicates after the cells were transferred to the nitrogen-deprived condition. Symbols: ○, PHB content; △, total dry weight of cells.

3. Results and Discussion

3.1 *PHB accumulation in cyanobacteria*
Strain MA19 accumulated PHB under nitrogen-deprived conditions (Figure 1). PHB content reached 20% of total cell dry weight after incubation for 11 days. Such cells were used to isolate PHB granules.

Table 1 Effect of sonication on recovery and purity of PHB granule

Sonication time (min)	10	20	30
Recovery of PHB (%)	51.7	94.0	92.4
Chlorophyll a (μ g/mg)	10.9	1.2	0

Recovery of PHB(%)=(amount of PHB in lower band)/(total amount of PHB in the sample).

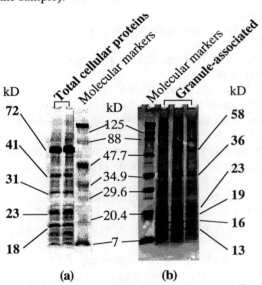

(a)   (b)

Figure 2.   Coomassie blue-stained SDS-PAGE of proteins from MA19.

3.2 *Percoll gradient centrifugation of PHB granules*
Centrifugation of crude extract in Percoll gradient resulted in formation of two major bands. The upper band (1.0 to 1.05 g/cm³) contained unbroken cells and membranes and the lower band (1.15 to 1.2 g/cm³) contained mainly PHB granules. Sonication time affected recovery and purity of PHB (Table 1). The presence of chlorophyll a probably indicates the association of PHB granules with thylakoid membrane. Longer sonication resulted in dissociation of chlorophyll a from a major PHB granule fraction. These results

reveal possible role of thylakoid membrane in granule formation, especially from the point of view of membrane association of PHB synthase previously reported[11].

3.3 PHB-granule-associated proteins in cyanobacteria

Composition of proteins solubilized from PHB granules isolated from MA19 was compared to that of total proteins from MA19 by SDS-PAGE. The main bands (58, 36, 23, 19, 16, 13 kD) found in the granule-associated proteins (Figure 2 (b)) were not identical to those (72, 41, 31, 23, 18 kD) in the total cellular proteins (Figure 2 (a)). These results suggest association of some specific proteins with PHB granules in MA19.

References

1. Carr, N.G. (1966) Biochem. Biophys. 120, 308-310.
2. Jensen, T.E. & Sicko, L.M. (1971) J. Bacteriol. 106, 683-686.
3. Campbell, J., Stevens, S.E.J. & Balkwill, D.L. (1982) J. Bacteriol. 149, 361-363.
4. Vincenzini, M., Sili, C., De Philippis, R., Ena, A. & Materassi, R. (1990) J. Bacteriol. 172, 2791-2792.
5. Wanner, G., Henkelmann, G., Schmidt, A. & Kost, H.-P. (1986) Z. Natürforsch 41c, 741-750.
6. Capon, R.J., Dunlop, R.W., Ghisalberti, E.L. & Jefferies, P.R. (1983) Phytochemistry 22, 1181-1184.
7. Stal, L.J. (1992) FEMS Microbiol. Rev. 103, 169-180.
8. Miyake, M., Erata, M. & Asada, Y. (1996) J. Ferment. Bioeng. 82, 516-518.
9. Arino, X., Ortega-Calvo, J.-J., Hernandez-Marine, M. & Saiz-Jimenez, C. (1995) Arch. Microbiol. 163, 447-453.
10. Miyake, M. & Asada, Y. (1995) Photosynthesis: from Light to Biosphere V, 403-406.
11. Miyake, M., Kataoka, K., Shirai, M. & Asada, Y. (1997) J. Bacteriol. 179, 5009-5013.
12. Fukui, T., Yoshimoto, A., Matsumoto, M., Hosokawa, S. & Saito, T. (1976) Arch. Microbiol. 110, 149-156.
13. Haywood, G.W., Anderson, A.J. & Dawes, E.A. (1989) FEMS Microbiol. Lett. 57, 1-6.
14. Gerngross, T.U., Snell, K. D., Peoples, O. P., Sinskey, A. J., Csuhai, E., Masamune, S., & Stubbe, J. (1994) Biochemistry 33, 9311-9320.
15. Liebergesell, M., Sonomoto, K., Madkour, M., Mayer, F. & Steinbüchel, A. (1994) Eur. J. Biochem. 226, 71-80.
16. Stuart, E.S., Lenz, R.W. & Fuller, R.C. (1995) Can. J. Microbiol. 41, 84-93.
17. Steinbüchel, A., Aerts, K., Babel, W., Föllner, C., Liebergesell, M., Madkour, M. H., Mayer, F., Pieper-Fürst, U., Pries, A., Valentin, H. E. & Wieczorek, R. (1995) Can. J. Microbiol. 41, 94-105.
18. Wieczorek, R., Steinbüchel, A. & Schmidt, B. (1996) FEMS Microbiol. Lett. 135, 23-30.
19. Tandeau de Marsac, N. & Houmard, J. (1988) Method. Enzymol. 167, 318-328.

PHOTOSYNTHESIS AND PRODUCTIVITY RESPONSE TO CHANGES IN BLUE/RED LIGHT RATIO IN *DUNALIELLA BARDAWIL* UNDER FIELD CONDITIONS.

Roni Ashkenazi, Dan Yakir and Amnon Yogev
Dept. of Environmental Sciences and Energy Research, Weizmann
Institute of Science, Rehovot, Israel

Key words: bioreactor, light acclimation, blue light, diurnal changes, green algae.

1. Introduction

Light serves both as the energy source for photosynthesis and as an external regulating signal and it can be a limiting factor in photosynthetic productivity. To effectively utilize available light, plants and green algae can regulate leaf morphology, pigments composition, structure and function of thylakoid membranes and the overall rates of photosynthesis. These adaptation mechanisms can be induced by changes either in light intensity or quality, or both, and they appear to be highly conserved in all photosynthetic organisms examined (1). When the rate of photons absorption exceeds the rate of utilization of excitation energy, light utilization efficiency decreases and damage to the photochemical apparatus can occur. It can also result with regulated processes leading to the harmless dissipation of excess excitation energy. Among the radiationless dissipation mechanisms is the de-epoxidation of violaxanthin to form zeaxanthin molecule (the "xanthophyll cycle" (2)). Another protection mechanism is the accumulation of ß-carotene in lipoidal globules in the interthylakoid space (3, 4), accompanied by the induction of Cbr synthesis ("Carotene biosynthesis related" protein, (5)), typical to the halotolerant green algae *Dunaliella Bardawil.*
Acclimation to changes in the light spectral characteristics is well known (6, 7). There is evidence that in higher plants, blue light leads to the production of "sun type" chloroplasts, with a higher chl a/chl b ratio and lower granal stacking (8, 9, 10). In algae, the existence of a similar response is not clear yet. We observed concomitant changes in sunlight spectral characteristics, algal growth rates and pigments composition through the diurnal and seasonal cycles. We hypothesized that it is possible to enhance photosynthetic productivity and pigments synthesis by manipulating the spectral composition and intensity of the sunlight under field conditions.

2. Methodology

Outdoor photobioreactors consisted of two concentric Pyrex glass tubes with diameter ratio of 1.5 (7.5/5cm) were developed. Surface to volume ratio of the bioreactor was 0.51 and average concentration factor of 1.2 times the ordinary insolation was achieved at the surface of inner tube. Spatial and temporal distribution of light was maximized within the bioreactor by keeping a 90⁰ angle between sunlight plan of radiation and bioreactor surface through daytime. The culture was located within the inner tube, continuously stirred and under pH and temperature control. Sunlight spectral

4159

G. Garab (ed.), Photosynthesis: Mechanisms and Effects, Vol. V, 4159–4162.

modification experiments were carried out by using optical filter solutions within the intertubular space. The biological material used was the unicellular green algae *Dunaliella bardawil* (11). A routine of a three days experiment of continuous culture of growing cells was established.A procedure of sampling once to twice a day in the first two days and then every 3-5 hr on the third day of experiment was carried out. All samples were used for cell counting, pigments determination (12) and one of the following analysis: Quantum yield of photosynthesis determined by the rate of O_2 evolution (using a Clark-type DW2/2 oxygen electrode unit, Hanzatech Instruments), the estimation of photosynthesis efficiency of PSII and the photosynthetic antennae size by fluorescence induction technique (13), Cbr analysis by SDS-Polyacrylamide Gel Electrophoresis and Western Blotting (5), or Xanthophyll epoxidation state determination by HPLC (after extracted from whole cells with ethanol:hexane (2:1,v/v) as described by (14)). The determination of the epoxidation state of xanthophylls was based on (2) .Sunlight spectrum and intensity were measured simultaneously with sampling the algae during the day, in order to look for a correlation between the change in light regime and physiological state of the algae as well as its photosynthetic production. Experiments were carried out throughout the year.

3. Results and Discussion

A linear correlation between the sunlight intensity and the ratio of blue/red light was observed during the natural daily cycle (Fig. 1).

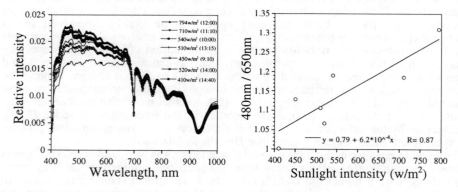

Figure 1: a. Typical Sunlight spectrums measured in Rehovot Israel, during 21 Dec. 97.
  b. Correlation between sunlight intensity and blue/red ratio.
Antennae size and photosynthetic efficiency were tightly related to the daily changes in the sunlight features. The antennae size and the photosynthetic efficiency measured on 11 Sep (Fig. 2) inversely correlated to both sunlight intensity and blue/red light ratio. Chl a/Chl b ratio reached a maximum value at noontime which was accompanied by minimal total chlorophyll concentration (not shown). The low photosynthetic efficiency at noontime (Fig 2) is consistent with the phenomenon of midday depression observed in higher plants (15).

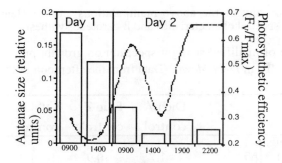

Figure 2: Changes in LHCII size (bars) and photosynthetic efficiency (line) through a daily cycle (10-11 Sep. 97).

A striking negative correlation was observed between changes in the Cbr concentration and xanthophyll epoxidation state during the daily cycle (Fig. 3). This supported the proposed function of Cbr in binding zeaxanthin to form a photoprotective complex within the light harvesting antennae (5).

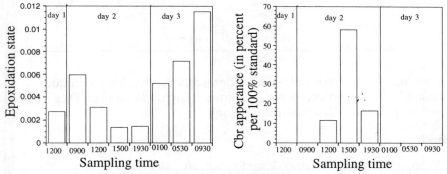

Figure 3: a. Epoxidation state of xanthophylls in *D. bardawil* during the daily cycle.
b. Accumulation of Cbr in *D. bardawil* during the daily cycle.

A clear diurnal pattern in cell division (not shown), pigments synthesis, antennae size, photochemical efficiency and Cbr accumulation was observed, corresponding to the diurnal cycle in sunlight spectrum and intensity.

Table. 1 shows the photosynthetic response to artificial manipulation of the sunlight spectrum. Subtraction of either blue (Potassium dichromate solution) or far red (CuSO$_4$ solution) light (and adjusting light intensities to equal in treatments and controls) resulted with increase in productivity as well as with photosynthetic response typical to low light. In both cases, total yield of ß-carotene (per ml) increased. Subtraction of part of the red light (with Methylene blue solution or blue zelofan) did not result with any significant effect on productivity. While cutting off light above 540nm (blue zelofan) induced partial high light response, reducing light in the range between 570-700 (Methylene Blue) did not cause a significant effect. ß-carotene production per cell could

4162

not be enhanced by any of the above manipulations and no clear correlation was discerned between spectral characteristics and ß-carotene synthesis.

Table 1: Influence of modified sunlight spectrum on growth, pigments concentration and quantum yield (Φ) in *D. bardawil*. (Data from the afternoon of the third experiment

	Dich./ control	Blue Z/ control	CuSO$_4$/ control	MB/ control
cells per ml. *10^5	1.7	0.9	1.4	1.1
ß-car. (pg/cell)	0.7	0.6	0.8	1.1
(µgr/ml)	1.2	0.5	1.1	1.2
Chl. (pg/cell)	1.7	1.0	1.5	0.8
(µgr/ml)	2.7	0.9	2.3	1.1
Chla /Chlb	0.9	1.4	0.7	0.9
Φ	*1.3	*1.1	**1.5	**0.7
antennae size	*3.4	*0.7	n.a	n.a

day).
   n.a = not analyzed ; * determined by Fluorescence induction technique.
** determined by Oxygen evolution technique.; (Dich = Potassium Dichromate, Blue Z = Blue Zelofan, MB = Methylene Blue).

References

1   Webb, R. M. Melis, A. (1995) Plant Physiol. 107, 885-893
2   Demmig-Adams, B. (1990) Biochemica et Biophysica Acta 1020
3   Ben-Amotz, A. Avron, M. (1989) J. Phycol. 25, 175-178
4   Ben-Amotz, A. Avron, M. (1983) Plant Physiol. 72, 593-597
5   Levy, H. Tal, T. Shaish, A. Zamir, A. (1993) Journal of Biol. Chem. 268, 20892-20896
6   Smith, H. Samson, G. Fork, D.C. (1993) Plant, Cell & Environment 16, 929-37
7   Morgan, D. C. Smith, H. (1976) Nature 262, 210-212
8   López-Juez, E. Hughes, M. J. G. (1995) Photochem. and Photobiol. 61, 106-111
9   Lichtenthaler, H. K. Buschmann, C. Rahmsdorf, U. (1980) in The Importance of blue light for the development of sun-Type chloroplasts. (S. Verlag, ed.)
10   Humbeck, K. Senger, H. (1984) in The Blue Light Factor in Sun and Shade Plant   Adaptation. (H. Senger, ed.) Springer-Verlag, Berlin Heidelberg
11   Ben Amotz, A. Katz, A. Avron, M. (1982) Journal of phycology 23, 176-181
12   Arnon, D. I. (1949) Plant Physiol 24, 1-15
13   Malkin, S. Armond, P. A. Mooney, H. A. Fork, D. C. (1981) Plant Physiol. 67
14   Shaish, A. Ben-Amotz, A. Avron, M. (1991) Journal of Phycol. 27, 652-656
15   Demmig-Adams, B. et al. (1989) Planta 177, 377-387

PRODUCTION OF OMEGA-3 FATTY ACIDS BY THE MICROALGA *MONODUS SUBTERRANEUS.*

Ping Ping Lam and Feng Chen
Fermentation laboratory, The Department of Botany,
The University of Hong Kong. Pokfulam Road, Hong Kong.

Key words: biomass, environment, microalgae, unsaturated fatty acids.

Introduction

The demand of producing omega-3 fatty acids from the microalgae have recently increased as recent developments indicating the nutritional and pharmaceutical importance of these long chain omega-3 fatty acids in the human diet. The nutritional importance of these compounds has been underscored by the fact that several countries, including the Scandinavian countries and Canada, have now established daily recommended dietary intakes values for omega-3 fatty acids. The British Nutrition Foundation has also published recommendations for dietary intake of these unsaturated fatty acids. As a result, markets for long chain omega-3 fatty acids are beginning to develop in the areas of health supplements and food enrichment for both the animal and human. PUFAs have demonstrated the therapeutic potential in the treatment of various diseases and disorders, including cardiovascular problems, a variety of cancers and inflammatory diseases (1). Some of the effects of omega-3 fatty acids may be explained in terms of their ability to alter the balance of prostaglandin and leukotriene eicosanoids, which mediate inflammatory and immune responses (2).

The current source of these omega-3 fatty acids are from fish and fish oils. However the production of these PUFAs were not sufficient as many fish can only produce little amount of them, so production cannot meet the demand of the people and the price of these PUFAs would therefore be very high. Also the PUFAs from this source have the problem of an undesirable odor, as well as the geographical and seasonal variations in quality. For the reason of these limitations, an alternative source of omega-3 fatty acids production was prompted. And the alternative source would be the microalgae as they were the primary producers of these PUFAs and it stimulated the interest of using microalgae as a source of these vital compounds. Food and pharmaceutical quality production can be enhanced both by the degree of process

G. Garab (ed.), Photosynthesis: Mechanisms and Effects, Vol. V, 4163–4166.
© 1998 *Kluwer Academic Publishers. Printed in the Netherlands.*

control and by the sterility achieved through a fermentation process, when compared to the production of these PUFAs from marine fish oil.

Monodus subterraneus is a microalga that has very high eicosapentaenoic acid content (3). So increasing the growth of it can enhance the production of this PUFA from the culture and therefore optimization of the growth of it will be determined in this experiment.

Materials and methods

Monodus subterraneus (UTEX 151) was obtained from the University of Texas Culture Collection (Austin, TX) and it was maintained on the nutrient agar slants as suggested by the UTEX. This nutrient agar slants contains 0.25g $NaNO_3$, 0.025g $CaCl_2 2H_2O$, 0.075g $MgSO_4 7H_2O$, 0.075g K_2HPO_4, 0.175g KH_2PO_4, 0.025g NaCl and 1g proteose peptone in one liter of distill water while 1 drop of $FeCl_3$ and 2 ml of trace element was optional. This proteose medium was only for the maintenance of the *Monodus* culture and the growth rate of the *Monodus* was very low. So the culture can be maintain for over 3 months and subculture of the *Monodus* will only necessary for every 3 months.

While for faster growth of the *Monodus*, it was cultivated on BG-11 medium as described by Iwamoto and Sato (3). One liter BG-11 medium contains 1.5g $NaNO_3$, 0.04g K_2HPO_4, 0.075g $MgSO_4 7H_2O$, 0.036g $CaCl_2 2H_2O$, 0.006g Citric Acid, 0.006g Ferric Ammonium Citrate, 0.001g Na_2-EDTA, 0.02g Na_2CO_3 and 1 ml trace element solution. And for the trace element solution, one liter will contains 2.86g H_3BO_3, 1.81g $MnCl_2 4H_2O$, 0.222g $ZnSO_4 7H_2O$, 0.39g $Na_2MoO_4 2H_2O$, 0.079g $CuSO_4 5H_2O$ and 0.0494g $Co(NO_3)_2 6H_2O$. There will be two set of cultures in this experiment, one set will changing the $NaNO_3$ concentration from 3g/L, 4.5g/L, 6g/L to 7.5g/L while keeping the glucose concentration at 40g/L that was in excess to eliminate the glucose limitation problem of growth of the microalgae. While the other set of cultures will be using the best $NaNO_3$ concentration from the previous experiment but changing the glucose concentration from 10g/L, 20g/L, 30g/L to 40g/L The cultures were grown in Erlenmeyer flasks. and the pH were adjusted to 8. The flasks were put into an incubator with the temperature kept at 30 °C , while the light intensity was kept at 2.5 klux. All the flasks were continuously shake with 150 rpm. Cultures were grown exponentially under the appropriate conditions for at least four days prior to the onset of the experiment.

Biomass dry weight determinations were performed by harvesting culture samples, filtering the cells through the GF/C filter paper (Whatman), washing twice with distilled water, and drying under 80 °C overnight to get a constant weight.

Result

The effect of nitrogen concentration in the culture was determined and showing in the FIG.1. It was showing that the nitrogen concentration was inversely proportional to the growth of the *Monodus*. The highest dry cell mass was 7.4g/L obtained at day 12 with the culture using 3g/L NaNO₃. However, when the nitrogen concentration increased from 3g/L, the growth of the other cultures become slower and the maximum dry cell mass obtained was a little bit lower.

While the effect of glucose concentration in the culture was showing in FIG.2 and the highest dry cell mass was nearly the same in all culture except for the lowest concentration. For the cultures with 20g/L, 30g/L and 40g/L glucose, the maximum dry cell mass were 7.4g/L in day 10, 11 and 12 respectively.

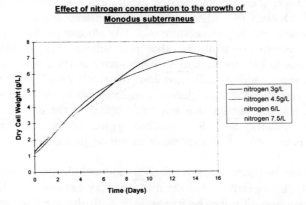

FIG.1. Growth curves of the cultures in different nitrogen concentration.

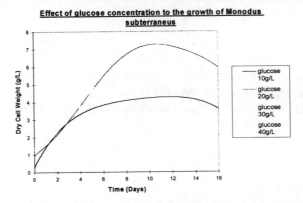

FIG.2. Growth curves of the cultures in different glucose concentration.

4166

Discussion

From the result in FIG.1, it was found that 3g/L $NaNO_3$ was the most optimum nitrogen concentration for the *Monodus subterraneus* to growth. While increasing the nitrogen concentration will slower the growth of the *Monodus* and also lower the cell concentration of the cultures. This result may be cause by the inhibition effect of the nitrogen concentration of the cultures. When the nitrogen concentration go beyond the optimum, there will be a toxic effect to the culture. And the higher the nitrogen concentration, the greater the toxic effect is and therefore the growth of the cultures become inversely related to nitrogen concentration when the concentration higher than optimum.

Based on the result from the above experiment, the optimum $NaNO_3$ concentration was determined and therefore the second experiment will only change the glucose concentration and keep the other components constant in all cultures. From the result, it was found that only the culture with 10g glucose per liter grew slower and got lower cell concentration while the other three cultures showed very similar result. And it may be due to the limitation of nitrogen source as there was only 3g/L $NaNO_3$ in the cultures and so the dry cell mass can only reach 7.4g/L in maximum. In this experiment, it was showing that glucose concentration in the culture can reach as high as 40g/L and did not causing any toxic effect to the culture as those three cultures grew similarly. So 20g/L glucose would be now the most optimum concentration as the nitrogen concentration cannot be increased.

In conclusion, the optimum nitrogen concentration and glucose concentration would be 3g/L and 20g/L respectively and the maximum dry cell mass would be 7.4g/L. In fact, the cell concentration can be increased if both the medium optimization and environmental conditions optimization were done. Other experiments have to be done more in the future.

References

1. Simopoulos, A.P., *J. Nutr.* 119:521 (1989).
2. Samuelsson, B., *Science* 220-568 (1983).
3. Iwamoto, H., and S. Sato, *J. Am. Oil Chem. Soc.* 63:434 (1986).

CHANGES IN THE PHOTOSYNTHETIC APPARATUS DURING ACCUMULATION OF SECONDARY CAROTENOIDS IN FLAGELLATES OF *HAEMATOCOCCUS PLUVIALIS*

C. Hagen, S. Schmidt & J. Müller
Institut für Allgemeine Botanik, Friedrich-Schiller-Universität Jena,
Am Planetarium 1, D-07743 Jena, Germany

Key words: astaxanthin, Chl fluorescence induction, cytochrome b/f(c), light acclimation, microalgae, O_2 evolution

1. Introduction

The unicellular volvocalean green alga *Haematococcus pluvialis* as one of the astaxanthin-synthesising organisms receives close attention from a biotechnological point of view [1]. In previous studies it was proposed that secondary carotenoid (SC) synthesis is related to the formation of resting cells [2]. However, SC accumulation was clearly shown in flagellates demonstrating this process as independent of senescence [3, 4].

Advanced physiological, biochemical and ultrastructural studies of the accumulation period are hindered by the extraordinary resistance of the thickened cell wall of the resting cells (aplanospores). To avoid this disadvantage, a batch cultivation system was established that results in accumulation of SC in flagellates within one week [4]. That allows us to characterise influences of the accumulation process on metabolic activities without interferences by formation of the resting state.

In this study, changes in photosynthetic activity of SC-accumulating flagellates were analysed by oxygen evolution and Chl fluorescence measurements. Additionally, changes in main components of the photosynthetic apparatus were quantified by gel electrophoresis and western blot analysis.

2. Procedure

Haematococcus pluvialis Flotow (no. 192.80, culture collection of the University of Göttingen, Germany; synonym: *H. lacustris* [Girod] Rostafinski) was grown autotrophically as described previously [4]. After 5 days precultivation under continuous white light ($25 \, \mu mol$ photons $m^{-2} s^{-1}$), logarithmically growing flagellates were transferred to nitrate-deprived medium and stronger light ($150 \, \mu mol$ photons $m^{-2} s^{-1}$ of white light) to induce accumulation of SC [4].

Light-induced oxygen evolution was measured after 10 min of dark adaptation (dark respiration measurement) by a Clark-type oxygen electrode (biolytik, Bochum, Germany) in 2 ml of the corresponding flagellate suspension concentrated by gentle centrifugation to 500000 cells ml^{-1} in a buffer containing fresh nitrate-deprived medium, 20 mM $NaHCO_3$ and 50 mM HEPES at pH 7.1 and 20 °C. Initial gross photosynthesis rate were determined between 1.5 to 2.5 min after onset of red actinic light provided by a metal halogenid lamp and a red long-pass filter ($\lambda_T > 600$). A new sample was prepared for each intensity of actinic irradiation. Characteristic parameters of photosynthesis-irradiance-plots (light-saturated photosynthesis rate: P_{max}; initial slope: α) were derived from least-square fits of the data to a model described in [5].

G. Garab (ed.), Photosynthesis: Mechanisms and Effects, Vol. V, 4167–4170.
© 1998 Kluwer Academic Publishers. Printed in the Netherlands.

Chlorophyll fluorescence measurements were performed using PAM 101-103 equipment (Walz, Effeltrich, Germany) in a thermostated cuvette (DW 2/2; Bachofer, Reutlingen, Germany) connected to the detection unit ED-101US/D (Walz) to increase sensitivity. For P-I-type experiments, samples were measured at 15 µg Chl ml^{-1} and 20 °C in the same buffer as for O_2 concentration measurements. After 10 min predarkening and detection of dark-adapted basic (F_o, red modulated light, 0.04 µmol photons m^{-2} s^{-1}) and maximal fluorescence (F_m, 400-ms flash of red light, 4300 µmol photons m^{-2} s^{-1}), the sample was exposed to increasing intensities of actinic irradiation, and actual efficiency of PS II ($F_{m'}$-F_t/$F_{m'}$) was probed by measurement of flash-induced $F_{m'}$ after 1 min of exposure to each light intensity step.

Total proteins were solubilised and separated by SDS-PAGE [6], and specific proteins were analysed by probing nitrocellulose electroblots with corresponding antibodies (RUBISCO-LSU/SSU and LHC-II were kindly provided by U. Johanningmeier, Halle; CYT F by J. Meurer, Jena). Quantification was done using densitometry of the immunoblots stained by alkaline phosphatase reaction, protein detemination using the Micro-Lowry protocol from Bio-Rad (München, Germany).

Cell counting, determination of total Chl and carotenoid content, pigment pattern analysis by HPLC and measurement of photon flux densities were described previously [4].

3. Results and Discussion

Green flagellates in the logarithmic phase of growth containing no traces of SC were transferred to a cultivation medium deficient in nitrate, and were exposed to stronger light conditions. Accumulation of SC started before cell number had reached a steady state after 2 to 3 days of exposure and reached a level of approximately 50 pg cell^{-1} at the 7th day. Cellular Chl content halved and the Chl a to Chl b ratio reached a maximum during the first two days (Table 1, for detailed information cf. [4]).

Table 1: Changes in the pigment pattern of SC-accumulating flagellates during exposure to mild stress conditions (nitrogen-deficiency, increased cultivation light intensity).

	Chl a/Chl b [mol mol^{-1}]	Neo	Lut	β-Car [mmol (mol Chl a)$^{-1}$]	Vio+Zea+Ant	De-epoxidation state [(Zea+0.5Ant)/(Vio+Zea+Ant), mol mol^{-1}]
start	2.536 (0.031)^{*1}	51.31 (0.18)	254.66 (1.93)	146.35 (1.20)	54.06 (1.07)	0.134 (0.005)
1 h	2.542 (0.085)	50.33 (0.77)	261.80 (2.33)	149.68 (0.89)	56.38 (1.64)	0.232 (0.012)
4 h	2.539 (0.054)	47.65 (1.03)	292.25 (3.74)	148.62 (5.23)	73.54 (1.36)	0.363 (0.011)
1 d	2.783 (0.010)	51.66 (0.58)	359.49 (1.36)	174.99 (0.91)	106.50 (1.66)	0.419 (0.006)
2 d	2.706 (0.042)	60.02 (0.60)	406.57 (6.53)	170.74 (3.23)	99.27 (0.65)	0.443 (0.007)
4 d	2.770 (0.027)	68.25 (1.71)	332.23 (3.95)	186.52 (6.62)	96.47 (2.82)	0.396 (0.006)
4 d low light^{*2}	2.499 (0.007)	46.43 (0.24)	251.86 (1.99)	156.14 (0.36)	36.43 (0.44)	0.006 (0.003)

^{*1}SE, n = 3-6 / ^{*2}5 µmol photons m^{-2} s^{-1} of white light

Adaptation of flagellates to the increased cultivation light intensity was detected by an increase in P_{max} and a lowered α already after 16 h of exposure (Fig. 1), dark respiration rate almost tripled during that period (Fig. 1, table).

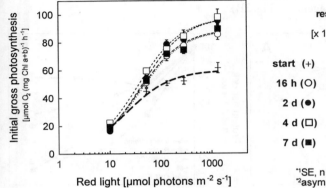

		Dark respiration rate [x 10^{-8} µmol O_2 cell^{-1} h^{-1}]	P_{max} [µmol O_2 (mg Chl)$^{-1}$ h^{-1}]	α [P x PAR]
start	(+)	12.40 (0.68)*1	59.77 (2.25)*2	2.75 (0.50)
16 h	(O)	30.88 (1.39)	88.83 (1.20)	2.25 (0.12)
2 d	(●)	22.22 (0.34)	98.20 (3.01)	2.30 (0.27)
4 d	(□)	23.14 (0.62)	97.81 (2.27)	2.71 (0.26)
7 d	(■)	21.76 (0.66)	89.47 (2.83)	2.38 (0.30)

*1SE, n = 24-95, 2 to 7 parallels of cultivation
*2asymptotic SE

Figure 1: P-I-characteristics of SC-accumulating flagellates.

Photosynthesis-irradiance-data obtained by Chl fluorescence measurements yielded only a small decrease of P_{max} in flagellates exposed for 4 or 7 days to nitrogen-deficiency and stronger light conditions (Fig. 2A). Green samples drawn before this mild stress exposure exhibited signs of photoinhibition at the highest actinic light intensity (Fig. 2A). Comparison between experiments using either red or blue actinic light demonstrated the shading function of SC (Fig. 2B, C).

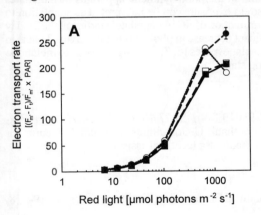

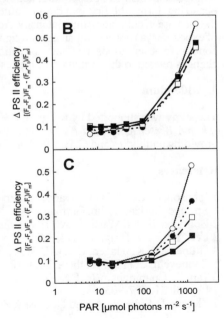

Figure 2: Changes in Chl fluorescence characteristics in SC-accumulating flagellates. Irradiation-induced decrease of open PS II efficiency was measured with red (B) or blue actinic light (C). Symbols: O, start; ●, 2 d; □, 4 d; ■, 7 d of exposure

4170

Efficiency of open PS II was reduced from 0.74 to 0.56 in a shock reaction to the increased cultivation light intensity during the first 3 h of exposure, then it reached a level of 0.68 which was almost stable for the entire SC accumulation period in flagellates (Fig. 3A). Nitrate-deprivation alone did not change PS II efficiency (Fig. 3A, closed circles).

In contrast to literature data [6], an extraordinary decrease in the level of cytochrome f was not detected in SC-accumulating flagellates (Fig. 3B).

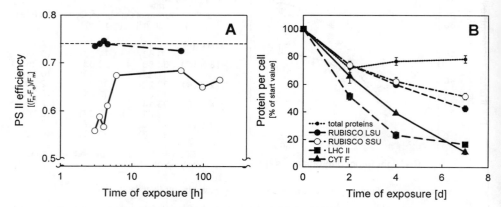

Figure 3: Changes in PS II efficiency (A) and in the protein pattern of the photosynthetic apparatus (B) in SC-accumulating flagellates. A. Data were obtained from suspensions exposed to nitrogen-deficiency and precultivation light ($\bullet$, 25 μmol photons m^{-2} s^{-1}) or 150 μmol photons m^{-2} s^{-1} (O).The dashed line marks the start value of 0.74.

Alltogether, changes in respiration activity and in the photosynthetic apparatus of flagellates occurring during the first two days of mild stress exposure can be interpreted as adaptation to the increase in cultivation light intensity. Accumulation of SC in flagellates starting at the 2nd day of exposure was not accompanied by a significant decrease in photosynthetic activity. That is in contrast to observations in SC-accumulating aplanospores [6, 7] and might be due to metabolic changes related to the formation of the resting state.

4. Addendum

The study was supported by the grant B301-96013 from the *Thüringer Ministerium für Forschung, Wissenschaft und Kultur*. The authors thank U. Johanningmeier and J. Meurer for providing antibodies and G. Bratfisch and M. Fiedler for technical assistance.

References

1 Johnson, E.A. (1995) in Advances in Biochemical Engineering (Fiechter A., ed.) pp. 119-178, Springer Verlag, Berlin, Germany
2 Boussiba, S. and Vonshak, A. (1991) Plant Cell Physiol. 32, 1077-1082
3 Lee, Y.-K. and Ding, S.-Y. (1994) J. Phycol. 30, 445-449
4 Grünewald, K., Hagen, C. and Braune, W. (1997) Eur. J. Phycol. 32, 387-392
5 Zonneveld, C. (1997) J. theor. Biol. 186, 381-388
6 Tan, S., Cunningham Jr., F.X., Youmans, M., Grabowski, B., Sun, Z. and Gantt, E. (1995) J. Phycol. 31, 897-905
7 Zlotnik, I., Sukenik, A. and Dubinsky, Z. (1993) J. Phycol. 29, 463-469

PHOTOINHIBITION AND ITS RECOVERY OF *SPIRULINA PLATENSIS* IN PHOTOAUTOTROPHIC AND MIXOTROPHIC CULTURES

Suk Man Cheung[1], Avigad Vonshak[2] and Feng Chen[1]
[1] Department of Botany, The University of Hong Kong, Hong Kong
[2] The laboratory for Microalgal Biotechnology, Jacob Blaustein Institute for Desert Research, Ben-Gurion University of the Negev, Sede-Boker Campus 84990, Israel

ABSTRACT
The inhibitory effect of high photon flux densities (PFD) on the growth of *Spirulina platensis* and its sequent recovery were investigated using photoautotrophic and mixotrophic cultures. The light-dependent O_2 evolution rate and variable chlorophyll fluorescence were measured after the cultures had been exposed to high PFD. Marked decreases in these two parameters of both cultures were found but the dropping rate and also the extent of mixotrophic cultures were slower and smaller. In adding chloramphenicol (CAP), an inhibitor of protein synthesis, no significant difference between the two cultures was seen. Both cultures were able to recover when placed in low PFD but photoautotrophic cultures recovered in a lower extent and a slower rate. The maximal photosynthetic rate (P_{max}) and light saturation value (E_k) of mixotrophic cultures were higher than those of the photoautotrophic cultures. Thus, mixotrophic cultures were less susceptible to high PFD stress. This can lead to a greater potential in using mixotrophic cultures for mass cultivation.

Keywords : D1 turnover, microalgae, O_2 evolution, photoinhibition

INTRODUCTION
Mixotrophic growth is characterized as an microorganism using not solely light as an energy source but also organic carbon source as well. *Spirulina* is used as a human food supplement for a long time (1, 2). They are rich in protein, water soluble vitamins, β-carotene and polyunsaturated fatty acids. This organism has been found to be able to utilize glucose as an organic substrate (3, 4). However, work on the photosynthetic characteristics of mixotrophic cultures is limited. It has been well documented that photoinhibition is one of the reasons that lower the production yield in large scale cultivation. This has been intensively investigated only in photoautotrophic cultures (5, 6) but not in mixotrophic cultures. Very little is known about the differences in photosynthetic response to light between the photoautotrophic and mixotrophic cultures. This study was undertaken to get more information on photoinhibition and recovery of mixotrophic culture of *Spirulina*, so that optimization of the productivity of algal cultures can be achieved.

G. Garab (ed.), Photosynthesis: Mechanisms and Effects, Vol. V, 4171–4174.
© *1998 Kluwer Academic Publishers. Printed in the Netherlands.*

CULTURE CONDITIONS

Spirulina platensis UTEX # 1926 (University of Texas Culture Collection) was grown in batch culture in Zarouk's medium (7) supplemented with or without 2 g L^{-1} glucose. The cells were grown in 250 mL Erlenmeyer flask containing 100 mL of nutrient medium. The alga was grown at 30°C under continuous illumination and shaking (130 rpm). Illumination was provided by cool white fluorescent lamps with a photon flux density (PFD) of 50 μmol photon $m^{-2}s^{-1}$.

RESULTS

When the P-I curve of the photoautotrophic cultures was compared to that of mixotrophic cultures, marked differences were found (Figure 1). The light-saturated photosynthetic rate, P_{max} and the saturation irradiance, E_k of mixotrophic culture are higher. This indicates that the photosynthetic capacity of mixotrophic culture is larger. This may conclude that mixotrophic cultures can utilize higher levels of PFD than photoautotrophic cultures and so less susceptible to photoinhibition

To further prove the decreased susceptibility of mixotrophic cultures to light, photoinhibition experiments were carried out. As shown in Figure 2, the photosynthetic activities of both cultures dropped significantly after exposing to high photon flux density (HPFD). The dropping rate and also the extent of mixotrophic cultures were slower and smaller. Therefore, it is believed that *Spirulina* cultures grown on a glucose supplemented medium are less sensitive to photoinhibition because of their higher photosynthetic capacity.

In order to find out which process (damage or recovery or both) in mixotrophic cultures is different from that in photoautotrophic cultures, another set of photoinhibition experiments was performed. In this case, an inhibitor of protein synthesis, chloramphenicol (CAP), , was added. After adding this, recovery of photoinhibited cells was prevented. As shown in Figure 3, there was no significant difference in the damage effects of photoinhibition in these two cultures. This may increase the possibility that the difference in the photosynthetic capacity of mixotrophic cultures is due to the recovery process.

To further confirm the difference in recovery responses, recovery experiments were carried out. After exposed to HPFD, the cells were placed to a dim light for recovery. As shown in Figure 4, it is found that the photoinhibited cells in mixotrophic cultures can recover faster. This significant difference may result in the larger photosynthetic capacity of mixotrophic cultures.

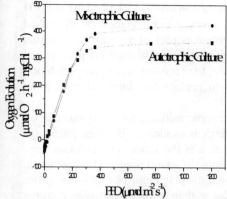

Fig. 1. Light Response Curves of *Spirulina platensis* in Two Cultures at 30°C

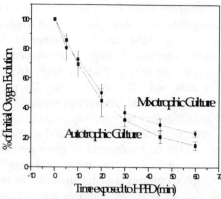

Fig. 2. Photoinhibition Effect on Oxygen Evolution of Two Different Cultures

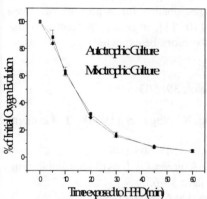

Fig. 3. Photoinhibition Effect in adding 100 mg L⁻¹ Chloramphenicol on Oxygen Evolution of Two Different Cultures

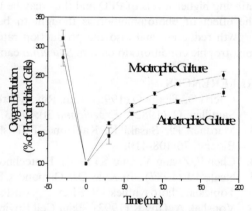

Fig. 4. Recovery of Oxygen Evolution in Photoinhibited Cells of Two Cultures

DISCUSSION

It is well known that the 32kDa apoprotein of the Q_B acceptor (D1 protein) in PS II is susceptible to be damaged by high irradiation during photoinhibition (8, 9). Ohad et al. (9) suggested that the turnover of D1 protein is part of a repair system that functions to replace damaged reaction centres with newly synthesized D1 protein and thus restore the normal PS II activity. In this case, the recovery from photoinhibition can be achieved. So, synthesizing D1 protein to replace the damaged ones is an important step in the recovery process.

Energy is required in protein synthesis. In mixotrophic cultures, more organic carbon substrate, i.e. glucose, is added and so more energy is available. Besides, glucose is known as the best substrate to generate energy as it is the fastest and efficient way. Therefore, it is easy to understand that the rate of synthesizing D1 protein in mixotrophic cultures is faster than that in photoautotrophic cultures. This leads to a result that mixotrophic cultures can recover faster than photoautotrophic cultures after photoinhibition. In adding chloramphenicol during photoinhibition experiment, the damage effects of both cultures were the same. This can further prove that the difference of the two cultures in recovery process is due to the synthesis of protein as chloramphenicol is a protein inhibitor that is used to block the repair process i.e. synthesis of D1 protein.

The faster repair process in photoinhibition can lead to a larger photosynthetic capacity of mixotrophic cultures. Therefore, the mixotrophic cultures are capable of utilizing higher levels of PFD and thus may be less susceptible to photoinhibition. As the effect of photoinhibition is thought to be responsible for a part of observed growth reduction and also the production rate (10, 11), a greater potential to use mixotrophic conditions to culture *Spirulina* can therefore be expected.

REFERENCES

1. Ciferri O, Tiboni O (1995) Ann. Rev. Microbiol. 39: 503-526.
2. Santillan, C (1982). *Experientia* 30:40-43.
3. Marquez FJ, Sasaki K, Kakizono T, Nishio N, Nagai S (1993) J. Ferment. Bioeng. 76:408-410.
4. Chen F, Zhang Y, Guo S (1996) Biotechnol. Lett. 18:603-608.
5. Neale PJ (1987) In Kyle DJ, Osmond CB, Arntzen CJ (eds), Photoinhibition: Topics in Photosynthesis 9. Elsevier, Amsterdam: 39-65.
6. Vonshak A, Guy R (1992) Plant Cell Environ. 15: 613-616.
7. Vonshak A (1987) Hydrobiologia. 151/152:75-77.
8. Kyle DJ, Ohad I, Arntzen CJ (1984) Proc. Natl. Acad. Physiol. 45:633-662.
9. Ohad, Kyle DJ, Arntzen CJ (1984) J. Cell Biol. 99:481-485.
10. Vonshak A, Richmond A (1985) Plant and soil. 89: 129-135.
11. Vonshak A, Guy R (1988) In *Algal biotechnology* (Stadler et al. (eds)), Elsevier Applied Science Publishers, London (1988).

ULTRASTRUCTURAL ASPECTS OF PHOTOAUTOTROPHIC *CHRYSANTHEMUM* CULTURE

Victoria Cristea[1], Francesca Dalla Vecchia[2], Constantin Crăciun[3] and Nicoletta La Rocca[2]
[1]Biological Research Institute, Republicii, 48, 3400 Cluj, Romania
[2]University of Padua, Biology Dept., G. Colombo, 3, 35121 Padua, Italy
[3]"Babeş -Bolyai" University, Clinicilor 5-7, 3400 Cluj, Romania

Key words: tissue culture, chloroplast, CO_2 concentration, electron microscopy, immuno-electronmicroscopy, in vitro photoautotrophy.

1. Introduction

In vitro photoautotrophic cell and tissue cultures represent a method to induce in aseptically developed cells, physiological processes as close as possible to those from nature. Unlike classical *in vitro* cultures, developed on culture media supplemented with an organic carbon source, photoautotrophic aseptic cultures develop on a medium lacking this source, which maintains heterotrophic nutrition, and usually requires a CO_2 supplemented atmosphere.

The effect of varied O_2 concentrations on the development and photosynthetic processes has so far been studied in photoautotrophic and photomixotrophic *Chrysanthemum morifolium* cultures (1, 2). The aim of this paper is to study photoautotrophic *Chrysanthemum morifolium* cultures as compared to classical ones or to those developed on varied sucrose concentrations and from the point of view of ultrastructure and functioning of the photosynthetic apparatus.

2. Material and Methods

Cuttings with 5-6 leaves, obtained from *Chrysanthemum morifolium* plantlets cultivated *in vitro*, on classical medium with 2% sucrose, were used. The culture medium contained α-naphthaleneacetic acid (NAA) 1 mg.l^{-1} and kinetin 0.5 mg.l^{-1}. The cuttings developed on four experimental variants: 1) 2% sucrose medium; 2) 0.3% sucrose medium; 3) and 4) sucrose free medium. The culture vessels of variants 1, 2 and 3 were closed with polyethylene film. The culture vessels of variant 4 were closed with suncaps (Sigma-transparent polypropylene film with 6.0 mm diameter 0.02 μm filter disc), and were supplemented with 2% CO_2, on the basis of the system proposed by Tichá (1996).

G. Garab (ed.), Photosynthesis: Mechanisms and Effects, Vol. V, 4175–4178.

For transmission electron microscopic (TEM) observations, the samples were processed as described by Rascio *et al.* (1994).

3. Result and Discussion

The ultrastructural aspects were followed in correlation with the data concerning the development and the photosynthetic processes of *in vitro* photoautotrophic *Chrysanthemum* plantlets, data detailed in another article (Cristea *et al.*, unpublished). Thus, the photoautotrophic plantlets present the highest values of shoot length, multiplication, leaves number, rhizogenesis, content of photosynthetic pigments, photosynthetic O_2 and respiratory O_2, as well as RUBPC-ase and PEPC-ase activity, compared to the other variants.

In our studies performed on *Chrysanthemum* plantlet leaves cultivated on 2% sucrose medium, as compared to the chloroplasts of a normal leaf, two situations occur. Either the chloroplasts have apparently normal thylakoids and grana, but most of the chloroplast volume is occupied by starch granules (Figure 1A). Or, the cells have less starch but have dilated thylakoids and increased interthylakoid spaces (Figure 1B). In the case of classical *in vitro* cultures, the presence of 2% sucrose in the culture medium, absolutely necessary if there are no gas exchanges with the exterior medium, explains the large quantity of starch from the chloroplasts. Only a few papers have studied the ultrastructural aspects of photoautotrophic plantlets cultivated *in vitro*. Thus, in *Rosa*, the chloroplast ultrastructure of plantlets classically cultivated *in vitro* corresponds to the ultrastructure of *in situ* developed chloroplasts. In other species, e.g. in *Prunus*, the chloroplast stroma is crossed by a network of disorganized thylakoid membranes (5).

The decrease to 0.3% of the sucrose concentration or its absence from the culture medium generates chloroplasts with an obviously affected thylakoid system, which is not organized into grana, has numerous plastoglobules and lacks starch (Figures 2 and 3). The same aspects were found in *Rosa* plantlets under similar conditions (6).

The 2% CO_2 supplementation of the atmosphere of the culture vessels which contain sucrose free medium allows the development of chloroplasts with normal thylakoids and well formed grana (Figures 4A and B). In this case, starch appears as a product of photosynthesis (Figure 4B), which also allows a superior development from all points of view. Today it is recognized that *in vitro* photoautotrophic cultures generally grow better at an increased CO_2 concentration, 0.5% or higher (7). The ultrastructure of chloroplasts of *in vitro* photoautotrophic *Chrysanthemum* plantlets is similar to that of normal leaves, which is also found in *Nicotiana tabacum* cell cultures (8, 9), *Chenopodium rubrum* (10) and in *Glycine max* (11).

Similar results were obtained also, at the *in vitro* photoautotrophic *Daucus carota* callus (12).

Immunocytochemical studies, using primary antibodies specific for PSI and PSII, show differences between photoautotrophic plantlets and those normally cultivated *in vitro*, less in the case of PSI and especially in the case of PSII. These results will be detailed in a future paper.

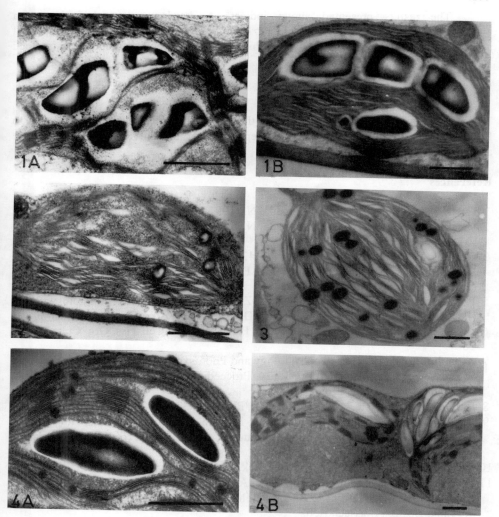

Figures 1-4. Ultrastructure of chloroplast from *Chrysanthemum morifolium* leaf cultivated *in vitro* under different experimental conditions. (1) On 2% sucrose medium – A) numerous starch granules are detected; B) chloroplasts with thylakoids and dilated interthylakoid spaces. (2) On 0.3% sucrose medium. (3) On 0% sucrose medium – note the damaged chloroplasts. (4A and B) On sucrose free medium and with 2% CO_2 – the normal chloroplast ultrastructure, similar to that of the *in situ* cl 'oroplast, supports the idea of its normal activity. Bars = 1 μm.

The results presented above indicate a development exclusively on a photosynthetic basis in the case of *in vitro* photoautotrophic plants, unlike classical aseptic culture. This *in vitro* photoautotrophic culture has been continuously maintained for 20 months.

Acknowledgements

This work was supported by a grant from the University of Padua. We are grateful to Prof. Nicoletta Rascio, from the Biology Department of this University for her valuable advice and help.

References

1 Tanaka, F., Watanabe, Y. and Shimada, N. (1991a) Plant Tiss. Cult. Lett. 8, 87-93
2 Tanaka, F., Watanabe, Y. and Shimada, N. (1991b) Environ. Control in Biology 29, 107-116
3 Tichá, I. (1996) Photosynthetica 32, 475-479
4 Rascio, N., Mariani, P., Dalla Vecchia, F., Zanchin, A., Pool, A. and Larcher, W. (1994) J. Plant Physiol. 144, 314-323
5 Pospíšilová, J., Čatský, J. and Šesták, Z. (1997) in Handbook of Photosynthesis (Pessarakli M., ed.) pp. 525-540, Marcel Dekker, New York - Basel - Hong Kong
6 Capellades, M., Lemeur, R. and Debergh, P. (1991) Plant Cell Tiss. Org. Cult. 25, 21-26
7 Widholm, J. M. (1995) in Ecophysiology and Photosynthetic *I. Vitro* Cultures (Carre F. and Chagvardieff P., eds.) pp. 15-24, Proceedings of the Intern. Symposium, 1-3 Decembre 1993, Aix en Provence, France
8 Brangeon, J. and Nato, A. (1981) Physiol. Plant. 53, 327-334
9 Takeda, S., Ida, K., Sato, F., Yamada, Y., Kaneko, Y. and Matsushima, H. (1992) in Research in Photosynthesis, Vol. I (Murata N., ed.) pp. 223-226, Kluwer Academic Publishers, The Netherlands
10 Hüsemann, W., Herzbech, H. and Robenek, H. (1984) Physiol. Plant. 53, 327-334
11 Horn, M.E. and Widholm, J.M. (1984) in Applications of Genetic Engineering to Crop Improvement (Collins G.B. and Petolino J.G., eds.) pp. 113-161, Nijhoff/Dr. W. Junk, Boston, Massachusetts
12 Cristea, V., Dalla Vecchia, F. and Crăciun, C. (1996) Giornale Botanico Italiano 130, 924-926

Exogenous electrical stimulation of tomato plants: Induction of proteinase inhibitor II gene expression and control of photosynthetic activity

Oliver Herde, Lothar Willmitzer and Joachim Fisahn
Max-Planck-Institut für Molekulare Pflanzenphysiologie, Karl-Liebknecht Strasse 25, D-14476 Golm, Germany

Key words: abiotic stress, fluorescence, gas exchange, membrane potential, signal transduction, xanthophyll cycle

1. Introduction

Electrical and hydraulic signals have been identified as physical signals that mediate the systemic expression of wound inducible genes (1,2). The appearance of variation potentials and fast transient voltage changes throughout most of the shoot has been reported following localized wounding by heating or burning (3). Furthermore, Wildon et al. (1) reported that mechanical wounding and localized burning generate electrical signals that are propagated through the plant thereby inducing Pin2 gene expression systemically.

Recently, we have demonstrated that injecting electrical current into leaves of tomato plants activated Pin2 gene expression in a local and systemic manner (4,5). This experiment corroborates the hypothesis that ionic movements are sufficient to induce Pin2 gene expression.

Furthermore, it could be shown that exogenous electrical stimulation is capable of regulating several photosynthetic parameters like assimilation- and transpiration rate, chlorophyll fluorescence and pigment composition (4,6).

2. Procedure

2.1 Current application
For current application a DC power supply was used and a voltage of 10 V was provided for 30 s during electric stimulation, see also Herde et al. (4).

2.2 Gas exchange measurements
Gas-exchange measurements were performed in an open system. All components of the measuring system were located inside a phytotron. Gas-exchange parameters (assimilation rate / transpiration rate) were measured by a compact-minicuvette system (Walz, Effeltrich, Germany). This commercially available gas exchange system comprises a binos100 infrared gas analyzer (Rosemount) and a climatized minicuvette

G. Garab (ed.), Photosynthesis: Mechanisms and Effects, Vol. V, 4179–4182.

(Walz, Effeltrich, Germany).
After the leaf was mounted in the measuring cuvette and the climate inside the cuvette was adjusted to ambient conditions the presence of a steady-state assimilation and transpiration was controlled for 30 min, see Herde et al. (4).

2.3 Chlorophyll fluorescence measurements

Chlorophyll fluorescence parameters q_P and q_N were measured at room temperature using a PAM 2000 chlorophyll fluorometer (Walz, Effeltrich, Germany). The q_P and q_N quenching of chlorophyll fluorescence represented values that were obtained after a steady state photosynthetic rate was reached (7). For all measurements, leaves remained attached to the whole plant and were mounted in a specially designed chamber. For details see Fisahn et al. (7).

2.4 Pigment analysis

Pigments were analyzed by high performance liquid chromatography (System Gold, programmable solvent module 126, programmable detector module 166, Beckmann Instruments, San Remon, CA, USA) by using a custom-made non-endcapped Zorbax ODS-column (Analysentechnik, Mainz, Germany), see also (7).

3. Results and Discussion

The effect of electric current on Pin2 gene expression and on the endogenous contents of ABA and JA in both wild type and ABA-deficient potato and tomato plants has been well described (4,5). Examination of Pin2 gene expression by Northern blot analysis revealed that application of electric current initiates the local and systemic accumulation of Pin2 mRNA, ABA and JA in wild type plants but not in ABA-deficient plants. These results indicate that electric current-induced Pin2 gene expression have some common steps which may require the presence of normal endogenous levels of ABA to trigger the later steps involved in gene activation. Furthermore, application of electric current lead to an accumulation of JA only in the wild type plants but not in ABA-deficient plants. These data strongly suggest that electrically-induced JA accumulation may require certain levels of ABA which exist in the wild type plants but not in the ABA-deficient plants, supporting the hypothesis that the site of action of JA is located after the site of action of ABA in the signal chain mediating the activation of Pin2 gene expression (5).

In a further series of experiments, participation of of both the action potential and the variation potential in Pin2 gene expression has been documented (6,8). In addition, decrease in turgor pressure could also contribute to the induction of Pin2 gene expression (6). The action potential is sufficient to slightly induce Pin2 gene expression, the variation potential, or the decrease in turgor pressure, seemed to induce a significant amplification of Pin2 gene expression (6). The correlation between wounding and loss of turgor has been already indicated indirectly by previous work, in which several stimuli led to a significant decrease in transpiration (4). In this work, mechanical injury induced a complex relaxation kinetics which was characterized by two time constants in the assimilation and transpiration rates, matching the time courses of the changes in the membrane potential. Modification of stomatal aperture detected by adaptations in the transpiration and assimilation rates is coupled to the flow of ions across the guard cell plasma membrane (9,10). Therefore, it could be suggested that this mechanism of turgor regulation is not only operative in stomatal guard cells but also in other cells of the plant. Moreover, turgor regulation and generation of action potentials seem to be closely associated with each other. In the ABA-deficient mutants of tomato, no changes in the membrane potential upon current application occurred. Therefore, in the absence of fast transient voltage changes, no adaptation in turgor pressure could be observed. In contrast, heat treatment induced the generation of a fast electrical transient although the endogenous ABA content was reduced. Consequently, turgor pressure decreased significantly within 3-4 min after stimulus application (6). In conclusion, Pin2 gene expression requires changes in the membrane potential or turgor pressure, respectively.

Tightly linked to the down-regulation of photosynthesis upon current application, measured by gas exchange, is the observation that the non-photochemical quenching starts to increase within 60 seconds after current application, reaching its maximum after approx. 30 minutes and maintained this level until 300 minutes after treatment (11). Similarly, the photchemical quenching started to decrease within 60 s after wounding, reaching its minimum after approx. 30 min and also maintained this level until 300 min after treatment (11). Because these reductions in photochemical quenching are correlated with decreases in the quantum efficiency of light-driven electron transport through PSII it has been proposed that these changes in photochemical quenching are related to the maintenance of a stoichiometric balance between reactions that consume and those that produce NADPH during steady state photosynthesis. A reduction of assimilation could restrict the rate of CO_2 fixation and consequently the rate of electron transport through PSII. Under these conditions the quantum efficiency of photosynthetic electron transport will decline and the degree of nonphotochemical quenching will rise to maintain stoichiometric balance.

Also important in this context, exogenous electrical stimulation is able to induce changes in the pigment composition of tomato plants. Five hours after electrical stimulation, total chlorophyll content remained almost unchanged, zeaxanthin content increased 5-6 fold, antheraxanthin content increased also, but to a minor extend, violaxanthin and neoxanthin on the other hand decrease significantly (11). Surprisingly, the ratio of the size of violaxanthin cycle pool compared to total chlorophyll content is not altered 5 h after electrical stimulation. As indicated by the pronounced increases of zeaxanthin and antheraxanthin content and the decrease of violaxanthin upon wounding, the deepoxidation state (DEPS) within the plant increases significantly upon current application. These changes in pigment composition after current application are very similar to those observed upon light- and heat stress. Usually, violaxanthin levels are rather high in moderately illuminated leaves, whereas levels of antheraxanthin and zeaxanthin are often negligible (12). On exposure of these leaves to high light, violaxanthin undergoes a two-step enzymatic deepoxidation to zeaxanthin via an intermediate, antheraxanthin. Therefore zeaxanthin accumulates at the direct expense of violaxanthin, although levels of antheraxanthin do not increase during this process (12). Therefore, light- or heat stress induce modifications on the level of pigment composition and chlorophyll fluorescence that are similar to those of current application.

Taken together, exogenous electrical stimulation, representing ionic movements in the apoplast, are sufficient to generate electrical signals which propagate throughout the whole plant. As a consequence, these signals are able to induce proteinase inhibitor II gene expression. Moreover, electrical stimulation produce characteristic changes in the gas exchange relaxation kinetics consisting of a fast and a slow time constant, similar to the action- and variation potential seen on the membrane potential level. Complementary, other photosynthetic parameters like chlorophyll fluorescence and pigment composition undergo specific alterations. In particular, non-photochemical quenching increases, deepoxydation of violaxanthin to zeaxanthin is strongly elevated and ABA-content of the leaves is raised.

References

1 Wildon, D. C., Thain, J. F., Minchin, P. E. H., Gubb, I. R., Reilly, A. J., Skipper, Y. D., Doherty, H. M., O´Donnell´, P. J. & Bowles, D. J. N. (1992) Nature 360, 62-65

2 Malone, M., Palumbo, L., Boari, F., Monteleone, M. & Jones, H. G. (1994) Plant Cell Environ. 17, 81-87

3 Pickard, B. G. (1973) Bot. Rev. 39, 172-201

4 Herde, O., Fuss, H., Peña-Cortés, H. & Fisahn, J. (1995) Plant Cell Physiol. 36, 737-742

5 Herde, O., Atzorn, R., Fisahn, J., Wasternack, C., Willmitzer, L. & Peña-Cortés, H. (1996) Plant Physiol. 112, 853-860

6 Herde, O., Peña-Cortés, H., Willmitzer, L. and Fisahn, J. (1998) Bot. Acta, in press

7 Fisahn, J., Kossmann, J., Matzke, G., Fuss, H., Bilger, W. & Willmitzer, L. (1995) Physiol. Plant. 95, 1-10

8 Herde, O., Peña-Cortés, H., Willmitzer, L. and Fisahn, J. (1998) Planta 206, 146-153

9 Raschke, K. & Hedrich, R. (1985) Planta 163, 105-118

10 Fromm, J. & Eschrich, W. (1993) J. Plant Physiol. 141, 673-680

11 Herde, O., Peña-Cortés, H., Fuss, H., Willmitzer, L. and Fisahn, J. (1998) Physiol. Plant., in press

12 Foyer, C.H., Dujardyn, M. & Lemoine, Y. (1989) Plant Physiol. Biochem. 27, 751-760

23. Artificial systems

CONGRESS ON PHOTOSYNTHESIS

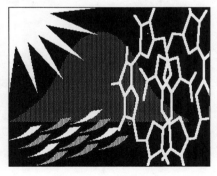

XIth INTERNATIONAL · BUDAPEST ✳ 1998

A MODEL FOR THE PRIMARY PHOTOCHEMISTRY OF PHOTOSYSTEM II

Dirk Bumann and Dieter Oesterhelt
Department of Membrane Biochemistry, Max Planck Institute for Biochemistry,
82152 Martinsried, Germany

Key words: Chl, electron transfer, modelling, O_2 evolution, P680, pheo,

Introduction

Photosystem II (PSII) catalyzes the light-induced oxidation of water. To this end, a chlorophyll cation with the highest redox potential of all biological systems is generated. A detailed knowledge of this process is crucial for an understanding of photosynthesis and might assist in the improvement of artificial photoconversion in solar cells.

Existing evidence: Three-dimensional structure

The minimal complex capable of the primary photochemistry is the photosystem II reaction center (PSII-RC) [1]. Two of its subunits (polypeptides D1 and D2) are homologous to the subunits L and M of the photosynthetic reaction center of purple bacteria (bRC) [2,3]. The pigment content and the general photochemistry of both complexes are very similar [4]. The three-dimensional structure of PSII-RC has not yet been solved but the structure of two bRCs is known to atomic resolution [5]. Both bRC structures are very similar despite considerable differences in the sequences which indicates a highly conserved three-dimensional structure of this complex.

Based on the close similarity of PSII-RC and bRC, structural models have been developed for PSII-RC [6-9]. Several critical predictions of these structural models have been verified with site-directed mutants [10-13] and spectroscopic methods [14-21] providing broad support for a close structural similarity between bRC and PSII.

G. Garab (ed.), Photosynthesis: Mechanisms and Effects, Vol. V, 4185–4191.
© 1998 *Kluwer Academic Publishers. Printed in the Netherlands.*

Existing evidence: Primary photochemistry

In bRC two bacteriochlorophylls (bchls) form a excitonically coupled dimer, the so-called special pair, which acts as the primary electron donor. After excitation the special pair reduces a bacteriopheophytin via a monomeric bchl [22]. The same general photochemistry occurs in PSII-RC [4]. An excited chlorophyll (chl) species acts as the primary donor and reduces a pheophytin. It seems reasonable to assume that homologous pigments have identical function, i.e. a chl dimer is the primary donor and a monomeric chl assists in electron transfer to the pheophytin.

However, despite a large number of spectroscopic studies the chl(s) that constitute the primary donor have not been identified. The interpretation of the experimental data is very difficult due to the presence of various pigments with broadly overlapping absorptions and the distortion and heterogeneity of the samples [23-27]. As a result, the kinetics of charge transfer [28-30], the molecular structure of the primary donor (monomer or weakly coupled dimer) [23-25, 31-37], the minimal pigment content of PSII-RC [26, 27, 38], and the assignment of certain regions in the Qy absorption band to specific pigments [23, 24, 27, 31, 37, 39-45] all remain highly controversial.

Accordingly, it is not possible to rule out that the photochemistry of PSII either is identical to that of bRC or rather different [24, 37, 44]. Among the possibilities is a monomeric primary donor that is homologous to the monomeric bchl of bRC [32, 46]. The oxidized donor could accept an electron from the redox-active tyrosine Z [13, 47] either directly or *via* chls homologous to the special pair of bRC.

Regardless of the favored model there is a general problem in understanding the primary photochemistry of PSII. To drive water oxidation, the redox potential of the primary donor has to be much higher (1.1 V [48]) than in bRC [49] that does not oxidize water. Due to its very high redox potential, the primary donor of PSII can oxidize other chls [50, 51]. According to the structural models, each of the 4 chls that are homologous to the bchls of bRC has at least one close neighbor. Hence, regardless of the actual assignment of the primary donor, the oxidation of other chls could be a rapid side reaction competing with the oxidation of water [41]. However, the quantum efficiency indicates that almost no side reactions occur in the photochemistry of normal PSII. How this is achieved, remains unclear.

Existing evidence: recent results from photoinhibition

Recent photoinhibition studies [52] showed that under certain conditions two chls of PSII are destroyed. Both chls are apparently essential for the electron transfer from Z to the primary donor, but they are distinct from the primary donor that is not affected. Both chls have Q_y transitions almost parallel to the membrane plane and are probably located close to the primary donor.

Model building

The orientation of the destroyable chls resembles that of the bchls constituting the special pair of bRC. The signal/noise ratio obtained [52] does not allow to decide whether the two chls are excitonically coupled but the overall bandwidth of the Q_y-transition rules out any strong coupling in PSII-RC [43]. If the destroyable chls are tentatively assigned to chls homologous to the special pair of bRC, the primary donor would be a chl homologous to one of the monomeric bchls (Fig. 1) as has been already proposed based on results obtained with electron paramagnetic spectroscopy [32]. Due to their location, the destroyable chls could critically influence the electron transfer from Z to the primary donor so that their destruction would block this electron transfer. While other possibilities exist only this model allows such a simple explanation of the photoinhibition data.

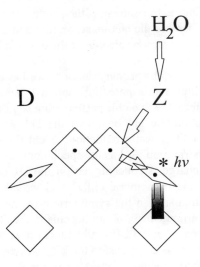

Fig. 1: Proposed model for the primary photochemistry of photosystem II of oxygenic photosynthesis. Squares represent pheophytins, dotted squares represent chlorophylls, the star depicts the monmeric primary donor, the shaded arrow indicates the primary charge seperation, the empty arrows indicate subsequent electron transfer reactions.

4188

A monomeric primary donor cation could oxidize the tyrosine Z either directly or *via* the chls that would function as intermediate donors [46]. It was inferred from recombination experiments that a state chl$^+$pheo$^-$ in PSII-RC has a separation distance similar to that of (special pair)$^+$pheo$^-$ in bRC [53] indicating that at some point the chls homologous to the special pair are oxidized. Moreover, the highly controversial spectroscopic evidence for PSII could be partly explained by the presence of two different chl cations that are involved in the primary photochemistry. Hence, experimental evidence supports the proposed role of the destroyable chls as intermediary electron donors (as it is shown in Fig. 1).

Chl cations involved in the primary photochemistry of PSII have a lifetime in the ns time range [47]. Such a comparably long lifetime could cause a high risk of charge recombination, especially if electron and hole are close to each other during all that time [24]. Therefore it is important which chl cation has this long lifetime. If the primary donor is the only chl cation and is directly reduced by Z, the charge separation would be very small for that whole lifetime. In contrast, if the primary donor quickly oxidizes the destroyable chls this would increase the charge separation and thereby lower the risk of recombination. Hence, functional reasons also favor the destroyable chls as intermediary donors (as it is shown in Fig. 1).

To drive the oxidation of water, the primary donor as well as the two destroyable chls would have to have rather high redox potentials above 1 V (compared to 0.83 V of chl$^+$/chl *in vitro* [54]). Therefore, the two chls participating as intermediary donors could oxidize the neighboring monomeric chl bound to the D2 polypeptide (chl$_{D2}$) which would lower the quantum yield of water oxidation. A chl dimer cation in PSII has a highly asymmetric charge distribution [36] as the special pair of many bacterial systems [55]. Possibly, the positive charge is concentrated on the D1 half of the dimer which would prevent an attack of the monomeric chlD2 and would also explain the faster oxidation of tyrosine Z as compared to the symmetric tyrosine D [13, 20]. The model would thus allow for a close arrangement of various chls in the PSII-RC without the risk of unproductive side reactions. Hence PSII could function as a water-splitting device without major structural changes as compared to the bRC structure.

In conclusion, the described model can most easily explain the existing evidence. Interestingly, recent evidence suggests the sequential occurrence of two different radical pairs [56] as it is predicted by the model described here. A definite elucidation of the primary photochemistry, however, requires to resolve the contradicting experimental results and to obtain more spectroscopic data about the participating species.

Acknowledgment

D.B. was supported by Fonds der Chemischen Industrie and Stiftung Volkswagenwerk in terms of a Kekulé scholarship. We thank P.Mathis, G.Renger, A.W. Rutherford, and A.Trebst for critically reading the manuscript and helpful comments.

References

1   Nanba, O. and Satoh, K. (1987) Proc. Natl. Acad. Sci. USA 84, 109-112

2   Michel, H. and Deisenhofer, J. (1986) in Photosynthsis III (Staehelin, L.A. and Arntzen, C.J. eds.), pp. 371-381, Springer, Berlin

3   Trebst, A. (1986) Z. Naturforsch. 41c, 240-245

4   Mathis, P. and Rutherford, A.W. (1987) in Photosynthesis (Amesz, J. ed.), New Comprehensive Biochemistry vol. 15, pp. 63-96

5   Ermler, U., Fritzsch, G., Buchanan, S.K. and Michel, H. (1994) Structure 2, 925-936

6   Michel, H. and Deisenhofer, J. (1988) Biochemistry 27, 1-7

7   Svensson, B., Vass, I., Cedergren, E. and Styring, S. (1990) EMBO J. 9, 2051- 2059

8   Ruffle, S.V., Donelly, D., Blundell, T.L. and Nugent, J.H.A. (1992) Photosynth. Res. 34, 287-300

9   Nugent, J.H., Ruffle, S.V. and Berry, M.C. (1993) Biochem. Soc. Trans. 21, 22-25

10 Ohad, N. and Hirschberg, J. (1992) Plant Cell 4, 273-282

11 Tommos C., Davidsson, L., Svensson, B., Madsen, C., Vermaas, W., Styring, S. (1993) Biochemistry 32, 5436-5441

12 Tang, X.-S., Chisholm, D.A., Dismukes, G.C., Brudvig, G.W. and Diner, B.A. (1993) Biochemistry 32, 13742-13748

13 Barry, B.A. (1995) Meth. Enzym. 258, 303-319

14 Ganago, I.B., Klimov, V.V., Ganago, A.O., Shuvalov, V.A. and Erokhin, Y.E. (1982) FEBS Lett. 140, 127-130

15 Breton, J. (1985) Biochim. Biophys. Acta 810, 235-245

16 Breton, J. (1990) in: Photosynthesis. Energy Conversion by Plants and Bacteria (Govindjee, ed), pp. 153-194, Academic Press, New York

17 Hirsh, D.J. Beck, W.F., Innes, J.B. and Brudvig, G.W. (1992) Biochemistry 31, 532-541

18 Kodera, Y., Takura, K. and Kawamori, A. (1992) Biochim. Biophys. Acta 1101, 23-32

19 Gilchrist, M.L., Ball, J.A., Randall, D.W. and Britt, R.D. (1995) Proc. Natl. Acad. Sci.USA 92, 9545-9549

20 Koulougliotis, D., Tang, X.-S., Diner, B.A. and Brudvig, G.W. (1995) Biochem. 34, 2850-2856

21 Mimuro, M., Tomo, T., Nishimura, Y., Yamazaki, I. and Satoh, K. (1995) Biochim. Biophys. Acta 1232, 81-88

22 Arlt T., Schmidt, S., Kaiser, W., Lauterwasser, C., Meyer, M., Scheer, H. and Zinth, W. (1993) Proc. Natl. Acad. Scie. USA 90, 11757-11761

23 Otte, S.C.M., van der Vos, R. and van Gorkom, H.J. (1992) J. Photochem. Photobiol. B 15, 5-14

24 Van der Vos, R., van Leeuwen, P.J., Braun, P. and Hoff, A.J. (1992) Biochim. Biophys. Acta 1140, 184-198

25 Seibert, M (1993) in: The Photosynthetic Reaction Center I (Deisenhofer, J. and

Norris, J.R. eds), pp. 319-356, Academic Press, New York

26 Pueyo, J.J., Moliner, E., Seibert, M. and Picorel, R. (1995) Biochemistry 34, 15214-15218

27 Kronermann, L. and Holzwarth, A.R. (1996) Biochemistry 35, 829-842

28 Durrant, J.R., Hastings, G., Joseph, D.M., Barber, J., Porter, G. and Klug, D.R. (1993) Biochemistry 32, 8259-8267

29 Wiederrecht, G.P., Seibert, M., Govindjee, Wasielewski, M.R. (1994) Proc. Natl. Acad. Sci. USA 91, 8995-9003

30 Giorgi, L.B., Nixon, B.J., Merry, S.A.P., Joseph, D.M., Durrant, J.R., de las Rivas, J., Barber, J., Porter, G. and Klug, D.R. (1996) J. Biol. Chem. 271, 2093-2101

31 Van Kan, P.J.M., Otte, S.C.M., Kleinherenbrink, F.A.M., Nieveen, M.C., Aartsma, T.J. and Van Gorkom, H. (1990) Biochim. Biophys. Acta 1020, 146-152

32 Van Mieghem, F.J.E., Satoh, K. and Rutherford, A.W. (1991) Biochim. Biophys. Acta 1058, 379-385

33 Noguchi, T., Inoue, Y. and Satoh, K. (1993) Biochemistry 32, 7186-7195

34 Allakhverdiev, S.I., Ahmed, A., Tajmir-Riahi, H.-A., Klimov, V.V. and Carpentier, R. (1994) FEBS Lett. 339, 151-154

35 Carbonera, D., Giacometti, G. and Agostini, G. (1994) FEBS Lett. 343, 200-204

36 Rigby, S.E.J., Nugent, J.H.A. and O'Malley, P.J. (1994) Biochemistry 33, 10043-10050

37 Schelvis, J.P.M., van Noort, P.I., Aartsma, T.J. and van Gorkom, H.J. (1994) Biochim. Biophys. Acta 1184, 242-250

38 Eijckelhoff C. and Dekker, J.P. (1995) Biochim. Biophys. Acta 1231, 21-28

39 Braun, P., Greeberg, B.M. and Scherz, A. (1990) Biochemistry 29, 10376-10387

40 Kwa, S.L.S., Newell, W.R., van Grondelle, R. and Dekker, J.P. (1992) Biochim. Biophys. Acta 1099, 193-202

41 Van Gorkom, H.J. and Schelvis, J.P.M. (1993) Photosynth. Res. 38, 297-301

42 Garlaschi, F.M., Zucchelli, G., Giavazzi, P. and Jennings, R.C. (1994) Photosynth. Res. 41, 465-473

43 Cattaneo, R., Zucchelli, G., Garlaschi, F.M., Finzi, L. and Jennings, R.C. (1995) Biochemistry 34, 15267-15275

44 Durrant, J.R., Klug, D.R., Kwa, S.L.S., van Grondelle, R., Porter, G., Dekker, J.P. (1995) Proc. Natl. Acad. Sci. USA 92, 4798-4802

45 Hillmann, B., Brettel, K., van Mieghem, F., Kamlowski, A. , Rutherford, A.W. and Schlodder, E. (1995) Biochemistry 34, 4814-4827

46 Van Mieghem, F.J.E. and Rutherford, A.W. (1994) Biochem. Soc. Trans. 21, 986-991

47 Gerken, S., Brettel, K., Schlodder, E. and Witt, H.T. (1988) FEBS Lett. 237, 69-75

48 Renger, G. (1993) Photosynth. Res. 38, 229-247

49 Lin, X., Murchison, H.A., Nagarajan, V., Parson, W.W., Allen, J.P. and Williams, J.C. (1994) Proc. Natl. Acad. Sci. USA 91, 10265-10269

50 Visser, J.M.W., Rijgersberg, C.P. and Gast, P. (1977) Biochim. Biophys. Acta 460,

36-46

51 Thompson, L.K. and Brudvig, G.W. (1988) Biochemistry 27, 6653-6658

52 Bumann, D. and Oesterhelt, D. (1995) Proc. Natl. Acad. Sci. USA 92, 12195-12199

53 Volk, M., Gilbert, M., Rousseau, G., Richter, M., Ogrodnik, A. and Michel- Beyerle, M.-E. (1993) FEBS Lett. 336, 357-362

54 Watanabe, T. and Kobayashi, M. (1991) in: Chlorophylls (Scheer, H. ed), pp. 287-316, CRC Press, Boca Raton

55 Rautter, J., Lendzian, F., Lubitz, W., Wang, S. and Allen, J.P. (1994) Biochemistry 33, 12077-12084

56 Müller, M.G., Hucke, M., Reus, M. and Holzwarth, A.R. (1996) J. Phys. Chem. 100, 9527-9536

USING SYNTHETIC MODEL SYSTEMS TO UNDERSTAND CHARGE SEPARATION AND SPIN DYNAMICS IN PHOTOSYNTHETIC REACTION CENTERS

Michael R. Wasielewski[a,b], Gary P. Wiederrecht[a], Walter A. Svec[a], Mark P. Niemczyk[a], Kobi Hasharoni[c], Tamar Galili[c], Ayelet Regev[c], and Haim Levanon[c]
[a]Chemistry Division, Argonne National Laboratory, Argonne, IL 60439 USA
[b]Department of Chemistry, Northwestern University, Evanston, IL 60208 USA
[c]Department of Physical Chemistry, The Hebrew University of Jerusalem, Jerusalem 91904, Israel

Key words: EPR, time-resolved spectroscopy, electron transfer

1. Introduction

Over the past 20 years many biomimetic systems have been synthesized to model the photosynthetic reaction center (RC).(1) The purpose of these modeling studies is to use relatively simple electron donor-acceptor molecules with restricted donor-acceptor distances and orientations to probe the electronic interactions between photosynthetic pigments that lead to efficient, long-lived charge separation in photosynthetic reaction centers. These assemblies are designed to mimic several key properties of the RC protein: 1) multi-step ET to increase the lifetime of the radical pair (RP) product, 2) high quantum yield, fast formation and slow recombination of the singlet-initiated charge separation, 3) temperature independent ET rates, and 4) spin polarization of the RP states. Most RC models fulfill only a subset of these criteria. Moreover, there is one key property of the RC primary photochemistry that has proven very difficult to mimic. It is the unique ability of the RP intermediate within the photosynthetic RC to yield, upon charge recombination, a triplet state that retains a memory of the precursor RP spin state, and exhibits the time-resolved EPR (TREPR) signal characteristic of this triplet state.

It is well recognized that singlet photochemistry governs charge separation in natural photosynthesis, while in model systems, both singlet and triplet channels are competitive and depend on several variables such as relative orientations, distances, and electronic couplings between the participating redox partners.(2) In addition, the solvent reorganization energy (λ_s) due to solvent dipoles reorienting around the ion pair is of prime importance for tuning the energy levels of the charge-separated states, such that the branching ratio of the electron transfer rates, i.e., singlet- or triplet-initiated routes, can be controlled. Thus, to achieve the goal of mimicking natural photosynthesis by biomimetic supramolecular systems, it is essential to develop experimental methods to explicitly determine the energy levels of radical ion pair (RP) states. We show that this goal can be achieved by blending together molecular architecture, solvent properties, and fast EPR detection of paramagnetic transients.

G. Garab (ed.), Photosynthesis: Mechanisms and Effects, Vol. V, 4193–4200.
© *1998 Kluwer Academic Publishers. Printed in the Netherlands.*

Recent studies of intramolecular electron transfer (IET) illustrate the unique properties of liquid crystals (LCs) that make them interesting solvents in which to study covalently linked donor-spacer-acceptor systems.(3-6) Most importantly, the IET rates in these solvents are reduced by several orders of magnitude (from ps to ns), permitting the observation of IET processes on sub-microsecond time scales using time-resolved electron paramagnetic resonance (TREPR) spectroscopy.(3) This reduction of IET rates is due to the nematic potential associated with the alignment of the LC molecules, which restricts the isotropic molecular reorientation found in conventional solvents.(3,7,8) These inherent properties of LCs permit tuning the RP states into a time domain where the paramagnetic transients can be monitored over a wide temperature range.(3,9)

The powerful approach of coupling TREPR with LC solvents has permitted the elucidation of photochemical mechanisms that otherwise could not be observed. In particular, this technique recently permitted the first observation of the special spin-polarized triplet excited state in a multi-component model system generated by the same mechanism of back electron transfer observed in natural photosynthesis. (10,11) It is noteworthy that this model system consists of non-porphyrinoid entities and further studies of multistep ET processes within porphyrinoid systems dissolved in LCs would be valuable.

2. Mimicry of the Radical Pair and Triplet States in Photosynthetic Reaction Centers with a Synthetic Model

An attractive feature of nematic liquid crystals for the study of anisotropic charge transfer is the "nematic potential" associated with the reduced reorientational capability of the liquid crystalline solvent molecules in the presence of an ion pair, as compared to isotropic solvents.(9) This effect is associated with a high activation energy for reorientational motion in an aligned solvent and dramatically slows the rate of photoinduced charge recombination. Since intramolecular charge transfer in isotropic solvents frequently occurs on picosecond time scales, the nematic potential permits a wider variety of experimental techniques to be utilized since charge transfer lifetimes can be increased by 3-4 orders of magnitude.(9) In addition to the nematic potential, a second critical role that liquid crystals can play in spectroscopic applications is the control of LC director alignment as a function of the sign of the magnetic susceptibility of the LC molecules. LCs with positive susceptibilities align parallel to the magnetic field, while those with negative susceptibilities align perpendicular to the magnetic field. This has a large effect in the polarization dynamics observed in EPR and can assist in the assignment of the observed photoinduced state.

In the example given here, supramolecular systems were synthesized that model the photosynthetic reaction center (RC), as illustrated in Figure 1. They were designed to mimic several key properties of the RC protein: (1) multistep electron transfer (ET) to increase the lifetime of the radical pair (RP) product; (2) high quantum yield, fast formation, and slow recombination of the singlet-initiated charge separation; (3) temperature independent ET rates; and (4) spin polarization of the RP states. Until this work, most RC models fulfilled only a subset of these criteria with very few reports employing time-resolved electron paramagnetic resonance spectroscopy (TREPR). We describe here TREPR results on a photosynthetic model system, **1**, in a nematic LC that does not contain the natural

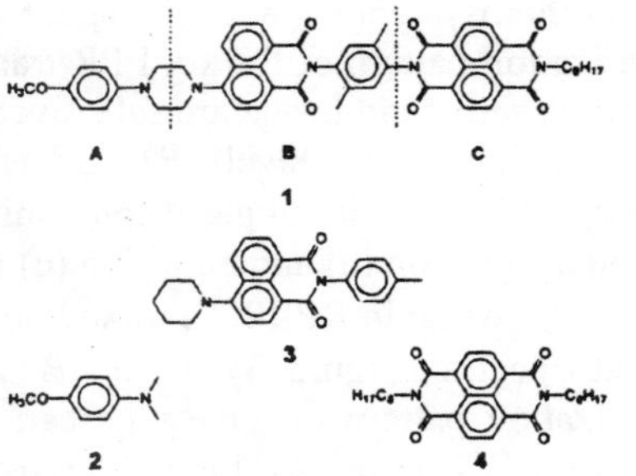

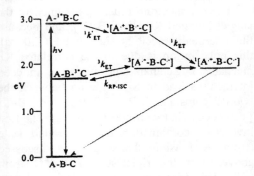

Figure 1. Supermolecule **1** and the control molecules are illustrated.

pigments, yet closely mimics the spin dynamics of triplet state formation found only in photosynthetic RCs. The development of donor-acceptor systems that mimic photosynthetic mechanisms, and that are at the same time electronically and structurally simpler than the natural photosynthetic pigments, may aid our understanding of the mechanistic details of the radical pair dynamics. The design of supermolecule **1** followed criteria established for promoting high quantum yield charge separation in glassy media.

Figure 2. Energy level diagram of all relevant IET and ISC routes. The energies were determined in toluene.

Transient optical absorption spectroscopy carried out on **1** in toluene determined the nature of the intermediates and the rate constants for intramolecular electron transfer (IET) between the electronic states given in the energy level diagram in Figure 2. Laser excitation at 295K and 420 nm selectively excites chromophore B. The lowest excited singlet state of B accepts an electron from A with $\tau = 8$ ps. A subsequent dark ET step with $\tau = 430$ ps forms the final RP, $^1[A^{\cdot+}\text{-}B\text{-}C^{\cdot-}]$, with a lifetime of 310 ns. Photoexcitation of **1** oriented in solid LC results in broad EPR spectra, at the two indicated orientations, with additional narrow lines superimposed at the center of the spectra (Figure 3). The line shapes of the broad spectra clearly suggest that they are due to a triplet state.

In all model systems reported thus far, triplet states observed by EPR are formed via a spin-orbit intersystem crossing (SO-ISC) mechanism. Another possible mechanism is radical pair intersystem crossing (RP-ISC), which forms a triplet upon recombination within the RP, and which has been observed by TREPR only in bacterial and in green plant photosystems I and II. These two mechanisms can be differentiated by the polarization pattern of the six EPR transitions at the canonical orientations. In SO-ISC, the three zero-field levels are selectively populated and this selectivity is carried over to the high field energy levels. RP-ISC is also selective, but acts directly on the high-field triplet sublevels via singlet-triplet mixing S-T$^\circ$ (or S-T$^{\pm 1}$). Thus, SO-ISC results in mixed absorption (a) and emission (e) lines within a particular EPR transition, i.e., T^i—T$^\circ$ (i=±1), while in RP-ISC a mixed polarization pattern is impossible. Inspection of the triplet spectra (Figure 3) in the B $\perp$ L and B $\parallel$ L orientations shows that the polarization pattern of a,e,e,a,a,e can only be attributed to a RP-ISC mechanism, as found for RC proteins. This unique triplet state is found to be localized on C, and the triplet EPR spectrum of **4** exhibits identical zero-field splitting parameters, with polarization pattern of a,a,a,e,e,e typical of SO-ISC.

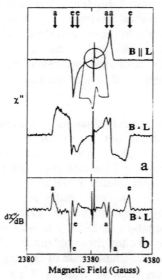

Figure 3. Direct detection CW-TREPR spectra of the triplet state A-B-^{3}*C in a LC, at two orientations of the LC director, L, taken 700 ns after the laser pulse (420 nm) at 150K (the same spectra were recorded at lower temperatures down to 100K). The narrow spectrum is an expansion of the RP signal, observed in the center of the triplet spectra at both orientations. At this temperature, the narrow spectrum is due to the primary ET route (Figure 2). (b) Numerical differentiation of the B$\perp$L spectrum.

Upon warming of the sample from 100 to 330 K, two observations are made. First, the triplet spectra are detectable only when the superimposed narrow line spectrum exhibits a time independent e,a pattern. This phenomenon occurs in the temperature range 100-240 K. Above 240 K the time-evolved narrow spectra exhibit phase inversion (Figure 4). The e,a spectrum must be assigned to the singlet-initiated RP, ^{1}[A$^{\cdot +}$-B-C$^{\cdot -}$], and because the intermediate RP is short-lived (430 ps), this polarization is due to the correlated RP mechanism. Zeeman and hyperfine induced interactions lead to S-T$^\circ$ mixing between the two RP states, i.e., ^{1}RP—^{3}RP; thus, ^{1}[A$^{\cdot +}$-B-C$^{\cdot -}$] can decay via two routes (Figure 2). The first involves recombination to the ground state (from ^{1}RP), while the second, as discussed above, forms the triplet state, A-B-^{3}*C (from ^{3}RP). This triplet state decays via a solvent-controlled secondary ET to repopulate ^{3}RP, or directly to the ground state. The existence of the secondary ET reaction, which is triplet initiated (^{3}k$_{ET}$), is evident from the phase change of the spectra in Figure 4. The singlet-initiated e,a pattern is observed at short times after the laser pulse evolves at later times into a triplet-initiated a,e pattern. The temperature dependence of the secondary ET is consistent with the phase transition temperatures of the LC:

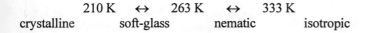

210 K ↔ 263 K ↔ 333 K

crystalline   soft-glass   nematic   isotropic

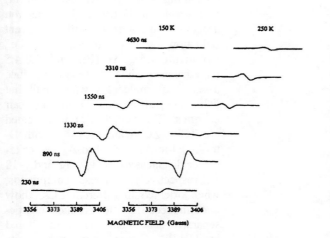

MAGNETIC FIELD (Gauss)

Figure 4. TREPR spectra of the spin-polarized RP, $A^{.+}\text{-}B\text{-}C^{.-}$, at two temperatures in the LC. The spectra are presented as a function of the time following the laser pulse (420 nm). Notice the signal phase change at 250 K. The g-factors of the (e,a) and (a,e) spectra are the same.

In the soft-glass regime, the solvent molecular motion is not frozen and is sufficient to allow ET to occur. Thus while the forward ET reaction seems to be independent of the LC phase (solid and fluid), the secondary ET becomes active only in the soft-glass and in the fluid nematic phases. The disappearance of the triplet spectrum, which occurs with the RP signal phase inversion, is probably due to a high quantum yield of the secondary (back)reaction, making $A\text{-}B\text{-}^3{}^*C$ short-lived.

The observation of this triplet state in **1** by TREPR demonstrates that most of the electronic states found in the primary photochemistry of photosynthetic RCs can be mimicked successfully in synthetic models interacting with LCs.

3. Determination of Energy Levels of Radical Pair States in Photosynthetic Models Oriented in Liquid Crystals through Time-Resolved Electron Paramagnetic Resonance

It is well recognized that singlet photochemistry governs charge separation in natural photosynthesis, while in model systems, both singlet and triplet channels are competitive and depend on several variables such as relative orientations, distances, and electronic couplings between the participating redox partners.(2) In addition, the solvent reorganization energy (λ_s) due to solvent dipoles reorienting around the ion pair is of prime importance for tuning the energy levels of the charge-separated states, such that the branching ratio of the electron transfer rates, i.e., singlet- or triplet-initiated routes, can be controlled. Thus, to achieve the goal of mimicking natural photosynthesis by biomimetic supramolecular systems, it is essential to develop experimental methods to explicitly determine the absolute values of the energy levels of RP states. We have shown that this goal can be achieved by blending together molecular architecture, solvent properties and fast EPR detection of paramagnetic transients.

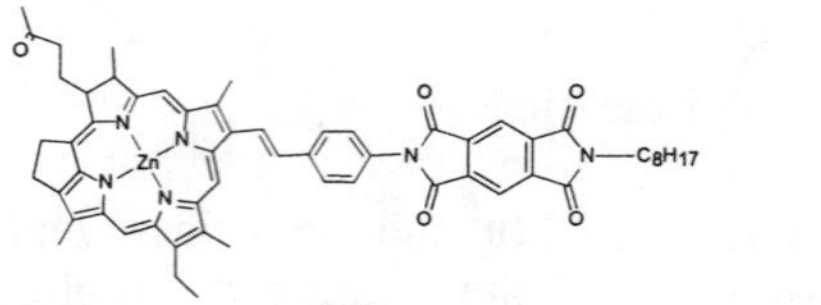

ZCPI

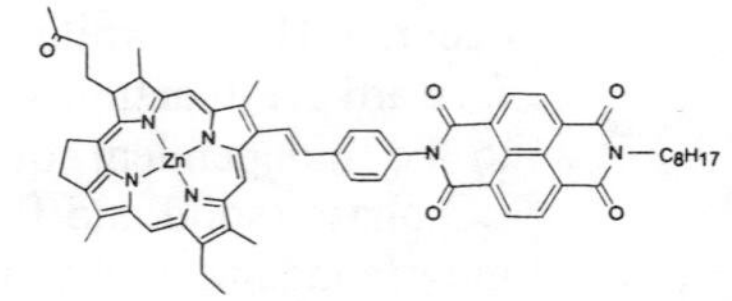

ZCNI

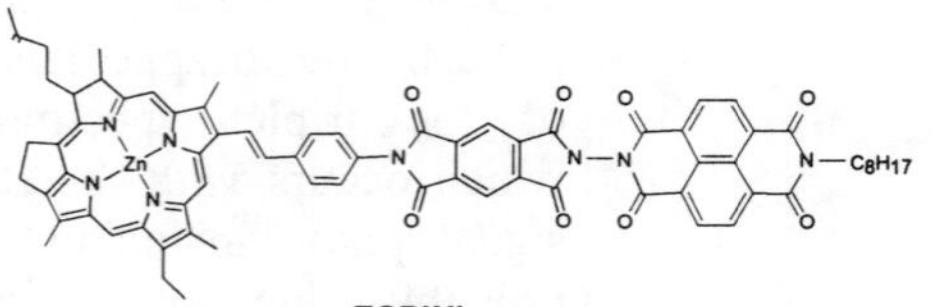

ZCPINI

Figure 5. Photosynthetic model systems are illustrated.

The present work deals with a series of covalently linked compounds (Figure 5) containing a chlorophyll-like (chlorin) electron donor, D (ZC). Two electron acceptors are used with different reduction potentials, i.e., pyromellitimide, A_1 (PI) and 1,8:4,5 naphthalenediimide, A_2 (NI) to produce a series of molecules with small but deliberate differences of the ion-pair energies. The compounds investigated are ZCPI, ZCNI and ZCPINI, with D-to-A_1, D-to-A_2 and D-to-A_2 center-to-center distances of ~ 11, ~ 11 and ~ 18 Å, respectively. These compounds, when oriented in different LCs, show photoinduced IET to produce charge-separated states that can be monitored by TREPR. The origin of such a state and the spin dynamics associated with it strongly depend not only on λ_s, but very substantially on the molecular architecture. Thus, while $ZC^{\cdot+}$- $PI^{\cdot-}$ and $ZC^{\cdot+}$- $NI^{\cdot-}$, with short donor-acceptor distances, exhibit triplet radical pair (TRP) spectra, i.e., $^3[D^{\cdot+}$- $A^{\cdot-}]$, the much longer donor-acceptor distance in the triad, $ZC^{\cdot+}$-PI-$NI^{\cdot-}$, results in a correlated radical pair (CRP) spectrum, i.e., $^{1,3}[D^{\cdot+}$- A_1- $A_2^{\cdot-}]$. These two types of spectra can be differentiated only by TREPR, via the dipolar and/or the exchange interactions, which strongly depend on the donor-acceptor distance. We further show that the spectral analysis in terms of the energy states scheme in Figure 6, illustrating the different routes of triplet and RP states production, permits an accurate assignment of the energies of the RPs in the different phases of the LC solvents. This is the first demonstration of RP energy level determination for short-lived RPs by tuning the solvent reorganization energy through the different phases of LCs.

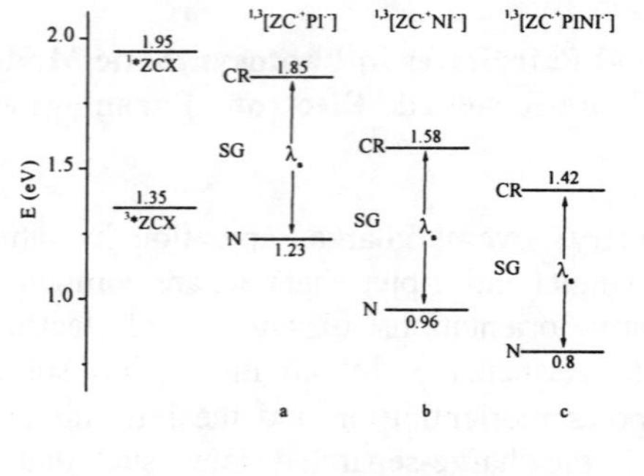

Figure 6. The temperature and phase dependent energy levels of the model systems are shown.

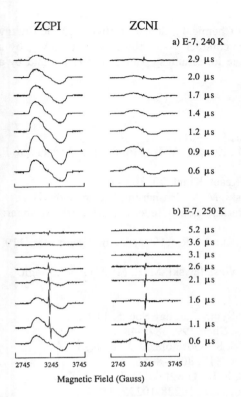

Figure 7. Direct-detection TREPR spectra (triplets and RPs), for different times after the laser pulse, of the photoexcited dyads in the soft glass of E-7 at a) 240K and (b) 250K.

While space constraints limit complete discussion of these results, Figure 7 illustrates the main theme of this work. The spectra clearly illustrate that the ZCPI radical pair has a lower driving force for charge separation because the radical pair is only evident at the higher temperature, whereas the radical pair of ZCNI is present at both temperatures. Furthermore, the spectra illustrate the large increase in the solvation ability of the soft glass, as only a 10 K increase in temperature produces a strong signal from the ZCPI radical pair. In fact, as the temperature is increased through the soft glass region, approximately 0.65eV of solvent reorganization energy is gained. Thus, by observing the appearance of the radical pair signals at different temperatures, and considering the difference in the redox potentials of the molecules shown in Figure 5, accurate assignment of the radical pair energy levels can be achieved. Furthermore, polarization considerations leads to the assignment of the dyads as triplet radical pairs, the triad displays the CRP mechanism.

4. Conclusions

Our current work in modeling reaction center dynamics has resulted in the observation of each major spin-dependent photochemical pathway that is observed in reaction centers. The development of new, simpler model systems has permitted us to probe deeply into the mechanistic issues that drive these dynamics. Based on these results we have returned to biomimetic chlorophyll-based electron donors to mimic these dynamics. Future studies will focus on the details of electronic structure and energetics of both the donor-acceptor molecules and their surrounding environment that dictate the mechanistic pathways and result in efficient photosynthetic charge separation.

5. Acknowledgment

This work was supported by the Division of Chemical Sciences, Office of Basic Energy Science, U.S. Department of Energy under contract W-31-109-Eng-38 (MRW), by a U.S.-Israel BSF grant (HL), by the Deutsche Forschungsgemeinschaft (HL), and by a Volkswagen grant (HL).

References

1 Wasielewski, M. R. (1992) Chem. Rev. 92, 435-461.
2 Joran, A. D.; Leland, B. A.; Felker, P. M.; Zewail, A. H.; Hopfield, J. J.; Dervan, P. B. (1987) Nature 327, 508-511.
3 Levanon, H.; Hasharoni, K. (1995) Prog. React. Kinet. 20, 309-46.
4 Greenfield, S. R.; Svec, W. A.; Wasielewski, M. R.; Hasharoni, K.; Levanon, H. *In The Reaction Center of Photosynthetic Bacteria*; Michel-Beyerle, M.-E., Ed.; Springer: Berlin, 1996; pp. 81-87.
5 Wiederrecht, G. P.; Svec, W. A.; Wasielewski, M. R.; (1997) J. Am. Chem. Soc. 119, 6199-6200.
6 van der Est, A.; Fuechsle, G.; Stehlik, D.; Wasielewski, M. R. (1997) Appl. Magn. Reson. 13, 317-35.
7 Meier, G.; Saupe, A. (1966) Mol. Cryst. 1, 515-525.
8 Martin, A. J.; Meier, G.; Saupe, A. (1971) Symp. Faraday Soc. 5, 119-125.
9 Hasharoni, K.; Levanon, H. (1995) J. Phys. Chem. 99, 4875-4878.
10 Hasharoni, K.; Levanon, H.; Greenfield, S. R.; Gosztola, D. J.; Svec, W. A.; Wasielewski, M. R. (1995) J. Am. Chem. Soc. 117, 8055-8056.
11 Hasharoni, K. .; Levanon, H.; Greenfield, S. R.; Gosztola, D. J.; Svec, W. A.; Wasielewski, M. R. (1996) J. Am. Chem. Soc. 118, 10228-10235.

SYNTHETIC PROTEIN MAQUETTE DESIGN FOR LIGHT ACTIVATED INTRAPROTEIN ELECTRON TRANSFER

Christopher C. Moser, R. Eryl Sharp, Brian R. Gibney,
Yasuhiro Isogai and P. Leslie Dutton

Johnson Research Foundation and Department of Biochemistry and Biophysics
University of Pennsylvania, Philadelphia, PA 19104

1. Introduction

One of the best ways to understand the natural engineering of electron transfer proteins is to construct them ourselves. We design and synthesize simplified *de novo* synthetic redox protein, called maquettes, which seek to abstract the features essential for electron transfer function from the large and complex electron transfer proteins found in nature. With these simplified systems we can explore which properties of protein redox centers are relatively easy to reproduce, and reflect a low level of engineering, and which are more specialized and reflect an element of significant naturally selected design. We can also appreciate the physical chemical features that are essential (rather than incidental) for bioenergetic functions, including light activated electron transfer.

We have chosen α-helical bundles as our design scaffold. These bundles are robust, readily tolerant to amino acid substitution, and well described experimentally [1]. Our sequences are based on an amino acid heptad repeat AEELLKK, which defines 2 turns of an amphipathic α-helix in which the polar groups of equal numbers of negatively charge glutamates and positively charged lysines are directed towards one face of the helix, and non-polar groups alanine and leucine are directed towards the opposite face. We extend the heptad repeat 3.5 times for 7 turns, comparable to the transmembrane helices found in many natural electron transfer proteins. This extra length, compared to commonly synthesized experimental peptides, discourages unraveling of the a helices and affords high stability to the α helical unit.

These helices spontaneously self-assemble in aqueous solution to form a 4 helix bundle with a hydrophobic interior. Interactions between charged groups of separate helices and the packing of interior residues define the geometry of the self assembly. Although a bundle can be formed with 4 independent and identical helices, we generally covalently link bundles by one of two methods. By beginning the helices with a Cys and several flexible Gly residues, pairs of 31 amino acid helices can be oxidized in the presence of atmospheric oxygen to form disulphide links. Alternatively, a full 62 amino acid helix-turn-helix sequence can be synthesized. Both are efficiently produced on continuous flow solid-phase synthesizers using the fluorenyl methoxy carbonyl (Fmoc)/'Bu strategy. Thus our four helix bundles are generally made of two identical di-helical subunits. Depending upon the choice of the internal residues, our bundles self-assemble into a syn- geometry with parallel helix-loop-helix subunits or an anti-geometry, with subunit loops at opposite ends of the bundle. Resolved NMR structures of bundles in solution show an overall two fold symmetry [2] confirmed by emerging x-ray crystal structures.

The *de novo* synthesis of this bundle design affords us a great deal of flexibility to engineer redox center sites and multiple redox center electron transfer proteins. The

G. Garab (ed.), Photosynthesis: Mechanisms and Effects, Vol. V, 4201–4206.

extraordinary stability of these bundles compared to most natural proteins (unfolding at temperatures typically above 100°C) allows us to modify the structures with relatively little concern for compromising protein folding. We can also expose these electron transfer proteins to greater environmental extremes (such as pH 4 or 10) while describing the redox center properties. Solid phase synthesis permits us to do a vast array of chemical modifications and attachments to the peptides in organic solvents while still attached to the synthetic resin.

2. Ligand Attachment of Redox Centers

There are several strategies we use to secure redox centers in these bundles. The simplest is to place histidines in the bundle interior to ligand to metal porphyrins such as hemes. Figure 1 shows a series of synthetic hemoproteins that we have constructed

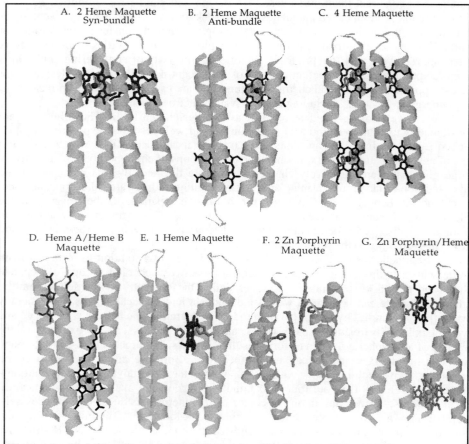

A. 2 Heme Maquette Syn-bundle   B. 2 Heme Maquette Anti-bundle   C. 4 Heme Maquette

D. Heme A/Heme B Maquette   E. 1 Heme Maquette   F. 2 Zn Porphyrin Maquette   G. Zn Porphyrin/Heme Maquette

Figure 1: Multi-heme maquette design using His ligation and self-assembly.

which bind heme added to aqueous solution. By placing histidines on each helix 10 residues from the middle of the loop, the helix-loop-helix subunits provide *bis*-His ligation for added heme, for 2 hemes per 4 helix bundle. Either a *syn*- or *anti*-configuration can be constructed, depending on the sequence in the hydrophobic interior (Figure 1A and B) [3, 4].

By placing additional histidines 24 residues from the middle of the loop (2 heptad repeats from the 10 position), we can provide *bis*-His ligation for up to 4 hemes [3]. Heme binds tightly to the 10 positions with 15 and 135 nM dissociation constants and to the 24 position with 3.5 and 22 μM dissociation constants, and display classical heme B visible spectra and EPR properties. 10 and 24 position hemes have distinguishable α bands at 558 and 560 nm, respectively. Heme redox midpoint potentials can be manipulated by placing charged arginines near the ends of the helices. Heme redox properties can also be adjusted by protonating or deprotonating glutamates near the hemes as a function of pH [5]; in other words, heme oxidation and reduction can be coupled to proton exchange. Depending on conditions, maquette-bound protoporphyrin IX displays a range of E_m values from -230 to 40 mV.

Because of the distinct binding constants of hemes to the histidines in multiheme synthetic proteins, and because heme dissociation is slow (days at room temperature) it is possible to bind different heme types within a single bundle. Figure 1C shows a mixed hemeA/hemeB maquette that we have constructed. Well separated α band maxima at 591 and 558 nm and heme midpoints of -65 and -200 mV means that intraprotein electron transfer is possible in this maquette. Indeed, we have observed steady state reduction of oxygen catalyzed by this protein when supplied with an exogenous reductant (Gibney et al., Biochemistry submitted).

Using a modified strategy of providing only one histidine per helix-turn-helix subunit allows us to create single heme maquettes in which added heme is *bis*-His ligated between protein subunits (Figure 1D) [6]. In this case, heme binding affinity is noticeably weaker (Kd 23 μM) than typical intra-subunit heme binding. However, using porphyrin which prefers 5-coordinate ligation such as Zn protoporphyrin IX, results in much tighter binding of two porphyrins per 4 helix bundle (Kd <1 and 10 nM). The binding of Zn porphyrins to maquettes introduces the ability to light activate electron transfer from the Zn porphyrin excited state, analogous to Mg chlorophylls.

A slightly different strategy places a histidine on the first helix 10 residues from the middle of the turn, and a second histidine on the second helix 24 residues from the middle of the turn. In the 4 helix bundle there are 4 histidines, but added heme can only *bis*-His ligate between protein subunits. By titrating in 1 heme per bundle, then 1 Zn porphyrin per bundle, a light activatable mixed porphyrin maquette has been constructed.

Redox centers can also be bound using the coordination chemistry of the amino acid Cys in helix-turn-helix peptides. FeS clusters formed from $FeCl_2$ and Na_2S added in solution spontaneously self-assemble with 4 Cys built into the loop region to generate a cubane 4Fe4S cluster with an Em_8 of -350 mV and EPR properties similar to the natural protein ferredoxin. By providing additional *bis*-His ligations, we have constructed FeS/Heme maquettes, also capable of intraprotein electron transfer [7].

3. Covalent Attachment of Light Activated Redox Centers

A distinctly different strategy of securing redox centers to the bundles is provided by directly covalently linking the redox center to unprotected residues while the peptide is still attached to the synthetic resin. Thus one of the 4 propionic acids of coproporphyrin links with a free amine at the end of a helix, which results in a 4 heme, 2 coproporphyrin bundle after the addition of heme (Figure 2A) [8]. The light activatable coproporphyrins spontaneously dimerize, forming a structure reminiscent of chlorophyll dimers in photosynthetic reaction centers. However, the placement of the light activatable unit at the end of the bundle is not an efficient design for stable intraprotein charge separation. A generic equation for calculating electron transfer rates is based upon basic non-adiabatic electron tunneling theory and a survey of natural electron tunneling proteins:

$$\log k_{et} = 15 - 0.6\,R - 3.1\,(\Delta G^{\circ} - \lambda)^2/\lambda$$

where the tunneling rate k_{et} is in units of s^{-1}, edge-to-edge distance R is in units of Å and the free energy (ΔG°) and reorganization energy (λ) of the electron transfer reaction are in units of eV [9, 10]. Using this relation, the 15 Å distance between the coproporphyrin dimer and the nearest heme assures that a 3 μsec light activated charge separation will be followed by an even more rapid 1 μsec charge recombination to the coproporphyrin ground state, well before the anticipated 800 μsec heme-10 to heme-24 electron transfer rate.

Securing a light activatable center within the bundle provides a much more attractive strategy. A Cys residue at an interior position half-way along the length of one helix of a helix-turn-helix peptide reacts with a brominated flavin to form a flavo-protein maquette with two flavins per bundle [11]. Burial of the flavin raises its average n=2 midpoint potential from -135 mV in solution to -95 mV. Light activation

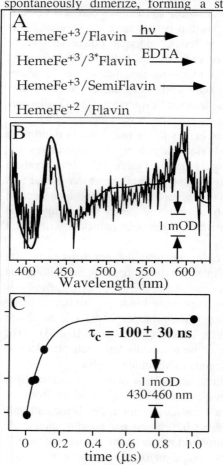

Figure 2: Flavin/Heme maquette electron transfer. **A.** Sequence of light activated reactions. **B.** Light induced absorbance changes on 1 μs after a laser flash and after continuous illumination (smooth curve). **C.** Spectral changes take about 100 ns.

of the flavin creates a long-lived triplet state (lifetime 20 μs). In the presence of an exogenous nitrogenous base as electron donor, such as EDTA, excited flavin extracts an electron to become a moderately good semiquinone reductant (Figure 3A).

Histidines incorporated into the synthetic flavo-protein sequence bind protoporphyrin IX to make a flavo-hemo maquette (Figure 2B). However, the expected oxidized flavin/semiquinone flavin redox potential of -130 mV is similar to that of bound protoporphyrin IX. A higher potential heme would facilitate light activated electron transfer. Although it is possible to raise the heme E_m by strategic placement of polar and charged residues, it is more straight-forwardly accomplished by binding oxo-meso-heme to the histidines. This inherently more oxidizing heme raises the heme E_m to -30 mV.

Light activation of the flavo-oxo-meso-heme maquette [11] produces the expected visible absorption changes (Soret bandshift and induction of the reduced 600 nm α band, Figure 3B) corresponding to intraprotein electron transfer reduction of the heme on a timescale of about 100 nsec (Figure 3C). This reaction is somewhat slower than the anticipated tunneling rate of 30 nsec based upon the generic tunneling equation and the modeled flavin to heme edge-to-edge distance of around 10 Å. Thus we expect intraprotein electron transfer could still be rate limited by the electron transfer between the light activated flavin and the EDTA in solution.

The flexibility of the *de novo* synthetic 4 helix bundle motif allows us to explore many designs within the light activated flavo-hemo-protein theme. Light activated electron transfer has been observed when the flavin has been attached to Cys placed at different ends of the helix. Light activation of reduced flavin creates an alternative strong electron donor in its own right. We intend to exploit the dark n=2 electron redox properties of the flavin as an electron acceptor analogous to the Q_B of photosynthetic reaction centers by using other light activatable groups, such as the Zn porphyrins, in combination with heme redox centers to make photosynthetic maquettes.

4. Lessons on redox protein engineering

Our exploration of maquettes has shown that a relatively low level of engineering is required to bind hemes in the hydrophobic interior of synthetic, and presumably natural, proteins. The driving force of *bis*-His ligation is apparently strong enough to recruit nearby histidines, even if the histidines are found on two separate subunits within a bundle, provided a sufficiently large hydrophobic interior satisfies the basic hydrophobic/hydrophillic partition requirements of the porphyrin ring. Single his ligation, as in the Zn porphyrins appears to be even easier. Manipulation of heme redox properties away from an approximately -150 mV average is possible, but requires more specific design. Midpoint potentials are more easily manipulated by changing the heme type. Light activatable redox centers can be self-assembled into proteins, although covalent linkage, a strategy used in some natural proteins, is convenient for the experimentalist. Incorporation of light-activatable centers within the protein interior enables rapid electron tunneling by virtue of bringing redox partners within the 4 to 14Å distance range seen in natural electron transfer proteins.

5. References

1. Bryson, J.W., S.F. Betz, H.S. Lu, D.J. Suich, H.X.X. Zhou, K.T. O'Neil, and W.F. DeGrado (1995) Science 270, 935-941.

2. Gibney, B.R., J.S. Johansson, F. Rabanal, J.J. Skalicky, A.J. Wand, and P.L Dutton (1997) Biochemistry 36, 2798-2806.

3. Robertson, D.E., R.S. Farid, C.C. Moser, J.L. Urbauer, S.E. Mulholland, R. Pidikiti, J.D. Lear, A.J. Wand, W.F. DeGrado, and P.L. Dutton (1994) Nature 368, 425-432.

4. Grosset, A.M., F. Rabanal, R.S. Farid, D.E. Robertson, D.L. Pilloud, W.F. DeGrado, and P.L. Dutton (1996) in Peptides: Chemistry, Structure, and Biology, (P.T.P. Kaumaya and R.S. Hodges, ed.), Vol. pp. 573-574.

5. Schifman, J., C.C. Moser, W. Kalsbeck, D.F. Bocian, and P.L. Dutton (1998) Biochemistry submitted.

6. Sharp, R.E., J.R. Diers, D.F. Bocian, and P.L. Dutton (1998) J. Amer. Chem. Soc 120, 7103-7104.

7. Gibney, B.R., S.E. Mulholland, F. Rabanal, and P.L. Dutton (1996) Proc. Natl. Acad. Sci. U.S.A. 93, 15041-15046.

8. Rabanal, F., W.F. DeGrado, and P.L. Dutton (1996) J. Amer. Chem. Soc. 118, 473-474.

9. Moser, C.C. and P.L. Dutton (1992) Biochim. Biophys. Acta 1101, 171-176.

10. Moser, C.C., J.M. Keske, K. Warncke, R.S. Farid, and P.L. Dutton (1992) Nature 355, 796-802.

11. Sharp, R.E., C.C. Moser, F. Rabanal, and P.L. Dutton (1998) Proc. Natl. Acad. Sci. in press.

Baceriochlorophyll-serine based photochemotherapy; type III PDT?

Avigdor Scherz[*1], Sharon Katz[1,2], Yahel Vakrat [1,2], Vlad Brumfeld[1],
Eva Gabelmann[1,2], Judith Zilberstein[2], Dieter Leupold[3], [4]James R. Norris,
Hugo Scheer[5], and Yoram Salomon [2].

[1] Dept. of Plant Sci. The Weizmann Inst. of Sci. Rehovot, 76100, Israel

[2] Dept. of Biol. Reg. The Weizmann Inst. of Sci. Rehovot, 76100, Israel

[3] Max-Born Institute, Berlin, Germany

[4] Dept. of Chemistry, University of Chicago, Illinois, USA

[5] Institute of Botany, Ludwig-Maximilians-University, Munchen

Keys words: Bacteriochlorophyll, cancer, photodynamic therapy, vascular bed,
paramagnetic resonance imaging, electron spin resonance, reactive oxygen species
Abbreviations: PDT = photodynamic therapy; Bchl-Ser = Bacteriochlorophyll-serine; MRI =
magnetic resonance imaging; ESR = electron spin resonance; DMPO = 5,5-dimethyl-1-pyrroline
N-oxide; TEMP = 3,3,5,5-tetramethyl-pyrroline n-oxide; ROS = reactive oxygen species

1. Introduction:

Photodynamic therapy (PDT) is a new modality of cancer therapy in which dye molecules (sensitizers), excited by light, generate highly reactive but shortly leaving chemicals within the tumor tissue. These reactive chemicals interact with vital cellular components and subsequently cause cell death (1).

PDT is thought to commence either with an electron (type I) or energy (type II) transfer from the relatively long-living triplet state of the sensitizer to a nearby oxygen molecule or other electron acceptors (2-5). The energetic requirements of the two processes, however, are quite different. Oxidation of sensitizers that are presently used in PDT requires ~2.5-3.0 eV (UV-VIS radiation), whereas their promotion into the triplet state, from which they can transfer energy to oxygen molecules, requires only ~1.1-1.5 eV (RED-NIR radiation). Since RED and NIR penetrate much deeper than UV-VIS radiation into animal tissue, PDT is currently promoted by type II processes. The photosynthetic pigments, chlorophylls, and bacteriochlorophylls are, however, exceptions. Their oxidation potential is ~0.8-1.0 and ~0.5-0.7 eV, respectively (in organic solvents), which means that excitation of Bchls into their lowest-excited singlet state (S^1, 1.3-1.6 eV) is sufficient to promote both electron and energy transfer to oxygen molecules by NIR radiation that can penetrate deeply into the treated tissue (1,6) The electron transfer process is of particular interest, since it usually results in the formation of the highly potent OH radicals.

In light of these considerations we have synthesized conjugates of Bchls and their metal derivatives with different residues, such as amino acids, peptides, and proteins that modified their hydrophylicity and affinity to tumor tissues (7-10). One of these reagents, Bacteriochlorophyll-serine (Bchl-Ser), has already been tested and found highly efficient against tumor cells in culture and as a vascular reagent in PDT of mice implanted with tumors (11-13). Recent studies have

G. Garab (ed.), Photosynthesis: Mechanisms and Effects, Vol. V, 4207–4212.
© 1998 *Kluwer Academic Publishers. Printed in the Netherlands.*

indicated that the highest photodynamic effect is achieved shortly after the Bchl-Ser is injected into the mice when most of the sensitizer is still in the blood and its cellular content is too small to confer cytophototoxicity. Hence, we began to study the photochemistry and physics of Bchl-Ser in aqueous and micellar solutions. Consequently, we determined that in both systems, Bchl-Ser derivative generates only hydroxyl radicals and anions. We therefore, propose that these reactive oxygen species (ROS) induced the observed damage to the tumor vascularization which ended in tumor necrosis and eradication.

2. Materials and methods

2.1 Animal tumor treatment

Preparation of Bchl-ser was recently described (Rosenbach-Belkin et al, 1996, Scherz et al, 1997-8). M2R mouse melanoma tumors were implanted in CD1 nude male mice and allowed grow to ~ 100-250 mm^3 before treatment. PDT protocol consisted of local or i.v. injection of 200-500 mg of Bchl-Ser into the mouse tail vein and immediate illumination of the tumor site (580-750 nm) by an xenon lamp (Medic-Light, Haifa, Israel or Biospec, Moscow, Russia) for 30 min (60-100 mW/cm^2). Histopathology and Magnetic Resonance Imaging (MRI) experiments were performed as previously described (13-15).

2.2. Detection of reactive oxygen species

Reactive oxygen species (ROS) were detected by electron spin resonance (ESR) spectroscopy using a Bruker ER 200 D-SRC spectrometer, equipped with a Bruker ER 040 XK-X-band microwave bridge that was set at 9.78 GHz. The samples were examined using a microwave power of 20 mW, modulation of 1 gauss, a field center at 3500 gauss, a scan range of 100 gauss, and a scan time of 200 sec. The samples were inserted into a flat quartz cell and illuminated inside the cavity using a LS3-PDT lamp (Biospec-Moscow) in which light below 720 and above 830 nm was filtered out. The light intensity was set at 27 mW/cm^2. Each sample was scanned 6 times.

The optical absorption of each sample was recorded (before and after measuring the ESR) using a Spectronic GENESYS (Milton Roy) spectrometer.

In order to trap $OH^\bullet$, $O_2^{\bullet-}$, and $CH_3^\bullet$, we used DMPO (D5766, Sigma). The DMPO (20 μl to 1 ml PBS) was vacuum-filtered twice through activated charcoal. TEMP (T9891, Sigma) was used in an attempt to trap singlet oxygen.

The participation of $O_2^{\bullet-}$ and H_2O_2 in the formation of $OH^\bullet$ was examined by adding 1100 units/ml of super-oxide-dismutase (SOD, s-251, Sigma) and 57 units/ml of catalase [c-9322 (16))] respectively, to the sample prior to illumination within the ESR cavity. Desferrioxamine (d-9533, 1 mM, Sigma) was also added to exclude the participation of ion metals.

For anaerobic measurements the samples were flushed with argon for 30 minutes and then transferred to the flat quartz cell.

2.3 Transient absorption spectroscopy of Bchl-Ser in aqueous solutions

Transient absorption at times shorter than 400 pico-seconds was examined using a mode-locked 510C Nd-YAG laser (Quantel, France) with a pulse duration of 35 pico-seconds that pumps a PD10 dye laser (Continuum, USA), and generates white light (470-870 nm) after passing through a 5-cm quartz cell filled with water. Excitation was finally performed with 765 nm pulses (14 pico-seconds, 120 mW). The transient optical absorption was monitored for samples in a continuous flow at different time intervals after excitation using an intensified DIDA system of Princeton Instruments. The flow rate was selected to prevent accumulation of photochemical products between consecutive measurements. Transient absorption at longer times was recorded at the Max Born Institute (MBI) in Berlin as previously described (17). Determination of ϕ_Δ, the quantum yield for singlet oxygen formation, was performed either continuously or by a time-resolved technique using a 380 nm excitation wavelength and perinaphthenone as a reference sensitizer (ϕ_Δ = 0.98 (18)). Specifically, the initial singlet oxygen emission intensity was measured as a function of the laser energy, where the slopes of the linear plots provided the relative quantum yields. This method allows separating background fluorescence on a nano-second time scale from the singlet oxygen emission, which decays in microseconds.

3. Results and discussion

3.1. Tumor response to Bchl-Ser-based PDT.

3.1.a Animal survival

We found that a single PD treatment with Bchl-Ser and light led to progressive destruction of the tumor, culminating in its complete eradication in 17 out of 20 animals treated by local intra-tumor drug injection and in 14 out of 16 animals treated by i.v. drug administration, showing an average rate of survival of about 85% with mice living for over 200 days. In control mice, untreated but irradiated or Bchl-Ser-treated and non irradiated, tumor growth progressed and the animals died within their expected life span (55 ± 5 days, n=20) (15).

3.1.b Effect of PDT on vascular function

Vascular damage caused by PDT, 1 and 24hr after treatment, was determined by MRI from signal enhancement time curves after i.v. administration of Gd-DTPA. We found that the T_2 relaxation time was substantially shortened in the tumor volume, indicating vascular permeation and hemorrhage formation within 1 hr after illumination. Eu-IgG accumulated in treated tumors to an extent of about 6-8 times higher than in untreated tumors in the same animal, 1 hr after treatment, supporting this. After 24 hr, the region exposed to PDT showed a total lack of Gd-DTPA uptake and a substantially lower accumulation of Eu-IgG. These results indicated complete vascular arrest in the treated tumor. During illumination, we observed light dependent oxygen depletion that culminated in complete oxygen arrest a few hours after treatment (12).

Histopathological examination of the tumor 24 hr after treatment showed that the blood vessels became extremely dilated and occluded by thrombi and hemorrhages that were characterized by erythrocyte infiltration. This microscopic analysis provided further support that the vascular bed of the tumor is a primary target of Bchl-Ser-based PDT, as proposed here. Necrosis was usually observed after 24 hr in the tumor zone, with no signs of damage to collateral areas that were simultaneously illuminated after i.v. drug injection. This high differentiation indicates that the response of the tumor vascular bed to the suggested Bchl-Ser-based PDT, is markedly different from the response of normal tissue, providing high selectivity to the suggested protocol.

It is reasonable to assume that at the time of illumination, the level of Bchl-ser in the exrtra-vascular part of the tumor is too low to confer cytophototoxicity. Furthermore, *in vitro* studies have shown that the maximal drug effect is achieved in serum containing cell cultures, only after four hours of incubation (11, Mazor et al, unpublished data). Therefore, the photodynamic action of Bchl-Ser that results in vascular damage and subsequent tumor necrosis, probably involves the cellular components of the blood and starts at the cell/plasma interface. These conclusions provide a rationale for examining the photocehemistry and physics of Bchl-Ser in aqueous solutions.

3.2 Generation of reactive oxygen species by NIR illumination of Bchl-Ser in aqueous solutions.
The ESR spectrum of Bchl-Ser in PBS solution prior to illumination under atmospheric conditions showed a single band with a half-width of 13.5 G and a g-factor of 2.0055 (Fig. 1). This is similar to previously reported signals of Bchl radicals (19) but it decayed during illumination or upon deaeration. The ESR spectrum of aerated solutions of Bchl-Ser and DMPO in PBS before and during illumination indicated extensive formation of OH radicals (Fig. 1). However, there was no indication of O_2^- radical generation or the participation of hydrogen peroxide, since neither catalase nor SOD had any effect on the intensity or shape of this spectrum. No OH· signal was observed after extensive deaeration. Adding TEMP to an irradiated Bchl-Ser in PBS solution did not yield an ESR triplet signal, which implies that singlet oxygen was not formed in any appreciable amount.

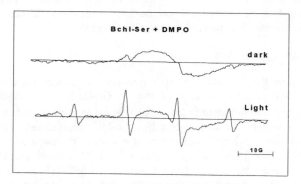

Figure 1. The ESR spectrum of Bchl-Ser+DMPO in aerated PBS solution

The differential optical absorption of Bchl-Ser in deaerated PBS solution after illumination showed an absorption increase at ~540 and 660 nm accompanied by bleaching of the Bchl-Ser Q_x transition at ~590 nm. The entire spectrum could be simulated well by a combination of 1 Bchl cation radical, 1 Bchl anion radical minus the absorption of two Bchls in that region. Some decay of the positive signals was observed at longer times (but still, within the first nano-second). Surprisingly, the spectrum was markedly different from the one expected for the formation of a Bchl triplet excited state. Illumination of Bchl-Ser in aerated PBS solutions resulted in a similar spectrum at time 0 followed by a gradual increase of the 540 nm band and a concomitant decrease of the 640 nm transition intensity (Fig. 2). The negative peak at ~600 nm shifted to the red and decreased in intensity at longer times (>100 picoseconds).

Time-resolved fluorescence spectroscopy of the Bchl-Ser/PBS solution showed no luminescence of singlet oxygen molecules at 1200-1300 nm. Some luminescence was observed for Bchl-Ser in chloroform. This observation is in agreement with the ESR data.

During illumination the pH of $\sim 10^{-5}$M Bchl-Ser/PBS solution increased by almost one unit.

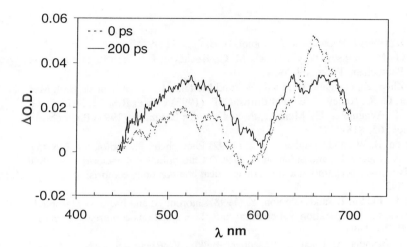

Figure 2. Differential absorption of Bchl-Ser in aerobic PBS solution at zero time and 200 picosecond after irradiation with a 35 picosecond pulses at 765 nm

4. Conclusions

In light of these observations we propose that illumination induces ultrafast electron transfer from one Bchl to another in loosely coupled dimers (such dimers were recently proposed by Katz [Ph.D. Thesis, Weizmann Institute]). The anion can reduce oxygen if present, starting a chain of events ending with the ejection of the hydroxyl radical and anion in a non-Fenton reaction [desferioxamine (d-9533, 1 mM, Sigma)], a well known iron chelator, had no effect on the radical formation excluding the participation of ion metals). In the presence of oxygen, some of the Bchl-Ser molecules may already be oxidized, causing to the observed dark signal. Since the radicals seem to be ejected directly from the excited sensitizer, their formation may fall into neither type I nor type II processes. Rather, it may represent a class by itself; a type III process. Although hydroxyl radical formation from bacteriochlorin in 20% DMSO/PBS solution was recently reported, this processes seemed to involve a classical Fenton reaction (20).

The treatment protocol as proposed here is novel and tailored to the properties of the water-soluble Bchl-Ser. The radicals formed probably interact with the blood cellular components, resulting in the aforementioned wave of damage. Hopefully, it opens the way for PDT of larger, solid tumors.

5. Acknowledgment: Y.S. is the incumbent of the Charles and Tillie Lubin Chair of Biochemical Endocrinology. This work was supported by Steba-Beheer NV; The Israel-German Foundation-GIF, grant # I-0497-140.02/96; The EC-TMR facility in MBI, and a generous gift from Mrs. Sharon Zuckerman of Toronto, Canada.

References

1   Moan, J. (1990) J. Photochem. Photobiol. B: Biol. 5, 521-524
2   Smith, G., McGimpsey, W. G., Lynch, M. C., Kochevar, I. E., and Redmond, R. W. (1994) Photochem. Photobiol. 59, 135-139
3   (1993) Zhiyi, Z., Nenghui, W., Oian, W., and Meifanm L., Free radicals Biol. Med. 14, 1-9
4   Buettner, G. R., Kelley, E. E., and Burns, C. P. (1993) Cancer Res. 53, 3670-3673
5   Hadjur, C., Wagnierer, G., Monnier, P., and van den Bergh, H. (1997) Photochem. Photobiol., 65, 818-827
6   Henderson, B. W., and Dougherty, T. J. (1992) Photochem. Photobiol. 55, 145-157
7   Scherz, A., Fiedor, L., and Salomon, Y. (1997) Chlorophyll and bacteriochlorophyll derivatives, their preparation and pharmaceutical compositions comprising them. I. US Patent #5,650,292
8   Scherz, A., Fiedor, L., and Salomon, Y. (1998)Chlorophyll and bacteriochlorophyll derivatives, their preparation and pharmaceutical compositions comprising them. II. US Patent #5,726,169
9   Gross, S., Brandis, A., Chen, L., Rosenbach-Belkin, V., Roehrs, S., Scherz, A., and Salomon, Y. (1997) Photochem. Photobiol. 66, 872-878
10   Fiedor, L., Rosenbach-Belkin, V., Sai, M., and Scherz, A. (1996) Plant Physiol. Biochem. 4, 393-398
11   Rosenbach-Belkin, R., Chen, L., Fiedor, L., Tregub, I., Pavlotsky, F., Brumfeld, V., Salomon, Y., and Scherz, A. (1996) Photochem. Photobiol. 64, 174-181
12   Zilberstein, J., Bromberg, A., Frantz, A., Rosenbach-Belkin, V., Kritzmann, A., Pfeferman, R., Salomon, Y., and Scherz, A. (1997) Photochem. Photobiol. 65, 1012- 1019
13   Zilberstein, J., Rosenbach-Belkin, V., Neeman, M., Bendel, P., Kohen, F., Gayer, B., Scherz, A., and Salomon, Y. (1998) in The 7th Biennial Congress International Photodynamic Association (Patrice, T., ed), Nantes, France
14   Abramovitch, R., Meir G., and Neeman, M. (1995) Cancer Res. 55, 1956-1962,.
15   Zilberstein, J., Rosenbach-Belkin, V., Neeman, M., Bendel, P., Kohen, F., Gayer, B., Scherz, A., and Salomon, Y (1998), Cancer Res. Submitted
16   Sigma. Sigma catalog , 246, (1998)
17   Stiel, H., Teuchner, K., Hild, M., Freyer, W., Leupold, D., Scherz, A., Noy, D., Simonin, I., Hartwich, G., and Scheer, H. (1997) J. Luminescence 72-74, 612-614
18   Stiel, H., Teuchner, K., Hild, M., Freyer, W., Leupold, D., Simonin, I., Hartwich, G., Scheer, H., Noy, D., and Scherz, A. (1998) J. Amer. Chem. Soc. Submitted
19   Lubitz, W. (1991) in Chlorophylls (Scheer, H., ed), pp. 903-944, CRC press
20   Hoebecke, M., Schuitmaker, H.J., Jannink, L.E.; Dubbelman, T.M.A.R., Jacob, A., and Van de Vorst, A. (1997) Photochem. Photobiol. 66, 502-508

Synthetic Four-Helix-Bundle Protein carrying 1 Or 2 Chlorophyll Derivatives

A. Struck[1], H. Snigula[2], H.-K. Rau[1], P. Hörth[1], H. Scheer[2] and W. Haehnel[1]

[1] Biologie II, University of Freiburg, Schänzlestr. 1, D-79104 Freiburg
[2] Botanisches Institut, University of München, Menzingerstr. 67, D-80638 München

Key words: synthetic proteins, chlorophyll, pigment-protein interactions, circular dichroism

1. Introduction

In recent years the structures of several (bacterio)chlorophyll proteins have elucidated the interactions between proteins and cofactors (1,2). It is therefore tempting to test our understanding by designing synthetic proteins with similar cofactors.

Based on recent success in designing heme proteins (3), a modular approach was taken, which relies on a synthesis of a four helix bundle. In order to reduce potential complicating, in current work, we introduced three structural modifications: a) to avoid the purely understood extended hydrophobic interactions of the chlorophylls we have removed the phytyl chain, b) to avoid a mixture of epimers, the 13^2 carbomethoxy group was removed and c) for a better positioning, the chlorophyll derivatives were bound covalent to the peptides.

The synthetic peptides were loaded with 1 or 2 chlorophyll a derivatives such, that they were principally capable of an edge to edge interaction, reminiscent of the B850, B870 pigments in LH2, LH1 and RC, respectively, from purple bacteria.

2. Methods

Pigments

The pigments were synthesized from chlorophyll a. Modifications involved demetallate with acetic acid (4), demethoxycarbonylate at $C13^2$ to Ppheid a by refluxing for 21 h in pyridine under argon (5), transesterification by refluxing for 1 h under argon with methanol/5% H_2SO_4 (2), ozonolysis of the 3-vinyl group (6), and remetallation with Ni or Zn.

Peptide Syntheses
Sequences
Mp-A: Ac-N-L-E-E-L-L-K-K-L-Q-E-A-L-E-K-A-Q-K-L-L-K(Mp)-NH₂
Mp-B: Mp-G-N-A-L-E-L-**H**-E-K-A-*l*-K-Q-**H**-E-E-L-*ll*-K-K-L-NH₂
Mp-C: Mp-G-N-A-L-E-L-L-E-K-A-L-K-Q-L-E-E-L-L-K-K-W

G. Garab (ed.), Photosynthesis: Mechanisms and Effects, Vol. V, 4213–4216.
© 1998 Kluwer Academic Publishers. Printed in the Netherlands.

Mp = 3-Maleimidopropionyl-, *I* and *II* are modified amino acids for the covalent linkage of 1 or 2 pigments

Template: T(Trt)(Acm)(Trt)(StBu): cyclo[C(Acm)A-C(Trt)P-G-C(StBu)A-C(Trt)P-G]

MOP: T[Sp-A][Sp-B][Sp-A][Sp-C]
Sp = 3-Succinimidopropionyl-

The helical peptides Mp-A, -B and -C were synthesized by solid phase peptide synthesis (SPPS) using standard Fmoc/tBu protection strategy. The templates were synthesized manually by SPPS on NovaSys-TGT resin and cyclized in solution as described (3). The chemoselective coupling of the maleimide group to the cysteine thiol group was performed in 3/2 (v/v) 0.2 M sodium phosphate, pH7/acetonitrile. The solution with a concentration of approximately 20 mM of each compound with an excess of the maleimide compound of 1.05 to 1.1 was stirred for 30 min at room temperature.

Circular dichroism (CD) and simultanous absorption spectra were recorded on a Mark V dichrograph (ISA/Jobin Yvon).

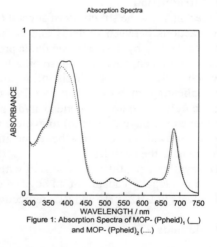

Absorption Spectra

Figure 1: Absorption Spectra of MOP- (Ppheid), (⎯) and MOP- (Ppheid)₂ (....)

3. Results

<u>Protein design</u>
The antiparallel four helix bundle protein MOP was prepared by linking the helices A-C in the pattern ABAC to a template with four Cys via maleimidopropionyl groups. These groups were attached to a N-terminal Gly or the side chain of a C-terminal Lys to control the orientation of the helices. The basic design with three different helices is similar to that of two modular proteins capable of light-induced electron transfer between metallocenters (7). The hydrophilic outer surface of the amphilic helices was designed to stabilize the helical structure. In the hydrophobic face of helix B, one or two cofactors were linked covalently at the lysins marked with *I* and *II* (see sequence in the method section). 3 amino acids downstream, on the same helix B, there are 2 Histdins for ligation with the central metal of the coupled chlorophylls. Based on this design, four different pigment binding synthetic proteins were synthesized:

1.) MOP-(Ppheid)₁, 2.) MOP-(Ppheid)₂, 3.) MOP-(Ni-Ppheid)₁, 4.) MOP-(Ni-Ppheid)₂
Ppheid = pyropheophorbide

All four synthetic pigment bound to the four helix bundle proteins were purified by HPLC and show the expected masses (ESI-MS).

Absorption spectroscopy

The absorption of the MOPs carrying 1 or 2 pigments are rather similar. The only significant differences are small redshifts and broadening of the Q_y bands in the doubly conjugated MOPs. Similar results were obtained with the free bases and the Zn-complexes. In Figure 1 MOP-(Ppheid)$_1$ shows an absorption maximum of the Q_Y band at 682 nm. In MOP- (Ppheid)$_2$, the Q_Y band shows a redshift of 5 nm towards 687 nm. A similar redshift is observable in the characteristic Q_X bands, from 514 nm to 519 nm and 548 nm to 553 nm, respectively. Remarkable is also the strong absorption decrease at 403 nm in MOP-(Ppheid)$_2$.

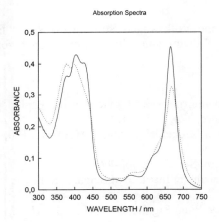
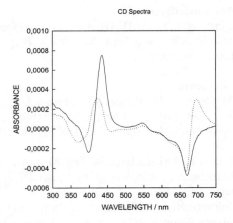

Figure 2: Absorption (left) and CD Spectra (right) of MOP-(Ni-Ppheid)$_1$ (——) and MOP-(Ni-Ppheid)$_2$ (·····).

The absorption spectra of MOP-(Ni-Ppheid)$_1$ and MOP-(Ni-Ppheid)$_2$ are shown on the left side of Figure 2. Again, as in the metal free MOP-Ppheid, there is a redshift of 4 nm within the Q_Y band from 664 nm (MOP-(Ni-Ppheid)$_1$) to 668 nm (MOP-(Ni-Ppheid)$_2$). Remarkable is the redshift broadening of the Q_Y absorption band. The Q_X absorption band shows similar characteristic red shifts. The Soret absorption shows, as in the metal free compounds, a decrease at 425 nm in the dimer. In addition there is an absorption increase at 376 nm.

CD Spectroscopy

The CD spectrum of the MOP carrying one chromophore shows only positive or negative bands with extrema close to the absorption maxima. By contrast, s-shaped features are present in the Q_Y and Soret regions, which are characteristic for interacting pigments. The zero crossings are to the red of the absorption maxima (Figure 2). There is an increase in the Q_Y absorption around 697 nm (s-shaped), which is absent in the MOP-(Ni-Ppheid)$_1$.

4. Conclusions

It is possible to bind 1 or 2 chlorophylls covalently in the hydrophobic pocket of a synthetic protein of the antiparallel four-helix-bundle type. The CD-spectra show the characteristics for monomeric pigments in the MOP carrying a single chlorophyll, and s-shaped pattern characteristics for dimeric pigments in MOP-(Ni-Ppheid)$_2$. From the lack of interaction signals in MOP-(Ni-Ppheid)$_1$ we conclude, that this interaction is intramolecular in MOP-(Ni-Ppheid)$_2$. The absorption spectra show for the 2 pigment compounds a for pigment interaction typical red shift. Similar experiments, like the one with Ni pigments, were done with Zn derivatives.

5. Acknowledgements

Support by the BMBF (Beo/22/0311152) and the Volkswagen Stiftung (to W.H.) is gratefully achknowledged. We thank Brigitte Wilhelm and Christa Reichenbach for technical assistance.

6. Literature

1. McDermott, G., Prince, S.M., Freer, A. A., Hawthornthwaite-Lawless, A.M., Papiz, M.Z., Cogdell, R.J. and Isaacs, N.W. (1995). Nature, 374, 517-521.
2. Deisenhofer, J., Epp, O., Miki, K., Huber, R. and Michel, H. (1984). J.Mol.Biol., 180, 385-398.
3. Rau, H.K. and Haehnel, W. (1998). J. Am. Chem. Soc., 120, 468-476.
4. Rosenbach-Belkin, V. (1988). Dissertation, Weizmann Institute of Science, Rehovot, Israel.
5. Chlorophylls. In: Handbook of Chromatography, Plant Pigments Vol. 1., Köst, H.-P., Ed., Boca Raton, CRC press, pp. 235-307.
6. Fischer-Drsek, R. (1995) Dissertation, Institute of Organic Chemistry, University of Fribourg, Switzerland.
7. Rau, H. K., De Jonge, N., Haehnel, W., (1998), PNAS, in press.

LIGHT-INDUCED ELECTRON TRANSFER BETWEEN A SUBSTITUTED TYROSINE AND [Ru(bpy)$_3$]$^{2+}$

Mark J. Burkitt, Yves Frapart, Ping Kenez, Licheng Sun[†], Markus Tamm[†], Ann Magnuson, Björn Åkermark[†] and Stenbjörn Styring

Biochemistry, Chemical Center, University of Lund, S- 221 00 Lund, Sweden; and [†]Royal Institute of Technology, Department of Organic Chemistry, S-100 44 Stockholm, Sweden

Key words: artificial systems, charge separation, EPR, free radicals, photochemistry

1. Introduction

In the Swedish Consortium for Artificial Photosynthesis[1], our aim is to develop sustainable systems of solar energy utilization based on synthetic supermolecules inspired by the natural photosynthetic apparatus.

In recent studies, we have shown that a tyrosine molecule covalently attached to [Ru(bpy)$_3$]$^{2+}$ undergoes rapid oxidation to a tyrosyl radical following photo-oxidation of the metal center in the presence of the electron acceptor [Co(NH$_3$)$_5$Cl]$^{2+}$ (1,2). Moreover, it has been shown that the inclusion of a synthetic Mn^{II}-Mn^{II} complex in this reaction system results in the rapid repair of the tyrosyl radical, accompanied by formation of the corresponding Mn^{II}-Mn^{III} complex (3). These reactions provide a functional mimic of the charge-separation reactions of Photosystem II. With the aim of improving the efficiency and extent of oxidizing equivalent accumulation on the manganese dimer, we are currently developing a generation of supermolecules in which the ruthenium, tyrosine and manganese units are components of a single supermolecule.

In the supermolecule **1b**, a (Mn^{II})$_2$ unit is bound to the tyrosine moiety of the tyrosine-[Ru(bpy)$_3$]$^{2+}$ dyad *via* nitrogen ligands from a pair of dipyridylamine (dpa) substituents and the tyrosine phenolate oxygen (Fig. 1). It was recently shown by laser flash-photolysis studies, involving the observation of [Ru(bpy)$_3$]$^{2+}$ recovery at 450 nm, that an electron is transferred rapidly from the tyrosine-dpa-(Mn^{II})$_2$ moiety of **1b** following photo-generation of [Ru(bpy)$_3$]$^{3+}$ (4). The purpose of the investigations described in this report was to use electron paramagnetic resonance (EPR) spectroscopy to determine whether or not the dpa-substituted tyrosine phenol moiety undergoes oxidation during this process, as described above for the simpler molecule in which electron transfer from the manganese dimer is intermolecular. We were particularly interested in the effects of the dpa substituents on phenoxyl radical formation, and therefore conducted the EPR studies reported here on the manganese-free ligand **1a** (Fig. 1).

---

[1] The Swedish Consortium for Artificial Photosynthesis comprizes members from the Department of Biochemistry, University of Lund; the Department of Physical Chemistry, University of Uppsala; and the Department of Organic Chemistry at the Royal Institute of Technology, Stockholm.

G. Garab (ed.), Photosynthesis: Mechanisms and Effects, Vol. V, 4217–4220.
© *1998 Kluwer Academic Publishers. Printed in the Netherlands.*

2. Materials and Methods

2.1 Synthesis
The synthesis of compounds **1a** and **1b** will be reported elsewhere.

2.2 Detection of a phenoxyl radical from **1a** in aqueous solution
To a saturated aqueous solution of $[Co(NH_3)_5Cl]^{2+}$ was added DMSO (10 %, v/v) and **1a** (2 mM). An EPR spectrum was recorded immediately upon the commencement of continuous laser-flashing (523 nm, 5 Hz, 1.5 W) into the cavity of a Bruker ESP 300 spectrometer from a GCR-200 Series Nd:Yag laser. Diphenylpicrylhydrazyl (in 10 % aqueous DMSO) was used as a g marker ($g = 2.0036$).

2.3 Detection of a phenoxyl radical from **1a** in non-aqueous solvent
The reaction mixture, containing **1a** (2 mM) and methyl viologen (MV.PF$_6$) in acetonitrile, was subjected to continuous laser-flashing at 5 Hz (as above) for one minute, after which an EPR spectrum of the resultant methyl viologen radical cation was recorded. A kinetic spectrum of underlying, transient species was then constructed from time-mode absorption measurements using field positions at which absorption by the methyl viologen radical is zero. Diphenylpicrylhydrazyl in acetonitrile was used as a g marker.

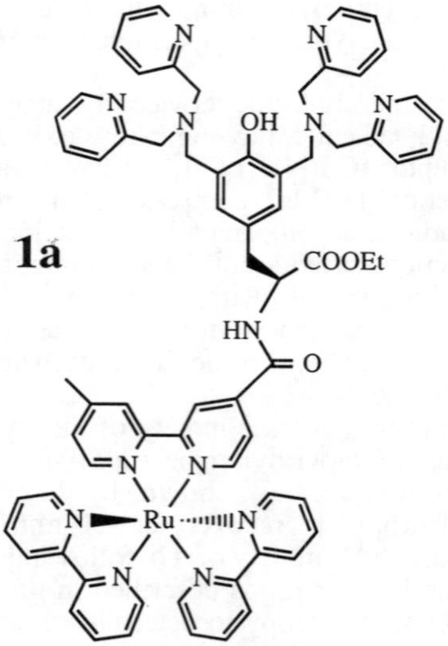

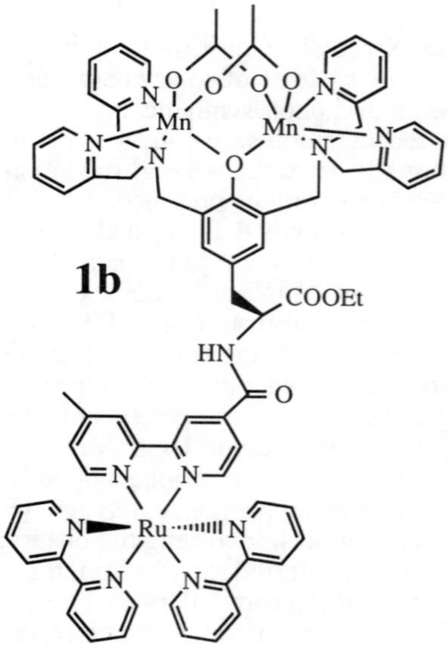

Figure 1. Structures of the molecules employed in this study.

3. Results and Discussion

When a solution of **1a** and methyl viologen in acetonitrile was subjected to continuous laser-flashes, the highly characteristic EPR spectrum of the MV$^+$ radical cation was observed (not shown). The detection of the methyl viologen radical demonstrates photo-induced electron transfer from **1a** to an external acceptor, methyl viologen, thereby confirming that the $[Ru(bpy)_3]^{2+}$ complex retains its photo-physical properties when incorporated into the supramolecular structure **1a**.

The presence of the strong EPR signal from MV^{+} precluded the straightforward detection of signals from any other radicals (*e.g.*, phenoxyl), therefore an alternative electron acceptor, $[Co(NH_3)_5Cl]^{2+}$, was employed in preliminary attempts to explore the possibility of radical formation following photo-oxidation at the $[Ru(bpy)_3]^{2+}$ site in **1a**.

The continuous, *in situ* laser-flashing of an aqueous solution of **1a** in the presence of $[Co(NH_3)_5Cl]^{2+}$ resulted in the generation of a phenoxyl radical, as evidenced by its EPR signal, having an isotropic g value of 2.0044 (5) (Fig. 2). The detection of this radical confirms that the ruthenium(III) generated in **1a** upon illumination is able to abstract an electron from the tyrosine substituent. Optical kinetic studies have confirmed that that this occurs *via* intramolecular electron transfer (4).

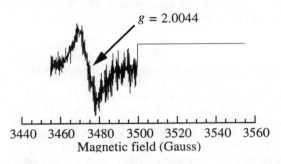

$g = 2.0044$

3440 3460 3480 3500 3520 3540 3560
Magnetic field (Gauss)

Figure 2. EPR spectrum of a neutral phenoxyl radical observed during the continuous laser-flashing of **1a** in an aqueous solvent system containing $[Co(NH_3)_5Cl]^{2+}$.

Complex **1b** is stable in acetonitrile, but loses manganese in water. Consequently, it was considered necessary - to be of value as a model compound for **1b** - to also investigate phenoxyl radical generation from **1a** in acetonitrile. Since the MV^{+} radical absorbs strongly at 600 nm, the PF_6 salt of this acceptor (MV. PF_6) has been employed in optical kinetic studies on the photochemistry of **1a** in acetonitrile (4). For these reasons, we also considered it necessary to demonstrate phenoxyl radical formation from **1a** in acetonitrile using MV. PF_6 as the acceptor. This was attempted using a kinetic approach, in which EPR absorptions were recorded (in time-mode) at magnetic fields at which the multi-line spectrum of the MV^{+} radical has no absorption, *i.e.* where its signal crosses the baseline. Using traces obtained at several such positions, a transient spectrum was constructed, revealing clearly the presence of an underlying signal. This signal was found to have a g-value of *ca.* 2.0029, suggesting it is from a protonated phenoxyl radical, $PhOH^{+}$ (5) (Fig. 3).

This finding indicates that the rapid deprotonation which accompanies (or precedes) electron transfer from the tyrosine phenolate moiety to the photo-generated Ru(III) in the aqueous solvent system, generating $PhO^{\cdot}$, does not occur in acetonitrile. This is consistent with the very low pK_a values of phenoxyl radicals, which have been observed previously in their protonated forms only in non-aqueous solvents (5).

4220

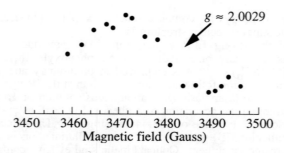

$g \approx 2.0029$

| | | | | | |
|3450|3460|3470|3480|3490|3500|

Magnetic field (Gauss)

Figure 3. Kinetic EPR spectrum attributed to a phenol radical cation, generated as a transient species following the application of a single laser-flash to **1a** in acetonitrile containing methyl viologen.

4. Conclusions

Taken together with complementary findings from optical spectroscopy (*4*), the present EPR studies confirm that $[Ru(bpy)_3]^{2+}$ retains its photo-physical properties when incorporated into the supramolecular structures **1a** and **1b**. The observation of electron transfer from the phenol moiety in **1a** to the photo-generated $[Ru(bpy)_3]^{3+}$ unit is crucial to the development of charge-separation systems to be used ultimately for the oxidation of water because it demonstrates that the recovery of $[Ru(bpy)_3]^{2+}$ is not *via* simple recombination. This biomimetic model of the light-induced charge separation reactions of Photosystem II will be now be developed further, with the aim of accumulating sufficient oxidizing equivalents on **1b** to bring about the oxidation of water: complementary cyclic voltammetry studies have shown that the redox potentials associated with the oxidation of **1b** are within the necessary range to achieve this goal (*4*).

5. Acknowledgments

This work was supported by grants from the Knut and Alice Wallenberg Foundation, the European TMR program (TMR network CT96-0031), the Nordic Energy Research Program and the Swedish Natural Science Research Council.

6. References

1. Sun, L., Hammarström, Norrby, T., Berglund, H., Davydov, R., Andersson, M., Börje, A., Korall, P., Philouze, C., Almgren, M., Styring, S., and Åkermark, B. (1997) *Chem. Comm.*, 607-608.
2. Sun, L. Berglund, H., Davydov, R., Norrby, T., Hammarström, L., Korall, P., Börje, A., Philouze, C., Berg, K., Tran, A., Andersson, M., Stenhagen, G., Mårtenson, J., Almgren, M., Styring, S. and Åkermark, B. (1997) *J. Am. Chem. Soc.* **119**, 6996-7004.
3. Magnuson, A., Frapart, Y., Horner, O., Åkermark, B., Sun, L., Girerd, J.-J. and Styring, S. (manuscript submitted for publication).
4. Hammarström *et al.*, to be published.
5. Dixon, W. T. and Murphy, D. (1976) *J. Chem. Soc., Faraday Trans. 2*, **72**, 1221-1230.

RADICAL MAQUETTES AS MODELS OF RADICAL ENZYMES

Cecilia Tommos,[1] Jack J. Skalicky,[2] A. Joshua Wand,[2] and P. Leslie Dutton[1]
[1]Johnson Research Foundation, Department of Biochemistry and Biophysics,
University of Pennsylvania, Philadelphia, PA 19104; [2]Department of Chemistry,
State University of New York, Buffalo, NY 14260

amino acid, design of proteins, electron transfer, free radicals, protein radicals, proton transfer

A most interesting development in recent years is the number of redox proteins that have been recognized to employ amino acids, including tyrosine, tryptophan, cysteine, glycine and derivative thereof, as redox cofactors [1]. The use of sidechains as endogenous cofactors raises the question as to how these amino acids specifically are targeted for redox chemistry. One theme emerging for the systems studied thus far is the presence of a metal site in the immediate vicinity of the redox-active sidechain that generates the oxidizing potential necessary to form the radical, which, in turn, reacts with the substrate. Photosystem II in oxygenic photosynthesis provides an exception as a light-activated chlorophyll complex, P680, oxidizes two nearby tyrosines, denoted Y_Z and Y_D, in which the former is directly involved in water oxidation [2,3]. Redox-active sidechains change their pK values dramatically upon oxidation and in several systems, coupled deprotonation and reprotonation reactions are part of the catalytic cycle. This suggests protein-mediated mechanisms to allow a directional flow of protons during

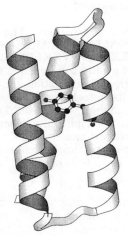

Figure 1. Model of α_3Y^1.

the redox reactions. In recent models for the water-oxidizing process, for example, redox-induced proton exchange at Y_Z has been proposed to serve as the device by which the product protons are delivered from the catalytic site and towards the thylakoid lumen [2-4]. In some enzymes, the radical is confined and directed towards local chemistry. In other systems, such as the ribonucleotide reductase enzymes involved in DNA synthesis, the initial radical is generated at a position remote from the site of catalysis and radical migration is involved in substrate turnover [1]. The high reactivity, combined with the ease by which protein radicals migrate, suggest that they must be under strict biological control so as not to damage the host organism. To delineate the general principles of protein radicals is of considerable interest and here we describe a novel approach both to challenge our current knowledge of these systems and, if successful, to serve as a model system for radical enzymes. Our aim is to design and construct a sidechain radical maquette - a simplified synthetic protein containing an amino acid that is targeted for redox chemistry. The design of the first generation of radical maquettes is aimed to satisfy four basic criteria. The synthetic protein should; i) contain well-defined secondary and tertiary structures, ii) contain a single cofactor located within the protein core and shielded from the bulk solution, and otherwise

G. Garab (ed.), Photosynthesis: Mechanisms and Effects, Vol. V, 4221–4224.

iii) be "redox inert" and not contain residues expected to have reduction potentials close to the chosen cofactor. Finally, besides the absorption by the cofactor, it should iv) have a small extinction coefficient at 280 nm. With this as a basis, two small globular proteins including a single tyrosine or tryptophan have been constructed. Figure 1 shows a working model for the maquettes, which amino-acid sequences only differ in residue 32, the position of the cofactor. The basic sequence used for the radical maquette system [5] was derived from the 3-helix bundle family of synthetic coiled-coil peptides developed by DeGrado and coworkers [6-8]. The sequence of the single-stranded helix-loop-helix-loop-helix peptide, designed α_3 [5,8], contains a total of 65 residues. Each helix is designed to contain 19 residues and the three helices are joined by two glycine loops each of four residues [5]. The core of the protein consists of six hydrophobic layers composed of 8 leucines, 3 isoleucines, 6 valines, and the aromatic sidechain. The exterior positions are occupied by 17 glutamates, 17 lysines, 2 arginines and 3 alanines, arranged to satisfy both intra- and inter-helical electrostatic interactions. In the following we describe the synthesis and characterization of the tyrosine and tryptophan radical maquettes, denoted α_3Y^1 and α_3W^1, respectively.

Experimental Section

Protein Synthesis. The α_3Y^1 and α_3W^1 peptide chains were as follows: Ac-R•VKALEEK•VKALEEK•VKAL-GGGG-R•IEELKKK•(Y/W)EELKKK•IEEL-GGGG-E•VKKVEEE•VKKLEEE•IKKL-*CONH$_2$*. They were assembled on a MilliGen 9050 solid-phase synthesizer by using standard Fmoc/^{t}Bu chemistry on a 0.2 mmol scale. Following N-terminus acetylation, the synthesized peptide was cleaved from the solid resin support and its sidechains concomitantly deprotected in a trifluoroacetic acid (TFA):1,2-ethanedithiol:water (90:8:2) mixture for 4.5 h. The resin was filtered off and the crude product precipitated in ether, dissolved in water with 0.1% TFA, lyophilized, and purified by reversed-phased HPLC on a C18 column with an acetonitrile:water:0.1% TFA gradient. The purity of the final products was assayed by analytical HPLC and their molecular masses, 7417 and 7440 Da for α_3Y^1 and α_3W^1, respectively, were verified by mass spectroscopy.

Characterization of Secondary and Tertiary Structures. The secondary structure and thermodynamic stability of the synthesized proteins were studied by CD and NMR spectroscopies. Figure 2 displays the far-UV spectrum of α_3Y^1, which shows the positive 192 nm and negative 208 and 222 nm features characteristic of an α-helical structure. The α_3W^1 system gave rise to an essentially identical spectrum (not shown). From the CD intensity at 222 nm, we estimate that the proteins are 65-70% α-helical [9]. In addition, numerous NH-NH NOESY correlations are observed that is consistent with α-helical secondary structures (not shown). The stability of the synthetic proteins at 25° C relative to their chemically denaturated states was studied by monitoring the reversible loss of their secondary structure, probed by the decrease in ellipticity at 222 nm, as a function of guanidine hydrochloride concentration. The α_3Y^1 and α_3W^1 proteins melt with sharp transitions displaying m values of 2.8 and a midpoint of unfolding at 1.9 and 2.2 M Gdn:HCl, respectively (not shown).

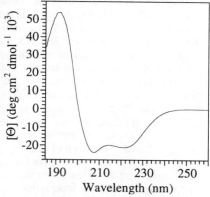

Figure 2. CD spectrum of α_3Y^1.

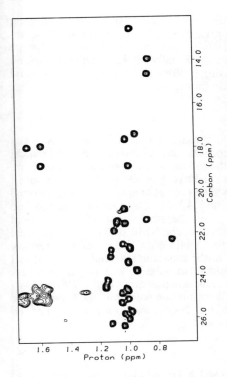

Assuming a two-state folded/unfolded event, we estimate the free energy of unfolding in the absence of denaturant to be 5.4 and 6.2 kcal/mol for α_3Y^1 and α_3W^1, respectively. These values are within the range observed for natural proteins of similar size [10]. The steep melting transitions indicate that the unfolding events are highly cooperative, which implies well-packed hydrophobic cores for these two systems. Consistent with this, Figure 3 shows a ^{13}C-HSQC NMR spectrum obtained on α_3Y^1, which reflects the magnetic environment of the methyl groups in the protein. The ^{1}H-^{13}C crosspeaks are well resolved, indicating that the methyl groups reside in unique magnetic surroundings and provides evidence that α_3Y^1 has an well-defined tertiary structure in solution. The magnetic environment in random coil or molten globule structures is more homogenous and would give rise to methyl correlations in a much narrower ^{1}H and ^{13}C chemical shift range than is observed in Figure 3. The presence of a single set of methyl ^{1}H-^{13}C correlations further suggests that the protein is populating a single conformer in solution.

Figure 3. ^{13}C-HSQC NMR spectrum of α_3Y^1.

Characterization of the Tyrosine and Tryptophan Sidechains. Fluorescence spectroscopy was performed to determine if the tryptophan in α_3W^1 is buried in the interior of the protein as designed. The dotted line in Figure 4 represents tryptophan in solution, which has an emission maximum at 360 nm. The solid line represents the fluorescence from α_3W^1 at neutral pH. The emission maximum is blueshifted to 324 nm, consistent with a tryptophan in a buried position [11]. Similar, measurements on α_3Y^1 established that the tyrosine is also shielded from solvent. Tyrosine has an absorption maximum at 275 nm, which, upon deprotonation of the phenol headgroup, redshifts to 293 nm. Titration of the tyrosine in α_3Y^1, monitored by the absorbance at 293 nm, reveals a pK of 11.3, which is 1.2 pH units higher as compared to free tyrosine in solution and is consistent with a position shielded from the bulk medium. In order to monitor the intactness of α_3Y^1 within the pH range studied, changes in protein folding was monitored by CD. The α_3Y^1 protein starts to lose its secondary structure concomitant with the de-protonation reaction. Most likely, these two events are coupled, that is, high pH disrupts the secondary and tertiary structures, the dielectric of the protein rises and allows the tyrosine to deprotonate. The creation of a charge, in turn, speeds up the unfolding process. Consistent with this notion,

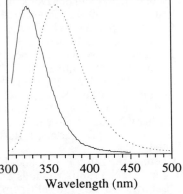

Figure 4. Fluorescence spectra from tryptophan (dotted line) in solution and (solid line) in α_3W^1.

α_3Y^1 has ~25% left of its α-helical structure at pH 13.5. In contrast, the α_3W^1 system, which cofactor is still protonated and neutral at this pH, has ~50% of its secondary structure still intact. This suggests that the pK shift of 1.2 pH units represents the lower limit for the protein-induced change in acidity of the buried tyrosine. Finally, NOESY data obtained on α_3Y^1 reveal dipolar couplings between the phenol ring protons and several nearby methyl groups, indicating a buried position of this sidechain (not shown).

Summary

A number of proteins have been recognized in which redox-active amino acids play key roles in catalysis. Here we describe the design and construction of synthetic proteins, radical maquettes, aimed to study the interplay between the radical and its protein environment. Initially, we have focused on the aromatic cofactors and two globular proteins including a single tyrosine or tryptophan have been developed. The radical maquettes are synthesized as 65 residue peptides with molecular weights around 7.4 kDa. Their are designed to fold into global structures containing three α-helices with the single aromatic amino acid positioned within the hydrophobic core. Circular dichroism and NMR studies on the purified proteins show that they form thermodynamically stable, well-defined secondary and tertiary structures. Spectral characteristics and pKa shifts of the aromatic sidechains suggest that they are located within the protein interior and shielded from the bulk phase, as designed. Voltammetry studies are in progress to measure the reduction potentials of the Y˙/Y and W˙/ W redox couples in the synthetic proteins relative to their solution potentials and to determine how these values change as a function of protein folding. These measurements are intended to guide in further development of the radical maquettes.

Acknowledgments.

This work was supported by NIH Grants GM41048 to P.L.D. and GM35940 to A.J.W, and NSRA grant GM18121 to J.J.S. C.T. gratefully acknowledges support from STINT. We thank Professor DeGrado for making avaible α_3 design information prior to publication.

References

1 Stubbe, J. and van der Donk, W. A (1998) Chem. Rev. 98, 705-762.
2 Britt, R. D. (1996) In Oxygenic Photosynthesis: The Light Reactions (Ort, D. R.; Yocum, C. F. eds.) pp.137-164, Kluwer Academic Publishers, Dordrecht, The Netherlands.
3 Tommos, C. and Babcock, G. T. (1998) Acc. Chem. Res. 31, 18-25.
4 Hoganson, C. W. and Babcock, G. T. (1997) Science 277, 1953-1956.
5 Johansson, J. S., Gibney, B. R., Skalicky, J. J., Wand, A. J., Dutton, P. L. (1998) J. Am. Chem. Soc. 120, 3881-3886.
6 Lovejoy, B., Choe, S., Cascio, D., McRorie, D. K., DeGrado, W. F., Eisenberg, D. (1993) Science 259, 1288-1293.
7 Ogihara, N. L., Weiss, M. S., DeGrado, W. F., Eisenberg, D. (1997) Prot. Sci. 6, 80-88.
8 Bryson, J. W., Desjarlais, J. R., Handel, T. M., DeGrado, W. F. (1997) Prot. Sci. 6, 1404-1414.
9 Chen, Y-H., Yang, J. T., Chau, K. H. (1974) Biochem. 13, 3350-3359.
10 Pace, C. N. (1986) Methods Enzymol. 131, 266-280.
11 Lakowicz, J. R. (1983) Principles of Fluorescence Spectroscopy, Plenum Press, New York, USA.

Axial Ligand Coordination and Photodissociation of Nickel Substituted Bacteriochlorophyll-a

Dror Noy[1], Vlad Brumfeld[1], Idan Ashur[1], Roie Yerushalmi[1], Hugo Scheer[2] and Avigdor Scherz[1*]
1. Dept. of Plant Sciences, Weizmann Inst. of Science, Rehovot 76100, Israel
2. Botany Dept., University of Munich, Menzinger Str. 67, D-80638 Munich, Germany

Key words: axial ligands, electronic structure, excited state lifetime, photochemistry, pigment binding, ultrafast spectroscopy.

1. Introduction

The study of metalloporphyrin-containing enzymes involves fundamental issues of chemistry such as catalysis, coordination chemistry, as well as charge and light-energy transfer at different time domains. Nickel porphyrins ([Ni]-Pors) have proved to be very useful biomimetic models in such studies both *in vitro* and in reconstituted proteins (1,2). Their special electronic properties, i.e. ultrashort excited-state lifetimes, reflect the interactions between the nearly degenerate frontier molecular orbitals (FMOs) of the porphyrin π-system and the d^8 electronic configuration of the central Ni(II) atom, which may be either low spin (doubly occupied d_{z^2} and empty $d_{x^2-y^2}$) or high spin (singly occupied d_{z^2} and $d_{x^2-y^2}$), depending on the state of coordination (3). At the same time, the degeneracy of the FMOs in D_{4h} porphyrins causes complicated and partially unresolved electronic spectra that may obscure any detailed interpretation of the [Ni]-Pors photophysics and photochemistry.

Metal-substitution in bacteriochlorophylls (BChls), as we have recently suggested (4,5), has opened the way to overcome these shortcoming. Their four non-degenerate excited configurations cause four excited singlet and four excited triplet states that can be followed separately. Our previous investigations have indicated that Nickel-substituted BChl ([Ni]-BChl) is the only metal derivative in which the central atom changes its coordination number from four to six in the presence of pyridine (Py). On the other hand, when incorporated into the accessory BChl sites of bacterial reaction centers (RCs), [Ni]-BChl was found to be coordinated only by a single histidine residue (6). Consequently we began to search for five-coordinated [Ni]-BChls *in vitro*. Having [Ni]-BChl in three states of coordination provided a unique opportunity to study the effects of axial ligands on its electronic structure, photochemistry, and photophysics. Specifically, we succeeded in resolving the optical absorption spectra of [Ni]-BChl in all three coordination states, in the presence of the strong axial ligands Py and imidazole (Im). In addition, we observed the ultrafst photodissociation of the ligands.

G. Garab (ed.), Photosynthesis: Mechanisms and Effects, Vol. V, 4225–4228.
© *1998 Kluwer Academic Publishers. Printed in the Netherlands.*

2. Materials and Methods

2.1 Experimental

[Ni]-BChl was prepared according to Hartwich et al. (4). Solutions of dry acetonitrile (AN) and Py (Aldrich) were kept in vials sealed with rubber septa and transferred via nitrogen-purged syringes. A 3M solution of Im (Sigma) in dry AN was prepared just before measuring. Absorption spectra of [Ni]-BChl in dry AN were recorded on a Cary 5 UV-VIS-IR double-beam spectrophotometer using the baseline correction mode in a standard 10-mm quartz cuvette sealed with a rubber septum and purged with nitrogen. Py or Im was added through the septum by using a microliter syringe. A pump-probe setup was used for time-resolved absorption spectroscopy. The sample was excited by using 690 nm, 14 ps, 90 mJ pulses generated by a Continuum PD10 dye laser that was pumped by a frequency-doubled Nd-YAG laser (532 nm, 38 ps). White light pulses (480-860 nm), generated by focusing the residual fundamental beam (1064 nm, 38 ps) of the Nd-YAG laser on a 5-cm pathlength cylindrical water cell, were used for probing. The pump-probe interval was controlled by an optical delay line. Spectra were detected by using a double diode array (DIDA 512), utilizing dedicated software (Winspec), all from Princeton Instruments. Pump pulse intensity was monitored by a power meter (Coherent LabMaster). The measurement accuracy was about 0.01 OD units. Solutions of [Ni]-BChl in 0.5 ml diethyl ether (DE) and pure Py were measured in a 0.2-mm quartz cuvette sealed with a Teflon stopper to prevent solvent evaporation. Py was added using a microliter pipettor.

2.2 Data analysis

Analysis of the digitized spectra was carried out with Matlab® software using custom programs as well as Matlab's standard optimization and statistics toolboxes.

Modeling factor analysis (7) was used for separating the highly overlapping spectra of the different coordination states of [Ni]-BChl. Our evolution model considered two equilibrium constants

$$[Ni]\text{-}BChl + L \leftrightarrows [Ni]\text{-}BChl.L \qquad\qquad K_1, \qquad (1)$$

$$[Ni]\text{-}BChl.L + L \leftrightarrows [Ni]\text{-}BChl.L_2 \qquad\qquad K_2,$$

where L = Py or Im. Thus, the mole fraction of each coordination state was a function of the axial ligand-binding constants and could be determined by searching for the maximum likelihood of K_1, and K_2, which best reproduced the experimental data. Confidence intervals for K_1 and K_2 were calculated according to O'Dea et al. (8).

The Q_y transition bands from the isolated spectra of the [Ni]-BChl species were symmetrized by reflecting their low energy half onto their high-energy side and fitted with a Voigt function(convolution of Gausssian and Lorentzian). The excited-state lifetimes were estimated from the full width at half maximum of the Lorentzian component of this function (9).

3. Results and Discussion

3.1 Spectral isolation of [Ni]-BChl coordination states

No clear characteristic band could be observed for [Ni]-BChl.L in AN/Py mixtures. However, the systematic red shift of the "isosbestic" point in the Q_x region indicated the evolution of [Ni]-BChl.Py with increasing concentrations of Py. When L = Im, the five-coordinated species caused a distinct characteristic band at 573 nm. Nevertheless, we were able resolve distinct absorption spectra related to [Ni]-BChl, [Ni]-BChl.L (Fig. 1) and [Ni]-BChl.L$_2$. Axial ligand-binding constants were found to be $3.28 < K_1 < 4.0$ and $36 < K_2 < 72.5$ for L = Py whereas for L = Im, we observed a larger K_1 but smaller K_2. The recovered spectra are characteristic of BChl in different coordination states, as recently suggested by Hartwich et al. (4) and Noy et al. (5).

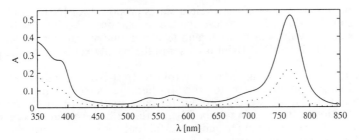

Figure 1. Spectrum of [Ni]-BChl in dry AN containing 49 mM Im (–) compared with that of pure [Ni]-BChl.Im isolated by factor analysis (····)

As expected, the Q_x transition was the most sensitive to axial coordination. It was red shifted from 534 nm in [Ni]-BChl to 571 and 573 nm 'in [Ni]-BChl.Py and [Ni]-BChl.Im, respectively and even further to 596 and 607 nm in [Ni]-BChl.Py$_2$ and [Ni]-BChl.Im$_2$, respectively. Note that the ligand identity has a minor but additive effect on the Q_x band position. The lowest energy electronic transition band, Q_y, was blue shifted in [Ni]-BChl.L and [Ni]-BChl.L$_2$ by a similar amount relative to [Ni]-BChl because of the $d_{x^2-y^2}$ metal orbital occupation (3). We have therefore assumed that both [Ni]-BChl.L and [Ni]-BChl.L$_2$ have a triplet ground state as suggested for five-coordinated Ni-protoporphyrin in Hemoglobin (1).

3.2 Excited state lifetimes from spectral lineshapes

The profile of the Q_y transition band of [Ni]-BChl appears to be strongly Lorentzian and could be fitted with a Voigt function (10). The same transition in [Ni]-BChl.L appears to be more Gaussian. Assuming that the Lorentzian character corresponds to natural line-broadening because of ultrafast relaxation of the singlet excited state, the excited state lifetimes appeared to depend on the ligand field, increasing from less than 40 fs in square-planar [Ni]-BChl to ~110 fs in square-pyramidal [Ni]-BChl.L.

3.3 Time-resolved absorption spectroscopy of [Ni]-BChl in the presence of pyridine

Picosecond absorption spectroscopy of [Ni]-BChl.Py$_2$ in pure Py and in DE with different Py concentrations indicated an exponential ground-state recovery with a time constant of 100 ps, both in pure Py and in DE containing Py. Nevertheless, the recovery was still incomplete even after 400 ps. However, bleaching of the Q_y and Q_x absorption bands of [Ni]-BChl.Py$_2$ at 763 and 596 nm, respectively, was accompanied by an absorption increase at 780 and 534 nm, corresponding to the Q$_y$ and Q$_x$ absorption bands of [Ni]-BChl, respectively. The transient spectra could be simulated by a linear combination of pure [Ni]-BChl, [Ni]-BChl.Py and [Ni]-BChl.Py$_2$ spectra. No bleaching from [Ni]-BChl in pure DE was detected, indicating that ground state recovery is faster than our time resolution.

Conclusions

The distinct absorption spectra of four-, five- and six-coordinated [Ni]-BChl were isolated from the spectra of [Ni]-BChl in dry AN containing Py or Im ligands. To the best of our knowledge, the absorption spectra of pure five-coordinated [Ni]-Por were never isolated, neither *in vivo* nor *in vitro*.

The mechanisms for the relaxation of excited states proposed for [Ni]-Por (11,12) were adapted with some modifications to describe the relaxation of excited [Ni]-BChl. This relaxation probably involves ultrafast ligand-to-metal charge transfer. However, since the highest occupied MO of [Ni]-BChl is closer to the metal excited state, the relaxation is faster than in [Ni]-Por.

The ultrafast radiationless transitions that follow [Ni]-BChl excitation result in a substantial dissipation of heat to the immediate surroundings. Moreover, photodissociation of axially ligated protein residues may result in strong but local vibrational excitation of the protein matrix. Therefore, we believe that RCs reconstituted with [Ni]-BChl can serve as a unique and powerful tool for studying protein vibrational dynamics and its effects on photoinduced electron transfer.

Acknowledgement

This study was supported by SFB grant #533 (project A6) and the Willstater-Avron-Minerva foundation for photosynthesis.

References

1 Shelnutt, J. A.; Alston, K.; Ho, J. Y.; Yu, N. T.; Yamamoto, T. and Rifkind, J. M. (1986) Biochemistry 25 620-627.
2 Alden, R. G.; Ondrias, M. R. and Shellnut, J. A. (1990) J. Am. Chem. Soc. 112 691-697.
3 Ake, R. L. and Gouterman, M. (1970) Theoret. Chim. Acta 17 408-416.
4 Hartwich, G.; Fiedor, L.; Simonin, I.; Cmiel, E.; Schafer, W.; Noy, D.; Scherz, A. and Scheer, H. (1998) J. Am. Chem. Soc. 120 3675-3683.
5 Noy, D.; Fiedor, L.; Hartwich, G.; Scheer, H. and Scherz, A. (1998) J. Am. Chem. Soc. 120 3684-3693.
6 Chen, L. X.; Wang, Z. Y.; Hartwich, G.; Katheder, I.; Scheer, H.; Scherz, A.; Montano, P. A. and Norris, J. R. (1995) Chem. Phys. Lett. 234 437-444.
7 Malinowski, E. R. (1991) Factor Analysis in Chemistry 2nd Ed. Wiley, New York .
8 O'Dea, J.; Osteryoung, J. and Lane, T. (1986) J. Phys. Chem. 90 2761-2764.
9 Hartwich, G.; Friese, M.; Scheer, H.; Ogrodnik, A. and Michel-Beyerle, M. E. (1995) Chem. Phys. 197 423-434.
10 Jansson, P. A. (1984) in Deconvolution with application in spectroscopy (Jansson, P. A. ed.) pp. 1-34 Academic Press, New York.
11 Rodriguez, J. and Holten, D. (1989) J. Phys. Chem. 91 3525-3531.
12 Rodriguez, J. and Holten, D. (1990) J. Phys. Chem. 92 5944-5950.

STABLE PHOTOSYNTHETIC COMPLEX OF *RHODOSPIRILLUM RUBRUM* IN REVERSE MICELLES

C. Obregón[1], A. Srivastava[1], A. Darszon[2] and Reto J. Strasser[1]
[1]Bioenegetics Laboratory, University of Geneva, CH-1254, Geneva, Switzerland, [2]Inst. Biotechnology, UNAM, Cuernavaca, Morelos, México

Key Words: Bacterial photosynthesis, BChl fluorescence induction, OJIP transient, phospholipid complex, purple bacteria, low water systems.

1. Introduction

Water plays a fundamental role in determining the reaction rates and the thermodynamics of biological systems. However, the way in which water influences catalysis, structure and organization are not completely understood. One approach to resolve this problem is through low water system (water-in-oil-microemulsions) (1,2). The activity of water soluble enzymes, and to a lesser extent membrane proteins, in reverse micelle solutions have been studied extensively (1,2). Reverse micelles allow the possibility of controlling the amount of water available to proteins, organelles and cells housed in their interior (1). Here we have shown how water influences the function and organization of photosynthetic complexes. Although the photosynthetic complexes of spinach thylakoid membranes, and green alga *C. reinhardtii* can be "solubilized" in various reverse micellar systems (3), only the bacterial photosynthetic complex (BPC) could be transferred either to HTS (Hexadecane-Tween-Span) or PLC (Phospolipid complex) reverse micelles in a stable functional form. Bacteriochlorophyll (BChl) fluorescence induction transients (OJIP) and photo-oxidation of the reaction center, P, measured as 820nm absorption changes (4), have been used to evaluate the photosynthetic capacity of the bacterial reaction center. Subsequently, we have studied the stability of BPC in PLC-liquid preparation and BPC in PLC after immobilizing them in resin. HTS-reverse micellar systems have been used to investigate the effect of water on the primary photosynthetic activity.

2. Procedure

Wild type cells of *Rhodospirillum rubrum* were grown anaerobically in M-medium with succinate as carbon source (5). BPC in PLC-Reverse Micelles: Soybean phospholipids were dissolved (10 mg) in isooctane (1 ml). *R. rubrum* cells, equivalent to about 1 mg protein in 70 μl, were added to 1 ml phospholipid reverse micelles. The mixture was sonicated for 15 min in a water bath sonicator (Bandelin Sonorex Super RK-102H) under dim light at 15-20°C. After sonication, 100 μl of ice cold 1 M $MgCl_2$ was added to the emulsion, which was then vortexed for 1-2 min. The preparation was further centrifuged for 3-5 min at 4000 RPM (Universal/K2S, Whitish, Tuttlingen, Germany), and the organic fraction containing the BPC in PLC was collected from the top of the tube. The heavy fraction of the BPC in PLC was collected after further centrifuging the PLC at 35 000 RPM (Kontron T-1080 with TFT 80.4 rotor). The heavy fraction of BPC

G. Garab (ed.), Photosynthesis: Mechanisms and Effects, Vol. V, 4229–4232.
© *1998 Kluwer Academic Publishers. Printed in the Netherlands.*

in PLC was used to test the water dependency of primary photosynthetic activity of the BPC in HTS-reverse micelles (Twin 85 (77 mM) and Span 80 (8.6 mM) in hexadecane). The BPC in PLC-reverse micelles were either used directly or after the solvent was evaporated with N_2 gas. The semisolid BPC in PLC-reverse micelles were also embedded in commercially available Araldite.

BChl fluorescence induction kinetics were measured with a fluorometer (Plant Efficiency Analyzer, PEA, Hansatech Ltd., King's Lynn, Norfolk, England) with 650 nm of 600 Wm^{-2} light intensity (6). BChl fluorescence signals were detected using a PIN photocell after passing through a 890 nm filter. Light induced absorption changes at 820 nm were measured with the Hansatech P700 measuring system in combination with the PEA head, replacing one 650 nm LED (light emitting diode) with a 820 nm LED. Absorption spectra were recorded by using Perkin-Elmer, Lambda 3, UV/VIS spectrophotometer.

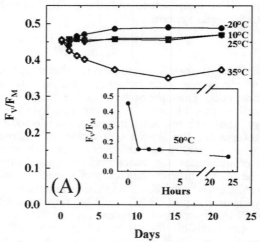

Figure 1. Effect of temperature on the stability of BPC in PLC probed by OJIP BChl fluorescence induction kinetics and the photooxidation of the reaction center, P, at 820 nm.

(A) The PLC containing BPC were kept at different temperatures in the liquid preparation in a small vial. Before measurement, the vials were kept at room temperature for 30 min. Insert shows F_V/F_M for the vial kept at 50°C. The x-axis for the main figure (A) is shown in days and the insert in hours.

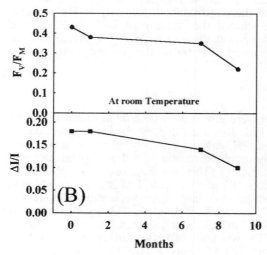

(B) The organic phase of the PLC containing BPC was evaporated under N_2 gas and the semisolid bacterial photosynthetic complex in PLC was then mixed with the resin (Araldite). An aliquot was dropped on a microscopic slide and left to dry for about 1 hr. Changes in the F_V/F_M ratio and photooxidation of the reaction center, P, measured as light-induced absorption changes, were measured over a period of months. At the beginning the room temperature was between 22 to 25°C, but in the later part of the experiment it had increased up to 32 to 35°C.

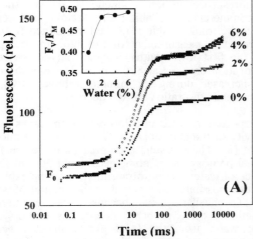

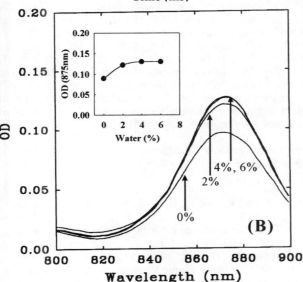

Figure 2: Effect of different amounts of water on the primary phochemical activity of BPC in PLC-reverse micelles (heavy fraction, as described in the procedure). Water concentrations in the HTS-reverse micelles were pre-adjusted and then BPC in PLC was added. Although they were stable for several hours, data shown here were obtained after 30 min of incubation.

Fig (A): Effect of water on the BChl fluorescence induction kinetics. BPC in PLC were transferred in HTS-reverse micelles and exposed to 600 W m^2 s^{-1} of red actinic light. The insert in (A) exhibits the effect of water on the maximum quantum yield of primary photochemistry, F_V/F_M

Fig (B): Effect of water on the absorption spectra of BPC in PLC after transferring them in HTS-reverse micelles. For all the samples, auto zero was adjusted at 800 nm. Insert in Fig. (B) shows the water dependent changes in OD at 875 nm.

3. Result and Discussion

Several membrane bound enzymes remain highly active in phospholipid reverse micelles (2). The bacterial photosynthetic complex in PLC was found to be active for several days depending on the temperature. The yield of primary photochemistry (F_V/F_M) of BPC in PLC, probed by OJIP BChl fluorescence induction kinetics (6), was found to be constant between -20°C to 25°C for days (Fig. 1A) and even for months (data not shown). However, they were found to be totally unstable above 45-50°C (insert in Fig. 1A). After evaporating the organic solvent from the PLC-reverse micelles containing BPC, the remaining complexes showed full fluorescence induction kinetics (data not shown). These BPCs were also embedded in commercially available resin, Araldite, and kept at room temperature. They were found to be totally stable (Fig. 1B), as judged from the OJIP induction kinetics as well as from the measurement of the 820 nm absorption changes (photooxidation of P). In the first few months, insignificant differences in F_V/F_M ratio and the light induced absorption changes at 820 nm ($\Delta I/I$) were observed. However, as the room temperature increased during the summer (above 30°C), the primary photosynthetic activity measured as F_V/F_M ratio and $\Delta I/I$ decreased.

Water movements to CF1 have been reported to occur in chloroplast membranes during photophosphorylation (7) and the activity of the chloroplast ATPase in reverse micelles has been found to be water dependent (8). These results prompted us to search for a biological system from which functional photosynthetic complexes could be transferred into reverse micelles where the influence of water on phototransduction can be explored. In PLC-reverse micelles there is very little possibility to manipulate the amount of water. We have chosen another system, the HTS micellar system, because it has been perhaps the most successful in housing functional cells, probably due to its hydrophobic-hydrophilic balance (9). Removal of water from the HTS-reverse micellar system containing BPC in PLC leads to a decrease of the variable fluorescence (Fig. 2A). However, even after the addition of a small amount (2%) of water, an increase in the variable fluorescence was observed. Insignificant differences in the F_V/F_M ratio were observed in between 2, 4 and 6% of water in HTS-reverse micellar system. The requirement of only a small amount of water on the intactness of the system is also obvious in the absorption spectra (Fig. 2B). In conclusion, we have shown that (1) the BPC can be transferred in functional form in both types of reverse micelles; (2) the BPC in PLC showed full activity of primary photochemistry for several months even at room temperature and a water dependent increase in the variable fluorescence; (3) the activity of primary photochemistry is preserved after immobilizing the BPC in PLC in resins.

Acknowledgment: Supported by the Swiss National Foundation (3100-046860.96 and 3100-052541.97) and Société Académique de Genève to AS and RJS.

References

1 Luisi, P.L., Giomini, M., Pileni, M.P. and Robinson, B.H. (1988) Biochim. Biophys. Acta 947, 209-246

2 Darszon, A and Shoshani L. (1992) in Biomolecules in Organic Solvents, (Gómez-Puyou, A., ed.) pp. 35-65, CRC Press, Orlando, FL

3 Srivastava A., Darszon, A. and Strasser R.J. (1997) *Chimia* 51, 443

4 Woodbury, N. W. and Allen, J.P. (1995) in Anoxygenic Photosynthetic Bacteria (Blankenship, R.E., Madigan, M.T. and Bauer, C.E., eds.) 527-557, Kluwer Academic Publishers, The Netherlands

5 Ghosh, R., Hardmeyer, A., Thoenen, I. and Bachofen, R. (1994) Appl. Environ. Microbiol. 60 1698-1700

6 Strasser, R.J. and Ghosh, R. (1995) in Photosynthesis: from Light to Biosphere (Mathis, P., ed.). Vol II, 915-918. Kluwer Academic Publisher, The Netherlands

7 Zolatareva, E.K., Gasparyan, M-E., Yaguzhinsky, L.S. (1990) *FEBS Lett.* 272, 184-186

8 Kernen, P., Degli-Agosti, R., Strasser, R.J. and Darszon, A. (1997) Biochim. Biophys. Acta 1321, 71-78.

9 Pfammatter, N., Famiglietti, N.M., Hochkoeppler, A. and Luisi, P.L. (1992) in Biomolecules in Organic Solvents (Gómez-Puyou, A., ed.) CRC Press, Orlando, FL.

NATURE OF CHLOROPHYLL *a* SELF-ASSEMBLY IN MIXED POLAR ENVIRONMENTS

Radka Vladkova
Institute of Biophysics, Bulgarian Academy of Sciences,
Acad. G.Bonchev Str. Bl. 21, 1113 Sofia, Bulgaria

Key words: aggregates, fluorescence, H-bond, model systems, pigment binding, antenna pigment organization

1. Introduction

Chlorophyll *a* (Chl *a*) and bacteriochlorophyll *c* are the main light-harvesting pigments of the antennae complexes in higher plants and the chlorosomes in green photosynthetic bacteria, respectively. These two light-harvesting complexes are typical example for the existence of two kinds of light collection strategies, based on whether the proteins play major role or not in the pigments organization (1). Almost nothing is known why these molecules organize so differently as antenna pigments, and especially, which properties of their native environments promote this different organization. Evaluation of the basic principles that govern the process of Chls aggregation in model systems mimicking the native microheterogeneous environment can make it possible to achieve a better understanding of the factors controlling Chls organization *in vivo*.

Despite the fact that the formation of Chl *a* aggregates in organic solvents-water mixtures have been well documented, there are only a few studies aiming at clarifying and quantifying the role of the medium on the process of Chl *a* aggregation (2,3). In these studies, only the physical properties of the medium have been taken into consideration. A possible way for quantitative characterization of all specific and non-specific solvation properties of the medium is to use some of the empirical solvent scales (4). Several empirical solvent parameters have already been applied in model studies on monomeric Chl *a* (5). In the present work, two empirical solvent scales are used. Based on this, it was possible to correlate the process of Chl monomer→oligomer transition, as well as the type of oligomer organization with the solvent parameters. The nature of Chl *a* self-assembly in polar solvent-water mixtures is discussed in the light of its similarity with the process of micellization/self-assembly of other amphiphilic molecules such as non-ionic surfactants and polar lipids in aqueous media.

2. Procedure

2.1 *Materials*
Chl *a* was purchased from Sigma and used without further purification. Methanol (MeOH), ethanol (EtOH), acetonitrile (ACN), acetone (ACE), tetrahydrofuran (THF), and pyridine (PYR) were freshly dried and distilled prior to use. Water was doubly

G. Garab (ed.), Photosynthesis: Mechanisms and Effects, Vol. V, 4233–4236.
© 1998 *Kluwer Academic Publishers. Printed in the Netherlands.*

distilled. For every series of solvent-water mixtures, equal amounts of a stock solution of Chl a in CCl_4 were put into test-tubes, and the solvent was evaporated with a stream of dry nitrogen. The thin films of Chl a were dissolved in the desired amount of solvent, and then, the required amount of water was added.

2.2 Methods

Absorption and fluorescence spectra were recorded on a Specord UV/VIS double beam spectrophotometer (Carl Zeiss, GDR) and a Perkin-Elmer MPF 44B spectrofluorometer, respectively. The Chl solutions were excited at 430 nm at right angles using 4 nm slits. Measurements were carried out in closed 1 cm quartz cuvettes at 25° C. The relative fluorescence quantum yields of Chl a in solvent-water mixtures versus that in the pure organic solvent were determined by dividing the ratio of the integrated fluorescence intensities (600-770 nm) of both samples by the ratio of their absorption at the excitation wavelength. The refractive index n was measured by an Abbe-type refractometer (Carl Zeiss Jena) at 25° C. The other parameters of solvent-water mixtures were taken from the literature or calculated as described in (6).

3. Results and Discussion

3.1. Properties of the medium promoting Chl a monomer → oligomer transition

Figure 1 shows the effect of water addition on the fluorescence quantum yield of Chl a in different solvent-water mixtures. These curves were used as a diagnostic tool for the degree of Chl monomer→oligomer transition.

The mole fraction of water giving rise to the mid-point value of the quantum yield ($f_{1/2}$) was determined for each solvent, and with respect to the $f_{1/2}$ values the solvents used arrange in the order: THF > PYR > ACE > ACN > EtOH > MeOH. The quantitative characterization of the different properties of the mixtures at $f_{1/2}$ water content is done by the values of the empirical parameters: the polarity parameter $E_T(30)$ from the single parametric scale of Reichardt and Dimroth (4), and the four parameters: polarity P, polarizability Y, electrophilicity E, and nucleophilicity B, from the approach of Koppel and Palm (7).

Table 1 shows that the values of $E_T(30)_{1/2}$, and those of $Y_{1/2}$ and $E_{1/2}$ parameters can be taken

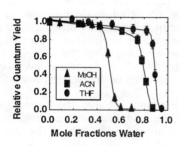

Figure 1. Relative fluorescence quantum yield of Chl a ((2.7 - 2.9 × 10^{-6} M) in different solvent-water mixtures.

Table 1. Mean values of the empirical solvent parameters for the different solvent-water mixtures at $f_{1/2}$ mole fraction of water: polarity parameter $E_T(30)_{1/2}$ (in kcal/mol), and Koppel and Palm's parameters: polarity $Y_{1/2} = (\varepsilon_{1/2} - 1)/(\varepsilon_{1/2} + 2)$, polarizability $P_{1/2} = (n_{1/2}^2 - 1)/(n_{1/2}^2 + 2)$, electrophilicity $E_{1/2}$, and nucleophilicity $B_{1/2}$.

Parameter	$E_T(30)_{1/2}$	$Y_{1/2}$	$P_{1/2}$	$E_{1/2}$	$B_{1/2}$
Weighted mean value	56.1 ± 0.4	0.9463 ± 0.0014	0.2133 ± 0.0068	18.7 ± 0.2	158.2 ± 10.0
Relative SD	1.6%	0.36%	7.8%	2.1%	15.4%
Sum of rel. deviations	3.2%	0.69%	13.5%	0.53%	34.5%

as equal within the experimental error ($\leq 4\%$) for all six mixtures. The parameter $E_T(30)$ is essentially a scale reflecting dipole-dipole interactions (function of the dielectric constant ε) with the addition of a hydrogen-bonding ability of the solvent (8). The non-specific polarity parameter Y also reflects the dipole-dipole (orientational) interactions, while the specific solvation parameter electrophilicity E is measuring the Lewis acidity (electrophilic solvating power, hydrogen-bond donating ability) (7). Based on the identical results obtained by applying two quantitative approaches, we conclude that the hydrogen bonding ability and dipole-dipole interaction of solvent-water mixture play most important role in promoting Chl self-assembly. The same properties of the medium are found to be important for the self-assembly of surfactants in aqueous media (9).

3.2 Properties of the medium determining the type of chlorophyll oligomers

In order to elucidate which properties of the medium are responsible for the type of aggregates formed, a comparison of the Chl absorption spectra at the onset and completion of the transition in different solvent-water mixtures was carried out (Fig. 2). At the onset of the transition, the most important difference is in the ratio $\varepsilon_s/\varepsilon_{s'}$, the intensity ratio of the main and first satellite Soret bands. This ratio is very similar in MeOH, EtOH, ACN, and ACE, while in PYR and THF it is considerably higher. Taking into account that Soret-band intensity is sensitive to the electrophilic properties of the solvents (5), the difference in this ratio reflects the higher degree of hydration of Chl a in the former mixtures, as compared to the latter. This result is important with respect to the understanding of the factors controlling the type of aggregates formed.

At the end of the transition, a profound difference is observed in the types of oligomers formed. MeOH, EtOH, ACN and ACE-water mixtures form Chl aggregates absorbing at 740-760 nm, while PYR and THF form aggregates absorbing at 667-670 nm.

Between the two kinds of polar solvents, clear difference in the solvent parameters P and B (7) is found. The polarizability parameter P is a function of the index of refraction n, and reflects the dipole-induced dipole interactions. The nucleophilicity parameter B measures the Lewis basicity (nucleophilic solvating power, electron pair donating ability) of the solvent. We conclude that the nucleophilic and polarizability properties of Chl environment are those that have essential role in modulating the type of aggregate organization. The higher coordination ability and polarizability of PYR and THF prevent the higher hydration of the monomeric Chl a at the onset of the transition, as well as the formation of 750 nm absorbing aggregates using the water molecules (10). These two properties of the solvent modulate the degree of Chl a hydration, and by this way, the

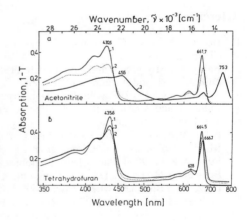

Figure 2. Absorption spectra of Chl a in ACN (a) and THF (b) at different mole fractions of water, f. In ACN, (1a): $f = 0$; (2a): $f = 0.75$ (50 vol%); (3a): $f = 0.87$ (70 vol%). In THF, (1b): $f = 0$; (2b): $f = 0.82$ (50 vol%); (3b): $f = 0.95$ (80 vol%); concentration of Chl a, 2.8×10^{-6} M.

hydrophilic/hydrophobic balance of the Chl molecule. The type and shape of the micelles formed by non-ionic surfactants and polar lipids depend on this quantity in a similar way (11,12).

Based on the present results, it is difficult to determine the exact Chl organization in both groups of solvent-water mixtures. However, the hyperchromism of Q_y band (13) in MeOH, EtOH, ACN, and ACE (Fig. 2), from one side, and the hypochromism of Q_y band in PYR and THF mixtures, from the other side, reflect the different strength of Chl macrocycle-macrocycle interactions. In the former group, these interactions are much more stronger than in the latter one. Additionally, the similarities of the absorption spectra with Chl aggregates studied by other methods (14,15) could be useful to propose the organization of the two types of Chl aggregates formed in polar solvent-water mixtures. It could be suggested that MeOH, EtOH, ACN and ACE-water mixtures give rise to Chl organization resembling inverse cylindrical micelles, while the aggregates in PYR and THF-water mixtures give rise to a form with different curvature compared to the inverse micelles.

3.3 Nature of Chl a self-assembly in polar solvents-water mixtures

The Chl *a* aggregation in polar solvents-water mixtures resembles the process of micellization/self-assembly of non-ionic surfactants and polar lipids in aqueous media based on the similarity in the factors found as important during the both processes: (i.) essential role of hydrogen-bonding ability and dipole-dipole interaction of the medium in promoting the self-assembly; (ii.) important role of the degree of hydration in determining the type and shape of the assemblies.

Such analogy was recently found (16) based on the similarity in the critical behavior during the Chl *a'* self-assembly in aqueous methanol.

References

1. Holzwarth, A.R., Griebenow, K. and Schaffner, K.(1992) J. Photochem. Photobiol. A: Chem. 65, 61-71.
2. Balny, C., Brody, S.S. and Hoa, G.H.B. (1969) Photochem. Photobiol. 9, 445-454.
3. Dijkmans, H. and Aghion, J. (1974) Plant & Cell Physiol. 15, 739-745.
4. Reichardt, C (1979) Solvent Effects in Organic Chemistry. Verlag Chemie, Weinheim
5. Renge, I. and Avarmaa, R. (1985) Photochem. Photobiol. 42, 253-260.
6. Vladkova, R., submitted.
7. Palm, V.A. (1977) The Fundamentals of the Quantitative Theory of Organic Reactions. Khimia Publishing House, Leningrad.
8. Suppan, P. (1990) J. Photochem. Photobiol. A: Chem. 50, 293-330.
9. Beesley, A.H., Evans, D.F. and Laughlin, R.G. (1988) J. Phys. Chem. 92, 791-793.
10. Katz, J.J., Bowman, M.K., Michalski, T.J. and Worcester, D.L. (1991) In Chlorophylls (Scheer, H., ed.), pp. 211-235. CRC Press, Boca Raton, FL.
11. Luzzati, V. (1968) In Biological Membranes, Physical Fact and Function (Chapman, D., ed.) pp. 71-123, Academic Press, New York.
12. Shinoda, K. (1983) Progr. Colloid & Polymer Sci. 68, 1-7.
13. Scherz, A. and Parson, W.W. (1984) Biochim. Biophys. Acta 766, 666-678.
14. Worcester, D.L., Michalski, T.J. and Katz, J.J. (1986) Proc. Natl. Acad. Sci. U.S.A. 83, 3791-3795.
15. Kadoshnikova, I.G. and Kiselev, B.A. (1979) Biofizika 24, 811-814.
16. Oba, T., Mimuro, M., Wang, Z.-Y., Nozawa,T., Yoshida, S. and Watanabe, T. (1997) J. Phys. Chem. B 101, 3261-3268.

CHLOROPHYLLa - PHOTOACTIVE SPECIES IN PHOTOVOLTAIC AND PHOTOELECTROCHEMICAL DEVICES

Tugulea G. Laura, Antohe Stefan, Dulcu Ileana
Faculty of Physics - University of Bucharest
P.O.Box MG-11, Bucharest-Magurele, 76900 Romania

Key words: electron transfer, electronic structure, light activation, modelling, photoconversion, photochemistry

1. Introduction

There has been much recent interest in photochemical and photoelectrochemical investigation of chlorophyll and related pigments. Different types of photovoltaic and photoelectrochemical devices using chlorophylls have been realized and high efficiencies for light photoconversion have been reported. As an extension of our previous work, done on photovoltaic devices with chlorophyll a, we realized photoelectrochemical devices using chlorophyll a species absorbing at 740 nm. The performed experiments on both electrochemical and photovoltaic devices, using P740 chlorophyll in thin films, lead to a comparative study and the explanation of chlorophyll a behaviour in the process of charge separation under illumination and of transfer of electrons.

2. Procedure

Chlorophyll a (Chla) was prepared from fresh spinach leaves by Strain&Svec method and checked for purity (absorption and fluorescence in VIS). The Chla a species: P740 (a water adduct of chlorophyll a, absorbing at 740nm) was obtained by dissolving the fresh prepared chlorophyll a in n-pentan and maintaining the solution at temperatures below -10°C. Thin films of chlorophyll a were prepared by electrodeposition. The SnO_2 coated glass, providing the transparent conducting substrate, was chemically cleaned to remove any residual grease. On these substrates a thin layer of Chla of various thickness was electrodeposited from the suspension/solution of Chla in n-pentan, the SnO_2 coated glass being the cathode and the electric field having an intensity of 1000V/cm. The film thickness was controlled by varying the molar concentration of Chla solution and/or by varying the electrodeposition time. The film thickness was estimated from the absorption spectra and from interferometric microscopy measurements. The optical absorption spectra of the films were obtained on a double beam UV - VIS spectrophotometer Lambda 2S Perkin Elmer.

4237

G. Garab (ed.), Photosynthesis: Mechanisms and Effects, Vol. V, 4237–4240.
© 1998 Kluwer Academic Publishers. Printed in the Netherlands.

2.1. *Photovoltaic measurements*

Photovoltaic devices of sandwich type: (SnO$_2$ electrode/P740 chlorophylla/Hg) have been prepared by depositing a Hg drop on top of P740 chlorophyll layer. The active area, defined by the Hg electrode, was 0.3 cm^2. The current - voltage characteristics under dc bias were measured using a stabilised power supply in connection with a helipot potentiometer, giving a resolution of 20 mV. Device current and voltage output measurements were made with a Philips microameter and a Keithley 614 electrometer, respectively. illumination of The devices were illuminated through the SnO$_2$ electrode, using a 650W halogen lamp. A monochromator (in the spectral range 350 - 800 nm) was used in order to obtain the spectral responses of the cells (action spectra). All measurements were performed in atmospheric air and at room temperature. The action spectra are corrected to equal photon density at each wavelengths and normalized (the peak at 800 nm is unity).

2.2. *Photoelectrochemical measurements*

The SnO$_2$ electrode coated with a thin film of P740 chlorophyll a represented the working electrode (area about 0.4 cm^2) of a three-electrode electrochemical cell, with a Pt counter electrode and a standard Ag/AgCl (saturated KCl in water) as reference. As support electrolyte, solutions of Na$_2$SO$_4$, of different ionic strengths, in phosphate buffers (pH range: 6.9 - 8) were used. A μAUTOLAB (Eco Chemie) electrochemical analyser and GPES software have been used to perform the experiments on the electrochemical cell. The working electrode was illuminated by monochromatic light (in the spectral range 350 - 800 nm) using a halogen lamp of 650 W. Action spectra have been obtained by plotting the photocurrents, measured for a given electrode voltage, at different wavelengths. All measurements were performed in atmospheric air and at room temperature. The action spectra are corrected to equal photon density at each wavelengths and normalized (the peak at 800 nm is unity).

3. Results and discussions

3.1. *Photovoltaic behaviour*

The dark current-voltage characteristics of the photovoltaic devices of sandwich type (SnO$_2$ electrode/P740 chlorophylla/Hg) exhibited small rectification and the explanation for this behaviour is the existence of two blocking contacts: a p-n junction at the interface (SnO$_2$ electrode/P740 chlorophyll a) and a Schottky barrier between chlorophylla and Hg. The photoresponse of the photovoltaic cell is the result of the balance between the two blocking contacts. This behaviour is also confirmed by the kinetics of short-circuit currents on illumination with light at different wavelenghts.

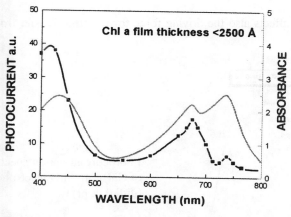

Figure 1
Photocurrent action spectra
of (SnO₂/P740 Chlala/Hg)
and absorption spectra of
Chl a on SnO₂ electrode

Figures 1 and 2 show the action spectra of short-circuit photocurrents for two sandwich cells, differing by the thickness of the chlorophyll a film. The action spectrum of the photovoltaic cell   (SnO₂ electrode/P740 chlorophylla/Hg) is matching the absorption spectrum of the chlorophyll a film if the thickness of the Chla film < 2500A (Figure 1), but is the inverse of the absorption spectrum for thick films (4500 A, Figure 2).

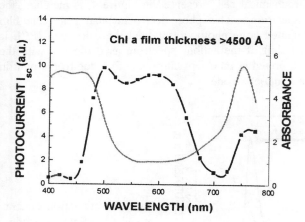

Figure 2
Photocurrent action spectra
of (SnO₂/P740 Chlala/Hg)
and absorption spectra of
Chl a on SnO₂ electrode

3.2. *Photoelectrochemical behaviour*

Generation of anodic photocurrents was interpreted schematically in terms of the electron donation from excited chlorophyll a to SnO₂ electrode, followed by the reduction of Chl a cation by the free electrons from electrolyte solution. The spectral behaviour of the photoelectrochemical cell was investigated and compared with the absorption spectrum of chlorophyll a film. Figure 3 presents the pH dependence of the photoresponse when using electrochemical cells with the same thickness of chl a film and electrolyte solutions with the same ionic strenght.   The value of 7.5 was found as optimum pH. The change in pH of the electrolyte causes a shift of the semiconductor electrode potential and therefore affects also the driving force for electron transfer from chlorophyll a to SnO_2electrode.

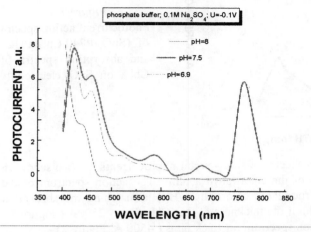

Figure3
Photocurrent action spectra of Chl a on SnO_2electrode (different pH)

The action spectrum of the electrochemical cell, presented in Figure 4, is matching the P740 chlorophyll a absorption spectrum. This photoelectrochemical cell (chla film thickness is less than 1000A) exhibited large photocurrents and the calculated IPCE is 1.2 % at 740 nm. The small red shift of the action spectrum as compared with the absorption spectrum could be interpreted as an electrochromic effect due to the built in field at the interface chlorophyll a / SnO_2electrode.

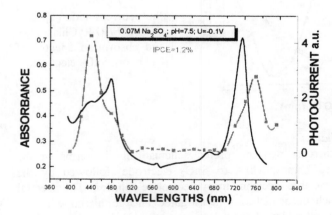

Figure4
Photocurrent action spectra of Chl a on SnO_2electrode (1000 A film thickness)

The behaviour of the photovoltaic and photoelectrochemical devices realized with P740 chlorophyll a in thin films lead to the conclusion that chlorophyll a is the photoactive component in both devices and the thickness of the P740 chlorophyll a film is the dominant factor in controlling the charge photogeneration and electron transfer.

PHOTOPOTENTIAL GENERATION IN GREEN BACTERIA CELLS AND CELL FRAGMENTS LOCATED IN AN ELECTROCHEMICAL CELL

Arkadiusz Ptak, Alina Dudkowiak and Danuta Frąckowiak

Institute of Physics, Poznań University of Technology, ul.Piotrowo 3, 60-965 Poznań, Poland

Key words: Bchl, Chl, chlorosomes, liquid crystals, model systems, oligomers.

1. Introduction

Illumination of the pigmented membranes of photosynthetic organisms generates the difference in electric potential between both surfaces of the membrane [1,2]. As a result, some ions can be transfered through the membranes more easily. Almost all pigmented elements of photosynthetic organisms such as chloroplasts [1,3], thylakoids [4] or chromatophores [5] located between two electrodes exhibit photovoltage generation.

In this work we have measured an effect of photovoltage generation in green bacteria cells and cell fragments located in a photoelectrochemical cell.

The green sulfur bacteria *Prosthecochloris aestuarii* contains antenna bodies - chlorosomes, predominantly containing bacteriochlorophyll (Bchl) *c*. The chlorosomes differ from other giant antenna bodies because their internal structure is predominantly generated by the aggregation of the Bchls and there is no significant involvement of protein [6]. The Bchls *c* in chlorosomes form large aggregated structures (oligomers) which are responsible for the spectral properties of the chlorosomes. The main role of these oligomeric structures is to harvest the light energy and to transfer it to the reaction center where it is used for the charge separation. We wanted to check if such oligomeric forms of the pigments were able to participate in the photovoltage generation in a photoelectrochemical cell (PhC).

2. Materials and Methods

The PhC used in our experiments consists of two transparent electrodes: both semiconducting (In_2O_3) or one semiconducting and the second - metallic (gold), separated by teflon distancers (thickness: 10-60μm). The investigated samples: green bacteria cells as well as model systems - photosynthetic pigments in nematic liquid crystal (LC) matrix, were located in the PhC. The electrical properties such as current-voltage characteristics, kinetics of photocurrent generation and photovoltage action spectra were measured. The details of the construction of the PhC as well as the methods of photoelectrical measurements were described previously [7,8].

G. Garab (ed.), Photosynthesis: Mechanisms and Effects, Vol. V, 4241–4244.
© *1998 Kluwer Academic Publishers. Printed in the Netherlands.*

3. Results and Discussion

The photovoltage action spectra shown in Fig.1B are recalculated on the same number of quanta (10^{13} quanta/cm^{2}s) reaching the surface of the sample. In the photovoltage action spectrum of the native sample the red absorption band located at about 750nm (Fig.1A) is not observed. The red band at about 675nm is formed as a result of the pigment desaggregation. The Soret band maximum of the action spectrum is located at 410-415nm, what strongly suggests the participation of bacteriopheophytin c. It is clear that even for the sample having high absorption in the oligomer region the "maximum of oligomers" is practically absent in the photovoltage action spectra.

In the suspension of chloroplasts or thylakoids the light gradient effect occurs because the front and back sides of the thylakoids are illuminated differently [1,9]. Similar effect for whole PhC is observed. The strongly illuminated semiconductor electrode gives higher negative potential than the opposite one.

In order to establish the molecular mechanisms of photovoltage generation some simpler models consisting of photosynthetic pigments in nematic liquid crystal located in the PhC were also investigated [7,8]. Nematic liquid crystal simulates fluid and oriented structure of molecular membrane. The kinetics of photocurrent generation under continous illumination for such a model system are slightly similar to these for bacteria fragments. The change of the

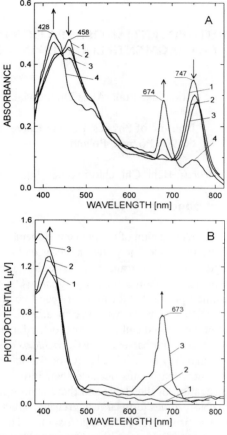

Fig.1.A. Absorption spectra of the samples: curve 1: native whole cells of bacteria; 2: bacteria fragments measured directly after sample preparation; 3: bacteria fragments measured 3-4 h after sample preparation; 4: bacteria fragments 24h after sample preparation. **B.** Photopotential action spectra of the same samples. All the samples located in the photoelectrochemical cell with two semiconducting (In$_2$O$_3$) electrodes.

photoelectrochemical cell thickness does not influence on the photocurrent amplitude. It suggests that only a thin layer of pigment molecules adsorbed on the electrodes takes part in the photopotential generation.

The kinetics of photocurrent generation in the time range from microseconds to

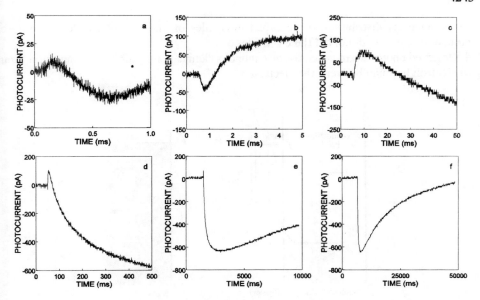

Fig.2a-f. Time evolution of the photocurrent signal generated by DCM dye laser (max.658nm) in the asymmetric cell with Chl *a* in nematic liquid crystal. Time ranges are given.

minutes for the model system illuminated by laser flashes were also measured [8]. These kinetics measured for the Chls in nematic liquid crystal located in the assymetrical PhC (with semiconductor and gold electrodes) are very complex (Fig.2). Much simpler kinetics were obtained in the same system for the solution of protonated synthetic dye (stilbazolium mero-cyanine). In this case the charge distribution in the volume of the cell prevents the ions migration.

4. Conclusions

1. Oligomers of Bchl *c* located in chlorosomes are inactive in photopotential generation on the electrodes of the PhC.

2. A thin layer of pigment molecules adsorbed on the semiconductor electrodes is responsible for photopotential generation. There is an injection of electrons from the first excited singlet state of adsorbed molecules into the conduction band of the semiconductor electrode. Proposed mechanism of the photovoltage generation is present in Fig.3.

3. Observed photovoltage (=difference between the light-generated electric potentials of the two electrodes of the PhC) on the symetrical PhC (with two semiconductor electrodes) can be due to two efects:

a) the light gradient effect → different number of quanta reaching the electrodes → different number of electrons injected into the electrode → difference between photopotentials on the electrodes → photovoltage;

b) assymetrical distribution of adsorbed molecules on the two electrodes → different number of electron donors at the electrodes.

4. Observed photocurrent generated by continous illumination is due to ion migration or/and to electron hopping between the electrodes.

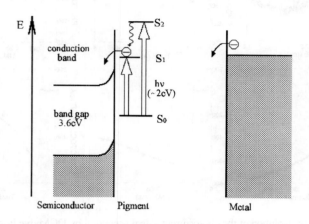

Fig.3. Scheme of energy states of the pigment and the electrodes.

References

1 Fowler, C.F. and Kok, B. (1974) Biochim. Biophys. Acta 357, 308-318

2 Dobek, A., Deprez, J., Paillotin, G., Leibl, W., Trissl, H.W. and Breton, J. (1990) Biochim. Biophys. Acta 1015, 313-321

3 Barabás, K., Váró, G. and Keszthelyi, L. (1994) J. Photochem. Photobiol. B Biol. 26, 37-44

4 Witt, H.T. (1975) in Bioenergetics of Photosynthesis (Govindjee, ed.) pp.493-554, Academic Press, New York - San Francisco - London

5 Trissl, H.-W., Kunze, U. and Junge, W. (1982) Biochim. Biophys. Acta 682, 364-377

6 Foidl, M., Golecki, J.R. and Oelze, J. (1994) Photosynth. Res. 41, 145

7 Ptak, A., Chrzumnicka, E., Dudkowiak, A. and Frąckowiak, D. (1996) J. Photochem. Photobiol. A Chem. 98, 159-163

8 Ptak, A., Der, A., Toth-Boconadi, R., Naser, N.S. and Frąckowiak, D. (1997) J. Photochem. Photobiol. A Chem. 104, 133-139

9 Paillotin, G., Dobek, A., Breton, J., Leibl, J. and Trissl, H.-W. (1993) Biophys. J. 66, 379-385

24. Emerging techniques

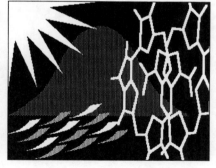

CONGRESS ON PHOTOSYNTHESIS

XIth INTERNATIONAL

BUDAPEST ✳ 1998

A NEW HIGH-SENSITIVITY 10-ns TIME-RESOLUTION SPECTROPHOTOMETRIC TECHNIQUE ADAPTED TO *IN VIVO* ANALYSIS OF THE PHOTOSYNTHETIC APPARATUS.

Pierre Joliot, Daniel Béal, Fabrice Rappaport.

CNRS UPR1261 IBPC, 13, rue P. et M. Curie 75005 Paris, France

1. Introduction

In the past years, we have developed a new type of spectrophotometer in which the absorption is sampled by short monochromatic flashes produced by a xenon flash tube (1-2). These instruments are double-beam spectrophotometers in which the ratio between the measuring and the reference signal is computed for each flash in order to correct for the energy fluctuations of the flashes (3 %). The time resolution is determined by the duration of the flashes (2 µs) and the light-induced spectral changes can be analyzed in a time window ranging from 2 µs to 10 s with a signal to noise ratio of about 10^5. Owing to the high sensitivity of these techniques and to the optical design which allows collection of a large fraction of the light scattered by the biological sample, these spectrophotometers are particularly well suited to the analysis of light induced spectral changes in isolated chloroplasts as well as in unicellular algae.

In the spectrophotometric technique presented here the monochromatic flashes are produced by an optical parametric oscillator (OPO) pumped by the third harmonic of Nd:Yag laser (6 ns duration). Continuously tunable flashes are obtained in a spectral domain ranging from 415 nm to 680 nm and from 730 nm to 2000 nm. Compared to the Xenon flash spectrophotometer, this technique presents several improvements: 1) The time resolution (10 ns) is improved 200-fold; 2) sample of optical density as large as 3 may be analyzed with an optimal signal to noise ratio, owing to the high energy output of the detecting flashes; 3) the spectral bandwidth of the detecting flashes is close to 1 nm, compared to 5 nm when using a large operture monochromator as in the Xenon flash technique.

2. Description of the apparatus

The Laser MOPO 710 (Spectra-Physics) delivers monochromatic flashes of 10 to 40 mJ depending upon the spectral range. The laser beam is split into two beams in order to illuminate both the measuring and the reference cuvettes. For each detecting, flash the

G. Garab (ed.), Photosynthesis: Mechanisms and Effects, Vol. V, 4247–4252.

ratio between the measuring and the reference signals is computed in order to correct for the large fluctuations of the energy of the OPO flashes (10 to 40 %). This requires the energy fluctuations of the measuring and detecting beams to be well correlated, a requisite that is not met when using a conventional beam splitter as a semi- transparent mirror.

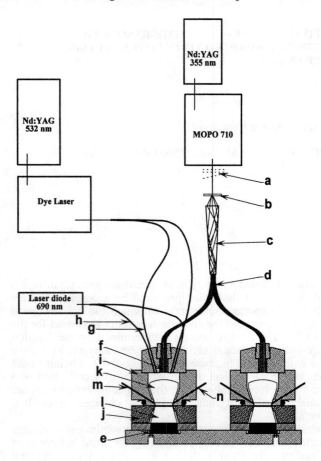

In the device shown in figure 1 the laser beam is scattered by quartz plates (b) with grounded surfaces. These plates are applied to the entrance of a trapezoidal light pipe (c) 50 mm long (entrance section 10 mm x 10 mm, output section 10 mm x 3 mm). At the output of this light pipe the light is collected on the common section of a Y-shaped bundle of optic fibers (d). These fibers are randomly distributed between the two branches of the Y-shaped light pipe providing the two beams which illuminate the measuring and the reference cuvettes, respectively. After computing the ratio N_m/N_r where N_m and N_r are the number of photoelectrons emitted by the measuring and reference photoreceptors respectively, the signal to noise ratio is mainly limited by the quantum noise and equal to ~ $\sqrt{N}$. Thus, ~2 x 10^{10} photoelectrons are required to achieve a signal to noise of 10^5. The corresponding noise level is 5 10^{-6} absorption units, a value close to the limit of our instrument. The light transmitted or scattered by the biological sample is collected over a wide solid angle by a conical light pipe (l) and illuminates two silicon photodiodes (e) of large surface (18 mm x 18 mm). It is worth pointing out that a photomultiplier cannot be used as a photoreceptor: first, the quantum yield of the photocathode of a photomultiplier is much lower than that of a silicon photodiode; second, most of the photocathodes cannot cope with the high peak current intensity (0.1 A) generated by the 6 ns flashes.

The actinic flashes are provided by a tunable dye laser pumped by the second harmonic of Nd:Yag laser (Spectra Physics). This laser beam illuminates a bundle of optic fibers (g) which form a crown around the bundle of fiber which delivers the detecting light. The actinic light is collected by a conical light pipe (k) which improves the homogeneity of the actinic illumination of the measuring cuvette. Continuous actinic illumination is provided by a laser diode (690 nm; maximum power 500 mW) which illuminates the entrance of a second bundle of optical fibers (h).

The photoreceptors (e) are protected from the actinic light by filters which absorb (colored filters) or reflect (interference filter) the actinic light.

The cuvettes shown in figure 1 are optimized for the analysis of unicellular cell suspensions. The 2.5 mm thick cuvette is a conical cavity in aluminum closed by the two conical light pipes (k and l). The diameter of the cuvette is 13 mm and 10 mm for the upper and lower faces respectively. The light scattered by the cell suspension is partly reflected by the polish aluminum wall of the cuvette and collected over a wide angle by the lower conical light pipe toward the large surface photodiodes (3.4 cm^2). A dialysis membrane can be interposed between the two aluminum blocks, thus dividing the cuvette into two compartments. In this case, the biological sample is introduced in the lower (0.5 mm thick) compartment. The cell suspension or a buffer in the presence or absence of a dialysis membrane respectively, is flowed in the upper compartment via channels m and n.

When the actinic effect of the detecting flash is negligible (e.g. in the case of ligand photodissociation) a cuvette of similar design than that shown in figure 1 but of smaller diameter (4 mm) is better suited. The small diameter results in the increase of the energy density input thereby allowing the study of low photochemical efficiency processes.

3. Treatment of the signals and command of the experiment.

The signal delivered by the two photodiodes (Hamamatsu S 3204-46) is integrated using a 4.7 nF capacitor. The time constant of discharge of the capacitor (200 μs) is much longer than the duration of flash. Therefore, the signals at the output of the two photodiodes are proportional to the number N of photoelectrons emitted during the flash. These signals are applied to the input of a differential amplifier (gain of 8). The output signal, which is proportional to the difference between the measuring (M) and the reference (R) is digitized using a 18 bit analog-digital converter (Analogic inc. ADC 5020). A second analog-digital converter digitizes the reference signal M. For each detecting flash the ratio (M-R) x 8 / R is computed. The signal M and R have to be adjusted to similar value (+/- 10 %) to avoid the saturation of the differential amplifier. The energy of the detecting flashes is most generally adjusted to obtain a signal of about 1 V. In these conditions the number of photoelectrons is close to 3×10^{10}, which results in a signal to noise ratio of about 10^5 (see above).When not using the differential amplifier, the noise is limited by the coding accuracy of the analog-digital converter.

The analog-digital conversions are triggered 2 μs before and 5 μs after each detecting flashes. Computing the difference between these numbers eliminates low frequency perturbations which may be superimposed to the signal. Light artifact associated with the actinic flash are also rejected by this procedure, provided that Δt is larger than 7 μs.

The pulses which control the Pockell cell of the actinic and detecting lasers are synchronized by two 10 Hz clocks. The detecting clock is shifted with an adjustable time delay (Δt) with respect to the actinic clock. The time delay can be adjusted from 0 to 60 ms with minimum steps of 5 ns by a computer driven program. The jitter between the actinic and detecting flashes is lower than 1 ns. Single or multiple detecting flashes can be selected using an electromechanical shutter in the cavity of the OPO laser. When measuring a flash induced absorption change, the absorption of the dark-adapted material is first sampled by 1 to 4 detecting flashes. The actinic flash is fired at a time (100 ms - Δt) after the last flash of the base line. Several detecting flashes are fired sequentially when analyzing kinetics in a time range larger than 100 ms. When using the dialysis membrane device, the settling of the biological material on the bottom of the lower compartment eliminates any long-term absorption drift. It is then possible to analyze slow light-induced absorption changes in a

4250

time range extending to more than 100 s. For Δt ranging from 0 to 100 ms, a single detecting flash can be fired after each actinic flash. To cover this time range, repetitive experiments involving several actinic flashes must be performed with various time delays. In the experiment shown in figures 2 and 3, the time delay is progressively incremented after each actinic flash from the lowest to the highest values (0 ns to 50 ms). Then, a second set of experiment is performed by varying Δt from the highest to the lowest values. The two series of experiments are then averaged. This procedure, entirely computer driven minimizes the consequences of an eventual drift in the properties of the biological material.

4. Examples of application of the spectrophotometric technique

In the case of oxygenic photosynthetic material (isolated chloroplasts, unicellular algae or leaves), the actinic flashes are tuned at 700 nm in order to improve the homogeneity of the actinic illumination. At this wavelength, high energy flashes (about 1-2 mJ) are required to saturate photosystem I and II reaction centers. Actinic light transmitted through the sample is blocked by a blue filter (Schott BG 39, 5mm thick).

4.1 Kinetics of the flash induced absorption changes measured at 518 nm in whole cells of Chlorella sorokiniana.

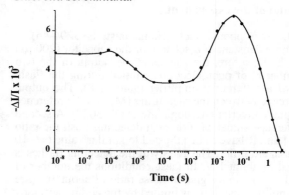

Experiments have been performed in anaerobic conditions with a PSII lacking strain of *Chlorella sorokiniana* (3). The algae are illuminated by saturating flashes 4.5 s apart. In figure 2 the absorption changes have been plotted on a logarithmic time scale. The absorption changes mainly result from the combination of the formation of carotenoid triplet (4) and of an electrochromic shift undergone by chlorophyll b and carotenoid induced by the formation of a transmembrane potential (5).

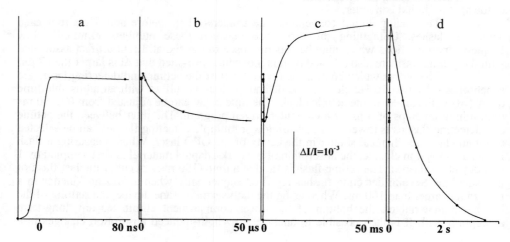

These absorption changes have been analyzed in a time window ranging from 10 ns to 4 s. i.e. more than 8 orders of magnitude. The total duration of the experiment is 2.5 min. The rising phase observed on the faster linear time scale (figure 3a) mainly reflects the time resolution of the method. For $\Delta t = 0$, about half of the photons of the detecting flashes are distributed before the actinic photons. Then, the amplitude of the signal is about half of its maximum value which is reached in less than 10 ns. A first relaxation ($t_{1/2}$ 8 μs) reflects the decay of the carotenoid triplet (figure 3b). Following this relaxation phase, the absorption changes are mainly associated with the electrochromic shift and the signal is proportional to the number of PS I charge separations induced by the actinic flash. The rising phase observed in the 1 to 50 ms time range (figure 3c) reflects transmembrane charge transfer (electron or proton) induced by the cytochrome b_6f turnover (6). The absorption decay ($t_{1/2} \sim$ s) reflects ion leak through the membrane or proton leak through the membrane ATP-synthase (figure 3d). The total duration of the experiment can be reduced when analyzing a single relaxation process by using a smaller number of time delays (about 5) chosen in the time range of the scrutinized process.

4.2 Flash induced absorption changes in the blue region of spectrum, measured in intact leaves.

In figure 4 are shown the photoinduced absorption changes measured at 432 nm and 435 nm. The time interval between actinic flashes is 2 s and the total duration of the experiment 1 min. In the blue region of the spectrum, the optical density is about 2.5 (about 0.3 % of the incident light is transmitted by the leaf). To keep the number N of photoelectrons close to its optimum value, the energy of the detecting flashes has to be increased to values close to 5 μJ, corresponding to about 10^{13} incident photons. Such high

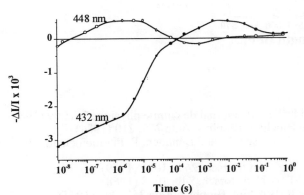

energy detecting flashes induce significant actinic effects (about 5 % of the reaction centers excited per flash). The actinic effect of the detecting flashes is however negligible in the green region of the spectrum (not shown).

The absorption changes measured at 432 nm are mainly associated with the photoxidation of PS II and PSI primary donors. The relaxation in the 10 μs range is mainly associated with the reduction of P_{700}.

4252

4.3 Time resolved spectra measured in whole cells of Rhodopseudomonas viridis.

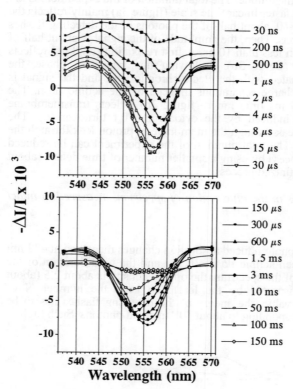

Reaction centers of *Rhodopseudomonas viridis* include a tetraheme cytochrome polypeptide. These cytochromes act as secondary donors to the oxidized primary donor P_{780}^+. The experiments have been performed in anaerobic conditions which induce the reduction of the two high potential hems (cyts 556 and 559). Reaction centers are excited by a saturating actinic flash at 1064 nm (the fundamental mode of the Nd:Yag laser). The time-resolved spectra show the sequential electron transfer from cyt559 to P^+ ($t_{1/2}$ = 200 ns) followed by the electron transfer from cyt556 to cyt559$^+$ ($t_{1/2}$ 2.5 µs). In a longer time range, one electron is transferred from a soluble cyt552 to cyt556 . These data are presented in more details in (7).

References

1. Joliot, P., Beal, D. and Frilley, B. Journal de chimie physique 77, 209 (1980).
2. Joliot, P. and Joliot, A. Biochim. Biophys. Acta 765, 210 (1984).
3. Lavergne, J., Delosme, R., Larsen, U. and Bennoun, P. Photobiochem. photobiophys. 8, 207 (1984).
4. Wolff, C. and Witt, H. T. Z. Naturforsch. 24 b, 1031 (1969).
5. Junge, W. and Witt, H. T. Z. Naturforsch. 24b, 1038 (1968).
6. Joliot, P. and Delosme, R. Biochim. Biophys. Acta 357, 267 (1974).
7. Rappaport, F., Béal, D., Verméglio, A. and Joliot, P. Photosynth. Res. 55, 317 (1998).

CHLOROPHYLL FLUORESCENCE:
NEW INSTRUMENTS FOR SPECIAL APPLICATIONS

Ulrich Schreiber
Julius-von-Sachs Institut für Biowissenschaften, Universität Würzburg,
Julius-von-Sachs Platz 2, D-97082 Würzburg, Germany

*Key words: guard-cell chloroplast, microalgae, microscopy, microfibre probe,
phytoplankton*

1. Introduction

During the past decade there has been remarkable progress in the understanding
and practical use of chlorophyll fluorescence in plant science (1,2,3). This progress has
resulted from fruitful interactions between three different research disciplines: basic
research (dealing with the dynamics of excitation transfer, photochemical charge
separation and electron transport), applied research (making use of fluorescence as an
noninvasive tool) and the development of new instruments and methodology, to measure
fluorescence and to extract the essential information from it. In particular, with the
introduction of pulse-amplitude-modulated (PAM) fluorometers and the saturation pulse
method of quenching analysis (4), chlorophyll fluorescence has gained widespread
applications. It provides manifold information at various levels of the complex process of
photosynthesis, starting from light absorption, energy transfer and primary energy
conversion, and ending with the export of assimilates from the chloroplast. Due to its
large signal amplitude, chlorophyll fluorescence traditionally has been a pioneering tool
in photoynthesis research. Making use of recent progress in optoelectronics and
microprocessor/computer technology, a new generation of chlorophyll fluorometers with
extreme sensitivity and selectivity has been developed, the essential features of which
will be outlined in the present communication.

2. Description of Instruments and Results

2.1 *Phytoplankton analysis by 4-wavelength excitation technique*

Since the early work of Lorenzen (5), chlorophyll fluorescence has become
increasingly important for assessment of phytoplankton mass and primary productivity
(6). Very sensitive techniques have been developed to measure chlorophyll content and
to analyze basic parameters of photosynthetic activity in natural surface waters down to
0.1 µg Chl/l (7, 8). The so far available instrumentation has been limited by the fact that
it cannot distinguish between different types of phytoplankton, like green algae, diatoms
and cyanobacteria. In principle, such distinction is possible on the basis of the specific
fluorescence excitation properties of differently pigmented phytoplankton groups (8, 9).
A practical version of such a device has been developed (Kolbowski and Schreiber,
unpublished) and recently has become generally available (PHYTO-PAM Phytoplankton
Analyzer, Heinz Walz GMBH, Effeltrich, Germany).

The PHYTO-PAM employs light-emitting-diodes (LED) to excite chlorophyll
fluorescence alternatingly by 10 µs light pulses at four different wavelengths (470, 535,

G. Garab (ed.), Photosynthesis: Mechanisms and Effects, Vol. V, 4253–4258.
© *1998 Kluwer Academic Publishers. Printed in the Netherlands.*

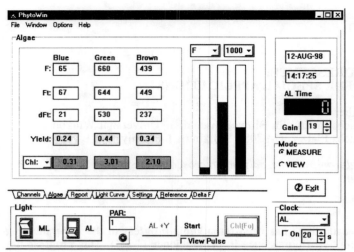

Figure 1
User surface of the
PHYTO-PAM with
display of „Algae"-
window showing
deconvoluted values
of fluorescence,
saturation pulse
induced dF, Yield,
and chlorophyll
concentrations for
cyanobacteria,
green algae and
diatoms in a river-
water sample.

620 and 650 nm). The fluorescence pulses are detected by a photomultiplier and amplified under microprocessor-control, resulting in 4 separate continuous signals (4 channels). The fluorometer is operated in conjunction with a Pentium-PC and special Windows software for data deconvolution and analysis. The fluorescence information is displayed in seven different "Windows" on the PC-monitor screen. Fig. 1 shows the user surface and screen-layout with the "Algae"-window being active. Data from a typical measurement, which characterizes the phytoplankton in a river-water sample (Main river) are displayed. The measurement involved a chlorophyll determination (Chl, in µg/l) on the basis of chlorophyll fluorescence yield in the quasi-dark state (Fo) and determination of the quantum yield of photosystem II with the help of a brief pulse of saturating light. When a saturation pulse is triggered, the momentary fluorescence yield (Ft) is sampled and the increase of fluorescence (dF=Fm-Ft) is determined. The quantum yield (Yield) corresponds to dF/Fm. The displayed data for cyanobacteria (Blue), green algae (Green) and diatoms (Brown) were calculated from the original 4-channel fluorescence data by an on-line deconvolution routine, based on previously stored "reference excitation spectra". Such "spectra", which consist of only four points at 470, 535, 620 and 650 nm, can be readily measured under "Reference" for any pure algae culture. The reliability of deconvolution depends on proper choice of reference spectra. However, in any case, the differences between cyanobacteria, green algae and diatoms are sufficiently large to allow at least a coarse differentiation, even if the particular species contained in a sample were not identified.

Fluorescence not only provides information on the content of phytoplankton, but also on its photosynthetic activity, making use of the saturation pulse method. In particular, the effective photosystem II quantum yield observed during continuous illumination is closely correlated with relative electron transport rate (10). Under "Light Curve", the PHYTO-PAM provides a routine for measuring light response curves, which give insight into the light saturation properties and photosynthetic capacity of the differently pigmented algal groups. In Fig.2 the Light Curves corresponding to the sample of Fig.1 are shown. Only the responses of green algae (3.0 µg Chl/l) and of diatoms (2.1 µg Chl/l) are displayed, as the content of cyanobacteria (0.3 µg Chl/l) was too low to give a satisfactory response with a single recording. In the given example, the light responses of green algae and diatoms were very similar, with I_k(green) = 222 and

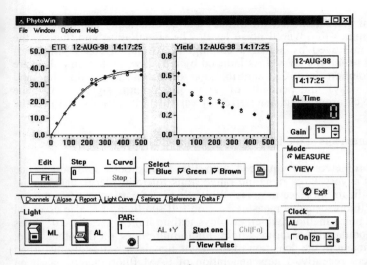

Figure 2
User surface of the
PHYTO-PAM in the
„Light Curve" mode
of operation.
Display of light
response curves of
relative electron
transport rate (ETR)
and of effective
quantum yield,
deconvoluted for
green algae and
diatoms in a river-
water sample (see
also Figure 1).

I_k(brown) = 212 μmol quanta/m^{2}s PAR. For comparison, it may be mentioned that a pure culture of *Ankistrodesmus* grown in artificial light was characterized by a much lower value of I_k = 38 μmol quanta/m^{2}s PAR (original data not shown).

While it is not surprising that, as in the example of Fig.2, green algae and diatoms, having experienced the same light conditions, display very similar light response curves, in other cases one may also encounter samples with different behaviour. An example is given in Fig. 3, which shows light response curves of green algae and diatoms in a water sample from Sydney Harbour. In contrast to the green algae, in this case the diatoms show an unusually low quantum yield in the dark, which rises in low light, before it declines again in parallel to the quantum yield of the green algae. This unusual behaviour of diatoms appears likely to reflect partial reduction of the photosystem II acceptor pool in the dark by electrons from reduced stroma components, possibly via NADPH-dehydrogenase, as previously observed in cyanobacteria (11). As a consequence, in the dark the diatoms may be in state 2 (12), which by illumination would be reversed into state 1.

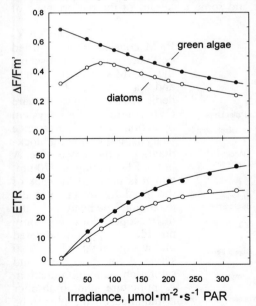

Figure 3 Light response curves of green algae and diatoms in a Sydney Harbour water sample.

With the present optical geometry of the PHYTO-PAM, which essentially has been previously described (9), in a single recording the detection limit is at 0.1 μg Chl/l. The limiting factor is a background signal corresponding to ca. 5 μg Chl/l. Although this background signal can be

4256

readily determined (from measurement with filtrate) and automatically subtracted from the total signal, it determines the overall noise level. The signal/noise and, hence, the detection limit can be improved by signal averaging. A special "Delta F"-mode of operation is provided, in which the ΔF-values induced by repetitive saturation pulses are averaged, with deconvolution into the contributions of green algae, diatoms and cyanobacteria. In this way, independent of any background signal, only the photosynthetically active chlorophyll (characterized by variable fluorescence, ΔF) can be assessed.

2.2 Ultrasensitive measurements at the level of single cells and chloroplasts

Standard PAM fluorometers are equipped with PIN-photodiodes as fluorescence detectors, which display a very large dynamic range, thus tolerating large background signals without any significant increase in noise. While this is of particular advantage in field applications, when fluorescence is measured e.g. in full sun light, it is also essential for use of the saturation pulse method, which employs actinic light pulses ca. 10^4 times more intense than the modulated measuring light. Photomultipliers, on the other hand, while being far more efficient in the detection of low light signals, are strongly disturbed by background light. Very recently, a new generation of PAM fluorometers has been developed (Schreiber and Gademann, unpublished) and now has become generally available, which makes use of the exceptional sensitivity of photomultipliers and still allows quenching analysis by the saturation pulse method (PAM-CONTROL, Universal Control Unit for Ultrasensitive Chlorophyll Fluorescence Measurements, in conjunction with various emitter-detector units, Heinz Walz GmbH). Here a short description of the so-called MICROSCOPY-PAM fluorometer and some examples of the performance of this new device will be presented.

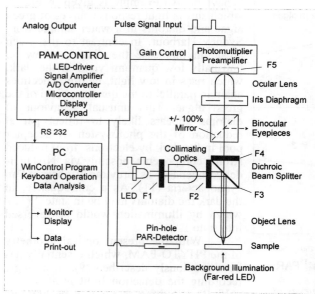

The MICROSCOPY-PAM consists of a modified epifluorescence microscope, the PAM-CONTROL unit and a Pentium-PC with dedicated Windows-software (WinControl) for system operation and fluorescence analysis. Fig.4 shows a block-diagram of the new device. A blue LED peaking at 470 nm, which is installed in place of the usual Xe-arc lamp in the excitation pathway of the microscope, does not only provide pulse-modulated measuring light, but also serves for actinic and saturation pulse illumination. A miniature photomultiplier is mounted on the phototube adaptor of the microscope. By visual inspection, with the help of an iris diaphragm the field of view can be narrowed

Figure 4 Block diagram of MICROSCOPY-PAM chlorophyll fluorometer. See text for further details.

down, such that the fluorescence characteristics of a particular, microscopically small object (as e.g. a single cell or chloroplast) can be selectively assessed.

In Fig.5 typical recordings of dark-light induction curves are displayed, which were measured with the MICROSCOPY-PAM. It is apparent that with a single cell of

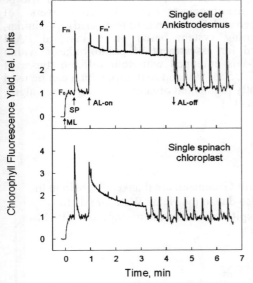

Figure 5 Typical recordings of dark-light induction curves with repetitive application of saturation pulses as measured with the MICROSCOPY-PAM.

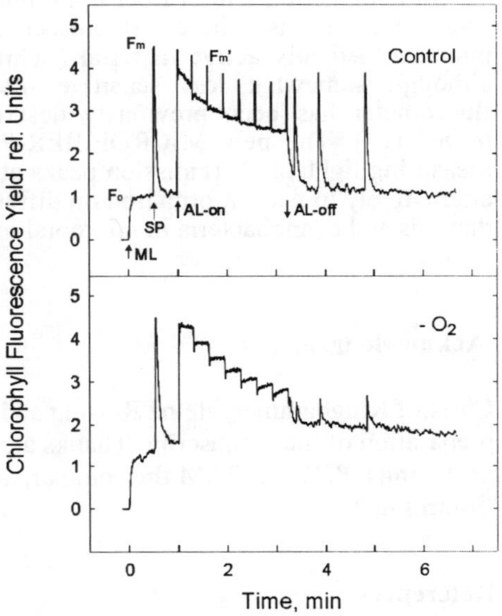

Figure 6 Dark-light induction curves with repetitive saturation pulses for quenching analysis measured on a single guard cell pair of Vicia faba in presence and absence of O_2.

Ankistrodesmus (3x12 µm) and a single spinach chloroplast (ϕ 5 µm) essentially the same fluorescence information can be gained as usually obtained with algae suspensions or whole leaves containing millions of cells. When only the measuring light (ML) is applied, minimal fluorescence yield, Fo, is monitored. Maximal fluorescence, Fm, is induced by a saturation pulse, SP. Following onset of actinic illumination (AL-on), repetitive saturation pulses are applied to assess maximal fluorescence yield, Fm', which is lowered with respect to Fm by nonphotochemical quenching. The WinControl software provides on-line calculation of effective quantum yield and quenching coefficients, which are displayed on the PC monitor screen (not shown). As with the PHYTO-PAM described above (see Fig.2), also a window for recording of light response curves is available (not shown).

An important practical application of the MICROSCOPY-PAM relates to the study of guard-cell photosynthesis. As shown in Fig.6, the fluorescence responses of *Vicia faba* guard cells display all the well-known features of a fully functional photosystem II and also of apparently normal electron transport activity. However, this electron transport does not necessarily reflect Calvin cycle activity and CO_2-fixation. This may be concluded from the fact that it is severely suppressed when molecular oxygen is removed. While further discussion of this interesting aspect would be out of scope of the present contribution, it may be assumed that O_2-dependent electron flow involving the Mehler-Ascorbate-Peroxidase cycle (13) plays a particular role in guard cell chloroplasts as an effective means of providing ATP for ion-pumping which is essential for stomata functioning.

In conjunction with a fiber-optic microprobe the PAM-CONTROL unit can also be used for ultrasensitive fluorescence measurements in different layers of photosynthetically active material (Schreiber and Gademann, unpublished). A similar, although somewhat less sensitive measuring system based on a standard PAM fluorometer has been previously described for assessment of photosynthesis within leaves (14). The new MICROFIBER-PAM is equipped with four different types of measuring light LEDs (emission peaks at 470, 530, 590 and 650 nm), which can be used alternatively to gain information on differently pigmented organisms, as e.g. green algae, diatoms and cyanobacteria in microbial mats (15).

Acknowledgements

Christof Klughammer, Heinz Reising and Rolf Gademann are thanked for help in the preparation of the manuscript. Thanks are also due to the Heinz Walz GmbH for providing a PHYTO-PAM fluorometer, an epifluorescence microscope and a PAM-Control unit.

References

1 van Kooten, O. and Snel, J. (1990) Photosynthesis Res. 25, 147-150
2 Schreiber, U., Bilger, W. and Neubauer, C. (1994) in Ecophysiology of Photosynthesis (Schulze, E.D. and Caldwell, M.M, eds.) pp. 49-70, Springer, Berlin
3 Dau, H. (1994) Photochem. Photobiol. 60, 1-23
4 Schreiber, U., Schliwa, U. and Bilger, W. (1986) Photosynth. Res. 10, 51-62
5 Lorenzen, C.J. (1966) Deep-Sea Research 13, 223-227
6 Falkowski, P.G. and Kolber, Z. (1995) Aust. J. Plant Physiol. 22, 341-355
7 Kolber, Z. and Falkowski, P.G. (1993) Limnology and Oceanography 38, 1646-1665
8 Schreiber, U., Schliwa, U. and Neubauer, C. (1993) Photosynth. Res. 36, 65-72
9 Kolbowski, J. and Schreiber, U. (1995) in Photosynthesis: from Light to Biosphere (Mathis, P. ed.) pp. 825-828, Kluwer Academic Publishers, Dordrecht, The Netherlands
10 Genty, B., Briantais, J.-M. and Baker, N.R. (1989) Biochim. Biophys. Acta 990, 87-92
11 Mi, H., Endo, T., Schreiber, U. Ogawa, T. and Asada, K. (1992) Plant Cell Physiol. 33, 1233-1237
12 Allen, J.F. (1991) Biochim. Biophys. Acta 1098, 275-335
13 Schreiber, U., Hormann, H., Asada, K. and Neubauer, C. (1995) in Photosynthesis: from Light to Biosphere (Mathis, P. ed.) pp.813-818, Kluwer Academic Publishers, Dordrecht, The Netherlands
14 Schreiber, U., Kühl, M., Klimant, I. and Reising, H. (1996) Photosynth. Res. 47, 103-109
15 Kühl, M. and Jorgensen, BB. (1992) Limnol. Oceanogr. 37, 1813-1823

TOPOGRAPHY OF LEAF CARBON METABOLISM AS ANALYZED BY CHLOROPHYLL a-FLUORESCENCE.

WEIS, E., MENG, Q.*, SIEBKE, K., LIPPERT, P. & ESFELD, P.
University of Münster, Institute of Botany, D-48148 Münster, Germany
*Shandong University, Taian, 271018 Shandong, PR China

1. Introduction

Leaf metabolism is highly flexible, depending on the physiological and developmental state. Often, metabolic activity is not uniformly distributed over the leaf area. In fully developed source leaves, mesophyll cells are specialized to photosynthesis, sucrose synthesis and export. Distribution of leaf metabolism is relatively uniform. In growing young leaves (sink leaves) the situation is more complex. Growth occurs over the entire leaf area, but cell division, expansion and differentiation occur pre-eminently in different zones. This is reflected in a complex and dynamic topography of leaf metabolism and photosynthetic activity. Growing cells are metabolically active. They do not export but retain metabolites. Metabolites feed and cause ‚sink‘ regulation of photosynthetic metabolism (e.g. [1]). But even in fully developed source leaves the topography of metabolism is not necessarily uniform. Leaves are major sites of pathogen attack, wounding or other stress, and plants react with local defense, which, in turn, induce redistribution of carbon fluxes and ‚sink‘ control of photosynthetic metabolism [2].

In this study we will demonstrate, that chlorophyll a-fluorescence can be used as a sensitive and non-invasive diagnostic tool to analyze the dynamic topography of leaf metabolism during development, pathogen interaction or other conditions which disturb the leaf's metabolism. Chlorophyll a-fluorescence imaging was recently introduced and used to derive images of fluorescence quenching and electron transport [3, 4, 5]. Under certain conditions, images of photosynthetic carbon fixation can be derived. We present examples of leaf development and pathogen interaction to demonstrate, that the topography of leaf metabolism can be highly complex and dynamic.

2. Materials and Methods

The computer based video camera system, connected with a gas exchange system was basically as described in [5] and modified as in [6]. Using equations from Genty and coworkers [7] pixel by Pixel calculation of fluorescence parameters and calibration of electron transport rates were carried out slightly modified as in [5]. ‘Assimilation Images’ were taken in saturating light, 450-680 ppm CO_2 and 2 % O_2, total-flux-images

4259

G. Garab (ed.), Photosynthesis: Mechanisms and Effects, Vol. V, 4259–4264.

in 450 ppm CO_2 and 21 % O_2. ‚Induction images' were calculated from 8-12 flux images taken during the induction period after a dark period (several hours) in 60 ppm CO_2 and 21 %O_2. Flux images were summed up and normalized to the image taken when electron transport was at its maximum. All experiments were carried out with attached leaves, placed in a gas exchange chamber. Elicitor was directly injected into the leaf mesophyll.

3. Results

Assimilation images in sink leaves. In figure 1, images of photosynthetic electron transport in a young developing leaf are displayed. The left image in figure 1 is taken in almost saturating light, 680 ppm CO_2 and 2 % O_2. Under such conditions, photorespiration is largely suppressed and electrons are mainly consumed by carbon fixation, i.e., the image may represent the topography of photosynthetic carbon fixation. In the top of the leaf, the assimilation rate is about as high as in an older, fully developed ‚source leaf'. The assimilation rate is markedly lower in the cell division and expansion zones at the leaf base and along developing major veins (differentiation zones). In 21 % O_2 (figure 1, right image) the overall flux is increased, due to electron consumption by photorespiration. In the top of the leaf (high assimilation zone) O_2-dependent increase is rather moderate, but the increase is substantial in the lower leaf zones. In the middle part of the leaf the overall flux is about as high as in low oxygen. Only a smaller area at the leaf base still exhibits very low flux. The experiment demonstrates that the expanding leaf can be subdivided into three metabolic zones which coincide with developmental zones: In the leaf top chloroplasts, photosynthetic metabolism and mesophyll anatomy are as fully developed as in a source leaf. Gas exchange between air and mesophyll is optimal and carbon fixation is high. In a second zone, which represents the expansion and differentiation zone, carbon fixation is low but the overall flux in 21 % O_2 is high. In this zone chloroplasts and photosynthetic metabolism have already developed, but entry of CO_2 is not optimal, because the cell and leaf anatomy has not yet developed: mesophyll cells have not yet fully expanded, chloroplast arrangement within the cells is still irregular (while it is highly organized in fully developed palisade cells) and the intercellular air space system has not yet developed (from microscopic studies, not shown). Due to non-optimal leaf anatomy, the diffusion resistance for CO_2 is high, and intracellular CO_2 concentration is too low to allow efficient carbon fixation. In this zone photosynthetic flux mainly represents photorespiration. In a third zone at the leaf base the electron flux is low, even in high O_2. In this cell division zone chloroplasts have not yet fully developed (not shown).

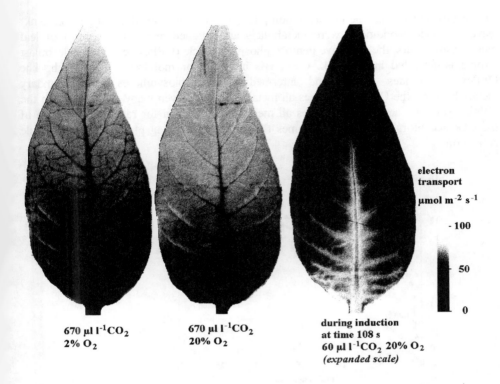

electron
transport
μmol m^{-2} s^{-1}

- 100

50

0

670 μl l^{-1}CO$_2$
2% O$_2$

670 μl l^{-1}CO$_2$
20% O$_2$

during induction
at time 108 s
60 μl l^{-1}CO$_2$ 20% O$_2$
(expanded scale)

Figure 1   *Electron transport of a young tobacco leaf. Left and middle: 350 μmol/m²/s light, right: 150 μmol/m²/s light after 14 hours of dark adaptation.*

Photosynthetic induction images. Secondary kinetics of photosynthesis, in particular, 'photosynthetic induction' reflect the overall metabolic situation in the chloroplast and, thus, is a valuable diagnostic tool to analyze the general metabolic state of a leaf. In source leaves or source areas of young leaves photosynthetic induction is slow and exhibits a lag phase. As an example, photosynthetic induction of a developed chickpea leaf is shown in figure 2 (control). Slow induction depends on the autocatalytical aspect of the reductive pentose phosphate cycle: starting from a low dark level of metabolites, initial regeneration of the CO_2-acceptor pool (RuBP) is slow and it requires time to build up an optimal metabolite concentration to support high assimilation rates. In sink mesophyll areas of young leaves or after 'sink' control of leaf metabolism, as, e.g., after induction of defense, photosynthetic induction is fast, even after overnight darkening. Figure 2 shows induction of chickpea leaflets inoculated with a fungal pathogen. Elicited leaf areas (induced 'sink' areas) co-exist with non-elicited areas. Photosynthetic induction is biphasic: the slow phase represents normal source areas, the rapid phase elicited leaf areas. By integration images taken during induction we can image the topography of 'fast induction' areas (see figure 3). We also see fast induction in the cell division and differentiation zones of young leaves (see right image in figure 1).

The metabolic basis for the fast induction phenomenon is explained as follows: in ‚sink' leaves or under conditions were metabolites are retained and ‚sink' control of leaf metabolism occurs, the oxidative pentose phosphate cycle (OPPcycle) in the chloroplast stroma is activated in the dark. Carbon is imported or mobilized from starch. The OPPcycle provides energy and intermediates for biosyntheses and secondary metabolism. As the OPPcycle shares all metabolites (and even many enzymes) with the Calvin cycle, it's activation provides all metabolites required for rapid regeneration of the CO_2 acceptors and thus, overcomes the ‚autocatalytic' lag phase for Calvin cycle activation.

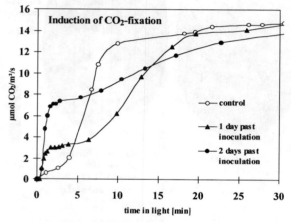

Figure 2   *Induction of photosynthesis in chickpea leaflets after a dark period of 10 to 14 hours (ca. 60 ppm CO_2 and 21 % O_2). One leaflet was infected with the fungal pathogen Ascochyta rabiei and induction was measured 1 and 2 days past inoculation (dpi). During this time the leaflet exhibits defense reactions, accompanied by induction of `sink' metabolism (involvement of increasing apoplastic invertase). 4 to 6 dpi necrotic spots appear.*

In an earlier study, more subtle metabolic differentiation at a much smaller scale around microveins could be visualize by imaging photosynthetic oscillations (Siebke & Weis 1996, Photosyn. Res 45,225-237).

Imaging leaf defense reaction. Figure 3 (A) shows an images of fast induction zones induced by local treatment with an elicitor. In treated areas the shikimic acid pathway and defense reactions are induced (e.g. phytoalexin synthesis (not shown, see [9])). High level of elicitor can induce local necroses. Similar symptoms are induced by fungal pathogens (see Fig. 2). Early defense reactions are accompanied by ‚sink' control of leaf carbon metabolism: assimilates are retained in the mesophyll, apoplastic invertase is induced and respiration increases (not shown). This situation can be visualized by ‚fast induction imaging. The slight reaction in the control leaflet, where water was injected instead of elicitor solution, is due to wounding. While most defense symptoms and necrosis formation remains strictly localized to the elicited area (indicated by circles), the induction of ‚sink' control as visualized by fast induction imaging, slowly spreads over the leaflet and eventually covers large parts of the leaf. Sometimes, these secondary defense areas exhibit low assimilation rates (electron transport in 450 ppm CO_2 and 2 % O_2) while total electron transport (in 21 % O_2) is still high (figure 3 B, C). Local necrosis formation is accompanied by a total collapse in photosynthetic activity (figure 3). Necrotic spots are surrounded by fast-induction-areas indicating secondary induction of defense-related ‚sink' metabolism (figure 3 left bottom leaflet). Secondary expansion of defense reactions within the leaf is characteristic and may be related to short distance SAR (systematically acquired resistance), mediated by signaling substances such as salicylic acid. Induction of apoplastic invertase and related metabolic changes has been discussed to be involved in SAR [10].

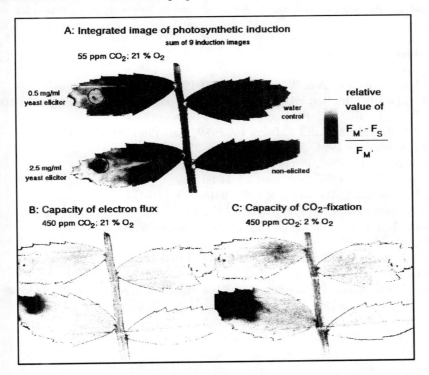

Figure 3 *Photosynthesis of chickpea leaflets ca. 15 hours past elicitation. A: Induction image of electron transport after a dark period of 12 hours in the first 10 min in light (60 ppm CO_2 and 21 % O_2). Steady state electron flux in 500 μmol/m²/s light and 450 ppm CO_2 of CO_2-fixation (C: 2 % O_2) and maximum electron transport (CO_2-fixation plus photo-respiration B: 21 % O_2).*

Fast photosynthetic induction is accompanied by fast induction of non-photochemical fluorescence quenching (NPQ). It can be explained as follows: onset of electron transport causes membrane energization and related ‚high energy quenching' [11]. In source leaves, quenching is low, due to the lag phase for electron flux. Therefore, under appropriate conditions, imaging of photosynthetic induction can be replaced by NPQ imaging. As NPQ imaging is easy to perform it could open the possibility to use this imaging for routine analysis, e.g., early detection of plant-pathogen interaction.

4. Conclusions

Chlorophyll *a*-fluorescence imaging provides valuable and non-invasive diagnostic tool to analyze the topography of photosynthesis and ‚sink' control of leaf metabolism. It is highly valuable to detect and follow early events in plant – pathogen interaction. As chl. *a*-fluorescence imaging is non-invasive, it can be used to follow dynamic processes in a single leaf.

5. Literature

1 Stitt M, von Schaewen A & Willmitzer L 1990, Planta 183, 40-50
2 Scholes J.D. 1992, In: Pest and pathogens: plant response in foliar attack, pp. 85-106 Ayres P.G. ed., BIOS Sci. Publ.
3 Daley P F, Raschke K, Ball J T, Berry J A 1989, Plant Phys. 90, 1233-1238
4 Genty B & Meyer S 1994, Aust. J. Plant Phys. 22, 277-284
5 Siebke K & Weis E 1995, Planta 196, 155-165
6 Jensen M, Siebke K 1997, Symbiosis 23, 183-196
7 Genty B, Briantais J M, Baker N R 1989, BBA 990, 110-124
8 Siebke & Weis 1996, Photosyn. Res. 45, 225-237
9 Daniel S, Tiemann K, Wittkampf U, Bless W, Hinderer W, Barz W 1990, Planta 182: 270-278
10 Herbers K, Meuwly P, Frommer W B, Métraux J-P, Sonnewald U, 1996, Plant Cell 8, 793-803
11 Krause G H & Weis E 1991, Annu. Rev. Plant Physil. Plant Mol. Biol. 42: 313-349

E-mail: weise@uni-muenster.de

AN INSTRUMENT FOR THE MEASUREMENT OF SUNLIGHT EXCITED PLANT FLUORESCENCE.

I. Moya, L. Camenen, G. Latouche, C. Mauxion, S. Evain and Z.G. Cerovic.
CNRS - LURE, Univ. Paris-Sud, BP34, 91898 ORSAY, FRANCE.

Key words : environment, light reflection, remote sensing, whole plant.

1. Introduction

In the last 20 years, several spectroscopic methods have been considered for characterizing the status of plant canopies and assessing their biomass and production. Among these methods, the use of the fluorescence signal emitted by plants under laser or daylight excitation has been the object of an intense research activity. Sun induced chlorophyll fluorescence in the upward light at sea surface has been considered since more than 20 years (1, 2). Following the successful application of airborne remote sensing (3, 4) ESA and NASA started up the development of ocean color sensors which includes appropriate channels for the detection of chlorophyll fluorescence. In this report we present attempts to follow the variations of the sun-induced chlorophyll fluorescence yield and reflectance at distance, to extend this approach to continental vegetation monitoring.

2. Procedure

2.1 *Fluorescence measurements using the Fraunhofer lines principle*

Under natural sunlight illumination the amount of chlorophyll fluorescence emitted by a leaf, which may represent up to 1% of the absorbed light in the visible part of the spectrum, is difficult to quantify because it is blurred by the reflected light. However, at certain wavelengths where the solar spectrum is attenuated (Fraunhofer lines), the fluorescence signal can be quantified. Fig. 1 compares a low resolution emission spectrum of the solar radiation at the sea level superimposed to the fluorescence emission and

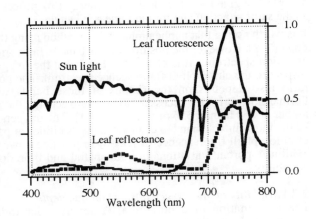

Fig. 1. Solar radiation at sea level compared to fluorescence and reflectance spectra of a green leaf.

G. Garab (ed.), Photosynthesis: Mechanisms and Effects, Vol. V, 4265–4270.

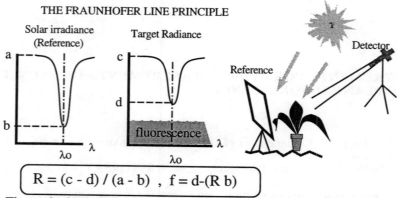

THE FRAUNHOFER LINE PRINCIPLE

$$R = (c - d) / (a - b) \ , \ f = d - (R \ b)$$

Fig 2. The method is based on the partial filling of the absorption band by the sun excited emission of the luminescent target. The calculation supposes that the wavelengths at which a (c) and b (d) are measured are sufficiently close to consider $R(\lambda)$ and $f(\lambda)$ constant.

reflectance spectra of a green leaf. Solar irradiance exhibit three main absorption bands in the red and near infra-red part : The Hα band at 656 nm is due to the hydrogen absorption by the solar atmosphere whereas two bands at 687 nm and 760 nm are due to the molecular oxygen absorption by the terrestrial atmosphere. These bands largely overlap the chlorophyll fluorescence emission spectrum of leaves. Therefore it is tempting to use these bands to monitor the chlorophyll fluorescence emission under daylight excitation by the method of the Fraunhofer line filling. This method was first introduced by McCord (5) and applied to fluorescence remote sensing by Plascyk (6). In short, this method compares the depth of the line in the solar irradiance to the depth of the line in the radiance of the target. Fig. 2 summarize the principle and shows related equations. The output parameters are the fluorescence contribution (**f**) to the light emitted by the target and the reflectance coefficient (**R**). R is defined as the ratio of the energy flux reflected by the sample in a given solid angle and the energy flux reflected by a white lambertian surface in the same situation and for the same solid angle.

It is worth noting that applications of this method using the Hα line have been successfully done (7, 8). However the position of this band is far from the maximum of the chlorophyll emission spectrum. It is clear from Fig 1 that the oxygen bands are better situated with respect to the chlorophyll fluorescence spectrum. The potential gain in the signal to noise ratio also depends on the shapes of these bands which are compared in Fig 3. Furthermore, it is also necessary to take into account the reflectance which is 5 to 10 times higher at 760 than at 656 or 687 nm (Fig 1). The oxygen absorbing band at 760 appears to be the most suitable band to evaluate the possibilities of passive measurements for detection of chlorophyll fluorescence of leaves. In practice however the oxygen absorption bands exhibit important variations in both spatial and time domains as a result of atmosphere variability.

2.2 *Variability of the oxygen absorption line depth.*
The illumination of a horizontal target by solar radiation at ground level may be decomposed into three components : (1) direct radiation, (2) diffused radiation by the atmosphere and (3) radiation diffused by the environment. Although direct radiation account for up to 85% of the total irradiance during a sunny day at noon, the diffused

:adiation cannot be neglected. For example up to 45% of the solar radiation is diffuse when the sky contains 4/8 of cumulus. In the case of completely cloudy sky all solar radiation is diffuse. The air mass crossed by light during its way through the atmosphere is usually represented by the air mass number **m**. The absorption of a clear atmosphere for a zenithal direction corresponds to the absorption of an uniform layer of ≈ 8.4 Km of air at the sea level pressure (**m** = 1). The depth and the shape of the oxygen absorption bands depends of the path length of solar radiation. The atmosphere path length changes with sun zenithal angle (Fig. 4) and with the type of atmosphere propagation. As a result **m** may vary from 0.5 to 2, depending on experimental conditions.

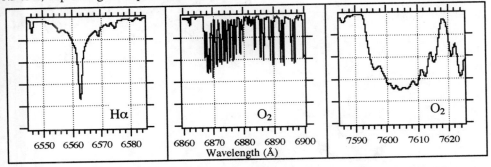

Wavelength (Å)

Fig 3. Spectra of the 3 bands of interest for monitoring chlorophyll fluorescence.

To ensure valid measurement attention should be paid to provide identical illumination to the target and the reference. In addition to the spatial constraints, there are also temporal constraints, because the depth of the oxygen absorption band may change suddenly due to light intensity variations (clouds). Therefore the **a**, **b**, **c** and **d** reflectance signals mentioned in Fig. 2 should be acquired simultaneously.

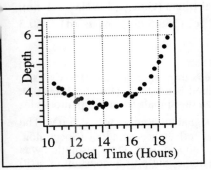

Fig. 4. Variation of the apparent depth of the 760 nm oxygen band with time at Orsay (48° 48' N , 2° 10' W) on June 4 1997.

2.3 Passive fluorescence detector at 760 nm

The system was aimed to follow fluorescence and reflectance changes as a function of natural daylight variations. For cost reasons commercially available interferential filters and monochannel detectors (photodiodes) were chosen. Two approaches are possible: to measure simultaneously the four signals (**a**, **b**, **c** and **d**) using four detectors or to measure them sequentially using only one detector. The former case implies to maintain correlation between the four detectors at less than $\approx 5.10^{-4}$, which seemed difficult to attain. The later case involves a mechanical system which exceeds our machining possibilities. We decided to build an intermediary system using two detectors to measure simultaneously **a** and **b** or **c** and **d** using a beam splitter (microscope cover). Fig. 5 shows a view of the instrument. As the **a** and **c** channel are detected on the reflection side, the difference between reflection and transmission of the beam splitter compensates approximately for the differences between the **a** (**b**) and **c** (**d**)

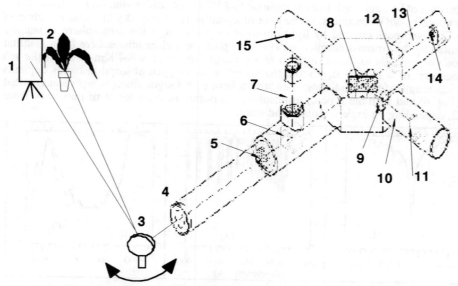

Fig. 5. Components of the passive detector at 760 nm. 1: Reference, 2: Target, 3: chopped mirror, 4: Field Stop, 5: High pass filter, 6: Polarization cube, 7: Finder, 8: Beam splitter, 9: Interference filter (758.5 nm), 10 (13): lenses, 11 (14): Photodiodes, 12: Interference filter (760.5 nm), 15: Beam dump.

signals. A chopped mirror alternates the field of view towards the reference or the target. To maintain the signals stable during several hours, both filters and detectors are located inside a heating jacket which maintains the temperature constant at 35 C. An RC filter has been introduced in the output circuit of each photodiode to exactly match the response time constant within $\pm 10^{-4}$. Two similar interferential filters (Omega, FWHM = 1 nm, transmission 70%, Φ = 25 mm) are used to define **a** (**c**) (758.5 nm) and **b** (**d**) (760.5 nm). The field of view of the system is $\approx 5°$.

2.4 *Calibration of the instrument*
The transmission of filters and the efficiency of detecting photodiodes (HUV 2000, EGG), were measured using the emission of a calibrated black body (Li-Cor 1800-02). Two digital voltmeters (HP 34401) were used to digitize the signals from the photodiodes with a resolution of 6 digits. In the absence of input signals the noise was < 5 μV. The reference signal was measured by pointing onto a flat reflectance standard (Spectralon, Labsphere). For the **a** signal up to 3 V was obtained in full sun light. Time resolution was limited by the chopped mirror to ≈ 1 Hz in the normal mode, but a time constant of ≈ 30 ms was achieved under stable sun illumination for kinetic measurements.

3. Results and discussion

3.1 *Measurements of variable fluorescence*
In order to test the accuracy of our passive detector we performed a concomitant measurement of a fluorescence induction curve with a fluorometer specially developed at LURE for kinetic chlorophyll fluorescence measurements on leaves at distance (FIPAM).

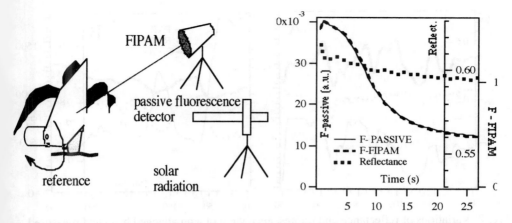

Fig. 6 Left: Set-up for the comparison of chlorophyll fluorescence transients obtained by the passive and active (FIPAM) method. Right: fluorescence induction transients recorded on a single leaf of *Prunus laurocerasus* . PAR $\approx$1000 µmoles photon m^{-2} s^{-1}.

A brief description of the principle and a typical application of the FIPAM is presented by Flexas et al. in this issue. Fig. 6 (left) shows the setting of the different components. A distance of $\approx$ 1m is sufficient to avoid interferences or shading between different parts of the set-up. In order to obtain a sudden dark to light transition, a leaf attached to the plant is maintained under the shade by the reference plate which can be moved rapidly by a stepping motor. After recording the reference signals, the reference is moved and the transient fluorescence signals are recorded simultaneously by the passive detector and by the FIPAM. The detection wavelength was the same for both detectors. After 30 s the reference is measured again. Due to mechanical limitations only the slow decreasing phase of the kinetics is observed. In Fig. 6 (right) the two fluorescence curves are plotted on different scales, but both starting from zero. So it is obvious that the two experiments are measuring the same fluorescence parameter.

3.2 *Passive fluorescence measurements at distance under variable sunlight*
In order to test the capacity of our system, we measured continuously during several hours on a single bean leaf attached to the plant. Due to clouds, solar radiation often changes between direct and diffuse, generating important PAR variations (200 to 1900 µmoles photon m^{-2} s^{-1}). Fig. 7 A shows a typical record after adaptation to moderate light. The relative stationary fluorescence yield (Fs) is obtained by dividing **f** by **a** which can be considered as a measurement of PAR. It is worth noting that Fs is positively correlated with PAR in this range of light intensity. After two hours of high light exposure, Fs becomes negatively correlated with PAR (Fig. 7 B) whereas the reflectance signal increases notably and exhibit huge changes which parallels Fs. Anti parallel correlation between PAR and Fs has been already reported Cerovic et al. (9) and Flexas et al. (this issue). It is interpreted as an accumulation of non photochemical quenching (NPQ), resulting from the high irradiance. NPQ relaxes when the light decreases. The threshold at which the anti parallel variations are observed is dependent of the type of plant, light acclimation and stomata aperture. A striking result is the observed variation of **R** which mimics the Fs variations when pronounced NPQ is occurring, although **R** is also varying under moderate light (Fig. 7 B). At variance, **R** is negatively correlated with fluorescence during the rise of

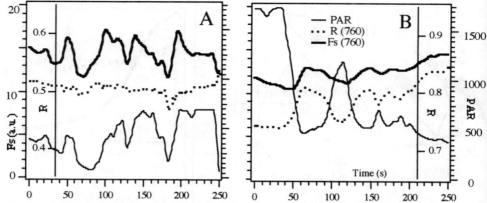

Fig. 7. Variations of reflectance and fluorescence yield of a an attached bean leaf measured with the passive fluorometer at 1 m of distance. A: 20 minutes after exposure to moderate light. B: After 2 hours of high light exposure. Conditions are kept constant during all the measurement.

the induction kinetics after dark adaptation (Fig. 6 Right) Variations of apparent reflectance around 750 nm after a shade to light transition have been already reported (10) and attributed to the chlorophyll fluorescence because reflectance variations were hardly detectable when superimposed with fluorescence variations.

In spite of the apparent complexity of **R** variations, this parameter is of great interest, as a complement of **f** for assessing several aspects of leaf or canopy function. The passive fluorometer described in this work demonstrate the possibility to measure Fs at distance with high accuracy. At the same time it provides the reflectance parameter **R** free of fluorescence contribution. Moreover, the possibility to perform the same type of measurement in two additional channels: at 687 nm (fluorescence) and 531 nm (reflectance changes in the green region) has been also tested with success in our laboratory.

4. References

1 Neville, R.A. and Gower, J.F.R. (1977) J. Geophys. Res. 82, 3487-3493
2 Morel, A. and Prieur, L. (1977) Limnology and Oceanography 22, 709-722
3 Doerffer, R. (1993) ICES mar. Sci. Symp. 197, 104-113
4 Gower, J.F.R. and Borstad, G.A. (1989) Adv. Space Res. 9, 461-465
5 McCord, T.B. (1967) J. Geophys. Res. 72, 2087-2097
6 Plascyk, J.A. (1975) Opt. Eng. 14, 339-346
7 Plascyk, J.A. and Gabriel, F.C. (1975) IEEE Trans. Instrum. Meas. 24, 306-313
8 Carter, G.A., Theisen, A.F. and Mitchell, R.J. (1990) Plant Cell Environ. 13, 79-83
9 Cerovic, Z.G., Goulas, Y., Gorbunov, M., Briantais, J.-M., Camenen, L. and Moya, I. (1996) Remote Sens. Environ. 58, 311-321
10 Gamon, J.A., Field, C.B., Bilger, W., Bjorkman, O., Fredeen, A.L. and Penuelas, J. (1990) Oecologia 85, 17

PATCH CLAMPING THE PHOTOSYNTHETIC MEMBRANE: A SENSITIVE TOOL TO STUDY CHLOROPLAST BIOENERGETICS

W.J. Vredenberg, J.H.A. Dassen and J.F.H. Snel
Dept Plant Physiol., Graduate school of Experimental Plant Sciences
Wageningen Agricultural University, Arboretumlaan 4
NL6703 BW Wageningen, the Netherlands

Key words: charge separation, Q-cycle, photocurrent, lumen conductance

1. Introduction

Electric parameters of the chloroplast have been shown to contain information on several bioenergetically relevant characteristics of the photosynthetic machinery (1,2). Single channel recording of the envelope of a pea chloroplast in an inside-out configuration has demonstrated a 50 pS anion channel to be involved in protein import into the organel (3). By applying single turnover saturating light flashes to a patched chloroplast in a so called whole thylakoid configuration the operation and actual performance of two distinguishable light-dependent current generators could be demonstrated (4).

Observations that the photocurrent kinetics were strongly modified at different holding potentials were indicative of light-induced conductance changes in the chloroplast resistance network (5,6). Evidence for the suggestion that these conductance changes arise primarily from alterations in the electric properties of lateral conductance phases along the thylakoid has come from patch-clamp experiments with chloroplasts in the presence of ammonium chloride (7).

This contribution illustrates some of the unique features of the patch clamp method for studying the bioenergetic performance and dynamic behavior of the photosynthetic membrane. The multi-phasic photocurrent profiles upon a train of multiple flashes show the following characteristics: (i) photocurrent generation originates from trans-thylakoid charge transfer accompanying reaction center (RC)- and Q-cycle turnover, (ii) a binary oscillation of the Q-cycle current generator with high activity in even numbered flashes, (iii) a 15-30% decrease in the amplitude of the RC-driven current in the second and following flashes of a flash train, concomitantly with an increase in the dark recovery time of the current. The decrease in amplitude and decay rate constant of the photocurrent in a double flash after dark adaptation are interpreted in terms of a change in the electric conductance of the thylakoid lumen. Data are interpreted to indicate a light control of the thylakoid lumen via a narrowing of the planar sheet-like structures by 1 to 3 single turnover flashes.

2. Procedure

Peperomia metallica plants were grown under a 12/12 hr photoperiod (irradiance of 1 W/m^2) at 25/20 °C day/night temperature and 60-70% relative humidity. Young fully expanded leaves were used for chloroplast isolation. Further details are given in refs (4,6). The microscope, patch-clamp amplifier, headstage and micromanipulator were mounted on a vibration-free table which permitted photocurrent recording from one chloroplasts up to several hours.

G. Garab (ed.), Photosynthesis: Mechanisms and Effects, Vol. V, 4271–4276.
© *1998 Kluwer Academic Publishers. Printed in the Netherlands.*

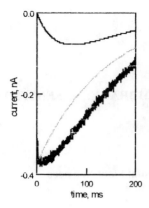

Single-turnover induced photocurrent (lower noisy curve) in a dark-adapted *Peperomia* chloroplast in the presence of 50 mM NH_4Cl and 0.6 mM methylviologen. The fit of the response (lower solid curve) is the sum of the photocurrent generated by reaction centers (I_{RC}, middle curves) and by secondary electron flow (I_Q, top curves) with (in pA)

$$I_{RC} = -380.\exp(-t/134),$$
$$I_Q = -200[1 - \exp(-t/65)].\exp(-t/134)$$

Photosensitive thylakoid area (S)

Pipette

Chloroplast

Figure 1

Calculation of thylakoid area (bleb diameter) associated with single turnover induced photocurrent (see top figure):

amplitude (I_{RC})	**380 pA**
decay time (τ_0)	**134 ms**

Number of charges ($I_{RC}*\tau_0$) through pipette

$5.1 \ 10^{-11}$ C ($= 3.2 \ 10^8$ electrons)

which (with $\alpha \sim 0.5$) is about 50% of charges generated by RC's. Thus number of charge-generating RC's

$$N_{RC} \sim 6.4 \ 10^8$$

This means, with 2 RC's per ETC and an approx. ETC density of $1. \ 10^{15} \ m^{-2}$, that the photocurrent (electrons)

are generated in a **thylakoid surface**

$$S = 3.2 \ 10^{-7} \ m^2$$

If this surface is considered as a sphere (bleb) than its diameter d is:

$$d = 320 \ \mu m$$

Diameter chloroplast

$$10\text{-}20 \ \mu m$$

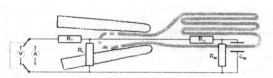

Electric equivalent scheme of the whole-thylakoid configuration. C_M membrane capacitance, R_M membrane resistance, R_A access resistance to the thylakoid membrane; R_T electrode tip resistance, R_L leak resistance. The two encircled symbols at the left indicate the configuration of the patch-clamp amplifier in the voltage- (A) or current clamp (V) mode. The present experiments were done in voltage-clamp mode. The fraction of the photocurrent flowing through R_M is denoted by α (see further text).

Patch-pipettes were prepared from borosilicate glass, fire-polished before use and filled with the chloroplast suspension medium. The electrode tip resistance R_T was about 10 MΩ. A whole thylakoid configuration was achieved by suction of a chloroplast on to the tip of the electrode. Usually the (5 min) dark adapted chloroplast was illuminated with a train of 8 (or more) successive Xe flashes (8 μs) with a dark time interval of 1s. In other experiments a train of 2 flashes was given to a dark adapted chloroplast with a dark time interval between flash pairs of 5 to 200 s. In some special cases the flash trains were preceeded by a preillumination light pulse of several seconds duration. Flashes were not wavelength-filtered.

The electric equivalent scheme of the chloroplast recording configuration (Fig. 1), the signal detection system as well as the details of curve analysis are the same as described elsewhere (4,6).

3. Results and Discussion

Fig. 1 shows an example of the photocurrent detected by an electrode sucted onto a *P. metallica* chloroplast in a 'whole chloroplast' configuration under voltage clamp ($V_h = 0$) at a ms time resolution upon a saturating single turnover light flash. As illustrated in detail elsewhere (4), each response can be fitted with two exponentials and is the sum of currents generated by fast charge separation in the reaction centers ($R1_{RC}$) and by secondary electron flow ($R1_Q$). The two current generators differ in actual number and rate constant for a single trans-membrane charge transfer (N_{RC}, k_{RC} and N_Q, k_Q, respectively), but are subject to the same diffusive dissipation process. The photocurrent response can be written as

$$I(t) = [I_{RC}(t)] + [I_Q(t)] = [I_{RC} + I_Q\{1 - \exp(-k_2.t)\}] * \exp(-k_0.t)$$

in which I_{RC}, I_Q are proportional to the actual number (N_{RC}, N_Q) of operational current generators that perform a single turnover, k_0 ($= 1/\tau_0$) the relaxation constant (time) of the photocurrent current decay and $k_2 = k_Q - k_0$, in which k_Q is the rate constant of a single charge transfer by a Q-cycle current generator. The area under the photocurrent traces [$I_{RC}(t)$] and [$I_Q(t)$] is, except for an attenuation factor $1-\alpha$, a direct measure of N_{RC} and N_Q, respectively; α is the fraction of the generated current that is branched through R_M and does not flow into the electrode. As in general R_T and R_L are small as compared to R_A (2,8), $\alpha = R_A/(R_A+R_M) = \tau_0 / \tau_M$, in which $\tau_M = R_M * C_M$ is the 'native' relaxation time of the thylakoid membrane, which can be determined independently from the decay of a flash-induced P515 response (9). It should be noted that in the present configuration with *P. metallica* chloroplasts in general 30 to 40% of the current through R_A leaks through R_L. Fig. 1 illustrates, in agreement with earlier conclusions (8,9) that the recorded photocurrents come from a thylakoid area which is many times larger than the area of the chloroplast envelope. In this case the current generating thylakoid area is about 10^3 higher than the chloroplast area. This value is even higher when the current 'loss' through the leak resistance R_L is taken into account.

Figure 2 shows the photocurrent profiles measured upon the first 4 (out of a train of 8) successive single turnover saturating flashes, separated by a dark interval of 1 s, given to a dark adapted chloroplast. There is a clear difference in the extent and kinetics between these responses. As compared to the photocurrent upon the first flash the extent of the fast initial rise of the current in the following flashes is lower whereas the dark recovery lasts longer. The difference in the initial current amplitude between the two flashes is 23%. The difference in dark recovery becomes more clear after a deconvolution of the responses into those of I_{RC} and I_Q.

Figure 3 shows the photocurrent parameters $I_{RC.n}$, $\tau_{0.n}$ and $N_{Q.n}$ estimated for the full train of 8 single turnover flashes (see fig. 2 for the current profile of the first 4 flashes) in dependence of the flash number n. It shows or alternatively contains the evidence for the

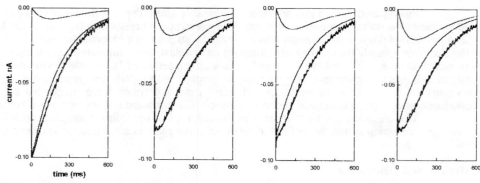

Figure 2. Single turnover induced photocurrents (lower noisy curve in each panel) in a 5 min dark adapted *Peperomia* chloroplast upon 4 single turnover saturating flashes given at an interval of 1 s. The fit of the response (lower solid curves) is the sum of the photocurrent generated by reaction centers (I_{RC}, middle curves) and by secondary electron flow (I_Q, top curves).

following conclusions. (i) An approx. 20-22% decrease in the amplitude I_{RC} of $R1_{RC}$ accomplished by 3 to 4 flashes; in general the amplitude in different chloroplasts is found to be constant after the 4th flash and. (ii) An increase in the dark relaxation time τ_0 of $R1_{RC}$, also

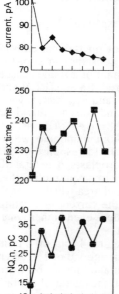

most pronounced in the first flashes, and anti-parallel with the photocurrent amplitude $I_{RC,n}$. (iii) No appreciable change, if at all in the number of charge separating active reaction centers $N_{RC,1}$ per flash as concluded from the areas covered by $I_{RC}(t)$. (iv) A significant binary oscillation in the number of charges $N_{Q,n}$ generated by $R1_Q$, with a low(er) transport in odd-numbered flashes. (v) The rate constant of Q-cycle charge generation k_Q is in the range between 7 and 20 s^{-1} and independent of the flash number. (vi) The average ratio $N_{Q,n}/N_{RC,1}$ for a pair of flashes at higher flash numbers is about 0.4. This is the first quantification (but see also ref. 4) of the oscillatory Q cycle in intact chloroplasts in terms of the number of

Figure 3 Photocurrent amplitude ($I_{RC,n}$), relaxation time ($\tau_{0,n}$) and number of charges transferred by Q-cycle ($N_{Q,n}$) as a function of the flash number (n) in a train of 8 flashes.

charges involved in successive single turnovers of the photosystems. A binary oscillation in the P515 response, ascribed to a Q-cycle has been reported before (10) and is now nicely confirmed by our photocurrent measurements. It should be added that in some chloroplasts no Q-cycle activity was found under our experimental conditions. The reason for this is unknown as yet; it may be related to an unfavorable local environment in the native chloroplast (e.g. pH, quinol concentration, redox potential).

The anti-parallel pattern of $I_{RC,n}$ and $\tau_{0,n}$ in the first 4 flashes with a lower value of $I_{RC,n}$ in the 2nd and following flashes is routinely observed in individual chloroplasts. Assuming the model in Fig. 1 is correct, this points to an increase in R_A after the first flash. Changes in either R_M or R_L would not have caused a change of this kind (8). The possibility that a light-induced change in C_M has occurred during the first flash also has been shown to be unlikely (1,6). Data collected for various chloroplasts indicate that the relative increase in R_A caused by a single turnover flash in a dark adapted chloroplast is in the range between 15 and 30%. Noteworthy is the relatively slow dark relaxation (in the time range of minutes) of these resistance changes within the thylakoid lumen under normal conditions,as illustrated in fig. 4. Here the difference between the photocurrent amplitudes upon a double flash given to a dark-adapted chloroplast ($I_{RC,1}$ - $I_{RC,2}$) is measured as a function of the darktime interval between the two successive flash pairs. The reversion to the initial dark-adapted state after one flash takes more than 120 s for the chloroplast used in this experiment.

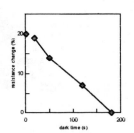

Figure 4 Dark recovery of change in lumen resistance after flash.

Replacement of K^+ by NH_4^+ in the chloroplast isolation medium was found not to affect the responsiveness of R_A upon a flash (7), although R_A was found to decrease upon continuous illumination in conjunction with an osmotic swelling of the thylakoid lumen. This energy-dependent decrease in R_A was concluded from the photocurrent profiles before and after preillumination of a chloroplast in the presence of the amine. This is shown in fig. 5. Control chloroplasts, i.e. with 'physiological' KCl did not show this behavior in prolonged energization.

Analyses like those shown in figs. 2 and 3 indicate that the changes in the photocurrent profile are not caused by changes in the number of reaction centers actually involved in the single turnover charge separation. Such change (11), if

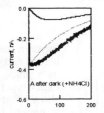
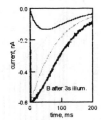

Figure 5. Flash-induced photo-current in *Peperomia* chloroplast 1s before (A) and 3s after a 4s illumi-nation in presence of 50 mM NH₄Cl

occurring at all under our conditions is negligible, i.e. the number of centers which are closed due to impaired or retarded reversion of the primary donor and/or acceptor of the RC ('inactive centers') is not increased significantly upon the flash. Thus the lower photocurrent amplitude accompanied by a retardation of the decay kinetics in the second and following flashes is mainly, if not completely caused by changes in a current-limiting series resistance (R_A) of the chloroplast network rather than by changes in the population of active reaction centers. We presume that the flash-induced increase in the chloroplast access resistance R_A originates mostly, if not completely from a change in thylakoid geometry. Electrostatic repulsion forces within the thylakoid lumen due to the negative surface charges of opposing membranes set limits to the distance of nearest approach, i.e. to the thickness of the sheet-like lumenal (and stromal) spaces in a granum. The positive charges created inside the lumen by the light driven trans-thylakoid charge separation will have a screening effect on these surface charges causing a decrease in the repulsive forces, leading to a narrowing and shrinkage of the lumen, and consequently to an increase in the lateral resistance of the lumen. A requirement of 1 to 3 turnovers of the reaction centers, present at a density of approx. 2.10^3 μm^{-2} for effective screening of the surface charge would indicate a lumenal surface charge density of about 0.08 $\mu C.cm^{-2}$. This charge density is about 10 times lower than that reported for the stroma-oriented thylakoid surface (12). In order to account for the high photon efficiency of the proposed surface charge screening one might speculate that the PS II density in the grana stacks is much higher than the RC density averaged over the whole thylakoid area (13).

Figure 6 summarizes our view on the low light requiring structural changes of the thylakoid organization pattern in the intact chloroplast. A change of this kind which alters the proximity of membrane proteins is likely to be accompanied by changes in energy transfer within and between LHC's Forthcoming research will focus on the possible link of these changes with

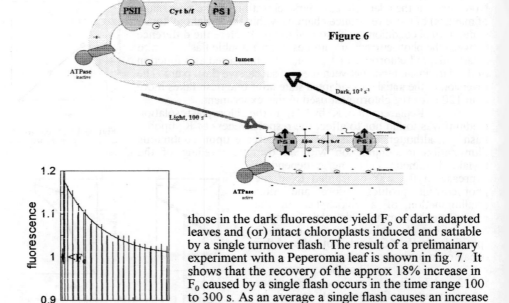

Figure 6

those in the dark fluorescence yield F_0 of dark adapted leaves and (or) intact chloroplasts induced and satiable by a single turnover flash. The result of a preliminary experiment with a Peperomia leaf is shown in fig. 7. It shows that the recovery of the approx 18% increase in F_0 caused by a single flash occurs in the time range 100 to 300 s. As an average a single flash causes an increase in F_0 of about 20% which reverses in the dark with a relaxation time of about 100 s.

Figure 7. Fluorescence yield (upward deflections) before and after a saturating pulse measured at 15 s intervals by low intensity modulated light in a XePAM. Continuous curve is the exponential fit of $\Delta F/F_0$ (17.8%)with relaxation time 100 s

Acknowledgments

We thank Alexander Bulychev, Tijmen van Voorthuysen and Paul van den Wijngaard for their contribution. This research was partly financed by a grant from the Netherlands Organization for Scientific Research (NWO)

References
1. Voorthuysen T van (1997) The electrical potential as a gauge of photosynthetic performance in plant chloroplasts. A patch-clamp study. PhD thesis. Wageningen Agricultural University , Wageningen
2.. Vredenberg WJ (1997). Bioelectrochem Bioenerg 44(1): 1-11
3. Wijngaard PWJ van den, Vredenberg WJ (1997) J. Biol. Chem. 272:29430-28433
4. Voorthuysen T van, Dassen JHA, Snel JFH and Vredenberg WJ (1996). Biochim Biophys Acta 1277: 226-236
5. Bulychev AA, Antonov VF and Schevchenko EV (1992). Biochim Biophys Acta 1099: 16-24
6. Voorthuysen T van, Bulychev AA, Snel JFH, Dassen, JHA and Vredenberg WJ (1997) Bioelectrochem Bioenerg 43: 41-49
7. Bulychev AA, Dassen JHA, Vredenberg WJ, Opanasenko VK and Semenova GA (1998). Bioelectrochem Bioenerg, in press
8. Vredenberg WJ, Dassen, JHA and Snel, JFH(1998) Photsynth. Res. In press
9. Vredenberg WJ, Bulychev AA, Dassen, JHA, Snel, JFH. and Voorthuysen T van (1995) Biochim Biophys Acta 1230: 77-80
10. Velthuys BR (1981). FEBS Letters 115: 167-170
11. Chylla RA and Whitmarsh J, (1989). Plant Physiol 90: 765-772
12. Schapendonk A.H.C.M., Hemrika-Wagner A.M., Theuvenet A.P.R., Wong Fong Sang H.W. Vredenberg W.J. and Kraayenhof R. (1980) Biochemistry 19: 1922-1927.
13. Trissl H-W and Wilhem C (1993). TIBS 18: 415-419

NON-UNIFORM CHLOROPHYLL FLUORESCENCE QUENCHING DURING PHOTOSYNTHETIC INDUCTION IN WILTING LEAVES, AS REVEALED BY A FIELD-PORTABLE IMAGING SYSTEM.

C. B. Osmond[1] , C. Büchen-Osmond[1] and P. F. Daley.[2]   [1] Research School of Biological Sciences, Institute of Advanced Studies, The Australian National University, Box 475, Canberra, ACT 2601, Australia.   [2] Environmental Restoration Division, Lawrence Livermore National Laboratory, PO Box 808/L528, Livermore, CA 94550, USA

Key words: Induction, leaf anatomy-physiology, photosystem 2, stomata, water stress.

1. Introduction

Although chlorophyll fluorescence imaging methods have been used to show patchy fluorescence quenching associated with patchy stomatal responses following application of ABA and other treatments (1, 2), the significance of patchiness in natural habitats, and whether it can be ascribed to stomatal responses and/or metabolic regulation (3, 4, 5), remains open to question (6, 7).   Better understanding of the areal distribution of photosynthetic activity is important to evaluate photosynthesis in natural habitats (8).   A field-portable instrument has been developed (9) to help determine whether these phenomena are species or environment specific, and whether they are of widespread occurrence.   Leaves of plants growing in mesic natural habitats often show transient wilting in bright light.   The preliminary data presented here show patchy fluorescence quenching may be widespread during wilting, and suggests that fluorescence imaging could be routinely used to augment photosynthetic measurements in the field.

2. Materials and methods

A field-portable fluorescence imaging system described (9) and evaluated (10) earlier, was used to examine turgid and wilting, detached and attached leaves of plants in air, in the laboratory and the field.   Detached leaf experiments were done with wilted plants growing in full sunlight, immediately outside the laboratory.   Leaves were placed in a leaf clip affixed to the imaging system and dark-adapted for 1-2 min.   Detached leaves of turgid plants were kept in a vial of water during observation.   In most experiments, fluorescence induction transients were excited using 500 μmol photons m^{-2} s^{-1} red light (663 nm) from the LEDs in the illuminator,and images of fluorescence were recorded on videotape at constant exposure.   The field-portable equipment, powered by rechargeable batteries, was set up on a tripod, and selected, attached leaves placed in the leaf clip on the imaging system.   Wilting and turgid leaves were examined on intact plants in full sunlight on mown meadows in the Botanic Gardens, in Darmstadt and Göttingen, Germany, where wilting of herbaceous monocotyledons and dicotyledons was common within 15 min of exposure to full sunlight.   Similar observations were made on herbs

G. Garab (ed.), Photosynthesis: Mechanisms and Effects, Vol. V, 4277–4280.
© 1998 *Kluwer Academic Publishers. Printed in the Netherlands.*

growing in light gaps in a beech forest near Arnsberg, Germany, when the light intensity increased from 60 to 1,500 μmol photons m^{-2} s^{-1} during the 30 min passage of sunflecks. Chlorophyll fluorescence images were analysed as described earlier (9), and leaf water content was found by expressing fresh weight of leaf discs at time of measurement as a percentage of fresh weight of discs floated on water over night.

3. Results and Discussion

The field-portable imaging system was used in its most simple configuration; tape recording of photosynthetic induction experiments in a survey mode. In detached, wilted leaves of *Potentilla*, vein-defined areas of slower quenching appeared in the first 20-40s of fluorescence transients (Fig. 1). At modest levels of water loss these patches faded within 120s, but persisted under greater stress. In most cases, but not all, rehydration of leaf discs excised from the areas observed led to restoration of uniform quenching, of the sort seen in controls (Fig. 1). Similar patterns were commonly observed in detached, wilted leaves of *Frageria*, but vein-associated slower quenching of fluorescence was observed in *Vicia* and *Hibiscus*. In *Oxalis*, random points of slow quenching lit up during otherwise uniform quenching of surrounding tissue. Quantitative analysis of the fluorescence quenching rate in the patches and surrounding tissues was done by measuring pixel values at six selected points in successive images (Fig. 2).

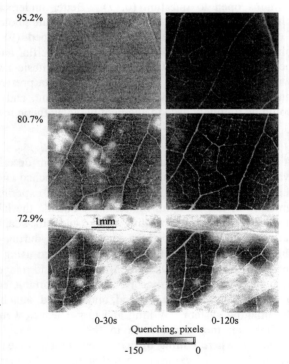

Figure 1: Quenching of chlorophyll fluorescence in *Potentilla* leaves from field grown plants that were allowed to wilt on the laboratory bench. The images shown were calculated by subtracting the images captured at 30s and at 120s from the initial image captured on illumination. Uniform quenching was found in turgid (top) controls with 95.2% water content, in contrast to the patchy areas of slow quenching in wilted leaves (middle) with 80.7% water content. The vein-defined patches persisted in the most stressed leaves (bottom) with only 72.9% water content.

Figure 2: Time course of change in fluorescence intensity in bulk tissue (●) and patches (O) as they appeared in detached leaves of herbs that wilted on exposure to full sunlight. *Oxalis* leaflets were folded when collected.

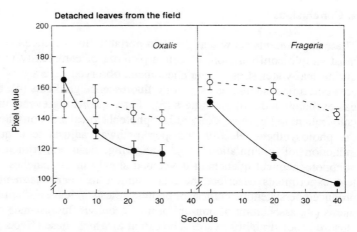

Detached leaves from the field

The patchiness of chlorophyll fluorescence quenching in wilting leaves was not an artefact of detached leaf experiments. Attached leaves of herbs, shrubs and grasses that wilted on exposure to full sunlight under field conditions in two successive European summers showed many variations of patchy quenching. Many Compositae on display in the Botanic Gardens Göttingen wilted on the first sunny morning after 3 weeks of rain, and patchiness was observed in *Chrysanthemum segetum, Helianthus annus, H. helianthoides, Heliopsis scarbara* and *Rudbeckia hirta*. Leaves of *Sanguisorba* in a meadow, that folded on each other after wilting, showed persistent broad areas of slow quenching along the mid-vein (data not shown). Bright patches in wilted *Taraxacum* and *Lamium* leaves faded within 30-50s, whereas pale-green areas between veins of *Geum* (possibly due to nutrient stress or infection) showed persistent high fluorescence (Fig.3), during a 30 min sunfleck.

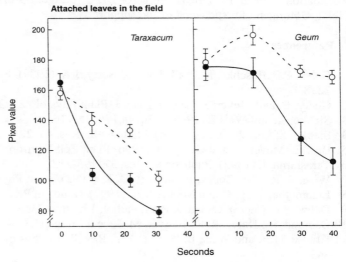

Attached leaves in the field

Figure 3: Time course of change in fluorescence intensity in bulk tissue (●) and patches (O) as they appeared in attached leaves of wilted plants in the field. *Geum* leaves growing in a light gap showed persistent high fluorescence in areas of intervein chlorosis.

4. Conclusions

These experiments show that the field-portable fluorescence imaging system, even in its most simple configuration, reveals a plethora of complexity in fluorescence quenching during leaf water stress. The phenomena observed here are of about the same scale in area and time response as oscillating fluorescence gradients (4), but are rather faster than gas exchange-induced patchiness (5). Early quenching events in fluorescence transients (subsequent to Fp) most likely reflect photochemical and non-photochemical processes as the photosynthetic electron transport pathway adjusts to a quasi steady-state during induction of assimilation (11). Saturating flash experiments (10) confirmed that nonphotochemical quenching developed slowly in the patches. Although the detached leaf experiments described here are similar in their treatment severity to those that produced persistent patchiness of photosynthesis in leaves of uprooted field-grown cotton plants (7), association of non-uniform chlorophyll fluorescence quenching with stomatal closure is not straightforward. The extent to which these effects on electron transport are a consequence of prior stomatal closure, or impaired metabolism due to water stress, is unclear. Perhaps they portray "memory effects" of localised variations carbon assimilation, sustained changed in ΔpH or adenylate status. One plausible indication that these areas of slow quenching may be associated with localised changes in water status comes from the large patches that form along the mid-rib when *Sanguisorba* leaflets fold on each other during wilting. Folding is presumably turgor activated and associated with local changes in osmotic relations that may also influence fluorescence transients.

5. Acknowledgments

The portable imaging system was developed with support of NSF grant BSR 891451, and the research was made possible by a Forschungspreis from the Alexander von Humboldt Foundation. The authors thank Professor Dr. U. Lüttge, Darmstadt and Professor Dr. H. Heldt, Göttingen, for sponsorship and support.

6. References

1 Daley, P.F, Raschke, K., Ball, J.T. and Berry, J.A. (1989). Plant Physiol. 90, 1233-1238.
2 Genty, B. and Meyer, S. (1994). Aust. J. Plant Physiol. 22, 277-284.
3 Siebke, K. and Weis E (1995a). Planta 196, 155-165.
4 Siebke, K. and Weis, E. (1995b). Photosynth. Res. 45, 225-237.
5 Bro, E., Meyer, S. and Genty, B. (1996) Plant Cell Environ. 19, 1349-1358.
6 Terashima, I. (1992). Photosynth. Res. 31, 195-212.
7 Wise, R. R., Orti-Lopez, A. and Ort, D. R. (1992). Plant Physiol. 100, 26-32.
8 Lichtenthaler, H. K. and Miehe, J. A. (1997) Trends in Plant Sciences 2, 316-320.
9 Daley, P.F. (1995). Canad. J. Plant Pathol. 17, 167-173.
10 Osmond, B., Daley, P. F., Badger, M. R. and Lüttge,U. (1998). Bot. Acta (in press).
11 Krause, G.H. and Weis, E. (1991). Ann. Rev. Plant Physiol. Plant Mol. Biol. 42, 313-349.

DIFFUSIVE AND NON-DIFFUSIVE LIMITATIONS OF HETEROGENEOUS PHOTOSYNTHESIS DETERMINED BY CHLOROPHYLL FLUORESCENCE IMAGING FOR *ROSA RUBIGINOSA* L. LEAVES.

Sylvie Meyer and Bernard Genty
Groupe Photosynthèse et Environnement, Laboratoire d'Ecologie Végétale,
Bât. 362, CNRS URA 2154, Université Paris Sud, Orsay 91405, France.
Fax (33 1) 69 41 72 38

Key words : abscisic acid, fluorescence imaging, gas exchange, leaf physiology, photosynthetic heterogeneity.

1. Introduction

Photosynthesis can be heterogeneously distributed over a leaf, especially during periods of environmental stress. Heterogeneity has usually been explained by patchy stomatal closure [1,2] but local metabolic limitations might also be involved, *e. g.*, during induction by light [3]. Discrimination between stomatal and metabolic limitations of photosynthesis has been usually done by measuring photosynthetic capacity under non-limiting CO_2 availability [2, 4]. We investigated the potential contribution of stomatal conductance and metabolism in determining heterogeneity of photosynthesis by exposing the leaf to short pulses of high CO_2 concentration while imaging the leaf chlorophyll fluorescence. Relationship between diffusive and metabolism dependent limitations was characterised for detached leaves of *Rosa rubiginosa* after ABA-feeding.

2. Procedure

2.1 CO_2 and H_2O gas exchange measurements and mapping of PSII photochemical yield, ($1-\Phi/\Phi_m$), using chlorophyll fluorescence imaging were simultaneously performed on detached leaves of *Rosa rubiginosa* L. (heterobaric and hypostomatous) as described in [5, 6]. Experiments were done at a subsaturating photon flux density of 650 µmol m^{-2} s^{-1} in air. Images of ($1-\Phi/\Phi_m$) were recorded just before and during a brief transition to 0.4% O_2, (i) before ABA addition, at 340 and 49 µmol mol^{-1} CO_2, (ii) after feeding the leaf with 10^{-4} M ABA, when steady-state was reached at 340 µmol mol^{-1} CO_2. Images were also recorded during a transition to 0.74% CO_2 for 30 to 90 s in 21% O_2 and during transition to 0.4% O_2. Gas exchange parameters were calculated according to [6, 7].

2.2 Total electron transport rate, J_t, was calculated from ($1-\Phi/\Phi_m$) according to [8,9] as:

$$J_t = 4I((1 - \Phi/\Phi_m) - d)/k \qquad (1)$$

where I is the incident photon flux density, k and d are the slope and the Y intercept of the linear relationship between ($1-\Phi/\Phi_m$) and the quantum yield of gross CO_2

4281

G. Garab (ed.), Photosynthesis: Mechanisms and Effects, Vol. V, 4281–4284.

4282

assimilation as measured under non photorespiratory conditions [5,8]. Since images of $(1-\Phi/\Phi_m)$ computed from fluorescence images were obtained under constant irradiance, they can also be seen as maps of the relative rate of linear electron transport.

3. Results and Discussion

3.1 Distribution of photosynthesis in ABA-fed leaves. Before ABA-feeding (Fig 1A), at steady-state photosynthesis in 0.4% O_2, the values of $(1-\Phi/\Phi_m)$ were uniformly distributed over most of the leaflet area. The frequency distribution was unimodal and largely symmetrical. During a brief exposure to 0.74% CO_2, $(1-\Phi/\Phi_m)$ increased (Fig. 1B) and the distribution of $(1-\Phi/\Phi_m)$ became much more uniform (Fig. 1B).

Figure 1. Images and frequency distributions of $(1-\Phi/\Phi_m)$ taken during transition in 0.4% O_2 before (A, C) and during (B, D) a transition to 0.74% CO_2 in control (A, B) and ABA-fed (C, D) leaf of Rosa. C O_2 mole fraction was 310-340 μmol mol^{-1}. 3-4 pixels represent one stoma and 10 epidermal cells. The bar indicates 1 cm. Pixel intensity was scaled using a 8 bit grey scale where black and white correspond to $(1-\Phi/\Phi_m)$ of 0 and 0.5, respectively (A, B, D) or 0 and 0.4, respectively (C). The class size of the frequency distributions was 0.01. Mean with SD are indicated. Stomatal conductance was 377 and 66 mmol H_2O m^{-2} s^{-1} for A and C, respectively, net CO_2 assimilation was 22.3 and 6.0 μmol m^{-2} s^{-1} for A, and C respectively.

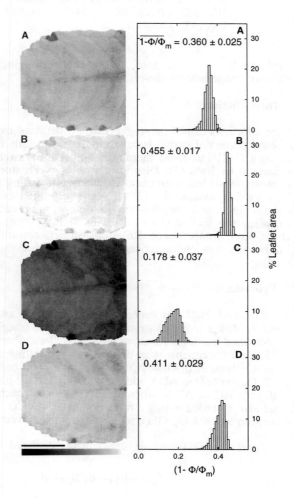

During ABA-feeding, ($1-\Phi/\Phi_m$) heterogeneously decreased according to large areas (Fig. 1C). The pattern of heterogeneity varied from experiment to experiment but the frequency distributions remained unimodal [see 6]. In contrast to the control distribution, the distribution of ABA-fed leaves often skewed towards low ($1-\Phi/\Phi_m$), with a larger coefficient of variation. During a brief transition to 0.74% CO_2, the heterogeneous distribution of ($1-\Phi/\Phi_m$) largely disappeared, but the distribution remained skewed towards left (Fig. 1D). Mean ($1-\Phi/\Phi_m$) reached a value between the means of ($1-\Phi/\Phi_m$) obtained in control before and during transition to 0.74% (Fig. 1A, B). This shows that the capacity of electron transport for CO_2 fixation remained high after stomatal closure.

The rapidity and the amplitude of the reversion of mean ($1-\Phi/\Phi_m$) during transition to high CO_2 (Figs. 1C, D) showed that ABA-feeding mainly involved an heterogeneous limitation of CO_2 supply due to stomatal closure. This has been already described in [6]. However, in [6], the reversion was complete which is not the case here where a higher irradiance has been used. The lack of full recovery indicated that ABA also induced a non-diffusive limitation of photosynthesis.

3.2 Non-diffusive limitation and its dependency on diffusive limitation of photosynthesis. By keeping control leaf near the CO_2 compensation point, we investigated the non-diffusive limitation due to long term CO_2 deficiency (Fig. 2). After a one hour CO_2 deficiency, ($1-\Phi/\Phi_m$) obtained before and during transition to 0.74% CO_2 decreased more than during ABA-treatment (Fig. 2). Moreover, whatever the treatment was, the lower ($1-\Phi/\Phi_m$) was before transition to 0.74% CO_2, the lower ($1-\Phi/\Phi_m$) was during transition to high CO_2 and this dependency was conserved for any site of the leaflet (data not shown). This suggests that electron transport capacity depended on electron transport rate at steady-state photosynthesis estimated just before transition to 0.74% CO_2.

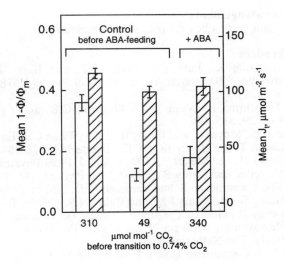

Figure 2. Changes of mean ($1-\Phi/\Phi_m$) in 0.4% O_2 in control and ABA-fed Rosa leaf before (open bar) and during transition to 0.74% CO_2 (hatched bar). Control images were taken in 310 and 49 (for 1h) μmol mol^{-1} CO_2. J_t was calculated following equation 1.

4284

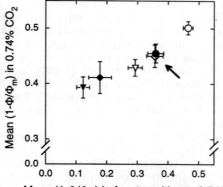

Figure 3. Relationship between mean $(1-\Phi/\Phi_m)$ obtained during transition to 0.74% CO_2 and mean $(1-\Phi/\Phi_m)$ obtained before this transition. O_2 concentration was 21% (open symbols) and 0.4% (closed symbols). Symbols correspond to the control leaf held in 310 (o, ●) and 1h in 49 $\mu mol\ mol^{-1}\ CO_2$ (◊, ♦), and to the ABA-fed leaf in 340 $\mu mol\ mol^{-1}\ CO_2$ (∇, ▼). The leaf was the same as in Fig. 1.

The dependency between mean $(1-\Phi/\Phi_m)$ obtained during transition to 0.74% CO_2 and mean $(1-\Phi/\Phi_m)$ obtained before this transition was the same in 0.4 and 21% O_2 (Fig. 3). Interestingly (Fig. 3, arrow), for a same value of $(1-\Phi/\Phi_m)$ obtained in 0.4 and 21% O_2, $(1-\Phi/\Phi_m)$ obtained during transition to high CO_2 was the same. This shows that the capacity of the electron transport rate to fix CO_2 depended on the total electron transport rate at steady-state photosynthesis but not on the partitioning of electron transport between carboxylation and oxygenation of RuBP. Rubisco deactivation may be involved in this non-diffusive limitation of photosynthesis during CO_2 deficiency or stomatal closure as predicted by [10]. In conclusion, ABA-feeding primarily involved a limitation of CO_2 diffusion mediated by heterogeneous stomatal closure which in turn induced non-uniformly distributed reduction of metabolic activities [see also 11,12].

Acknowledgements
We thank P. Leadley for reading the manuscript.

References
[1] Raschke K., Patzke J., Daley P.F. and Berry J.A. (1990). In current Res. in Photosynth., (Baltscheffsky M. ed.), vol IV, 573-578, Kluwer Academic Publishers, Netherlands
[2] Terashima I., Wong S-C., Osmond C.B. and Farquhar G.D. (1988). Plant Cell Physiol. 29, 385-394
[3] Bro E., Meyer S. and Genty B.(1996). Plant Cell Environ. 19, 1349-1358.
[4] Graan T. and Boyer JS. (1990). Planta 181, 378-384
[5] Genty B. and Meyer S. (1995). Aus. J. Plant Physiol. 22, 277-284
[6] Meyer S. and Genty B. (1998) Plant Physiol. 116: 947-957
[7] Von Caemmerer S. and Farquhar G.D. (1981). Planta 156, 199-206.
[8] Genty B., Briantais J-M. and Baker N.R. (1989). Biochim. Biophys. Acta 990, 87-92
[9] Genty B., Goulas Y, Dimon ., Peltier G., Briantais J.M. and Moya I. (1993). In Res. in Photosynth., (Murata N. ed.), vol. IV, 603-610, Kluwer Academic Publishers, Dodrecht, The Netherlands
[10] Von Caemmerer S. and Edmondson D. L. (1986). Aus. J. Plant Physiol. 13, 669-688
[11] Vassey T. L., Quick W. P., Sharkey T. D. and Stitt M. (1991) Physiol. Plant., 81, 37-44

RETRIEVAL OF THE ACTUALLY EMITTED CHLOROPHYLL FLUORESCENCE OF LEAVES

C. Buschmann[1], A.A. Gitelson[2], H.K. Lichtenthaler[1]

[1] Botanical Institute, University of Karlsruhe, D-76128 Karlsruhe, Germany
[2] Remote Sensing Laboratory, J. Blaustein Institute for Desert Research,
Ben-Gurion University of the Negev, Sede Boker Campus, 84990, Israel

Key words: in vivo spectroscopy, leaf anatomy, light reflection, optical properties

1. Introduction

The Chlorophyll (Chl) fluorescence of green leaves measured at ambient temperature is characterized by maxima near 685 and 735 nm (1). When, during leaf development the Chl content of the leaf increases, the ratio between the short wavelength maximum F685 and the long wavelength maximum F735 decreases strongly with an exponential function (1, 2, 3). In particular the red Chl fluorescence F685 decreases due to the re-absorption of the actually emitted fluorescence on its way to the leaf surface where it is sensed. This is due to the fact that the spectrum of fluorescence emission overlaps with the absorption spectrum of Chls (peak near 680 nm (3)). Thus re-absorption strongly changes the spectral shape of the actually emitted fluorescence inside the leaf. Here we present the actually emitted Chl fluorescence spectra retrieved by taking into account the absorption inside the leaf as deduced from the measurement of transmittance and reflectance spectra

2. Procedure

2.1 Plant material

Leaves with different Chl content (Tab. 1) were collected in June from a tall plane tree (*Platanus acerifolia*) growing on the campus of the University of Karlsruhe. A very young, thick and small leaf (No. 1) was compared with a fully expanded leaf of medium Chl content (No. 2) and an older, smaller leaf with a high Chl content (No. 3).

Table 1: Characteristics of the three leaves of a plane tree studied.

Leaf	No. 1: low Chl	No 2: medium Chl	No. 3: high Chl
Chl content	225 mg m^{-2}	315 mg m^{-2}	550 mg m^{-2}
State of development	very young	young, fully expanded	fully developed
Leaf tissue structure	soft	soft - solid	solid
Leaf area	15.0 cm^2	142.0 cm^2	67.8 cm^2
Leaf thickness	0.86 mm	0.23 mm	0,2 mm

G. Garab (ed.), Photosynthesis: Mechanisms and Effects, Vol. V, 4285–4288.

2.2 *Measurements of spectra*

Various optical *in vivo* spectra of intact leaves were measured between 650 and 800 nm in steps of 0.5 nm. The spectra were always taken at the same spot of the leaf from the upper and the lower leaf side. For reflectance and transmittance spectra a spectrometer with integrating sphere (UV2101PC, Shimadzu, Kyoto/Japan) was used. Chl fluorescence emission spectra were taken at room temperature with a spectrofluorometer (LS52, Perkin-Elmer, Überlingen/Germany) exciting at 430 nm.

2.3 *Retrieval of the actually emitted Chl fluorescence spectra*

For the retrieval of the actual fluorescence (for details see (4)) first the absorption A of the leaf was calculated from the reflectance R and the transmittance T of the leaf at each wavelength (A, R and T in %): $A = 100 - (R + T)$

Then the fluorescence F measured outside the leaf (leaf surface) was corrected for the absorption A of the leaf resulting in the retrieved fluorescence f as actually emitted inside the leaf: $f = F / (100 - A)$

3. Results and Discussion

3.1 *Reflectance and Transmittance Spectra*

The *in vivo* reflectance spectrum of a green leaf is characterized by a low signal between 650 and 700 nm and a subsequent rise towards longer wavelengths (Fig. 1). A similar shape can be found in the *in vivo* transmittance. In the spectral range between 650 and 750 nm both reflectance and transmittance are mainly determined by the absorption characteristics of the Chls. Above 750 nm, much of the radiation is reflected or transmitted, only a small amount of radiation is retained within the leaf due to scattering as influenced by the leaf structure.

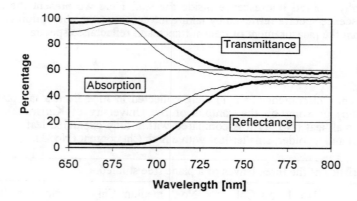

Figure 1. In vivo spectra of reflectance and transmittance of leaves of a plane tree (*Platanus acerifolia*). Transmittance spectra are shown with a reversed scale (minimum at the 100% line) in order to demonstrate the absorption given as the difference between reflectance and transmittance. Contrasted are the tissue with the highest (leaf 3, upper side: bold line) and the lowest Chl content (leaf 1, lower side: thin line).

3.2 Measured Fluorescence Emission Spectra

The emission spectrum of the red to far-red Chl fluorescence is characterized by maxima near 685 and 735 nm (Fig. 2, Tab. 2). The red maximum F685 overlaps with the absorption maximum (cf. Fig. 1). Thus, red Chl fluorescence emitted deeper inside the leaf is partly re-absorbed by the Chls on its way to the leaf surface. As compared to the originally emitted Chl fluorescence, the measured Chl fluorescence F685 possesses a much lower maximum, whereas the far-red maximum F735 is only little affected (practically no overlapping with absorption band of Chl). As a consequence, the fluorescence ratio F685/F735 decreases with increasing Chl content of that leaf side facing the detector (Table 2) and the position of the red maximum shifts to longer wavelengths.

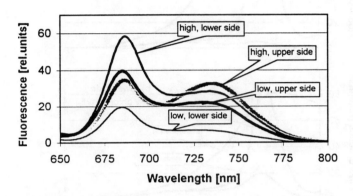

Figure 2. Measured *in vivo* Chl fluorescence emission spectra of plane leaves with low and high Chl content (cf. Tab. 1) measured from the upper and lower leaf side.

Table 2. Chl fluorescence characteristics of the three plane leaves (cf. Tab. 1).

	low Chl		medium Chl		high Chl	
	lower side	upper side	lower side	upper side	lower side	upper side
Emission maximum [nm]						
Measured: Fmax	685.0	685.0	686.5	686.5	686.5	686.0
Retrieved: fmax	683.5	684.0	683.0	682.5	682.0	682.0
Fluorescence ratios						
Measured: Fmax/F735	2.99	1.80	2.03	1.28	2.05	1.06
Retrieved: fmax/f735	19.4	27.6	13.5	16.6	13.2	14.8

3.3 Retrieved Fluorescence Spectra

By the calculation procedure given above, the originally emitted Chl fluorescence inside the leaf was retrieved (Fig. 3). This retrieved or actual Chl fluorescence f685 is about 10 times higher than the measured Chl fluorescence F685 at the leaf surface. In the actual retrieved fluorescence spectrum the red maximum f685 becomes dominant and the long wavelength far-red fluorescence f735 is reduced to a small shoulder. The position of the red maximum is slightly shifted towards shorter wavelengths (Table 2).

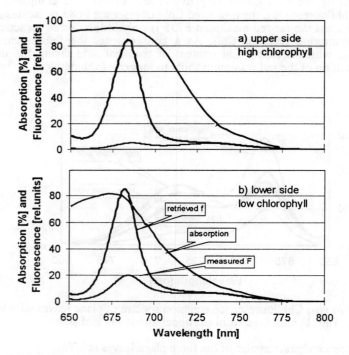

Figure 3. Retrieved and measured Chl fluorescence emission spectrum as well as spectrum of absorbed radiation of a leaf of a plan tree. Contrasted are the tissue with a) the highest (leaf 3, upper leaf side) and b) the lowest Chl content (leaf 1, lower leaf side).

4. Conclusion

The actually emitted Chl fluorescence of leaves can be retrieved by the aid of absorption spectra. The originally emitted actual red Chl fluorescence f685 is ca. 10 times higher than the measured Chl fluorescence F685. The shape of the actually emitted fluorescence spectrum resembles that of the fluorescence emission of a Chl *a* in solution.

References

1 Lichtenthaler, H.K., and Rinderle, U. (1988) CRC Critic. Rev. Anal. Chem. 19, 29-85
2 Buschmann, C. (1981) in Photosynthesis, Vol. 5 (Akoyunoglou, G., ed.) pp 417-427, Balaban Intern. Sci. Serv., Philadelphia
3 Buschmann, C., and Lichtenthaler, H.K. (1998) J. Plant Physiol. 152, 297-314
4 Gitelson, A.A., Buschmann, C., and Lichtenthaler, H.K. (1998) J. Plant Physiol. 152, 283-296

FLUORESCENCE IMAGING OF PHOTOSYNTHETIC ACTIVITY IN STRESSED PLANTS VIA CHLOROPHYLL FLUORESCENCE.

Oliver Wenzel and Hartmut K. Lichtenthaler
Botanical Institute II, University of Karlsruhe, 76128 Karlsruhe, Germany

Key words: Chl fluorescence induction, R_{Fd}-values, fluorescence imaging

1. Introduction

Various stressors such as ozone and highly reactive oxygen species ($^1O_2$, $O_2{}^-$, OH, H_2O_2), air pollution gases (sulfur dioxide, NOX etc.), drought, heat and high-light stress as well as increased levels of UV-radiation will affect the chemical and pigment composition of leaves, their photosynthetic performance and modify their optical and fluorescence properties (1). Therefore stress effects in plants are detectable via changed reflectance and fluorescence emission spectra (1, 2). Green plants illuminated with UV-radiation, blue or green light emit a red (690 nm) and far-red (740 nm) Chl fluorescence (3-5). A stress-induced decrease in chlorophyll content can be detected by a significant increase of the Chl fluorescence ratio red/far-red (F690/F740) (3, 4, 6) and a decrease in photosynthetic activity by a decline of the variable Chl fluorescence ratio R_{Fd} (3, 4). The Laser-Induced Fluorescence Imaging System (LI-FIS) allows to determine fluorescence images and fluorescence ratio images permitting an early detection of pigment changes in the photosynthetic apparatus and a decline of the photosynthetic function (4, 7). Gradients from the leaf rim to the center part of the leaf can only be seen by fluorescence imaging (7, 8).
Here we describe a new compact FIS equipped with a UV-Flash Light (FL-FIS) and demonstrate the use of both FIS devices for imaging of the photosynthetic activity.

2. Procedure

2.1 R_{Fd}-values and photosynthetic activity
Green pre-darkened (15 min) leaves show upon illumination Chl fluorescence induction kinetics known as Kautsky effect (4, 9). The Chl fluorescence rises to a maximum (F_m) within 200 ms and then slowly declines to a much lower value within 4 or 5 min, the steady state F_s. The ratio of the fluorescence decrease F_d (from F_m to F_s) to the steady state fluorescence F_s ratio (F_d/F_s), also known as variable Chl fluorescence ratio R_{Fd}, is correlated to the potential photosynthetic capacity of leaves (6, 10).

G. Garab (ed.), Photosynthesis: Mechanisms and Effects, Vol. V, 4289–4292.

2.2 *Fluorescence imaging*

The new compact Karlsruhe FL-FIS (Fig. 1): It is based on a less expensive construction than the hitherto applied Karlsruhe/Strasbourg LI-FIS. The Nd-YAG laser was replaced by a UV-containing white pulsed light source (Xenon flashing lamp, manufactured by the TU Budapest, Hungary). The CCD-camera was substituted by a simple, cheaper model (Camille camera, manufactured by PHOTONETICS, Germany). In order to enable fast changes of the filters for images in the four fluorescence bands (440, 520, 690, 740 nm) a filter wheel was placed in front of the CCD-camera which is controlled via a RS 232 interface from the image software on the computer (PC). The already existing image software for controlling and synchronizing the fluorescence measurements was replaced by an all inclusive new development (ARP-Camille) of PHOTONETICS. The new software allows a centralized control of all FIS components by the PC. It controls the camera and the wheel with fluorescence filters, the trigger-system to synchronize the gated image intensifier of the CCD-camera and the Xenon flashing lamp, the accumulation of images and evaluation of results. The whole FIS instrumentation was mounted on an optical bench in front of an sample holder to guarantee a fixed angle between Xenon flashing lamp and CCD-camera in order to avoid deviations between the illuminated sample and the area which is imaged by the camera. An excitation beam expander was placed in front of the flashing lamp together with a holder for the interference band pass filter (DUG 11, SCHOTT, Germany) reducing the light of the Xenon lamp to the spectral UV-A range for fluorescence excitation. The advantage of the LI-FIS or FL-FIS instruments is that they simultaneously record the Chl fluorescence signatures of several hundred pixels over the whole leaf area, or even of several leaves and plants (7, 8).

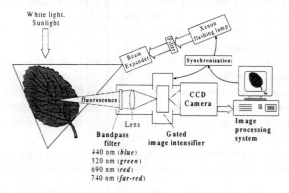

Figure 1: Setup of the Karlsruhe Fluorescence Imaging System for Plants.

Via computer-aided data processing one obtains false color images not only of the Chl fluorescence intensity, but also of the Chl fluorescence ratios of whole leaves. In contrast to point data measurements of conventional spectrofluorometers, the Chl fluorescence gradients between the margins and the central leaf vein, or between tip and petiole of leaves are visualized (4, 8). In fact, early stress events do not affect the whole leaf area at once but initially only parts of it. Thus their monitoring permits a very early stress detection in plants at such a stage where countermeasures can still be taken to overcome the stress-induced changes in order to avoid severe damage to the plants (1, 7).

3. Results and Discussion

The photosynthetic activity of leaves and the uptake of the herbicide diuron (DCMU), which blocks photosynthetic electron transport and photosynthetic quantum conversion by binding to the D1 protein of photosystem II, can easily be studied by changes in Chl fluorescence (4, 11). In this analysis, the red chlorophyll fluorescence (F690) at F_m (after 1 s) and at F_s (reached after 5 min illumination with white light) is recorded via the FIS. In the control leaf, the high Chl fluorescence yield at F_m decreased evenly over the whole leaf area to F_s (Fig. 2). In contrast, in the leaf of the *Digitalis* plant, which had been treated with 10^{-5} M diuron (via the roots), F_m only decreased to F_s in those leaf parts that were still free of the herbicide. Thus the herbicide uptake is visualized in the fluorescence images by a failure of the Chl fluorescence to decline from F_m to F_s. This decline in Chl fluorescence from F_m to F_s can be quantified via the variable Chl fluorescence decrease ratio R_{Fd} (6).

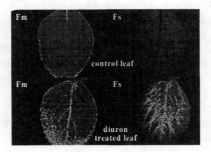

Figure 2: Red chlorophyll fluorescence (F690) at F_m and F_s of a *Digitalis* leaf with and without diuron treatment.

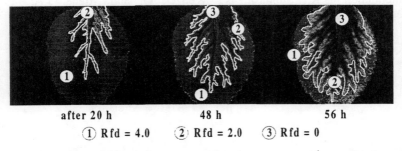

Figure 3. Decrease and loss in photosynthetic function with increasing uptake time of diuron (10^{-5} M) indicated by the decline of R_{Fd} values measured at 690 nm.

Images of R_{Fd} ratios (Fig. 3) of a green attached *Digitalis* leaf indicating the loss of photosynthetic activity with increasing uptake time of diuron (10^{-5} M) applied via the roots. The decline of R_{Fd} values from 4.0 at the beginning of the experiment to zero starts from the middle leaf vein and then moves over the whole leaf area.

4. Conclusion

Fluorescence imaging with the Karlsruhe FL-FIS or the Karlsruhe/Strasbourg LI-FIS represents a superior means of detecting early stress shown here for the decrease of photosynthetic activity (decline in R_{Fd}-values). This non-invasive technique is far more precise than point data measurements of conventional spectrofluorometers. The two fluorescence imaging systems work already in the near distance of up to 10 m and will be further developed for remote sensing of the state of health of terrestrial vegetation.

Acknowledgements: The authors would like to thank Drs. M. Sowinska and M. Lang for taking the fluorescence images.

References

1 Lichtenthaler, H.K. (1996) J. Plant Physiol. 148, 4-14

2 Lichtenthaler, H.K, Gitelson, A. and Lang, M. (1996) J. Plant Physiol. 148, 483-493

3 Lichtenthaler, H.K. and Rinderle, U. (1988) CRC Critical Rev. Anal. Chem. 19, Suppl. I, 29-85

4 Lichtenthaler, H.K. and Miehé, J.A. (1997) Trends in Plant Sciences 2, 316-320

5 Buschmann, C. and Lichtenthaler, H.K. (1998) J. Plant Physiol. 152, 297-314

6 Babani, F. and Lichtenthaler, H.K. (1996) J. Plant Physiol. 148, 555-566

7 Lichtenthaler, H.K., Lang, M., Sowinska, M., Heisel, F. and Miehé, J.A. (1996) J. Plant Physiol. 148, 599-612

8 Lang, M., Lichtenthaler, H.K., Sowinska, M., Summ, P. and Heisel, F. (1994) Botanica Acta, 107, 230-236

9 Kautsky, H. and Hirsch, A. (1931) Naturwissenschaften 19, 964

10 Tuba, Z., Lichtenthaler, H.K., Csintalan, Z., Nagy, Z. and Szente, K. (1994) Planta 192, 414-420

11 Lichtenthaler, H.K., Subhash, N., Wenzel, O. and Miehé, J.A. (1997) In: Procceedings of the IGARSS'97, Singapore, IEEE/USA, 1799-1801

SCREENING OF *ARABIDOPSIS* MUTANTS LACKING DOWN-REGULATION OF PHOTOSYSTEM II USING AN IMAGING SYSTEM OF CHLOROPHYLL FLUORESCENCE

[1]Shikanai T., [1]Shimizu K., [2]Endo T. and [1]Hashimoto T.
[1]Graduate School of Biological Sciences, Nara Institute of Science and Technology, Ikoma, Nara, 630-0101 Japan. [2]Division of Applied Life Sciences, Graduate School of Agriculture, Kyoto University, Sakyo, Kyoto, 606-8502 Japan.

Key Words: Chlorophyll fluorescence, Mutant, *Arabidopsis*

Introduction

Since absorption of excess light energy leads to depression of photosynthetic activities, it is important to regulate the light harvesting in higher plants (1). Plants have evolved complex mechanisms to dissipate excess light energy by modulating photosynthetic electron flow (2). To dissect the regulation mechanisms molecular biologically and physiologically, we have screened *Arabidopsis* mutants lacking down-regulation of photosystem (PS) II by using a chlorophyll fluorescence imaging system.

Chlorophyll fluorescence emitted from PS II reflects the state of photosynthetic electron flow. It can be used to screen mutants lacking activities utilizing or dissipating absorbed light energy, which lead to a high-chlorophyll fluorescent phenotype (3-5). Most of the high-chlorophyll fluorescence mutants are defective in components of the main pathway of photosynthesis and are expected to be seedling-lethal under photoautotrophic growth conditions. To concentrate mutants concerning auxiliary and regulatory mechanisms of photosynthesis, we performed screening using mutagenized M2 seedlings grown photoautotrophically on soil. 59 high-chlorophyll fluorescence mutants were divided into six groups according to patterns of a pulse amplitude modulated (PAM) chlorophyll fluorometer.

Materials and methods

Plant Materials
Arabidopsis thariana M2 seeds (ecotype Landsberg *erecta* and Columbia) mutagenized by ethyl methanesulfonate (EMS) were purchased from Lehle Seeds (Round Rock, TX). M2 seeds (ecotype WS) mutazenized by T-DNA insertion (6) were provided by Arabidopsis Biological Resource Center. For mutant screening, seeds were sown on soil and were incubated for two weeks at 23°C under the light conditions (60 $\mu Em^{-2}sec^{-1}$, 16hr light /8hr dark cycles).

Chlorophyll fluorescence imaging
Mutant screening was performed by using a fluorescence imaging system essentially as

4293

G. Garab (ed.), Photosynthesis: Mechanisms and Effects, Vol. V, 4293–4296.
© 1998 *Kluwer Academic Publishers. Printed in the Netherlands.*

reported by Niyogi et al (7). The system consists of an actinic light source (providing white light at 300 $\mu Em^{-2}sec^{-1}$), a CCD camera equipped with a far-red light transmitting filter (>680nm) and a computer for image analysis. Chlorophyll fluorescence was monitored by using the actinic light source with a filter transmitting light less than 590 nm.

Prior to screening seedlings were dark-adapted for 20 min. Fluorescence image was recorded before (F1) and after (F2) the actinic light illumination for 2 min. Seedlings with less quenching of the fluorescence (F1 -F2) were further characterized by a PAM chlorophyll fluorometer.

Chlorophyll fluorescence measurements
In vivo chlorophyll fluorescence was detected with a PAM fluorometer (Walz, Effekyrich, Germany). The maximum yield of chlorophyll fluorescence was induced by an 800-ms saturating pulses of white light prior to and during the actinic light illumination (100 and 370 $\mu Em^{-2}sec^{-1}$ for 4.5 min).

Results and Discussion

Arabidopsis mutants that exhibited a reduced quenching of chlorophyll fluorescence were identified using a chlorophyll fluorescence imaging system. Figure 1 shows a representative pattern of screening. Although dark-adapted leaves emitted high chlorophyll fluorescence (Fig. 1, 0 min), it was quenched during the actinic light illumination due to both utilization and dissipation of light energy (Fig. 1, 2 min). Putative mutants emitting high chlorophyll fluorescence after the actinic illumination were further characterized by a PAM chlorophyll fluorometer under the two light conditions (100 and 370 $\mu Em^{-2}sec^{-1}$). Approximately 21,000 M_2 seeds by EMS (13,000 seeds of the ecotype Landsberg *erecta* and 8,000 seeds of the ecotype Columbia) mutagenized by EMS were screened resulting in the identification of 38 putative mutants. Approximately 30,000 seeds mutagenized by T-DNA insertion (ecotype WS) were also screened resulting in the identification of 23 putative mutants. Mutants showed divergent visible phenotypes including changes in color and size of leaves. Although the growth were reduced generally, they can grow photoautotrophically and set seeds normally.

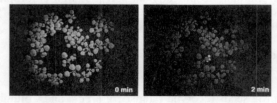

Fig. 1 Screening of mutants with less quenching of chlorophyll fluorescence. Chlorophyll fluorescence image was monitored before (0 min) and after (2 min) the actinic light illumination.

According to the chlorophyll fluorescence patterns, mutants were classified into six groups (Table 1).

Group 1 The mutants of this group showed slightly higher steady-state chlorophyll fluorescence (Fs) than wild type (Fig. 2A and B). In many mutants abnormality in chlorophyll fluorescence patterns were more evident under the higher light conditions (370 $\mu Em^{-2}sec^{-1}$). Mutants with higher Fs exhibited more reduced Fv/Fm, suggesting that photoinhibition may have occurred as secondary effects of mutations.

Group 2 This group represents mutants exhibiting greatly reduced nonphotochemical quenching (Fig. 2C). Mutations may be defective in pH-dependent energy dissipation. Although npq1 lacking violaxanthin deepoxidase grows as well as the wild type (7), growth rate of some group 2 mutants was reduced. Fv/Fm values of mutants were similar to that of the wild type (0.74-0.76).

Group 3 The mutants of this group are characterized by extremely high Fo, which leads to reduction in Fv/Fm ranging from 0.07 to 0.67 (Fig. 2D) . Meurer et al. reported that mutants with Fv/Fm below 0.5 exhibited reduced level of polypeptides composing PS II (3). Possibly mutations in this group are related to PS II. In many mutants, Fs levels dropped below Fo after 0.5-2 min of fluorescence induction. Extremely high Fo and an increase in Fo quenching can be explained by interruption of energy transfer from the peripheral antenna to the reaction center.

Group 4 Only one mutant (LE17-11) belongs to this group. Although fluorescence induction patterns were similar to those of group 2, nonphotochemical quenching was much more reduced (Fig. 2E). Under moderate light condition (370 $\mu Em^{-2}sec^{-1}$) saturating pulses of white light induced transient reduction of fluorescence, indicating that electron carriers of the intersystem were overreduced. Fv/Fm was rather similar to that of the wild type (0.72).

Group 5 The mutants of this group showed reduced quenching of fluorescence especially at the low light intensity (100 $\mu Em^{-2}sec^{-1}$) (Fig. 2F). Notably, the quenching in this group was accelerated by saturating white-light pulses.

Group 6 In this group Fs is extremely high (Fig. 2G). Although a great reduction in nonphotochemical quenching is similar to that of groups 2 and 4, Fv/Fm is reduced in many mutants. Transient reduction of fluorescence by saturating white-light pulses indicated that electron carriers in the intersystem are overreduced even at the low light intensity (100 $\mu Em^{-2}sec^{-1}$).

Table 1 Classification of 59 mutants into six groups.

ecotype	mutant number	Fv/Fm*	ecotype	mutant number	Fv/Fm
Group 1			*Group 4*		
Ler **(EMS)	3	0.76 - 0.78	Ler (EMS)	1	0.72
Col ***(EMS)	3	0.68 - 0.73			
WS (T-DNA)	8	0.57 - 0.76			
Group 2			*Group 5*		
Ler (EMS)	2	0.76	Ler (EMS)	2	0.59, 0.68
Col (EMS)	4	0.75 - 0.76	Col (EMS)	1	0.75
WS (T-DNA)	1	0.74			
Group 3			*Group 6*		
Ler (EMS)	6	0.17 - 0.64	Col (EMS)	11	0.35 - 0.78
Col (EMS)	4	0.07 - 0.55	WS (T-DNA)	1	0.59
WS (T-DNA)	12	0.20 - 0.66			

* Fv/Fm of wild type (Landsberg *erecta*) is 0.75.
** Landsberg *erecta*; *** Columbia

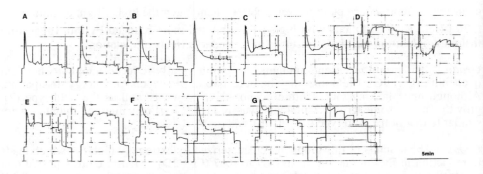

Fig. 2 Chlorophyll fluorescence patterns of the wild type and mutants. Mutants LE20-7-1 (B, group 1), LE17-8 (C, group 2), LE16-7 (D, group 3), LE17-11 (E, group 4), LE18-18 (F, group 5) and WT40-6-3 (G, group 6) represent each group, as well as Landsberg erecta (A, wild type). Chlorophyll fluorescence was induced under the actinic light illumination at 100 $\mu Em^{-2}sec^{-1}$ (left pattern) and 370 $\mu Em^{-2}sec^{-1}$ (right pattern) for 4.5 min. Saturating white-light pulses were supplied every 1 min.

Imaging of chlorophyll fluorescence is a powerful tool to screen the photosynthesis mutants. Approximately 0.2% of EMS-mutagenized lines and 0.07% of T-DNA insertion lines showed various kinds of abnormalities in chlorophyll fluorescence patterns. Although we isolated mutants which can grow under photoautotrophic conditions, leaky mutations may be directly associated with the components of the main photosynthetic pathway. To select mutants concerning the modulation of the electron flow from the mutant pool, mutants are exposed to environmental stresses, high light and low CO_2 concentration. Adaptation to the stresses will be evaluated by monitoring the change in chlorophyll fluorescence and photosynthetic activities.

Acknowledgments

This work was supported by the Japan Society for the Promotion of Science Grant No. JSPS-RFTF97R16001.

References

(1) Asada (1994) in Photoinhibition of photosynthesis BIOS Scientific Publishers, Oxford, UK pp129-142.
(2) Horton P, Ruban AV and Wlters RG (1996) Ann Rev Plant Physiol Plant Mol Biol 47, 655-684
(3) Somerville CR (1986) Annu Rev Plant physiol 37, 467-507.
(4) Meurer J, Meierhoff K and Westhoff P (1996) Planta 198, 385-396.
(5) Niyogi KK, Grossman and Björkman O (1998) Plant Cell 10, 1121-1134.
(6) Feldman KA (1991) Plant J 1,71-82.
(7) Niyogi KK, Grossman and Björkman O (1997) Plant Cell 9, 1369-1380

FAST-RESPONSE DOUBLE-MODULATION FLUOROMETER.

Kaftan, D.[1], Trtílek, M.[2], Kroon, B.[3] and Nedbal, L.[1]
[1]NRC-GCC, Institute of Microbiology, Opatovický mlýn, CZ-37981 Třeboň, Czechia, [2]P.S.Instruments, Koláčkova 39, CZ-62100 Brno, Czechia, [3]A.Wegener Institute, Am Handelshafen 12, D-27570 Bremerhaven, Germany
Key words: fluorescence, photosystem II, heterogeneity, instrumentation, S-states, Q_a

1. Introduction

Single-turnover saturating flashes of light are used to induce synchronous operation of Photosystem II reaction centers. The flash must be so short that the frequency of double-turnovers is negligible and must be powerful enough to turnover all reaction centers. The flashes produced by Xe flashlamps suffer from a flash-to-flash variability and from a tail extending frequently tens of microseconds after the trigger. We have shown earlier that it is feasible to use light-emitting diodes (LED) to generate square-shaped single-turnover flashes that saturate the Q_A reduction in cyanobacteria (1, 2). Here, we demonstrate that the same technique works also for green plants that have smaller PSII effective antenna cross-section than cyanobacteria.

The double-modulation fluorometer does not only generate single-turnover flashes from LED arrays but can also measure fluorescence induction <u>during</u> the flash (Flash Fluorescence Induction, FFI). The duration of the flash is much shorter than the Q_A^- reoxidation time (3) - and, similarly to the induction measured in presence of DCMU, the fluorescence kinetics detected on such a short time-scale reflects primarily accumulation of Q_A^-. The flash fluorescence induction (FFI) allows to assess the Photosystem II antenna heterogeneity without poisoning the organism by the Q_B-site blocking herbicides.

In parallel to the Flash Fluorescence Induction, the double-modulation fluorometer also allows to measure Q_A^- reoxidation kinetics, transients associated with S-state transitions, Kautsky effect and other fluorescence phenomena.

2. Procedure

The fast-response double-modulation fluorometer was constructed by a modification of the FL-100 fluorometer (P.S.Instruments, Brno, Czechia; www.psi.cz). The instrument consists of PC with modified FluorWin software, microprocessor controlled Control Unit and Optical System. The minimum sampling period was lowered to 100 ns at the 12 bit resolution in the channel 1 and kept 10 μs / 16 bit in the channel 2. The Optical System was modified to accelerate the high-current switching to ca. 400 ns response time. All other features including programmable timing, duration and energy for individual actinic or measuring flashes are preserved from the earlier version (1).

G. Garab (ed.), Photosynthesis: Mechanisms and Effects, Vol. V, 4297–4300.
© 1998 *Kluwer Academic Publishers. Printed in the Netherlands.*

The measuring flashes are generated by two sets of 7 orange LEDs positioned on the opposite sides of the sample in a 10 x 10 mm fluorescence cuvette. The duration of the measuring flashes can be set by software in the range 500 ns to 35 μs (usually 3-5 μs). Typical irradiance during a measuring flash is 10^3 μE.m^{-2}.s^{-1}.

Actinic flashes are generated by two sets of 48 red or orange LEDs. The duration range is 500 ns to 90 μs. The maximal incident irradiance during an actinic flash is ca. 200.000 μE.m^{-2}.s^{-1}.

Actinic light is generated by two sets of 12 blue or orange LEDs. The duration of the irradiance may be set from 500 ns to hours. Typical irradiance is 10^2 μE.m^{-2}.s^{-1}. Additional set of 7 far red LEDs (735 nm) can excite preferentially Photosystem I.

The experimental organism was green alga *Scenedesmus quadricauda*.

3. Results and discussion.

Flash Fluorescence Induction.

The fluorescence induction of the green alga *Scenedesmus quadricauda* measured during a 90 μs long actinic flash is shown in Fig.1. The incident irradiance during the flash was 208.000 μE.m^{-2}.s^{-1}. The sampling period was 100 ns.

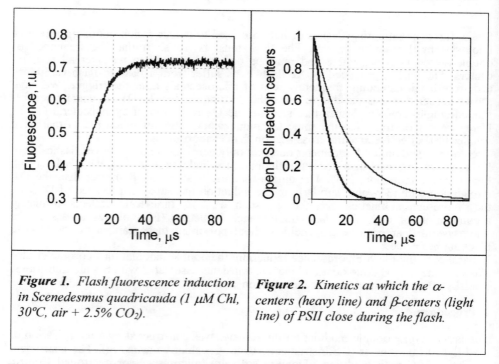

Figure 1. *Flash fluorescence induction in Scenedesmus quadricauda (1 μM Chl, 30°C, air + 2.5% CO₂).*

Figure 2. *Kinetics at which the α-centers (heavy line) and β-centers (light line) of PSII close during the flash.*

The curve was deconvoluted into contributions from α- and β-centers (reviewed in (4)) and the resulting kinetics at which the centers close during the flash is shown in Fig.2. The measurement was done in absence of any Q_B-site blocking herbicide.

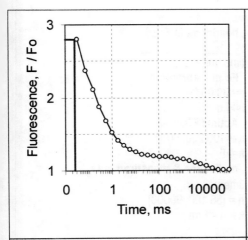

; **header defining experimental constants**
MeasuringFlash= 3 µs
ActinicFlash = 25 µs

.....

A_FlashVoltage=100
M_FlashVoltage=50

.....

;_____

; **timing sequence**
time unit µs
<0>=>F
[56,100..100s]=>fm

Figure 3. *The fluorescence transient reflects the Q_A^- reoxidation after a 25 µs saturating flash. First measuring point comes 56 µs after the actinic flash trigger and 31 µs after the end of the flash. The solid square represents the off-edge of the flash.*

Partial experimental protocol (computer code). *At the time 0 µs, a flash (F) lasting 25 µs is fired. Measuring flashes (f, 3 µs) and measurements (m) are executed in logarithmically timed series: first 56 µs, second 100 µs and last 100 s after the F-trigger. There are 4 measurements per decade (100,178,316,562,1000...).*

Double modulation technique.

In contrast to the Flash Fluorescence Induction, most experimental protocols use non-periodic timing of measurements and flashes in the double modulation technique. The capacity of the fluorometer is demonstrated in experiments monitoring the Q_A^- reoxidation (Fig.3) and the fluorescence transients reflecting the advancement of the S-states in a series of single-turnover flashes (Fig.4).

Kautsky effect and fluorescence induction in presence of DCMU can be measured with an experimental protocol similar to the one used in Fig.3. Actinic flash (F) is replaced by continuous actinic light. The measuring flashes and measurements are spaced logarithmically (ca. 10-20 / decade). The irradiance period, usually seconds to minutes, and voltage used to generate light are again defined in the protocol. Adding actinic flash routine anytime during the experiment can provide data for quenching analysis based on single-turnover flashes or on long actinic flashes similar to those generated by PAM fluorometer.

Fluorescence transient occurring in organisms of oxygenic photosynthesis exposed to several single-turnover saturating flashes is shown in Fig.4.

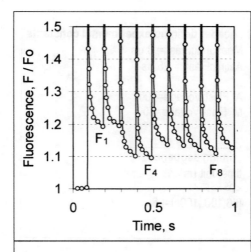

Figure 4. *The fluorescence transient elicited by a series of 9 saturating flashes spaced 100 ms apart. The fluorescence F_n measured 99 ms after n^{th} flash reflects the S-state of the oxygen-evolving complex.*	**Partial experimental protocol.** *Fo is measured at 30, 60 and 90 ms. Nine actinic flashes (F) are fired 100 ms apart starting at 100 ms. Measuring flashes & measurements (fm) are executed in logarithmically timed series: first 56 μs, second 100 μs and last 99 ms after each F-trigger.*

The data were obtained with a PIN diode as a detector. Typical chlorophyll concentrations in algal suspensions were 1-6 μM Chl$a$+b. Using a photomultiplier Hamamatsu H6780-01 and modified optical compartment, fluorescence kinetics were obtained with concentrations down to 6 nM Chl$a$+c in *Phaeodactylum tricornutum* (Maria A. van Leeuwe, unpublished).

4. Acknowledgements

The research was supported by grant from the European Commission INCO-COPERNICUS project IC15 CT96 0105.

5. References

1 Nedbal, L. and Trtilek, M. (1995) in Photosynthesis: From Light to Biosphere, Vol. 5 (Mathis P., ed.) pp.813-816, Kluwer Academic Publishers
2 Trtilek, M., Kramer, D. M., Koblizek, M. and Nedbal, L. (1997) J. Lumin. 72-74, 597-599
3 Bowes, J. M. and Crofts, A. R. (1980) Biochim. Biophys. Acta 590, 373-384
4 Laverne, J. and Briantais, J. M. (1996) in Oxygenic Photosynthesis: The Light Reactions (Ort, D. R. and Yocum, C. F., ed.) pp.265-287, Kluwer Academic Publishers, Dodrecht, The Netherlands

RAPID DEPTH-PROFILING OF THE DISTRIBUTION OF 'SPECTRAL GROUPS' OF MICROALGAE IN LAKES, RIVERS AND IN THE SEA.

The method and a newly-developed submersible instrument which utilizes excitation of chlorophyll fluorescence in five distinct wavelength ranges

M. Beutler[1], K. H. Wiltshire[2], B. Meyer[2], C. Moldaenke[3] and H. Dau[4]
[1]c/o Prof. U.-P. Hansen, Inst. für Angewandte Physik, Univ. Kiel, Olshausenstr. 40, D-24098 Kiel, Germany
[2]MPI für Limnologie, Postfach 165, D-24302 Plön, Germany
[3]bbe Moldaenke, Schauenburgerstr. 16, D-24118 Kiel, Germany
[4]FB Biologie, Univ. Marburg, Lahnberge, D-35032 Marburg, Germany

Key words: aquatic ecosystems, bioproductivity, biotechnology, environmental stress, global climate changes, phytoplankton

1. Introduction

The differentiated assessment of the phytoplankton distribution is a prerequisite for a qualified estimate on the rate of primary production by phytoplankton and on its dependence on environmental factors. Also, supervision of the phytoplankton can facilitate the early identification of an unusual or stressed status of the aquatic ecosystem (e.g., algal blooms, toxic substances, oxygen deficit, etc.).

The presently available methods for determination of the phytoplankton distribution in waters often lack the spatial and temporal resolution required to obtain a thorough understanding of the role of phytoplankton in aquatic ecosystem. The typical delay between sampling a water volume and obtaining the final results mostly excludes the use of conventional methods for supervision or monitoring tasks. Furthermore, presently established methods are often highly costly in terms of manpower.

In this contribution we describe a methodical approach (Beutler, 1998; a similar approach was proposed by Kolbowski and Schreiber, 1995) and a newly-developed submersible instrument which enables the discrimination of four spectral groups of microalgae with high spatial and temporal resolution (and without delay between the *in-situ* measurement and the final result).

2. Results and Discussion

In oxygenic photosynthesis, the chlorophyll (Chl) fluorescence measured around 685 nm is predominantly emitted by the Chl of the Photosystem II (PS II) antenna system which consists of the evolutionarily conserved Chl a-core antenna and species-dependent peripheral antennae (Fig. 1). The Chl fluorescence intensity is reasonably well described by

$$F(\lambda_{ML}) = N_{PSII}\, c_{instr}\, I_{ML} \{ A_{peri}(\lambda_{ML}) + A_{core}(\lambda_{ML}) \}\, \Phi^F \qquad (1)$$

with: N_{PSII}, number of PS II in the sample volume; c_{instr}, a fluorometer-dependent constant; I_{ML}, intensity of the measuring light; $\Phi^F = k_F/\{k_{th}+k_F+k_P\}$, fluorescence yield.

G. Garab (ed.), Photosynthesis: Mechanisms and Effects, Vol. V, 4301–4304.

4302

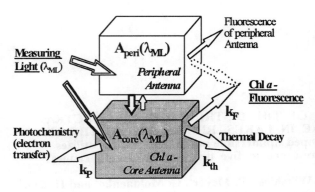

Figure 1. General model for PS II antenna systems.

$A_{peri}(\lambda_{ML})$ and $A_{core}(\lambda_{ML})$ denote the absorption cross-section of the peripheral and core antenna, respectively. The exciton dynamics in the PS II core antenna are characterized by rapid exciton equilibration (Dau, 1994). The details of the exciton dynamics in the peripheral antennae (e.g., the phycobilisome system of cyanobacteria) are, in the present context, of minor importance and therefore not considered.

According to Eq. 1, $A_{peri}(\lambda_{ML})$ and $A_{core}(\lambda_{ML})$ determine the Chl-fluorescence excitation spectrum. The color of a photosynthetic organism is determined (or at least influenced) by $A_{peri}(\lambda_{ML})$. Furthermore, the color of algae is a major taxonomic criterion. Consequently, various taxonomic groups differ significantly in their fluorescence excitation spectrum. We call species groups characterized by similar fluorescence excitation spectra ' spectral groups of microalgae ' (see Fig. 2).

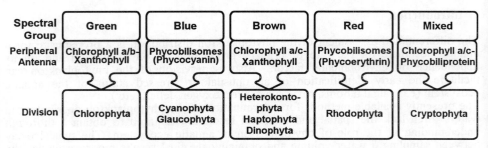

Figure 2. Spectral groups of microalgae. Note that 'Brown' does not label *Phaeophyta*. Because microalgae of the *Rhodophyta* group are of little importance with respect to intended investigations, this spectral group is not considered in the following.

We measured fluorescence excitation spectra for several species of each spectral group using five distinct excitation wavelengths (Beutler et al., in prep.). The mean excitation probabilities per Chl a-concentration (in instrument dependent units) are shown in Fig. 3. Using the 'norm-spectra' shown in Fig. 3, it becomes possible to determine quantitatively the algae population distribution (meaning the Chl a-concentration per spectral alga group) within the sample volume of a submersible instrument (Beutler et al., in prep.). It is worth mentioning that the Chl a-concentration per spectral algal group is a particularly useful parameter because it is closely related to the rate of primary production per water volume (unpublished results; the relation of cell numbers to the rate of primary production is less direct).

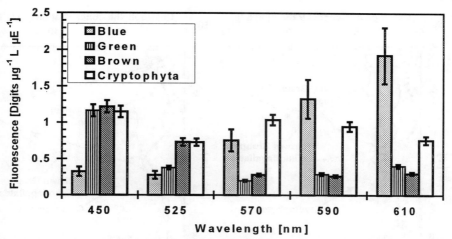

Figure 3. Fluorescence excitation probabilities for four spectral algal groups. The error bars represent the species-dependent differences in the excitation spectrum of an individual spectral group.

We developed a submersible probe with the following characteristics: robust windows and stainless-steel housing (l = 45 cm, $\varnothing$ = 14 cm) suitable for water-depths up to 100 m; continuous monitoring of the submersion depth by means of an integrated pressure sensor; five-color excitation using long-living light-emitting diodes (LED); rectangular detection at the PS II-fluorescence emission peak (685 nm) using an optimized optical bandpass-filter combination and a robust, red-sensitive miniature photomultiplier; microprocessor-control of LED-pulse sequences and data acquisition; high sensitivity and dynamic range enables measurement of fluorescence excitation spectra at extremely low Chl concentrations; high temporal resolution; continuous monitoring of the light attenuation enables the use of an attenuation correction; storage of data in the submersible probe or on-line data transfer (via RS 485 intersection) to a personal computer (laptop PC) on board of, e.g., a research vessel; 'intelligent' data evaluation algorithm (Beutler et al., in prep.) and user-friendly visualization software for personal computers. The submersible probe described above facilitates rapid depth-profiling of the phytoplankton population distribution; an example is shown in Fig. 4.

In conclusion, the presented methical approach in combination with the newly-developed submersible instrument facilitates assessment of algal population distribution with presently unsurpassed spatial and temporal resolution. For many investigations, the resolution of 3 or 4 algal groups should be sufficient. The accuracy of the algal-group differentiation is limited by the species-dependent variability within an individual algal group (see Fig. 3) and by the influence of environmental factors on the fluorescence yield (e.g., an inaccuracy of about 10% due to photoinhibitory fluorescence quenching is possible, data not shown). The presented experimental approach could become an important new tool in aquatic ecology and for supervision of aquatic resources. Presently not feasible investigations (or supervision tasks) may become manageable.

4304

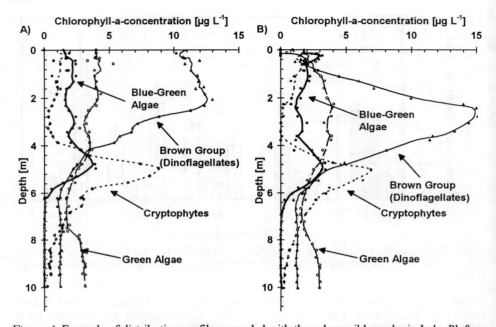

Figure 4. Example of distribution profiles recorded with the submersible probe in Lake Plußsee (Northern Germany) at 04.08.98 showing vertical migration of dinoflagellates. Both measurements were made at the same location at 9:30 *am* and 2:00 *pm*. The phytoplankton consisted of dinoflagellates (*Ceratium* spp.), chlorophyta (*Phacotus* sp.), blue-green algae/cyanobacteria (*Microcystis* spp., *Anabaena* spp.) and cryptophyta (*Cryptomonas* spp.). Dinoflagellates were dominating. At 9:30 *am* most of the dinoflagellates are situated at the surface (0 m - 2 m) of the lake. A maximum for cryptophyta was found at circa 5 metres.
At 2:00 *pm* the dinoflagellates were moving downwards in water layers with higher nutrient concentrations. Their maximum concentration could be found at 3 m depth while the other algal groups did not migrate.

Acknowledgements. We are grateful to the engineers and technicians of *bbe Moldaenke* for their relentless (and successful) efforts to construct a robust and user-friendly instrument. We thank Prof. U.-P. Hansen (FB Physik, Univ. Kiel) for scientific advice, and we acknowledge the contributions of A. Möller, D. Lohse and M. Meyerhöfer to various *in-vivo* and *in-situ* tests carried out in lakes, rivers and in the seas of Northern Germany and in Korea.

References
1 Beutler M. (1998) Diploma thesis, Mathematisch-Naturwissenschaftliche Fakultät, Christian-Albrechts-Universität zu Kiel.
2 Kolbowski J. and Schreiber U. (1995) in Photosynthesis: From Light to Biopshere (Mathis P. ed.), 825-828.
3 Dau (1994) Photochem. Photobiol. 60, 1-23.
4 Beutler M., Wiltshire K., Moldaenke C. and Dau H. ms in preparation.

CONTINUOUS CHLOROPHYLL FLUORESCENCE AND GAS EXCHANGE MEASUREMENTS AS A NEW APPROACH TO STUDY WATER STRESS EFFECTS ON LEAF PHOTOSYNTHESIS

Jaume Flexas[1], Jean-Marie Briantais[2], Zoran Cerovic[2], Hipólito Medrano[1] and Ismael Moya[2]

1.Instituto Mediterráneo de Estudios Avanzados – Universitat de les Illes Balears (UIB-CSIC), Departament de Biologia Ambiental, Carretera Valldemossa Km. 7,5; 07071 Palma de Mallorca, Baleares, Spain. 2.Laboratoire pour l'Utilisation du Rayonnement Electromagnétique (LURE), Centre Universitaire Paris-Sud, B.P. 34; 91898 Orsay Cedex, France.

Key words: grapevines, drought stress, photosynthesis, electron transport, energy dissipation, remote sensing.

1. Introduction

Water stress leads to limitations in leaf carbon assimilation, which involves an excess of excitation energy at the level of photosystems reaction centres promoting down-regulation mechanisms of electron transport (1, 2). Thus, parallel decreases in both CO_2 assimilation (A) and electron transport rate (ETR) can be expected and, in consequence, it seems possible to detect water stress from chlorophyll fluorescence measurements. Moreover, it has been reported that water stress induces marked effects on steady-state chlorophyll fluorescence (Fs) daily pattern (3,4). In water-stressed grapevines significant reduction of ETR has been found, although such reduction was lower than that of A, suggesting an increase in alternative ways for electron consumption (4).

Recent development of new equipment by I. Moya group makes it possible to measure continuously and at distance the most commonly used chlorophyll fluorescence parameters. The purpose of this work was to apply this new equipment, joint with a continuous recording of gas exchange, to study the water stress effects on leaf photosynthesis.

2. Procedure

One-year-old plants of *Vitis vinifera* (L.) were grown in pots inside a greenhouse at LURE, Orsay (France). Pots were irrigated periodically until starting measurements, (August 1996). Water stress was induced withholding watering from the first day of measurements. Photosynthetic performance (both fluorescence and gas-exchange) was followed continuously during the following 10 days in a single leaf. The Li-6400 (Li-Cor Inc., USA) chamber was placed carefully to maintain natural leaf position. CO_2

G. Garab (ed.), Photosynthesis: Mechanisms and Effects, Vol. V, 4305–4308.
© 1998 Kluwer Academic Publishers. Printed in the Netherlands.

concentration inside the chamber was maintained at about 360 μmol mol^{-1}. The CO_2 assimilation (A), stomatal conductance (g), as well as photosynthetic photon flux density (PPFD) and air and leaf temperatures were recorded. Environmental heterogeneity was present because sunny and cloudy days. Glasshouse structure also caused temporally shade over the leaf. PPFD achieved midday peaks of 1200 μmol photon m^{-2} s^{-1} during sunny days. A new fluorometer (FIPAM) designed and built at LURE (Orsay, France) by I. Moya was used for continuously recording fluorescence parameters at 0.5 m from the leaf. Fluorescence measurements were made in the same piece of leaf used by the Li-6400 chamber by shooting through its transparent window. The measured leaf remained apparently undisturbed after measurements.

3. Results and Discussion

3.1 Effects of water stress on the diurnal time course of chlorophyll fluorescence and gas-exchange

Under well-watered conditions, diurnal time course of ETR (fig.1A) followed the diurnal pattern of irradiance. The rate of A also followed the same pattern during the morning. However, from midday values, a progressive decrease in A was recorded, due to midday stomata closure (not shown). Such decrease was not accompanied by ETR decreases, thus, the ratio ETR/A increased during the afternoon. Such afternoon imbalance between ETR and A has already been reported for some C_3 plants, including grapevines (4, 5), and it has been associated with relative increase in the photorespiration rate. This increase seems to be enough to protect from photochemical damage (6), as pre-dawn values of Fv/Fm were completely recovered at the end of the day, after about one hour of darkness (not shown). In contrast, under drought conditions, diurnal patterns of A and ETR were clearly different during the entire day (fig. 1B). A was near zero during most of the day, but ETR was maintained high. These results suggest that under water-stress there is an increase of electron transport to final acceptors different than CO_2, such as O_2 (4, 5).

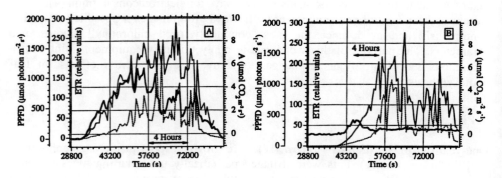

Figure 1. Diurnal time courses in sunny days of ETR (continuous thin line), A (continuous thick line) and PPFD (dotted line), under irrigation (A, 9th August 1996) and drought (B, 17th August 1996).

3.2 Effects of progressive soil drying on stomatal conductance, CO_2 assimilation and electron transport rate

Progressive soil drying was followed by different degrees of reduction in A, g and ETR (fig. 3). Averaged values of those parameters were obtained from measurements taken at 200 μmol photon m^{-2} s^{-1} to include comparable data of both sunny and cloudy days, for the full period of water stress development (from 8th August to 17th).

Stomata closure was the earliest response to soil drying. Decrease of CO_2 assimilation followed similar pattern as g, but with some delay. In consequence, intrinsic water-use-efficiency progressively increased with water stress. ETR was maintained more or less constant within a narrow range of values during the whole period. Only the values corresponding to afternoon of the last days of the drought cycle were clearly decreased (not shown). Thus, a progressive increase in the ratio ETR/A was observed with soil drying. As seen when examining diurnal time courses, these results show again that water stress does not cause major inhibition of the photochemical mechanism (2, 4), and that electron transport to O_2 should be increased (2, 4, 5).

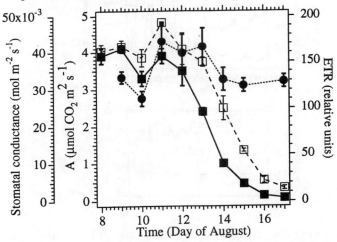

Figure 2. Averaged values ± standard error of ETR (circles), A (open squares) and g (filled squares) at 200 μmol photon m^{-2} s^{-1}, along the days of water stress development.

3.3 Interest of fluorescence parameters for water stress assessment

It is not possible to estimate the rate of CO_2 assimilation only from chlorophyll fluorescence measurements in grapevines. However, chlorophyll fluorescence seems to be a very useful tool for water stress detection, especially the diurnal response of Fs to light intensity. Figure 3 shows details (periods of 4 hours) of drought-associated change in the Fs response to PPFD, for different degrees of water stress. Under irrigated conditions, there was a positive correlation between Fs and irradiance. However, only 5 days after, with mild water stress, the positive correlation was maintained at low light intensities (below 250 μmol photon m^{-2} s^{-1}), but at higher light intensities the correlation was inverse. Three days after, the negative correlation was found even at low light

4308

intensities. The water stress-associated inverse correlation between Fs and light intensity is a characteristic signal of water stress, which might be related with strong increase of the non-photochemical quenching (3, 4, 7), and it could be used in early detection to prevent severe water stress. FIPAM capacity for Fs detection at distance is therefore a promising tool for ecophysiological and practical purposes.

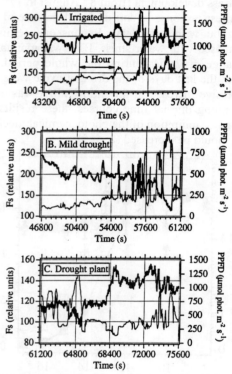

Figure 3. Relation between Fs (continuous line) and PPFD (dotted line). Detail of four hours periods. A. Well-watered plant, corresponding to 9/8/96. B. Mild water stress situation, 14/8/96. C. Severe water stress situation, 17/8/96.

References

1 Cornic, G. (1994) in Photoinhibition of Photosynthesis: from molecular mechanisms to the field (Baker, N.R. and. Bowyer, J.R, eds.) pp. 297-313, Bios Scientific Publishers, Oxford

2 Foyer,C., Furbank,R., Harbinson, J. and Horton, P.(1990) Photosynth. Res. 25, 83-100

3 Cerovic, Z.G., Goulas, Y., Gorbunov, M., Briantais, J.-M., Camenen, L. and Moya, I. (1996) Remote Sens. Environ. 58, 311-321

4 Flexas, J., Escalona, J.M. and Medrano, H. (1998) Plant Cell Environ (in press)

5 Lal, A., Ku, M.S.B. and Edwards, G.E. (1996) Photosynth. Res. 49, 57-69

6 Kozaki, A. and Takeba, G. (1996) Nature 384, 557-560

7 Flexas, J., Escalona, J.M., Cifre, J. and Medrano, H. (1998) (this issue)

PHOTOSYNTHETIC INDUCTION IN IRON-DEFICIENT SUGAR BEET LEAVES: A TIME-RESOLVED, LASER-INDUCED CHLOROPHYLL FLUORESCENCE STUDY

F. Morales[1], R. Belkhodja[1], Y. Goulas[2], J. Abadía[1] & I. Moya[2]
[1]CSIC, Aula Dei Experimental Station, Apdo. 202, 50080 Zaragoza, Spain
[2]LURE - CNRS, Bat 209 D, 91405 Orsay, France

Key words: chl fluorescence induction, iron deficiency, laser spectroscopy, remote sensing, time-resolved spectroscopy.

1. Introduction

Chlorophyll fluorescence has become a useful, non-destructive and non-intrusive tool in photosynthesis, plant physiology and early detection of stress conditions in plants (1-2). The use of Chl fluorescence has been extended to remote sensing of terrestrial vegetation (3, and references therein). Specific LIDAR (light detection and ranging) apparatus have been developed for the remote sensing of Chl fluorescence of terrestrial vegetation (4-7). These devices use laser sources and gated light detection to discriminate changes in Chl fluorescence yield from scattered and reflected solar radiation, and also from Chl fluorescence excited by sunlight. This approach is similar to that used in the pulse amplitude modulation (PAM) technique applied in near-contact fluorimeters (8-9). The main limitation of both LIDARs and PAM-fluorimeters in remote sensing applications is that they measure signal amplitude (an extensive parameter), which depends on distance and also on the degree of atmospheric light transmission. This problem can be solved by performing simultaneous measurements at two wavelengths and using fluorescence ratios as signatures (5, 10). An alternative approach is to measure fluorescence lifetime, which is an intensive parameter (11). This approach was used successfully for terrestrial vegetation (3) with a τ-LIDAR that uses picosecond laser pulses (6-7).

The main aim of this work was to characterize the photosynthetic induction of sugar beet leaves affected by Fe deficiency by using the τ-LIDAR, an apparatus that could monitor stress conditions at distance by means of photosynthesis-related remote sensors. We compared the results obtained using a τ-LIDAR (6-7) and a modified PAM-fluorimeter (3) to investigate the photosynthetic induction of sugar beet leaves affected by Fe deficiency. This allowed us to perform simultaneously and compare under natural conditions measurements of Chl fluorescence lifetimes (made with the τ-LIDAR) and Chl fluorescence yields (made with the modified PAM-fluorimeter).

2. Procedure

2.1 *Plant material*

Sugar beet (*Beta vulgaris* L. cv. Monohil, Hilleshög, Landskröna, Sweden) was grown in a growth chamber in half-Hoagland nutrient solution with or without iron. Plants were grown at a PPFD of 350 μmol photons m^{-2} s^{-1} PAR. They grew at a temperature of 20 °C (day) and 15 °C (night), 80% relative humidity and a photoperiod of 16 h light/8 h dark.

G. Garab (ed.), Photosynthesis: Mechanisms and Effects, Vol. V, 4309–4312.
© *1998 Kluwer Academic Publishers. Printed in the Netherlands.*

2.2 Methods

Chlorophyll concentration per area was estimated non-destructively by using a SPAD-502 Minolta device. For calibration of the apparatus, 40 leaf disks across all the Chl range used were first measured with the SPAD, then frozen in liquid N_2, extracted as described previously (12-13) and the extracts analyzed spectrophotometrically according to (14).

Measurements of Chl fluorescence yields were performed using a modified PAM-fluorimeter (3). The adaptation was based on a emission-detection unit, consisting of a laser diode (Philips CQL840/D, Eindhoven, The Netherlands) as excitation source ($\lambda = 635$ nm), a Fresnel lens (0.15 m of diameter), and a photodiode (S3590-01, Hamamatsu, Japan) protected by a red highpass filter (RG-665, Schott, France). The unit permitted to measure Chl fluorescence yields from a distance of 0.5 to 1 m using the PAM principle and electronics (PAM-101, Walz, Effeltrich, Germany). Data were automatically collected by a computer every 15 s by means of a home-made acquisition program.

The operation principle of the τ-LIDAR has been described previously (6-7). Excitation light was provided by a tripled mode-locked Nd-YAG laser (35 ps pulse duration of 1 mJ at 335 nm) (Quanta System, Milano, Italy) from a distance of 15 m, which produced a spot 6 cm in diameter. Light coming from the leaf surface (emitted Chl fluorescence plus reflected light) were collected through a Fresnel lens (diameter 0.38 m) placed in the laser beam axis, and detected with a high speed crossed field photomultiplier (Sylvania, Model 502). Chlorophyll fluorescence and the backscattered signal were selected with a red highpass (RG-665, Schott, France) and a 335 nm interference filter (Oriel, Stratford, Connecticut), respectively. Current pulse from the photomultiplier output was digitized with a high bandwidth transient analyzer (Tektronix SCD 1000, 1-GHz bandwidth).

The mean Chl fluorescence lifetime was estimated using the barycenter lifetime (τ_{bar}), that can be estimated easily on line during the experiment. Backscattered signals from 8 pulses were averaged and compared to the average of 8 Chl fluorescence signals. This sequence was repeated every 2 min, and τ_{bar} was calculated on line as described previously (3). The deconvolution correctness was judged as described previously (15).

Measurements were performed in the greenhouse with intact plants. One leaf was attached to a stand and exposed to radiation from the τ-LIDAR, the modified PAM-fluorimeter and a light-supplementing projector (350 μmol photons m^{-2} s^{-1} PAR).

3. Result and Discussion

Light was suddenly increased from 40-50 (natural sunlight in the greenhouse at leaf level) to 350 μmol photons m^{-2} s^{-1} PAR by switching on a slide projector. Under natural greenhouse light, control and Fe-deficient leaves had similar Chl fluorescence yields (approximately 1.0 relative units; Fig. 1). However, τ_{bar} was higher in Fe-deficient leaves than in controls; control, moderately and severely Fe-deficient leaves had τ_{bar} values of approximately 0.35, 0.45, and 0.50 ns (Fig. 1A, B and C), respectively. Switching on the projector caused a transient increase in Chl fluorescence yield that was also accompanied by a transient increase in τ_{bar} values. It should be mentioned that maximum fluorescence values are not depicted in Fig. 1 because the sampling rate (1 data every 15 s) was too small to detect the peak maximum. At steady-state photosynthesis, control and Fe-deficient leaves had Chl fluorescence yields of approximately 1.0 and 1.2 relative units. The τ_{bar} values increased with Fe deficiency; control, moderately and severely Fe-deficient leaves showed τ_{bar} values of approximately 0.45, 0.70 and 0.75 ns, respectively. After switching off the projector, Chl fluorescence yields and τ_{bar} values became similar to those found at the beginning of the experiment.

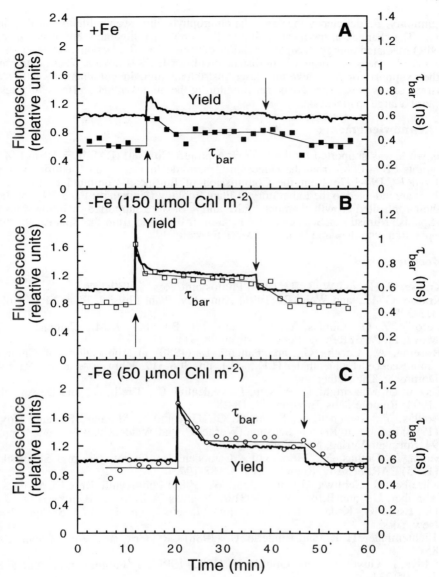

Figure 1. Changes in mean chlorophyll fluorescence lifetime (τ_{bar}, in ns) and in chlorophyll fluorescence yield during adaptation to a sudden increase in light intensity from 40-50 (natural sunlight in the greenhouse at leaf level) to 350 μmol photons m^{-2} s^{-1} PAR. Control (400 μmol Chl m^{-2}; A), moderately (150 μmol Chl m^{-2}; B) and severely (50 μmol Chl m^{-2}; C) Fe-deficient leaves were used. Light was increased by switching on a slide projector. Arrows indicate the switching on and off of the projector.

In summary, Fe deficiency increases the chlorophyll fluorescence lifetime in illuminated leaves. The changes in the mean chlorophyll fluorescence lifetime seem to be associated to NPQ-mediated energy dissipation and/or closure of PSII reaction centers (16). This effect can be measured from a long distance with a τ-LIDAR device. These data provide further support for the view that laser instrumentation, in combination with remote sensors, offers new perspectives for monitoring the photosynthetic effects of stress on plants at a large spatial scale.

4. Acknowledgments

This work was supported by the EUREKA project No. 380 (LASFLEUR) to I.M. and by grants PB94-0086 from the Dirección General de Investigación Científica y Técnica and AIR3-CT94-1973 from the Commission of European Communities to J.A. The authors are indebted to the Laboratory of Plant Ecology (University Paris XI) for the use of their controlled growth chambers. F.M. and R.B. were supported by fellowships from the Spanish Ministry of Science and Education (MEC) and from the Spanish Institute of Cooperation with the Arab World (ICMA), respectively.

References

1 Krause, G. H., and Weis, E. (1984) Photosynth. Res. 5, 139-157
2 Krause, G. H., and Weis, E. (1991) Ann. Rev. Plant Physiol. Plant Mol. Biol. 42, 313-349
3 Cerovic, Z. G., Goulas, Y., Gorbunov, M., Briantais, J.-M., Camenen, L., and Moya, I. (1996) Remote Sens. Environ. 58, 311-321
4 Rosema, A., Cecchi, G., and Pantani, L. (1988) in Applications of Chlorophyll Fluorescence, (Lichtenthaler H.K., ed.), pp. 306-317. Kluwer Academic Publishers, Dordrecht, The Netherlands
5 Cecchi, G., Mazzinghi, P., Pantani, L., Valentini, R., Tirelli, D., and De Angelis, P. (1994) Remote Sens. Environ. 47, 18-28
6 Goulas, Y., Camenen, L., Schmuck, G., Guyot, G., Morales, F., and Moya, I. (1994) in Laser in Remote Sensing, (Werner C. and Waidelicheds W., eds.), pp. 89-94. Springer-Verlag, New York
7 Moya, I., Goulas, Y., Morales, F., Camenen, L., Guyot, G., and Schmuck, G. (1995) EARSeL Adv. Remote Sens. 3, 188-197
8 Schreiber, U., Schliwa, U., and Bilger, W. (1986) Photosynth. Res. 10, 51-62
9 Schreiber, U., and Bilger, W. (1993) in Progress in Botany, (Behnke H.D., Lüttge U., Esser K., Kadereit J.W and Runge M., eds.), pp. 151-173. Springer-Verlag, New York
10 Lichtenthaler, H. K., and Rinderle, U. (1988) CRC Crit. Rev. Anal. Chem. 19, 29-85
11 Moya, I., Guyot, G., and Goulas, Y. (1992) ISPRS J. Photogramm. Remote Sens. 47, 205-231
12 Abadía, J., and Abadía (1993) in Iron Chelation in Plants and Soil Microorganisms, (Barton L.L. and Hemming L., eds.), pp. 327-343. Academic Press Inc., New York
13 Morales, F., Abadía, A., Belkhodja, R., and Abadía, J. (1994) Plant Cell Environ. 17, 1153-1160
14 Lichtenthaler, H. K. (1987) Meth. Enzymol. 148, 350-382
15 Grinwald, A., and Steinberg, I. (1974) Anal. Biochem. 59, 583-598
16 Morales, F., Belkhodja, R., Goulas, Y., and Abadía, J., and Moya, I. (1998) Submitted

CHLOROPHYLL *a* FLUORESCENCE SENSES WATER AND NaCl FLUXES ACROSS PLASMA MEMBRANES OF CYANOBACTERIUM *Synechococcus* sp. PCC7942

Stamatakis, K., Ladas, N., Alygizaki-Zorba, A., Papageorgiou, G. C.,
NRC Demokritos, Institute of Biology, Athens, Greece 153 10

Key words: Chl a fluorescence induction, salt stress, cell volume, whole cells, quenching.

1. Introduction

We shall demonstrate that chlorophyll (Chl) *a* fluorescence senses NaCl-induced volume changes of *Synechococcus* sp. PCC7942 cells and responds to them quantitatively (1, 2). The cells import Na^+ and Cl^- passively (2-4) and export Na^+ in exchange for protons against ΔpH *via* Na^+/H^+ antiporters located in plasma membranes (PM). The antiporters utilize the pH difference (outside acidic) that respiratory electron transport and a P-type ATPase generate across PM (4-6). NaCl concentration upshocks induce osmotic space changes (3) which are reported by changes in Chl *a* fluorescence (1, 2). Osmotic volume changes may also occur during light/dark-acclimative transitions of cyanobacteria.

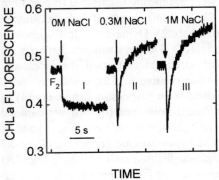

Fig 1. Time courses of Chl *a* fluorescence of dark-acclimated *Synechococcus* after addition (arrows) of NaCl-free and NaCl-containing medium to achieve the indicated end concentrations (dilution 1:4).

G. Garab (ed.), Photosynthesis: Mechanisms and Effects, Vol. V, 4313–4316.

2. Procedure

Synechococcus was cultured as in (1). The culture medium (BG11 plus 20 mM Hepes-NaOH, pH 7.5) was used for assays and for sorbitol solutions of defined osmolality. Cell suspensions are designated as hyper- or hypo-osmotic by reference to turgor threshold of cells (maximal suspension osmolality at which cells maintain turgor). Chl a fluorescence was excited with the modulated excitation (650 nm, $\Delta\lambda$= 25nm, 1.6 kHz, 1μs flashes) of a PAM fluorometer that detects and displays synchronous fluorescence signals. Cell suspensions (18μg Chl a/ml; 10μM DCMU) were either dark-acclimated (4 min darkness) prior to fluorometry or light-acclimated with actinic light (3.5 mE m^{-2} s^{-1}; Corning filter CS 3-67). The modulated excitation (0.07μE m^{-2}s^{-1}) could not acclimate the cells to light. Suspension osmolalities, turgor thresholds and packed cell volumes were measured as in (3).

3. Result and discussion

Injection of NaCl-free medium (Fig 1, arrow) suppressed fluorescence from F_2 to a lower level because of sample dilution (1:4; trace I). Injection of NaCl (final 0.3 M, trace II and 1.0 M, trace III, arrows) triggered remarkably different events. First, fluorescence dropped to lower level ($t_{1/2} \leq 10$ ms; ref. 2) and then it rose to $F_2' > F_2$ ($t_{1/2} \approx 200$ ms). These fluorescence changes resemble closely osmotic volume changes of *Synechococcus* after NaCl concentration upshocks (3). The initial cell volume decrease due to hyper-osmotic shock by NaCl is reported by a synchronous Chl a fluorescence decrease and the subsequent cell volume increase, due to NaCl and water influx is reported by a fluorescence increase.

Cells perceive as osmotica solutes that are excluded or pass the PM slowly. NaCl is not osmoticum for *Synechococcus* sp. PCC 7942, sorbitol is (1,2). Proof is given in Fig. 2 which shows that microhematocrit packed cell volumes (sum of intercellular space, the osmotic cell space and non-osmotic cell space) were practically insensitive to NaCl but not to sorbitol. Hyper-osmotic media quench Chl a fluorescence of light-acclimated (state 1) but not of dark-acclimated cyanobacteria (state 2). Hyper-osmotic suspensions of dark- acclimated cells cannot be acclimated to light(1). This may suggest that cells are liable to hyper-osmotic contraction in the light-acclimated state but not in the dark-acclimated state, or that light-acclimated cells are swollen relative to dark-acclimated cells. The hypothesis is supported by the similar dependencies on suspension osmolality (a) of hyper-osmotic fluorescence quenching of light-acclimated cells and (b) of hyper-osmotic volume suppression (1,2).The Chl a fluorescence of light-acclimated cells (Fig. 3) and of NaCl-swollen dark-acclimated cells (B) is subject to hyper-osmotic quenching. The normalized fluorescence quenching magnitudes of light-acclimated cells ($\Delta F\ F_2^{-1}$, Fig. 4A) and of dark-acclimated cells ($\delta F\ F_2^{/-1}$, Fig. 4B; definitions in Fig. 3 legend) relate quantitatively to medium osmolality (contributed by sorbitol).

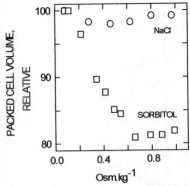

Fig 2. Packed cell volumes of *Synechococcus* sp. PCC7942 cells as a function of suspension osmolality (adjusted either with sorbitol or NaCl).

In the presence of NaCl, the turgor threshold was nearly 2.5 times larger than in its absence. Cytoplasmic osmolality increased after NaCl uptake by cells. Agents that block respiratory electron transport, H^+-ATPase activity, or the collapse $\Delta pH + \Delta \psi$ across the PM suppress the active Na^+ efflux but not the passive NaCl influx (4-6). At such conditions, internal Na^+ rises and cell volumes increase. In Fig. 5 we examined whether Chl *a* fluorescence can report these changes. *Synechococcus* cells were either cultured under light (normal cells) or they were deprived of light for the last 16 hours (starved cells). NaCl-induced Chl *a* fluorescence kinetics were recorded as in Fig. 1. Starved cells, which sustained higher cytoplamic levels of Na^+ and experienced larger volume increases

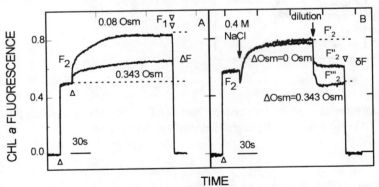

TIME

Fig 3. (A) Suspension osmolality affects on the light-induced Chl *a* fluorescence rise in *Synechococcus* (osmoticum: sorbitol). (B) Chl *a* fluorescence kinetics of dark-acclimated *Synechococcus* cells after NaCl addition and injection of sorbitol-free ($\Delta Osm = 0$) or sorbitol-containing medium ($\Delta Osm = 0.343$). Δ, V, on/off times of weak modulated excitation. Δ, V, on/off times of actinic light. Symbols are defined on the figure.

4316

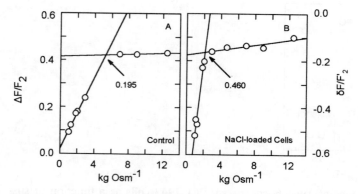

Fig. 4. Hyper-osmotic Chl *a* fluorescence quenching as a function of reciprocal suspension osmolality in *Synechococcus* (A). Light-acclimated cells. (B). NaCl-treated, dark-acclimated cells.

were also characterized by a more pronounced rise of Chl *a* fluorescence.

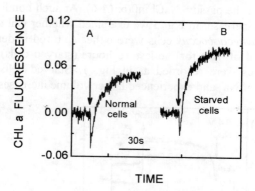

Fig 5. Chl *a* fluorescence kinetics induced by NaCl ($\Delta C = 0.4$ M) in dark-acclimated *Synechococcus*. (A) Normal cells. (B) Light-deprived (starved) cells.

References

1 Papageorgiou, GC, Alygizaki-Zorba, A. (1997) Biochim. Biophys. Acta. 1335, 1-4
2 Papageorgiou, GC, Alygizaki-Zorba, A., Ladas, N., Murata, N. (1998). Physiol. Plantarum (in press)
3 Blumwald, E., Melhorn, R. J. and Packer, L. (1983) Plant Physiol. 73, 377-380
4 Nitschmann, W. H., Packer, L. (1992) Arch. Biochem. Biophys. 204, 347-352
5 Paschinger, H. (1977) Arch. Microbiol. 113, 285-291
6 Fresneau C., Riviere, M.E., Arrio, B. (1993) Arch. Biochem. Biophys. 306, 254-260

THE ENERGETIC CONNECTIVITY OF PSII CENTRES IN HIGHER PLANTS PROBED IN VIVO BY THE FAST FLUORESCENCE RISE O-J-I-P AND NUMERICAL SIMULATIONS

Alexandrina D. Stirbet[1,2], Alaka Srivastava[1] and Reto J. Strasser[1]
[1]Bioenergetics Laboratory, University of Geneva, CH-1254 Geneva, Switzerland
[2]Biophysics Department, University of Bucharest, R-76900 Bucharest, Romania

Key words: Chl fluorescence induction, photosystem 2, grouping

1. Introduction

The fluorescence intensity of chlorophyll in plants, algae and cyanobacteriae depends on the state of the PSII reaction centres [1]. In terms of the quencher theory of Duysens & Sweers [2], the fast polyphasic fluorescence rise from a minimum F_0 to a maximum value F_P $(= F_M$ in saturating light) is mainly due to the photoreduction of the fluorescence quencher, the primary acceptor Q_A to its nonquenching form Q_A^-. The shape of the induction curve is also influenced by the excitation energy transfer among PSII units, commonly denoted as PSII connectivity [3] or grouping [4]. The quantitative relation between the relative variable chlorophyll fluorescence, $V(t) = [F(t)-F_0]/(F_M-F_0)$, and the fraction of PSII with reduced Q_A (closed centres), $B(t)$, has an hyperbolic form and is described by the well-known equation [4,5] :

$$V(t) = \frac{B(t)}{1+C[1-B(t)]}, \quad (1) \qquad \text{with} \qquad C = p\frac{F_{as} - F_0}{F_0} = p\frac{F_M - F_0}{F_0}V_{as}, \quad (2)$$

where C is the parameter for the curvature of the hyperbola, p is the overall probability of connectivity between the PSII units, F_{as} is the fluorescence value used in the normalisation of the fluorescence curve (asymptote), and V_{as} is the relative variable fluorescence at the asymptote. An identical equation, based on the radical pair model has also been derived [6]. All formulae have been established for single turn over situations (e.g., in presence of DCMU), where Q_A is reduced only once and the fluorescence transient has an O to J shape, with $F_J \approx F_M$. Eq. (1) shows that for unconnected photosystems (so called separate packages, with $p = 0$ and therefore, $C = 0$) the relative variable fluorescence is identical to the fraction of closed reaction centres of PSII, $B_E(t)$, and the kinetics of the fluorescence transient becomes exponential:

$$V_E(t) = B_E(t) = 1 - e^{-k(t-t_0)}, \quad (3).$$ The exponential constant k can be calculated if a

curve $V_E(t)$ can be derived: $k = \dfrac{1}{t-t_0}\ln\dfrac{F_{as} - F_0}{F_{as} - F_E(t)}, \quad (4).$

In this work we will present a tentative method to calculate p, the overall probability of connectivity between the PSII units (overall grouping probability), from the experimental O-J-I-P fluorescence induction curve [7] measured in vivo under physiological conditions

G. Garab (ed.), Photosynthesis: Mechanisms and Effects, Vol. V, 4317–4320.
© 1998 Kluwer Academic Publishers. Printed in the Netherlands.

on dark-adapted pea leaves. We will compare the value of p of a control leaf with that obtained with a DCMU-treated leaf.

2. Material and Methods

Experiments were done with mature intact pea leaves. In DCMU treatment, a droplet of 500 µl of DCMU solution (170 µM in distilled water) was added on the axial side of the leaf and kept in the dark for 24 hours. Fluorescence induction was performed on control and DCMU-treated leaves at room temperature using a shutter-less fluorometer (Plant Efficiency Analyser built by Hansatech Ltd.). The dark-adapted samples were illuminated one second with continuous light (600 Wm^{-2}; peak at 650 nm). The fluorescence signals were detected after passing through a long-pass filter (50% transmission at 720 nm). The first reliable point of the transient is measured at $t_0 = 0.05$ ms after the onset of illumination, and was taken as F_0.

3. Result and Discussion

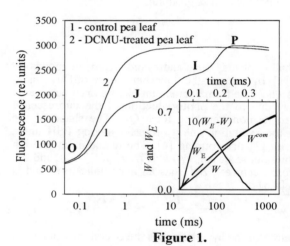

Figure 1.

Fig.1 shows the fast fluorescence induction kinetics of a control pea leaf (curve 1) and of a DCMU-treated pea leaf (curve 2) measured in high light, and presented in a logarithmic time scale. It can be observed that after our DCMU treatment (without alcohol and dry leaf surface) the F_M value is almost the same for both samples. Under high light, typically over 500 Wm^{-2}, two intermediary steps designated as F_J (the "J" level; at ~2 ms) and F_I (the "I" level; at ~30 ms) normally appear in the transient [7] (see curve 1 in Fig. 1). It has been shown [8] that the shape of the O-J part (t ≤ 2 ms) of a normal transient measured *in vivo* is proportional to the O-J

rise in the presence of DCMU. Therefore, we normalise both transients at V_J:

$$W(t) = \frac{V(t)}{V_J} = \frac{F(t) - F_0}{F_J - F_0}, \quad (5).$$

With separate package units (single turn over condition), these kinetics become

exponential: $W_E(t) = B_E^W(t) = \dfrac{V_E(t)}{V_J(t)} = \dfrac{F_E(t) - F_0}{F_J - F_0} = 1 - e^{-k(t-t_0)}, \quad (6).$

In the **inset** of Fig. 1 the sigmoidicity of the experimental curve $W(t)$ can be seen; this has been related to the energetic connectivity among PSII centres. In order to measure the degree of PSII connectivity (i.e., the probability p from the relation (2)) we have constructed an exponential curve with an asymptote $V_{as} = V_J$ (see the dashed curve W_E in the inset of Fig.1), corresponding to a hypothetical unconnected PSII population which undergoes a single reduction of Q_A to Q_A^-. Both curves, $W(t)$ and $W_E(t)$, have one common point $W^{com} = W_E^{com}$ beside the origin. From the relation (6), the value of the rate constant k of the exponential fluorescence rise can be calculated by replacing $F_E(t)$ by the common experimental point F_{com}:

$$k = \frac{1}{t_{com} - t_0} \ln \frac{F_J - F_0}{F_J - F_{com}} = -\frac{1}{t_{com} - t_0} \ln(1 - W^{com}), \quad (7).$$

The fluorescence rise of the unconnected PSII system can be calculated as:

$$W_E(t) = 1 - e^{-k(t-t_0)} = 1 - \left(\frac{F_J - F_{com}}{F_J - F_0}\right)^{\frac{t-t_0}{t_{com}-t_0}} = 1 - \left(1 - W^{com}\right)^{\frac{t-t_0}{t_{com}-t_0}} = B_E^W(t), \quad (8).$$

Fig. 2 shows the exponential constant k of the control (solid curve) and DCMU-treated

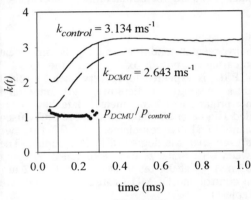

Figure 2.

(dashed curve) leaves as a function of time of the common point W^{com}. It appears that k is quite constant in an interval starting from 0.3 ms to 0.9 ms. We choose 0.3 ms to calculate k by the common point (t^{com}, W^{com}). Knowing k, one can calculate the hypothetical exponential curve $W_E(t)$ at any time according to the Eq.(6). The difference between the two curves, $W_E - W = \Delta W$ (see the inset of Fig. 1) shows the effect of energetic connectivity (grouping) between photosynthetic units on the fluorescence transient. At the time corresponding to the maximal value of ΔW, we have the best resolution to calculate the hyperbola parameter C of

Eq. (1). The relative variable fluorescence $V(t)$ and $V_E(t)$ or $W(t)$ and $W_E(t)$ are expressed as functions of the fraction of closed reaction centres. For the small time interval, until 0.3 ms, we can write $B(t) = B_E(t) = V_E(t) = W_E(t)V_{as} = W_E(t)V_J, \quad (9).$

The curvature constant C of the hyperbola can be derived successively from Eqs. (1),(9),(3),(6) and (2), as:

$$C = \frac{B - V}{V(1-B)} = \frac{B_E - V}{V(1-B_E)} = \frac{V_E - V}{V(1-V_E)} = \frac{W_E - W}{W(1-W_EV_J)} = \frac{F_M - F_0}{F_0}V_J p, \quad (10).$$

We calculate W_E according to Eq. (6) by setting the time at the position of highest sensitivity, $t = 0.1$ ms. Finally, the overall probability of the connectivity between PSII units, p value, was calculated using the relation (2), where $F_{as} = F_J$, the fluorescence value used in the normalisation of the $W(t)$ and $W^{DCMU}(t)$ curves.

Table 1.

ms	F_0 0.05	F 0.1	F^{com} 0.3	F_J 2	F_M max	V_J 2	W 0.1	W^{com} 0.3	W_E 0.1	k 0.3	C	p
control	604	730	1283	1854	2992	0.523	0.101	0.543	0.145	3.13	0.47	0.23
DCMU	626	777	1696	2839	2963	0.947	0.068	0.483	0.124	2.64	0.92	0.26

Table 1 (on the previous page) shows the estimation of the constant k, the curvature constant C, and the probability p for energy transfer among PSII units, derived from experimental fluorescence O-J-I-P transient of control and DCMU-treated pea leaves. The method to estimate in vivo the probability of energy transfer between photosynthetic units (grouping) can be modified in several ways. For example, the PEA instrument allows us to measure very accurately the initial slope dF/dt_0 of the fluorescence transient O-J-I-P. Therefore, the slopes dV/dt_0, dW/dt_0 and dW_E/dt_0 are available. After differentiation of Eq. (1) for $t = 0$ we get:

$$\frac{dV}{dt_0} = \frac{dB}{dt_0}\frac{1}{1+C}, \quad (10) \quad \text{or as described above:} \quad \frac{dW}{dt_0} = \frac{dW_E}{dt_0}\frac{1}{1+C} \text{, and } C \text{ can be}$$

calculated replacing W and W_E by Eqs. (6) and (10) using the optimal time, $t = 0.1$ ms.

4. Conclusions

PSII function is highly sensitive to the environmental conditions (e.g., pollutants, heat and water stress, excess light, and increasing carbon-di-oxide), and the O-J-I-P fluorescence induction curve, a monitor of PSII, is expected to be very useful in plant biology research, and related areas such as ecology, horticulture, agronomy, and biotechnology. Specific processing and interpretation of experimental data have been proposed in our laboratory (i.e., the so called JIP-Test) and have been successfully used in the analysis of the physiology of the plants [7,8]. To complement the JIP-Test, we propose here a way to measure and compare, *in vivo*, the degree of PSII grouping. The overall probability of the connectivity between PSII units, p, can be obtained by using only few points of the transient: F_0 (at 0.05 ms) , F at 0.1 ms, F at 0.3 ms, F_J at 2 ms, and F_M. We have tried our new method on control and DCMU-treated leaves, and we have obtained well correlated values (see Table 1).

Acknowledgments. This work was supported by Swiss National Foundation - Grant No. 3100-046860.96/1 and Societé Académique de Genève.

REFERENCES
1. Govindjee, (1995) Aust. J. Plant Physiol. 22:131-60.
2. Duysens, L.N.M. and Sweers, H.E. (1963) In Studies on Microalgae and Photosynthetic Bacteria, Japanese Society of Plant Physiologists, ed., pp. 353-372. Tokyo: University of Tokyo Press.
3. Joliot, A. and Joliot, P. (1964) C.R.Acad. Sci.Paris 258:4622-4625.
4. Strasser, R.J. (1978) In Chloroplast Development (Akoyunoglou, G. et al., eds.) pp. 513-524. Holland: Elsevier.
5. Strasser, R.J. (1981) In Structure and Molecular Organization of the Photosynthetic Apparatus (Akoyunoglou, G., ed.) pp. 727-737. Philadelphia: Balaban Int.Sci.Ser.
6. Lavergne, J. and Trissl, H.-W. (1995) Biophys. J. 68:2474-2492.
7. Strasser, R.J., Srivastava, A. and Govindjee, (1995) Photochem. Photobiol. 61:32-42.
8. Strasser, B.J. and Strasser, R.J. (1995) In Photosynthesis: from Light to Biosphere (Mathis, P., ed.) pp. 977-980. Kluwer Academic Publishers.
9. Stirbet, A.D., Govindjee, Strasser, B.J. and Strasser, R.J. (1998) J.Theor.Biol. 193:131-151
10. Krüger, H.J., Tsimilli-Michael, M. and Strasser, R.J. (1997) Physiol. Plantarum 101: 265-277.

ACTIVITY AND HETEROGENEITY OF PSII PROBED IN VIVO BY THE CHLOROPHYLL *a* FLUORESCENCE RISE O-(K)-J-I-P

Reto J. Strasser[1] and Merope Tsimilli-Michael[1,2]
[1]Bioenergetics Laboratory, University of Geneva, CH-1254, Jussy-Geneva, Switzerland, [2]Cyprus Ministry of Education and Culture.

Key words: adaptation, electron transport, energetic connectivity, environmental stress, reaction centres, structural changes

1. Introduction

The photosynthetic organisms undergo in nature perpetual state changes in order to adapt to a perpetually changing environment (1). The environmental changes are provoked by many different factors, natural and physical/chemical, as well as any combination of any of them. Therefore, and moreover considering the huge heterogeneity of the photosynthetic material in nature, a plethora of survival strategies is employed. However, a survival strategy is a combination of adaptive processes, each regulating structural features which consequently determine function (1). Therefore, the plethora of macroscopically different strategies might rather arise from many possible combinations of much fewer adaptive processes. A certain combination may be specific for a certain behavioural type.

The question is, therefore, whether and how the different possible *structural/functional* features can be expressed in measurable parameters which can, furthermore, be distinguished from one another even when several of them undergo simultaneous changes upon a multiple environmental change. This question poses the necessity of criteria which should be specific for each possible structural/functional change, defined in the frame of a theoretical model and approachable through a suitable experimental procedure. Even more, since the aim is to get an access to the understanding of the behaviour of the photosynthetic material not only *in vivo* but also *in situ*, the experimental procedure must be non-invasive, easily handled and quickly conducted in order to serve *screening purposes*. Procedures utilising the Chl *a* fluorescence signals can fulfil these requirements when used either in techniques imaging big areas of photosynthetic material, or for measuring the fast fluorescence kinetics (in 1 s) which provides a rapid way of screening small, but many, samples of photosynthetic material.

In our laboratory we have established a rapid screening procedure, the JIP-test (2), by which the fast polyphasic Chl *a* fluorescence rise O-J-I-P is analysed, leading to the calculation of several structural and functional parameters that offer a quantification of the PSII behaviour (2-4). These parameters, all referring to time zero (onset of the fluorescence induction) are (a) the specific (per reaction centre, RC) energy fluxes for absorption (ABS/RC), trapping (TR_0/RC) and electron transport (ET_0/RC), (b) the corresponding phenomenological (per active cross section, CS) fluxes, ABS/CS, TR_0/CS and ET_0/CS, and (c) the flux ratios or yields, i.e. the maximum quantum yield of primary photochemistry (φ_{Po}), the efficiency (ψ_0) with which a trapped exciton can move an electron into the electron transport chain further than Q_A^-, and the quantum yield (φ_{Eo}) of ET.

G. Garab (ed.), Photosynthesis: Mechanisms and Effects, Vol. V, 4321–4324.
© 1998 *Kluwer Academic Publishers. Printed in the Netherlands.*

In this paper we extend the JIP-test in order to tackle the question (A) of rate constants and light dependency of electron transport and (B) of PSII heterogeneity. The latter is also linked to adaptation, in the sense that adaptive processes appear to create and regulate different types of heterogeneity of the PSII structure and function. We here focus on heterogeneity concerning (a) the donor side, (b) the acceptor side, (c) the inactive (quenching sinks) vs. active RCs and (d) the connectivity (grouping) of the PSII units.

2. Experimental procedure

The Chl a fluorescence transients are measured and recorded (for 1 s) by the PEA fluorimeter (Plant Efficiency Analyser PEA, built by Hansatech Instruments Ltd. King's Lynn Norfolk, PE 30 4NE, GB) with a 12-bit resolution and a 12-bit signal resolution (5).

3. The extension of the JIP-test

3.1 Electron transport

So far, the ET activity is calculated by the JIP-test only at time zero (ET_0) (for definitions see Introduction), by utilising the initial slope M_0 of the fluorescence induction curve (FIC), which is proportional to $[(TR_0/RC)-(ET_0/RC)]$, and the initial slope $M_{0,DCMU}$ of the FIC that the sample exhibits when poisoned with DCMU, and which is proportional to TR_0/RC. However, in the JIP-test the value of $M_{0,DCMU}$ is derived without DCMU, as it has been simulated (2) by M_0/V_J, where V_J the relative variable fluorescence at 2 ms (J-step). Hence $\psi_0 = (ET_0)/(TR_0)$ is calculated.

When the estimation of the ET activity is required, especially in relation with productivity, the assumption usually made is that it is equal to the TR activity. However, this is only true when $[Q_A^-]$ is constant. In general, ET activity is expected to be equal to $[Q_A^-].k_{ab}$, where k_{ab} the rate constant for the electron transfer to Q_B. Still, this does not explain why at time zero, though $[Q_A^-]_0 = 0$, the slope M_0 is smaller than $M_{0,DCMU}$. We here tackle this contradiction as following: At any time the TR_t flux is partially used to create an ET flux and partially to build up a potential energy by accumulating redox equivalents Q_A^-, which concomitantly provoke an additional ET flux. The general equation is:

$$(dB/dt) = (1-B).(TR_0/RC) - [(1-B).(TR_0/RC).\psi + B. k_{ab}]$$

which at t=0 gives: $(dB/dt)_0 = (TR_0/RC).(1-\psi_0)$. The probability ψ, defined by a constellation of rate constants, can be assumed as constant (equal ψ_0) throughout the FIC (1 s), during which structural changes do not occur. The expression in square brackets gives the ET_t/RC, and the expression $(1-B).(TR_0/RC)$ gives the TR_t/RC. Thereafter, and assuming B = V (no connectivity), we get:

$$ET_t/RC = (1-V_t).(M_0).(1-V_J)/V_J + V_t.k_{ab}$$

Applying this equation for the J-step (2 ms) where $dV/dt = 0$ and thus $ET_J/RC = TR_J/RC$, the k_{ab} is determined, provided that the fluorescence transient is induced by a saturating light. Therefore, we get an access to another structural parameter of the sample. Moreover, under steady state conditions we get: $[Q_A^-/Q_{A, total}]_S = B_S = (ABS/RC) / [K + (ABS/RC)]$, which is the hyperbolic light saturation curve, with $K = k_{ab}/[(1-\psi). \varphi_{Po}]$.

3.2 Heterogeneity

(a) Donor side: When a photosynthetic organism is exposed to strong heat stress, the oxygen evolving capacity decreases due to a blockage between the oxygen evolving complex (OEC) and PSII, before tyrosine Yz. Under the same conditions, an additional step, denoted as K-step appears at 300 μs in the fluorescence transient (now O-K-J-I-P) (6). It has been shown that the appearance of the K-step and the increase of its amplitude go in parallel with the damping of the period four oscillations of the Chl a fluorescence yield (7).

Therefore, the O-K-J-I-P can be used as a tool for the evaluation of the heterogeneity at the donor side of PSII. A quantification of the O-K-J part can be given by the expression $V_K/(V_K+V_J)$, where V_K and V_J the relative variable fluorescence at the K- and J-step. The comparison of inhibited with non-inhibited samples with respect to the $V_K/(V_K+V_J)$ value can provide, as a first approximation, a measure of the fraction of non-functioning OECs.

(b) Acceptor side: The structure of the PSII complex can also show an heterogeneity concerning the Q_A-Q_B binding. Adaptation processes can also regulate this heterogeneity, altering the fraction of the non-Q_B binding RCs. The O-J-I-P transient can provide by numerical simulations an estimation of this fraction (8). We here present another test which, for the moment, appears easier and faster for this evaluation. For this test, the fluorescence measurement (denoted as 1st hit) conducted after a dark period long enough to ensure the reopening of all RCs, is followed by a second measurement (2nd hit). The duration of the dark interval between the two hits is experimentally defined as the time required for the fast reopening of the RCs, assuming that those RCs that remain reduced after this time are the non-Q_B binding. Once the determination of the suitable for each organism dark interval is done, the test comprising of the two hits can rapidly screen many samples. The utilised values are those at the extrema of the two successive transients, here denoted as F_0', F_M' (1st hit) and F_0'', F_M'' (2nd hit). F_M'' is often slightly lower than F_M', because the 1st hit triggers already a state change. Therefore, the value of F_0'' has to be corrected by normalising F_M' on F_M''. The corrected value, $F_0''.(F_M'/F_M'')$, is then treated as referring to a point on the 1st hit transient and the relative variable fluorescence at this point, V_0'', is calculated. Assuming no connectivity, V_0'' is equal to the fraction of closed, i.e. of non-Q_B binding, RCs. Hence:

$$[\text{non-}Q_B] / [(Q_B \text{ binding}) + (\text{non-}Q_B)] = [F_0''.(F_M'/F_M'') - F_0'] / [F_M' - F_0'] = 1 - \varphi_{Po}''/\varphi_{Po}'.$$

(c) Active/inactive reaction centres: The maximum quantum yield of primary photochemistry, $\varphi_{Po} = 1-(F_0/F_M)$, widely used for the comparison of photosynthetic material, refers to the whole measured sample which might be heterogeneous. It gives therefore an average value of all the PSII units in the sample. A possible heterogeneity is the one concerning the existence of RCs which have been inactivated, in the sense of being transformed to quenching sinks (9,3,10). The fluorescence yield of the corresponding units is equal to that of open units. The JIP-test has the advantage of providing a calculation of TR_0/RC, i.e. the trapping flux per active RC. Thereafter, and since $\varphi_{Po} = TR_0/ABS$, the quantity ABS/RC can be calculated, which expresses the total absorption divided by the total active RCs and, therefore, gives a measure of the average antenna size. In most of the cases we have studied, we witnessed that the increase of ABS/RC resulting after the photosynthetic organism is exposed to light or mild heat stress, is not due to absorption changes (as tested by reflectance measurements). Thus, the increase of ABS/RC indicates the decrease of active RCs by the transformation of a fraction of them to quenching sinks. For the quantification of this fraction, the sample is compared with a control sample (subscript c) having all RCs active:

$$\frac{(RC_{active})}{(RC_{total})} = \frac{(RC_{active})}{(RC_{active})_c} = \frac{(ABS/RC)_c}{(ABS/RC)} = \frac{(TR_0/RC)_c}{(TR_0/RC)} \cdot \frac{(\varphi_{Po})}{(\varphi_{Po})_c} = \frac{(M_0/V_J)_c}{(M_0/V_J)} \cdot \frac{[1-(F_0/F_M)]}{[1-(F_0/F_M)]_c}$$

Moreover, having an access to TR_0/RC, we can follow possible structural changes in the units with active RCs, as the TR flux is defined by $k_P/(k_P+k_N)$, where k_P and k_N the photochemical and nonphotochemical deexcitation rate constant respectively. Furthermore, as the F_0 value is defined by $k_F/(k_P+k_N)$ and k_F is assumed to remain constant, we can conclude from the comparison of the TR_0/RC and F_0 changes, how each of the two rate constants is affected. However, it is interesting to note that in the stress cases we have studied, we mostly observed that the employed adaptive strategies regulate much more pronouncedly the heterogeneity of active/inactive RCs than the rate constants of the still active units.

(d) Unit-unit connectivity: The regulation of the energetic connectivity of the PSII units seems to be also an adaptive strategy of the photosynthetic apparatus. Connectivity (grouping) is revealed as a sigmoidicity in the FIC or, at least as a deviation from the exponential towards a sigmoidal shape (11,12). Connectivity (grouping) is quantified by the according probability, p_G (12), and is therefore defined in a continuum from zero to one. Efforts to measure p_G were dealing so far with DCMU-poisoned or by low temperature (77 K) blocked photosynthetic material, where the FIC reflects pure photochemical events. In our laboratory we developed a procedure to calculate p_G from the FIC of samples in the absence of DCMU, *in vivo* and *in situ*, by numerical simulations of the O-J part of the fluorescence transient (13). With the derived equations the probability p_G is calculated from the experimental values $F_{50\mu s}$, $F_{100\mu s}$, $F_{300\mu s}$, F_{2ms} and F_M, as following:

$$p_G = \frac{W_E - W}{W(1 - W_E V_J)} \cdot \frac{F_0}{(F_M - F_0)V_J}$$

where: $V_J = (F_{2ms} - F_{50\mu s}) / (F_M - F_{50\mu s})$ $\qquad$ $W = (F_{100\mu s} - F_{50\mu s}) / (F_{2ms} - F_{50\mu s})$

$\qquad$ $W_E = 1 - [(F_{2ms} - F_{300\mu s}) / (F_{2ms} - F_{50\mu s})]^{1/5}$

4. Conclusion

In this paper we have described how the fluorescence transient O-(K)-J-I-P can provide an access to information concerning structural parameters defining electron transport and to the evaluation of different types of heterogeneity of the PSII structure and function. The JIP-test, thus extended, becomes now an even more useful tool for studies of stress and stress adaptation of the photosynthetic organisms, *in vivo* and *in situ*.

Acknowledgements: Swiss National Foundation (SNF 31-46.860.96/ 31-52.541.97) and Soc. Acad. Genève to R.J.S. Cyprus Ministry of Education and Culture to M.T.-M.

References

1 Tsimilli-Michael, M., Krüger, G.H.J. and Strasser, R.J. (1996) Archs. Sci. Genève, 49, 173-203
2 Strasser B.J. and Strasser, R.J. (1995) in Photosynthesis: From Light to Biosphere, (Mathis P., ed.) Vol. V, pp.977-980, Kluwer Academic Publishers, The Netherlands
3 Krüger, G.H.J., Tsimilli-Michael, M. and Strasser, R.J. (1997) Physiol. Plant. 101, 265-277
4 Strasser, R.J., Srivastava, A. and Tsimilli-Michael, M. (1999) in Probing Photosynthesis: Mechanism, Regulation and Adaptation (Yunus M., Pathre U. and Mohanty P., eds.), Taylor and Francis, London, UK (in press)
5 Strasser, R.J., Srivastava, A. and Govindjee (1995) Photochem. Photobiol. 61, 32-42
6 Srivastava, A., Guisse, B. Greppin H. and Strasser, R.J. (1997) Biochim. Biophys. Acta 1320, 95-106
7 Strasser, B. (1997) Photosynth. Res. 52, 147-155
8 Strasser, R.J. and Stirbet, A.D. (1998) Mathematics and Computers in Simulation (in press)
9 Krause, G.H. (1988) Physiol. Plant. 74, 566-574
10 Tsimilli-Michael, M., Pêcheux, M. and Strasser, R.J. (1998) Archs. Sci. Genève (in press)
11 Joliot, A. and Joliot, P. (1964) C. R. Acad. Sci. Paris 258, 4622-4625
12 Strasser, R.J. (1981) in Photosynthesis III. Structure and Molecular Organisation of the Photosynthetic Apparatus (Akoyunoglou G., ed.) pp.727-737, Balaban International Science Services, Philadelphia, Pa.
13 Stirbet, A.D. and Strasser, R.J. (these Proceedings)

OSCILLATIONS OF THE CHLOROPHYLL *a* FLUORESCENCE RELATED TO THE S-STATES OF THE OXYGEN EVOLVING COMPLEX

Bruno J Strasser and Reto J Strasser
Laboratory of Bioenergetics, University of Geneva,
CH - 1254 Jussy, Switzerland (email: strasser@bioen.unige.ch)

Key words: Chl *a* fluorescence induction, S-states, OJIP transients, unicellular algae, P680, primary donor

1.Introduction
The chlorophyll *a* fluorescence transient is one of the most used tools in photosynthesis research (1) to probe the photosystem II (PSII) reactions. When using high actinic light (typically $600 Wm^{-2}$), the fluorescence transient shows a polyphasic rise (2,3) called O-J-I-P transient. The step J occurs around 2 ms, and the step I around 30 ms. The O-J rise has been attributed to a partial closure of the RCs, the J-I to the closure of the remaining centers, and the I-P rise to the removal of plastoquinone quenching (3,4). However, it has been shown previously that the S-state of the OEC plays also a crucial role in the fluorescence rise (*see e.g.* 2).

We have modified a commercially available fluorimeter (PEA from Hansatech), in order to incorporate a single turnover flashing device. We have investigated fluorescence transients after different number of single turnover flashes (rate 1 Hz), in order to set the OEC in a specific S-state. Long dark adaptation brings the OEC in the S_1 state. Then, each single turn-over flash "pumps" the system in the next S-state.

Using a shutterless system, and strong continuous actinic light ($600 Wm^{-2}$), we can measure the initial level of fluorescence within microseconds after the onset of the light, and therefore have a reliable measurement of F_o. With a continuous measurement ($10\mu s$ resolution), we can resolve the initial slope of the fluorescence rise, which gives us a measure of the primary rate constant for photochemistry (5). From these investigations, we have gained an increased understanding of the role of the OEC in PSII photochemistry and its effect on fast fluorescence kinetics. Furthermore, the proposed method can now be used for various *in vivo* investigations to monitor the OEC activity (*see e.g.* Govindjee et al., *these proceedings*.)

2. Material and Methods
The unicellular green algae *Scenedesmus obliquus* was grown autotrophically. Chl *a* fluorescence transients were measured by a PEA fluorimeter (Hansatech Ltd., King's Lynn, Northfolk, UK), single turnover flashes were provided by a xenon flash lamp (4).

3. Results and Discussion
The fluorescence transients after different number of single turn-over flashes are presented in Fig 1. Only the first phase of the transient, the O-J phase, is markedly affected by the flash number. The fluorescence transients, normalized to the dark adapted

G. Garab (ed.), Photosynthesis: Mechanisms and Effects, Vol. V, 4325–4328.
© 1998 Kluwer Academic Publishers. Printed in the Netherlands.

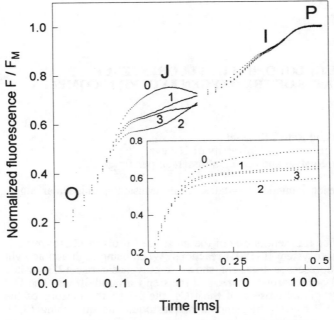

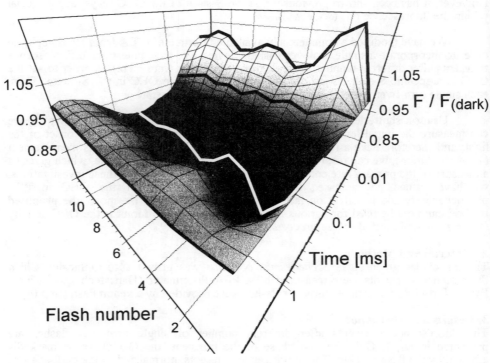

Figure 1 (left):
Fluorescence induction kinetics (O-J-I-P) after different number of single turnover flashes. The preilluminating flash number is labeled next to the transient. The F_M level was unaffected by the flash number.

Figure 2 (bottom):
Fluorescence induction kinetics (O-J-I-P) as a function of time and preilluminating flash number. All transients are normalized to the dark adapted transient.

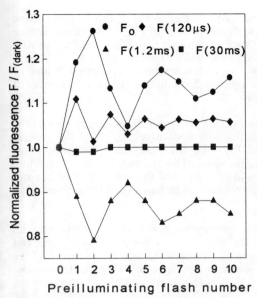

Figure 3:
Flash number dependence of the fluorescence intensity at various times of the transients.

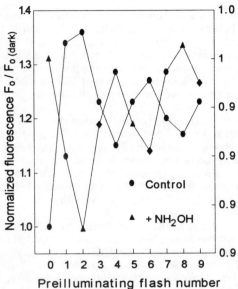

Figure 4:
Effect of 40μM hydroxylamine on the F_o oscillation pattern. In the presence of hydroxylmine, it is shifted by 2 flashes.

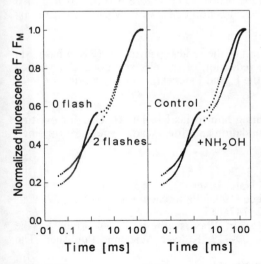

Figure 5:
Fluorescence transients after 2 flashes (*left*), or after 40μM hydroxylamine treatment (*right*).

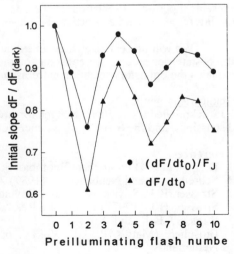

Figure 6:
The initial slope of the fluorescence transients (dF / dt_0) as a function of flash number.

state, are presented as a function of time and flash number in Fig 2. In Fig 3, we show the flash number dependence of the initial fluorescence intensity F_o, of the fluorescence intensity at 120 µs, at J (1.2 ms), and at I (30ms). F_o shows a period four oscillation with a maximum on flash 2, 6 and 10, as it has been observed earlier (6). The J level shows a period 4 oscillation in the opposite direction, *i.e.* with a maximum on flash 4, 8 and 12 (2). The fluorescence level at 120 µs shows a period 2 oscillation. Thus with a series of transients, we can measure all three oscillations.

In order to investigated if the observed period four oscillations are related to the OEC or not, and are of the same origin as the well known oscillation of the oxygen release oscillation (7), we have inactivated the OEC by milimolar concentration of hydroxylamine, which is known to inactivate the OEC. This treatment resulted also in a complete suppression of oxygen evolution, and of the period four oscillations (data not shown). It confirms that a functional OEC is necessary in order to be able to observe the period four fluorescence oscillation.

A further confirmation comes from the micromolar concentration treatment with hydroxylamine which have been proposed to "set back" the oxygen clock by 2 steps, and therefore cause a 2 flash delay in the oxygen release pattern (8). Figure 4 shows that the F_o oscillation pattern is also shifted by 2 flashes in the presence of 40µM hydroxylmine. The fluorescence transients measured after 2 flashes (S_3 state), and after hydroxylamine treatment (0 flash) are presented in Fig 5. The first phase of the transients is alike suggesting the similar nature of the two states.

Lastly, we measured the initial slope (dF/dt_0, measured between 50 and 100µs) of the fluorescence transients after different number of flashes (6). This slope is closely related to the primary rate constant for photochemistry (7). Figure 6 shows that it also presents a period 4 oscillation, in the opposite direction of F_o, even when normalized by the fluorescence yield attained at the J level.

We will present elsewhere an extension of the radical pair model (RRP) based on the S-state dependent rate constant for the reduction of the electron donor Y_z^+, which explains the opposite oscillation patterns at the F_o and F_J level (*paper in preparation*).

Acknowledgment
BJS acknowledges Prof. H. Senger for inviting him to his lab, Dr. H. Dau for exciting discussions, and the Swiss Natl. Science Foundation and Soc. Acad. Genève for funding.

References
1. Dau, H. (1994) Photochem Photobiol 60, 1-23
2. Schreiber, U. and Neubauer, C. (1987) Z Naturforsch 42c, 1255-1264
3. Strasser, R.J., Srivastava, A. and Govindjee (1995) Photochem Photobiol 61, 32-42
4. Strasser, B.J. (1997) Photosynth. Res. 52, 147-155
5. Strasser, B.J. and Strasser, R.J. (1995) In Photosynthesis: From Light to the Biosphere (Mathis, P., ed.) Vol.IV, pp.909-912, Kulwer Academic Publisher, The Netherlands.
6. Lavergne, J. and Lecci, E. (1993) Photosynth. Res. 35, 323-343
7. Joliot, P., Joliot, A., Bouges, B. and Barbieri, G. (1971), Photochem. Photobiol. 14, 287-305
8. Bouges-Bocquet, B. (1980) Biochim. Biophys. Acta 594, 85-103

INHIBITION OF CHLORORESPIRATION DOES NOT AFFECT THE F_0 IN *PHAEODACTYLUM TRICORNUTUM*

C. Geel, W. Groen-Versluis and J.F.H. Snel
Wageningen Agricultural University, Laboratory for Plant Physiology,
Arboretumlaan 4, NL-6703 BD Wageningen, Netherlands

Keywords: Chl fluorescence induction, electron transport, marine algae, nonphotochemical quenching, productivity, state transitions,

1. Introduction

The minimal fluorescence, *i.e.* the fluorescence of open PSII centres, can be used to estimate PSII excitation in a phytoplankton sample. As the minimal fluorescence in the light adapted state, F_0', is not easily measured in the light, we used the minimal fluorescence in the dark adapted state, F_0, to estimate PSII excitation in the light (1). This method gave a good relationship between PSII electron flow and carbon assimilation at low and medium irradiances. Several factors might affect the relationship between F_0 and F_0' and therefore limit the use F_0 for estimation of PSII absorption cross section: state transitions from state II to state I, changes in spill-over from PSII to PSI or chlororespiration resulting in energy dependent fluorescence quenching in the dark (2). The maximal value of F_M in *P. tricornutum* and several other species at low light instead of in the dark (3) indicates that these processes do occur. A transition from state II to state I would imply that F_0' at low light is higher than F_0. Changes in spill-over from PSII to PSI would affect both F_0 and F_M and would cause F_0' at low light to be higher than the F_0 measured in the dark. Energy dependent quenching caused by chlororespiration would lower the F_V/F_M and the apparent F_0 in the dark adapted state, but the latter effect might be masked by dark reduction of Q_A by chlororespiration. We have applied conditions which modulate chlororespiration and studied the effects on the apparent F_0 in *P. tricornutum*, a species relatively well examined with respect to regulation of light harvesting and chlororespiration.

2. Materials and Methods

Growth of *Phaeodactylum tricornutum*, measurement of dissolved oxygen, chlorophyll fluorescence and the determination of fluorescence parameters have been described (3). Chlorophyll fluorescence was measured with the Xe-PAM fluorometer. F_0 is defined as the minimal fluorescence observed in the presence of measuring light only. F_0' is the minimal fluorescence observed the light adapted state. Experiments were done at 20 °C and oxygen saturation unless indicated otherwise. The oxygen concentration in the algal samples was lowered by injecting small amounts of glucose into the cuvette containing glucose oxidase. Antimycin A, CCCP and nigericin were added to the samples 5 minutes before the start of the measurement. Final concentrations (in µM) were: antimycin A: 50; CCCP: 5; nigericin: 5. The final ethanol concentration never exceeded 1 %. Samples measured at 5 and 10 °C were adapted to that temperature for 30 minutes in the dark prior to measurement. Far red light intensity was measured in W m^{-2} and converted to photon flux density units.

G. Garab (ed.), Photosynthesis: Mechanisms and Effects, Vol. V, 4329–4332.

Table 1. Effects of temperature, oxygen concentration, antimycin A, CCCP and nigericin on F_0, F_M and F_V/F_M of *P. tricornutum*. Data expressed as a percentage of the control at 20 °C and oxygen content equivalent to 100 % air saturation. The F_V/F_M values of the control samples were between 0.64 and 0.70. n represents the number of observations.

condition		n	F_0 ($\pm$std)	F_M ($\pm$ std)	F_V/F_M ($\pm$ std)
temperature:	10 °C	10	102 ± 3	114 ± 3	106 ± 1
	5 °C	10	109 ± 4	121 ± 4	106 ± 1
oxygen:	22 % O_2	4	94 ± 1	102 ± 1	105 ± 1
	4 % O_2	3	89 ± 1*	91 ± 2*	101 ± 1
inhibitor:	antimycin A	16	108 ± 18	104 ± 9	98 ± 6
	CCCP	15	95 ±13	105 ± 11	104 ± 3
	nigericin	11	105 ± 16	108 ± 13	101 ± 2

* decrease not reversible by readdition of oxygen

3. Results

Geel *et al.* (3) showed that F_V/F_M in *P. tricornutum* increased with decreasing temperature and suggested that this increase might be interpreted as inhibition of chlororespiration at low temperature. Table 1 shows that at 10 °C F_V/F_M is higher due to an increase of F_M. This temperature effect might be explained by the removal of chlororespiration but as well by a change in energy distribution via spill-over. Lowering the temperature to 5 °C led to an increase of F_0 and F_M, but not to a further increase in F_V/F_M. The increase of F_0 and F_M is not easily explained by a light-induced removal of energy dependent quenching as that should result in an increase of F_V/F_M as well. The increase of both F_0 and F_M is better explained by a decrease in spill-over from PSII to PSI or by a state transition from state I to state II.

Involvement of chlororespiration and its effect on both F_0, F_M and F_V/F_M in the temperature experiments was examined by inhibition of chlororespiration. At low oxygen concentrations chlororespiration should be low (4). F_0 was slightly decreased and F_M was slightly increased at 22% oxygen content, leading to a small increase of F_V/F_M (Fig.1, Table 1). Below an oxygen content of 22% F_M decreased. At 4% oxygen content both F_0 and F_M where smaller than in air- saturated water, but F_V/F_M was about the same (Fig. 1, Table 1). In contrast to F_0, F_M did not readily relax after readdition of oxygen.

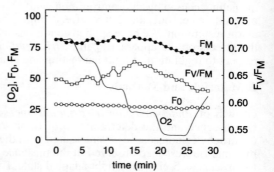

Figure 1 Effect of oxygen concentration on F_0, F_M and F_V/F_M.

At 10 °C a lowering of the oxygen content did not cause an effect on F_V/F_M (data not shown). This suggests that chlororespiration, if present, does not affect fluorescence at 10 °C.

Antimycin A, an inhibitor of chlororespiration (5) and the uncouplers CCCP and nigericin were added to dark adapted samples of *P. tricornutum* to dissipate the hypothetical proton gradient in the dark. Antimycin A, CCCP and nigericin gave only small effects on F_0, F_M and F_V/F_M (Table 1) which, in view of the relatively large standard deviation, should be considered insignificant. Therefore we conclude that in the dark adapted state the proton gradient, if present at all, is not large enough to induce energy dependent quenching.

Ting and Owens (2) found a small but significant increase of both F_0 and F_M using anaerobiosis, antimycin A, nigericin and CCCP. Their experiments, however, were done after preillumination with far red light (704 nm, 10 W m^{-2}). Figure 2 shows the dependency of F_0, F_M and ϕ_{PSII} on the photon flux density of the far red light. F_0 slightly decreased in low far red light and remains constant at higher photon flux densities. F_M and ϕ_{PSII} both show a small increase at low photon flux densities and a decrease at higher photon flux densities. The maximum of F_M was found at about 35 µmol m^{-2} s^{-1} far red light. This increase of F_M might be

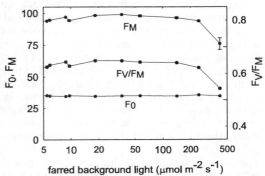

Figure 2 The effect of far red background light on F_0, F_M and F_V/F_M.

explained both by a state II to state I transition or by the elimination of energy dependent fluorescence quenching between darkness and 35 µmol m^{-2} s^{-1}. At 65 µmol m^{-2} s^{-1} of far red light, which is approximately the photon flux density used by Ting and Owens (3), we found a slightly higher value of both F_M and ϕ_{PSII} compared to the value measured in darkness. The decrease of F_M at higher photon flux densities is probably due to energy dependent fluorescence quenching. The proton gradient required for the energy dependent quenching might be caused by PS I driven cyclic electron flow, or by linear electron flow driven by far red light delivered to the PSII reaction centre as well.

Figure 3 shows the effect of a 30s far red light pulse (245 µmol m^{-2} s^{-1}) on the fluorescence transients of dark adapted *P. tricornutum* cells. The far red light induces an immediate increase of fluorescence. When the far red light is turned off, the fluorescence yield shows a transient increase followed by a slow relaxation, indicating a transient reduction of the PQ-pool related to unequal energy distribution between PSI and PSII. This change in energy distribution in favour of PSII resembles a state II to state I transition. As the duration of the far red light pulse is only 30s, protein phosphorylation of LHC or PSII components is not very likely. The fact that a state transition is observed at all, implies that the unbalance in energy distribution between PSII and PSI causes the very low excitation light (integrated PFD: 4.3 nmol m^{-2} s^{-1}) itself to induce a significant reduction of the PQ-pool in *P.*

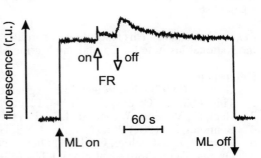

Figure 3 Effect of far red light on the fluorescence yield. ML: measuring light; FR: far red light.

tricornutum. The observation (1) that at these very low PFD's the F_V/F_M is light dependent, supports this interpretation.

4. Discussion

The proportional changes in F_0 and F_M (F_V/F_M = constant) induced by nigericin, antimycin A and CCCP indicate that nonphotochemical quenching is absent in the dark adapted state and that the changes in F_0 and F_M must be caused by changes in PSII cross section or changes in spillover. An increase in F_M might involve a transition from state II in the dark to state I in the light. If the transition were regulated via redox-regulated protein phosphorylation, the PQ-pool would have to be more reduced in the dark than in the light, which is not apparent from the data in Fig.2. The timescale of the transition (seconds rather than minutes, Fig.3) is much faster than the timescale of state transitions in higher plants. A mechanism involving regulation of energy distribution in *P. tricornutum* by light dependent ionic redistribution, as suggested by Owens (6), might be more compatible with the rate of these changes. A decrease of the amount of spill-over between darkness and low light would imply that in the dark adapted state energy distribution towards PS I is favoured.

The data show that under our conditions chlororespiratory activity was insignificant or did not significantly affect the redox state of Q_A and the F_V/F_M. Our experiments at very low excitation light using the sensitive Xe-PAM fluorometer indicate that the F_V/F_M of *P. tricornutum* is sensitive to even low light as the F_V/F_M increased from 0.65 to 0.70 upon decreasing the measuring light from 17 nmol m^{-2} s^{-1} to 2 nmol m^{-2} s^{-1} (1). It is therefore very well possible that in earlier work on chlororespiration in *P. tricornutum* measuring light, or far red light used to oxidize Q_A^- prior to the measurements, may have induced the energy dependent quenching attributed to chlororespiratory activity. In our experiments a dark adaptation period of only 30 min was used after harvesting of the algae. This period may have been too short to induce the chlororespiratory phenomena described by Feild et al (7).

Conclusions

It is concluded that in *P. tricornutum* the F_0, measured after 30 min. dark adaptation, is not significantly affected by photochemical or nonphotochemical quenching related to chlororespiration. Adaptation to light may involve state transitions or changes in spill-over leading to higher minimal fluorescence levels in the light adapted state. More insight in the relation between nonphotochemical quenching and F_0 quenching will help to reduce the error in estimating the PSII excitation in the light adapted state from F_0 measured in the dark adapted state.

Acknowledgements This research was partially financed the Cornelus Lely foundation and the National Institute for Coastal and Marine Management.

References
1 Geel, C. (1997) Diss. Wageningen Agricultural University, ISBN 90-5485-789-7
2 Ting, C.S. and Owens, T.G. (1993) Plant Physiol. 101, 1323-1330
3 Geel, C., Versluis. W. and Snel, J.F.H. (1997) Photosynth. Res. 51, 121-134
4 Bennoun, P. (1982) Proc. Nat. Acad. Sci. USA 72, 4352-4356
5 Ravenel, J. and Peltier, G. (1991) Photosynth. Res. 28, 141-148
6 Owens, T.G. (1986) Plant Physiol. 80, 739-746
7 Feild, T.S., Nedbal, L. and Ort, D.R. (1998) Plant Physiol. 116, 1209-1218

STUDY OF FLUORESCENCE PHENOMENA IN GREENING BARLEY LEAVES.

Philippe Juneau and Radovan Popovic
University of Québec in Montréal, Chemistry Dept./TOXEN

Key words: Chl fluorescence induction; Greening; Light harvesting complexes

1. Introduction

Photosystem II chlorophyll a fluorescence in fully developed green leaves is dependent to the oxidoreduction state of its electron transport carriers and related processes (1, 2). It is widely accepted, that the variable fluorescence induction kinetics is a representative indicator of the photosynthetic electron transport in intact leaf. After the rapid rise of the fluorescence yield induced by continuous illumination, the fluorescence level is lowered by the competition between the photochemical reaction and the processes not directly related to the redox state of Q_A, known as non-photochemical quenching phenomena (Q_N) (3). Beside the fact that Q_N can be related to the photoinhibition of photosynthesis (Q_I) (4), it was found that the major part of Q_N in green plants is dependent to the interthylakoids Δ pH, so called energy-dependent quenching value (Q_E) (5). However, another part of Q_N (named as Q_T) is also attributed to the phosphorylation and uncoupling of some peripheral PSII light harvesting complexes (LHCIIb) (4).

The slow fluorescence induction kinetics in greening leaves should be dependent to the development state of photosynthetic apparatus. During the greening process, the LHCII development takes place where the number of chlorophyll molecules associated with each reaction centre was increased and consequently the energy transfer from PSII to PSI became evident (6).

Since photochemical and non-photochemical energy dissipation is highly dependent to LHCII, PSII associated electron transport and related proton gradient, it is of interest to understand how structural and functional aspects of developing photosynthetic apparatus are involved in the fluorescence energy dissipation processes. In the present report, we investigated the change of modulated fluorescence parameters, in order to understand the strong fluorescence quenching effect during the first 2 minutes of illumination of greening etiolated barley leaves (7).

G. Garab (ed.), Photosynthesis: Mechanisms and Effects, Vol. V, 4333–4336.
© 1998 Kluwer Academic Publishers. Printed in the Netherlands.

2. Materials and Methods

Barley seedlings (*Hordeum vulgare* L. cv. Sophie) were grown for 6 days at 22°C in darkness on vermiculite soak water and then exposed to 3, 6, 12 or 24 hours of continuous white light (100 $\mu E \cdot m^{-2} \cdot s^{-1}$). Etiolated greening leaves were treated with 125 mM NaF, 0,1 mM DBMIB or 600 μM DCMU during 1, 2.5 or 3 hours respectively. Modulated fluorescence measurements were done by using a PAM fluorimeter (MFMS/2S, Hanzatech) as describe (3). The modulated, the actinic and the flash light intensity were 2 $\mu E \cdot m^{-2} \cdot s^{-1}$, 225 $\mu E \cdot m^{-2} \cdot s^{-1}$ and 800 $\mu E \cdot m^{-2} \cdot s^{-1}$ respectively. Fluorescence parameters, ϕm, $\phi' m$, Q_P and Q_N, were evaluated as describe earlier (3).

3. Result and Discussion

The modulated fluorescence kinetics of barley greening leaves was changed at different stage of the greening process, were, during the first 12 hours of greening, all the variable fluorescence yield induced by actinic light was entirely quench (figure 1). When the dark adapted greening leaves were exposed to actinic light, the fluorescence yield decreased rapidly below the initial Fo level reaching in 1 min F_{MIN} level. However, the fluorescence yield was increased to Fo level again after 2 minutes of continuous illumination. The pronounced quenching effect seen during the first 2 minutes of illumination was the biggest at 6 hours of greening. Q_N value was decreased from 0.9 to 0.5 during 24 hours of greening and approached to the value for fully green leaves (0.4) (data not shown). Since the complexity of greening process, were the PSII and PSI proteins and pigments synthesis and their functional link into photosynthetic electron transport take place (8, 9), interpretation of Q_N became also a matter of complexity.

During the greening process the most evident undergoing change is related to the synthesis of LHCII and their functional association with the photosynthetic electron transport process (6, 10). It seems that the maximum of non-photochemical quenching effect during the first 2 minutes of illumination is a response to the functional properties of LHCII. We supposed that the protein phosphorylation process may takes an important place as a fluorescence quencher. Indeed, when 6 hours greening leaves were treated with NaF, we noticed that the decay of variable fluorescence after Fp rise is much slower compared to the non-treated leaves (figure 2). Since NaF is a strong inhibitor of the phosphatase and therefore suppress the reassociation of LHCIIb with the PSII (11), we interpret that the change of the fluorescence quenching effect in NaF treated leaves is resulted by modification of LHCIIb functions.

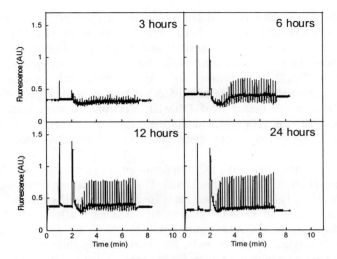

Figure 1: Fluorescence kinetics of barley greening leaves at different stage of the greening process. For more details see Materials and Methods.

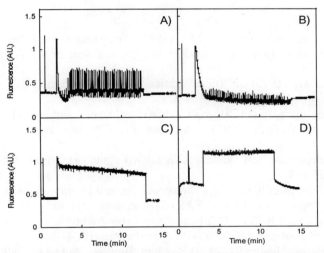

Figure 2: Effect of 1 NaF (B), DBMIB (C) and DCMU (D) on the fluorescence patern of 6 hours greening barley leaves. A) represents the fluorescence emission of the non-treated leaves. For further details see Materials and Methods.

The LHCII-b phosphorylation is regulated by the redox state of the plastoquinones, since when PQ is in a reduce state the phosphorylation can occur (12, 13). When the DBMIB was used to block electron transport and to maintain PQ in a reduce state, LHCIIb phosphorylation is undergoing followed by their uncoupling with PSII reaction centres (12, 13). Therefore, we may interpret that rapid decrease of fluorescence after the maximum rise (figure 2c) is link to the LHCIIb dissociation from PSII. Since, in the presence of DBMIB, electron transport communication between PSII an PSI was inhibited, the further very slow fluorescence quenching phenomena was probably induced by cyclic electron transport associated with PSII (14).

However, in the presence of DCMU, PQ was maintained in oxidised state, therefore it prevents the activation of the phosphokinase (13), and consequently the LHCIIb uncoupling with active PSII reaction centres. Therefore, as well expected under this conditions, no any decay of variable fluorescence was seen after the initial rise of fluorescence induced by actinic light (figure 2d).

The LHCIIb phosphorylation appears to be link to the strong fluorescence quenching of the Fo value seen under high light intensity in greening leaves.

4. References

1) Duysens L.N.M. and Sweers H.E. (1963) In Studies on Microalgae and Photosynthetic Bacteria. Jpn. Soc. Plant Physiol, Univ. Tokyo Press, Tokyo, pp 353-372

2) Papargeorgiou G. (1975) In Bioenergetics of photosynthesis. Govindjee (ed), Academic Press, New York, USA, pp 319-371

3) Schreiber U., Schilwa U. and Bilger W. (1986) Photosynth. Res. 10: 51-62

4) Krause G.H. and Weis E. (1991) Annu. Rev. Plant Physiol. Plant Mol. Biol. 42: 313-349

5) Horton P., Ruban A.V. and Walters R.G. (1996) Annu. Rev. Plant Physiol. Plant Mol. Biol. 47: 655-684

6) Baker N.R. and Strasser R.J. (1982) Photobiochem. and Photobiophys. 4: 265-273

7) Juneau P. and Popovic R (1997) ACFAS 65th Congress, Canada

8) Baker N.R. and Leech R.M. (1977) Plant Physiol. 60: 640-644

9) Chitnis P.R. and Thornber J.P. (1988) Photosynth. Res. 16: 41-63

10) Wellburn A.R. and Hampp R. (1979) Biochim. Biophys. Acta 547: 3800-397

11) Bennett J. (1980) Eur. J. Biochem. 104: 85-89

12) Allen J.F., Bennett J., Steinback K.E. and Arntzen C.J. (1981) Nature 291: 25-29

13) Allen J.F. (1995) Physiologia Plantarum 93: 196-205

14) Arnon D.I. and Tang G.M.S. (1988) Proc. Nat. Acad. Sc. USA 85: 9524-9528

PORTABLE PLANT HEALTH MEASUREMENT SYSTEM

Nejat Aksoy

Abyss Research Inc.
580 Broadway Suite 408, New York, NY 10012
nejat@photosynthesis.org

KeyWords: Chl fluorescence induction, photosystem 1, photosystem 2, Qa, S metabolism, whole leaf

1. Introduction

This system is designed to assist diagnosis of the plant health globally. The system is formed by portable plant health measurement devices connected to a diagnosis and analysis center through a flexible information network. A flexible network is formed so that users from the remote areas as well as internet are able to use the system. The hardware and software is designed in an open technology for easier upgrades. Portable plant health measurement instrument is a networkable leaf flash spectrophotometer capable of measuring Qa, Electrochromy, P700, Fluorescence , S fluorescence , reflectance spectra, temperature, humidity and image of the leaf with GPS information. The instrument is an extended version of the Leaf Flash Spectrometer (M.K.Saygin, TUBITAK, Teknopark, Gebze, Turkey). The network and intelligent user interface options of the system can be used by any commercially or user designed instrument.

The user interface for the instrument developed in a leveled way so that a farmer with a limited technical knowledge will be able to use it efficiently. An expert software system is able to make decisions on the health of the measured leaf by using the central system data. The portable instrument will assist the user to measure the plants and give the final results of the measurement together with relevant information such as suggestions, weather information, local news etc. depending on the geographical location.

For the determination of the photosynthetic condition and health of the plants several measurement techniques are developed (1, 2) as well as the known techniques are integrated in measurement process. The system is designed to be an open system so that the new techniques, user defined changes, other instruments can easily be integrated into the device. The techniques enable us to get information not only on the overall

G. Garab (ed.), Photosynthesis: Mechanisms and Effects, Vol. V, 4337–4340.
© 1998 *Kluwer Academic Publishers. Printed in the Netherlands.*

photosynthetic efficiency of the plant but on different parts of the photosynthetic machinery. The reason of the stress or damage can also be estimated. Collecting the photosynthetic measurements together with the image of the leaf and the geographic coordinates gives an opportunity to understand the leaf photosynthesis better. Since plant type and location information will be collected a global plant health monitoring and diagnosis system can be formed by incorporation of the satellite data.

2. Methods and Discussion

2.1. Hardware and the software

Design is based on a fast pentium based upgradable embedded controller. A virtual reality port is also added to the instrument so that a wearable screen is an option for the user. For integration with user designed or other instruments all universal ports are added to the device. A fast ethernet network card and a 56 K modem card are also integrated into the device. Optical section of the instrument is based on miniature optics technology. The leaf image retrieval section is based on a high resolution CCD imager . The spectrophotometer is again a miniature fiber optic CCD based one. A GPS card is also incorporated into the device however the user is able to use its own GPS devices. In the development of the software for the instrument and the network we used object-oriented technology. The embedded controller operating system is tightly integrated with the network. Smalltalk language and object oriented database with integrated fuzzy logic algorithms creates an intelligent, leveled user interface. The instrument operating system, data collection and management, internet and network connectivity, Image analysis, analysis of data, transfer of information from weather services and databases are all integrated tightly.

2.3. Measurement techniques

2.3.1 S fluorescence

As a direct application of flash spectroscopy to leaves, it was shown that the fluorescence level of dark adapted leaves, measured 800 ms after saturating single turnover flashes, oscillates with a period of four (2). Figure 1.A shows the experimental measurement of the fluorescence level of a dark-adapted spinach leaf during a train of saturating single-turnover flashes.

2.3.2 Qa

Figure 1.C shows simultaneous measurement of Qa from a spinach leaf measured in repetition mode with 10 additions and 1 s dark time at 533.9 and 544 nm.

2.3.3 P700

A method to measure P700 redox changes is to observe broadband increase in absorbance caused by the p700+ cation radical at 810-830 nm. In this wavelength region chlorophyll fluorescence is not excited and emission is small. Extent of P700 oxidation in intact leaves during steady state illumination by the measurement of 820 nm absorbance changes was reported by Weiss(3). A typical P700 signal of a hedera leaf is in figure 1.B (4)

2.3.4 Electrochromy

Figure 1.D shows a typical electrochromy signal from a hedera helix leaf. Ten signals are averaged with 300 ms dark time between each.

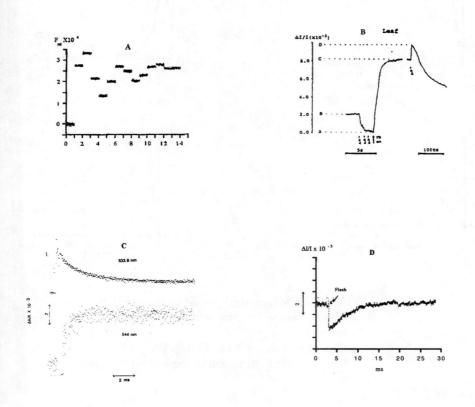

FIG. 1.A.Oscillations of the fluorescence of a dark adapted leaf by flashes
      B. P700 signal from Hedera Helix
      C. Qa signal from a hedera Helix leaf at 533.9 and 544 nm s.
      D. Electrochromy signal measured at 523 nm on the Hedera Helix

2.3.9 Study of signals with different stress treatments

4340

Water Splitting enzyme S is the most affected part of the photosynthetic chain from the stress. Usually, first damage occurs on this enzyme. Measurement of the quaternary oscillation from the enzyme S gives information about all electron transport chain. Any disfunctioning component of the chain will distort the signals.

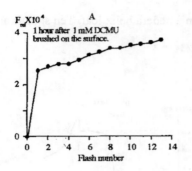
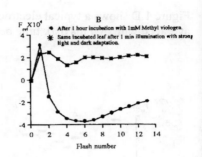
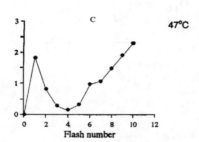
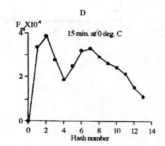

FIG. 2.A. Effect of DCMU on the S signal on Hedera Helix
B. Effect of Methyl violegen with dark adaptation and after light treatment
C. Effect of heat treatment at 47 deg. C
D. Effect of 0 deg C treatment of the leaf for 15 minutes.

REFERENCES

1)Aksoy, N., Uz, M., and Saygin, O (in preperation)
2)Aksoy, N, (1995) PhD Thesis, Bogazici University
3) Krause, G.H., Weis, E. (1988) Applications of Chlorophyll Fluorescence, pp. 3-11. H.K.Lichtenthaler ed., Kluwer Academic Publishers.
4) Uz, M., Saygin, O (1994) Photosynth.Res. 40, 175-179.

FLOW CYTOMETRY OF PROTOPLASTS FROM C4 PLANTS

[1]Pfündel E. and [2]Meister A., [1]Universität Würzburg, Botanik II, D-97082 Würzburg, Germany (pfuendel@botanik.uni-wuerzburg.de). [2]Institut für Pflanzengenetik und Kulturpflanzenforschung, D-06466 Gatersleben, Germany (meister@ipk-gatersleben.de).

Key words: fluorescence, maize, optical properties, protoplasts

1. Introduction

In leaves from species with the C4 dicarboxylic acid pathway of photosynthesis, two different chloroplast types exist: the mesophyll (MES) and the bundle sheath (BS) chloroplasts. In most C4 plants, MES and BS chloroplasts differ in their PSII/PSI ratios (1) and, hence, exhibit different chlorophyll fluorescence intensities at wavelengths smaller than 700 nm (R) relative to those at wavelengths longer than 710 nm (FR). Earlier, we analysed by flow cytometry the chloroplasts from C4 species by measuring the R and the FR fluorescence (2,3). Flow cytometric sorting yielded essentially pure MES and BS chloroplasts. Fluorescence spectroscopy of the pure chloroplast fractions revealed that in BS chloroplasts from many NADP-malic enzyme-type C4 plants, PSII is present in significant amounts and that its excitation energy is efficiently transferred to PSI.

Our previous studies demonstrated that pure MES and BS chloroplasts were needed to understand the light energy distribution in C4 leaves. Similarly, pure preparations of MES and BS protoplasts would aid analysis of the differentiation of the two cell types at both the genetic and biochemical levels. Here, we apply the optical conditions used for chloroplast sorting to analyse the protoplasts from NADP-malic enzyme C4 plants. The results reveal that chlorophyll fluorescence reabsorption in protoplasts blurs the spectroscopic differences between MES and BS chloroplasts and prevents sorting of the two protoplasts types. We suggest in future flow cytometry of C4 protoplasts that the short-wavelength spectral window should be confined to wavelengths at which chlorophyll fluorescence reabsorption is reduced.

2. Materials and methods

Plant growth conditions and leaf digestion was as described in (2) and (4), respectively. Under our conditions, leaf digestion for 3 h did not dissolve the bundle sheath strands in *Zea mays* and, hence, resulted in mostly MES protoplasts. By contrast, bundle sheath strands from C4 species of the genus *Flaveria* were enzymatically digested and yielded both MES and BS protoplasts. Protoplasts released from the leaves were filtered through a 80 μm nylon mesh and sedimented in the two different centrifugation steps to enrich MES and BS protoplasts as described in (4). The two pellets were resuspended and combined. The protoplast suspension was further purified by using a pH 7.0 Dextran/KCl-Mannitol gradient (4). Protoplasts collected from the interphase between the two gradient components were analysed. Chloroplasts were prepared from protoplasts by sedimenting

G. Garab (ed.), Photosynthesis: Mechanisms and Effects, Vol. V, 4341–4344.
© 1998 *Kluwer Academic Publishers. Printed in the Netherlands.*

the protoplasts followed by resuspension in distilled water and vigorous stirring. Flow cytometry was performed with a FACStar[Plus] sorting flow cytometer (Becton Dickinson, San Jose, CA). In the flow cytometer, a string of single particles is generated by injecting the protoplast suspension into a fluid stream which passes through a nozzle. Excitation at 488 nm of chlorophyll fluorescence occurs below the nozzle tip. Fluorescence, at right angle to the excitation beam, was separated into the R and FR spectral windows. All flow cytometrical conditions were as described in (2) except the nozzle diameter was increased to 200 μm and the sheath fluid pressure was reduced to 5 p.s.i. Fluorescence emission spectra were measured in a LS 50B fluorimeter (Perkin Elmer, Buckinghamshire, England) with 440 nm as the excitation wavelength (2).

3. Results and discussion

Figure 1A shows the flow cytometrical analysis of MES protoplasts from maize. The dots represent the red fluorescence (F_R) of individual particles plotted against the corresponding far red fluorescence (F_{FR}). We observed a region of particles exhibiting high fluorescence intensity, and one region exhibiting low fluorescence. The strongly fluorescing particles disappeared after the protoplasts were distorted (Fig. 1B). Hence, the strongly fluorescing particles correspond to protoplasts and the weakly fluorescing particles to chloroplasts. Our data confirm other results (5) that demonstrate the presence of significant numbers of chloroplasts in protoplast preparations. One important factor that produces chloroplasts from protoplasts is the mechanical stress exerted on protoplasts while passing the flow cytometer.

We characterised the fluorescence properties of protoplasts and chloroplasts by calculating the quotient of F_{FR}/F_R. The frequency distributions of F_{FR}/F_R show that F_{FR}/F_R in chloroplasts is roughly 80% of that in protoplasts (Fig. 1C). Figure 1D depicts the F_{FR}/F_R distribution of chloroplasts which were prepared by exposing protoplasts to low osmotic conditions. Under these conditions the F_{FR}/F_R in chloroplasts is further decreased and dropped to approximately 70% of the protoplast value.

So far, we demonstrated that the ratio of short-wavelength to long-wavelength fluorescence intensities differs

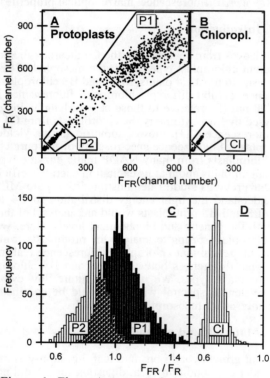

Figure 1: Flow cytometry of maize MES protoplasts. **A**: dot plot of protoplasts. The solid line-polynoms, P1 and P2, indicate the regions analysed in the frequency distribution diagrams in **C**. **B**: dot plot of chloroplasts prepared from protoplasts. **D**: frequency distribution diagram of the region Cl in panel **B**.

between protoplasts and chloroplasts. To gather information on the absolute fluorescence differences, we used fluorescence spectroscopy. Figure 2 shows fluorescence emission spectra of protoplasts and chloroplasts suspensions. The chlorophyll a + b concentration in the two suspensions was 1 μM. Chloroplasts were prepared by suspending protoplasts in distilled water. We concluded from Fig. 1 that the low osmotic conditions increase the fluorescence changes occurring on the transition from protoplasts to chloroplasts. Therefore, it is likely that osmotic effects also amplify the fluorescence changes in Fig. 2. In any case, Figure 2 clearly demonstrates that the peak at 680 nm in chloroplasts is significantly increased compared to the corresponding protoplast peak, where only minor differences were detected in the FR spectral region. This means that the lowered ratio of F_{FR}/F_R in chloroplasts, as established by flow cytometry (Fig. 1C), is caused by increased F_R emission from chloroplasts.

We believe that fluorescence reabsorption modulates the F_R emission. Our assumption is based on the fact that chlorophyll absorbs in the R spectral region but not in the FR spectral region. Therefore, reabsorption primarily affects the F_R signal. Fluorescence absorption is related to the average distance between the origin of a fluorescence quantum within an absorbing domain and the surface of the domain. The optical path lengths within a chloroplast are much smaller than those within a protoplast. Hence, we explain the relatively high F_R emission from chloroplasts by their relatively small particle size.

We then asked if flow cytometry can detect MES and BS protoplasts in C4 protoplast isolations. We addressed this question by comparing the flow cytometrical patterns of protoplasts from *Flaveria australasica*, a C4 species, with that of protoplasts from *Flaveria pringlei*, a C3 plant (Fig. 3). In comparison to *F. pringlei*, *F. australasica* exhibited a broader F_{FR}/F_R frequency distribution and the histogram peak was shifted to increased F_{FR}/F_R values. In NADP-malic enzyme-C4 species, the BS protoplasts are expected to show higher F_{FR}/F_R values than MES protoplasts because of their increased PSI/PSII stoichiometry. Therefore, the presence of BS protoplasts in the preparation from *F. australasica* explains well the different F_{FR}/F_R distribution curves of the C3 and the C4 plant. Flow cytometry, however, did not detect two dis-

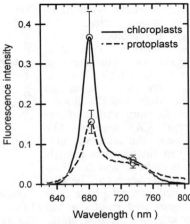

Figure 2: Fluorescence emission spectra at room temperature of protoplasts (dashed line) and of chloroplasts (solid line) from *Flaveria bidentis*. Spectra are not normalised and represent the mean of 6 protoplast or 4 chloroplast samples. Standard deviations of the short-wavelength maximum and of the fluorescence value at 735 nm are indicated as bars. Chlorophyll a + b concentration was 1 μM for protoplast and for chloroplast suspensions, respectively.

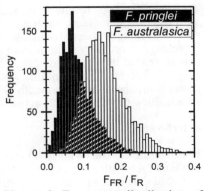

Figure. 3: Frequency distribution of F_{FR}/F_R of protoplasts from *Flaveria pringlei*, a C3 plant, and of *Flaveria australasica*, a C4 plant.

tinct particle populations that would indicate MES and BS protoplasts in *F. australasica*. This contrasts with our previous work in which flow cytometry clearly identified the MES and BS chloroplasts from various *Flaveria* C4 species (3). Hence, the relatively uniform fluorescence properties of MES and BS chloroplasts seems to be distorted in protoplasts to a degree that results in the overlap of the two protoplast populations. We suspect the reasons for the fluorescence behaviour are that protoplasts of different size are created by the mechanical stress in the flow cytometer, and

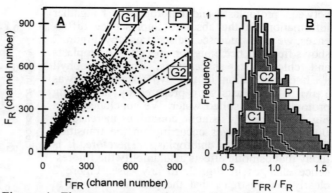

Figure 4: Flow cytometry of protoplasts and chloroplasts from *Flaveria palmeri*. <u>A</u>: dot plot of protoplast preparation. the polynom drawn as dashed line marks the area P that was used to calculate the frequency distribution shown in <u>B</u>. The solid line polynoms indicate the gates, G1 and G2, used for protoplast sorting. The chloroplasts released from the sorted G1 protoplasts are denoted C1, and those derived from G2 protoplasts are called C2. Chloroplasts were flow cytometrically analysed. The results from chloroplast analyses are depicted as histograms in <u>B</u>.

that the acceleration force during hydrodynamic focusing in the nozzle can create locally increased chloroplast concentrations within many protoplasts. Both phenomena are likely to affect the protoplast's optical properties which modulate fluorescence reabsorption and, consequently, the F_R signal. Thus, reabsorption artefacts appear to be the major obstacle to differentiating between MES and BS protoplasts. Finally, we tested if MES and BS protoplasts are enriched in the extreme regions of a dot plot (Fig. 4). The distribution of F_{FR}/F_R in chloroplasts derived from protoplasts with extremely high or extremely low F_{FR}/F_R overlap. This indicates that sorting from the edge region of the dot plot is inefficient.

To summarise, our results show that reabsorption of chlorophyll fluorescence interferes with the flow cytometric sorting of MES and BS protoplasts from C4 plants. Within our short-wavelength spectral window, fluorescence reabsorption decreases with increasing wavelengths: on the transition from protoplasts to chloroplasts, the fluorescence intensity was increased by a factor of 2.4 at 680 nm, the corresponding value was 1.4 at 700 nm (Fig. 2). Therefore, to minimise reabsorption artefacts during future flow cytometry of protoplasts, the short-wavelength window should be confined to higher wavelengths at which chlorophyll reabsorption is reduced.

4. References

1 Edwards, G. and Walker, D. (1983) C₃, C₄: mechanisms, and cellular and environmental regulation, of photosynthesis. Blackwell Scientific Publications, Oxford, UK
2 Pfündel, E. and Meister, A. (1996) Cytometry 23, 97-105
3 Pfündel, E., Nagel, E. and Meister, A. (1996) Plant Phys. 112, 1055-1070
4 Boinski, J.J., Wang, J.-L., Xu, P., Hotchkiss, T. and Berry, J.O. (1993) Plant Mol. Biol. 22, 397-410
5 Galbraith, D.W. (1994) Meth. Cell Biol. 42, 539-561

A STUDY OF THYLAKOID MEMBRANE ARCHITECTURE BY CONFOCAL MICROSCOPY

Monaz Mehta and Christa Critchley
Department of Botany, The University of Queensland,
Queensland 4072, Australia

Key words: fluorescence imaging, chloroplast, bundle sheath chloroplasts, membrane structure, spinach, stroma membranes

1. Introduction

Literature on chloroplast structure from the past 50 years is dominated by electron microscopic studies which involve the fixation of the chloroplasts by harsh chemical processes. Electron micrographs disclose the chloroplast as being enclosed by a double membrane and traversed by a complex membrane system called the thylakoids (1). In mesophyll chloroplasts this membrane system appears spatially segregated into appressed regions known as grana and non-appressed areas called stroma thylakoids. The thylakoid membranes contain large amounts of chlorophyll pigments which fluoresce in the orange-red wavelengths (2) and exist in protein-pigment complexes which are embedded within the membrane, called photosystems. These photosystems convert captured light energy into electrical current.

It is generally believed that grana regions are enriched in photosystem II (PS II) and that photosystem I (PS I) is thought to be aggregated in the stroma thylakoids, with a small percentage occurring in the grana margins (3). At 20°C PS II has a peak fluorescence at 685 nm, with PS I emitting a much weaker signal at 730 - 740 nm (2).

Ultrastructural studies of live, photosynthetically active chloroplasts in the past 50 years are the exception rather than the rule (4-9). The primary objective of this study was to observe chloroplasts of different species that had not undergone treatment by harsh chemical fixatives and stains which might alter chloroplast shape and internal membrane architecture.

2. Materials and Methods

Confocal scanning laser microscopy (CSLM) is a laser based light microscopy technique with the laser providing a light source that is bright, monochromatic and has a high degree of spatial and temporal coherence (10). Focusing the condenser and objective onto the same point in a specimen results in any stray light from planes other than the focal plane, to be cut out. When this advantage is coupled with extremely

G. Garab (ed.), Photosynthesis: Mechanisms and Effects, Vol. V, 4345–4348.

narrow apertures in the light path, a highly resolved image of a very thin (optical) section can be obtained.

For our confocal microscope studies, we used a Biorad MRC-600 CSLM, with a krypton-argon laser at 488 nm and barrier filters of 560 and 770 nm long pass (LP) for fluorescence, based on the premise that the latter would primarily pass PS I fluorescence. Images were taken either at a single focal depth and in a single scan or as averages of eight one second scans.

To observe thylakoid membrane arrangement in chloroplasts from *Pisum sativum* and *Spathiphyllum sp.* var. Symphony 200 were isolated according to Spencer and Unt's (5) protocol. Bundle sheath chloroplasts were always observed within their cells. Bundle sheath strands were mechanically isolated from leaves of *Sorghum bicolor*, *Saccharum officinalis*, *Zea mays* and *Panicum miliaceum* to Agostino *et al.* (11). To improve resolution of bundle sheath chloroplasts, the thick bundle sheath cell walls were partially digested by incubating the strands in medium containing 0.2% pectinase (*Rhizopus sp.*) and 2.0 % cellulase (*Aspergillus niger*) (12) for approximately 30 minutes. Some C_3 chloroplasts were treated with 5mM DCMU to enhance PS II fluorescence (13).

3. Results and Discussion

In chloroplasts of *Spathiphyllum sp.* and *Pisum sativum* grana were seen in confocal images taken parallel to the membrane plane. These grana emerged as regular, circular fluorescent shapes (Fig. 1, left), evenly distributed throughout the chloroplasts, with some appearing in different focal planes.

The most notable feature of mesophyll chloroplast thylakoid structure observed with the confocal microscope was the lack of fluorescing membrane connecting the grana regions. Chloroplasts imaged using two different emission filters appeared identical, although the fluorescence signal coming through the 770 LP filter was much lower (not shown).

A comparison of such granal C_3 mesophyll chloroplasts with bundle sheath chloroplasts of different C_4 species was made. Agranal *Saccharum officinalis* bundle sheath chloroplasts fluoresced considerably less and homogeneously and appeared not to have any grana (Fig. 1, right). *Sorghum bicolor* BSC, another agranal type (15), also emitted dim, homogeneous fluorescence, while *Zea mays* BSC, for whom biochemical and EM evidence suggests rudimentary grana, showed uniform fluorescence studded with a small number of brightly fluorescing grana (not shown). In contrast, *Panicum* BSC resembled mesophyll chloroplasts, with large numbers of discretely fluorescing grana. In none of our many images taken of C_3 and C_4 mesophyll chloroplasts did we see connecting membranes between the grana.

Because agranal BSC are deficient in PSII activity, chlorophyll fluorescence from sorghum and sugarcane BSC must come from PSI. We therefore suggest that the image (Fig.1, right) must derive essentially from stroma type thylakoids. Given that the fluorescence from agranal bundle sheath chloroplasts as seen in sugarcane and sorghum, could only come from PSI, we surmised that the confocal microscope should have been able to detect some PSI fluorescence in stromal membranes, even if the signal was low. A light microscopic study by Jungers and Doutreligne (17) of *Pellionia* and *Pilea* stem chloroplasts found green pigment partitioned exclusively in the grana, observed inter-granal spaces were non-coloured and that existing starch grains were visible through these regions. They also noted that grana in *Pellionia* had a tendency to fuse into strings during the course of an observation, a feature we also observed many times.

Recent studies of live chloroplasts labelled with green fluorescent protein have confirmed observations of live plastids made by Wildman (16) about interconnections and exchange of material between chloroplasts in a cell (9). These results appear to contradict the traditional idea about plastids being enclosed by a double membrane. Our own observations of isolated *Spathiphyllum sp.* chloroplasts in isotonic buffer (5) disclosed plastids which were extremely cup shaped, also in contradiction with the hemispherical chloroplasts often seen in the EM (unpublished data).

We believe it is important to re-investigate the true nature of thylakoid membrane architecture in alive chloroplasts. Are there any stromal membranes connecting grana, and if so, are they green? Are the non-appressed membrane regions as rich in PSI as they are currently believed to be?

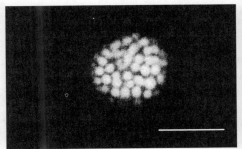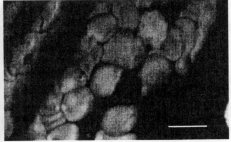

Figure 1. Confocal microscope images of isolated Spathiphyllum sp. chloroplast (left) and of chloroplasts in an intact bundle sheath strand of Saccharum officinalis (right). Excitation with 488 nm and 560 nm emission filter. Scale bars = 5 μm.

The confocal images raise questions about the nature of the thylakoid membrane in live chloroplasts. Do the thylakoids truly occur as seen in EM pictures? Is the membrane spatially segregated into laterally heterogeneous regions where the bulk of PSI occurs in the stroma membranes, and almost all PSII exists in the grana. Do the grana appress only under experimental conditions (17)?

Confocal imaging with a blue excitation laser combined with a 770 LP emission filter may still detect the "tail-end" of PSII (granal) fluorescence, which may have masked any low PSI (stroma thylakoid) fluorescence. When chloroplasts were treated with sodium dithionite, a PSI inhibitor, confocal images of the chloroplasts showed the same membrane configuration as untreated chloroplasts (unpublished data).

Further experimentation will aim to verify the location of PSI on the thylakoid membrane using techniques which will least interfere with the functional integrity of the chloroplasts.

Acknowledgements

This work was supported by an Australian Research Council Large Grant to CC. The authors wish to thank Mr Colin Macqueen of the Special Research Centre for Vision, Touch & Hearing, UQ, for help with the confocal microscopy.

References

1 Lichtenthaler, H.K., Buschmann, C., Döll, M., Fietz, H.J-., Bach, T., Kozel, U., Meier, D. and Rahmsdorf, U. (1981) Photosynthesis Research 2, 115-141
2 Govindjee (1995) Aust. J. Plant Physiol. 22, 131-160
3 Anderson, J. and Andersson, B. (1980) Biochim. Biophys. Acta 593, 427-440
4 Spencer, D. and Wildman, S. (1962) Aust. J. Biol. Sci. 15, 599-610
5 Spencer, D. and Unt, H. (1964) Aust. J. Biol. Sci. 18, 197-210
6 Wildman, S. G., Jope, C.A. and Atchison, B.A. (1980) Bot. Gaz. 141, 24-36
7 Punnett, T. (1967) J. Cell Biol. 35, 108a
8 van Spronsen, E.A., Sarafis, V., Brakenhoff, G.J., van der Voort, H.T.M. and Nanninga, N. (1989) Protoplasma 148, 8-14
9 Köhler, R.H., Cao, J., Zipfel, W.R., Webb, W.W. and Hanson, M. (1997) Science 276, 2039-2042
10 Inoué, S. (1995) In Handbook of biological confocal microscopy (Ed. J.B. Pawley) pp 1-17, Plenum Press.
11 Agostino, A., Furbank, R.T. and Hatch, M.D. (1989) Aust. J. Plant Physiol. 16, 279-290
12 Jenkins, C.L.D and Boag, S. (1985) Plant Physiol. 79, 84-89
13 Trebst, A. (1980) In Methods in Enzymology pp 675-715, Academic Press, Inc.
14 Edwards, G. and Walker, D.A. (1983) In C_3, C_4: mechanisms, and cellular and environmental regulation, of photosynthesis pp 299-325, Blackwell Scientific Publications
15 Downton, W.J.S., Berry, J.A. and Tregunna, E.B. (1970) Z. Pflanzenphysiol. 63, 194-199
16 Wildman, S. (1964) Organelles in living plant cells: their appearance and behaviour in natural speed photography. University of California Extension Media Center
17 Jungers, V. and Doutreligne, J. (1943) Cellule. 49, 409-419

SPECTRAL AND TIME-RESOLVED ANALYSIS OF LIGHT-INDUCED CHANGES OF NADPH FLUORESCENCE IN CHLOROPLASTS

G. Latouche, F. Montagnini, Z.G. Cerovic and I. Moya.
CNRS-LURE, Universite Paris-Sud, 91405 Orsay Cedex, France.

Key words : fluorescence lifetime, modelling, optical properties, photochemistry, reconstituted chloroplasts, redox potential.

1. Introduction

The use of NADPH fluorescence to estimate the redox state of NADP *in vivo*, or in intact isolated chloroplasts, is a very attractive method. Compare to usual quantitative techniques which use isolation procedures and enzymatic cycling (e.g. 1), fluorescence is immediate and non-destructive. But NADPH is not the only blue fluorophore excited by UV light present in chloroplasts and even less in leaves, therefore an important work has been done to investigate the blue-green fluorescence (BGF) of leaves (2) and isolated chloroplasts (3). The present report is the direct continuation of these investigations. Intact isolated chloroplasts exhibit a light-induced increase of BGF. This increase is only due to the variation of NADPH (3). In chloroplasts, an important part of the NADP is bound to proteins (4) and the binding to proteins increases the lifetime and the quantum yield of NADPH fluorescence (5). We performed decay and decay-associated spectra (DAS) in the dark and under light to determine how these populations of NADPH having different surroundings are involved in the light-induced increase of chloroplasts BGF. In addition, the emitted NADPH fluorescence being strongly reabsorbed by photosynthetic pigments (especially chlorophylls) (3), we tried to recover the NADPH concentration in intact chloroplasts from their BGF by the use of a model based on Beer-Lambert's Law.

2. Materials and Methods

Highly intact chloroplasts from young pea (*Pisum sativum* L.) leaves were isolated according to the procedure described in (6). Broken chloroplasts were prepared at the same time as described in (7), and chloroplast extracts (CE), stroma fraction, were obtained from intact chloroplasts as in (8). Reconstituted chloroplasts were prepared in the measuring cell by mixing intact chloroplasts broken in the cell by hypo-osmotic shock and CE. The final stroma/thylakoïd ratio was 11 times greater than in intact chloroplasts. Fluorescence spectra and decays were recorded on the Super-ACO synchrotron in Orsay (9). The set-up was modified for measurements on standard quartz cell (1 cm optical path) in front face configuration, and a laser diode (656 nm) was added to provide actinic photosynthetic light (intensity: 50 μE m^{-2} s^{-1}) without interfering with fluorescence measurements. The pulse-modulated fluorimeter used to test chloroplasts activity and the model, was a new version of the device described in (7). The pulsed excitation light (1 μs duration) is delivered by a xenon flash lamp, and the blue-green fluorescence was measured with a photomultiplier based detector insensitive to continuous light.

4349

G. Garab (ed.), Photosynthesis: Mechanisms and Effects, Vol. V, 4349–4352.

	kinetic component										mean lifetime
	very short (C1)		short (C2)		medium (C3)		long (C4)		very long (C5)		
	τ (ns)	f (%)	τ (ns)	f (%)	τ (ns)	f (%)	τ (ns)	f (%)	τ (ns)	f (%)	tm (ns)
NADPH	0.31	45.0 (1.7)	0.54	51.6 (1.7)	2.3	3.4 (0.2)	—	—	—	—	0.50 (0.01)
Rec. chloro.	0.16		0.52		1.34		3.77		10.98		
Dark		9.7 (2.9)		18.6 (4.2)		31.2 (2.7)		25.3 (2.0)		15.2 (3.9)	3.15 (0.46)
Light		12.1 (1.2)		22.4 (5.5)		31.9 (2.1)		24.7 (2.2)		8.8 (2.8)	2.46 (0.37)

Table 1. Lifetimes (τ) and fractional intensities (f) of the different kinetic components of fluorescence decays of NADPH and of reconstituted chloroplasts (Rec. chloro.). During dark to light transition chloroplasts fluorescence was doubled. In brackets are the standard deviation of the results that were all obtained by global deconvolution of several fluorescence decay. The emission wavelength was 456 nm. NADPH was in an Hepes buffer solution (pH = 7.9), and excitation wavelength was 340 nm. For reconstituted chloroplasts the excitation wavelength was between 340 and 350 nm.

3. Results and Discussion

All decays and DAS of chloroplasts BGF presented here were measured on reconstituted chloroplasts. Given the high complexity of these fluorescence decays, an important variation of light-induced BGF was needed to visualize the contribution of the different lifetime components to the light-induced changes. Reconstituted chloroplasts showed a much larger increase in light-induced BGF than intact chloroplasts because there was a larger NADPH/photosynthetic pigments ratio. As can be seen in table 1, there were no significant variations in the fractional intensities of the kinetics components of chloroplasts fluorescence in the dark compared to light, except for the very long component (C_5). This indicates that all these components participate to the light-induced increase of BGF, except C_5, and explains the decrease of the mean lifetime (τ_m) under actinic light. There are several hypothesis to explain the observed light-induced increase of chloroplasts BGF. First, it can be the result of an increased proportion of NADPH bound to protein over free NADPH. Second it could be a simple reduction of $NADP^+$ to NADPH with three possibilities: the form bound to proteins is preferentially reduced; the free form is preferentially reduced; both forms are almost equally reduced. Our results on fluorescence decay of NADPH (table 1), very close to the ones previously obtained (10), allowed us to conclude that the free form of NADPH appears in the very short (C_1) and the short (C_2) kinetic components of chloroplasts BGF. According to the literature, the fluorescence of NADPH bound to proteins (2.9 ns and 6.9 ns) (11) would contribute to the medium (C_3) and the long (C_4) kinetic components. An increase in the bound/free NADPH ratio (first hypothesis above) would lead to an increase in τ_m and in the fractional intensities of C_3 and C_4 under actinic light, which is not the case. Therefore, the second hypothesis is correct. Moreover, both forms of NADPH were equally reduced because neither the fractional intensities of the kinetic components containing the fluorescence of free NADPH (C_1, C_2), nor the ones containing the fluorescence of bound NADPH (C_3, C_4) increased under light. In order to confirm and complete these results, excitation spectra and DAS were measured in the dark and under actinic light (Fig. 1), and the spectra for each kinetic component were calculated. In the total excitation spectra, there is a large peak at 288 nm which is due to the contribution of the tail of the protein fluorescence attaining the emission wavelength (456 nm). The major contribution to this

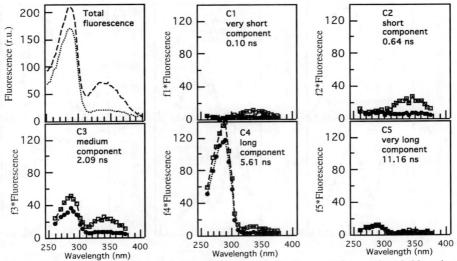

Figure 1. Decay-associated excitation spectra of the blue-green fluorescence of reconstituted chloroplasts in the dark and under actinic light (50 μE m^{-2} s^{-1}). The spectra obtained in the dark are plotted with short dotted lines and closed circles, and those obtained under light with long dotted lines and open squares.

protein fluorescence comes from C$_4$. The medium component contributes also for a small part and a very small contribution comes from C$_5$. This is in agreement with the lifetime of protein fluorescence reported in the literature (12). From the spectra presented in Fig. 1, light-induced difference spectra for each kinetic component were calculated (Fig. 2). The difference spectrum of total fluorescence shows a peak at 340 nm which corresponds to the maximum of excitation of NADPH. The smaller peak at 288 nm (Fig. 2) can be interpreted as a "transfer" from proteins to bound NADPH that can occur as a direct energy transfer (13) or as a reabsorption by NADPH of protein fluorescence. C$_1$ and C$_2$ contribute for a large part to the 340 nm peak, confirming the participation of free NADPH to the light-induced increase of chloroplasts BGF. Unlike C$_1$ and C$_2$ that contribute only to the 340 nm peak, C$_3$ and C$_4$ are also responsible for the 288 nm peak. The contribution of C$_3$ and C$_4$ to the 340 nm peak is due to the direct fluorescence of the bound NADPH. The contribution of C$_3$ to the 288 nm peak is equal to its contribution to the 340 nm peak and can be interpreted as a transfer of excitation energy from proteins to bound NADPH (13). The contribution of C$_4$ to the 288 nm peak is much larger than its contribution to the 340 nm peak, therefore it couldn't be due to an energy transfer. It can come from the re-emission at 456 nm of the light energy first absorbed by proteins, fluoresced at 350 nm (maximum fluorescence emission of proteins) and that was reabsorbed by NADPH. With these DAS we have confirmed that the light-induced increase of chloroplasts BGF is due to the reduction of both free and bound NADPH.

In order to use NADPH fluorescence to quantify NADPH concentration and its variations in intact chloroplasts we adapted a simple model based on Beer-Lambert's Law that was developed in biotechnology to estimate the biomass by the measurements of NADPH fluorescence (14). To test this model and obtain numerical values of its different parameters, experiments were performed in which NADPH concentration and photosynthetic pigment concentration (provided as washed broken chloroplasts) were varied independently (data not shown). The results confirm that for small concentrations of NADPH (i. e. under 10 μM), which is the case for intact chloroplasts in solution,

4352

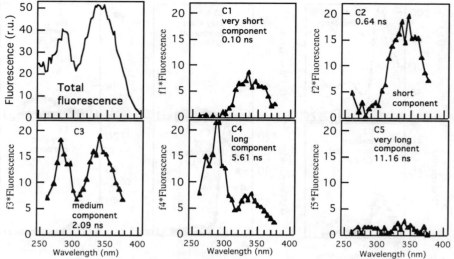

Figure 2. Decay-associated spectra of the light-induced increase of blue-green fluorescence of reconstituted chloroplasts. They were obtained from the results presented in Fig. 1 by difference.

NADPH fluorescence depends linearly on NADPH concentration, but the proportionality is strongly affected by the concentration of photosynthetic pigments. By applying the model with the obtained parameters to intact chloroplasts we were able to make an estimation of NADPH concentration inside the chloroplasts. We found 0.7 mM NADPH under light, which is not far from the previously reported measurements for spinach (0.9 mM) (1). We confirmed also that during the dark to light transition the concentration of NADPH in chloroplasts was doubled.

The presented results confirm that NADPH fluorescence can be used for non-destructive continuous and quantitative monitoring of the light-induced changes of NADP redox state in chloroplasts.

References

1 Takahama, U., Shimizu-Takahama, M. and Heber, U. (1981) Biochim. Biophys. Acta 637, 530-539
2 Buschmann, C. and Lichtentahler, H.K. (1998) J. Plant Physiol. 152, 297-314
3 Cerovic, Z.G., Langrand, E., Latouche, G., Morales, F. and Moya, I. (1998) Photosynth. Res. 56, 291-301
4 Usuda, H. (1988) Plant Physiol. 88, 1461-1468
5 Salmon, J.-M. (1977) J. Chim. Phys. 2, 239-245
6 Cerovic, Z.G. and Plesnicar, M. (1984) Biochem. J. 223, 543-545
7 Cerovic, Z.G., Bergher, M., Goulas, Y., Tosti, S. and Moya, I. (1993) Photosynth. Res. 36, 193-204
8 Lilley, R.M. and Walker, D.A. (1974) Biochim. Biophys. Acta 368, 269-278
9 Cerovic, Z.G., Morales, F. and Moya, I. (1994) Biochim. Biophys. Acta 1188, 58-68
10 Visser, A.J.W.G. and VanHoek, A. (1981) Photochem. Photobiol. 33, 35-40
11 Gafni, A. and Brand, L. (1976) Biochemistry 15, 3165-3170
12 Wolfbeis, O.S. (1985) in Molecular Luminescence Spectroscopy. Methods and Applications: Part 1 (Schulman, S.G., ed.), pp. 167-370, John Wiley & Sons, New York.
13 Velick, S.F. (1961) in Light and Life (McElroy, W.D., Glass, B., ed.), pp. 108-143, Johns Hopkins Press, Baltimore, Maryland.
14 Wang, N.S. and Simmons, M.B. (1991) Biotech. Bioeng. 38, 1292-1301

THERMOLUMINESCENCE OF PLANT LEAVES .
Instrumental and experimental aspects.

Ducruet J.M., Toulouse A. and Roman M.
SBE, Bât 532, INRA/CEA Saclay 91191 Gif-sur-Yvette cedex France

Keywords : higher plants, leaf physiology, temperature stress.

1. Introduction

Chlorophyll luminescence (delayed fluorescence) originates from photosystem-II (PS-II) and results from the recombination of charges pairs created by a previous illumination. Thermoluminescence (TL) is a particular technique to study luminescence emission, which consists in (i) cooling the sample before or immediately after an illumination at a temperature sufficiently low to practically block those light-emitting back-reactions which are investigated (ii) revealing by a linear temperature increase the different types of recombining charge pairs, which correspond to elementary TL bands.
Although (thermo)luminescence has been used since many years for investigating PS-II charge stabilisation in thylakoids, sub-chloroplast particles or frozen leaves or algae (1,2), applications to stress studies have been hindered by both instrumental and experimental problems. A major limiting factor has been a cumbersome and costly instrumentation, a problem which can now be solved. We describe here a compact and unexpensive TL device. Furthermore, (thermo)luminescence emissions from leaf tissues may exhibit sharp discrepancies compared to those observed from isolated thylakoids, which reflects the influence of the surrounding cell medium and the physiological state of the leaf.

2. Instrumentation

The TL-measuring device, derived from that briefly described (3,4), is formed by a TL measuring head (Fig.1), a power supply for temperature regulation and a computer with an interface.

2.1 *Sample holder and temperature regulation*
The temperature regulation of the sample is done by a single-stage 4 X 4 cm Peltier plate (Marlow DT1089 or DT12-8-01LS), sandwiched between a water-cooled brass block and a stacking made of a 0.05 mm thin copper film, a thermally insulating silicone rubber and a Celeron plate. Temperature is monitored by an Omega CO1-K flat thermocouple inserted under the copper film. The temperature is regulated by a calculated variable voltage driving a power amplifier and by a TTL-triggered current-inverter for

G. Garab (ed.), Photosynthesis: Mechanisms and Effects, Vol. V, 4353–4356.
© 1998 Kluwer Academic Publishers. Printed in the Netherlands.

heating/cooling. With a 20°C water flow, a linear gradient can be achieved from -15°C to 80°C. The measuring cuvette is formed by a Φ 25 mm cylindrical well bored into the silicone and Celeron plates, the sample (leaf disc or filter) with 100 µl water is softly pressed on the copper film by a Pyrex window and a ring. For high-temperature TL (20° to 150°C), a similar sample holder is used, with a Thermocoax heater instead of a Peltier.

2.2 *Luminescence excitation and detection*
A 5-arms PAM-Walz light guide is placed in front of the sample for illuminations, then the TL measuring head is laterally moved front to the detector.
Luminescence is detected through a RG 665 filter by a compact Hamamatsu H 5701-50 red-extended photomultiplier, with built-in high voltage converter (power supply ± 15 volts) and amplifier. The detector is located at 2.5 cm atop of the sample. An ultra-weak blue LED illuminating the sample through a 480 nm filter can be triggered in order to determine the F_0 fluorescence and TL band intensities on the same sample.

2.3 *Computer interfacing*
Interfacing is performed by either a plugged-in Advantech PCL818 card, a parallel port Nat. Instr. DAQ-Pad1200 or a PCMCIA NI DAQ-Card1200. Temperature and TL are recorded through Analog to Digital converting channels. A power-amplified Digital to Analog converter regulates the current through the Peltier or the Thermocoax heater. Four Digital Output TTL channels trigger, respectively, the heating/cooling current inverter, a xenon flash, an Light source and the pulsed blue LED. Signal acquisition and analysis are done with a dedicated software (5).

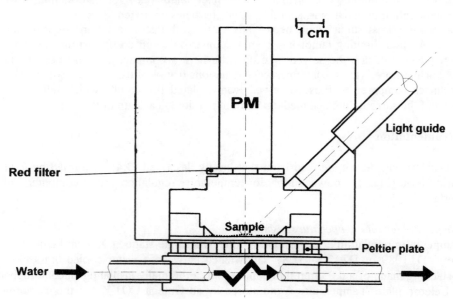

Figure 1. Sectional view of the TL measuring head. PM : photomultiplier tube.

3. thermoluminescence properties of leaves

In isolated thylakoids, flash sequences induce a B band ($S_2S_3Q_B^-$ recombination). It is centred near 35°C at neutral pH and splits into B1 (S3) and B2 (S2) bands when the pH is lowered (1,2). Fig. 2 shows that TL signals from leaves behave differently.

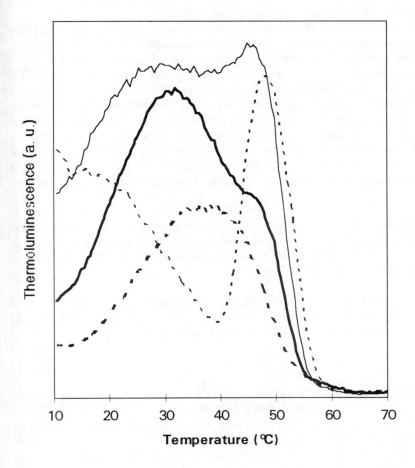

Figure 2. TL from cucumber leaf discs, after illumination at 10°C. Thick line: 1 flash, thin line: 3 flashes, thick dots: 3 flashes + nigericin, thin dots: 30 s far-red light (PAM 10)

Following flash sequences, a new band peaks at 47°C after 3 flashes (Fig, 2) or 2 flashes (not shown) and appears as a shoulder after 1 flash. The intensity of this band is highly dependent on plant species and growth conditions. It can be more reproducibly induced by a far-red illumination (3) and it is suppressed by uncouplers (Fig. 2) or by freezing.

This TL band also observed in intact chloroplasts or algal cells (not shown) corresponds to a luminescence afterglow (6) resulting from the back-transfer of electrons towards the initially silent $S_2S_3Q_B$ centers (7). This back-transfer requires both a reducing pool and a proton gradient. The flash-induced 'afterglow' TL band has been shown to reflect the reduction state of the photosynthetic electron transport chain in CAM-inducible plants (8). It should be noticed that this phenomenon makes unreliable the decompositions of luminescence decays into exponential phases for leaf material.

Fig.2 also shows that only the B band in the presence of nigericin peaks at 38°C and that the B band in the absence of uncoupler is slightly downshifted to 31°C after 1 flash, to 28°C after 3F and to below 20°C after far-red. This downshift can be ascribed to a low lumen pH which unstabilises the S_2 (1F) state and, to a geater extent, the S_3 state (3F). Far-red light generates S_2/S_3 and induces a stronger lumen acidification (3).

Freezing the sample produces TL signals more similar to those observed in thylakoids. However, in some plant species, freezing a leaf fragment below -5°C strongly distorts the B band. In a liquid N2-cooled set-up (5), freezing at -40°C or -20°C induces a sharp band near 15°C instead of 38°C in maize or at 10°C in grapevine (not shown). The origin of this freezing artefact not observed with thylakoids is unknown, although it might be tentatively explained by a release of vacuolar compounds following membrane disruption.

A better understanding of the TL properties of unfrozen cells or leaf tissues would provide an informative non-invasive probe of the photosynthetic metabolism. For such applications, TL apparatuses transportable close to the experimental fields would allow freshly collected samples to be studied.

Acknowledgements : we thank M. Hamang, L. Liagre and G. Lecointe for their contribution to building the TL devices. M. Roman is supported by N.I.L.P.R.P., Lasers Depart., Bucharest, Romania and by a grant from MENRT.

References

1 Sane, P.V. and Rutherford, A.W. (1986) in Light emission in plants and bacteria (Govindjee, Amesz J. and Fork D.C. eds) pp 329-360, Academic Press, New-York

2 Vass, I. and Inoue, Y. (1992) in The Photosystems : Structure, function and molecular biology (Barber J., ed .) pp 259-294, Elsevier, Amsterdam

3 Miranda, T. and Ducruet, J.M. (1995) Plant Physiol. Biochem. 33, 689-699

4 Ducruet, J.M. (1995) in Proceed. Photosynthesis and Remote Sensing, Montpellier (Guyot G., ed.) pp 93-96, INRA, Paris

5 Ducruet J.M., and Miranda, T. (1992) Photosynth. Res. 33, 15-27

6 Bertsch, W.F. and Azzi, J.R. (1965) Biochim. Biophys. Acta 94, 15-26.

7 Sundblad, L.G., Schröder, W.P., and Akerlund, H.E. (1988) . Biophys. Acta 973, 47-52.

8 Krieger, A., Bolte, S., Dietz, K.J. and Ducruet, J.M. (1998) Planta 205, 587-594

MEASURING P700 ABSORBANCE CHANGES IN THE NEAR INFRARED SPECTRAL REGION WITH A DUAL WAVELENGTH PULSE MODULATION SYSTEM

Christof Klughammer and Ulrich Schreiber
Julius-von-Sachs Institut für Biowissenschaften, Universität Würzburg,
Julius-von-Sachs Platz 2, D-97082 Würzburg, Germany

Key words: chloroplast, ferredoxin, P700, Photosystem I, transient absorption spectroscopy

1. Introduction

Monitoring the oxidation state of the primary donor P700 of photosystem I (PS I) via near infra red absorption spectroscopy has been providing valuable information on PSI-driven photosynthetic electron transport *in vivo* (1, 2, 3). Redox changes of P700 are preferentially detected at a single wavelength band in the region of the $P700^+$ cation radical absorption band peaking at 820 nm. NIR-measuring light > 800 nm is not absorbed by chlorophyll, does not generate chlorophyll fluorescence and has no actinic effect on photosynthesis of green plants. It can be applied at high intensity with the advantage of a high S/N ratio. In order to discriminate the measuring light from continuous background light the former usually is modulated. Modulation up to MHz frequencies can be easily achieved by infra-red emitting diodes (IRED). Such IREDs display bandwidths of 30-50 nm which is suitable for the broad $P700^+$ absorption band. Single wavelength systems based on a modulated IRED (e.g. ED-800T, Walz, Effeltrich) have been successfully used in monitoring light induced P700 redox changes in intact leaves (1, 2). Superimposed less wavelength-specific changes (as scattering changes) in most cases are slow and may be separated from $P700^+$ by application of saturation pulses (4) or by switching off the actinic light from time to time and observing the rapid signal relaxation. However, the low selectivity of the single wavelength measurement becomes a major problem with suspensions of isolated chloroplasts or algae. Unspecific optical changes caused by shrinking, swelling and settling of the chloroplasts or cells lead to serious distortions of the much smaller $P700^+$ signal, and artefacts caused by chemical additions and sample stirring preclude the observation of bathochromic absorption changes.

In order to overcome the principal problems of single wavelength $P700^+$ measurements we have developed a new dual wavelength emitter detector unit which combines with a standard PAM Chlorophyll Fluorometer (PAM 101, Walz, Effeltrich) to a dual wavelength pulse modulated spectrophotometer with 100 µs time resolution. The system detects strictly differential absorbance changes (810 nm minus 860 nm) and, therefore, is selective for absorbance changes caused by P700. Scattering changes and stirring noise in chloroplast suspensions are considerably decreased as compared to the conventional single beam method. Measurements of chemically induced P700 redox changes in stirred suspensions of chloroplasts or algae become feasible with the new system. Examples of application using chloroplast suspensions and intact leaves are presented.

G. Garab (ed.), Photosynthesis: Mechanisms and Effects, Vol. V, 4357–4360.
© *1998 Kluwer Academic Publishers. Printed in the Netherlands.*

2. Materials and Methods

Fig. 1 shows the experimental set-up for a dual wavelength measurement of chemical- and light-induced P700 redox changes on a stirred chloroplast suspension. The emitter pulses of the PAM 101 chlorophyll fluorometer (Walz, Effeltrich) trigger the driver of infrared emitting diodes (IREDs) at a frequency of 100 kHz. For each trigger pulse the driver produces two sequential measuring light pulses in the dual wavelength emitter (E), a sample (810 nm, 30 nm HBW) and a reference (860 nm, 40 nm HBW) light pulse, respectively. The IRED's are focused by a perspex cone on one arm of the multibranched fiberoptics. Another arm of the fiberoptics guides measuring light from the sample back to a photodiode detector (D). Only the difference signal of the two pulse signals is amplified by the AC-coupled preamplifier and fed into the detector entry of the PAM 101. The difference signal can be manually balanced to zero and then amplified with high gain.

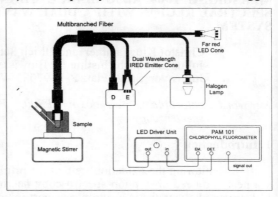

Figure 1. Experimental set-up of the dual wavelength measurement of P700 redox changes in stirred chloroplast suspensions.

3. Result and Discussion

3.1 Improvement of the $P700^+$ signal quality by the dual wavelength method

In Fig. 2 an example is presented showing the improvement of the signal quality by the dual wavelength method on light- and chemically induced absorption changes in rapidly stirred suspensions of intact chloroplasts. In case of the dual wavelength measurement (Fig. 2A) a light induced bleaching signal (810-860 nm) caused by reduction of PSI acceptors (3) and a distinct P700 oxidation after addition of 0.5 mM nitrite (open arrow) are clearly visible on top of the background stirring noise (2×10^{-4}). In contrast, when the same experiment was repeated with the standard ED-800-T single wavelength unit (Fig. 2B) the light induced bleaching was obscured by the large stirring noise (2×10^{-3}) and by a slow wavelength-unspecific apparent absorption increase. The P700 oxidation signal caused by addition of nitrite can hardly be assessed against the background of the unspecific change and the stirring noise.

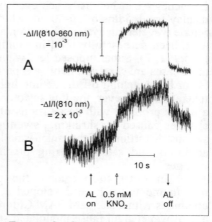

Figure 2. Comparison of dual (A) and single (B) wavelength $P700^+$ measurements on suspensions of rapidly stirred intact chloroplasts (60 µg chl/ml). Nitrite was added 10 s after actinic light on (250 µE/m²s).

3.2 *Ferricyanide induced oxidation of P700*

Single wavelength measurements suffer particularly from slow signal drifts. As shown in Fig.3, the high selectivity for differential absorption changes in the NIR spectral region and the signal stability of the new system even allows the observation of the extremely slow ferricanide induced oxidation of P700. After addition of 3 mM ferricyanide to a stirred sample of osmotically ruptured chloroplasts a rapid differential absorption increase with an amplitude of 10^{-3} is observed followed by the slow P700 oxidation change within 5-10 minutes. Application of a 800 ms light pulse leads to complete reduction of P700, and the chemical oxidation of P700 starts again after a lag time of about 3 minutes. The lag time may be taken as a measure of e⁻-equivalents stored at the donor side of PSI. The rapid absorption increase was not reversed by the light pulse.

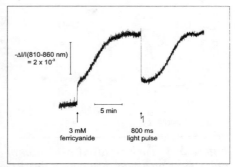

Figure 3. Oxidation of P700 by 3 mM of ferricyanide and its reduction by a light pulse (3000 $\mu E/m^2 s$) in ruptured chloroplasts (100 μg chl/ml) in presence of 0.2 μM nonactin and 0.2 μM nigericin.

3.3 *Binding of ferricyanide to ferredoxin binding sites*

The rapid differential absorption change upon ferricyanide addition was studied in some more detail as shown in Fig. 4. The change with a half rise time of 600 ms was not caused by an artefact but turned out to be sensitive to the presence of ferredoxin. It could be completely suppressed by addition of 20 μM of this PSI acceptor prior to ferricyanide addition. Almost the same amplitude of the change was observed when adding ferrocyanide instead of ferricyanide, whereas no change was found with intact chloroplasts (not shown). Hence, the observed change was not due to an oxidation process but must be induced by binding of the ferro/ferricyanide anion to a ferredoxin binding site at the thylakoid surface. The detection of such small chemically induced absorption changes may provide valuable information on protein binding sites or inhibitor binding.

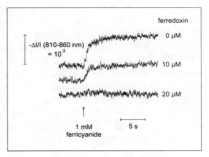

Figure 4. Kinetics of ferricyanide binding to ferredoxin binding sites of thylakoids. Various concentrations of ferredoxin were added 2 min before the addition of ferricyanide as indicated.

3.4 *Probing the reduction of PSI acceptors by time resolved oxidation of P700 in an intact leaf induced by a rapid 30000 $\mu E/m^2 s$ dark light transient*

In the following example of application, advantage is taken of the high time resolution of the system and its effective discrimination against background light as a consequence of the 100 kHz pulse modulation of the measuring light. The time resolved

4360

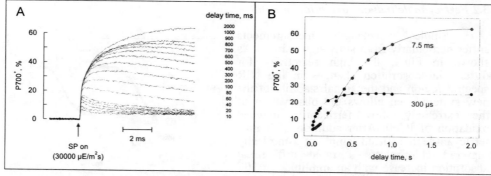

*Figure 5 **A**. Time resolved oxidation of P700 induced by a strong red light saturation pulse (SP) in a leaf disc which has been preilluminated by a 300 ms white light pulse (6000 µE/m²s). The delay time between preillumination pulse and onset of SP is varied.* ***B*** *Plot of P700⁺ amplitudes 300 µs and 7.5 ms after onset of the saturation pulse.*

oxidation of P700 in a leaf disc by a 30000 µE/m²s saturation pulse with steep light-on characteristic is demonstrated. In Fig. 5A the P700 oxidation kinetics during the first 8 ms after light on are depicted. In all cases P700 was fully reduced before light-on. The redox state of PSI acceptors was varied by a 300 ms preillumination light pulse (6000 µE/m²s) given on a dark adapted leaf disc shortly before and by variation of the delay time between the end of the preillumination and the onset of the saturation pulse. Depending on the delay time, a rapid P700 oxidation phase with an amplitude up to 25% of total P700 and a rise time of less than 300 µs is observed, followed by a slower re-reduction phase at delay times < 300 ms or by a slower oxidation phase at delay times > 300 ms. We interpret the polyphasic rise of P700 oxidation to reflect the redox state of various PSI acceptors which quite slowly reoxidise after a light pulse applied to dark adapted samples of leaves or chloroplasts (3,5). The amplitude of the rapid oxidation phase may be proportional to the oxidation state of the secondary PSI acceptor A_1 (5) whereas the amplitude of the slow P700 redox change may reflect oxidation of the iron-sulfur centers within PSI and/or of ferredoxin. If so, a plot of the P700 amplitudes taken at 300 µs and at 7-8 ms versus delay time would show the reoxidation kinetics of A_1^- and of FeS-centers, respectively, after the 300 ms preillumination pulse (Fig. 5B).

References

1 Harbinson, J. & Woodward, F.I. (1987) Plant, Cell Environ. 10, 131-140
2 Schreiber, U., Klughammer, C. and Neubauer, C. (1988) Z. Naturforsch. 43c, 686-698
3 Klughammer, C. and Schreiber, U. (1991) Z. Naturforsch. 46c, 233-244
4 Klughammer, C., Schreiber, U. (1994) Planta 192, 261-286
5 Klughammer, C., Pace, R.J. (1997) BBA 1318, 133-144

POSSIBILITIES OF NEW SYSTEM FOR LD/CD MEASUREMENTS IN PHOTOSYNTHESIS RESEARCH

Durchan M., Vácha F., Motejl M., and Štys D.
Laboratory of Biomembranes, University of South Bohemia,
Branišovská 31, CZ-37005 České Budějovice, Czech Republic

Key words: circular dichroism, fluorescence, LHC II, light scattering, linear dichroism, photosystem 2, reaction centers

1. Introduction

A new system based on JASCO J-715 spectropolarimeter was built up for extended linear/circular dichroism (LD/CD) measurements in photosynthesis research. Our device with accessories makes possible the simultaneous measurement of fluorescence-detected LD/CD (FDCD), or light-scattering. We are able synchronously to detect the light-induced changes in LD/CD spectra. The system is equipped with liquid-helium cryostat for measurements down to 1.5 K.

In this work, we present the first results demonstrating the possibilities of our equipment. The main objects for our tests were the light-harvesting chlorophyll *a/b* protein complex (LHC II), the reaction centre of Photosystem II (RC of PS II), and the subchloroplast particles enriched in PS II (BBY particles).

2. Material and Methods

2.1 Measuring system

The block diagram of optical system is shown in Figure 1. The main unit is the model J-715/150-S standard outfit (JASCO Corp., Japan). A xenon lamp (150 W) is used as the light source. The double monochromator uses crystal prisms that have different axial orientations, so the light is not only monochromated, but also linearly polarized. Piezo-elastic modulator works at 50 kHz. The beam has a low divergence, so the low-temperature measurements can be carried out in cryostat (Static Optistat Bath with ITC 503 temperature controller, Oxford Instr., U.K.) located at the back of sample compartment. The sample compartment is equipped with constant temperature water inlet/outlet port. In the case of room temperatures, the thermostated cell holder can be used; the adapter plate for cryostat can be removed, and the standard detector housing can be mounted closely to the sample compartment. Besides standard photomultiplier, we can connect the additional high sensitivity PM (R3896, Hamamatsu, Japan), or diode VIS/NIR detection accessory. The electrical system of spectropolarimeter has two channels as external input terminal. The light-induced changes in spectra can be recorded by synchronous detection with additional light source Fiber-Lite PL800 plus fibre optics (Dolan-Jenner, USA), light chopper model 651 (EG&G Instr., USA) through the additional lock-in amplifier.

G. Garab (ed.), Photosynthesis: Mechanisms and Effects, Vol. V, 4361–4364.
© 1998 *Kluwer Academic Publishers. Printed in the Netherlands.*

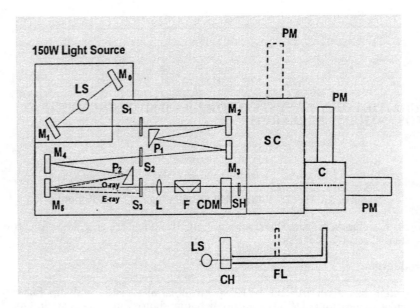

Figure 1. Block Diagram of Optical System

M mirrors, LS light sources, S slits, P prisms, L lens, F filter, CDM modulator, SH shutter, PM photomultiplier tube, SC sample compartment, C cryostat, CH light chopper, FL fibre optics

2.2 Material

The PS II-enriched membrane fragments were isolated from pea according to Berthold et al. (1). The LHC II trimers were prepared according to Burke et al. (2), the RCs of PS II according to Nanba and Satoh (3). Low-temperature spectra were measured at samples diluted in 60 % glycerol, samples for room-temperature ones were diluted in water.

3. Results and Discussion

3.1 Fluorescence-detected CD

Figure 2 illustrated the CD spectra of LHC II trimers, and RC of PS II at 77 K. Spectra correspond closely to data previously published (e.g. in (4)) with main bands in the red, and several intensive bands in the Soret, blue region. More interesting results seem to be FDCD spectra. As you can see (Fig. 2A), FDCD spectra of LHC II trimers include some bands which are not present or are not so significant in absorption CD spectra. Especially studying carotenoids in LHC II seems to be possible by FDCD (the band about 506 nm).

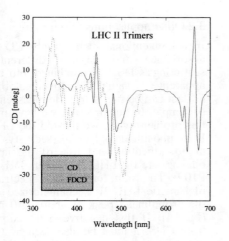

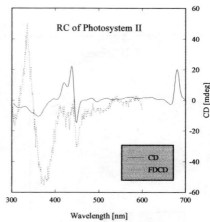

(FD)CD of RCs (Fig. 2B) where only six chlorophylls *a*, two pheophytins and two β-carotenes are present is more simple.

Figure 2. CD and FDCD at 77 K

Fluorescence was detected by RG645 filter (λ > 645 nm)

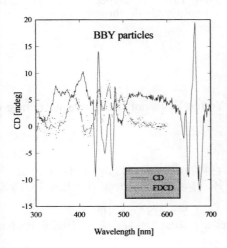

The low-temperature CD spectra of the BBY particles are depicted in Figure 2C. The absorption CD spectrum of BBY displays identically the same position of bands as LHC II trimers in Soret region, i.e. positive bands at 430, 443, 468, 481 nm, and negative ones at 436, 458, 474, and broad 487-97 nm band, but interestingly, in the FDCD spectrum some of them are lost. This is fate of 436 or carotenoid 497 nm bands. In FDCD of BBY, the (-) 451 nm band which is one of the main FDCD bands of RC of PS II is clearly seen.

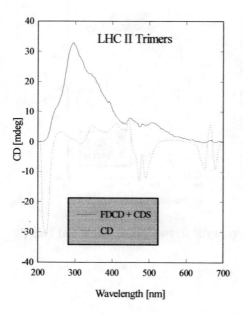

CD [mdeg]

LHC II Trimers

FDCD + CDS
CD

Wavelength [nm]

Figure 3. FDCD and circular differential scattering of LHC II trimers at 277 K

3.2 Other applications of our system

In a conventional dichrograph, CD of absorbance and circular differential scattering (CDS) are combined into the apparent CD signal. Moreover, another artifact can be raised from luminescence of sample. In our practise, we have met this problem when we titrated the LHC II by phosphatidylglycerol (see: Štys et al. Identification and role of the lipid binding site on LHC II, poster SY1-2-P109). The room temperature spectrum of trimeric LHC II is characterized by the pronounced peak at about 230 nm (Fig. 4). If this negative band comes from CDS, it must be also seen by the second detector in perpendicular direction. No such a negative signal was detected. In the second channel, we can clearly examine the FDCD of the aromatic aminoacid residues (290, and 340 nm) as well as the FDCD bands of chlorophylls, but no such a negative band. From this result, we can presume that the 230 nm CD band is not connected neither with the aromatic aminoacids in the LHC II polypeptides nor with the photosynthetic chromophores.

Acknowledgements

This work was supported by the Ministry of Youth, Education and Physical Training of the Czech Republic (grant project No. VS96085).

References

1 Berthold, D.A., Babcock, G.T. and Yocum, C.F. (1981) FEBS Lett. 134, 231-234
2 Burke, J.J., Ditto, C.L. and Arntzen, C.J. (1978) Arch. Biochem. Biophys. 187, 252-263
3 Nanba, O. and Satoh, K. (1987) Proc. Natl. Acad. Sci. USA 84, 109-112
4 Hemelrijk, P.W., Kwa, S.L.S., van Grondelle, R. and Dekker, J.P. (1992) Biochim. Biophys. Acta 1098, 159-166

PHOTOSELECTION EFFECTS IN EPR-DETECTED TRIPLET STATES OF PHOTOSYNTHETIC PIGMENT MOLECULES.

I.I. Proskuryakov, I.B. Klenina[*], I.V. Borovykh, P. Gast and A.J. Hoff, Huygens Laboratory, Leiden University, P.O.Box 9504, 2300 RA Leiden, The Netherlands, and [*]Institute of Basic Biological Problems, Pushchino 142292, Russia

Key words: bacterial reaction center, D1/D2-cyt b-559, EPR, polarised spectroscopy, time-resolved spectroscopy, triplet states

1. Introduction

Photosynthetic reaction centers (RCs) transform the energy of light absorbed by pigment molecules into chemical energy. RCs of bacteria are best characterised both functionally and structurally. In them, excitation of the primary donor P (dimer of bacteriochlorophylls) leads to a sequence of electron-transfer reactions to intermediate acceptor (bacteriopheophytin) and then to a quinone primary acceptor. When electron transfer past the intermediate acceptor is blocked, reverse electron transfer to oxidised P occurs, populating the triplet state of P (at T<77 K with almost-unity yield). Though the triplet states are the side products, their properties make them excellent probes of the RC photochemistry (for review see [1, 2]). Magnetophotoselection, MPS, or EPR detection of triplet states excited with polarised light, proved to be a powerful technique for the studies of the triplet states. Usually MPS is performed utilising cw EPR detection [1]. First application of time-resolved direct-detection EPR (DD-EPR) to the study of a synthetic aromatic molecule was reported in [3]. DD-EPR is advantageous in such studies, for it is essentially free from a drawback of cw EPR MPS, the latter's strong dependence on spin-lattice relaxation. This enables investigations in a wide temperature range. Here we report on the first observation of DD-EPR MPS of the primary donor triplet state of *Rb. sphaeroides* R26 RCs and D1/D2-cyt b-559 complexes of plant photosystem II.

2. Materials and Methods

Primary donor triplet state was studied in *Rb. sphaeroides* R26 reaction centers with prereduced or removed primary acceptor, and in D1/D2-cyt b-559 reaction center complexes of plant photosystem II. Quinone-depleted RCs were prepared as in [4]. D1/D1 complexes were isolated according to [5]. Prereduction of the quinone acceptor in Q_A-containing RCs was performed by freezing the sample in the light in the presence of 10 mM sodium ascorbate. A typical EPR sample contained about 66% v/v glycerol.

G. Garab (ed.), Photosynthesis: Mechanisms and Effects, Vol. V, 4365–4368.
© *1998 Kluwer Academic Publishers. Printed in the Netherlands.*

As the excitation light sources an optical parametric oscillator (for the range 750 - 920 nm) or a dye laser (oscillating at 670 - 720 nm) were used. For their pumping, a Continuum Surelite I laser was utilised. This resulted in excitation pulses of ca. 5 mJ with 4 ns duration. The OPO pulse was attenuated to ca. 0.5 mJ by means of a neutral density filter. The output of the lasers is vertically polarised. To rotate the polarisation plane of the exciting light, a system of two mirrors was utilised.

Time-resolved direct detection EPR measurements were performed on a home-built X-band spectrometer equipped with a Varian optical transmission cavity TE_{102}. The EPR signal was recorded using boxcar integration, sampling the output of the microwave mixer from 0.2 to 1.5 μs after the laser flash.

Spectral simulation of the triplet signals was performed within the assumption of selective T_o population in high magnetic field, and in angular integration photoselection with light polarised parallel or perpendicular to the magnetic field was taken into account.

3. Results and Discussion

3.1. Rb.sphaeroides reaction centers
Figure 1 shows the effect of 90° rotation of the excitation light polarisation plane on the ^3P spectrum of *Rb. sphaeroides* R26 RCs with reduced Q_A. Simulation of the spectra provides the values of the polar angles of $\mathbf{Q}_y$ relative to the triplet axes system of the primary donor, $\theta = 66°$, $\varphi = 62°$. These values qualitatively agree with the values obtained from LD-ADMR (reviewed in [6]). The existing difference between the two sets may be ascribed to the lower temperature in ADMR measurements, i.e., 1.5 K. Considerable deviation of $\mathbf{Q}_y$ from the x-y triplet axes plane most probably is related to the incomplete C_2 - symmetry of the primary donor dimer. Simulation of the ^3P spectra of *Rb. sphaeroides*

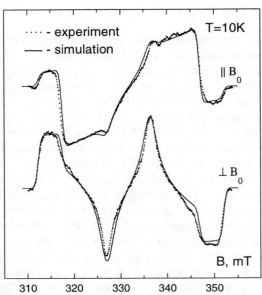

Figure 1. Experimental and simulated spectra of ^3P. The spectra were recorded under 900 nm excitation with polarisation plane parallel and perpendicular to the magnetic field.

R26 RCs with Q_A removed yields similar θ, φ values. These spectra do not change in the temperature range 10 - 100 K (illustrated in Fig. 2 for T = 10 - 40 K). In contrast, the ^3P spectra of Q_A - reduced RCs undergo a change in the shape between 10 and 30 K (Fig. 2), which then remains constant up to 100 K. This temperature dependence can not be simulated by variation of Q_y orientation. It relates to the initial polarisation of the spectrum, for the shape does not change when signal sampling period is changed in the range 0.2 - 2 μs. In the interval 10 - 30 K the relaxation rate of the $Q_A\,Fe^{2+}$ complex experiences a drastic increase. We thus ascribe the observed temperature dependence of the ^3P spectra in Q_A - reduced RCs to a deviation of the triplet state population from pure S - T_0 due to complex spin evolution in the precursor radical pair caused by interaction with the fast-relaxing complex.

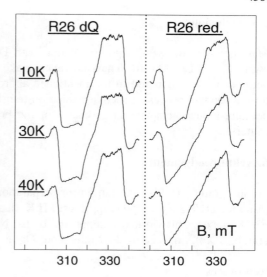

Figure 2. Temperature dependence of the ^3P spectra obtained with excitation polarisation plane parallel to the magnetic field at 900 nm.

3.2. D1/D2-cyt b-559 complexes

Photoselection effects are much weaker in the case of D1/D2-cyt b-559 complexes excited into the 680 absorption band. This fact may be due to the presence of several spectrally - overlapping pigment molecules connected by energy transfer. To diminish the effect of accessory chlorophylls absorbing at shorter wavelengths, excitation at 685 nm was utilised, and the spectra recorded with polarisation perpendicular and parallel to the magnetic field were subtracted. Figure 3 shows that such difference spectra are strongly dependent on temperature. As the D1/D2-cyt b-559 complexes do not contain Fe^{2+}, the cause of the dependence is different from that in Q_A - reduced bacterial RCs. We assume that in the case of D1/D2 complexes the reason is the formation at 10 K of a 'trapping state' which does not transfer energy to the primary donor. At 30 K energy transfer is feasible, changing the orientation of the effective Q_y in the triplet axes system.

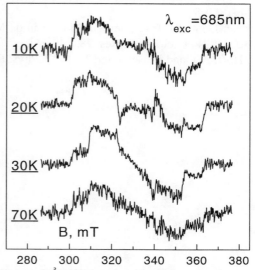

Figure 3. $^3P680$ 'perpendicular' - minus - 'parallel' spectra at several temperatures.

5. Conclusions

Measurements of magnetophotoselection with DD-EPR provides information similar to cw EPR and LD-ADMR. They, however, make possible measurements in a wider temperature range compared to the mentioned approaches. This enables studies of temperature-dependent processes such as excitation transfer, triplet hopping, and of possible temperature-induced changes in geometry of the primary donor of photosynthetic reaction centers.

6. Acknowledgement

We are grateful to Ms. S.J. Jansen for preparations of bacterial RCs, and to Ms. M. Germano for D1/D2-cyt b-559 complexes. I.B.K. thanks INTAS for travel support (grant No. 93-2849-ext). This work was supported by the Netherlands Foundation for Chemical Research (SON), financed by the Netherlands Organisation for Scientific Research (NWO).

References

1 Budil, D.E. and Thurnauer M.C. (1991) Biochim. Biophys. Acta 1057, 1-41
2 Angerhofer, A. (1991) in Chlorophylls (Scheer H., ed.) pp.945-991, CRC Press, Boca Raton
3 Ikoma, T., Akiyama, K., Tero-Kubota, S. and Ikegami, Y. (1991) J. Phys. Chem. 95, 7119-7121
4 Okamura, M.Y., Isaakson, R.A. and Feher G. (1975) Proc. Natl. Acad. Sci. USA 72, 3491-3495
5 van Leeuwen, P.J., Nieveen, M.C., van de Meent, E.J., Dekker, J.P. and van Gorkom, H.J. (1991) Photosynth. Res. 28, 149-153
6 Hoff, A.J. and Deisenhofer, J. (1997) Phys. Reports 287, 1-247

PHOTOSELECTION IN ESP SPECTRA OF PHOTOSYNTHETIC REACTION CENTERS

I.V. Borovykh[a], I.B. Klenina[b], P. Gast[a], A.J. Hoff[a] and I.I. Proskuryakov[a,b]
[a]Department of Biophysics, Huygens Laboratory, P. O. Box 9504, 2300 RA Leiden, The Netherland
[b]Institute of Basic Problems of Biology, RAS, Pushchino, 142292, Russia

Key words: bacterial reaction center, EPR, time - resolved spectroscopy, polarised spectroscopy, radical pair, *R. sphaeroides* R26.

1. Introduction
The first relatively stable stage of photoinduced electron transfer in reaction centers (RCs) of many photosynthetic bacteria is $P^+Q_A^-$, where P is the so - called primary donor (dimer of bacteriochlorophyll molecules), and Q_A - primary acceptor, a molecule of quinone. As early as 1977 [1], it was reported that when the normally present magnetic interaction between Q_A^- and paramagnetic Fe^{2+} - ion is disrupted, an electron spin - polarised (ESP) photoinduced signal may be detected using time - resolved EPR. This signal was subsequently attributed to the state $P^+Q_A^-$, which since 1987 [2] is considered as a spin - correlated radical pair (SCRP). Simulation of experimental ESP spectra of $P^+Q_A^-$ using the SCRP model, will provide information on the structural organisation of RC and on interactions between its cofactors. Recently we observed that the shape of the ESP signal of $P^+Q_A^-$ in Zn^{2+} - substituted RCs of *Rb. sphaeroides* R26 strongly depends on the wavelength of the excitation laser flash. This dependence arises due to the effect of photoselection, i.e., selective excitation of certain orientations of RCs in the sample by plane - polarised laser light. When taking photoselection into account, the SCRP model satisfactorily describes the quite different ESP signals for light polarised parallel and perpendicular to the magnetic field of the spectrometer.

2. Materials and Methods
RCs of *Rb. sphaeroides* R26 were isolated as described in [3]. The Fe^{2+} - ion was removed according to [4,5] and replaced by diamagnetic Zn^{2+} as described in [6]. As the excitation light source we used a Continuum Surelite I laser pumping an optical parameter oscillator (OPO). The excitation wavelength was varied from 560 to 1050 nm. The OPO light is known to be 95% vertically plane - polarised. To rotate the polarisation plane to horizontal, a system of two mirrors was used. The pulsewidth was about 4 ns at output power of 5 mJ. Time - resolved direct detection (DD) EPR measurements were performed on a home - built X - band spectrometer as described in [7]. The EPR signal was recorded using a boxcar integrator. The integration period was 0.1 - 1.5 µs after laser flash. The measurements reported here were performed at 30 K. Spectral simulation was done using the treatment developed by Hore et al. [2] for spin - correlated radical pairs. The anisotropy of Q_A^- **g** - tensor and linewidth, but not that of P^+, was accounted for.

4369

G. Garab (ed.), Photosynthesis: Mechanisms and Effects, Vol. V, 4369–4372.
© 1998 *Kluwer Academic Publishers. Printed in the Netherlands.*

3. Results and Discussion

Figure 1 shows the excitation wavelength dependence of the $P^+Q_A^-$ ESP spectra obtained with vertically - polarised light (polarisation perpendicular to the magnetic field). It was

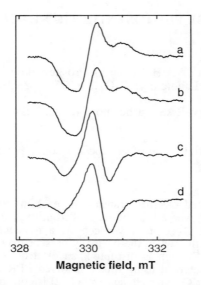

Figure 1. Direct-detection EPR spectra of the state $P^+Q_A^-$ in Zn - substituted reaction centers of *Rb. sphaeroides* R26 recorded at different excitation wavelengths: (a) - 900 nm, (b) - 850 nm, (c) - 810 nm, (d) - 790 nm. Other experimental conditions are described in the text. The spectra are normalised to the same maximum amplitude. Here and further absorption of micro-waves corresponds to the positive direction.

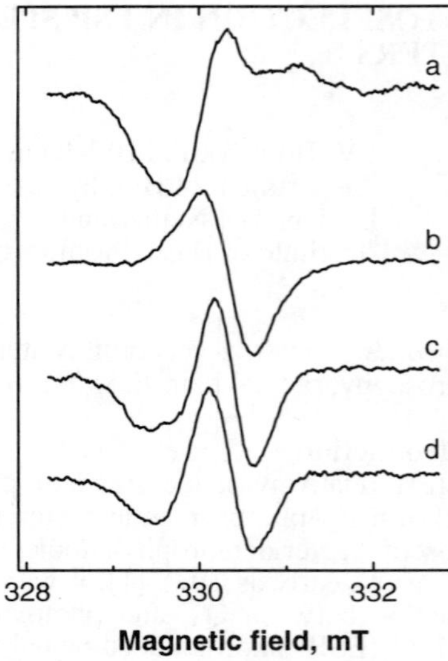

Figure 2. Normalised spectra of the state $P^+Q_A^-$ measured with 900 nm (a, b) and 820 nm (c, d) excitation with light polarised perpendicular (a, c) and parallel (b, d) to the magnetic field. Differences of curve (a) and the spectrum of Fig. 1, a illustrate the sensitivity to exact experimental conditions.

observed that the $P^+Q_A^-$ spectrum does not change when excited within the absorption band of the primary donor, and shows high degree of wavelength dependence within the bands of the monomeric bacteriochlorophylls (BChl). Three important points should be kept in mind when rationalising the experimental data: the $P^+Q_A^-$ EPR properties are anisotropic, the OPO light is 95% vertically polarised, and at changing excitation wavelength, varying ratios of P and monomeric BChl molecules of RCs with different transition dipole moments are excited. Thus, the reason for the observed spectral dependence may be the phenomenon of photoselection. This was proved in the experiments with rotating the excitation light polarisation plane parallel and perpendicular to the magnetic field of the spectrometer (Fig. 2). Rotation of polarisation plane results in drastic changes of the ESP signal of $P^+Q_A^-$ produced by exciting into the Q_y - 900 nm (Fig. 2 a,b and Fig. 3, left part) and Q_x - 610 nm (Fig. 3, right part) bands of the primary donor P, while a smaller effect is present under 820 nm light (Fig. 2 c,d). A much lower degree of photoselection when exciting into the 800 nm band is the result of overlapping absorption of two monomeric BChl molecules and probably, of P, with different orientations of their optical transition dipoles. The experimental spectra with excitation wavelength 900 nm and 610 nm were simulated using the SCRP approach [2], taking into account the photoselection effect. It was introduced by multiplying the spectra calculated for a given orientation of RC relative to the magnetic field by a factor proportional to the intensity of polarised light absorption. The results of simulation (Fig. 3) were obtained with the following values of polar angles relative to the Q_A^- **g** - tensor frame:

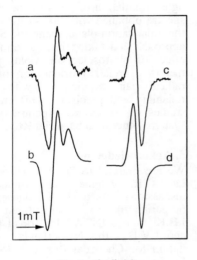 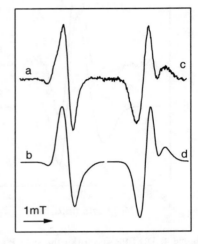

Figure 3. Experimental (a,c) and calculated (b,d) EPR spectra of $P^+Q_A^-$. Experimental spectra are recorded at excitation wavelength 900 nm (left) and 610 nm (right) with the polarisation plane perpendicular (a) and parallel (c) to the magnetic field. The spectra are arbitrarily normalised.

$\theta = 59°$, $\varphi = 51°$ for the $P^+Q_A^-$ dipolar axis, and $\theta = 7°$, $\varphi = 125°$ for the $\mathbf{Q_y}$ and $\theta = 7°$, $\varphi = 41°$ for the $\mathbf{Q_x}$ optical transition moment of the primary donor. The orientations of the dipolar axis and of the $\mathbf{Q_y}$ and $\mathbf{Q_x}$ calculated from these parameters differ from that determined from the X-ray structure [8] by 21° for dipolar axis, and by 25° and 3° for $\mathbf{Q_y}$ and $\mathbf{Q_x}$, respectively. Finally, we note that by carefully selecting the excitation wavelength, using glycerol - free samples, and placing a depolarising lightguide in front of the EPR cavity, it is possible to record a spectrum, which does not demonstrate any dependence on the orientation of the polarisation plane of the exciting light (Fig. 4). This non - photoselected spectrum is simulated quite well with the SCRP model for isotropic excitation. Photoselection leads to the dependence of the ESP signal of $P^+Q_A^-$ on excitation wavelength, which explains conflicting results obtained in different laboratories. The analysis of this dependence provides information on the orientation of the effective transition dipole when several pigments are excited, which is important for

4372

better understanding the structure of the RC absorption spectrum. Magnetophotoselection of the $P^+Q_A^-$ signal provide data on the structure of non - crystallised photosynthetic reaction centers.

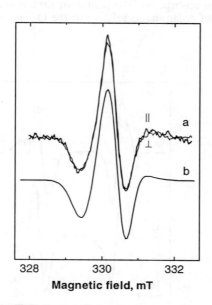

Figure 4. DD-EPR spectra of the state $P^+Q_A^-$ with light polarisation plane parallel and perpendicular to the magnetic field (a), and their simulation with the same set of parameters as used for calculating the spectra of Fig. 3 b,d (for more details see text). Experimental spectra are normalised in amplitude. Excitation wavelength 820 nm.

4. Conclusions

The observed wavelength dependence of the spin - polarised signal of $P^+Q_A^-$ in Zn^{2+} - substituted reaction centers of *Rb. sphaeroides* R26 is attributed to the photoselection effect of plane - polarised excitation laser light. Simulation of pairs of the $P^+Q_A^-$ spectra obtained with 90° - rotation of the polarisation plane, using the SCRP approach and taking into account the effect of photoselection, yields parameters of relative orientation of the radical pair components with high reliability and precision. This enables studies on structural organisation of systems other than bacterial RCs.

5. Acknowledgement

We are grateful to Dr. P.J. Hore and Dr. C.R. Timmel for stimulating discussions, and to Ms. S.J. Jansen for preparing the Zn^{2+} - substituted RCs I.B.K. thanks INTAS for travel grant and I.V.B - the Netherlands' Foundation for Chemical Research (SON) for financial support.

References

[1]. A.J. Hoff, P. Gast, J.C. Romijn, (1977), FEBS Lett. 73, pp. 185 - 190

[2]. P.J. Hore, D.A. Hunter, C.D. Mckie, A.J. Hoff, (1987) Chem. Phys. Lett. 137, pp. 495 - 500 and G.L. Closs, M.D.E. Forbes and J.R. Norris, (1987), J. Phys. Chem. 91, 3592 - 3599

[3]. G. Feher and M.Y. Okamura, in: The photosynthetic bacteria, eds. R.K. Clayton and W.R. Sistrom (Plenum Press, New York, 1978)

[4]. J.S. van den Brink, R.J. Hulsebosch, P. Gast, P.J. Hore and A.J. Hoff, Biochemistry (1994), 33, pp. 13668 - 13677

[5]. D.M. Tiede and P.L. Dutton, (1981), BBA, 637, pp. 278 - 290

[6]. R.J. Debus, G. Feher and M.Y. Okamura, (1986),Biochemistry, 25, pp. 2276 - 2287

[7]. M.K. Bosch, Ph.D. Thesis, Leiden University, Netherlands, (1995)

[8]. U. Ermler, G. Fritsch, S.K. Buchanan and H. Michel, (1994), Structure 2, pp. 925 - 936

NANOSECOND PROTEIN DYNAMICS OF THE RC AND LH COMPLEXES AS MEASURED BY COHERENT INELASTIC NEUTRON SPIN-ECHO SPECTROSCOPY.

Gall A.[1,2], Seguin J.[1], Bellissent-Funel M-C.[2] and Robert B.[1]

[1] *Section de Biophysique des Protéines et des Membranes,* DBCM/CEA & URA2096/CNRS, C.E. Saclay, 91191 Gif-sur-Yvette Cedex, France.

[2] *Laboratoire Léon Brillouin* (CEA-CNRS), CEA-Saclay, 91191 Gif-sur-Yvette Cedex, France.

Key words: bacterial reaction centre, light harvesting complexes, membrane protein, modelling, time-resolved spectroscopy.

1.Introduction

In light of the x-ray crystal structures of the photochemical reaction center (RC) and its peripheral light-harvesting (LH2) complex, and the neutron diffraction experiments on their accompanying detergent rings (1, R.J.Cogdell, personal communication), we have purified these proteins in order to measure and compare their coherent dynamics in the nanosecond time-domain using inelastic neutron spin-echo (NSE) spectroscopy (2). We chose to investigate the RC and LH2 as they are both integral membrane proteins, have x-ray crystal structures with resolutions superior than 2.5 Å (3,4), and contain pigment-cofactors that serve as molecular markers which monitor protein structure and conformation. A further advantage in comparing the dynamics of these two proteins is that, although having near identical numbers of amino acids (ca. 850), they have widely differing tertiary and quaternary structures. In summary, LH2 consists of 18 independent transmembrane spanning α-helices located within 9 identical α/β-heterodimer subunits that forms a ring-like structure in the membrane. However, the RC has three much larger sub-units (L,M & H) with less overall symmetry, and is more globular-like. For indepth reviews of the RC and LH2 structures see refs. (3) and (4).

Inelastic neutron spectroscopy exploits the isotopic specific coherent neutron scattering crossections (σ^{coh}), incoherent neutron scattering crossections (σ^{inc}), neutron absorption scattering crossections (σ^{abs}) and the energy transfer, $\hbar\omega$, resulting from neutron-protein

G. Garab (ed.), Photosynthesis: Mechanisms and Effects, Vol. V, 4373–4376.

interactions. The two principle quantities measured in inelastic experiments are: (i) the energy transfer, $\hbar\omega$, between the initial, E_0, and final, E, energies of the neutron

$$\hbar\omega = E - E_0 = \frac{\hbar^2}{2m}\left(k^2 - k_0^2\right)$$

where k_0 and k are the corresponding wavevector moduli, and (ii) the scattering vector modulus, $\vec{Q}$, corresponding to the momentum transfer: $\hbar\vec{Q} = \hbar\vec{k} - \hbar\vec{k}_0$. In NSE spectroscopy, the final polarisation of the scattered neutrons, $P(Q,H)$, is measured as a function of the applied magnetic field, H. The magnetic field is converted to a time variable by a Larmor precession equation resulting in a window of Fourier time measurements (5). $P(Q,H)$ is proportional to the real part of the intermediate scattering function, $I(Q,t)$, where $I(Q,t)$ is the signal intensity at time, t:

$$I(Q,t) \propto \int \cos(\omega t).S(Q,\omega)d\omega$$

If the dynamic structure factor, $S(Q,\omega)$, has a lorentzian shape then its Fourier transform, $I(Q,t)$, can be expressed as an exponential decay, $\exp^{(-t/\tau)}$, where $1/\tau = D_{eff}Q^2$, D_{eff} is the effective diffusion coefficient. For a detailed synopsis of NSE methodology see (5), and references therein.

In this work, we present inelastic neutron spin-echo data showing that it is feasible to obtain information on coherent nanosecond dynamics for membrane proteins where only labile hydrogen-deuterium exchange occurs. This opens up the way for computer simulations on the slow collective breathing-actions of membrane proteins.

2. Procedure

The RC and LH2 complexes were purified based on the protocols used for x-ray crystallography trials; polypeptide compositions were verified by SDS-PAGE and protein integrity by absorption spectroscopy. Deuterium-exchanged proteins (40mg/ml) were isolated and equilibrated in a fully deuterated buffer (10mM D_{11}-Tris.Cl, pD 8.2) containing detergent (RC, D_{28}-n-Octylb-D-Glucopyranoside; LH2, n-Dodecyl-N,N-Dimethylamine-N-oxide). NSE data were collected using the spectrometer MESS (G3.2, *Orphée Reactor*, CEA-Saclay). Spectra were recorded at 288K at Q=0.06 Å^{-1} (λ=6 Å) and at Q=0.20 Å^{-1} (λ=5 Å) giving a maximum Fourier time of 17ns. Reduced background scattering levels in the sample chamber were provided by helium gas as it has reduced σ^{scat} and σ^{abs} than ordinary air. Graphite was used to determine the resolution of the spectrometer and to normalise the protein NSE echoes.

3. Results and Discussion

The polypeptide composition as revealed by SDS-PAGE, and room-temperature absorption spectroscopy taken after the NSE experiment confirm that the purified proteins were not denatured by the experimental conditions (data not shown).

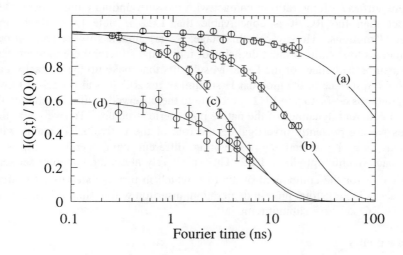

Figure 1. The Q-dependence of normalised NSE signals for (a) RC, Q=0.06 Å^{-1}, τ=138 ns (b) LH2, Q=0.06 Å^{-1}, τ=17.4 ns (c) RC, Q=0.20 Å^{-1}, τ=4.7 ns (d) LH2, Q=0.20 Å^{-1}, τ=10.2 ns. T=288K. The solid lines represent the best-fits found for the intermediate scattering functions using single exponential decay functions, see text.

Q-dependence on Protein dynamics: Shown in figure 1 are the intermediate scattering functions for the two proteins. It is evident that the nanosecond global motions of the RC and LH2 complexes are somewhat different. Contributions from the detergent microemulsions can be discounted as no NSE echoes were observed for the detergent-buffer samples for either value of Q (data not shown). Remembering that $1/\tau = D_{eff}Q^2$, at Q=0.06 Å^{-1} the effective diffusion coefficient (D_{eff}) of LH2 is at least 5 fold than that for the RC. While at Q=0.20 Å^{-1} the RC (τ≈5 ns) appears to be more dynamically fluid than LH2 (τ≈10 ns) within our time window of 0.25-17 ns. Moreover, as indicated by extrapolation to t=0, at Q=0.20 Å^{-1} LH2 contains significantly higher levels of sub-nanosecond dynamics (i.e. NSE echo ≈ 0.6, figure 1d) than the RC for the same scattering vector (figure 1c). Therefore, we conclude that the effective diffusion coefficient is both dependent on protein structure and scattering vector.

The influence of tertiary/quaternary structure on protein dynamics: In this work we have shown that it is possible to measure coherent nanosecond protein dynamics, observe differences based on membrane protein tertiary/quaternary structures and not solely on the numbers of amino acids present.

When compared to the RC, the numerous single transmembrane-spanning polypeptides belonging to LH2, in conjuction with its oily interior (6), result in a structure that has a larger protein-surface/volume ratio in contact with the surrounding environment. Being less compact than the RC it is conceivable that LH2 is able to facilitate more subnanosecond motions. However at this stage, we are unable to decisively allocate specific dynamics to specific protein domains or individual structural components, such as the transmembrane helices of the L and M subunits that make up the co-factor cage, or the hydrophilic surface of the globular H-subunit of the RC. It is also unclear to what extent the numerous co-factors in LH2 contribute to the observed motions, and how they affect overall coherent dynamics of the native nonomeric structure. To overcome these uncertainties we are presently investigating the role of the individual domains within these proteins, e.g. by selectively deuterating different components, such as the monomeric bacteriochlorophylls in LH2. This selectively alters the coherent scattering crossection (σ^{coh}) for the component of interest, which in turn isolates its NSE signal. Other approaches include total protein deuteration, insertion within lipid bilayers, and in the case of the RC selective subunit removal.

Acknowledgements

A.G. is grateful to the Royal Society (London) for an ESEP Fellowship. The authors would like to express their gratitude to Dr. A. Brûlet (LLB) and Mr. O. Choquet (LLB) for assistance during NSE data acquisition.

References

1 Roth, M., Arnoux, B., Ducruix, A., and Reiss-Husson, F. (1991). Biochemistry. 30, 9403-9413

2 Mezei, F. (1972) Z.Physik 255, 146-1604

3 Ermler, E., Fritzsch, G., Buchanan, S.K., and Michel, H. (1994). Structure 2, 925-936

4 Cogdell, R.J., Isaacs N.W., Freer, A.A., Arrelano, J., Howard.T.D., Papiz, M.Z., Hawthornthwaite-Lawless, A.M. and Prince, S.M., (1997). Prog. Biophys. molec. Biol., 68, 1-27

5 Lecture Notes in Physics, Vol. 128: Neutron Spin Echo. (1980) (Mezei, F., ed.), pp.1-253 Springer-Verlag, Berlin, Germany

6 Freer, A.A., Prince, S.M., Sauer, K., Papiz, M.Z., Hawthornthwaite-Lawless, A.M., McDermott, G., Cogdell, R.J. and Isaacs N.W. (1996). Stucture, 4, 449-462

SEPARATION AND CHARACTERISATION OF WHEAT PROTEIN PSII BY HIGH RESOLUTION CHROMATOGRAPHY TECHNIQUES

[1]Zolla L., [2]Huber C. G. ,[1]Timperio A.M., [1]Bianchetti M. and [3]Corradini D.
[1]Department of Environmental Science,   University of Tuscia, Viterbo, Italy
[2]Institute Analytical Chemistry, Leopold-Franzens University, Innsbruck, Austria
[3]Chromatography Institute, CNR Montelibretti, Roma, Italy

Keywords: *HPLC, mass spectrometry, LHC II, photosystem II, chromatography, environmental stresses.*

1. Introduction

Up to now the available methods for separation and identification of antenna proteins are both expensive and technically demanding while traditional approaches by SDS-PAGE are not only cumbersome but also rather ineffective for evaluating differences in the relative quantity of each components unless time consuming antibody titration is used.   It seems, thus, necessary to develop a rapid and sensitive method to identify and determine quantitatively the several components of two photosystems. With this aim in mind, the electrophoretic migration behaviour of three closely related hydrophobic intrinsic membrane proteins of the spinach photosystem II light-harvesting complex (LHC II) was investigated in free solution capillary electrophoresis with running electrolyte solutions containing either anionic, zwitterionic or non-ionic detergents and by reversed phase HPLC ([1];[2]) In this study we present an HPLC method for the resolution of both the protein components major and minor antenna of wheat upon direct injection of BBY particles into the column. The resolution of these proteins directly in the solubilized thylakoid membranes allows estimating with accuracy the relative subunit stoichiometry of photosystem II.

2. Procedure

The experiments were carried out on two different liquid chromatograph units: a Beckman (Fullerton, CA, USA) System Gold system and a Perkin Elmer (Norwalk, CT, USA) Model 200 C system having a Mode l785 A UV detector, and a Model LC 240 Florescence detectors connected in series. Samples were loaded onto the column by a Model 7125NS-005 Rheodyne (Cotati, CA, USA) sample injection valve with either a 50-μLor a 20 μL sample loop.
The experiments were performed using two Vydac (The Separation Group, Hesperia,

G. Garab (ed.), Photosynthesis: Mechanisms and Effects, Vol. V, 4377–4380.
© 1998 Kluwer Academic Publishers. Printed in the Netherlands.

CA, USA) Protein C-4 columns of either 250 x 4.6 mm I.D. or 250 x 10 mm I.D., both containing 5-μm porous butyl silica. All solutions were filtered through a Millipore (Milan, Italy) type FH 0.5-μm membrane filter and degassed by bubbling with helium before use.

SDS-PAGE electrophoresis

The fractions collected from the semi-preparative chromatographic separation were lyophilised, dissolved in 120 mM TRIS/HCl pH 8.45 buffer, containing 120 mM DTT, 5 M urea and 4% (w/v) SDS, and then analysed by SDS-PAGE urea, according to the method reported by Shagger [3].

HPLC

The Vydac C-4 columns were pre-equilibrated with 38% (v/v) aqueous acetonitrile solution containing 0.1% (v/v) TFA and samples were eluted by gradient consisted of a first linear gradient from 38 to 55.4% (v/v) acetonitrile in 22 min, followed by 3 min isocratic elution with the eluent containing 55.4% acetonitrile, followed by a second gradient segment from 55.4 to 61.8% (v/v) acetonitrile in 8 min and by a third gradient segment from 61.8 to 95% acetonitrile in 1 min. The last gradient segment up to 95% acetonitrile was used in order to ensure that eventual hydrophobic contaminants of the PS II antenna system were eluted from the column.

Mass spectroscopy

The HPLC-ES-MS experiments were carried out with a Model Rheos 400 low-pressure gradient micro pump (Flux Instruments, Karlskoga, Sweden), a Model S4100 column oven (Sykam, Gilching, Germany), a Rheodyne Model 7520 microinjector equipped with a 0.5-μL sample loop, a Model Linear UV-VIS 200 variable wavelength detector (Linear Instruments, Fremont, CA, USA) equipped with a 3mm/1.2-μL flow cell, and a PC-based data system from Gynkotek (Germering, Germany). ESI-MS ([4]) was performed on a Finnigan MAT TSQ 7000 triple quadrupole mass spectrometer (Finnigan MAT, San Jose, CA, and USA) equipped with the electrospray ion source.

2. Results and Discussion

Photosystem II membranes were isolated from thylakoid membranes of wheat leaves by Triton X-100 extraction as reported in the experimental section.. This material, corresponding to the BBY preparation, was then subjected to sucrose-gradient ultracentrifugation in order to isolate the protein components of the minor and major PS II antenna system as well as of the reaction-center complexes. The sucrose gradient ultracentrifugation produced 3 different green bands the second of which (S2) contained a mixture of the major and minor PS II antenna system, whereas the third green band (S3) contained mainly the protein components of the major PS II antenna system.

The optimal separation of the protein components of the PS II antenna system was obtained by a multisteps gradient which was slightly different in shape on the two HPLC units employed to perform this study, as reported in the Material and Methods section.

Figure 1A reports the chromatogram obtained upon injection of the material harvested from the second band of the sucrose-gradient ultracentrifugation, containing a mixture of the protein components of both the major and the minor PS II antenna system. Figure1B shows that upon direct injection of BBY to the column the protein components of the PS II, the major and minor antenna system are well resolved without interference from the other proteins components of PS II. Moreover the presence of the latter does not affect resolution and retention times .

The identification of the protein components of the major antenna complex of PS II resolved by RP-HPLC was performed by various SDS-PAGE methods, different in buffer composition, followed by either Coomassie and silver staining, and by mass spectroscopy.

The fractions collected from semi-separative chromatographic separation were lyophilized and then analyzed by SDS-PAGE urea, according to the method reported by Schagger [3]. Since. in TRIS/tricine electrophoresis, antenna proteins migrate in the following order of increasing mobility: CP29>CP26>LHCII>CP24 and, within LHCII, Lhcb1>Lhcb2>Lhcb3, from the results obtained by SDS-PAGE (data not shown) we are able to assign a name to each peak: peak 1 does not contain protein components; peak 2, CP24 (Lhcb6); peak 3, the Lhcb2 component of LHCII; peak 4, Lhcb1 components of LHCII; peak 5, Lhcb3; peak 6, CP29 (Lhcb4); peak 7, CP26 (Lhcb5).

In order to confirm the above assignation, we have performed a molecular mass determination of proteins contained in each peak by the combined use of a microbore HPLC column coupled on-line to a mass spectrometer equipped with electrospray ion source (ESI-MS).

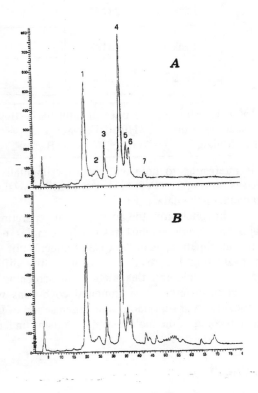

Figure 1: RP-HPLC separation of the second band of sucrose gradient (panel A) and BBY from wheat (panel B). See the text for details.

Table 1 reports the molecular masses obtained by the deconvolution of the ESI-MS spectra in comparison with the values expected on the basis of the gene and protein sequences available on the databanks.

	MOLECULAR MASS (MEASURED)	MOLECULAR MASSES (Expected) from OTHER SPECIES
Peak 2	22800	22813 (S)
Peak 3	24675	24834 (T)
Peak 4	24827	25026(M) 25041(P) 24969(M) 24906(P) (P)
Peak 5	24288	24285 (B) 24308 (T)
Peak 6	28382	27804 (B)
Peak 7	27100	26607 (M) 27642 (T)

Table 1: Molecular masses of light harvesting proteins present in the BBY of wheat, determined by HPLC-ESI-MS
M= Maize; T=Tomato; B=Barley ; P=Petunia; S= Spinach

It is interesting to note that the experimentally obtained values of the molecular masses are in excellent agreement with the values expected on the basis of the gene available on the databanks.

In conclusion, the possibility of separating all protein components of the PSII major and minor antenna system in samples not subjected to the sucrose-gradient ultracentrifugation, as in the case of injection of BBY, is expected to be advantageous for evaluating the relative content of the different protein components of PSII. Moreover employing this method in screening photosynthetic mutants and plants adapted to different environmental conditions will be useful on the elucidation of composition and supramolecular organisation of LHC II and will possibly increase the understanding of the molecular mechanisms underlying the physiological adaptations.

References

[1]   Zolla, L., Bianchetti, M., Timperio, A.M., Mugnozza, G.S. and Corradini, D. (1996) Electrophoresis 17, 1597-601.
[2]   Zolla, L., Bianchetti, M., Timperio, A.M. and Corradini, D. (1997) J. Chromatography A 196, 142-150
[3]   Schagger H. and von Jagow G. (1987) Anal. Biochem. 166, 368-379
[4]   Rotham, L.D. (1996) Anal. Chem. 68, 587R.

IMAGING THE NATIVE COVALENT STATE OF THYLAKOID PROTEINS BY ELECTROSPRAY-IONIZATION MASS SPECTROMETRY

Julian P. Whitelegge, K.F. Faull, C.B. Gundersen and S.M. Gómez
Department of Chemistry & Biochemistry, Pasarow Mass Spectrometry
Laboratory, University of California, Los Angeles, CA 90095, USA.
jpw@chem.ucla.edu

Key words: phosphorylation D1 D2 CP43 CP47 palmitoylation

1. Introduction

The ideal analysis of any protein includes a mass spectrum of the intact molecule to define the native covalent state and its heterogeneity. A versatile procedure for effective electrospray-ionization mass spectrometry (ESI-MS) of intact intrinsic membrane proteins purified using reverse-phase chromatography in aqueous formic acid/isopropanol has been developed (1). The spectra of spinach D1 and D2 as well as bacteriorhodopsin achieved mass measurements that were within 0.01 % of calculated theoretical values, setting a benchmark standard for analysis of such molecules (1). Typically, genome data needs manipulation before agreement with calculated masses is achieved. DNA sequencing errors, post-transcriptional, post-translational modifications as well as protein damage must all be considered. Spectra frequently reveal lesser quantities of other molecular species that can usually be equated with covalently modified sub-populations of the dominant proteins. Here we present spectra of the larger Photosystem 2 (PS2) subunits from pea illustrating the accuracy and resolution afforded by ESI-MS.

2. Procedure

2.1 Sample Preparation
Leaves from two week old greenhouse grown pea (*Pisum sativum* v. Alaska) plants were used for preparation of PS2 enriched membranes (2). Samples were prepared by acetone precipitation prior to dissolution in 60 % HCOOH.

2.2 Reverse-phase chromatography
Reverse-phase chromatography was performed as previously described (1,2). The poly(styrene-divinylbenzene) coplymer (Polymer Lab.s PLRP/S; 2.1 x 150 mm; 5 μm x 300 Å) stationary phase was eluted at a flow rate of 100 μl/minute. Smaller thylakoid polypeptides were eluted with a standard TFA, H_2O, CH_3CN gradient. To elute the larger PS2 polypeptides the column was equilibrated in 95 %A, 5 %B, prior to linear gradient elution to 100 % B over 55 minutes (A, 60% HCOOH; B, 2-propanol).

2.3 Electrospray-ionization Mass Spectrometry
Mass spectra were recorded on a Perkin Elmer Sciex API III triple-quadrupole mass spectrometer with an Ionspray™ source. The instrument was scanned with a step size of 0.3 in all experiments. The computations of measured protein molecular weight were performed using MacSpec 3.3 software (Sciex). Calculated molecular weights were generated using PeptideMap 2.1 (Sciex).

G. Garab (ed.), Photosynthesis: Mechanisms and Effects, Vol. V, 4381–4384.
© 1998 *Kluwer Academic Publishers. Printed in the Netherlands.*

3. Results and Discussion

The psbA, psbB, psbC and psbD gene products from pea thylakoids have been imaged using ESI-MS. The masses measured using ESI-MS were compared to masses calculated for the pea polypeptides based upon genome data and previously described post-translational modifications.

Table I. Masses of Larger Pea Photosystem 2 Polypeptides by ESI-MS.

Polypeptide (gene)	Molecular Mass (Measured[1])	Molecular Mass (Calculated[2])	Molecular Mass (difference)
D1 (psbA)[3]	38037.0	38033.6	+ 3.4
D2 (psbD)[4]	39443.6	39479.5 (39437.5)	- 35.9 (+ 6.1)
CP43 (psbC)[5]	50217.8	50204.9	+ 12.9
CP47 (psbB)[6]	55912.0	55902.0	+ 10.0

1. The average of two independent measurements is presented. Mass accuracy of 0.01 % is typically achieved under the conditions used. 2. Average masses based upon natural isotopic abundance. 3. D1 was translated from pea chloroplast psbA [Oishi, K.K., Shapiro, D.R., Tewari, K.K. (1984) Mol Cell Biol 4, 2556-2563] and processed at the N-terminus by Met1 removal and N-acetylation [Michel, H., Hunt, D.F., Shabanowitz, J., Bennett, J. (1988) J Biol Chem 263, 1123-1130] and at the C-terminus by cleavage after Ala344 [Takahashi, Y., Nakane, H., Kojima, H., Satoh, K. (1990) Plant Cell Physiol 31, 273-280]. 4. D2 was translated from pea chloroplast psbD [Bookjans, G.B., Stummann, B.M., Rasmussen, O.F., Henningsen, K.W. (1986) Plant Mol Biol 6, 359-366] and processed at the N-terminus by Met1 removal with N-acetylation [Michel et al, 1988]. Non-acetylated form in parentheses. 5. CP43 was translated from pea psbC [Rasmussen, O.F., Bookjans, G., Stutmann, B.M. and Henningsen, K.W. (1984) Plant Mol. Biol. 3, 191-199] and processed at the N-terminus removing amino acids 1-14 prior to N-acetylation [Michel et al., 1988]. 6. CP47 was translated from pea chloroplast [Lehmbeck, J., Stummann, B.M. and Henningsen, K.W. (1989) Physiol.Plant. 76, 57-64]. Met1 and Gly2 were removed prior to N-acetylation to obtain the closest possible mass match.

The measured mass of D1 was within 0.01 % of the value calculated based upon the pea psbA sequence with post-translational processing as for the spinach protein. The masses of the other three proteins did not fall within 0.01 % of calculated values and it therefore necessary to consider the origin of the difference. Since our measurements usually achieve 0.01 % accuracy, as shown for D1, we infer incorrect mass calculation based on the wrong primary structure assignments for the other three proteins. The largest difference was observed for D2 which measured 35.9 Da smaller than the calculated value. Agreement is closer if the N-terminus is not acetylated though an internal sequence difference could also account for the difference. Small sequencing errors are quite common and post-transcriptional modification of RNA results in primary sequence modification in some thylakoid proteins (3). Consequently, the primary structure of D2 requires more detailed investigation to localize and define the site(s) of deviation. The measured masses of CP43 and CP47 were higher than calculated values. One possibility for the difference is partial methionine oxidation on the experimental protein. The larger the protein the more methionine residues it will tend to contain thereby raising the probability that a single molecule will carry at least one oxygen on methionine. Thus the mass centroid may be shifted upward. If 16 Da are added to the calculated masses for each of CP43 and CP47 a much closer agreement is observed with the measured values. The molecular weight spectrum of CP47 (Figure 1B) indeed shows evidence for such a centroid shift in the form of the shoulder on the left-hand side of the peak. Oxidation may explain the somewhat elevated mass measured for D2 in a previous study (4). Increasing machine resolution, by decreasing step size for example, will allow distinction of non-oxidized from oxidized species in order to resolve this problem.

Figure 1. Electrospray-ionization MassSpectrum of the CP47 apoprotein.
A. Mass spectrum of CP47. ESI-MS leads to the generation of a collection of multiply charged ions varying in the number of charges carried. B. Molecular weight reconstruction of uncharged CP47. Y-axis is Relative Intensity (%).

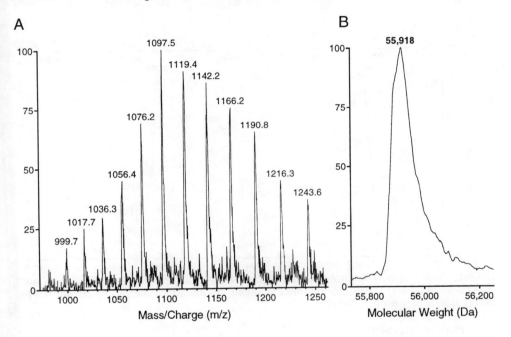

The resolution afforded by ESI-MS provides an excellent means of observing molecular heterogeneity of the proteins. CP47 (Figure 1) showed little heterogeneity when compared to the other proteins (Figure 2), all of which showed evidence of a sub-population of molecules 80 Da larger than the most abundant species. This mass difference corresponds to a single phosphorylation which is known to occur at the N-terminus of D1, D2 and CP43 (see Table I). In another experiment using spinach reaction centers (2), a small population of D1 was observed to be 238 Da larger than the main species. This most probably corresponds to the palmitoylated form of the protein (5) indicating that this modification is internal and not associated with the excised C-terminus.

Figure 2. Molecular Heterogeneity of the larger PS2 Polypeptides.
Molecular weight reconstructions of PS2 polypeptides. Molecular weight (Da) is on the X-axis. A. Pea D1. B. Pea D2. C. CP43. D. Spinach D1.

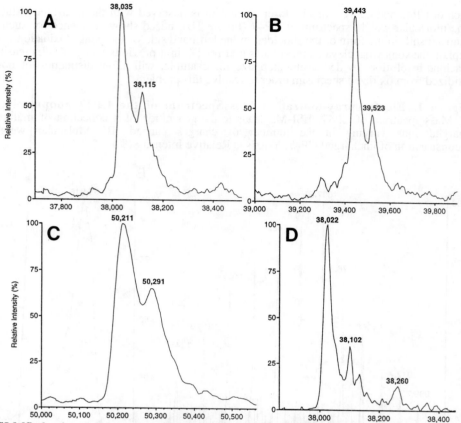

ESI-MS clearly provides a powerful means for the physiologist and biochemist to monitor phosphorylation and relate it to the other covalent modifications that occur during the life of reaction-center proteins *in vivo* (6).

4. Acknowledgments.

We thank the NIH and the W.M. Keck Foundation for their financial support. This work was presented with assistance from the Katherine E Steinbeck Memorial Fund.

References

1 Whitelegge, J.P., Jewess, P., Pickering, M.G., Gerrish, C., Camilleri, P. and Bowyer, J.R. (1992) Eur. J. Biochem. 207, 1077-1084
2. Whitelegge, J.P., Gundersen, C.B. and Faull, K.F. (1998) Protein Sci. 7, 1423-1430.
3. Bock, R., Hagemann, R., Kossel, H., Kudla, J.(1993) Mol.Gen.Genet. 240, 238-44.
4. Sharma, J., Panico, M., Barber, J., Morris, H. (1997) J. Biol. Chem. 272, 33153-7.
5. Mattoo, A.K. and Edelman, M. (1987) Proc. Natl. Acad. Sci. USA 84, 1497-1501.
6. Whitelegge, J.P. (1996) in The Handbook of Photosynthesis (Pessarakli, M., ed.) pp. 241-256, Marcell Dekker, New York, USA.

25. Educational aspects

CONGRESS ON PHOTOSYNTHESIS

XIth INTERNATIONAL

BUDAPEST ✳ 1998

PHOTOSYNTHESIS AND THE WORLD WIDE WEB

Larry Orr[1] and Govindjee[2]
[1]Arizona State University, Tempe, AZ 85287-1604; [2]University of Illinois,
265 Morrill Hall, 505 S Goodwin Avenue, Urbana, IL 61801-3707, USA
(E-mail:gov@uiuc.edu)

1. Introduction

The *World Wide Web* is an important resource for public awareness and for educating all citizens of the world including its political leaders, granting agencies, pre-university and university students, as well as teachers and researchers in all fields of photosynthesis. The Web began in 1989 when Tim Berners-Lee proposed a hypertext project as a means for scientists at the European Particle Physics Laboratory, CERN (Conseil Européen pour Recherche Nucleaire), located in Geneva, Switzerland, to exchange data with other scientists around the world (1,2) who were linked by a plethora of wide-area networks using a newly standardized communications protocol, TCP/IP (Transmission Control Protocol/Internet Protocol). The various wide-area networks multiplied and developed into what we now know as the Internet, not a single network at all, but a matrix of networks with numerous redundant connections (3). At first, only scientific data was exchanged via Web servers on the Internet, though it soon became possible for persons on one computer hooked to the Internet to access the information located on another computer that acted as a Web server. All that was needed was an Internet connection and the URL (Universal Resource Locator) address of the server with the information. It was also possible to create text documents containing embedded URL addresses called hypertext links. By selecting a link via keyboard command or, more commonly, by a mouse click, the server serving as host for the text document would route the user to the server that actually contained the material referenced in the link. This material was usually displayed as lines of plain text. In 1993, the National Center for Supercomputing Applications (NCSA) at the University of Illinois at Urbana-Champaign released *Mosaic*, allowing Web text documents to be displayed graphically on computer monitors (4).

The scientific community embraced Mosaic and other simple Web readers, called web browsers. The capabilities of the web browsers were quickly grasped by most universities and many companies. The Web was suddenly accessible by anyone with an Internet connection and a browser. Information in the form of text, pictures, sound, and movies became common as the developers of Mosaic went on to form Netscape Communications, Inc., which produced its commercial product, *Navigator*. Navigator was a hit and the Web exploded into the "entity" we know today. Microsoft's *Internet Explorer* soon followed and the general public became involved in the Web in a big way. Commercial corporations were quick to establish web sites advertising their products or posting information that the public might want. Universities and research organizations began expanding their small informational sites to larger sites containing massive amounts of information, including personnel directories, general and specific information for all facets of the organizations, course catalogs, registration forms, and on-line classes.

G. Garab (ed.), Photosynthesis: Mechanisms and Effects, Vol. V, 4387–4392.
© 1998 *Kluwer Academic Publishers. Printed in the Netherlands.*

Photosynthesis web sites began appearing in late 1994. Some were large sites devoted to covering the entire field of photosynthesis and related scientific areas. Others were sites containing information about specific areas of study such as reaction centers or photosystems. Individual investigators also began developing their own web sites filled with information about their interests and laboratories. On-line tutorials and entire courses began to appear. In this review, we will present relevant information on the availability of information on the World Wide Web under five major headings: (1) Group Sites; (2) Individual Researchers; (3) Educational Sites; (4) Books and Journals; (5) Other Sites. An expanded version of this paper (including a major section on "Searching the Web") is available on the Web at: http://www.life.uiuc.edu/govindjee/photoweb/ as well as at http://photoscience.la.asu.edu/photosyn/photoweb/. Due to length restrictions, it is impossible to include every worthy web site in this review. Thus, we will highlight a few of the most well known sites in each category. Other sites will be covered in the Web version of this review. Our sincere apologies to anyone whose site we have overlooked.

2. Group Sites

One of the earliest photosynthesis sites, the Arizona State University (ASU) Photosynthesis Center site, appeared in 1995 and is maintained by one of us (L.O.). This award-winning site, http://photoscience.la.asu.edu/photosyn/, is very comprehensive, showcasing not only the work of the Center, but also providing original material and numerous links to other sites devoted to individual and group photosynthesis research and to sites of interest to researchers, educators, students and the general public. The University of Illinois at Urbana-Champaign (UIUC) hosts several important sites. A site by one of us (G) (http://www.life.uiuc.edu/govindjee/) is much more than an individual site as it includes information on a variety of topics, including course web pages, a major tutorial/essay on "The Photosynthetic Process" (by J. Whitmarsh and Govindjee), movies, photos and comments on Robert Emerson, Eugene Rabinowitch, William Arnold, Lou Duysens and Stacy French, as well as a tutorial on "Photosynthesis and Time" and other teaching materials such as slides. UIUC is also the home of a wonderful and highly educational site of A.R. Crofts (http://ahab.life.uiuc.edu/index.html; we will return to this site again). The Photosynthesis Research Unit, affiliated with the USDA Agricultural Research Service, covers many areas of photosynthesis crop research at http://www.life.uiuc.edu/pru/homepage/text.html. The Theoretical Biophysics Group at UIUC presents an excellent site (http://www.ks.uiuc.edu/Research/psu/psu.html) devoted to "Light-Harvesting in Bacterial Photosynthesis." The Plant Cell Biology group at Lund University, Sweden, http://plantcell.lu.se/, contains much useful information and links as well as interesting pages on "Light, Time and Micro-Organisms," imaging chlorophyll fluorescence and much more. The Institute of Crystallography at the Free University of Berlin has an important site devoted to X-ray structure analysis of Photosystem I, http://userpage.chemie.fu-berlin.de/fb_chemie/ikr/ag/Saenger/phosys/. The Biophysics group at Leiden University hosts an important site devoted to "Photophysical Processes in Photosynthetic Reaction Centers" (http://www.biophys.LeidenUniv.nl/Research/RCs/) and a second one on "Energy and Electron Transfer in Photosynthetic Membranes" (http://www.biophys.LeidenUniv.nl/Research/PMs/); however, in August 1998, the data were of 1995 vintage. The Avron-Wilstätter Center for Photosynthesis Research at the Weizmann Institute of Science has a large site, devoted to many areas of research, http://weizmann.ac.il/publications/Scientific_Activities/1995/wilstatter_ctr.html.

3. Individual Researcher's Sites

There are too many good individual sites to list them all; the following outstanding sites are mentioned in alphabetical order by last name. Additionally, some of the Educational Sites discussed in Section 4 are also individual sites. Our web version will list many more sites and expand on some of those mentioned here. Once again, our apologies to those whose sites have been left out.

Parag Chitnis has an excellent site at http://www.public.iastate.edu/~chitnis/ that describes many areas of his research.Christa Critchley's site is at http:// www. botany.uq. edu/critchley.htm/. Antony Crofts' site serves as an example of an excellent research site, http://ahab.life.uiuc.edu/index.html. It contains a great deal of personal information, plus a very well done lab home page. Also, there are pages for the bc-complex, Photosystem II, fluorescence quenching, and a course on Biological Energy Conversion (see below for more detailed information). Devens Gust has just expanded his site, http://photoscience.la.asu.edu/photosyn/faculty/gust/index.htm, and now includes information on research into artificial photosynthesis and molecular electronics. Richard Malkin discusses his work with Photosystem I at his site at http: //mollie. berkeley. edu /Faculty/facultymembers/Malkin.html. Another great site is Jonathan Marder's, http://indycc1.agri.huji.ac.il/~marder/, which contains many pages of non-typical information as well as an excellent tutorial on the photosynthetic reaction center. At http://www-server.bcc.ac.uk/biology/conrad.htm, Conrad Mullineaux describes his work on phycobilisomes and also on the mobility of light-harvesting systems. Richard Sayre's site, http://www.biosci.ohio-state.edu/~rsayre/, is under construction but already contains information on Photosystem II, bioremediation using *Chlamydomonas*, and research involving the cassava plant. Tom Sharkey's site is at http://www.wisc.edu/botany/sharkey.html. An interesting and useful site is that of Vladimir Shinkarev that has links to US granting agencies, search engines, data bases, publishers and current content searches, etc: http://130.126.50.91/vlad/. Wim Vermaas has developed a very large section devoted to his course in Genetic Engineering and Society at http://photoscience.la.asu.edu/photosyn/faculty/vermaas.html. The site of Neal Woodbury, http://www.public.asu.edu/~laserweb/woodbury/woodbury.htm, includes information on his lab's study of molecular dynamics and mechanism in protein mediated chemical reactions, as well as information on the developing field of biophotonics (optical biomolecular devices), and extensive notes and exams for courses he teaches.

We note that Melvin Calvin's obituary, as well as his Nobel Prize speech, can be found at: http://www.lbl.gov/Science-Articles/Archive/Melvin-Calvin-obit.html

4. Educational Sites

The ASU Center has developed an extensive list of links to many educational sites with comments on the educational level needed to understand the information contained therein, http://photoscience.la.asu.edu/photosyn/education/learn.html. An excellent overview of the entire photosynthesis process is contained in the on-line paper by J. Whitmarsh and Govindjee, http://www.life.uiuc.edu/govindjee/paper/gov.html. Another excellent site is the MIT Hypertextbook on biology, with its chapter on photosynthesis, http://esg-www.mit.edu:8001/esgbio/ps/psdir.html. The site at http://pluto.clinch.edu/ home/jrb/public_html/photo.html contains "Photosynthesis: An Introductory Internet Tour" and has links to information about early work (even includes photographs of

Aristiotle, van Helmont and others) that led to knowledge about the photosynthetic process. Wim Vermaas has written an introduction to photosynthesis aimed at the general public that is also available on-line at the site http://photoscience.la.asu.edu/photosyn/education/photointro.html. Devens Gust has a page devoted to the question, "Why Study Photosynthesis?" explaining why photosynthesis studies are important to many areas not immediately identified with photosynthesis (http://photoscience.la.asu.edu/photosyn/study.html). The University of Illinois has created introductory sites and links for photosynthesis at http://www.life.uiuc.edu/bio100/lectures/sp98/lects/07s98-photosyn.html and at http://ampere.scale.uiuc.edu/pb102/lectures/07photosynthesis102.html. The section on the Virtual Chloroplast, http://ampere.scale.uiuc.edu/pb102/07/virtchlor.html is wonderful. One of us (G) has slides that are accessible for direct downloading, or can be viewed by using a password (to be obtained by writing to him) for his spring 1998 Biology121 course; there is also a page devoted to the Kautsky Effect, http://www.life.uiuc.edu/govindjee/kautsky.html, that contains a nice short movie of the effect; with the help of M. Lexa, G has developed a graphical site for "Photosynthesis and Time" (http://ampere.scale.uiuc.edu/~m-lexa/gov) which uses a click-on program as well as a movie to illustrate the step by step reactions taking place in the chloroplast thylakoid membrane. Robert Blankenship has a good discussion of photosynthetic antennas and reaction centers at http://photoscience.la.asu.edu/photosyn/education/antenna.html. John Allen has created "Light, Time and Micro-Organisms" (http://plantcell.lu.se/LTM/default.html) to demonstrate the pt_s scale of early events. Michael Lynch, a graduate student at University College Dublin, has created an interesting site that illustrates how to build a fluorimeter (http://www.ucd.ie/~app-phys/michael/research.html) and also has a nice review of the photosynthesis process. Harry Frank has a good page on carotenoids, http://www.lib.uconn.edu/chemistry/faculty/frank.htm. Another site with much information on pigments is http://www.ucmp.berkeley.edu/glossary/gloss3/pigments.html, hosted by the University of California, Berkeley. T. Murphy has a site containing the manual for a plant physiology laboratory, http://www-plb.ucdavis.edu/courses/plb111L/, which contains many preparation techniques. John Whitmarsh has a page on "Electron Transport and Energy Transduction," http://www.life.uiuc.edu/pru/labs/whitmarsh/chapter7/contents.html, which is taken from the book, Photosynthesis: A Comprehensive Treatise, edited by A. S. Raghavendra (see below). John Marder's page (mentioned earlier) contains a very nice tutorial on reaction centers, http://indycc1.agri.huji.ac.il/~marder/rc_view/.

John Cheeseman has created a commercial photosynthesis teaching program, http://www.life.uiuc.edu/cheeseman/JC.software.html. Another commercial program, Photosyn Assistant, is designed to help in the interpretation, comparison and modeling of photosynthesis, http://www.scientific.force9.co.uk/photosyn.htm. Ross Koning has created a couple of pages of interest for a plant physiology class he teaches. The light reactions are discussed at http://koning.ecsu.ctstateu.edu/Plant_Physiology/LightRxns.html, and the Calvin cycle at http://koning.ecsu.ctstateu.edu/Plant_Physiology/Calvin.html. Another botany course from the University of Alberta covers the light reactions at the site http://gause.biology.ualberta.ca/courses.hp/bot240/lecover11.html and carbon fixation at http://gause.biology.ualberta.ca/courses.hp/bot240/lecover12.html.

A. Crofts has several pages of information and links developed for a biophysics course he teaches. Two of the more relevant pages are the Introduction (http://arc-gen1.life.uiuc.edu/Bioph354/lect1.html), and Introduction to Photosynthesis (http://arc-gen1.life.uiuc.edu/Bioph354/lect20.html). Also see http://www.life.uiuc.edu/govindjee/photoweb/2books.html for a list of single or two-authored books.

5. Books and Journals

Although books and journals about photosynthesis have not been placed on-line for economic reasons, there are web sites that discuss them and commercial sites that sell them or list their contents and, sometimes, their abstracts. Publishers often include the Tables of Content for their journals and other information, such as Instructions to Authors.

The most current set of *books* on photosynthesis and related matters is the Advances in Photosynthesis series being published by Kluwer Academic, with one of us (G) serving as its Series Editor. Descriptions and ordering information can be found at two sites: the publisher (http://www.wkap.nl/series.htm/AIPH), and at http://photoscience.la.asu.edu/ photosyn/books/advances.html. This series currently contains 7 volumes, with the 2 most recent volumes specifically published in time for display at this Congress: Vol. 6, "Lipids in Photosynthesis: Structure, Function and Genetics," edited by P.-A. Siegenthaler and N. Murata; and Vol. 7, "The Molecular Biology of Chloroplasts and Mitochondria in *Chlamydomonas*", edited by J.-D. Rochaix, M. Goldschmidt-Clermont and S. Merchant. Several other volumes are currently in various stages of production.

"Aquatic Photosynthesis," by Paul G. Falkowski (Brookhaven National Lab) and John Raven (University of Dundee) is found at http://www.blackwell-science.com/~cgilib/ bookpage.bin?File=3245 (1997, Blackwell Science).

"Plant Biochemistry and Molecular Biology," by Hans-Walter Heldt (Institute of Plant Biochemistry, Göttingen) (with the collaboration of Fiona Heldt) is found at the site http://www1.oup.co.uk/bin/readcat?Version=900797103&title=Plant+Biochemistry+and +Molecular+Biology&TOB=209439&H1=185927&H2=209114&H3=209115&H4=209 205&count=1&style=full (1997, Oxford University Press).

"Photophysics of Photosynthesis. Structure and Spectroscopy of Reaction Centers of Purple Bacteria," by A. J. Hoff and J. Deisenhofer. In: Physics Reports, 287 (1997)

"Photosynthesis: A Comprehensive Treatise", edited by A. S. Raghavendra, is found at http://www.cup.org/Titles/57/052157000X.html (1997,Cambridge Press).

"Photosynthetic Unit and Photosystems–History of Research and Current Views (Relationship of Structure and Function), by A. Wild and R. Ball (of Germany). More information can be found at http://www.life.uiuc.edu/ govindjee/photoweb/4books.html (1997, Buckuys Publishers).

Additional links to older volumes and books intended for young readers and the general public can be found at http://photoscience.la.asu.edu/photosyn/books.html.

Some selected *Journals* are listed below:
Australian Journal of Plant Physiology: http://www.publish.csiro.au/journals/ajpp/
Biochimica et Biophysica Acta: http://www.elsevier.nl/inca/homepage/sah/bba/menu.sht
Biochemistry: http://pubs.acs.org/journals/bichaw/index.html
Biophysical Journal: http://www.biophysj.org/
Journal of American Chemical Society: http://pubs.acs.org/journals/jacsat/index.html
Journal of Biological Chemistry: http://www.jbc.org/
Journal of Photochemistry and Photobiology (JPP):

http://www.kumc.edu/ASP/ PAPHome/pap_home.html
Journal of Physical Chemistry: http://pubs.acs.org/journals/jpchax/index.html
Nature: http://www.nature.com/
Photochemistry and Photobiology: http://www.kumc.edu/ASP/PAPHomepap/home. html
Photosynthesis Research: http://www.wkap.nl/journalhome.htm/0166-8595
Plant Molecular Biology: http://www.wkap.nl/journalhome.htm/0167-4412
Science: http://www.sciencemag.org/

6. Other Sites

6.1 Societies and Organizations: Some selected ones are listed below:

The International Society of Photosynthesis Research:
   http://www.life.uiuc.edu/ plantbio/ispr/.
American Chemical Society: http://www.acs.org/
ASP--American Society for Photobiology: http://www.kumc.edu/POL/
ASPP--American Society of Plant Physiologists: http://www.aspp.org/
Biophysical Society: http://www.biophysics.org/biophys/society/biohome.htm
Japanese Society of Plant Physiologists: http://www.nacos.com/jspp/jspp01.html
Phycological Society of America: http://jupiter.phy.ohiou.edu/psa/

6.2 Related Sites. The following sites contain information of interest to researchers
   studying photosynthesis:

Chlamydomonas Genetics Center: http://www.botany.duke.edu/dcmb/chlamy.htm
Cyanosite: http://www-cyanosite.bio.purdue.edu/index.html
CyBib: http://www-cyanosite.bio.purdue.edu/cybib/cybibhome.html.

The ASU Photosynthesis Center site also maintains links to a number of commercial
vendors providing equipment and software: http://photoscience.la.asu.edu/photosyn/
links.html.

We recommend that that those starting searches for photosynthesis web sites should first
use search engines such as HotBot (http://www.hotbot.com) and AltaVista
(http://www.altavista.digital.com), utilizing the advanced searching features that those
programs offer, and search strings in those and other search engines. For more
information we refer you to the section on Searching the Web in the expanded version of
this paper, available on-line at http://www.life.uiuc.edu/govindjee/photoweb/ and also at
http://photoscience.la.asu.edu/photosyn/photoweb/.

References

1 Bernes-Lee, T., Cailliau, R., Geoff J.-F. and Pollermann, B. (1992) Electronic
   Networking 2: 52-58
2 Williams, B. (1996) The World Wide Web for Teachers, p. 14. IDG Books, Foster
   City, CA
3 Kroll, E. (1992) The Whole Internet. User's Guide & Catalog, pp 11-18. O'Reilly and
   Associates, Sebastopol, CA
4 Engst, A.C. (1994) Internet Starter Kit, p. 90. Hayden Books, Indianapolis

PROBLEMS WITH ANSWERS FROM PHOTOSYNTHESIS: A WAY TO TEACH BIOPHYSICS

Péter Maróti
Department of Biophysics, József Attila University, Szeged
Egyetem utca 2. Szeged, Hungary H-6722

Key words: bacterial reaction center, bioenergetics, charge separation, coupling/uncoupling, electron transfer, environment

We have for some decades engaged in providing biophysics lectures and laboratory practicals for undergraduate and graduate students majoring in physics, biology or medicine. Biophysics courses generally involve the teaching of considerable number of quantitative theoretical and practical concepts and the time available is strictly limited. The presentation of a crucial equation or application to the class is often prefaced by a remark such as "it may be shown that...". The student is then left either to accept it without any deeper insight and a sense of the limitations, or to digest the concept by seeking out the derivation or the solution of a biophysics problem, often from educationally unsatisfactory research papers which abound in phrases such as "it follows that...". In many cases such points are obvious only to the authors and the referees.

In our biophysics courses, we prefer to use (homework) problems throughout the semester as a study guide. Many students are pleased to have a greater opportunity to apply the concepts of biophysics. Very recently a textbook was published which presents more than 250 current problems from modern biophysics and related fields of application, together with detailed solutions [1]. The problems emphasize concepts and information that are essential for an understanding of the fundamental physical processes in life sciences, including photosynthesis and medical science. The book is divided into 11 chapters and the topics follow the sequence of dimensions and diversity of the living world. It starts with the basic principles relating to the energetics of the living world, and moves on to problems from the microworld (biophysics of molecules and membranes) and the macroworld (biomechanics, biophysics of organs, radiation and the environment). The subsequent chapters are devoted to problems concerning the application of different experimental methods in biophysics, in photosyntheis and in medicine (diagnostic and therapeutic methods and medical imaging). The problems in biostatistics help the reader to understand and to digest the concepts and the methods of evaluation of experimental data.

G. Garab (ed.), Photosynthesis: Mechanisms and Effects, Vol. V, 4393–4396.
© 1998 *Kluwer Academic Publishers. Printed in the Netherlands.*

The book is written for undergraduate and graduate students, with a view to their improving their problem-solving ability in physics. The reader is faced with the great challenge of finding solutions to the problems, but at the same time his or her knowledge of important concepts and relations is reinforced. The treatment of the problems is straightforward and well-documented, and their digestion does not usually demand any special background knowledge. The solutions provide full discussions of the problems and are well separated from the problems themselves. As the level of difficulty of the problems covers a wide range, both beginners and advanced readers will find pleasure in the book. I hope that the book will become an essential in the education of physicists and biologists, as well as being invaluable for medical and pharmaceutical students interested in the application of physical principles to problems in biology and medicine.

The natural sciences, including primarily physics, chemistry and biology, have made substantial contributions to the identification and analysis of biophysical problems. The boundary between the disciplines is diffuse and therefore the methods used to solve biophysics problems are widely used by other disciplines as well. Photosynthesis as a complex process of practically unlimited spatial and temporal dimensions has been the target of research initiated by different disciplines. Many of these topics will cover essential parts of the subjects of biophysics. Below, a couple of problems were selected from different fields of photosynthesis to demonstrate how these problems with their solutions will contribute to improve the student's problem-solving ability. Because of the space limitation, only the problems will be presented.

Bioenergetics

1. The photosynthetic green alga *Chlorella fusca* is to be cultivated in thermostated aqueous solution through which carbon dioxide (CO_2) gas is bubbled. Stationary growth requires the transport of 0.1 L CO_2 gas per minute through the culture. The cells will grow under atmospheric pressure ($1 \cdot 10^5$ Pa) and room temperature (25 °C). The CO_2 will be taken from a CO_2 gas cylinder with a volume of 20 L at a temperature of 20 °C. What is the minimum pressure required in the cylinder to ensure the necessary gas supply continuously for one week (7 days)? The gas in the high-pressure cylinder should be regarded as a real (not perfect) gas (a van der Waals gas with $a = 0.366$ Pa·m^6·mol^{-2} and $b = 4.29 \cdot 10^{-5}$ m^3·mol^{-1} in the van der Waals equation).

2. Consider the redox couple ubiquinone/ubiquinol (Q/QH_2), which plays a key role in many energy-converting living systems (e.g. photosynthesis). Determine the degree of reduction of the Q/QH_2 couple as a function of the actual redox potential at room temperature ($T = 293$ K) and pH = 7.0. Describe the pH-dependence of the Gibbs energy (redox midpoint potential, E_m) of the couple. E_m is 60 mV at pH 7.0.

3. In the photosynthetic reaction center protein, the light-induced electron passes from a primary quinone molecule (Q_A) to a secondary quinone molecule (Q_B). The (interquinone) electron transfer is sensed by the pK change in an adjacent protonatable residue of an amino acid: its pK values are $pK_{A^-} = 9.8$ and $pK_{B^-} = 11.3$ if the electron is on Q_A or on Q_B, respectively. (The protonatable group is closer to Q_B than to Q_A.) To

examine the role of protonation, determine the pH-dependence of the (observed) Gibbs energy of the interquinone electron transfer.

4. The redox dye diaminodurene (DAD) can be reduced by $n = 2$ electrons and can bind 2 electrophilic ligands (H^+-ions) with (proton) dissociation constants of K_1^O and K_2^O in its oxidized and K_1^R and K_2^R in its reduced (electron-rich) forms (Fig. 1).

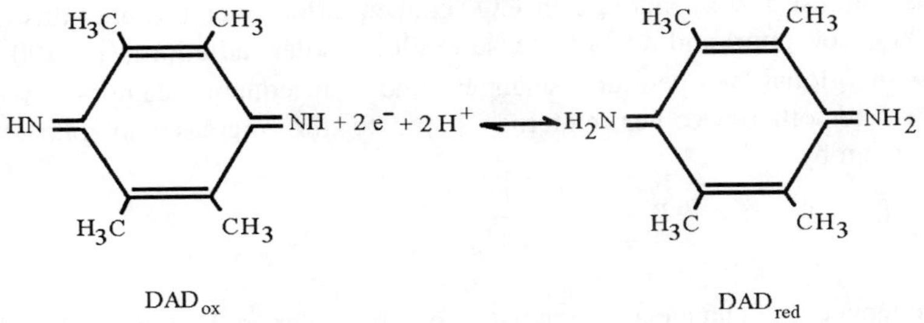

$$DAD_{ox} \qquad NH + 2\,e^- + 2\,H^+ \longleftarrow H_2N \qquad DAD_{red}$$

Figure 1

What is the pH-dependence of the midpoint redox potential (E_m) of the dye? The midpoint redox potential of the oxidized form at very high pH ($>> pK_1^R$) is $E_m(O) = 170$ mV and the pK values are $pK_1^R = 10.0$, $pK_2^R = 6.0$, $pK_1^O = 2.0$ and $pK_2^O = 1.0$. Discuss the special cases.

Global photosynthesis and the environment

5. Assume that 1% of the surface of the Earth of radius $R = 6.8 \cdot 10^6$ m is covered by green plants. The saturating rate of photosynthetic oxygen evolution from a leaf is $P_{max} = 20\ \mu mol \cdot m^{-2} \cdot s^{-1}$. At the present rate, how much time is required to exchange the entire atmospheric content of oxygen? Estimate the global net productivity of terrestrial plants per year.

6. The continuous illumination of a leaf of a green plant by monochromatic light of wavelength $\lambda = 680$ nm and intensity $I = 14$ W·m^{-2} will maintain an oxygen evolution (or CO_2 fixation) rate of $P = 7.6\ \mu mol \cdot m^{-2} \cdot s^{-1}$ at saturating CO_2. What is the quantum requirement of oxygen evolution (CO_2 fixation) in this experiment and how is it related to the theoretical minimum? What is the overall photosynthetic efficiency of energy transduction if the combustion energy of carbohydrate $C_6H_{12}O_6$ is $\Delta H = 2,822$ kJ·mol^{-1}. Discuss the role of different losses that might diminish the actual efficiency of photosynthesis under field conditions.

7. What wavelength of light is sufficient for the photochemical reaction of activation energy $w = 280$ kJ/mol? In this simple case, monochromatic light is considered effective.

8. The intensity of solar radiation on the Earth's surface is $I = 520$ W/m^2. (a) What energy is absorbed in $t = 5$ h by spherical tree canopy with a diameter $d = 20$ m, which is able to absorb 85% of the incident radiation? (b) Under such conditions, $m = 2.7$ kg of glucose is synthesized in the tree. The molecular weight of glucose is $M = 180$ g/mol, and photosynthetic glucose production requires $w_s = 2.8$ MJ/mol energy. What percentage of the absorbed energy is utilized for glucose production?

9. The uncurbed growth of the human population and the exponential expansion of industrial activity can cause global climate changes on the Earth. Moderate model calculations suggest a 20% increase in CO_2 concentration over the preindustrial level ($C_0 = 0.03\% = 300$ ppm) and a 6% increase in global solar radiation ($I_0 = 400$ W·m^{-2}), together with global temperature changes and consequent changes in rainfall, evaporation and other weather patterns. Estimate the increase in photosynthetic production given by

$$P = \frac{\alpha I \cdot \tau C}{\alpha I + \tau C}$$

as a consequence of changes in intensity of the solar radiation I and the CO_2 concentration C ($\alpha = 1 \cdot 10^{-8}$ kg CO_2·J^{-1} and $\tau = 2 \cdot 10^{-3}$ m·s^{-1}).

10. An algal biomass is cultivated in a well-stirred pool of area A and depth d under an incident light intensity I_0. The rate of photosynthetic activity per unit mass of the biomass is a hyperbolic function of the actual light intensity I:

$$P = \frac{\alpha I \cdot P_{max}}{\alpha I + P_{max}}$$

where α is a constant and P_{max} is the maximum rate of photosynthetic activity measured at saturating light intensity. Determine the rate of photosynthetic production of the entire biomass of the pool as a function of the concentration of the algae. At what concentration is it recommended to harvest the biomass? The optical extinction coefficient of the algae at the wavelength of illumination is ε.

Reference

1 Maróti, P., Berkes L. and Tölgyesi F. (1998) Biophysics Problems: A Textbook with Answers pp 492, Akadémiai Kiadó, Budapest, Hungary

Acknowledgements

The work was supported by Pro Renovanda Cultura Hungariae, Hungarian Ministry of Education (MKM FKFP 1288/97, AMFK 043/98), Hungarian Science Foundation (OTKA 17362/95) and International Human Frontier Science Program (RG-329/95M).

Keyword index[§]

AUTHOR INDEX

1

lii